Robert H. Barth

Associate Professor of Zoology
University of Texas at Austin

Robert E. Broshears

Vanderbilt University

THE INVERTEBRATE WORLD

SAUNDERS COLLEGE PUBLISHING

Philadelphia New York Chicago
San Francisco Montreal Toronto
London Sydney Tokyo Mexico City
Rio de Janeiro Madrid

Address orders to:
383 Madison Avenue
New York, NY 10017

Address editorial correspondence to:
West Washington Square
Philadelphia, PA 19105

This book was set in Souvenir Light by Trade Composition Co.
The editors were Kendall Getman, Michael Brown, Lyn Peters, Celine Keating, Lloyd Black, Priscilla Estes, Patrice Smith, Willoughby Broderick, Maryanne Miller, and Don Reisman.
The art & design director was Richard L. Moore.
The cover design was done by Richard L. Moore.
The artwork was drawn by Vantage Art.
The production manager was Tom O'Connor.
This book was printed by Fairfield Graphics

Cover credit: Cover photo ©1981 Ken Kasper. Subject: The orange spiral structures are part of a serpulid tubeworm.

THE INVERTEBRATE WORLD ISBN 0-03-013276-2

Library of Congress catalog card number 80-53916.

1234 144 987654321

CBS COLLEGE PUBLISHING
Saunders College Publishing
Holt, Rinehart and Winston
The Dryden Press

Dedicated to
Sidney J. Townsley

Acknowledgements

We are grateful for the guidance and support of the many people who contributed to the making of this book. Invertebrate specialists around the world gave generously of their time and knowledge in reviewing the manuscript. Our friends and families contributed in many ways, sustaining us through the writing with their understanding and humor. Finally, we thank the publishers for their patience and professionalism.

Individual chapters were reviewed by the following specialists:

CHAPTER	
Protozoans	Stuart Krassner, UC-Irvine; John O. Corliss, U. of Maryland; E.L. Powers, U. of Texas-Austin
Sponges	George Hechtel, SUNY at Stony Brook
Coelenterates	R.D. Campbell, UC-Irvine; David Krupp, U. of Hawaii
Flatworms	E. Ruffin Jones, U. of Florida
Pseudocoelomates	R.A. Rohde, U. of Massachusetts
Annelids	R.P. Dales, Bedford College, U. of London
Mollusks	Kenneth J. Boss, Harvard University; Ruth D. Turner, Harvard University
Intro to Arthropods	Howard Schneiderman, UC-Irvine
Chelicerates	Herbert W. Levi, Harvard University; Jonathan Reiskind, U. of Florida
Crustaceans	Linda Mantel, CCNY
Myriapods	Richard L. Hoffmann, Radford College
Insects	Howard Schneiderman, UC-Irvine
Lophophorates	Susan Cummings, UC-Irvine
Echinoderms	David Pawson, Smithsonian Inst.
Other Deuterostomes	Gary Freeman, Univ. of Texas-Austin

In addition, Janice Moore, Univ. of New Mexico, reviewed our sections on the parasitic helminths, and Milton Fingerman, of Tulane University, contributed a review of the complete final manuscript.

These reviews have ensured a more complete and accurate text than would otherwise have been possible. The omissions and errors that remain are clearly our own responsibility, and we encourage readers to bring them to our attention.

The Invertebrate World represents a joint effort of the staffs at Holt, Rinehart and Winston and Saunders College Publishing. We are particularly grateful to Lyn J. Peters, who served faithfully as our developmental editor. Special thanks are also due to Don Reisman, whose efforts during the production phases added handsomely to the book's final form.

Preface

> Bring forth with thee every living thing that is with thee, of all flesh, both of fowl, and of cattle, and of every creeping thing that creepeth upon the earth; that they may breed abundantly in the earth, and be fruitful, and multiply upon the earth.
>
> — Genesis

> Races came and went, species passed away, but ever new species arose, more lovely, or equally lovely, always surpassing wonder. The fountainhead was incorruptible and unsearchable. It had no limits. It could bring forth miracles, create utter new races and new species in its hour, new forms of consciousness, new forms of body, new units of being. To be man was as nothing compared to the possibilities of the creative mystery.
>
> — D. H. Lawrence

We live in an invertebrate world. Although often less conspicuous than their vertebrate cousins, invertebrates nevertheless are pervasive in each principal biological environment. Invertebrates account for roughly 97% of all known animal species; it is to them, these "creeping things," that the world belongs.

As a 97% approximation of all animal biology, invertebrate zoology presents problems of management. Due to the expanding biological curriculum, most colleges and universities can offer no more than a one semester survey of the invertebrate phyla, an unfortunate situation that compounds the tasks of students, professors, and text writers alike. For so much biology to be presented in so little time, a considerable condensation of material is necessary.

We have written *The Invertebrate World* in response to the need for a manageable text to accompany a one semester course on the invertebrates. Material has been carefully selected to present a unified, comprehensive yet streamlined discussion of these animals. The introductory chapter is a brief portrait of the physical and biological world, its principal environments and their particular demands on living things. Invertebrates have evolved a variety of organizational schemes, or body plans, which meet these demands. With a focus on the basic elements of each body plan we then describe the various invertebrate phyla, utilizing example organisms that are best-known and/or most representative of their groups. In so doing we provide the reader with a picture of the major themes of invertebrate evolution. After emphasizing its general plan, we describe the diversity within each invertebrate phylum. In choosing the traditional phylum-by-phylum approach, we were persuaded by the belief that it will thus be easier for the reader to focus primarily on the biology of whole animals. This approach seems most appropriate for an introductory invertebrate zoology course, perhaps particularly so in that most courses in the modern biological curriculum are approached in a comparative life process manner. As the most successful of the various schemes of invertebrate organization, the major phyla receive greater emphasis, reflecting our belief that in a short course the student's time is spent most profitably in studying the means of major biological success. Less successful phyla are included not only for the sake of completeness but also for the instruction offered by their adaptive limitations.

Stress is placed on the structural, physiological, and behavioral adaptations of organisms to their particular environment. Morphological descriptions are kept to a minimum in the belief that anatomy is best learned in the laboratory from the animals themselves with the aid of an appropriate lab manual. Taxonomy, the traditional bugaboo of introductory invertebrate courses, receives limited emphasis. In consideration of the capacity of any student's memory in a short course, we emphasize taxonomic organization only to the extent that it serves as a framework for the biological information and directly enhances the evolutionary theme. A concluding chapter considers invertebrate phylogenetic theories, summarizing earlier discussions of evolutionary relationships among the phyla.

As a book for a one-semester course, *The Invertebrate World* has a limited capacity for depth of discussion. For those whose interests lead them beyond the scope of this book, there is a list of references at the end of each chapter. We are cognizant of the difficulty presented to the student by the overwhelming vocabulary of invertebrate zoology; accordingly, there is a glossary covering essential terminology. In the hope that a good figure will often be worth its thousand words, photographs and drawings are used extensively. The book is written with the assumption that the reader has had some exposure to biological science, specifically that provided by an introductory course in general biology.

Finally, we hope that *The Invertebrate World* is a book that can be enjoyed and one that may instill in its readers an increasing appreciation of invertebrate organisms. Invertebrates, like humans, are expressions of Lawrence's "creative mystery." Coping with essentially the same biological situations, they are our companions in the world. Our lives are enriched as we wonder at them. It is wonder that guides and motivates science. And it is to the invertebrates, more than to humanity, more than to elephants or eagles, that a world of wonder ultimately belongs.

We invite comments from all readers.

Table of Contents

1

Of Organisms and Environments —An Introduction

Our planet is distinguished by the presence of living things. These living things are mostly animals without backbones. Biologists have described over a quarter-million species of plants and nearly four times as many animals. Of the latter, a mere 3% have vertebrae, while the remaining 97% represent a vast range of invertebrate types. We have relatively greater knowledge of the smaller group, the vertebrates, within which we count our own species. Invertebrate animals are a much more diverse assemblage about which much is known and a great deal unknown. Yet, by most measures, invertebrates collectively represent the most successful manifestations of organic evolution. This book offers a short account of their world.

A central fact of life is change. Molecules, organisms, and environments evolve continuously. A molecule of oxygen in the air today may become part of a marine crustacean tomorrow; next week that crustacean may be digested by a squid, the squid itself fated to die years from now on the ocean floor, passing its elements on to yet other creatures. Marine arthropods, stranded on the swampy edges of receding Paleozoic seas, have given rise over millions of years to the insects, invertebrate masters of the terrestrial environment and by some measures the most successful of all living things. The millennia have brought changes also to sinking South Pacific islands, leaving behind rings of coral and their network of quiet life—the sponges, bryozoans, and echinoderms of the reef. The molecules, organisms, and environments of the invertebrate world evolve with each passing day and epoch.

Because organisms and environments participate inseparably in the evolutionary process, our appreciation of the invertebrates is limited unless we understand something of the biology of environments. Animals inhabit four general types of environments: the sea, fresh water, land, and the bodies of other organisms. Each environmental type has advantages and disadvantages for life; each makes special demands and grants special privileges to the creatures that it harbors. An invertebrate growing up in the sea is quite a different animal from one that parasitizes the body of a terrestrial snail. To understand how and why these creatures differ and, moreover, to comprehend the entire biological world and learn why it is largely an invertebrate world, we begin with a brief study of environments.

The Environments of the Invertebrate World

THE SEA

The largest and oldest of animal environments is the sea. Seventy percent of the earth's surface lies beneath the oceans, and in terms of volume, the proportion is even more impressive. The greater depth of the sea compared to land elevation places over 99% of the biosphere under marine waters. In salt water, primitive macromolecules organized over 3 billion years ago into the first living organisms. Over time, virtually all invertebrate groups began in the sea. Subsequently, many animals migrated into other environments. Yet, except for the insects, the sea remains the home of the majority of modern invertebrates. The oldest, largest, and in many ways the most benign of animal environments, the sea is, at least for noninsects, also the most popular invertebrate habitat.

Why is the sea such a favorable place to live? This question can be approached from several viewpoints, but all answers relate to the fact that the marine environment represents the very cradle of life. Life began in the sea because the marine environment so readily provided the water, salts, gases, light intensities, etc., that the life process requires. In a very real sense, the invertebrate world is just what the sea ordered! Thus we find that the protoplasm and tissue fluids of many invertebrates are of a composition similar to that of sea water (see Table 1.1). Accordingly, the marine environment is osmotically favorable to its inhabitants. Moreover, the vast size of the oceans fosters a uniformity of environmental factors, including temperature. Marine invertebrates thus have minimal needs for thermal and osmotic regulation. Their biology tends to be regulated less vigilantly than that of freshwater and terrestrial organisms, and as a group, marine invertebrates are among the slowest metabolizers.

The marine environment is subdivided into several regions, and marine invertebrates are categorized according to these zones. Generally, the subenvironments of the sea are the intertidal zone, the continental shelf and slope, the pelagic region, and the abyssal plain (Fig. 1.1).

The **intertidal zone** refers to the coastal area lying

TABLE 1.1 Ionic Composition of Sea Water and the Body Fluids of Humans and Certain Invertebrates*

	Na^+	K^+	Ca^{2+}	Mg^{2+}	Cl^-	SO_4^{2-}
Human	145 (100)	5.1 (3.5)	2.5 (1.7)	1.2 (0.83)	103 (71)	2.5 (1.7)
Hydrophilus (insect)	119 (100)	13 (11)	1.1 (0.93)	20 (17)	40 (34)	0.14 (0.13)
Lobster	465 (100)	8.6 (1.9)	10.5 (2.3)	4.8 (1.0)	498 (110)	10 (2.2)
Venus (clam)	438 (100)	7.4 (1.7)	9.5 (2.2)	25 (5.7)	514 (120)	26 (5.9)
Sea cucumber	420 (100)	9.7 (2.3)	9.3 (2.2)	50 (12)	487 (120)	30 (7.2)
Sea water	417 (100)	9.1 (2.2)	9.4 (2.3)	50 (12)	483 (120)	30 (7.2)

*Upper numbers are expressed in millimoles (mM) per liter, lower numbers are relative, expressed in terms of 100 units of Na^+. (From A.G. Loewy and P. Siekevitz, *Cell Structure and Function*, 1966, Holt, Rinehart and Winston.)

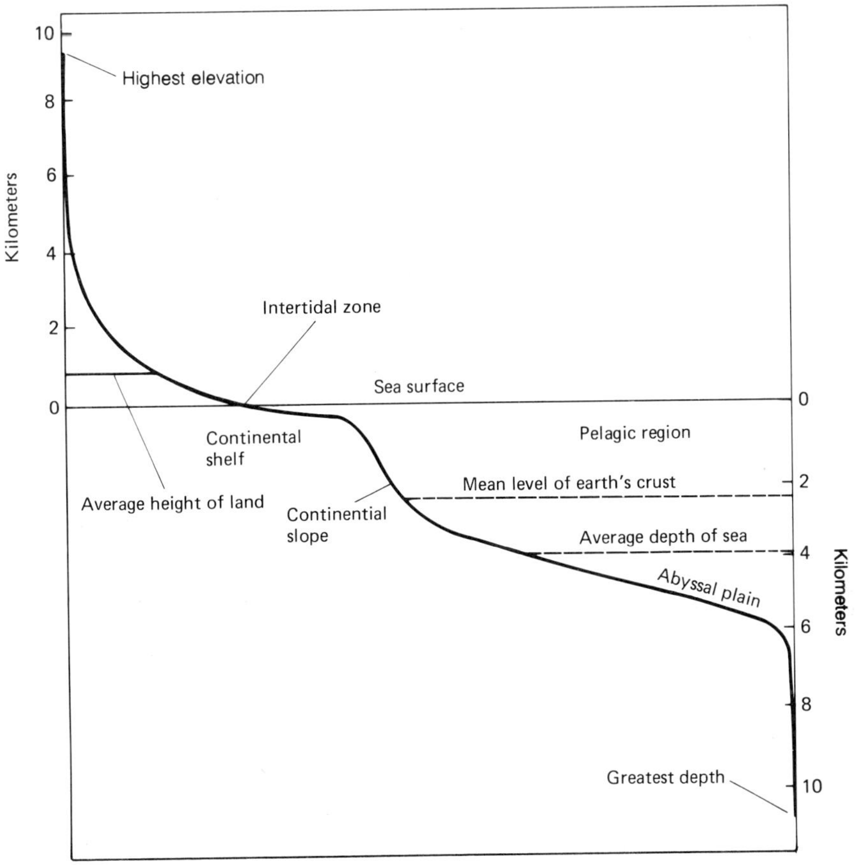

Figure 1.1 Regions or subenvironments of the oceans. (From A.S. Pearse, *Animal Ecology*, 1939, McGraw-Hill.)

between the high- and low-tide marks. This zone supports a wealth of invertebrate life and, because of its accessibility, has been well explored by both biologists and laypersons. Intertidal forms range from sand-burrowing worms to the varied and fascinating community of the tide pool. Throughout this subenvironment, animals are segregated in part according to their tolerance of drying air. Invertebrates most sensitive to desiccation live near the low-tide mark, while desiccation-resistant species are found on the rocks near the high-tide limit. The latter include various snails and barnacles whose hard shells shelter them from the dryness that occurs at each low tide.

The **continental shelf** is a shallow platform extending outward from the coast. It averages about 50 kilometers (km) in width, but variations from a few to several hundred kilometers exist. The continental shelf is a particularly rich region, where the biological advantages of the marine environment are supplemented by the runoff of nutritive organic and mineral substances from land. Most invertebrate groups are represented on the shelf, which is also a prize fishing ground. In addition, its shallow waters allow extensive light penetration and thus support active plant life. The continental shelf slopes to a depth of 200 meters (m), where it drops steeply to the abyssal plain.

The **abyssal plain** is the deep ocean floor. There, comparatively few invertebrates live in a world of constant cold, darkness, and intense hydrostatic pressure. The sea floor averages 4000 m in depth (compared to an average land elevation of 850 m), but some trenches plunge an awesome 10 km below the water surface.

The **pelagic region** refers to the open sea. There swimming and floating invertebrates lead mobile, unattached lives.

Marine organisms are divided into three groups according to their habitats and life-styles: the plankton, the nekton, and the benthos. The organisms that float near the surface of the sea constitute the marine **plankton.** Planktonic organisms include both plants **(phytoplankton)** and animals **(zooplankton)**. Because of its dissolved bicarbonate salts, marine water holds 50 times more carbon dioxide than fresh water does. The abundance of this important raw material for photosynthesis contributes to the sea's ability to support a rich phytoplankton. Indeed, marine phytoplankton account for a majority of the earth's photosynthetic activities. These plants are the primary producers of the open sea, and on them the food chains of most marine organisms are based. The many zooplanktonic forms fall into two subcategories: holopelagic types, such as jellyfish, copepods, and marine protozoans, which spend their entire lives adrift; and the larval forms of bottom-dwelling species, such as echinoderms, decapod crustaceans, and mollusks. These temporary planktonic forms settle to the bottom when larval development is completed.

Nekton refers to swimming animals. Relatively few invertebrates are included in this group, which is dominated by fish and mammals. Indeed, only the cephalopods (squid, cuttlefish, and octopods) are well represented among the invertebrate nekton.

Most marine invertebrate species belong to the **benthos.** This group is composed of bottom-dwelling animals and includes burrowers and crawlers as well as many sessile forms.

FRESH WATER

The freshwater environment is younger and far less uniform than the sea. Accounting for less than 1% of the earth's surface moisture, fresh water is distributed throughout many systems of lakes, ponds, swamps, rivers, and streams. Many lakes and streams are products of the last Ice Age, but even the older, slow rivers are young compared to the ocean. In such a geologically short time, life has barely begun to adapt to these environments.

Compared with the sea, freshwater systems are unstable because of their small size. The relatively small amount of water involved in such systems increases their vulnerability to climatic influences, local geographic disturbances, and fouling by human or natural forces. Lakes freeze over in winter and ponds dry out in summer; dams cripple the great rivers, and avalanches alter the paths of mountain streams.

The fertility of freshwater systems is highly variable. The amount of life supported by any freshwater environment depends on currents, local substratum conditions, and the size and age of the water body. In general, older, larger bodies of water flowing slowly over fertile soils contain the most living organisms. Lakes, then, are more fertile than rivers or ponds. Slow rivers in turn support more life than small, cascading streams. The continuous currents of many freshwater systems militate against the formation of stable biological communities. Moreover, the continual addition of relatively sterile water by atmospheric precipitation also discourages plant and animal life.

Clearly, invertebrates that inhabit fresh water have some difficulties not encountered by their marine cousins. Foremost is the problem of osmoregulation. The environment of freshwater animals is decidedly hyposmotic and must be kept at bay. Osmoregulation is a rigorous, energy-consuming activity in all freshwater forms and accounts for the higher metabolic rate of these animals as compared to marine types. Freshwater organisms also must survive interruptions in favorable environmental conditions. Freezing, drying, and fouling are frequent threats to freshwater invertebrates; hence their life cycles often provide a special emergency stage. When environmental disaster strikes, the typical freshwater invertebrate enters this dormant, resistant stage until suitable conditions return.

Given the difficulties of the freshwater environment, we might wonder why invertebrates colonized it. The best answer to this question lies in some general principles of evolution. Organisms and environments change continuously. Isolated parts of the sea very gradually became fresh water; their flora and fauna either adapted to this change or perished. Estuaries and marshes connecting marine with freshwater environments also served as pathways to freshwater life. Plants surely made the transition first, creating a habitat in which animal life was possible. The ensuing competition among marine forms contributed to selection for adaptation to alternative life styles. The first freshwater organisms, probably bottom dwellers with low metabolic rates, benefitted from their transition to a new, originally less competitive environment. Other invertebrates followed.

LAND

The terrestrial environment is perhaps the most demanding of all habitats. Because biological reactions take place solely in aqueous solution, terrestrial organisms must maintain an aqueous internal environment amid the dry external one. Under these conditions, constant regulation is necessary to prevent water loss. This rigorous regulation is expressed in terms of behavior, osmoregulatory activities, and/or water-tight external coverings. Many essential resources are irregularly distributed on land; hence land animals must search at length for food and shelter. The terrestrial environment is thus a stimulating as well as a difficult habitat, and the animals that live on land are among the most metabolically active.

Air is about one eight-hundredths as dense as water. Some movements are easier in this rarefied medium, but gravity is a major problem. The buoyant qualities of water allow some aquatic invertebrates to reach considerable size, but the absence of supporting vertebrae severely limits the size of terrestrial invertebrates.

A great advantage of the terrestrial environment is its abundance of oxygen. Water contains relatively limited amounts of dissolved oxygen and gradually loses the gas with increasing temperature and salinity (Fig. 1.2). Maximal concentrations reach only 10 milliliters (ml)/liter at 0°C. But the atmosphere is 21% oxygen; hence terrestrial animals are well supplied for their increased metabolic needs.

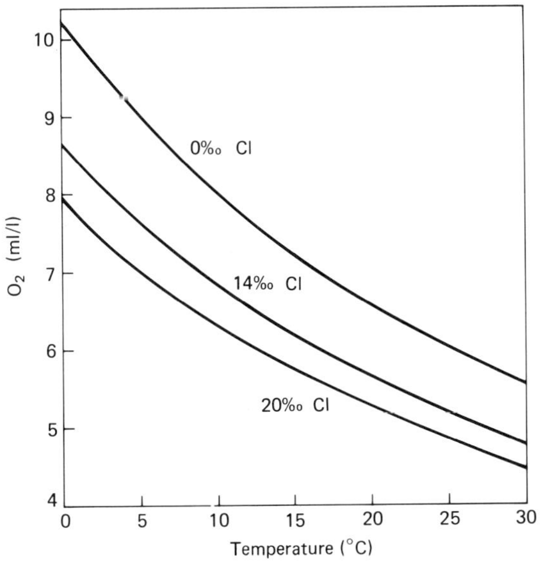

Figure 1.2 Amounts of oxygen dissolved in water of selected chlorinities as a function of temperature. (From O. Kinne, Ed., *Marine Ecology: A Comprehensive Integrated Treatise on Life in Oceans and Coastal Waters, Vol. 1, Part 1*, 1970, Wiley-Interscience.)

Except for several classes of arthropods, invertebrates have not adjusted well to life on land. The rigors of the terrestrial environment—the drying air and scarcity of food—demand a structural and physiological sophistication beyond the adaptive capacity of most invertebrate body plans. Accordingly, many invertebrate air breathers are secretive creatures that live under rocks and vegetation and in moist soils. The insects, however, are true masters of the land. Their diversity of types, representing 75% of all described animal species, accounts for the numerical superiority of terrestrial invertebrate forms.

THE PARASITIC ENVIRONMENT

A large number of invertebrates live as parasites on or within the bodies of other animals and plants. New parasitic species are identified all the time, and some biologists suspect that they may represent a majority of all animals. In certain respects, the parasitic environment is a most favorable one. Host organisms maintain appropriate environmental conditions for the parasites, including the provision of foodstuffs; however, the lack of continuity in the parasitic habitat is a major drawback. An established parasite may seem quite successfully adapted, but the ultimate survival of each species depends on the ability of succeeding generations to find new host environments.

Various subenvironments define several types of parasitic invertebrates. **Ectoparasites** attach to their hosts externally; **endoparasites** occupy the host interior. Endoparasitic subtypes include body cavity or **coelozoic** parasites, tissue or **histozoic** parasites, and intracellular or **cytozoic** parasites.

Some of the qualities of parasitic life are outlined in a special essay (see Chap. Six); further discussion of this specialized environment is reserved for the accounts of the major parasitic groups.

The Unity of the Invertebrate World

A second fact of life is the basic unity of living things. Every animal exhibits the same basic systems: first, **maintenance systems**—the structures and processes that define the boundaries of the living unit and maintain order within it. Maintenance systems involve size, form, and trafficking of materials between an organism and its environment. Second, **activity systems** ultimately involve the behavior of the animal. Such systems gather information from the environment, interpret it, and then organize and execute responses. Third, **continuity systems** effect the growth and reproduction of similar organisms and thus ensure the continuance of the species.

MAINTENANCE SYSTEMS

Every living system is highly organized. For its biological functions to proceed normally, an organism must maintain a series of well-defined relationships among its internal components and between itself and its environment. However, according to the Second Law of Thermodynamics, all systems tend toward a state of maximum disorder. Thus, in seeking to maintain order, the maintenance systems of all organisms are pitted against one of the great principles of the universe. Maintenance systems include aspects of general morphology—body form, size, and symmetry. Morphology is associated with the ways in which animals meet other maintenance responsibilities: nutrition and digestion, respiration, circulation, osmoregulation, and excretion.

General Morphology

Invertebrates are of many shapes and sizes, varying from a microscopic radiolarian sphere to the streamlined, 20 m missile of the giant squid. Size and shape affect biological functions, and the relationship between size and surface is central to all maintenance systems. The surface area of an organism is an expression of the square of its length, while overall size (volume) is a cubed function of the same dimension. Thus, as an animal increases in length, the ratio between its surface area and its volume decreases. Small invertebrates have large surface areas compared to their volumes; large animals have comparatively smaller surfaces. The biological implications of this relationship are far-reaching. Larger organisms maintain more cells and tissues in locations far removed from the external environment. The transport of nutrients, wastes, and respiratory gases between the environment and internal areas presents major maintenance problems. On the other hand, a larger size has its advantages. Generally, larger animals are more resistant to environmental changes and less vulnerable to trauma and predation. Moreover, the isolation of internal cells (provided they are adequately serviced) provides opportunities for tissue specialization. Cells that need not deal directly with the external environment may differentiate for very specific functions. Such differentiation and division of labor among specialized cell and tissue types are central to the evolution of all multicellular organisms.

Another key feature of animal form is symmetry. Most invertebrates exhibit radial or bilateral symmetry, although asymmetrical and intermediate forms exist. Radial symmetry is basically a passive organization. Multiple planes can be drawn through the center point or axis of a radially symmetrical animal, each dividing the organism into two mirror images. Body parts are arranged concentrically along these axes (Fig. 1.3). Radial symmetry is considered passive because it is suited primarily to a **sessile** life-style. Sessile invertebrates, those spending their lives attached to a substratum, are limited in their ability to make whole body responses to environmental stimuli. Their radial symmetry allows them to monitor and respond equally well in all directions. Bilateral symmetry, on the other hand, is a more active organization. In bilateral animals, only one plane divides the body into two mirror images; this plane runs from the dorsal to the ventral surface along the longitudinal axis (Fig. 1.4). Body parts are arranged singly along this axis or in pairs on either side. Bilateral symmetry is displayed by more

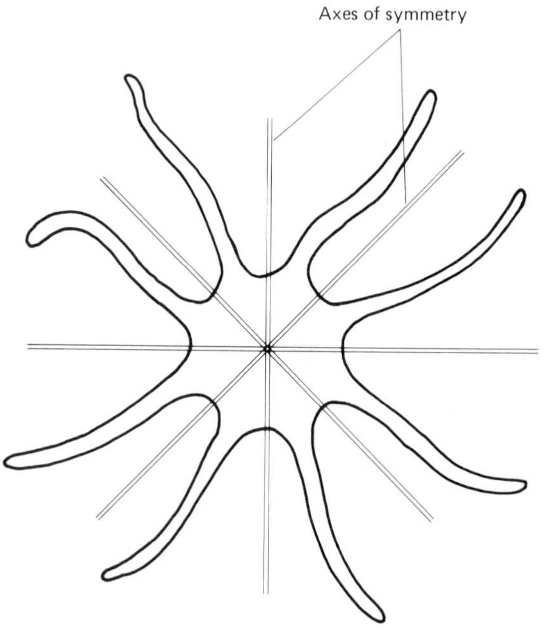

Figure 1.3 Radial symmetry. Any plane passing through the oral-aboral central axis divides the animal into equal halves.

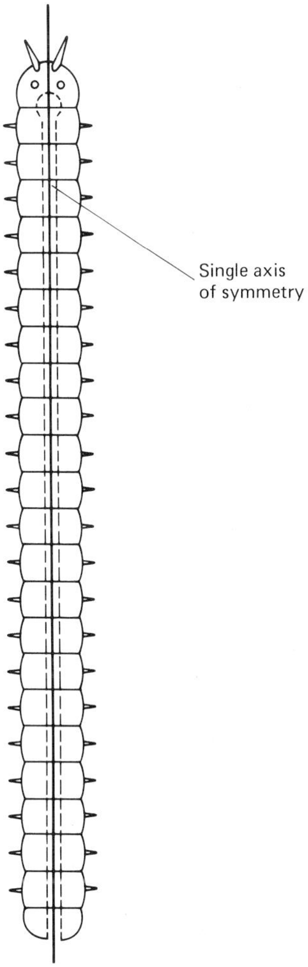

Figure 1.4 Bilateral symmetry. A single plane running dorso-ventrally along the longitudinal axis divides the animal into equal right and left halves.

active animals, as it organizes all locomotor machinery to produce movement in a single, forward direction.

Nutrition and Digestion

Animals obtain their energy by consuming other living things or once-living things. Diets and methods of food acquisition are highly variable among the invertebrates, and nutritional factors influence the entire biology of each species. Foodstuffs are of animal, plant, or bacterial origin. They may be ingested in solid or liquid form, swallowed whole or reduced first by enzymes and/or lacerative structures. Methods of food acquisition include predation, parasitic feeding, scavenging, detritus feeding, and filter feeding. Rapacious predators actively seize living prey, which they kill and eat. Usually these animals are relatively large, fast, and/or strong, at least with respect to their prey. Parasites are not unlike predators, but typically, they are too small to kill their food. Instead, parasites feed without causing the death of their prey, then called the host. Scavengers are animals that eat dead and decaying organisms. Detritus feeding refers to the consumption of nonliving organic particles in association with microorganisms. Detritus is found in terrestrial soils, beneath freshwater bodies, and on the ocean floor. Finally, filter feeding is a means by which many relatively passive invertebrates obtain food. These always aquatic, often sessile animals strain planktonic food and/or detritus from surrounding waters. Tentacles, mucous glands, and ciliary tracts are common structures of filter feeders.

Detritus is particularly abundant on the ocean floor, where it supports a diverse invertebrate fauna. Filter feeding is limited to aquatic and, particularly, to marine forms; only in these environments does the abundance of food permit such a passive food acquisition technique.

Digestion converts foodstuffs into forms usable by the feeding individual. This process normally includes the breakdown of food by enzymes into molecules that can be absorbed and incorporated into the feeder. Most animals have an organ system for digestion; this system usually includes food-gathering structures (ciliary nets, teeth, claws, etc.), an oral opening, a muscular swallowing area (the pharynx), an enzyme-rich area (the stomach and intestine), an absorption area (the intestine), and an anus. Most, but by no means all, invertebrates possess these structures.

Respiration

Gas exchange is central to the release of organic energy in most animals. Some invertebrates can obtain energy from their foodstuffs without oxygen, but these animals constitute a minority. Anaerobic metabolism is always less efficient than its aerobic counterpart and leaves organic acids as cumbersome waste products. Aerobic metabolism has advantages: not only does it release energy more efficiently, but its waste product, carbon dioxide, is more disposable. Accordingly, the respiration of most animals involves the uptake of oxygen and the removal of carbon dioxide.

Oxygen always is absorbed in a dissolved state. Aquatic organisms primarily contact oxygen as a dissolved gas, and thus have little trouble with its form. Difficulties may occur, however, concerning its availability. Water seldom contains even 1% dissolved oxygen, and increased temperature and salinity rapidly reduce its gas-holding capacities (see Fig. 1.2). Fresh water is particularly vulnerable in this respect. In freezing weather, surface ice may isolate a freshwater body from atmospheric oxygen. Moreover, the limited size of freshwater systems allows them to heat rapidly, thus diminishing their oxygen supply.

Terrestrial invertebrates have a different oxygen problem. The oxygen supply is plentiful on land, as air consistently contains 21%, but there the difficulty lies in absorbing the gas in dissolved form. A moist respiratory surface is required, but such a surface is threatened continuously by desiccation. If its respiratory surface is not secluded in a cavity or tube with limited outside exposure, an animal is forced to minimize its direct contact with the air. Among invertebrates, only insects, certain other arthropods, and some terrestrial snails possess a well-protected respiratory organ. Other terrestrial invertebrates are confined to moist soils or other shaded, humid regions, where the threat of desiccation is minimal.

Circulation

Respiratory gases, nutrients, and other biological materials must be distributed properly throughout an organism. The extent to which special circulatory systems are involved in this distribution is largely a function of animal size. Diffusion distributes biological materials at a rather slow but steady rate, and in many very small invertebrates this process provides adequately for all circulatory needs. However, most larger animals require a system of tubes through which a circulatory medium flows. Biological materials are shuttled through these circulatory tubes and thus are distributed with a greater efficiency than could be effected by diffusion alone. Gas transport is the most critical function of the circulatory system. Frequently, special respiratory pigments (such as hemoglobin) bind to oxygen and expedite its transport by the circulatory medium.

Osmoregulation

The control of water is another important biological process. Life is waged best when protoplasm and tissue fluids maintain an optimum osmotic constitution. We emphasized that protoplasm and tissue fluids are somewhat similar to sea water. Accordingly, marine invertebrates experience comparatively few osmoregulatory problems. Freshwater organisms, on the other hand, are threatened constantly by the influx of water from their hyposmotic environment. Animals of inland waters usually possess some osmoregulatory system for expelling water from their bodies. Typically, this system includes a network of tubules that gather excess water and transport it to a muscular pump for expulsion. Terrestrial organisms have the opposite problem: their osmoregulatory systems must retain water in their bodies. Most land animals also bear water-resistant coverings, such as the cuticles of insects and the shells of terrestrial snails.

Excretion

The excretory function is related to osmoregulation. Metabolism always produces unusable end products, notably the nitrogenous wastes of protein breakdown. These nitrogenous wastes can be eliminated in one of three major forms, the simplest but most toxic of which is ammonia. Urea, a compound formed from ammonia and carbon dioxide, requires energy for its production but is less toxic. Finally, uric acid is a complicated molecule of low toxicity. Uric acid is energetically expensive to produce, but offers the great advantage of being nearly insoluble in water; hence its excretion involves the loss of very little body fluid. The form in which nitrogenous wastes are excreted is largely a function of the osmoregulatory situation of the animal involved. Again, environment is the dominant factor. Aquatic organisms, particularly the smaller ones, eliminate ammonia. This highly soluble waste diffuses away rapidly, its toxicity quickly reduced by dilution in the surrounding waters. Larger aquatic animals and some terrestrial ones cannot tolerate ammonia; they tend to produce and excrete the less toxic urea. Meanwhile, most terrestrial invertebrates eliminate their nitrogenous wastes as insoluble uric acid. Water is simply too precious on land to be used for diluting ammoniacal wastes.

ACTIVITY SYSTEMS

Activity systems comprise those structures and functions by which an animal, as a whole organism, responds to its world. They include a means for environmental surveillance, information processing and coordinated, locomotor response. These activities are carried out by nerves, muscles, and sense organs.

Nerves

Living systems require considerable coordination. They must communicate information within their boundaries in order to integrate the activities of their various parts and so behave as whole organisms. The most common communication systems are composed of nerve cells, elongated cells along whose processes electrical impulses are propagated. Biological information is encoded in these impulses and thus messages are communicated throughout the nervous system. This electrical phenomenon is the crux of animal coordination. Only two basic messages are conveyed by the system: an impulse and an absence of impulse. Yet, because of the potential complexity of temporal and spatial impulse patterns, these simple messages communicate elaborate information. Nerve impulses occur with different frequencies and travel along different combinations of nerve cells; thus an animal can code different types of information.

The layout of the nervous system is critical. In the most primitive invertebrates, the system consists merely of nerve cells arranged in a diffuse network, or nerve net. Adequate stimulation of the net at any one point results in the propagation of impulses in all directions. Nerve cells are separated by junctions called synapses, across which information is sent in the form of chemical transmitters. Although in most nerve nets, synapses permit impulse transmission in both directions, synaptic systems also provide the possibility of unidirectional conduction. Impulse transmission can be limited to a single direction by restricting the chemical transmitters to one side of the nerve cell junction. By such an arrangement, impulses arriving from only one direction can be communicated across the synapse. Synapses may be inhibitory (blocking impulse transmission) as well as excitatory—both types are important for the execution of coordinated responses.

Animal evolution has witnessed a decided trend toward the formation of central nervous systems. Within a central nervous system, large numbers of concentrated nerve cells evaluate the information flowing through the organism. A well-developed central system increases an animal's capacity to cope with complex informational patterns and, likewise, its ability to organize complex responses. The development of the central nervous system is one of the most important and fascinating processes in invertebrate evolution.

The communication of electrical impulses throughout the body via the nervous system is vital, but equally important is the processing of this information by the nerves. Basically, their job is to instruct the muscles.

Muscles

Muscles are the primary effectors for most invertebrates. Muscular movement depends on the nerve-stimulated activity of actin and myosin filaments. Arranged in parallel rows, these filaments produce muscular contractions by sliding past one another in response to nervous input.

Two fundamental properties of muscles greatly influence animal design. First, muscles always exert force on a skeletal system to which they are anchored; and second, muscles do not elongate actively.

Invertebrates definitively lack the bony endoskeletons to which vertebrates attach their muscles. Among the invertebrates, shells and exoskeletons are the principal hard parts and serve as the sites of muscular anchorage. Many invertebrates, however, have absolutely no hard parts. Their muscles may pull against tough connective tissues or thin cuticles, but with such pliant skeletal elements, muscular efficiency suffers. Accordingly, we might expect invertebrates without hard parts to be small and relatively slow creatures.

There is, however, a rather efficient type of invertebrate skeleton that employs no hard parts. This **hydrostatic skeleton** features a contained body of fluid. Hydrostatic skeletons exploit the fact that water easily adapts to various container shapes, but cannot be compressed in volume. In invertebrates with hydrostatic skeletons, fluid is contained within a closed, usually tube-shaped cavity surrounded by two or more layers of muscles (Fig. 1.5) arranged at right angles to one another, commonly in longitudinal and circular patterns. Contraction of the circular muscles causes the contained fluid to exert pressure against the end walls of the cavity, elongating the cavity and stretching the longitudinal muscles in the process. Longitudinal contractions have the opposite effect and result in a shorter, wider cavity and stretched circular muscles. By alternately contracting these muscle layers, invertebrates with hydrostatic skeletons manage general body movements, including creeping, crawling, and swimming.

Muscles cannot increase in length by their own active processes; rather they must be stretched by other,

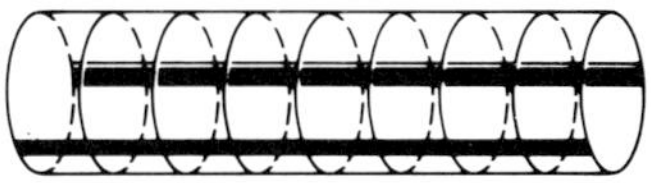

Figure 1.5 A hydrostatic skeleton features an enclosed volume of fluid with surrounding circular and longitudinal muscle antagonists. Contraction of circular muscles produces a thinning and lengthening of the body while contraction of the antagonistic longitudinal muscles produces a thickening and shortening of the body. (From R.B. Clark, In *The Lower Metazoa*, E.C. Dougherty, Ed., 1963, University of California.)

so-called antagonistic muscles. Thus, for both hard and hydrostatic skeletons, at least two sets of muscles are usually required. In the arthropod, flexor and extensor muscles extending across an appendage joint are antagonistic to one another (Fig. 1.6). In invertebrates with hydrostatic skeletons, circular and longitudinal muscle layers likewise antagonize each other (Fig. 1.5). Muscles form the bulk of many animals, and their antagonistic relationship, as well as the general association between muscles and skeletal systems, is a strong determining factor in invertebrate design.

Muscles respond to information transmitted by nerves. But how does the nervous system gain the information needed to instruct the muscles? To answer this question, we describe the sense organs.

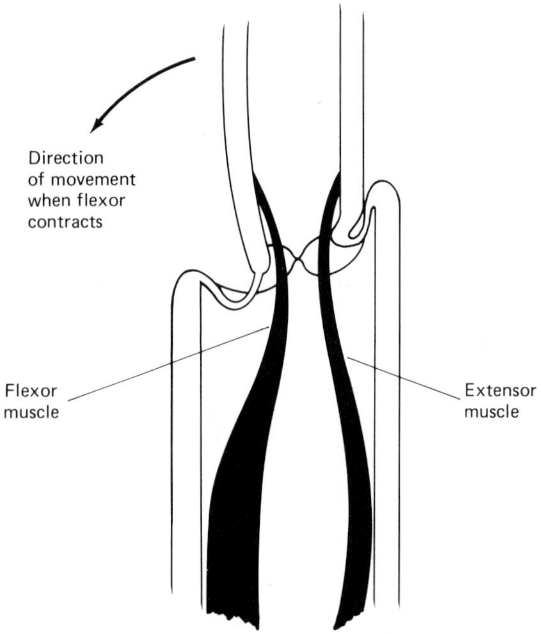

Figure 1.6 Diagram of an arthropod limb joint sectioned at right angles to the axis of articulation showing the antagonistic flexor and extensor muscles. (From W.D. Russell-Hunter, *A Biology of Higher Invertebrates*, 1969, Macmillan.)

Sense Organs

Animals live in a world of solid objects, chemicals, light, sound, and gravity. To respond to such a world, they must determine the quality, quantity, and distribution of its elements. To that end, animals are equipped with sense organs—structures absorbing certain types of energy from the environment and transducing that energy into nerve activity.

Sense organs vary widely in form and complexity. Single cells with tactile sensitive hairs are among the simplest and most common sensory structures, while the image-forming eyes of advanced invertebrates rank among the supreme accomplishments of organic evolution. Chemosensitive cells, cells sensitive to sound waves, and equilibrium centers called statocysts, are among the many sensory devices of the invertebrates.

Sensory equipment reflects an animal's environment and life style. For example, established adult parasites find all their biological needs at close hand; consequently, their sense organs are poorly developed. On the other hand, active invertebrates, particularly the nektonic and terrestrial forms, must search for their food, shelter, and mates. Accordingly, these animals are well endowed with sensory devices. Bilaterally symmetrical animals typically move with one end constantly forward, and this forward end often bears a collection of sense organs. Such a concentration of sense organs and associated information-processing nervous tissue in bilaterally symmetrical animals is called **cephalization** and represents one of the major trends in invertebrate evolution.

Behavior

The product of nervous, muscular, and sensory activities is behavior. Using the information received by sense organs and interpreted and transmitted by nerves, antagonistic muscles act against skeletal systems to produce movement—and movement is central to behavior. From the sea anemone feeding with its tentacles to the male octopus maneuvering for copulation, all behavior involves motion. Behavior is an animal changing the spatial relationship between itself and its world. That change results in its feeding, reproducing, or eluding capture or in any of the countless acts necessary for survival.

CONTINUITY SYSTEMS

Continuity systems include the structures and functions by which animals ensure the perpetuation of their species, among which are sexual and asexual reproduction. A variety of life cycles is displayed, often featuring one or more developmental forms.

Sexual Reproduction

Sexual reproduction is the most widespread means of species perpetuation. Such reproduction typically involves the union of a sperm produced by a male reproductive system with an egg from a female system. The methods by which gametes are united include some of the most interesting events in the invertebrate world.

Sperm-producing systems and egg-producing systems usually are located in separate male and female individuals. However, both systems can occur in a single animal—a condition known as **hermaphroditism.** Hermaphroditic invertebrates are common in environments with sparse populations. In such places, meetings between two members of the same species occur infrequently; hermaphroditism ensures the capacity for sexual union between any pair of individuals. For the same reason, sessile invertebrates are often hermaphroditic. Contact between the gametes of these organisms is a statistically risky affair, as it is difficult for sessile animals to maneuver into sexually facilitating positions. Their dual production, usually in enormous numbers, of sperm and eggs ensures an adequate fertilization percentage.

Although possible, self-fertilization by hermaphrodites is rare. The evolutionary advantage of cross-fertilization in producing new genetic combinations normally outweighs the relative convenience of self-fertilization. Many hermaphroditic forms discourage self-fertilization by releasing sperm well in advance of egg maturation. Such staggered gamete production is called **protandry.** Also, some hermaphrodites require a copulating partner for sperm ejaculation.

Development

During the period between zygote formation and adulthood, the environment quite pronouncedly influences the pattern of development. In the supportive environment of the sea, immature stages often lead free-swimming lives, completely unsheltered by their parents. Marine larvae are abundant indeed and contribute to the diversity of planktonic life. The austerity of fresh water prevents such independence by young invertebrates. Limited food supplies, overpowering currents, and osmotic and thermal threats severely restrict the number of larval forms in the freshwater environment. Accordingly, freshwater invertebrates often are brooded by their parents and do not assume an independent existence until they have developed adult characters. Because freshwater environments are so variable, many of the invertebrates inhabiting them exhibit special resistant stages in their life cycles. At the onset of adverse conditions—a drying pond in summer or a winter freeze—such invertebrates assume resistant forms, which persist until favorable conditions return.

The terrestrial environment is so demanding that most land animals are born well developed. Alternatively, adults may return to aquatic habitats to reproduce. Notable exceptions, however, include the insects. Insects have evolved a highly specialized metamorphic life cycle in which larval forms themselves are well adapted to land. Finally, the parasitic environment supports some very complex life cycles, including larval forms that locate a host for the succeeding adult generation.

Asexual Reproduction

Asexual reproduction remains an important element in the continuity of the invertebrate world. Many organisms that normally reproduce by sexual means retain some capacity for asexuality. Asexual reproduction is most common among primitive forms, particularly the Protozoa. It is also a common means for rapidly increasing a population under optimum environmental conditions. Many invertebrates in temperate-zone fresh waters reproduce asexually during the comfortable summer months. A large population is established, which then turns to sexual reproduction during the fall and winter. Asexuality is common also among colonial invertebrates and is responsible for the large size of marine colonies such as the corals. Sessile forms may rely on asexual reproduction when isolation prevents a sexual union.

Asexual reproduction usually involves one of three methods—binary fission, budding, or fragmentation. In binary fission, the "mother" organism divides into two "daughter" individuals of equal size. In budding, the mother organism produces a single, and much smaller, daughter. Finally, fragmentation refers to the simultaneous division of the mother animal into many daughter individuals.

Regeneration, a curious phenomenon related to asexual reproduction, refers to the ability of an organism to replace major body parts by the growth and differentiation of healing tissues. This process is evidenced most strikingly in invertebrates that regenerate an entire organism from a relatively small piece of an old one. Regeneration and asexual reproduction are twin capacities, consistently occurring in the same animal groups.

The Diversity of the Invertebrate World

With the exception of the basic qualities shared by all living things, invertebrates fail to demonstrate a single trait in common. Rather, they are characterized by what they are not. They are *not* vertebrates; thus these animals, which dominate the biological world, are clumped into one vast and heterogeneous assemblage. From protozoans to sea stars, from squid to insects, invertebrates represent the essential diversity of animal life itself. Here we only begin to describe the invertebrates, mentioning their basic types and structural, functional, and behavioral diversities.

All living things are classified into groups according to biological similarities. Each principal group represents a unique approach to life, a basic plan which, modified by the various organisms in the group, solves their biological problems of maintenance, activity, and continuity. Each of these groups is called a **phylum.** Opinions vary as to the number of invertebrate phyla, and our account of 32 separate phyla admittedly is subject to debate. All phyla are subdivided into classes, the classes into orders, and the orders into families. Intermediate taxonomic levels, such as subphyla, subclass, and subfamily, are common in certain taxonomic groups. Families are composed of genera; a genus consists of its member **species.** Species represent populations of similar organisms naturally capable of mating and producing fertile offspring. Of the six taxonomic categories, only the phylum and the species purport to be biologically defined units. The other four represent arbitrary assemblages of organisms on a scale of pyramiding similarities.

Considerable use of the terms "primitive," "lower," "advanced," and "higher" is inescapable in describing the invertebrate world. These are convenient but unfortunate words, which can mislead easily. If we assume that life arose during a fixed geological period, it is true that all organisms today share an equally long evolutionary history. None stopped evolving, thus remaining at primitive levels, while others continued evolution and so attained advanced status. Rather, animal evolution witnessed the retention of ancestral characters by some organisms and the development of new characters by others. The terms lower and primitive are reserved simply for those organisms most reminiscent of the ancestral type; advanced and higher species represent more divergent and specialized forms.

Related to this distinction is the concept of biological success. Commonly we speak of insects and mollusks as very successful groups, while horseshoe crabs and brachiopods are described as relative failures. In this context, biological success refers to species abundance and ecological diversity. Yet, in a very real sense, *all* living groups are profoundly successful, for they have survived the rigors of natural selection for millions of years. Members of smaller and/or more primitive groups may occupy only a few of the many possible ecological niches, but their very existence testifies to their success.

Animal evolution is a great puzzle. Behind the organisms and phyletic organizations of the modern animal kingdom lie millions of years of organic evolution—mutations and natural selection, gene flow, migrations, and speciation. Today's animals represent only a dynamic living moment in the evolutionary history of their world. Our clues to the rest of that history are few and difficult to analyze. Unlike the vertebrates, most invertebrate phyla left very few fossils; thus we are unable to study their ancient forms in stone. Comparative anatomy, physiology, biochemistry, and embryology are our best tools in the attempt to reconstruct the invertebrate past. By comparing the biologies of contemporary, related species, we can conjecture the form of their common ancestor. Embryology may furnish the most direct clues. The old adage, "ontogeny recapitulates phylogeny," no longer is accepted literally, but the activities of embryos are often related to the evolutionary background of their species and, if properly interpreted, can advance our understanding of their ancestry.

Barring some unforeseeable breakthrough in our information-gathering techniques, the entire story of invertebrate evolution will never be told. That story is one of the most fascinating philosophical puzzles for the twentieth century biologist, and probably will remain so for the biologists of centuries to come.

The Study of the Invertebrate World

Invertebrates are fundamentally involved in all aspects of our planet's ecology; essentially, the study of invertebrates is the study of life on earth. From the insects that pollinate our crops to the lobster on the dinner table, invertebrates are a vital part of the human experience. On a more profound level, every invertebrate participates in the web of life on our planet; they, more than we, built this web and its prosperity depends foremost on invertebrate, not vertebrate, health.

Career biologists would never dispute the pervading influence of invertebrates. Much of our knowledge of basic biological processes was attained by experimenting with these organisms, and invertebrates continue to play a most active role in many fields of biological research. Crustaceans and mollusks, for example, are instrumental in neurophysiological studies. Experiments on the giant axons of the latter have contributed greatly to our basic understanding of nervous conduction. The fruit fly has served in countless genetic studies, while echinoderms have helped unlock many of the secrets of embryonic development.

Invertebrates merit study for their own sakes also, as marvelous expressions of the evolution of their world. To them, more than to humans, elephants, or eagles, the world belongs. And it is with a special sense of wonder and respect that we write their story.

for Further Reading

The invertebrate biology literature is enormous and ever expanding, far too large to be cited exhaustively in a single book. The references we include merely sample the thousands of existing books, monographs, review articles, and original research papers. Hopefully, they will facilitate access to literature on topics of interest. The best introduction to the literature on a particular group of animals or a specific physiological or ecological topic is likely in a recent monograph or review article. For this reason, we include as many monographs and review articles as possible in our lists for further reading. Beyond this, *Biological Abstracts* and *Zoological Record* are two general bibliographic sources for biologists. Both provide listings of articles, papers, monographs, and books on biological topics. The former provides brief summaries of its entries; the latter, however, is historically more complete. These two sources are updated regularly, and thus will provide ready access to the recent literature after the specific references cited in this book are out of date.

Each chapter in this book concludes with suggestions for further reading. Specific reference citations are annotated if the content is not clearly indicated by the title. A number of general works are listed at the conclusion of this chapter. To conserve space, these works, with few exceptions, are not cited again, although clearly they represent valuable sources for most of the subsequent chapters also. Hence students interested in a specific group of animals or in a particular topic should return to this list very early in their literature search. We restricted the reference citations in subsequent chapters to works dealing mainly with the subject matter of the chapter. Perusal of these lists reveals that much of the original research literature on invertebrates appears in various scientific journals. These journals are so numerous, and so few of them are restricted to invertebrates alone, that we did not attempt a journal listing. A few journals dealing exclusively with research on particular groups of animals are mentioned in the appropriate chapter bibliographies.

Single-Volume Textbooks on Invertebrate Biology

Alexander, R. M. *The Invertebrates.* Cambridge University Press, New York and London, 1979.

Barnes, R. D. *Invertebrate Zoology.* 4th ed., Saunders College Publishing, Philadelphia, 1980. (A comprehensive text)

Barrington, E. J. W. *Invertebrate Structure and Function.* 2nd ed., Halsted Press, New York, 1979. (A comparative treatment of invertebrate animals, organ system; an excellent follow-up to a phylum-by-phylum survey course)

Buchsbaum, R. *Animals without Backbones.* 2nd ed., University of Chicago Press, Chicago, 1976. (An elementary text, well illustrated)

Gardiner, M. *The Biology of Invertebrates.* McGraw-Hill, New York, 1972. (A comparative treatment of invertebrate animals, organ system by organ system; and excellent follow-up to a phylum-by-phylum survey course)

Hickman, C. P. *Biology of the Invertebrates.* 2nd ed., C. V. Mosby, St. Louis, 1973.

Laverack, M. S., and Dando, J. *Essential Invertebrate Zoology.* 2nd ed., Halsted Press, New York, 1979.

Marshall, A. J., and Williams, W. D. *Textbook of Zoology: Invertebrates.* American Elsevier, New York, 1972. (A text emphasizing the classical taxonomic and morphological approaches; examples drawn mainly from the Australian and Indo-Pacific faunas)

Meglitsch, P. A. *Invertebrate Zoology.* 2nd ed., Oxford University Press, New York, 1972.

Multivolume Treatises on Invertebrate Groups

Bronn, H. G. (Ed.) *Klassen und Ordnungen des Tierreichs.* C. F. Winter, Leipzig and Heidelberg, 1866– (Many volumes; series not completed. Perhaps the most detailed classical zoological treatment of the animal kingdom)

Grassé, P. P. (Ed.) *Traité de Zoologie.* Masson et Cie, Paris, 1948. (Covers the entire animal kingdom; series nearly complete. The French counterpart to *Klassen und Ordnungen des Tierreichs*)

Hyman, L. H. *The Invertebrates.* McGraw-Hill, New York, 1940– (Six volumes completed. The most recent advanced treatment in English. Individual volumes are cited in subsequent chapters.)

Kaestner, A. *Invertebrate Zoology.* Interscience, New York, 1967–1970. (An English translation of a well-known three-volume German work on invertebrates; lophophorates and echinoderms not included)

Moore, R. C. (Ed.) *Treatise on Invertebrate Paleontology.* Geological Society of America and University of Kansas Press, Lawrence, 1952–(Many volumes; series incomplete. A detailed treatment of fossil invertebrates)

General Works on Invertebrate Morphology, Physiology, Ecology, or Behavior

Beklemishev, W. N. *Principles of Comparative Anatomy of Invertebrates.* University of Chicago Press, Chicago, 1969. (English translation of the classic Russian morphological text; two volumes)

Brusca, G. J. *General Patterns of Invertebrate Development.* Mad River Press, Eureka, CA, 1975.

Bullock, J. H., and Horridge, G. A. *Structure and Function of the Nervous System of Invertebrates.* W. H. Freeman, San Francisco, 1965. (Two volumes)

Carthy, J. D., and Newell, G. E. (Eds.) *Invertebrate Receptors.* Academic Press, New York, 1968.

Corning, W. C., Dyal, J. A., and Willows, A. O. D. (Eds.) *Invertebrate Learning, Vols. I and II.* Plenum Press, New York, 1973.

Eltringham, S. K. *Life in Mud and Sand.* Crane, Russak, New York, 1971. (The ecology of mud and sand habitats)

Florkin, N., and Scheer, B. T. (Eds.) *Chemical Zoology, Vols. I-VIII.* Academic Press, New York and London, 1967–1974.

Fretter, V., and Graham, A. *A Functional Anatomy of Invertebrates.* Academic Press, London, 1976.

Giese, A. C., and Pearse, J. S. (Eds.) *Reproduction of Marine Invertebrates. Vols. I-V,* Academic Press, New York, 1974–1978.

Hardy, A. C. *The Open Sea.* Houghton Mifflin, Boston, 1956. (A delightful introduction to the study of plankton)

Highnam, K. C., and Hill, L. *The Comparative Endocrinology of the Invertebrates.* 2nd ed., University Park Press, Baltimore, 1977.

Jennings, H. S. *Behavior of the Lower Organisms.* Indiana University Press, Bloomington, 1976.

Kume, M., and Dan, K. *Invertebrate Embryology.* Clearing House for Federal Scientific and Technical Information, Springfield, Va., 1968.

MacGinitie, G. E., and MacGinitie, N. *Natural History of Marine Animals.* 2nd ed., McGraw-Hill, New York, 1968.

Mill, P. J. *Respiration in the Invertebrates.* St. Martin's Press, London, 1972.

Newell, R. C. *Biology of Intertidal Animals.* 3rd ed., Marine Ecological Surveys Ltd., Faversham, Kent, U.K. 1979.

Nicol, J. A. C. *Biology of Marine Animals.* 2nd ed., Interscience, New York, 1969.

Prosser, C. L. (Ed.) *Comparative Animal Physiology.* 3rd ed., W. B. Saunders, Philadelphia, 1973.

Ricketts, E. F., Calvin, J., and Hedgepeth, J. W. *Between Pacific Tides.* 4th ed., Stanford University Press, Stanford, California, 1968.

Russell-Hunter, W. D. *A Life of Invertebrates.* MacMillan, New York, 1979. (Emphasizes functional morphology)

Salanki, J. *Neurobiology of Invertebrates.* Akademiai Riado, Budapest, 1973.

Stancyk, S. E. *Reproductive Ecology of Marine Invertebrates.* University of South Carolina Press, Columbia, 1979.

Stephenson, T. A., and Stephenson, A. *Life Between Tidemarks on Rocky Shores.* W. H. Freeman, San Francisco, 1972. (Systematic coverage of the rocky intertidal fauna throughout the world)

Vernberg, W. B., and Vernberg, F. J. *Environmental Physiology of Marine Animals.* Springer-Verlag, New York, 1972.

Wells, M. *Lower Animals.* McGraw-Hill, New York, 1968. (A brief, but delightful account of invertebrate activity systems)

Identification Guides

Brusca, R. C. *Common Intertidal Invertebrates of the Gulf of California.* 2nd ed., University of Arizona Press, Tuscon, 1980.

Edmondson, W. T., Ward, H. B., and Whipple, G. C. (Eds.) *Freshwater Biology.* 2nd ed., John Wiley and Sons, New York, 1959. (A guide to the identification of freshwater organisms)

Gosner, K. L. *Guide to the Identification of Marine and Estuarine Invertebrates.* Interscience, New York, 1971. (Keys to the identification of marine invertebrates from Cape Hatteras to New England)

Light, S. F., Smith, R. I., Abbott, D. P., and Weesner, F. M. *Intertidal Invertebrates of the Central California Coast.* (R. I. Smith and J. T. Carlton, Eds.), 3rd ed., University of California Press, Berkeley, 1975.

Kozloff, E. N. *Keys to the Marine Invertebrates of Puget Sound, the San Juan Archipelago and Adjacent Regions.* University of Washington Press, Seattle, 1974.

Pennak, R. W. *Fresh-Water Invertebrates of the United States.* 2nd ed., Wiley-Interscience, New York, 1978.

Smith, R. I. (Ed.) *Keys to Marine Invertebrates of the Woods Hole Region.* Contribution No. 11. Systematics-Ecology Program. Marine Biology Laboratories, Woods Hole, Mass., 1964.

Smith, D. L. *A Guide to Marine Coastal Plankton and Marine Invertebrate Larvae.* Kendall Hunt Publishing Co., Dubuque, Iowa, 1977.

Voss, G. L. *Seashore Life of Florida and the Caribbean. E. A. Seemann Publishing Co. Miami, 1976.*

Laboratory Manuals

Beck, D. E., and Braithwaite, L. F. *Invertebrate Zoology Laboratory Workbook.* 3rd ed., Burgess, Minneapolis, 1968.

Dales, R. P. (Ed.) *Practical Invertebrate Zoology.* University of Washington Press, Seattle, 1970.

Freeman, W. H., and Bracegirdle, B. *An Atlas of Invertebrate Structure.* Heinemann Educational Books, London, 1971.

Lincoln, R. J. and Sheals, J. G. *Invertebrate Animals: Collection and Preservation.* British Museum (Natural History), London, 1979.

Sherman, I. W., and Sherman, V. G. *The Invertebrates: Function and Form: A Laboratory Guide.* 2nd ed., Macmillan, New York, 1976.

2

The Protozoans

A most diverse group of organisms comprises the phylum Protozoa. These "first animals" of the invertebrate world are characterized by their unicellularity. While cells are the basic building units of all higher creatures, cells in protozoans represent whole organisms in themselves. Each protozoan cell fulfills all the maintenance, activity, and continuity responsibilities of a total living animal. Unicellularity does not diminish the complexity of protozoan forms; rather, the phylum contains some of the most highly evolved and sophisticated cells in the animal world.

Protozoan evolution features the continuing differentiation and specialization of subcellular organelles. Also important are changes in metabolic capabilities and environmental responsiveness. This extensive infracellular evolution has produced a variety of protozoan forms. Indeed, unicellular organisms vary so greatly that their phyletic unity is universally disputed. One of the more serious problems with a uniphyletic scheme is the inclusion among the Protozoa of both plantlike and animal-like forms. Several alternative classification schemes exist, including some dividing the protozoans into several distinct phyla. A currently popular scheme establishes three lower kingdoms distinct from both plant and animal groups. Of fundamental importance in this scheme is the distinction between **prokaryotic** and **eukaryotic** cells. The first living cells were probably prokaryotic; that is, they lacked membrane-enclosed organelles in the cytoplasm and, hence, true nuclei, mitochondria, and plastids. Modern prokaryotes, including the bacteria and the blue–green algae, make up the Monera kingdom. All other cells, from those of protozoans to those of higher plants and animals, are eukaryotic, with the familiar assortment of membrane-bound organelles. A second kingdom comprises the unicellular organisms. Called the Protista, this kingdom encompasses the protozoans and the various algal groups (except blue-green algae). A third kingdom is reserved for the fungi.

Multiple kingdoms help clarify the evolutionary relationships among lower organisms. However, solely for instructional purposes, we retain here the traditional concept of a single protozoan phylum, while expressing an awareness of its inherent difficulties and emphasizing the diversity of the individual groups.

In our classification, these groups include four subphyla. (For a revised classification of protozoans, see Levine, et al., in the list of readings at the end of the chapter.) The most primitive is the Sarcomastigophora, an extremely heterogeneous taxon whose adult members are characterized by pseudopodia and/or flagella. Mastigophorans are the predominantly flagellated forms, including many plantlike organisms. Sarcodines include the pseudopodous amoebas, foraminiferans, heliozoans, and radiolarians. The second and third subphyla represent two wholly parasitic groups historically known as sporozoans. These protozoans are separated now into subphylum Sporozoa and subphylum Cnidospora, both characterized by highly resistant, reproductive spores in their complex parasitic life cycles. The fourth subphylum is the Ciliophora, whose single class features the ciliate protozoans. Ciliates are distinguished by hairlike locomotor structures called cilia and by multiple nuclei of two distinct types. Protozoologists specializing in the study of these animals generally rank the Ciliophora as a phylum.

Protozoans are cosmopolitan in distribution, living wherever adequate moisture is available. They are most abundant in aquatic habitats, particularly in marine plankton, but terrestrial forms can be numerous in damp soils. The current census describes over 30,000 living species.

Protozoan Unity

The overwhelming fact of life for protozoans is their unicellularity. The external world lies just one plasma membrane away. To deal with that preeminently near environment, and moreover to conduct all of the essential life processes, protozoans rely only on the structures of a single cell. Their biology is a lesson in the marvelous potentiality of a single cell.

MAINTENANCE SYSTEMS

General Morphology

As we pointed out, protozoans can be compared with the basic units of multicellular organisms, consisting of a mass of cytoplasm and organelles bounded by a plasma membrane. Their organelles include the familiar nuclei, mitochondria, ribosomes, endoplasmic reticulum, and Golgi apparatus. A general account of cell biology is beyond the scope of this book; thus discussion of these organelles will be minimal. We concentrate here on the ways in which protozoan cells function as whole organisms.

Protozoans are naturally quite small animals, although they qualify as some of the largest known cells. Most species are microscopic, measuring less than 100 micrometers (μm) in greatest dimension. A few protozoans, from 1 to 3 millimeters (mm) in length or diameter, are visible to the naked eye; and one known fossil foraminiferan measured an incredible 19 centimeters (cm) across! All types of symmetry are seen in this phylum, from the perfectly spherical radiolarians (Fig. 2.1) to the rather amorphous amoebas (Fig. 2.2).

The protozoan surface is bounded by a typical plasma membrane, which besides defining the structural limits of the protozoan, represents the intermediary between the organism and its environment. All environmental influences on a protozoan either affect the plasma membrane directly or exert their effects after passing through it. In controlling the traffic across its surface, the plasma membrane demonstrates differential permeability. Some substances (e.g., water and ammonia) diffuse freely across the membrane, while others (e.g., certain salts and large molecules) are barred passage. Additionally, the membrane may

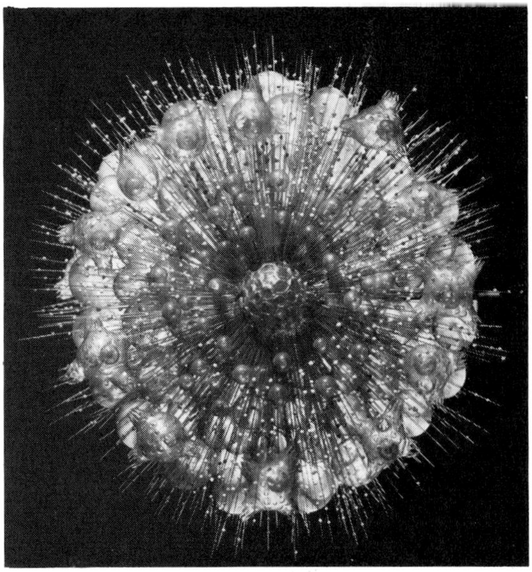

Figure 2.1 Photograph of a glass model of a radiolarian (*Trypanosphaera regina*). (Courtesy of the American Museum of Natural History.)

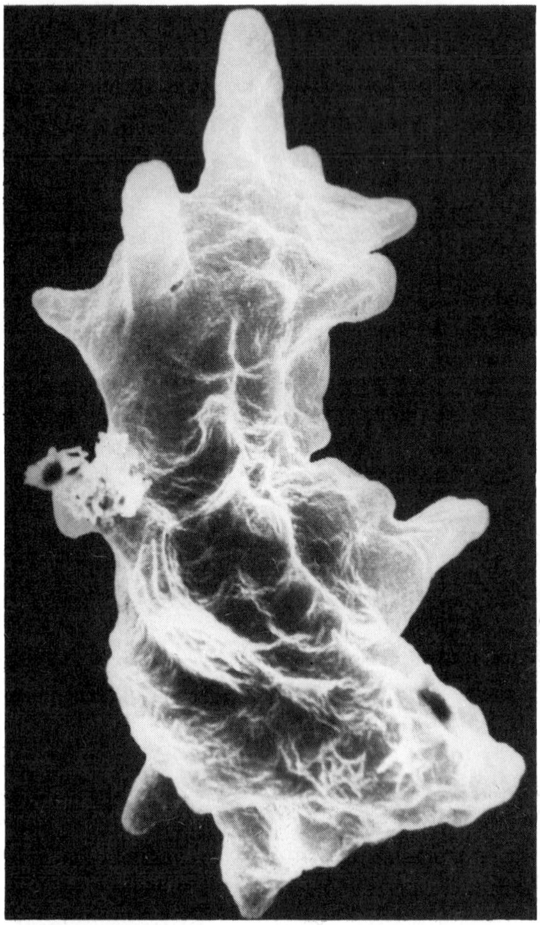

Figure 2.2 Scanning electron micrograph of *Amoeba proteus*. (Photo by Eugene B. Small and Donald S. Marzalek; courtesy of G.A. Antipa.)

transport nondiffusible materials into the cell by active processes.

The plasma membrane may be reinforced by a variety of structures. Such structures include external shells (commonly called **tests**) as well as elaborations of the plasma membrane itself, which form a heavy membranous sheath called a **pellicle.** Tests of calcium carbonate or silica (Fig. 2.3) may be secreted from within the cell; alternatively, the membrane's outer surface may be coated with sticky mucus to which foreign debris clings as a protective patchwork cover (Fig. 2.4). Pellicles, commonly associated with the ciliates, represent complex plasma membranes. These membranes occur in multiple layers and usually incorporate several organelles. Tests and pellicles are common among most freshwater and terrestrial protozoans, as they afford the increased protection needed by inhabitants of these environments. Such structures, however, seldom cover the entire surface of a protozoan because a feeding site must remain exposed.

The plasma membrane bounds the cytoplasm, the aqueous medium of all living cells. Cytoplasm exists in one of two colloidal phases. In the **sol phase,** various subcellular particles are in suspension, and the cytoplasm is relatively fluid and granular. In the **gel phase,** colloidal particles combine to form a clear gelatinous cytoplasm. Cytoplasm in the gel phase borders the plasma membrane and is called **ectoplasm.** Cytoplasm in the central area of the cell is commonly in the sol phase and is referred to as **endoplasm.** Protozoan cytoplasm can alternate between the sol and gel phases, and such transitions are implicated in some types of cellular movement.

Directly involved in the movements of many protozoans are fibrous structures known as **flagella** and **cilia.** Flagella and cilia are hairlike projections from the cell surface, effecting movement by undulating or stroking. These locomotor organelles are very common among protozoans and appear within higher animal phyla as well. They resemble each other structurally, and a cilium may be regarded as a rather specialized flagellum. The principal difference between the two organelles is that cilia tend to be shorter, more numerous, and better organized both structurally and functionally.

Figure 2.3 Scanning electron micrograph of the test of *Euglypha rotunda*. (From R.H. Hedley, Bull. Brit. Museum of Natural History (Zoology), 1973, 25:121.)

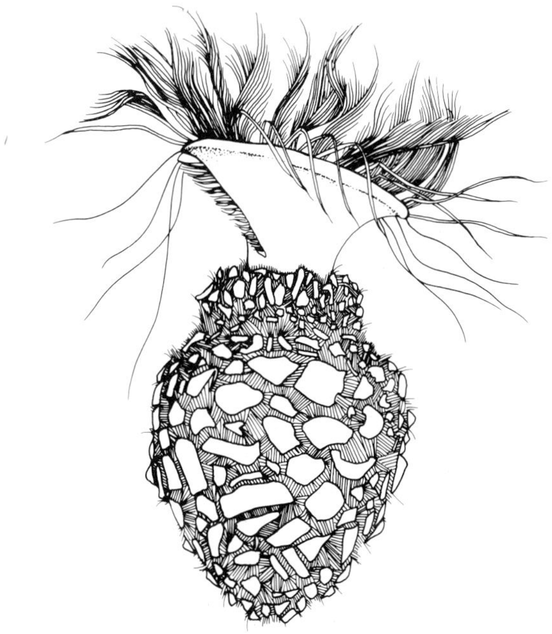

Figure 2.4 *Tintinnopsis ventricosa,* showing particulate matter clinging to the sticky outer surface forming a protective covering. (From J.O. Corliss, *The Ciliated Protozoa: Characterization, Classification, and Guide to the Literature,* 1979, Pergamon.)

Flagella and cilia arise from a **basal body** (or **kinetosome**) located in the ectoplasm adjacent to the plasma membrane. There is also evidence that basal bodies contain DNA and perhaps have some powers of self-replication. A highly ordered bundle of **microtubules** originates from the basal body and projects outward to form the central axis of the flagellum or cilium. Microtubules are tubes whose walls are composed of chains of protein molecules. Tubulin, a common protein within these chains, has an amino acid sequence similar to that of actin, a typical protein of true muscle fibers. The microtubular bundle, or **axoneme,** consists of two central tubules encircled by nine double tubules (Fig. 2.5). The axoneme is surrounded by a plasma membrane continuous with that of the entire organism.

Flagella and cilia move by sliding or contracting their microtubules. Such actions produce the undulations and/or oar strokes that propel so many protozoans and, in some cases, also serve ventilating and food-gathering purposes. Often flagella occur singly; although even in multiflagellate forms, these organelles act independently. Cilia, on the other hand, are well organized both structurally and functionally. Typically

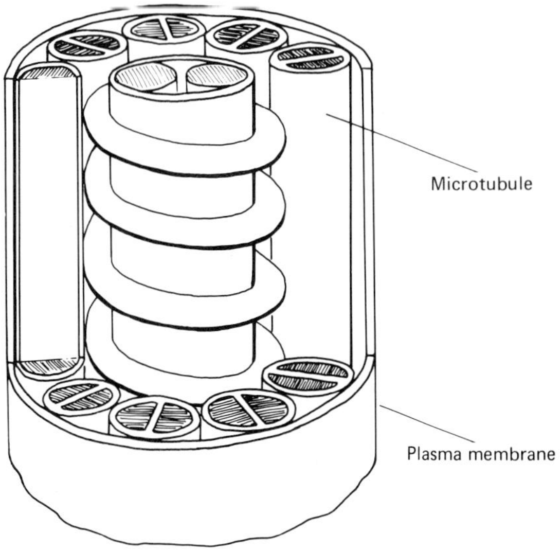

Figure 2.5 A diagrammatic reconstruction illustrating the ultrastructure of a cilium. (From N.J. Berrill, *Growth, Development and Pattern,* 1961, W.H. Freeman.)

arranged in longitudinal rows, the cilia are interconnected by a sophisticated network called the **infraciliature.** Each ciliary basal body issues a fine fibril, which joins a bundle of other fibrils coursing parallel to the body surface (Fig. 2.6). This bundle of fibrils is called a **kinetodesma;** the longitudinal unit of joined cilia, basal bodies, and kinetodesmata constitutes a **kinety,** the major unit of the infraciliature. The infraciliature is a structural feature of the pellicle and accounts for the complexity of this layer among ciliate protozoans.

Every protozoan has at least one nucleus, and many species have two or more of these organelles. Protozoan nuclei carry genetic information and direct protein synthesis much as the nuclei of higher organisms do. However, protozoan nuclei tend to diversify more in form and function. Sponge nuclei and squid nuclei are rather similar, except, of course, in their DNA content; indeed, all higher animals contain cells with nuclei of a basically standard type. But, among the protozoans, nuclei assume a variety of numbers and forms. They may be specialized for different roles. Ciliates, for example, have two nuclear types: a polyploid macronucleus for maintenance and a diploid micronucleus for continuity. Nuclear diversity is yet another example of the intensive subcellular evolution characteristic of protozoans.

Two other organelles allowing the protozoan to

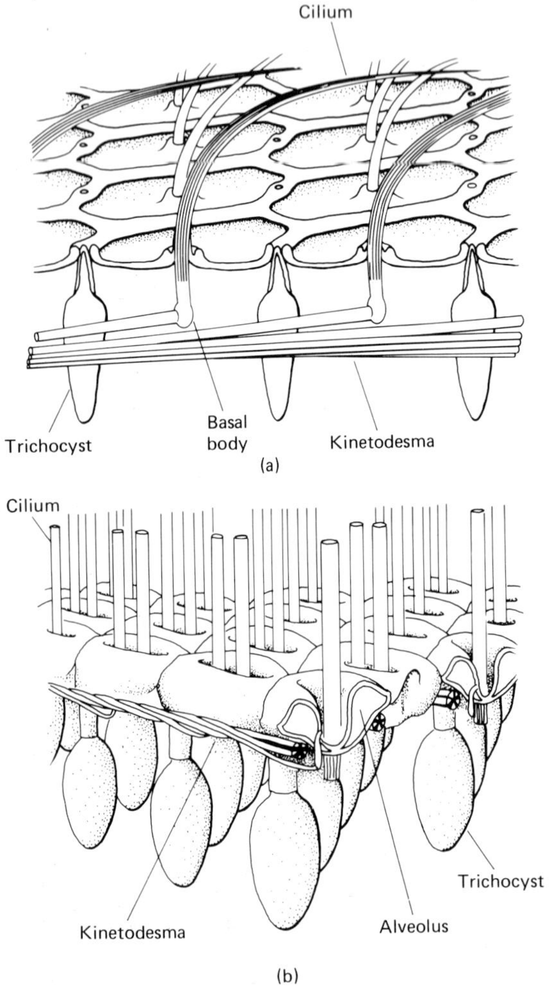

Figure 2.6 Two diagrams illustrating the ultrastructure of the infraciliature of ciliated protozoa. (*a*) Grell's interpretation of pellicular and subpellicular elements in *Paramecium.* Note the cilia, trichocysts, and the relationship between the basal bodies and the kinetodesma; (*b*) A composite drawing of the pellicular and subpellicular organelles of *Paramecium bursaria* as interpreted by Ehret and Powers. Note the twisted nature of the kinetodesma and the relationship of the pellicular subunits to each other and to the cilia, which are paired in this species. (From Corliss, *The Ciliated Protozoa: a* after K.G. Grell, *Protozoologie*, 1956, Pergamon; *b* after C.F. Ehret and E.L. Powers, Int. Rev. Cytol., 1959, *8*:97.)

function as a whole organism are the food vacuole and the contractile vacuole. These structures are specialized for digestive and osmoregulatory purposes, respectively, and are discussed subsequently in an account of these maintenance systems.

Nutrition and Digestion

All active cells in the invertebrate world have the same basic nutritional requirements: carbohydrates, proteins, fats, various salts and minerals, and a few vitamins. In higher animals, food acquisition and digestive systems gather foodstuffs from the environment and convert them into nutrient forms usable by individual cells. Protozoans, meanwhile, have just one cell for the acquisition, digestion, and utilization of food. One cell's ability to perform all of these functions is limited, yet protozoans manage very well. In doing so, they demonstrate some basic principles of animal nutrition.

The raw materials a protozoan (or any organism) must obtain from its environment depend not only on its biological needs but also on its synthetic abilities. Organisms capable of manufacturing carbohydrates from carbon dioxide and water do not need to eat other living things. Creatures that synthesize food from nonbiological sources are labeled **autotrophic.** Most autotrophic organisms utilize solar energy for their synthetic activities and, accordingly, are called **photosynthetic** or **photoautotrophic.** Photosynthesis is a definitive process of plants, but many flagellates traditionally claimed by zoologists for the animal kingdom engage in photosynthetic activities. A few flagellates obtain some of their energy by oxidizing nonorganic compounds. Such nutrition, autotrophic but not photosynthetic, is called **chemosynthesis.**

Photosynthesis and chemosynthesis are not widespread methods of nutrition among protozoans; indeed, few protozoans are strictly photoautotrophic and none is exclusively chemoautotrophic. Most unicellular animals acquire their food in organic form and then convert it to usable materials. This basically animal scheme is called **heterotrophic** nutrition. Heterotrophic methods include the capture and ingestion of other organisms in essentially solid form (**holozoic** nutrition) and the uptake of organic materials dissolved or suspended in a liquid (**saprozoic** nutrition). Even among autotrophic flagellates, a considerable capacity for heterotrophic nutrition exists. Many photoautotrophs switch to heterotrophy in the dark; other species depend on combinations of autotrophic and heterotrophic methods. The list of organic nutrients required by these organisms is not lengthy. Generally, such forms possess considerable synthetic abilities, much more extensive than those of ciliates and other strict heterotrophs. The line between autotrophy and heterotrophy is less defined than was once supposed, and apparently these primitive forms have not perfected the photosynthetic process to the extent seen in green plants. This conclusion supports recent hypotheses that the earliest living organisms were saprozoic heterotrophs rather than autotrophs.

Holozoic and saprozoic nutrition involves the formation of food vacuoles by **phagocytosis** or **pinocytosis**. For either of these processes to occur, some feeding part of the plasma membrane must be exposed to the environment. Upon stimulation from an appropriate source, the local membrane distorts to surround potential food. Full enclosure of the food produces a food vacuole which then detaches from the membrane boundary. This procedure is basic to both phagocytosis and pinocytosis. The only apparent difference between the two processes is that the former involves wider membrane and cell distortions than does the latter. Phagocytosis is associated with holozoic nutrition, as it allows the ingestion of larger food particles, while pinocytosis properly describes the formation of small food vacuoles (often called vesicles) in saprozoic organisms.

Often, food vacuoles are formed at a persistent **cytostome,** or mouth, of the protozoan cell. Cytostomes are particularly characteristic of species with well-developed pellicles. Usually the feeding site is located in a recessed surface area or vestibule where the protective covering can be interrupted safely. Cilia or flagella may be associated with the cytostome, where they establish food-gathering currents (Fig. 2.7).

After its formation, a food vacuole circulates within the endoplasm and its contents are digested. Digestion is aided by enzymes contributed from lysosomes, membrane-bound organelles that fuse with the food vacuole. Typically, digestion occurs in two phases. During the first phase, the contents of the food vacuole become increasingly acidic and the vacuole itself decreases in size. This phase is associated with the death and hydrolysis of prey tissues. The second digestive phase features a gradually enlarging and more alkaline food vacuole. Digested nutrients are absorbed from the vacuole either by diffusion or by a secondary pinocytosis. Undigestible wastes are egested from the cell when the shrunken vacuole merges with the plasma membrane. In protozoans with permanent tests or pellicles, egestion occurs at a fixed locus called the anal pore or **cytoproct.**

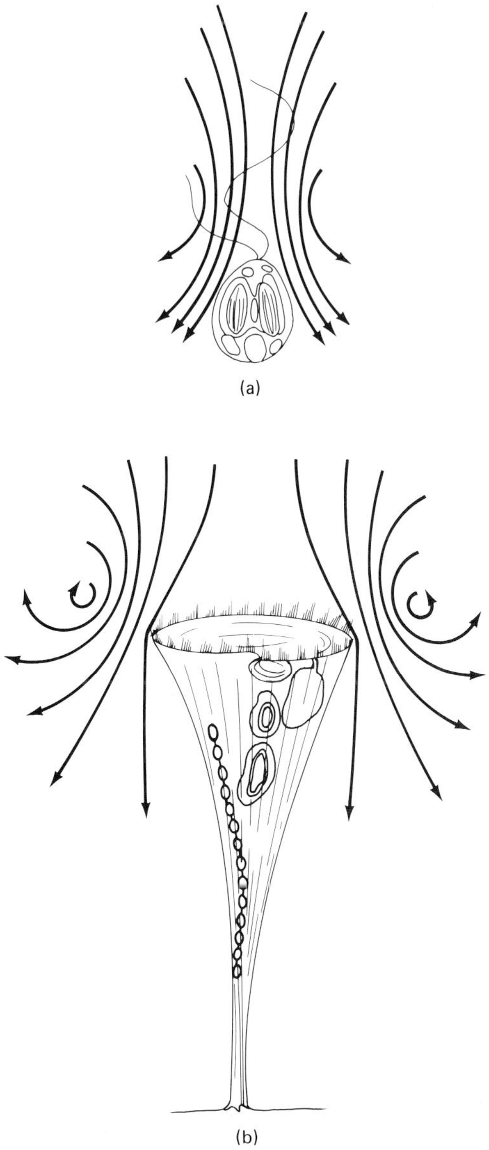

Figure 2.7 Water currents bringing in food are created by flagella in (*a*) the chrysomonad, *Ochromonas*, and by the coordinated activity of cilia in (*b*) the ciliate, *Stentor*. (From M.A. Sleigh, *The Biology of Protozoa*, 1973, Edward Arnold.)

Circulation and Respiration

No specialized organelles for circulation or respiration have been described in the Protozoa. Physical diffusion and simple **cyclosis,** or cytoplasmic streaming, sufficiently distribute materials inside the cell. Most protozoans are too small to have internal transport problems. Respiratory exchange and circulation could present difficulties for the very largest members of the phylum; however, these animals are invariably elongate and/or flattened in form and their extensive surface areas mollify any increased transport demands.

The majority of protozoans are aerobic, depending on oxygen as the final hydrogen acceptor in their energy-producing reactions. Anaerobic metabolism, however, is also common in the phylum. Many species apparently can tolerate low oxygen supplies, although they do not thrive under such conditions. Facultative anaerobiosis is common among freshwater protozoans, whose environment can be depleted of oxygen quite precipitously. The phylum's obligate anaerobes are mostly endoparasitic forms.

Excretion and Osmoregulation

Excretion, like circulation and respiration, is a biological problem directly related to animal size. The excretion of nitrogenous wastes presents no problem to single-celled animals; their small size allows simple diffusion to suffice. Ammonia is their dominant waste product.

Osmoregulation, however, is a maintenance function less related to size. Every animal cell, whether representing an entire protozoan organism or a small unit of a larger beast, must maintain a favorable water balance with its environment. The principal subcellular organelle associated with osmoregulation in protozoans is the **contractile vacuole.** Contractile vacuoles are bounded by a plasma membrane and commonly are located in the ectoplasm near the cell surface. They occur singly or in small numbers. In protozoans without hard coverings, these organelles are usually transitory and may form when several smaller vacuoles coalesce. In protozoans with tests or pellicles, the contractile vacuoles are associated with a permanent surrounding network of canals (Fig. 2.8a). In either case, the organelle enlarges as it slowly fills with excess water. It then collapses suddenly, expelling its contents to the outside (Fig. 2.8b). Cytoplasmic pressure probably causes the vacuolar collapse in some protozoans, but in a few ciliates contractile fibrils (similar to those present in flagella and cilia) have been implicated.

Contractile vacuoles are developed most extensively among freshwater species. These protozoans must do more osmotic work because of their hypotonic environment. Small size is an additional handicap to all protozoans, because their high surface-to-volume ratio causes a maximum vulnerability to osmotic pressure. Indeed, a contractile vacuole in a small, freshwater protozoan may contract every few seconds and expel a volume of water equal to the body volume in only a few minutes. By comparison, marine protozoans generally possess small contractile vacuoles that discharge only a few times per hour, and these organelles may be entirely missing from endoparasitic species.

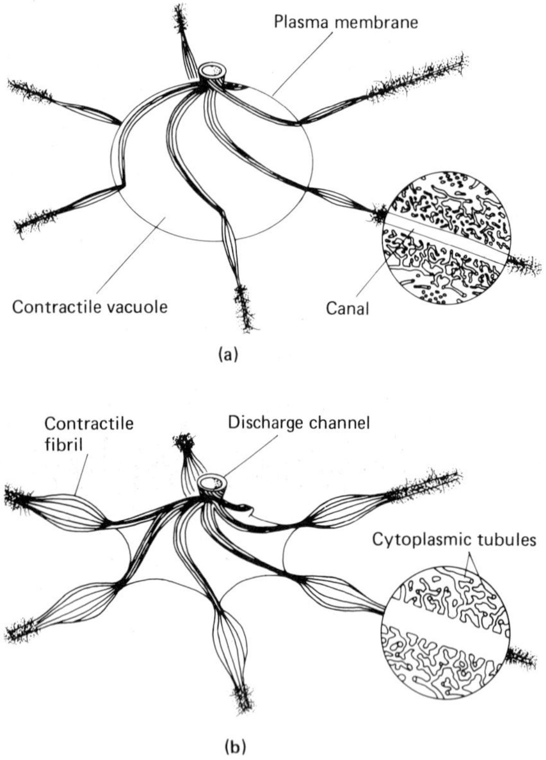

Figure 2.8 A three-dimensional representation of the contractile vacuole of *Paramecium aurelia* shown (*a*) as it fills and (*b*) just after its sudden collapse expelling its contents through the discharge channel to the outside. (From A. Jurand and G.G. Selman, *The Anatomy of Paramecium Aurelia*, 1969, Macmillan.)

ACTIVITY SYSTEMS

Within their own small worlds, protozoans can be highly active animals. Nonetheless, they are small creatures whose unicellularity precludes the existence of true nerves, muscles, or sense organs. Certain subcellular organelles, however, serve some of the same functions in protozoans as do the nerves, muscles, and sense organs in higher animals.

Nothing resembling an information-conducting system has been described in the protozoans. (A once-attractive suggestion that the infraciliature of the ciliates serves as a conducting system is without con-

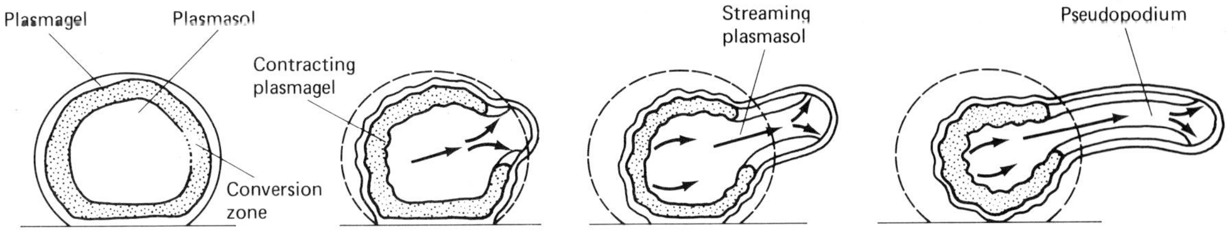

Figure 2.9 A diagram illustrating the "rear contraction theory" of pseudopod formation by an amoeba. Arrows indicate direction of flow of the plasmasol. (For further explanation see text.) (From T.L. John & E.C. Bovee, Physiological Reviews, 1969, *40*:830.)

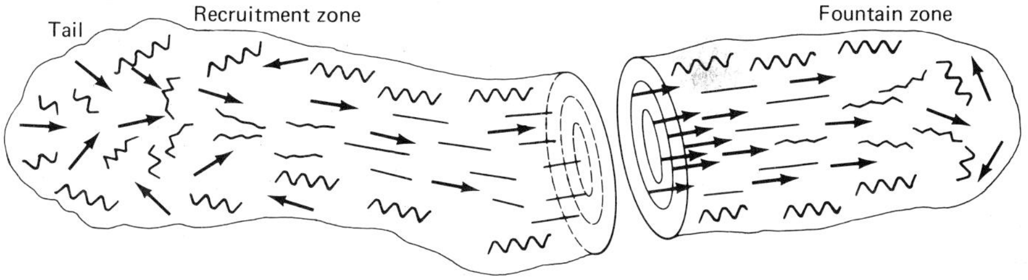

Figure 2.10 A diagram illustrating the "frontal contraction theory" of pseudopod formation. (For explanation see text.) (From R.D. Allen, Scientific American, 1962, *206* (2):112.)

vincing support.) With the exception of simple photoreceptors in certain phytoflagellates, distinct sensory organelles are also absent. Of course, true protozoans do not require the complex intercellular communication systems common to higher animals. Environmental stimuli simply impinge on exposed areas of the cell and effect general changes in cell metabolism. Typically, the cytoplasm is continuous throughout the body of a protozoan, and this living, circulating medium provides internal communication.

Research on protozoan movement focuses on the three main locomotor organelles in the phylum: pseudopodia, cilia, and flagella. These organelles form the basis of much of the phylum's taxonomy and are reviewed as we consider each of the major protozoan groups. At this point, however, we offer only a general account of their activities.

From the time amoebas were first discovered, their movements have fascinated and bewildered biologists. These protozoans apparently move by transporting cytoplasm into advancing extensions of their plasma membrane. These extensions of the cell are called **pseudopodia,** or "false feet." Numerous theories have been advanced to explain pseudopodial movement, and today at least two views merit wide support: the "rear contraction theory" and the "frontal contraction theory." Both theories recognize the reversible colloidal states of cytoplasm, but they differ concerning the source of the pressures that cause pseudopodial movement.

The rear contraction theory argues that the contraction of plasmagel causes plasmasol to stream across the cell and flow into a pseudopodium formed directly opposite the point of plasmagel contraction (Fig. 2.9). The rear contraction theory thus depicts plasmasol as an essentially passive substance, streaming only in response to external pressure. The frontal contraction theory describes a more active role for the plasmasol. This theory proposes that contractions occur in the advancing pseudopodial tip itself. According to this theory, plasmagel forms a tube along the entire length of the pseudopodium (Fig. 2.10). Plasmasol streams through this tube until it reaches the pseudopodial tip; there the fluid cytoplasm establishes a "fountain zone" through which it flows before streaming back along the edge of the gelatinous tube. The plasmasol then contracts as it is incorporated into the tube. Such contraction pulls more plasmasol down the center of the pseudopodium and through the fountain zone, thus continuously increasing pseudopodial length. The frontal contraction theory argues for a tensile plasmasol, one that pulls itself along rather than streaming passively according to external pressure.

Related to the frontal contraction theory are other

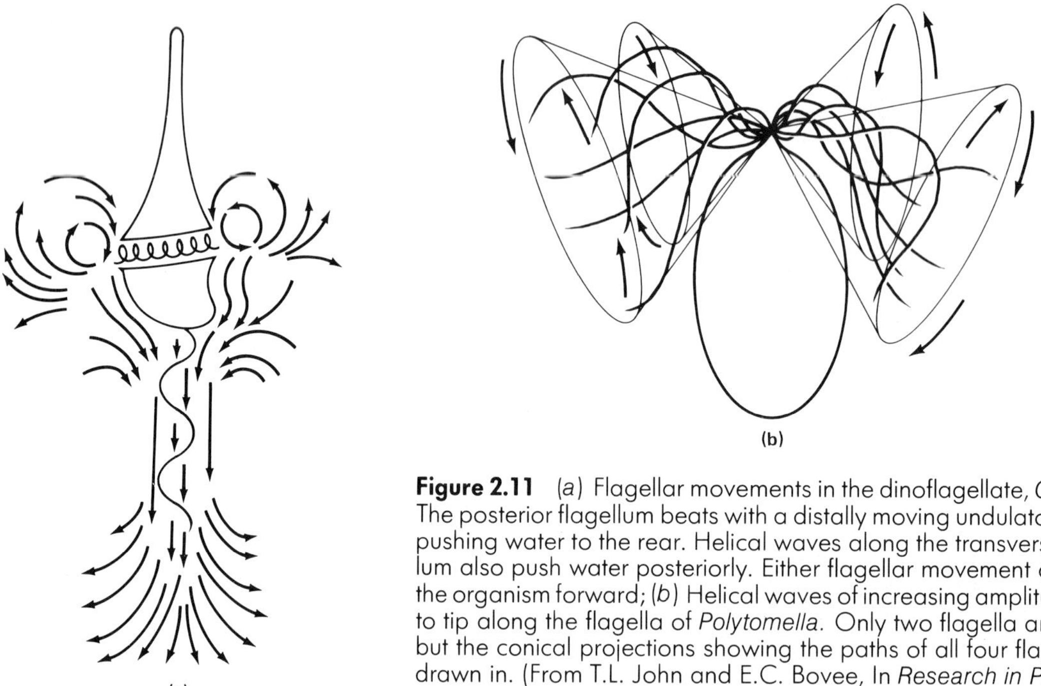

(a) (b)

Figure 2.11 (*a*) Flagellar movements in the dinoflagellate, *Ceratium.* The posterior flagellum beats with a distally moving undulatory wave, pushing water to the rear. Helical waves along the transverse flagellum also push water posteriorly. Either flagellar movement can drive the organism forward; (*b*) Helical waves of increasing amplitude base to tip along the flagella of *Polytomella.* Only two flagella are shown but the conical projections showing the paths of all four flagella are drawn in. (From T.L. John and E.C. Bovee, In *Research in Protozoology, Vol. 1,* T.T. Chen, Ed., 1967, Pergamon.)

models based on a tensile plasmasol. These theories note the existence of proteinaceous fibrils throughout the cytoplasm and suggest that the contraction or sliding of such structures causes amoeboid movement. Another family of theories has been generated from electrical studies demonstrating that polarizing stimuli affect pseudopodial activity. Obviously, amoeboid movement must be researched further. The final resolution of the question probably will recognize more than one model of amoeboid movement, a reflection of the considerable diversity of this heterogeneous phylum at all structural and functional levels.

The morphology of cilia and flagella has been described. These hairlike structures propel a majority of protozoans and are such effective organelles that they occur repeatedly throughout the higher animal phyla. (Indeed, human sperm cells represent a flagellated stage in our own life cycle.) On a functional level, it is relatively easy to distinguish between flagella and cilia. Flagella beat with a standing undulatory motion (Fig. 2.11a). In most cases, waves originate from the flagellar base and pass outward along the shaft; oppositely directed impulses are rare, but do occur. The entire undulation either is confined to a single plane or describes a helix (Fig. 2.11b). Flagellar action exerts forces on the surrounding medium in the direction of wave propagation. **Mastigonemes** (Fig. 2.12), tiny lateral projections along the shaft, increase the surface area of the organelle and thus improve its capacity to push against the environment. Mastigonemes influence the direction of water flow along the flagellum and thereby steer the organism.

In contrast to flagellar undulations, ciliary movements are of a stroking nature. Indeed, cilia often are compared to the oars of a rowboat. The shaft stiffens through a quick, effective stroke and then bends as a slower recovery stroke brings the organelle to its original position (Fig. 2.13). The recovery stroke typically is counterclockwise and not in the same plane as the effective stroke. Ciliary movement involves sliding microtubules within the axoneme. According to a widely accepted theory, peripheral tubules form linkages with each other. These tubules maintain a constant length, but because of their cross-linkages, any sliding causes the cilium to bend (Fig. 2.14). An organized program of active sliding throughout the length of the cilium produces the effective stroke. The recovery stroke occurs when reversed sliding is limited to a localized region which gradually ascends the shaft.

Cilia also are distinguished from flagella by their coordination. Flagella seldom occur in structural series and are quite independent functionally. Cilia, on the other hand, are organized both structurally and functionally, and their coordinated strokes produce faster

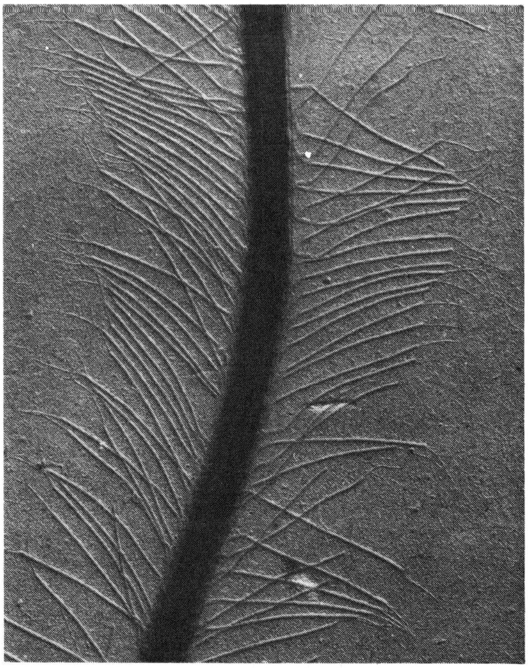

Figure 2.12 A micrograph showing mastigonemes along the flagellum of *Paraphysomonas cylicophora* (a chrysomonad). (From B.S.C. Leadbeater, Norweg. J. Bot., 1974, *19*:179.)

and more precisely controlled locomotor patterns. Effective strokes occur in sequence along each kinety (Fig. 2.15), and are directed in a slightly diagonal, posterior direction. Such an arrangement causes a ciliate to rotate as it moves along a spiral path (Fig. 2.16). Although the search for intracellular control systems continues, there is only limited evidence that the infraciliature or the cytoplasm plays a major role in ciliary coordination. More probable is the theory that cilia beat in organized patterns simply because they are stimulated simultaneously and/or because the cilia themselves stimulate each other mechanically, a so-called hydrodynamic linkage occurring between adjacent cilia.

Protozoans respond to a number of environmental stimuli and in some ways pursue an optimal relationship with their environment. Most of their activity is by trial and error. A protozoan responds to light, chemicals, temperature, and touch; it turns away from unpleasant encounters with these environmental stimuli and continues in any direction of favorable stimulation. In this manner a protozoan moving through an area of gradient qualities eventually locates where the pH, temperature, light, and chemical composition are most supportive.

Most protozoans respond negatively to touch. Upon contacting a solid object, a flagellate or ciliate reverses the beating of its locomotor organelles and swims away in an opposite direction. Amoebas, however, behave differently; typically, their pseudopodia cling to solid objects. Light, provided it is not too intense, is attractive to most protozoans. Photosynthetic forms orient to receive an optimal supply of solar energy. Chemical stimulation plays an important role in food acquisition and habitat selection. Also, endoparasitic protozoans probably rely heavily on chemical clues as they migrate through the host body in search of a specific domicile.

Amoebas seldom progress more than a few micrometers per second. Flagellates are roughly 10 times as fast. The swiftest protozoans are the ciliates, whose speed averages 100 times that of the typical amoeba.

Attempts to demonstrate classical conditioning and habituation in protozoans have been moderately successful. Habituation, the gradual "learning" not to respond to a repeated stimulus, is indicated by several experiments, although this and other quasilearning phenomena are viewed skeptically by many protozoologists. Unicellular organisms present special problems as subjects for behavioral studies because all of their bodily resources tend to be involved in each of their biological functions. The gentlest stimulation of a protozoan necessarily influences its entire body. It is thus difficult to resolve the sensory, integrative, and motor elements of protozoan activity and to isolate these functions from other biological systems. Because of their small size and our relatively gross instruments of measure, analysis of the finer points of protozoan responsiveness is, at this point, unattainable.

CONTINUITY SYSTEMS

Most cells and organisms reproduce their own kind. While cellular reproduction and organismal reproduction are related but separate tasks for multicellular animals, these two processes are identical among the protozoans. In no other systems do the protozoans display more diversity than in their means of reproduction. In activities ranging from the simple mitotic fission of amoebas to the complex nuclear exchanges of ciliate conjugation, this phylum demonstrates the potentiality of cellular reproduction.

Protozoan reproduction includes both asexual and sexual schemes. Asexual methods include fission and budding, two processes that never involve the mixing of nuclear material from two parent organisms. A single asexually reproducing protozoan divides into two or more daughter cells, each receiving an essen-

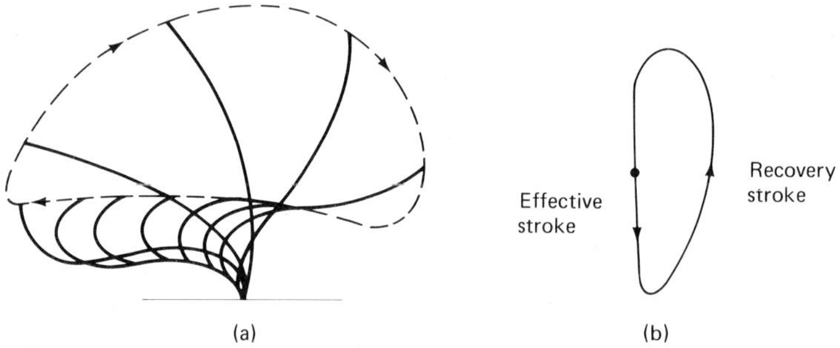

Figure 2.13 Ciliary movement. (*a*) A ciliary stroke as seen from the side. The cilium is straightened during the left to right effective stroke and bent during the recovery stroke; (*b*) The path followed by the tip of a cilium during a stroke cycle, showing the typical counter-clockwise loop. (From Sleigh, *The Biology of Protozoa*.)

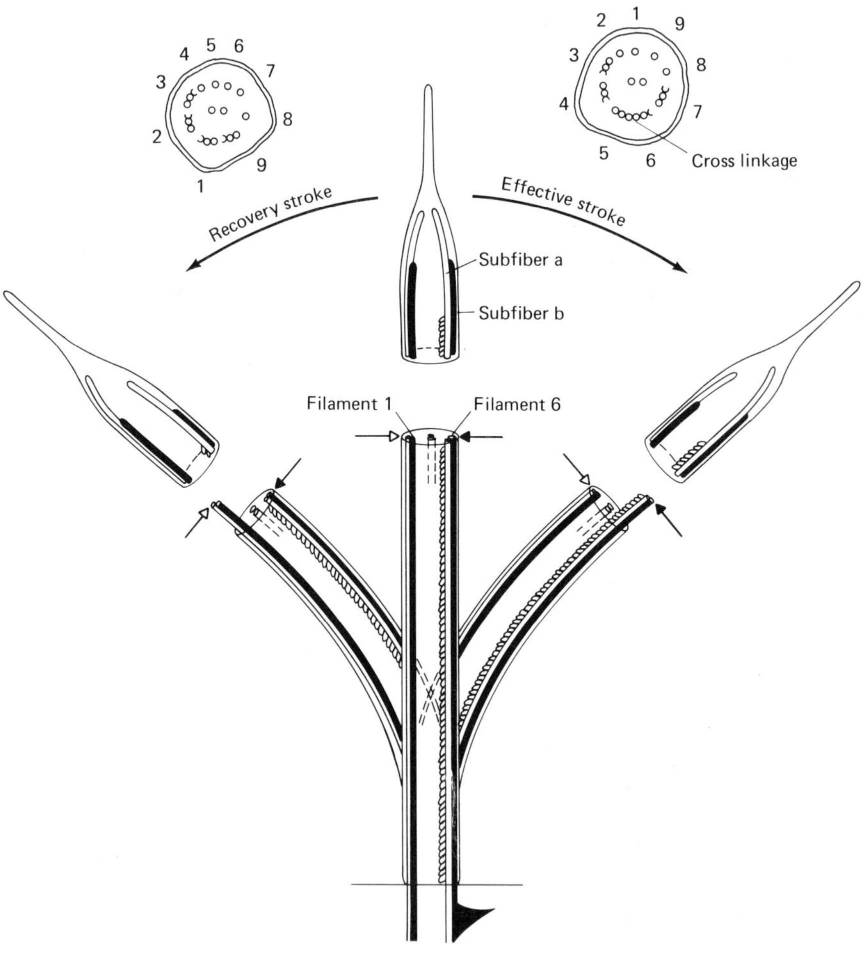

Figure 2.14 The sliding filament hypothesis of ciliary motility. Behavior of two double filaments (1 and 6) is illustrated when a cilium is bent to either side of a straight position. (For additional discussion, see text.) (From P. Satir, J. Cell Biol., 1968, *39*:79.)

tially identical supply of the nuclear material of the parent organism. Sexuality, the alternative to fission and budding, takes several forms in the Protozoa. Sexual reproduction always implies the fusion of two units carrying nuclear information; among protozoans, these units may be whole cells or mere nuclei.

Asexual Reproduction

Asexual reproduction usually is regarded as the most primitive form of animal proliferation. Its various methods are found at some stage in the life cycle of all protozoans, but asexuality is most common among

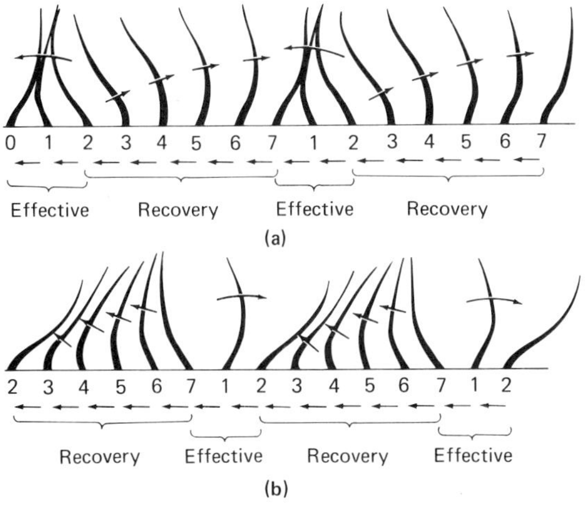

Figure 2.15 A diagram showing two types of metachronal rhythms of ciliary beating. In (*a*), the direction of the effective stroke corresponds to that of the metachronal waves. In (*b*), the two directions are opposite. (From K.G. Grell, *Protozoologie*, 1973, Springer-Verlag.)

the primitive orders of the Sarcomastigophora. Indeed, there are many amoebas and flagellates that reproduce solely by asexual means.

Binary fission is the most common method of asexual reproduction. Typically, a mitotic division of the nucleus precedes a cytoplasmic constriction that creates two complete and identical daughter cells (Fig. 2.17a). Binary fission is thus essentially the same process as mitotic cell division among metazoans. The process is by no means a simple affair and characteristically involves several stages. A preparatory stage precedes active fission. Preparations may include the dedifferentiation of some structures and/or the replication of certain subcellular organelles. Often, structures that occur singly in the adult protozoan must be duplicated to ensure their inclusion in each daughter cell. Oral structures, basal bodies, and protective tests commonly undergo prefission replication. When fission finally occurs, the cytoplasmic division may separate two already highly differentiated organisms (Fig. 2.17b). During a postfission period, the daughter cells differentiate and grow into mature protozoans.

Multiple fission is also common among protozoans. This method involves multiple nuclear and cytoplasmic divisions within a single parent structure and results in the simultaneous release of many daughter cells. Multiple fission is an asexual means of rapidly increasing a population and is associated with strategic points in the life cycle of many protozoan parasites. The sporozoans and cnidosporans, in particular, reproduce in this way. Multiple fission is also a common method of reproduction among the foraminiferans and radiolarians.

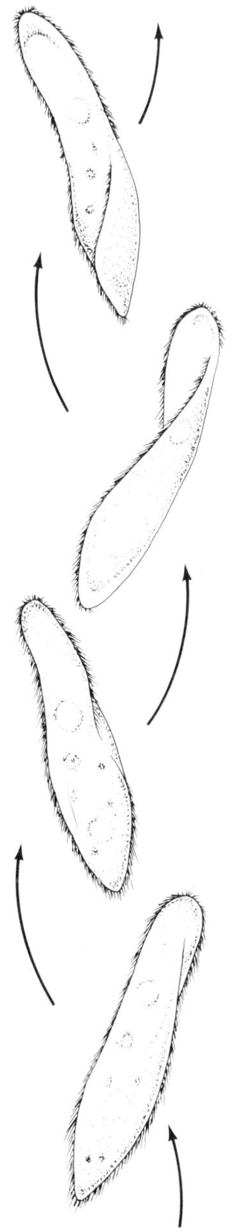

Figure 2.16 *Paramecium* rotates as it moves along a spiral path. (From H. Curtis, *The Marvelous Animals*, 1968, The Natural History Press: Doubleday.)

Finally, budding is an asexual process producing two protozoans of different size. A smaller, or daughter, cell forms and detaches from a larger mother cell. Budding occurs most frequently in sessile protozoans with short-lived, free-swimming immature stages. The

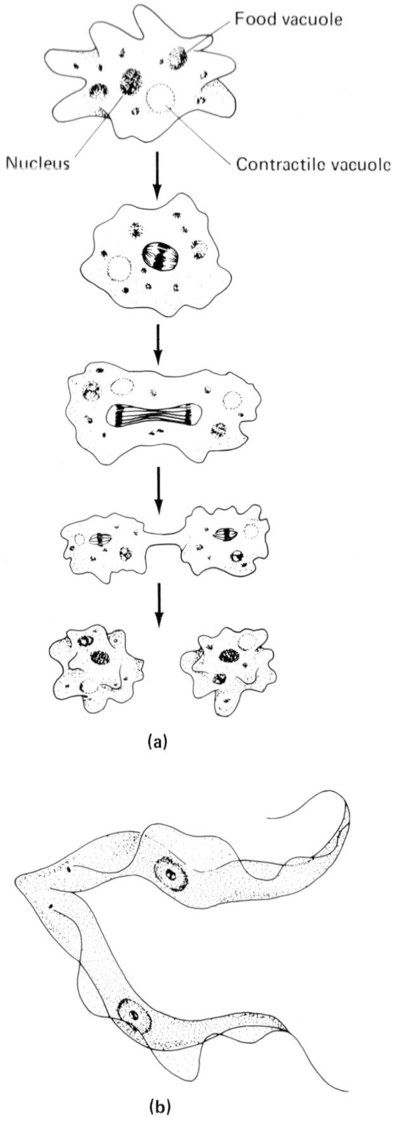

Figure 2.17 (*a*) Binary fission in *Amoeba;* (*b*) Binary fission in *Trypanosoma brucei.* (*b* from D.L. Mackinnon and R.S.J. Hawes, *An Introduction to the Study of Protozoa*, 1961, Oxford University.)

suctorian ciliates, for example, are sessile as adults, but bud off daughter cells that actively swim before settling and maturing. Implied in such a life cycle is a metamorphosis by which the swimming, immature form differentiates into a sessile adult.

Sexual Reproduction

Despite their asexual tendencies, protozoans have developed a wide range of mechanisms for sexual reproduction. Sexuality effects a reorganization of nuclear material. The products of a sexual union differ from the parent forms and thus represent unique experiments on which the forces of organic evolution may operate. Sexual reproduction involves two basic activities: meiosis, or the production of reproductive units containing a haploid number of chromosomes; and fertilization, in which a diploid condition is restored by the union of two haploid units.

Among the protozoans, a diversity of meiotic patterns produces several types of haploid units. Whole protozoans may differentiate into sex cells called gametes. Such cells are called **isogametes** when all are identical in appearance; the term **anisogametes** refers to gametes of two distinct types. Commonly, one gamete of an anisogametic pair is smaller and more motile than the other. This smaller gamete is regarded as a male prototype; the larger gamete is considered female. Even among isogametes, there are usually strain types; thus all conspecific isogametes cannot fuse with one another. Often a "+" and "−" strain are identified, and fertilization is only possible between individuals of opposite strains. Mating strains discourage sexuality between closely related individuals, while promoting the mixing of genetic material from dissimilar organisms.

While meiosis immediately precedes gametogenesis and fertilization in most protozoan life cycles (as in those of multicellular animals), some protozoan groups, particularly among the plantlike flagellates, display an opposite arrangement. These protozoans are haploid throughout most of their lives. The fusion of haploid gametes produces a transitory, diploid zygote, but meiosis quickly restores the dominant haploid condition.

Sexuality among ciliate protozoans is peculiar and includes the phenomena of **conjugation** and **autogamy.** Conjugation involves the pairing of two ciliates and the establishment of a cytoplasmic bridge between them. Complicated nuclear reorganizations follow, including meiosis and the exchange of chromosomes between the conjugating pair. The two ciliates then separate and swim away, each with a revised nuclear constitution. Conjugation thus does not directly increase cell number; however, the process does bestow the evolutionary advantages of sexuality on each of its participants. Moreover, conjugation has a considerable rejuvenating effect on ciliate populations. Studies indicate that these protozoans can reproduce asexually (by transverse, binary fission) only for a limited number of generations. Eventually, conjugation must occur if the population is to prosper, whereupon another series of asexual generations may follow.

Autogamy is related to conjugation in that similar nuclear reorganizations take place, but only a single organism is involved. Autogamy is regarded as a sexual event because it includes meiosis and a nuclear union. Like conjugation, autogamy can have stimulating effects on populations stymied by prolonged periods of asexuality.

Ciliates are not the only protozoans whose life cycles combine sexual and asexual modes of reproduction. Alternations of sexual and asexual phases are general in the phylum and also appear in numerous metazoans.

Another aspect of protozoan life cycles is **encystment.** Protozoans subjected to severe environmental stress may encyst in thick protective envelopes where they remain until favorable conditions return. Terrestrial and freshwater forms are especially adept at encystment, as their environments are subject to sudden and threatening changes. Encysted protozoans are well adapted for travel and may be carried long distances by winds, water currents, or other animals. Parasitic protozoans, notably the sporozoans and cnidosporans, often survive the difficult journey between hosts while in an encysted stage. Finally, encystment may play a supportive role in reproduction. Meiosis, fertilization, fission, or early development may involve such extreme physiological upheavals that a protozoan requires a temporary respite. It may encyst to concentrate on reproduction alone.

Protozoan Diversity

The protozoans represent the most diverse group called a phylum in this book. The variety of structural and functional plans presented by unicellular organisms is extensive; indeed, their only common feature is unicellularity. Grouping all protozoans in one phylum is comparable to lumping all multicellular invertebrates together. Both schemes betray a gross lack of appreciation for the considerable diversity of the individuals involved. Within their single cells, protozoans carry on life activities in many and marvelous ways, and an admittedly simplified taxonomic approach must not blind us to their variety.

Approximately 30,000 living protozoan species have been described, and due to their small size, thousands of undiscovered species undoubtedly exist. Adaptive radiation in this large and heterogeneous phylum centers on locomotor organelles, nuclear morphology, reproductive formats, and modes of nutrition. Locomotor organelles refer to the pseudopodia, cilia, and flagella already described. These structures provide keys to the general taxonomy of the phylum. The first subphylum, the Sarcomastigophora, includes the amoeboid and flagellated protozoans. Amoeboid forms are classified further according to several pseudopodial types. The Sporozoa and Cnidospora, the second and third protozoan subphyla, lack adult locomotor organelles, whereas the fourth group, the Ciliophora, possesses cilia. Reproductive schemes and nuclear morphologies are extremely varied throughout all groups. Asexual reproduction is common in the Sarcomastigophora, although sexuality is exhibited by many of its members. Sporozoans and cnidosporans alternate asexual and sexual phases in their rather complex life cycles, which also feature encysted stages called spores. Finally, the Ciliophora exhibit conjugation; members of this subphylum are distinguished further by their possession of two nuclear types. Nutrition is the last major system on which protozoan diversity centers. Within the mastigophorans, we distinguish two main groups, the autotrophic phytoflagellates and the heterotrophic zooflagellates. The Sporozoa and Cnidospora feed parasitically. And, finally, specializations of the oral ciliature for feeding purposes are the basis for much of the taxonomy of the Ciliophora.

SUBPHYLUM SARCOMASTIGOPHORA

The Sarcomastigophora is the largest protozoan subphylum and is thought to include the most primitive unicellular forms. This group contains amoeboid and flagellated animals, as well as some flagellated organisms described properly as neither animals nor plants. Asexual reproduction is general in this subphylum, and many forms reproduce solely by asexual means. To the extent that sexuality exists, it involves the union of haploid cells. Spores are unknown in the Sarcomastigophora. Sarcodines (the amoebas) and mastigophoreans (the flagellates) may be uninucleate or multinucleate, but when they occur, multiple nuclei are always monomorphic. The relationship between amoeboid and flagellated types is often quite remote, but the groups are placed in a common subphylum because of the existence of intermediate species possessing both flagella and pseudopodia at different points in their life cycles.

Class Mastigophorea

The mastigophoreans are probably the most primitive invertebrates. Indeed, the presence of photosynthetic individuals in this group suggests that these organisms

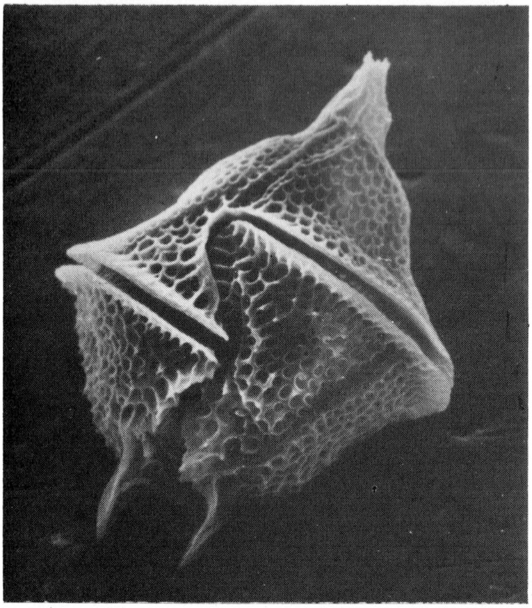

Figure 2.18 Scanning electron micrograph of *Gonyaulax digitale* showing the girdle and the sulcus. (Photo courtesy of J.D. Dodge, Department of Botany, Royal Holloway College, University of London.)

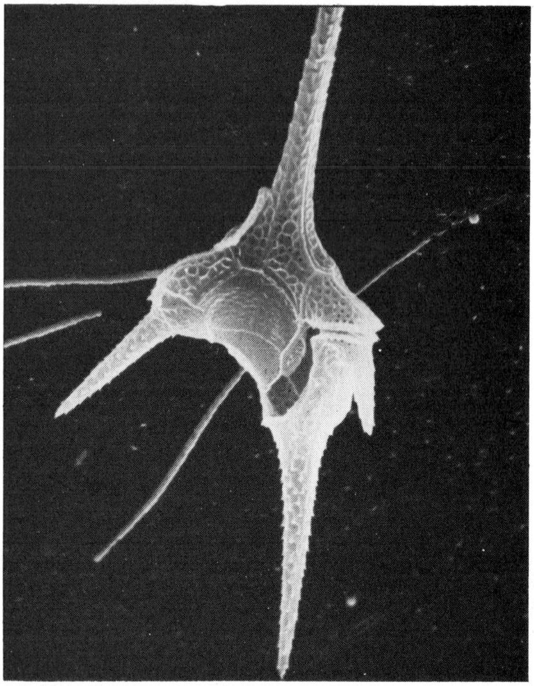

Figure 2.19 Scanning electron micrograph of *Ceratium hirudinella.* (From J. Phycol., 1970, *6*:137. Photo courtesy of J.D. Dodge.)

originated when animal and plant forms were not totally distinct. The class is distinguished by one or more locomotor flagella in each adult. Mastigophoreans traditionally are separated into two subgroups, the plantlike phytoflagellates and the more animal-like zooflagellates, although this division may have no real phyletic significance.

THE PHYTOFLAGELLATES. The phytoflagellates are ancient forms that figured prominently in the origins of both the animal and the plant kingdoms. These flagellated organisms move like animals, but the presence of photosynthetic pigments in many species and their general reliance on autotrophic nutrition are definite plantlike qualities. We include the phytoflagellates here because of their animal affinities, but botanists claim them as well.

Phytoflagellates commonly possess only one or two flagella, and their photosynthetic pigments are restricted in most cases to organelles called **chromoplasts.** Five important orders make up this group: dinoflagellates, cryptomonads, chrysomonads, euglenoids, and volvocids. Additional smaller orders exist but are not discussed here.

Dinoflagellates are abundant in the plankton of marine and fresh waters. These protozoans are distinguished by two sculptured grooves, one transverse (the **girdle**) and one longitudinal (the **sulcus**), along their body surfaces (Fig. 2.18). A flagellum occupies each groove. The dinoflagellate body is surrounded by a cellulose envelope, either simple and continuous or consisting of separate hardened plates. Dinoflagellate envelopes may be sculptured into elaborate spines, fins, and other projections (Fig. 2.19). Typically, reproduction is by binary fission. The outer envelope ruptures along predetermined longitudinal sutures, and each daughter cell then reforms its missing half (Fig. 2.20). Multiple fission has been described in a few species. An encysted stage is very common in the dinoflagellate life cycle and indeed may be a dominant phase. Encysted dinoflagellates are said to be in **palmella** form. Palmella forms have lost their flagella and usually drift as relatively undifferentiated cells still cloaked in their original cellulose (Fig. 2.21). Multiple fission may occur during encystment. Many dinoflagellates in palmella form engage in symbiosis with other protozoans, anemones, corals (see Chap. Five, Special Essay) and clams. Symbiotic dinoflagellates are called **zooxanthellae.** *Ceratium* (Fig. 2.19) is a common freshwater dinoflagellate, and *Noctiluca* (Fig. 2.22) is a large luminiscent form which

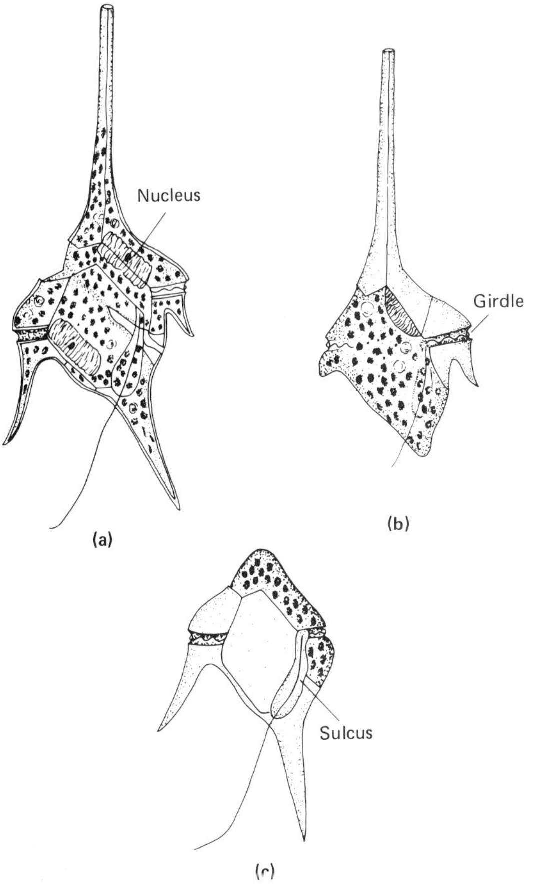

Figure 2.20 Binary fission in *Ceratium hirudinella*. (*a*) Division in progress; (*b*) and (*c*) The two daughter cells after division. (From Grell, *Protozoologie*.)

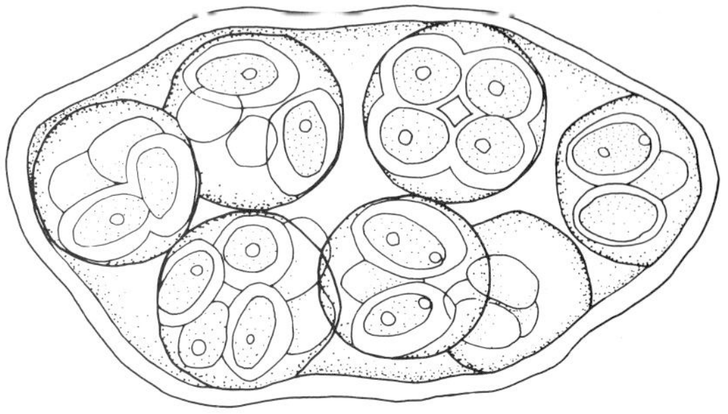

Figure 2.21 *Chlamydomonas*, a volvocid, in the palmella stage. (From L.H. Hyman, *The Invertebrates: Protozoa through Ctenophora, Vol. I*, 1940, McGraw-Hill.)

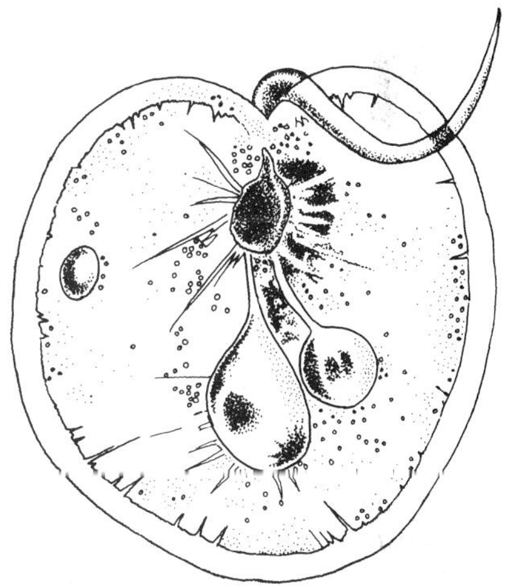

Figure 2.22 *Noctiluca*, a large bioluminescent marine dinoflagellate. (From J.L. Sumich, *Biology of Marine Life*, William C. Brown.)

contributes significantly to marine plankton. *Gymnodinium* and *Gonyaulax* (Fig. 2.18) are marine genera notorious for their explosive population increases, called "blooms." Blooms sometimes cause the dreaded "red tides" of coastal waters, which involve concentrations as great as 40 million individual dinoflagellates per milliliter of sea water. Such concentrations color the water reddish-brown and, because of the dinoflagellates' toxic waste products, can result in the wholesale death of marine life. The precise causes of blooms remain unknown.

The cryptomonads and the chrysomonads are related phytoflagellate groups. These protozoans display yellow or brownish chromoplasts (the color comes from pigments other than chlorophyll) and an elaborate asexuality. Cryptomonads typically bear a pair of flagella in association with an anterior vestibule (Fig. 2.23). This unwalled reservoir is the site of food vacuole formation as well as contractile vacuole discharge. All other surfaces of the cryptomonad body are covered with a cellulose wall. Current descriptions of reproduction in this order are limited to asexual schemes. Longitudinal binary fission is the rule, but encystment is common and may precede elaborate developmental programs, as in the chrysomonads described below.

Chrysomonads lack the vestibule of the cryptomonads and commonly have two flagella of unequal length. The longer flagellum is directed anteriorly and the shorter posteriorly; the latter may be absent. Chrysomonads are often naked, although some species secrete an external case called a **lorica** (Fig. 2.24). This phytoflagellate order is extremely diverse

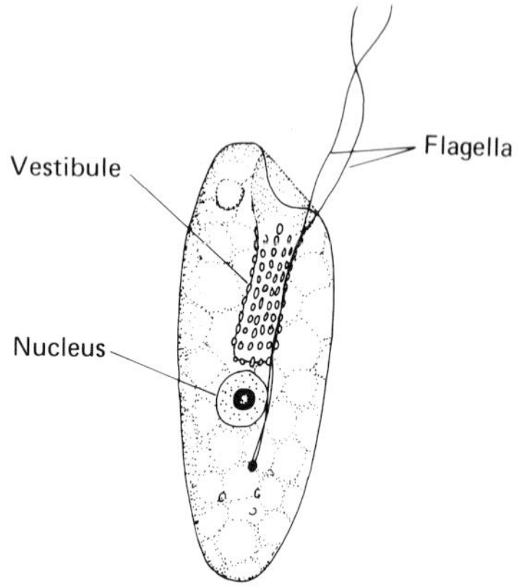

Figure 2.23 A cryptomonad, *Chilomonas paramecium.* (From Grell, *Protozoologie.*)

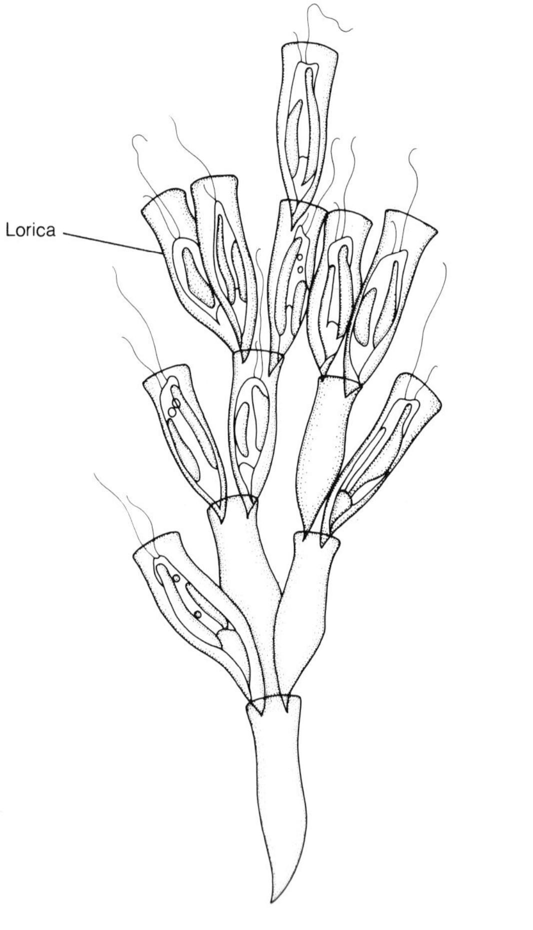

Figure 2.24 A colonial chrysomonad, *Dinobryon sertularia.* (From Grell, *Protozoologie.*)

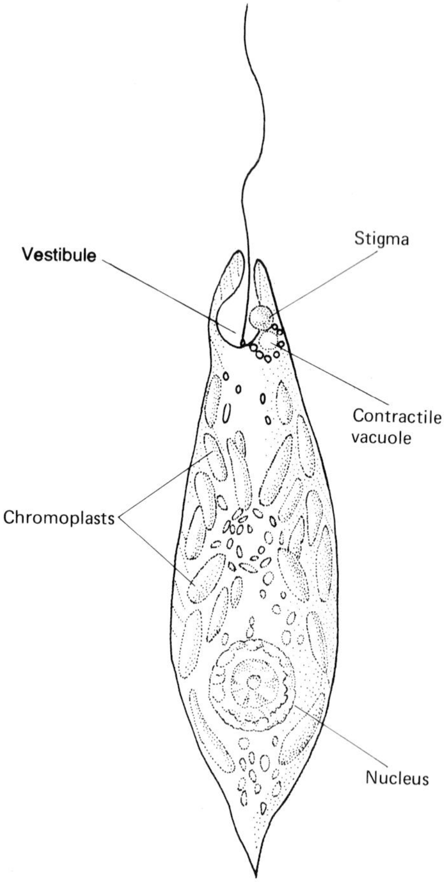

Figure 2.25 *Euglena viridis.* (From Grell, *Protozoologie.*)

and features sessile and free-swimming members, colonies as well as individuals, and even a few partially amoeboid forms. Some chrysomonads are holozoic, but the group as a whole is overwhelmingly plantlike. Reproduction is by longitudinal binary fission, which follows preparatory divisions of the nucleus, chromoplast, and flagellar apparatus. Encystment is general in this order, whose cysts are constructed of silica in contrast to the cellulose of cryptomonad cysts. The loss of flagella and the general metabolic curtailment associated with precystic activity often signal the advent of developmental changes. Encysted chrysomonads may secrete a gelatinous matrix in which future generations of daughter cells become embedded. If flagella are retained, these chrysomonads form motile colonies. Nonflagellated colonies consist of individuals in the palmella stage.

The euglenoids comprise the fourth important order of phytoflagellates. They are somewhat larger proto-

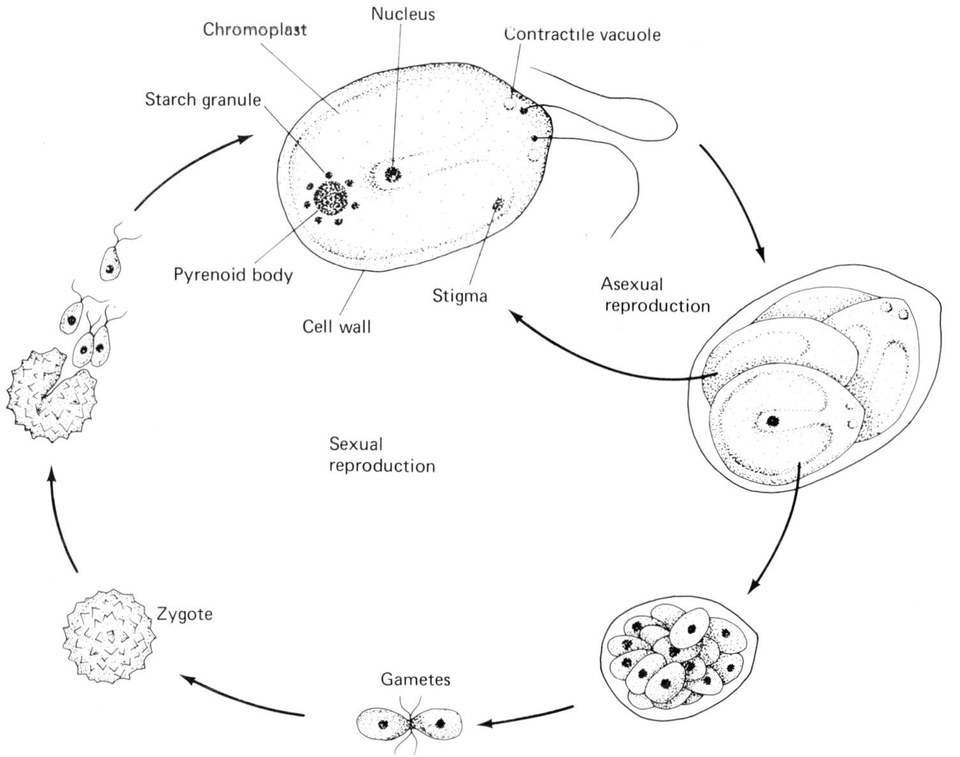

Figure 2.26 Life cycle of *Chlamydomonas.* (For explanation see text.) (From D.E. Beck and L.F. Braithwaite, *Invertebrate Zoology Laboratory Workbook*, 3rd ed., 1968, Burgess.)

zoans, whose pliable, usually cylindrical cell bears one or two flagella. As in the cryptomonads, these flagella often are associated with an anterior vestibule. Green euglenoids, including the well-known *Euglena* (Fig. 2.25), possess a photoreceptive red eyespot or **stigma,** which allows them to orient toward sunlight during photosynthesis. Colorless euglenoids rely solely on saprozoic and/or holozoic modes of nutrition. *Peranema* is a common colorless form. Reproduction in this order is primarily asexual and may occur during free-swimming or encysted stages. Palmella forms are known. Rarely, euglenoids reproduce by sexual means, usually involving the union of isogametes.

The volvocids are a large group of advanced phytoflagellates. These are generally small, rigid organisms bounded by a cellulose wall and possessing one, two, or sometimes four flagella. Volvocids are always green and photosynthetic, and possess photoreceptive stigmata. An outstanding character of this order is its tendency toward a colonial organization of increasing complexity. Sexuality is well developed in such colonial forms and thereby marks its only real success among the phytoflagellates.

Chlamydomonas (Fig. 2.26) is a frequently studied, solitary volvocid that displays both asexual and sexual reproduction. Its asexuality involves binary or multiple fission, and complete daughter cells are formed within the cellulose confines of the parent organism (Fig. 2.26). Subsequent rupture of the parental envelope releases the new generation. Sexuality in *Chlamydomonas* involves the union of gametes which are essentially identical to vegetative cells (Fig. 2.26). Isogametes or anisogametes (differing slightly in size) may occur, but mating strain restrictions are always in effect. A thick-walled zygote is formed and undergoes meiosis, producing four new individuals.

Colonial formation by volvocids reflects in part the pattern of cell division in this order. Daughter cells tend to remain together for a variable time following mitosis, and most volvocid colonies merely represent a permanent association of cells of like parentage. In *Gonium* (Fig. 2.27) and other small, primitive colonies, all individuals are identical in structure and function. Sexuality involves isogametes, and all members are potent. *Volvox* (Fig. 2.28) is a more advanced volvocid colony. Its spheroid colonies represent "superprotozoans," orga-

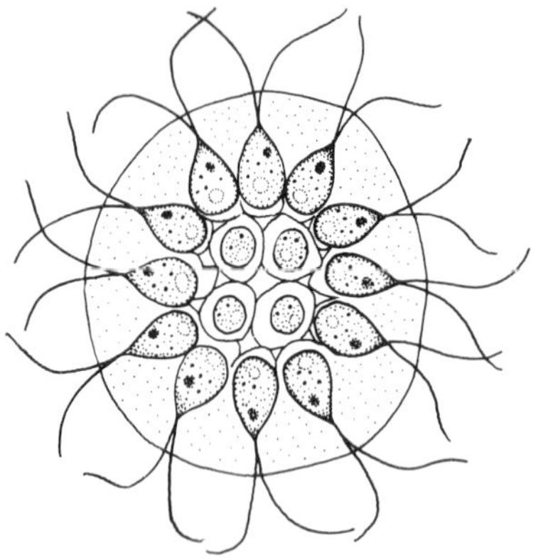

Figure 2.27 A colony of *Gonium pectorale*. (From Sleigh, *The Biology of Protozoa*.)

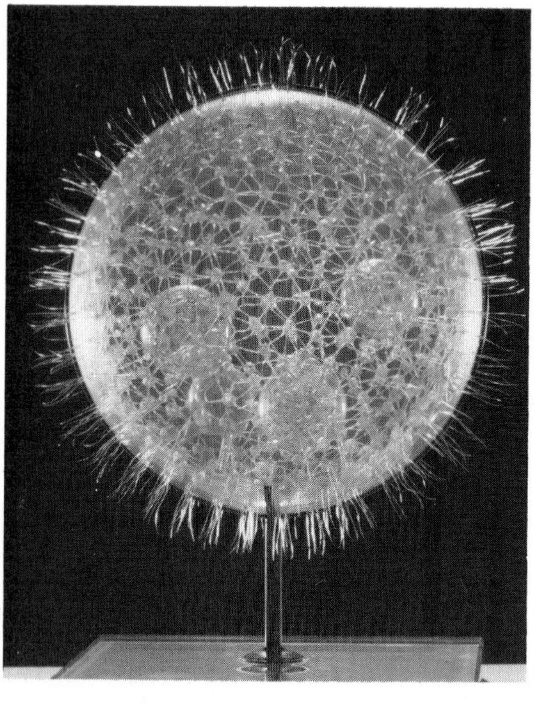

Figure 2.28 Photograph of a glass model of a *Volvox* colony. (Courtesy of the American Museum of Natural History.)

nizations of cells that cooperate for mutual benefit. Polarization exists in a *Volvox* colony, as its anterior members are stronger swimmers and have better-developed stigmata. Sexual potency is limited to posterior cells, which produce highly differentiated anisogametes. Intracolonial fertilization is followed by a complex program of early development. The zygote formed undergoes multiple cell divisions which produce a daughter colony (Fig. 2.29). Originally, all flagella are directed toward the hollow center of their colonial sphere. Then a remarkable inversion occurs, whereby the young *Volvox* colony turns inside out. The resulting externally flagellated spheroid ruptures through the parent colony to begin its independent life.

Volvox and related colonies are the objects of evolutionary speculation. Some scientists suggest that these volvocids represent the evolutionary link between the protozoan and metazoan worlds. Indeed, some argue that *Volvox* is not a protozoan at all! The generally accepted rule is that a colony remains protozoan if its nonreproductive members are all of one type. By this guideline, the classification of *Volvox* is debatable, but comparative studies with other volvocids do support its designation as a protozoan. The possible significance of volvocids in the evolution of protozoans and metazoans is discussed in a later section.

THE ZOOFLAGELLATES. The second general group of flagellated protozoans includes the colorless forms not closely related to the phytoflagellates. These unicellular organisms are unquestionably animals and are as diverse as their plantlike counterparts. Many subgroups have been established, based primarily on differences in number and position of flagella and surrounding structures. We review only four of the more important orders: choanoflagellates, kinetoplastids (trypanosomes), metamonads, and rhizomastigids.

The choanoflagellates are easily recognized protozoans: freshwater organisms with a single flagellum surrounded at its base by a protoplasmic collar (Fig. 2.30). Most species are sessile, and a few form colonies whose individuals are united by a common stalk (Fig. 2.30c) or a gelatinous matrix (Fig. 2.30b). Undulations of the choanoflagellate flagellum draw currents through the collar, where food particles adhere. Trapped nutrients are passed to the small cell body for digestion within food vacuoles. Often, choanoflagellates are compared with the choanocytes of sponges (see Chap. Four), and some theories derive the sponges from these zooflagellates.

Kinetoplastids are a very diverse group of zooflagellates whose best-known members include the parasitic trypanosomes. Trypanosomes commonly infect humans and other vertebrates, and two highly pathogenic species, *Trypanosoma gambiense* and *T. rhodesiense*, are responsible for African sleeping sickness. These parasites proliferate asexually within the digestive tract of tsetse flies, and infective stages are transmitted to humans via the bite of these blood-suck-

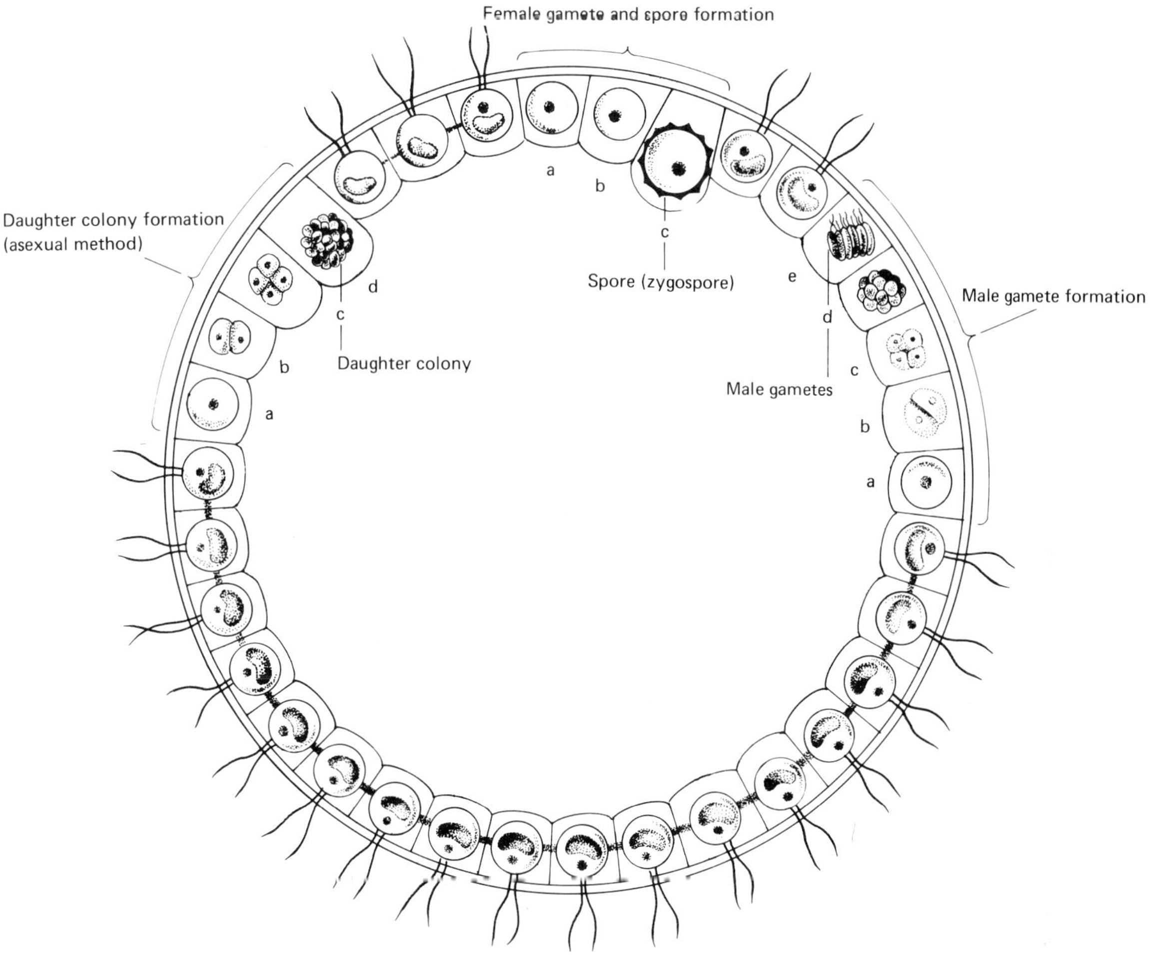

Figure 2.29 A diagram illustrating sexual and asexual reproduction in *Volvox.* (From Beck and Braithwaite, *Invertebrate Zoology Laboratory Workbook*.)

ing insects. Depending on the species, a trypanosome may assume up to four different body forms (Fig. 2.31) during its complex life cycle. This group is recognized most easily by its **kinetoplast,** a large body containing almost one-fourth of the animal's DNA. Also characteristic of these protozoans is the flagellum-bearing undulating membrane of the infective stage (Fig. 2.32). Other trypanosome parasites of humans include *T. cruzi* of South and Central America, the causative agent of Chagas' disease. The genus *Leishmania* is widely distributed in the tropics and causes "oriental sore." Many trypanosomes are insect gut parasites, and it is thought that parasitization of vertebrate bloodstreams represents a secondary specialization. Accidental transmission to vertebrates by blood-sucking insects presumably paved the way for this evolutionary development. Plant parasites also have been described in this group.

The metamonads are a large group of zooflagellates with many symbiotic members. Metamonads can be arranged in a series of groups (formerly considered separate orders) of increasing structural complexity. Trichomonads are among the simpler forms. *Trichomonas vaginalis* (Fig. 2.33a), a parasite of the human genital tract, has four anterior flagella and a recurrent flagellum with an undulating membrane. Also characteristic of trichomonads is the **axostyle,** a bundle of microtubules extending the length of the body. Increased structural complexity among metamonads involves a multiplication of the organel-

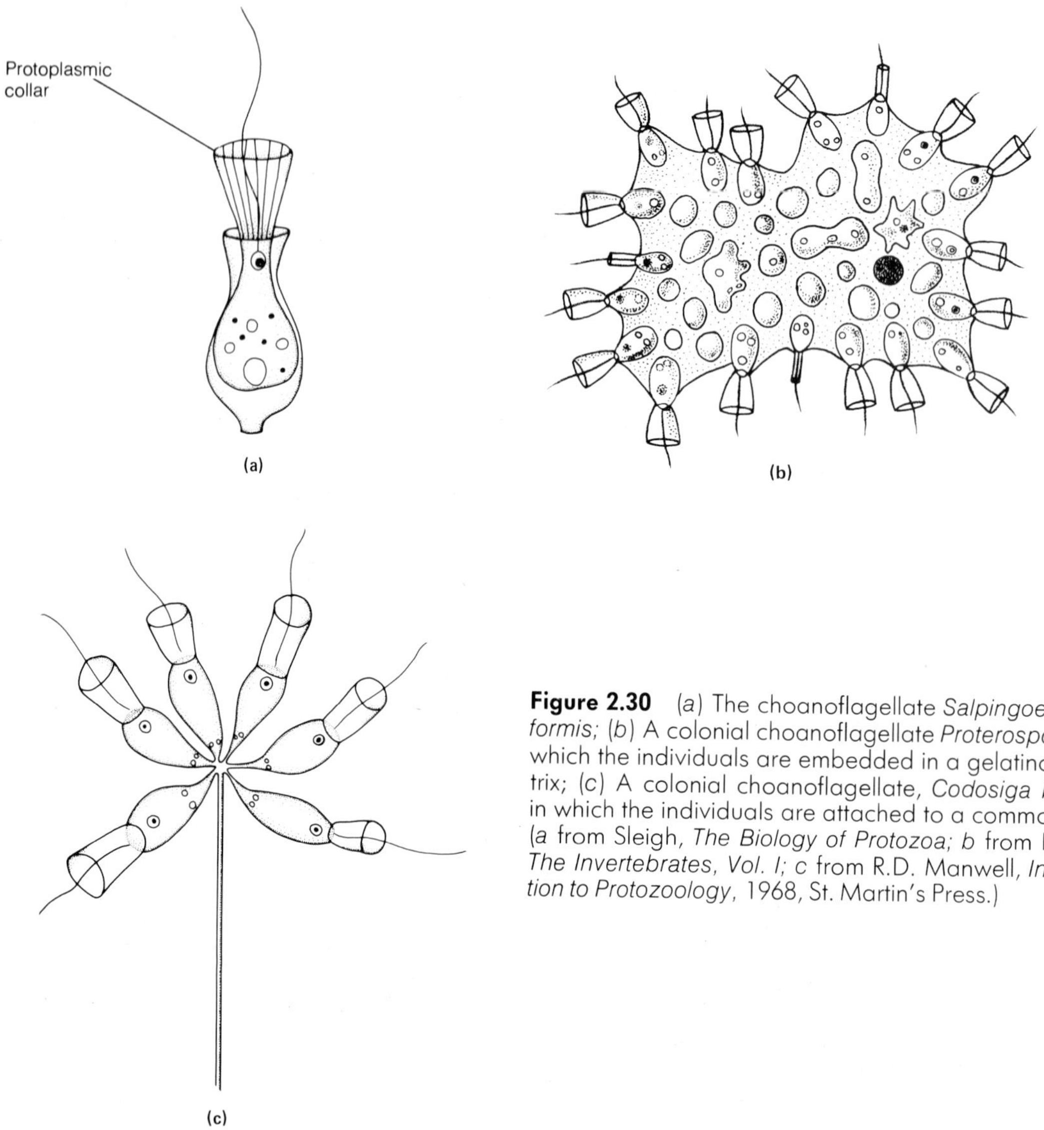

Figure 2.30 (*a*) The choanoflagellate *Salpingoeca fusiformis;* (*b*) A colonial choanoflagellate *Proterospongia* in which the individuals are embedded in a gelatinous matrix; (*c*) A colonial choanoflagellate, *Codosiga botrytis,* in which the individuals are attached to a common stalk. (*a* from Sleigh, *The Biology of Protozoa; b* from Hyman, *The Invertebrates, Vol. I; c* from R.D. Manwell, *Introduction to Protozoology*, 1968, St. Martin's Press.)

les composing the flagellar apparatus. For example, the vertebrate gut parasite *Giardia* (Fig. 2.33b) possesses two nuclei and two sets of four flagella.

This multiplication of organelles climaxes with the hypermastigids, the most sophisticated metamonads and some of the most highly evolved protozoans. They possess large numbers of flagella in anterior rows or tufts (Fig. 2.33c). This order is wholly commensal or mutualistic in wood roaches, termites, and ruminant mammals. Hypermastigids occur in large numbers in the host gut, where they may provide the enzyme necessary for cellulose digestion. Thus termites and roaches tap an abundant but seldom exploited food source, while the hypermastigids are furnished a home. The obligate nature of the relationship is demonstrated when termites are defaunated by subjection to high oxygen concentrations. Hypermastigids are anaerobes and are killed by intense exposure to oxygen; the termites also are doomed, as they cannot digest their wooden food, even though they continue to eat normally.

Hypermastigids reproduce asexually within their symbiotic partners, but if the association is with an insect, sexuality may be induced when the host molts. Apparently, ecdysone, the insect molting hormone, stimulates hypermastigid gamete formation. Encysted

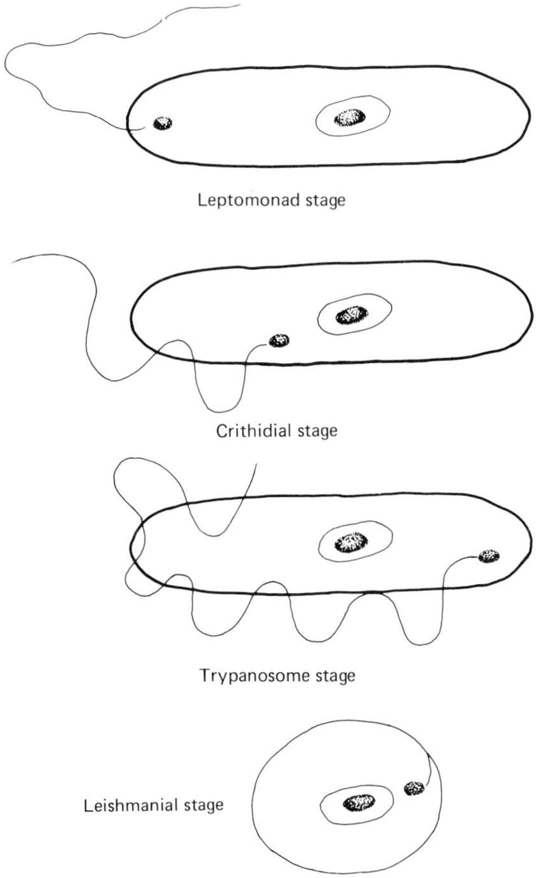

Figure 2.31 The four body forms which may be exhibited by a trypanosome during its life cycle.

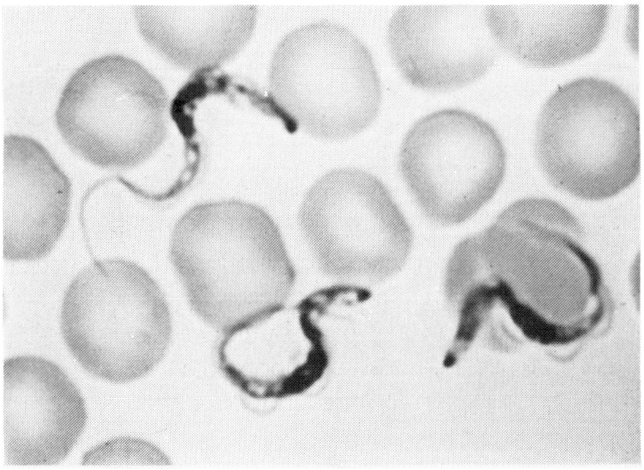

Figure 2.32 A photograph of *Trypanosoma rhodensiense* among human red blood cells. Note the flagellum with its undulating membrane. (Courtesy of Carolina Biological Supply.)

zygotes are shed with the feces and await consumption by another insect host. *Trichonympha* (Fig. 2.33c) is a representative hypermastigid genus.

The rhizomastigids are the last zooflagellate order under consideration and represent a living link to the amoeboid members of the Sarcomastigophora. To wit, the rhizomastigids are protozoans permanently possessing both flagella and pseudopodia. They are relatively simple organisms, primarily inhabiting fresh water. Some members are extremely dangerous pathogens capable of causing amoebic meningitis. *Mastigamoeba* (Fig. 2.34) is a well-known rhizomastigid genus.

Class Sarcodinea

The sarcodines compose the second main division of the subphylum Sarcomastigophora. These protozoans are distinguished by their possession of pseudopodia and lack of flagella in the adult stage. Their reproduction is primarily by binary or multiple fission, although sexuality has been described in some species. Various subgroups are characterized by pseudopodial types and by special protective structures. Five major orders are recognized: the naked and the testate amoebas of the subclass Rhizopodia, the foraminiferans of the subclass Granuloreticulosia, and the radiolarians and heliozoans of the subclass Actinopodia.

Four general types of pseudopodium are recognized. The **lobopodium** (Fig. 2.35a) is a relatively unstructured blunt projection of the cell containing both ectoplasm and endoplasm. The naked amoebas have lobopodia, and it was primarily with this pseudopodial type that our discussion of amoeboid movement was concerned. Testate amoebas possess lobopodia or **filopodia** (Fig. 2.35b). The latter are slender, pointed pseudopodia containing mostly ectoplasm. Another type is the **reticulopodium** (Fig. 2.35c), which consists of a network of anastomosing filopodia. Reticulopodia are best known in the foraminiferans. The final pseudopodial type is the **axopodium** (Fig. 2.35d), a slender, ectoplasmic locomotor structure reinforced by a central, longitudinal axis of spiraling microtubules (Fig. 2.36). Because of this axis, axopodia are more permanent than the other pseudopodia and are characteristic of radiolarians and heliozoans.

THE AMOEBAS. Naked amoebas include the familiar amoeboid protozoans of the genera *Amoeba* (Fig. 2.2) and *Chaos* (also known as *Pelomyxa*). Structurally, they are simple animals, but their predominantly freshwater and parasitic range suggests a considerable

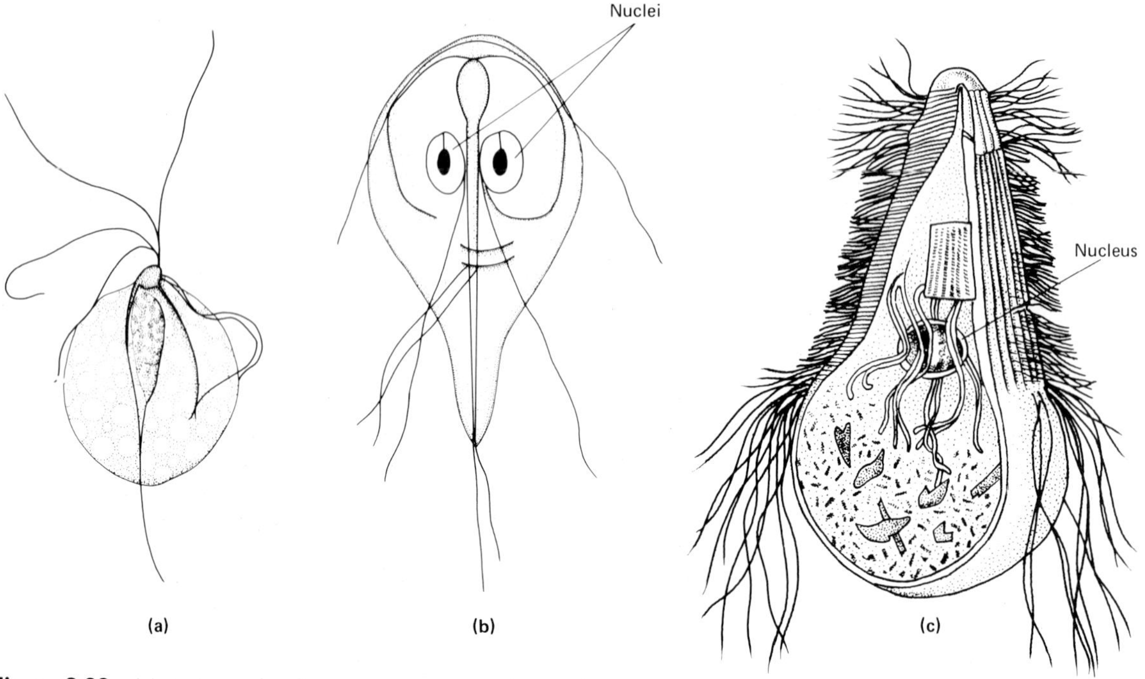

Figure 2.33 Metamonads of increasing structural complexity. (*a*) *Trichomonas vaginalis;* (*b*) *Giardia enterica;* (*c*) *Trichonympha,* from the gut of a termite. (*a* from Grell, *Protozoologie; b* and *c* from Hyman, *The Invertebrates, Vol. I.*)

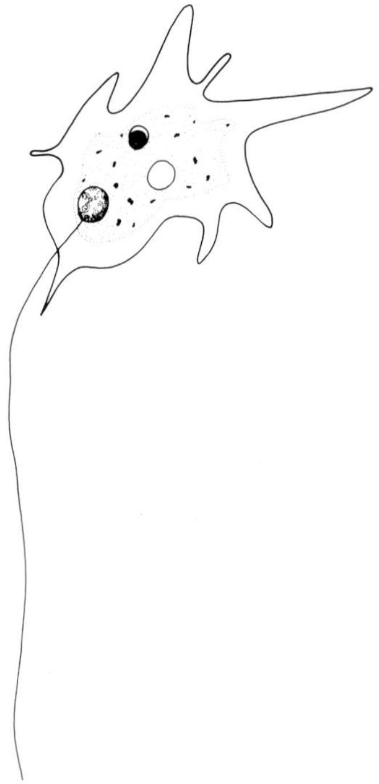

Figure 2.34 *Mastigamoeba,* a form showing both pseudopodia and a flagellum. (From Hyman, *The Invertebrates, Vol. I,.*)

physiological sophistication. The naked amoebas are rather large protozoans, and multiple nuclei are common among them. *Chaos* (Fig. 2.35a), for example, may attain a diameter of 5 mm, and its nuclei can number in the hundreds. Reproduction in this order appears to be exclusively asexual. As is so common among freshwater and parasitic protozoans, encystment is frequent. *Entamoeba histolytica* is responsible for amoebic dysentery in humans. This amoeboid parasite emerges when its cyst is consumed accidentally. When present in large numbers, the mature stage can be quite destructive, producing massive lesions in the wall of the large intestine. Curiously, there are no mitochondria in this species.

The testate amoebas live within a single chambered test. This test may comprise chitinoid or siliceous plates, as in *Arcella* (Fig. 2.37a), and *Euglypha* or may consist of cemented foreign particles, as in the sandy tests of *Difflugia* (Fig. 2.37b). A single opening allows the amoeba to extend its pseudopodia (either lobopodia or filopodia) for feeding and locomotion. In some species, reproduction is by total binary fission, and each daughter cell reconstitutes the missing half of its test. If the test is too thick and rigid to participate directly in the fission process, budding is common, and the daughter cell remains attached until its new test is formed (Fig. 2.38). The group is confined

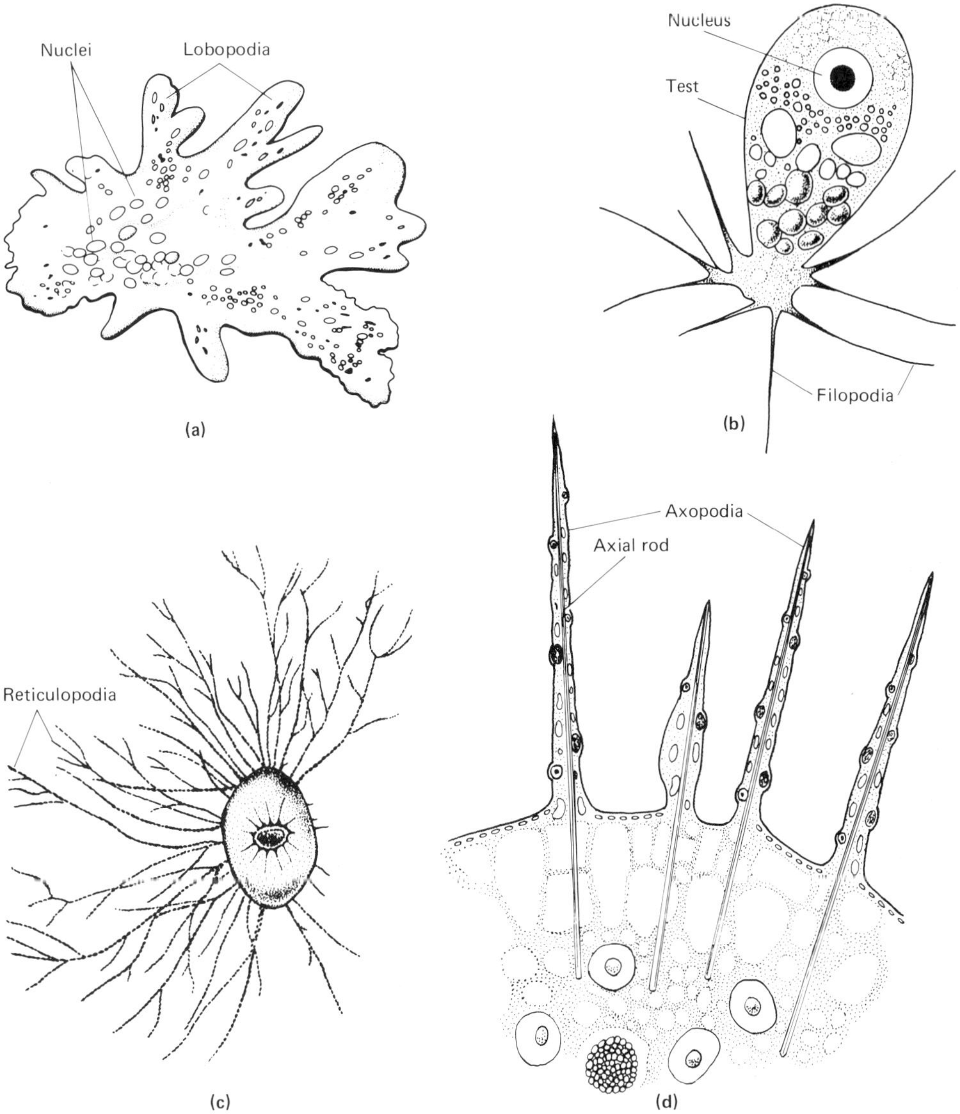

Figure 2.35 (*a*) *Chaos carolinensis* showing lobopodia; (*b*) *Chlamydophrys* showing filopodia; (*c*) *Gromia*, a foraminiferan, showing reticulopodia; (*d*) A portion of a heliozoan, *Actinosphaerium eichorni*, showing axopodia. (*b* and *c* from Hyman, *The Invertebrates, Vol. I*; *d* from Sleigh, *The Biology of Protozoa*.)

largely to fresh water, but some testate amoebas occupy semiterrestrial habitats, often in association with mosses.

THE FORAMINIFERANS. Foraminiferans are a sarcodine order whose members also possess tests. Their tests differ however, from those of the amoebas in that they are divided into a series of chambers, only the last of which opens to the outside (Fig. 2.39). Through numerous perforations in the test, cytoplasm emerges to form a reticulopodial net (Fig. 2.35c). This net serves chiefly in food capture. Nearly anything that contacts the adhesive pseudopodia—from organic particles and unicellular algae to small crustaceans—may serve as food. Foraminiferans are primarily marine, and their abandoned calcareous tests constitute significant geological deposits on the ocean floor. The association of foraminiferan deposits with oil pockets has kindled extensive paleontological interest in the group. Indeed, most of the 20,000 known fossil

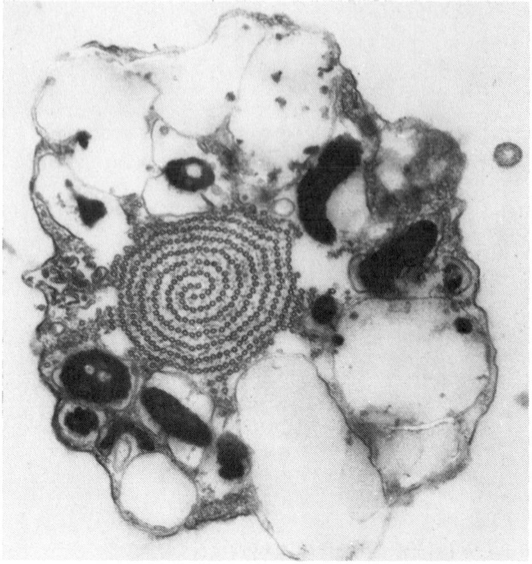

Figure 2.36 An electronmicrograph showing a cross section through an axopodium of *Actinosphaerium.* Note the central axis of spiraling microtubules. (From Sleigh, *The Biology of Protozoa.* Photo courtesy of Dr. A.C. MacDonald.)

protozoans are foraminiferans. About half of all described protozoans, living and fossil, are members of this order.

Reproduction among the foraminiferans is a complicated process involving a life cycle of alternating sexual and asexual stages (Fig. 2.40). The entire scheme has been described for relatively few species; thus the account given here is not necessarily general for the order. The asexual stage is large and multinucleate, distinguished further by the small initial chamber of its test. Multiple fission produces smaller, uninucleate forms whose first test chambers are considerably larger. These asexual products are responsible for gamete production. Gametes may be released or, perhaps more typically, retained within a fertilization cyst. Zygotes emerge from the cyst and develop into asexually reproducing forms. The cycle then begins anew.

Globigerina (Fig. 2.41) is a planktonic foraminiferan whose long calcareous spicules improve flotation. Most foraminiferans, however, inhabit the sea floor in shallower waters.

THE RADIOLARIANS. Radiolarians are spheroidal, largely marine planktonic sarcodines and are among the most beautiful protozoans. This order is characterized by near-universal symmetry, axopodia, and an elaborate skeletal structure. A central membranous

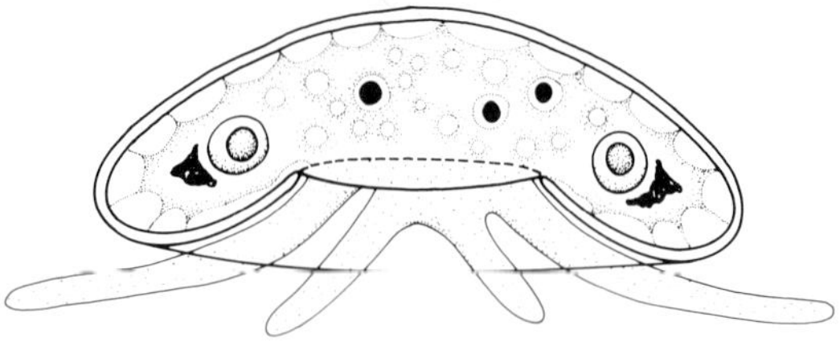

(a)

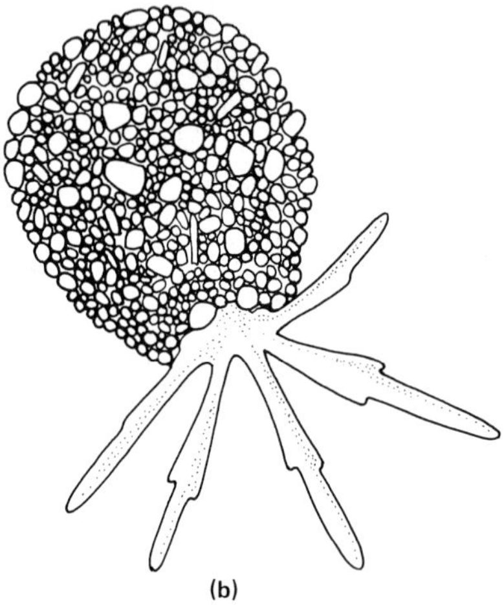

(b)

Figure 2.37 (*a*) *Arcella vulgaris* optical section and side view combined; (*b*) *Difflugia* showing the test of sand grains. (*a* from Grell, *Protozoologie*; *b* from Hyman, *The Invertebrates, Vol. I.*)

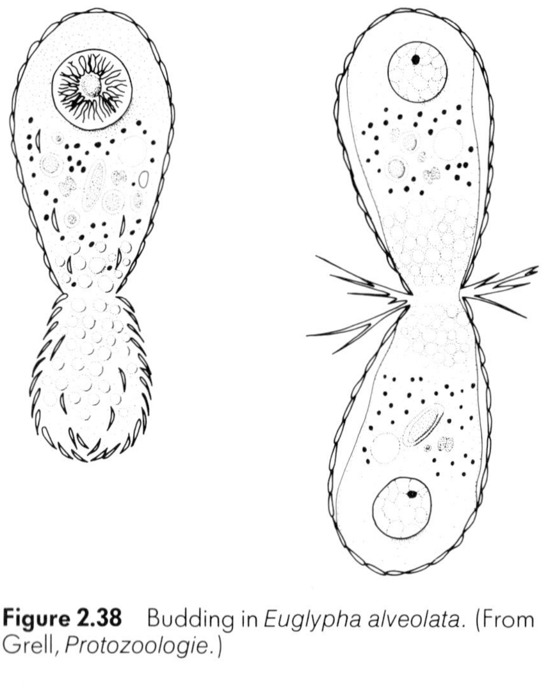

Figure 2.38 Budding in *Euglypha alveolata.* (From Grell, *Protozoologie.*)

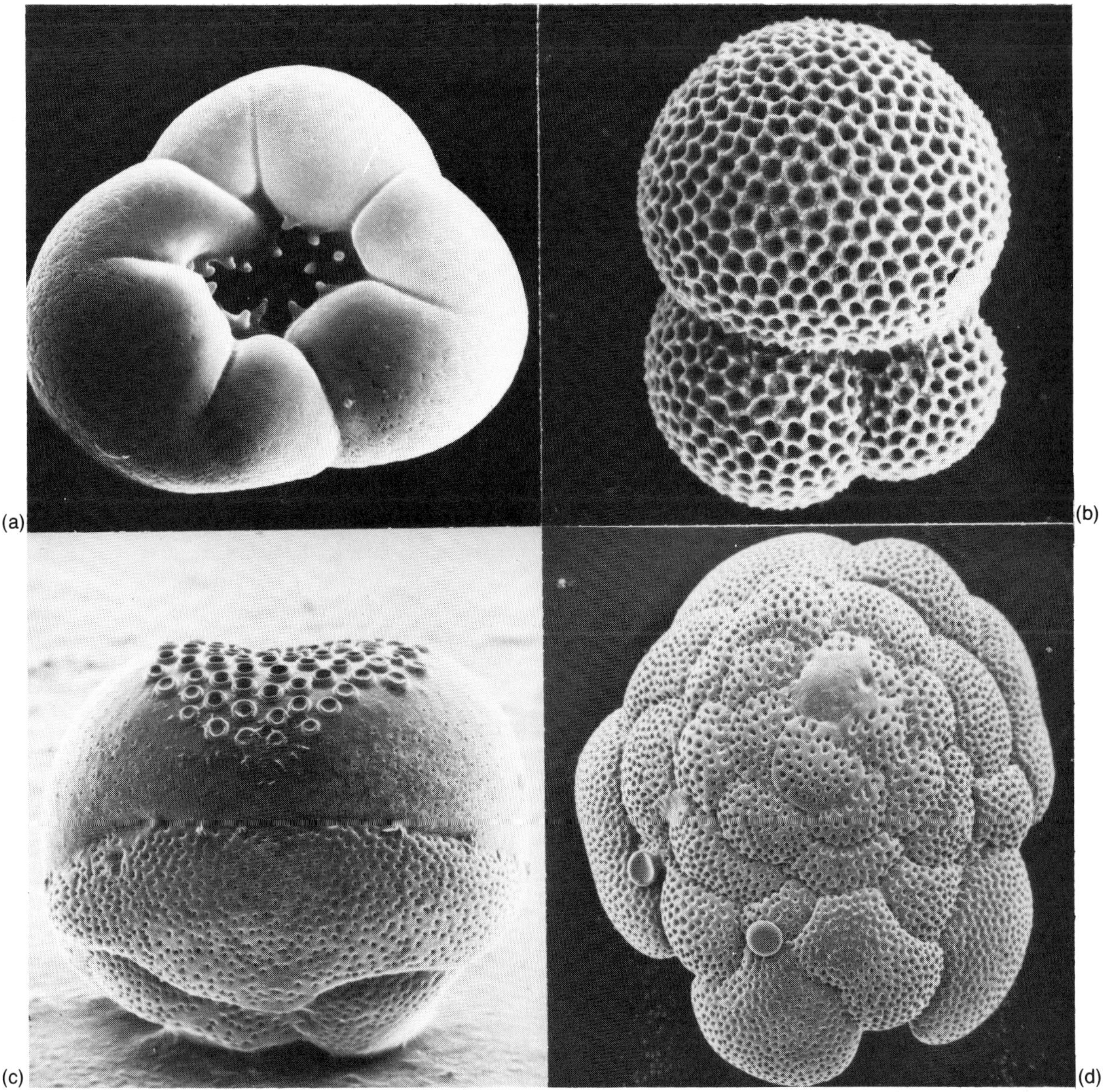

Figure 2.39 Scanning electron micrographs of foraminiferan tests. (*a*) *Rotaliella heterocaryotica*, X1400; (*b*) *Globigerinoides sacculifer*, X105; (*c* and *d*) *Tretomphalus bulloides*; (*c*) Sexual form, X120; (*d*) Asexual form, X80. (From Grell, *Protozoologie*.)

capsule divides the radiolarian cell into two regions (Figs. 2.1 and 2.42a). Within the capsule, the cytoplasm is granular and contains one or more nuclei. Outside the capsule is a frothy, highly vacuolated region called the **calymma.** Long axopodia responsible for food entrapment originate from the calymma or the capsular membrane. Captured food is passed into the calymma for digestion within food vacuoles, and nutrients may be transported to the innermost cytoplasm through perforations in the central capsule. Large vacuoles in the calymma are apparently flotation devices. During sea storms these vacuoles discharge their low-density fluids, causing the radiolarian to sink. At a more stable depth, the vacuoles reform and the animal ascends to surface waters. The radiolarian calymma is a habitat for zooxanthellae. The latter may

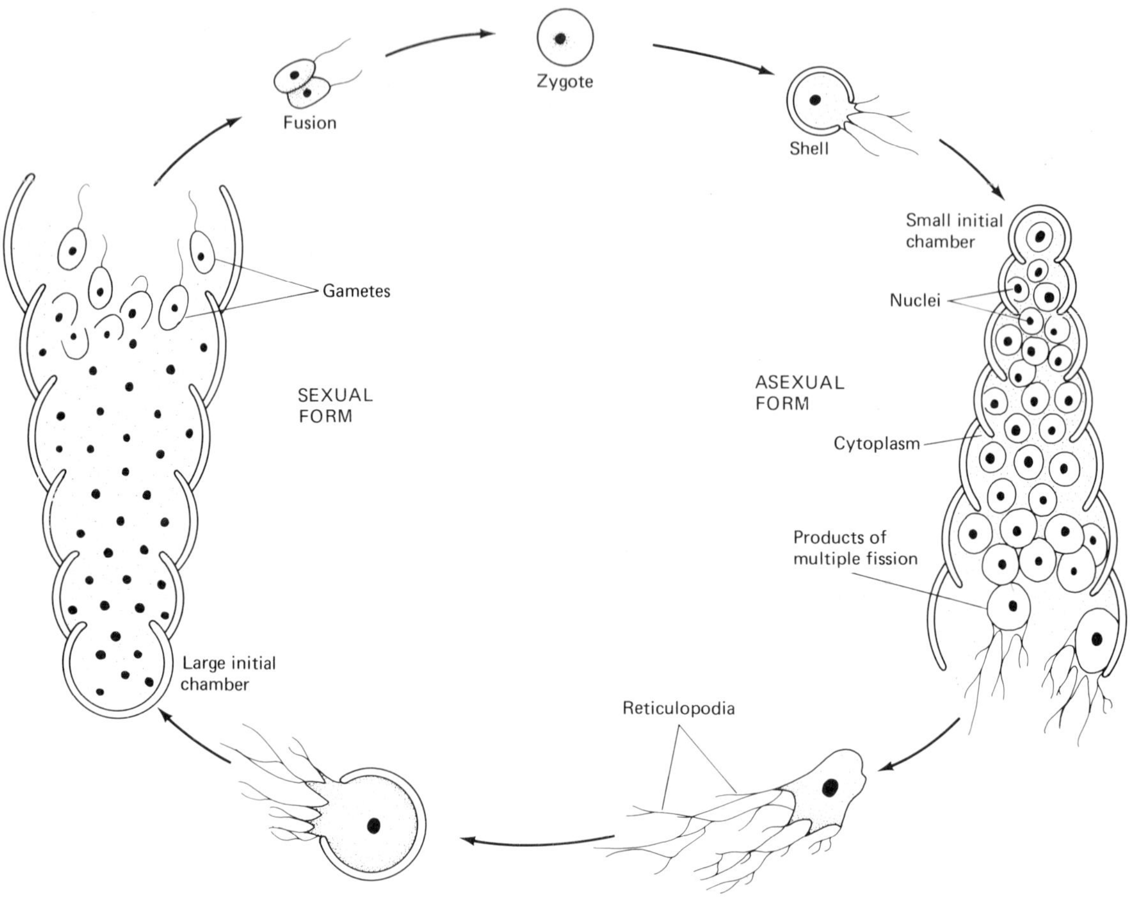

Figure 2.40 The life cycle of a foraminiferan showing alternating sexual and asexual stages. (From Beck and Braithwaite, *Invertebrate Zool. Laboratory Workbook.*)

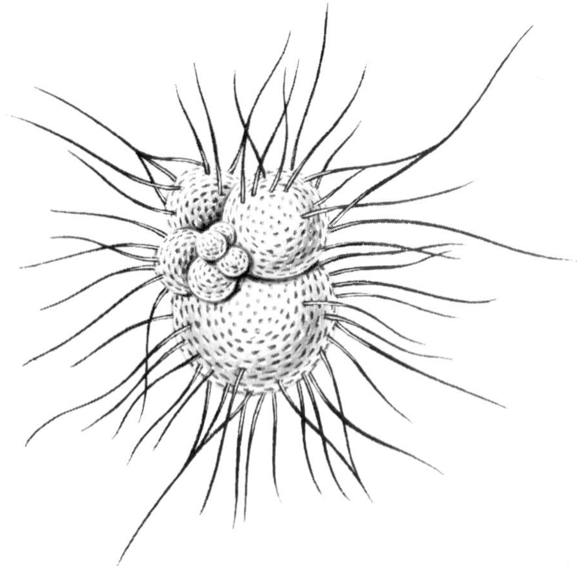

Figure 2.41 Globigerina, a planktonic foraminiferan. (From Hyman, *The Invertebrates, Vol. I.*)

help radiolarians to survive prolonged periods of starvation.

The radiolarian skeleton is the glory of the order, and many species create truly exquisite structures (Figs. 2.41 and 2.42). These siliceous skeletons are organized in concentric branches. In young animals they are confined to the calymma, but growth may bring some skeletal parts within the central capsule. Indeed, in some rapidly growing forms, the central elements of the skeleton eventually may occupy the nucleus itself.

Many aspects of radiolarian biology are poorly investigated because of the pelagic distribution of the group. Their reproduction is not well described. Binary fission and budding are common, and sexuality apparently occurs. Their sexual reproduction may involve flagellated gametes, but the details remain unclear.

THE HELIOZOANS. The last sarcodine order is reminiscent of the radiolarians, but its members are adapted primarily for life in fresh water. Heliozoans

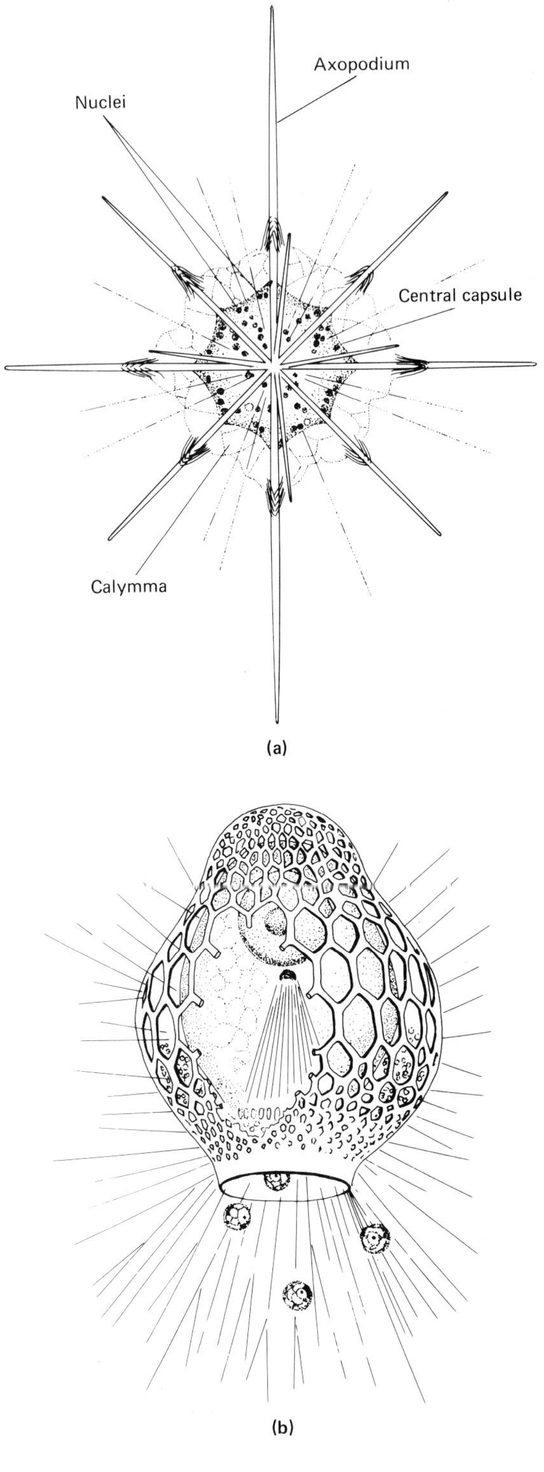

Figure 2.42 (*a*) *Acanthometron elasticum* showing the division of the cytoplasm into the two regions typical of radiolarians; (*b*) *Cyrtocalpis urceolus*. (From Grell, *Protozoologie*.

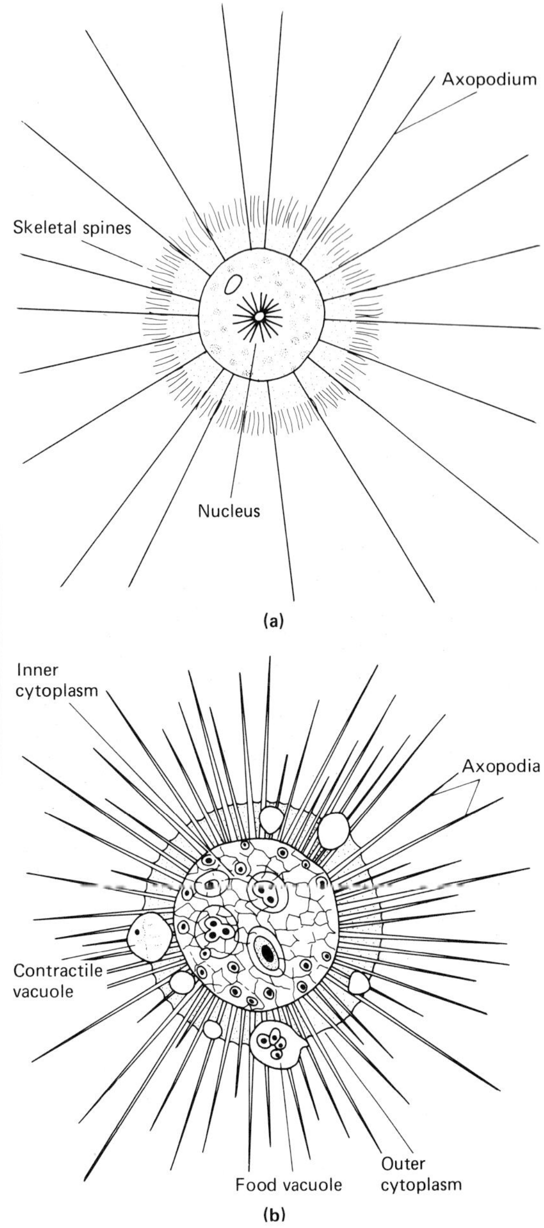

Figure 2.43 (*a*) *Heterophrys myriapoda*, a uninucleate heliozoan; (*b*) *Actinosphaerium eichorni*, a multinucleate heliozoan. (*a* from R.P. Hall, *Protozoology*, 1953, Prentice Hall, Inc.; *b* From F. Doflin, *Lehrbuch der Protozoenkunde*, 1949, Verlag von Gustav Fischer Jena.)

have distinct inner and outer cytoplasmic regions, but no central capsule intervenes (Fig. 2.43). A skeleton may be present, but is less elaborate than those of radiolarians. Long axopodia project a considerable length from the body; their axial tubules may originate

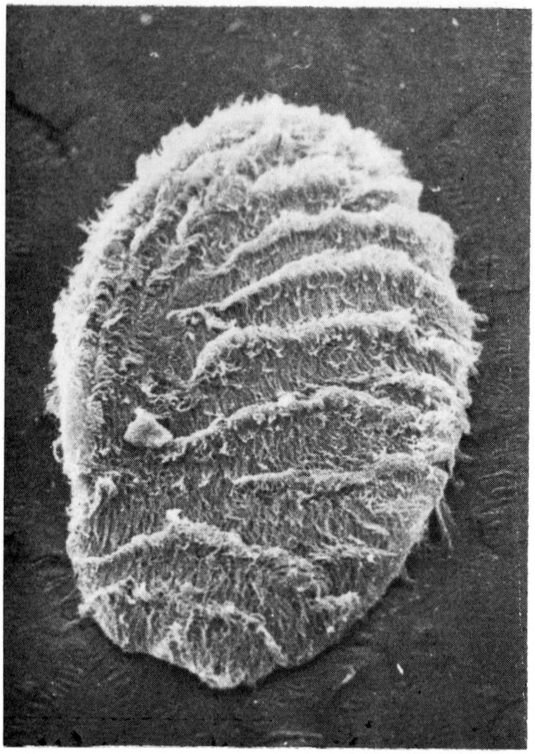

Figure 2.44 Scanning electron micrograph of *Opalina ranorum* showing the characteristic pattern of ciliary beating. (From S.L. Tamm, Proc. Roy. Soc. London, 1970, *175*:219–233. Photo couresy of Dr. S.L. Tamm.)

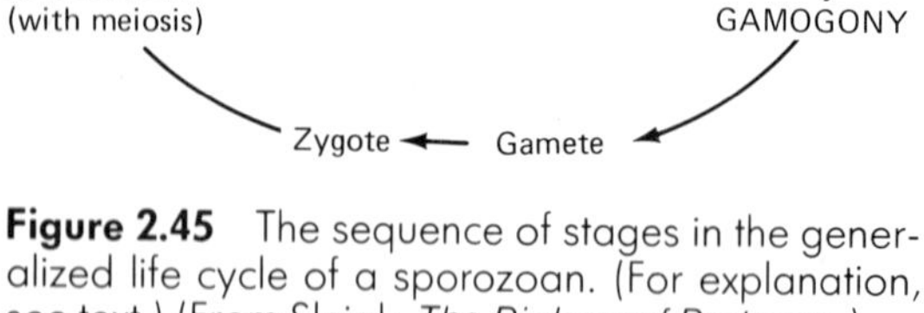

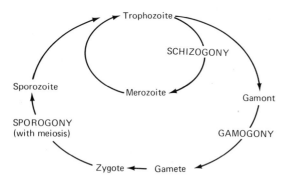

Figure 2.45 The sequence of stages in the generalized life cycle of a sporozoan. (For explanation, see text.) (From Sleigh, *The Biology of Protozoa*.)

from one or more nuclei within the central cytoplasm. Heliozoan axopodia add buoyancy to these primarily floating creatures and play a central role in food capture. When several axopodia contact a large particle of food, the axial filaments are resorbed and

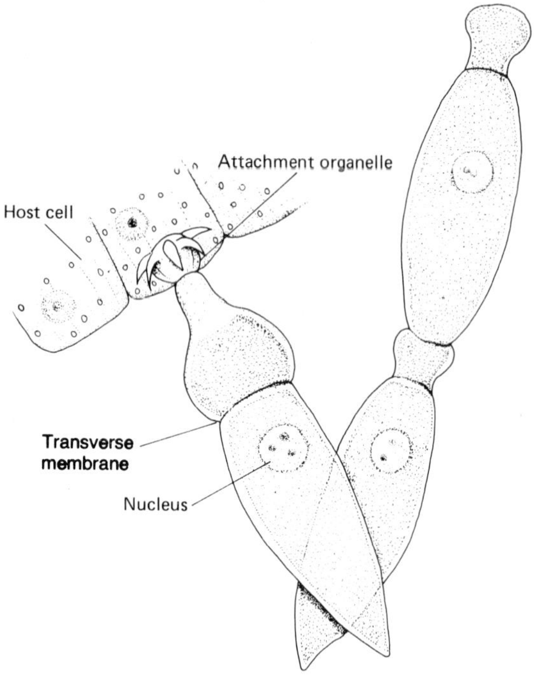

Figure 2.46 Diagram showing the anatomical characteristics of gregarines. On the right are two gregarines in syzygy. (From Beck and Braithwaite, *Invertebrate Zoology Laboratory Workbook.*)

phagocytosis occurs. The outer cytoplasmic region is highly vacuolated, as in the radiolarians, and is the site of digestive activity.

Some heliozoans employ their axopodia in walking over hard substrata. The axial filaments of these organelles slide in relation to one another and thus contract or extend the axopodium. The contraction of forwardly located axopodia and the lengthening of those to the rear effects a slow, rolling motion of the cell.

Under favorable environmental conditions, heliozoan reproduction is by binary fission. Food shortages may initiate sexuality. A starved heliozoan may lose its axopodia, encyst, and undergo gametogenesis. Fertilization occurs within the cyst, and when favorable environmental conditions are restored, a young heliozoan emerges.

Heterophrys (Fig. 2.43a), *Actinosphaerium* (Fig. 2.43b), and *Actinophrys* are representative heliozoan genera.

Class Opalinatea

Members of this small class of multinucleate sarcomastigophorans superficially resemble ciliates and

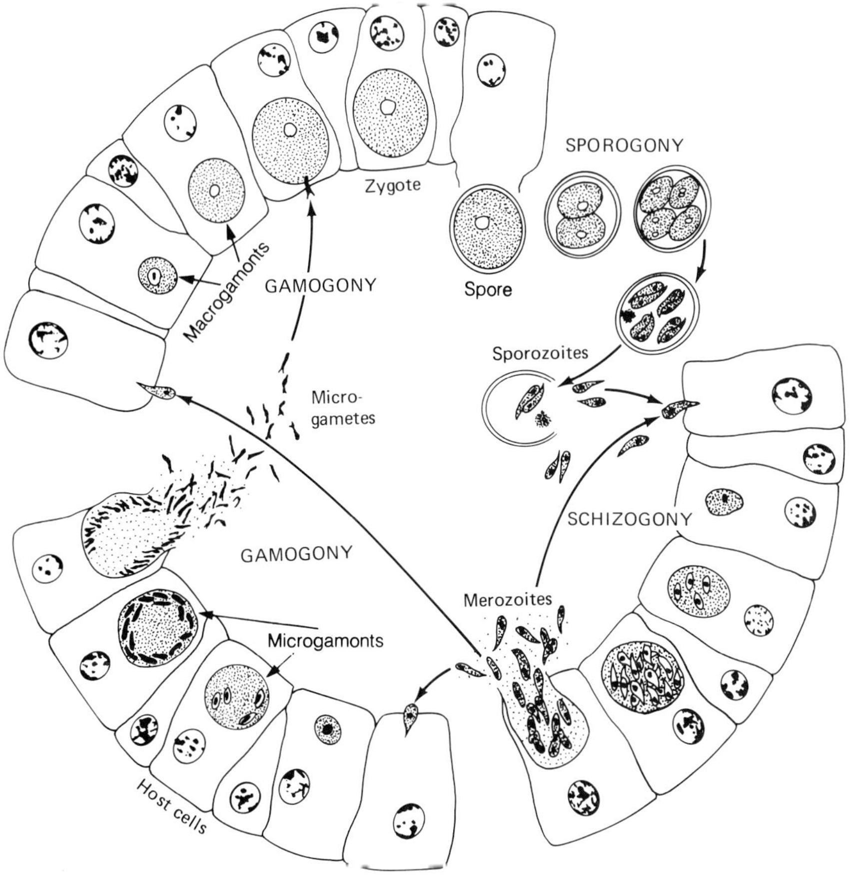

Figure 2.47 The life cycle of the coccidian, *Eimeria.* (For further explanation, see text.) (From E.R. Noble and G.A. Noble, *Parasitology*, 3rd ed., 1972, Lea and Febiger.)

were once classified as such because of their multiple rows of cilia. However, opalinids differ markedly from ciliates in their lack of differentiated macro- and micronuclei and in their sexual reproduction, which involves syngamy rather than conjugation. In asexual reproduction, fission occurs between rows of locomotor organelles (as in flagellates) rather than across rows (as in ciliates). This and certain other morphological features place opalinids taxonomically nearer the flagellates than the ciliates.

Opalinids are commensals living in the large intestines of frogs and toads. In *Opalina ranarum* (Fig. 2.44) the life cycle involves asexual reproduction in adult frogs and sexual reproduction in tadpoles.

SUBPHYLUM SPOROZOA

The wholly endoparasitic Sporozoa represents the second protozoan subphylum. Its 4000 or more species are characterized by a very complex life cycle involving both asexual and sexual reproduction. With few exceptions, the sporozoan life cycle features an encysted stage, the **spore**, which houses the infective form of the parasite. A wide range of invertebrates and vertebrates, including humans, serve as hosts for these protozoans.

Structurally, sporozoans are rather simple animals, a quality characteristic of many highly evolved parasitic forms. With the exception of the microgametes of some species, no locomotor organelles appear in the subphylum and other subcellular structures are not well described. Sporozoan diversity focuses instead upon reproduction and life cycle.

A generalized, sporozoan life cycle is depicted schematically in Figure 2.45. Typically, spores contain a variable number of small, infective **sporozoites,** which are released when the spore enters a new host. Sporozoites often penetrate individual cells. Growth produces a **trophozoite** which may proliferate asexually. This asexual reproduction is called **schizogony**

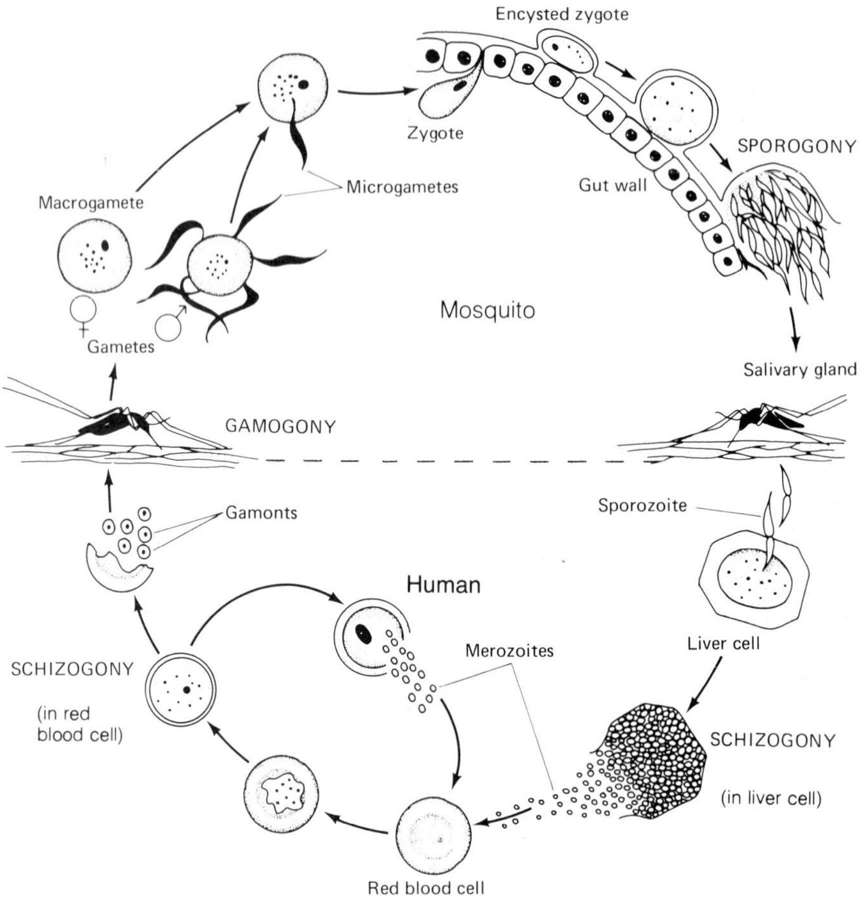

Figure 2.48 The life cycle of *Plasmodium vivax*, a coccidian causing human malaria. (For further explanation, see text.) (From H. Curtis, *Biology*, 2nd ed., 1977, Worth.)

and its products are labeled **merozoites.** The merozoites continually infect new host cells. Eventually, trophozoites or merozoites mature into sexual forms called **gamonts.** The term **gamogony** describes their maturation, which includes the transformation of the gamont into one or more gametes. Both isogametes and anisogametes have been described in this subphylum. Fertilization forms a zygote which may encyst before undergoing yet another reproductive act, **sporogony.** Sporogony, which includes meiotic activities by the zygote, produces the next generation of sporozoites. Sporozoites commonly are confined within a protective envelope or spore, but in a few species sporeless life cycles are known (see below).

Thus the sporozoan life cycle is complex, involving schizogony, or infective asexual reproduction, gamogony, or sexual maturation and gamete production, and sporogony, the production of spores. Two important sporozoan groups are recognized, both belonging to the class Telosporea: the gregarines and the coccidians.

THE GREGARINES. Gregarines are larger sporozoans whose evolutionary history as parasites may be comparatively short. They seem less highly adapted than the coccidians for parasitic life. Gregarine size dictates that these protozoans spend much of their lives within the body cavities and extracellular spaces of their hosts. In addition, the tremendous reproductive potential usually associated with successful parasitism is rather limited in this class, as schizogony is lacking in many species. Gregarine hosts include several invertebrate groups, notably annelids and insects.

The gregarine body is generally elongate, and a transverse membrane may define separate anterior and posterior regions (Fig. 2.46). The anterior region usually bears an attachment organelle, a prerequisite structure for extracellular parasites. *Gregarina* is common in the digestive tract of many insects. Spores consumed with food liberate sporozoites within the insect gut, and growth to the trophozoite stage follows. There is no schizogony. Rather, a unique gregarine process called **syzygy** unites two older trophozoites. Such paired forms mature into gamonts and encyst within a common envelope. The production of isogametes follows, and many zygotes are formed within the original cyst. Each zygote undergoes sporogony to produce a single spore with its resident sporozoites. In the passed feces of the parental host, these spores and their infective stages await a new generation of the life cycle.

Thus, for the gregarines of this representative genus, reproduction is primarily sexual, and numerical increases result from the multiple production of gametes and sporozoites. Other gregarine types include forms which undergo schizogony; also, some gregarines are intracellular parasites during the early trophozoite stage. However, these features are much more typical of the second major sporozoan group, the coccidians.

THE COCCIDIANS. Coccidians are smaller and more specialized than the gregarines. This sporozoan group parasitizes higher invertebrates and vertebrates. Their vertebrate hosts include humans, as this class contains *Plasmodium*, the causative organism of malaria. Coccidians are intracellular parasites that leave their host cells only as infective stages. Schizogony is common in this class, and multiple cycles of asexual reproduction greatly increase the infective population. Two coccidian genera are considered here: *Eimeria*, often a tissue parasite of vertebrates; and *Plasmodium*, infecting blood cells during a complicated life cycle involving two hosts.

Several species of *Eimeria* (Fig. 2.47) are parasites in the epithelia of the internal organs of birds and mammals. Sporozoites invade epithelial cells and grow into trophozoites. Schizogony produces numerous merozoites, which in turn infect more cells. Schizogony may continue for several cycles, as each merozoite generation gives rise to another of exponentially increased number. In time, merozoites mature into gamonts of two types. **Macrogamonts** produce single, large, nonmotile macrogametes; by repeated divisions, smaller **microgamonts** give rise to many biflagellated microgametes. Microgametes leave the host cell and enter macrogametes for fertilization. The resulting zygote encysts and undergoes sporogony. Departure in host feces often precedes sporogony, which is completed within the zygotic cyst and typically produces a single spore with eight sporozoites. Thus these coccidians depend primarily on asexual means to bolster their populations; sexuality in this genus involves anisogametes and does not directly increase cell number.

Coccidians that parasitize blood cells represent the major protozoan threat to human health. Particularly dangerous are certain species of *Plasmodium* whose infections produce malaria. *Plasmodium* belongs to a special group of coccidians whose life cycles exploit two hosts and lack the spores normally characteristic of their subphylum. *Plasmodium* sporozoites are introduced directly into the human blood stream when an infected mosquito takes a blood meal (Fig. 2.48). These immature stages enter liver cells, grow into trophozoites, and begin schizogony. More liver cells may be infected with each merozoite generation, but after at least one schizogony cycle in the liver, red blood cells are the primary targets. Schizogony continues within these blood cells through several generations, producing a *Plasmodium* population of enormous proportions. Typically, there is some synchrony of asexual activity, as many red blood cells simultaneously rupture and release a new merozoite generation. Synchronous blood cell ruptures also release toxins which cause the paroxysmal chills and fever symptomatic of malaria. In time, some merozoites halt schizogony and differentiate into gamonts, which in turn become macrogametes and microgametes. Gamonts remain quiescent within red blood cells until they are consumed by a feeding mosquito. Within that insect's stomach, the gametes mature and form zygotes. Zygotes bore through the gut wall and then attach to the peritoneum, where they encyst and undergo sporogony to form naked sporozoites. No spores are displayed in *Plasmodium*. Sporozoites migrate to the salivary glands of their insect host, where they await another human invasion. Thus *Plasmodium* undergoes schizogony and early gamogony within its human host, and gamogony completion and a modified sporogony without spore formation within an insect vector.

SUBPHYLUM CNIDOSPORA

Cnidosporans are a parasitic group that earlier was classified among the sporozoans but now is considered a distinct subphylum. Cnidosporans share certain traits with the sporozoans; sporelike stages are common in the life cycle and continuity systems are comparably complex. But this subphylum differs in the details of its life cycle and in the presence of peculiar

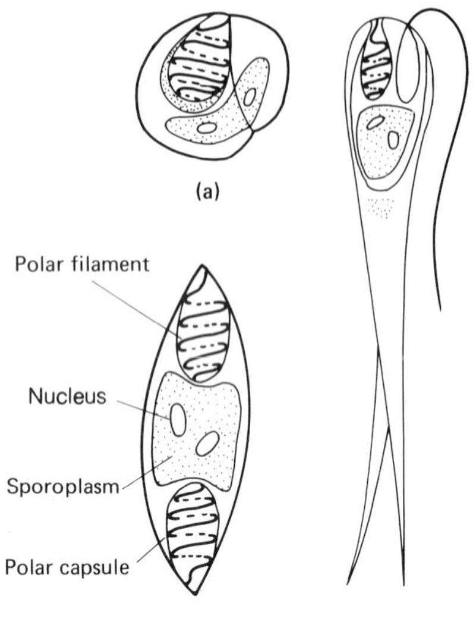

Figure 2.49 Myxosporidian spores. (*a*) *Unicapsulian muscularis* from halibut; (*b*) *Myxidium lieberkuhni* from pike; (*c*) *Henneguya* sp. from a freshwater fish showing one polar filament extruded. (From Sleigh, *The Biology of Protozoa*.)

organelles called **polar filaments.** Two classes of cnidosporans are recognized, the Myxosporidea and the Microsporidea.

THE MYXOSPORIDEA. Myxosporidians include the larger cnidosporans, most of which are tissue and cavity parasites of fish. Figure 2.49 depicts the distinctive spores of this class. The spore wall is divided into two valves meeting in a suture; within these valves, one or more **polar capsules** contain up to four spirally coiled polar filaments. Also present is a central, living **sporoplasm.** The sporoplasm may form two separate, uninucleate masses or a single, binucleate unit.

When the myxosporidian spore is ingested by a host, it migrates to an appropriate cavity or tissue, whereupon the polar filaments are extruded for anchorage. The sporoplasmic nuclei fuse, and the sporoplasm emerges in amoeboid form (Fig. 2.50). This nuclear fusion may be considered a sexual act and probably contributes to genetic variability within the class. The cnidosporan invades accessible areas, undergoing repeated schizogony to increase its infective potential. Distinct daughter cells may be formed, but asexual division may involve nuclei alone, in which case a large (up to several millimeters) plasmodial parasite is formed. In time, some nuclei and their surrounding cytoplasm encyst and undergo sporogony. Sporogony in this class is intriguing in that it features a multicellular stage which behaves in a peculiarly metazoan manner. Mitotic divisions create a six-celled unit within the cyst. Two of these cells become the valves of the future spore, two become its polar capsules, and the remaining two cells constitute the sporoplasm. Such developmental differentiation is unparalleled among the protozoans and has fueled controversy as to the exact nature of this group. Some students disqualify the myxosporidians from membership in the Protozoa, and few taxonomists agree completely on their proper place in the invertebrate world.

THE MICROSPORIDEA. Microsporidians are intracellular parasites in almost all animal groups. Insects are the most common hosts. This second cnidosporan class has very tiny spores, some of which are bacterial in size. Their life cycle involves schizogony and sporogony but no sexual activity of any kind. Microsporidian spores possess only one polar filament.

The genus *Nosema,* containing destructive parasites of honey bees and silk moths, has received major attention. Its spores house a single sporoplasm around which the polar filament is coiled (Fig. 2.51). Polar capsules are absent. When the spore is ingested, the hollow filament everts and the sporoplasm emerges through it to invade a host cell. Schizogony proceeds until the host cell is filled with parasites; sporogony follows. Microsporidians leave the host cell only as spores. Shed with the feces, these spores remain dormant until eaten by an appropriate host.

SUBPHYLUM CILIOPHORA

The fourth and final protozoan subphylum is the Ciliophora. Only one class, the Ciliata, is represented and all of its members are called ciliates. Ciliates are a large group—over 7000 species have been described—and a radiating evolution has generated many ciliate forms and life-styles. There are solitary and colonial species, and both symbiotic and free-living forms. Many species are free-swimming; some are sessile. Aquatic environments are favored, but ciliates live also in moist soils and parasitic habitats. Despite such diversity, however, the Ciliophora probably is the most unified protozoan subphylum. Its unifying characters include a complex living pellicle, nuclear dimorphism, and a unique sexuality.

The ciliate pellicle (Fig. 2.6) is a well-ordered external covering of rather consistent architecture throughout the class. This living envelope comprises a series of membranes and intramembranal spaces, the entire

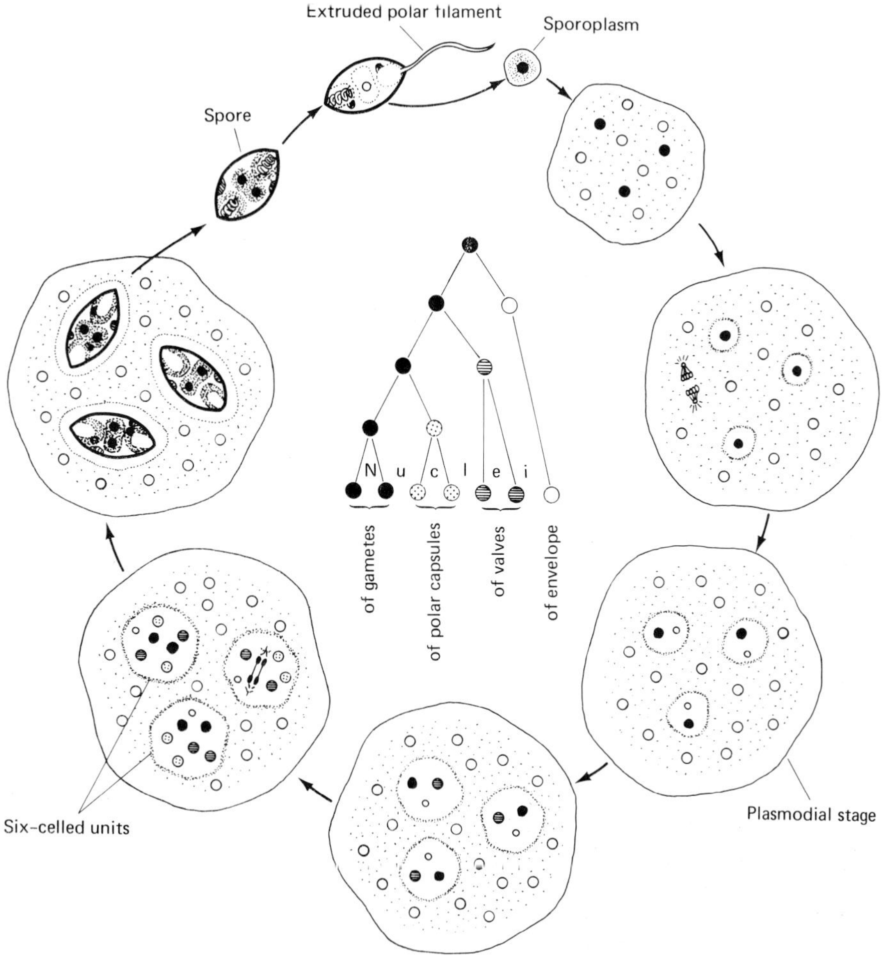

Figure 2.50 Life cycle of a myxosporidian. (For additional explanation, see text.) (From Grell, *Protozoologie*.)

infraciliature, and numerous organelles called **trichocysts.** An outer plasma membrane is continuous over the body surface and covers each projecting cilium. Beneath this surface, additional membranes delineate an array of spaces called **alveoli.** The infraciliature is located among the alveoli; commonly, two of these spaces separate adjacent kineties. Just beneath the infraciliature are the **trichocysts.** These organelles are distributed generally over the body surface in many ciliates, but in some species they may be restricted to the oral region. Trichocysts discharge long striated shafts tipped with barbs which may contain toxins (Fig. 2.52). Their discharge serves offensive and defensive functions and may be instrumental in prey capture and anchorage. The cytoplasm abutting the pellicle is usually quite dense and provides anchorage for the trichocysts. In addition, this ectoplasm may contain **myonemes,** large contractile bundles of microfilaments (Figs. 2.62 and 2.65).

Ciliate nuclei occur in variable numbers, but two distinct nuclear types are displayed by virtually every member of the subphylum: the large, typically polyploid **macronucleus,** which serves vegetative functions; and the **micronucleus,** which is considerably smaller and diploid and is responsible for continuity. Asexual reproduction in this group always involves a mitotic division of the micronucleus, while the macronucleus often simply pinches in two. Less commonly, the macronucleus may degenerate altogether before fission and then reform from the micronucleus in each daughter cell. Ciliate fission typically occurs along a transverse plane; that is, fission divides the

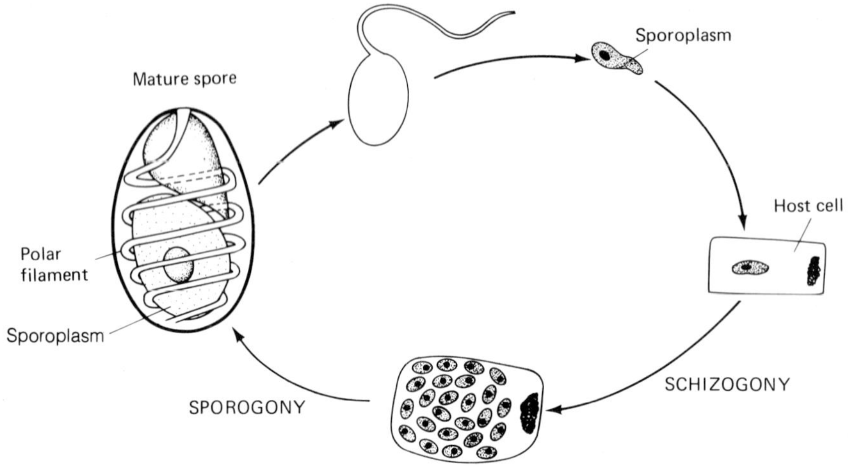

Figure 2.51 The life cycle of the microsporidian, *Nosema*. (For further explanation, see text.) (From Sleigh, *The Biology of Protozoa*.)

Figure 2.52 An electron micrograph showing discharged trichocyst tips from *Paramecium*, X14,000. (From M.A. Jakus, *Bio. Bull*, 1946, *91*:141. Photo courtesy of Dr. Jakus.)

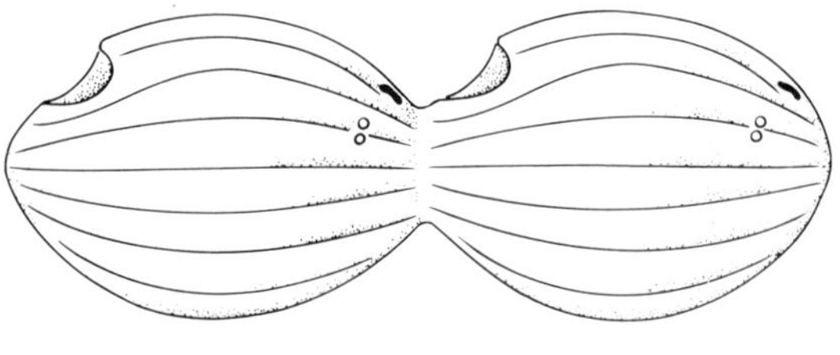

Figure 2.53 *Tetrahymena* in a late stage of fission, the dividing line cutting across the kineties. (From Corliss, *The Ciliated Protozoa*.)

animal across its kineties and is never longitudinal as in the flagellates (Fig. 2.53). The apparent advantage of nuclear dimorphism is that it liberates the macronucleus from reproductive chores. Polyploidy in this larger nucleus may generate a chromosome number as high as 13,000 n, an impracticality for any nucleus concerned with the genetic integrity of its species. Such an awesome DNA store is prolific in the RNA transcription necessary for daily activities, while the micronucleus can concentrate exclusively on reproduction.

Ciliate sexual reproduction is unique. Micronuclei alone serve as gametes and are exchanged between two individuals during a process known as **conjugation.** Conjugation is initiated when, under appropriate physiological conditions, two compatible ciliates fashion a cytoplasmic bridge between their cells (Fig. 2.54). Nuclear reorganization follows and varies depending on the number of each nuclear type present. The conjugation of a ciliate like the familiar *Paramecium caudatum*, which has one macronucleus and one micronucleus, is representative and relatively simple to follow (Fig. 2.55). The macronucleus of a conjugating *P. caudatum* disintegrates. A two-step meiotic division by the micronucleus produces four haploid nuclei, three of which are resorbed by the cytoplasm. The remaining haploid micronucleus divides once mitotically. One of the mitotic products in each cell then migrates across the cytoplasmic bridge and fuses with the stationary nucleus of its partner. The two ciliates then separate. Following conjugation, each animal has a genetically reconstituted micronucleus, which by a series of mitotic divisions and differentiations (in some cases accompanied by cell divisions), restores the original number of each nuclear type. Conjugation per se does not increase the ciliate population, but does allow the genetic recombination

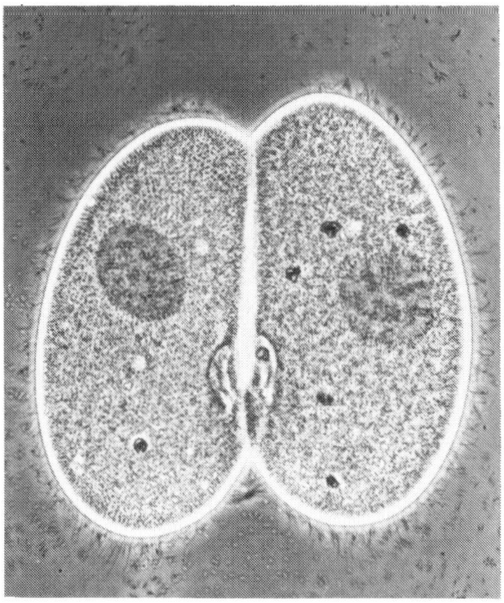

Figure 2.54 Photomicrograph of a pair of *Paramecium trichium* in conjugation. (Photo by D. Ammerman, from Grell, *Protozoologie*.)

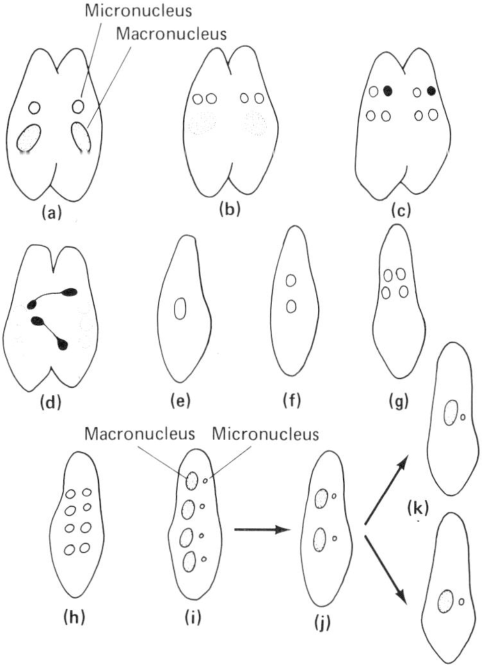

Figure 2.55 The sequence of nuclear events during and just after conjugation in *Paramecium caudatum*. (a–d) Conjugation; (e–k) Post-conjugation events in one of the exconjugants. The other conjugant would show the same pattern. (For further explanation, see text.) (From P.A. Meglitsch, *Invertebrate Zoology*, 2nd ed., 1972, Oxford University.)

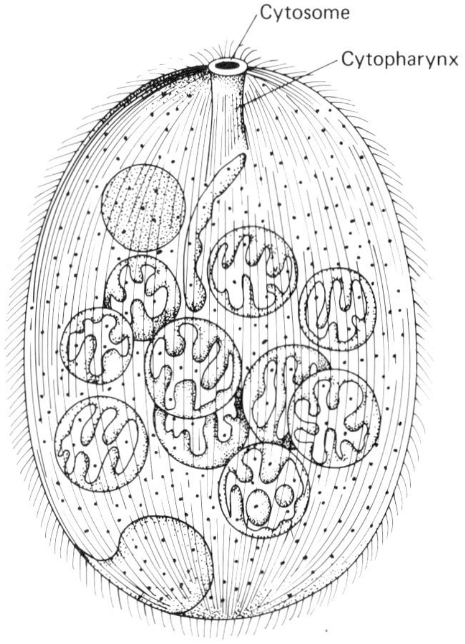

Figure 2.56 *Prorodon teres* showing the uniform ciliature and anterior cytopharynx. (From Grell, *Protozoologie*.)

stimulating the evolution of sexual creatures. Asexual divisions normally follow conjugation quite closely in time; but again, reproductive programs vary with the individual species.

Another essentially sexual form of ciliate reproduction is called **autogamy.** Autogamy resembles conjugation in general terms of nuclear activity, but involves only a single individual. A lone ciliate engaged in autogamy proceeds to a stage at which it contains two identical haploid micronuclei; these two nuclei then fuse, producing a diploid (and homozygous) individual. Apparently, autogamy has at least some of the same salutory effects as conjugation. A ciliate culture may thrive only for a limited number of asexual generations. Conjugation or autogamy then becomes essential to the continued growth of the culture.

Often, mating types occur among ciliates. Individuals bear sexual labels and may conjugate only with a conspecific member of an appropriate type. Sexual identity probably is localized in the pellicle and may be discriminated during the formation of the cytoplasmic bridge. Mating types prevent the inbreeding of closely related individuals and thus enhance genetic variability within a ciliate population.

Diversity within the Ciliophora concerns the adaptive radiation of the ciliature, particularly in the oral regions. Four subclasses of the Ciliata represent four different trends in ciliary evolution.

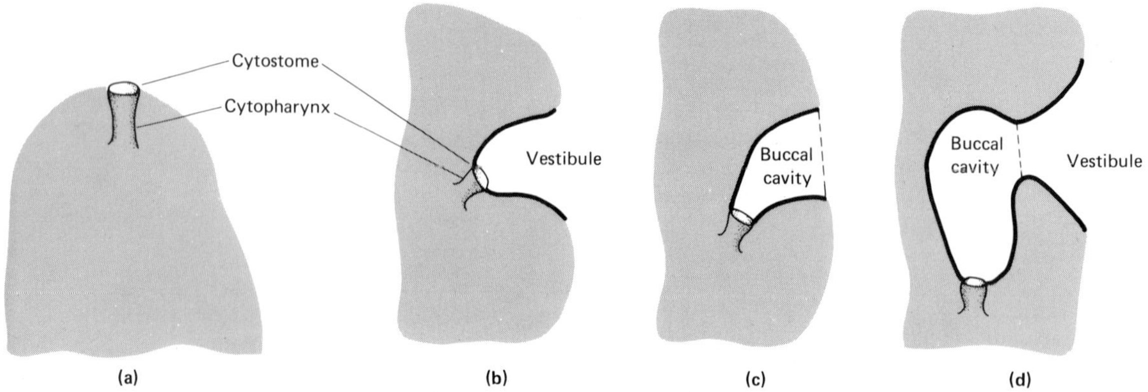

Figure 2.57 A diagram illustrating evolutionary trends in the oral region of holotrichous ciliates. (For explanation, see text.) (From Corliss, *The Ciliated Protozoa*.)

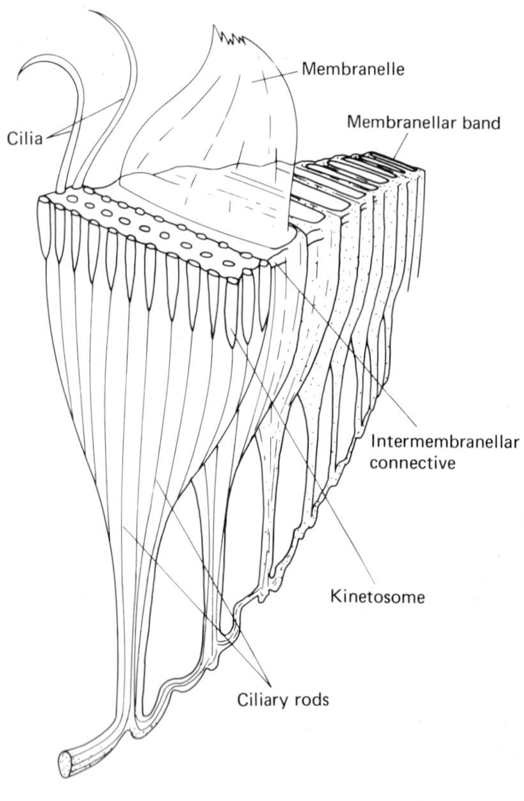

Figure 2.58 A drawing showing the structure of membranelles as viewed through an electron microscope. Each membranelle possesses up to 3 rows of cilia with 20 to 30 cilia per row. (From I.W. Sherman and V.G. Sherman, *The Invertebrates: Function and Form: A Laboratory Guide*, 2nd ed., 1976, Macmillan.)

THE HOLOTRICHIA. Holotrichs are the most primitive ciliates. Lower genera like *Prorodon* (Fig. 2.56) possess a uniform ciliature. The anterior cytostome or mouth usually opens into a short cytopharynx terminating within the endoplasm (Fig. 2.57a). A decided evolutionary trend witnesses the displacement of the holotrich mouth toward one side of the body and the development of a preoral chamber. If the preoral chamber is unciliated or it its ciliature is simply continuous with that of the general body, the chamber is called a **vestibule** (Fig. 2.57b). Many holotrichs possess a preoral chamber whose cilia are organized into compound organelles; such chambers are called **buccal cavities** (Fig. 2.57c). Some advanced holotrichs such as *Paramecium* display both a buccal cavity and a vestibule (Fig. 2.57d). Ciliary organelles bordering the buccal cavity are of two general types: the **membranelle** and the **undulating membrane.** A membranelle (Fig. 2.58) comprises up to three short rows of cilia. Undulating membranes (Fig. 2.59) represent cilia fused in a single, longer row. *Tetrahymena* (Fig. 2.60) is a holotrich genus displaying a typical arrangement of these buccal organelles. Three membranelles are present on one side of the buccal cavity of this ciliate, and a single, undulating membrane flanks the other side.

Most free-living holotrichs are holozoic, and both raptorial and ciliary feeders are described. Figure 2.61 depicts the photogenic *Didinium* as it prepares to engulf a *Paramecium. Tetrahymena* uses a different technique of food acquisition. Members of this genus are primarily bacterial feeders, culling their food from water currents established by the metachronal beating of their buccal ciliature. In every case, holotrichs form food vacuoles at the innermost point of the cytostome or cytopharynx.

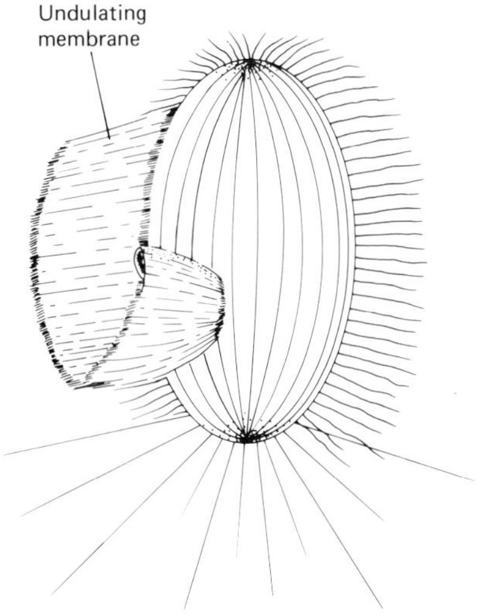

Figure 2.59 *Pleuronema*, showing a well-developed undulating membrane. (From Corliss, *The Ciliated Protozoa*.)

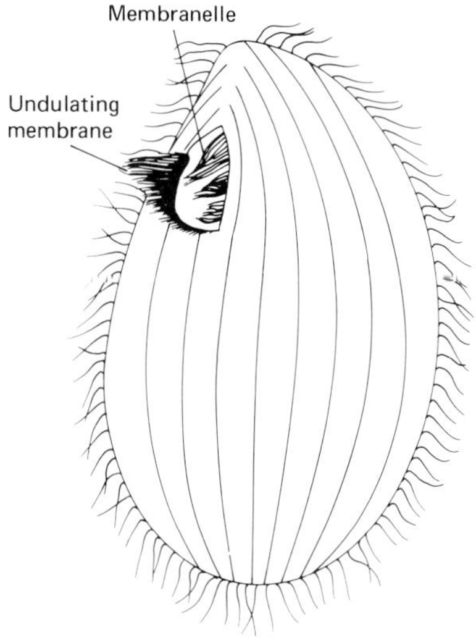

Figure 2.60 *Tetrahymena*, showing an arrangement of buccal organelles typical of holotrichs. (From Corliss, *The Ciliated Protozoa*.)

THE PERITRICHIA. The peritrichs are ciliates whose buccal cavity is known as an **infundibulum.** *Vorticella* (Fig. 2.62) is a well-known, representative genus. The infundibulum is deep and tubular, actually quite funnel shaped, and is bordered on either side by a spiraling row of cilia. Called **polykineties,** these ciliary rows are the peritrich counterparts of the membranelles and undulating membrane mentioned above. The polykineties form a ciliary whorl at the animal's anterior end and then descend along the infundibulum in a counterclockwise spiral.

Peritrich body form differs from the holotrich norm; the differences are attributable primarily to the sessile life styles pursued by most members of the group. Body ciliature is reduced considerably, and many species possess a long, contractile stalk. Contractions of the stalk are effected by myonemes located beneath the pellicle (Fig. 2.62). Other pellicular structures, including the entire infraciliature, remain intact despite the extreme reduction in body ciliation. This persistence of the infraciliature among the peritrichs and other specialized ciliates suggests the relatively recent appearance of minimally ciliated forms. Further evidence of a fully ciliated ancestor is seen in *Vorticella* "larvae." Produced by budding, these immature forms move under the power of a posterior circlet of cilia.

Vorticella undergoes conjugation, but its sessile condition is associated with certain logistical changes. Two distinct conjugating forms occur. Macroconjugants remain stationary, while smaller microconjugants break loose from their stalks and swim. Microconjugants contact the larger forms (Fig. 2.62) and the characteristic nuclear exchanges occur. The microconjugant never releases, however; a spent "male" peritrich simply withers and dies.

Peritrichs often form colonies with two or more individuals sharing a common stalk (Fig. 2.63a). Other colonial patterns occur, including groups of stalkless peritrichs embedded in a gelatinous mass (Fig. 2.63b).

THE SPIROTRICHIA. The spirotrichs possess the most sophisticated oral ciliature in the subphylum. Typically, a long zone of membranelles spirals into the buccal cavity in a clockwise direction (Fig. 2.64). (Recall that peritrichs display a counterclockwise spiral.) Spirotrichs usually retain their body ciliature, although cilia often are organized into groups called **cirri** (Fig. 2.64). Spirotrichs are very common in aquatic environments, and the group includes some of the largest known ciliates.

Stentor (Fig. 2.65) is a representative spirotrich genus. Its species may measure well over a millimeter in length and therefore number among the relatively few protozoans visible to the naked human eye. *Stentor*'s buccal field is situated apically on its often trumpet-shaped body. Its macronucleus has a peculiar shape, long and constricted periodically so as to resemble a string of beads. Such nuclear morphology is shared by another common spirotrich, *Spirostomum* (Fig. 2.66). This long and slender ciliate has a very large oral zone, and its compound ciliary organelles

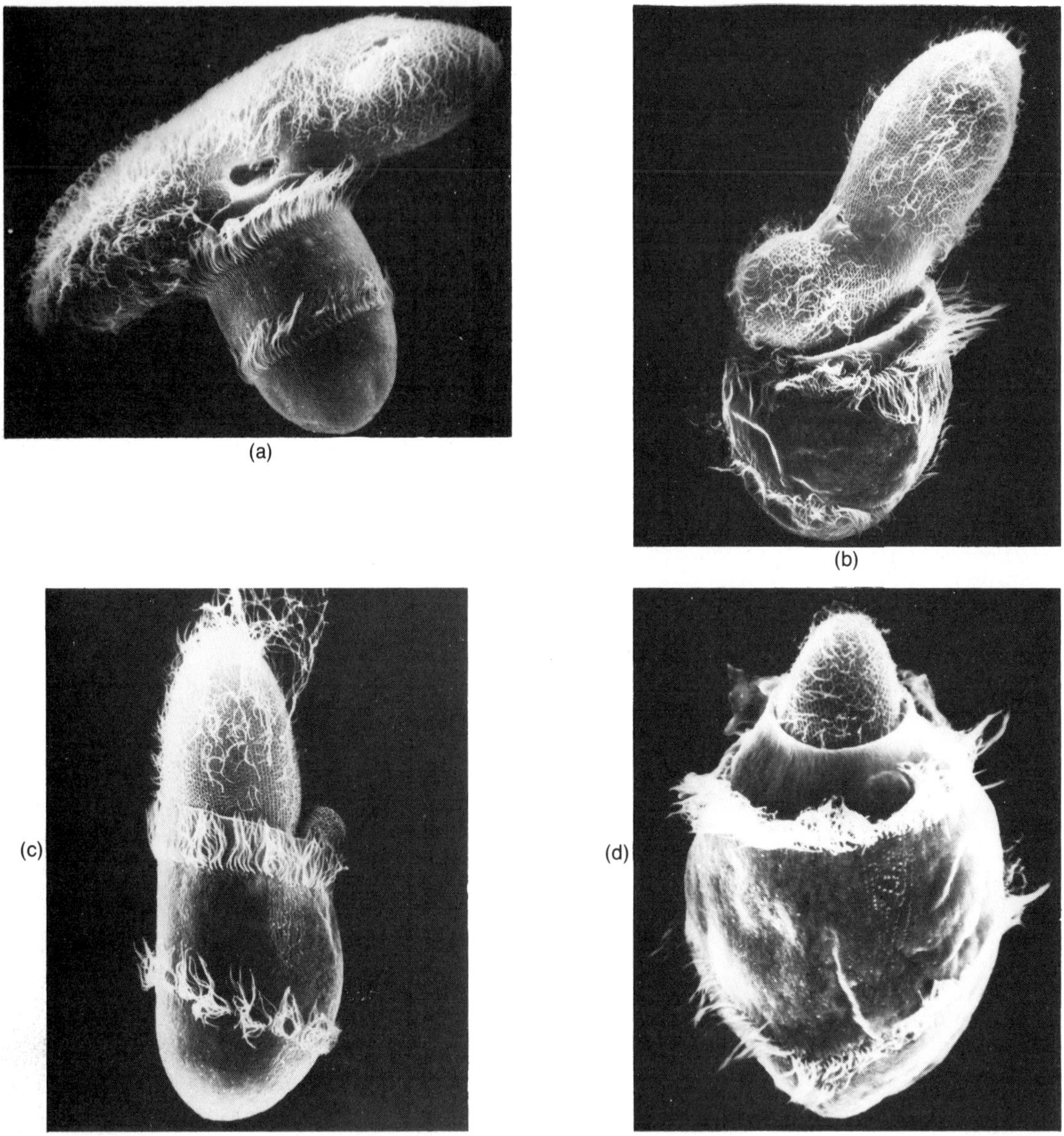

Figure 2.61 (*a–d*) Scanning electron micrographs showing *Didinium nasutium* feeding on *Paramecium multimicronucleatum*. (*a* and *b*) Early ingestion, X1400 and X1500, respectively; (*c*) Mid-ingestion, X1300; (*d*) Late ingestion, X1800. (From H. Wessenberg and G.A. Antipa, J. Protozool., 1970, *17:250*. Photos courtesy of G.A. Antipa.)

extend lengthwise along the body surface. Both *Stentor* and *Spirostomum* are freshwater organisms.

Another group of spirotrichs is primarily marine, and its members possess a test composed of cemented foreign particles. *Tintinnopsis* (Fig. 2.4) is one of the few freshwater forms included among these testate ciliates, whose oral ciliature is often quite sophisticated.

Euplotes (Fig. 2.64a) is a specialized spirotrich genus. These ciliates and their close relatives are dorsoventrally flattened, and their ciliature is reduced and highly modified. Typically, all body cilia are gathered into cirri, which may assume locomotor functions, allowing the animal to creep over the substratum.

Most spirotrichs are ciliary feeders, harvesting small organisms and organic particles from surrounding

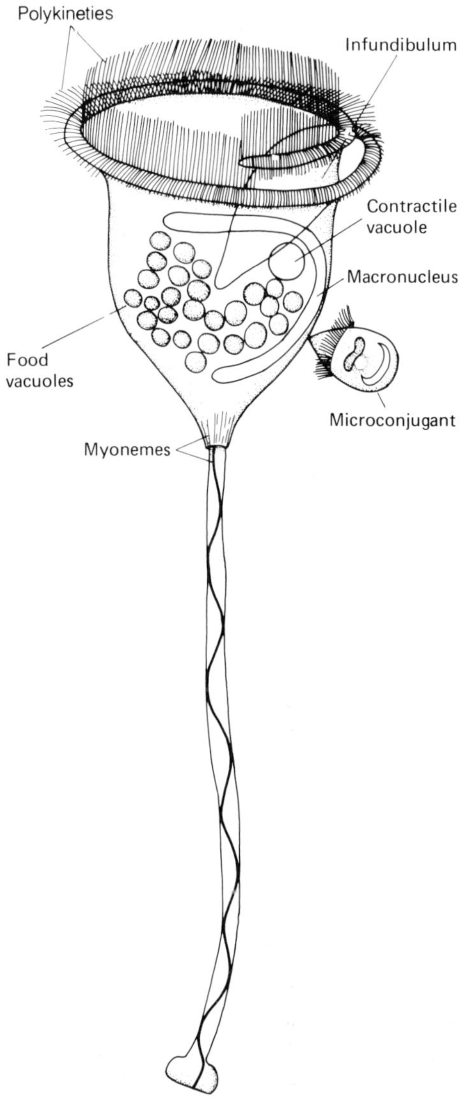

Figure 2.62 *Vorticella*, a peritrichous ciliate. Note the microconjugant attached to the stalked macroconjugant. (From Hyman, *The Invertebrates, Vol. I.*)

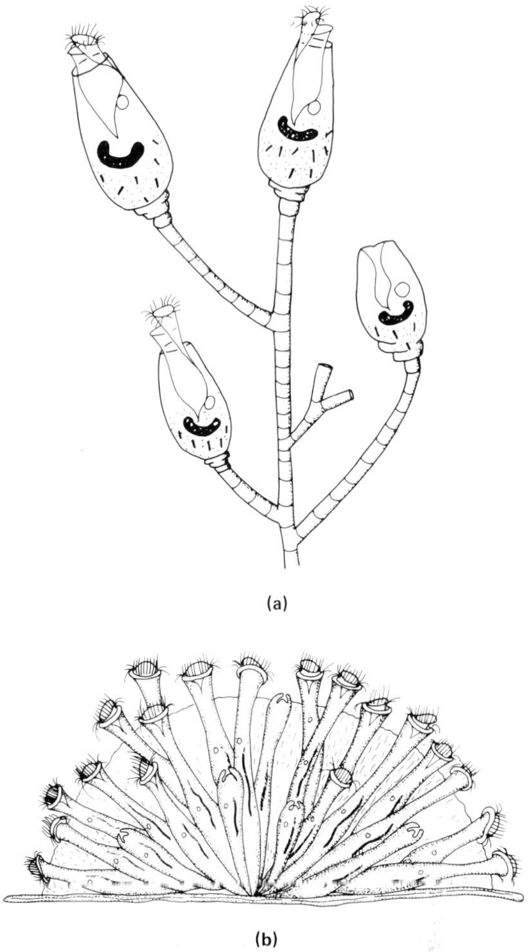

Figure 2.63 (*a*) *Opercularia ramosa*, a colonial peritrich with a non-contractile stalk; (*b*) *Ophrydium sessile*, a colonial peritrich with stalkless individuals embedded in a gelatinous mass. (From Sleigh, *The Biology of Protozoa.*)

waters. Their sophisticated buccal ciliature is highly effective in producing feeding currents.

THE SUCTORIA. Suctorians are unusual in that, as adults, they are totally devoid of cilia. These protozoans are normally sessile and are often stalked. *Ephelota* (Fig. 2.67) and *Acineta* (Fig. 2.68) are common genera. Suctorians capture prey with tentacles which originate as tubular outgrowths of the plasma membrane. Some tentacles are pointed and prehensile, serving as prey-capturing devices; others are knobbed, hollow tubes through which prey cytoplasm is sucked into the cell proper. Just how this sucking force is generated remains unknown, but most hypotheses include an active role by the microfilaments that reinforce the tentacular walls.

While obviously distinct from other members of the subphylum, suctorians are undoubtedly ciliates. They display dimorphic nuclei, and neighboring individuals may conjugate. Additionally, ciliated daughter cells bud from their suctorian parent and swim actively before settlement and metamorphosis to the adult form (Fig. 2.68).

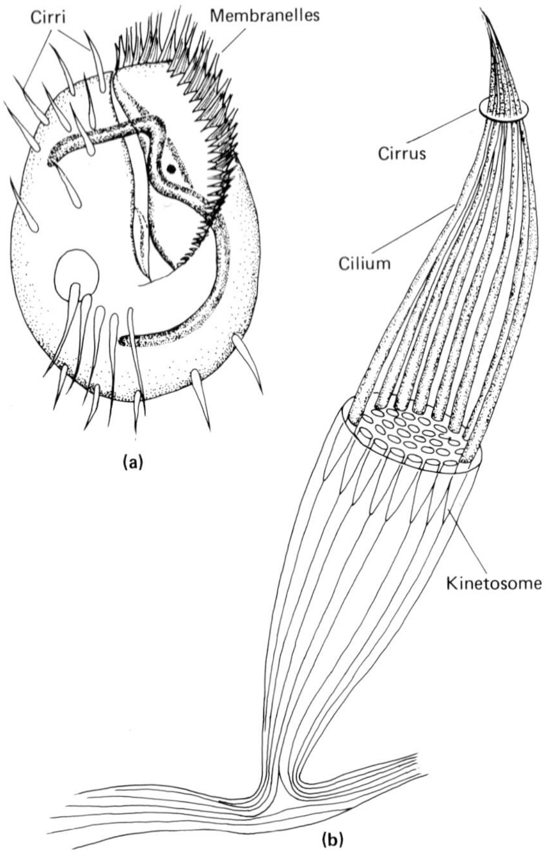

Figure 2.64 (*a*) *Euplotes patella*, a cirri bearing spirotrich; (*b*) Diagrammatic reconstruction of a cirrus showing structure as revealed by the electron microscope. A cirrus consists of 4 to 6 rows of cilia with approximately 6 cilia per row. (*a* from Sleigh, *The Biology of Protozoa*; *b* from Sherman and Sherman, *The Invertebrates: Function and Form.*)

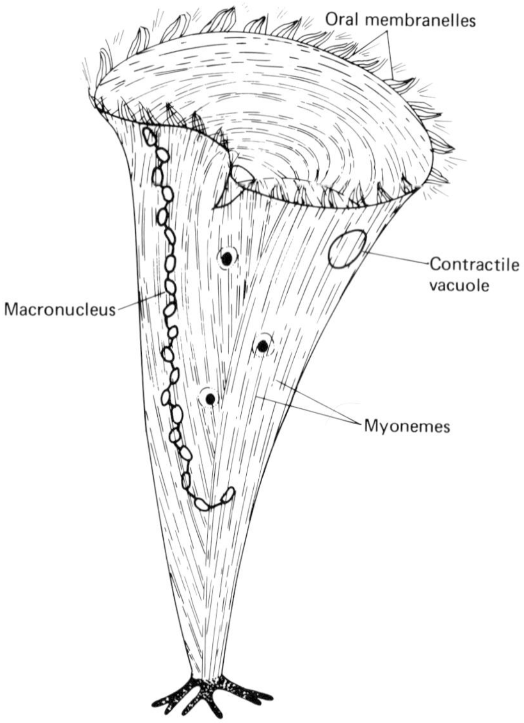

Figure 2.65 *Stentor coeruleus*, a common freshwater spirotrich. (From Sleigh, *The Biology of Protozoa.*)

Protozoan Evolution

Protozoan evolution was and is a momentous affair. Few areas of study address themselves more directly to the historical beginnings of the animal world. The evolution of protozoans involves three major areas: the origin of the phylum, the relationship between its various groups, and the protozoan relationship with the metazoan world.

The origin of unicellular organisms was a landmark in biological history. Certainly, life had existed for some time before protozoans appeared; indeed, some biologists believe that the evolution of cellular life from noncellular, organic precursors required a time period equal to that which has elapsed since protozoans first appeared. Precellular evolution is a fascinating and speculative biochemical story that lies outside the scope of this book; it involved countless experiments with biological organization. We will never know all the kingdoms and subkingdoms of precellular life that may have appeared and perished. But somewhere, perhaps one or two billion years ago, the mutual ancestors of blue–green algae, bacteria, and modern plants and animals gave rise to cells. A single origin or multiple origins may have occurred. Eukaryotic cells were organized far more effectively than anything previously seen in the biological world. Their membranes, well-defined nuclei, and other organelles allowed an unprecedented efficiency. Indeed, most of the history of life on earth reflects the continuing evolution of this basic biological unit.

The first animals probably were saprozoic forms that sipped the organic soup of the primeval ocean. As early cellular life became subject to an oxidizing rather than a reducing atmosphere, photosynthetic

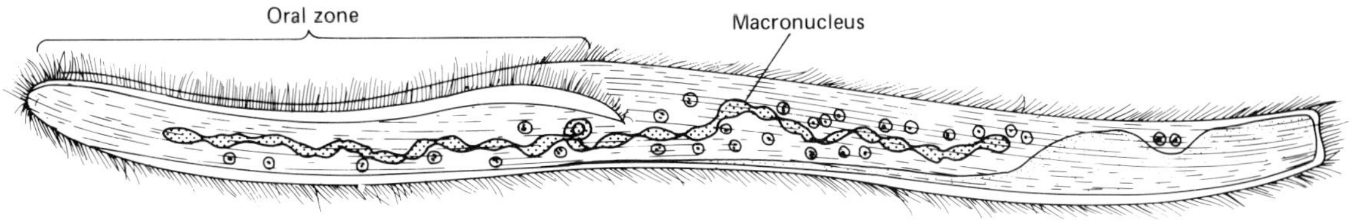

Figure 2.66 *Spirostomum ambiguum.* (From Grell, *Protozoologie.*)

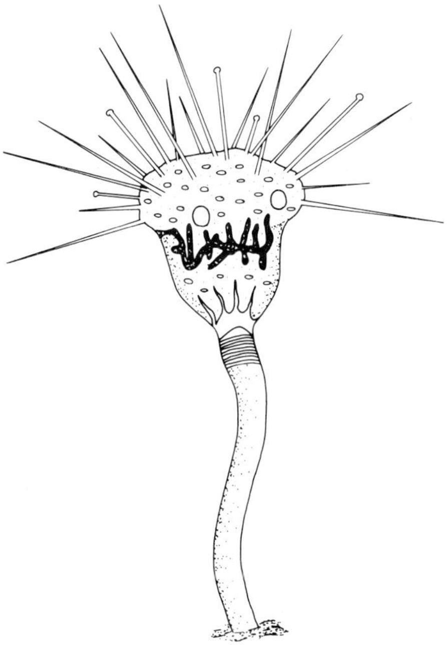

Figure 2.67 *Ephelota gemmipara*, a marine suctorian. Both pointed and knobbed tentacles are present. (From Sleigh, *The Biology of Protozoa.*)

forms developed and flourished. Heterotrophic and autotrophic types were not distinct at that time, and it is premature to assign these ancient organisms to either the plant or the animal kingdom. Phytoflagellates are recognized by most biologists as the most primitive protozoans, and these creatures apparently figure in the ancestry of both major kingdoms. At present, the relationships among the principal protozoan groups remain obscure. The distinct possibility that cellularity has multiple origins further compounds the task of the protozoologist. Moreover, the protozoans have not provided us with a particularly helpful fossil record. Only foraminiferans and radiolarians are well documented in Precambrian deposits, but even these fossils shed little light on intergroup affinities.

Figure 2.69 displays some proposed lines of evolutionary relationship within this phylum. None is proven or, at our current level of knowledge, even provable. Such conjectures are simply all that we can manage. Phytoflagellates theoretically gave rise to zooflagellates and possibly to sarcodines. Alternatively, sarcodines may be derived from zooflagellates. The sporozoans and cnidosporans probably diverged from separate zooflagellate stocks. Ciliates, usually considered the most highly evolved protozoans, likewise are traced to a zooflagellate origin.

The last major question regarding protozoan evolution involves the relationship between unicellular and multicellular animals. Numerous theories have been proposed on the matter, and two have emerged with considerable support. The first theory argues that metazoans arose from colonial flagellates. According to this view, individual cells within these colonies gradually subordinated themselves to the welfare of the colony as a whole. Differentiation specialized member cells for unique functions and left them dependent on each other for ultimate survival. In this manner, the colony itself became a single living animal. This so-called **colonial theory** is consistent with the position that coelenterates and sponges (probably independently evolved from zooflagellates) are the most primitive metazoan organisms. These multicellular invertebrates are organized along simple patterns and share a basically radial symmetry with many flagellate colonies.

A somewhat less popular theory proposes the origin of metazoans from multinucleate protozoans. This theory holds that ciliatelike forms experienced an adaptive compartmentalization of their multinucleate cytoplasm. Numerous subordinate cells were formed, each complete with its own nucleus and specialized for a certain function in the life of the now multicellular organism. This **plasmodial theory** points toward the lower flatworms as the most primitive metazoans. Such flatworms are bilaterally symmetrical and possess cilia and multinucleate tissues. Strict adherents of the plasmodial theory then argue that coelenterates and other metazoans evolved from these worms.

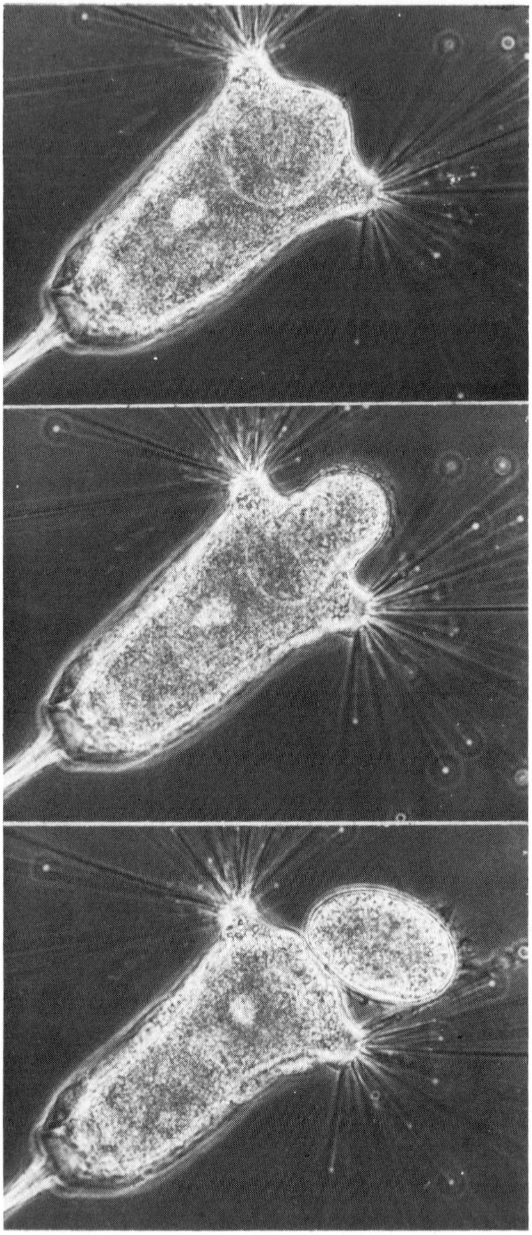

Figure 2.68 The suctorian *Acineta tuberosa.* These photographs illustrate the budding process. (From Grell, *Protozoologie.*)

The colonial theory and the plasmodial theory represent two perspectives from which most ideas of evolution in the lower invertebrate world proceed. We return to these theories as we encounter the metazoans with which they deal. Chapter 18 demonstrates further the central role played by protozoan evolution in the phylogeny of the invertebrate world.

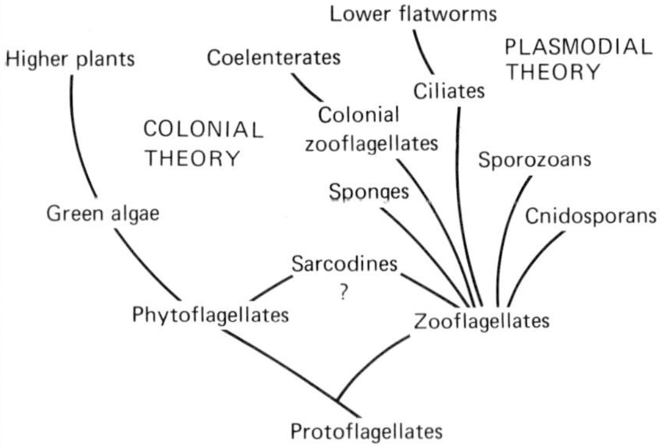

Figure 2.69 Possible evolutionary relationships within the phylum Protozoa and between unicellular and multicellular organisms.

Bioluminescence

Light plays an important role in the life of nearly all organisms. It is an orienting factor for circadian and annual rhythms, and its stimulation of visual organs provides many animals (ourselves included) with their major source of worldly information. The primary biological value of light is in photosynthesis, where its raw energy input sustains living things. Light figures in biological systems in yet another way. A surprisingly large and curious assortment of animals produce their own light—a phenomenon known as bioluminescence.

In recent years, chemists have made considerable progress in understanding the molecular basis of bioluminescence. Most of their work, however, lies outside the scope of this book. Suffice it to say that an enzyme acts on a substrate in such a way that an electron is elevated to a higher energy level. When that electron returns to its original level, energy is released in the form of cold light. Thus bioluminescence involves the conversion of chemical energy into light energy, much as photosynthesis involves an opposite conversion. The enzymes catalyzing bioluminescent reactions are called **luciferases;** the substances on which they act are called **luciferins.** Luciferases and luciferins vary widely throughout the animal world and may be rather specific to each organism.

The capacity for bioluminescence is distributed rather randomly. Bacteria, fungi, protozoans, and nearly all metazoan phyla have at least a few bioluminescent species. The phenomenon is particularly common among protozoans, coelenterates, ctenophores, squid, amphipods, fish, and certain well-known insects. Sponges, annelids, sedentary mollusks, arthropods, and echinoderms also have their luminescent forms. The freshwater environment harbors almost no light-producing species, while in the ocean the process is most common among planktonic and deep-sea forms.

Bioluminescence can be categorized in several ways. In many animals, light is produced not by the organism itself but by symbiotic bacteria. Often these animals can be identified because their bacterial luminescence is continuous, whereas other creatures produce light in flashes after mechanical, chemical, or nervous stimulation. Bioluminescence may be extracellular or intracellular. In extracellular types, secretions containing a luciferin and luciferase are mixed either outside the body or within intercellular fluids. In intracellular luminescence, the raw materials are contained within subcellular granules. Light-producing cells, commonly known as **photocytes,** may be housed within an organ called a **photophore.** Photophores often bear accessory structures, including lenses, filters, screens, and reflecting layers. Animals with well-developed photophores exercise fine control over their light emission.

Why certain bacteria, fungi, and protozoans luminesce is not understood. Among higher animals, light production serves several functions, including defense, predation, and communication. Sudden flashes of light may startle a would-be predator, and certain squid secrete luminescent clouds to obscure themselves from view. Some worms glow from their less vulnerable end, making that end more likely to be taken by a predator. Luminescence by deep-sea predators may aid their search for food, and a few cave-dwelling insects attract prey with their lights. Communication, usually of a sexual nature, is perhaps the most sophisticated use of bioluminescence. Fireflies signal to their mates with distinctive flash patterns, and spawning couples of certain marine worms glow as they release their gametes for external fertilization.

Bioluminescence almost surely did not arise as a means of defense, predation, or communication. These higher behavioral levels barely apply to protozoans, let alone bacteria and fungi. Biochemical studies have led to the speculation that bioluminescence is a very ancient phenomenon, perhaps common to most organisms in primeval times. Primevally, the earth had a reducing atmosphere, and oxygen may even have been toxic to the first living things. Bioluminescence could have arisen as a side product of biochemical pathways which evolved primarily to remove oxygen from these organisms. This hypothesis is supported by the fact that modern luciferins are associated with cellular respiration. Today that respiration usually includes the reduction of oxygen to form water, a process releasing energy for adenosine 5′-triphosphate (ATP) production. Ancient organisms may have reduced oxygen simply to expel it from their systems, and the energy from this reduction could have been released as light. With the perfection of oxidative metabolism, bioluminescence slowly disappeared from most species. It remains in a few groups, such as bacteria, fungi, and protozoans, where it appears to be a vestigial trait. Also, those organisms whose light-producing abilities became adapted to specific problems—such as defense, predation, and communication—remained bioluminescent. This scheme accounts well for the widespread but sporadic display of bioluminescence in the modern invertebrate world.

for Further Reading

The literature on Protozoa is vast. We are forced, therefore, to be very selective and have listed only a few classic references, together with recent books and review articles and a few original research papers on topics of current interest. The *Journal of Protozoology* is devoted exclusively to reports of research on Protozoa.

Allen, R. D. Comparative aspects of amoeboid movement. *Acta Protozool.* 7:291–299, 1970.

Baker, J. R. *Parasitic Protozoa.* Hutchinson, London, 1969.

Bamforth, S. S. Terrestrial protozoa. *J. Protozool.* 27: 33–36, 1980.

Bardele, C. F. A microtubule model for ingestion and transport in the suctorian tentacle. *Z. Zellforsch.* 130: 219–242, 1972.

Brehm, P. and Eckert, R. An electrophysiological study of the regulation of ciliary beating frequency in *Paramecium. J. Physiol.* 282: 557–568, 1978.

Buetow, D. E. *The Biology of Euglena.* Academic Press, New York, 1968. (Two volumes)

Chen, T. T. *Research in Protozoology.* Pergammon Press, Oxford, 1967–1972. (Includes detailed reviews of virtually all aspects of protozoan biology; four volumes)

Cheung, A. Ciliary activity of Stationary in *Opalina. Acta. Protozool.* 17: 153–162, 1978.

Corliss, J. O. *The Ciliated Protozoa: Characterization, Classification, and Guide to the Literature.* 2nd ed., Pergammon Press, New York, 1979.

Corliss, J. O. The ciliate Protozoa and other organisms: Some unresolved questions of major phylogenetic significance. *Am. Zool.* 12: 739–753, 1972.

Corning, W. C., Dyal, J. A., and Willows, A. O. D. (Eds.) *Invertebrate Learning, Vol. 1.: Protozoans through Annelids.* Plenum Press, New York/London, 1973.

Cullen, K. J. and Allen, R. D. A laser microbeam study of amoeboid movement. *Exp. Cell. Res.* 128: 353–362, 1980.

Dogiel, V. A. *General Protozoology.* 2nd ed., University Press, Oxford, 1965. (A text emphasizing the biology of Protozoa rather than the taxonomy. Revised by J. I. Poljanskij and E. M. Chesin)

Dougherty, E.C. (Ed.) *The Lower Metazoa, Comparative Biology and Phylogeny.* University of California Press, Berkeley and Los Angeles, 1963.

Eckert, R. Bioelectric control of ciliary activity. *Science.* 176: 473–481, 1972.

Ehret, C. F., and Powers, E. L. The cell surface of *Paramecium. Int. Rev. Cytol.* 8: 97-133, 1959.

Elliot, A. M. (Ed.) *The Biology of Tetrahymena.* Dowden, Hutchinson, and Ross, Stroudsburg, Pa., 1973.

Fenchel, T. Suspension feeding in ciliated protozoa: structure and function of feeding organelles. *Arch. Protistenkd.* 123: 239–260, 1980.

Garnham, P. C. C. *Malaria Parasites and Other Haemosporidia.* Blackwell, Oxford, 1966.

Grell, K. G. *Protozoology.* Springer, Berlin, 1973. (English translation of a classic German text. A well-balanced treatment beautifully illustrated)

Grell, K. G. Cytogenetic systems and evolution in foraminifera. *J. Foraminiferal Res.* 9: 1–13, 1979.

Hansen, H. J. Test structure and evolution in the Foraminifera. *Lethaia.* 12: 173–182, 1979.

Hawes, R. S. J. The emergence of asexuality in Protozoa. *Quart. Rev. Biol.* 38: 234–242, 1963.

Hedley, R. H. and Adams, C. G. (Eds.) *Foraminifera.* Academic Press, New York, 1974.

Hill, D. L. (Ed.) *The Biochemistry and Physiology of Tetrahymena.* Academic Press, New York and London, 1972.

Hoare, C. A. *The Trypanosomes of Mammals: A Zoological Monograph.* Blackwell, Oxford, 1972.

Hyman, L. H. *The Invertebrates, Vol. I: Protozoa through Ctenophora.* pp. 44–232, McGraw-Hill, New York, 1940. (A concise treatment of Protozoa, although now quite out of date. The Chapter "Retrospect" in Vol. 5 (1959) surveys work on Protozoa from 1938 to 1958)

Jahn, T. L., Bovee, E. C. and Jahn, F. F. *How to Know the Protozoa.* W. C. Brown, Dubuque, Iowa, 1979.

Jeon, K. W. (Eds.) *The Biology of Amoeba.* Academic Press, New York, 1973.

Jones, A. R. *The Ciliates.* St. Martin's Press, New York, 1974.

Jurand, A., and Selman, G. G. *The Anatomy of Paramecium aurelia.* Macmillan, London, 1969.

Kidder, G. W. (Ed.) *Chemical Zoology, Vol. I: Protozoa.* Academic Press, New York, 1967. (Series edited by M. Florkin and B. T. Scheer. Fourteen extensive review articles, mostly dealing with the physiology and biochemistry of Protozoa)

Kudo, R. R. *Protozoology.* 5th ed., Charles C. Thomas, Springfield, Ill., 1966. (A classical protozoology text emphasizing taxonomy)

Leedale, G. F. *Euglenoid Flagellates.* Prentice-Hall, Englewood Cliffs, NJ, 1967.

Leith, A. Variability of the limited life span state in amoeba. *Exp. Cell. Res.* 127: 261–268, 1980.

Levine, N. D., Corliss, J. D., et al. A newly revised classification of the Protozoa. *J. Protozool. 27*: 37 58, 1980.

Levine, N. D. and Ivens, V. The coccidia of secetivores. *Rev. Ivet. Parasitol.* 39: 261—298, 1979.

Levine, N. D. and Ivens, V. *The Coccidian Parasites of Ruminants.* University of Illinois Press, Urbana, 1970.

Mackinnon, D. L., and Hawes, R. S. J. *An Introduction to the Study of Protozoa.* Oxford University Press, London, 1961. (A species by species approach emphasizing representative forms. Contains much practical information)

Manwell, R. D. *Introduction to Protozoology.* Dover Publications, New York, 1968.

Murray, J. W. *An Atlas of British Recent Foraminiferids.* Heinemann, London, 1971.

Murray, J. W. *Distribution and Ecology of Living Benthic Foraminiferids.* Crane, Russak, New York, 1973.

Nigrini, C. and Moore, T. C. *A Guide to Modern Radiolarians.* Cushman Foundation for Foraminiferal Research, Washington, D.C., 1979.

Noble, E. R., and Noble, G. A. *Parasitology.* 3rd ed., Lea and Febiger, Philadelphia, 1972. (Contains an extensive treatment of parasitic protozoans found in humans)

Ogden, C. G. An ultrastructural study of division in *Euglypha. Protistologica* 15: 541—556, 1979.

Pal, R. A. The osmoregulatory system of the amoeba *Acanthamoeba castellami. J. Exp. Biol.* 57: 55076, 1972.

Parducz, B. Ciliary movement and coordination in Ciliates. *Int. Rev. Cytol.* 21: 91–128, 1967.

Patterson, D. J. Organization and classification of the protozoan *Actinophyrs sol. Microbios.* 26: 165–207, 1980.

Paulin, J. J., Henk, W. and Steiner, A. Surface manifestations of ciliate morphogenesis: Regeneration in *Stentor* and suctorian budding revisted. *Scanning Electron Microsc.* 527–532, 1980.

Pitelka, D. R. Ciliate ultrastructure: Some problems in cell biology. *J. Protozool.* 17: 1–10, 1970.

Pohley, H. J., Dornhaus, R. and Thomas, B. The amoebo-flagellate transformation, a system-theoretical approach. *Biosystems.* 10: 349—360, 1978.

Read, C. P. *Parasitism and Symbiosis,* Ronald Press, New York, 1970. (A text emphasizing host-parasite relationships)

Rudzinska, M. A. The fine structure of malaria parasites. *Int. Rev. Cytol.* 25: 151–199, 1969.

Ruffolo, J. J. Feeding apparatus of the ciliate protozoan *Euplotes eurystomus. Scanning Electron Microsc.* 533–536, 1980.

Sandon, H. *Essays on Protozoology.* Hutchinson, London, 1963.

Sarjeant, W. A. S. *Fossil and Living Dinoflagellates.* Academic Press, London, 1974.

Satir, P. Studies on cilia. III. Further studies on the cilium tip and a sliding filament model of ciliary motility. *J. Cell Biol.* 39: 77–94, 1968.

Scholtyseck, E. *Fine Structure of Parasitic Protozoa.* Springer-Verlag, Berlin and New York, 1979.

Seravin, L. N. Mechanisms and coordination of cellular locomotion. *Comp. Physiol. Biochem.* 4: 37–111, 1971.

Seravin, L. N. and Gerassimova, Z. P. A new macrosystem of ciliates. *Acta Protozool.* 17: 399–418, 1978.

Seravin, L. N. and Orlovskaja, E. E. Feeding behavior of unicellular anumals: I. The main role of chemoreception in the food choice of carnivorous protozoa. *Acta Protozool.* 16: 309–332, 1978.

Sleigh, M. A. Cilia. *Endeavour.* 30: 11–17, 1971.

Sleigh, M. A. *The Biology of Protozoa.* Edward Arnold, London, 1973. (A brief but excellent treatment of protozoan biology, well illustrated and up to date)

Sleigh, M. A. *Cilia and Flagella.* Academic Press, London, 1973.

Small, E. B., and Marszalek, D. S. Scanning electron microscopy of fixed, frozen, and dried Protozoa. *Science.* 163: 1064–1065, 1969.

Sonneborn, T. M. Breeding systems, reproductive methods and species problems in Protozoa. In *The Species Problem* (E. Mayr, Ed.), Orno, New York, 1974.

Stout, J. D., and Heal, O. W. Protozoa. *Soil Biology* (N. A. Burges and F. Row, Eds.), pp. 149–195. Academic Press, London, 1967.

Tait, A. Evidence for diploidy and mating in trypanosome. *Nature.* 287: 536–538, 1980.

Tamm, S. L. A scaning electron microscope study. *J. Cell Biol.* 55: 250–255, 1972.

Van Houten, J. Two mechanisms of chemotaxis in *Paramecium. J. Comp. Physiol. A Sens. Neural Behav. Physiol.* 127: 167–174, 1978.

Vickerman, K., and Cox, F. E. G. *The Protozoa.* John Murray, London, 1967.

Wagtendonk, W. J. van (Ed.) *Paramecium—A current Survey.* Elsevier, Amsterdam/London/New York, 1974.

Warton, A. and Honigberg, B. M. Structure of trichomonads as revealed by scanning electron microscopy. *J. Protozool.* 26: 56–62, 1979.

Whittaker, R. H. New concepts of kingdoms of organisms. *Science.* 163: 150—159, 1969.

3

Introduction to the Metazoans

The vast majority of invertebrate cells are organized to form multicellular organisms. Multicellular animals, known collectively as metazoans, enjoy numerous advantages over their protozoan forebears. Their multicellular constitutions allow a greater size and specialization of body parts. Metazoans can organize entire cells—or entire tissues of cells or entire organs of tissues—into biological machinery far beyond the capability of protozoans. Protozoans certainly have succeeded with their limited raw materials, but only the metazoans had the means to become the truly dominant members of the invertebrate world.

The Early Embryogeny of Metazoans

Most metazoans, and certainly all sexual ones, begin their life cycle as a single cell, typically a fertilized egg. This single cell then undergoes a complicated embryogeny, or growth and differentiation to a multicellular being. As discussed in Chapter One, this embryogeny may furnish clues to the ancestry of the metazoan involved.

Typically, an early period of rapid cleavage forms a hollow sphere of cells called the **blastula.** Next, certain cells are transported into the center of the sphere by a process known as **gastrulation,** which occurs in several different ways. One method is *invagination* and involves the inward migration of cells from the so-called vegetal pole into the **blastocoel** or central cavity of the blastula (Fig. 3.1a). Gastrulation by invagination commonly forms a hollow two-layered structure. The central cavity of the resulting **gastrula** is called an **archenteron** and represents the future gut of the animal. Its exterior opening is the **blastopore.** Because of their yolk content, the vegetal pole cells of some blastulas are larger than the animal pole cells (Fig. 3.1b). In this case, gastrulation occurs by **epiboly,** a process in which the more rapidly dividing animal pole cells expand around the vegetal pole cells, displacing them into the blastocoel (Fig. 3.1b). Gastrulas produced by epiboly possess a blastopore and at least a small archenteron. Other gastrulation

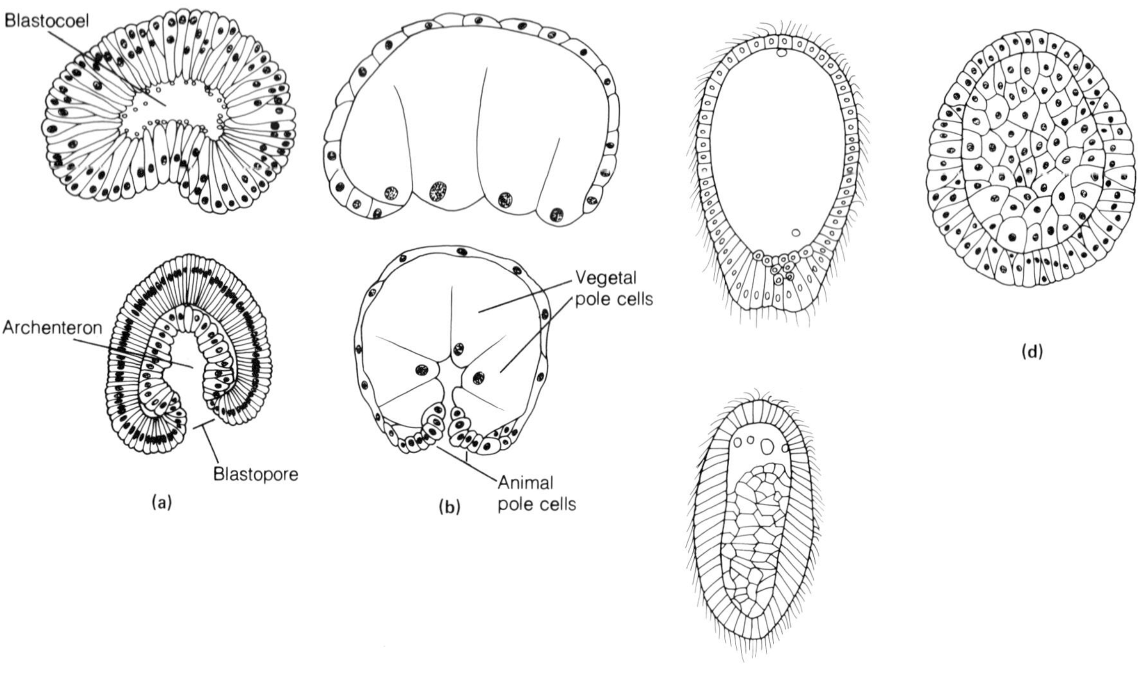

Figure 3.1 Patterns of gastrulation. (*a*) Gastrulation by invagination. Two stages in the gastrulation of *Terebratulina* (Brachiopoda); (*b*) Gastrulation by epiboly. Two stages in the gastrulation of *Crepidula* (Mollusca, Gastropoda); (*c*) Gastrulation by unipolar ingression. Two stages in the gastrulation of *Aequorea* (Coelenterata, Hydrozoa); (*d*) Gastrulation by delamination. The stereogastrula of *Clava* (Coelenterata, Hydrozoa). (From A. Richards, *Outline of Comparative Embryology*, 1931, John Wiley.)

methods include ingression and delamination. Gastrulation by **ingression** occurs when surface cells migrate into and fill the blastocoel (Fig. 3.1c). Ingression may be unipolar or multipolar (occurring at several points on the blastula surface). Gastrulation by **delamination** results from tangential cell divisions isolating an inner portion of the blastula (Fig. 3.1d). Gastrulas produced by ingression and delamination lack both the blastopore and the archenteron. Being solid, they are known as **stereogastrulas.**

After gastrulation, two groups of embryonic cells are distinguishable. The first group covers the external surface of the gastrula; these cells represent the future **ectoderm,** the germ layer that forms the outer body covering as well as the nervous and sensory systems. The second cell group lies within the gastrula; these cells constitute another germ layer, the **endoderm,** which produces the digestive system. A third germ layer, the **mesoderm,** is present in most metazoans. The mesoderm forms the muscles, the reproductive structures, and the major elements of most respiratory, circulatory, excretory, and osmoregulatory systems. The origin of the mesoderm involves rather complex processes. Mesoderm formation differs among invertebrate types and is described subsequently as we encounter some of the grouping schemes for the metazoan phyla.

Metazoan Organization and Classification

During cellular and germ layer differentiation, multicellular embryos become subject to the metabolic consequences of metazoan life. We remarked earlier that protozoan cells are limited in size because of the problems of internal transport. Circulatory and respiratory needs likewise limit the distance at which metazoan cells can function away from an external surface. Multicellularity does protect the internal cells

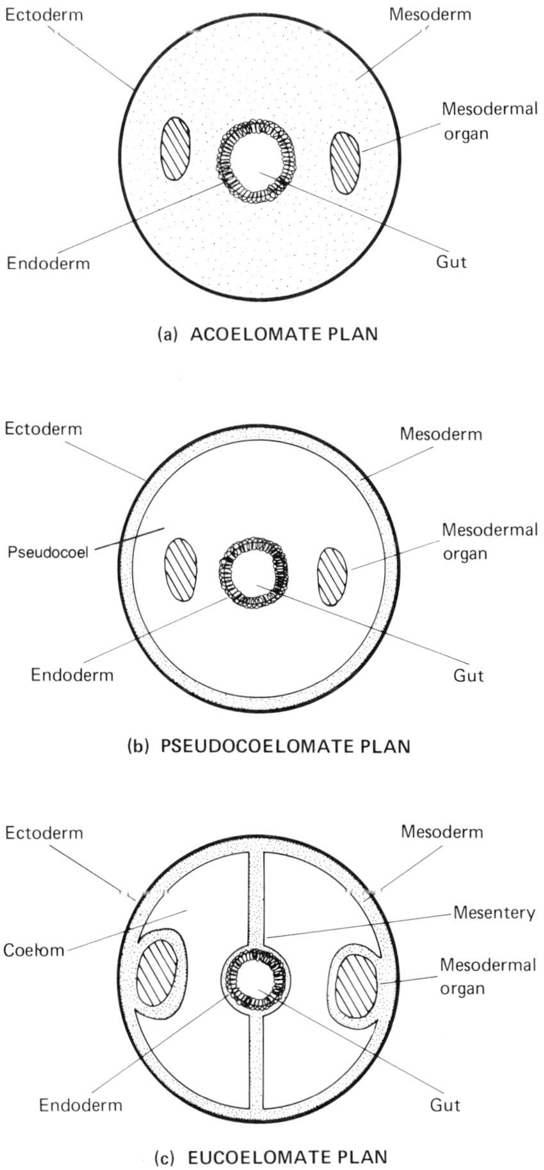

Figure 3.2 Metazoan body plans in cross section showing secondary body cavities. (*a*) The acoelomate plan with no secondary body cavity; (*b*) The pseudocoelomate plan featuring a pseudocoel; (*c*) The eucoelomate plan featuring a true coelom.

of metazoans, but those cells nonetheless must be supplied with nutrients and oxygen. Accordingly, metazoan organizational patterns reflect adaptive compromises between the advantages and liabilities of increasing size. A common feature of these organizational patterns is the body cavity referring to a hollow internal region completely separate from the gut. Different types of body cavities exist and form the basis of a system of metazoan classification.

The metazoan phyla are divided into a number of different groups, distinguished by body symmetry; degree of subordination of cells, tissues, and organs; existence and type of body cavities; and details of certain embryogenic processes.

SYMMETRY

In Chapter One, we discussed radial and bilateral symmetry as two major organizational patterns. Radial symmetry is a passive condition associated most often with sessile life-styles. Bilateral symmetry is displayed by more active invertebrates. Two metazoan phyla, the Coelenterata and the Ctenophora, display primary radial symmetry and thus compose the superphylum Radiata. With the exception of the sponges, all other metazoan groups demonstrate primary bilateral symmetry; they are all members of the superphylum Bilateria. Sponges are odd metazoans in many ways and their symmetry, although perhaps more radial than bilateral, is not well described by either of the two terms. Indeed, the sponges are so odd that they frequently are classified as a separate division of the Metazoa, the Parazoa. All other metazoans belong to the Eumetazoa.

SUBORDINATION OF CELLS

In metazoans, the welfare of individual cells is subordinate to the needs of the whole animal. Several levels of subordination may be involved. To some extent, the physiological independence of each component cell is preserved in sponges; thus we may say that sponges express a **cellular level of organization.** In coelenterates and ctenophores, groups of similar cells are organized structurally and functionally into tissues. Accordingly, we say that these radiate animals demonstrate a **tissue level of organization.** Finally, the remaining metazoans possess organs and organ systems composed of various tissues. These higher animals display an **organ-system level of organization.**

BODY CAVITIES

Metazoans cannot realize their size potential if all their cells are packed together. A mass of continuous cells would have virtually the same metabolic problems as a single cell of comparable size. Because of transport problems, in all but the tiniest metazoans, cells are organized around internal spaces. Such spaces include the respiratory and circulatory vessels of higher invertebrates. In terms of general metazoan organization, however, the body cavity represents a major adaptation of multicellular animals to their internal transport needs. The body cavity commonly surrounds the digestive tract, creating a "tube-within-a-tube" architecture (Fig. 3.2b,c), which is general among higher metazoans.

Sponges and the radiate phyla have no nondigestive body cavities; thus only the Bilateria is involved in a classification of cavity types. The flatworms and nemertines also lack a specialized body cavity (Fig. 3.2a); hence, these bilateral worms are termed **acoelomates.** The nematodes, rotifers, and several minor phyla possess a cavity between their endodermal and mesodermal tissues (Fig. 3.2b). These invertebrates are called **pseudocoelomates.** Their body cavity, the **pseudocoel,** may represent a persistent blastocoel. The pseudocoelomate gut lacks mesodermal components; except for anterior and posterior attachments, it lies freely within the fluid-filled pseudocoel. The annelids, mollusks, arthropods, lophophorates, echinoderms, and chordates possess a true **coelom** (Fig. 3.2c), a body cavity located wholly within mesodermal tissues. A layer of mesoderm, the **peritoneum,** surrounds the gut and also lines the body wall. The peritoneum thus forms a continuous border for the coelom; it joins above and below the gut in double layers or **mesenteries** which attach the gut to the body wall. Metazoans possessing a true coelom are called **eucoelomates** and unquestionably dominate the invertebrate world.

Eucoelomates and, to a lesser extent, the pseudocoelomates enjoy distinct advantages over those lower invertebrates lacking body cavities. These spaces not only ameliorate internal transport problems but also provide room for the differentiation and growth of internal organs. A fluid-filled coelom or pseudocoel also may function as a hydrostatic skeleton. Finally, these cavities decrease the bulk of higher metazoans without sacrificing body size or strength. Thus locomotion is rendered more efficient.

EUCOELOMATE EMBRYOGENY

Several aspects of metazoan embryogeny contribute to our understanding of the relationships among the many phyla involved. We give primary consideration to the details of the early embryogeny of the eucoelomates, as these processes distinguish two major branches of the higher invertebrate world: the *Protostomia* and the *Deuterostomia.* Protostomes include annelids, mollusks, arthropods, and related minor groups. These pnyla represent a majority of all metazoans and thus establish the protostomate assemblage as the major evolutionary line in the animal kingdom. Some authorities also include the acoelomates and pseudocoelomates in the protostomate line on the basis of somewhat limited evidence. However, the distinctions between protostomes and deuterostomes are most apparent in the eucoelomate groups. The deuterostomes include the echinoderms, chordates, and related minor phyla. Protostomes and deuterostomes differ in their cleavage styles, in their mouth formation, and in the origin of their mesoderm and coelom. Lophophorates display somewhat intermediate characteristics and thus do not fit clearly into either the Protostomia or the Deuterostomia.

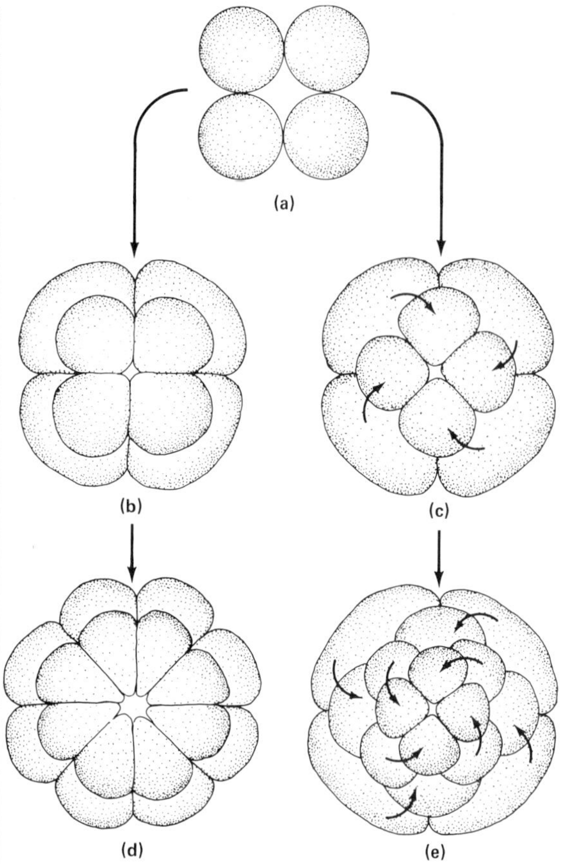

Figure 3.3 A comparison of radial and spiral cleavage. (*a*) 4-celled stage, identical in both radial and spiral cleavage; (*b* and *d*) 8 and 16-celled stages, radial cleavage; (*c* and *e*) 8 and 16-celled stages, spiral cleavage. (After Richards, *Outline of Comparative Embryology.*)

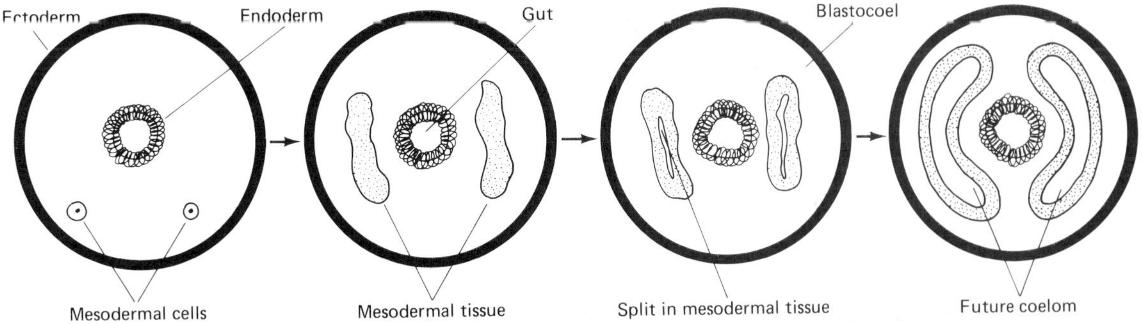

Figure 3.4 Coelom formation by schizocoely. A split within each mesodermal zone widens to become the coelom.

Cleavage Styles

Initial cleavage in both protostomes and deuterostomes proceeds identically to the four-celled stage (Fig. 3.3a). The next mitotic cycle, which produces an eight-celled embryo, distinguishes the two groups. In deuterostomes, the cleavage plane is perpendicular to the polar axis of the embryo, so that the second quartet of cells lies immediately above the first quartet (Fig. 3.3b). Such a design is called **radial cleavage.** Protostomes display **spiral cleavage,** in which the cleavage plane producing the second quartet of cells is oriented obliquely to the embryo's polar axis (Fig. 3.3c). Each cell thus lies in a furrow created by the two cells above or beneath it. Radial and spiral cleavage continue through numerous mitotic cycles (Fig. 3.3d, e).

Another cleavage factor distinguishes the protostomes from the deuterostomes: protostomate cleavage tends to be determinate. Determinate cleavage refers to the fact that, as early as the four-celled stage, the fate of each embryonic cell is fixed. If one of these first four cells is isolated, that cell has the potential to produce only one-fourth of the protostomate animal. In contrast, deuterostomate cleavage tends to be indeterminate. A cell isolated from an early deuterostomate embryo maintains the potential to develop into an entire organism.

Mouth Formation

Two distinct sites of mouth formation further delineate the two eucoelomate groups. Protostomes form their mouths from the blastopore or at a site very near it. Thus, as the word protostome indicates, the mouth is the first-formed opening of the gut. Deuterostomate mouths always are formed at a considerable distance from the blastopore; indeed, the deuterostomate blastopore is often the forerunner of the anus. In these animals, the mouth is the second-formed gut opening.

Origin of the Mesoderm and the Coelom

The last major distinction between the protostomes and the deuterostomes focuses on the origin of the mesoderm and its coelom. Consistent with the determinate cleavage of the group, all protostomate mesoderm arises from a single, mesodermally fated cell. This cell divides to form two cells, each of which proliferates a zone of mesodermal tissue along its side of the embryo (Fig. 3.4). Subsequently, a split develops within each mesodermal zone. This split widens into a cavity which becomes the protostomate coelom. Such coelomic formation is called **schizocoely,** and therefore protostomes may be called **schizocoelomates.**

Deuterostomes display a different pattern of mesoderm formation involving paired evaginations of the archenteron (Fig. 3.5). Evaginations of the wall of this primitive gut expand and eventually separate as lateral pouches. The pouch walls represent the future mesoderm of the organism, while the pouch cavity itself is the forerunner of the deuterostomate coelom. The deuterostomate method of coelom formation is called **enterocoely,** and these animals sometimes are called **enterocoelomates.**

Typically, both protostomate and deuterostomate coeloms expand to fill and obliterate the blastocoel. Eucoelomate mesoderm abuts both the ectoderm of the body wall and the endoderm of the gut, and the tube-within-a-tube structure is thereby complete.

The biology of the various metazoan groups forms the subject matter of succeeding chapters. A more detailed version of the classification scheme outlined in this chapter may be found in the Appendix.

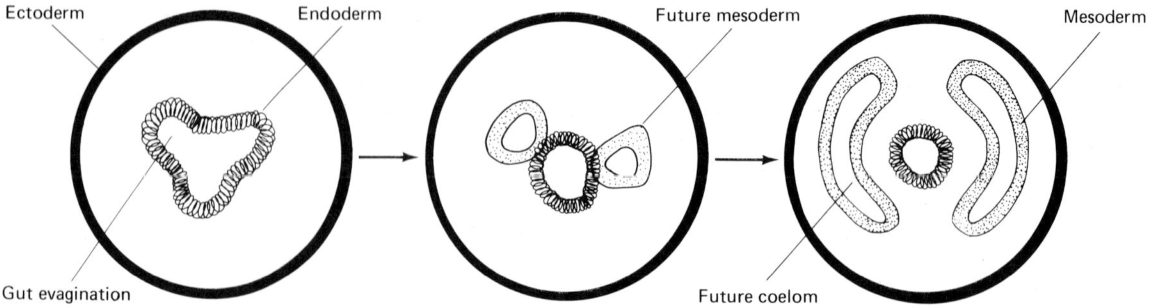

Figure 3.5 Coelom formation by enterocoely. Lateral pouches develop from the gut, separate, and later expand to form the coelom.

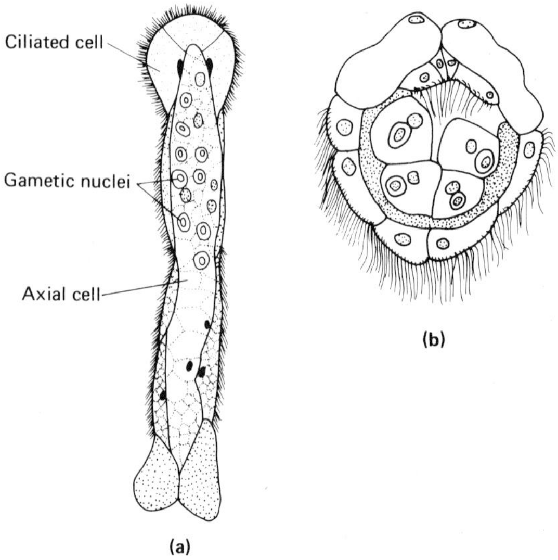

Figure 3.6 (*a*) The nematogen stage of a dicyemid mesozoan (*Pseudicyema truncatum*); (*b*) Infusoriform larva of a dicyemid mesozoan. (From L.H. Hyman, *The Invertebrates, Vol. I: From Protozoa through Ctenophora*, 1940, McGraw-Hill.)

The Enigmatic Mesozoa

The Mesozoa comprises approximately 50 species of tiny parasitic multicellular animals which are very difficult to classify. Structurally, mesozoans are very simple, but they exhibit the complex life cycles typical of parasites. Two groups exist, the dicyemids and the orthonectids.

Dicyemid mesozoans parasitize the excretory organs of squid and octopods. The **nematogen** stage is commonly found in young hosts. This wormlike animal is 1 to 7 mm long and consists of an axial cell surrounded by a layer of ciliated cells (Fig. 3.6a). The anteriormost cells function in attachment, while the axial cell asexually produces vermiform larvae. These larvae increase the infective population. As the host matures, nematogens become **rhombogens.** Rhombogens are structurally similar to nematogens but produce different larvae, known as **infusoriform** larvae (Fig. 3.6b). Actually infusoriform larvae are produced sexually within **infusorigens.** The latter may be highly reduced hermaphroditic individuals occupying the axial cell of the parent rhombogen. Infusoriform larvae exit with the host's urine and probably enter some bottom-dwelling invertebrate which serves as an intermediate host. This hypothetical intermediate host remains unknown, as does the means by which the parasite reenters the primary host.

Orthonectids are uncommon parasites of a variety of marine invertebrates, including flatworms, nemertines, polychaetes, mollusks and echinoderms. The small wormlike adults are diecious and live unattached within the host. Their body consists of an internal mass of eggs or sperm surrounded by a layer of ciliated cells (Fig. 3.7a, b). When sexually mature, the adults leave their host and mate as free-living forms. Zygotes develop within females and give rise to larvae (Fig. 3.7c) rather like the infusoriform larvae of dicyemids. These larvae leave the parent and soon become infective. Within the tissue spaces of a new host, each larva forms a multinucleated amoeboid mass (Fig. 3.7d) which produces males and females by asexual means.

Zoologists dispute the correct phylogenetic position of the Mesozoa. Basically, there are two opposing views. One holds that mesozoans are very degenerate flatworms. According to the second view, the mesozoans represent an offshoot from the early metazoans and thus have a long independent history. Certain resemblances to parasitic flatworms in both life cycle and structure support the first view, but these similarities may result from evolutionary convergence. (Superficial resemblances between unrelated parasites are common: See Chap. 6, Special Essay.) The blastula- or stereogastrulalike appearance of mesozoans supports the notion that they are truly primitive multicellular animals that arose independently from protozoans. This evidence, however, also is subject to other interpretations. In the absence of more definitive information, the relationship of the mesozoans to other metazoans remains an enigma.

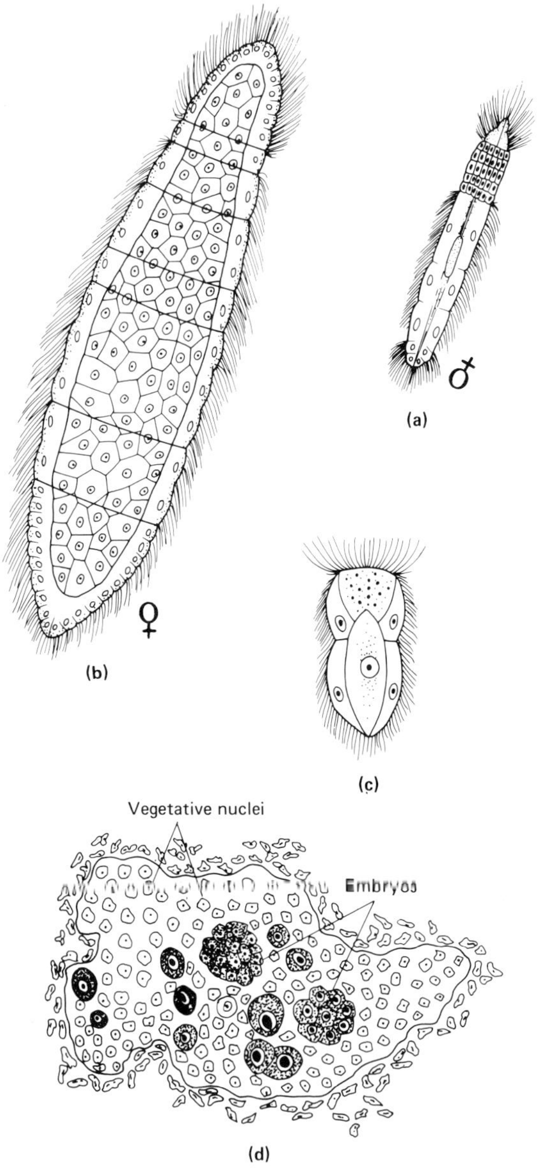

Figure 3.7 (*a* and *b*) Adult male and female of *Rhopalura giardii*, an orthonectid mesozoan parasite of brittle stars; (*c*) Larva of an orthonectid mesozoan, *Microcyema vespa;* (*d*) A male-producing amoeboid mass derived from a larva of *Rhopalura ophiocomae*, an orthonectid parasite of brittle stars. (*a* and *b* from S.F. Harmer and A.E. Shipley, Eds, *Cambridge Natural History, Vol. II*, 1901; *c* and *d* from Hyman, *The Invertebrates, Vol. I*.)

for Further Reading

Anderson, D. T. *Embryology and Phylogeny in Annelids and Arthropods.* Pergammon Press, Oxford and New York, 1973.

Balinsky, B. V. *An Introduction to Embryology.* 4th ed., Saunders, Philadelphia, 1975.

Brusca, G. J. *General Pattern of Invertebrate Development.* Mad River Press, Eureka, CA, 1975.

Ebert, J. D., and Sussex, I. M. *Interacting Systems in Development.* Holt, Rinehart and Winston, New York, 1970. (An excellent text emphasizing cell interactions at the molecular level)

Greenberg, M. J. Ancestors, embryos and symmetry. *Syst. Zool.* 8: 212–221, 1959.

Hyman, L. H. *The Invertebrates, Vol. II: Platyhelminthes and Rhynchocoela.* McGraw-Hill, New York, 1951. (The introduction discusses invertebrate body symmetry, cavities and cleavage patterns.)

Lapan, E. A. and Morowitz, H. The mesozoa. *Sci. Am.* 227: 94–101.

Lash, J., and Whittaker, J. R. (Eds.) *Concepts of Development.* Sinaver Associates, Stamford, Conn., 1974. (An advanced text with chapters written by a team of experts.)

Mathews, W. W. *Atlas of Descriptive Embryology.* Macmillan, New York, 1975. (The first three chapters include photographs of insect, nematode, and echinoderm development.)

McConnaughey, B. H. The Mesozoa. *The Lower Metazoa* (E. C. Dougherty, Ed.), pp. 151–168, University of California Press, Berkeley, 1963.

McKenzie, J. *An Introduction to Developmental Biology.* Blackwell, Oxford, 1976.

Reverberi, G. (Ed.) *Experimental Embryology of Marine and Freshwater Invertebrates.* North-Holland Publishing Co., Amsterdam and London, 1971.

Stunkard, H. W. The life history and systematic relations of the Mesozoa. *Quart. Rev. Biol. 29:* 230—244, 1954.

Stunkard, H. W. Clarification of taxonomy in Mesozoa. *Syst. Zool.* 21: 210–214, 1972. (Includes a useful literature review)

Szebenyi, E. S. *Atlas of Developmental Embryology.* Fairleigh Dickinson University Press, Rutherford, NJ, 1977.

4

The Sponges

The sponges occupy a unique position in the invertebrate world. Known as the phylum Porifera, these animals are somewhat intermediate between metazoans and colonial protozoans. They have no organs, no well-defined tissues, and no mouths or digestive cavities; and among their cells, there is little metazoan-style coordination. To emphasize this intermediate condition, some biologists classify the phylum in a distinct subdivision of the Metazoa, the Parazoa.

The way of life for sponges is radically different from that for other animals. Indeed, sponges do not look or behave like most animals at all. Externally, they have few typically animal features, a property which long caused confusion about their identity. It was not suggested until 1765 that the sponges might be animals, and only in the last century did this idea gain complete acceptance.

Adult sponges are sessile animals whose bodies house cell-lined channels through which water circulates (Fig. 4.1). Their external surfaces bear many small pores and one or more large pores. Water enters a sponge through its small pores, called **ostia,** and after circulating through the channel network, exits through a large pore, the **osculum.** Flagellated collar cells called **choanocytes** (Fig. 4.2) line some of the water channels. Choanocytes power the channel currents and exchange materials with circulating waters. The sponge is covered externally by epidermal cells called **pinacocytes.** Between pinacocytes and choanocytes is a middle layer, or **mesoglea,** composed of wandering amoeboid cells and various skeletal elements.

Mesogleal skeletal elements form the basis of sponge taxonomy. Four classes are recognized: the Calcarea, the Hexactinellida, and the Demospongiae, and the Sclerospongiae. Calcareous sponges have skeletons constructed from small calcium carbonate spikes, or **spicules.** Hexactinellida, the glass sponges, are a highly evolved deep-water group with skeletons composed of siliceous spicules organized in regular networks. Demospongiae, by far the largest group of sponges, possess **spongin** fibers and/or siliceous spicules. (Spongin is a protein peculiar to sponges and is related to the collagen of vertebrate connective tissue.) The Sclerospongiae are a small, specialized, tropical reef-dwelling group of sponges.

Most sponges live in shallow, marine waters, where they attach to rocks, submerged timbers, plants, other animals, and similar substrata. Some occur at great depths in the ocean, and a few species have invaded fresh water.

Sponge Unity

MAINTENANCE SYSTEMS

General Morphology

Schematically, the simplest sponge is a tube with lateral ostia and a single apical osculum (Fig. 4.1). Externally, it is covered with pinacocytes—flat undifferentiated cells with some contractile properties. The central cavity, or **spongocoel,** is lined with choanocytes. Choanocytes have sievelike collars composed of slender, linear pseudopodia connected by fine microfibrils (Fig. 4.2). Each cell has a single flagellum within its collar. A gelatinous matrix, or **mesoglea,** between the two cell layers houses skeletal elements as well as several types of wandering cells called **amoebocytes.**

This schematic sponge is radially symmetrical. Such symmetry is probably the primitive condition in the phylum, but most modern sponges tend toward asymmetry. A lack of symmetry reflects the way of life for sponges. Active, aggressive animals have heads and streamlined bilateral bodies, while most sessile creatures exhibit the omnidirectional alertness that radial symmetry provides. But sponge behavior involves very little except water circulation; thus the symmetry of these animals is less ordered. Indeed, body form often is determined by local water and substratum conditions and not by a species-specific mechanism. Sponges of the same species often assume different shapes under different environmental conditions, and the same environment may cause different species to assume similar body forms. More central to species identity is the sponge's internal morphology, including its skeletal architecture and the arrangement of its water channels.

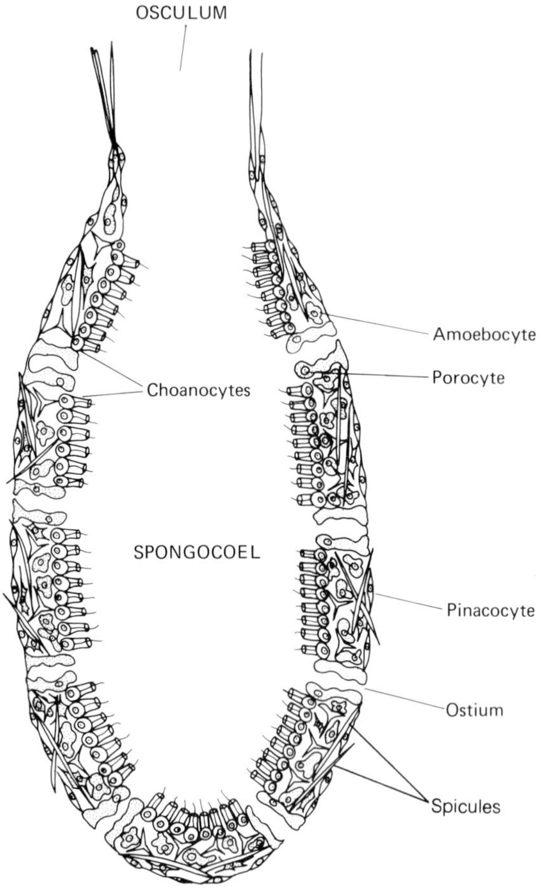

Figure 4.1 A longitudinal section through a simple asconoid sponge. (From L.H. Hyman, *The Invertebrates, Vol. I: Protozoa through Ctenophora,* 1940, McGraw-Hill.)

A sponge's most critical surface is the interface between its choanocyte collars and internally circulating waters. There most vital physiological processes take place, including food uptake, respiration, and excretion, as well as certain reproductive activities. Efficient performance of these tasks depends on two main factors: an adequate rate of flagella-powered water circulation and intimate contact between choanocyte collars and the water medium. Internal water volume increases with the cube of water channel diameters, while choanocyte surface area remains a squared function of the same measure. Therefore, small-diameter channels provide the greatest density of stroking flagella and filtering collars. Sponge evolution witnesses a decided trend in this direction—in the plumbing of advanced sponges, small pipes are the rule.

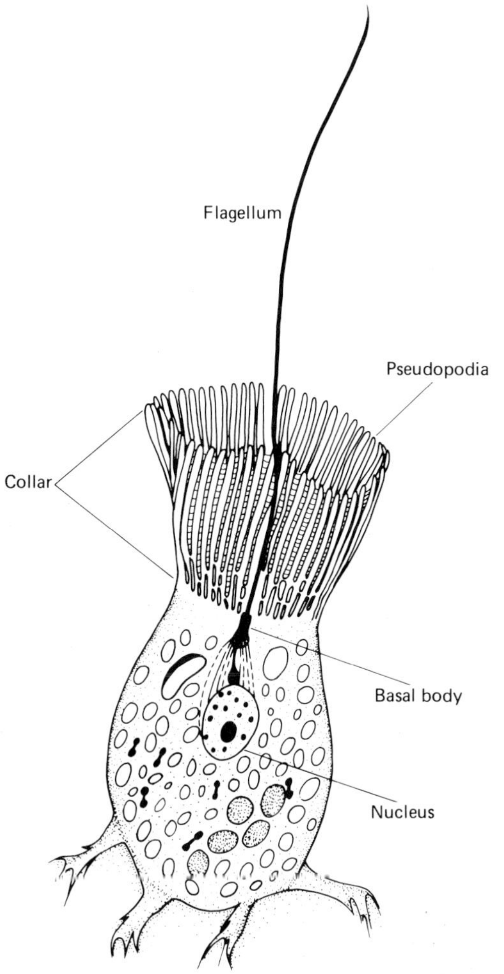

Figure 4.2 A drawing of a choanocyte based on an electronmicrograph. (From R. Rasmont, Ann. des. Sciences, Naturelles, Zool., 1959, *1*(2):253.)

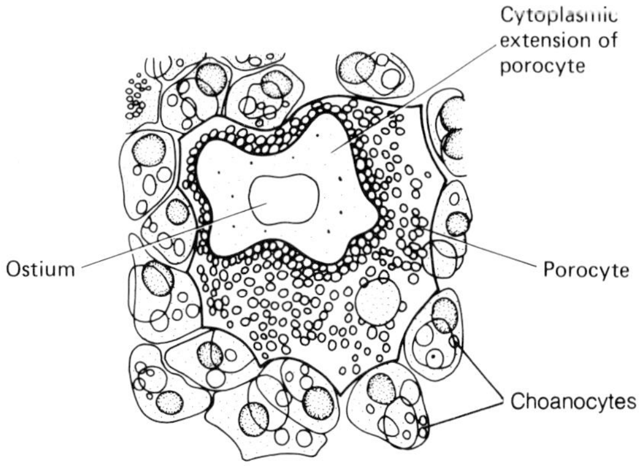

Figure 4.3 A porocyte of *Leucosolenia* surrounded by choanocytes sectioned through the cell bodies. (From Hyman, *The Invertebrates, Vol. I.*)

Three general types of water channel systems exist, although they tend to grade into each other: the asconoid, the syconoid, and the leuconoid. These are descriptive and not taxonomic terms; thus their use in formal sponge classification is limited.

An **asconoid** sponge has a very simple water channel system. Our schematic sponge (Fig. 4.1) has basic asconoid characters—a body plan built around a single spongocoel lined internally with choanocytes (Fig. 4.4a). The external surface is covered with pinacocytes and porocytes (Fig. 4.3). A **porocyte** is a tubular cell with a pore, the ostium, passing through it. Water enters an asconoid sponge through its ostia-bearing porocytes, passes through the spongocoel, and is expelled from a single apical osculum. Many asconoid sponges are colonial. They are also quite small, as their body plan limits intimate contact between choanocytes and internal water. Even with the largest choanocytes in the phylum, asconoid sponges have a relatively slow rate of water flow; their metabolic rate is correspondingly low.

The **syconoid** condition is easily derived from the asconoid plan. Basically, a series of body wall foldings create blind side pockets, or **radial canals,** along the spongocoel (Fig. 4.1b). Porocytes are absent, as water enters the radial canals through small intercellular spaces called **prosopyles** (Fig. 4.4b). Choanocytes are limited to the radial canals; pinacocytes line the remainder of the spongocoel. In most syconoid sponges, a cortex of pinacocytes and mesogleal material surrounds the radial canals externally (Fig. 4.4c). Within this cortex, **incurrent canals** open to the outside world via **dermal ostia.** In such syconoid sponges, water flow is as follows: dermal ostia, incurrent canals, prosopyles, radial canals, spongocoel, osculum. In contrast to asconoid sponges, syconoid sponges are usually solitary.

The most elaborate plumbing system is found in **leuconoid** sponges, where the spongocoel is reorganized as an intricate system of chambers and canals branching through a considerably bulkier sponge body (Fig. 4.4d). The external cortex is also more elaborate, as subdermal spaces precede the incurrent canals. Choanocytes are confined to small chambers, where their contact with circulating water is very intimate. A small pore called an **apopyle** opens from each **choanocyte chamber** into an excurrent water channel system. That system comprises a network of

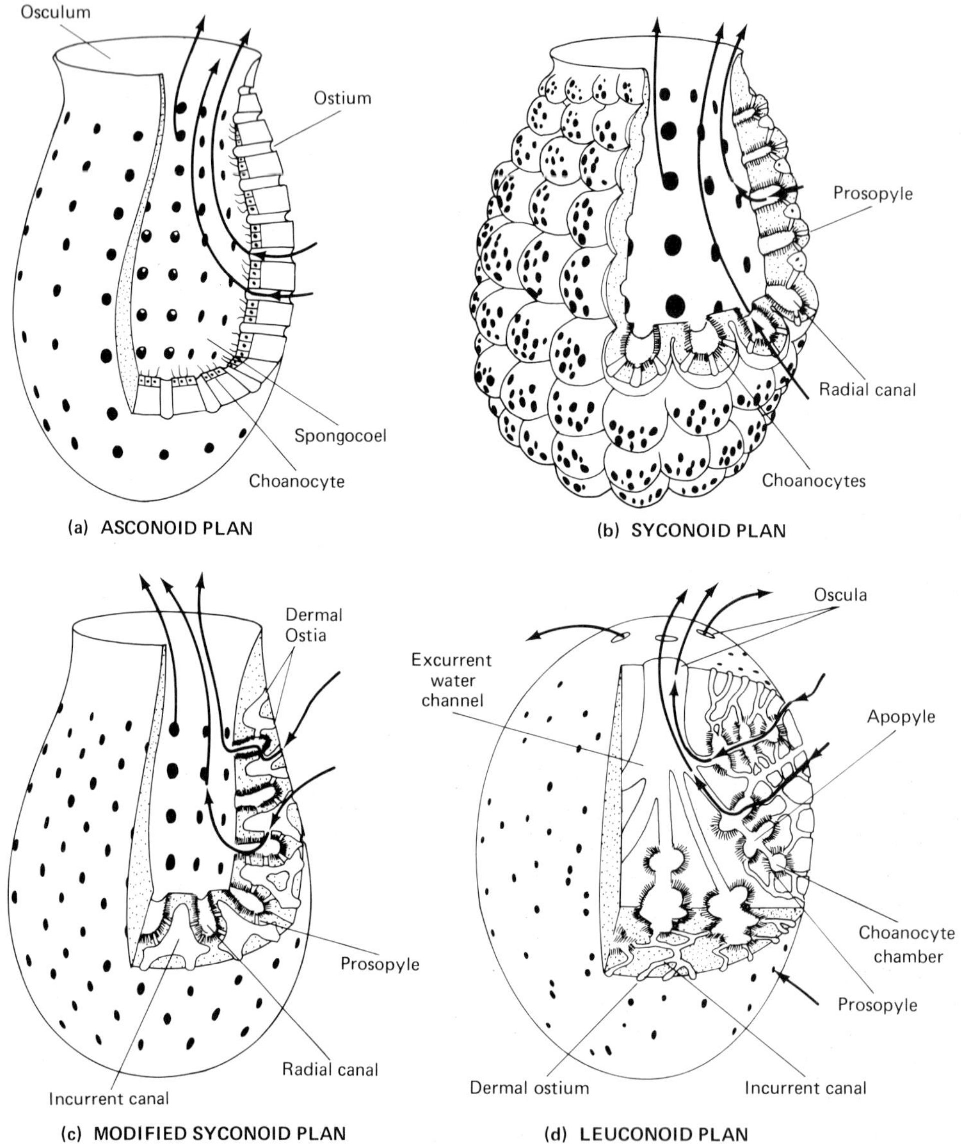

Figure 4.4 A diagramatic representation of the four basic sponge body plans showing the pathway of water flow through each. (*a*) Asconoid plan; (*b*) Syconoid plan; (*c*) Modified syconoid plan; (*d*) Leuconoid plan. (For further explanation see text.) (From F.M. Bayer and H.B. Owre, *The Free-Living Lower Invertebrates*, 1968, Macmillan.)

channels that gradually increase in size until they reach an osculum. Water flow in leuconoid sponges thus follows this route: dermal ostia, subdermal spaces, incurrent canals, prosopyles, choanocyte chambers, apopyles, water channels, larger water channels, osculum.

The leuconoid plan continues a trend established by syconoid sponges. The original spongocoel is increasingly compartmentalized, thus generating more internal surface area. Most important, leuconoid choanocytes sieve virtually every drop of water passing through the sponge body. Their physiological efficiency allows leuconoid sponges to attain a size and diversity unparalleled in the phylum.

Leuconoid sponges tend to form very large colonies with many oscula. In such colonies, boundaries be-

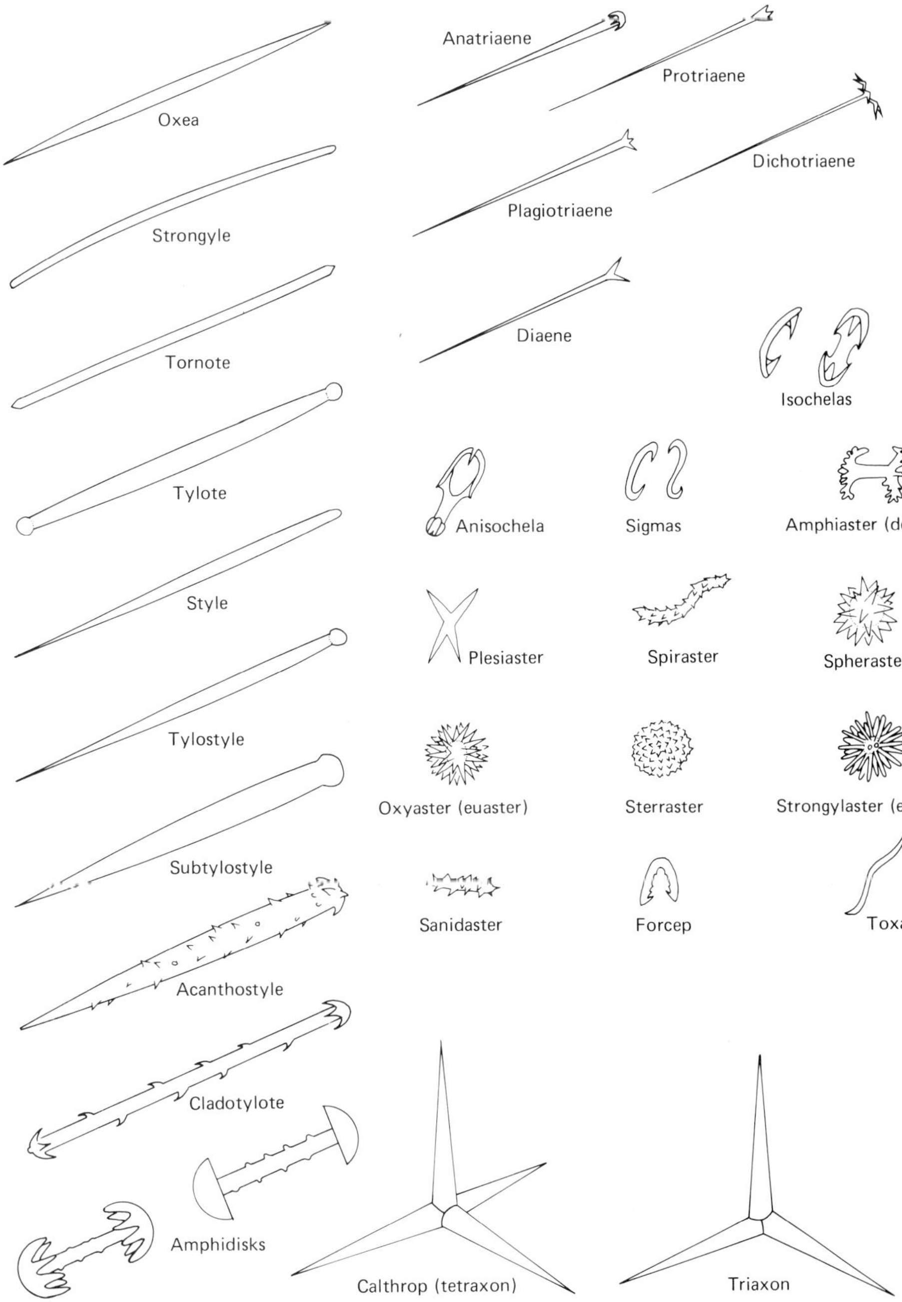

Figure 4.5 A variety of spicule types found in sponges. (From D.E. Beck and L.F. Braithwaite, *Invertebrate Zoology Laboratory Workbook*, 3rd ed., 1968, Burgess.)

tween individuals are quite obscure. Some researchers consider a single osculum and its associated channels and cells as the sponge individual. Given the intricacy of most leuconoid forms, however, such individuals defy isolation as physiological and structural units. Moreover, the relative independence of sponge cells further complicates a definition of the individual sponge.

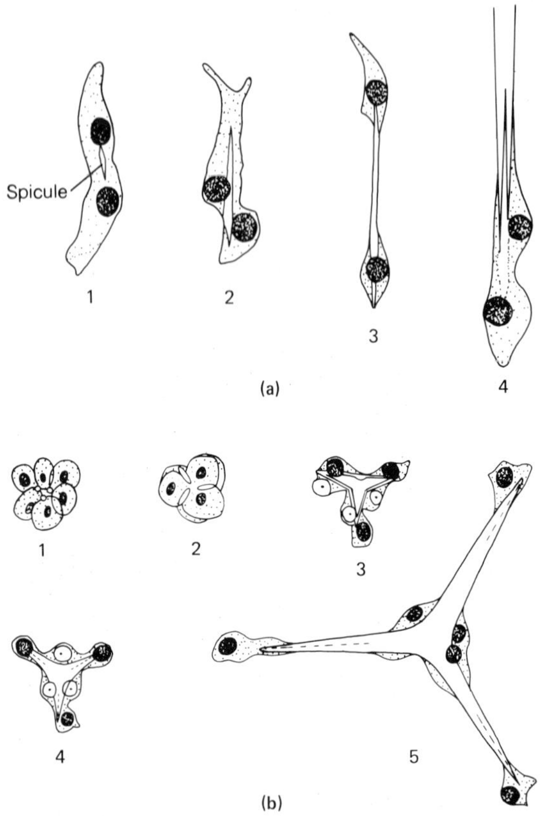

Figure 4.6 (*a*) Secretion of a monaxon spicule: (*1*) Spicule starting between two nuclei. (*2* and *3*) Spicule lengthening and nuclei separating into two cells. (*4*) Completion of spicule tip; (*b*) Formation of a triradiate spicule: (*1* and *2*) Three rays starting as granules in a cluster of 6 cells and then lengthening. (*3*) Three rays united. (*4*) Founder and thickener cells at work. (*5*) A late stage in spicule formation. (From Hyman, *The Invertebrates, Vol. I.*)

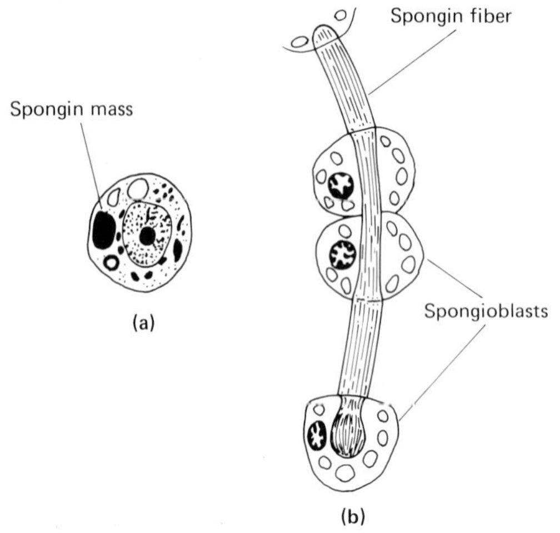

Figure 4.7 Formation of a spongin fiber. (*a*) Spongioblast containing a mass of spongin; (*b*) Spongioblasts in series secreting a spongin fiber. (From Hyman, *The Invertebrates, Vol. I.*)

In contrast to their other features, sponge skeletal elements are rigidly organized. Sponge taxonomic systems emphasize skeletal details; thus, as is so common in invertebrate zoology, an elaborate terminology has developed. We mention here the basic features and matters of physiological importance. This terminology is so extensive, however, that the bulk of sponge skeletal description is reserved for advanced texts.

Skeletal elements support and lend form to the sponge body. Their sharp ends may surface externally, to the discouragement of would-be predators. Sponge skeletons consist of numerous calcareous spicules, siliceous spicules, or spongin fibers, or of some combination of the latter two. Debris is an important component in some groups. Spicules assume many shapes and sizes. They may be straight and needlelike or curved, pronged, or hooked; some are spherical with star- or burrlike projections, while others have highly irregular shapes. Figure 4.5 displays several spicule types along with their terminology. We defer further description of skeletal structure to the discussion of sponge diversity.

Scleroblasts, a type of wandering cell in the sponge mesoglea, secrete spicules by a process related to cell division (Fig. 4.6a). Spicule production begins within a single scleroblast undergoing mitosis. The potential spicule forms along a spindle fiber. As mitosis is completed, one daughter cell (the founder cell) continues to originate new spicule length, moving away from the other daughter cell as the spicule grows. The second daughter cell (the thickener cell) gradually follows the first and thickens the new spicule. Both daughter scleroblasts eventually exit from the youngest end of the spicule. This process is similar whether calcium carbonate or silica is the raw material. Spicules with more than one axis are produced by multiple teams of scleroblasts (Fig. 4.6b). Spongin, a meshwork of protein fibers, is manufactured inside wandering cells called **spongioblasts.** Longer fibers are formed by spongioblasts working in series (Fig. 4.7).

In a phylum known for the noncoordination of its cells, the formation of sponge skeletal elements represents an extraordinary degree of intercellular cooperation. It has been suggested that scleroblasts meet at random to form multirayed spicules, but in many cases, spicule arrangement is too regular to support this hypothesis. Alternatively, mechanical forces (e.g., water flow) may orient the scleroblasts. The regulation of spicule shape is also poorly understood. The regularities of crystallization offer only a partial explanation, although deviations from crystal growth patterns might result from organic material contained within the spicule. (Siliceous spicules, for example, contain an organic axis surrounded by an organic sheath.) Another important influence on spicule shape is the effect exerted on founder cell movements by the surrounding cell layers. Currently, however, the various factors controlling the association and division of scleroblasts, the sites of spicule formation, and the production of specialized spicules remain a mystery.

Amoebocytes transport nutrients and, possibly, gametes through the mesogleal layer. Other wandering cells store food; some contain the pigments which grant brilliant coloring to their species. Histological studies suggest that a curious cell type, the **collencyte** (Fig. 4.8), may perform nervelike functions. Finally, **archeocytes** are totipotent, differentiating into other wandering cells according to the sponge's needs. Archeocytes also are involved in gamete production.

Nutrition and Digestion

Sponges eat and digest somewhat like protozoans. Bacteria, small planktonic organisms, and organic detritus are trapped by the fibrous meshwork of the choanocyte collar and transferred to the cell body (Fig. 4.2). Digestion, which is completely intracellular, may

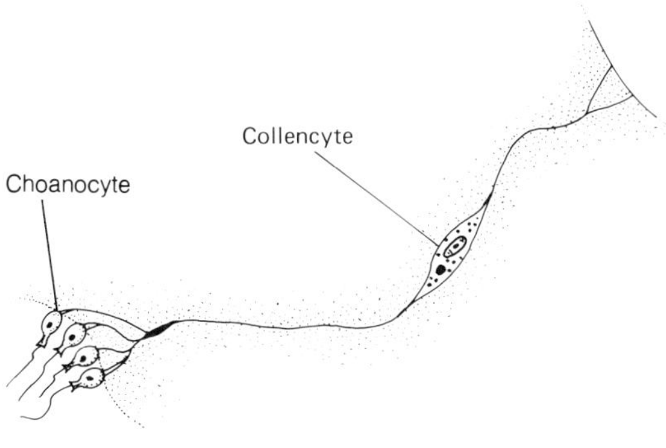

Figure 4.8 A collencyte within the mesoglea of a sponge. (From Tuzet and Pavans de Ceccatty. Compte Rendu de L'Academie des Sciences, Paris, 1953, 237:2342.)

begin within the choanocytes, but nutrients are rapidly transferred to amoebocytes. Within food vacuoles in these wandering cells, the common acid-to-alkaline transition takes place. The products of digestion are distributed throughout the sponge body by amoebocytes, and undigestible materials are egested into the water canal system and eventually expelled through the osculum.

Virtually all circulating water in leuconoid sponges passes through a choanocyte collar. The distance between collar pseudopodia averages 0.1 μm, while the prosopyles have an average width of 5 μm. Hence, such sponges are suited for straining particles in the 0.1- to 5-μm range. Although choanocytes are responsible for most sponge ingestion, particles too large to enter the prosopyles can be ingested by amoebocytes. Additionally, particles too large to enter the dermal ostia occasionally are phagocytosed by pinacocytes. Sponges are primarily particulate feeders, but choanocytes may pinocytose dissolved proteins. Many sponges possess large numbers of unicellular algae and bacteria, either in the mesoglea or within amoebocytes or archeocytes. Whether these organisms contribute to sponge nutrition is unknown (but see Special Essay, Chap. 5, on mutualism between algae and corals).

Circulation, Respiration, Excretion, and Osmoregulation

Because virtually all sponge cells are at a surface, diffusion and continuous circulation of water amply provide for the trafficking in materials central to the life process. Food and minerals are brought in; digestive wastes are pumped out. Respiratory gases are exchanged; nitrogenous wastes in the form of ammonia are expelled. Sponges have an understandably low metabolic rate, a factor contributing to the ease by which these maintenance activities are performed.

The sponge water current is produced by the uncoordinated beating of choanocyte flagella. In the leuconoid sponge, water enters each choanocyte chamber via two or three prosopyles and departs via a single apopyle. Because the lumen of the apopyle is larger than that of all the prosopyles combined, water enters the chamber faster than it leaves. Overall, the water speed through the incurrent canals is about twice that through the excurrent channels. Each choanocyte flagellum beats towards the apopyle; hence there is a pressure increase from prosopyle to apopyle. The lowered pressure on the prosopylar side of the chamber sucks water and suspended food particles into the sponge. The increased pressure in the

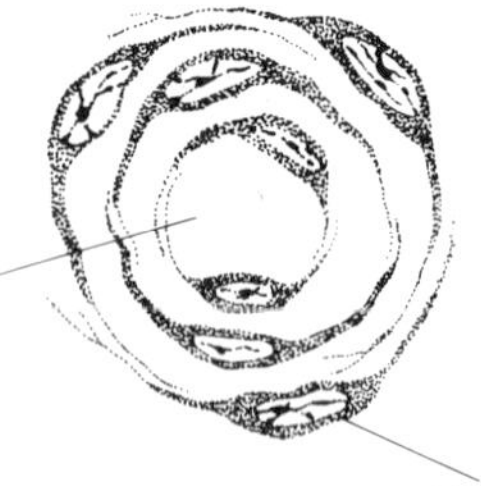

Figure 4.9 Myocytes surrounding a prosopyle. (From A. Dendy, Quart. J. Micros. Sci., 1893, *132*:159.)

excurrent channels ensures that water and waste products are carried well away from the sponge. Large volumes of water pass through a sponge. For example, the freshwater *Spongilla* may pump 70 times its body volume each hour.

Sponges are primarily aerobic, but they typically extract a low percentage of oxygen from the water passing through their canal system. Oxygen uptake increases, however, after a period of osculum closure, suggesting an oxygen debt as a result of enforced anaerobiosis. Freshwater sponges osmoregulate on a cellular level through the formation of contractile vacuoles.

ACTIVITY SYSTEMS

Sponges are not highly coordinated animals. They have no nervous system, no muscles, no sense organs, and correspondingly little behavior.

Communicative functions have been suggested for collencytes (Fig. 4.8), but physiological studies fail to support this idea. If collencytes do transmit information, experiments indicate that they do not use electrical impulses in the process. Pinacocytes are generally contractile, although they lack the sliding filaments associated with muscle cells. Long pinacocytes curve around prosopyles and oscula (Fig. 4.9); such cells are called **myocytes** and their contraction regulates the size of these openings. Although the spongocoel is an inner fluid-filled space, it does not constitute a hydrostatic skeleton. It has no organized body wall musculature. Moreover, the spongocoel–water channel system is so full of holes that no controllable body of water could be enclosed.

With such limited effectors and in the probable absence of coordinating mechanisms, sponge behavior is limited. In response to gross stimuli—removal from water, buffeting, or extreme temperature change—sponges contract generally. Stimulation at one point can have effects at a distance. However, any distant responses inevitably arrive very slowly, and hours may elapse before the entire sponge completes a reaction. These responses likely involve simple cell-to-cell communication. A domino effect may occur, in which the contraction of one cell stimulates a similar action in adjacent cells. Local responses also might be felt through the changes they effect in water circulation. A contracted pore causes pressure changes which could be felt at some distance.

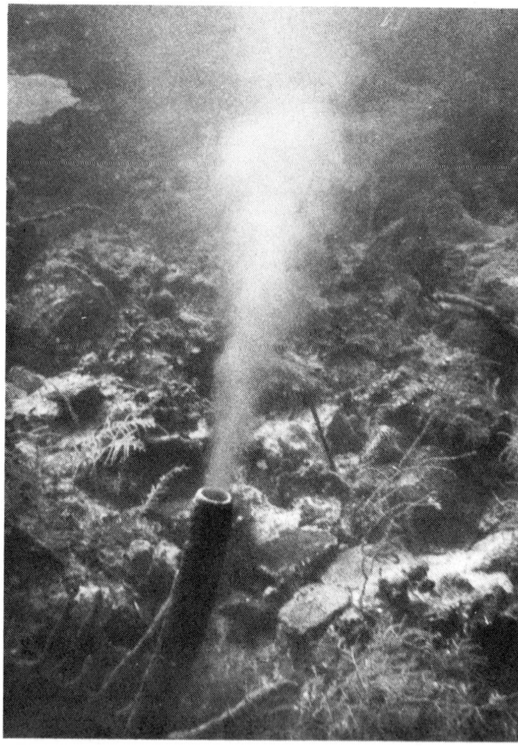

Figure 4.10 Release of sperm through the osculum of a large tubular West Indian sponge, *Verongia acheri*. (From H.M. Reiswig, Science, 1970, *170*:539.)

The feasibility of cell-to-cell communication in sponges reduces our need to search for more specialized conducting systems. Also arguing against such specialized systems is the limited coordination in these animals on every level. Choanocyte flagella in the same chamber beat with no integrated rhythm, and their activity may be uncoordinated with the sponge's osculum. Under stress conditions, the osculum may close even though choanocytes continue to draw water. In time, a sponge can become so bloated by trapped water that it explodes.

Of course, sponges do not require much overt be-

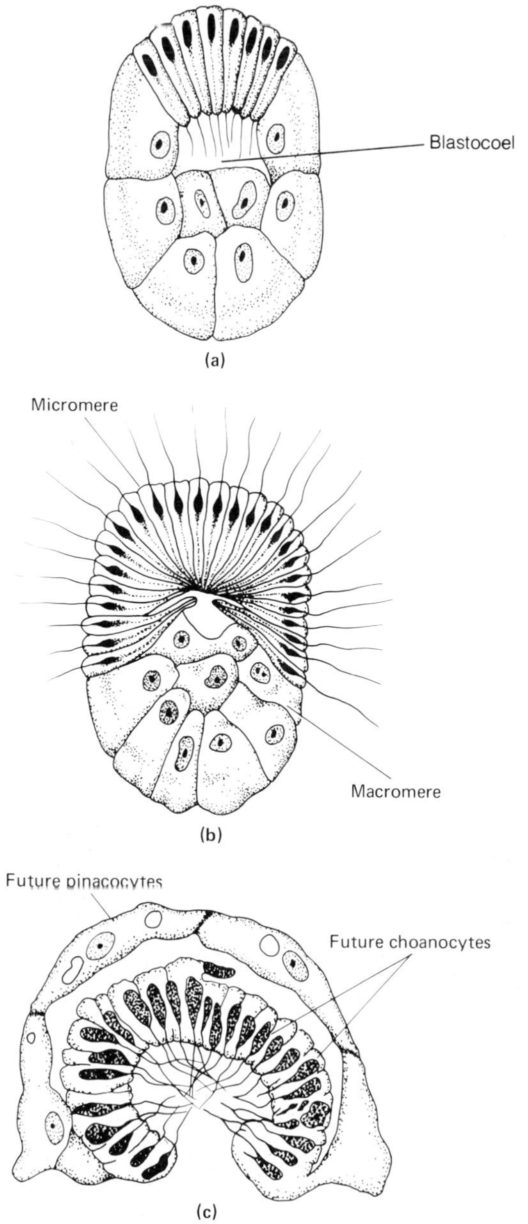

Figure 4.11 Development in calcarean sponges. (*a*) Blastula stage showing 8 macromeres and 8 flagellated micromeres with flagella extending into the blastocoel; (*b*) Inversion yields an externally flagellated amphiblastula larva; (c) Amphiblastula larva after gastrulation and attachment. (From Hyman, *The Invertebrates, Vol. I.*)

havior. Filter feeding precludes any necessity to pursue prey; water movements, rather than sponge movements, are the order of the day. Sponges have very few predators, due in part to their skeletons but apparently also to a supreme unsavoriness. Sponges are prickly, and many evidently do not taste good; most animals leave them alone.

CONTINUITY SYSTEMS

Reproduction and Development

Sponges reproduce both sexually and asexually. Gametes probably develop from archeocytes, but choanocytes may be involved in sperm production. Sperm are released into the water channel system and reach the outside world through an osculum (Fig. 4.10); upon entering another sponge's ostia, the sperm are taken up by choanocytes. The collar cell with a sperm detaches from the canal wall and moves into the mesoglea. When a suitable egg is encountered, the choanocyte fuses with it and fertilization occurs. This peculiar form of sperm transfer has been described for the calcarean sponges and apparently is general in the phylum. However, amoebocytes may "chaperone" sperm in some Demospongiae.

Most sponges are hermaphroditic, but cross-fertilization is ensured by the nonsynchronous maturation of eggs and sperm in a single individual. Typically, sperm mature before eggs do. This condition is called **protandry** and is widespread among hermaphroditic species.

Embryogeny is one of the most curious aspects of sponge biology. Developmental patterns vary widely, and the following account is descriptive only of some better-studied calcareans. In these sponges, a fertilized egg divides within the parental mesoglea to the 16-cell blastula stage. This blastula has eight macromeres and eight flagellated micromeres; flagella extend into the central blastocoel (Fig. 4.11a). Macromeres then stop dividing, but the smaller flagellated cells continue to divide. Next the flagellated micromeres migrate through an opening between the macromeres. In a process without parallel among other metazoans, the entire sponge embryo turns inside out. The resulting, externally flagellated larva is called an **amphiblastula** (Fig. 4.11b). The amphiblastula enters an excurrent water channel and exits from the sponge to assume a brief, free-swimming existence. Before or after the larva attaches to a substratum, gastrulation takes place. Here again, sponge embryogeny is unique: it is the *micromeres* that gastrulate. Inside the young organism, micromeres become choanocytes, amoebocytes, and archeocytes; macromeres differentiate into pinacocytes, porocytes, and scleroblasts. These developmental fates are opposite to those ob-

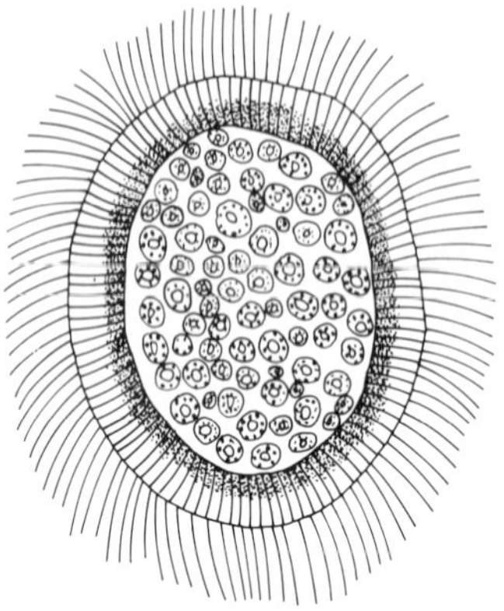

Figure 4.12 Parenchymella larva of a leuconoid sponge. (From Hyman, *The Invertebrates, Vol. I.*)

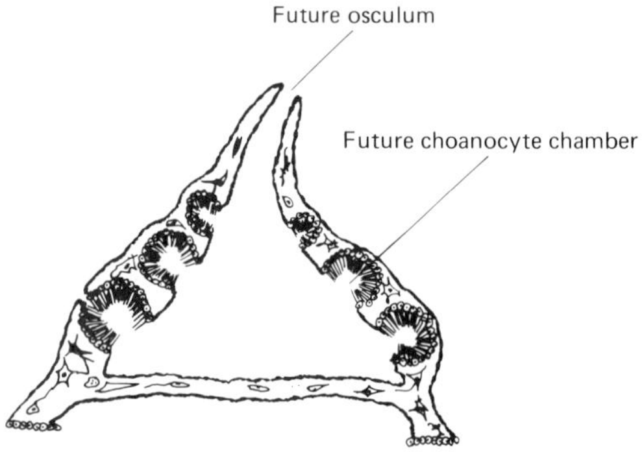

Figure 4.13 A rhagon, an immature stage in the development of many leuconoid sponges. (From Hyman, *The Invertebrates, Vol. I.*)

served in the rest of the invertebrate world. In other metazoans, macromeres are the source of internal (endodermal) cells, while micromeres differentiate very early into epidermal tissues. The sponge amphiblastula attaches at its blastopore. The blastocoel becomes a spongocoel and an osculum opens at its free end (Fig. 4.11c).

The preceding events produce an asconoid sponge. Syconoid types develop if the body walls become folded. Leuconoid sponges exhibit a different larval form called a **parenchymella** (Fig. 4.12). These larvae are completely bounded by flagellated cells except for a small area at the macromere pole. Internally, they are filled with archeocytes. Upon larval attachment to the substratum, the flagellated cells migrate into the center of the organism; there they form choanocytes and amoebocytes. Development may recapitulate asconoid and syconoid stages, or the leuconoid form may arise via an immature stage known as a **rhagon** (Fig. 4.13). The rhagon has a very thick body wall that hollows out to form choanocyte chambers and water channels.

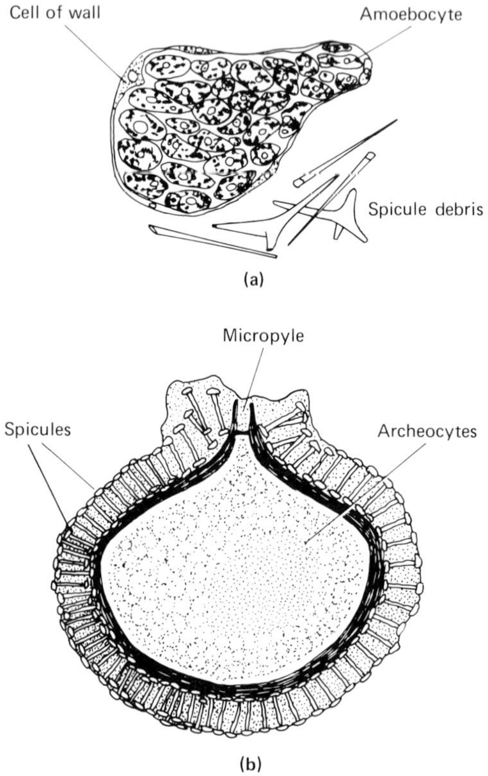

Figure 4.14 Asexual reproduction in sponges. (*a*) A reduction body of a calcarean sponge; (*b*) A gemmule of a freshwater sponge. (From Hyman, *The Invertebrates, Vol. I.*)

Random settling with heavy mortality in unfavorable areas largely determines sponge distribution. In some species, however, larval behavior may play a major role. As the sponge larva settles, there is usually a creeping phase of several hours duration during which the larva may exhibit phototactic and geotactic responses. Such settling larvae display more behavior than sponges do at any other time in their lives.

Asexual reproduction by budding occurs when a portion of sponge mass isolates itself from the parent organism. The new individual may remain attached, as in colonial sponges, or may detach for an independent life.

Certain asexual methods more peculiar to the sponges involve specialized groups of cells. During

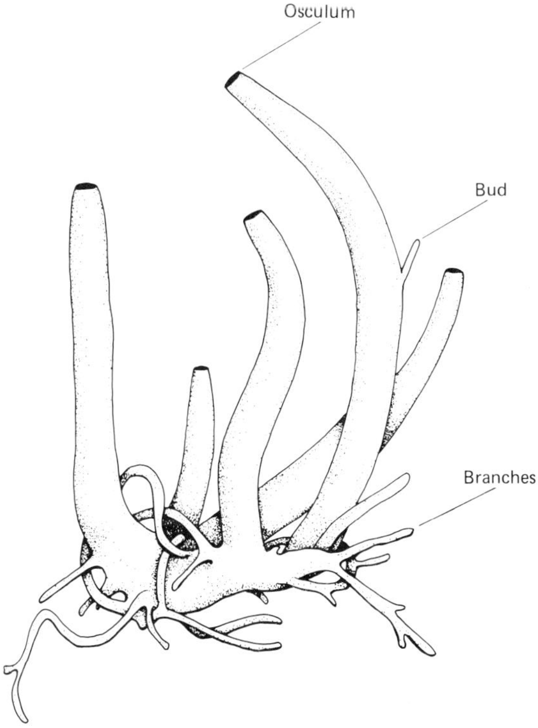

Figure 4.15 A portion of a colony of *Leucosolenia*, an asconoid sponge. (From Beck and Braithwaite, *Invertebrate Zoology Lab Workbook.*)

unfavorable periods, some sponges disintegrate into **reduction bodies** (Fig. 4.14a)—clusters of amoebocytes bounded by a thick wall. When better environmental conditions return, these clusters germinate into young sponges. **Gemmules** are formed by a few marine species, but are most characteristic of freshwater sponges. Similar to reduction bodies, gemmules comprise a mass of food-filled archeocytes surrounded by columnar membranes and spicule secreting cells (Fig. 4.14b). Germinating cells emerge through a pore called the **micropyle.** In certain marine sponges, gemmules produce larvae which are virtually indistinguishable from those formed by a sexual union. However, larvae from freshwater gemmules develop directly into small leucons. The archeocytes emerge from the micropyle in three waves. The first two differentiate into pinacocytes and choanocytes; the last wave provides archeocytes.

Motile sperm and free-swimming larvae help disperse sponge populations. Gemmules and reduction bodies contribute to a lesser extent, but their primary ecological value lies elsewhere. Sessile organisms like sponges cannot migrate from their original environments. They have no way to escape a suddenly fouled home or the cyclical austerity of temperate-zone fresh waters. Gemmules and reduction bodies allow sponges to survive such problem periods and return to

(a)

(b)

Figure 4.16 (*a*) A cluster of calcareous sponges, *Leucosolenia complicata* from a wharf piling. (*b*) A single syconoid sponge, *Sycon coronatum.* (*a* courtesy of D.P. Wilson; *b* courtesy of Ward's Natural Science Establishment.)

life as usual, if and when better conditions are restored. Gemmules are the form in which most freshwater sponges overwinter.

The stimuli triggering gemmule formation are unknown. Some gemmules exhibit a true **diapause,** similar in many respects to that of insects. Diapause is a state of developmental arrest; exposure to low temperatures is required before developmental competency can be restored. It is thus an important adaptation for overwintering. In most nondiapausing species, the gemmules remain within the parent body until freezing temperatures occur. The parent sponge secretes a substance preventing premature germination.

The independence of individual sponge cells and the capacity of these animals for asexual reproduction suggest that sponges are capable of regeneration. A remarkable experiment demonstrates the extensiveness of their regenerative capacities. A sponge body can be pressed through a silk cloth and separated into cells and cell clusters. These strained cells become amoeboid, contact each other, and form growing clusters. When enough cells of each type are gathered, a new sponge individual is formed.

A Review

To conclude our discussion of unity within this phylum, we review the general characteristics of sponges. Basically, sponges are:

(1) primitive metazoans retaining some protozoan character;

(2) animals without organs, definite tissue layers, mouths, or digestive cavities;

(3) sessile animals whose bodies house choanocyte-lined channels through which water circulates;

(4) invertebrates without nerves, muscles, or sense organs;

(5) mostly protandrous hermaphrodites whose embryogeny features a unique inversion process;

(6) a phylum whose taxonomy is based primarily on skeletal elements composed of calcium carbonate, silica, and/or spongin.

Sponge Diversity

A phylum as physiologically primitive as the sponges has a limited capacity for adaptive experimentation. Accordingly, adaptive radiation in sponges focuses on two major fronts: the development of more elaborate water channel systems, and the construction of diverse skeletons. The asconoid, syconoid, and leuconoid systems have been described. Further variations on these water channel themes are influenced more by local environmental conditions than by genetic means. Skeletal elements provide the primary basis for sponge taxonomy.

CLASS CALCAREA

Calcareous sponges are distinguished by spicules of calcium carbonate. Calcareans represent the smaller, simpler members of the phylum and are usually restricted to shallow coastal waters. This class includes all of the asconoid and syconoid sponges as well as a few leuconoid forms. Individual members are never more than 10 cm in height. Their calcareous spicules are of similar size and include one-, three-, and four-pronged types. Arranged according to species-specific structural patterns, these spicules provide form and protection for the sponge. Some project externally, and a collar of protruding spicules commonly surrounds the osculum. *Leucosolenia* and *Sycon* (Figs. 4.15 and 4.16) represent the basic asconoid and syconoid styles, respectively.

CLASS HEXACTINELLIDA

Members of this class, commonly known as glass sponges, possess the phylum's most beautiful skeletons. Composed primarily of six-pointed siliceous spicules, these skeletons are characteristically vase shaped and radially symmetrical. Fused spicules form a latticework that may be very delicate in appearance, as in Venus's flower basket (*Euplectella;* Fig. 4.17).

The hexactinellid body centers around a spacious spongocoel. Flagellated canals or chambers extend radially from the spongocoel in a syconoid or primitively leuconoid arrangement. A sieve plate of fused spicules usually covers the osculum. Pinacocytes are entirely absent from the class; external surfaces are covered instead by a syncytium of amoebocyte pseudopodia.

The glass sponges are the most highly individualized members of the phylum. They are usually 10 to 30 cm high, but individuals up to a full meter in height have been reported. Hexactinellids are deep-water sponges. They are most common in tropical waters at depths of 450 to 900 m, although some species range down to 5 km.

An interesting commensal relationship occurs between *Euplectella* and some shrimp species. A sexual pair of small, young shrimp enter a sponge and reach adult size in the spongocoel, where the oscu-

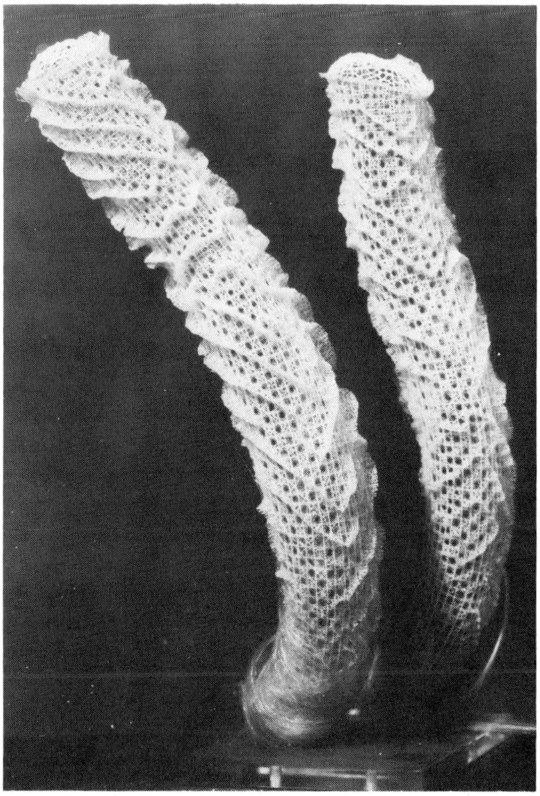

Figure 4.17 Skeletons of Venus's flower-basket, *Euplectella*, a hexactinellid sponge. (Photo courtesy of the American Museum of Natural History.)

Figure 4.18 Spongin skeleton of a commercial bath sponge (*Spongia*) from the Mediterranean. The large openings are the oscula. (Photo by Betty Barnes, courtesy of W.B. Saunders Company.)

lar sieve prevents their escape. Feeding on plankton, the shrimp remain together until death.

CLASS DEMOSPONGIAE

About 95% of all sponge species belong to the Demospongiae, the most advanced class in the phylum. All members of this class have leuconoid body forms. Their skeletons are constructed from siliceous spicules, spongin fibers, or some combination of the two. When both are present, the spicules are usually embedded within spongin fibers. Siliceous spicules occurring alone are never six-pointed, as in the glass sponges. The few genera without skeletal elements are believed to be primitive.

Demospongiae are widely distributed in the world's oceans. Some, like the tropical loggerhead sponges, reach considerable size: a few have been measured at more than a cubic meter. All types of growth patterns are seen; colors, too, are variable, because of the pigmented granules within special amoebocytes. The commercial bath sponges, whose skeletons are composed exclusively of spongin fibers and debris (Fig. 4.18), belong to this class, as do all freshwater sponges (Fig. 4.19). Sponges have invaded freshwater habitats perhaps more than once, but never very successfully.

Several Demospongiae are involved in biological relationships with other invertebrates. The boring sponges (e.g., *Cliona*) attack calcium carbonate surfaces such as corals and mollusk shells; they decompose the mineral areas and grow into the resulting holes. Other sponges grow on old mollusk shells inhabited by hermit crabs. In this symbiosis, the sponge receives a substratum, while the hermit crab benefits from increased camouflage. Eventually, the mollusk shell may erode, leaving the crab to occupy a portion of the sponge body itself. Some true crabs camouflage themselves by holding a piece of sponge over their backs. In all these relationships, the general nonedibility of sponges is an additional factor in the crab's protection.

CLASS SCLEROSPONGIAE

This fourth recently discovered class of coralline sponges contains six species from Jamaican coral reefs. They possess a leuconoid body form but differ from other leuconoid sponges in that they have an internal skeleton of spongin and siliceous spicules with an outer covering of calcium carbonate.

Sponge Evolution

Sponges probably arose at some point far back in Precambrian time when a group of protozoan flagellates adapted along an evolutionary cul de sac. The

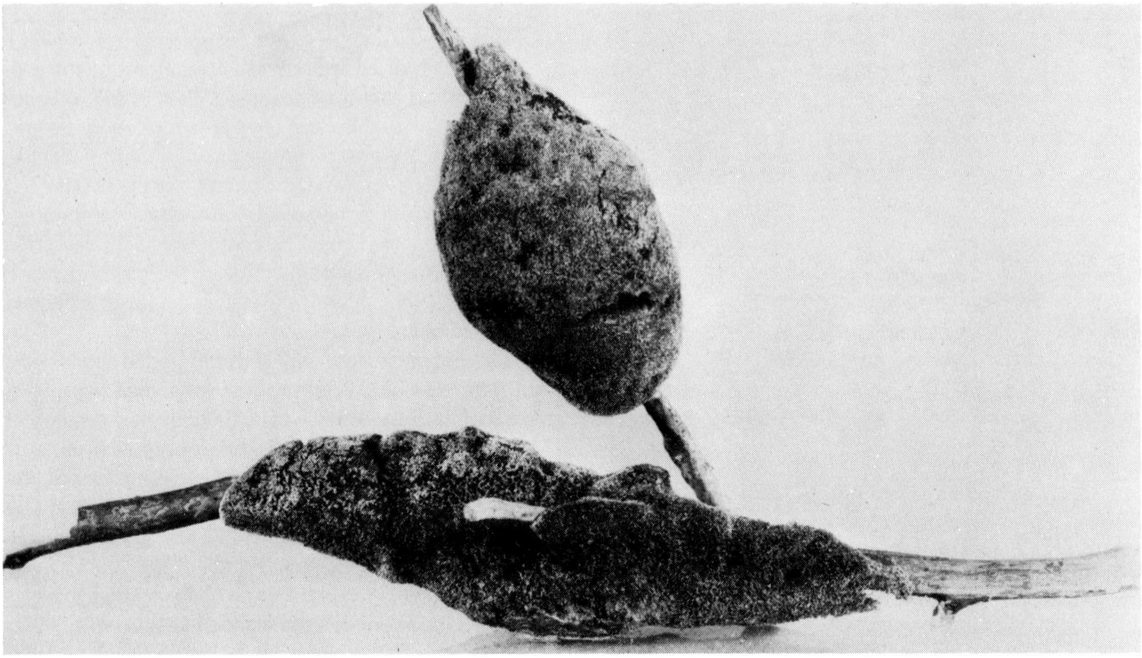

Figure 4.19 Freshwater sponges (*Spongilla*) growing attached to sticks. (Courtesy of the American Museum of Natural History.)

theory that sponges represent a dead end in invertebrate evolution is supported on several counts. No other animal group's principal opening to the outside world is excurrent rather than incurrent. Sponge embryogeny is bizarre. The true tissues and digestive cavities that typify other metazoans are absent, while choanocytes—very rarely present in other groups—dominate sponge physiology. The inclination is to propose an independent origin for the sponges, arising from a distinct stock of colonial flagellates or at least diverging from the rest of the metazoan line at a very early date.

Hopefully, the similarity between sponge choanocytes and the collar cells of choanoflagellates has not been overlooked. Sponges may have originated from these protozoans; however, choanoflagellates are primarily a freshwater group and have nothing corresponding to the embryonic stages of sponges. The phytomonad *Volvox* undergoes a developmental inversion like that of sponges, suggesting a common ancestor. *Volvox,* however, is photosynthetic. A common ancestry is still possible, but the loss of photosynthetic ability—a great sacrifice for any living thing—would have been required early in the process. For the present, sponge origin, like that of the other metazoans, remains a matter of speculation.

Whatever their origin, sponges have made an important transition to multicellular life. Their pinacocyte and choanocyte cell groupings represent primitive levels of tissue formation. Yet sponges have not fully exploited the many opportunities of metazoan life, and we might wonder what advantages they have achieved over protozoans. Perhaps their most important asset is increased size. Larger size affords an organism a better chance in dealing with challenging environments. Gametes and embryos, in particular, can be protected within multicellular clusters. Moreover, all cells need not deal on all sides with the external world. Adaptive possibilities thus exist for limiting the responsibilities of individual cells; specialization may occur, as well as cooperation among cells. Sponges retain some protozoan traits; thus these possibilities are not well developed in this phylum.

for Further Reading

Ayling, A. L. Patterns of sexuality, asexual reproduction, and recruitment in some subtidal marine Demospongiae. *Biol. Bull.* 158: 271–282, 1980

Bayer, F. M., and Owre, H. B. *The Free-Living Lower Invertebrates.* pp. 1–24. Macmillan, New York, 1968. (A useful general account of the Porifera, nicely illustrated)

Bidder, G. P. The relationship of the form of a sponge to currents. *Quart. J. Microsc. Sci.* 67: 292–323, 1923.

Berquist, P. R. *Sponges.* Hutchinson and Co., London, 1978.

Berquist, P. R. The ordinal and subclass classification of the Demospongiae (Porifera): Appraisal of the present arrangement and proposal of a new order. *N. Z. J. Zool.* 7: 1–6, 1980.

Brauer, E. B. Osmoregulation in the freshwater sponge. *Spongilla lacustris. J. Exp. Zool.* 192:181–192, 1975.

Bullock, T. H., and Horridge, G. A. *Structure and Function of the Nervous System of Invertebrates, Vol. I.* pp. 450–453, W. H. Freeman, San Francisco, 1965.

Elvin, D. W. Seasonal growth and reproduction of an intertidal sponge, *Haliclona permollis. Biol. Bull.* 151: 108–125, 1976.

Fell, P. E. Porifera. In *Reproduction of Marine Invertebrates, Vol. I.* (Giese, A. C. and Pearse, J. S., eds.), pp. 51–132. Academic Press, New York, 1974.

Finks, R. M. The evolution and ecologic history of sponges during Paleozoic times. *Symp. Zool. Soc. London.* 23: 3–22, 1970.

Fjerdingstad, E. J. The ultrastructure of choanocyte collars in *Spongilla lacustris. Z. Zellforsch.* 53: 645–657, 1961.

Florkin, M, and Scheer, B. T. *Chemical Zoology, Vol. II: Porifera, Coelenterata, and Platyhelminthes,* pp. 1–77. Academic Press, New York, 1968. (Six chapters on selected topics relating to sponge biology and physiology)

Fry, W. G. (Ed.) *The Biology of The Porifera,* Symposia of the Zoological Society of London, No. 25. Academic Press, New York, 1970. (A collection of various aspects of sponge biology including the Sclerospongiae, the new fourth class of sponges; the proceedings of a symposium held in in September 1968)

Garrone, R. *Frontiers of Matrix Biology, Vol. 5* Phylogeny of connective tissue. Morphological aspects and biosynthesis of sponge intercellular matrix. S. Karger, Basel, Switzerland, 1978.

Harrison, F. W., and Cowden, R. R. (Eds.) *Aspects of Sponge Biology,* Academic Press, N.Y., 1976. (A symposium volume containing articles on a variety of topics concerning the biology of sponges

Hartman, W. D., and Reiswig, H. M. The individuality of sponges. In *Animal Colonies:* Development and Function through Time (R. S. Boardman, et al., Eds.) pp. 567—584. Hutchinson, and Ross, Stroudsburg, PA, 1973.

Hartman, W. D., Wendt, J. W., and Wiedenmayer, F. *Living and Fossil Sponges: Notes for a Short Course,* University of Miami Press, Miami, 1980.

Humphreys, T. Species specific aggregation of disassociated sponge cells. *Nature* (London) 228: 685–686, 1970.

Hyman, L. H. *The Invertebrates, Vol. I: Protozoa through Ctenophora,* pp. 284–364, McGraw-Hill, New York, 1940. (The standard reference work on the biology of sponges; includes an extensive bibliography. The chapter "Retrospect" in Vol. V summarizes investigations on sponges from 1938 to 1958)

Johnson, M. F. Significance of life history studies of calcareous sponges for species determination. *Bull. Mar. Sci.* 28: 570–574, 1978.

Johnson, M. F. Studies on the reproductive cycles of the calcareous sponges *Clathrina coriaceae* and *C. blanca. Mar. Biol.* 50: 73–80, 1979.

Jones, W. E. Is there a nervous system in sponges? *Biol. Rev.* 1–50, 1962.

Jorgensen, C. B. *The Biology of Suspension Feeding.* Pergamon Press, New York, 1966.

Levi, C. Gastrulation and larval phylogeny in sponges. In *The Lower Metazoa.* (E. C. Dougherty, Ed.). pp. 375–382. University of California Press, Berkeley, 1963.

Pennak, R. W. *Fresh-Water Invertebrates of The United States.* 2nd ed., pp. 80–98, Wiley-Interscience

Reiswig, H. M. Particle feeding in natural populations of three marine demosponges. *Biol. Bull.* 141: 568–591, 1971.

Reiswig, H. M. Coloniality among the Porifera. In *Animal Colonies: Development and Function through Time* (R. S. Boardman, et al., Eds.), pp. 549—565, Dowden, Hutchinson, and Ross, Stroudsburg, PA, 1973.

Reiswig, H. M. Water transport, respiration and energetics of three tropical marine sponges. *J. Exp. Mar. Biol. Ecol.* 14: 231–249, 1974.

Reiswig, H. M. The aquiferous systems of three marine Demosporagiae. *J. Morphol.* 145: 493–502, 1975.

Reiswig, H. M. Discharge of gametes by sponges of the Caribbean Sea. *Mem. Soc. Cienc. Nat. La Salle.* 36: 183–192, 1976.

Simpson, T. L. and Gilbert, J. J. Gemmulation, gemmule hatching and sexual reproduction in freshwater sponges: The life cycle of *Spongilla lacustris* and *Tubella pennsylvanica. Trans. Am. Microsc. Soc.* 92: 422–433, 1973.

Tuzet, O. The phylogeny of sponges. In *The Lower Metazoa* (E. C. Dougherty, Ed.), pp. 129–148, University of California Press, Berkeley, 1963.

Vogel, S. Evidence for one-way valves in the water flow system of sponges. *J. Exp. Biol.* 76: 137–148, 1978.

Wells, H. W., Wells, M. J., and Gray, I. E. Ecology of sponges in Hatteras Harbor, North Carolina. *Ecology* 45: 752–767, 1964.

Wiedenmayer, F. *Shallow-water Sponges of the Western Bahamas.* Birkhaeuser Verlag, Basel, Switzerland, 1977.

Wilkinson, C. R. Skeletal morphology of coral reef sponges: A scanning electron microscope study. *Aust. J. Mar. Freshw. Res.* 30: 793–802, 1979.

5

The Radiate Animals: Coelenterates and Ctenophores

Some of the most beautiful invertebrates are the radially symmetrical coelenterates and ctenophores. Comprising about 9000 living species, these twin phyla include jellyfish, sea anemones, corals, comb jellies, and the freshwater hydras. The floral beauty of many coelenterates led Renaissance scholars to classify them as marine flowers. By the 1700s they were recognized as animals, but only in the last century were these organisms distinguished from such diverse groups as echinoderms, sponges, and bryozoans.

The radiate phyla possess two distinctly metazoan features making them far more complex than sponges: a **gastrovascular cavity** and the organization of cells into functional tissues. The gastrovascular cavity, or **coelenteron,** is an internal space for the digestion and distribution of nutrients. A much greater range of food sizes is available to an animal with such a cavity, and entire new feeding styles are possible. The coelenteron is the only body cavity in these animals; its single opening forms a mouth. The cells lining this cavity constitute a functional tissue. Externally, coelenterates and ctenophores are covered by another tissue layer, and they possess organized contractile and nervous elements as well. Thus, the radiate phyla display a tissue level of organization.

Coelenterates and ctenophores display primary radial or biradial symmetry. This condition differs from the universal symmetry of a sphere in that a single axis joins the definite upper and lower ends of the animal. Body parts are arranged concentrically about this axis. In coelenterates and ctenophores, the gastrovascular cavity surrounds the central axis, with the mouth located at its upper end. In coelenterates, shifts toward biradial or radiobilateral symmetry occur when alterations in mouth structure create polarities perpendicular to the oral–aboral axis (Fig. 5.1).

Members of the Coelenterata (also called the Cnidaria) and Ctenophora are exclusively aquatic and, with few exceptions, are limited to the marine

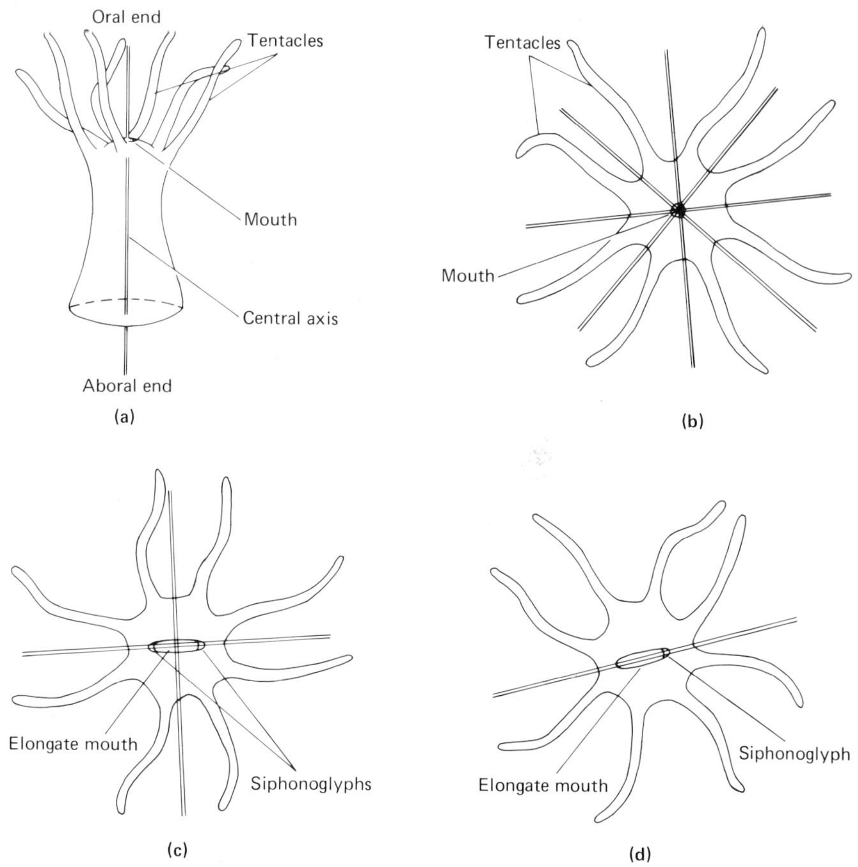

Figure 5.1 Coelenterate symmetries. (*a*) Side view of a radially symmetrical coelenterate polyp showing the oral-aboral central axis; (*b*) In a radially symmetrical animal innumerable oral-aboral planes passing through the central axis will yield equal halves; (*c*) In a biradially symmetrical animal only two oral-aboral planes (at right angles to each other) will yield equal halves; (*d*) In a radiobilaterally symmetrical animal only a single oral-aboral plane will yield equal halves. (From D.E. Beck and L.F. Braithwaite, *Invertebrate Zoology Laboratory Workbook*, 3rd ed., 1968, Burgess.)

environment. Coelenterates are by far the more abundant group, totaling about 8900 species.

The Coelenterates

The schematic coelenterate might be likened to a flexible bottle with the cork pulled (Fig. 5.2a). Its outer and inner surfaces are covered by tissue layers. Between these surface sheets lies a third layer, the **mesoglea,** which varies from a noncellular membrane to a thick fibrous meshwork with cellular components. Tentacles surround the uncorked mouth and contain extensions of the three body layers. The tentacles are armed with stinging cell organelles called **nematocysts.** With bizarre exceptions (see page 261), nematocysts are peculiar to the coelenterates and are important taxonomic tools in some groups.

Although some coelenterates are quite large, their structural simplicity is always preserved. The phylum's maintenance systems remain rudimentary. The primitive nervous and muscular elements of coelenterates have been the focus of basic research because they represent the simplest examples of such systems. These systems make something like real behavior possible. Continuity systems in these animals are diverse and complex. The basic life cycle scheme, on which there are numerous variations, features the alternation of a motile, sexually reproducing jellyfish, or **medusa,**

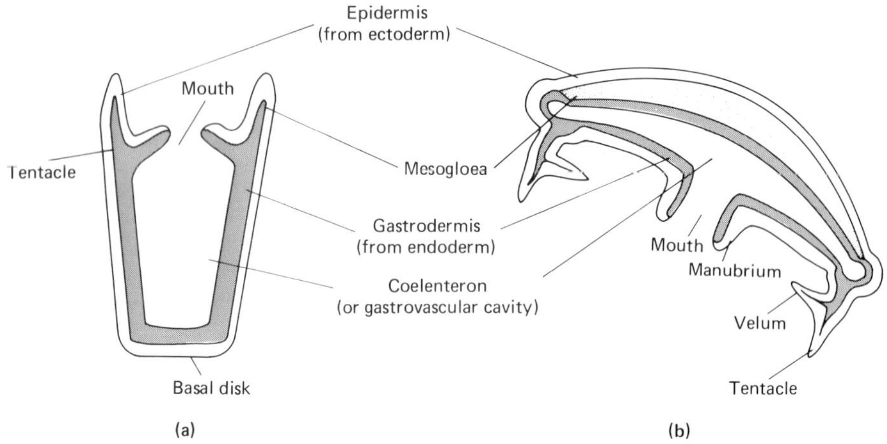

Figure 5.2 The two basic body forms exhibited by coelenterates. (*a*) The sessile polyp; (*b*) The motile medusa. (Modified from W.D. Russell-Hunter, *A Biology of Lower Invertebrates*, 1968, Macmillan.)

with a stationary, asexually reproducing **polyp.** Polyps and medusae may be extremely modified for their respective life-styles. One stage may dominate the life cycle, even to the extent that the other is eliminated completely. The polypoid stage is often colonial, composed of asexually proliferated individuals. Some colonial coelenterates feature both polypoid and medusoid members within a single colony.

At first glance, the body forms of the typical polyp and medusa may not indicate their intimate relationship. A polyp corresponds well to our bottle simile, but the average medusa is bell or umbrella shaped (Fig. 5.2b). Yet, with reference to the three-layered body wall of all coelenterates, their morphologies are easily derived from one another. Basic tissue layers do correspond; the major structural difference is that the medusa has a much bulkier, gelatinous mesoglea. Because of this increased bulk, the medusa is drawn out radially and compressed along its oral–aboral axis. A medusa's mouth is located at the end of a mouth tube, or **manubrium,** suspended from the lower, or subumbrellar, surface of the bell; tentacles extend from the bell margin. Accordingly, the association of tentacles and mouth is not as intimate as among polypoid forms. Adhered to the substratum by their pedal disks, polyps orient with their mouths upward; medusae swim mouths down.

The alternation of polypoid and medusoid forms is an example of **temporal polymorphism.** Each form, or morph, represents a stage within one generation of the life cycle. In many colonial coelenterates, different morphs exist simultaneously and are specialized for particular functions.

Adaptive radiation in this phylum features increases in the structural complexity of the coelenteron and considerable variations in the life cycle scheme. Three large classes are recognized: the Hydrozoa, the Scyphozoa, and the Anthozoa. Hydrozoans include the simpler forms; their coelenterons are relatively unmodified, and both polyps and medusae commonly participate in the life cycle. This class contains the freshwater hydras and many colonial forms, including the dreaded Portuguese man-of-war. Scyphozoans, including most large jellyfish, are a relatively unified group in which the medusa is dominant. The most complex gastrovascular cavities are found in the anthozoans, a class in which the medusoid form is totally absent. Sea anemones and corals are anthozoans.

Coelenterate Unity

MAINTENANCE SYSTEMS

General Morphology

The epidermis of coelenterates comprises ectodermally derived cells modified for supportive, sensory, defensive, and locomotor functions. An internal cell layer is modified for digestion and distribution. Termed the **gastrodermis,** this internal layer is endodermal in origin. Between the epidermis and the gastrodermis is the mesoglea, a structureless gel in the

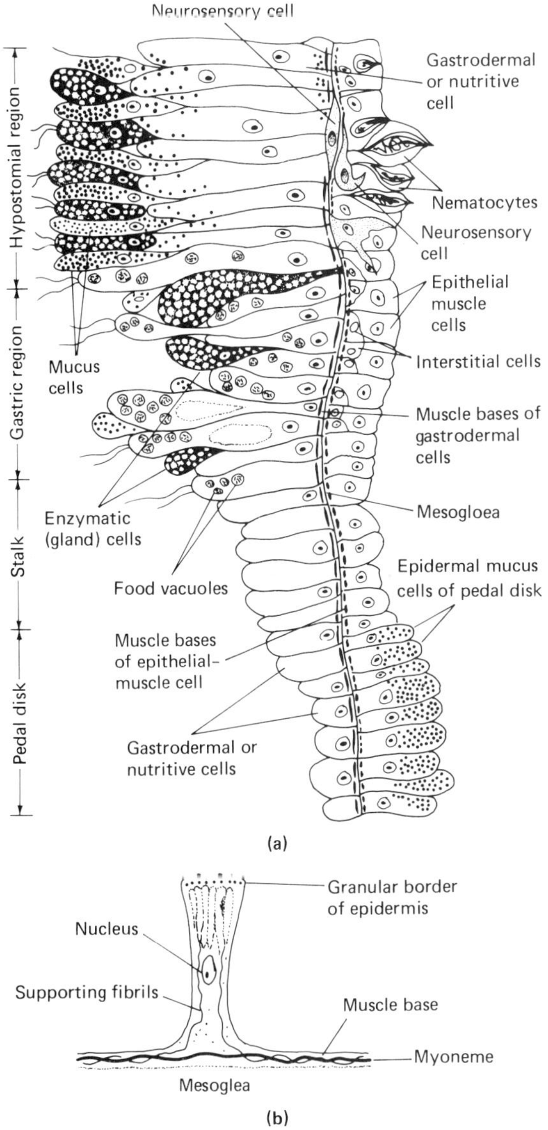

Figure 5.3 *Hydra.* (*a*) Section through the body wall; (*b*) Epithelial muscle cell. (*a* from I.W. Sherman and V.G. Sherman, *The Invertebrates: Function and Form: A Laboratory Manual*, 2nd ed., 1976, Macmillan; *b*: Hyman, *The Invertebrates, Vol. I: Protozoa through Ctenophora, 1940,* McGraw-Hill.)

simplest coelenterates, but somewhat cellular and fibrous in more complex forms.

The epidermis may be ciliated or flagellated, and its cell membranes may be obscure. Five basic cell types are found: epithelial cells, often containing contractile fibers; gland cells; neurosensory cells; nematocytes; and interstitial cells (Fig. 5.3a).

Epithelial cells, often called epithelial–muscle cells because of their contractile elements, compose most of the epidermis and serve in the support, body wall definition, and general protection of the animal. They are usually columnar or cuboidal, but their shape varies throughout the phylum. Epithelial cells make fibrous or amoeboid connections with the mesogleal layer. Contractile fibers within these cells extend longitudinally along the mesoglea (Fig. 5.3b). Such fibers represent the chief longitudinal muscle elements in the simpler hydrozoans, while in more complex coelenterates, epidermal muscles separate as an independent tissue layer.

Gland cells are plentiful in the epidermis, particularly in the oral and pedal disks. Some of these cells manufacture a mucus used to glue the coelenterate to a solid surface or to entangle prey. Other epidermal gland cells secret a protective casing around some polypoid morphs. This casing varies from a transparent, chitinoid sheath to the calcium carbonate skeletons of the corals.

Neurosensory cells in coelenterates represent the simplest nervous elements in the invertebrate world. These epidermal cells lie near the mesoglea, and their processes extend between the epithelial layer and the muscle fibers (Fig. 5.3a). By synaptic connections, these neurons form a loose, conducting **nerve net.** Sensory sites occur when sensitive cell processes are exposed at the body surface (Fig. 5.3a).

Nematocytes are a fourth epidermal cell type. Nematocysts, the thread weapons diagnostic of the coelenterates, are enclosed within these ovoid cells. Distributed throughout the epidermis, nematocytes are most abundant in the oral and tentacular areas, where they may be organized into units called batteries. Further description of these remarkable devices is reserved for our discussion of coelenterate activity systems.

The last epidermal type is the **interstitial cell,** an undifferentiated formative cell from which the other types in the epidermis are derived. These usually solitary cells are wedged between the bases of epithelial–muscle cells (Fig. 5.3a). As they differentiate into another cell type, these cells migrate to an appropriate place in the body wall. Small clusters of cells called **nematoblasts** are in transition between interstitial cells and nematocytes. In all sexually reproducing coelenterates, interstitial cells are responsible for gamete production.

Thus these five cell types compose the outer layer of a coelenterate. In primitive hydrozoans, this epidermis consists of a single cell layer. With the increasing development of distinct muscle tissue and an increased number of neural elements, higher coelenterates display an epidermis virtually three cells thick: the outer epithelial cells, nematocytes, and gland cells; the nerve net; and the muscle tissue against the mesoglea.

The cells lining the gastrovascular cavity resemble those of the epidermis. Nutritive cells, similar to epithelial cells and often with comparable contractile elements, dominate the internal wall. Flagella on the expanded distal ends of these cells churn the contents of the coelenteron as food is engulfed into vacuoles. Pleated membranes increase the surface area of these nutritive cells, and the entire gastrovascular wall may be folded or otherwise partitioned, thus enhancing its absorptive potential. Compartmentalized coelenterons are characteristic of the anthozoans. Also, in anthozoans the oral ectoderm indents to form a gullet, or **stomodeum** (Fig. 5.39).

Gland cells secreting mucus and digestive enzymes are abundant in the oral region. A gastrodermal nerve net is present, while contractile elements within the gastrodermal cells form a strong circular musculature. Interstitial cells are rare. Nematocytes are common in the oral region, and in anthozoans they occur along the free margins of the septa, which compartmentalize the coelenteron.

The mesogleal layer may be thought of as variously active glue. In simple hydrozoans it is little more than an extra membrane between the epidermal and gastrodermal layers, while in more advanced coelenterates it serves a variety of more specific functions. The mesoglea plays a skeletal role and may anchor muscle fibers. In medusae, it is greatly expanded and defines the bell shape. There is some evidence that epidermal and gastrodermal nerve nets communicate through mesogleal neurons. In any case, this gelatinous layer can be a pathway for wandering interstitial cells, and the products of digestion must traverse it to reach the epidermis.

Nutrition and Digestion

Nutritional activities are rather uniform throughout this phylum. Coelenterates are carnivorous and more or less raptorial in their acquisition of food. Planktonic invertebrates and even small fish are ushered through a remarkably expandable mouth into the coelenteron, where digestion is initiated extracellularly by gland cell secretions. These secretions contain enzymes (largely proteases) which reduce edible tissues to a size suitable for phagocytosis. Digestion is completed intracellularly within food vacuoles. Undigestible wastes are expelled through the mouth. In many coelenterates, there seems to be regional specialization within the coelenteron, whereby some areas are particularly active in enzyme secretion and others are adapted solely to food uptake.

Respiration, Excretion, Circulation, and Osmoregulation

Most coelenterate cells lie at or very near a surface. This structural simplicity precludes the need for elabo-

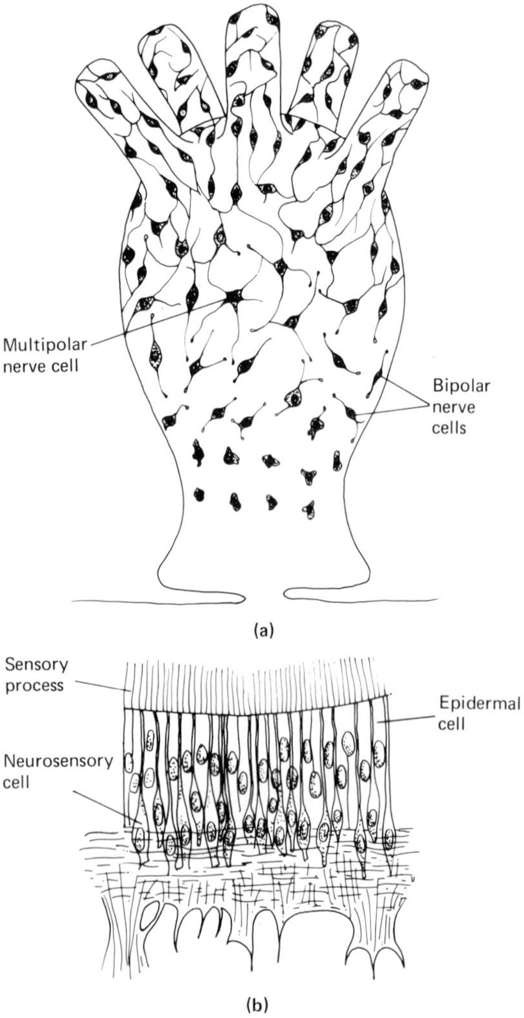

Figure 5.4 (*a*) A bud of *Hydra* showing the formation of the nerve net; (*b*) Neurosensory cells in a sensory pit of a rhopalium, a compound sense organ of scyphozoan medusae. (*a* from J.H. Bullock and G.A. Horridge, *Structure and Function in the Nervous System of Invertebrates, Vol. I*, 1965, W.H. Freeman, after C.H. McConnell, Quart. J. Micros. Sci., 1933, 75:495; *b* from Hyman, *The Invertebrates, Vol. I*.)

rate internal transport systems. Simple diffusion and the stirrings of gastrodermal flagella provide an adequate distribution of nutrients and respiratory gases. Nitrogenous wastes in the form of ammonia likewise diffuse away. Coelenterates live mainly in clean water with a high oxygen content. Their respiration is strictly aerobic. The respiratory rates of these animals are generally low, because of their small amount of organic material relative to their volume. The anthozoans are the most cellularized coelenterates and correspondingly have the highest respiratory rates.

Osmoregulation presents little problem for the marine members of the phylum; their body fluids are essentially in osmotic harmony with the sea. Lacking osmoregulatory mechanisms, coelenterates live only in regions within the limits of their salinity tolerance. Most have a narrow tolerance and thus are restricted to habitats possessing normal ocean salinities. Because of the nature of their habitat, freshwater forms are threatened with internal flooding. A lowered osmotic concentration within their bodies helps to alleviate this threat. Also, *Hydra* pumps a hyposmotic fluid out through its mouth. Apparently, its entire gastrovascular cavity serves as a "contractile vacuole."

ACTIVITY SYSTEMS

Muscles and Nerves

Coelenterates are the most primitive animals with distinct activity systems. Their nervous and muscular features were outlined in the histology discussion, and their relative simplicity leaves little more to say about organization. Longitudinal fibers in the tentacles and epidermal walls often are collected into strong muscles. Circular muscle fibers are well developed in the gastrodermis, particularly near the oral and pedal disks. Sphincter muscles surround the mouth. In anthozoans, the gastrodermal musculature includes strong longitudinal fibers within the septal walls, as well as circular muscles. Finally, all free-swimming medusae have powerful circular muscles in the subumbrellar region of the bell margin. These swimming muscles are an example of epithelially derived fibers completely independent of the epidermis.

Nerve cells are organized in diffuse nerve nets, but since they are never highly concentrated, nothing like a central nervous system is formed (Fig. 5.4a). Most nerve cells are multipolar, forming a number of terminal branches, although this phylum witnesses an increase in bipolar cells among its advanced members. Synaptic contacts link neighboring nerve cells as in higher animals, but in most coelenterates impulse conduction is not polarized. Thus, nerve cells regularly conduct impulses in both directions across the synapse. This situation is due to the presence of synaptic vesicles on both sides of the nerve junction; appropriate stimulation of either side releases the chemicals of nervous transmission from these vesicles, thus conducting an impulse across the synapse. In contrast, the nerve cells of higher phyla display unidirectional conduction because synaptic vesicles are limited to one side of the junction. Coelenterate nerve cells also differ from those of higher organisms in having naked fibers; the Schwann cells that commonly surround the axons of advanced forms are completely absent from this phylum.

Any sufficient stimulation of the nerve net spreads out evenly across it, much as ripples emanate from a stone thrown into water. The extent to which impulses spread across a nerve net depends on the initial strength of the stimulus. Most coelenterates summate stimulus information through time and space. A stimulus too weak to elicit a response by itself may be sufficient if repeated often enough or if administered simultaneously at multiple sites on the animal. These twin phenomena, termed **temporal** and **spatial summation,** respectively, also occur in higher animals.

The nerve net is denser in the active oral regions of coelenterates. Indeed, in many medusae a distinct nerve ring is associated with the subumbrellar swimming muscles (Fig. 5.5). While not remotely approaching a central nervous system, such densities do foreshadow the increasing concentration of nerves in the evolution of higher animals.

Coelenterates also possess cytoplasmic conducting systems. These systems are much more diffuse than the nerve nets, and their conduction rates are far slower. Epidermal cells and muscle elements have been implicated in nonnervous conduction. Some nonneural systems are called mechanical conducting systems, because the excitation experienced by one element is conveyed by physical means to adjoining elements. (Recall that a similar coordinating mechanism was proposed for sponges.) Nonneural systems may represent various stages in the evolution of cells specialized for conduction. They are largely absent from the anthozoans, whose nervous systems are the best developed in the phylum.

Sense Organs

Sense organs are not well developed in this phylum as a whole. Characteristic of their sessile existence,

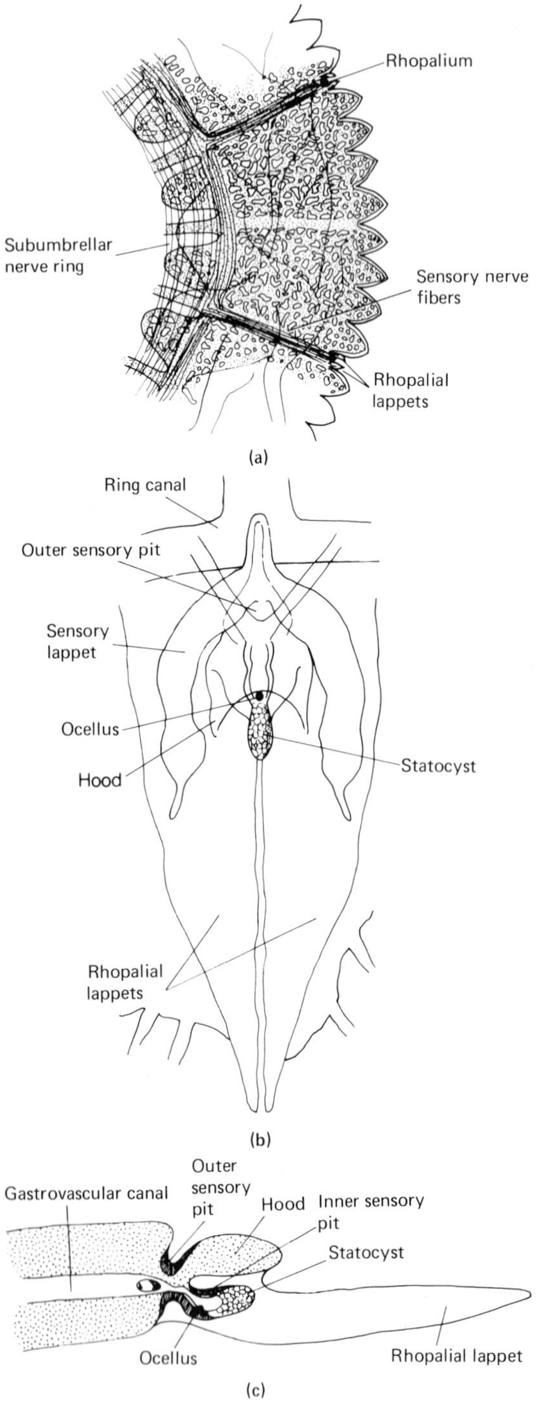

Figure 5.5 The rhopalium of Scyphozoa. (*a*) Section of the bell margin of *Rhizostoma* showing the location of rhopalia and the associated nerve plexus including the subumbrellar nerve ring; (*b*) Aboral view of a rhopalium of *Aurelia*; (*c*) Diagram of a radial section through a rhopalium. (From Hyman, *The Invertebrates, Vol. I.*)

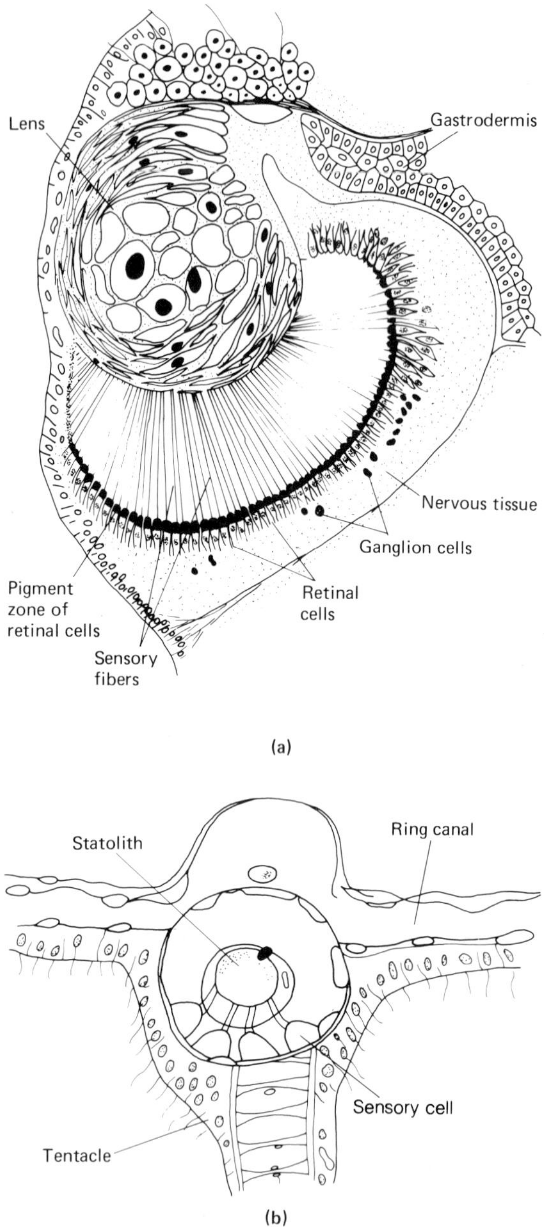

Figure 5.6 (*a*) Section through an ocellus of *Carybdea*, a scyphozoan; (*b*) Section through a statocyst of a medusae of *Obelia*, a hydrozoan. (From Hyman, *The Invertebrates, Vol. I.*)

polyps have rudimentary sensory structures, consisting primarily of neurosensory cells whose long processes are exposed to the environment (Fig. 5.4b). At the point of exposure, a bristle, or some such sensitive structure, responds to environmental conditions (touch, temperature, chemicals) and relays information to the nerve net. Consistent with their more active

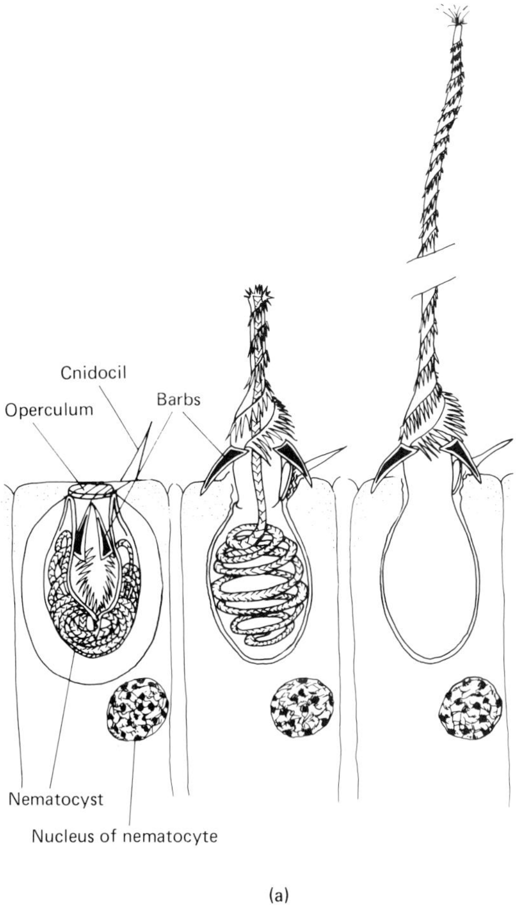

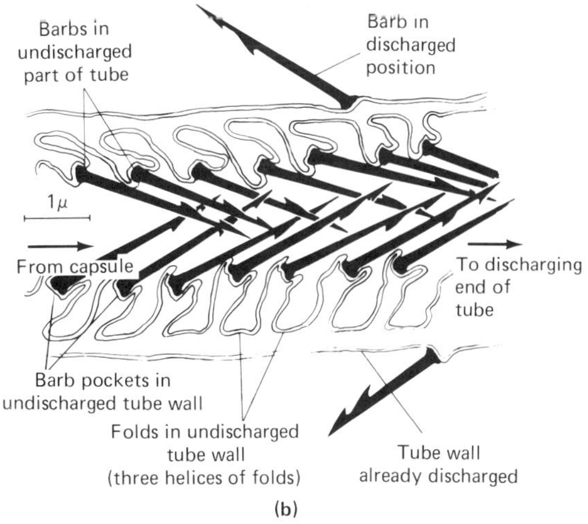

Figure 5.7 (*a*) A nematocyst in the process of discharge. Left to right: undischarged nematocyst; partially discharged nematocyst; completely discharged nematocyst; (*b*) Diagrammatic longitudinal section through a partially discharged nematocyst thread, based on electron micrographs. Barbs are mounted along the spirally folded inner wall of the thread, which discharges by turning inside out. (*a* from M. Wells, *Lower Animals*, 1968, McGraw-Hill; *b* from Russell-Hunter, *A Biology of Lower Invertebrates.*)

life-styles, medusae possess specialized sensory structures. Eight compound sense organs, termed rhopalia (Fig. 5.5), occupy the bell margins of scyphozoan medusae. Typically, each **rhopalium** contains: (1) two sensory pits, which may be chemoreceptive; (2) a photoreceptive **ocellus;** (3) two sensory flaps (lappets), which may be tactile; and (4) an equilibrium center called a **statocyst.** Hydrozoan medusae commonly possess ocelli and/or statocysts.

The ocellus and the statocyst are primitive sense organs. An ocellus is a rudimentary eye comprising pigmented photosensitive cells (Fig. 5.6a). Most visual organs in advanced phyla are essentially variations on the ocellus theme. Statocysts feature a small fluid-filled chamber into which tactile hairs protrude. A granule (called a statolith) occupies this chamber and differentially stimulates the hairs when gravitational forces cause it to move (Fig. 5.6b). Interpretation of this hair stimulation allows the animal to maintain proper balance.

It is noteworthy that the various sense organs described are the only true organs in coelenterates.

Nematocysts: A Coelenterate Specialty

Coelenterates possess a remarkable cell organelle specialized for a number of environmentally related activities. This organelle, the **nematocyst,** is contained within an ovoid cell, the **nematocyte.** The nematocyst proper consists of a coiled, spirally pleated thread turned inside out and encapsulated in a double-walled chamber (Fig. 5.7a). This chamber is covered by a porthole lid, or **operculum.** Extending from the nematocyte is the **cnidocil,** a bristled structure triggering nematocyst discharge (Fig. 5.7a). A full understanding of that discharge has not been achieved, but osmotic pressure apparently is involved. According to the most acceptable theory, high osmotic pressure develops inside an undischarged nematocyst capsule, due in part to the impermeability of the capsule wall. Upon appropriate stimulation of the cnidocil, this impermeability breaks down and an osmotic gradient immediately forms. Water rushes along this gradient into the nematocyst capsule. Rapidly swelled by the increase in water volume, the nematocyst thread

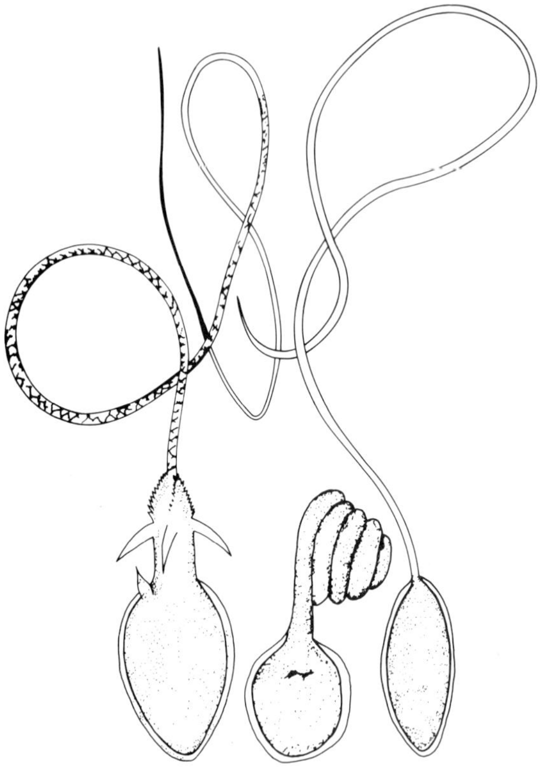

Figure 5.8 Three basic nematocyst types: penetrant, volvent, and glutinant. (From Sherman and Sherman, *The Invertebrates: Function and Form.*)

lashes outward, everting and unfolding its spiraled pleats (Fig. 5.7b). Sharp spines may emerge from the everting, spiraling tip like swirling knife blades; not meager weapons, these can penetrate tough crustacean cuticles. In some cases, pressure exerted on the nematocyst capsule by surrounding muscle elements also influences discharge. In view of the diversity of nematocyst types, we perhaps should not expect the same discharge mechanism to pertain to all.

Nearly 20 nematocyst types have been identified, with the greatest variety found in the Hydrozoa. A coelenterate species usually has several different types. Nematocyst types fall into three major groups: (1) the penetrants, sharply pointed threads that puncture their targets and inject crippling neurotoxins from their open tips; (2) the volvents, nematocysts that wrap around and entangle prey; and (3) the glutinants, whose sticky secretions fasten an organism to the substratum (Fig. 5.8).

Nematocysts function primarily in food capture, but they also provide the principal means of protection for jellyfish and unencased polyps. Evidently, they guard well, because these soft and otherwise vulnerable creatures are preyed upon by very few animals. In fact, their nematocyst artillery is so highly respected that most predators leave them completely alone. Some jellyfish encounters with swimmers have been fatal, thus including these coelenterates among the very few invertebrates individually deadly to humans.

Nematocyst discharge is an example of an independent effector system. The physiological condition of the coelenterate does have a mediating influence, but firing ultimately occurs in response to stimuli received by the cell itself. There is no intervention by nervous elements. Frequently, a dual source of stimulation of the cell is required. In some coelenterates, certain types of nematocysts may not discharge upon tactile stimulation with a clean glass rod, but will discharge if touched by a rod previously dipped in crustacean extract. Clearly, both chemical and tactile input are necessary here. In nematocysts employed in adhesion during locomotion, discharge is inhibited by crustacean extract.

Behavior

A coelenterate can react to its environment as a whole organism, responding in a neuromuscularly coordinated manner to environmental stimuli. The coelenteron of polypoid morphs performs excellently as a **hydrostatic skeleton.** Filled with water and pinched shut by oral sphincter muscles, it becomes a stiff tube manipulated by body wall muscles. Accordingly, polyps can lean, squat, twist, and nod from side to side. Solitary polyps may creep slowly along the substratum by graded contractions of circular muscles in the pedal disk. Others paddle with their muscular tentacles or perform somersaults, flipping from tentacles to pedal disk and back. Such active behavior, rare among polyps, is typical of medusoid coelenterates. Contraction of the strong circular muscles of its subumbrella rapidly expels water from the lower skirts of the medusa, jetting the animal away. These muscles are stretched by the elasticity of the thick mesogleal layer. Medusae exercise limited directional control over their water jet. Controlled movements are confined largely to the vertical plane, and horizontal travel is dominated by sea currents.

Coelenterate tentacles capture potential food by a combination of gross entanglement and nematocyst discharge. Chemical stimuli emanating from potential prey help stimulate discharge of the nematocyst types employed in prey capture. The movement of the tentacles toward the coelenteron and the opening of the mouth also depend on chemical stimuli from the food.

Specific chemical releasers for this feeding response have been identified in several coelenterate species. Tracts of cilia assist the tentacles in bringing food against the muscular swallowing mouth. Subsequent gross body movements may churn the contents of the coelenteron and, once extracellular digestion is complete, body wall contractions force undigested materials out the opened mouth.

Like most animals, coelenterates in distress demonstrate a maximum amount of muscular activity. Medusae swim vigorously. Polyps at first contract the body grossly, withdrawing the tentacles and tightly sealing the vital oral region. If the polyp has some hard protective casing, withdrawal is normally sufficient to survive an attack. On the other hand, an unencased polyp must resort to desperate measures. Threatened by potential predators such as starfish, some sea anemones compact themselves and then detach abruptly from the substratum. Thus they bail out of the situation, abandoning their fate to the currents and seafloor geography.

The more complex nerve nets of anthozoans have made them the focus of a number of studies. Certain experiments with sea anemones, in particular, demonstrate how the nervous machinery determines the behavioral repertoire of the animal.

A little game played with sea anemone tentacles reveals some of the neuromuscular properties previously discussed. Weak electrical or mechanical stimulation of a single tentacle elicits movements by that tentacle alone. Slightly increased stimulation draws neighboring tentacles and a portion of the oral disk into the response. More irritation causes the tentacles to grasp the stimulating object and attempt to ingest it, while major force applied to that same single tentacle may result in a complete contraction of the anemone.

Anemones demonstrate temporal and spatial summation of noncontinuous stimuli; hence the aforementioned increases in stimulus strength can be increases in frequency as well as in intensity. The anemone nervous system thus broadcasts impulses in proportion to their total strength. Responses appear in jerks or waves, as stimulus strength gradually pushes nervous excitation across more distant synapses. Because individual nerve fibers respond in the all-or-none fashion of higher organisms, these synaptic transmissions are critical to a widening response. Because of the small diameter of all coelenterate neurons, nerve impulses and associated responses are slower than those of the larger neurons in higher phyla.

Individual muscles require an external force to restore their resting length. For the sea anemone, these forces involve the mesoglea and the coelenteron. The mesoglea is highly elastic and contains a diagonal lattice of fibers. Deformed during contraction, this mesogleal lattice gradually returns to its normal position when muscles relax. Meanwhile, the beating of stomodeal cilia draws water back into the coelenteron, eventually restoring its volume. The mouth remains open, as during the water-expelling contraction, by virtue of contracted radial muscles in the oral disk.

Controversy surrounds the question of autonomous behavior in sea anemones. Autonomous behavior is that inspired from within the nervous system itself and is hence fundamentally distinct from a simple reflex response. Anemones do creep slowly, and a few species can swim by sustained rhythmic contractions of the body musculature. Time-lapse photography reveals certain slow, spontaneous movements—the column may elongate or contract, sway from side to side, or undergo peristaltic movements. While these activities may represent autonomous behavior, some biologists believe that they are merely delayed digestive responses or reactions to unobserved stimuli.

Another controversy centers on whether anemones can learn. Sea anemones gradually fail to respond to simulated prey, such as extract-soaked paper, and eventually "tune out" constant irritation, such as a dripping water tap. While these phenomena may represent behavioral changes due to experience—a basic definition of learning—it is not clear that such changes are recorded in the nerve net itself. The sensory powers of the animal simply may have fatigued. Whatever the final conclusion, sea anemones display a remarkable variety of responses, especially for animals be-

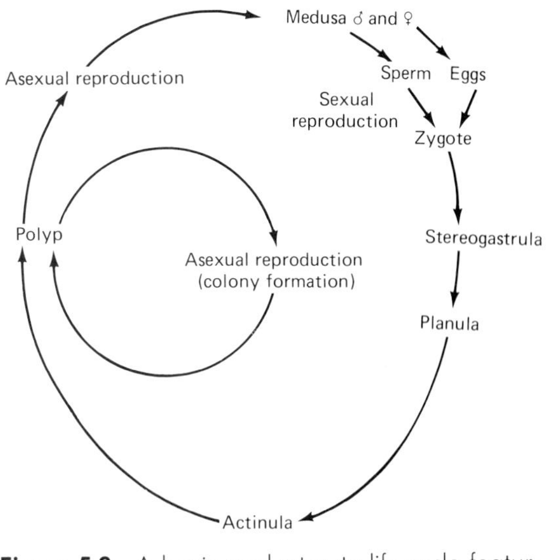

Figure 5.9 A basic coelenterate life cycle featuring both polyp and medusa stages.

Figure 5.10 Photograph of a budding *Hydra*. (Courtesy of the American Museum of Natural History.)

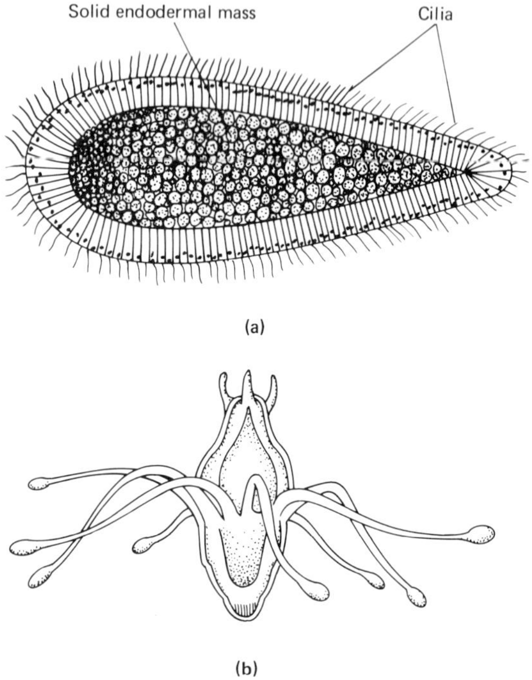

Figure 5.11 (*a*) A planula larva of a hydrozoan; (*b*) An actinula stage of a colonial hydroid. (*a* from Sherman and Sherman, *The Invertebrates: Function and Form*.)

having without the benefits of a central nervous system.

CONTINUITY SYSTEMS

Reproduction and Development

Coelenterate life cycles typically feature distinct individual organisms and both sexual and asexual reproduction. Genital systems (gonads) are never complex and consist only of clustered gamete-producing cells. We describe here a basic life cycle (Fig. 5.9) and some common variations. Variations are often so extreme, however, that much of the life cycle discussion must be deferred to the description of coelenterate diversity.

Asexual reproduction by polyps leads to the growth of the common colonial forms. These colonies exhibit a number of growth patterns. Often, individuals are proliferated along a distinct common stem. Composed of the same cellular layers as the polyps themselves, this stem allows a continuous gastrovascular cavity throughout the colony. Stem growth may be linear or arborescent, or colonies may spread in crusty sheets. Polyps also may bud from the basal epidermis of existent individuals, a method particularly common among solitary forms (Fig. 5.10). In every case, interstitial cells are involved in these reproductive processes.

Eventually, the asexual product is not a polyp, but a medusa. This morph separates to assume a swimming and/or planktonic existence. Medusae are commonly diecious, producing only eggs or only sperm. Gametes usually are released to the open sea, but female medusae may retain their eggs for internal fertilization. After fertilization, highly variable cleavage patterns result in a hollow blastula. Gastrulation, usually by uni- or multipolar ingression, produces a ciliated, ovoid **stereogastrula.** With minimal modification, the latter becomes a solid free-living larva known as a **planula** (Fig. 5.11a). After a brief planktonic existence, the planula larva settles on a suitable substratum and dif-

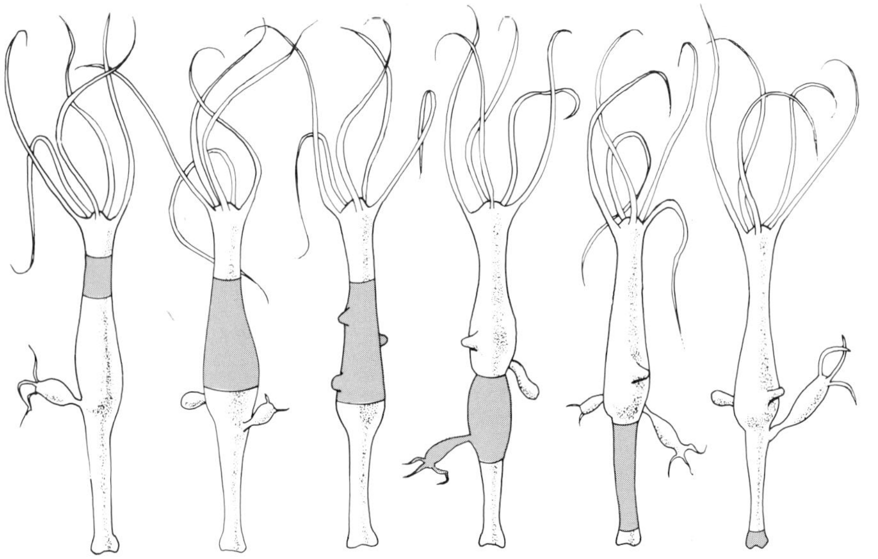

Figure 5.12 A stained piece of alien *Hydra* tissue is grafted into place in an unstained individual. As the animal grows the grafted tissue moves toward the base, successively becoming part of the gastric region, budding zone, stalk, and pedal disk. (From P.A. Meglitsch, *Invertebrate Zoology*, 2nd ed., 1972, Oxford University; after Brien and Renier Decoen.)

ferentiates into an immature polypoid called an **actinula** (Fig. 5.11b). The actinula grows into a mature polyp, and the cycle is completed.

Numerous variations in this cycle exist, and most feature a suppression of the medusoid, polypoid, or planula stage. Medusae may remain attached to colonial polyps as gamete-producing buds. Some polyps even lack these sexual buds and possess their own internal gonads. (The latter case holds for all anthozoans.) Meanwhile, those life cycles dominated by the medusoid stage may feature a tiny, solitary polyp whose primary function is to bud off medusae. In extreme cases, the asexual, budding stage may be completely absent, with the planula developing directly into a sexual medusa. Whether formed in a medusa or a polyp that has assumed sexual functions, the zygote may be protected during its early development. All or part of the planula stage may be passed within capsules released by the parent or within the parent organism itself (Fig. 5.20). Freed at some later stage, such offspring are better prepared for independent life. Schemes suppressing the free-living larval stage are particularly advantageous for freshwater forms, whose environment presents limited opportunities for planktonic existence. Variations in the life cycles of coelenterates reflect differences in the selective forces affecting each species. In some cases, there may be strong selection for protection of offspring, even at the expense of dispersal. Here we may expect suppression of the free-swimming medusa and the planktonic larval stage. In other cases, selection for dispersal is stronger and the medusa stage shares or even dominates the life cycle.

An alternation of motile, sexually reproducing morphs with a sessile, asexual stage provides the coelenterates with certain advantages. Sexual reproduction allows the recombination of genetic material central to the evolutionary process. Asexual reproduction by stationary forms facilitates a rapid exploitation of any especially favorable environments.

Coelenterates have high regenerative powers, as do most animals able to reproduce asexually. Early experiments with two hydrozoans (a *Hydra* species and a colonial hydroid) demonstrate the diversity of regenerative growth in the phylum. Cylindrical cross sections cut from a hydra develop into whole animals. The original polarity of the animal is retained, as new tentacles sprout from the oral end and a new pedal disk forms at the aboral end. Apparently, a metabolic gradient is in effect; the metabolism of intact animals is progressively quicker toward the oral region, and severed sections seem to "remember" that.

If new tissue is grafted into the column of a living hydra, it is gradually displaced toward the base of the

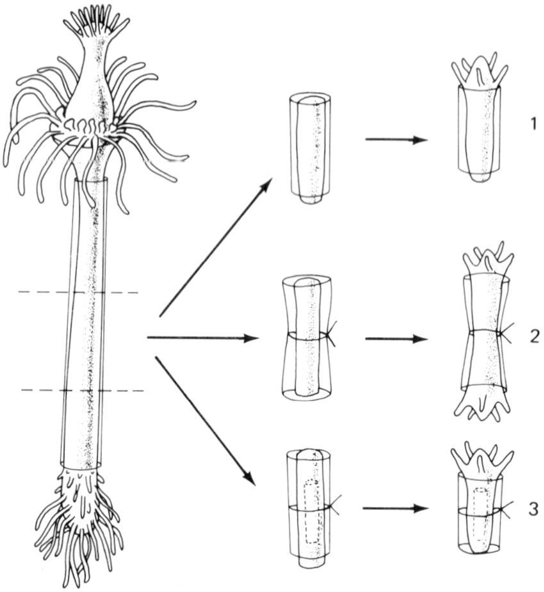

Figure 5.13 Regeneration experiments on *Tubularia.* (1) A cut piece of stem forms a polyp at the original oral end; (2) A ligature around the piece of cut stem allows the formation of polyps at both ends; (3) Insertion of a capillary tube in the ligated region prevents the formation of the aboral polyp. (From Meglitsch, *Invertebrate Zoology.*)

intact animal. The alien tissue traverses each cell zone and eventually is eliminated from the stem (Fig. 5.12). Such downward movement is caused by the continuous production of new cells from an apical growth zone. This new tissue spreads down the column and upward into the tentacles, differentiating into other cell types as required. All hydra cells are thus continuously renewed, and in a matter of weeks an entirely new animal exists.

Regeneration in the colonial hydroid *Tubularia* follows an interesting pattern (Fig. 5.13). (1) A cut stem without polyps buds at the end nearest its original polyp, but no new polyps develop if a polyp remains at one end of the cut stem. (2) If a bare stem is tied off in the center, polyps develop at both ends. (3) If a capillary tube is inserted to join two previously ligated ends, a polyp forms only at the end nearest the original polyp. Apparently, some substance spreads out from the polyp end of each stem and, distributed by gastrovascular channels, inhibits the formation of polyps elsewhere.

The remarkable regenerative powers of hydras may be demonstrated in yet another way. Dissociated hydra cells can be centrifuged into clumps; these cell clumps then reorganize into one or more small hydras. Such regeneration from mere clumps of cells is reminiscent of sponge cell reaggregation.

A Review

Before discussing adaptive radiation in this phylum, we review here the common coelenterate features that form the basis of their continuing evolution. Basically, coelenterates are:

(1) radially symmetrical animals at the tissue level of organization;

(2) composed of three body layers—a cellular epidermis, a mesoglea, and a cellular gastrodermis;

(3) endowed with a single body cavity, the coelenteron, and nematocyst-bearing tentacles;

(4) lacking in specialized excretory, osmoregulatory, circulatory, and respiratory systems;

(5) the most primitive animals with true muscular and nervous systems;

(6) polymorphic animals which regularly reproduce by sexual and asexual means.

Coelenterate Diversity

Despite their simple organizational scheme, coelenterates display an extraordinary diversity, due in part to their tendency to form colonies. In surveying the adaptive radiation of this phylum, we discuss each of its three classes in turn.

CLASS HYDROZOA

There are 2700 species of hydrozoan coelenterates. Most members of the class have both polypoid and medusoid stages in their life cycles, although some exist only as polyps and others only as medusae. Hydrozoans differ from other coelenterates in possessing a noncellular mesoglea and a simple gastrovascular cavity without partitions. The gonads are derived from the epidermis. In hydrozoan medusae, a thin circular flap, the **velum,** extends inward from the margin of the bell, obscuring a portion of the subumbrellar surface (Figs. 5.2 and 5.18). The velum contains a band of circular muscle that aids the subumbrellar muscles in swimming.

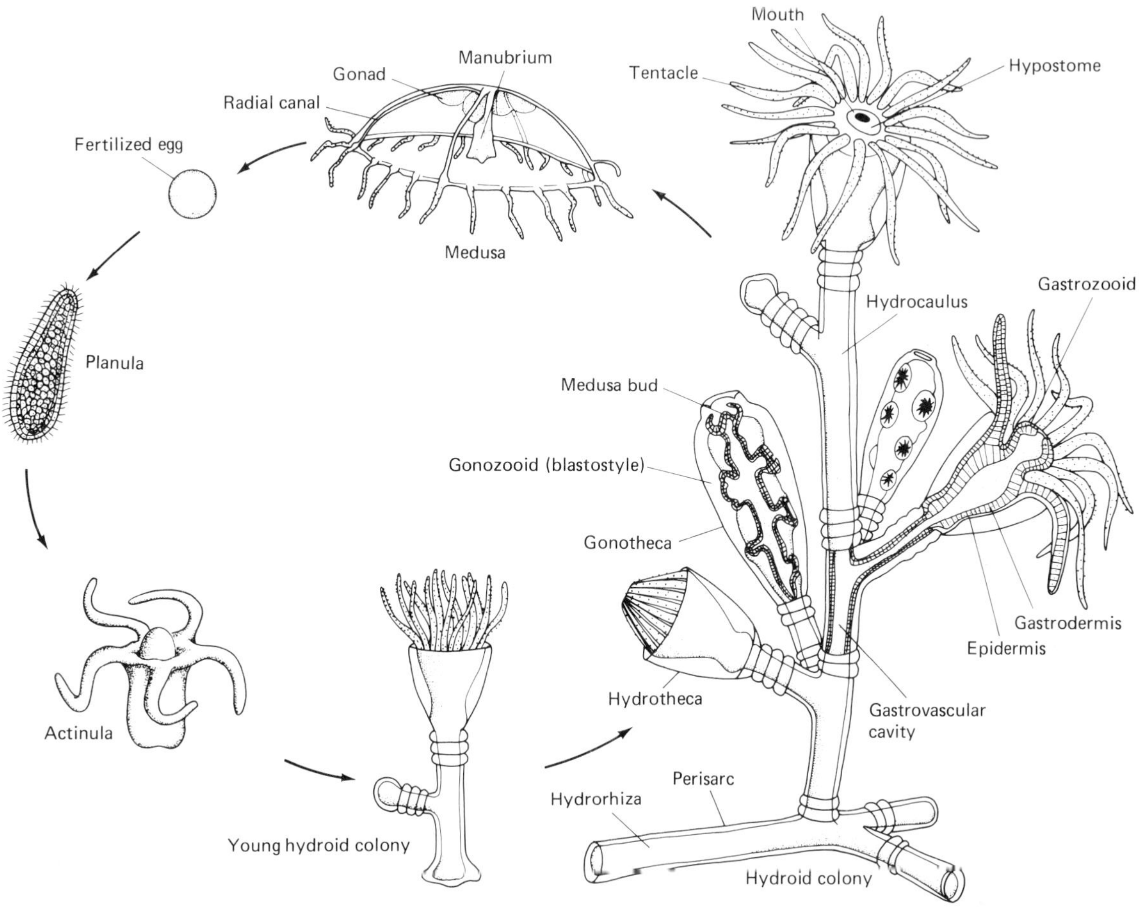

Figure 5.14 Life cycle of *Obelia*, a calyptoblast hydroid. Structure of an *Obelia* colony at the right. (Modified from R.D. Barnes, *Invertebrate Zoology*, 4th ed., 1980, Saunders College Publishing.)

The class Hydrozoa contains two major orders, the Hydroida and the Siphonophora, and three minor orders, the Trachylina, the Hydrocorallina, and the Chondrophora.

The Hydroida

Perhaps the most characteristic coelenterate life form is a sessile, branched colony composed of many specialized polyps (termed **zooids**). This life form is best exemplified by the colonial hydroids. Colonial hydroids are most abundant in shallow marine waters near the shore, where they are attached to hard substrata. Their colonies vary from small inconspicuous tufts less than 2 cm long to large shrubby growths almost a meter high and containing many thousands of polyps. Individual polyps may be colorless and indiscernible to the naked eye or brilliantly colored flowerlike structures 2 to 3 cm in diameter.

A well-known colonial hydroid is *Obelia* (Fig. 5.14), which occurs commonly as a whitish or brownish growth on algae and rocks. *Obelia* is anchored to the substratum by a branching rootlike structure, the **hydrorhiza.** A vertical axis, the **hydrocaulus,** supports lateral branches; these branches bear terminal zooids, the large majority of which are vase-shaped feeding polyps known as **gastrozooids.** Each gastrozooid possesses an oral ring of 24 tentacles and is enclosed in a transparent glassy cup, the **hydrotheca.** Less numerous are the long cylindrical **gonozooids.** These morphs are enclosed in transparent cases known as **gonothecae** and bear lateral offshoots varying in form depending on their development. Such offshoots give rise to free-swimming, sexual medusae.

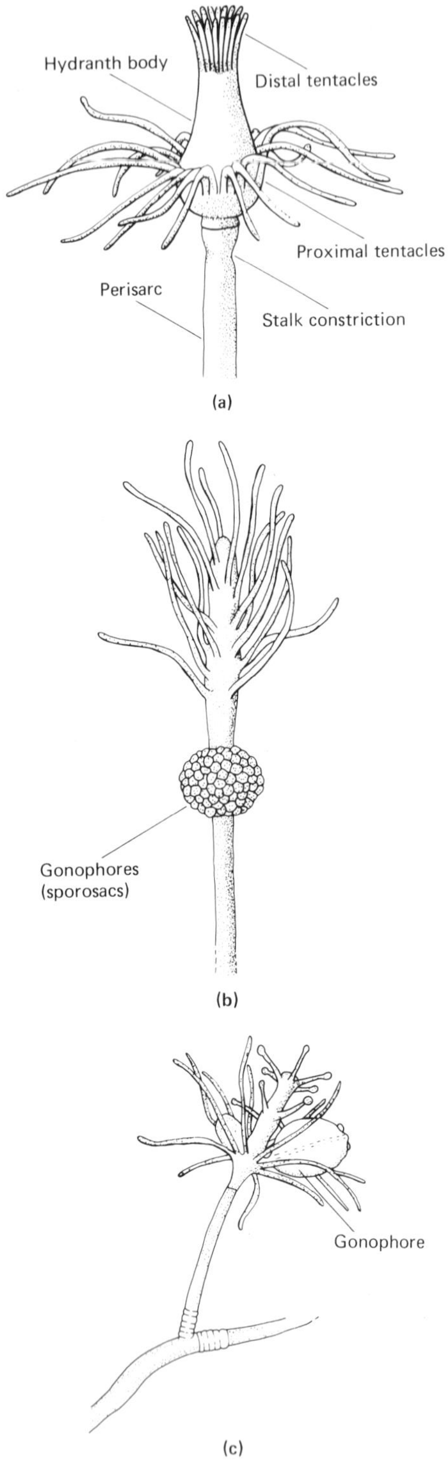

Figure 5.15 Polyps of colonial hydroids. (a) *Tubularia*; (b) *Clava* with sporosacs; (c) *Pennaria* with gonophores. (From Sherman and Sherman, *The Invertebrates: Function and Form.*)

The hollow stems uniting the colony's zooids share the three body layers—an epidermis, a thin mesoglea, and a gastrodermis surrounding an extension of the coelenteron. These stem structures are known as the **coenosarc.** The epidermis of the stem secretes a thin, protective chitinoid tube or **perisarc.**

Hydroids are divided into two groups according to their perisarc. The perisarc is restricted to the stems in athecate hydroids (suborder Gymnoblastea), whereas in thecate hydroids (suborder Calyptoblastea) the perisarc also surrounds individual zooids, as in *Obelia.*

Tentacle arrangements vary on the gastrozooids of different hydroids. *Obelia* has a single whorl around an oral cone known as the **hypostome.** *Tubularia* possesses a pair of whorls, one around the mouth and one around the base of the polyp. *Clava* features a scattered irregular arrangement, while *Pennaria* combines a basal whorl with scattered tentacles (Figs. 5.14 and 5.15).

Continuous throughout the colony, the coelenteron is lined with flagellated gastrodermal cells. The flagella create currents that distribute materials to all parts of the colony. In this manner, nonfeeding zooids receive nutrients. Individual zooids are flexed by epithelial–muscle cells. The stems themselves are immobile because of the rigid perisarc.

Most hydroid colonies include gastrozooids and gonozooids. The sexual medusoid stage may be produced on well-developed polyps with mouths and tentacles or upon polyps that have lost these structures. Gonozooids of the latter type, properly referred to as **blastostyles,** are characteristic of *Obelia* (Fig. 5.14). Other morphs may be present, such as the **dactylozooids.** These mouthless, defensive polyps represent single extensible tentacles loaded with nematocysts (e.g., *Hydractinia*; Fig. 5.16a).

Hydroid colonies vary in form according to their patterns of growth and branching. Some colonies are encrusting mats of hydrorhizae that give rise directly to individual polyps. Certain *Hydractinia* (Fig. 5.16b), commensals on mollusk shells inhabited by hermit crabs, form such flat colonies. However, most colonial hydroids have a main stem which produces lateral polyp-bearing branches. New polyps may bud from the tip of the stem, as in **sympodial** colonies, or may form at a growth zone located just below the apical polyp, as in **monopodial** colonies (Fig. 5.17). In either monopodial or sympodial colonies, branching may occur in a single plane, producing a plumelike colony, or in all planes, giving rise to a bushy or shrubby growth.

Thecate and athecate hydroids differ in their free-swimming medusae. Thecate hydroids produce flattened, saucer-shaped **leptomedusae** (e.g., *Obelia;* Fig. 5.14). In such medusae, gonadal buds associated with

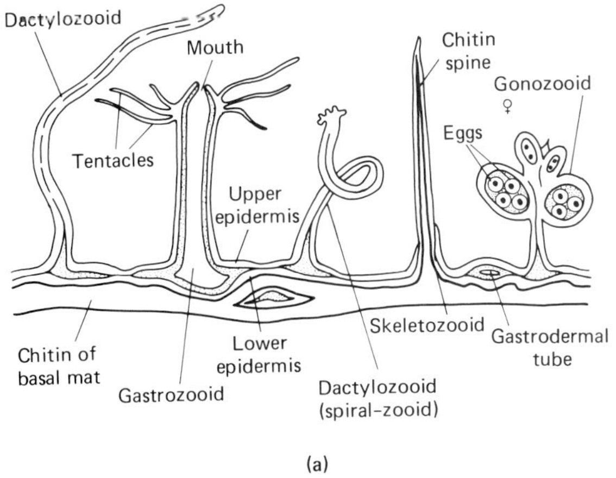

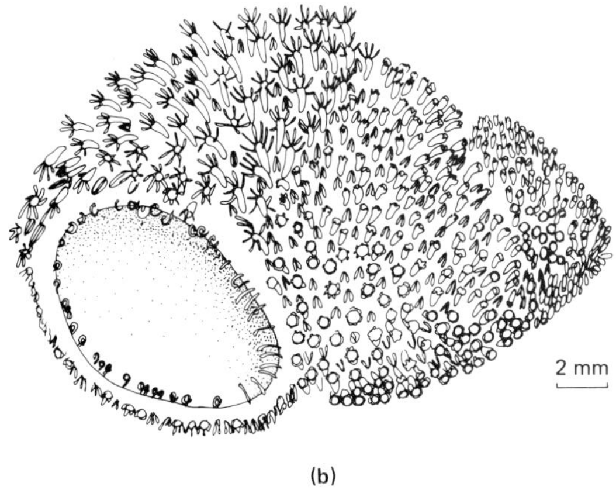

Figure 5.16 (*a*) Diagram of a portion of a colony of *Hydractinia* showing the several different types of polyps present; (*b*) A colony of *Hydractinia* living on a snail shell. (*a* from W.D. Russell-Hunter, *A Biology of Lower Invertebrates; b* from R.K. Josephson, J. Exp. Biol., 1961, *38*:561.)

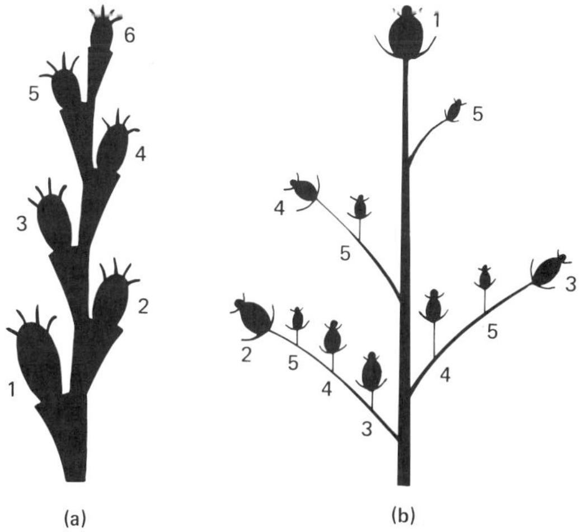

Figure 5.17 Schematic representation of two types of hydroid colony growth. (*a*) Sympodial growth; (*b*) Monopodial growth. Polyp age is indicated by relative polyp size, and by numbering the polyp generation with the original polyp numbered 1. (From Meglitsch, *Invertebrate Zoology*.)

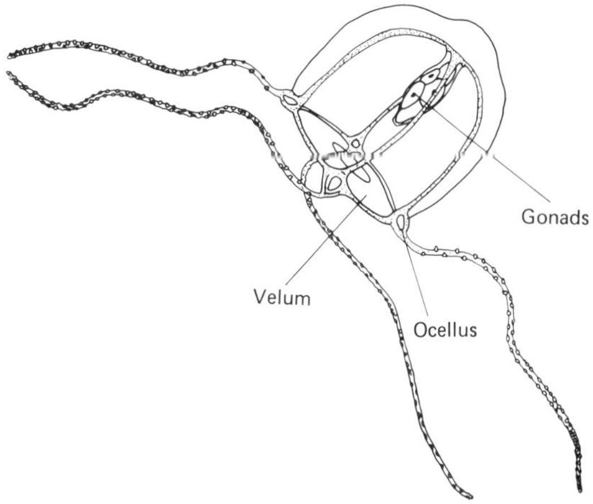

Figure 5.18 The anthomedusa of an athecate (gymnoblast) hydroid, *Sarsia*.

four radial canals project from the subumbrellar surface. The velum is poorly developed. Marginal sense organs include statocysts and, sometimes, ocelli. In the **anthomedusae** of athecate hydroids, the arched umbrella produces a characteristic bell or miter shape (Fig. 5.18). Its walls are thick because of an expansion of the exumbrellar mesoglea. The velum is well developed. Other characteristics of anthomedusae are gonad development on the manubrium and an absence of statocysts. Marginal sense organs usually include ocelli.

Hydroid medusae are usually very small. Some are barely visible to the naked eye, but the largest *(Aequorea)* may be 40 cm in diameter. Most hydroid medusae are transparent except for the gonads, manubrium, and marginal sense organs, which may be brightly colored. Many are luminescent.

Hydroids such as *Obelia* and *Bougainvillea* exhibit the temporal polymorphism typical of coelenterate life cycles. Polyps dominate the life cycle. A fixed polyploid colony usually survives for extended periods, producing ephemeral free-swimming medusae when conditions are favorable. In the evolution of both phylogenetic lines, there has been a pronounced reduction of the medusoid phase. For example, in many athecate

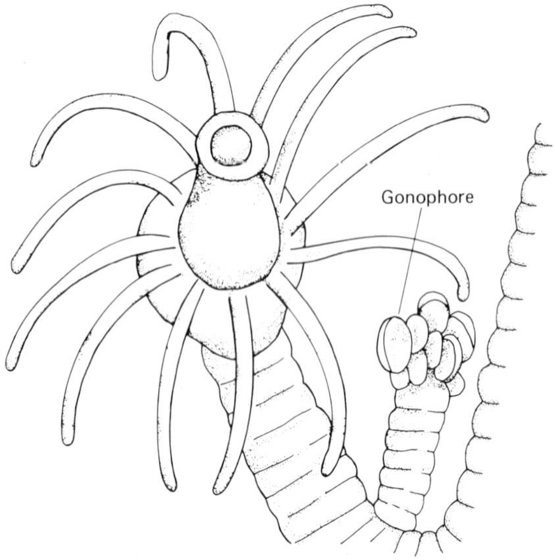

Figure 5.19 A portion of a colony of *Eudendrium* showing gonophores produced on blastostyles. (From Meglitsch, *Invertebrate Zoology*.)

hydroids, medusoid buds are retained only as sessile **gonophores** on the colony, and there is no free-living jellyfish stage at all. Gonophores may be produced on blastostyles (gonozooids) (e.g., *Eudendrium;* Fig. 5.19) or on gastrozooids (e.g., *Tubularia;* Fig. 5.20). Gametes mature within the gonophore. In some hydroids, fertilization is external and the zygote develops into a free-swimming planula larva. In others, fertilization is internal and fertilized eggs are retained within the female gonophore for varying periods of time. In *Tubularia,* an extreme case, young are not released until the actinula stage. Gonophores usually retain some vestiges of medusoid structure, but all traces may disappear, as in the gonophores of *Clava* (Fig. 5.15b). These highly modified gonophores, termed **sporosacs,** contain little more than gonads. A similar medusoid reduction is seen among some thecate hydroids.

Related to the athecate hydroids are the familiar freshwater hydras. These hydroids exist only as solitary polyps; budded individuals always separate and no colonies are formed. Hydras have no skeleton; individuals can locomote, but often make temporary pedal attachments to the substratum. There is no medusa in the life cycle of these freshwater animals, gonads simply appear on the polyploid column when conditions are favorable (Fig. 5.21). Hydras commonly reproduce by asexual budding throughout the spring and summer (Fig. 5.10). In the fall, sexuality is induced, perhaps by cooler temperatures and the progressive shortening of day length. Fertilized eggs are surrounded by protective coats that allow them to overwinter. Upon the return of warmer temperatures in the spring, development is initiated and young hydras hatch without having passed through a planktonic planula stage.

Hydras thus asexually increase their population during the favorable spring and summer months and sexually produce resistant stages for overwintering. Such a life cycle closely parallels that of other freshwater invertebrates, including sponges, rotifers, and certain crustaceans.

Because of their specializations for life in fresh water, hydras are not typical coelenterates or even typical hydrozoans. Yet, because of their availability, they are often studied in zoology courses as if they were.

The Trachylina

The order Trachylina is a primitive group of hydrozoans in which the polyploid stage is reduced or absent and the medusoid stage dominates. Trachyline medusae resemble those of hydroids; they are of small or moderate size and have a velum. They differ chiefly in that their marginal sense organs are modified tentacles derived in part from the gastrodermis and in that they often develop directly from the egg without an intervening polypoid phase. Most trachyline medusae are pelagic and are seldom encountered except in the open sea. Two notable exceptions (Fig. 5.22) are: *Gonionemus,* a shallow-water, bottom-dwelling form that clings to algae and other objects by adhesive suckers; and *Craspedacusta,* a small freshwater medusa that occurs sporadically in various parts of the world, particularly in artificial bodies of water.

The Hydrocorallina

Members of this group display a massive calcareous skeleton secreted by the coenosarc and hence superficially resemble the anthozoan stony corals. Examples are *Stylaster* and the well-known stinging coral, *Millepora* (Fig. 5.23); the latter possesses very toxic nematocysts. Colonies may exist as irregular lumpy masses or upright plates divided into antlerlike branches up to 60 cm tall (Fig. 5.23a). The skeleton is perforated by minute pores of two sizes. The larger gastropores contain gastrozooids. Each of these morphs is surrounded by a ring of smaller dactylopores containing dactylozooids (Fig. 5.23b). *Millepora* produces free-swimming but rather degenerate

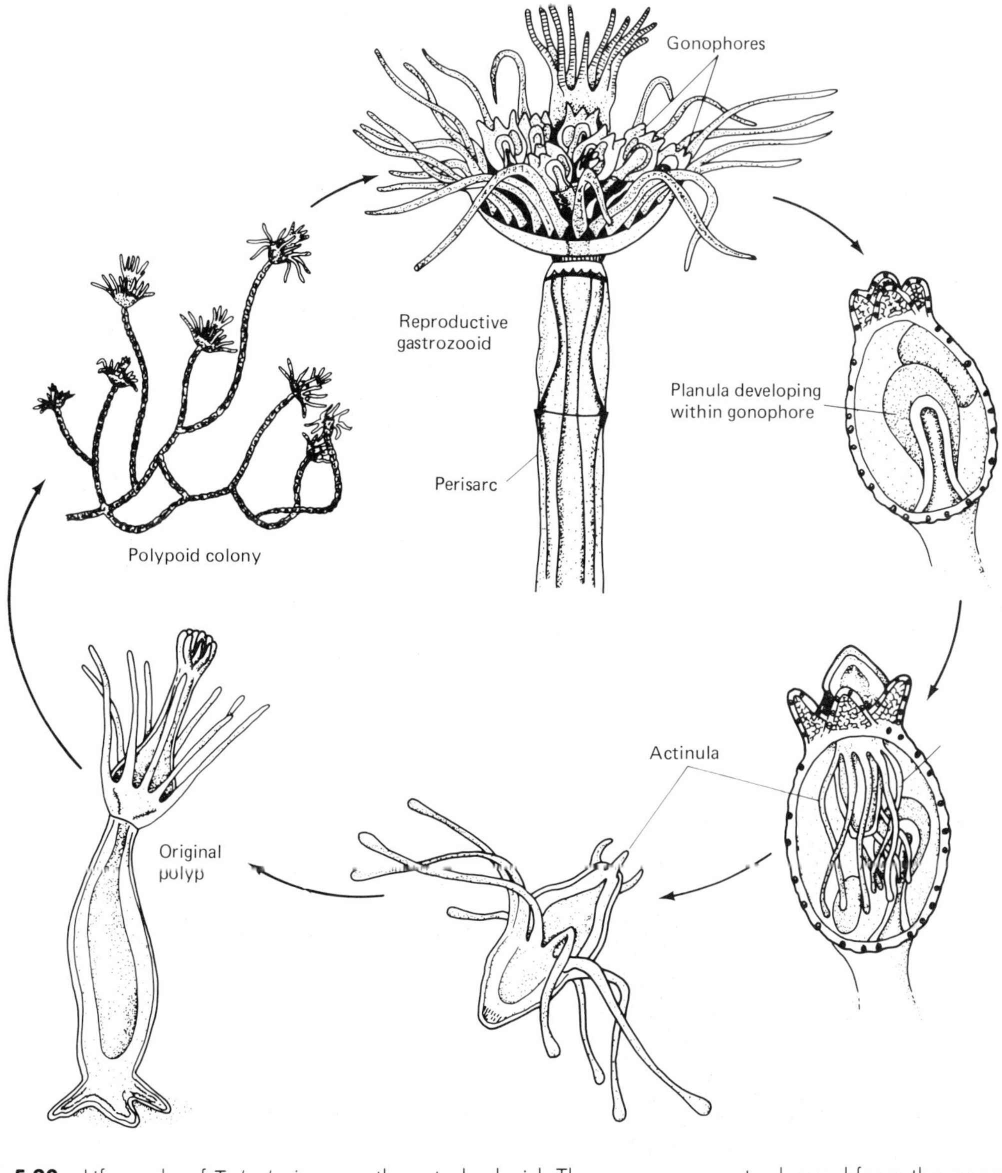

Figure 5.20 Life cycle of *Tubularia,* an athecate hydroid. The young are not released from the gonophores until the actinula stage.

medusae. The attached gonophores of *Stylaster* essentially are sporosacs, which release the young as planulae. Hydrozoan corals are most abundant on coral reefs in the Indo-Pacific region, but do occur elsewhere, as along the west coast of North America.

The Siphonophora

Among the most remarkable coelenterates are the siphonophores. These pelagic, swimming, or floating hydrozoan colonies exhibit more polymorphism than any other coelenterates. Best known of the siphonophores, not only because it is commonly washed ashore on ocean beaches but also because of its capacity to sting violently, is *Physalia,* the Portuguese man-of-war (Fig. 5.24).

Each siphonophore colony comprises numerous polypoid and medusoid individuals of different types (Fig. 5.25). Polypoid morphs are specialized for feeding, prey capture and defense, and for the production

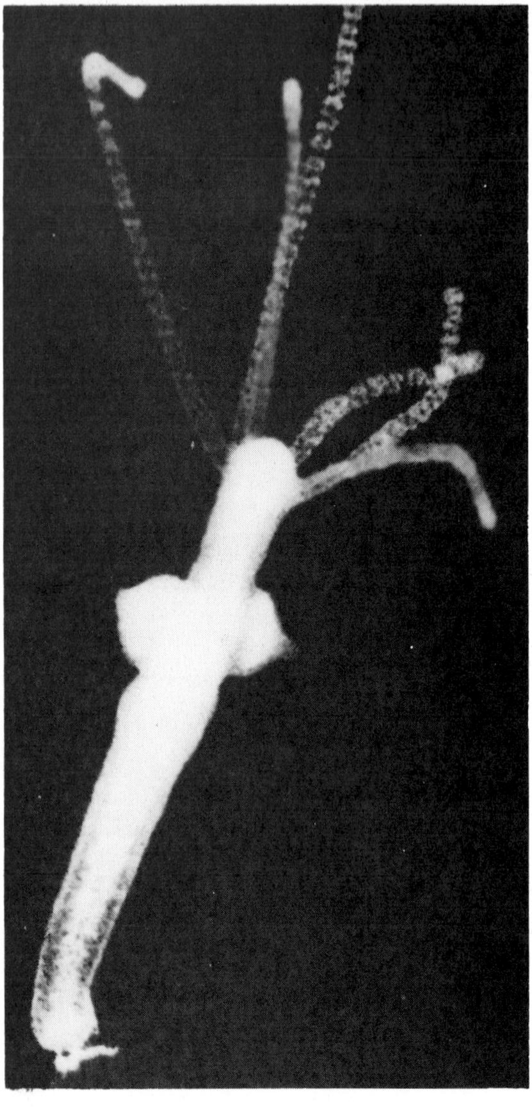

Figure 5.21 Photograph of *Hydra* with spermaries. (Courtesy of Carolina Biology Supply.)

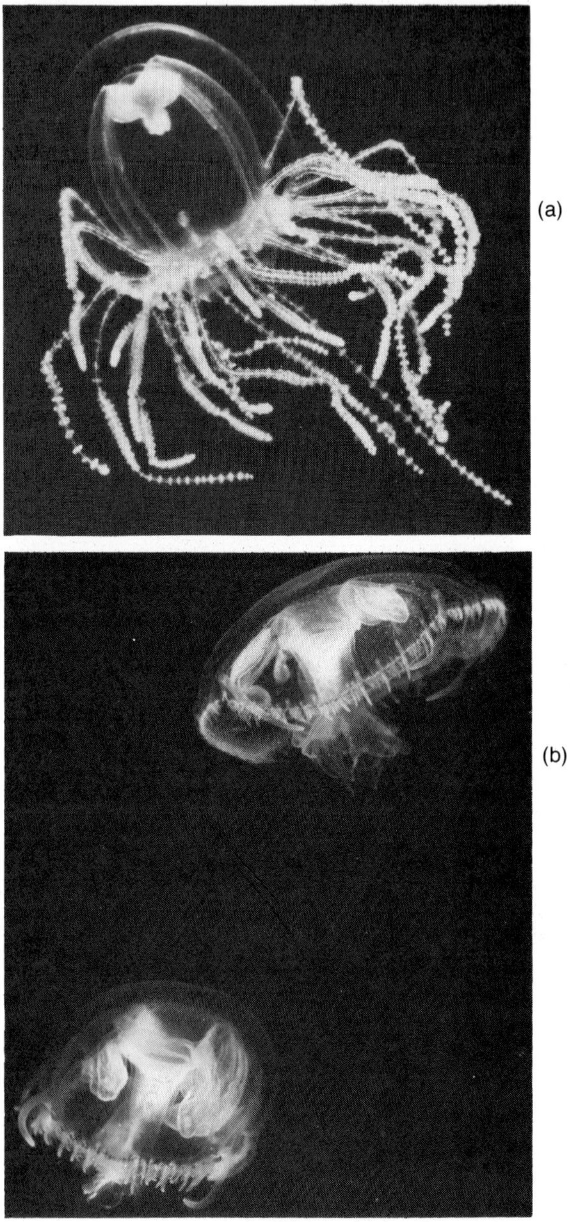

Figure 5.22 (*a*) Photograph of *Gonionemus*, a trachyline medusa; (*b*) *Craspedacusta sowerbyi*, a cosmopolitan freshwater trachyline medusa of sporadic occurrence. (*a* courtesy of D.P. Wilson; *b* courtesy of Grant Heilman.)

of reproductive morphs. Here we refer to the familiar gastrozooids, dactylozooids, and gonozooids, respectively. Siphonophore gastrozooids have a mouth and a single, long, branched tentacle. Dactylozooids lack a mouth, and their single, very long, unbranched tentacle is studded with batteries of toxic nematocysts. These tentacles are highly extensible and may reach a length of 10 m in *Physalia*. Gonozooids are often highly branched, bearing large numbers of gonophores.

Medusoid morphs are concerned with reproduction and locomotion. Gonophores, although never released, usually possess typical medusoid structures in addition to gonads. Swimming bells or **nectophores** possess a medusoid bell with a velum, but lack a manubrium and tentacles (Fig. 5.26). Nectophores are

(a)

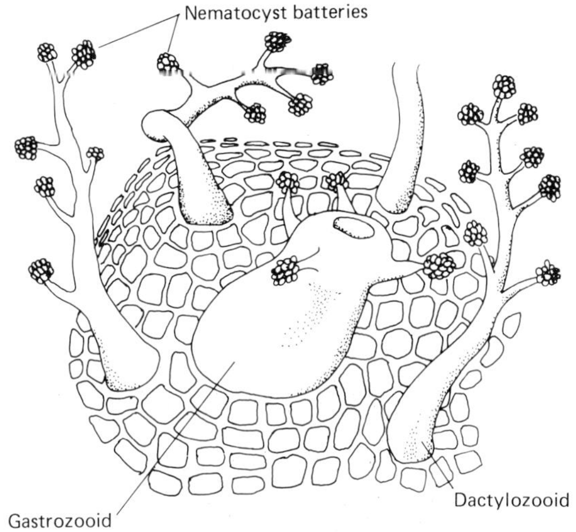

(b)

Figure 5.23 (*a*) A colony of stinging coral, *Millepora;* (*b*) A *Millepora* gastrozooid surrounded by a ring of dactylozooids. A whole colony comprises a large number of these units. (*a* from L. Agassiz, Mem. Mus. Comp. Zool., 1880, 7(1):1; *b* from H.N. Moseley, H.M.S. Challenger Reports, Zool., 1880 2(7):248.)

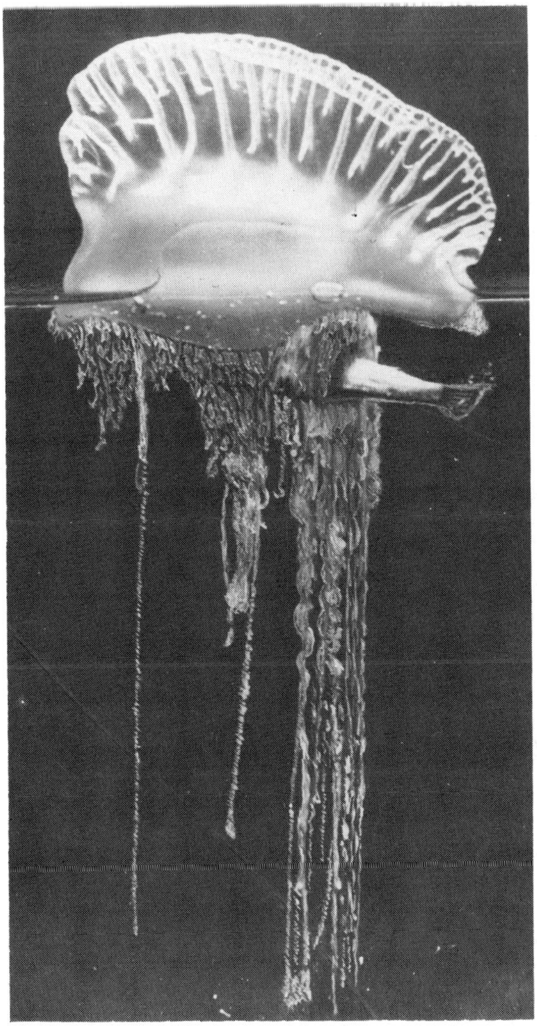

Figure 5.24 *Physalia,* the Portuguese-man-of-war, a well known siphonophore. (Photo courtesy of N.Y. Zoological Society.)

always highly muscular and their contractions propel the colony through the water. Curious structures called bracts or **phyllozooids** are well supplied with nematocysts and are presumably protective in function. Another siphonophore structure, probably derived from an aboral invagination of the larva, is the gas float or **pneumatophore** (Figs. 5.26c and 5.27). The pneumatophore is a double-walled chamber lined with a chitinoid epidermal secretion; a gas gland is located at its base. This gland secretes a gas with a composition approximating that of air, although in *Physalia* it contains a high concentration of carbon monoxide. The pneumatophore is best developed in

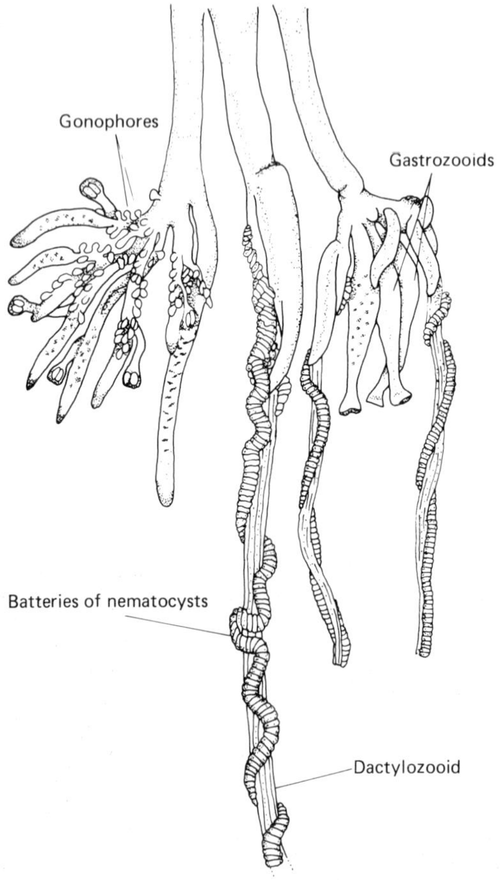

Figure 5.25 A portion of a *Physalia* colony showing several of the morphs characteristic of this colonial species. (From Beck & Braithwaite, *Invertebrate Zoology Laboratory Workbook*)

Physalia, where it forms an ovoid bladder up to 30 cm long with an erectile crest (Fig. 5.24). The muscular wall of the float permits changes in shape and the lateral rolling that keeps it moist. The float maintains *Physalia* at the surface; in many other siphonophores, smaller floats keep the colony at a particular depth. Recent investigations indicate that siphonophores may be chiefly responsible for the "deep-scattering layer" that has been detected by echo sounders in all oceans at depths of 300 to 800 m. Concentrations of these animals as great as 300 per 1000 cubic meters (m^3) have been observed at such depths. The small gas-filled floats of these siphonophores seem to be perfect resonators.

Three groups of siphonophores are defined on the basis of colony structure: (1) those with swimming bells but no float; (2) those with a small float and a long train of swimming bells; and (3) those with a large float and no bells. A colony of the first type is composed of one or more bells and a tubular stem bearing zooids organized into cormidia (Fig. 5.26a). Each **cormidium** includes a phyllozooid, a gastrozooid, a gonozooid, and a dactylozooid (Fig. 5.26b). A budding zone just below the nectophores produces additional cormidia. Cormidia may break loose from the parent colony and live independently. Colonies of the second type have a long chain of cormidia below their nectophores (Fig. 5.26c). *Physalia* is an example of the third type of colony (Fig. 5.24). Here a disk beneath the float contains the gas gland and the common gastrovascular cavity. Cormidia are suspended from the underside of the disk, each consisting of a gastrozooid, a gonozooid with both male and female gonophores (an unusual arrangement in coelenterates), and a dactylozooid with an extremely long tentacle. After capturing prey (often fish of considerable size), dactylozooid tentacles contract to about 10 cm in length, thus drawing the prey to the mouths of the gastrozooids. These movements require considerable coordination among members of the colony.

The Chondrophora

Members of this order, once grouped with the Siphonophora, form even more highly organized and integrated colonies. Indeed, some coelenterate specialists view chondrophorans as single individuals rather than colonial forms. These animals possess a many-chambered chitinoid pneumatophore which may sport a thin triangular sail (e.g., *Velella;* Fig. 5.27). Beneath the pneumatophore is a large central gastrozooid which supplies the nutritive needs of the whole colony. Concentric rings of other zooids surround the central gastrozooid. Innermost are gonozooids which release ephemeral free-swimming medusae; outside these morphs is a peripheral ring of dactylozooids.

In colonial hydrozoans, each individual zooid is comparable to a single hydra. Such zooids are differentiated both morphologically and physiologically to serve specific roles in the colony. In siphonophores and chondrophorans, particularly, individual zooids subordinate to the overall welfare of the colony. From the physiological point of view, the colony assumes the characteristics of a single individual and its zooids those of organs.

Both siphonophores and chondrophorans are animals of the open sea. As with many oceanic forms of life, their biology is not well known. Revealing their

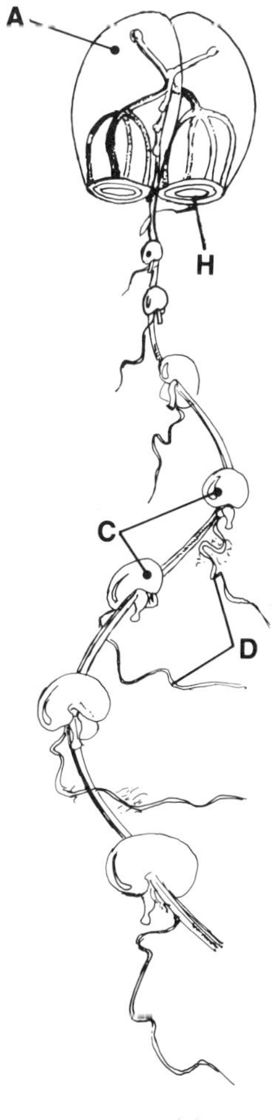

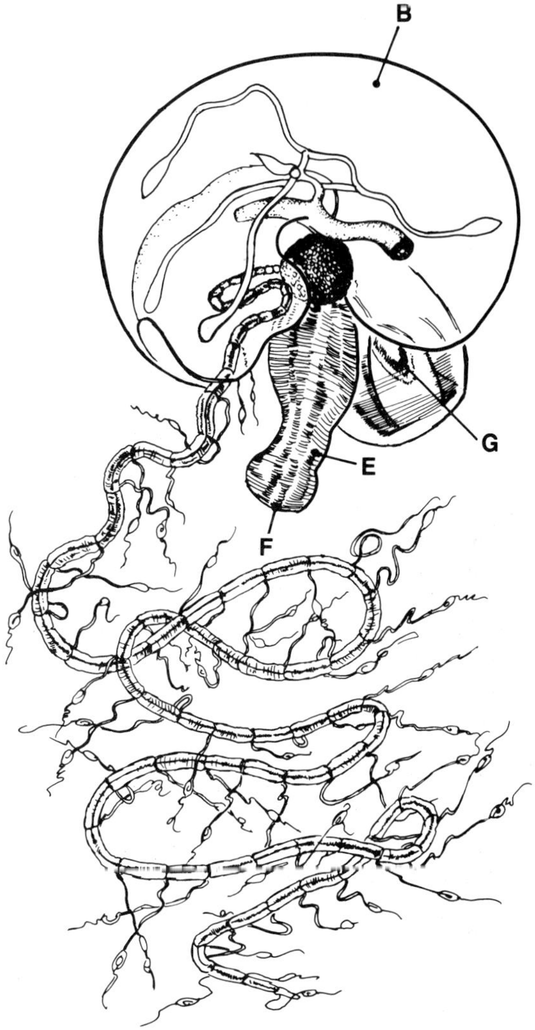

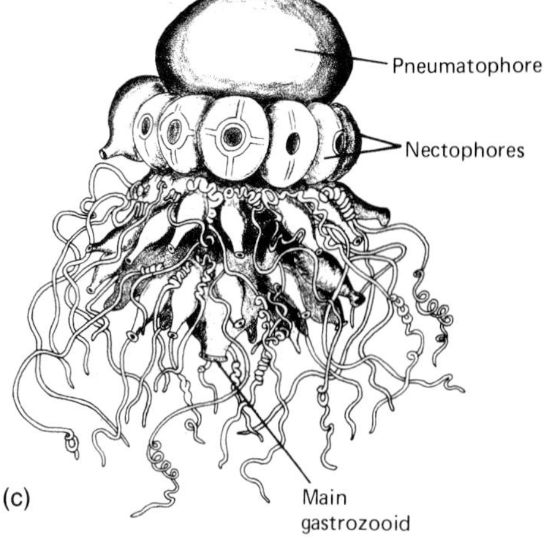

Figure 5.26 (*a*) *Praya*, an example of a siphonophore colony with swimming bells but no float; (*b*) A single cormidium from a *Praya* colony; (*c*) *Stephalia*, an example of a colony with a float and numerous swimming bells. (*a* from J.W. Fewkes, Bull. Mus. Comp. Zool., 1880, *6*: 127; *b* from E. Haeckel, H.M.S. Challenger Reports, Zoology, 1888, *28* (77); *c* from Hyman, *The Invertebrates, Vol. I.*)

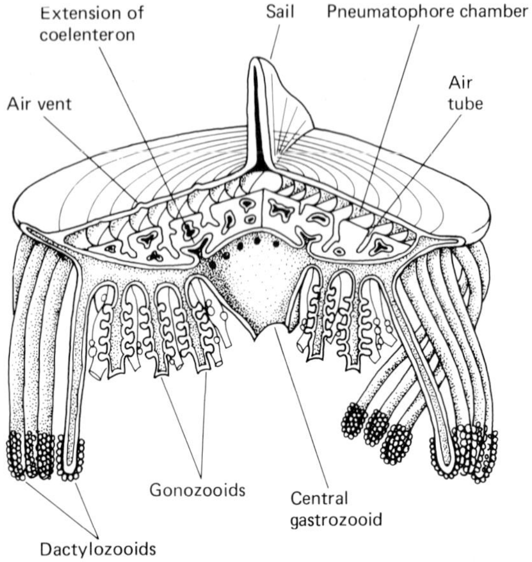

Figure 5.27 Cut-away diagram of *Velella* a chondrophoran. (Modified from Delage et Hérouard, *Traité de Zoologie Concrète*, 1896, Schiecher Freres.)

secrets will challenge the ingenuity of future generations of marine biologists. At present, however, they number among the more mysterious denizens of the invertebrate world.

CLASS SCYPHOZOA

Scyphozoans are the coelenterates most commonly called jellyfish. Free-swimming medusae are the dominant morphs in this class. The polypoid stage is reduced considerably and in some species is entirely absent. With 200 species, the scyphozoans compose the smallest class in this phylum, but their size (large by coelenterate standards) and the frequently dangerous stings of their nematocysts have made them a group well known to the layperson. Exclusively marine in distribution, scyphozoans are found from polar to equatorial seas and from surface waters to depths of 3000 m.

Scyphozoans differ from hydrozoan medusae in several ways. Typically, scyphozoans are larger, averaging 20 to 30 cm in diameter. The giant jellyfish *Cyanea capillata* may measure 2 m across, and its 800 tentacles can extend downward 30 to 60 m, distinguishing it as the largest individual coelenterate. The larger size of scyphozoans is attributable primarily to a bulky, partially cellularized mesoglea. This mesoglea, often of an extremely stiff gelatinous material, defines the form of the scyphozoan bell, which varies from a tall dome to a flat, broad disk (Figs. 5.28 and 5.33). Tentacles fringe the bell margin, which often is notched to form **lappets.** Between the lappets are tentacles and rhopalia; the latter are the principal sensory centers in these animals. Scyphozoans never possess the subumbrellar velum characteristic of hydrozoan medusae. Also, in contrast to the hydrozoans, scyphozoan gonads are gastrodermal, not epidermal, in origin.

A distinct tetramerous symmetry is evidenced in this class. Typically, the mouth tube, or **manubrium,** is produced as four oral arms dangling from the center of the subumbrella (Fig. 5.28). These oral arms form the sides of a square mouth. In some scyphozoans, the coelenteron comprises a central stomach area and four symmetrically arranged pouches separated by septa. From these pouches, an elaborate radial canal system lined with ciliated or flagellated cells extends outward to the bell margin, where a ring canal is often present (Fig. 5.29). Through such canals, nutrients and other biological materials are distributed throughout the animal. This canal system is well-developed in advanced scyphozoans, but such types commonly lack septa and thus have no distinct gastric pouches. Gastrodermal gonads are located in the coelenteron pouches when present; in scyphozoans lacking pouches, the horseshoe-shaped gonads are confined to the stomach floor. Tetramerous symmetry is also apparent in the arrangement of tentacles and rhopalia; there are four or, more often, multiples of four of each of these structures.

Because scyphozoans are larger coelenterates, they have more internal transport problems than do most hydrozoans. The radial canal system distributes biological materials. External openings from this system may serve an excretory function. Scyphozoans often possess four symmetrically arranged subumbrellar funnels (Fig. 5.28). Extending into the bulky mesoglea, these depressions probably facilitate respiratory exchange.

Their larger size necessitates that scyphozoans eat more. These animals enjoy a varied diet, ranging from planktonic organisms to small or medium-sized fish. Typically, a scyphozoan traps prey by descending upon it. A strong, circular subumbrellar musculature propels the coelenterate vertically; subsequently, as the animal sinks, prey is entangled by the oral arms and/or the bell tentacles. In plankton-feeding species, tracts of cilia and numerous mucous glands on the bell and tentacles together bring food to the oral arms and ultimately to the mouth. Within the coelenteron, nematocysts situated on gastric filaments (Fig. 5.28)

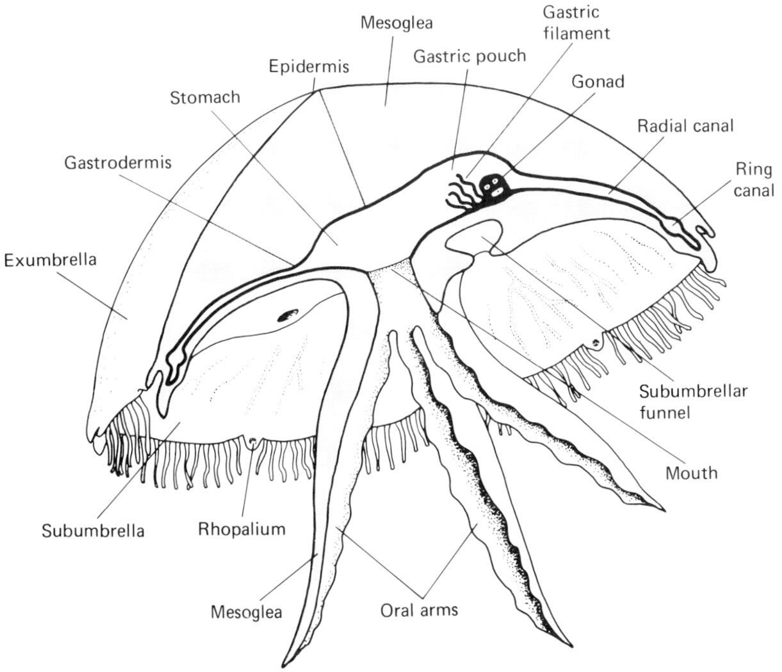

Figure 5.28 A cut-away side view showing the anatomy of a scyphozoan medusa. (Modified from Barnes, *Invertebrate Zoology.*)

subdue larger prey, and digestion proceeds as in hydrozoans. Scyphozoans are negatively phototrophic, and most feed in surface waters at night.

With few exceptions, scyphozoans are diecious. Gametes exit from the mouth, but eggs often are retained by the oral arms, where fertilization and brooding may occur. Radial, indeterminate cleavage yields a hollow blastula, which develops into a typical planula larva. After settling, the planula forms a polypoid **scyphistoma** (Fig. 5.30). Scyphistomae are small, ranging from a few millimeters to several centimeters in height. Under favorable conditions, the scyphistoma asexually produces others of its kind. In a less supportive environment, it buds off immature, medusoid forms called **ephyrae** (Fig. 5.31). Ephyrae are produced either singly or in groups by transverse fission at the oral end of the scyphistoma, a process called **strobilation.** Almost microscopic in size, ephyrae grow and mature into adult scyphozoans. Scyphistomae may live for several years, periodically undergoing strobilation. Despite their size and complexity, adult medusae are much shorter-lived.

Exceptions to this generalized life cycle occur among pelagic scyphozoans, whose habitat is lacking in substrata. Such jellyfish have undergone an even further reduction of the polypoid stage. The scyphistoma may develop within the adult coelenteron in these species or may be completely absent.

Four scyphozoan orders deserve brief mention. The Stauromedusae are an aberrant group combining polypoid and medusoid forms in a single individual. Typically, their eight-sided medusoid bell is supported by a polypoid column. These scyphozoans are sessile and completely immobile. *Haliclystus* (Fig. 5.32), "the clown," is a representative genus. Cubomedusae are an order of warm-water jellyfish whose bells are distinctly cuboidal. These animals have four large tentacles or groups of tentacles extending from the corners of their bell. A marginal nerve ring connects sensory rhopalia, a linkage unique in the class. Cubomedusae include some of the most dangerous scyphozoans. Nematocyst stings by several species of sea wasps can cause severe damage to people. In Australia alone, at least 50 human fatalities are attributed to these coelenterates. *Chironex* (Fig. 5.33) and *Chiropsalmus* are among the better-known and more feared genera.

The largest and most familiar scyphozoans belong to the order Semaeostomae. Bell shape in this group varies considerably, and the oral arms are usually quite conspicuous. A cruciform stomach lacks septa, but a well-developed system of radial canals extends to the bell margin. *Aurelia* (Fig. 5.34), the scyphozoan type most commonly studied in the introductory laboratory, is a member of the Semaeostomae. The giant *Cyanea* of Arctic waters is also included here, as is *Chrysaora*, the highly poisonous sea nettle (Fig.

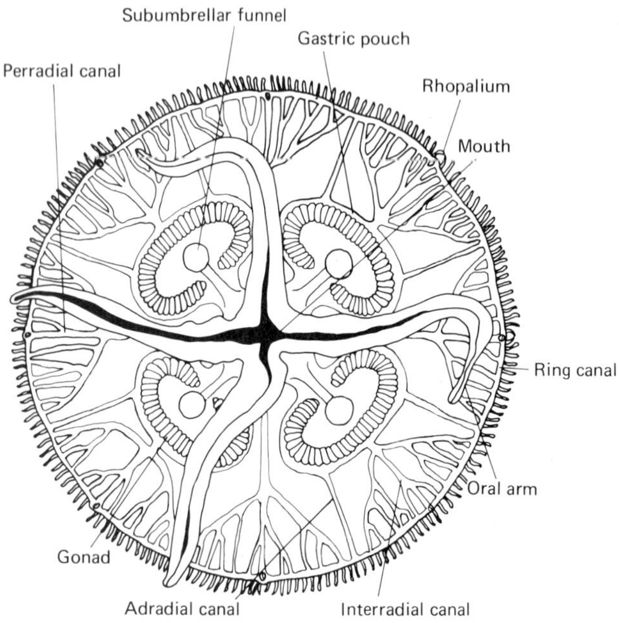

Figure 5.29 Oral or subumbrellar view of a scyphozoan medusa (*Aurelia*) featuring the gastrovascular canal system. (From Beck and Braithwaite, *Invertebrate Zoology Laboratory Workbook.*)

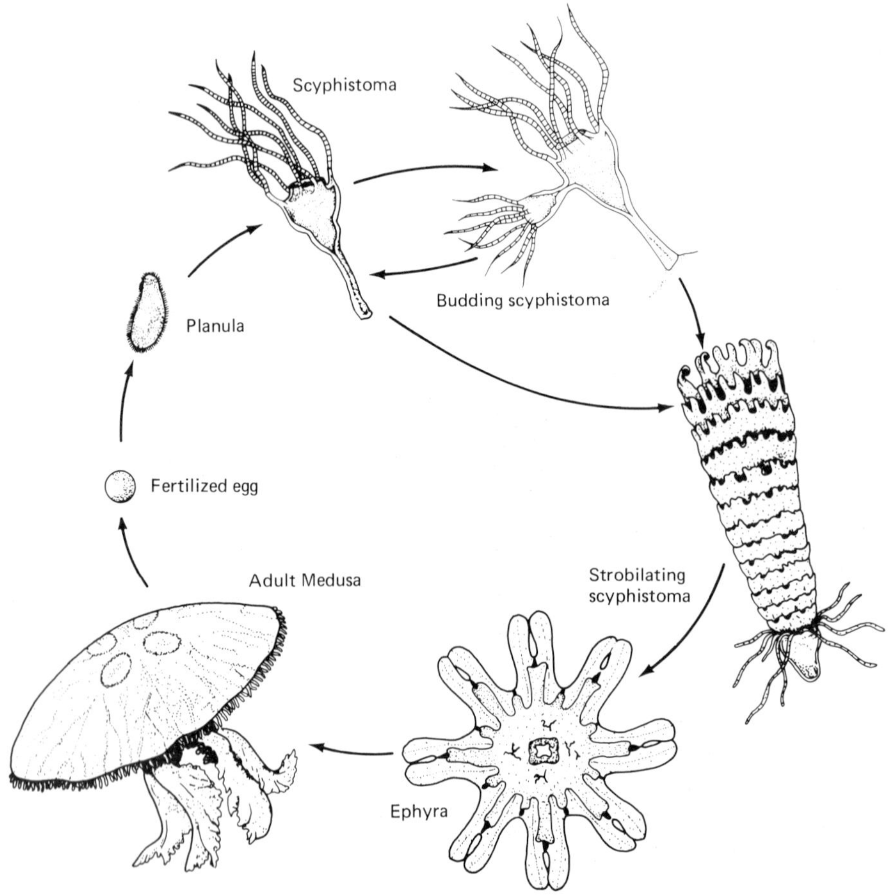

Figure 5.30 Life cycle of *Aurelia*, a scyphozoan featuring a scyphistoma stage. (Modified from Barnes, *Invertebrate Zoology.*)

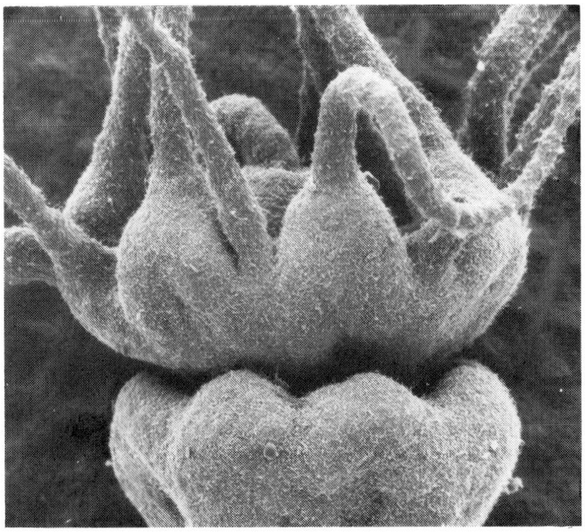

Figure 5.31 Scanning electron micrograph of a strobilating scyphistoma of *Aurelia.* (Photo courtesy of D.B. Spangenberg and Wm. Kuenning.)

Figure 5.32 *Haliclystus,* a sessile stalked stauromedusa. (Photo courtesy of D.P. Wilson.)

5.35). *Pelagia* represents one of several open-sea genera whose polypoid stage has been totally suppressed. Semaeostomes live in all oceans.

A fourth order, the Rhizostomae, includes those scyphozoans specialized as filter feeders. In these jellyfish, the frilled edges of the oral arms have fused over the mouth. The manubrium houses a complex series of canals opening to the outside through numerous oral pores. Through these pores collected plankton enter the gastrovascular cavity. Most rhizostomes are confined to warm, shallow waters. The bottom-dwelling *Cassiopeia* (Fig. 5.36) and the stiff spheroidal *Stomolophus,* or "cabbage head", are representative genera.

CLASS ANTHOZOA

Corals and sea anemones are the featured organisms of the Anthozoa, the third and largest class of coelenterates. Numbering about 6000 species, anthozoans include solitary, sessile individuals as well as colonies. Their life cycle is dominated by a sexually and/or asexually reproducing polyp; medusae are entirely absent from this class.

In general appearance, an anthozoan resembles a hydrozoan polyp. In both, a body column supports an oral ring of tentacles. Anthozoans, however, are usually larger and sturdier because of their thick, fibrous, and partially cellularized mesoglea. Further, the entire column wall is invaginated at the anthozoan mouth to form an ectodermally lined tubular pharynx. This stomodeal structure typically extends at least one-third of the way to the pedal disk and is ovoid in cross section. At one or both sides of the stomodeum, a ciliated groove or **siphonoglyph** may be present (Fig. 5.37). Siphonoglyphs alter the symmetry of the anthozoans. When two siphonoglyphs are present, two planes of symmetry exist and the animal is said to be biradially organized. The presence of a single siphonoglyph allows only one symmetrical axis; in such anthozoans, symmetry is technically (but not functionally), bilateral. The pharynx is supported by **septa,** gastrodermal and mesogleal ingrowths of the body wall. Some septa merge with the pharynx. Called complete septa, these ingrowths usually contain strong, longitudinal retractor muscles. Other septa, either too short to contact the pharynx or located below the stomodeal invagination, are called incomplete septa; they terminate freely within the gastrovascular cavity. Complete and incomplete septa thus divide the anthozoan coelenteron into compartments and greatly increase its surface area. In contrast to hydrozoans, anthozoans (like scyphozoans) have gastrodermally derived gonads. These gonads are located along the septa (Fig. 5.39). Skeletons, secreted externally by the epidermis or formed internally from mesogleal cells, are common in the class.

Anthozoan taxonomy reflects variations in the form and arrangement of tentacles, siphonoglyphs, septa, and skeletal elements. Two subclasses are recognized: the Zoantharia, which includes sea anemones and the stony corals; and the Alcyonaria, which includes the soft corals, sea fans, and sea pansies.

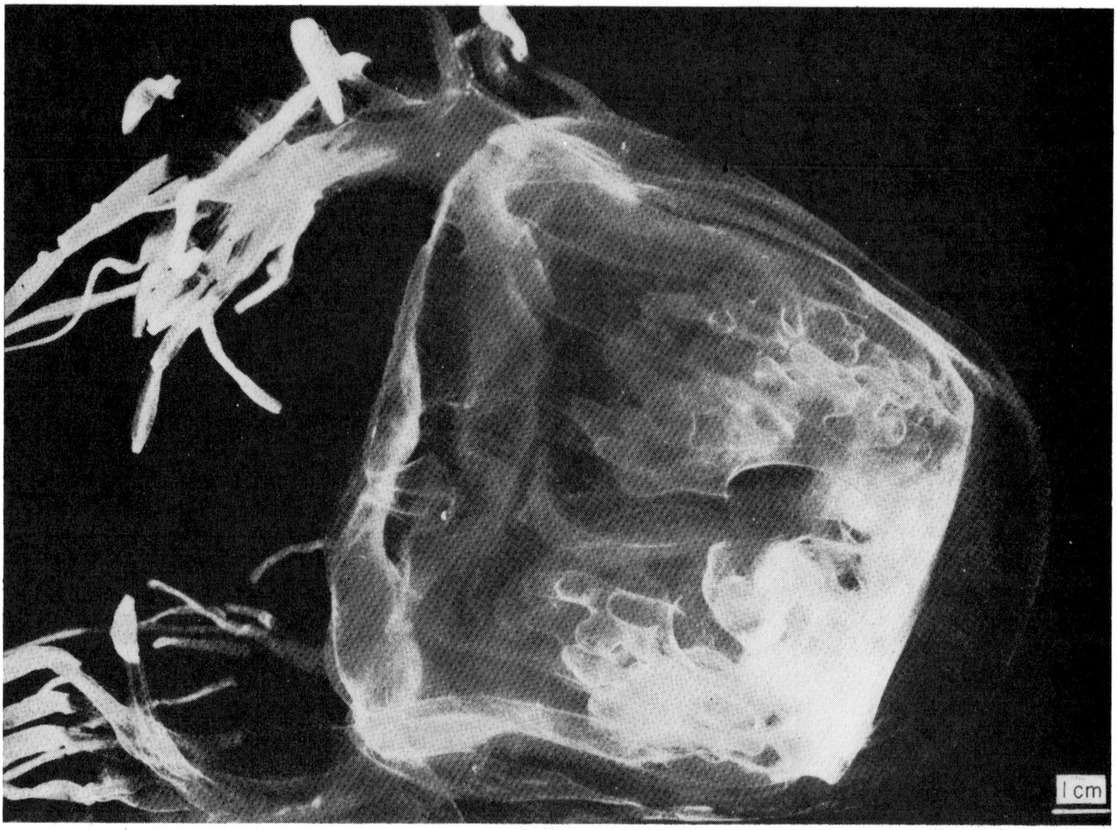

Figure 5.33 The sea wasp, *Chironex fleckeri*, a cubomedusa. (From Barnes, *Invertebrate Zoology*.)

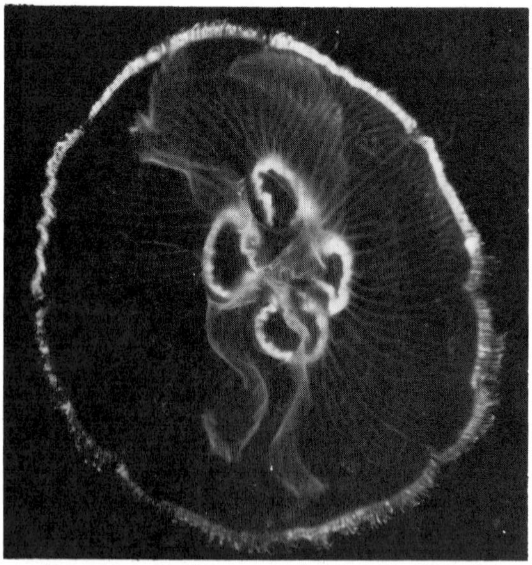

Figure 5.34 *Aurelia*, a semaeostome scyphozoan. (Photo courtesy of D.P. Wilson.)

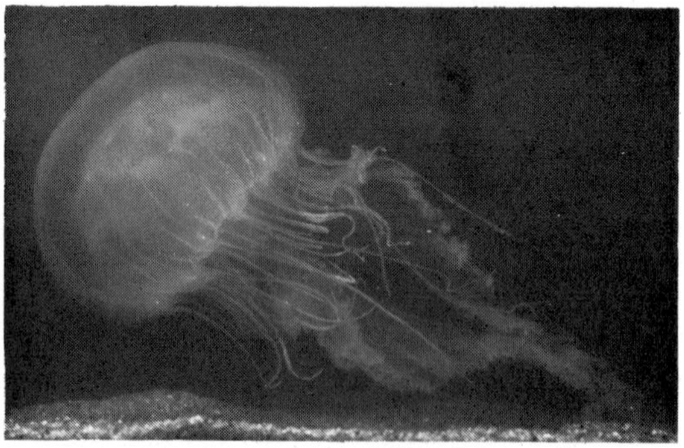

Figure 5.35 The sea nettle, *Chrysaora*, a semaeostome. (Photo courtesy of Grant Heilman.)

Subclass Zoantharia (Hexacorallia)

Zoantharians constitute the more diverse subclass. Typically, their tentacles and septa are arranged in patterns of six (Fig. 5.37a), accounting for the group's alternate name, the Hexacorallia. The sea anemones

Figure 5.36 A specimen of *Cassiopeia*, a rhizostome scyphozoan lying in its characteristic feeding position on the bottom with the oral surface upwards. (From Fritz Goro, Life Magazine.)

and the stony corals, the two principal zoantharian groups, are discussed separately.

The ancient Greeks thought that sea anemones (order Actiniaria) were marine flowers. The group does contain some strikingly beautiful animals whose colorful whorls of tentacles often create a floral impression (Fig. 5.38). Sea anemones attach to hard substrata in tide pools and below the low-tide mark along rocky coastlines throughout the world. They are common inhabitants of coral reefs and are exploited by hermit crabs in camouflaging the latter's mollusk shell homes.

Structurally, sea anemones consist of a stout column with an aboral pedal disk and one or more oral rings of tentacles (Fig. 5.39). The tentacles are simple and tubular and surround the slitlike mouth. A stomodeum extends from the mouth into the coelenteron, where it is supported by hexamerously arranged septa. Typically, sea anemones possess six pairs of complete septa (termed primaries) and one or more cycles of incomplete septa. Incomplete septa are located between pairs of primary septa and are termed secondaries, tertiaries, etc. (Fig. 5.37a). Usually two siphonoglyphs are present; the pairs of primary septa associated with these grooves are known as **directives.**

The free margins of septa often are modified as trilobed **septal filaments.** The middle lobe possesses nematocytes and enzymatic gland cells, while ciliated cells dominate the two lateral lobes (Fig. 5.40). The middle lobes may be drawn out basally into long threads called **acontia** (Fig. 5.39). When the animal contracts, the acontia may be thrust through the **cinclides,** small lateral pores in the body wall which permit the exit of water from the gastrovascular cavity. Acontial threads probably have a defensive function. In addition to their role in the subjugation and digestion of prey, the septal filaments may be sites of absorption and intracellular digestion

The siphonoglyphs, aided by the ciliated lobes of the septal filaments, circulate gastrovascular water which transports nutrients and dissolved gases throughout the animal. Perforations in the complete septa (Fig. 5.39) ensure adequate flow through the upper compartments of the coelenteron. The restoration of upright body posture after contraction depends on water imported along the siphonoglyph.

Sea anemones are mostly solitary creatures, although asexual reproduction by budding may create, at least temporarily, an apparent colonial organization (Fig. 5.41). Asexuality in these animals also includes longitudinal fission and pedal laceration. In the latter process, an anemone detaches abruptly from the substratum, tearing away from a portion of its adhered pedal disk. The disk fragment then regenerates a new animal. Anemones produced by asexual means may lack a regular arrangement of septa.

Sexual reproduction is more common, and the group contains both hermaphroditic and diecious species. Protandry is the rule among hermaphroditic anemones. Some species brood their developing young within the coelenteron, but in others, both fertilization and development occur in the sea. Typically, a planula larva differentiates from a hollow blastula. A stomodeal invagination occurs at the blastopore, while septa grow inward from the column wall. The planula attaches and sprouts tentacles, and the next generation's anemone is formed.

The second major group of zoantharians includes the reef-building stony corals (order Scleractinia). These animals are quite similar to anemones; the major difference is their hard calcareous skeleton. Other distinctions include smaller polyp sizes, a colonial organization in most species, and the absence of siphonoglyphs.

Epidermal cells of the lower column and basal disk secrete calcium carbonate around each polyp. These secretions form a skeletal cup, or **theca** (Fig. 5.42). On the bottom of the theca, several radially arranged partitions are deposited between folds in the aboral body wall (Fig. 5.42). The skeleton of a single polyp is called a **corallite** (Fig. 5.43). In colonial corals, all polyps are interconnected by lateral extensions of the body wall. This connective sheet contains both upper and lower epidermal and gastrodermal layers, as well as an extension of the gastrovascular cavity. The lower

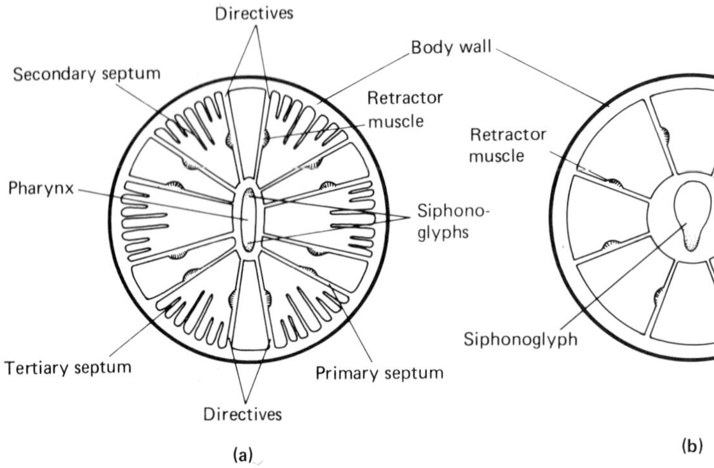

Figure 5.37 Diagrammatic cross sections through anthozoan polyps at the level of the stomodeum. (*a*) A zoantharian; (*b*) An alcyonarian.

Figure 5.38 Photograph of a sea anemone. (Courtesy of S. Arthur Reed.)

epidermis secretes a calcareous skeleton between adjacent corallites. Thus a coral colony is covered completely by living tissue, and the coelenterons of all members are continuous.

Sexual reproduction similar to that of anemones creates a single polyp, which founds a new colony through budding. The colonial form reflects the budding pattern. Budding may occur from the connective sheet or from the polyps themselves. Arborescent patterns yield colonies such as the staghorn *Acropora* (Fig. 5.44a), while incomplete lateral budding produces the convolutions characteristic of brain corals (Fig. 5.44b). Over 200 coral species contribute to some large reefs, and their varied growth patterns

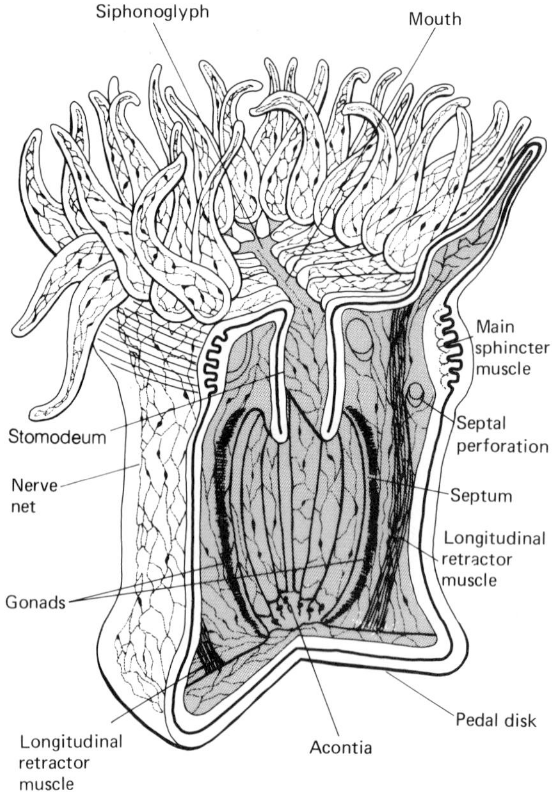

Figure 5.39 Cut-away side view diagram illustrating the anatomy of a sea anemone, *Calliactis*. (From Wells, *Lower Animals*.)

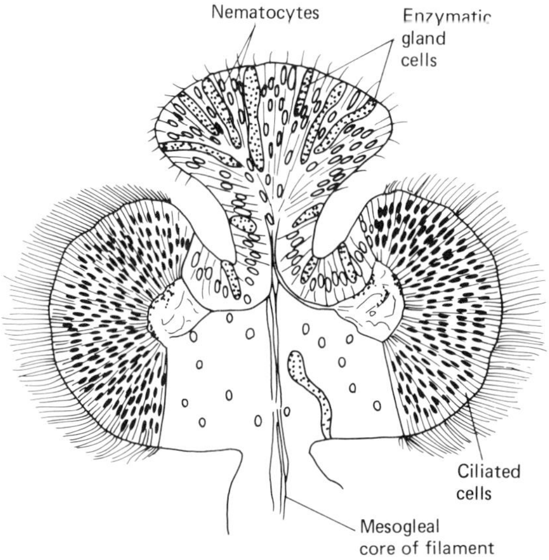

Figure 5.40 Section through a septal filament of a sea anemone, *Metridium*, showing histological structure. (From Hyman, *The Invertebrates, Vol. I.*)

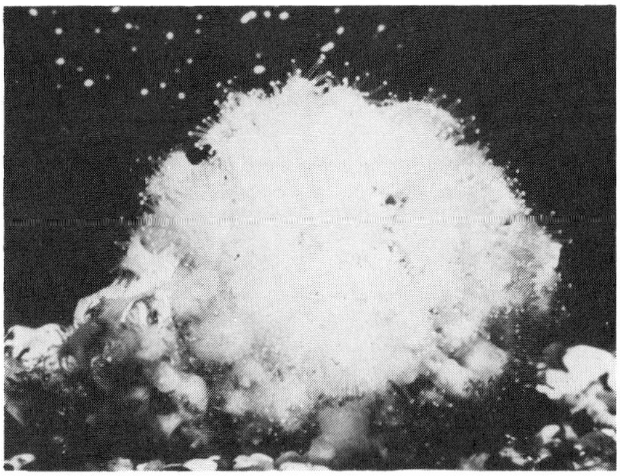

Figure 5.41 Photograph of a "colony" (actually a cluster) of sea anemones, *Corynactis californica*. (Courtesy of Ward's Natural Science Establishment.)

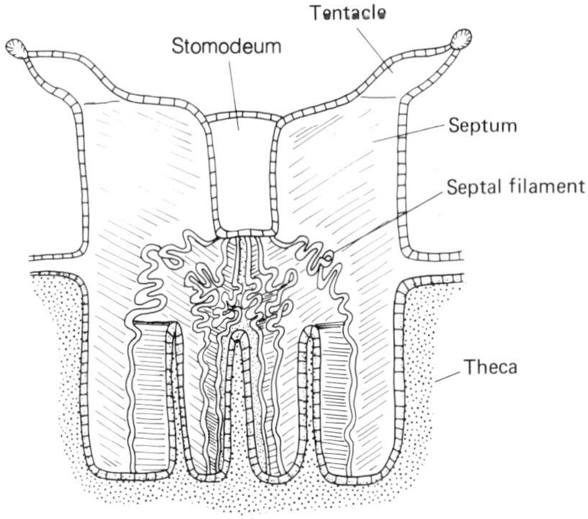

Figure 5.42 Diagram of a longitudinal section through a coral polyp in its skeletal cup. (From Hyman, *The Invertebrates, Vol. I.*)

Figure 5.43 Photograph showing the skeletal cups (corallites) of individual coral polyps. (Courtesy of S. Arthur Reed.)

create a fascinating world of shapes and textures. The coral reef is not only the home of anthozoan polyps, but an entire community to which most invertebrate phyla contribute members. A mosaic of sponges, hydroids, echinoderms, tunicates, polychaetes, mollusks, crustaceans, and fish thrives in the reef environment.

Reef-building corals are strictly tropical in distribution. They thrive at temperatures between 23 and 25°C and are healthiest at depths of less than 30 m. Corals survive at greater depths, but do not build reefs because of the light requirements of their mutualistic zooxanthellae (see Special Essay, this chapter). Three general types of coral reefs are described. Simple fringing reefs are continuous with the shore (Fig. 5.45a). Barrier reefs are separated from the shore by a lagoon channel (Fig. 5.45a). The Great Barrier Reef of Australia is the most famous of these coral formations. It extends 2000 km along the coast of Queensland, from which it is separated by a channel 50 m deep and up to 200 km wide. The third reef type is the atoll—circular coral islands not associated with an existing coastal area (Fig. 5.45b). Atolls often rise abruptly from the sea in regions where the water is many hundreds of meters deep. The circular reef surrounds a central lagoon.

(a)

(b)

(c)
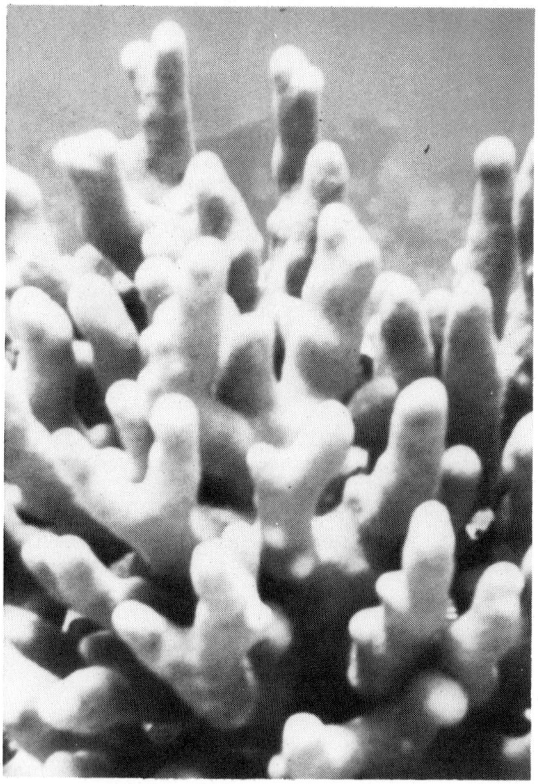

Figure 5.44 Colonies of coral. (*a*) Branches of *Acropora*, staghorn coral; (*b*) Brain coral, *Diploria*; (*c*) Finger coral, *Porites*. (Photos courtesy of Ward's Natural Science Establishment.)

Because reefs are built only in shallow waters, fringing reefs can be explained easily; they simply grow out from shore until the water becomes too deep. But an explanation for barrier reef and atoll growth has been a chronic problem for biologists. Two theories have survived over a century of debate. The first was formulated by Charles Darwin during his historic voyage on the *HMS Beagle* from 1831 to 1836. Darwin proposed that atolls originated in fringing reefs which once surrounded volcanic islands. As one of these islands gradually subsided into the ocean, its coral reef gradually grew upward. Thus the reef continued to occupy a constant depth. When the outlying portions of the island sunk, a barrier reef surrounded all or part of the remaining land. Total subsidence of the island left a ring of coral, an atoll (Fig. 5.45b).

Darwin's theory of atoll formation remains very popular. Indeed, cores sunk several hundred meters through existing atolls reveal that the reef consists of skeletons from shallow-water species. A few deeply sunk cores have contracted volcanic bedrock. Evidently, shallow-water corals have been growing upward for a long time under the influence of shifting

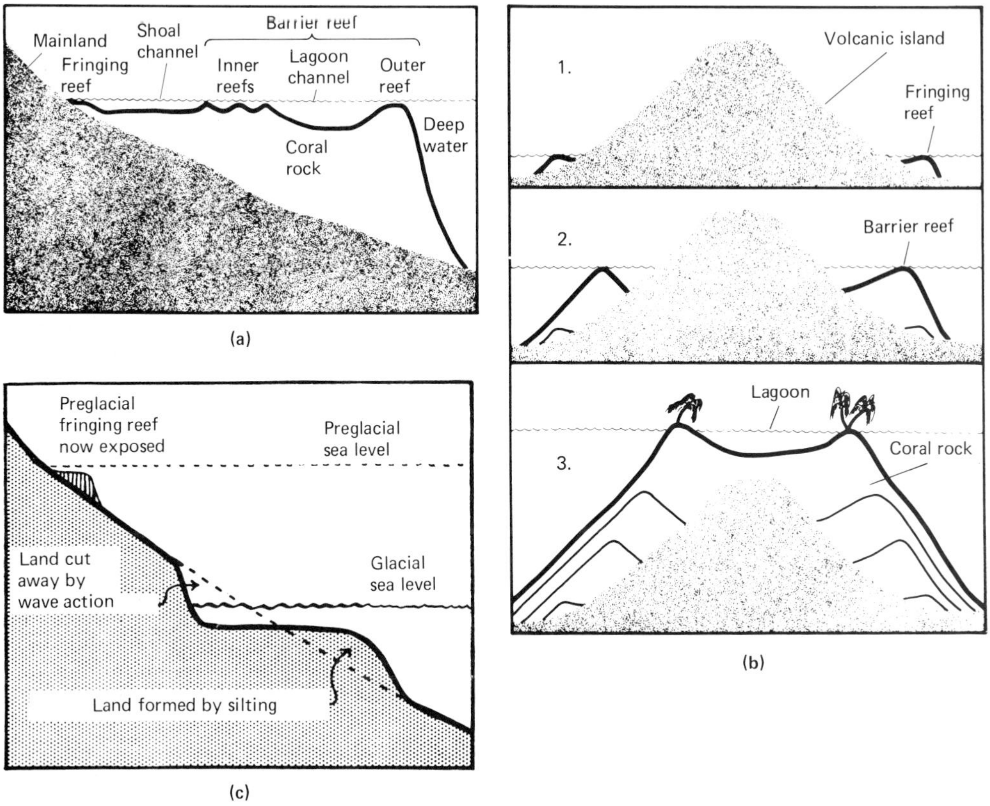

Figure 5.45 (*a*) Diagram showing position of two basic reef types relative to the shore, a fringing reef and a barrier reef; (*b*) Diagram showing the formation of fringing reefs, barrier reefs and atolls as the land sinks beneath the water according to Darwin's theory. (1) Fringing reef, (2) Barrier reef (3) Atoll; (*c*) Diagram showing the formation of a platform by wave action during the last Ice Age, an important feature of Daly's theory of reef formation. (From F.G. Walton Smith, *Atlantic Reef Corals*, 1971, University of Miami.)

land and water levels. This basic Darwinian idea is at the heart of almost all theories of atoll formation. One other theory has gained considerable support and, together with Darwin's contributions, may explain the formation of most coral reefs. According to Reginald A. Daly, the last Ice Age played a major role in coral formation. The great polar and temperate-zone glaciers of that period tied up such large volumes of water that tropical seas sank as much as 50 m. Most previously formed reefs were killed by exposure to air. Meanwhile, wave action against newly exposed coastlines established broad flat platforms extending out into the sea (Fig. 5.45c). As the Ice Age ended, coral growth on these platforms kept pace with the rising waters. The fact that many atoll lagoons and barrier reef channels are approximately 50 m deep is cited as evidence in support of Daly's theory.

Of the several smaller orders of Zoantharia, the order Ceriantharia, or burrowing anemones, is noteworthy. Burrowing anemones are solitary forms with a tall columnar body aborally rounded and buried in sand or mud. Tentacles are arranged in two circles on the oral disk, one at the margin and one around the mouth. There is a single siphonoglyph, and all septa are complete. Ceriantharid burrows are often lined with protective mucus and agglutinated sand and bottom debris. *Cerianthus* (Fig. 5.46) is a well-known member of this group.

Subclass Alcyonaria (Octocorallia)

The second anthozoan subclass is distinguished by octamerous symmetry. Alcyonarian polyps always have eight tentacles, and their eight septa are all com-

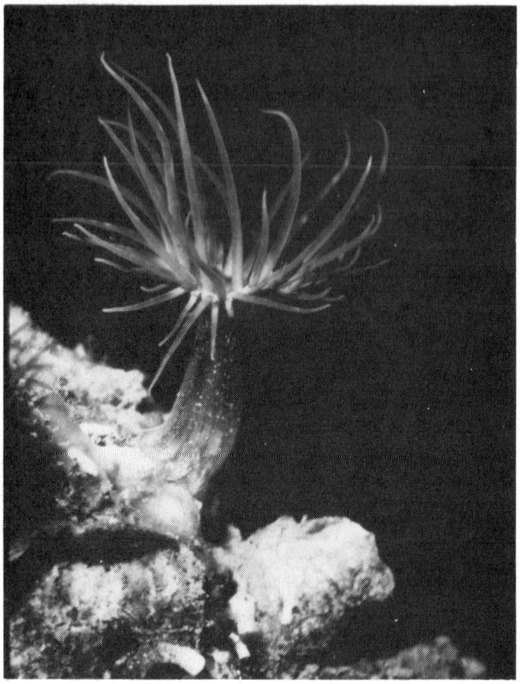

Figure 5.46 Photograph of *Cerianthus*, a burrowing anemone. (Courtesy of S. Arthur Reed.)

plete (Fig. 5.37b). This organization suggests an alternative name for the subclass, the Octocorallia. One siphonoglyph is the rule. The arrangement of longitudinal retractor muscles on the septa differs from the zoantharian arrangement. Also alcyonarian polyp morphology is much less diverse than that of zoantharian polyps. Alcyonarians include a number of colonial forms known best by their common names: the sea pansies, the sea fans, the sea pens, the sea whips, the soft corals, and the organ-pipe corals.

The tentacles of alcyonarian polyps bear featherlike projections called **pinnules** (Fig. 5.47). These pinnules cull food from surrounding waters. Alcyonarians are almost wholly colonial, but their organization differs markedly from that of stony corals, particularly with regard to skeletal elements. Alcyonarian colonies are covered by **coenenchyme,** a soft tissue in which the individual polyps are housed. Coenenchyme is mostly mesogleal in composition, but does feature gastrovascular tubes called **solenia** (Fig. 5.47) and a thin epidermis. Solenia are narrow extensions of the coelenteron of each polyp, and through such tubes the entire colony is united. New polyps are budded from the solenia. Amoebocytes within the coenenchyme secrete an internal skeleton composed of either calcareous spicules or a horny material. Thus the alcyonarian skeleton is contained completely within the living tissue of the colony and is not secreted externally, as in the Zoantharia.

As in the stony corals, alcyonarian growth patterns are quite diverse. The skeleton of the familiar gorgonians (order Gorgonacea, including the sea fans and sea whips) features an axial rod composed of a horny material called **gorgonin.** This rod is surrounded by a coenenchyme cylinder in which the polyps are imbedded. The overall organization of the axial rods determines the colony's appearance. Sea fans have a flattened lattice form (Fig. 5.48), while sea whips resemble plastic-coated wires (Fig. 5.49). Another alcyonarian group (order Stolonifera) includes the organ-pipe coral *Tubipora* (Fig. 5.50). A superficial resemblance to organ pipes occurs in this genus, whose parallel polyps are housed within tiers of cylindrical tubes. Finally, colonies of sea pens and sea pansies (order Pennatulacea; Fig. 5.51) are founded by a large, anchored polyp. Secondary, often dimorphic polyps develop on the upper portion of the primary polyp, which is secured in soft substrata.

Although alcyonarians enjoy worldwide distribution, they are most abundant and conspicuous in warmer waters and are particularly prominent in coral reef communities.

Coelenterate Evolution

Coelenterate evolution is a controversial topic which involves the debate on the origin of the Metazoa. Two schools of thought exist concerning how the coelenterates fit into metazoan history. One asserts that they have degenerated from the acoelomates. This position argues that the anthozoans are the most primitive members of the phylum. As evidence, supporters cite the "remnants" of bilateral symmetry in the group and claim that radial symmetry in scyphozoans and hydrozoans has evolved in association with sessile and/or planktonic life-styles. The alternative theory proposes that coelenterates arose directly from protozoan stock and argues that primary radial symmetry is original in the phylum and that the hydrozoans are the most primitive coelenterate class.

Certain observations support the latter theory. Proceeding from the hydrozoans to the anthozoans, we observe a progressively more complex nervous organization and the gradual cellularization of the mesogleal layer. These processes are consistent with a

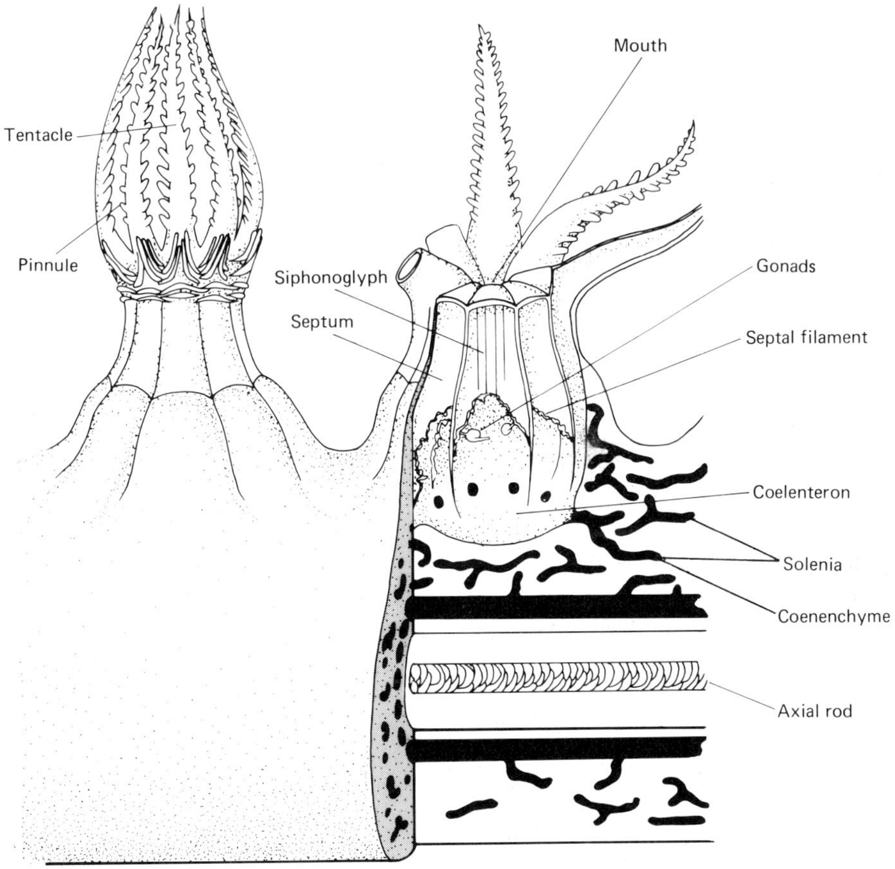

Figure 5.47 A cut-away view of an alcyonarian colony illustrating characteristic structural features. (From R.C. Moore, Ed., *Treatise on Invertebrate Paleontology, Part F; Coelenterata,* 1956, Geological Society.)

progressively evolving line and suggest that the anthozoans are the most advanced coelenterate class. Theories on the evolution of the Metazoa are discussed more fully in Chapter 18.

We may wonder whether the polyp or the medusa is the phylum's original form. Again, biologists argue for both sides, but prevailing opinion favors an ancestral medusa. The fact that coelenterate zygotes form planula larvae suggests that the phylum arose from a creeping, ciliated planuloid organism. The development of tentacles would have produced an actinulalike, free-swimming form, the primeval medusa. The trachyline hydrozoans may represent the most primitive coelenterates, as their life cycle is dominated by such a simple medusoid stage. Later in coelenterate evolution, asexual reproduction by the actinula larva could have led to a distinct polypoid stage. According to one theory, the polyp represents an extended larval form specialized for asexual reproduction. With the establishment of the polyp, a gradual suppression of the medusa began. Various stages in that suppression are evidenced in the Hydroidea, while medusae have been phased out totally in the Anthozoa. The exact relationships among the coelenterate classes and orders remain speculative, but scyphozoans and anthozoans apparently diverged very early from the hydrozoan line. The opposite view, that the polyp is the original form, is held by a few coelenterate specialists. According to this view, the medusa is a specialized dispersal form evolved independently in both the Hydrozoa and the Scyphozoa.

Coelenterates are fascinating and beautiful animals and are certainly successful members of the invertebrate world. Yet they remain at the **tissue level of organization** and probably did not give rise directly to the higher animal phyla. Restrictions on their evolu-

(a)

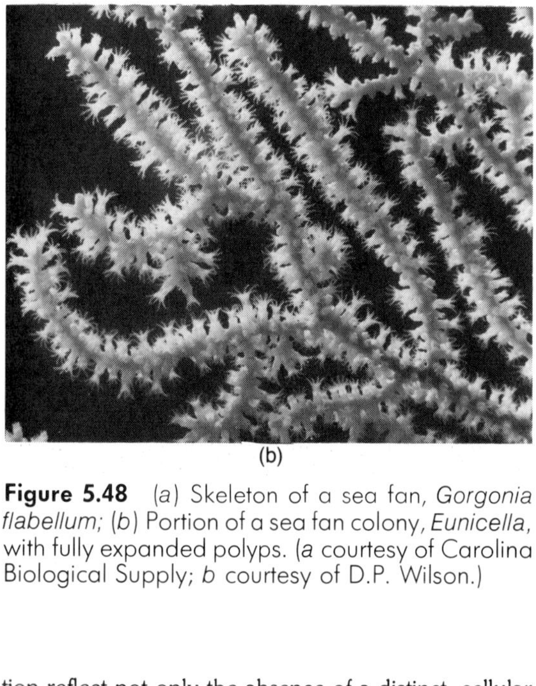

(b)

Figure 5.48 (*a*) Skeleton of a sea fan, *Gorgonia flabellum*; (*b*) Portion of a sea fan colony, *Eunicella*, with fully expanded polyps. (*a* courtesy of Carolina Biological Supply; *b* courtesy of D.P. Wilson.)

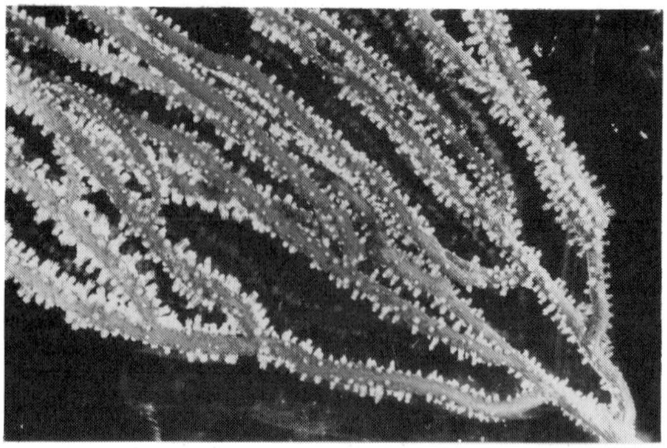

Figure 5.49 A sea whip, *Leptogorgia virgulata*, another member of the order Gorgonacea. The polyps are white while the skeleton ranges from purple to scarlet, orange or yellow. (Photo courtesy of Jack Rudlow.)

tion reflect not only the absence of a distinct, cellular mesoderm, but also their basic radial symmetry. When an animal group retains radial symmetry, any new adaptational structure must be repeated along each line of symmetry. Such repeated structures apparently are difficult to evolve.

The Ctenophores

The second radiate phylum, the Ctenophora, represents a small group of marine organisms commonly known as comb jellies. These animals are distributed throughout the world's oceans. Only 90 species have been identified.

Comb jellies or ctenophores are related to the coelenterates and sometimes are classified with them as a subphylum. Ctenophores exhibit the biradial symmetry seen in certain anthozoans, and a coelenteron is their only body cavity. Histological similarities include the presence of epidermal and gastrodermal layers separated by a gelatinous mesoglea, the latter often forming the bulk of the animal. However, the ctenophores have had a long and independent evolution, during which several distinctly ctenophoran characters have developed. Foremost among these are eight meridional rows of ciliated plates and a distinctive developmental form, the **cydippid** larva. Ctenophores are monomorphic; alternating polypoid and medusoid stages are unknown in the phylum.

Ctenophores display a variety of shapes and sizes, but the typical body form is ovoid or spheroidal. *Pleurobrachia*, the sea walnut (Fig. 5.52), represents a typical body plan. Two poles are defined: the oral pole bearing the mouth and the aboral pole topped with a complex **apical sense organ.** From this apical organ, eight radially arranged **comb rows** extend over the body surface toward the oral pole. Each comb row comprises many parallel plates of long, partially fused cilia (Fig. 5.53). Also, at opposite sides of the ctenophoran body, two long tentacles emerge from paired, ciliated pits (Fig. 5.52 and Fig. 5.55). Called tentacular sheaths, these pits represent invaginations of the epidermis. The positioning of the two

Figure 5.50 Skeleton of an organ pipe coral, *Tubipora musica*, (Order Stolonifera). (Courtesy of Carolina Biological Supply.)

tentacles, together with the slitlike mouth, imparts biradial symmetry to the body.

Ctenophoran tentacles have a muscular core (allowing contraction into the tentacular sheath) and an epidermis containing adhesive cells called **colloblasts** (Fig. 5.54a). Colloblasts comprise a hemispherical mass of secretory granules which is attached to the core of the tentacle by a spiral filament coiled around a straight filament. The straight filament represents the highly modified nucleus of the colloblast cell. At least one ctenophore, *Euchlora rubra*, possesses nematocysts. However, for most members of the phylum, colloblasts assume nematocystlike functions in prey capture and defense.

Ctenophores trap zooplankton with the sticky colloblasts of their tentacles. The food-laden tentacles then are wiped across the mouth. The mouth opens into a flattened, tubular pharynx, or stomodeum, where gland cell secretions initiate extracellular digestion. The resulting nutrient broth is transported directly to the stomach. The ctenophoran stomach gives rise to an elaborate system of gastrovascular canals (Fig. 5.55). Two pharyngeal canals recurve and lie parallel to the long axis of the pharynx. Two transverse canals pass at right angles to the pharyngeal canals and shortly divide into three branches. The middle branch on each side leads to the base of the tentacular sheath. Lateral branches, two on each side, bifurcate to form a grand total of eight canals, which join eight meridional canals, one beneath each of the eight comb rows. Finally, an aboral canal passes from the stomach to

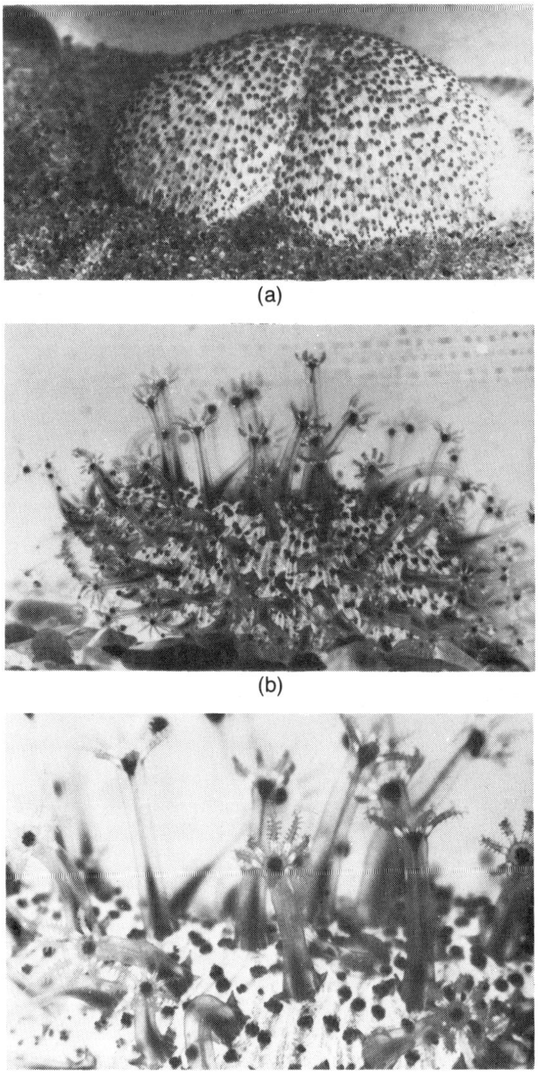

Figure 5.51 The sea pansy. *Renilla* (Order Pennatulacea). (*a*) Whole colony, polyps contracted; (*b*) Whole colony, polyps expanded; (*c*) Close-up of expanded polyps showing the eight feathery tentacles characteristic of alcyonarians. (Photos courtesy of Ward's Natural Science Establishment.)

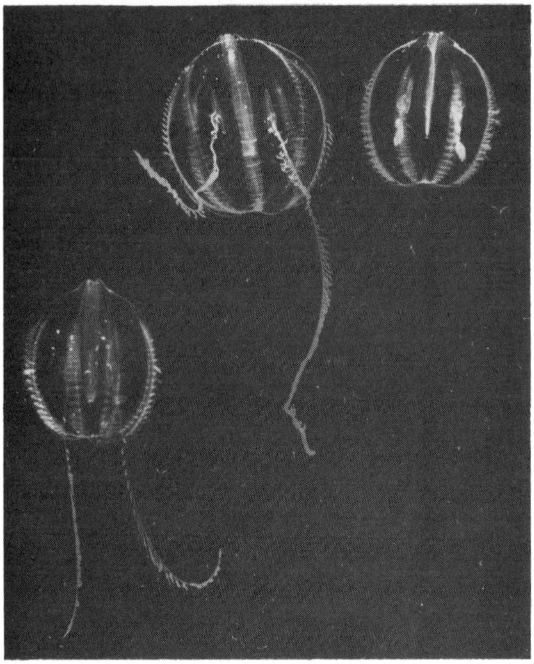

Figure 5.52 Photograph of *Pleurobrachia*, the sea walnut. (Courtesy of D.P. Wilson.)

the aboral pole, where it divides into four short canals, two ending blindly and two opening to the outside via anal pores near the apical sense organ. Within this gastrodermal canal system, absorption and intracellular digestion occur. Like the gastrovascular cavity of coelenterates, the system provides for internal transport and the canal walls are major respiratory surfaces.

Ciliated structures called **cell rosettes** (Fig. 5.54b) surround pores leading from the gastrodermal canals into the mesoglea. Cell rosettes may control traffic between these two areas, and some workers believe that such control is excretory and/or osmoregulatory in nature.

The ctenophoran nervous system comprises a subepidermal plexus which has condensed from a diffuse nerve net. It is thus comparable to the system found in higher coelenterates. Nervous elements are well developed below the apical sense organ and beneath each of the comb rows. The comb rows are responsible for the locomotion of these animals. Their cilia beat in an organized fashion to propel the ctenophore along. The apical sense organ is a balancing center apparently coordinating the comb rows. It consists of a mass of calcareous particles (the statolith) supported by four long tufts of cilia acting as springs (Fig. 5.53). From each spring, a pair of ciliated grooves pass outward and connect with the two comb rows of that quadrant. Tilting of the animal causes the statolith to press more heavily on one of the ciliary springs. The

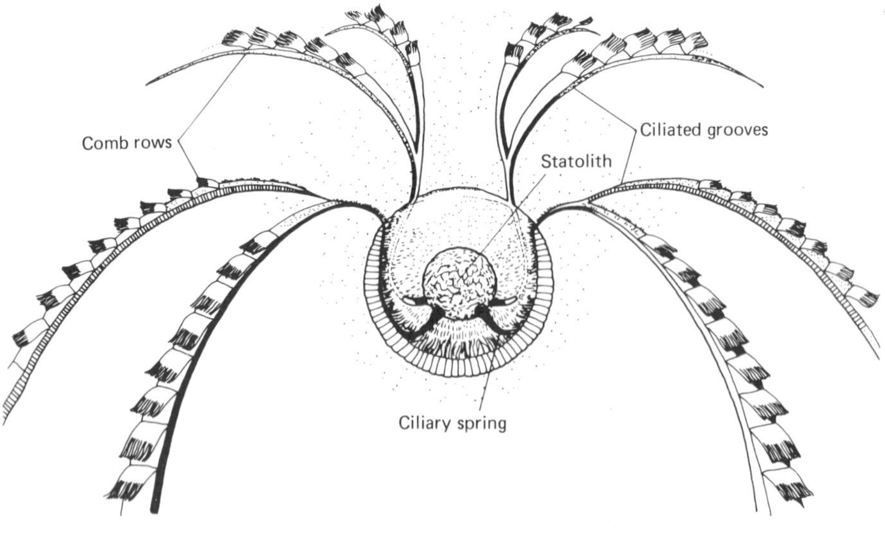

Figure 5.53 Diagram of the aboral sense organ of a ctenophore from which the eight comb rows take their origin. (From F.M. Bayer and H.B. Owre, *The Free-Living Lower Invertebrates*, 1968, Macmillan.)

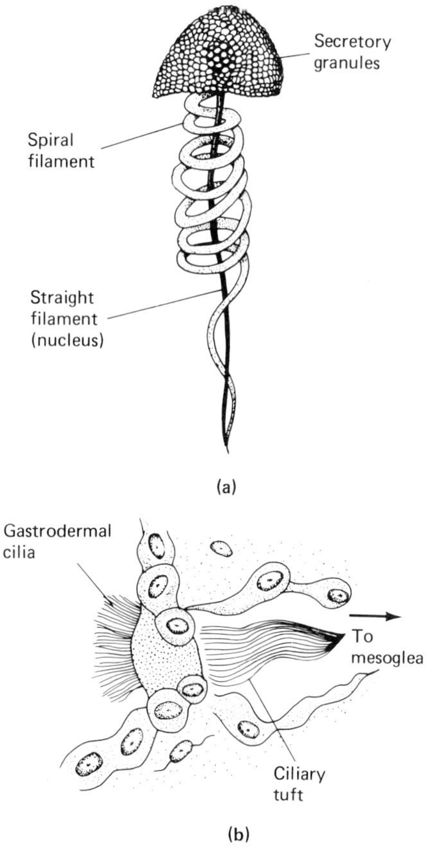

Figure 5.54 (*a*) A colloblast from a ctenophore tentacle. (*b*) A cell rosette from *Coeloplana*. (*a* from Bayer and Owre, *The Free-Living Lower Invertebrates; b* from Hyman, *The Invertebrates, Vol. I.*)

resulting stimulus is transmitted via the ciliated grooves to the adjoining comb rows, eliciting the vigorous beating of their ciliated plates. Such activity restores upright posture. Removal of the apical sense organ does not stop ciliary beating, but does prevent coordination among the comb rows.

Ctenophoran muscles differ from their coelenterate counterparts in that the former arise from amoeboid cells in the mesoglea. The ctenophoran musculature is contained largely within the mesoglea and is developed best around the mouth and pharynx. Nervous elements are associated with these muscles. The ctenophoran mesoglea is thus a more active layer than that of most coelenterates.

Luminescence is a widespread phenomenon among ctenophores. This curious biological process

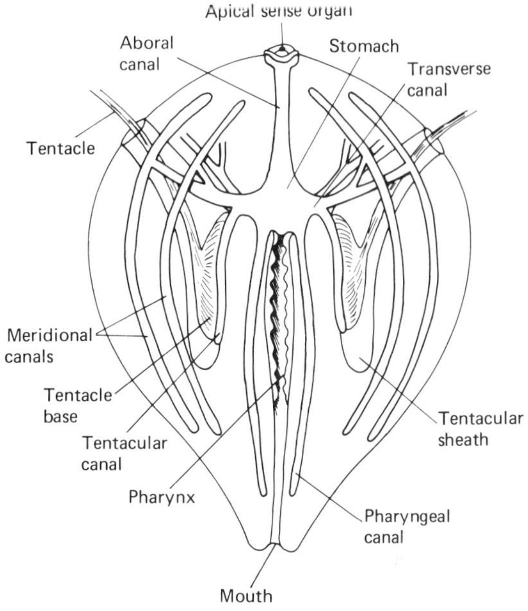

Figure 5.55 A diagram illustrating the gastrovascular canal system of a cydippid ctenophore. (From Hyman, *The Invertebrates, Vol. I.*)

occurs along the meridional canals. Since most ctenophores are transparent, color and light displays are particularly conspicuous. However, the adaptive significance of such displays remains obscure. (See Chap. Two, Special Essay.)

Ctenophores are hermaphrodites. Gonads are located in the walls of their meridional canals. Usually gametes exit through the mouth, but in some species short sperm ducts may lead to the aboral surface. Fertilization usually takes place in the sea. Cleavage is determinate (recall that coelenterate cleavage is indeterminate), and a solid blastula and gastrula are formed. A free-swimming larva called a *cydippid* develops (Fig. 5.57). The pharynx and tentacular sheaths arise from ectodermal invaginations, while four ciliated bands appear on the body surface. The latter divide to produce the eight comb rows typical of the phylum. Some ctenophores can become sexually mature as larvae under appropriate conditions. The gonads later regress and reappear in the adult stage.

With the exception of a few very specialized bottom-dwelling forms, asexual reproduction seems to be absent from the phylum. However, a capacity for regeneration is well developed; thus ctenophores are an exception to the general rule that asexual repro-

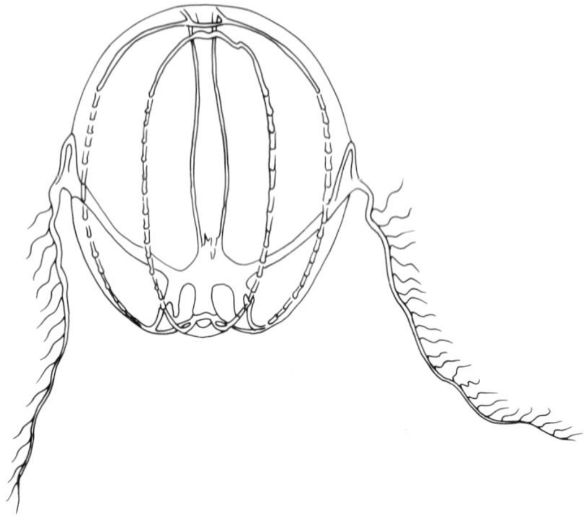

Figure 5.56 A cydippid larva of a ctenophore, *Bolinopsis*.

duction and regenerative capacity occur in the same animals.

Most ctenophores are planktonic in the open sea or inshore waters. Some genera have modified the basic body form exhibited by *Pleurobrachia*. *Mnemiopsis* (Fig. 5.57a), a common form along the Eastern and Gulf Coasts of North America, possesses two large oral lobes and four other ciliated projections around the mouth. Some ribbon-shaped ctenophores are flattened along the axis joining the tentacular sheaths. These animals include *Cestum*, known as Venus's girdle, and *Velamen* (Fig. 5.57b). *Cestum* may attain a length of 1 meter and swims with graceful undulations. Other ctenophores, notably *Coeloplana* (Fig. 5.57c), are flattened orally–aborally and have assumed a bottom-crawling existence. Superficial resemblances between *Coeloplana* and some turbellarian flatworms have stimulated speculation that the acoelomate animals arose from crawling ctenophores. However, *Coeloplana* and related genera are specialized ctenophores (many are commensals with alcyonarians and echinoderms) and thus are poor candidates for the ancestors of a primary line in animal evolution.

All ctenophores mentioned thus far are members of the class Tentaculata. A much smaller class, the class Nuda, includes ctenophores without tentacles. *Beroe* (Fig. 5.57d), a large, flattened, bell-shaped form with an enormous mouth and pharynx, is a well-known member of this class.

The Ctenophora probably diverged long ago from the ancestral medusoid coelenterate. Ctenophoran similarities with hydrozoan medusae further encourage the view that the latter are the oldest coelenterate form. However, to confuse the issue, similarities between ctenophore and anemone embryos suggest an affinity between the Ctenophora and the Anthozoa. In view of the differences between ctenophores and coelenterates, specifically those involving early embryonic development, we must conclude that the two phyla have had a long, independent evolutionary history. With our present knowledge, it is impossible to determine the point of origin of ctenophores from coelenterates with any degree of certainty.

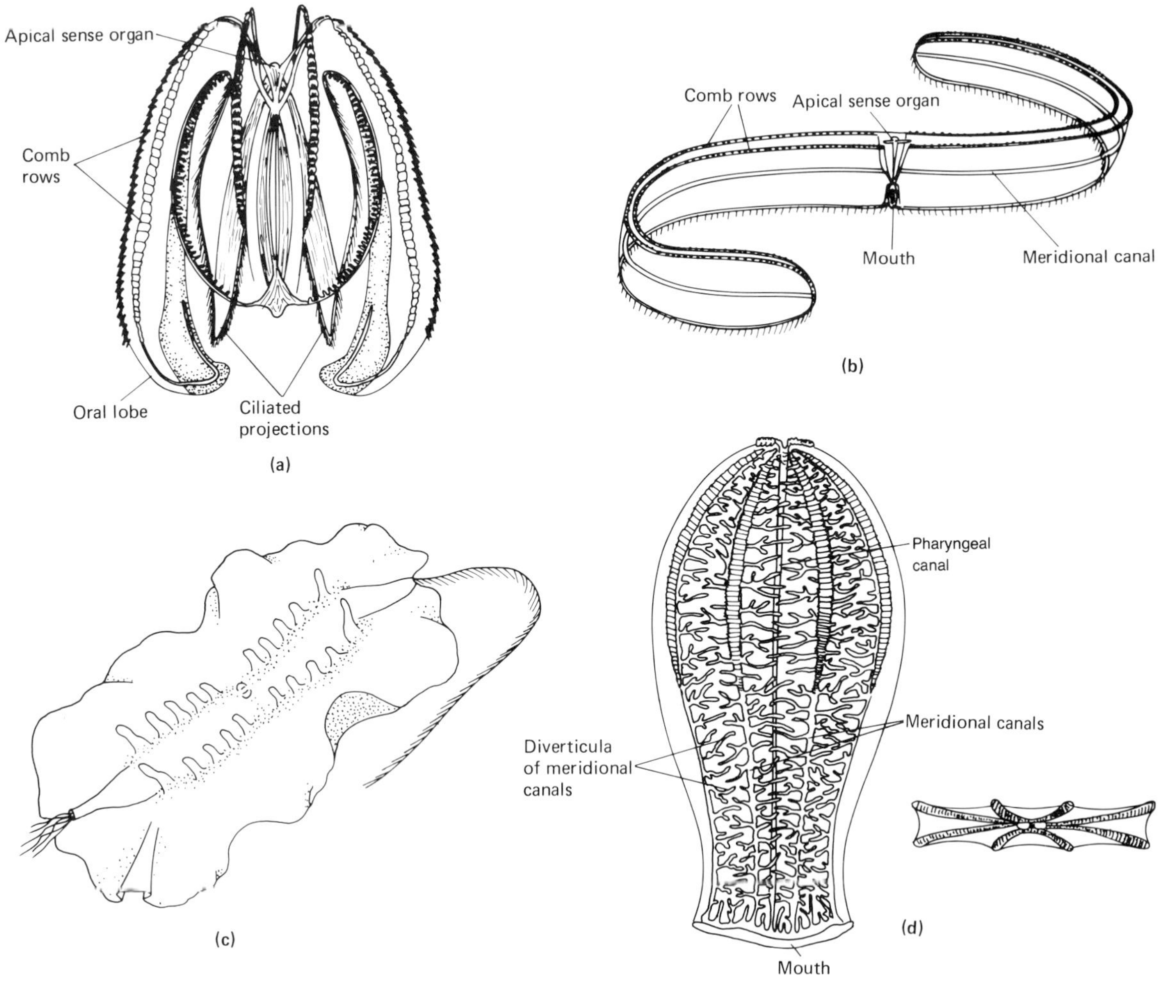

Figure 5.57 (*a*) *Mnemiopsis* (Order Lobata); (*b*) *Velamen* (Order Cestida); (*c*) *Coeloplana* (Order Platyctenea); (*d*) *Beroe*, an atentaculate ctenophore. *Right:* aboral view. *Left:* lateral view. (*a*, *b* and *d* from Hyman, *The Invertebrates*, *Vol. I*; *c* from C. Dawydoff, Arch. Zoo. Exp. et Gen., 1938, *80*:125.)

Mutualism: A Coelenterate Example

Invertebrates relate to each other in many different ways. Predators and parasites seek other animals for food and/or residence, while prey and host organisms avoid these would-be exploiters. Within their own species, animals may seek companionship and sex, but many membrs of the invertebrate world simply ignore each other. Meanwhile true cooperation between individuals of different species is surprisingly common.

Two animals of different species can pursue a relationship directly and simultaneously benefiting both of them. Such a relationship is called mutualism and was encountered earlier between termites and *Trichonympha* and between sponges and hermit crabs. Among coelenterates, mutualistic relationships are abundant and include associations between sea anemones and hermit crabs and between corals and many reef organisms. One of the most intimate and fascinating of all mutualistic relationships involves the reef-building corals and the photosynthetic, unicellular organisms called zooxanthellae.

Zooanthellae have been identified as the palmella stage of certain dinoflagellates. One species, ***Gymnodinium microadriaticum,*** resides within the gastrodermis of reef-forming polyps, where it accrues obvious benefits of shelter and support. The advantage to the coral is somewhat less obvious. Certain biological molecules (small carbohydrates and amino acids) pass from the zooxanthellae to the coral and are probably important nutritionally to many coral species. However, the suggestion that the polyps consume entire zooanthellae is discouraged because of the carnivorous habits of these coelenterates. Indeed, if starved in the dark, polyps often discharge their mutualistic partners. Evidently, light and/or a well-nourished coral is vital to the relationship. Another proposal, that zooxanthellae "pay their rent" with photosynthetically produced oxygen, also is not supported. Induced variations of 50 to 122% in local oxygen pressure fail to influence coral metabolism; moreover, polyps stripped of their zooxanthellae do not develop oxygen debts. Apparently, these coelenterates are well adapted to oxygen levels in the sea and need no auxiliary source.

Another theory is more tenable. A distinct correlation exists between the conditions favoring photosynthesis by zooxanthellae and reef building by corals. Both processes thrive only in the euphotic, upper 30 m of the ocean. The deposition of calcium carbonate occurs 10 times faster by day than by night, and a cloudy sky can reduce the rate of reef formation by half. The stimulative effect of light is absent from all corals lacking zooxanthellae, and such corals do not form reefs. It appears that the contribution of the zooxanthellae is made in light (i.e., while most actively engaged in photosynthesis) and is related directly to reef formation.

Reef formation by the deposition of calcium carbonate is thought to involve the following reactions:

$$Ca^{2+} + 2HCO_3^- \rightleftarrows Ca\,(HCO_3)_2 \rightleftarrows CaCO_3 + H_2CO_3$$

$$H_2CO_3 \uparrow\downarrow H_2O + CO_2\uparrow$$

Calcium and bicarbonate ions from sea water are converted into the calcium carbonate of the reef. As in all equilibrium reactions, the rate of this process is controlled in part by the removal of the end products. Here the zooxanthellae play a critical role, as their continuous uptake of carbon dioxide for photosynthesis shifts the equilibrium of the reaction to the right. Zooxanthellae also may aid in recycling phosphates and nitrogenous wastes within the reef. This recycling possibly contributes to the ability of many corals to survive in nutrient-poor water.

Thus coral and zooxanthellae live together and contribute to each other's welfare. The invertebrate world encompasses many such mutualistic associations. Following the dictates of natural selection, it is thus a world not only of predation and parasitism but also of cooperation between species for their mutual benefit.

for Further Reading

Arai, M. N. and Brinckmann-Voss, A. Hydromedysae of British Columbia and Puget Sound, Canada. *Can. Bull. Fish. Aquat. Sci.* 204: 1–192, 1980.

Bayer, E. M., and Owre, H. B. *The Free-Living Lower Invertebrates.* pp. 25–143, Macmillan, New York, 1968. (A useful account of coelenterates and ctenophores with very fine illustrations)

Benos, D. J., and Prusch, R. D. Osmoregulation in freshwater *Hydra. Comp. Biochem. Physiol.* 43A: 165–171, 1972.

Biggs, D. C. Field studies of fishing, feeding and digestion in siphonophores. *Mar. Behav. Physiol.* 4: 261–274, 1977.

Burnet, A. L. (Ed.) *Biology Of Hydra.* Academic Press, New York, 1973. (A collection of papers on many aspects of *Hydra* research)

Cairns, S. Guide to the commoner shallow-water gorgonians of Florida, the Gulf of Mexico, and the Caribbean region. Sea Grant Field Guide Series #6, University of Miami, 1976.

Campbell, R. D. Tissue dynamics of steady state growth in *Hydra littoralis III.* Behavior of specific cell types during tissue movements. J. Exp. Zool. 164: 379–391, 1967.

Campbell, R. D. Cnidaria. In *Reproduction in Marine Invertebrates, Vol. I.* (Giese, A. C. and Pearse, J. S., eds.) pp. 133–200, Academic Press, New York, 1974.

Carlgren, D. A Survey of the Ptychodactiaria. Corallimorpharia and Actiniaria. *K. Sven. Vetenskapsakad.* Handl. Ser. 4 1L No. 1, 1949. (A monograph on the sea anemones of the world)

Carre, D. Hypothesis on the mechanism of cnidocyst discharge. *Eur. J. Cell. Biol.* 20: 265–271, 1980.

Carre, C. and Carre, C. Triggering and control of cnidocyst discharge. *Mar. Behav. Physiol.* 7: 109–117, 1980.

Chamberlain, J. A. Mechanical properties of coral skeleton: Compressive strength and its adaptive significance. *Paleobiology.* 4: 419–435, 1978.

Devaney, D. M. and Eldgredge, L. G. (Eds.) *Reef and Shore Fauna of Hawaii, Vol. I.* Bishop Museum Press, Honolulu, 1977.

Florkin, M., and Scheer, B. T. (Eds.) *Chemical Zoology, Vol. II: Porifera, Coelenterata, and Platyhelminthes.* pp. 81–284, Academic Press, New York, 1968. (Six chapters on selected topics relating to coelenterate biology and physiology. The chapter by H. M. Lenhoff on the feeding response, digestion, and nutrition of coelenterates is particularly noteworthy.)

Franc, J. M. Organization and function of ctneophore colloblasts: an ultrastructural study. *Biol. Bull.* 155: 527–541, 1978.

Fraser, C. *Hydroids of the Atlantic Coast of North America.* University of Toronto Press, Toronto, 1954. (A monograph on the American North Atlantic hydroids)

Gierer, A. Hydra as a model for the development of biological form. *Sci. Am.* 231 (6): 44–54, 1974.

Hand, C. On the origin and phylogeny of the coelenterates. *Syst. Zool.* 8: 191–202, 1959.

Horridge, G. A. Relations between nerves and cilia in ctenophores. *Am. Zool.* 5: 357–375, 1965.

Hyman, L. H. *The Invertebrates, Vol. I: Protozoa through Ctenophora.* pp. 365–696, McGraw-Hill, New York, 1940. (The standard reference work on the biology of coelenterates and ctenophores; includes an extensive bibliography. The chapter "Retrospect" in Vol. V summarizes investigations on coelenterates and ctenophores from 1938–1958 and includes current ideas concerning invertebrate phylogeny.)

Jeffrey, S. W., and Haxo, F. T. Photosynthetic pigments of symbiotic dinoflagellates (zooxanthellae) from corals and clams. *Biol. Bull. Woods Hole* 135: 149–165, 1968.

Jennison, B. L. Gametogenesis and reproductive cycles in the sea anemone *Anthopleura elegantissima. Can. J. Zool.* 57: 403–411, 1979.

Jones, O. A., and Endeau, R. (Eds.) *Biology and Geology of Coral Reefs.* Academic Press, New York, 1973. (A massive reference work in four volumes on the biology and geology of coral reefs; papers by a number of specialists)

Kaestner, A. *Invertebrate Zoology, Vol. I.* pp. 43–153, Wiley Interscience, New York, 1967.

Kramp, P. L. Synopsis of the medusae of the world. *J. Mar. Biol. Assoc.* (U.K.) 40: 1–469, 1961. (Brief descriptions of all known hydromedusae and scyphomedusae; extensive bibliography)

Lane, C. E. The Portuguese man-of-war. *Sci. Am.* 202: 158–168, 1960.

Lang, J. Interspecific aggression by scleractinian corals; Why the race is not only to the swift. *Bull. Mar. Sci.* 23: 260–279, 1973.

Lenhoff, H. M., and Loomis, W. L. F. (Eds.) *The Biology of Hydra and of Some Other Coelenterates.* University of Miami Press, Coral Gables, Fla., 1961. (A collection of 24 papers presented at a symposium on the physiology and ultrastructure of *Hydra* and other coelenterates with particularly good coverage of nematocyst structure and function, feeding, and nutrition in *Hydra,* and development and regeneration in *Hydra* and other coelenterates.)

Lenhoff, H. M., Muscatine, L., and Davis, L. V. (Eds.) *Experimental Coelenterate Biology.* University of Hawaii Press, Honolulu, 1971. (A collection of 25 papers representing the results of research at the Hawaii Institute of Marine Biology. Topics covered include: growth and development, feeding behavior, food transport and metabolism, endosymbiosis (mutualism) with algae and calcification.)

Lentz, T. L. Hydra: Induction of supernumerary heads by isolated neurosecretory granules. *Science* 150: 633–635, 1965.

Lewis, D. H., and Smithe, D. C. The autotrophic nutrition of symbiotic marine coelenterates with special reference to hermatypic corals. *Proc. Roy. Soc. London Ser. B* 178: 11–129, 1971.

Lewis, J. B. and Price, W. S. Feeding mechanisms and feeding strategies of Atlantic reef corals. *J. Zool.* 176: 527–544, 1975.

Lubbock, R. Chemical recognition and nematocyte excitation in sea anemone. *J. Exp. Biol.* 83: 283–292, 1979.

Mackie, G. O. (Ed.) *Coelenterate Ecology and Behavior,* Plenum Publishing Corporation, New York, 1976. (A series of original research papers and reviews from the 3rd International Symposium on Coelenterate Biology. Emphasizes ecology, reproductive biology, functional morphology, and behavioral physiology.)

Miller, R. L., and Wyttenbach, C. R. The developmental biology of the Cnidaria. *Am. Zool.* 14: 437–866, 1974. (Proceedings of a symposium held in Washington, D.C., in December 1972. Up-to-date coverage of such topics as gametogenesis, fertilization, embyogenesis, cellular activities during growth and regeneration, morphogenetic aspects of growth in Hydra and colonial hydroids, growth and regeneration in scyphozoans)

Muscatine, L., and Lenhoff, H. M. (Eds.) *Coelenterate Biology: Reviews and New Perspectives.* Academic Press, New York, 1974.

Muscatine, L., Pool, R. R., and Trench, R. K. Symbiosis of algae and invertebrates: aspects of the symbiont surface and the host-symbiont interface. *Trans. Am. Microsc. Soc.* 94: 450–469, 1975.

Newell, N. D. The evolution of reefs. *Sci. Am.* 226: 54–65, 1972.

Oliver, W. A. The relationship of the scleractinian corals to the rugose corals. *Paleobiology* 6: 146–160, 1980.

Pennak, R. W. *Fresh-Water Invertebrates of the United States,* 2nd ed., pp. 99–113, Wiley-Interscience, New York, 1979. (A treatment of freshwater hydrozoans with a key to North American species)

Porter, J. W. Autotrophy, heterotrophy, and resource partioning in Caribbean reef-building corals. *Am. Nat.* 110: 731–742, 1976.

Rees, W. J. (Eds.) *The Cnidaria and their Evolution.* Symposia of the Zoological Society of London, No. 16. Academic Press, New York, 1966. (A collection of 17 papers on various aspects of coelenterate biology. Eight papers deal with coelenterate evolution and phylogeny.)

Reeve, M. R., Walter, M. A., and Ikeda, T. Laboratory studies of ingestion and food utilization in lobate and tentaculate ctenophores. *Limnol. Oceanogr.* 23: 740–751, 1978.

Rose, S. M. *Regeneration.* Appleton-Century-Crofts, New York, 1970. (One chapter is devoted to regeneration in coelenterates.)

Ross, D. M. Behavioral and ecological relationships between sea anemones and other invertebrates. *Annu. Rev. Mar. Biol. Oceanogr.* 5: 291–316, 1970.

Schaller, H. and Gierer, A. Distribution of the head activating substance in *Hydra* and its localization in membranous particles in nerve cells. *J. Embryol. Exp. Morphol.* 29: 39–52, 1973.

Sebens, K. P. The regulation of asexual reproduction and indeterminate body size in the sea anemone *Anthopleura elegantissima. Biol. Bull.* 158: 370–381, 1980.

Singla, C. L. Statocysts of hydromedusae. *Cell. Tiss. Res.* 158: 391–407, 1975.

Stimson, J. S. Mode and timing of reproduction in some common hermatypic corals of Hawaii and Enewetak. *Mar. Biol.* 48: 173–184, 1978.

Stoddart, D. E. Ecology and morphology of recent coral reefs. *Biol. Rev.* 44: 433–498, 1969.

Taylor, D. L. The cellular interactions of algal-invertebrate symbiosis. *Adv. Mar. Biol.* 11: 1–56, 1973.

West, D. A. The epithelio-muscular cell of hydra: Its fine structure, three-dimensional architecture and relationship to morphogenesis, *Tissue Cell.* 10: 629–646, 1978.

West, D. A. Symbiotic zoanthids of Puerto Rico. *Bull. Mar. Sci.* 29: 253–271, 1979.

Westfall, J. A. Ultrastructural evidence for neuromuscular systems in coelenterates. *Am. Zool.* 13: 237–246, 1973.

Westfall, J. A., Yamataka, S., and Enos, P. D. Ultrastructural evidence of polarized synapses in the nerve net of *Hydra. J. Cell Biol.* 51: 318–323, 1971.

Yonge, C. M. Ecology and physiology of reef building corals. In *Perspectives in Marine Biology.* (A. A. Buzzati-Traverso, Ed.), pp. 117–135. University of California Press, Berkeley, 1958.

Yonge, C. M. The biology of coral reefs. *Adv. Mar. Biol.* 1: 209–260, 1963.

Yonge, C. M. Living corals. *Proc. Roy. Soc. London Ser. B* 169: 329–344, 1968.

6

The Acoelomate Animals: Flatworms and Nemertines

Radial symmetry is not consistent with an active life-style. Active life-styles are associated with bilateral symmetry, and since natural selection often favors active, aggressive creatures, most members of the animal world are bilateral. Indeed, except for some protozoans, the sponges, and the radiate phyla, all invertebrates display primary bilateral symmetry. These animals may be grouped into a superphylum, the Bilateria.

Bilateral symmetry has evolved in conjunction with increased mobility. The machinery of locomotion (nervous, muscular, sensory, and skeletal) became far more efficient when adaptations occurred to produce movement in one general direction. Muscle tissues proliferated and body form became more streamlined. Moreover, the anterior end developed neurosensory organs due to selective pressures generated by the active animal's expanding need for information. This anterior concentration of sensory and nervous tissues describes **cephalization,** or head formation. The result, a mobile beast with a head, is basically descriptive of most animals today.

All members of the Bilateria are triploblastic; that is, they possess three embryonically formed germ layers. Ectoderm and endoderm are present, as in the radiates. The germ layer unique to bilateral animals is mesoderm. This intermediate germ layer forms muscles and organ systems for circulation, respiration, excretion, osmoregulation, and reproduction.

This chapter is concerned with the flatworms and nemertines, the most primitive animals in the Bilateria. These are generally small, secretive creatures, many of which are ocean bottom dwellers. Flatworms are the first metazoan group in which parasitism is common, and two classes, the flukes and the tapeworms, have achieved great success with the parasitic way of life. Nemertines are a minor group of long, fragile worms with an eversible proboscis. Flatworms and nemertines are distinguished from all other bilateral phyla in possessing no secondary body cavity. Lacking a coelom, these animals are referred to as acoelomate. Their mesoderm is a solid, cellular layer between the ectoderm and the endoderm; such a solid mesoderm is called a **parenchyma.** Embedded within this parenchyma are true reproductive and excretory–osmoregulatory organ systems. Mesodermal muscles are well developed in both phyla, and the nemertines alone possess a rudimentary circulatory system. The acoelomates thus demonstrate an organ-system level of organization. This organization is shared by all higher invertebrates.

Despite their evolutionary advances, free-living acoelomates enjoy only limited success. Their limitations are usually attributed to metabolic problems associated with a solid mesoderm (*see* the following). It has been suggested that the high frequency of parasitism among flatworms reflects an adaptive "awareness" of these metabolic problems and an attempt to live under a different set of physiological rules.

The Flatworms

Members of the phylum Platyhelminthes, commonly known as flatworms, are the most primitive creatures in the Bilateria. About 15,000 species are described. As implied by their common name, most flatworms are dorsoventrally thin animals. Some, such as land planarians (Fig. 6.1), are long and vermiform with a distinct head set off by lateral projections called auricles; other flatworms are oval or leaf-shaped and have no clearly marked head (Fig. 6.2). Flatworms are the most primitive animals with true organ systems. They possess a specialized excretory-osmoregulatory organ and a complex, typically hermaphroditic reproductive system. The simple digestive tract lacks an anus and is reminiscent of that observed among anthozoan coelenterates. Mesodermal muscles as well as a submuscular nervous system are usually well developed. Compared with most radiate animals, flatworms are quite active.

The phylum is rather diverse, largely because of its parasitic forms. Three classes are recognized: the Turbellaria, the Trematoda or flukes, and the Cestoda, commonly known as tapeworms. Turbellarians include all the free-living flatworms and are certainly the most primitive members of the phylum. They are largely marine in distribution and usually crawl along the bottom ooze, under rocks or on seaweed. Some members are pelagic, and others (e.g., the planarians) have invaded fresh water, where they are always bottom dwellers. Although a few groups have colonized the terrestrial environment, they are confined largely to moist forest floors in tropical and subtropical areas. Ancient turbellarians almost certainly gave rise to trematodes and cestodes, all of which are parasitic.

Parasitism is a specialized condition requiring extraordinary adaptations in body form, physiology, and behavior. The Special Essay concluding this chapter discusses these adaptations and should contribute to an understanding of the considerable divergence displayed by parasitic flatworms. Flukes and tapeworms are, in fact, so highly specialized that they no longer possess many typically flatworm characters. For this reason, our discussion of flatworm unity is limited to the turbellarians. While turbellarians are themselves a diverse group, a description of these free-living worms outlines the basic plan of the phylum. Our discussion of platyhelminth diversity then describes the alterations made in this basic turbellarian plan as it was adapted to parasitic life-styles.

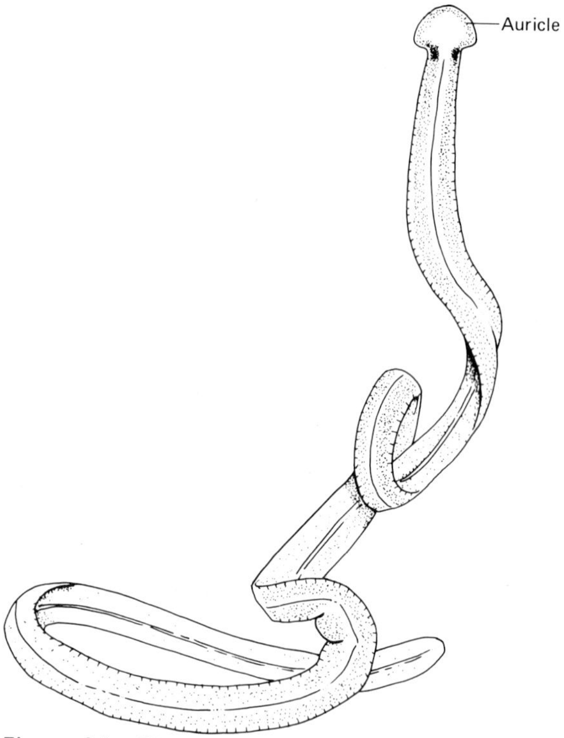

Figure 6.1 *Bipalium kewense*, a planarian found in greenhouses and damp localities in mild climates. Yellowish olive, with dark longitudinal stripes, it may reach a length of 35 cm. (From L.H. Hyman, *The Invertebrates, Vol. II: Platyhelminthes and Rhynchocoela*, 1951, McGraw-Hill.)

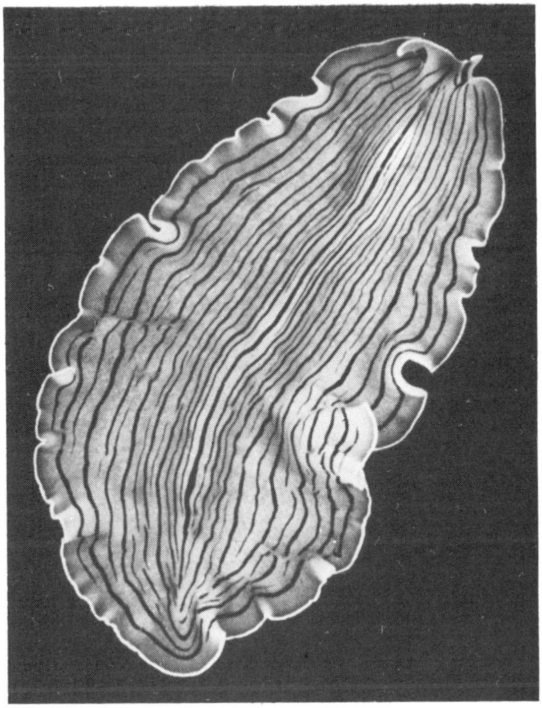

Figure 6.2 A marine polyclad flatworm, *Prostheceraeus vittatus*, about 3 cm long. (Photo courtesy of D.P. Wilson.)

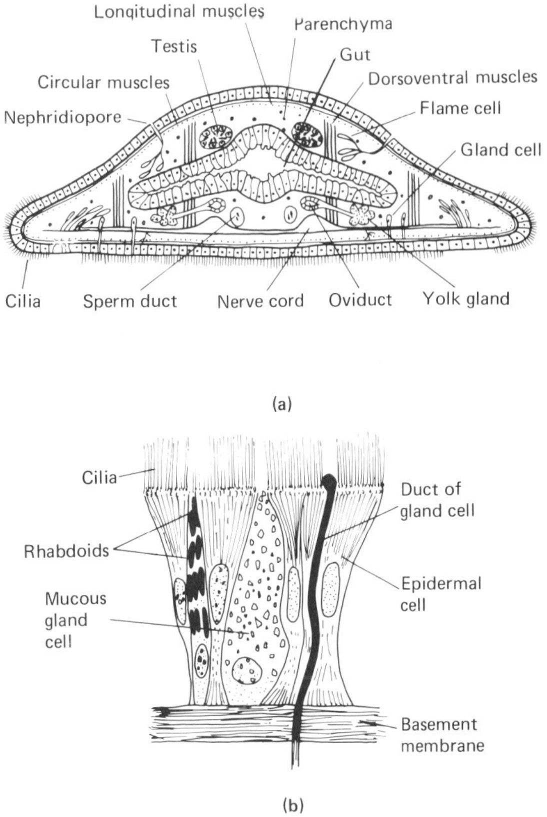

Figure 6.3 (*a*) A diagrammatic cross section through a generalized turbellarian showing body wall layers and major internal organs; (*b*) A section through the epidermis of a polyclad turbellarian. (*a* from R. Buchsbaum, *Animals Without Backbones: An Introduction to the Invertebrates*, rev. ed., 1976, University of Chicago; *b* from Hyman, *The Invertebrates, Vol. II.*)

Flatworm Unity

MAINTENANCE SYSTEMS

General Morphology.

In schematic fashion, Figure 6.3a depicts a cross section through a generalized turbellarian. The epidermis is ciliated primitively and may form a syncytium. Cilia are best developed ventrally and are limited to that surface in many larger species. The external surface often bears suckers or other adhesive organs. Large gland cells secrete mucus along which the worm glides. Such gland cells may be located in the mesoderm, in which case their long, narrow ducts lead to the epidermal surface (Fig. 6.3b). These ducts may project slightly from the epidermis, where in cooperation with the sticky secretions of other gland cells, they help the worm adhere to the substratum. Rod-shaped epidermal bodies known as **rhabdoids** (Fig. 6.3b) are peculiar to free-living flatworms. Rhabdoids are secreted by epidermal gland cells, but their function remains unknown. One theory holds that rhabdoids produce a capsule that, upon release, secretes a protective film over the flatworm's body. Some authorities have suggested a defensive function for these structures.

A noncellular basement membrane divides the epidermis from the mesodermal tissues (Fig. 6.4). Immediately beneath this membrane are circular, diagonal, and longitudinal muscles (Fig. 6.4), some of which may be anchored to the basement membrane itself. Below these muscle layers are well-developed longitudinal nerve cords. These cords are arranged

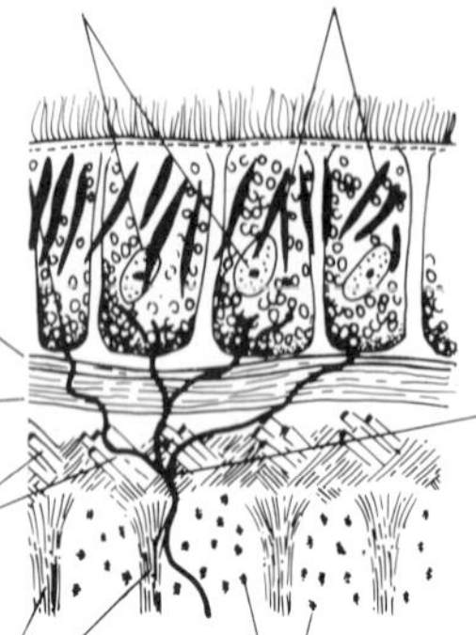

Figure 6.4 A section through the body wall of the terrestrial triclad *Geoplana* showing the epidermis and the arrangement of the muscle layers. From L. von Graff, *Monographie der Turbellarien*, Leipzig, 1882.)

radially in many species, but a marked trend toward a bilateral arrangement occurs in higher turbellarians (Fig. 6.8). In primitive members of the class, some muscular and nervous elements persist in association with the epidermis. Muscle fibers may originate from epidermal cells, as in the lower coelenterates, and vestiges of a coelenterate-type nerve net remain even in advanced turbellarians. To some specialists, these conditions suggest a close evolutionary relationship between turbellarians and coelenterates.

The rest of the mesoderm comprises complex parenchymal muscles and tightly packed cells (Figs. 6.3a and 6.4). Small intercellular spaces do exist, however, and some wandering amoebocytes may be present. The parenchymal musculature is developed best in larger flatworms. Muscle fibers are arranged in diagonal, dorsoventral, transverse, and longitudinal patterns. Differentiated among the mesodermal mass is the reproductive system, a complicated and highly variable apparatus featuring, in nearly every case, both egg- and sperm-producing structures. Also located in the mesodermal layer (although probably of ectodermal origin) is the protonephridial system, an excretory—osmoregulatory complex of ciliated bulbs and tubules. Free-living flatworms possess a simple digestive system composed of a ventral mouth, a pharynx, and a blind, often branched, intestine; there is seldom a permanent anus.

Digestion and Nutrition

Most turbellarians are carnivorous, feeding on plankton, small crustaceans, and other flatworms. Cadavers as well as live prey are common food items. A muscular pharynx (Fig. 6.5) is often protrusible through the ventral mouth, where it can seize an edible object. Paralyzing, and perhaps digesting, enzymes are secreted to subdue prey; even a muscular penis may struggle against particularly recalcitrant foodstuffs! Food is pulled through the mouth by the retraction of the pharynx and is distributed throughout the blind intestine. The importance of extracellular versus intracellular digestion varies with the species. However, the intestine does not possess separate enzyme-secreting and absorbing areas, indicating that extracellular digestion in the gut is minimal. Nutritive cells, which compose the single-layered intestinal wall, phagocytose small food particles. Digestion is completed in food vacuoles within these cells. In the absence of an anus, undigested wastes are expelled through the mouth.

The various turbellarian orders display trends toward increasingly complex pharynges and branched intestines. Three pharyngeal types exist (Fig. 6.5): The *simple* pharynx is merely a ciliated invagination of the epidermis. In more complex forms, this invagination is muscularized and protrusible. A *bulbous* pharynx features an eversible bulb suspended in a small pharyngeal cavity. The most complex turbellarian pharynx is *plicate*. A plicate pharynx consists of a long folded or ruffled tube protrusible through the mouth. At its base, it is continuous with the general body layers. This advanced pharynx is a true organ comprising nervous, glandular, and muscular tissues.

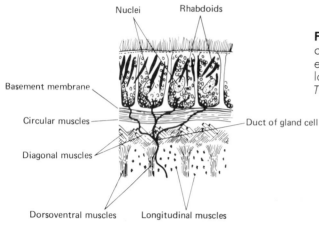

The very primitive marine acoels have no permanent digestive cavity. In this group, a short, simple pharynx opens into a digestive zone (Fig. 6.6a) consisting of nutritive tissues, which may be syncytial in some species. Temporary digestive cavities form around food particles as phagocytosis occurs. In some cases, nutritive cells may emerge through the acoel mouth to surround foodstuffs and usher them into the

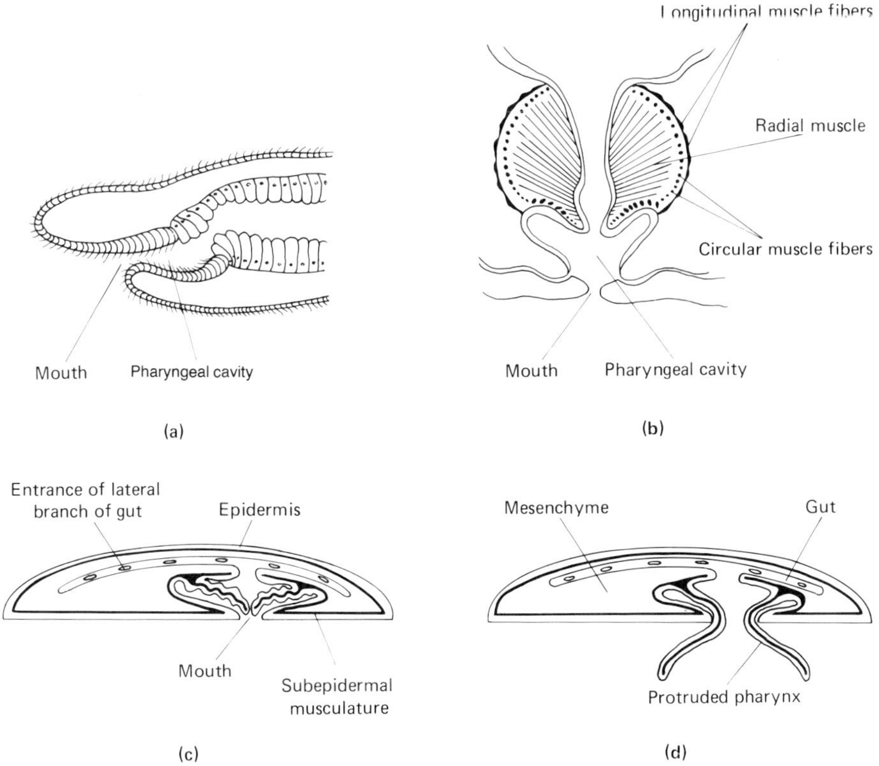

Figure 6.5 Turbellarian pharyngeal types. (*a*) Simple; (*b*) Bulbous; (*c* and *d*) Plicate: retracted in (*c*) and protruded in (*d*) (*a* from L. von Graff, *Monographie der Turbellarien*, *b* from L. von Graff, In *Klassen und Ordnungen des Tierreichs*, H.G. Bronn, Ed., *Vol. IV*, 1904, Vermes, *c* and *d* from J.B. Jennings, Biol. Bull. 1957, *12*:63.)

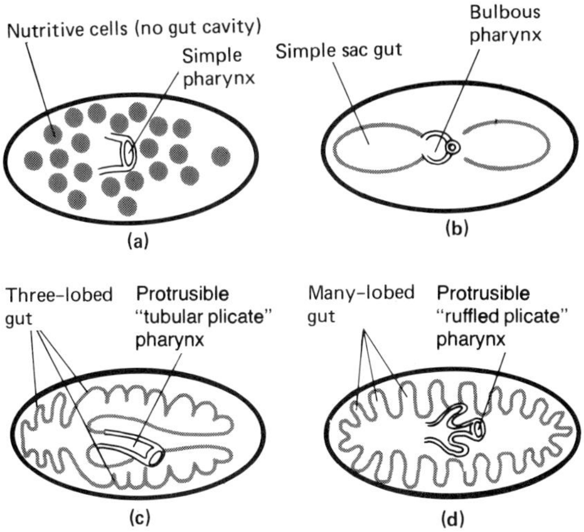

Figure 6.6 Turbellarian digestive cavities. (*a*) Acoel; (*b*) Neorhabdocoel; (*c*) Triclad type; (*d*) Polyclad. (From W.D. Russell-Hunter, *A Biology of Lower Invertebrates*, 1968, Macmillan,)

body. All other turbellarians have permanent digestive cavities. Several orders of smaller worms possess a simple blind tube, but in larger turbellarians digestive cavities are quite complex. These cavities are branched in various patterns (Fig. 6.6) that greatly increase their surface area and ensure that sites of digestion and absorption are not far removed from any body tissues. This adaptive branching compensates in part for the lack of a circulatory system in these animals and thus permits some size increase.

Freshwater planarians resist starvation by metabolizing their own digestive and parenchymal tissues. Individuals may survive a 99% reduction of their original size.

Respiration and Circulation

With their possession of a relatively solid mesodermal layer, the flatworms are subject to certain metabolic problems. These difficulties are central to flatworm biology and severely limit the potential size of individual members of the phylum. Recall that coelenterates occasionally attain large size because they restrict most of their living tissues to surface areas. Mesoglea, which has little or no metabolic requirements, constitutes the bulk of these larger coelenterates. However, a cellularized mesoderm has substantial metabolic needs that, in turn, direct an increase in metabolic rate. Planarians, for example, have a per-gram oxygen consumption 10 times that of coelenterates.

Triploblastic animals can service their mesodermal tissues in at least one of two ways. If they are flat organisms with all tissues close to an external surface, diffusion plays the major role in distributing metabolic materials. Alternatively, these materials can be transported by the coelom and the respiratory, circulatory, excretory, and osmoregulatory organ systems. Basically, flatworms exhibit the first method. Although their oxygen consumption is higher, they are essentially no better equipped for gas transport than are the coelenterates. Special organs for circulation and respiration are never present. However, the phylum does introduce an excretory–osmoregulatory system.

Excretion and Osmoregulation

Flatworms display the invertebrate world's most primitive organ for the elimination of water from the body. This organ, the **protonephridium**, is the forerunner of the advanced water control structures (kidneys) of higher animals. Basically, a protonephridium consists of a large central cell connected by a tubule to a pore at the body surface. The central cell has one to many blind, hollow, ciliated projections (Fig. 6.7a). These projections are called **flame bulbs** because of the flickering motion of their cilia; the central cell is termed a **flame cell**. The protonephridial tubule is formed from a series of cells resembling sponge porocytes. Apparently, water diffuses across the flame bulbs and into the flame cell lumen, where ciliary action establishes an excurrent flow. Quickened by muscular contractions in the tissues surrounding the tubule, this current flows to the **nephridiopore**, where water and dissolved wastes are voided. Organized groups of protonephridia constitute a protonephridial system (Fig. 6.7b).

Protonephridia are lacking in the marine acoels and are generally less developed in other marine turbellarians, while these organs attain their highest complexity among freshwater forms. This situation suggests that the primary function of the protonephridial system is osmoregulatory rather than excretory. Ni-

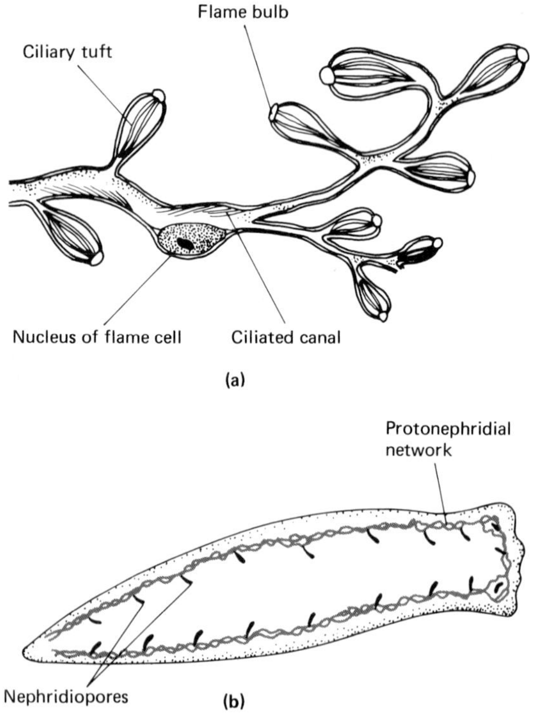

Figure 6.7 (*a*) Diagram of a flame cell with multiple flame bulbs; (*b*) Diagram of the protonephridial system of a triclad turbellarian. (*a* from C. Reisinger, Zoologischer Anzeiger, 1923, *56*:205; *b* from J. Wilhelm, Zeit. f. Wissenschaftliche Zoologie, 1906, *80*:544.)

trogenous wastes in the form of ammonia can be voided by surface diffusion in these thin animals; however, some ammonia leaves with the water expelled through a nephridiopore. Osmoregulation, on the other hand, is a more complex problem, especially for freshwater forms. Moreover, the marine turbellarian's composition is slightly less salty than sea water; hence these animals also have osmoregulatory needs. Flatworms avoid brackish situations where salinities fluctuate most sharply and osmotic control problems are most intense.

ACTIVITY SYSTEMS

The key themes of flatworm evolution, the establishment of bilateral symmetry and the development of mesodermal organ systems, are related directly to the active life-styles pioneered by this phylum. Cephalization, the development of neurosensory and muscular systems of increasing size and complexity, and a more elaborate behavioral repertoire represent major advances by the flatworms over the radiate animals. Higher invertebrate phyla have evolved activity systems along the lines introduced by Platyhelminthes.

Muscles

Free-living flatworms have a well-developed musculature. Although in some primitive acoels contractile elements persist in association with the epidermis, in all higher turbellarians muscles form distinct subepidermal and parenchymal tissue layers. Typically, subepidermal muscles include an outer circular layer located immediately beneath the basement membrane, an inner layer of longitudinal fibers, and an intermediate network of diagonal fibers (Fig. 6.4). Diagonal muscles form a crisscross lattice. These fibers, as well as those in the longitudinal layer, often are collected into distinct bands. Parenchymal musculature features a maze of longitudinal, transverse, and dorsoventral fibers. Both subepidermal and parenchymal muscles are well organized in advanced turbellarians, accounting for a mobility unparalleled by lower invertebrates.

These several muscle layers, as well as the dense structure of the entire mesoderm, create a stiff turgor in turbellarians. Hard skeletal elements are absent from the class, but this turgor allows the entire body to function as a skeleton. Local muscular forces are transmitted through the dense body tissues themselves. Contracting muscles distort the animal temporarily, but antagonistic fibers and general body resilience rapidly restore normal posture.

Nerves

Among the free-living flatworms, nervous development from a radial, anemonelike nerve net to a truly bilateral system can be seen. Nervous system development in the group, and indeed in all higher animals, is dominated by two themes: (1) a concentration of associative neurons in the anterior end to form a brain; and (2) the consolidation of peripheral neurons into a few large nerve cords.

The acoels possess a coelenterate-type nerve net (Fig. 6.8a). This diffuse, radial net persists subepidermally in many higher turbellarians, but in most members of the class, it forms a submuscular plexus. All advanced turbellarians have consolidated their nervous elements into larger, more distinct units. Initially, the submuscular plexus is condensed into radial pairs of longitudinal nerve cords (Fig. 6.8b). These cords are interconnected by cross-strands, or commissures. Shifts toward bilateralism occur when the number of nerve cords is reduced. Characteristically, the two ventral cords emerge as the primary longitudinal conductors for the animal, and the ventral commissures become quite prominent (Fig. 6.8c). The resultant ladder-type nervous system is the prototype for the major continuing line in invertebrate nervous evolution.

Concomitantly, an equally important evolutionary change occurred in the anterior ends of flatworms. In primitive forms, a concentration of nervous tissue developed there, usually in association with a statocyst (Fig. 6.8a). Such an anterior nervous mass is termed a brain. In a primitive turbellarian, where the statocyst–neuron complex is quite tiny, the term is premature at best; but in an advanced flatworm, enough nervous tissues are gathered to produce the most primitive brain in the invertebrate world (Fig. 6.8c).

Flatworms have multipolar and bipolar nerve cells (as do coelenterates), and there is little morphological differentiation between axons and dendrites. Nonetheless, unidirectional conduction does occur because of the limited distribution of synaptic vesicles. Conduction is much more efficient than in the radiate phyla, as neurons are arranged for sensory, associative, and motor roles. Faster and more direct nervous transmission also results from the organization of flatworm neurons into large nerve cords.

Sense Organs

An animal's need for precise information about its environment is largely a function of the extent to

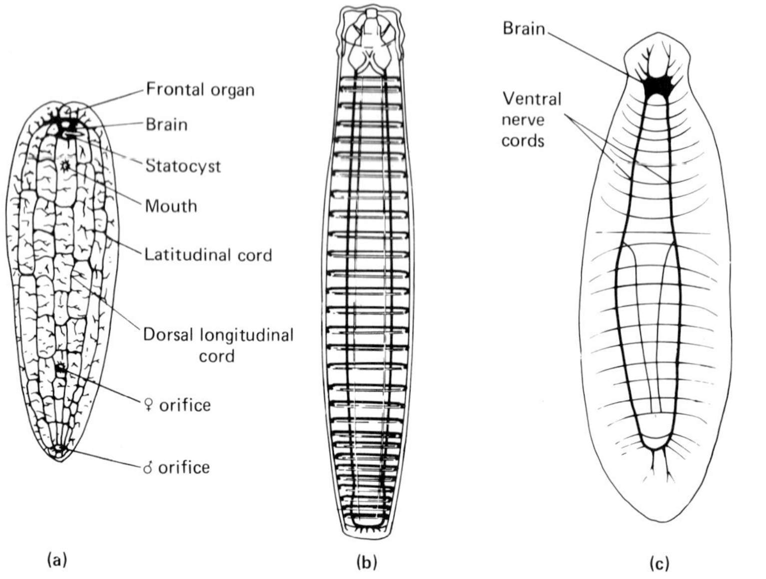

Figure 6.8 Turbellarian nervous systems. (*a*) The nerve net of *Convoluta*, an acoel showing an early stage in the formation of a brain at the anterior end; (*b*) The nervous system of *Bothrioplana* featuring radially arranged pairs of longitudinal nerve cords; (*c*) The nervous system of *Procerodes*, an advanced triclad nervous system featuring a pair of ventral nerve cords with prominent ventral commissures. (From T.H. Bullock and G.A. Horridge, *Structure and Function in the Nervous Systems of Invertebrates, Vol. I*, 1965, W.H. Freeman.)

which it must search for its biological needs. Active, bilateral invertebrates seek out food and sexual partners and fight or flee when confronted with danger. For these tasks, such animals need good sense organs. Moreover, on the premise that it is more profitable to know where one is going than where one has been, good sense organs are necessary at the anterior end in association with a processing center or brain. Free-living flatworms are no exception, and their collection of sense organs represents a considerable advancement over the sensory mechansims of the radiate animals.

Tactile and chemoreceptive cells are well distributed on the turbellarian body (Fig. 6.9a). Tactile cells possess terminal bristles surfacing through the epidermis; they are most concentrated in anterior regions and along the body margins. Paired ciliated grooves on the heads of many aquatic turbellarians contain a chemoreceptive complex. Ciliary action provides these sensory centers with a continuously circulating sample of external fluids; moreover, their bilateral arrangement allows flatworms to compare the directional qualities of important chemicals (e.g., meat juices or poisons) and to orient toward or away from them.

Turbellarian eyes comprise a group of light-sensitive cells lying within pigment cell cups (Fig. 6.9b). In some species, a thin layer of epidermal cells covers the pigment cups. Behavioral studies show that flatworms discriminate fairly well among light intensities. The bilateral arrangement of the eyes allows turbellarians to determine the direction of illumination and to orient toward it. The orienting response normally involves withdrawal, as most flatworms prefer the dark. Photoreceptive cells may be scattered throughout the epidermis of planarians, as these animals avoid light even after their eyes have been removed. The presence of photosensory elements in the epidermis underscores the importance these worms place on detecting and avoiding light. Flatworms are thin, moist animals, and those living on land cannot survive prolonged exposure to the desiccating sun.

Statocysts are present in most swimming turbellarians, but absent in crawling forms like the planarians. Planarians, and presumably other crawling flatworms, are positively **thigmotactic** on their ventral sides and negatively thigmotactic elsewhere. These conditions mean that the animals preferably maintain ventral contact with the substratum. If placed on its back, a planarian always attempts to turn over and regain ventral contact. Gravity apparently plays no role in this behavior.

Behavior

Turbellarians are the most primitive metazoans regularly moving about in search of their vital needs. Unlike the passive sponges and coelenterates, they pursue prey, flee from predators and noxious stimuli, and seek shelter and sex.

Most turbellarian behavior is slow and, even as worms go, rather undramatic. Small species move by

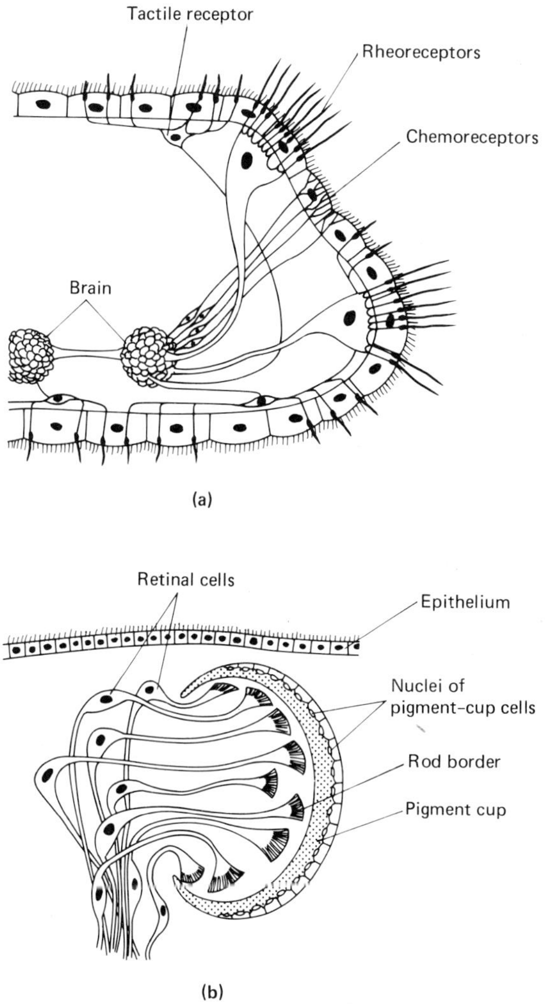

Figure 6.9 (*a*) A cross section through the anterior end of *Mesostoma*, a neorhabdocoel turbellarian showing the location of sensory receptors; (*b*) Section through an inverse pigment cup ocellus of a triclad turbellarian. (*a* from J. von Seler, Zeit. f. Morphologie und Okologie der Tiere, 1930, *18*:786; *b* from R. Hesse, Zeit. f. Wissenschaftliche Zoologie, 1897, *62*:527.)

ciliary action along the slime trails laid down by their mucous glands. Ciliary power, however, is insufficient to propel larger flatworms. (Such an attempt would be rather like trying to row an ocean liner.) Large aquatic turbellarians wriggle and squirm through the water, demonstrating the increasing importance of muscles in locomotion. Alternating contractions of longitudinal and circular muscles along the body account for the forward movements of nonswimming worms. Initially, circular muscles contract and cause the worm to become long and thin. By adhering to the substratum at an anterior point and then contracting its longitudinal muscles, the flatworm "inches" forward. A planarian can lift its anterior end and move its head from side to side. This swaying movement is produced by alternating contractions of longitudinal muscles on the right and left sides of the head. Meanwhile, the body is stiffened by the contraction of diagonal muscles.

Turbellarians can detect food at distances too great for determining in which direction the source material lies. A distant, decaying crayfish, for example, might stimulate chemoreceptors evenly on both sides of the worm. Knowing that food is in the area, but unable to pinpoint its exact location, a turbellarian may resort to **klinokinesis.** Klinokinesis involves sustained forward motion, provided the stimulation remains at a high level. If the overall stimulus intensity drops, the worm turns at random until a strong signal is restored. Forward motion is then resumed, and the turbellarian turns again only if the stimulation decreases. In this way, the search area gradually diminishes and the hungry worm eventually has its meal.

Turbellarians do or do not learn, depending on which experimental results we accept to support any given definition of learning. At the flatworm level, learning is difficult to study and conclusions tend to reflect rather personalized data interpretations. A wealth of experimental information exists. Two of the most interesting observations follow. Normally strongly negative to light, turbellarians may remain in illuminated areas if electric shocks have punished their previous attempts to move into darker zones. Planarians may favor one side of a Y-shaped maze if, on earlier trials, entry to the opposite side resulted in similar electric jolts. Both of these classical conditioning experiments suggest learning, but problems arise when results cannot be duplicated regularly. Identical experiments with different species under varying conditions of temperature and previous handling yield an array of conflicting data. Moreover, minor changes in experimental design usually disrupt supposedly learned behavioral patterns. (However, changes that seem "minor" to the experimenter might be very significant to the worm!) These problems notwithstanding, flatworms are the most primitive animals on which learning studies are feasible and will play a continuing role as biologists probe the mysteries of learning in the invertebrate world.

CONTINUITY SYSTEMS

To begin our description of flatworm reproduction, we reiterate that the current discussion of platyhelminth unity describes only the free-living turbellarians. The assumption of parasitism always entails many modifications in maintenance and activity systems. However, if a species is to survive the transition to parasitic life, the most fundamental alterations of its continuity systems are necessary. Accordingly, the trematodes and cestodes display considerable divergence from the general flatworm scheme described here.

Reproduction and Development

Almost without exception, flatworms are hermaphroditic and their dual sexual systems are usually very complex. This situation contrasts sharply with the relative simplicity of all other organ systems in the phylum. Reproductive complexity has developed within the phylum itself, as primitive turbellarians possess only rudimentary structures. Acoels have no testes; their sperm are produced along longitudinal bands of mesoderm. These sperm migrate through parenchymal spaces to a copulatory organ, the first penis we have encountered in the invertebrate world. Eggs also are produced *sans* gonads and, since primitively nothing like a vagina is present, impregnation commonly occurs by hypodermic injection. The penis of each worm pierces the body wall of its mate and deposits sperm in the latter's parenchyma. Sperm meet eggs and fertilization follows. Fertilized eggs may exit via the mouth, but more commonly they are liberated by rupture of the body wall, resulting in the death of the parent.

Higher flatworms have improved considerably on the acoel scheme. Extensive evolutionary experimentation with the reproductive organ system has produced considerable diversity, even within the various turbellarian groups. The following account is generally descriptive of these advanced systems. Moreover, the parasitic classes probably evolved from worms with this generalized reproductive system.

The generalized male system comprises testes, connective tubules, and a copulatory organ (Fig. 6.10a). The testes, which may be single, paired, or numerous, are drained by small tubules which communicate with larger tubes, the **vasa deferentia.** The latter transport sperm to the **male atrium.** The male atrium is a small cavity housing the copulatory organ. This organ is called a penis when it is protrusible, but in those species with an *eversible* copulatory organ, the proper term is a **cirrus.** Preceding the penis or cirrus, each vas deferens may be dilated as a sperm storage area or **seminal vesicle.** Adjacent prostatic glands often supply seminal fluids.

The female tract (Fig. 6.10a) includes one or more ovaries. Oviducts join the ovaries with a **genital atrium.** Here the copulatory organ of another flatworm is inserted. Incoming sperm are stored either in the **copulatory bursa,** a side chamber of the genital atrium, or in a special dilation of the oviduct called a **seminal receptacle.** When eggs mature, sperm leave their storage area and migrate into the oviducts, where fertilization occurs.

Turbellarian taxonomy keys on the yolk situation of each subgroup. The primitive acoels and the marine polyclads, generally recognized as lower turbellarians, have no yolk glands; they simply incorporate yolk into their eggs during oogenesis. Higher turbellarians include the neorhabdocoels and the triclads. In these worms, the female tract displays two functional areas. The **germarium** produces yolkless eggs, while the **vitellarium** is responsible for yolk cell production. Yolk cells later secrete a hard shell which surrounds them and one or more of the yolkless eggs. Fertilized eggs may be stored in yet another side chamber of the genital atrium, the uterus.

Turbellarians may have two genital atria, one for each sexual system. Often, however, a common atrium services both male and female tracts. Cross-fertilization through mutual copulation and insemination is the rule among these hermaphrodites (Fig. 6.10b). The necessity for copulation to stimulate ejaculation discourages self-fertilization, as does protandry. In the same animal, sperm mature before eggs do, so union between them is unlikely. A further consequence of protandry is the necessity for sperm storage facilities, such as the copulatory bursa and the seminal receptacle.

Eggs are laid in gelatinous masses or are fixed singly to the substratum by a sticky secretion from atrial glands. Development depends on zygotic yolk content. The lower orders, whose eggs themselves contain yolk, undergo spiral, determinate cleavage. In higher turbellarians (the neorhabdocoels and the triclads), however, embryonic development must accommodate external yolk cells. Although clearly derived from the spiral format of their primitive cousins, the cleavage pattern in both neorhabdocoels and triclads has been modified almost beyond recognition.

In most turbellarians, development is direct, and within a few days to a few weeks following fertilization, a small baby flatworm hatches and crawls or swims away. Some freshwater species lay thin summer eggs but much thicker overwintering eggs, in which embryonic development may require several months. Polyclads, a wholly marine group, include the only turbellarians with a free-swimming larval stage. This stage, a **Müller's** or **Götte's larva** (Fig. 6.11a),

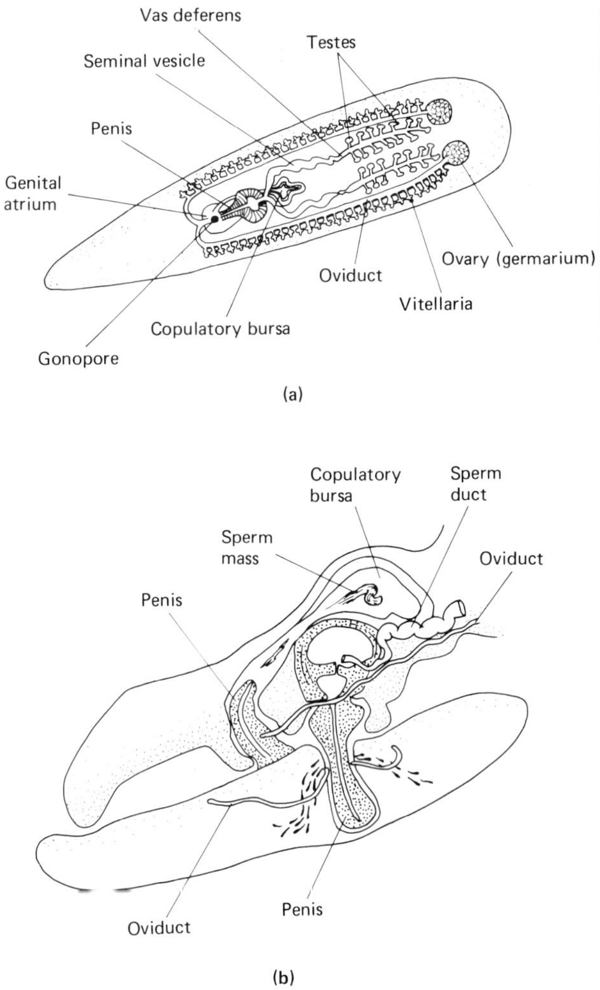

Figure 6.10 (*a*) Diagram of the hermaphroditic reproductive system of a triclad turbellarian; (*b*) Section through a copulating pair of *Dugesia gonocephela* (Turbellaria, Tricladida) showing the inserion of the penis of each pair member into genital atrium of the other. (*a* From P. Steinman, Revue Suisse de Zoologie, 1911, *19*(7):75; *b* from Hyman, *The Invertebrates, Vol. II.)*

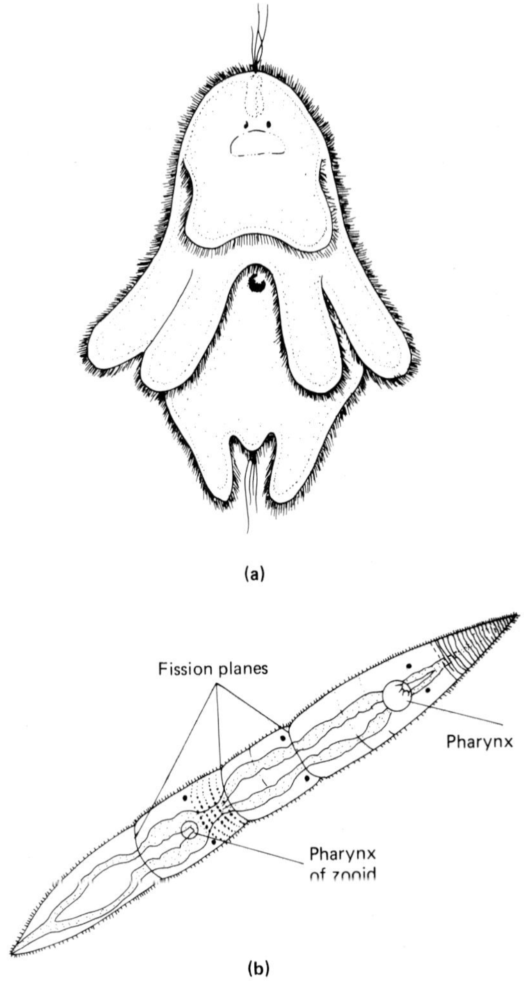

Figure 6.11 (*a*) Müller's larva of *Planocera multitentaculata*, a polyclad turbellarian; (*b*) Asexual reproduction in *Alaurina*, a catenulid turbellarian, producing a chain of zooids by transverse fission. (*a* from K. Kato, *Jap. Journal of Zool.*, *8*:537; 1940, *b* from A.V. Ivanoff, In *The Atlas of Invertebrates of the Far Eastern Seas of the U.S.S.R.*, E.N. Pavlovskii, Ed., Academiia nauk, U.S.S.R.)

undergoes a brief, planktonic existence before metamorphosing into an adult worm.

Asexual Reproduction and Regeneration

Some, but by no means all, turbellarians can reproduce asexually and can regenerate lost body parts. As is generally true concerning these powers, a worm capable of one usually is proficient in both. Some species proliferate asexually by forming zooids (Fig. 6.11b), chains of daughter flatworms produced by transverse fission. As each zooid matures, it separates from the chain and pursues an independent life. Many planarians reproduce by transverse fission without zooids. Instead, the original worm stretches along its main axis and snaps into two or more pieces. Each piece then regenerates its missing parts.

Planarian regeneration patterns are reminiscent of those observed in the coelenterate *Hydra* (see Chap. 5). In transversely sliced sections, a distinct metabolic gradient exists, ensuring the maintenance of the original anterior—posterior orientation in regenerated forms. Cross sections excised from the anterior end of a planarian are always more active metabolically than sections cut from the rear of the animal. Anterior slices regenerate faster, and smaller initial sections from this region can ensure full development of a new worm.

A Review

Before discussing diversity within the phylum, where the adoption of parasitism has caused many changes in the basic body plan, we review here the general characteristics of flatworms. Basically, free-living flatworms (turbellarians) are:

(1) bilaterally symmetrical, triploblastic, acoelomate worms;

(2) animals at the organ system level of organization;

(3) the most primitive creatures to have undergone cephalization;

(4) worms with a simple digestive tract lacking an anus;

(5) innovators of the protonephridial system;

(6) metazoans that have pioneered active, aggressive life styles;

(7) hermaphroditic animals with highly complex reproductive systems.

Flatworm Diversity

As demonstrated by the turbellarians, the flatworm body plan has had limited success. Free-living flatworms display a rather low profile at every level, and perhaps not surprisingly, the other two classes of Platyhelminthes have undergone a different adaptive route. In these classes, the Trematoda and the Cestoda, the phylum's basic plan has been modified to support parasitic life-styles. With parasitism, some considerable success has come to the flatworms. Of 15,000 known species of Platyhelminthes, 11,000 are parasitic trematodes and cestodes. Moreover, the expanding search for new species promises to favor parasitic forms even more heavily in the future census of these worms.

CLASS TREMATODA

Approximately 7000 species of trematodes, also known as flukes, are parasites with a wide range of vertebrate and invertebrate hosts. In general appearance, these parasitic flatworms clearly reveal their common heritage with the turbellarians (Fig. 6.12). Body form varies, as in the free-living group, but compared with most turbellarians, flukes are stocky and more oval in shape and have no clearly delimited head. Trematodes are covered externally by an unciliated **tegument,** a layer quite unlike the turbellarian epidermis. This tegument, long referred to as a "nonliving cuticle," is cytoplasmic and thus very much alive; it originates from long extensions of cells located deep within the parenchyma (Fig. 6.13). Strong attachment organs surround an anterior mouth and occur midventrally or at the posterior end (Figs. 6.14 and 6.15). The digestive system consists of the anterior mouth, a short pharynx and esophagus, and two intestinal tubes extending along each side of the worm (Fig. 6.14). Small intestinal diverticula may exist and are highly branched in some species.

Turbellarian muscular, nervous, and protonephridial systems have remained essentially intact in trematodes. Thus nervous and muscular systems do not show the degeneration so often associated with parasitism. This situation reflects the relatively simple nervous organization in the phylum. Also, among trematodes, the development of muscles for attachment and the greater use of muscles for movement (in the absence of cilia) maintain selective pressures on the nervous system. Subtegumental muscles are well developed, but parenchymal fibers may be degenerate. Sucking attachment organs are always highly muscularized and well-endowed nervously. Sense organs have degenerated somewhat among adult endoparasites, but ectoparasites and the free-living larvae of endoparasitic trematodes usually possess rudimentary eyes as well as tactile and chemoreceptors.

The protonephridial system remains functional in the class. Characteristically, two large longitudinal tubes collect the drainage from numerous flame cells. After coursing posteriorly along the main body axis, these tubes may loop and terminate at paired nephridiopores near the mouth. Immediately preceding the nephridiopores, a muscularized distention of each tube may form a primitive bladder. Alternatively, the protonephridial system may conclude posteriorly with the union of the two longitudinal tubes into a single, medial bladder and one excretory pore (Fig. 6.14). Trematodes retain a protonephridial system because of their osmotic needs. The largely anaerobic

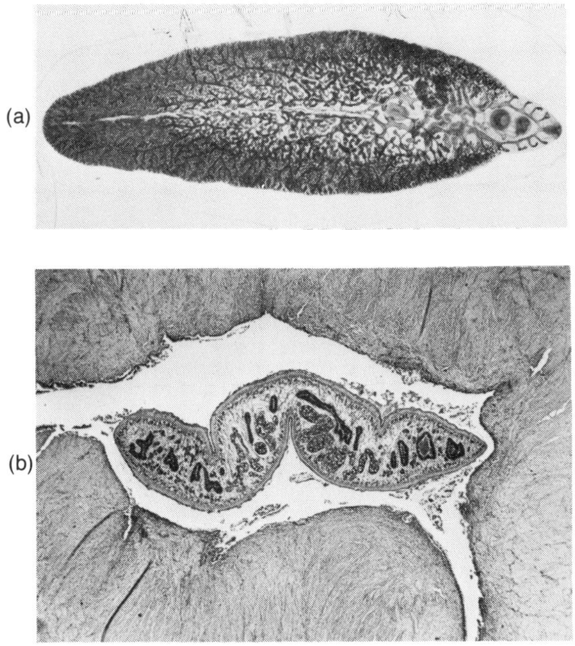

Figure 6.12 (*a*) Photograph of a stained whole-mount preparation of the sheep liver fluke, *Fasciola hepatica* (Trematoda, Digeneal); (*b*) Section of *Fasciola hepatica* in situ in sheep liver. (Photos courtesy of Turtox/Cambosco.)

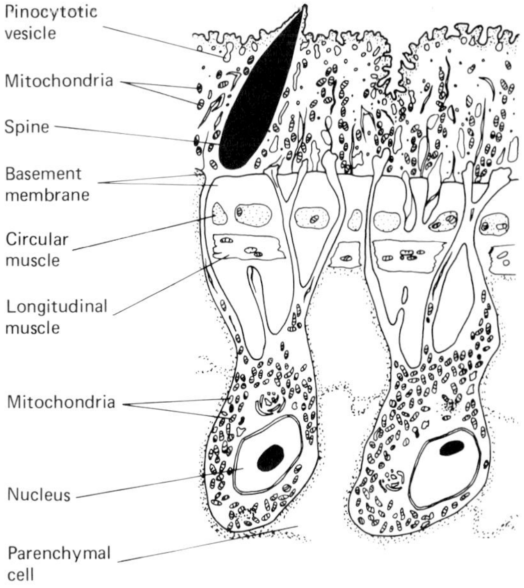

Figure 6.13 Drawing of a section through the tegument of the sheep liver fluke, *Fasciola hepatica*, based on electron micrographs. (From L.T. Threadgold, Quart. J. Micros. Sci., 1963, *104*:505.)

metabolism of these parasites produces high concentrations of organic acids in the body. Consequently, these worms are not in osmotic equilibrium with host fluids and must rid themselves of constantly inflowing water.

Trematode reproductive systems are rather uniform (Fig. 6.14). Typically, two sperm ductules originate from a single pair of testes. These ductules join to form a solitary vas deferens which transports sperm to the genital pore, the site of the copulatory organ. There a seminal vesicle stores sperm and prostatic glands add semen. The female system centers around a small central chamber, the **ootype** (Fig. 6.14). Ducts join the ootype with other reproductive structures—a single ovary, paired yolk glands, a seminal receptacle, Mehlis' glands, and the uterus. Hermaphroditism is general in this class and cross-fertilization occurs by mutual insemination. Each male copulatory organ ejaculates into the genital pore of its mate, and sperm are dispatched through the uterus to the ootype. Protandry militates against self-fertilization and requires a seminal receptacle in the female system, where sperm await the maturation of eggs. Ripe eggs descend from the ovary to the ootype, where they are fertilized. Yolk glands add nutrients and a capsule-forming material to the young zygotes, which then are ushered into the uterus and eventually expelled through the genital pore. **Mehlis' gland** secretions may lubricate uterine passageways as fertilized eggs are discharged. Egg production in the Trematoda is prolific; outputs up to 100,000 times those of free-living turbellarians have been observed.

Adaptive radiation in this class is reflected in the parasitizing specialties of its members. Two factors central to parasitic existence play major roles in trematode adaptation and are keys to their taxonomy. These factors are the life cycle and the attachment organ. Three orders are recognized: the Monogenea, the Digenea, and the Aspidobothrea.

Order Monogenea

The Monogenea are a small group of parasitic flatworms with only one host in their life cycle. They are largely ectoparasitic on the skin or gills of marine and freshwater fish; amphibians, turtles, and aquatic invertebrates also serve as hosts. A few monogenetic trematodes are endoparasites, usually living in the

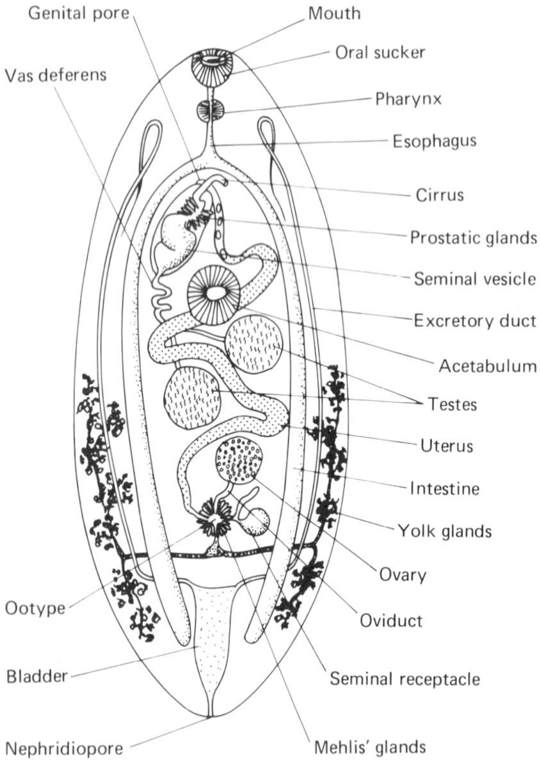

Figure 6.14 Drawing of a generalized trematode showing major features of internal anatomy. (From A.C. Chandler, *Introduction to Parasitology*, 1945, John Wiley.)

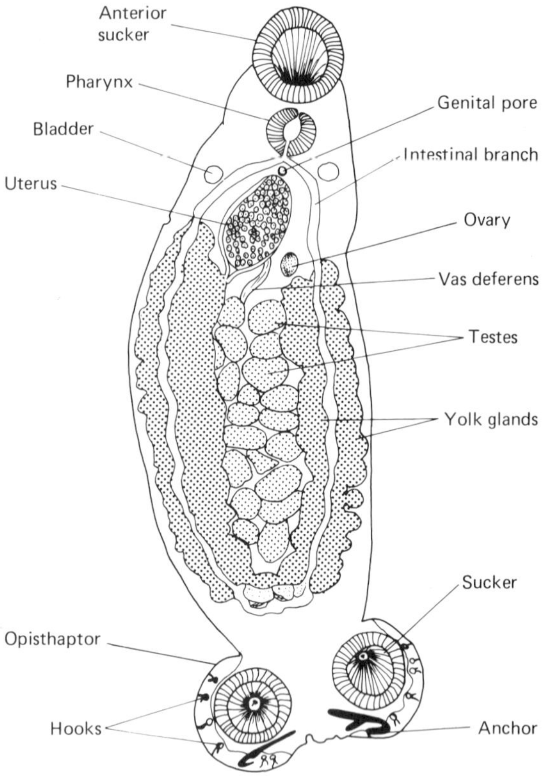

Figure 6.15 A diagram of a monogenetic trematode, *Sphyramura*, showing major structural features. (From Hyman, *The Invertebrates, Vol. II.*)

mouth, lungs, or bladder. The aforementioned habitats are relatively accessible from the outside, and their exploitation presents minor adaptational problems for this essentially ectoparasitic order.

The second outstanding character of the Monogenea is the **opisthaptor** (Fig. 6.15), a posterior attachment organ. The opisthaptor comprises muscularized suckers and a battery of sclerotized hooks and anchors. Smaller, anterior suckers are associated with the monogeneid mouth. Also characteristic of the order are paired, anterior nephridiopores.

Monogenetic trematodes present the simplest life cycle in this class. Eggs are released through the genital pore and immediately attach to the parental host by a long thread. After hatching, an **onchomiracidium** larva (Fig. 6.16) must locate its own host within a few hours or it will die. A precociously developed opisthaptor fastens the young larva to its new host, where growth and sexual maturation occur.

Monogeneids display a number of life cycle schemes, but space permits an account only of the following, well-known species. *Polystoma integerrimum* is parasitic on the gills and in the bladder of frogs and toads. Egg production by this parasite is stimulated when the amphibian host returns to water to breed. This sexual synchronization apparently is effected by the host's hormonal system. Amphibian and parasite eggs are released simultaneously, and the latter's onchomiracidia attach to tadpole gills. Those larvae parasitizing young tadpoles undergo rapid sexual maturation and produce another larval generation within a few weeks. When host metamorphosis occurs, these monogeneids slip through the gut and establish permanent residence in the host bladder. There sexual maturation requires about 3 years.

Order Digenea

The Digenea are a large group of endoparasitic trematodes. These flatworms have a complex life cycle with supernumerary larval stages and two to four

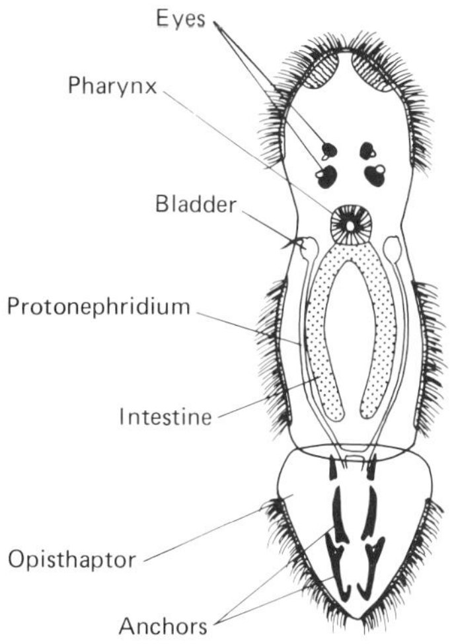

Figure 6.16 The onchomiracidium larva of a monogenetic trematode. *Benedenia.* (From Hyman, *The Invertebrates Vol. II.*)

different hosts. Adult digeneids parasitize all classes of vertebrates, usually in the digestive tract and its derivatives. Motile larvae are parasitic on aquatic vertebrates and invertebrates, commonly mollusks. Humans and domestic animals are not spared, as the order contains the notorious liver and blood flukes.

Attachment structures include an oral sucker and a similar ventral organ called the **acetabulum** (Fig. 6.14). The protonephridial system features many flame cells and a single, posterior nephridiopore. The description of their basic life cycle scheme is followed by specific examples.

The life cycle in Digenea begins with one of the millions of eggs laid by a sexual, usually hermaphroditic adult (Fig. 6.17). Because adults normally inhabit a vertebrate digestive tract, eggs are released most often in host feces. Upon contact with water, each egg hatches into a ciliated, free-swimming **miracidium.** Miracidia dart about until they contact the first intermediate host, commonly a snail; they penetrate the mollusk and establish residency in its digestive gland. Each miracidium then metamorphoses into a germinal sac called a **sporocyst.** Within the sporocyst, balls of embryonic cells give rise to numerous, vermiform rediae. Each **redia** contains primordia of adult organ systems, but most of its body is occupied by embryonic units destined to form yet another larval stage. In this next stage, the **cercaria,** adult systems are more evident. Cercariae have a flukelike form with a muscular tail for swimming. These larvae leave the snail and become free-living; they seek out the next intermediate host, often an arthropod or a fish, and encyst within its muscles. Encystment in muscle tissue involves loss of the cercarial tail and produces a dormant form called a **metacercaria.** When the second intermediate host is eaten by the primary host, adult metamorphosis occurs.

Thus a typical digenetic trematode has seven distinct stages in its life cycle: egg, miracidium, sporocyst, redia, cercaria, metacercaria, and adult. Only the miracidium and the cercaria, however, are independent larval forms. Sporocysts and rediae are nonmotile embryonic stages that asexually increase the larval population. Such larval proliferation is called **polyembryony.** Polyembryony reflects the recurring need of these parasites to find hosts. Although egg production is prolific, few miracidia successfully colonize the first intermediate host. Polyembryony replenishes the larval population so that similar decimations during the transition to each subsequent host can be withstood. To ensure completion of the circuitous journey to the primary host, each successful miracidium may give rise to 200,000 cercariae!

The Chinese liver fluke, *Opisthorchis sinensis* (Fig. 6.17), is a typical digeneid. As an adult, it parasitizes humans; larval stages infect snails and fish. The sheep liver fluke, *Fasciola hepatica* (Fig. 6.12) has only one intermediate host, an aquatic snail. Cercariae leave the snail and swim to shoreline vegetation. Here metacercariae encyst on plants later to be eaten by grazing sheep and cattle. Adult metamorphosis follows ingestion, and the mature fluke resides in the bile duct of its mammalian host.

Leucochloridium macrostomum has a fascinating mechanism to ensure its continuity. Its single intermediate host is a land snail; its primary host, a songbird. An apparently fateful situation exists in that the fluke's bird host does not eat snails, but rather feeds on insect larvae. *Leucochloridium macrostomum* counters brilliantly by disguising its snail host. Yellow sporocysts with bright green stripes vibrate vigorously within the snail's translucent tentacles (Fig. 6.18). Colorful and conspicuous, these pulsating striped tentacles resemble insect larvae. In addition, the parasite upsets the snail's behavior, causing the mollusk not to avoid light. These factors attract and deceive the songbird, which readily eats the tentacle–sporocyst. Subsequent developmental stages are passed in the bird's body, with the adult fluke completing sexual maturation in the intestine. Eggs are passed in the feces and, if eaten by a snail, hatch to begin another generation.

The blood flukes of the genus *Schistosoma* are very destructive flatworm parasites of humans. These tre-

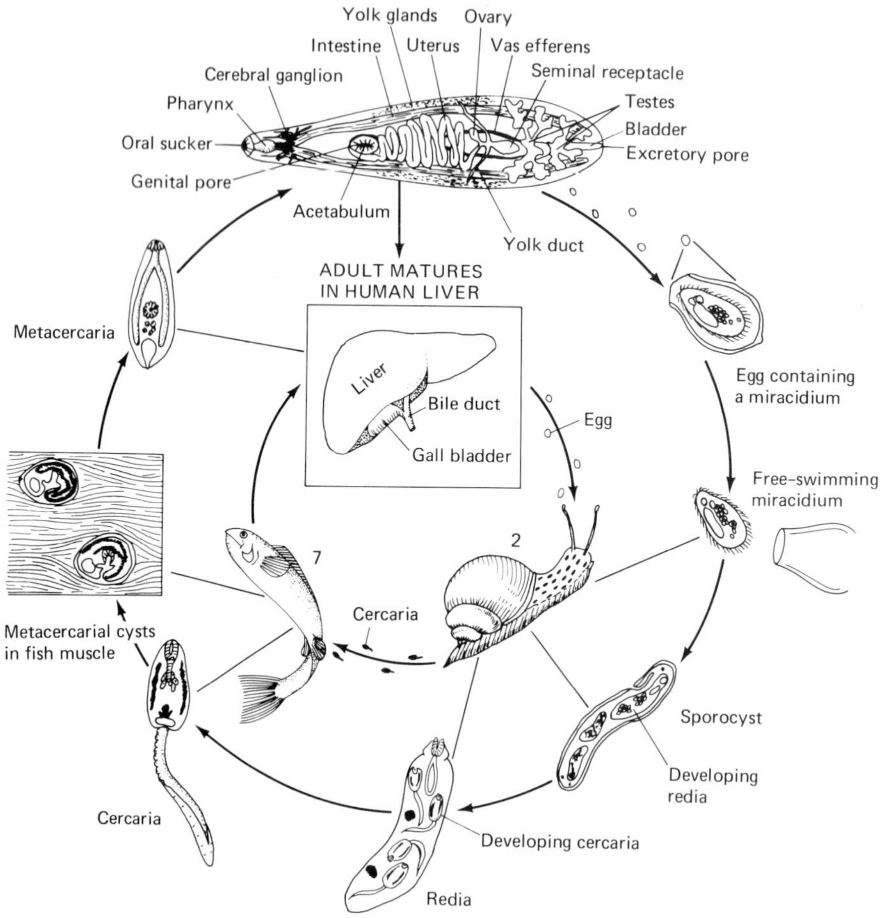

Figure 6.17 The human liver fluke, *Opisthorchis sinensis*, exhibits a life cycle quite typical of digenetic trematodes. (For description, see text.) (From C.P. Hickman, Sr., et al., *Integrated Principles of Zoology*, 5th ed., 1974, Mosby.)

matodes cause schistosomiasis, a debilitating disease that destroys the tissues of the lungs, liver, intestine, and/or bladder. Blood flukes are unusual flatworms in that separate male and female individuals exist. The male is the bulkier of the two, and the long but slender female actually resides with a longitudinal groove on his ventral surface (Fig. 6.19). Adults inhabit the veins of the intestind. Eggs are released into the blood stream and, depending on the species, migrate either to the large intestine or to the bladder. After boring through the wall of one of these organs, the eggs exit with the feces or urine. In contact with water, a miracidium hatches; if the miracidium enters an appropriate snail, polyembryony produces sporocysts and then cercariae. (There is no redial stage in blood flukes.) The cercariae, which have forked tails, swim away; after burrowing through the bare skin of a human host, they are carried by the blood stream to the lungs and liver. Adult development begins in these host organs, and maturation accompanies final adult settlement in the intestinal veins. The destructiveness of schistosomiasis is caused less by the adult fluke than by its developmental stages. Boring eggs and growing cercariae inflame and may destroy the tissues through which they pass. Death may result from this disease, which ranks with malaria and hookworm as one of the three greatest parasitic scourges of humans.

Order Aspidobothrea

A third trematode order shares common features with both the monogeneids and the digeneids. These

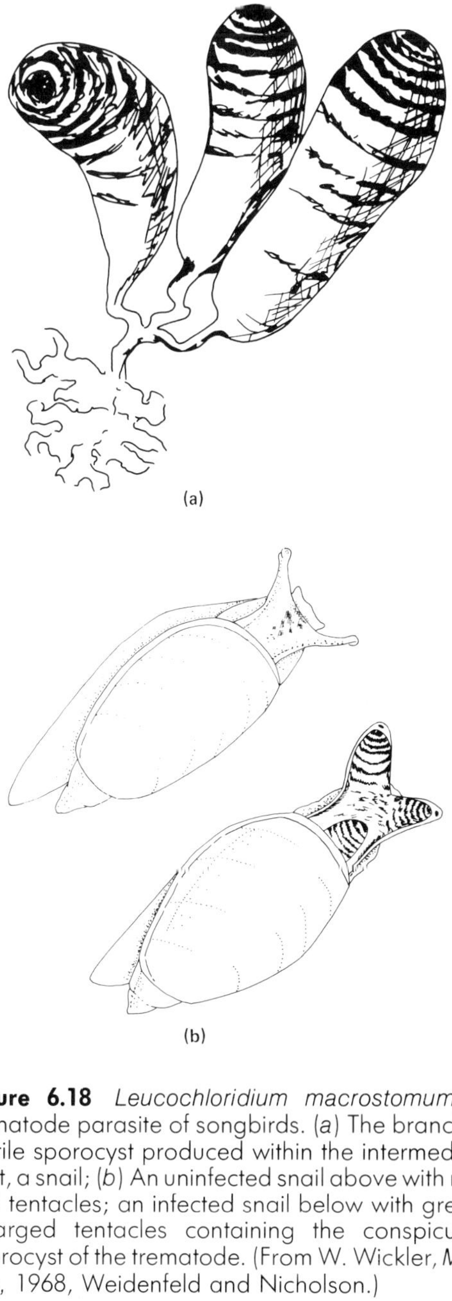

Figure 6.18 *Leucochloridium macrostomum*, a trematode parasite of songbirds. (*a*) The branched motile sporocyst produced within the intermediate host, a snail; (*b*) An uninfected snail above with normal tentacles; an infected snail below with greatly enlarged tentacles containing the conspicuous sporocyst of the trematode. (From W. Wickler, *Mimicry*, 1968, Weidenfeld and Nicholson.)

trematodes are mostly endoparasitic in fish, turtles, and mollusks. The distinguishing character of the order is an enormous sucker, usually called an opisthaptor, dominating the ventral surface (Fig. 6.20).

CLASS CESTODA

Most highly specialized of all flatworm classes, the Cestoda includes some of the most sophisticated parasites in the invertebrate world. About 4000 species are described. Adult cestodes are endoparasites in vertebrates. A complex life cycle features one or two intermediate hosts; larval hosts include invertebrates as well as vertebrates. In terms of reputation and number of species, the class is dominated by the notorious tapeworms of the subclass Eucestoda.

Subclass Eucestoda

Eucestodes are tapeworms: long, thin endoparasites in vertebrates, including humans. Representing the vast majority of cestodes, they appear very different from all other flatworms. The tapeworm body consists of a head, or **scolex,** a short neck region, and a long **strobila** composed of units called **proglottids** (Fig. 6.21). Whether these units constitute metameres is a matter of definition. A **metamere** is one of a series of homologous body segments, such as those of annelids and arthropods (see Chaps. 8 and 10). In cestodes, new proglottids are formed by continual transverse divisions in the neck region, and maturation and growth occur posteriorly. Thus the oldest and largest proglottids occupy the tapeworm's posterior. This situation excepts the strobila from most definitions of metamerism, which stipulate that new segments be formed exclusively at the posterior end.

The scolex bears the holdfast organs (Figs. 6.21 and 6.22). These hooks and/or suckers attach the animal to the lining of the host's gut. The scolex itself may be embedded in host tissues. Holding power is increased further by elaborations of the cestode tegument (Fig. 6.23), an outer covering similar to that found in trematodes. This tegument bears numerous extensions, called **microtriches,** which interlock with the microvilli of the vertebrate intestine.

Microtriches also increase the tegumental surface area. In the absence of a mouth and digestive system, cestode food acquisition depends on absorption across the body surface. Active transport and/or pinocytosis by the tegument is implied, and any increase in the tegumental surface area translates as increased food-gathering ability.

Tapeworms resemble most endoparasites in being facultative anaerobes. Oxidative enzymes are present,

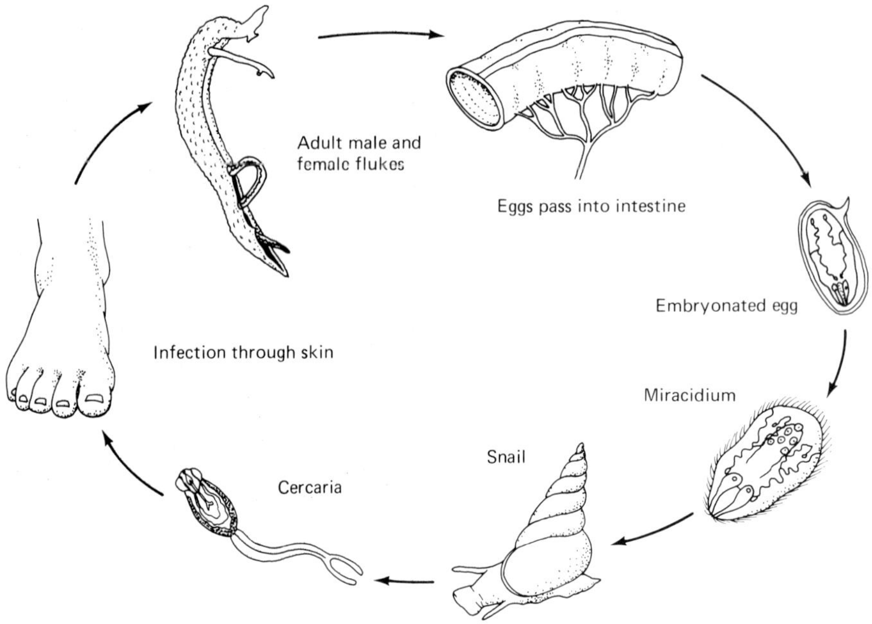

Figure 6.19 The life cycle of a human blood fluke, *Schistosoma mansoni.* (For description, see text.) (From Barnes, *Invertebrate Zoology*, 4th ed., 1980, Saunders College Publishing.)

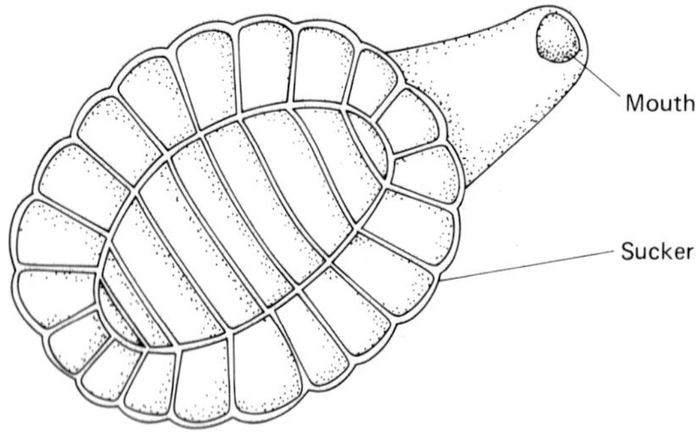

Figure 6.20 Ventral view of *Cotylaspis*, an aspidobothrian trematode. (From Hyman, *The Invertebrates*, Vol. II, 1951, McGraw-Hill.)

but in the frequent scarcity or absence of oxygen, tapeworms often rely on anaerobic metabolism.

As in the trematodes (and for similar reasons), nervous, muscular, and protonephridial systems in tapeworms retain considerable turbellarian character. Two longitudinal nerve trunks originate from the scolex and extend throughout the strobila. Ringed commissures join these trunks in each proglottid, and a connective ring in the scolex constitutes a small brain. Cestodes lack specialized sense organs, although sensory nerve endings populate the body surface. Both subtegumental and parenchymal muscles are well developed; parenchymal fibers are particularly strong. Large longitudinal muscles in this layer extend through the strobila without regard to proglottid boundaries. A protonephridial system comprises two pairs of tubules coursing longitudinally through the body (Fig. 6.21). Tubule pairs are located dorsolaterally and ventrolaterally, and the ventral pair is connected by a transverse duct in each proglottid. Numerous branch canals link the flame cells with these ducts. A pair of nephridiopores is located in the terminal proglottid.

Adult tapeworms direct much of their energy to egg production. Each proglottid houses a complete reproductive system (Fig. 6.21). With few exceptions, tapeworms are hermaphroditic, and their dual sexual

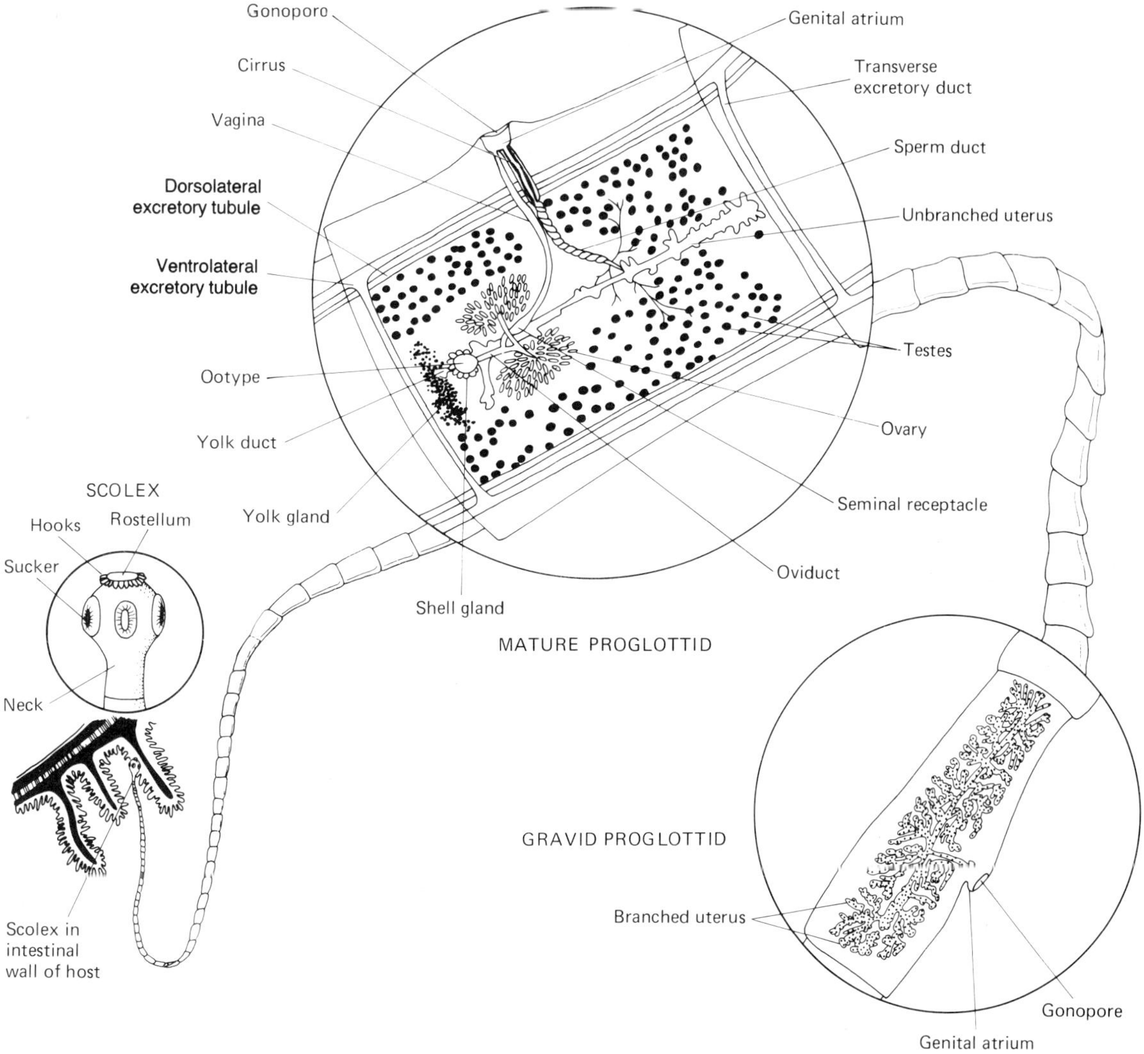

Figure 6.21 The pork tapeworm, *Taenia solium.* Inserts show enlargements of the scolex, a mature proglottid and a gravid proglottid. (Modified from C.A. Villee, *Biology*, 7th ed., 1977 W.B. Saunders, and W.H. Freeman and B. Bracegirdle, *An Atlas of Invertebrate Structure*, 1971, Heinemann.)

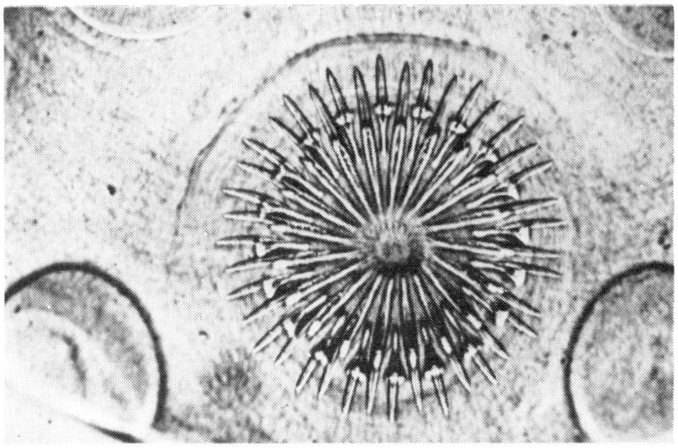

Figure 6.22 Rostellar view of a tapeworm scolex showing the hooks and suckers by which the worm attaches to the gut lining of the host. (Photo courtesy of Carolina Biological Supply.)

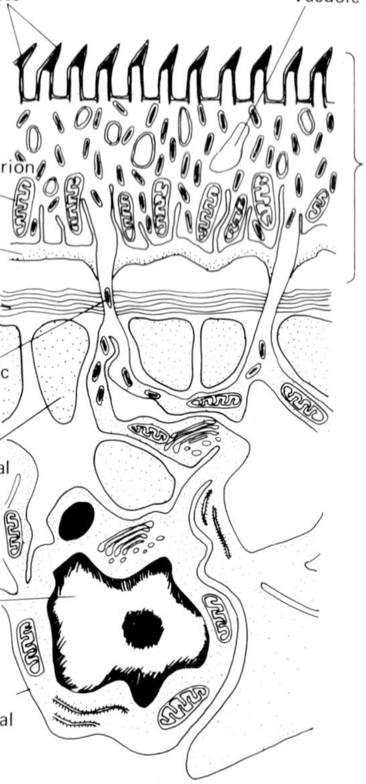

Figure 6.23 Drawing of a section through the tegument of a tapeworm based on electron micrographs. (From F. Beguin, Zeit. f. Zellforsch. Mikrosk. Anat., 1966, *72*:30.)

tracts resemble those of trematodes. In the male system, multiple scattered testes are connected by fine ducts to a copulatory complex consisting of a cirrus, prostatic glands, and a seminal vesicle. The female system features a central ootype with contributing ducts from the ovary, seminal receptacle, and yolk glands. A shell gland surrounds the ootype itself. Uterine arrangement differs, however, from the typical situation in trematodes. In most tapeworms, the uterus is a blind sac not communicating directly with the genital pore. A separate vaginal tube, part of which forms the seminal receptacle, connects the ootype with the body surface.

Anterior proglottids bear only the rudiments of reproductive organs. Sexual systems develop with age, and only the older, posterior units of the strobila possess fully mature organs. In another flatworm example of protandry, male structures always develop earlier than female ones. Cross-fertilization by mutual insemination between two individuals is typical of the subclass, but proglottids within the same worm may have intercourse for a limited number of generations. Some tapeworms release fertilized eggs on a continuous schedule. More commonly, however, eggs are stored in the branched uteri of posterior proglottids. These units, called **gravid** proglottids, bulge because of their egg load (Fig. 6.21). Gravid proglottids are shed one by one from the posterior end of the strobila, and these sacs of eggs exit the primary host with the feces.

The individual tapeworm egg faces extremely limited survival odds. It not only must develop through a circuitous life cycle, usually involving two or more hosts, but must succeed in most cases without polyembryony. In the absence of polyembryonic stages, one tapeworm egg—potentially just one tapeworm adult—does not represent a substantial investment in the future of the species. However, tapeworms produce mammoth numbers of eggs. Even the trematodes are surpassed in this regard, for in some tapeworm species, yields of up to a million eggs per day are common.

Larval stages locate a proper vertebrate host for the next adult generation. Immature forms can reach this primary host by infecting a link in its food chain. Lacking the trematode ability to bore into intermediate hosts, tapeworm larvae benefit from the indiscriminate appetites of these organisms. The developing parasite is passed through a food chain climaxing in the primary vertebrate host. Life cycles in the subclass thus depend on the eating habits of the vertebrate host, the eating habits of that host's prey, and so on, back to the parasite's first appearance in the food chain. As vertebrate food chains are highly variable, so too are the life cycles of tapeworms. Two examples barely introduce the variety of developmental patterns exhibited by these parasites.

One typical life cycle is demonstrated by *Dibothriocephalus latus* (Fig. 6.24), a tapeworm of fish-eating mammals. A ciliated larva called a **coracidium** hatches from the egg and is eaten by the first intermediate host, a copepod crustacean. Within its copepod host, the coracidium develops into a **procercoid.** If the copepod is eaten by the next intermediate host, a fish, the procercoid encysts in the latter's muscles as a **plerocercoid.** The plerocercoid larva resembles a young tapeworm, but lacks a strobila. If the infected fish is eaten by an appropriate mammal, strobilation occurs and the adult tapeworm establishes itself in the mammalian gut. There a single individual with as many as 4000 proglottids may attain a total length of 18 m.

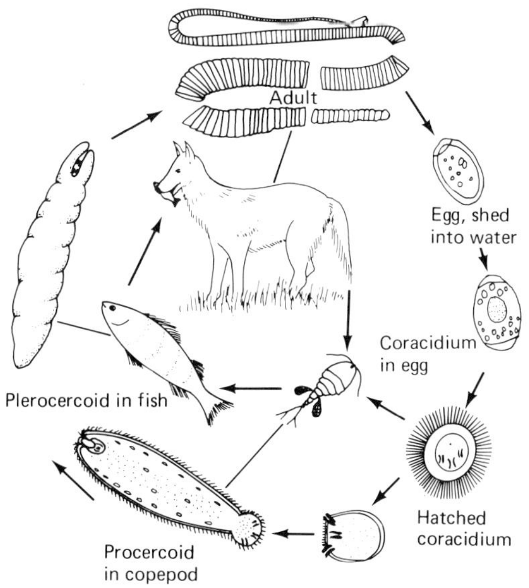

Figure 6.24 Life cycle of *Dibothriocephalus latus*, a tapeworm of fish-eating mammals, including humans. (For description see text.) (From C.P. Hickman, *Biology of the Invertebrates*, 1967, Mosby.)

The pork tapeworm of the human gut, *Taenia solium* (Fig. 6.21), is an advanced parasite with a simplified life cycle. Eggs are passed in human feces and do not hatch unless eaten by the single intermediate host, usually a pig. The first developmental stage is an **oncosphere,** a larva resembling a coracidium without its cilia. Oncospheres burrow through the pig's intestinal wall and enter the bloodstream, where they are transported to striated muscles. There each oncosphere develops into a **cysticercus** larva, also known as a bladder worm (Fig. 6.25). The cysticercus has an invaginated scolex and encysts in the pig's muscles. When a human eats infected, undercooked pork, the scolex evaginates and the tapeworm attaches to the lining of the small intestine, where it completes adult development.

Subclass Cestodaria

About 15 species of cestodes that are not tapeworms comprise the subclass Cestodaria. These flatworms are endoparasites in primitive fish and share some traits with the trematodes. They are nonstrobilated and possess only a single, hermaphroditic reproductive system; suckers are sometimes present (Fig. 6.26). Cestode affinities include the lack of a digestive system and the presence of similar larval forms.

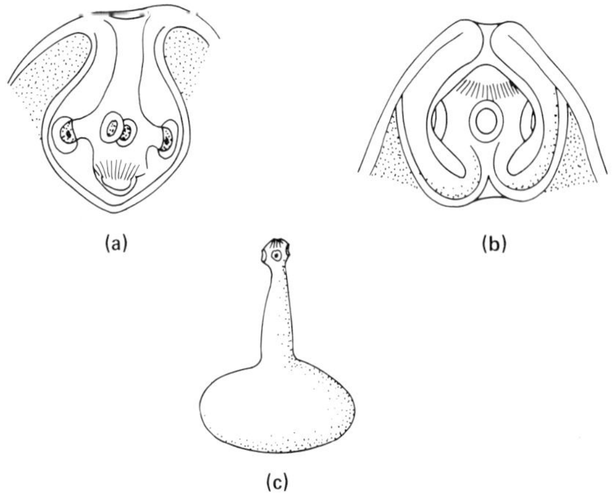

Figure 6.25 The cysticercus larva or bladder worm of *Taenia solium.* (*a*) The invaginated scolex; (*b*) The scolex partially evaginated; (*c*) The fully evaginated scolex of the cysticercus larva just before attachment to the intestinal lining of the host. (From T.J. Parker and W.A. Haswell, *A Textbook of Zoology, Vol. I*, 1957, Macmillan.)

Flatworm Evolution

The evolution of Platyhelminthes is an important subject with a history of spirited debate. Not only the origin of the Bilateria, but also the evolution of two major parasitic groups are involved. A number of theories have been advanced concerning the genesis of the phylum, but space permits discussion only of the two most widely accepted proposals.

The first proposal argues for the evolution of acoel turbellarians from a radial ancestor, the planula larva of Coelenterata. Recall that acoels are considered the most primitive flatworms by many specialists. They lack true reproductive and protonephridial systems, they have no gut, and their nervous organization is rudimentary. Acoel simplicity encourages comparison with the planula larva. Both are small marine organisms that move by ciliary action. In the common absence of a gut lumen, both groups are filled with a solid central mass of nutrient cells. To reconstruct the supposed transition from a planula stage to a primitive flatworm, it would remain primarily for the coelenterate larva to develop creeping locomotor patterns. Resultant selective pressures would be directed toward the development of bilateral symmetry, with accom-

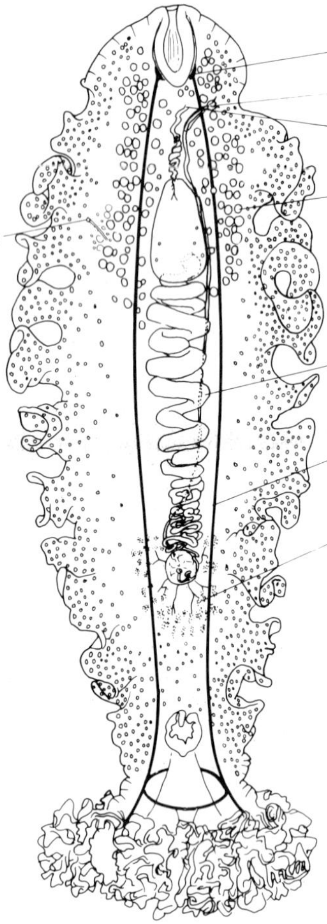

Figure 6.26 A diagram showing the major structural features of the cestodarian, *Gryocotyle fimbriata.* (From Hyman, *The Invertebrates, Vol. II.)*

panying cephalization and muscular, nervous, and sensory elaborations. As various organ systems were perfected, the higher turbellarian orders would appear.

A second theory proposes that acoel flatworms arose directly from protozoan stock. This theory, first encountered in the discussion of protozoan evolution (see Chap. 2), describes multinucleate ciliates as the immediate ancestors of the acoelomates. Cited in support of this proposal are the syncytial condition of some acoel tissues and the abundance of body cilia in lower flatworms. Both the planuloid and the ciliate theories of acoelomate origin are discussed in the concluding chapter.

Neorhabdocoels are likely candidates as the turbellarian ancestors of both parasitic classes. Several neorhabdocoel families are commensal or ectoparasitic on mollusks and echinoderms, and certain species bear a strong structural resemblance to trematode rediae. Although both probably stem from this same turbellarian order, trematodes and cestodes apparently had completely independent origins.

Parasitic flatworms can undergo incredibly complex life cycles. The difficulty of completing a single trematode or cestode generation is staggering, and questions arise as to what advantages these parasites accrue from such devious ways and how those ways came to be. There are no clear answers, but one theory suggests that the life cycles of modern parasites recapitulate their evolutionary histories. According to this theory, intermediate hosts originally were primary hosts; as advanced and presumably more profitably parasitized creatures evolved, adult trematodes and cestodes adapted to feed on them, while larval forms remained parasitic on the original hosts. Perhaps supporting this theory is the fact that certain neorhabdocoels, the alleged ancestors of flukes and tapeworms, are parasitic on mollusks. Recall that mollusks are frequent intermediate hosts of trematodes.

Certain flatworm characters may have facilitated the assumption of parasitism by so many members of the phylum. Indeed, on several counts, platyhelminths were *preadapted* for a parasitic way of life. Flatworms are thin, wriggly animals with a very generalized body form. They are supple organisms, capable of accommodating themselves to a number of physical habitats. Adhesive organs often are used in the locomotion of free-living species, and the adaptation of these organs for parasitic attachment would not appear difficult. Finally, the introduction of a true organ system for sexual reproduction minimizes the continuity problems that all parasites face.

By most measures of biological success, platyhelminths certainly set no records, but the phylum exhibits a sound body plan and, perhaps more noteworthy, introduces several important features in invertebrate evolution. Flatworm innovations include bilateral symmetry, a true mesoderm, the protonephridial system, a brain, and a true reproductive organ system. However, flatworms do have numerous problems, most of which relate to their acoelomate condition. Perhaps no other single factor preadapted the phylum to parasitism as much as the extensive metabolic problems presented by its rather solid mesoderm. As with the protozoans, the sponges, and the radiates, widespread free-living success also eludes the flatworms. Parasitic environments have become their refuge in the invertebrate world.

The Nemertines

The nemertines are unsegmented, acoelomate animals with a history of many names. They have been classified with an array of invertebrate groups, including planarians, leeches, nematodes, echiurans, and hemichordates. Their current status as an independent phylum is slightly more than a century old. Today the phylum is known as Nemertina, Nemertea, or Rhynchocoela, and its members are commonly called proboscis worms, ribbon worms, or nemerteans. Because they are so often compared with the more numerous and, phylogenetically and ecologically, more important flatworms, nemertines are often remembered only as "those other acoelomates."

The relatively small number of nemertine species—less than 600 have been identified—may overshadow the adaptive accomplishments of these worms. Structurally, they are more complex than flatworms, and several innovations enable them to achieve a larger body size than most platyhelminths. These innovations include a tubular alimentary canal with a mouth and anus and a closed circulatory system. Nemertines are also distinguished by their unique structure—a long, eversible proboscis apparatus contained within a special cavity, the **rhynchocoel**.

Although there are a few freshwater and terrestrial species, most nemertines live in the marine intertidal zone. They are found in all the oceans of the world, living under rocks and seaweed or burrowing in muddy sand. Some species live in selfconstructed, mucus-filled tubes, and a few are pelagic. Like flatworms, nemertines are long, thin organisms with cylindrical to ribbon-shaped bodies (Fig. 6.27). Their round or heart-shaped anterior does not constitute a distinct head because it does not contain the brain. Characteristically longer than flatworms, nemertines may attain lengths of up to 2 m.

Nemertine Unity

MAINTENANCE SYSTEMS

General Morphology

Nemertines resemble flatworms, but are more highly organized. A simple, ciliated epidermis overlies a dermis of connective tissues (Fig. 6.28). Numerous mucous gland cells are present in the epidermis or, less often, in the dermis. Ducts from these single-celled glands open on the body surface, which is well lubricated. Below the dermis are alternating longitudinal and circular muscle layers. Nemertine muscles are larger, stronger, and more ordered than those of flatworms. The subdermal musulature extends almost to the gut, from which it is separated by the remainder of the parenchyma. Dorsoventral muscles are often present in this narrow region. The gut wall itself is simple and turbellarian in character, rarely possessing any musculature of its own.

The most distinctive nemertine structure is the proboscis apparatus (Figs. 6.28b and 6.29). This multipurpose organ consists of a hollow tube lying in the rhynchocoel, a blind cavity completely separate from the gut. The proboscis-rhynchocoel complex opens through a proboscis pore above and in front of the mouth. The rhynchocoel wall is highly muscularized. The contraction of this wall exerts pressure on fluids within the rhynchocoel; this pressure causes the pro-

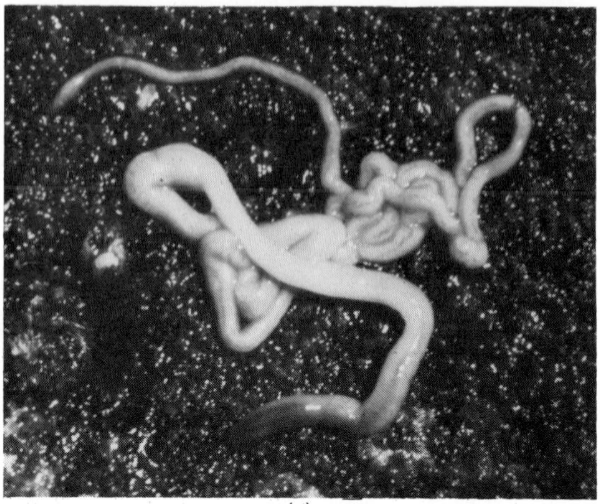

(a)

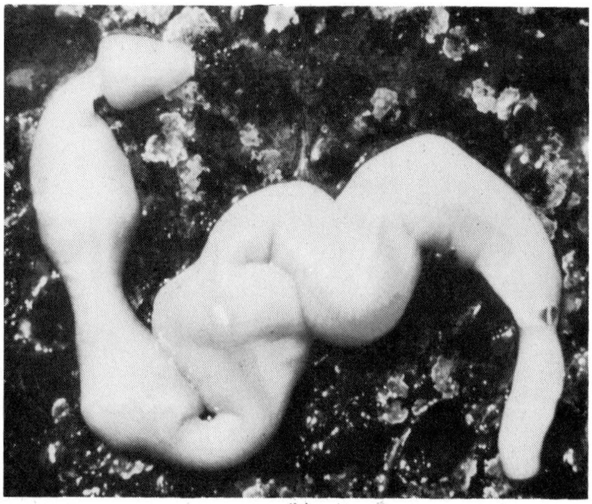

(b)

Figure 6.27 Two photos of nemertine worms. (*a*) *Amphiporus imparispinosus;* (*b*) *Amphiporus bimaculatus.* (Courtesy of Ward's Natural Science Establishment.)

boscis to evert and shoot outward through the proboscis pore (Fig. 6.29b). Typically, the proboscis is coiled before it fires and, when extended, may be much longer than the nemertine itself. A barb or stylet (Fig. 6.29) may be present distally. After extension, the nemertine withdraws the proboscis apparatus with its proboscis retractor muscle, a flexible set of fibers joining the tube to the rear of the rhynchocoel (Fig. 6.29). The proboscis may be fired so vigorously that these retractor muscles rupture. In such an extreme case, the proboscis apparatus is lost, but a new one can be regenerated.

The rhynchocoel, a body cavity totally contained within the mesoderm, fulfills some definitions of a true coelom. But the rhynchocoel does not surround endodermal tissues; nor does it provide space for the elaboration of mesodermal organ systems. It is thus in no way homologous with the secondary body cavities of coelomate animals. The rhynchocoel is merely an adjunct of the proboscis apparatus.

The proboscis may be used for defense, for locomotion, or for burrowing. Its primary function, however, is in prey capture.

Digestion and Nutrition

Nemertines are carnivores and scavengers, feeding primarily on annelids and small crustaceans. They seize food with their proboscis apparatus, which coils around the prey and exudes a sticky fluid. Nemertines with stylets on their proboscis tips may puncture and inject toxins into their victims. After capture, food is swallowed whole.

The nemertine digestive system represents one of the phylum's major adaptive improvements over the Platyhelminthes. The system comprises an anterior ventral mouth, a foregut, a long diverticulated intestine, and an anus (Fig. 6.30). Food is sucked through the mouth and into the foregut, where prey may be killed by acid secretions. Digestion occurs in the intestine, where glandular and phagocytotic cells are involved in the extracellular and intracellular breakdown of nutrients. Intestinal cilia move materials through the gut.

All of these characteristics are very turbellarian. The special feature of the nemertine digestive system is its anus. This second opening allows a one-way passage of nutrients through the gut. Eating and elimination can occur simultaneously, while food can be subjected sequentially to different enzymes. Gut structures can become adapted for precise functions at each sequential stage in digestion, although this possibility is minimally realized by the nemertines. The nemertine anus thus permits more orderly and efficient digestion than is possible among turbellarians.

Circulation and Respiration

Nemertines are the only acoelomates with a true circulatory system. Their system is a closed one, as blood remains within the walled confines of blood vessels and does not bathe the tissues directly. In its simplest form, the nemertine circulatory system comprises two vessels, one on either side of the gut, with a series of connecting blood spaces in the head and around the anus (Fig. 6.30). In some larger species, there is a middorsal vessel, and numerous connectives surround the gut and/or rhynchocoel (Fig. 6.30). Although the

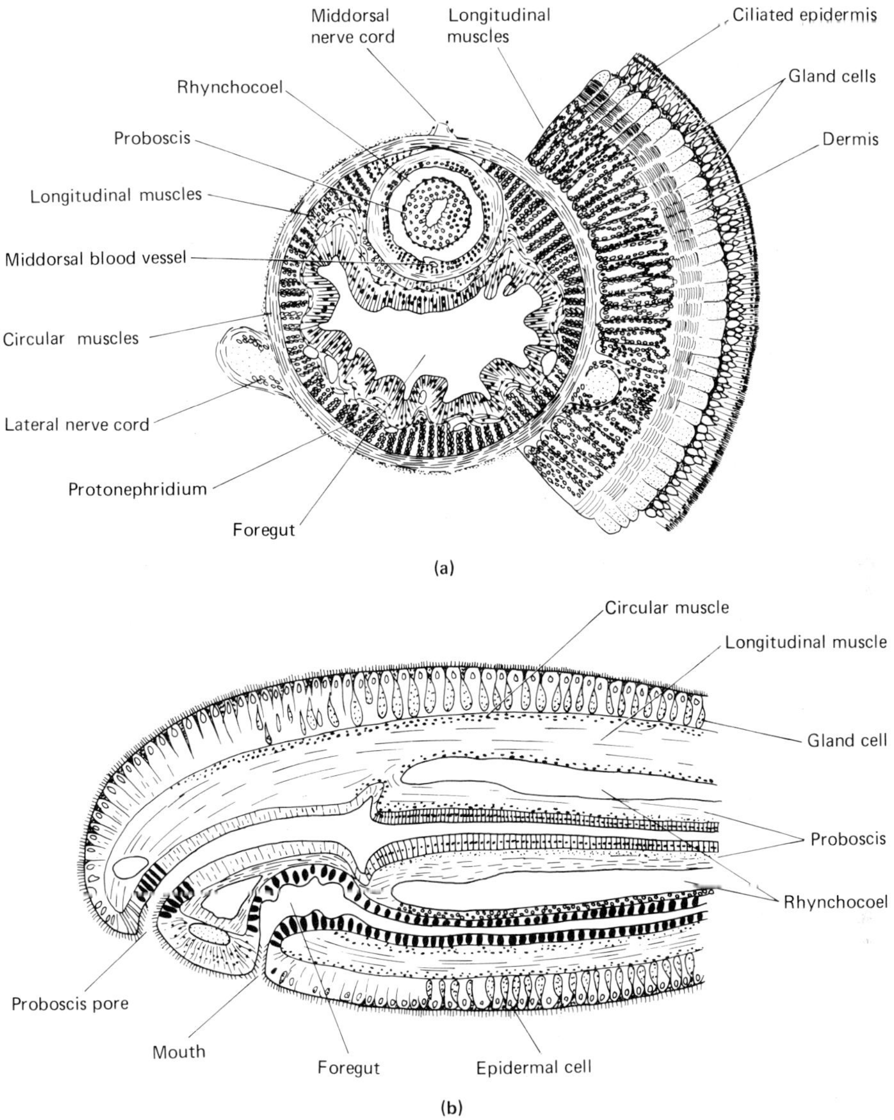

Figure 6.28 (*a*) Cross section through the anterior end of a nemertine worm, *Lineus*; (*b*) Longitudinal section through the anterior end of *Procarinina*, a nemertine with a separate mouth and proboscis pore. (From Hyman, *The Invertebrates, Vol. II.*)

major blood vessels are contractile, there are no organized routes of circulation. Blood merely sloshes as a result of local contractions, and flow reversals are frequent. Some species have blood pigments which may aid in oxygen transport, but most nemertine blood is colorless. In those species with pigmented blood, pigments always occur within corpuscles, an unusual occurrence outside the phylum Chordata.

Gas exchange in these thin animals usually involves simple diffusion. Some of the larger nemertines with well–vascularized foregut walls rhythmically swallow water, apparently for respiratory purposes. Such action may represent the adaptive beginnings of a respiratory system.

Excretion and Osmoregulation

A protonephridial system comprises flame cells, tubules, and nephridiopores, much as in flatworms. Primitively, there is one pair of protonephridia with

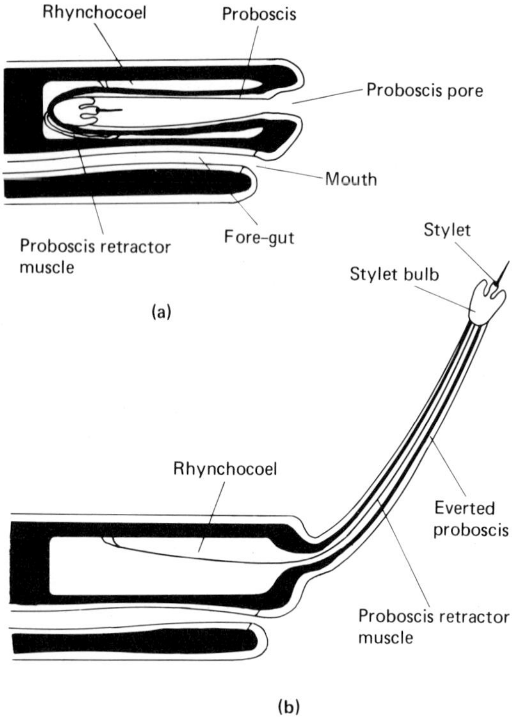

Figure 6.29 A diagrammatic representation of the proboscis apparatus of a nemertine worm. (*a*) Proboscis, retracted; (*b*) Proboscis everted and extended. (From Russell-Hunter, *A Biology of Lower Invertebrates*.)

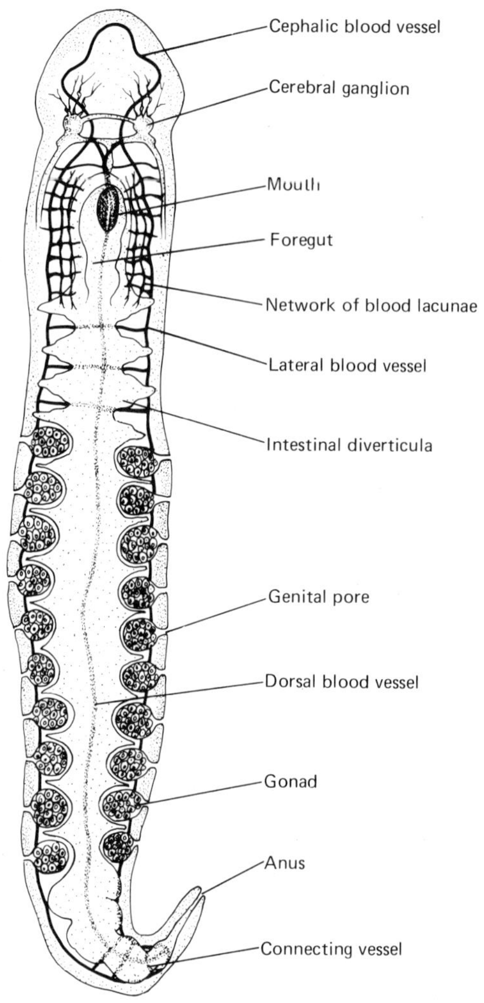

Figure 6.30 A diagram showing the internal anatomy of a generalized nemertine worm. (From L. Joubin, In *Faune Francais*, R. Blanchard and J. de Guerne, Eds., Soc d'Editions Scientifiques, 1893, Paris.)

nephridiopores at each side of the foregut. Frequently the system is much more complex and, as in the triclad flatworms, numerous nephridiopores may exist. In most nemertines, these elements are associated with the circulatory system (Fig. 6.31). Flame cells may be bathed directly in blood, strongly indicating a genuine excretory function. In addition, the protonephridial system handles osmoregulatory needs.

ACTIVITY SYSTEMS

Muscles and Nerves

The primary effector organ of nemertines is the proboscis apparatus. Otherwise, the phylum's musculature is basically flatwormlike, but more powerful. Muscle arrangements in the body wall vary among subgroups, indicating considerable adaptive experimentation with the locomotor system. However, the various arrangements are all mechanically less efficient than the highly stereotyped pattern found in annelid worms (see Chap. Eight). Reductions in the density of parenchymal tissues and corresponding increases in the volume of intercellular fluids improve the efficiency of the nemertine body skeleton (as compared to that of flatworms) and thus enhance locomotor capability. As a consequence, some nemertines can burrow in soft substrata—something platyhelminths cannot do.

The nervous system likewise resembles that of advanced turbellarians, although the major longitudinal

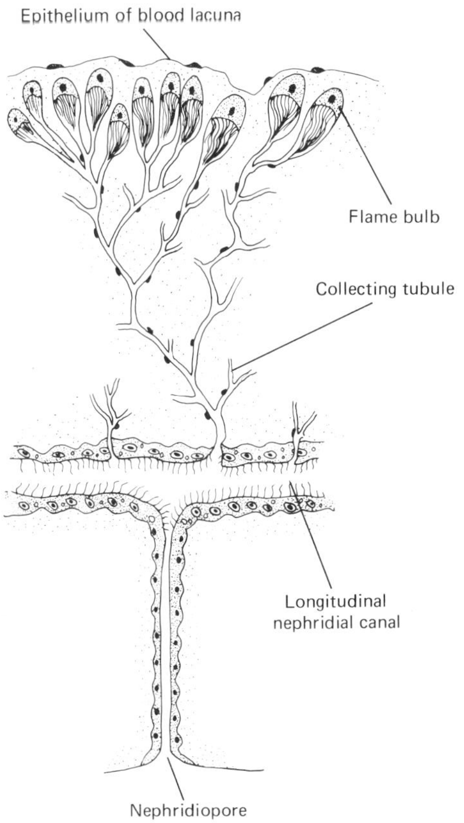

Figure 6.31 A complex nemertine protonephridium closely associated with the circulatory system and emptying into a nephridial canal. (From W.R. Coe, Trans. Conn. Acad. Arts and Sciences, 1943, 35:129.)

nerve trunks are lateral rather than ventral (Fig. 6.32). A middorsal cord is also prominent in many species. The nemertine brain is large for an acoelomate and consists of four lobes in a ring around the rhynchocoel.

Sense Organs

Nemertines and turbellarians have similar sensory structures. Tactile bristles are exposed through the epidermis, and pigment cup eyes occur in the anterior region. Anterior, ciliated grooves or slits are richly supplied with tactile sites and chemoreceptors. Other chemosensitive structures include the frontal organ, a ciliated pit found at the anterior tip of the animal, and the lateral organs, similar pits situated near the excretory pores. Statocysts are found in a few nemertines. Characteristic of the phylum is a pair of dorsal **cerebral organs** (Fig. 6.33), consisting of ciliated pits whose sensory and glandular cells associate with brain neurons. Food juices stimulate ciliary action in these deep pits, suggesting a chemoreceptive function for the organs.

Behavior

As might be expected, nemertines behave much like flatworms. Larger species swim or crawl by peristaltic contractions, while smaller worms glide under ciliary power along the slime trails laid down by their mucous glands. An extended proboscis apparatus may grasp a stationary object and then pull the nemertine forward as it retracts. In burrowing, the organ is thrust into the substratum and then forced deeper by whole-body contractions. Most nemertines avoid light and, like

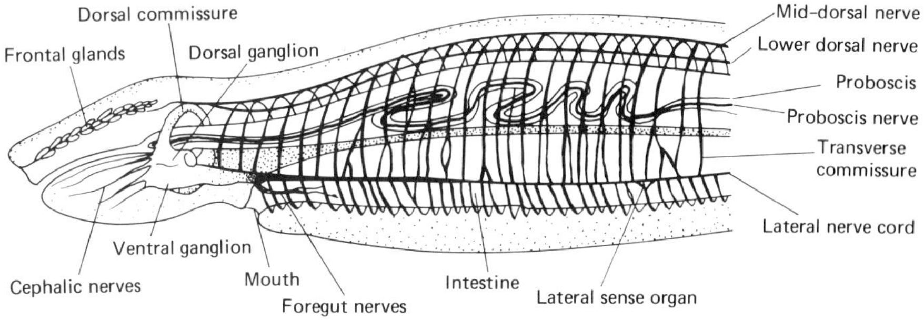

Figure 6.32 A lateral view of the anterior end of the nemertine *Tubulanus* showing the nervous system. (From O. Burger, *Fauna und Flora des Golfes von Neapel*, 1895.)

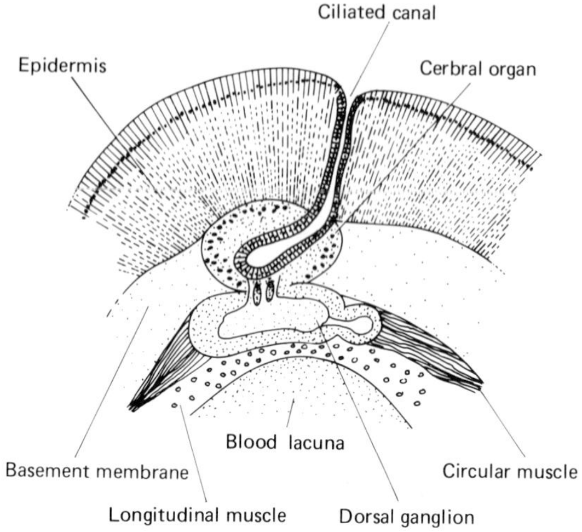

Figure 6.33 A dorsal cerebral organ of a nemertine worm. (From W.R. Coe, Trans. Conn. Acad. Arts and Sciences, 1943, *35*:129.)

flatworms, are strongly thigmotactic. Food is detected largely by chemical means, and in some species such detection is possible over distances of several centimeters.

CONTINUITY SYSTEMS

Reproduction and Development

Sexually, nemertines do not follow the flatworm plan. As a rule, proboscis worms are diecious, although the sexes are not readily distinguished by external features. Both males and females display simple, multiple gonads arising from parenchymal cells between the intestinal diverticula (Fig. 6.30). As each gonad sac matures, a small duct grows from it to the external surface, and gametes are released for external fertilization. Many nemertines enhance their reproductive potential when a male and female worm occupy a common burrow during the spawning period or when males discharge their sperm while crawling over female bodies.

Spiral, determinate cleavage produces a hollow, ciliated blastula. Development may be direct with no larval stages, in which case a solid gastrula is formed. Further differentiation and flattening with the development of bilateral symmetry precede the hatching of a young, ciliated nemertine, which superficially resembles a neorhabdocoel turbellarian. In those species with indirect development, a hollow gastrula differentiates into a helmet-shaped **pilidium** larva (Fig. 6.34). Similar to the Müller's larva of polyclad flatworms, the pilidium experiences a brief, free-swimming existence and then undergoes a complex metamorphosis into a young nemertine.

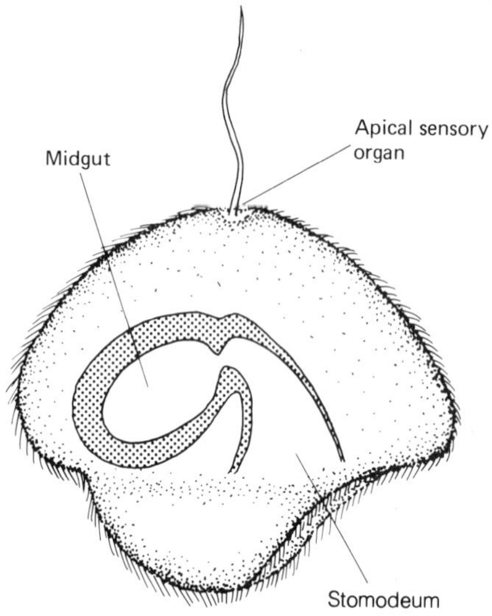

Figure 6.34 The pilidium larva of the nemertine *Cerebratulus*. (From Hyman, *The Invertebrates, Vol. II.)*

Regeneration and Asexual Reproduction

The related phenomena of regeneration and asexual reproduction are not as common among the Nemertina as among the Platyhelminthes. Nemertines are very delicate animals, and even careful handling can fracture them into many pieces. This fragility is a great problem in studying all phases of their biology. Yet, while nemertines are easily fragmented, only an anterior piece containing the brain usually regenerates into a whole animal. This condition suggests the existence of more centralized nervous and hormonal controls, representing another area in which this phylum is more specialized than the Platyhelminthes. In a few genera, notably *Lineus*, regeneration from posterior, brainless fragments is possible provided they contain a piece of nerve cord. Indeed, some species of *Lineus* regularly reproduce asexually by fragmenting into numerous small pieces, each of which then forms a daughter nemertine.

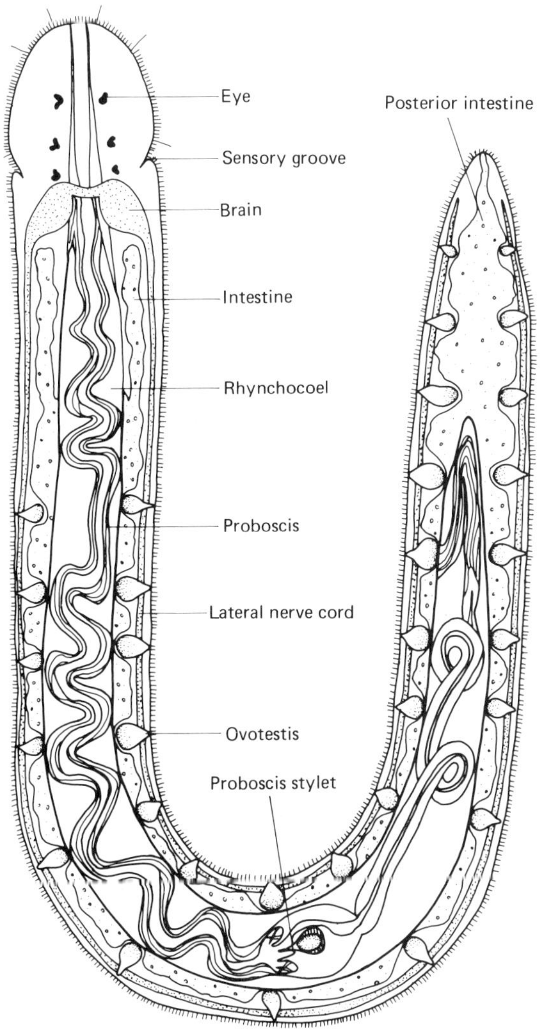

Figure 6.35 A diagram showing the general anatamy of *Prostoma rubrum*, a freshwater nemertine worm. (From R.W. Pennak, *Fresh-Water Invertebrates of the United States*, 2nd ed., 1978, Wiley-Interscience.)

Nemertine Diversity

The nemertines are a fairly uniform phylum. Anatomical variations are minor and include various arrangements of the musculature, the protonephridial and nervous systems, and the circulatory vessels. Two classes are recognized. Class Anopla includes the nemertines whose proboscises lack barbs; these marine worms are considered the more primitive class, and some species display the pilidium larva. *Tubulanus, Cephalothrix, Cerebratulus,* and *Lineus* and representative members. The second class of nemertines is the Enopla, a group with barbed proboscises and direct development. The single freshwater genus *Prostoma* (Fig. 6.35) and the subtropical terrestrial *Genonemertes* are members of this class. The many marine representatives include *Amphiphorus* and *Malacobdella* (species of which are commensals on mollusks).

Nemertine Evolution

Almost certainly, the nemertines originated from turbellarians, probably from the neorhabdocoels. Relative to their flatworm ancestors, they have attained a more complex and powerful musculature, a versatile proboscis apparatus, a circulatory system, and a one-way gut. These innovations may be responsible for the greater relative size of nemertines and the absence of parasitic species.

Yet, in species number and ecological diversity, nemertines are far less successful than platyhelminths. They have achieved length without segmentation or the development of a body cavity, and they lack an efficient hydrostatic skeleton. Like their flatworm cousins, nemertines ultimately are limited by their acoelomate condition.

Parasitism

Predators are animals that kill and eat other animals, generally smaller and weaker than themselves. We do not always like predators and their killings, but their life-style is accepted as an often cruel but necessary part of the animal world. Certain large predators, such as lions and eagles, even enjoy some cultural status. Provided their prey is not a competitively popular little beast (say, a "defenseless" baby antelope or sister's pet rabbit), these killers are held in quite high esteem. Men name their social clubs after them; armies and football teams march under their images; kings commission their immortalization in stone. In contrast, parasites, which are rather small creatures, regularly attack animals much larger than themselves, and no well-adapted parasite will kill the animal on which it feeds. But what might pass for bravery on the one hand and charity on the other fails to merit parasites even the slightest praise. Instead, they are culturally despised animals.

Parasites live and feed on, and at the expense of, "host" organisms. At first inspection, the parasitic life does not seem difficult, and certainly is anything but heroic. An established parasite clings to its host's surface or lives somewhere within the host's body and simply appropriates a share of local nutrients. In comfortable surroundings, the parasite seldom finds reason to move. But becoming *established* in or on a host is an awesome task. Indeed, parasitism presents some of the greatest adaptational challenges ever met by living things. In adapting to a number of common problems, parasites of different taxa exhibit considerable evolutionary convergence.

The two basic problems of parasitism are locating a specific host and maintaining residency there. Relative to the size of most parasitic creatures, the world is a very large place and contains few of the particular organisms that a given parasite may infest. In the parasitic world, there is no more arduous task than the initial one—the mere but vital location of a host. Once a potential host is found, the parasite sets up housekeeping. It must withstand immune and digestive reactions by the host and any other attempts to dislodge it. Finally, the parasite must extract a living wage from its host, ideally without causing too much damage.

It seems then that parasites have two quite contradictory needs: one, to roam and search; the other, to hold fast against all pressures. Typically, this dilemma is resolved by the assignment of opposing tasks to different stages of a complex life cycle. Motile larvae locate hosts, while the adult stage is adapted to full-time, stationary parasitism.

Adult parasites are often sessile animals, whose sensory, nervous, and muscular systems have degenerated. They may grasp their hosts with special hooking or sucking organs. Their digestive system often is reduced and may be entirely absent, in which case absorption occurs over the general body surface. This surface often is reinforced by a cuticle protecting the parasite from digestive or immune responses by the host. Ideally, however, a parasite triggers no immune response. A well-adapted parasite is one that the host accepts as its own kind. This compatibility serves both parasite and host. The host is spared the pain and/or death that parasitic infection can cause, while the parasite enjoys a healthy home.

Enter problem two. Lest our adult parasite die without issue, it must reproduce. Unfortunately, infection by multiple generations soon fouls the home environment, dooming host and parasites alike. Further, even under optimal conditions, the host's life span is limited. The only real solution is that the offspring leave to seek their own habitats in the host-scarce external world.

Parasitic offspring face all but impossible odds against finding a suitable host. However, on the principle that the near impossible becomes probable if tried often enough, parasitic parents provide quite adequately for their posterity. All of the metabolic energies spared by degenerate maintenance and activity systems are massed to support truly prolific reproductive organs. Many adult parasites sexually produce thousands upon thousands of eggs. Motile larvae hatch, and their great search is begun. Multiple larval stages are common, and asexual reproduction at any or all of these stages can exponentially increase the larval population. Intermediate larval hosts are the frequent sites of these reproductive events. When the final or primary host is located, the parasite matures sexually and the life cycle is complete.

But most parasitic larvae perish without finding a host, even an intermediate one, and infinitesimally few survive the journey to adulthood. The parasite's life is difficult, but of course, so is the life of the predator—be it lion, squid, or amoeba. Aside from human judgments, parasitism is, in this sense, parallel to predation. It is, at least, a means of surviving in a highly competitive animal world.

for Further Reading

Arai, H. P. (Ed.) *Biology of the Tapeworm Hymenolepsis diminuta.* Academic Press, New York, 1981.

Ax, O. Relationships and phylogeny of the Turbellaria. In *The Lower Metazoa.* (E. C. Dougherty, Ed.), pp. 191–224. University of California Press, Berkeley, 1963.

Bayer, F. M., and Owre, H. B. *The Free-Living Lower Invertebrates.* pp. 144–204. Macmillan, New York, 1968. (A nicely illustrated account of the free-living acoelomates. Coverage of the nemertines is particularly good.)

Clark, R. B. *Dynamics in Metazoan Evolution.* Clarendon Press, Oxford, 1964.

Combes, C., et al. The world atlas of cercariae. *Mem. Mus. Natl. Hist. Nat. Ser. A. Zool.* 115: 1-235, 1980.

Darlington, J. T. and Chandler, C. M. A survey of the planarians of Arkansas. *Southwest Nat.* 24: 141–148, 1979.

Erasmus, D. A. *The Biology of Trematodes.* Crane, Russak, New York, 1973.

Florkin, M., and Scheer, B. T. *Chemical Zoology, Vol. II.* Academic Press, New York, 1968. (In the seven chapters on platyhelminths, nutrition and digestion, intermediary metabolism, respiration, chemical aspects of ecology, and growth development, and cultural methods for parasitic flatworms are reviewed.)

Frandsen, F. Discussion of the relationships between *Schistosoma* and their intermediate hosts, assessment of the degree of host-parasite compatibility and evaluation of schistosome taxonomy. *Z. Parasitenkd.* 58: 275–296, 1979.

Gibson, R. *Nemerteans.* Hutchinson University Library, London, 1972. (The most recent and most detailed treatment of nemertine biology.)

Gibson, R. and Moore, J. Freshwater nemertines. *Zool. Linn. Soc.* 58: 177–218, 1976.

Hyman, L. H. *The Invertebrates, Vol. II: Platyhelminthes and Rhynchocoela.* McGraw-Hill, New York, 1951.

Heitkamp, U. The reproductive biology of *Mesostoma ehrenbergii. Hydrobiologica.* 55: 21–32, 1977.

Henley, C. Platyhelminthes (Turbellaria). In *Reproduction of Marine Invertebrates, Vol. I* (Giese, A. C. and Pearse, J. S., eds.), pp. 267–343. Academic Press, New York, 1974.

Koopowitz, H., Keenan, L., and Bernardo, K. Primitive nervous systems: Electrophysiology of inhibitory events in flatworm nerve cords. *J. Neurobiol.* 10: 383–396, 1979.

Lambert, A. Study of the larval stages of Monogenea of fishes. *Ann. Parasitol. Hum. Comp.* 53: 551–560, 1978.

Lebedev. B. I. Some aspects of monogenean existence. *Folia Parasitol.* 25: 131–136, 1978.

Martin, G. G. Ciliary gliding in lower invertebrates. *Zoomorphologie* 91: 249–262, 1978.

Minelli, A. A taxonomic review of the terrestrial planarians of Europe, *Bull. Zool. 44: 399–*420, 1978.

Moraczewski, J. Asexual reproduction and regeneration of *Catenula. Zoomorphologie* 88: 65–80, 1977.

Mueller, J. F. Helminth life cycles. *Am. Zool.* 5: 131–139, 1965.

Noble, E. R., and Noble, G. A. *Parasitology,* 4th ed., Lea and Febiger, Philadelphia, 1976.

Pantelouris, E. M. *The Common Liver Fluke.* Pergamon Press, Oxford, 1965. (A detailed account of the biology of *Fasciola hepatica,* including host pathology and treatment and control aspects)

Prusch, R. D. Osmotic and ionic relationships in the freshwater flatworm. *Dugesia dorocephal Comp. Biochem. Physiol.* 54A: 287–290, 1976.

Read, C. P. *Parasitism and Symbiology.* Ronald Press, New York, 1970. (A text stressing host-parasite relationships)

Riser, H. W. Nemertinea. In *Reproduction of Marine Invertebrates, Vol. I.* (Giese, A. C. and Pearse, J. S., eds.), pp. 359–389. Academic Press, New York, 1974.

Riser, N. W., and Morse, M. P. (Eds.) *Biology of the Turbellaria.* McGraw-Hill, New York, 1974.

Roberts, T. M., Ward, S., and Chernin, E. Behavioral responses of *Schistosome mansoni* miracidia in concentration gradients of snail-conditioned water. J. Parasitol. 65: 41–49, 1979.

Roe, P. Life history and predator-prey interactions of the nemertean *Paranemertes peregrina. Biol. Bull.* 150: 80–106, 1976.

Rose, S. M. *Regeneration.* Appleton-Century-Crofts, New York, 1970. (One chapter offers a good introduction to studies on flatworm regeneration)

Smyth, J. D. *The Physiology of Cestodes.* Oliver and Boyd, Edinburgh, 1966.

Smyth, J. D. *The Physiology of Trematodes.* Oliver and Boyd, Edinburgh, 1966.

Thorpe, W. H., and Davenport, D. (Eds.) *Learning and Associated Phenomena in Invertebrates, Animal Behavior Supplement.* Bailliere, Tindall and Cassell, London, 1964. (Includes three papers and a special discussion on the controversial topic of learning in planarians)

Tyler, S. Distinctive features of cilia in metazoans and their significance for systematics. *Tissue Cell.* 11: 385–400, 1979.

Wardle, R., McLeod, J. A., and Radinovsky, S. *Advances in the Zoology of Tapeworms.* University of Minnesota Press, Minneapolis, 1974.

Wilson, R. A., and L. A. Webster. Protonephridia. *Biol. Rev.* 49: 127–160, 1974.

7

The Pseudocoelomates

A body cavity represents the major adaptive solution to the metabolic problems inherent in the acoelomate plan. Replacing the parenchyma with a fluid-filled space establishes a tube-within-a-tube construction. The outer tube, or body wall, can move without disturbing the inner tube, comprising the digestive tract and the other internal organ systems. Between these tubes, a body cavity filled with an osmotically favorable fluid serves several functions. It provides a circulating medium for the distribution of metabolic materials. The alimentary canal and other organ systems can expand in size with minimal interference from surrounding tissues; and the fluid-filled cavity itself serves as a hydrostatic skeleton *par excellence.* The adaptive value of a body cavity is demonstrated by the fact that all higher animals are organized according to the tube-within-a-tube plan.

There are two basic types of body cavities: the true coelom and the pseudocoel. Most successful is the true coelom. The true coelom is a cavity completely within the mesoderm and lined throughout with a special mesodermal tissue called **peritoneum.** The **pseudocoel** is not a mesodermal cavity at all, but rather a space between the ectodermal and mesodermal tissues of the body wall and the endodermal gut.

Animals with pseudocoels are the subject of this chapter. The pseudocoelomates (as they are called) are mostly small, vermiform animals living in aquatic or parasitic habitats. Nematodes and rotifers are the most successful pseudocoelomates. Several smaller groups complete the list: gastrotrichs, kinorhynchs, nematomorphs, acanthocephalans, and possibly gnathostomulids. Acanthocephalans and gnathostomulids are recognized as distinct phyla. However, debate continues concerning the phylogenetic relationships among the other five pseudocoelomate groups.

The Aschelminthes

Nematodes, rotifers, gastrotrichs, kinorhynchs, and nematomorphs have certain features in common and collectively are referred to as the Aschelminthes. Besides being pseudocoelomates, they are all rather elongate and vermiform with little or no cephalization. Aschelminths possess an outer resistant cuticle. Their subcuticular hypodermis is often syncytial and bounds the body wall musculature, which is not stratified and typically lacks circular fibers. Adhesive glands are characteristic of many species. Respiratory and circulatory organs are absent from the group, but an excretory system is present, usually as protonephridia. The nervous system is simple, with a circumenteric ring and associated ganglia giving rise to a few large, longitudinal nerve cords. The straight and tubular gut includes a prominent pharynx and a subterminal anus.

Most aschelminths are diecious, and their simple reproductive system may terminate in common with the gut, thus forming a cloaca. Reproduction by **parthenogenesis** (i.e., without fertilization) is widespread in these animals. The strange phenomenon of cell or nuclear constancy is characteristic of some aschelminths. At some point during their development, mitosis ceases; thus these organisms are limited to a constant number of cells (or nuclei in the case of syncytial tissues).

Aschelminth classification is controversial. Some researchers argue for a close relationship between the groups and so consider them a single phylum of five classes. Other investigators discount such intimacy and rank each of the five groups as a separate phylum. Still other workers, seeking to arrange the aschelminths into two or three principal groups, consider them as classes temporarily without phylum assignment. This latter approach has some merit, but we suggest that any temporary classification be on the phylum level. Accordingly, in this text, "Aschelminthes" is a term without official taxonomic position; it refers tentatively to five separate but related phyla.

The Nematodes

The Nematoda, or roundworms, represent the most successful aschelminth phylum. Indeed, with 10,000 described species, they outnumber all other pseudocoelomates combined. They are ecologically among the most diverse invertebrates, being found in virtually every environment. Although widely distributed, nematodes remain basically aquatic animals, living in the ocean, fresh water, the bodies of other organisms, or the water film surrounding terrestrial soil particles. Nematodes are overwhelmingly abundant. They are possibly the most numerous metazoan organisms of the ocean floor, as estimates of their population density on mucky sea bottoms run into the billions per hectare. A single rotting apple was found to house 90,000 individuals of several species, while 3 to 9 billion of these worms have been counted per hectare in fertile farmlands throughout the world.

Nematode Unity

We might suppose that a large and ecologically diverse phylum like the nematodes would offer unending specializations in morphology, physiology, and behavior. However, nematodes are a surprisingly uniform lot, a factor making them exceedingly difficult to classify. They are characteristically small, unsegmented pseudocoelomates. Most species measure less than a millimeter in length, although some parasitic forms are much longer. Classically vermiform, nematodes are pointed at both ends and have no distinct head (Fig. 7.1). Covered by a cuticle, the external surface is often featureless.

It is not specialization, but rather the lack of it, that makes the phylum so successful. As we shall see, nematodes are very tolerant creatures which adapt readily to various environmental conditions. As do flatworms, nematodes maintain a generalized body plan and a low visibility. However, by virtue of their pseudocoel, nematodes exhibit a more efficient physiology and higher locomotor potential.

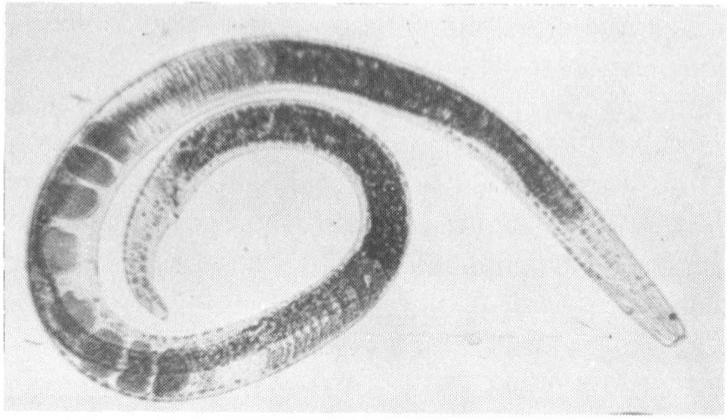

Figure 7.1 Photograph of a marine nematode. (Courtesy of Ward's Natural Science Establishment.)

MAINTENANCE SYSTEMS

General Morphology

Nematode morphology is relatively simple. These worms express the basic tube-within-a-tube plan. The outer tube is the body wall, consisting of a cuticle, a hypodermis, and longitudinal muscles. Nervous and excretory elements typically are embedded within this outer tube. Surrounded by the pseudocoel, the inner tube includes the alimentary canal and a relatively simple reproductive system.

The nematode cuticle comprises three layers of hardened protein (collagen) and organic and inorganic fibers. This tough, protective sheath may be sculptured into spines, ornaments, and rings, falsely suggesting segmentation (Fig. 7.2). The nematode cuticle is not inert; it is permeable to water, certain ions, and respiratory gases, but impermeable to a wide array of noxious substances. Crosswise fibers convey an elasticity to the cuticle.

Beneath the cuticle is a hypodermis. This layer is thickened middorsally, midventrally, and midlaterally into four longitudinal ridges (Fig. 7.3). In smaller nematodes, the hypodermis is cellular and all its nuclei are confined to the ridges; larger nematodes have a syncytial hypodermis with many scattered nuclei. Major longitudinal nerve trunks occupy the dorsal and ventral ridges of the hypodermis, and the lateral ridges of many species house excretory canals. A hypodermal adhesive gland opens at the posterior tip of the body in most free-living nematodes.

The hypodermal ridges divide the body wall musculature into four separate fields (Fig. 7.3). This musculature is unusual in that it includes only longitudinal fibers. Nematode muscles exist as individual units and do not form a stratified tissue. In another departure from the invertebrate norm, these muscles connect cytoplasmically with the longitudinal nerve trunks. Thin extensions from each muscle cell terminate on either the dorsal or the ventral nerve cord.

The nematode pseudocoel is spacious and filled with fluid. Free floating cells are lacking, but fixed cells of uncertain function may occur on the cavity's walls.

Suspended within the pseudocoel is the alimentary canal. The nematode digestive system comprises an anterior mouth, a short buccal cavity, a muscular and glandular pharynx, an intestine, a rectum, and a subterminal anus (Fig. 7.4). The mouth, buccal cavity and pharynx are lined with cuticle and thus constitute a stomodeum; the cuticular walls of the rectum define a proctodeum. Surrounding the mouth is a radial complex of lips and sensory projections. Typically, six cuticular lips, each with an outer and an inner papilla, encircle the oral aperture (Fig. 7.5). Nematode lips display numerous variations, including cuticular elaborations to form teeth, reductions in lip number, and the addition of papillae. Lip symmetry, however, remains radial, a condition which, to some researchers, suggests a sessile ancestry for these worms.

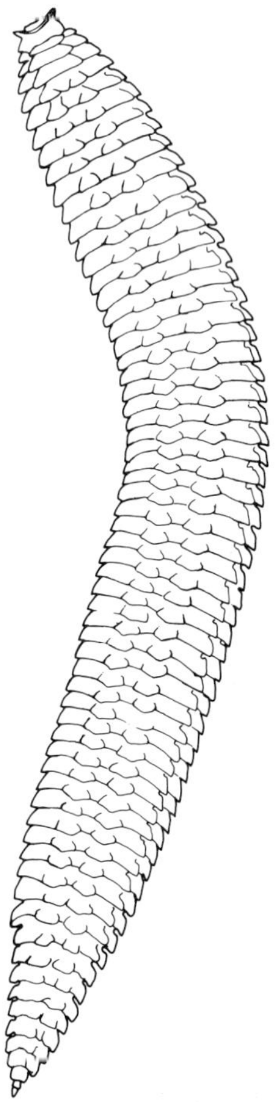

Figure 7.2 The sculptured cuticle of the nematode, *Circonema.* The cuticular rings falsely suggest segmentation. (From L.H. Hyman, *The Invertebrates, Vol. III: Acanthocephala, Aschelminthes and Entoprocta*, 1951, McGraw-Hill.)

Digestion and Nutrition

Consistent with their ecological diversity, nematodes enjoy a wide range of diets. They feed on plants and animals, both as predators and as parasites. Many nematodes are scavengers, while others feed entirely

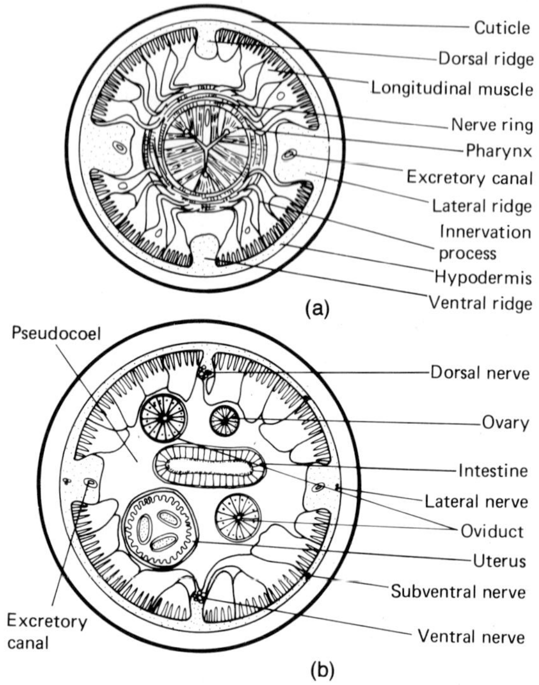

Figure 7.3 Transverse sections through a generalized nematode. (*a*) Pharyngeal region; (*b*) Intestinal region. (From D.L. Lee, *The Physiology of Nematodes*, 1965, Oliver and Boyd.)

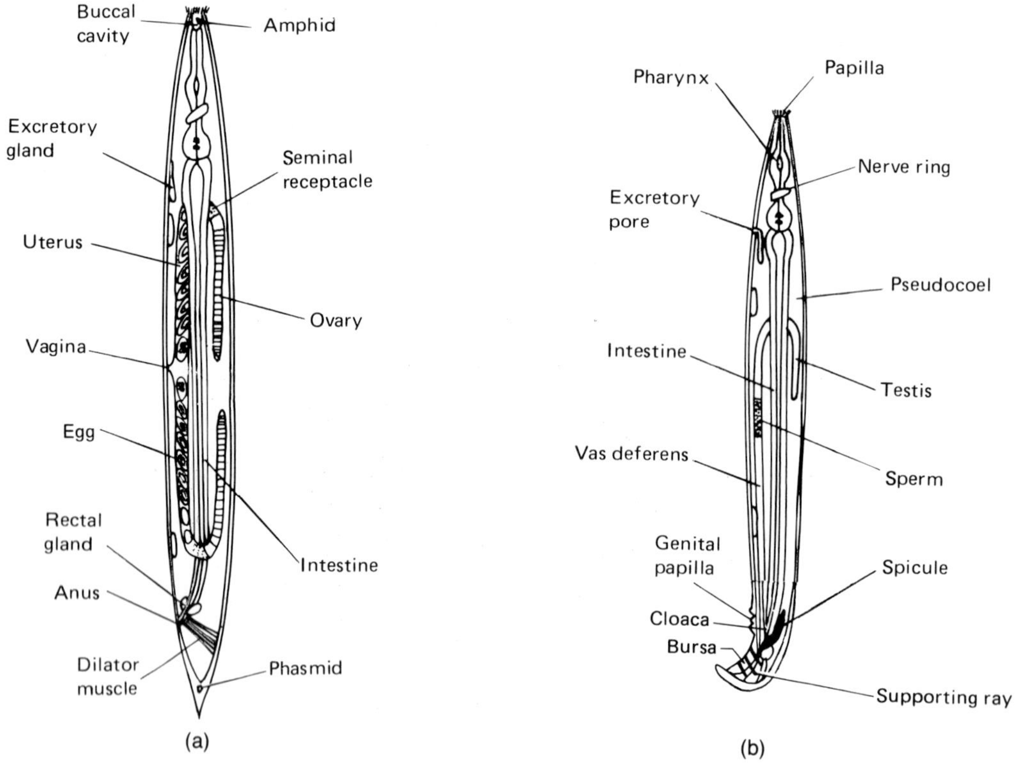

Figure 7.4 Morphology of a generalized nematode in lateral view. (*a*) Female; (*b*) Male. (From Lee, *The Physiology of Nematodes*.)

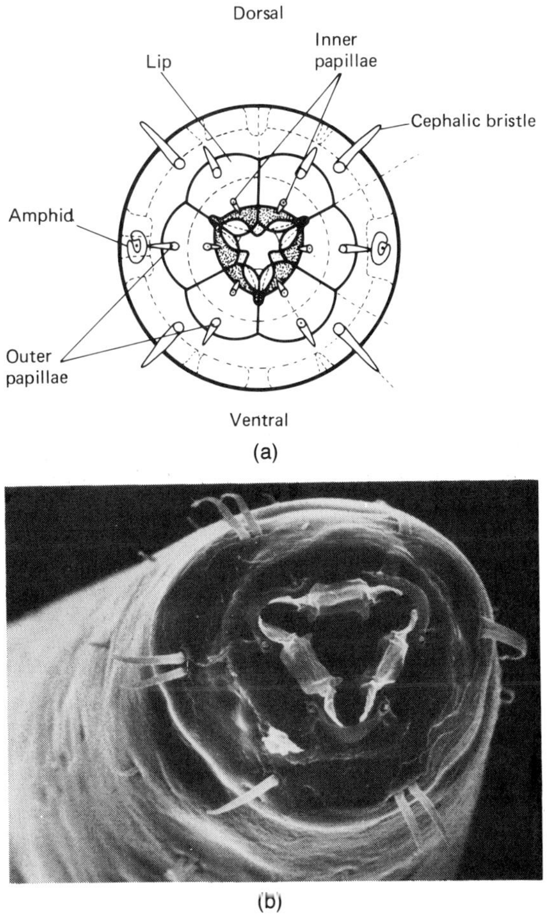

Figure 7.5 (*a*) Diagram of the anterior end of a generalized nematode showing the radial arrangement of lips and cephalic sensory structures. The central triradiate figure represents the pharynx; (*b*) A scanning electron micrograph of the head of *Enoplus* showing the triradiate mouth armed with three strong teeth, six minute inner labial papillae, and ten cephalic setae. (*a* from Hyman, *The Invertebrates, Vol. III*; *b* from W.L. Nicholas, *The Biology of Free-Living Nematodes*, 1975, Oxford University.)

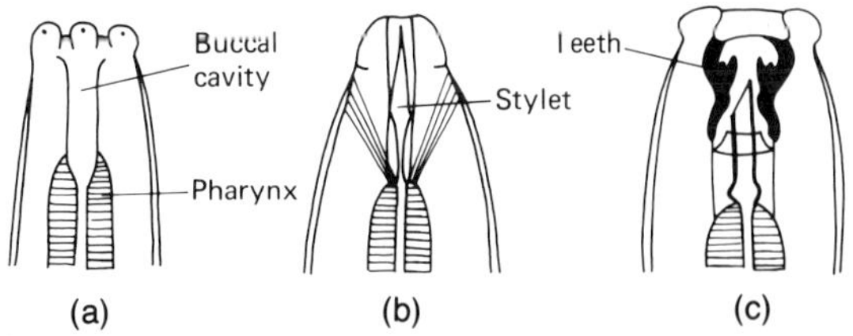

Figure 7.6 Diagrams of the anterior ends of three nematodes exhibiting three different feeding styles. (*a*) *Rhabditis*, feeding on microorganisms; (*b*) *Tylenchus*, a plant parasite; (*c*) *Actinolaimus*, a predator. (From Lee, *The Physiology of Nematodes*.)

on the microorganisms found in feces and mud. Although the phylum utilizes these varied nutritional techniques, individual species often have very restrictive diets.

The nematode mouth opens into the buccal cavity, where a long hollow stylet may be present (Fig. 7.6). This stylet can be extended through the mouth to pierce plant or animal prey and may form a tube for sucking in nutrients. The suction-producing organ is the pharynx (Figs. 7.3a and 7.6). Always highly muscularized and glandular, this organ controls the passage of food into the intestine, from which it is separated by a valve. The nematode intestine, a simple tube of glandular and absorptive cells, is usually collapsed because of the turgor pressure of the pseudocoelomic fluid. Food must be forced into the gut; thus a pharyngeal–intestinal valve and a strong pharyngeal pump are necessary. Between the intestine and the rectum, a second valve prevents pseudocoelomic pressure from squeezing nutrients out of the gut before they can be absorbed. The rectum leads to a subterminal, ventral anus (Fig. 7.4).

The nematode mouth bites, snatches, or sucks up foodstuffs, which are pumped by the pharynx into the intestine. Enzymes from pharyngeal and intestinal glands initiate digestion, which is almost entirely extracellular. As food passes posteriorly, gland cells in the intestinal wall give way to absorptive cells. The latter contain fine microvillar projections that dramatically increase the surface area, thus facilitating food uptake. Digested nutrients are passed into the pseudocoel for distribution throughout the body.

Nutritional habits determine any further specializations of the nematode digestive system. The hard parts of the mouth and buccal cavity are reduced in bacteria-feeding nematodes, while small herbivorous species often possess a hollow stylet for puncturing and extracting nutrients from plant cells. Some carnivorous species possess numerous teeth within the buccal cavity; others use a stylet to suck animal tissues, while blood-eating nematodes may secrete anticoagulants to ensure a steady flow of food from their prey.

Respiration and Circulation

Nematodes have no special structures for respiratory exchange or for the distribution of materials within their bodies. Apparently, the sloshing of pseudocoelomic fluids provides adequate internal transport. Meanwhile, the small size of these worms permits gas exchange by simple diffusion.

Many nematodes live in muddy soils, in ice-covered lakes, in the intestinal tracts of vertebrates, or in other environments where oxygen supplies are minimal. This situation suggests that these animals are capable of anaerobic metabolism. Many nematodes are indeed facultative anaerobes, surviving low-oxygen conditions (albeit with lowered metabolic activity), but quickly transferring to oxidative metabolism when oxygen supplies increase.

Excretion and Osmoregulation

Nematodes are an exception among the aschelminth phyla in that they have no protonephridial system. Their excretion and osmoregulation apparently are handled by a unique system primitively involving one or two unicellular structures, the **renette glands.**

Primitive nematodes possess single or paired renette glands suspended in the pseudocoel and connecting through a short duct to an anterior, ventral pore (Fig. 7.7a). The exact function of renette glands remains unknown in the primitive nematodes possessing them. Evidence from embryology and comparative morphology, however, indicates that similar single-celled glands have given rise to a system of excretory and osmoregulatory canals. Figure 7.7 depicts the renette glands and/or excretory canals of several nematode species and demonstrates a probable evolutionary sequence that could have transformed renette gland precursors into an excretory canal system. In advanced nematodes, these excretory canals course longitudinally through the lateral ridge of the hypodermis (Fig. 7.3).

An organism's osmoregulatory needs depend on its environment. Thus marine nematodes, whose pseudocoelomic fluid has a constitution not unlike sea water, have simple osmoregulatory systems. On the other hand, freshwater nematodes maintain well developed excretory canals which steadily remove excess water from the pseudocoel. Nematodes living in moist soil must possess particularly effective osmoregulatory systems, as their osmotic environment changes dramatically with each rainfall and warm sunny day.

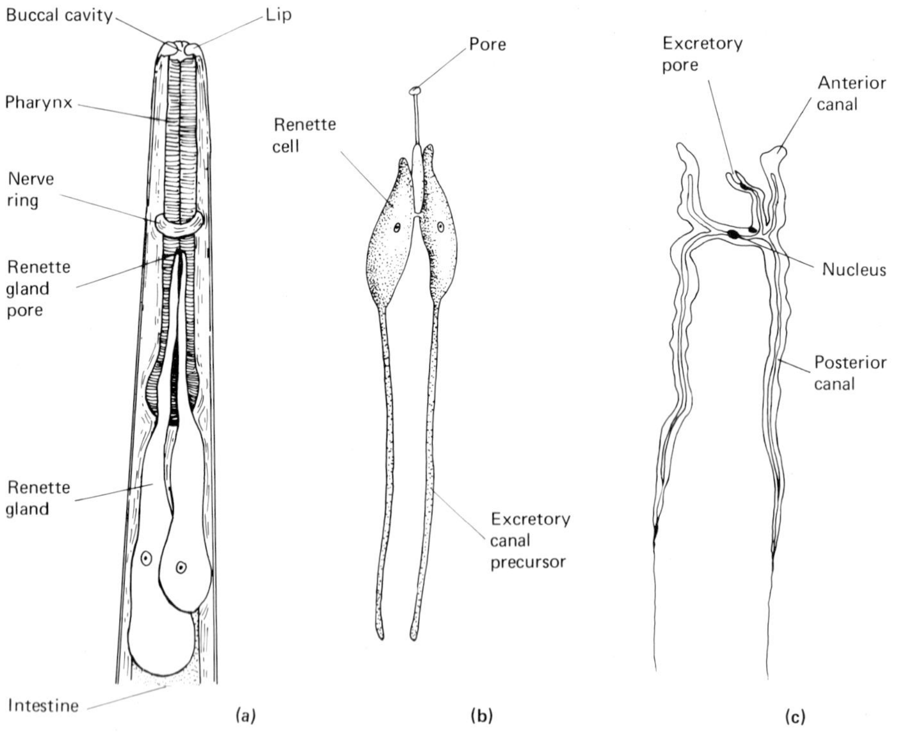

Figure 7.7 (*a*) Anterior end of *Rhabdias*, a nematode with two single-celled renette glands; (*b*) A juvenile *Ancylostoma* with posterior canals developing from the renette cells; (c) Excretory system of *Camallanus* featuring a complete set of excretory canals: two anterior and two posterior. (From Hyman, *The Invertebrates, Vol. III.*)

As mentioned previously, nematodes are a relatively tolerant lot. Perhaps in no other maintenance system is nematode flexibility as evident as in osmoregulation. Many species tolerate a wide range of osmotic pressures; their pseudocoelomic fluids simply harmonize with external conditions. For example, many parasitic nematodes in the hemocoel of insects survive great osmotic fluctuations whenever the host molts. Some nematodes living in soil, in moss, or in certain other plants can withstand considerable desiccation, entering into a state of greatly reduced metabolic activity known as **cryptobiosis** (See Special Essay, this chapter). In those nematodes maintaining constant conditions in the pseudocoel, some auxiliary control systems are effective. The cuticle itself is permeable to water entering the body, but not to water flowing in the opposite direction. Its differential permeability to various ions resists the diffusion of certain elements from the external environment. Also, the nematode gut may absorb undesirable ions and excess water and expel them through the anus.

Major excretory products in the phylum include ammonia, urea, and the organic acids produced by anaerobic metabolism. Under conditions of osmotic stress, some nematodes regulate the form in which their nitrogenous wastes are released. When water is plentiful, toxic ammonia is flushed from the system; but if water must be conserved, less toxic urea is the favored waste. *Ascaris,* a common parasite of humans usually excretes 69% of its nitrogenous wastes as ammonia and 7% as urea. However, threatened by water loss, this parasite conserves body fluids by excreting 52% urea and only 27% ammonia.

ACTIVITY SYSTEMS

Compared to flatworms, nematodes move a little faster and at less metabolic expense. Their advantage over the acoelomates is a function of their pseudocoel. Yet, in overview, nematodes are not very active animals, and improved as they may be in structure and physiology, their behavioral repertoire is almost as limited as that of the acoelomate phyla.

Muscles

Nematodes have peculiar muscles. Most cylindrical animals (e.g., anemones, flatworms, nemertines, annelids) have a stratified body wall musculature comprising alternating layers of circular and longitudinal fibers. But nematode muscles are not organized into layers, and circular fibers are entirely lacking. In the body wall, muscle cells are arranged in longitudinal series to form long cords. Each muscle cell makes its own cytoplasmic connection with the dorsal or ventral nerve cord. This relationship between muscle and nerve is unique, as nerve fibers characteristically penetrate muscles and not vice versa.

In addition to their body wall musculature, nematodes possess a highly muscularized pharynx (Fig. 7.3). Valves at either end of the intestine are controlled by short radial muscles, but the intestinal wall itself is not muscularized.

Nerves

The nematode nervous system structurally resembles that of higher turbellarians. Its circumpharyngeal nerve ring may be called a brain (Fig. 7.8). This ring gives rise to at least four ganglia (located dorsally, ventrally, and left and right laterally), which in turn dispatch nerves to the body. From the lateral ganglia, several short nerves lead to the lips and oral sense organs. Dorsal and ventral ganglia give rise to longitudinal nerve cords. The ventral cord, which may be double, is by far the more prominent and terminates in an anal ganglion. The ventral cord contains both sensory and motor fibers, while the dorsal cord comprises only motor elements. Two smaller longitudinal cords arise from the lateral ganglia and run posteriorly. These lateral cords are sensory.

On first inspection, the nematode nervous system does not seem more advanced than the acoelomate plan. However, nematodes have developed bipolar neurons. Along with other aschelminths, they are the most primitive animals possessing morphologically distinct axons and dendrites. In addition, their organization of nerves into separate motor and sensory units represents an important adaptation for the efficient, one-way conduction of nerve impulses.

Sense Organs

Judged by their sense organs, nematodes probably perceive the world much as flatworms do. Their major sense organs include sensory projections from the lips and the amphids. Labial bristles appear to be simple tactile receptors, larger but basically similar to those found scattered over the cuticular surface. **Amphids** are paired anterior pits located on or just outside the lip complex (Fig. 7.5). At the base of each pit, numerous gland and nerve cells communicate with the brain's lateral ganglia via an amphid nerve (Figs. 7.8, 7.9). Amphids probably are chemoreceptive and thus may be analogous to the ciliated grooves observed in the acoelomate phyla. The electron microscope reveals that the sensory processes of amphid cells are actually modified cilia. Prior to this discovery, cilia

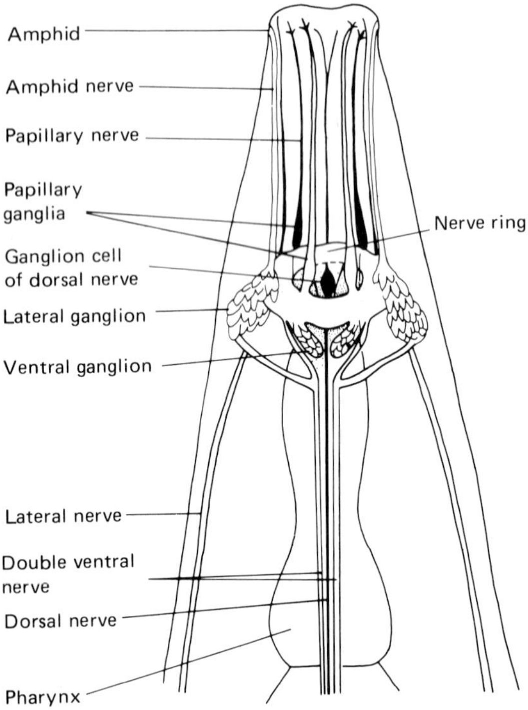

Figure 7.8 A diagram of the anterior part of the nervous system of a nematode (*Cephalobellus*). (Modified from Hyman, *The Invertebrates, Vol. III.*)

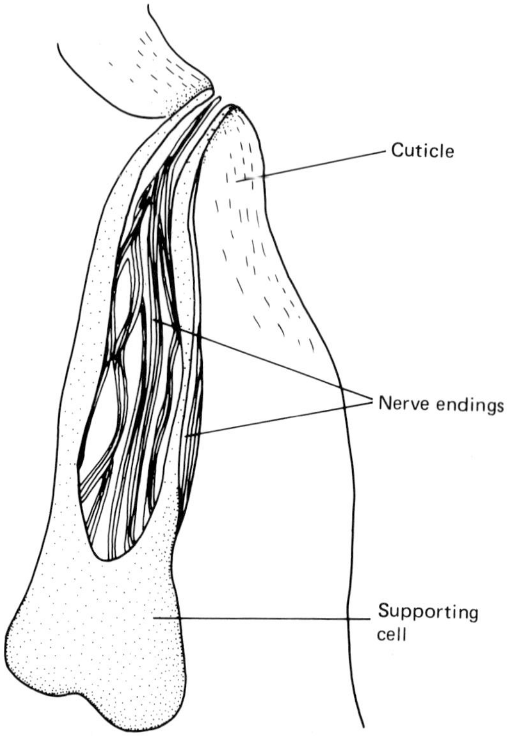

Figure 7.9 Drawing of an amphid from *Ascaris*. (From Hyman, *The Invertebrates, Vol. III.*)

were thought to be absent from nematodes. Amphids are best developed in free-living, marine species and are reduced and/or modified in parasitic forms. If all of these organs were originally chemoreceptors, some nematode parasites have altered their function; the amphids of certain zooparasitic species release an anticoagulant facilitating the uptake of host fluids.

Free-living nematodes sometimes have a pair of simple, pigment cup eyes with lenses derived from the cuticle. Other sense organs include the **phasmids**, a pair of unicellular glands in the tail region of some nematodes. Their exact function is unknown, but phasmids may be chemosensory and thus the posterior equivalent of amphids. Phasmids are best developed in zooparasitic forms.

Amphids and phasmids swell when nematodes are placed under osmotic stress. There is therefore some speculation that these organs help control the concentration of ions in the pseudocoelomic fluid.

Behavior

Nematodes are small, secretive creatures whose behavioral repertoire is decidedly limited. Most of them wander about in mud, finding food and mates in their own quiet way. However, their pseudocoel helps nematodes perform these functions a little more efficiently than flatworms can.

The fluid in the pseudocoel is under pressure and keeps the nematode body taut. In *Ascaris*, a large parasitic nematode, internal fluid pressure averages 70 cm H_2O, but can rise to 400 cm H_2O during locomotion. (By contrast, the internal fluid pressure of earthworms ranges from 3 to 30 cm H_2O.) By passing dorsoventral waves posteriorly along the body, a nematode slowly undulates forward (Fig. 7.10). In the absence of circular muscles, the elastic cuticle antagonizes the longitudinal contractions of the body wall. Nematodes can thrash themselves vigorously, but their contractions are often uncoordinated and, as a result, they have limited powers of locomotion. Nematodes swimming in open water are particularly inefficient. Terrestrial forms make more progress by pushing against solid particles or against the surface film of water surrounding soil. The phylum's locomotor style is best suited for life in such viscous media and represents a preadaptation for parasitic existence.

Nematodes respond to light, temperature, chemicals, and touch. Positive responses to light may help these worms find food at the surface of a body of water, while negative reactions shelter them from the drying sun. Attraction to heat facilitates host location among nematode parasites of mammals and birds. Chemicals play a major role in nematode behavior, as they provide information on the availability of food and, quite possibly, sexual partners.

CONTINUITY SYSTEMS

Many schemes of reproduction and life cycle have been described for the nematodes. Most species are diecious, with readily distinguishable sexes. Males are typically smaller than females; the former have curved tails and their copulatory structures are often visible externally (Fig. 7.4). Some nematodes, however, are hermaphroditic, while others use parthenogenetic modes of reproduction. Eggs produced by any of these methods develop on varying schedules and may hatch as embryos or as juvenile worms.

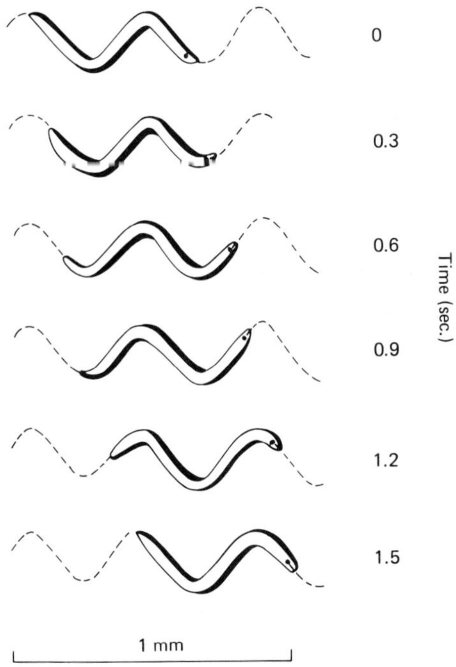

Figure 7.10 Wave formation along the body of the nematode, *Aphelenchoides ritzemabosi*, moving in a thin film of water. The posterior edge of each wave pushes against the water film, which exerts an equal but opposite thrust on the nematode. (From Lee, *The Physiology of Nematodes*.)

Reproduction and Development

Regardless of the mode of reproduction, nematode gonads are rather uniform. In males, a large germinal cell gives rise to ameboid sperm, which descend through a gradually widening, often spiraling testis. The base of the testis forms a seminal vesicle, which communicates with the rectum through a short ejaculatory duct. Within this muscularized duct, prostatic glands secrete fluids for sperm sustenance. In male nematodes, the rectum is called a cloaca because it forms part of both the digestive and the reproductive systems. Cloacal pouches house cuticularized spicules protrusible through the anus (Fig. 7.11).

While the male system involves only one testis, two ovaries are the rule in females. In design, each female gonad resembles a testis. Formed from a distal ovarian cell, eggs descend through a spiraling ovary, and a simple oviduct leads into one of the paired uteri (Fig. 7.4). Uteri join at a short vagina that passes through the cuticle to a midventral gonopore. Thus the female system has a separate outlet and is completely independent of the digestive tract; there is no cloaca in female nematodes.

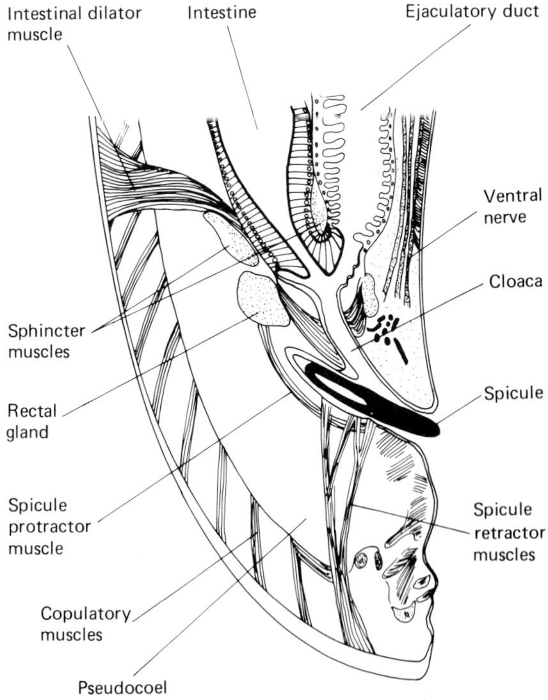

Figure 7.11 A longitudinal section through the posterior tip of a male *Ascaris* showing the cloaca and associated structures. (From Hyman, *The Invertebrates, Vol. III*.)

During copulation, the male pries the female gonopore open with his cloacal spicules, and his ejaculatory duct pumps sperm into the uteri. The distal end of each uterus is modified as a seminal receptacle, and it is here that fertilization occurs. Development may be completed in the uterus, in which case free-living offspring hatch from the mother nematode. More commonly, however, a major part of embryonic development occurs after the eggs have been laid.

Cleavage is determinate, but neither spiral nor radial patterns are evident. A hollow blastula gastrulates by epiboly. As development proceeds, the blastocoel persists to form the pseudocoel. Very early in development, future germinal cells are segregated from those destined to form somatic tissues. Later, as embryonic development is almost completed, cell division stops in all somatic tissues and is never resumed. Accordingly, except for the reproductive system, each nematode organ is limited to a certain number of cells. Cell numbers are specific to each species (Table 7.1).

Among terrestrial nematodes, there is a fairly high incidence of hermaphroditism and parthenogenesis. Hermaphroditic species resemble females, and their gonads typically produce first sperm, then eggs. Sperm are stored, and self-fertilization follows. Parthenogenesis is less common, and most primarily parthenogenetic populations include male members. A curious sort of parthenogenetic reproduction occurs in *Mesorhabditis belari*. Eggs destined to become females are never fertilized in this species, but nevertheless require the presence of sperm before development can begin.

Even among nematodes with a conventional sexuality, the male form may be reduced. For example, in the rat parasite *Trichosmoides crassicauda*, the minute males permanently reside within the female nematode's vagina.

Asexual reproduction and regeneration are practically unknown in the phylum. These conditions are expected among animals which, as adults, lose the capacity to produce new somatic cells.

Nematodes hatch as juvenile worms. Essentially, the young possess adult structures, but their reproductive systems are immature. Since nematodes maintain a constant number of somatic cells, postembryonic growth is limited to cell enlargement. Nevertheless, they can increase considerably in size. To accomodate growth, the cuticle is molted four times during the life cycle. Little is known of their molting physiology, but the process at least superficially resembles that of arthropods. After a new cuticle is secreted under the old, the latter is partially resorbed and then shed. The entire cuticle is molted, including the linings of the buccal cavity, pharynx, vagina, and rectum–cloaca. As in arthropods, the molting process apparently is controlled hormonally. A neurohormone from neurosecretory cells in the nerve ring stimulates the synthesis of certain molting fluid constituents by excretory cells. Nematode molting fluid digests the base of the old cuticle, thus separating it from the new cuticle. Some expansion or growth of the cuticle itself is possible, as most nematodes increase in size even after the final molt. Molting programs vary among species, but the first molt usually is passed in the egg before hatching. Subsequent stages in the life cycle may be separated by molts. Many parasitic nematodes arrest the molting process, utilizing the extra protection afforded by the temporary presence of a double cuticle when they undertake their journey between hosts.

TABLE 7.1

Tissue	**Number of cells**
Esophageal ganglia	161
Dorsal nerve, in head	$2 \times 6 =$ 12
Ventral nerve, in head	$2 \times 7 =$ 14
Ventral nerve	64
Lateral chords	$2 \times 14 =$ 28
Excretory organ	1
Body-wall musculature	64
Gut:	
esophagus, corpus	35
isthmus	0
bulb	24
esophago–intestinal valve	5
intestine	18
rectum	20
Connective tissue	16
	Total 464

Constant cell numbers in organs of the nematode *Turbatrix aceti*. (From Nicholas, *The Biology of Free-Living Nematodes*.)

A Review

In reviewing the important features of nematodes, we draw attention to many of the chief characteristics of pseudocoelomates and, more particularly, aschelminths. Basically, nematodes are:

(1) the largest group of animals bearing a secondary body cavity of the type known as a pseudocoel;
(2) cylindrical worms, exhibiting a remarkably uniform tube-within-a-tube plan of construction;
(3) worms possessing a simple tubular gut from

mouth to anus, with a muscular pumping pharynx as its major specialization;
(4) pseudocoelomates with a unique excretory system based on renette cells or tubules derived from them;
(5) the most primitive animals with morphologically distinct dendritic and axonic nerve processes;
(6) worms which, though well equipped with a hydrostatic skeleton, have limited powers of locomotion due to the absence of circular muscles;
(7) diecious animals with exclusively sexual reproduction and a species specific constancy in the numbers of cells comprising each organ;
(8) pseudocoelomates whose general features have preadapted them for a wide variety of habitats.

Nematode Diversity

Morphologically, nematodes are not a very diverse phylum. Most of these worms are, in fact, so similar in appearance that their taxonomy ranks among the most painstaking studies in invertebrate zoology. Nevertheless, nematologists report dozens of new species each year. The number of described types has more than doubled in the last few decades, and the total number of nematode species has been estimated to be as high as half a million. Such estimates are based on the apparent numerical superiority of free-living species and on the fact that virtually every thoroughly examined vertebrate hosts at least one and often many more unique parasitic nematodes. Cattle, for example, support 50 to 70 nematode types, while a recent count lists over 30 species as parasites of humans.

A highly unstable classification system currently recognizes up to a score of major nematode groups. Rather than defining these groups, we discuss nematode diversity in terms of the phylum's best-studied character, its feeding habits. Nematodes display virtually every mode of nutritional relationship. In no other group is it so instructive to trace the evolutionary progression between free-living and parasitic feeding styles.

Free-living nematodes include deposit feeders as well as raptorial forms; and, with the exception of ectoparasites on animals, all modes of parasitism have been reported in the phylum. Nematodes that consume decaying organic matter are particularly preadapted to parasitic life, and relatives of these free-living, terrestrial forms constitute the phylum's major zooparasitic groups.

Fecal and decaying organic matter is an environment of relatively low oxygen supply, high temperature, and osmotic instability. Such conditions are experienced by many parasites; hence a free-living adaptation to them implies some suitability for parasitic life. In addition, feces and dead organisms constitute small, noncontinuous habitats; thus nematodes dependent on them encounter transport problems comparable to those of parasitic animals. Like parasites, feces-ingesting and/or scavenging nematodes produce a large number of eggs, and dormant stages are common in their life cycles.

Nematodes of the genus *Rhabditis* are transported by insects from one feces mass to another. Within this genus, there is an evolutionary progression between species that merely ride the insect (e.g., *R. coarctata*) and those that are permanent, parasitic passengers (e.g., *R. insectivora*).

Within the order Tylenchida plant parasitic species probably evolved from fungal-feeding forms. Many free-living tylenchids eat fungi that live on plant roots. It would have been a relatively minor adaptive step for similar nematodes to become root-feeders. Because most nematodes are minute in comparison to most plant roots, the transition from simple foraging to a parasitic relationship with a single root also would not be difficult. Most likely, root-feeding nematodes adapted first as ectoparasites, then as endoparasites. In these worms, a sharp stylet often punctures and sucks out root fluids (Fig. 7.6b). Juvenile stages may form a small gall in the root tissues where they feed and mature. The next generation of juveniles migrates to other plants. Herbivorous insects may be exploited in this migration. Juvenile phytoparasitic tylenchids attach to a feeding insect and so secure passage to a new plant host. As in *Rhabditis*, such a transport relationship may lead ultimately to the parasitism of the insect.

The variety of nematode feeding styles and parasitic life cycles is almost endless, and the interested student should consult the chapter bibliography. Because of restrictions on space, we conclude our discussion of nematode diversity with an account of four of the phylum's more important parasites of humans.

Ascaris lumbricoides, an endoparasite of the human gut, is perhaps the best known nematode (Fig. 7.12). It is a very large species; the longer, female forms range up to 30 cm. Thousands of eggs are passed in host feces and must then be ingested for infection to occur. After hatching in the human digestive tract, juvenile worms pierce the intestinal wall and enter the hepatic portal system. Eventually they are transported by the blood stream to the lungs, where they enter the air passageways and are coughed up and then swal-

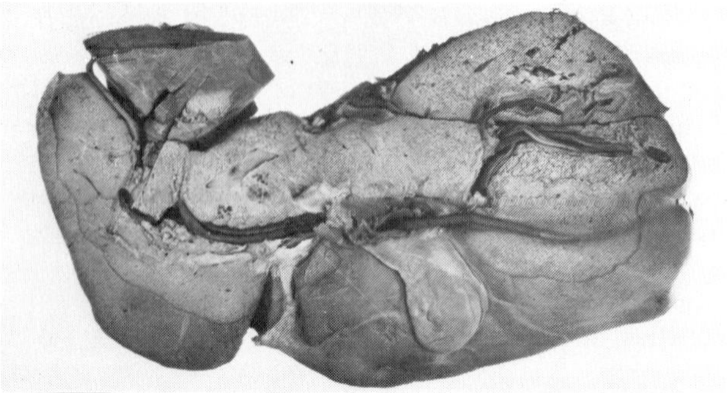

Figure 7.12 *Ascaris* in human liver tissue. (Courtesy of the United States Army Medical Museum.)

lowed. After returning to the intestine, the worms mature and attach to the gut wall, where in small numbers they cause no great harm. Large numbers, however, may clog the intestinal lumen. *Ascaris* produces enzyme inhibitors that protect against human digestive enzymes, a common defense among gut parasites. Because of the inadequate sewage treatment in most parts of the world and the near-indestructibility of its eggs, *Ascaris* is a very common animal. In 1947, it was estimated that this nematode parasitized 30% of the world's human population.

Hookworms of the genus *Ancylostoma* are the most destructive nematode parasites of man and also cause considerable damage to other vertebrate groups. Like *Ascaris*, adult hookworms are residents of the digestive tract, and eggs exit with host feces. But hookworm juveniles hatch outside the host's body; when immature stages are trod upon by a suitable vertebrate, they bore through exposed skin, migrate through the blood to the lungs, and are coughed up and then swallowed. Hookworm mouths have sharp cutting teeth which bite tightly into the intestinal wall of the host (Fig. 7.13). Unlike *Ascaris*, hookworms do not feed on the contents of the intestine. Rather, they are blood eaters, and a massive infestation of these parasites can produce severe anemia.

Trichinella spiralis is the causative agent of trichinosis. The tiny adult worm lives in the intestinal wall and is ovoviviparous. Larvae are transported by the blood to striated muscles, where they form calcified cysts (Fig. 7.14). If large numbers of worms are present, the resultant tissue destruction causes pain and stiffness. Transmission to another host occurs only if flesh containing the encysted worms is ingested by a potential host. In rats and pigs, this parasite can have a one host life cycle. In man, it has a two-host cycle, with the worm entering the human host in insufficiently cooked pork.

Filarial worms, the causative agents of elephantiasis, always have two animal hosts. Adults occupy the lym-

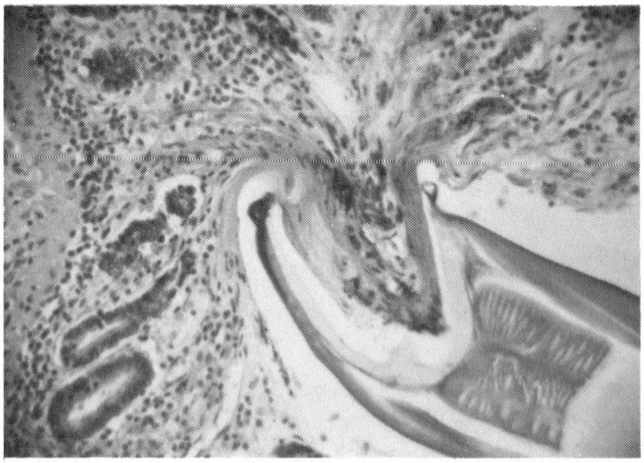

Figure 7.13 Section through a hookworm attached to the intestinal wall. The sharp teeth grasp the lining of the intestine firmly. (Photo courtesy of the United States Army Medical Museum.)

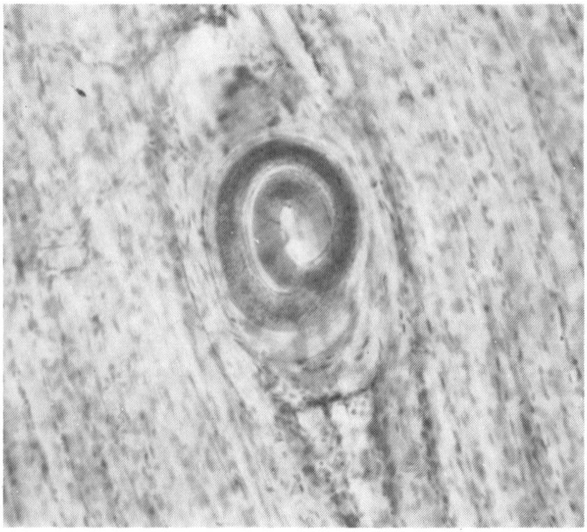

Figure 7.14 Larvae of *Trichinella spiralis* encysted in striated muscle of the host. (Courtesy of Carolina Biological Supply.)

phatic glands of vertebrates, and the ovoviviparously produced juveniles are released into the blood stream. Called **microfilariae,** these young nematodes migrate below the surface of the skin, where they are taken up by blood-sucking insects, commonly mosquitoes. Interestingly, microfilarial migrations toward the skin surface exhibit a circadian periodicity coinciding with the feeding activity of the mosquito. Growth proceeds within the intermediate host, and the filarial worms reenter a primary host during a subsequent feeding by the insect. Several filarial worm species parasitize man. Severe infection may obstruct the lymph vessels, causing the gross swelling of certain body tissues, a condition known as elephantiasis.

The Rotifers

The second largest aschelminth phylum consists of very small, aquatic pseudocoelomates called rotifers. Rotifers are abundant in fresh water the world over and are known commonly as wheel animals. This name describes a revolving wheel of cilia at the anterior end of most species. Typically, these creatures are less than half a millimeter long. Rotifers are thus on a size level comparable to that of larger protozoans, a factor which, combined with their prominent anterior ciliation, caused their early classification among the Ciliata. Subsequently, rotifers were mistakenly allied with annelids and arthropods before their final definition as a distinct aschelminth group. Numbering nearly 2000 species, the Rotifera rank with the nematodes as one of the two major pseudocoelomate phyla.

Rotifer Unity

MAINTENANCE SYSTEMS

General Morphology

The elongate, cylindrical body of the typical rotifer (Fig. 7.15) is composed of three major parts: a truncated anterior end, a large central trunk, and a terminal foot. The anterior end feartures the **corona**, a ciliary complex of varying organization. Sense organs are also present, and the central mouth lies within or at the base of the corona. Housing the rotifer's major

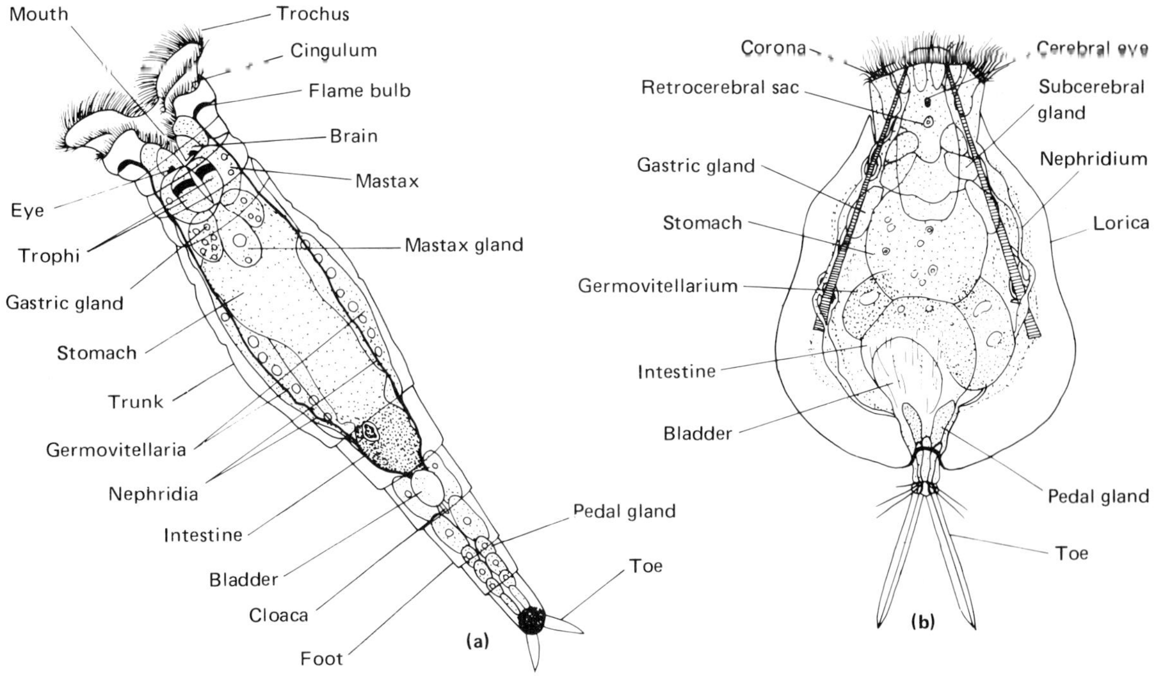

Figure 7.15 Two rotifers. (*a*) *Philodina roseola;* (*b*) A species of *Euchlanis.*

organ systems, the trunk itself is encased within a sculptured cuticle. Cuticular ornaments include rings and spines similar to those found among nematodes. Often the cuticle is greatly thickened to form an armored case called the **lorica** (Fig. 7.15b). Beneath the cuticle is the hypodermis, a syncytial layer with a constant number of nuclei. Subhypodermal muscles complete the trunk wall. The cuticle extends over the narrow, tapering foot. Frequently, this terminal region comprises telescoping sections that allow the foot to be retracted into the trunk (Fig. 7.15a). At the base of the foot are one to four projections called toes.

The corona is the most prominent feature of the phylum. Primitively, this organ probably consisted of a buccal field of cilia surrounding a ventral mouth. From this ventral field, a ciliated band extended around the apical end of the primitive rotifer (Fig. 7.16a). Perhaps the buccal field represented a vestige of ventral ciliation in a creeping ancestral rotifer. The corona has evolved in diverse ways, and today many ciliary patterns exist. Coronal cilia are often fused into cirri, membranelles, and other specialized structures.

Reductions and specializations in both the original buccal field and the apical band are common. In *Collotheca* (Fig. 7.16b) the corona is derived entirely from the buccal field, which has been modified as a funnel. Cilia are restricted to the interior of the funnel, and the mouth is located at its center. This buccal funnel is bordered by numerous spines and bristles that may be derived from cilia. Alternatively, the original buccal field is reduced and the apical, ciliary band dominates the coronal structure. Prominent cilia on the upper and lower borders of this band may form two ciliary wheels (Fig. 7.16c). The upper wheel is the **trochus,** and the lower wheel is called the **cingulum.** In the rotifers with this double-wheel arrangement, the mouth lies between the trochus and the cingulum.

Another coronal style involves two separate disks derived from the trochus (Fig. 7.15a). These ciliated wheels occur in common rotifer species and are largely responsible for the phylum's name. The two trochal disks occupy a retractable pedestal rimmed by the cingulum. The mouth lies between the cingulum and the double trochus.

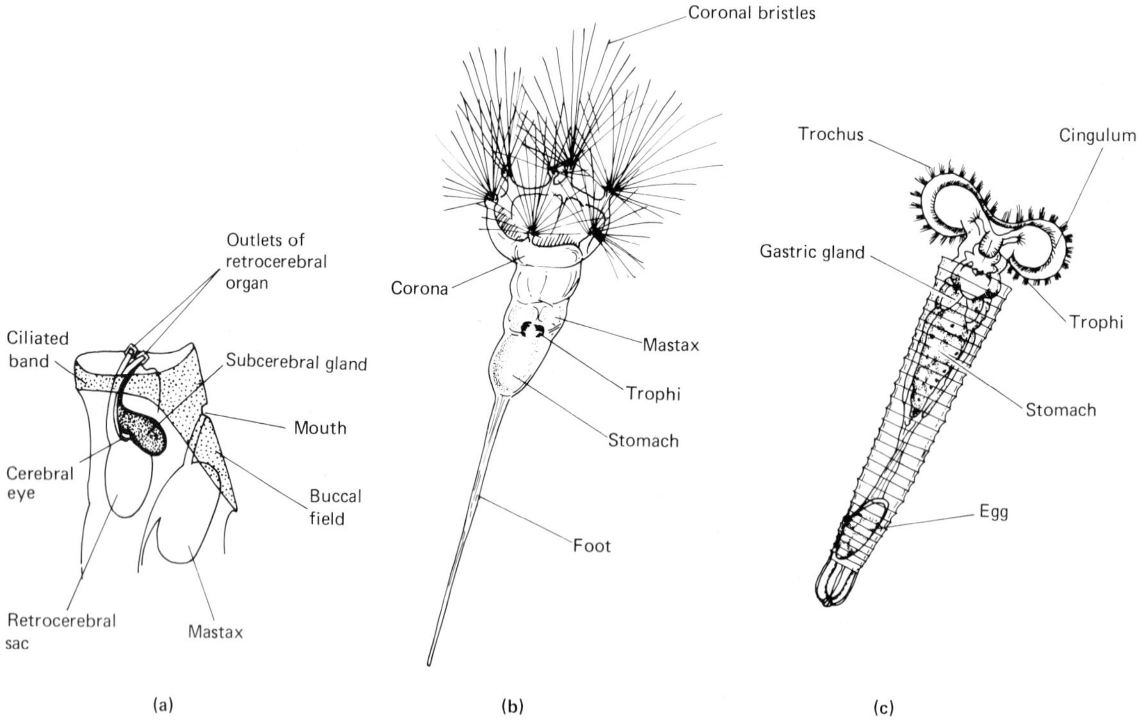

Figure 7.16 (*a*) Schematic drawing showing the relation of the corona to adjacent structures in a primitive rotifer; (*b*) The sessile rotifer *Collotheca* with a funnel-shaped corona derived from the buccal field; (*c*) *Limnias*, a rotifer in which the corona derives chiefly from the apical ciliated band. (*a* and *b* from Hyman, *The Invertebrates, Vol. III.*; c From W.T. Edmondson, Ed., *Freshwater Biology*, 1959, John Wiley.)

Digestion and Nutrition

Rotifers possess a complete digestive tract (Fig. 7.15). The mouth opens into a large, specialized pharynx called the **mastax.** The mastax is equipped with strong, masticating jaws or **trophi.** The many types of trophi are a key to the taxonomy of the phylum and can be correlated with diet (Fig. 7.17). Saliva from mastax glands lubricates food particles as they are fragmented and passed to the stomach. At the stomach–esophageal junction, gastric glands discharge their digestive enzymes. The stomach is the site of most digestion and absorption. Posteriorly, a ciliated intestine leads to a cloaca, and the digestive system terminates at a dorsal anus between the trunk and the foot.

Rotifers are small animals, and their diets have a distinctly protozoan character. Suspension feeders as well as active carnivores exist. The former extract minute organisms and organic particles from the water circulating through their coronas. Rotifers with trochal disks are well adapted for suspension feeding; their two ciliated wheels rotate in opposite directions, channeling food particles to the cingulum and ultimately to the mouth. The trophi of suspension feeders are adapted for grinding (Fig. 7.17a). Raptorial rotifers capture protozoans and small metazoans (including other rotifers) by one of two methods. Those with a coronal funnel, like *Collotheca* (Fig. 7.16b), commonly trap their prey. These rotifers wait until an unwitting animal enters the buccal funnel and then seal the funnel with long spines. Other raptorial feeders rely on a protrusible mastax with pincer- or tonglike trophi to grasp prey (Fig. 7.17b).

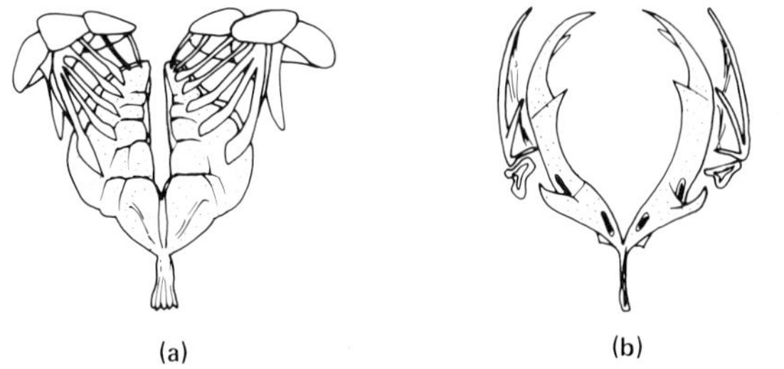

Figure 7.17 Two types of rotifer trophi, both anterior views. (*a*) The trophi of *Epiphanes senta*, modified for grinding; (*b*) The trophi of *Asplanchna*, modified for grasping prey. (From R.W. Pennak, *Fresh-Water Invertebrates of the United States*, 2nd ed., 1978, Wiley-Interscience.)

Respiration and Circulation

Rotifers, like other aschelminths, have no specialized respiratory or circulatory systems. They are small animals, and simple diffusion aided by the mixing of pseudocoelomic fluids provides adequate internal transport.

Excretion and Osmoregulation

Rotifers possess a simple protonephridial system (Fig. 7.15). Originating anteriorly in one to several flame bulbs, two long collecting tubules extend through the pseudocoel on either side of the body. These tubules fuse posteriorly and may form a muscular bladder, which communicates with the cloaca (Fig. 7.15). Alternatively, the protonephridial system may empty into the alimentary canal at the stomach–intestinal junction, in which case a muscularized cloaca serves as a bladder. The rotifer bladder regularly pumps water from the protonephridial system. A contraction rate of up to four times per minute indicates the volume of water that must be voided by these primarily freshwater animals. Rotifers are small enough for diffusion to meet their excretory needs; thus their protonephridial system is chiefly osmoregulatory in function.

In another example of cryptobiosis, some rotifers exhibit an incredible capacity to survive desiccation. Many live in temporary ponds or among moist plants like mosses which may dry out during certain seasons. Well adapted to such environments, these species can endure prolonged desiccation, returning to active life when supplied with water. One investigator reports the revival of a desiccated rotifer from dried moss almost 60 years old! Some metabolic activity is maintained in dormant forms (they may be killed by oxygen deprivation), but exactly how dried rotifers survive remains a puzzle (see the Special Essay, this chapter).

ACTIVITY SYSTEMS

Rotifer nerves, muscles, and sense organs follow the basic plan of the aschelminth phyla. However, the frequent use of anterior cilia in locomotion marks a major difference between rotifers and nematodes, while the small size of rotifers encourages behavioral comparisons with ciliate protozoans.

Muscles and Nerves

Rotifers have both circular and longitudinal muscles in the body wall. As in nematodes, these muscles are not stratified, but rather form distinct fibrous bands (Fig. 7.18). Occasionally, fibers connect the body wall with the gut. The rotifer nervous system is comparable to

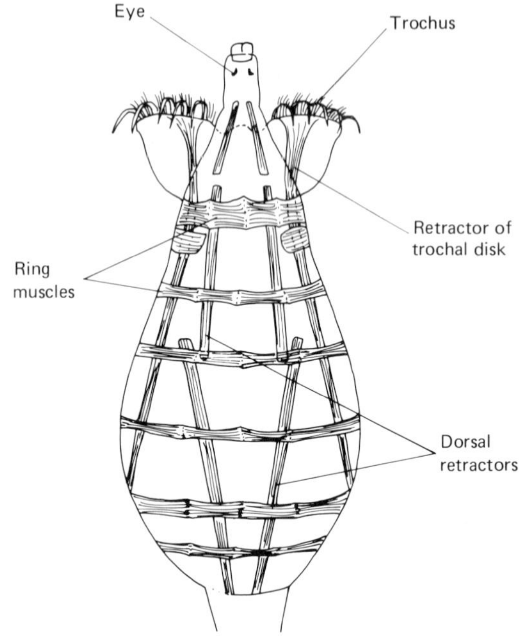

Figure 7.18 Dorsal view of *Rotaria* showing the arrangement of the body wall musculature. (From Hyman, *The Invertebrates, Vol. III.*)

that of nematodes. A cerebral ganglion or brain lies above the mastax and gives rise to the main body nerves. In some cases, these nerves include a pair of ventrolateral, longitudinal trunks; in others, the longitudinal trunks are ventral and dorsal in position. The trunks in turn give rise to ganglia and associated nerves communicating with the foot and other body structures. Motor and sensory nerves are distinct in the phylum. The mastax is well innervated, as are the sense organs of the anterior end.

Sense Organs

Rotifer sense organs include the sensory bristles and ciliated pits common to many lower invertebrates. These structures are developed particularly well in the coronal area. Some rotifers also possess one to five simple ocelli, which may include an eye on the dorsal surface of the brain.

A **retrocerebral organ** lies above the brain. This structure consists of a pair of glands and a single medial sac opening through ducts onto the apical ciliary field (Fig. 7.16a). The retrocerebral organ may be homologous with the frontal organ of acoelomates, but its exact function in rotifers remains unknown.

Behavior

Most rotifers are fairly active animals, certainly much more so than nematodes. Some species swim, others crawl, while many rotifers can do both. Typically, a rotifer swims by retracting its foot and propelling itself along under the power of coronal cilia. While swimming, rotifers loop and spiral like ciliate protozoans. Crawling involves minute steps with the toes or, more commonly, leechlike movements. The latter technique requires stretching the body, contacting the substratum with the anterior end, and then releasing and advancing the toe. Crawling is aided by adhesive secretions of the pedal glands.

Several rotifer groups are sessile. Accordingly, their behavior is limited to bending the trunk–foot and to coronal and mastax activity in food capture.

CONTINUITY SYSTEMS

Rotifers are sexual animals, but females dominate most reproductive activities. Indeed, male rotifers are always tiny and somewhat degenerate creatures; in some species, they have never been reported. Parthenogenesis is widespread in the phylum.

Reproduction

The female reproductive tract comprises one or two germovitellaria (Fig. 7.15). A **germovitellarium** is an ovary combined with a yolk gland. A small and specific number of ovarian nuclei are present, and each gives rise to a single egg. Eggs pass through a short oviduct to the cloaca. Reproductive structures in the male include a single testis and its adjoining, ciliated sperm duct (Fig. 7.19). Rotifer males have a degenerate digestive tract and thus no cloaca; the sperm duct has a discrete gonopore. Prostatic glands add seminal fluids to the sperm as the latter enter a copulatory organ modified from the gonopore wall. Copulation is commonly by hypodermic injection. Sperm may be of two types, one specilized for penetrating the female's cuticle and the other for fertilization.

Development

Developing eggs secrete their own protective shells and membranes. Prior to hatching, they are fixed to various stationary objects or to the body of the female. A few rotifers brood their eggs internally. A strongly determinate cleavage isolates cells destined to form the reproductive system at a very early stage in embryogenesis. In character with other aschelminth

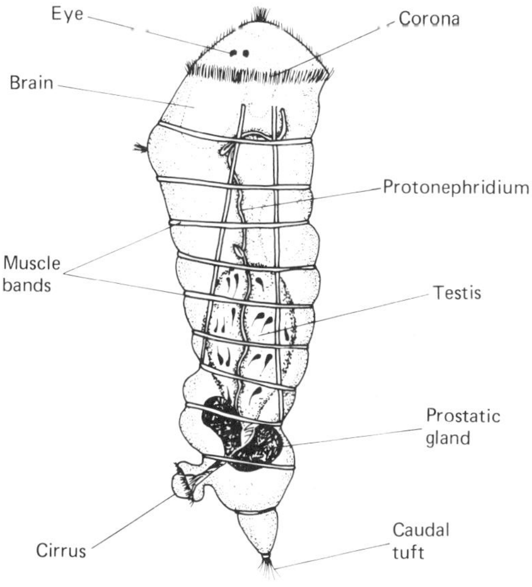

Figure 7.19 Male of *Collotheca* showing reproductive structures. (From Hyman, *The Invertebrates, Vol. III.*)

groups, nuclear division ceases during late development, and adult rotifers possess a species-specific number of cells in each organ system.

Males hatch as mature organisms, while females require only a few days to attain reproductive competence and full adult size. Their rapid maturity reflects the short lives of rotifers. With the exception of dormant eggs and durable desiccated forms, individual animals seldom live more than a few weeks.

Parthenogenesis is common among rotifers and, in some groups, is probably the sole means of reproduction. Certain species combine conventional sexual and parthenogenetic methods as an adaptation to an inconstant environment. In one such scheme (Fig. 7.20), all reproduction is parthenogenetic during the comfortable summer months. Only diploid, or **amictic,** females hatch from thin-shelled eggs. With the advent of autumn weather, haploid, or **mictic,** eggs are produced and, when unfertilized, hatch into male rotifers. A buildup of males in the population makes fertilization possible, and fertilized mictic eggs form very thick shells. These thick-shelled eggs give rise to the amictic females of the next season. Unlike all other eggs, they do not hatch within a few days, but rather remain dormant for up to several months, typically until springtime. This reproductive strategy is characteristic of rotifers living in temporary ponds and streams. Various environmental factors induce mictic

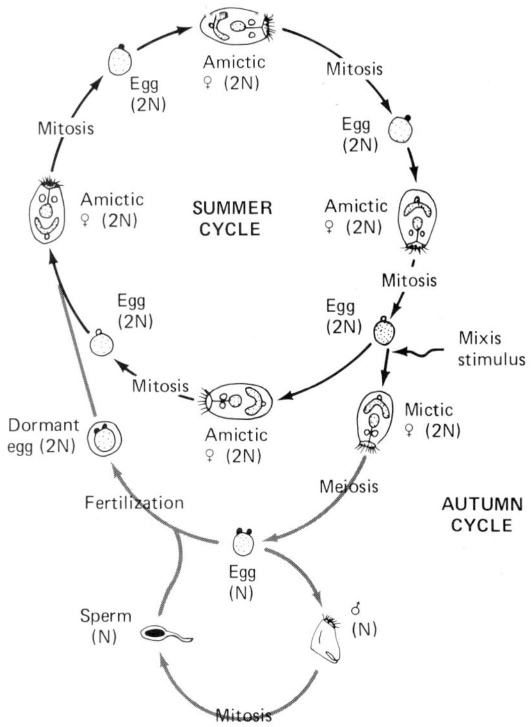

Figure 7.20 Life cycle of a monogonont rotifer showing the alternation of parthenogenetic and conventional bisexual reproduction. (From C.W. Birky, J. Exp. Zool., 1964, 155:273.)

egg production. These factors include high population densities, dietary deficiencies, and changes in the photoperiod. The relative importance of each factor seems to vary among species, and details of the control mechanism remain obscure.

Rotifer Diversity

Diversity among the rotifers involves modifications of the corona and mastax, various reproductive schemes, and locomotor habits. There is also an incidence of ectoparasitism in the phylum. Three classes are recognized: the Monogononta, the Bdelloidea, and the Seisonacea.

CLASS MONOGONONTA

This class includes the sessile rotifers as well as a few swimming forms. Monogonont rotifers are frequently predatory, and some species operate funnel traps (Fig.

7.16b) or protrusible mastaxes with grasping trophi (Fig. 7.17b). Males are generally small and degenerate; females possess a single germovitellarium. An alternation of reproductive methods featuring amictic and mictic eggs is observed in the class. Common monogonont rotifers include *Collotheca* (Fig. 7.16b), *Euchlanis* (Fig. 7.15b), and *Asplanchna*.

CLASS BDELLOIDEA

Bdelloid rotifers are perhaps the best-known members of the phylum. Their anterior end is retractable and bears the wheel-like corona. Bdelloid coronas comprise two trochal disks above a single cingulum (Fig. 7.15a). The class consists primarily of suspension feeders whose mastaxes are adapted for grinding. Bdelloids are swimming and creeping rotifers. Parthenogenesis is the rule in this class; indeed, no males have ever been identified. Females commonly possess two germovitellaria. *Philodina* (Fig. 7.15a) and *Rotaria* (Fig. 7.18) are commonly studied bdelloid genera.

CLASS SEISONACEA

This final group of rotifers contains two marine species that are epizoic on certain crustaceans. The body is elongate, and the reduced corona is modified for attachment. *Seison* (Fig. 7.21) is the only genus.

Like nematodes, rotifers are a peculiar group of animals presenting many fascinating biological questions: Why does mitosis stop during embryogenesis? What is the advantage of cell constancy? How do syncytial tissues differentiate? What mechanism controls the production of amictic versus mictic eggs? How can certain rotifers survive prolonged desiccation? These and many additional questions await the investigations of future biologists.

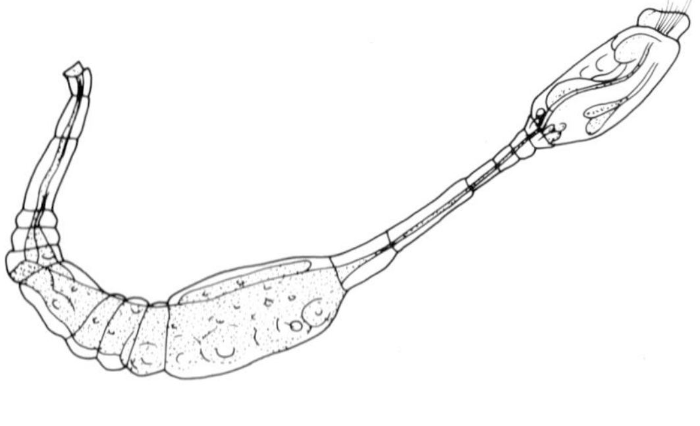

Figure 7.21 *Seison annulatus*, a member of the rotifer class Seisonacea. (From H.B. Ward and G.C. Whipple, *Freshwater Biology*, 1945, John Wiley.)

The aschelminth phyla include three smaller groups: the Gastrotricha, the Kinorhyncha, and the Nematomorpha. Totaling fewer than 750 species, these minor phyla have not been investigated thoroughly, and our current knowledge is limited primarily to anatomical descriptions. Accordingly, discussion of their biology will be brief.

The Gastrotricha

Gastrotrichs resemble rotifers in being small, aquatic aschelminths encased in a cuticle. However, gastrotrichs have no corona or mastax, and their cuticle is often sculptured into overlapped scales and long spines (Fig. 7.22). Cilia in diverse patterns often occupy the flattened ventral surface. These cilia are used in the group's gliding locomotion. Beneath the cuticle is a syncytial hypodermis, and a subhypodermal musculature similar to that found in rotifers borders a narrow pseudocoel. The gastrotrich digestive system (Fig. 7.23) features a muscular but jawless pharynx,

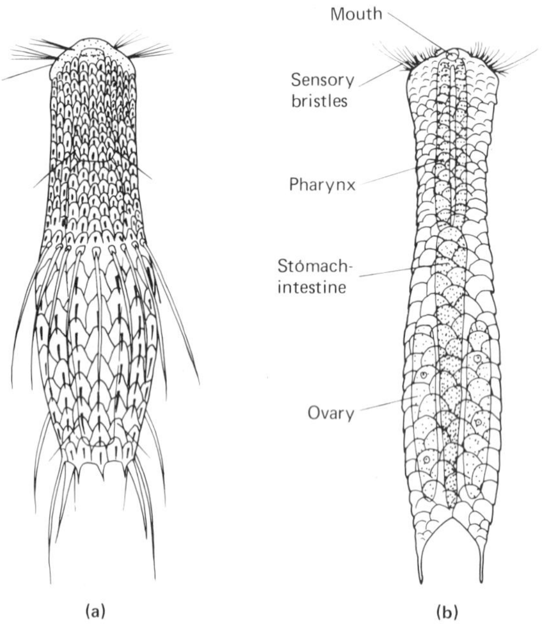

Figure 7.22 (*a*) A species of *Chaetonotus* (Gastrotricha) with a cuticle sculptured into scales and spines of various sizes; (*b*) *Lepidodermella*, a gastrotrich covered with spineless scales. (From Hyman, *The Invertebrates, Vol. III*.)

which sucks small organisms and organic particles through the mouth. Cilia surrounding the mouth aid in food capture. A protonephridial system exists in freshwater gastrotrichs, but is absent from most marine forms. Each of its paired tubules drains a single flame bulb and terminates in a ventrolateral nephridiopore (Fig. 7.23).

Two pharyngeal ganglia give rise to a pair of ventral, ganglionated cords (Fig. 7.23). Other short nerves innervate a typically aschelminth sensory complex composed of anterior bristles, ciliated pits, and eyespots on the surface of the brain.

In contrast to most pseudocoelomates, gastrotrichs are hermaphroditic, but the male system is so reduced in many freshwater species as to render them parthenogenetic. Ovaries and testes are simple sacs emptying through separate gonopores. Hatching as nearly mature organisms, gastrotrichs seldom live more than a few weeks. Populations may prepare for winter in rotiferan fashion by laying dormant eggs.

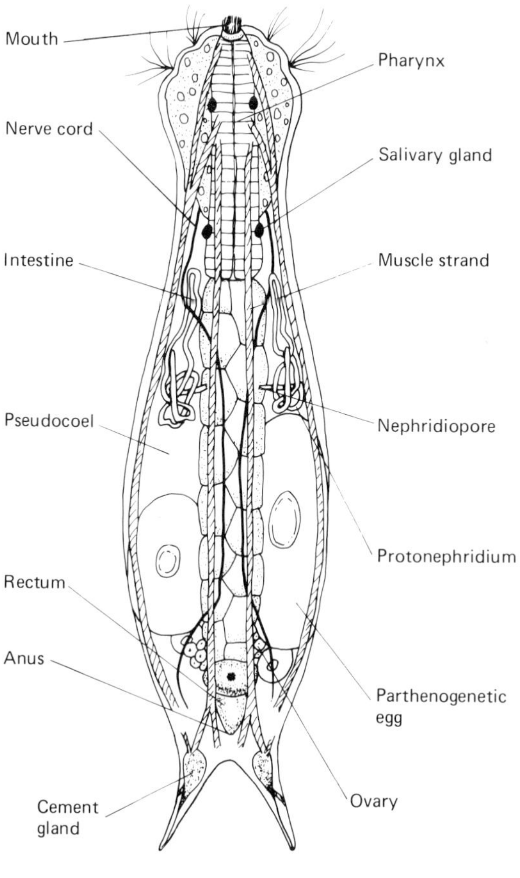

Figure 7.23 Ventral view of the internal anatomy of a typical gastrotrich. (From Pennak, *Fresh-Water Invertebrates of the United States.*)

The phylum contains both freshwater and marine species, most of which live in the interstitial spaces of detritus and bottom sediments. Others are found on the surfaces of submerged plants and animals. Although not nearly as abundant as rotifers, gastrotrichs are quite common in many freshwater environments.

The Kinorhyncha

The kinorhynchs are a marine phylum whose members burrow in muddy bottoms. Measuring up to a full millimeter in length, they are a little larger than rotifers and gastrotrichs. The most striking feature of kinorhynchs is their cuticle, which is divided into 13 distinct sections called **zonites** (Fig. 7.24). The first zonite represents a retractable head region, the second is a neck area, and the posterior 11 sections compose the body trunk. The cuticle surrounding each zonite comprises a dorsal spine-bearing plate and two or three ventral plates; these plates are joined by flexible cuticle. No external cilia are present. Cuticular zonation is reflected within the body wall, where a syncytial hypodermis and longitudinal musculature are serially arranged.

Kinorhynchs are mostly deposit feeders, and their digestive system recalls the nematode pattern (Fig. 7.24); its most distinguishing feature is a protrusible, spiny, oral cone. (The name Kinorhyncha means "movable snout.") The protonephridial complex is very similar to that found in gastrotrichs.

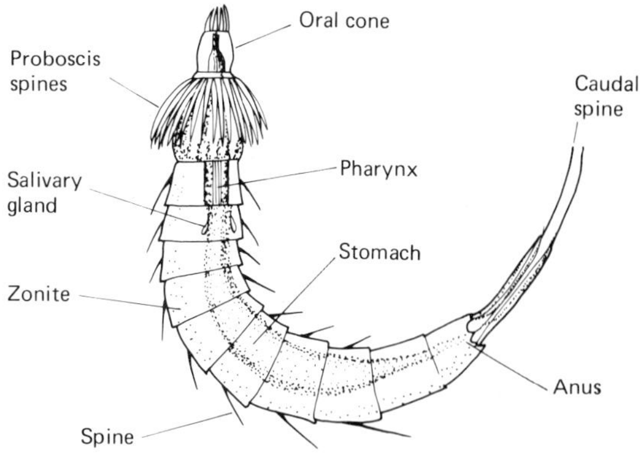

Figure 7.24 The basis anatomical features of an adult kinorhynch. (From M. Hartog, In *Cambridge Natural History, Vol. II*, S.F. Harmer and A.E. Shipley, Eds., 1959, Macmillan.)

The nervous system is associated closely with the hypodermis, where it exhibits some serial organization. A ventral nerve cord arises from a circumpharyngeal brain and connects ganglia in each zonite.

Kinorhynchs are diecious, but sexes are similar in size and general body form. Each sex has a pair of simple gonads and gonopores. Upon hatching, young kinorhynchs are quite immature, an uncharacteristic condition among the aschelminth phyla (Fig. 7.25). A series of molts is required before all the zonites appear and before most organ systems are developed fully.

The Nematomorpha

Nematomorphs are commonly called horsehair worms, a name referring to their long, thin bodies (Fig. 7.26). These aschelminths measure up to 1 meter in length, but seldom are more than 1 mm in diameter. Such hairlike morphology encouraged the medieval belief that nematomorphs grew spontaneously from horsehairs. Adult hairworms are free-living in marine or fresh water or in damp soil; their larvae parasitize arthropods.

The nematomorph body wall is typically aschelminth. It consists of a thick outer cuticle, a cellular hypodermis, and a subhypodermal musculature limited to longitudinal fibers. Often the pseudocoel is much reduced by cellular elements. The digestive tract is degenerate in the phylum; indeed, adults do not feed, while parasitic developmental stages absorb food across their body surfaces. Protonephridia are wholly absent. The nervous system resembles the kinorhynch plan, in that there is a ventral nerve cord within the hypodermis.

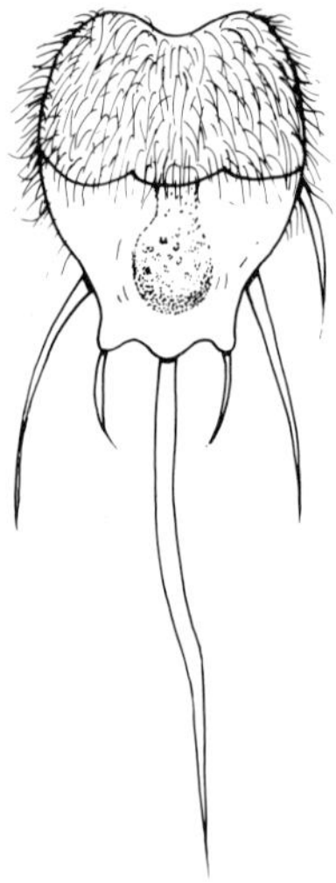

Figure 7.25 The first larval stage of a kinorhynch, genus, *Echinoderella*. (From Hyman, *The Invertebrates, Vol. III.*)

Nematomorphs are diecious and have a simple reproductive system terminating in a cloaca. Copulation occurs in water. Young hairworms pursue a brief, free-swimming existence before they parasitize shoreline insects or other arthropods. They grow, molting several times within the hemocoel of their host. Upon maturity, adult nematomorphs emerge from the host. However, they do not feed, and only the males swim or crawl actively.

The Other Pseudocoelomates

Our discussion of the pseudocoelomates ends with a brief look at two phyla not included in the aschelminth group: the Acanthocephala and the Gnathostomulida.

The Acanthocephala

With approximately 650 described species, the acanthocephalans rank third in number among the pseudocoelomate phyla. All adult members are endoparasites of vertebrates, while larval stages parasitize arthropods. Acanthocephalans are generally

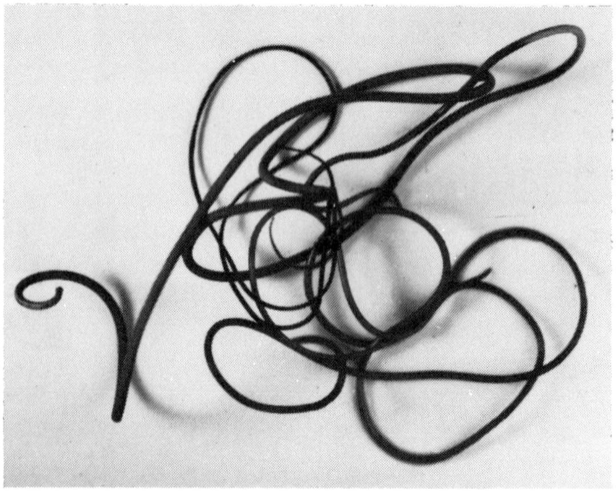

Figure 7.26 Three female horsehair worms (*Gordius*). (From Pennak, *Fresh-Water Invertebrates of the United States.*)

vermiform animals with a spiny body comprising an anterior proboscis, a short neck, and a large trunk (Fig. 7.27a). Most individuals measure only a centimeter or two, but some species are half a meter in length.

The most prominent feature of the phylum is the proboscis. This anterior organ bears sharp spines and hooks which anchor the parasite to its host. Retractor muscles withdraw the proboscis into the trunk; pro-

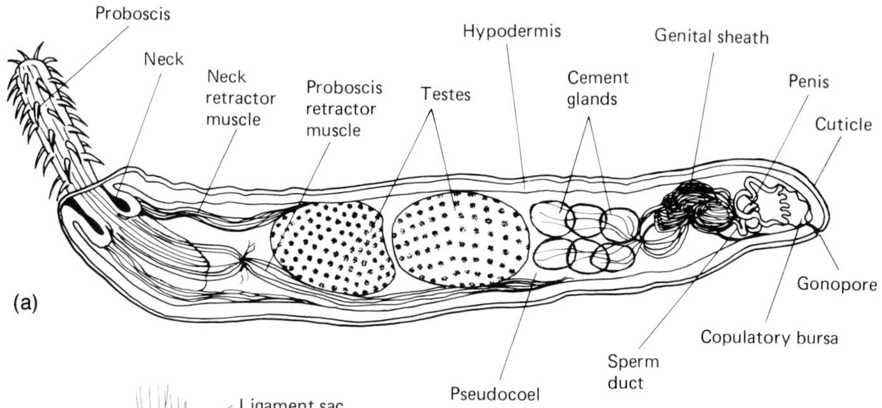

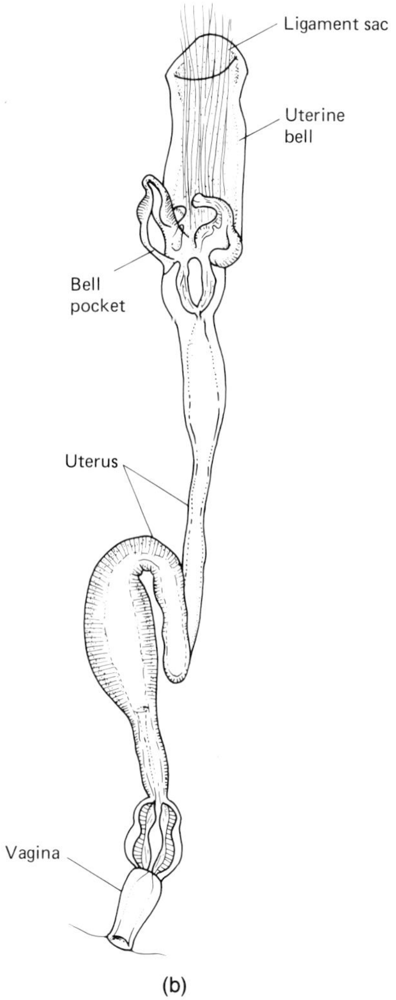

Figure 7.27 (*a*) A diagram of *Acanthocephalus* showing prominent structural features. (*b*) Female reproductive system of an acanthocephalan, *Bolbosoma*. (From Hyman, *The Invertebrates*, *Vol. III*.)

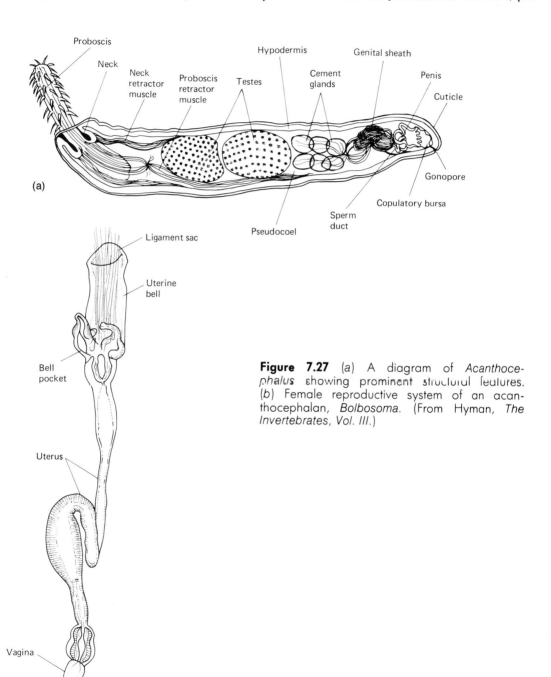

boscis extension involves the hydrostatic pressure of pseudocoelomic fluids.

The acanthocephalan body wall is unique among the pseudocoelomates. A thin epicuticle overlies a porous cuticle and a thick, fibrous hypodermis. The hypodermis is a syncytium housing an extensive net work of fluid-filled channels (Fig. 7.28). These channels communicate throughout the wall and open onto the body surface at small cuticular pores. A basement membrane separates the hypodermis from the musculature, which contains both circular and longitudinal fibers.

There is no digestive system in this phylum. Acanthocephalans apparently absorb nutrients across the body wall, a process involving the fluid-filled channels. These channels may distribute absorbed nutrients. A protonephridial system has been reported in a single acanthocephalan order, but appears to be lacking in most members of the phylum. A ventral, anterior ganglion or brain gives rise to longitudinal body nerves. Sense organs are rare, a common occurrence in parasitic groups, and may be limited to a few tactile sites and chemoreceptors.

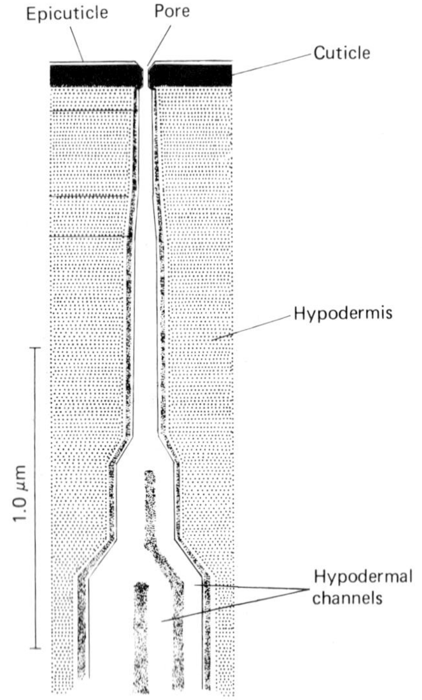

Figure 7.28 Diagrammatic reconstruction of a portion of the body wall of an acanthocephalan, *Polymorphus minutus*. (From D.W.T. Crompton, *An Ecological Approach to Acanthcephalan Physiology*, 1970, Cambridge University.)

Acanthocephalans are diecious (Fig. 7.27a, b), and although males are commonly smaller than females, both sexes occur in equal numbers. The male system consists of paired testes, a sperm duct, and a penis. The penis is protrusible through the copulatory bursa, a well-innervated pocket at the end of the trunk. The acanthocephalan male grasps a female at her gonopore, ejaculates, and then plugs the female orifice with secretions from his cement glands. Sperm migrate through the female's vagina and uterus to the **uterine bell.** This unique structure controls an entrance to the pseudocoel, where eggs, produced with egg balls, await fertilization. In young females, egg balls may define an ovary within a ligament sac, but the latter frequently ruptures and releases its balls into the pseudocoel. Within the pseudocoel, development proceeds to the first larval stage, the **acanthor** (Fig. 7.29). This immature form has precocious anterior hooks. It secretes three to four protective membranes and shells about itself and then passes through the uterine bell. The uterine bell admits only mature, shelled acanthors into the uterus; all earlier developmental forms are returned to the pseudocoel. Acanthors leave the adult body and exit from their parent's host with the latter's feces. Highly resistant to environmental stress, they remain dormant until eaten by certain insects or crustaceans. Development resumes within the hemocoel of

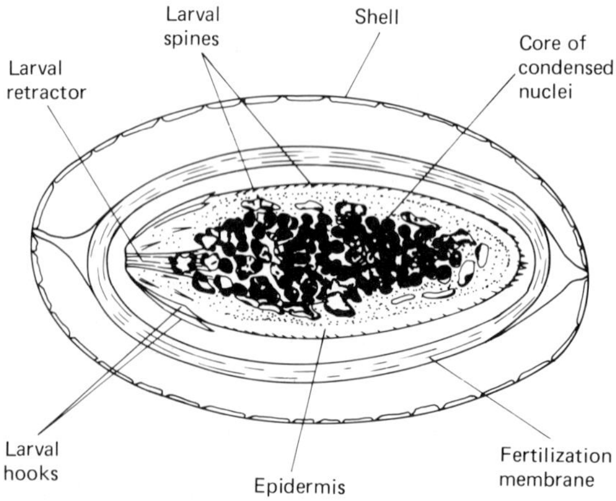

Figure 7.29 An acanthor, the first larval stage of an acanthocephalan (*Macracanthorhyncus hirudinaceus*). (From Hyman, *The Invertebrates, Vol. III.*)

the intermediate host, and the parasite progresses almost to adulthood before encysting. Ingestion by the primary vertebrate host reactivates the acanthocephalan, which matures sexually and lodges in the intestinal tract, attaching by its spiny proboscis.

Egg production is prolific in these parasites. Each female adult produces millions of ova, one or two of which may complete the long and circuitous life cycle.

Acanthocephalans successfully parasitize all vertebrate classes. Most species are parasitic on fish, but birds and mammals are frequent hosts as well. Almost 1500 acanthocephalans have been counted in the intestine of a single marine bird. *Macracanthorhynchus hirudinaceus* is a common parasitic threat to domestic pigs.

The Gnathostomulida

Gnathostomulids are so inconspicuous that they remained undescribed until 1956. Now more than 80 species of this most recently discovered animal phylum are known. Gnathostomulids are minute worms living interstitially in marine sands. They are found in most parts of the world, and some species are remarkably abundant.

Gnathostomulids are cylindrical animals less than 1 mm long, with an anterior head region, a trunk, and a tapering posterior (Fig. 7.30a). Each epidermal cell bears a cilium. Ciliary action produces a gliding locomotion. Several paired groups of longitudinal muscle fibers also contribute to body movement.

The gut resembles that of the smaller flatworms in being a simple tubular sac without an anus. The mouth opens ventrally behind the head and features comblike lips and a pair of toothed lateral jaws (Fig. 7.30b). Bacteria and fungi are the primary food. From their occurrence in environments containing little or no oxygen, we may conclude that gnathostomulids are facultative anaerobes. A simple epidermal nervous system is present. Sense organs are restricted to sensory bristles and ciliated pits.

Most gnathostomulids are hermaphroditic. The female system comprises a single ovary with a sperm storage sac; the male system features paired testes and a rather nematode-like copulatory organ that can be protruded through a posterior gonopore (Fig. 7.30a). Mutual impregnation and oviposition take place through the body wall. Direct development follows spiral cleavage. Some gnathostomulid life cycles feature an alternation of feeding, nonsexual stages and nonfeeding, sexual stages.

Phylogenetically, the Gnathostomulida are quite puzzling. They resemble flatworms in a number of basic features, yet possess structures, particularly the jaws, reminiscent of certain aschelminths. Gnathostomulids are probably related to these lower Bilateria and are described here for comparative purposes.

Pseudocoelomate Evolution

Little of consequence can be said concerning evolutionary relationships among the pseudocoelomate phyla. These soft-bodied animals have left very few fossils, and the biology of the smaller groups is too little known to argue convincingly in favor of any single phylogeny. However, in the absence of hard evidence, speculations abound. Based on similarities

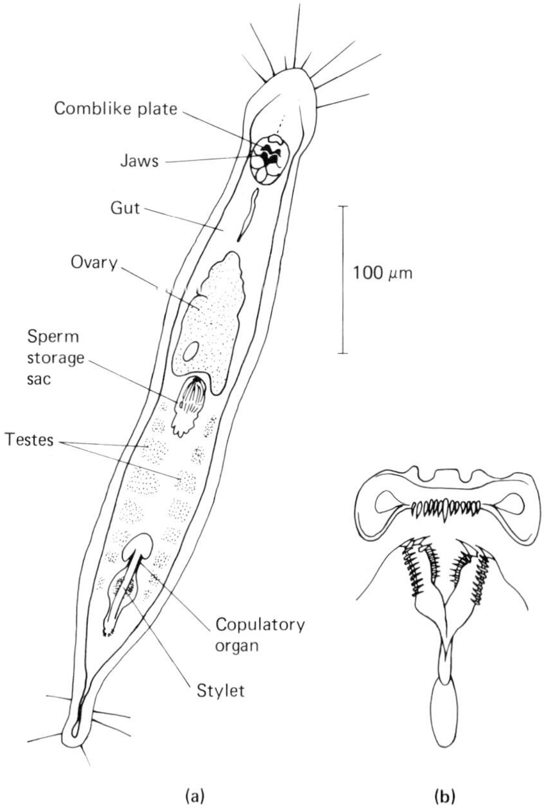

Figure 7.30 (*a*) Diagram of *Gnathostomula jenneri* showing major anatomical features; (*b*) The comb-like lip and toothed lateral jaws of the gnathostomulid *Austrognathia riedii*. (From W. Sterrer, Syst., Zool., 1972, *21*:151.)

in the digestive tract, one popular classification system allies the nematodes with the gastrotrichs. Alternatively, nematodes have been placed in a single phylum with the nematomorphs and the acanthocephalans. Other workers have classified the rotifers with the acanthocephalans and the nematomorphs with the gastrotrichs. Meanwhile, some arguments associate acanthocephalans and cestodes, two endoparasitic groups displaying similar developmental programs.

Clearly, the entire subject is muddled. We simply do not know how all these pseudocoelomates are related. However, we may assume that they arose from acoelomate stock and that more than a single origin was involved. One theory relates the evolutionary story as follows. Early metazoan evolution witnessed numerous experiments with secondary body cavities. Greatest success came with the perfection of the true coelom displayed by the most advanced animal phyla. Meanwhile, other adaptive lines produced invertebrates with "false coeloms." Thus, pseudocoelomates likely represent a polyphyletic group that is off the main line of animal evolution. These organisms have progressed beyond their acoelomate origins, but have not evolved the special type of body cavity that dictates major success in the invertebrate world.

At the beginning of this chapter, we outlined the advantages that pseudocoelomates enjoy over acoelomate animals. Their internal organs are independent of the body wall; these organs are bathed in fluid and can expand with minimal physical interference. Also, the pseudocoel provides internal transport routes and functions as a hydrostatic skeleton. Yet pseudocoelomates are not the most numerous invertebrates; thus, we should examine their limitations, particularly with reference to the eucoelomates.

Pseudocoelomic fluids exert considerable turgor pressure against the body wall and gut. In nematodes, this pressure is 10 times as great as that measured within the eucoelomic compartments of an earthworm. Such pressure collapses the gut and reproductive sacs; hence energy-consuming valves and pumps must move food and gametes through these organs. Digestive efficiency is limited further by the absence of a gut musculature; pseudocoelomates cannot move food by peristalsis. Also, the small size of pseudocoelomates prevents them from becoming dominant animals. The absence of a specialized circulatory system is a major factor inhibiting the evolution of large pseudocoelomates, and the factors mentioned above are certainly contributory. Small size allows the pseudocoel to provide internal transport functions; thus the pseudocoelomates have remained small.

Cryptobiosis

We humans have always been fascinated with death and the notion of an afterlife. That fascination extends from the necrolatry of ancient Egypt into modern biology, where we explore the mysteries of aging and the possibility of suspended animation during long-distance space travel. In the latter area, certain invertebrates are far ahead of us. Although they have not overcome the inevitability of aging and death, these organisms do survive long periods in a state that can barely be called living. This death-defying state is known as cryptobiosis and is exhibited by many species of nematodes, rotifers, and tardigrades.

There are many amazing accounts of the survival capacities of cryptobiotic invertebrates. When water was added to museum specimens of moss dried out over a century ago, rotifers and tardigrades were revived (although they died after a few minutes). Rotifers and tardigrades also have survived exposure to temperature extremes ranging from 150 to 0.008°C above absolute zero. Strong vacuums and oxygen deprivation fail to kill certain cryptobiotic nematodes, and these worms apparently can lose all their body water and still be revived. Electron bombardment within a scanning electron microscope is not immediately lethal to tardigrades; these creatures also can withstand X-irradiation 1000 times stronger than that required to kill a human being.

How do these animals survive such punishment? What selective forces have molded these remarkable talents? Unfortunately, the cryptobiotic condition is not well understood. However, the following observations suggest some of the factors at work. When the cryptobiotic state is initiated by desiccation (as is usually the case in nature), the organism undergoes certain postural adjustments. Nematodes spiral their bodies, and populations gather in large clumps. Individuals in the middle of these aggregations dry out very slowly. Rotifers and tardigrades assume a barrel-shaped form that minimizes their exposed surface area. With such postures, these animals lose water at a rate 1000 times slower than the norm. By retarding their desiccation, they may buy time to complete certain physiological adjustments necessary for cryptobiotic survival.

Studies reveal that oxygen consumption continues at a very low level in cryptobiotic organisms. Surprisingly, however, the survival rate may increase when these animals are kept under conditions of extreme desiccation and oxygen deprivation. This phenomenon suggests that cryptobiosis is most effective when metabolism is extremely low. Indeed, some researchers speculate that ideally cryptobiotic organisms suspend all metabolic activity. According to their theories, even the slightest metabolism during cryptobiosis may have an "eroding" effect on the organism.

If cryptobiosis involves the cessation of metabolism, some of our basic views on living systems must be altered. Life generally is described in terms of activity or process rather than structure; moreover, metabolism is considered essential to sustaining the organization of living systems. However, cryptobiosis suggests that structure alone may distinguish living and nonliving states. Cryptobiotic organisms may maintain a biologically sound organization without metabolism. How they accomplish this feat is not understood, but experiments with bacteria offer a possible clue. Certain bacteria cannot survive air-drying; however, if they are soaked in a sugar solution and then dried, significant numbers of individuals can be revived. Perhaps absorbed carbohydrate molecules in some way replace water in the insulation and support of vital structures. Similar events may occur in nematodes, rotifers, and tardigrades before they enter the cryptobiotic state.

The adaptive value of cryptobiosis is simpler to explain. Typically, cryptobiotic animals inhabit quite ephemeral, semiterrestrial environments, such as damp mosses and liverworts. To survive in such places, they must withstand frequent and unpredictable periods of desiccation, freezing, heating, and oxygen fluctuation. Cryptobiosis allows them to defy death during these problematic times and to resume life as usual when favorable conditions are restored. In this manner, some individuals may be remarkably long-lived. For example, it is estimated that the average tardigrade lives about 1 year in a continuously supportive environment. However, lengthy cryptobiotic intervals may extend the life span to 60 years or more.

for Further Reading

Behme, R., *et al., Biology of Nematodes: Current Studies.* MSS Information, New York, 1972. (A collection of 31 short papers on nematode biochemistry, and parasitism)

Bird, A. F. *The Structure of Nematodes.* Academic Press, New York, 1971. (A detailed account of nematode structure; includes treatment of methods used to prepare nematodes for study)

Crofton, H. D. *Nematodes.* Hutchinson University Library, London, 1966.

Croll, N. A. *The Behaviour of Nematodes.* Edward Arnold, London, 1970. (A concise treatment of all aspects of nematode behavior, with emphasis on movement and responsiveness to stimuli)

Crompton, D. W. T. *An Ecological Approach to Acanthocephalan Physiology.* Cambridge University Press, Cambridge, 1970.

Crowe, N. H., and Maden, K. A. Anhydrobiosis in tardigrades and nematodes. *Trans. Am. Micros. Soc.* 93: 513–524, 1974.

D'Houdt, J. L. Gastroicha. *Annu. Rev. Oceanogr. Mar. Biol.* 9: 141–192, 1971.

Ferris, V. R., Ferris, J. M., and Tjepkema, J. P. Genera of freshwater nematodes (Nematoda of eastern North America) Biota of Freshwater Ecosystem Ident. Man. 10. EPA, U.S. Govt. Printing Office, 1973.

Florkin, M., and Scheer, B. T. *Chemical Zoology. Vol. III, Section II: Nematoda and Acanthocephala.* Academic Press, New York, 1969. (Thirteen review articles give detailed coverage of nematode and acanthocephalan physiology and biochemistry. A single paper discusses other pseudocoelomate groups.)

Giese, A. C., and Pearse, J. S. (Eds.) *Reproduction of Marine Invertebrates Vol. I. Acoelomate and Pseudocoelomate Metazoans.* Academic Press, New York, 1974.

Gilbert, J. J. Dormancy in rotifers. *Trans. Am. Micros. Soc.* 93: 490–513, 1974.

Goodey, T. *Soil and Freshwater Nematodes,* 2nd ed. John Wiley and Sons, New York, 1963.

Higgins, R. P. A historical overview of Kinorhynch research: In Hullings, N. C. (Ed): Proceedings of the First International Conference on Meiofauna. *Smithsonian Contributions to Zoology,* 76: 25–31, 1971.

Hope, W. D., and Murphy, D. G. A taxonomic hierarchy and checklist of the genera and higher taxa of marine nematodes. *Smithsonian Contrib. Zool.* 137: 1–101, 1972.

Hyman, L. H. *The Invertebrates. Vol. III: Acanthocephala, Aschelminthes and Entroprocta.* McGraw-Hill, New York, 1951. (The classic treatise on the pseudocoelomate groups. The chapter "Retrospect" in Vol. V(1959) covers research on pseudocoelomates from 1950 to 1958.)

Lee, D. L. and Atkinson, H. J. *Physiology of Nematodes,* 2nd ed. Columbia University Press, New York, 1977.

Levine, N. D. *Nematode Parasites of Domestic Animals and Man.* Burhess, Minneapolis, 1968. (An exhaustive treatment of zooparasitic nematodes of medical and veterinary importance)

Mai, W. F. and Lyon, H. H. *Pictorial Key to the Plant-Parasitic Nematodes,* 4th ed. Comstock Publishing Associates, Cornell University Press, Ithaca, New York, 1975.

Nicholas, W. L. The biology of Acanthocephala. *Adv. Parsitol* 11: 671–706, 1973.

Nicholas, W. L. *The Biology of Free-Living Nematodes.* Clarendon Press, Oxford, 1975.

Paramonov, A. A. *Plant-Parasitic Nematodes, Vol. I.* Program for Scientific Translations, Jerusalem, 1968. (Treats mainly the morphological, ecological, and taxonomic aspects of phytoparasitic nematodes; translated from Russian.)

Riedl, R. J. Gnathostumlida from America. *Science* 163: 445–452, 1969.

Rieger, G. E., and Reiger, R. M. Comparative fine structure study of the gastrotrich cuticle evolution within the Aschelminthes. *Z. Zool. Syst. Evolutionsforsch* 15: 81–124, 1977.

Schaefer, C. Nematode radiation. *Syst. Zool.* 20: 77–78, 1971.

Sterrer, W. Systematics and evolution within the Gnathostomulida. *Syst. Zool.* 21: 151–173, 1972.

Tarjan, A. C., Esser, R. P., and Chang, S. L. An illustrated key to nematodes found in freshwater. *J. Water Pollut. Control Fed.* 49: 2318–2337, 1977.

Whitfield, P. J. Phylogenetic affinites of Acanthocephala: An assessment of ultrastructural evidence. *Parasitology* 63: 49–58, 1971.

Yeats, G. W. Feeding types and feeding groups in plant and soil nematodes. *Pedobiologia* 11: 173–179, 1971.

Zuckerman, B. M., Mai, W. F., and Rohde, R.A. (Eds.) *Plant Parasitic Nematodes Vol. I: Morphology, Anatomy, Taxonomy, and Ecology. Vol. II: Cytogenetics, Host-Parasite Interactions, and Physiology.* Academic Press, New York, 1971. (A collection of 22 review papers covering most aspects of the biology of phytoparasitic nematodes)

8

The Annelids

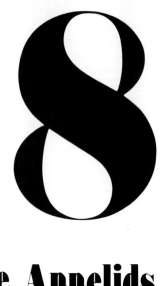

Some of the more advanced worms in the invertebrate world belong to the phylum Annelida. Compared with the flatworms and roundworms, annelids are generally larger and always more complex animals and demonstrate more diverse approaches to wormlike life. Under 9000 annelid species have been described—a number less than that described for Platyhelminthes or Nematoda. However, as most annelid groups lack parasitic members, we perhaps should not expect them to compete in species number with the largely parasitic lower phyla. Rather, annelids display a diversity of free-living forms, varying from terrestrial and freshwater burrowers to crawlers and tube dwellers of the marine benthos and even to some pelagic forms. Annelid successes are attributed primarily to their increased locomotor powers, which in turn depend on two structural characters which the phylum introduces: the true **coelom** and **metamerism**.

The basic structure of the true coelom was described in Chapter 3. The schematic diagram of a eucoelomate animal (Fig. 3.3c) conforms remarkably well to the generalized view of an annelid (Fig. 8.1). A spacious secondary body cavity lies entirely within the mesoderm; it is bordered to the outside by a parietal peritoneum adjacent to the body wall musculature, and its inner border forms a visceral peritoneum against the gut. We considered previously the advantages of a secondary body cavity: its use as a hydrostatic skeleton and internal transport system, the space it provides for the potential growth of mesodermal organs, and its reduced metabolic demands compared to those of the acoelomate parenchyma. The pseudocoels of nematodes and their relatives may provide these advantages, but not nearly to the extent that the true coeloms of annelids do. The entirely mesodermal location of the annelid coelom allows a maximum expansion of mesodermal organ systems. Muscles surround the annelid gut and aid in the movement of food through the digestive tract. Other

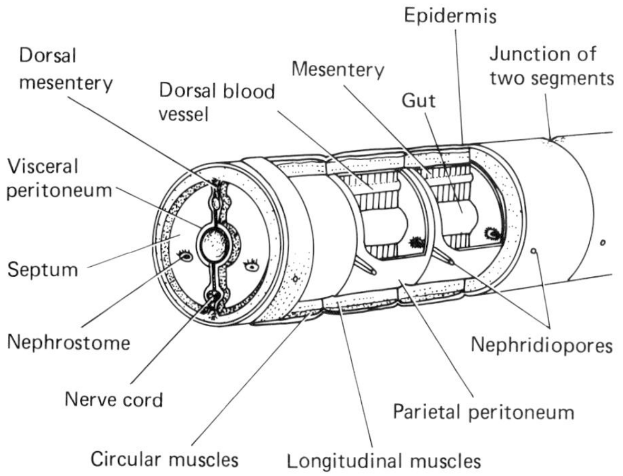

Figure 8.1 A generalized annelid worm. One segment is shown in cross section and portions of the body wall have been removed from three segments to reveal certain internal structures and the layers of the body wall. (From A. Kaestner, *Invertebrate Zoology, Vol. I*, 1967, Wiley-Interscience.)

mesodermal systems service digestive organs more efficiently than is possible in pseudocoelomate animals. Peritoneal mesenteries suspend the gut within the true coelom and further stabilize it against the disruptive movements of the body wall. Nerves and blood vessels within these mesenteries also isolate themselves from body wall activities.

Perhaps the most significant advantage of the coelom involves increased locomotor potential. The annelid coelom acts as a hydrostatic skeleton, much as the pseudocoel does; but the introduction of the second annelid specialty, metamerism, allows a skeletal efficiency unparalleled by lower worms. Metamerism—the segmentation of mesodermal body parts—involves a serial repetition of body segments along an anterior–posterior axis. Ideally, each segment in the metameric sequence possesses a full complement of mesodermal organs: muscles, circulatory and respiratory structures, osmoregulatory–excretory organs, and gonads. The nervous system (largely an ectodermal phenomenon) also displays a metameric organization.

Neighboring segments are separated internally by double layers of peritoneum, called **septa** (Fig. 8.1). Septa define the physical limits of each annelid segment and compartmentalize the coelom. In its capacity as a hydrostatic skeleton, the annelid coelom thus presents a series of discrete, individually controllable units. Segmental muscular forces can be applied with great precision, and such localized control of movement tremendously increases locomotor efficiency. Indeed, it is primarily their increased locomotor powers that enable annelids to exploit more habitats than the lower worms. Annelids can burrow—a talent undeveloped in flatworms and roundworms—and their crawling and swimming abilities are seldom matched by the lower Bilateria.

Three major annelid classes are recognized: the Polychaeta, the Oligochaeta, and the Hirudinea. Polychaetes are largely marine in distribution and include benthic crawlers and burrowers, tube dwellers, and a few swimming types. They are the most primitive class. Oligochaetes have adapted to freshwater and terrestrial life. The common earthworms are members of this class. Finally, the leeches of the Hirudinea are a primarily freshwater group specialized for predation and parasitism.

Annelid Unity

With the annelids, we begin our discussion of invertebrates whose evolutionary history has entailed considerable adaptive radiation and specialization of maintenance, activity, and continuity systems. Such a history tends to obscure the basic unity of a phylum and complicates a discussion of its general characters. Polychaetes constitute the primitive annelid class, but these marine worms have undergone a long and radiating evolution of their own. Oligochaetes display specializations for freshwater and terrestrial life, and leeches too have diverged in many ways from the main annelid line. Indeed, it becomes difficult to speak of the "main annelid line" at all. Accordingly, parts of our discussion of annelid unity reach back in biological time to the earliest appearance of segmented, eucoelomate worms. In considering hypothetical ancestral annelids, we better appreciate how this great worm phylum came to be and on what ancient characters its present diversity is based.

MAINTENANCE SYSTEMS

General Morphology

The study of annelid form involves a lesson in the mechanics of animal movement. With the notable exception of the arthropods, most active invertebrates employ sets of antagonistic muscles working against a hydrostatic skeleton. The most efficient arrangement of these muscles places one set at right angles to the other; commonly, one set is longitudinal and the other circular. In such an arrangement, the contraction of one muscle set maximally stretches the opposite set. For an animal to possess extensive longitudinal and circular muscles, it must be long and cylindrical; thus, this is the prevailing annelid body form.

The long and cylindrical body of the ancestral annelid probably comprised a series of identical segments (Fig. 8.2). Only the extreme anterior and posterior ends did not participate in the metameric sequence. The anterior presegmental tip, called the **prostomium,** was largely sensory in nature. The **pygidium,** or postsegmental end, bore the anus. The mouth was contained ventrally on the first anterior segment; this modified segment, called the **peristomium,** often fused with the prostomium and one or more other anterior segments to constitute a head. Still conspicuous in most modern annelids, the prostomium, peristomium, and pygidium are derived from

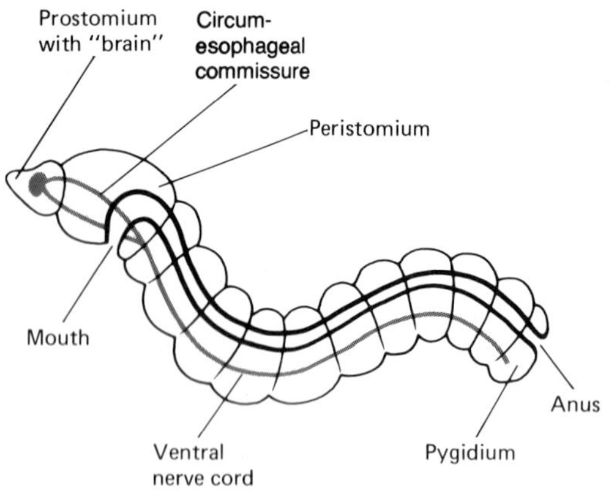

Figure 8.2 Greatly simplified schematic of a hypothetical ancestral annelid. (Modified from W.D. Russel-Hunter, *A Biology of Lower Invertebrates*, 1968, Macmillan.)

prominent larval structures. Segmentation arises in the annelid larva at a growth zone just anterior to the pygidium; here, segments continue to be produced throughout the growth period.

Figure 8.1 depicts a generalized view of an annelid worm. This section describes nearly every postperistomial segment in the hypothetical ancestral annelid. The body wall comprises a thin outer cuticle and epidermis, beneath which lie distinct circular and longitudinal muscle layers. The annelid wall is altogether a rather thin structure, a fact accounting for the frequent simplicity of other maintenance systems. A peritoneum separates this external cylinder from the coelom. The fluid-filled coelom surrounds the gut, which is suspended by dorsal and ventral mesenteries formed from double layers of peritoneum. The gut is typically straight and tubular; in an order reversed from that of the body wall, its circular muscles lie below its longitudinal ones.

Although modern annelids display many structural modifications, they remain remarkably faithful to the proposed ancestral form. One of their most common external modifications involves paired lateral or ventrolateral appendages in each segment (Fig. 8.13). Called **parapodia,** these appendages are conspicuous among the polychaetes. They provide traction during locomotion and may be adapted secondarily for respiration. The muscles of the body wall often are modified to manipulate these appendages (Fig. 8.13).

A marked trend in annelid evolution witnesses a reduction in rigid metamerism. Various segments and groups of segments have become increasingly specialized. In the more active annelid lines, anterior segments have become ever more involved in environmental surveillance and sensory analysis. Elaborate food-gathering structures have evolved in the anterior regions of many worms, particularly in tube-dwelling marine species which feed by filtering mechanisms. Certain midbody segments often have become the sole sites of reproduction, and in the oligochaetes and leeches, distinct reproductive organ systems have developed. Other mesodermal organs may be limited in distribution. These trends have transformed the annelids from animals with rather repetitive bodies to highly integrated worms with quite sophisticated organ systems.

Nutrition and Digestion

The ancestral annelids probably crawled over and through shallow marine bottoms, where they found a wealth of organic material. As the first animals capable of continuous burrowing, they could exploit the previously little-used food supply beneath the ocean floor. The earliest annelids either consumed the substratum whole, eventually digesting out its nutritive elements, or culled food from water currents above their burrows or tubes. Other species employed their burrows as hunting lodges, foraging for prey and then returning to eat in protective comfort.

An extensive radiation in feeding styles occurred early in annelid evolution and necessarily influenced the biology of the digestive system. The basic features of the annelid gut can be described here, but a description of its structural and functional specializations and an account of the varied diets of these worms must await a consideration of representative genera from each of the three classes.

The generalized annelid digestive system is straight and tubular with a peristomial mouth and a pygidial anus (Fig. 8.2). The system is not segmented, although it is related to the coelom and the body wall by metameric septa. Embryologically, the gut is divisible into three parts: a stomodeal foregut, a midgut, and a proctodeal hindgut. Both foregut and hindgut are derived from invaginated ectoderm and thus may be lined with cuticle. The midgut is endodermal.

The annelid foregut is the most varied area in the system. Its elaborations for food gathering and preliminary food processing reflect the diverse feeding habits of the phylum. Generally, the region contains the mouth, a buccal cavity, a muscular proboscis–pharyngeal complex, and the esophagus (Fig. 8.3). The proboscis is often protrusible through the mouth, whereupon it may snatch foodstuffs and pull them back into the worm. Many burrowing forms use their proboscis to excavate an underground home. In its protruded state, the proboscis may expose jaws and/or teeth which add considerable bite to the worm's feeding thrusts (Fig. 8.18). Such hard structures are formed from sclerotized cuticle. Contraction of the body wall increases fluid pressure within the enlarged coelomic compartment surrounding the proboscis, thus thrusting the organ outward through the mouth. Retractor muscles link the outer longitudinal fibers of the proboscis with the body wall; their contraction returns the protruded proboscis to its resting position.

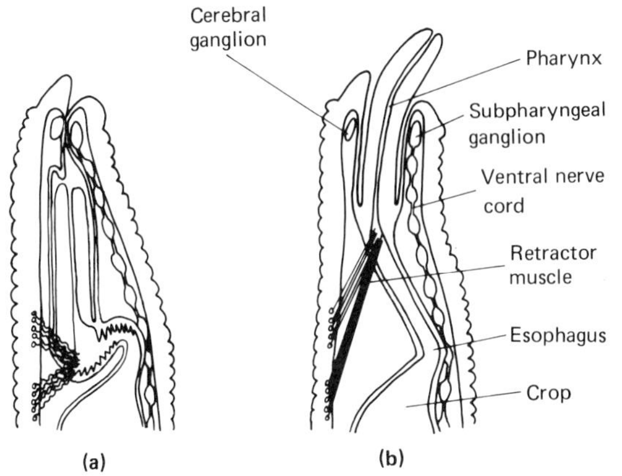

Figure 8.3 A sagittal section through the anterior end of a leech, *Glossiphonia complanata*. (*a*) Proboscis extended; (*b*) Proboscis retracted. (From H. Harant and P.P. Grassé, In *Traité de Zoologie, Vol. V*, P.P. Grassé, Ed., 1959, Masson et Cie.)

A narrow tube leads to the rear of the proboscis, which commonly is muscularized to form a pharynx (Fig. 8.3). Strong circular muscles within the pharyngeal wall are antagonistic with radial fibers attached to the main body cylinder; together these muscles constitute a powerful pharyngeal pump. Such a pump sucks captured food into the mouth and shuttles it down to the esophagus. Meanwhile, mucous glands in the pharynx moisten and soften foodstuffs. Esophageal dilations may form a food storage area called the **crop** and/or a separate area called the **gizzard** (Fig. 8.29a). Gizzard walls contain jaws of sclerotized cuticle, which further grind and mash food before its entry to the midgut.

Glandular secretions of the foregut soften and moisten incoming food. Some digestive enzymes may originate from pharyngeal or esophageal glands. Special **calciferous glands** associated with the esophagus take up calcium and other inorganic ions from the blood of detritus- and soil-feeding annelids. Calcite secretions from these glands then pass innocuously through the digestive tract and are removed at the anus.

The annelid midgut is relatively undifferentiated. It consists of a long stomach–intestine for the digestion and absorption of nutrients. The ancestral midgut was probably quite uniform in structure and function. However, adaptive trends toward sequentialism—the serial processing of nutrients—have produced a specific enzymatic and triturating stomach and a distinct absorptive intestine. Digestion is primarily extracellular. Septa inhibit extensive branching or coiling of the midgut and thus may contribute to the evolution of longer bodies among the annelids. The long intestine has segmental diverticula in some species, but sheer length, as well as some infolding of the dorsal intestinal wall, is the organ's primary adaptation for increasing its absorptive area. However, some shorter annelids, including the leeches (Fig. 8.39b), possess intestinal ceca with large absorptive surfaces. Certain annelids move small particles through the intestine by ciliary action; often such cilia occupy a ventral groove along the length of the midgut. Meanwhile, burrowing worms transport heavy volumes of soil and detritus through the gut by peristaltic contractions of the intestinal musculature.

The intestine grades imperceptibly into the hindgut, and the anus is located on the pygidium (but see the exception among leeches).

Circulation and Respiration

Annelids possess two systems for the internal transport of nutrients, wastes, and respiratory gases. These systems are a coelom and a closed blood vessel network. The first annelids probably relied on the coelom for most of their transport needs. As longer animals, they especially required a mechanism for shuttling materials from the intestinal sites of absorption to the extreme ends of their bodies. Oxygen also had to be transported from the body surface to the interior muscles and other organs. The ancestral coelom functioned well in these respects, and provided that annelids remained small and relatively sedentary, no other system was needed. But certain evolutionary trends contributed to the selective pressures that shaped a specialized blood vessel system. First, as annelids increased in size, their coelomic transport potential lagged behind their growing circulatory needs. Second, the compartmentalization of the coelom, while enhancing the worms' locomotor powers, impeded the longitudinal flow of their coelomic fluid. As annelids with metameric coeloms developed more efficient locomotor techniques, they became more active, burned more oxygen, and further increased in size. All of these factors militated for a specialized and more efficient blood vessel system.

The major annelid blood vessel is a large dorsal tube through which blood flows in an anterior direction (Figs. 8.1 and 8.4). Lateral connectives in each segment join a smaller ventral vessel, where blood flow is typically posterior. Sections of the dorsal tube and/or the anteriormost connectives may be highly contractile and thus function as "hearts." Secondary vessels connect with subepidermal and visceral sinuses, as well as with a small network surrounding the ventral neurons. The subepidermal sinuses are a major site of respiratory exchange. Visceral sinuses receive gut nutrients and begin their distribution through the annelid body. Secondary blood vessels also are associated with the metanephridia (Fig. 8.5.)

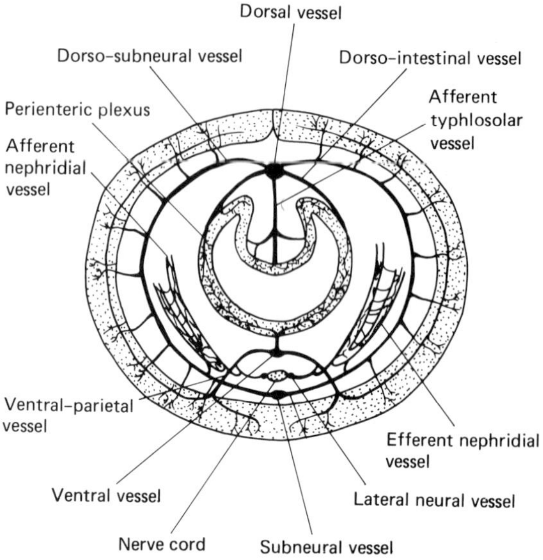

Figure 8.4 Cross section through an annelid body segment showing the major segmental blood vessels. (From C.A. Edwards and J.R. Lofty, *Biology of Earthworms*, 2nd ed., 1977, Chapman and Hall.)

Respiratory pigments are common in annelids and include hemoglobin and, in a few families, chlorocruorin. Hemerythrin is present in one genus. These large molecules are dissolved in the blood plasma. In some worms, the coelomic fluid contains respiratory pigments, in which case they exist as smaller molecules within corpuscles, a situation characteristic of vertebrate hemoglobin. Respiratory pigments play a varying role, depending on the species and the environmental conditions. In most annelids, oxygen consumption (i.e., metabolic and behavioral activity) is determined in part by the local supply of the gas, and respiratory pigments may become critical if oxygen concentrations drop. Terrestrial annelids usually enjoy an abundance of oxygen; in these worms, respiratory pigments increase the carrying capacity of the well-developed blood vessel systems. However, when heavy rains waterlog soils, thus sharply reducing their oxygen supply, most earthworms must crawl to the surface.

Another annelid respiratory adaptation involves epidermal gills. Thin-walled and highly vascularized, such gills present large surface areas to the environment. Gas exchange with the surrounding medium may be facilitated by ventilation currents. Polychaetes especially are noted for their gills (Figs. 8.20 and 8.21).

However, notwithstanding such gills, annelids rely essentially on oxygen uptake across the general epidermal surface. This reliance necessitates the continual lubrication of that surface, especially in terrestrial and burrowing forms; very importantly, it also militates against the evolution of protective insulation. Moreover, such general epidermal exchange can never be as efficient as gas exchange within well-structured internal gills, lungs, or similar respiratory pockets. Hence, their respiratory technique limits the ecological diversity of the annelid worms.

Osmoregulation and Excretion

The coelom combines with a specialized organ to handle osmoregulation and excretion in annelids. The specialized organ in this case is the **metanephridium** (or simply the nephridium). Basically, this organ is an elaborate protonephridium whose open internal end services a fluid-filled, secondary body cavity. Ancestral annelids likely had protonephridial systems, and flame cells persist in some contemporary, albeit primitive, members of the phylum. But metanephridia probably evolved early, in concert with the appearance of the true coelom.

The coelom may serve as a cesspool into which the body wall and gut empty their wastes. Metanephridia then selectively drain the coelom, resorbing any usable materials and fluids while passing nitrogenous wastes and excess water.

A single metanephridium consists of a ciliated funnel, the **nephrostome,** and a long coiled tubule terminating in a nephridiopore (Fig. 8.5). Typically, these organs are arranged in metameric pairs. Paired nephrostomes are mounted on the septa. They open into the coelomic compartment immediately anterior to the compartment housing their associated tubules and nephridiopores (Fig. 8.1). Elaborate blood capillaries surround the nephridia of most advanced annelids. This situation facilitates an efficient exchange of materials between the circulatory and the osmoregulatory–excretory systems. In some species, a relatively few highly vascularized metanephridia service the entire worm. This breakdown in metameric organization typifies the evolution of several annelid systems.

As we should expect, metanephridia are well developed in most freshwater and terrestrial annelids. These worms incur a maximum of osmotic stress, and considerable osmoregulatory work is vital to their health. Annelids in brackish situations may display a physiological program combining varying salinity tolerance with vigorous metanephridial action. Certain

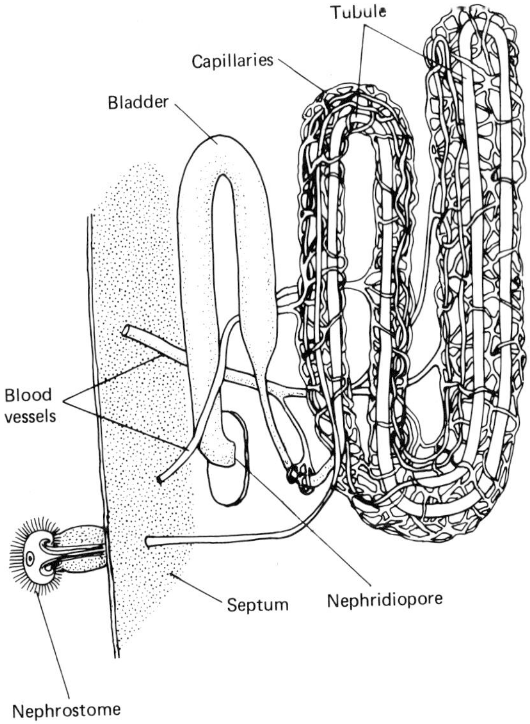

Figure 8.5 Diagram of an earthworm metanephridium. (From T.I. Storer, et al., *General Zoology*, 1972, McGraw-Hill.)

land forms possess a metanephridial system that greatly improves water retention. Their metanephridia do not terminate in surface nephridiopores, but rather discharge into the digestive tract. Subsequent water resorption by the intestine produces a very dry waste.

Ammonia is the dominant excretory component of most annelid urine. Urea and other complex nitrogenous wastes are present in smaller amounts, and their relative concentrations increase with the worm's need to conserve water. For example, earthworms placed in water excrete ammonia almost exclusively; returned to their native soil, they produce a urea-dominated urine.

In many annelids, yellow or brownish clusters of **chloragogen** tissue are associated with the intestinal wall (Fig. 8.29b). In addition to other roles, this tissue apparently has excretory functions. Chloragogen is a station of intermediary metabolism and probably is involved in the deamination of proteins (with the formation of ammonia and/or urea), the synthesis of respiratory pigments, and the storage of fats and starches. (Such responsibilities make the tissue the approximate physiological counterpart of the vertebrate liver.) Once filled with nitrogenous wastes, bits of chloragogen tissue break loose into the coelom, where, in association with coelomic amoebocytes, they may remain throughout the life of the worm.

ACTIVITY SYSTEMS

The success of the Annelida is attributable largely to the phylum's segmentation of both a true coelom and the body wall musculature. By increasing locomotor efficiency, such segmentation enables annelids to exploit new habitats and new, often more active, ways of life. As the coelom and body wall musculature evolved in these worms, a parallel evolution in nervous equipment allowed the coordinated movements of the segmental, hydrostatic units. Annelids that became adapted for active life styles also acquired prominent prostomial sense organs.

Muscles

The ancestral annelids possessed the basic wormlike arrangement of muscles already outlined (Fig. 8.1). When annelids first appeared, alternating layers of circular and longitudinal muscles were probably nothing new in the invertebrate world. However, a wholly mesodermal coelomic skeleton against which these antagonistic layers could operate was innovative. Contraction of the circular muscles generates coelomic fluid pressure, which can be exerted only in a longitudinal direction. When circulars contract, longitudinal muscles are stretched by this fluid pressure, and the worm elongates. Subsequent contraction of longitudinal muscles with a relaxation of circular ones causes the annelid to shorten in length and thicken in diameter. Given some device to prevent slippage, alternating longitudinal and circular contractions produce forwardly "inching" movements (Fig. 8.6). This classical locomotor technique can be observed in most annelids. Typically, muscular contractions occur as peristaltic waves beginning behind the prostomium and passing in a posterior direction.

The ancestral annelids probably originated from turbellarianlike creatures on the ocean floor. Their origin involved selective pressures favoring increased size and locomotor efficiency. Truly cylindrical animals can maintain only a relatively small surface area in continuous contact with the substratum. This inefficient situation is overcome, however, if the animals burrow. With traction possible on all sides, burrowing annelids could better exploit their tubular skeletons. The first annelids probably were not strong enough to sustain continuous tunneling and most likely lived in semipermanent burrows. Selective pressure on muscular development continued even in these forms, however, as peristaltic contractions were used to irrigate their burrows for respiratory and feeding purposes.

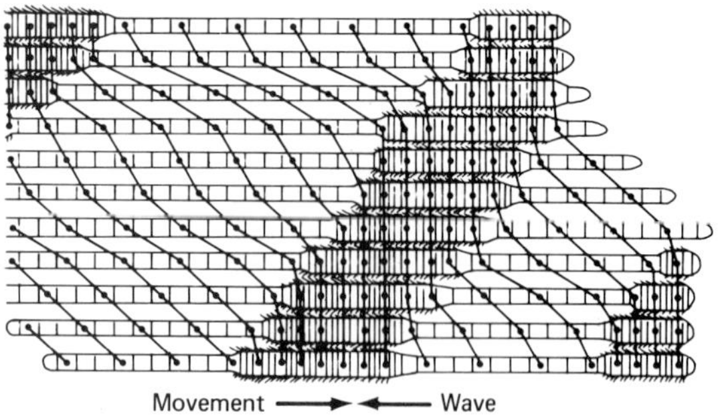

Figure 8.6 Alternating waves of contraction move posteriorly along the animal. Contraction of circular muscles causes elongation of a segment; contraction of longitudinal muscles causes shortening. Slippage is prevented by extension of the setae, and by traction against the sides of the burrow. (From M. Wells, *Lower Animals*, 1968, McGraw-Hill.)

Annelids display other locomotor techniques. Some polychaetes exploit the antagonistic relationship of longitudinal muscles on opposite sides of their bodies. Alternate contractions of these muscles throw the body into a series of lateral folds, and with traction supplied by prominent lateral appendages (parapodia), these annelids wiggle along (Fig. 8.14). Circular muscles are reduced in such worms and may be replaced with lateral fibers extending into the parapodia. Some other polychaetes actually move by parapodial stepping. In these forms, the lateral appendages can raise the main trunk above the substratum and advance in stepwise fashion. Parapodial steps occur in waves passing in alternating phases along each side of the body. Finally, certain polychaetes swim by undulating in the dorsoventral plane.

Nerves

The development of a powerful annelid musculature occurred concomitantly with adaptive trends in the phylum's nervous system. Muscular activity of all segments must be coordinated to produce proper, overall movements of the worm. A nervous organization well suited for elongate animals features longitudinal trunks communicating segmentally with the body wall musculature. The annelid nervous system is

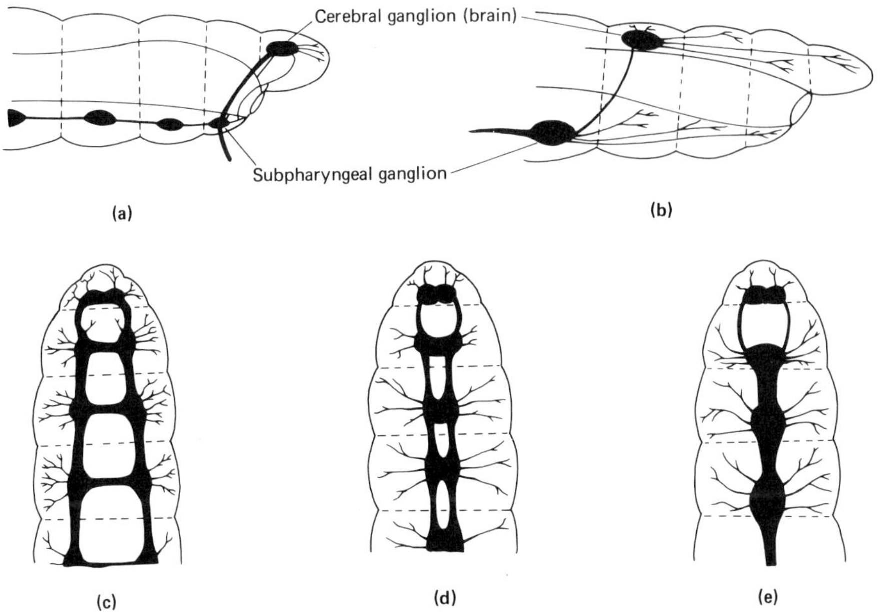

Figure 8.7 Evolutionary trends in the organization of the annelid nervous system. (*a*) In polychaetes the brain remains in the prostomium; (*b*) In oligochaetes and leeches the brain is located posteriorly; (*c*) The double ventral nerve cord and paired segmental ganglia of a primitive annelid; (*d*) An intermediate condition showing fusion of the segmental ganglia with retention of the double ventral nerve cord; (*e*) The single ventral nerve cord and single segmental ganglia of an advanced annelid. (From P.A. Meglitsch, *Invertebrate Zoology*, 2nd ed., 1972, Oxford University.)

built precisely in this way. A dorsal, cerebral ganglion or brain is located in the prostomium or in one of the first few anterior segments (Fig. 8.7 a,b). The brain receives input from nerves associated with the sense organs of the head, and its size reflects the development of these sensory structures. Circumenteric commissures connect the brain with a subesophageal or subpharyngeal ganglion, from which the major ventral, longitudinal nerve cords arise. Often, the gut is innervated by fibers originating from the circumenteric connectives. The evolution of the major ventral nerves in annelids represents a continuation of the trend established by the turbellarians (see Fig. 6.8). Primitively, two distinct cords are present (Fig. 8.7c); each links paired segmental ganglia. Cosegmental ganglia are connected by a lateral commissure, and each ganglion gives rise to secondary nerves. These nerves relay information to and from the gut and body wall. Sensory input from epidermal and muscular receptors associated with the vestiges of a subepidermal nerve plexus also enters the nerve cord at each ganglion. The nerve cords are intramuscular in lower annelids, but in more advanced forms, the cords have gradually been displaced to a protected position adjacent to the peritoneum. Another conspicuous adaptive trend leads to the consolidation of the ventral nerves. Advanced annelids display the fusion of segmental ganglia (Fig. 8.7d) and/or the union of both ventral tracts into a single nerve cord (Fig. 8.7e).

Most annelids possess giant nerve fibers. Composed of a single giant cell or several syncytial units, these large neurons conduct impulses at a rate over 1000 times faster than that of smaller, conventional neurons. Giant fibers rapidly transmit emergency information; they are the neuronal "hot lines" not only of annelids but of most larger and/or elongate invertebrates. Giant fibers mediate the escape responses of tube-dwelling and burrowing worms. Their discharge overrides any other information being transmitted through the system and results in the rapid and virtually simultaneous contraction of all longitudinal muscles of the body wall.

Although centralization is evident in the ventral nerve cord, the annelid nervous system apparently is not commanded to any great extent by the brain. Local reflex arcs apparently coordinate the muscular activity of each segment and prevent antagonistic fibers from contracting simultaneously. Peristaltic waves are perpetuated by the sequential activation of intersegmental reflexes. The ventral nerve cord is involved in all these reflex actions. However, it appears to relay very general information, and considerable peripheral discrimination is indicated in both sensory and motor situations. The brain rarely seems to know exactly what much of its body is doing, and its removal does not inhibit basic locomotion. The brain is important, however, in coordinating an annelid's behavior with the external world as perceived by its prostomial sense organs.

Sense Organs

Annelid sense organs include eyes, tactile sites and chemoreceptors, nuchal organs, and statocysts. These structures are quite variable among the three classes, and within the Polychaeta alone, considerable diversity is observed. This situation should be expected, however, because of the diversity of life-styles displayed within the phylum and particularly within its marine class. Actively crawling and swimming polychaetes possess relatively sophisticated sensory gear, while sedentary and less active forms display more rudimentary equipment. Most oligochaetes, for example, live in the rather dull world of terrestrial soil and lake bottom mud, and their sense organs are quite reduced. Leeches meanwhile pursue more active life-styles and so have keen sensory structures. But leeches originated from oligochaete stock and thus have no immediate ancestral foundation in sense organ biology. Accordingly, they have evolved unique sense organs that compare in quality with those of the polychaetes.

Annelid eyes display the widest diversity. The most active polychaetes possess cuplike prostomial eyes (Fig. 8.8a). Typically, such eyes are associated with pigments and a supporting cuticular cornea. A lens is secreted in some species, and focusing by muscular contractions and fluid pressure may allow image formation. In addition to prostomial eyes, polychaetes possess numerous photoreceptive centers scattered over the body surface. These so-called dermal eyes are the only light-sensitive structures in oligochaetes (Fig. 8.8b). Finally, leech eyes apparently are derived from oligochaete-like dermal eyes; typically, they consist of a vertical cluster of light-sensitive cells packaged in a pigment sheath (Fig. 8.8c).

Tactile cells are scattered widely over the annelid surface, but are concentrated around the mouth and particularly on the palps and tentacles of the peristomium and prostomium. These sensory units consist of a distal process projecting through the cuticle and a neuronal base communicating with the subepidermal nerve plexus (Fig. 8.9). Similar sensory cells, particularly those located about the mouth, may be chemoreceptive. However, the most conspicuous structures that may involve chemoreception are the **nuchal organs**. Located laterally on the prostomium, these paired ciliated pits are vital to the feeding behavior of the worm. At the base of each nuchal organ, neurons and gland cells are congregated, much as in the cerebral organs of nemertines.

Statocysts provide orienting mechanisms for many burrowing polychaetes. Removal of the statocysts from certain species results in an inability to fashion a proper burrow.

Behavior

Annelids crawl and burrow, some swim, and others lie in tubes. Relative to sponges, coelenterates, and the lower worms, they are fairly active animals; relative to most arthropods and the vertebrates, however, annelids are really quite slow. Indeed, in their behavior, as well as in many other aspects of their biology, annelids display a rather intermediate level between the lower and higher phyla of the invertebrate world.

As a rule, annelids are attracted by dim light but repelled by strong illumination. Apparently, light that stimulates dermal receptors alone is attractive to an annelid. Intervention by the brain may reverse such a response. Prostomial eyes, with their immediate cerebral innervation, may mediate negative reactions to light. If the prostomium and the brain are removed by surgery, the worm body will approach light of any intensity.

Tube-dwelling annelids, which expose their anterior ends for feeding, are particularly sensitive to changes in light intensity. A sudden major darkening signals the potential approach of a predator and causes the worm to retract rapidly into its tube. Such retraction follows the firing of the giant nerve fibers.

Except for the few swimming forms, annelids are strongly thigmotactic and thus favor postures producing maximum body contact with the substratum.

Many burrowing and tube-dwelling annelids display an innate periodicity in such vital activities as defeca-

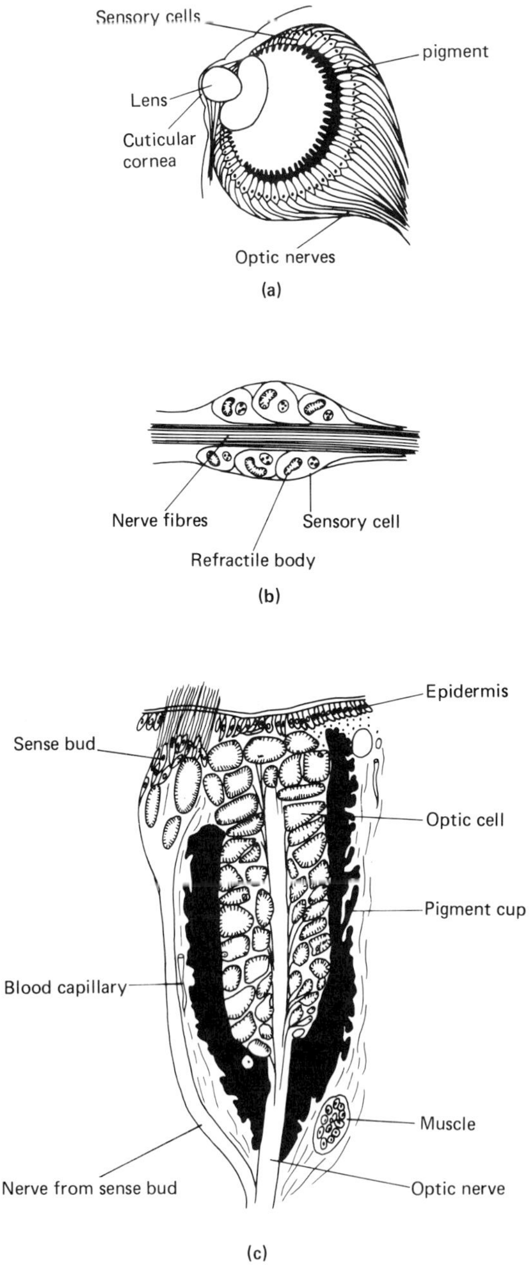

Figure 8.8 (*a*) The potentially image-forming eye of *Alciopa*, a planktonic polychaete; (*b*) A simple light sensitive structure, the so-called dermal eye of an earthworm; (*c*) Vertical section through the simple eye of *Hirudo*, a leech. A group of epidermal sensory cells known as a sense bud is also present. (*a* and *b* from Wells, *Lower Animals*, *c* from K.H. Mann, *Leeches*, 1962, Pergamon.)

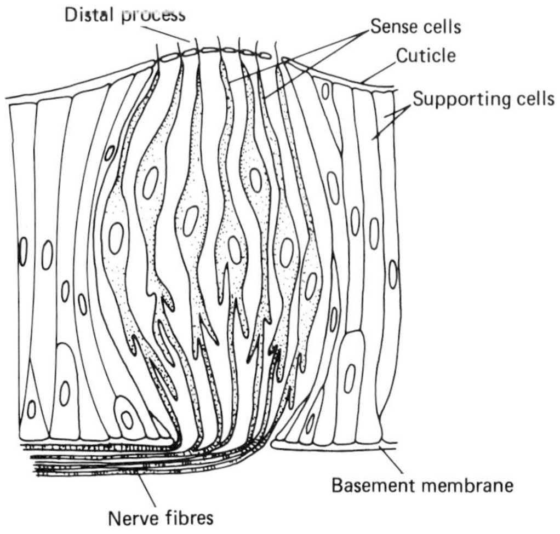

Figure 8.9 Diagram of a vertical section through an annelid epidermal sense organ. (From Edwards and Lofty, *Biology of Earthworms.*)

tion and the irrigation of their tubes and burrows. Some internal pacemaker regularly signals the worm when it is time to irrigate or to discharge undigested wastes. Researchers continue to investigate the nature of such mechanisms.

Finally, experiments on annelid learning have demonstrated the ability of these worms to formulate classical conditioning associations. Both earthworms and some errant polychaetes have performed well in maze experiments. These annelids can be taught to enter one side of a T-maze if attempts to enter the other side have been punished with electrical shocks. A hundred or so trials are sufficient to elicit the proper response in approximately three-fourths of the animals tested; memories extend over a 2- or 3-day period. Such data do not indicate that annelids are very intelligent. However, they are the lowest invertebrates in which classical conditioning can be demonstrated conclusively.

CONTINUITY SYSTEMS

Annelid continuity involves a variety of reproductive and developmental plans. Sexuality, hermaphroditism, asexuality, and regeneration are all well established in the phylum. The annelids undergo a typically protostomate embryogeny, which is distinguished further by the presence of a **trochophore** larva. The trochophore larva is common to both annelids and mollusks and represents an important focus for debate on the evolutionary kinships among annelids, mollusks and other invertebrate groups.

Reproduction and Development

The ancestral annelids were diecious and possessed paired gonads in each body segment. Most likely, these gonads were temporarily differentiated areas of the peritoneum and did not represent true organs. Gametes were released into the coelomic fluid, where their maturation was completed. Ripe gametes passed through special gonoducts or through the metanephridia, and external fertilization took place in the surrounding sea waters.

In today's primitive annelids, spiral determinate cleavage and gastrulation produce an embryo that rapidly develops into a trochophore (Fig. 8.10). The modern trochophore is a planktonic, top-shaped larva with an equatorial ciliary ring called the **prototroch.** An apical plate bears an additional tuft of cilia. Just below the prototroch, in the area of the original blastopore, a mouth and stomodeum form. A proctodeal anus, commonly surrounded by a small ciliary ring called the **telotroch**, opens at the lower extremity of the larva. An endodermal U-shaped gut connects the mouth and stomodeum with the anus. Next, a growth zone develops between the prototroch and the telotroch (Fig. 8.11). Here a massive proliferation of mesodermal tissue occurs, filling the original blastocoel. Paired mesodermal somites arise; these units represent the developing segments of the young worm, and their continuous serial deposition produces the animal's elongation (Fig. 8.11b). The mouth area, the prototroch, and the apical plate represent the presegmental prostomium, an area that diminishes in relative size as the young annelid grows. The central nervous system arises from the apical plate. The area surrounded by the telotroch corresponds to the future pygidium. With increased size, the postlarval stage settles to the substratum, where it begins its life as a segmented worm.

The polychaetes have modified the primitive system very little. These worms still feature separate sexes, and many species retain the fertility of each segment. However, a number of specialized polychaetes, especially the tube dwellers, possess only a few germinal segments. Also, gonoducts often merge with metanephridia, producing only one type of excurrent opening in each segment. Modern polychaete embryogeny is essentially faithful to the trochophore story just related.

The reproductive systems of oligochaetes and leeches are very highly specialized, displaying little to recall the ancestral condition. Remember that continuity systems within the Platyhelminthes were altered dramatically when flatworms adapted to parasitic life. Annelid evolution witnessed comparable changes when the phylum's reproductive and development systems were subjected to the selective

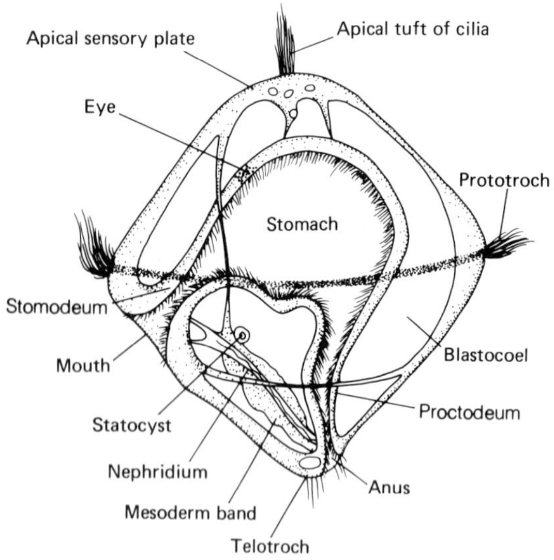

Figure 8.10 The trochophore larva of a polychaete worm. (From L.H. Hyman, *The Invertebrates, Vol. II: Platyhelminthes and Rhynchocoela* 1951, McGraw-Hill.)

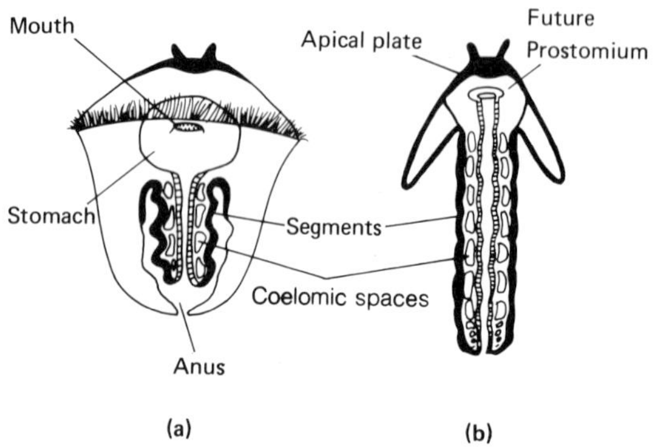

Figure 8.11 Post-trochophore development in a polychaete worm. (*a*) Initiation of mesodermal growth between prototroch and telotroch; (*b*) Serial deposition of segments along the anterior-posterior axis. The younger segments are located at the posterior end. (From C. Dawydoff, *In Traité de Zoologie, Vol. V*, Grassé, Ed.)

pressures of freshwater and terrestrial environments. The necessity of protecting both gametes and embryos in these harsher zones led to hermaphroditism, copulation, cocoon formation, and the complete suppression of an independent trochophore stage. These specializations, as well as some peculiarities in polychaete sexuality, are described upon considering each of the three classes in more detail.

Excepting the leeches, the annelids have considerable regenerative powers. They can reform lost structures, although the anterior is regenerated less readily than the trunk segments are. Some species even autotomize certain body parts if the whole animal can thereby be saved.

When regeneration is preceded by a regulated fragmentation, it becomes an important reproductive process. Asexuality occurs in some polychaetes and oligochaetes and includes budding as well as fragmentation and regeneration. Budding patterns include the formation of lateral fission zones along the body trunk (Fig. 8.12a). In these zones, daughter individuals differentiate to varying degrees and then separate from the mother worm. In some species, longitudinal chains of budding worms (then called zooids) may be formed (Fig. 8.12b).

Leeches alone among the annelid classes are incapable of either regeneration or asexual forms of reproduction.

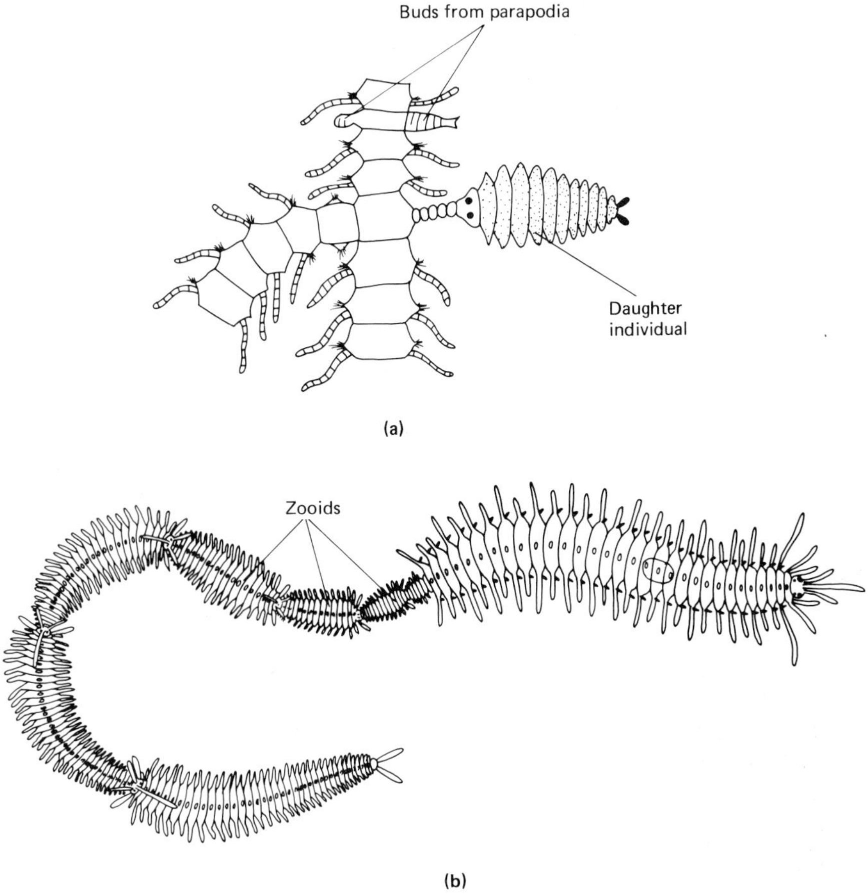

Figure 8.12 Asexual reproduction in annelids. (*a*) A portion of the polychaete *Syllis* showing budding of reproductive individuals from parapodia and also branching of the asexual stock; (*b*) *Autolytus purpureomaculatus*, exhibiting a longitudinal chain of budded worms (zooids) from the posterior tip. (*a* from R.W. Hegner and V.G. Engemann, *Invertebrate Zoology*, 1968, Macmillan; *b* from P. Fauvel, In *Traité de Zoologie, Vol. V*, Grassé, Ed.)

A Review

A radiating evolution has produced a diverse phylum Annelida, but some ancestral characters persist to unify the biologies of these worms. Basically, annelids are:

(1) segmented, eucoelomate worms;
(2) elongate animals with an anterior prostomium and a posterior pygidium;
(3) worms with a relatively straight, mouth-to-anus digestive tract;
(4) animals with a closed circulatory system dominated by a dorsal blood vessel;
(5) animals with excretory—osmoregulatory organs called metanephridia;
(6) worms with a brain, a major subesophageal ganglion, and a ventral nerve cord;
(7) the most primitive invertebrates in which learning definitely can be shown;
(8) protostomate animals, primitively displaying a trochophore larva.

Annelid Diversity

CLASS POLYCHAETA

The largest and most diverse annelid class is the Polychaeta, a marine group containing about 5300 species. Polychaetes are distinguished from other annelids by their possession of paired, lateral appendages, called parapodia (Fig. 8.13). These fleshy structures participate in the basic metamerism of the polychaete body and usually bear many projecting, chitinous bristles called **setae** (or chaetae; hence the class name). Generally, polychaetes are well-cephalized annelids. Most species are diecious and have many fertile segments. An independent trochophore larva is usually present in the life cycle.

But beyond these unifying characters, the class displays considerable diversity. Almost 40 families have been identified, but efforts to consolidate these groups at the order level have not been successful. To minimize taxonomic confusion, we limit our treatment of the class to a description of several well-known and representative genera. These polychaete genera may be placed into two nontaxonomic groups, the Errantia and the Sedentaria. The Errantia include the active (errant) polychaetes; the Sedentaria are less active, sedentary forms. While these groups represent artificial, nonphylogenetic divisions of the class, they nevertheless provide an instructive framework about which to organize a discussion of polychaete diversity.

The Errantia

Errant polychaetes are the more active members of the class—the swimmers, the crawlers, the continuous burrowers, and a few foraging tube dwellers. Such worms exhibit a persistent metamerism and considerable cephalization. The prostomium commonly bears palps, tentacles, and eyes. Well-developed parapodia in errant polychaetes are reinforced with skeletal rods called **acicula** (Fig. 8.13). The errant proboscis–pharynx is generally eversible and jawed. Representative genera include *Nereis, Aphrodita,* and *Glycera*.

Nereis (Fig. 8.14) often is cited as a "primitive polychaete." The genus does indeed display a number of ancestral characters: a very persistent metamerism, a well-developed head, and prominent, unmodified parapodia. These worms burrow in sand or mud near the tide mark and emerge at night to feed on a variety of small plants and animals. A few species are detritus eaters and one is commensal in hermit crab shells. The following discussion applies specifically to *Nereis virens*, the common clamworm, but most of the characters described are general throughout the genus.

The nereid head comprises the prostomium and the peristomium, the former dorsal to the latter (Fig. 8.15a). Prostomial sense organs include short paired tentacles and two stubby palps, as well as four dorsal eyes with gelatinous lenses. Modified peristomial parapodia form four pairs of longer dorsal tentacles. A powerful jawed proboscis is eversible through the mouth.

The parapodia of the nereid trunk demonstrate the classical architecture of these appendages. Indeed, the often bizarre parapodial forms of other polychaetes probably evolved from nereidlike precursors. Typically, these appendages present two lobes, a dorsolateral **notopodium** and a ventrolateral **neuropodium** (Fig. 8.13). Each lobe is supported by one or more acicula. Distally on both lobes are many chitinous setae extending from invaginated sacs. These setae are secreted from single cells within the distal sacs. Fingerlike projections or **cirri** are also common at the extreme upper and lower ends of the parapodial lobes.

Nereids employ their parapodia primarily in locomotion, and their body wall musculature is organized for the manipulation of these appendages.

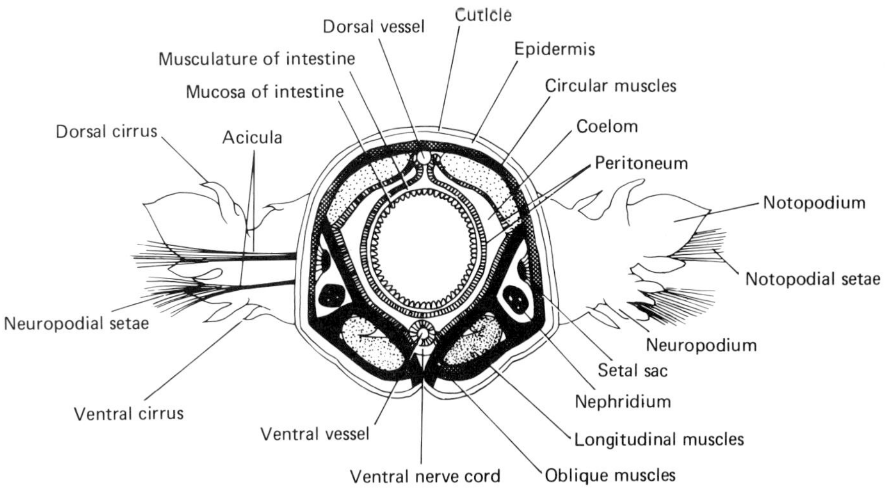

Figure 8.13 Cross section through a segment of a polychaete (*Nereis*) showing the paired lateral parapodia. The parapodium on the left shows the internal skeletal supports (the acicula); that on the right shows the external surface. (Modified from Kaestner, *Invertebrate Zoology, Vol. I*, Wiley-Interscience, 1967, and I.W. Sherman and V.G. Sherman, *The Invertebrates: Function and Form: A Laboratory Guide*, 2nd ed., 1976, Macmillan.)

Circular muscles are reduced, while oblique fibers connect the acicula with the ventral floor of the trunk (Fig. 8.13). Powerful longitudinal muscles on either side of the body antagonize each other during the wiggling movement of the worm, while the oblique muscles produce the paddling of the parapodia.

The broad flat surfaces of nereid parapodia are highly vascularized and represent the principal sites of respiratory exchange. Moreover, these appendages form true respiratory gills in many other polychaetes. Such gills often are derived from the dorsal cirri.

The nereid digestive system includes a mouth, an eversible proboscis and pharynx, and an esophagus associated with paired digestive glands (Fig. 8.15b). The system continues with a relatively straight and undifferentiated stomach—intestine, which concludes at the anus.

Metanephridia occupy every nereid segment except the peristomium. These osmoregulatory—excretory organs also transport gametes after the latter mature within the coelom. *Nereis* has no distinct gonads, and gametes simply differentiate in typical polychaete

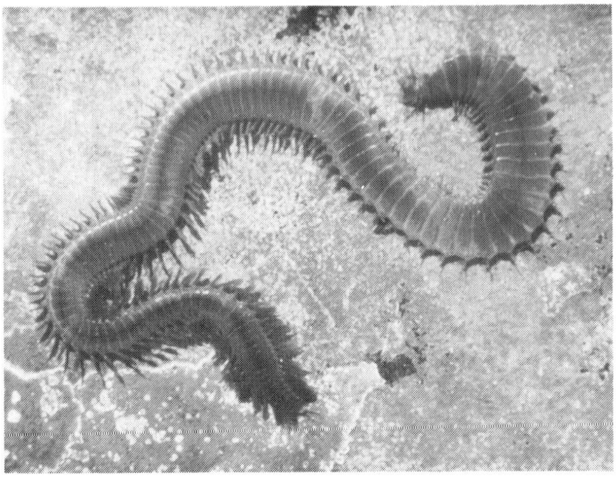

Figure 8.14 *Nereis vexillosa*, a clamworm of the Pacific coast of North America, and a close relative of the Atlantic coast clamworm, *Nereis virens*. (Photo courtesy of Ward's Natural Science Establishment.)

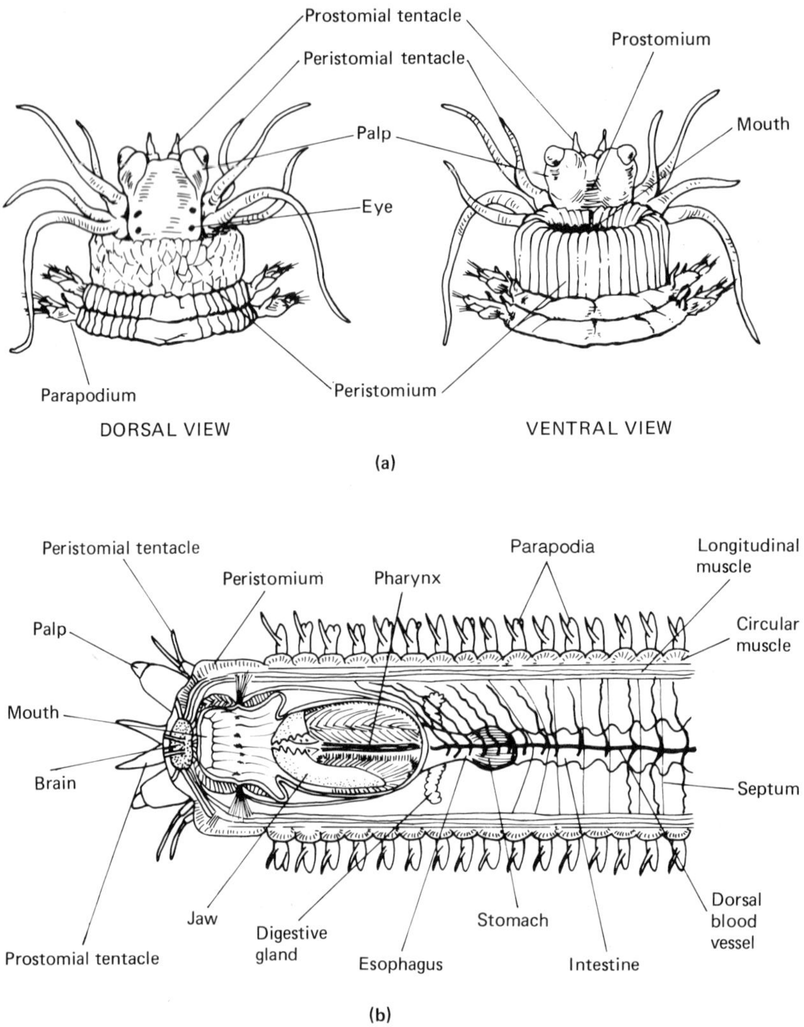

Figure 8.15 *Nereis virens.* (*a*) Dorsal and ventral views of the anterior end; (*b*) Dorsal dissection of the anterior portion. (*a* from Kaestner, *Invertebrate Zoology, Vol. I*; *b* from W.B. Benham, In *Cambridge Natural History, Vol. II*, S.F. Harmer and A.E. Shipley, Eds., 1959, Macmillan.)

fashion from the septal peritoneum. Sexes are separate, and a distinct periodicity in sexual maturation occurs. (See the following discussion of epitoky.)

Aphrodita (Fig. 8.16), commonly known as the sea mouse, is an errant polychaete whose parapodial elements are highly developed. Large and muscular neuropodia lift the main body of the worm from the substratum during their effective, stepping stroke. The notopodia, meanwhile, are reduced and extremely modified. Flat, segmental plates, probably representing specialized notopodial cirri, extend horizontally over the worm's dorsal surface. These plates, called **elytra**, establish a shallow respiratory channel above the animal and enable *Aphrodita* to ventilate its dorsal surface even when burrowing. The furry appearance of *Aphrodita* is attributable to its numerous setae.

Glycera (Fig. 8.17) is an active burrower. This errant polychaete fashions a gallery of underground tunnels, from which it emerges to seize prey. Its structural modifications represent adaptations for a burrowing, rapacious life-style.

Parapodia are reduced in size in *Glycera*, and the head region is pointed and streamlined. Both of these conditions facilitate the worm's movement through

Figure 8.16 The sea mouse, *Aphrodita aculeata* (dorsal view). The animal may be as much as 17 cm long and 7 cm wide. (Photo courtesy of D.P. Wilson.)

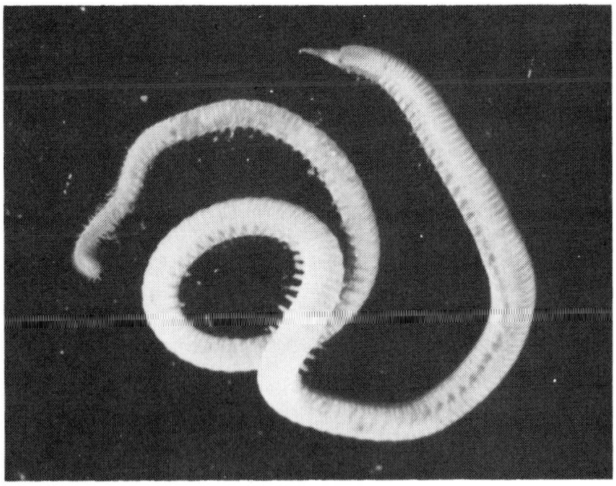

Figure 8.17 *Glycera americana*, a burrowing errant polychaete. (Photo courtesy of Ward's Natural Science Establishment.)

the marine floor. A rather long and conical prostomium may be the site of pressure receptors. Changes in water pressure above the burrow openings inform the worm of potential feeding opportunities above. The worm inches to a strategic opening and then rapidly lashes out to seize prey. The latter includes small crustaceans and other invertebrates of proper size, whose seizure is effected by an elaborate proboscis (Fig. 8.18). This large and powerfully eversible organ occupies the first 20 segments of the body. Septa are absent from this region, which therefore constitutes a single hydrostatic compartment. Contractions of the body wall musculature generate fluid pressures that force the proboscis out through the mouth. The organ is everted in the process, baring its four-jawed tip. Each of these jaws is serviced by a poison gland, making *Glycera* one of the most deadly hunters in its phylum.

Before concluding our discussion of the errant polychaetes, we call attention to a remarkable reproductive phenomenon associated with several errant families. That phenomenon is **epitoky** and involves the timely asexual formation of swarming, sexually reproducing individuals.

Epitokous polychaetes spend most of their lives on the marine floor in shallow waters. They feed and grow, but remain immature sexually; such forms are called **atokes**. Sexual maturation occurs only when an atoke produces an **epitoke** either by budding or by direct transformation. The sexually potent epitokes differ from atokes in several secondary characters. Commonly, they possess better photoreceptors and their posterior segments are larger to accommodate accumulating gametes (Fig. 8.19). Digestive organs are degenerate. Epitokes swim rather than crawl, and their parapodia and segmental musculatures are reorganized accordingly. More efficient, oar-shaped setae are also common.

Upon rising in swarms from the ocean floor, sexual epitokes congregate in surface waters. The ensuing reproductive spectacle ranks among the most orgiastic in the invertebrate world. Males usually perform a

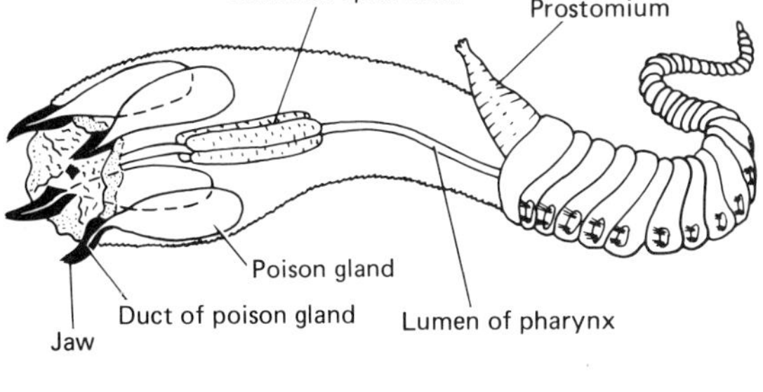

Figure 8.18 A schematic of *Glycera* emphasizing the everted proboscis apparatus used in prey capture. (From R.D. Barnes, *Invertebrate Zoology*, 4th ed., 1980, Saunders College Publishing.)

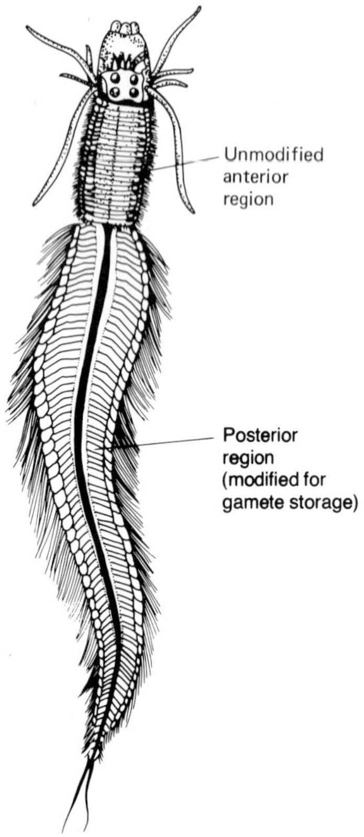

Figure 8.19 Male epitoke of *Nereis irrorata.* (From Fauvel, In *Traité de Zoologie, Vol. V,* Grassé, Ed.)

corkscrew dance about their female partners. Apparently, a female pheromone stimulates sperm release from the dancing male, and the presence of sperm causes eggs to be shed. Gamete release usually requires the rupture of the body wall and the immediate death of both adult worms. Fertilization then occurs externally.

In *Platynereis* (a worm in the same errant family as *Nereis*) unfertilized eggs can survive only a half-minute exposure to sea water, but a bizarre mechanism ensures their viability. The male worm inserts his anus into the mouth of the female and deposits sperm into her body. The digestive tracts of both sexes have degenerated; hence transmission occurs directly from the male to the female coelom. Fertilization takes place within the female, and resistant zygotes are shed immediately through her anus.

Epitokous swarming obviously involves synchronism. Sexual individuals must form within a limited time, and their mating ascents must occur simultaneously. Experimental evidence indicates that neurosecretory cells within the cerebral ganglia are responsible for epitokous periodicity. These cells apparently secrete a hormone inhibiting epitoky, as removal of the ganglia from immature worms initiates precocious epitoke formation. Reimplantation of the immature ganglia restores inhibition, but transplants from the cerebral ganglia of epitokes do not. Evidently, the inhibitory hormone is absent from the sexual forms.

In most species under investigation, swarming is related to the lunar cycle. The critical environmental factor appears to be the cyclical fluctuation in moonlight, as individual species swarm on a characteristic day of the lunar calendar. Samoan palolo worms, for example, swarm in October or November just after the third-quarter moon, while the Atlantic palolo mates in July during the first and third quarters of the lunar cycle. The Atlantic polychaetes swim to the surface a few hours before dawn, dance, and rupture at sunrise.

The Sedentaria

The second informal division of the Polychaeta comprises its more sedentary types. Such marine annelids live in permanent burrows or tubes, which they rarely, if ever, leave. In association with their sedentary ways and nearly constant orientation toward the environment, these polychaetes display regional specializations of the body. The phylum's ancestral metamerism is diminished in favor of specialized, supersegmental areas for food acquisition, osmoregulation–excretion, respiration, and reproduction.

Sedentary polychaetes have small parapodia in which acicula are never present. A protrusible proboscis is rare in these worms. Heads are reduced, as are prostomial sense organs, but elaborate anterior tentacles may collect food. Epidermal gills are common. Metanephridia and gonads usually are concentrated in distinct body regions. Genera representative of the Sedentaria include *Arenicola, Amphitrite,* and *Sabella. Chaetopterus,* meanwhile, is too fascinating a polychaete to omit from our discussion.

Arenicola (Fig. 8.20), the common lugworm of shallow and intertidal marine waters, is a typical sedentary polychaete. It lives in a permanent L-shaped burrow with its head toward the burrow's blind end. Its head is reduced considerably (a typical burrowing adaptation), and parapodia are conspicuous only as a few notopodial gills in the midbody and anterior regions.

Arenicola draws a respiratory–feeding current by peristaltic contractions of its body wall musculature. Water enters the open end of the burrow and passes from tail to head over the worm's body. Water then percolates vertically through the sand and the mud at the blind burrow end. Such a current aerates the gills while concentrating food particles in front of the mouth. Food suspended in the incoming water is trapped at the burrow's blind end and can be consumed promptly by the worm. Experiments indicate, however, that *Arenicola*'s foodstuffs are most concentrated near the surface of the substratum, and upward thrusts by the mouth are of primary importance in food gathering. Such thrusts cause nutrient-rich surface mud and sand to sink toward the polychaete's mouth and are responsible for the funnel-shaped depressions associated with lug-worm populations. Feeding thrusts involve an eversible but jawless proboscis. Such a proboscis is a rarity among sedentary polychaetes; in *Arenicola*, it is manipulated by retractor muscles and by three powerful septal diaphragms. The latter are highly muscularized septa which control the hydrostatic compartments surrounding the proboscis. Proboscis mucus traps particles of an appropriate size, which the worm then consumes. The lugworm's alimentary tract features a very glandular stomach and an intestine specialized for absorption. Amoebocytes take up nutrients from the intestinal mucosa; these cells apparently are the sites of final digestion. Defecation occurs at regular intervals when the worm backs its posterior to the burrow opening and piles its castings.

Lugworms are fairly tolerant of fluctuations in osmotic pressures. Over a certain range, their blood conforms to prevailing environmental conditions; a blood freezing point range of −0.29 to −1.72°C indicates this considerable tolerance. Such osmotic conformity represents an adaptation to life in coastal areas, where irregular water runoffs from land cause instability in local fluid composition.

The reproductive system in lugworms is typical of most sedentary polychaetes. Gamete production is limited to six genital segments. The nephridia in these segments function solely in the transport of gametes to the outside.

Arenicola is an influential agent on the geology of the coastal bottom. These polychaetes parallel the earthworms in terms of their soil-mixing activities. In areas with large lugworm populations, the entire upper 50 to 60 cm of the seabed is ploughed through within a 2- or 3-year period.

Most sedentary polychaetes are less active than *Arenicola*. *Amphitrite* (Fig. 8.21), for example, is a marine annelid that remains relatively still within a tight-fitting burrow. In adapting to such a life-style, *Amphitrite* displays the characteristic features of the Sedentaria. Its parapodia and prostomial sense organs are reduced, and the body has undergone extensive regional specialization.

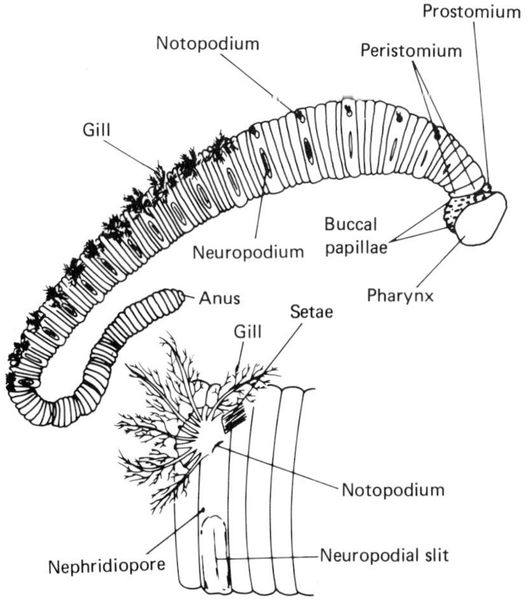

Figure 8.20 External features of *Arenicola*, the lugworm, with a close up of one of the gills modified from the notopodial lobe of a parapodium. (From Sherman and Sherman, *The Invertebrates: Function and Form.*)

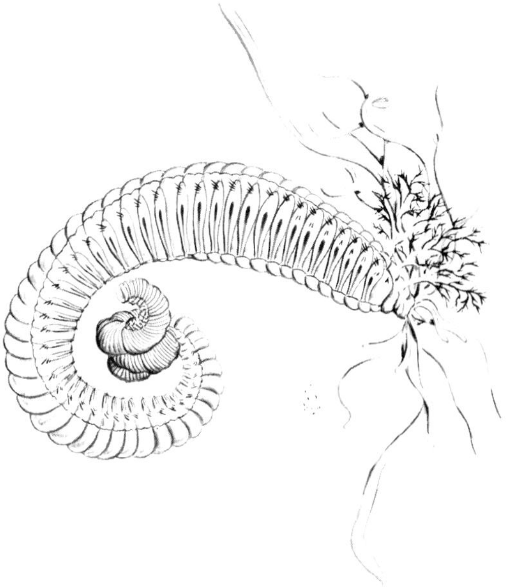

Figure 8.21 *Amphitrite*. (From F.A. Brown, Selected *Invertebrate Types*, 1950, John Wiley.)

The prostomium bears long, contractile tentacles that collect food and provide respiratory surfaces. Gas exchange also occurs across gills located dorsally on the anterior three segments. Such gills and tentacles ensure adequate respiration despite the worm's isolation within its tight burrow. Peristaltic movements of the body wall draw a tail-to-head current through the burrow, thus irrigating the respiratory surfaces.

Amphitrite is an indirect deposit feeder. Its tentacles stretch across the substratum and may creep along by ciliary action. Cilia border tentacular grooves along which microscopic food particles are transported. A copious supply of mucus lubricates these feeding tracts. Food accumulates at the tentacular bases and enters the mouth when each tentacle is wiped across the ciliated, peristomial lip. The digestive canal consists of a small, noneversible pharynx, an esophagus, a stomach, and a long intestine. Peristaltic contractions move foodstuffs through the anterior region of the gut, but intestinal transport depends on ciliary action within a long, ventral groove. The absorptive area of the intestine is increased by a slight coiling of the organ.

Sabella (Fig. 8.22) is a fan worm, a sedentary polychaete which dwells in a permanent tube constructed from mucus-bound sand grains. Like most fan worms, it displays a colorful crown of feathery prostomial tentacles. These tentacles, usually called radioles, extend from the open end of the worm's sand tube, where they collect both food and building materials. Cilia occupy the radiole surfaces, and their organized beating draws a water current up through the crown. Tentacular mucus traps suspended particles which are passed along radiole grooves toward the mouth. These particles are sorted according to size. Commonly, the largest particles are rejected, the smallest are eaten, and carefully selected medium-sized sand grains are incorporated into the sabellid tube. Tube construction involves the mixing of sand grains with mucus secreted from the walls of sand storage sacs. The latter are located ventrally just below the mouth (Fig. 8.23). From these sacs, a moist, supple string of mucus and sand is conveyed to a peristomial collar. This collar folds over the upper rim of the tube, where it molds new building material. As the mucus and sand are presented to the peristomial mold, the worm rotates, applying the new material evenly to its perimeter. When a fan worm is threatened, the radioles contract and the peristomial collar may fold over the anterior regions.

Other mucous glands occur in *Sabella*. Most prominent are the ventral, metameric glandular shields whose mucus is deposited along the length of the tube. Such mucus minimizes sand irritation of the rotating body trunk.

Other sabellid systems are predictable for sedentary polychaetes. True metanephridia are limited to a single enlarged pair at the anterior end. In fertile segments of the animal, specialized nephridia remain only for the transport of gametes. Fan worms are well adapted to one of the major problems of life within a tube—defecation. While many polychaetes risk

Figure 8.22 *Sabellastarki magnificans.* (Courtesy of Roger Klocek, Shedd Aquarium.)

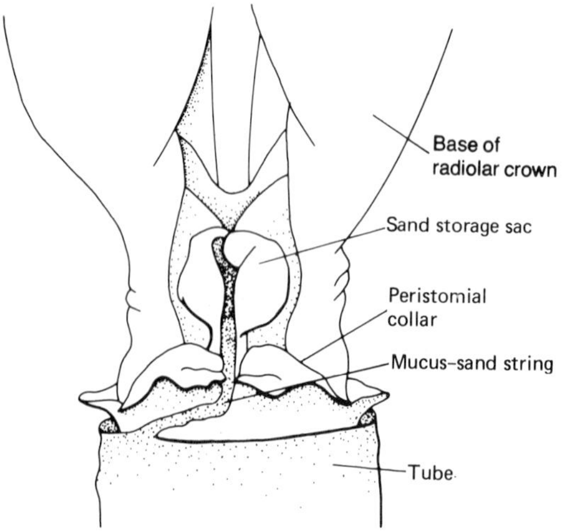

Figure 8.23 Ventral view of the base of the crown of radioles and the anterior end of the tube in *Sabella pavonina* showing the method by which the string of mucus and sand is applied to the edge of the tube. (From J.A.C. Nicol, *The Biology of Marine Animals*, 1967, John Wiley.)

periodic exposure outside their tubes during this process, *Sabella* discharges its undigested wastes without ever leaving its home. A ciliated midventral groove extends from the anus to the thoracic region, where it curves dorsally and terminates at the open end of the tube. Feces are packaged in mucus and then removed along this groove.

Other polychaetes also are known as fan worms. These annelids include close relatives of *Sabella*, as well as numerous types that live in tubes of calcium carbonate. The latter group, called serpulids (Fig. 8.24), possesses two large calciferous glands beneath the peristomial collar. These glands secrete calcium carbonate, which the collar molds onto the serpulid tube. Serpulid tubes may be anchored to rocks, shells, or other hard substrata; thus these fan worms exploit habitats normally excluding annelids. Usually, one of the serpulid radioles is modified as an operculum. When the feathery crown is retracted, this operculum protectively seals the open end of the tube.

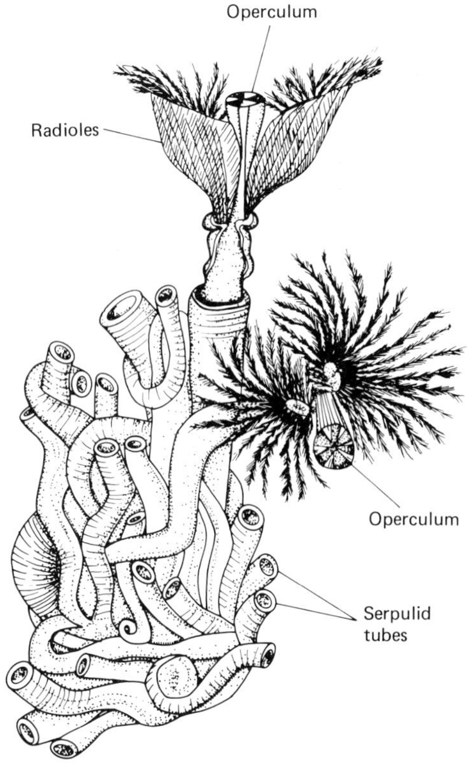

Figure 8.24 A group of calcareous tubes of *Serpula vermicularis*. Animals are protruding from the mouths of two of the tubes. (From Benham, In *Cambridge Natural History, Vol. II*, Harmer and Shipley, Eds.)

We conclude our survey of polychaete types with a discussion of one of the most bizarrely formed denizens of the invertebrate world. *Chaetopterus* (Fig. 8.25) is a sedentary marine annelid, which lives on mud flats within a permanent, horseshoe-shaped tube with openings at both ends. Its extremely modified body demonstrates the extent to which a polychaete can be specialized anatomically when it maintains a constant orientation to its environment.

The notopodia of the 14th, 15th, and 16th segments of *Chaetopterus* are fused as undulating, semicircular fans. Fan action draws water in an anterior-to-posterior direction over the worm. The 12th segment bears notopodial "wings" whose epithelia are ciliated and richly supplied with mucous glands. These broad wings control the local passage of the feeding current. As that current passes the 12th segment, large volumes of mucus are secreted by the notopodial wings. Current pressure forms a mucous sac extending posteriorly from the wings. This sac filters edible particles from the passing waters. Between the wings and the semicircular fans, a food cup projects from the worm's dorsal surface. This food cup harvests the mucous bag, gathering its contents into a compact food pellet. The pellet is transported along a ciliated, middorsal groove to the mouth for swallowing.

One other aspect of *Chaetopterus* biology is as amazing as its anatomy. For such a regionally

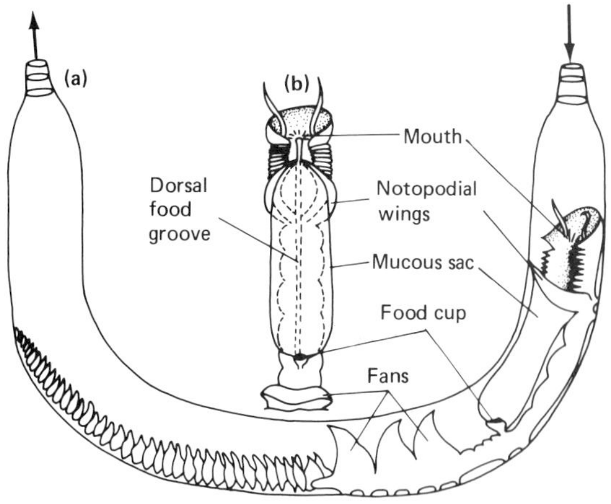

Figure 8.25 The parchment worm, *Chaetopterus variopedatus*. (*a*) Diagram of the worm feeding within its tube. Arrows show the direction of water flow through the tube; (*b*) Anterior portion of the worm, dorsal view. (From G.E. MacGinitie and N. MacGinitie, *Natural History of Marine Animals*, 1968, McGraw-Hill.)

specialized animal, *Chaetopterus* displays an incredible regenerative capacity. Experiments indicate that this polychaete can reform completely from any one of its first 14 segments!

Other Polychaetes

The genera described above reflect the diversity of the Polychaeta. However, still other life-styles exist within the class. The pelagic *Alciopa* (Fig. 8.26) represents a group of free-swimming, marine polychaetes. These worms commonly possess large eyes with distinct lenses. Other polychaetes make their homes by boring into mollusk shells, although the boring method remains undescribed.

Certain very small polychaetes live between sand grains and other bottom particles. These interstitial worms include most of the archiannelids. Archiannelids (Fig. 8.27) are reduced polychaetes; they have neither parapodia nor setae, and their segmentation is usually indistinct. Their epidermis may be ciliated. Archiannelids (as their name implies) originally were thought to represent the ancestral annelid type. Current opinion, however, considers their features to be derived rather than ancestral. The group may have a diverse origin, and similarities among its members could represent convergent adaptations to interstitial life.

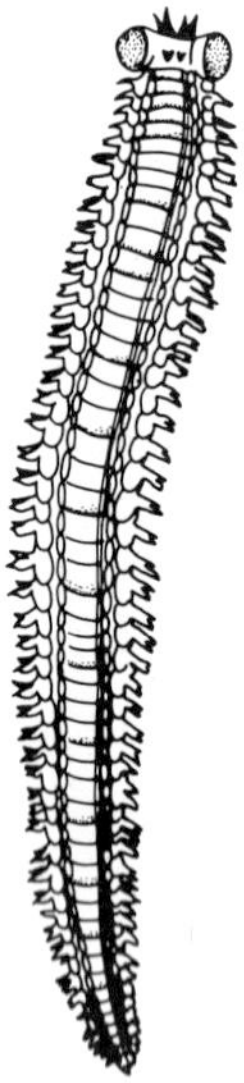

Figure 8.26 *Alciopa reynaudi*, a pelagic polychaete with large well-developed eyes. (From Fauvel, In *Traité de Zoologie, Vol. V*, Grassé, Ed.)

CLASS OLIGOCHAETA

The second of the three major annelid classes is the Oligochaeta, a group of segmented worms adapted for life in terrestrial and freshwater environments. Terrestrial forms constitute a majority of the 3100 known species and include many familiar earthworms. Life on land poses great problems for the annelid plan, and terrestrial adaptation has involved alterations in several of the phylum's ancestral characters. Many maintenance problems, however, have been circumvented by limitations on the terrestrial range. Most oligochaetes simply remain in soils with ample supplies of moisture, oxygen, and food. In addition, an adaptive background in burrowing has provided well for their success on land. In their continuity systems, however, oligochaetes have departed quite dramatically from the ancestral plan. Adult worms may be able to cope with the greater demands of the land environment, but gametes and developmental stages are far more vulnerable. The ancestral condition featuring external fertilization, unprotected embryos, and free-swimming larvae simply will not work for the oligochaetes. Accordingly, these annelids exhibit methods of protected gametic transfer and environmental isolation of the developmental stages. These processes address the hermaphroditism of all oligochaetes, their copulatory techniques, and their production of a cocoon by glandular secretions from a distinctive surface zone, the **clitellum** (Fig. 8.28).

Oligochaetes lack parapodia. Setae remain, although inconspicuously, in dorsolateral and ventrolateral positions, recalling the traditional location of the notopodia and neuropodia. (The relatively small number of these bristles in most oligochaetes accounts for the name of the class.) The prostomium lacks sensory appendages and the oligochaete head is much reduced. Recall that blunt heads and reduced parapodia are common among burrowing polychaetes; the occurrence of these two phenomena in the present class also reflects adaptation for a burrowing life-style. In their external appearance,

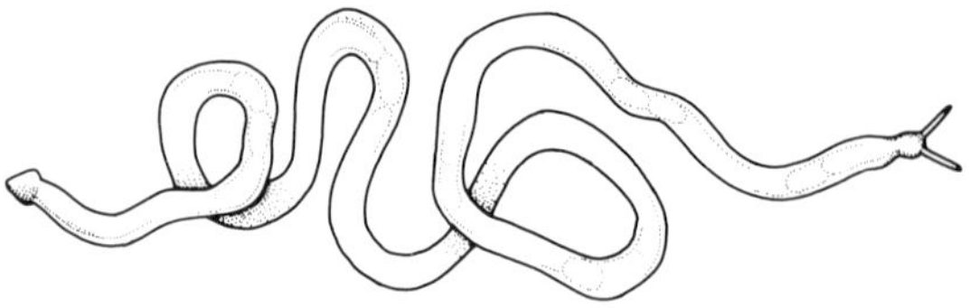

Figure 8.27 *Polygordius neapolitanus*, an archiannelid. (From P. de Beauchamp, In *Traité de Zoologie, Vol. V*, Grassé, Ed.)

oligochaetes display much less diversity than do polychaetes. The general absence of appendages and gills and a well-preserved trunk segmentation produce a fairly uniform body type throughout the group. Only the glandular clitellum, variously located over several segments in the anterior half of the body, breaks the surface monotony of these worms. Internally, the closed circulatory system, the metanephridia, and the nervous and muscular machinery retain strongly metameric characters. The foregut has undergone some specialization, but the most striking reorganization is observed in the oligochaete reproductive system. Oligochaetes have both ovaries and testes in one or more segments; these paired, dual gonads are serviced by a complex system of ducts leading to distinct male and female genital pores. The considerable diversity of genital structure forms the basis of most systems of oligochaete taxonomy.

Several oligochaete orders have been identified. However, paralleling our account of the Polychaeta, we discuss the Oligochaeta in terms of its two general divisions, the terrestrial forms and the freshwater groups.

The Terrestrial Oligochaetes

The oligochaetes that live on land are perhaps the most familiar worms in the invertebrate world. They

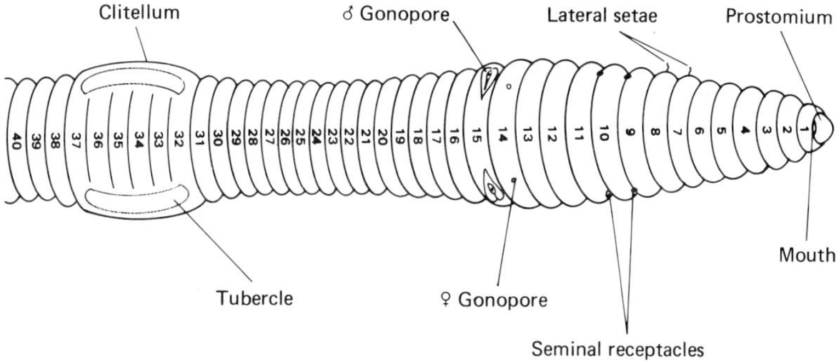

(a)

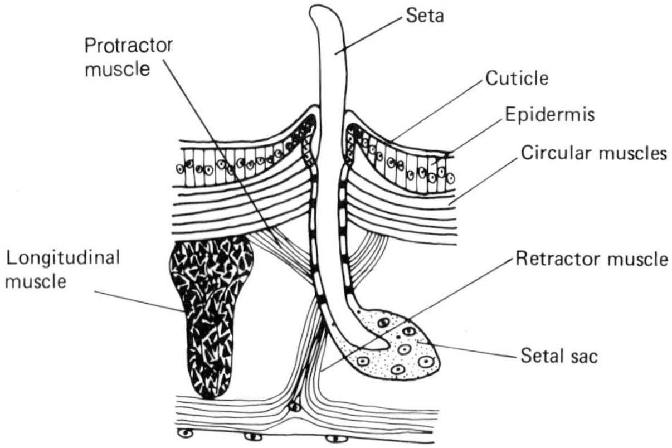

(b)

Figure 8.28 (*a*) Ventral view of the anterior third of an earthworm, *Lumbricus terrestris*. (Segments numbered); (*b*) Schematic transverse section through the body wall of an oligochaete showing setal musculature. (*a* from Kaestner, Invertebrate Zoology, *Vol. I*; *b* from M. Avel, In *Traité de Zoologie*, *Vol. V*, Grassé, Ed.)

are burrowers in moist, organically rich soils. These annelids are larger than the average polychaete: One report from South Africa describes an earthworm 6.78 m long and a full 8 cm in diameter! Most species, however, are a more moderate 10 to 30 centimeters in length.

The common earthworm *Lumbricus* (Fig. 8.28a) is familiar to most beginning students. This "night crawler" is a relatively sophisticated oligochaete, but nevertheless represents the terrestrial members of its class. *Lumbricus* is long and cylindrical, typically with over 100 segments. Metamerism is strongly evident and few external features are conspicuous. The head is reduced to a small, prostomial lip protruding over the peristomium. No sensory appendages are present. Four pairs of setae occupy every segment behind the peristomium; these pairs are located on both sides of the body tube in dorsolateral and ventrolateral positions (Fig. 8.29b). Protractor and retractor muscles (Fig. 8.28b) articulate these chitinous bristles, which are important for traction during locomotion. The most conspicuous external structure is the clitellum, a saddle-shaped glandular swelling over the 31st or 32nd to the 37th segment of mature worms (Fig. 8.28a). Numerous pores communicate between internal regions and the outside world. These pores include the segmental outlets of the metanephridial system, dorsal coelomic openings, and well-defined entrances and exits for the worm's reproductive facilities.

Night crawlers are herbivorous scavengers, feeding primarily on dead leaves and other decaying plant matter. Typically, they emerge from their burrows at night to eat. The posterior tip of *Lumbricus* usually remains in the burrow, so that rapid retraction (recall giant axons) is possible when predators threaten. The thick muscular pharynx (Fig. 8.29a) is somewhat protrusible, and food is sucked through the mouth by a pharyngeal pump. Pharyngeal glands moisten incoming food with mucus and also release a protease. The pharynx leads to a long, narrow esophagus. The earthworm esophagus is modified regionally as a thin-walled storage compartment called the crop and a cuticular and thick-walled grinding chamber called the gizzard. From the gizzard, food is passed into a long and relatively undifferentiated intestine. The earthworm intestine is distinguished by a prominent infolding of its dorsal wall, called the **typhlosole** (Fig. 8.29b). The typhlosole increases the surface area available for absorption of digested nutrients.

Earthworm respiratory and circulatory systems are faithful to the ancestral pattern. A closed circulatory network with a large dorsal blood vessel, a ventral vessel, and connective capillaries provides an efficient system of internal transport. Five prominent esophageal loops join the dorsal and ventral vessels in their segments (Fig. 8.30). These connectives, commonly called "hearts," regulate blood pressure in the system. Hemoglobin dissolved in the blood plasma transports approximately 40% of the worm's oxygen. Oxygen uptake is general over the moist body surface and is enhanced by a dense capillary network just below the cuticle.

The earthworm cuticle is lubricated by mucous gland secretions and by the controlled release of coelomic fluid through sphinctered dorsal pores. The cuticle must remain moist at all times because it serves as the respiratory surface. This condition sets a major limitation on the range of oligochaetes. Land obviously is the driest habitat, and the maintenance of a moist external surface amid such aridity is problematic. Earthworms manage by living in the soil and by burrowing deeper when surface regions dry out. Some types even estivate or hibernate when the environment undergoes seasonal severities.

Lumbricus has a highly regular metanephridial system comprising paired nephrostomes, highly coiled nephridial tubules, and nephriopores in every segment except the peristomium. Lengthy excretory tubules effectively resorb salts and allow earthworms to adjust to fluctuating osmotic conditions. As in most terrestrial oligochaetes, urea is the primary nitrogenous waste, although ammonia is excreted in variable quantities depending on environmental conditions. Earthworms survive prolonged submersion in water by excreting a copious, hypotonic urine. *Lumbricus* often emerges from the ground during heavy rains. Such behavior is not caused by osmotic problems, but rather by the lowered oxygen concentration of saturated soils.

Lumbricus displays a rather concentrated nervous system. Typical of most oligochaetes, the ventral nerve cord is singular and houses giant fibers. The brain occupies the third segment above the pharynx (Fig. 8.7b). Each segmental ganglion gives rise to three pairs of lateral nerves. These nerves contain both sensory and motor elements; sensory fibers may innervate immediately anterior and posterior segments as well as the local one.

Lumbricus has a well-developed body wall musculature with strong circular and longitudinal layers. During locomotion, these muscle layers act antagonistically against coelomic compartments. Characteristically, alternating circular and longitudinal pulses originate behind the peristomium and pass posteriorly along the body trunk. When the worm elongates, posterior setae extend to prevent slippage. These bris-

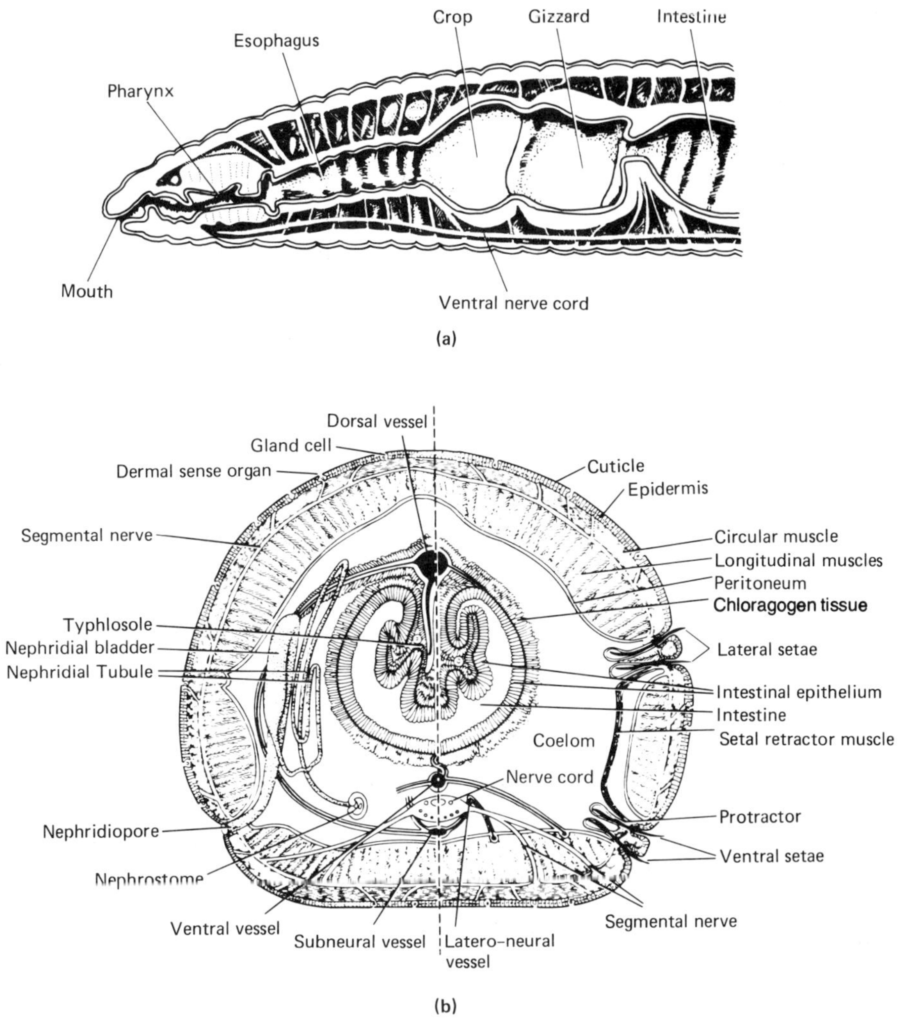

Figure 8.29 Internal anatomy of the earthworm, *Lumbricus terrestris*. (*a*) Sagittal section through the anterior digestive tract; (*b*) Cross section through the intestinal region. A nephridium is shown on the left side; setae are shown on the right. (*a* from Avel, In *Traité de Zoologie, Vol. V, Grassé*, Ed.; (*b*) from Sherman and Sherman, *The Invertebrates: Function and Form*.)

tles are retracted when anterior longitudinal muscles pull their segments forward. Earthworms burrow by thrusting their anterior ends into cracks and consuming the soil within their path. Soil is processed through the gut, where usable organic components are digested out. Mucus-coated feces are smeared against the sides of the burrow. Earthworm activity is invaluable to soil health, as it promotes drainage, aeration, and vertical mixing of soil types. Indeed, burrowing by earthworms represents one of the major causative factors in the evolution of modern soil itself.

Like most oligochaetes, *Lumbricus* lacks elaborate sense organs. No distinct eyes are present, although dermal photoreceptors are scattered over the body surface and are quite dense at anterior and posterior ends. Earthworms are attracted by weak light, but repulsed by light of stronger intensities. Numerous free nerve endings projecting through the cuticle respond to touch, chemicals, and other environmental stimuli.

The hermaphroditic reproductive system of this worm is typical of its class (Fig. 8.31). Male elements include two pairs of testes, one in the 10th segment

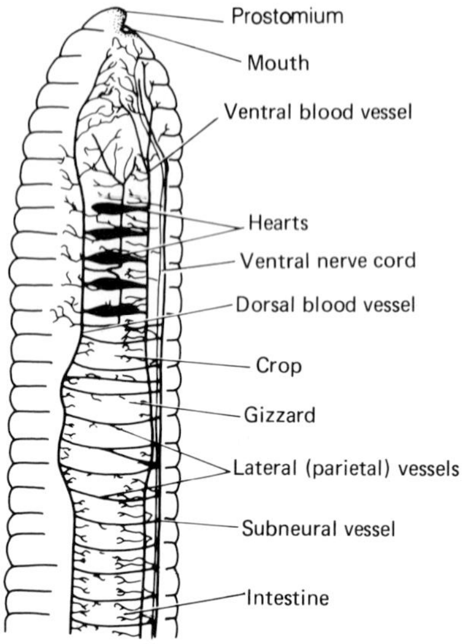

Figure 8.30 Lateral view of the anterior part of *Lumbricus terrestris* featuring the principal vessels of the circulatory system. (From D.E. Beck and L.F. Braithwaite, *Invertebrate Zoology Laboratory Workbook*, 3rd. ed., 1968, Burgess.)

and one in the 11th. Three pairs of seminal vesicles are present, one each in the 9th, 11th, and 12th segments. Nephridialike vasa deferentia drain the two segments containing the testes; these ductules merge posteriorly on each side and exit through a single pair of male pores on the 15th segment. At the posterior ends of the vasa deferentia, glandular tissue may constitute a prostate. The female system features only one pair of ovaries in the 13th segment. Paired ovisacs project posteriorly from the rear septa of the ovarian segment and are drained by oviducts. The oviducts terminate in paired genital pores on segment 14. Two pairs of seminal receptacles are located in the 9th and 10th segments.

Compared with those of the polychaetes, the reproductive facilities just described are complex indeed. All oligochaetes have similar systems, with taxonomically important differences in the number and segmental placement of the gonads and their accessory structures.

Gametes are produced within the gonads, but may complete their development in the associated coelomic compartment or inside the seminal vesicles and ovisacs. The latter are storage quarters for mature sperm and eggs, respectively. When two earthworms meet under appropriate conditions, mutual copulation occurs. Most oligochaetes orient in a head-to-tail fashion (Fig. 8.32) so that the male genital pores of

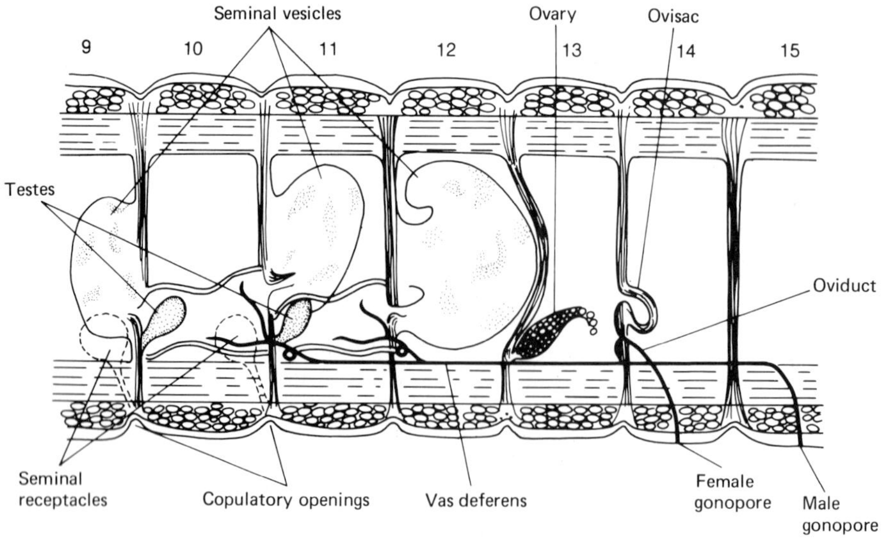

Figure 8.31 Longitudinal section through the reproductive segments of an earthworm. (Numbers correspond to specific body segments; see text.) (From Avel, In *Traité de Zoologie, Vol. V*, Grassé, Ed.)

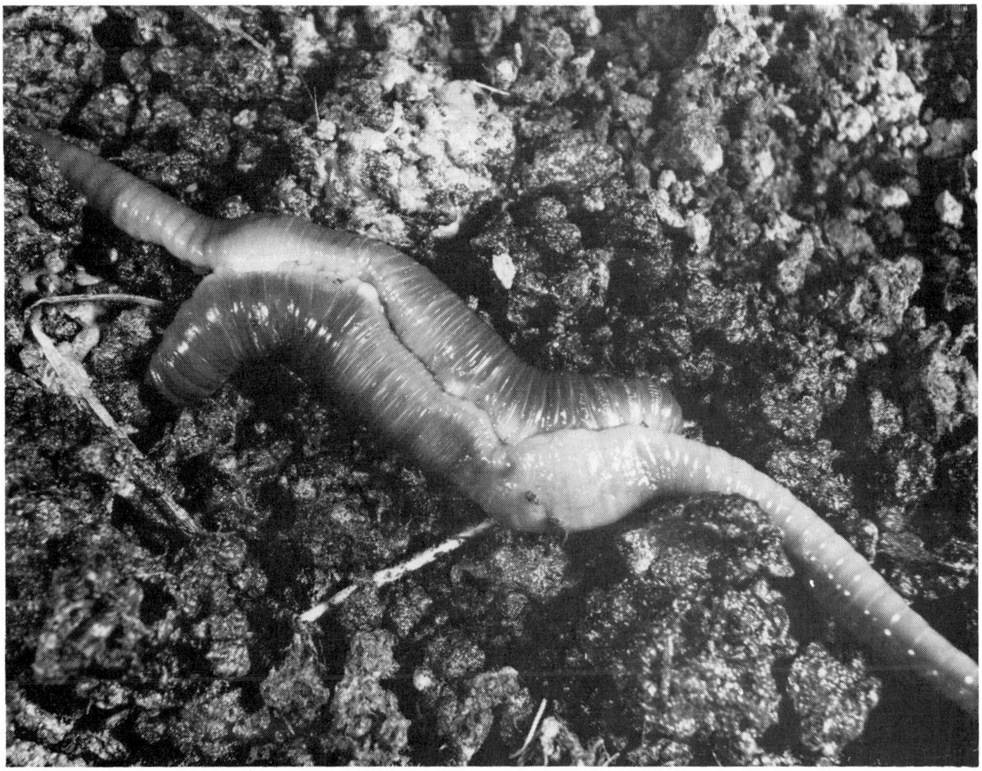

Figure 8.32 Two earthworms in copula. (Photo courtesy of Grant Heilman.)

each worm lie adjacent to the openings of the seminal receptacles of its partner. Mucous secretions from unicellular glands near the outer surface of the clitellum (Fig. 8.33) glue the two oligochaetes together. Special grasping setae intensify this copulatory embrace, as does a swelled lip around the male pore. Moreover, some oligochaetes even have a short penis. However, in *Lumbricus,* mating individuals never achieve mutual and simultaneous male pore/seminal receptacle contact. The clitellum of each worm adheres to its mate in the region of the latter's seminal receptacles. *Lumbricus* then fashions a longitudinal, ventral groove from the male pores of each worm to the seminal receptacle openings of the other. This groove is produced by muscular contractions along two parallel ridges. Bound together by copious mucous secretions, each worm deposits semen into the valley between the two ridges; muscular pulses then push the gametes along the groove and into the seminal receptacles of the mate (Fig. 8.34). *Lumbricus* requires 2 to 3 hours to complete this sperm transfer.

Several days following copulation, *Lumbricus* secretes a cocoon into which it deposits eggs for fertilization and development. First, the clitellum secretes a mucous coat around the anterior third of the worm. Second, unicellular glands situated below the mucus-secreting glands of the clitellum produce the cylindrical chitinoid walls of the cocoon (Fig. 8.33). At this point, the cocoon also surrounds the clitellum. A third set of glandular cells, lying deep within the clitellum, secretes nutrient albumin into the cocoon. When completed, the cocoon slips off the worm, passing anteriorly over the female genital pores and the openings of the seminal receptacles. Eggs and sperm are deposited inside the cocoon, where fertilization occurs. Once freed from the parent worm, the ends of the cocoon seal tightly and development begins.

Embryogeny in *Lumbricus* describes a basic protostomate pattern, although the classical spiral cleavage demonstrated by many polychaetes has been modified. The independent trochophore stage is suppressed entirely. Typically, only one worm per cocoon develops fully, emerging from 2 to 3 weeks following fertilization.

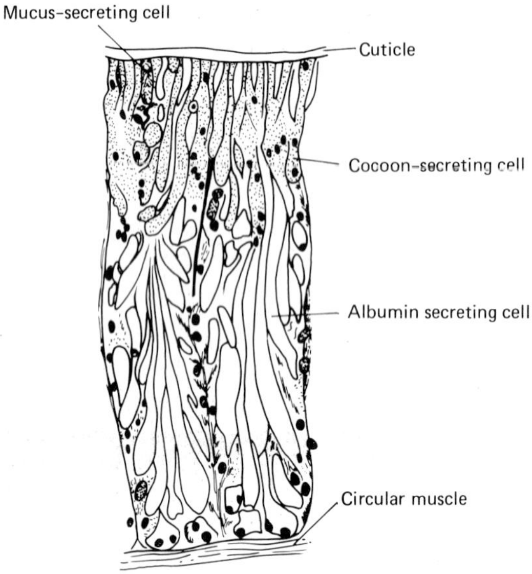

Figure 8.33 Section through the clitellum of *Lumbricus terrestris* showing the various specialized glandular epithelial cells. (From Avel, In *Traité de Zoologie*, Grassé, Ed.)

The Freshwater Oligochaetes

Many oligochaetes are aquatic, living mostly on the shallow bottoms of lakes and ponds. Others live among water vegetation, and a few thrive at considerable depths in the mud beneath large lakes. Freshwater oligochaetes are generally smaller than their terrestrial cousins and display several adaptations for aquatic life. Their setae commonly are longer, as some species employ them in swimming or food collection. Most freshwater oligochaetes, however, gather their food with a mucus-coated, eversible proboscis. The worms apply this sticky, muscular organ to the substratum, where food particles adhere to it. The proboscis then is withdrawn into the mouth. The digestive tract is similar to that of earthworms, although there is no typhlosole and the gizzard is reduced or absent.

Most freshwater oligochaetes can withstand oxygen shortages. However, respiratory requirements do limit most species to bottoms not more than 1 m below the water surface. These annelids possess hemoglobin that saturates at low oxygen concentrations, and they tolerate low respiratory rates. True gills are rarely present. Some worms that live on deep bottoms extend their posterior ends from their burrows and stir the surrounding water to increase ventilation (Fig. 8.35).

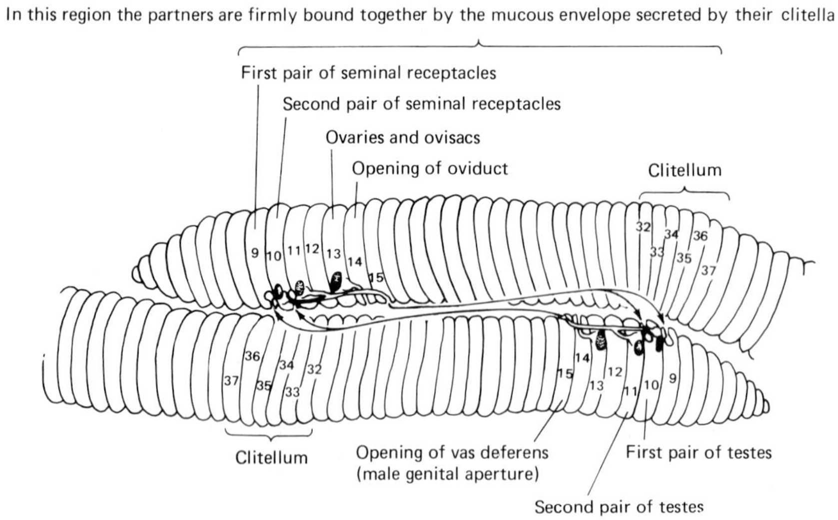

Figure 8.34 Schematic view of copulating earthworms. Arrows indicate the path of sperm from the male genital aperture of one worm to the seminal receptacles of its mate. (From F. Schaller, *Soil Animals*, 1968, University of Michigan.)

In freshwater oligochaetes, well-developed metanephridia excrete a copious and decidedly hypotonic urine. Ammonia is the predominant nitrogenous waste.

Sexual reproduction by aquatic oligochaetes follows the earthworm pattern. Asexual reproduction, however, is more widely spread among aquatic groups. Such reproduction involves transverse fission, and usually several daughter worms are formed during each fissional program. Regeneration may follow complete fission, but more often differentiation occurs before the daughter individuals separate. Separation of predifferentiated daughter worms usually occurs at one or more well-defined fission zones along the body trunk.

CLASS HIRUDINEA

The leeches of the class Hirudinea are the most specialized annelids. They are adapted primarily to life in slow, fresh water, where some species are predators and most are blood-sucking parasites. Some have invaded land, although mostly in humid tropical areas; a few marine species also exist. Leeches are never as small as many polychaetes and oligochaetes. The tiniest species are at least 1 cm long; an average size is 2 to 5 cm. Giant Amazonian leeches may be 30 cm long. With 500 known species, the group constitutes the smallest of the major annelid classes.

Leeches resemble the oligochaetes in several features, and it is assumed that they diverged (at least once) from primitive oligochaete stock. Characters common to the two classes include a lack of parapodia, reduced cephalization, a clitellum, and complex hermaphroditic reproductive systems. The Hirudinea, however, possess numerous distinctive features. The leech body is highly uniform and lacks setae. It is dorsoventrally flattened and commonly tapers to an anterior point (Fig. 8.36). Prominent anterior and posterior suckers are present. All leeches possess 33 or 34 body segments, although superficial

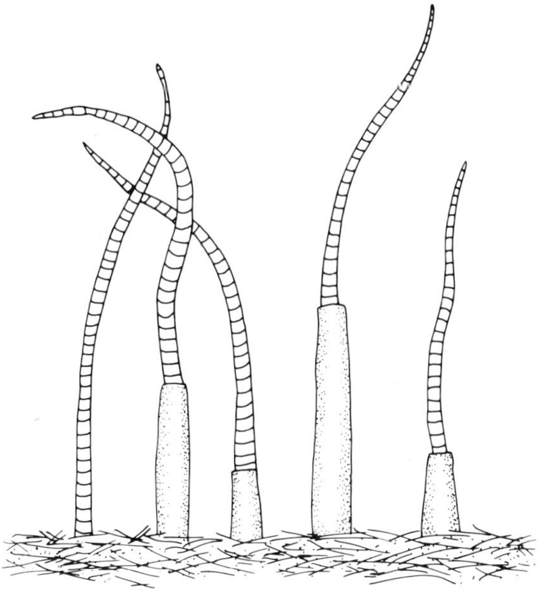

Figure 8.35 *Tubifex*, a typical tube-dwelling freshwater oligochaete. (From R.W. Pennak, *Fresh-Water Invertebrates of the United States*, 2nd ed., 1978, Wiley-Interscience.)

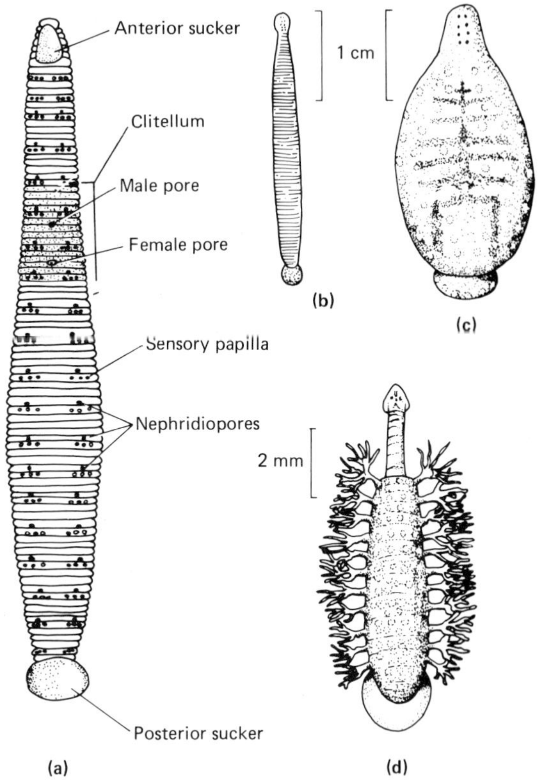

Figure 8.36 (*a*) Diagrammatic ventral view of the medicinal leech, *Hirudo medicinalis;* (*b*) A starved *Theromyzon tessulatum;* (*c*) A recently fed leech of the same species; (*d*) *Ozobranchus jantzeanus*, a leech with paired lateral gills. (From Mann, *Leeches.*)

rings called **annuli** obscure external segmentation. Internally, however, nervous and metanephridial systems maintain a regular metamerism and allow segments to be counted. The head region contains the leech eyes and mouth; the latter is circled by the anterior sucker. The trunk includes a distinct clitellum, and several posterior segments are fused into a powerful sucker. This sucker is highly muscularized with concentric rings of circular fibers and is turned ventrally.

In leeches, the body wall is more complex than in other annelids (Fig. 8.37). There is a thick dermal layer and oblique muscles occur between the conventional longitudinal and circular fibers. Laboratory dissections of leeches are difficult because of the volume and toughness of the dermis and the oblique musculature. Dorsoventral muscle bands contribute to the flattened form of these worms.

Internally, the leeches diverge even more from the annelid norm. There are no septa and the coelom itself is reduced beyond recognition. Connective tissue from the body wall has invaded the secondary body cavity, reorganizing the region into a complex system of interconnected sinuses and canals (Fig. 8.37 and 8.38). Chloragogen and loose mesenchyme called **botryoidal tissue** are also present in this region. While some leeches retain a closed blood vessel system, the coelomic network assumes internal transport responsibilities in most members of the class. Coelomic fluids carry respiratory pigments, and the sides of coelomic canals may contract to stimulate circulation. Very few leeches have gills. Respiratory gas exchange occurs over the general body surface. Many species ventilate this surface by attaching with the posterior sucker and then undulating the body.

Size largely determines whether a leech is a predator or a parasite (see Chap. 6, Special Essay). If these annelids are larger than their living food sources and if they can kill and eat their prey whole, then they are predators. Predatory leeches eat other worms, snails, and insect larvae. On the other hand, if leeches attack animals much larger than themselves and eat only part of those animals at a time, they are parasites. Leeches of this sort are always ectoparasites, and vertebrates are their common hosts. Parasitic leeches may remain on their hosts indefinitely, detaching only for reproductive purposes. Alternatively, they may take a brief meal and then detach until hunger recurs.

A parasitic leech attaches to its host with the anterior sucker, punctures its victim's skin, and then sucks out food (commonly blood). Most frequently, puncture involves triradiate cuticular jaws around the mouth

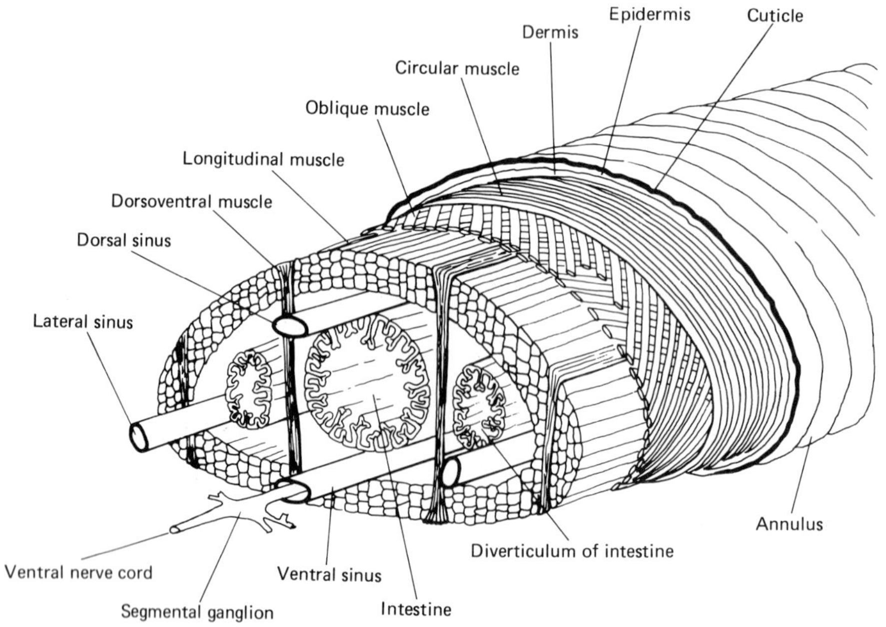

Figure 8.37 A cut-away diagram of the body of a leech showing the intestine and its diverticula, the ventral nerve cord, the chief coelomic sinuses, and the various layers of the body wall. (From J.G. Nicholls and D. Van Essen, Sci. Am, 1974, *230* (1):38.)

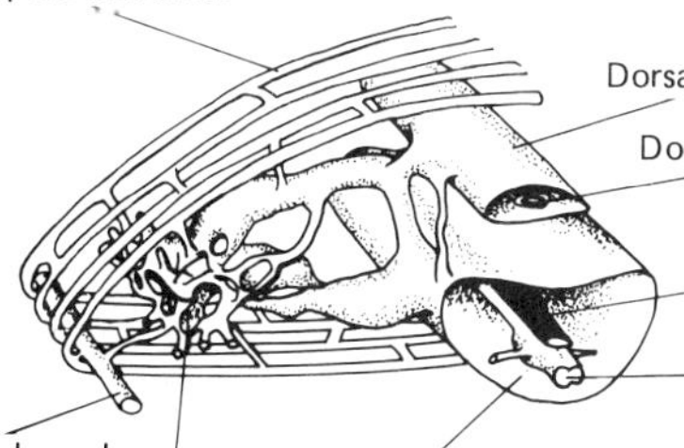

Figure 8.38 The arrangement of coelomic sinuses and canals of a leech, *Glossiphonia complanata*. (From F.E. Beddard, In *Cambridge Natural History*, *Vol. II*, Harmer and Shipley, Eds., 1959.)

(Fig. 8.39a). In some leeches, however, a proboscis is forced through the host's skin (Fig. 8.39b). A muscular, pumping pharynx creates the suction that draws out food. Salivary glands between the pharynx and the body wall discharge an anticoagulant called **hirudin** into the buccal cavity. A short esophagus precedes the leech stomach or crop. The latter often bears one to many lateral ceca, where food is stored prior to digestion. Such ceca may compensate for the relatively short length of many leeches, especially those which take huge meals (Fig. 8.36b, c). For example, the medicinal leech *Hirudo* consumes up to 10 times its body weight at one feeding. Lateral ceca accommodate this voluminous intake. Digestion is quite slow in leeches, a phenomenon at least partly attributable to the lack of endopeptidases in this class. Endopeptidases are enzymes that split large protein molecules at strategic internal bonds. Leeches possess only ectopeptidases; their protein digestion occurs only by the sequential removal of peripheral molecular groups and is therefore much slower. Because of their large meals and slow digestive action, many leeches need to feed only once every several months.

The leech stomach is separated from a tubular intestine by a pyloric sphincter. The intestine, which may have its own lateral ceca, extends to a short rectum and the anus. The latter occupies a dorsal position immediately in front of the posterior sucker (Fig. 8.39a, b).

Leech metanephridia occur in 10 to 17 metameric pairs in the middle third of the body. They excrete ammoniacal wastes and remove excess fluids from the voluminous blood meals. In addition, chloragogen and botryoidal tissues absorb larger waste particles.

Leeches have unique locomotor patterns: They crawl by a looping action involving the alternate contraction of all circular and all longitudinal muscles (Fig. 8.40). This action features: (A) attachment of the posterior sucker; (B) a wave of circular muscle contraction, which elongates the worm; (C) attachment of the anterior sucker with detachment of the posterior organ; and (D) a wave of longitudinal muscle contraction, which advances the rear body. During such locomotion, the general posture is maintained by contractions of the oblique musculature. The leeches' unique locomotion involves the entire body as a single hydrostatic unit. The gradual adoption of this technique removed selective pressures maintaining coelomic metamerism and thus contributed to the reduction of the leech coelom.

Many leeches swim. Swimming involves maximal flattening by the contraction of dorsoventral muscles and the passing of longitudinal waves along the worm's length.

The leech nervous system features a major ganglionic collar around the pharynx in the fifth or sixth body segment. This collar represents the brain, the circumenteric connectives, the subpharyngeal ganglion, and the fused ventral ganglia of the first few body segments. Such a nervous concentration reflects the integrated locomotion of these worms.

Consistent with their life-styles, most leeches possess sense organs comparable to those of the most active polychaetes. However, because leeches arose from oligochaete stock, their sense organs are quite different from those of the marine class. This condition illustrates the general principle that an organ once lost in an evolutionary progression is not re-evolved in its original form.

Leech sensory sites are scattered generally over the epidermis. Usually these sites are simply naked nerve endings extending through the cuticle. Sense organs include eyes and sensory papillae. Leech eyes comprise clusters of photosensitive cells individually not unlike those generally distributed over the trunk (Fig. 8.8c). Associated with pigment cups, these clusters occur dorsally on the first 2 to 10 segments. Sensory papillae represent tracts of sensory cells whose terminal bristles project through the cuticle (Fig. 8.8c).

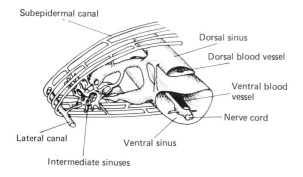

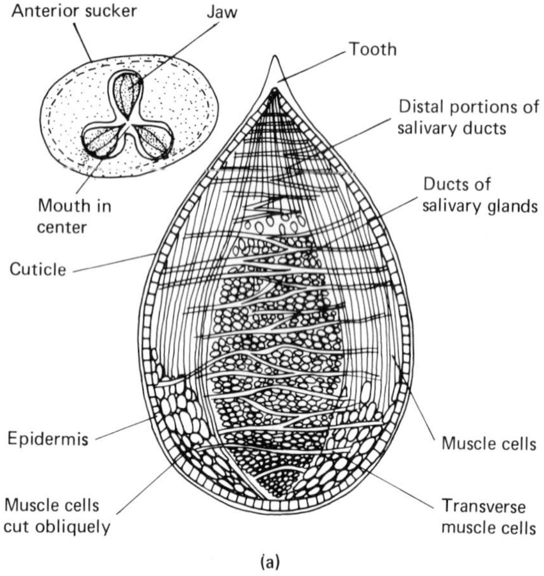

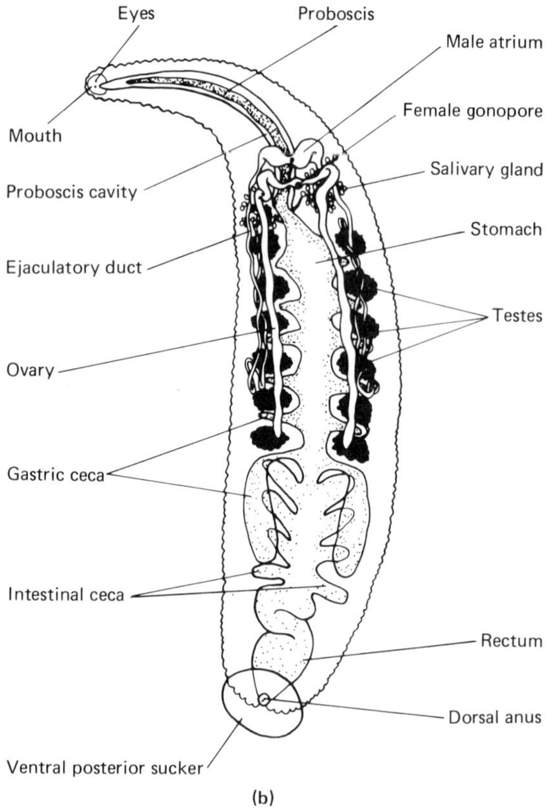

Figure 8.39 (*a*) Cross section through the jaws of *Hirudo medicinalis*, with an enlarged section through one of the jaws; (*b*) The digestive and reproductive tracts of the leech, *Glossiphonia*. (From Beck and Braithwaite, *Invertebrate Zoology Laboratory Workbook*.)

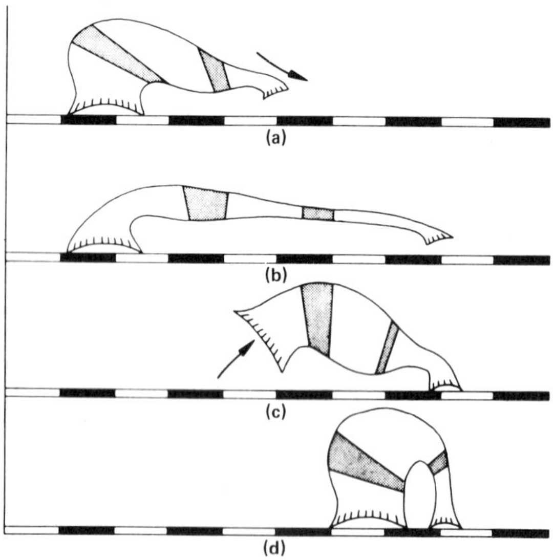

Figure 8.40 Locomotion sequence in a leech. (*a* and *b*) The leech extends its anterior end by contracting its circular muscles, making its body long and thin; (*c*) Attachment of the anterior sucker is followed by detachment of the posterior sucker and contraction of the longitudinal muscles, making the leech's body short and thick and drawing its posterior end forward; (*d*) The posterior sucker is set down near the anterior sucker and the process is repeated. The leech is a fluid filled bag of unchanging volume and the circular and longitudinal muscles act antagonistically around this single hydrostatic skeletal unit. (From Russell-Hunter, *A Biology of Lower Invertebrates*.)

These tracts may occur dorsally in rows or may surround the leech body. Their function is uncertain.

Leech responsiveness focuses on the location of prey/hosts. Photonegativity is general in this class, but some hungry leeches respond positively to light. Hungry leeches also are attracted to juices associated with their food sources. Water vibrations and shadows stimulate other leeches, particularly those parasitic on fish. Meanwhile, parasites on warm-blooded animals are sensitive to temperatures appropriate to their hosts.

Sexual reproduction among the leeches recalls the process in oligochaetes. Hermaphroditic systems feature distinct gonads, gonoducts, and gonopores; a glandular clitellum is present as well.

The male system (Fig. 8.39b) comprises four or more paired testes located segmentally behind the clitellum. Each testicular cavity represents a specialized remnant of the coelom. A short vas efferens connects each testis with a single vas deferens on either side of the body. The vasa deferentia grade into ejaculatory ducts before merging anteriorly at a single medial atrium. The atrial wall is muscularized and glandular. In some leeches, the male atrium is a site of **spermatophore** production. A spermatophore is a protective sperm package that is transferred from one individual to another. In leeches that do not produce spermatophores, a distal part of the atrial wall forms a penis. A single male gonopore opens onto the ventral surface of the 10th segment.

The simpler female system (Fig. 8.39b) includes a single pair of ovaries. Each ovary is surrounded by an ovisac; the ovisac cavity is a coelomic remnant. Oviducts extend anteriorly to a single, medial vagina and its ventral gonopore on the 11th segment. Thus the gonopores from both sexual systems are singular and occur within the clitellar region.

Impregnation in leeches follows one of two patterns, depending on whether or not a penis is present. In those groups possessing the male copulatory organ, sperm transfer occurs much as in the oligochaetes. Two worms orient head to tail with their ventral surfaces pressed tightly together. The two clitellar regions adhere, allowing immediate contact between each animal's male gonopore and its partner's female gonopore. The penis of each worm is everted into the female pore of its mate, and sperm are ejaculated into the vagina.

Leeches without a penis produce spermatophores within the male atrium. Their copulatory embrace involves the mutual adherence of ventral suckers and an intertwining of the body trunks. Ventral clitellar regions are brought together. The spermatophores then are expelled through the male gonopore by muscular contractions of the atrial wall; these packages of sperm actually penetrate the integument of the mate. Penetration results from the expulsive force and perhaps also from cytolytic properties of the spermatophore itself. Released into the coelomic sinuses and canals, sperm migrate to the vagina.

In both penis-bearing and spermatophore-producing leeches, the vagina serves as a sperm storage center. Fertilization occurs internally.

Leeches lay their eggs a few days or even several months after copulation. In most members of the class, the clitellum secretes a cocoon similar to that of the oligochaetes. The leech cocoon is stocked with albumin from clitellar glands, and fertilized eggs enter when the cocoon passes over the female gonopore. Most leeches simply deposit their cocoons in safe places, but a few types brood their young. In such species, the cocoon adheres to the ventral surface of the parent worm; after hatching, the baby leeches attach themselves directly to the parent with their suckers.

Leech embryogeny resembles the oligochaete plan. The trochophore stage is suppressed, and small leeches emerge from their cocoons a few weeks after deposition. Leeches commonly require an entire year to mature sexually. Reproduction tends to follow annual rhythms, with most matings occurring in the spring or summer months.

Unlike the other annelid classes, the Hirudinea exhibit neither asexuality nor regenerative powers.

OTHER ANNELIDS

Two other annelid groups may be considered the fourth and fifth classes of the phylum. These groups, the Branchiobdellida and the Acanthobdellida, display similarities to both oligochaetes and leeches.

The branchiobdellids (Fig. 8.41) are small annelids with 14 or 15 body segments. Possessing anterior and posterior suckers, they are always epizoic on freshwater crayfish. Setae are absent. Although little of their biology is described, the branchiobdellids clearly are specialized for epizoic life. The Acanthobdellida encompasses a single species, *Acanthobdella peledina*, which is parasitic on salmon and trout. This unique annelid possesses 27 annulated segments. The first five segments bear setae, while the last four are fused as a posterior sucker. There is no anterior sucker. *Acanthobdella* thus displays a mixture of oligochaete-and leechlike characters, and with the

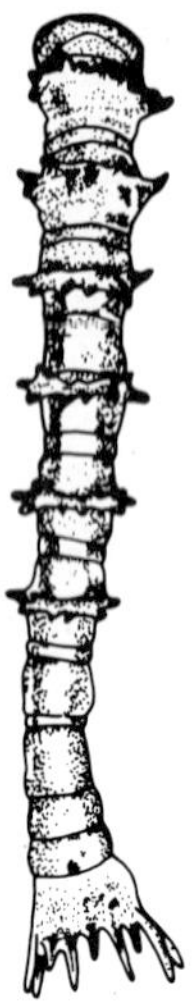

Figure 8.41 A branchiobdellid, *Stephanodrilus cirratus*. (From Avel, In *Traité de Zoologie, Vol. V*, Grassé, Ed.)

branchiobdellids, this species underscores the close evolutionary relationship between the two major nonmarine classes.

Annelid Evolution

Annelids are both simple and complex animals. On the one hand, they are a phylum of only moderate size and of predominantly marine distribution; they are soft-bodied worms with rather primitive respiratory and locomotor techniques. On the other hand, annelids are quite advanced compared with the invertebrates discussed in preceding chapters. They are eucoelomate and have introduced metamerism as a major structural phenomenon. Annelid digestive, circulatory, and excretory–osmoregulatory systems are often quite sophisticated; their nervous equipment likewise displays a relatively higher level of complexity. With the annelids, we begin to leave the "lower invertebrates" and to study the invertebrate world's major phylogenetic line, a line extending into the great arthropod classes.

The annelids present a number of interesting and highly speculative evolutionary questions. How did these worms arise? What is the origin of the true coelom? What is the origin of metamerism? What is the relationship between the Annelida and the Platyhelminthes, the Coelenterata, the Mollusca, and the Arthropoda? These questions are speculative indeed and only speculative answers are possible, both now and in the foreseeable future. However, asking them and thinking about their answers applies our faculties to the central phenomena of animal history. Such questions render the study of annelid evolution one of the most unifying disciplines in our study of the invertebrate world.

The evolution of both the coelom and metamerism are intertwined in most modern phylogenetic schemes. While many theories have been offered, only two are widely supported: the gonocoel—pseudometamerism theory and the enterocoel—cyclomerism theory. These theories, however, represent radically different views.

The **gonocoel–pseudometamerism theory** traces the ancestry of the annelids to early turbellarians. Recall that some flatworms display a serial repetition of gut diverticula, protonephridia, neural ganglia, and gonads. Such an arrangement is called **pseudometamerism.** The gonocoel–pseudometamerism theory suggests that the coelom originated when gonadal cavities expanded in volume, creating repeated coelomic compartments along the turbellarianlike animal (Fig. 8.42). Gametes still were produced from the compartmental walls (the septa of primitive annelids), and the body wall musculature and other organ systems organized around these units. The gonocoel–pseudometamerism theory is quite popular but suffers from two major difficulties. A gonadal origin lacks embryological support. (Coelomic origin occurs by enterocoely or schizocoely, not by "gonocoely.") Second, the theory suggests that the body wall musculature was among the last systems to assume a metameric organization. In view of the well-supported opinion that the primary advantage of a segmented coelom lies in locomotion, such late muscular involvement seems unlikely. Relieved of its locomotor responsibilities, the metameric coelom always regresses (see the leeches and, later, the arthropods). It is doubtful indeed whether adequate selective pressure for coelom formation could be generated if locomotor considerations were not at the forefront.

The **enterocoel–cyclometamerism theory** describes a coelomic origin from separated gastric pouches of the Anthozoa. Metamerism theoretically developed when these separated pouches further divided into serially arranged, paired compartments along a longitudinal axis (Fig. 8.43). The derivation of the coelom from a gastric cavity is consistent with its enterocoelous formation in deuterostomate animals. However, such embryological evidence provides the

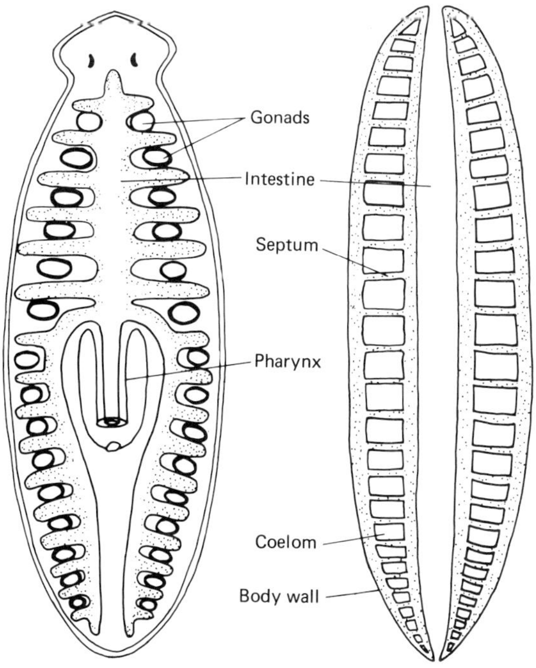

Figure 8.42 Diagram illustrating the gonocoel-pseudometamerism theory of coelom formation. (*a*) A turbellarian with gonads alternating with intestinal branches; (*b*) An annelid with gonads expanded to form cuboidal coelomic sacs that fuse to body and intestinal walls (From Hyman, *The Invertebrates*, *Vol. II.*)

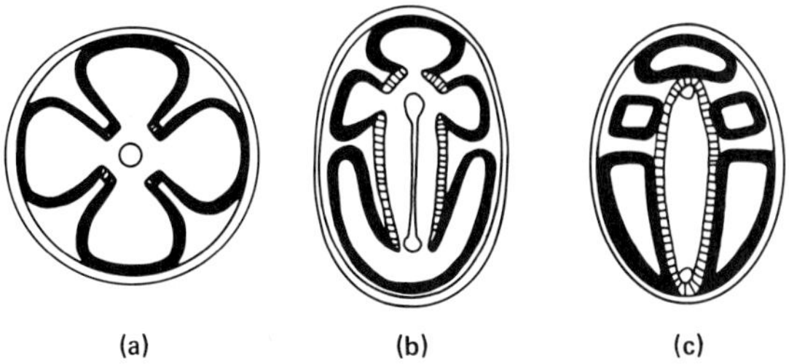

Figure 8.43 Diagrams illustrating the enterocoel-cyclometamerism theory of coelom formation. (*a*) A coelenterate with four gastric pouches and a mouth; (*b*) Division of the mouth to form mouth and anus, setting up a longitudinal axis; (*c*) Separation of the gastric pouches to form coelomic cavities. (From W.D. Hartman, In *The Lower Metazoa*, E.C. Dougherty, Ed., 1963, University of California.)

theory's only strong support. The enterocoel—cyclometamerism theory necessarily is linked to the school of evolutionary thought placing the anthozoans at the evolutionary base of the Metazoa (see Chap. 5). The enterocoel—cyclometamerism theory thus suffers from the same difficulties.

Regarding the relationships between the annelid groups, we restate the generally accepted positions. The Polychaeta represent the primitive class. The Oligochaeta diverged from ancestral forms either before or immediately after modern polychaete characters were established. The Hirudinea clearly are allied with the Oligochaeta. Finally, the Branchiobdellida and the Acanthobdellida are distinct groups whose mixed characters underscore the kinship between the two nonmarine classes.

Relationships between the Annelida and the "higher invertebrates" are rather clear in outline. Some alliance with the Mollusca is indicated by the presence of trochophore larvae in both phyla. Segmentation and other morphological and physiological phenomena support an especially close relationship with the Arthropoda. Further discussion of these relationships is reserved for the chapters dealing with these largest invertebrate phyla.

On leaving the Annelida, we emphasize both the advantages and the disadvantages of their eucoelomate, metameric plan. Annelid locomotor powers are unmatched by the lower invertebrates, yet an annelid's need to maintain a soft flexible body increases its vulnerability to attack. Respiration across the general body surface militates against the evolution of a protective external cover. Finally, while metamerism allows a more efficient locomotion, it nevertheless may encumber the development of specialized organs in these worms. If true metamerism is maintained, all new mesodermal developments must be repeated in each segment. Thus the Annelida, like the radiate animals, may experience a certain evolutionary inertia. Among other factors, that inertia helps maintain their status as simple yet complex members of the invertebrate world.

Other Protostomes

Before discussing the Mollusca, we pause to consider four minor phyla within the protostomate lineage. A common ancestry with the early annelids has been suggested for each of these groups. However, their exact position in the invertebrate world remains undetermined.

PHYLUM POGONOPHORA

The Pogonophora is one of the most recently discovered phyla. This group now consists of about 80 species of marine worms that live in tubes on the deeper continental slopes. Because of their late discovery, the difficulty of observing live specimens, and their rather fragile bodies, pogonophorans remain poorly described.

The long (10- to 85-cm) pogonophoran body exhibits three distinct zones (Fig. 8.44): a prosoma, the dominant trunk, and an opisthosoma. The prosoma comprises an anterior cephalic lobe and a rear glandular region involved in tube formation. One to many dozens of tentacles extend from the cephalic lobe and support numerous side branches or pinnules, as well as ciliary tracts. The truck region, which contributes most of the length of the worm, is studded with papillae. About halfway along its length are two girdles of short setae. The terminal region of the pogonophoran body has been discovered only recently. This area, the opisthosoma, is distinctly segmented and bears paired setae in the annelid fashion. The opisthosoma apparently functions in anchorage. It tends to separate from the trunk when the worm is pulled from its tube, a property accounting for its delayed discovery.

Each body region has its own coelomic compartment. Tentacles receive extensions of the prosomal coelom. The secondary cavity of the opisthosoma is partitioned by intersegmental septa.

Pogonophoran nutrition is a curious process. No mouth or digestive tract has ever been described. Apparently these animals absorb food across their body surface. Microscopic examination of the pogonophoran cuticle reveals ubiquitous microvilli, and the tentacles appear to be specialized for nutrient uptake. The tentacles may collect and transport food particles along their ciliated, mucus-coated surfaces. These surfaces are supplied with enzymatic gland cells, which would make extracellular digestion possible. Pogonophorans possess an annelidlike, closed circulatory system. Blood vessels irrigate the tentacles, whose surfaces probably are active in respiratory exchange. Other organ systems remain undescribed.

Pogonophorans have separate sexes; a pair of gonads occupies the trunk. Reproductive and developmental physiology remain under investigation, but preliminary reports indicate that the opisthosoma is defined by segmental mesoderm arising from teloblast cells. This observation, as well as the similarities between annelid and pogonophoran setae, argues for some intimate relationship between the two phyla. Prior to the discovery of the opisthosoma, po-

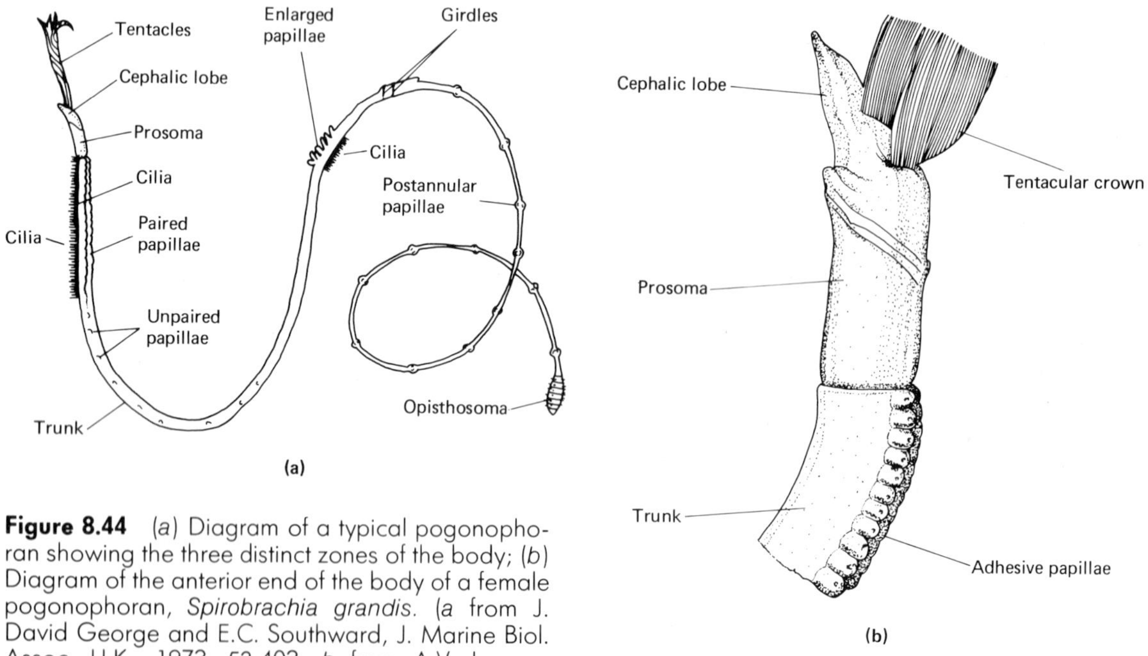

Figure 8.44 (*a*) Diagram of a typical pogonophoran showing the three distinct zones of the body; (*b*) Diagram of the anterior end of the body of a female pogonophoran, *Spirobrachia grandis*. (*a* from J. David George and E.C. Southward, J. Marine Biol. Assoc. U.K., 1973, 53:403; *b* from A.V. Ivanov, *Pogonophora*, 1963, Academic Press.)

gonophorans were allied with the deuterostomes on the basis of their tentacles, coelomic compartments, and superficial resemblance to hemichordates.

PHYLUM ECHIURA

Another small phylum of marine worms also appears to be related to annelids. These unsegmented animals are called echiurans. Approximately 100 species have been identified on shallow ocean bottoms, where they occupy sandy burrows or crannies in rock and coral deposits. A few live beneath deeper water.

The echiuran body ranges from a few millimeters to a meter or more in length. The common *Urechis caupo* (Fig. 8.45) averages about 20 cm. A cylindrical trunk houses the major organ systems; at its anterior end is a thin-walled cephalic lobe. This lobe, usually called the proboscis, may be homologous with the annelid prostomium. In some species, the proboscis constitutes over half the length of the worm, but in *Echiurus* (Fig. 8.46a), the trunk is the longer region. The basal edges of the proboscis enroll to form a gutter along the ventral surface. Meanwhile, the distal tip of the lobe is often broadly expanded (Fig. 8.46b). The echiuran body wall is reminiscent of that of the annelids.

An echiuran eats the detritus that adheres to the mucus-coated ventral surface of its proboscis. Proboscis cilia channel foodstuffs through the ventral gutter and into the mouth. A muscular pharynx then pumps the food into the long and extensively coiled digestive tract (Fig. 8.46a), which features an esophagus, a gizzard, perhaps a stomach, and a lengthy intestine. A short rectum leads to the posterior anus.

A closed circulatory system, very like that of the annelids, is found in the Echiura. A prominent dorsal vessel communicates through muscularized periintestinal connectives with a ventral blood vessel. Amoebocytes are present in the blood, but no respiratory pigments have been found. Echiuran respiration involves simple diffusion across the general body surface, especially across the proboscis. Some hemoglobin has been identified within cells populating the coelomic fluid. Excretory problems are met by paired metanephridia. In some species, only a single pair of metanephridia is present, but other echiurans may possess hundreds of these organs.

The echiuran musculature is organized along annelid patterns. In addition to the body wall muscles, there are smaller fibers that control the movements of a pair of large chitinous setae. The latter originate within sacs on the anterior ventral surface of the trunk (Fig. 8.46a). Echiurans also possess an annelidlike nervous system. A reduced cerebral complex occupies the proboscis and circumesophageal connectives meet a solitary ventral nerve cord. Specialized sense organs are lacking.

Echiuran sexes are separate. *Bonellia* is unusual in that males are as little as one-thousandth the size of females. These males develop from larvae that enter the body of an adult female, where their dwarf condition is induced by a female hormone. Larvae that

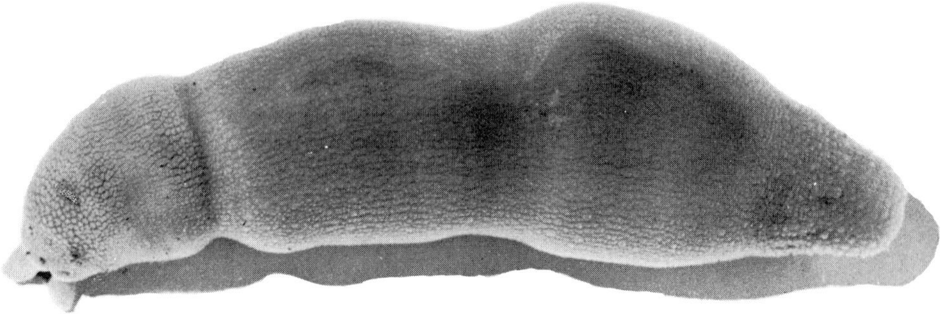

Figure 8.45 The echiuran worm *Urechis caupo*. (Photo from E.F. Ricketts, *et al.*, *Between Pacific Tides*, 1968, Stanford University.)

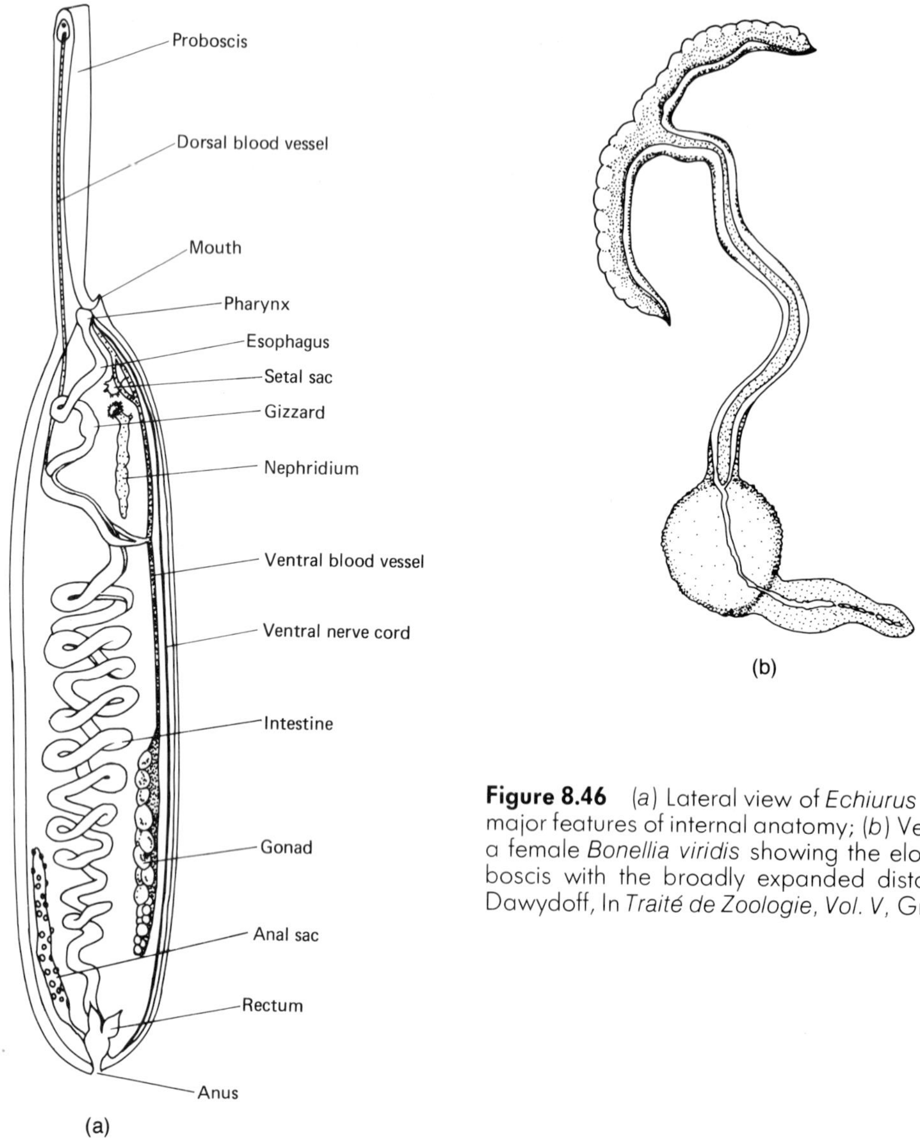

Figure 8.46 (*a*) Lateral view of *Echiurus* showing the major features of internal anatomy; (*b*) Ventral view of a female *Bonellia viridis* showing the elongated proboscis with the broadly expanded distal tip. (From Dawydoff, In *Traité de Zoologie, Vol. V*, Grassé, Ed.)

simply settle on the substratum develop into females. Echiuran gonads consist merely of specialized mesenteries that release immature sex cells into the coelom. After gametogenesis is completed, sperm and eggs exit through the nephridia and are fertilized in the sea. A typically protostomate embryogeny produces a trochophore larva. At this point, the young echiuran undergoes a brief metameric phase. Segmented coelomic pouches appear, and the nerve cord exhibits some metameric character. These phenomena are transitory however, and the adult worm is unsegmented. It has been estimated that some echiurans live as long as 60 years.

Because of this metameric interlude and the basic structure of their circulatory and nervous systems, setae, and body wall, echiurans appear to be closely allied with the annelids. Some common ancestry with the early polychaetes, perhaps stemming from early in the period during which true metamerism evolved, seems likely.

PHYLUM SIPUNCULA

The relationship between the Annelida and a third minor phylum, the Sipuncula, is less well defined. Sipunculans, commonly known as peanut worms, lead rather sedentary lives within protective homes on the ocean floor. Their domiciles include mud and sand burrows, abandoned mollusk shells or polychaete tubes, and natural rock crevices. A few sipunculans bore into limestone. About 350 species have been identified.

A large trunk and a narrow anterior region called the **introvert** define the cylindrical body (Fig. 8.47a). The introvert houses the worm's head and its distal end supports a ring of frilled tentacles. The surface of the introvert often is ornamented with spines, in contrast to the usually rather smooth trunk. The sipunculan body wall resembles that of the annelid, and there is a single spacious coelom. Overall body lengths vary from a few millimeters to almost a full meter.

Sipunculans feed on the organic material that their tentacles capture. These extendible organs collect suspended and/or deposited matter; mucous surfaces trap foodstuffs, which ciliary currents usher back to the mouth. The entire introvert may be withdrawn during ingestion. Other means of food procurement include direct ingestion of mud by actively burrowing forms and predation on polychaetes. The digestive tract comprises an esophagus and a long spirally coiled intestine (Fig. 8.47b). The intestine describes a U-shaped loop within the trunk; a rectum rises to the anterior, middorsal anus.

Sipunculans lack specialized circulatory and respiratory structures. These maintenance responsibilities are handled in part by the coelomic fluid, where hemerythrin is found within numerous corpuscles. Excretory work is done by a single pair of nephridia. Also involved in excretion are specialized cell complexes called **urns.** These complexes arise from peritoneum and then detach to roam freely through the coelomic fluid. They collect solid wastes before their eventual deposition within the body wall or removal via the nephridia.

Activity systems include the annelid-style musculature of the body wall, introvert retractor muscles, and an independent hydraulic system that operates the tentacles. The introvert is expanded by coelomic fluid pressure established by the contraction of trunk muscles. Retractor muscles, attached to the esophagus

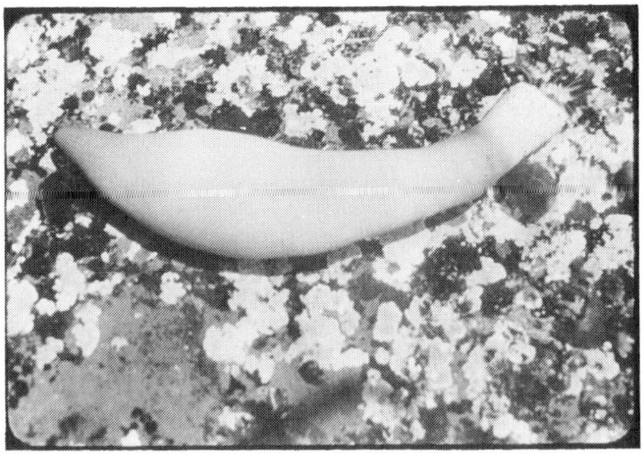

(a)

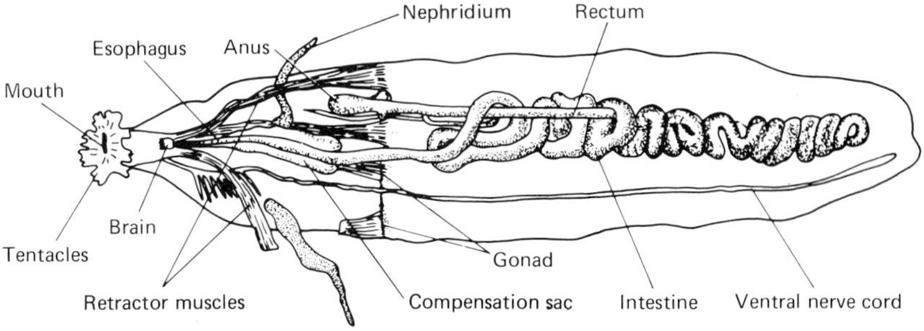

(b)

Figure 8.47 Sipunculan worms. (*a*) *Dendrostomium pyroides*. Note the introvert with its crown of branched tentacles at the anterior end; (*b*) Dissection of *Sipunculus nudus*, showing major internal structures. (*a* Photo courtesy of Ward's Natural Science Establishment; *b* from Hyman, *The Invertebrates, Vol. V*, McGraw-Hill, 1959.)

and anchored in the body wall, withdraw the organ. The tentacular system consists of a network of canals joining the hollow cores of the tentacles. These canals communicate with one or two compensation sacs. The degree of muscular contraction in the sac wall determines the amount of water within the tentacle cores and thus the degree of tentacular extension.

The sipunculan nervous system features a supraesophageal brain, paired connectives, and a ventral nerve cord (Fig. 8.47b). Sensory structures include batteries of sensitive cells along the surface of the introvert. **Nuchal organs,** paired ciliated pits thought to be chemoreceptive, are located dorsally on the introverts of some species. Other sipunculans possess paired pigment cup ocelli within the brain.

Continuity systems are rather simple. Immature gametes elaborated from the peritoneum complete development within the coelom. After discharge through the nephridia, they are fertilized in the sea. A protostomate embryogeny leads to the formation of a trochophore in some sipunculans, but in other species, development is direct. In any case, no metameric stages are observed.

Some relationship between the sipunculans and the annelids is suggested by their similarities in body wall structure and embryogeny. The absence of segmentation in the present phylum indicates that the sipunculans may have separated from the main annelid line before metamerism was established. The curious tentacular hydraulic system of these animals led to the recent hypothesis that sipunculans are related to the lophophorates and the echinoderms.

PHYLUM PRIAPULIDA

Phylogenetic uncertainties also surround the Priapulida, a protostomate phylum with only eight member species. These unsegmented animals once were included among the aschelminthes, but now occupy an as yet undefined branch within the eucoelomate line of invertebrate evolution. Priapulids are small creatures—their cylindrical bodies seldom exceed 5 cm in length—and are widely distributed on the coastal bottoms of colder seas. Most burrow in littoral mud and sand, but some live among coralline deposits and there is one tube-dwelling species.

The priapulid body is shaped somewhat like a cucumber (Fig. 8.48a). A protrusible proboscis defines the anterior third of the animal and is followed by a warty trunk region. Circular ridges deliniate superficial segments along the trunk. Variously shaped caudal appendages may extend from the posterior body. The body wall comprises an outer chitinoid cuticle, a thin

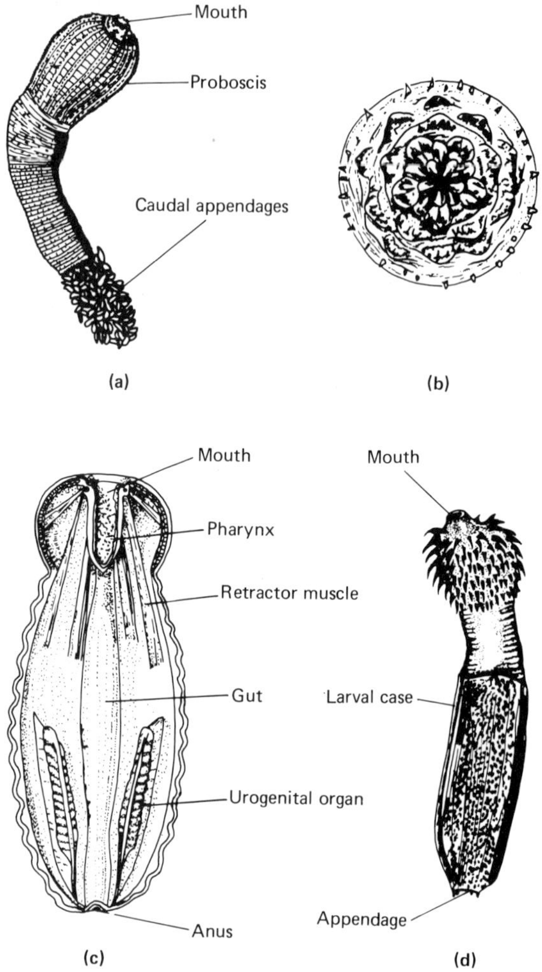

Figure 8.48 Priapulid worms. (*a*) External view of *Priapulus caudatus;* (*b*) Frontal view of a priapulid showing the arrangement of spines around a circular mouth; (*c*) Internal anatomy of *Halicryptus;* (*d*) Dorsal view of *Halicryptus* larva. (*a* from A.E. Shipley, In *Cambridge Natural History, Vol. II,* Harmer and Shipley, Eds., 1959, *b, c,* and *d* from Dawydoff, In *Traité de Zoologie, Vol. V,* Grassé, Ed.)

epithelium, circular and longitudinal muscle lyers, and a cellular peritoneum. Peritoneal mesenteries cross the coelom to support the internal organs.

Priapulids are carnivores and feed primarily on polychaetes. Proboscis papillae probably are active in food detection. Food capture involves the extension of the proboscis and the eversion of curved spines surrounding the mouth proper (Fig. 8.48a, b). These spines are applied to prey, which then is pulled into a muscular pharynx. The pharyngeal cuticle is modified as teeth. The rest of the digestive system consists simply of a long intestinal tube, a short rectum, and an anus at the trunk posterior end (Fig. 8.48c).

Priapulids are small, marine animals whose circulatory and respiratory needs are met without specialized equipment. Their coelomic fluid contains amoebocytes, and in one species, hemerythrin has been identified. The vesicles attached to the caudal appendages of certain priapulids may be involved in respiration. Two urogenital organs are present, one on either side of the digestive tract (Fig. 8.48c). Their central tubules, which also receive gonadal and protonephridial products, discharge through paired nephridiopores at the trunk posterior.

The priapulid nervous network is largely an epidermal phenomenon. The limited central elements include a circumpharyngeal ring and a single ventral nerve cord. Sense organs are rudimentary; there are no eyes, but proboscis and trunk papillae are probably important sensory sites. Priapulids burrow periodically. Alternate contractions of the body wall's circular and longitudinal fibers thrust the proboscis, and the animal moves forward by peristalsis.

Priapulids are diecious. Their paired gonads discharge through the protonephridial tubules. Details of fertilization and embryogeny (including the origin of the coelom) are not well researched. Eventually, a tiny larva develops, whose characters resemble those of the aschelminthes. A rotiferlike lorica surrounds the central body, and footlike appendages may be present at the posterior end (Fig. 8.48d). Spines may ring the mouth in the gastrotrich fashion. After an extended larval life, the lorica is shed. Continued growth requires periodic molting of the cuticle.

Priapulids thus demonstrate certain aschelminth affinities, although the likelihood of evolutionary convergence is often voiced. The cellularity of the peritoneum argues for the eucoelomate identity of the group, but confirmation from embryological sources is needed. If they are indeed eucoelomates, priapulids appear to rank among the most primitive animals of that description. Possibly, they represent an early experiment with the eucoelomate body plan—one that has had limited success in competition with the segmented worms.

for Further Reading

Anderson, D. T. *Embryology and Phylogeny in Annelids and Arthropods.* Pergamon Press, Oxford, 1973.

Banse, K. Sabellidae (Polychaeta) principally from the northeast Pacific Ocean. *J. Fish. Res. Board. Can.* 36: 869–882, 1979.

Baskin, D. G. Neurosecretion and the endocrinology of nereid polychaetes. *Am. Zool.* 16: 107–124, 1976.

Berrill, N. J. Induced segmental reorganization in sabellid worms. *J. Embryol. Exp. Morphol.* 47: 85–96, 1978.

Brinkhurst, R. O. Freshwater Oligochaeta in Canada. *Can. J. Zool.* 56: 2166–2175, 1978.

Brinkhurst, R. O. and Cook, D. G. (Eds.). *Aquatic Oligochaete Biology.* Plenum Press, New York and London, 1980.

Brinkhurst, R. O. and Jamieson, B. G. *Aquatic Oligochaeta of the World.* Toronto University Press, Toronto, 1972.

Clark, R. B. *Dynamics in Metazoan Evolution.* Clarendon Press, Oxford, 1964.

Clark, R. B. Systematics and phylogeny: Annelida, Echiura, Sipuncula. In *Chemical Zoology,* (M. Florkin and B. T. Scheer, Eds.) Vol. IV, pp. 1–68. Academic Press, New York, 1969.

Clark, R. B. and Olive, P. J. W. Recent advances in polychaete endocrinology and reproductive biology. *Ann. Rev. Oceanogr. Mar. Biol.* 11: 176–222, 1973.

Dales, R. P. *Annelids.* Hutchinson University Library, London, 1967,

Dales, R. P. and Peter, G. A synopsis of the pelagic polychaeta. *J. Nat. Hist.* 6: 55–92, 1972.

Edwards, C. A. and Lofty, J. R. *Biology of Earthworms,* 2nd ed. Chapman and Hall, London, 1977.

Fauchald, K. Polychaete phylogeny: a problem in protostome evolution. *Syst. Zool.* 23: 493–506, 1975.

Fauchald, K. *The Polychaete Worms. Definitions and Keys to the Orders, Families, and Genera.* Nat. Hist. Mus. of Los Angelos Co., Sci. Ser. 28: 1–190, 1977.

Fauchald, K. and Jumars, P. A. The diet of worms: a study of polychaete feeding guilds. *Ann. Rev. Oceanogr. Mar. Biol.* 17: 193–284, 1979.

Fitzharris, T. P. Regeneration in sabellid annelids. *Am. Zool.* 16: 593–616, 1976.

Florkin, M. and Scheer, B. T. (Eds.) *Chemical Zoology, Vol. IV: Annelida, Echiura, Sipuncula.* Academic Press, New York, 1969.

Gardener, S. L. Errant polychaete annelids from North Carolina. *J. Elisha Mitchell Sci. Soc.* 91: 77–220, 1975.

Goodnight, D. J., Hartman, O., and Moore, J. P. Oligochaeta, Polychaeta, and Hirudinea. In *Freshwater Biology* (W. T. Edmonson, H. B. Ward, and G. C. Whipple, Eds.), 2nd ed., pp. 522–557. John Wiley and Sons, New York, 1959.

Hartman, O. *Literature of the Polychaetous Annelids. Vol. I: Bibliography.* Privately published, 1959; *Vol. II: Catalogue of the Polychaetous Annelids of the World.* Allan Hancock Foundation, Paper No. 23, 1965.

Hermans, C. O. The systematic position of the Archiannelida. *Syst. Zool.* 18: 85–102, 1969.

Holt, T. C. The Branchiobdellida: Epizootic annelids. *Biologist* 50: 79–94, 1968.

Jacobsen, B. H. The feeding of the lugworm, *Arenicola marina. Ophelia* 4: 91–109, 1967.

Kaster, J. L. The reproductive biology of *Tubifex tubifex. Am. Midl. Nat.* 104: 364–366, 1980.

Klemm, D. J. *Freshwater Leeches of North America.* Biota of Freshwater Ecosystems Identification Manual No. 8, Envir. Prot. Agency, U.S. Printing Office, 1972.

Laverack M. S. *The Physiology of Earthworms.* Pergamon Press, New York, 1963.

Mangum, C. Respiratory physiology in annelids. *Am. Sci.* 58: 641–647, 1970.

Mill, P. J. (Ed.) *Physiology of Annelids.* Academic Press, London, 1978.

Mann, K. H. *Leeches (Hirudinea), Their Structure, Physiology, Ecology, and Embryology,* Pergamon Press, New York, 1962.

Moment, G. B. and Johnson, J. E., Jr. The structure and distribution of external sense organs in newly hatched and mature earthworms. *J. Morphol.* 159: 1–16, 1979.

Nicholls, J. G. and Van Essen, D. The nervous system of the leech. *Sci. Am.* 230: 38–48, 1974.

Olive, P. J. W. and Bentley, M. G. Hormonal control of oogenesis, ovulation and spawning in the annual reproductive cycle of the polychaete, *Nephthys hombergii. Int. J. Invertebr. Reprod.* 2: 205–222, 1980.

Orrhage, L. Strucczture and homologies of the anterior end of the polychaete families, Sabellidae and Serpulidae. *Zoomorphologie* 96: 113–168, 1980.

Pennak, R. W. *Fresh-Water Invertebrates of the United States,* 2nd ed., pp. 275–317, Wiley-Interscience, New York, 1978.

Reish, D. J. and Fauchald, K. (Eds.) *Essays on Polychaetous Annelids.* Allan Hancock Foundation, Los Angeles, 1977.

Rice M. E. Larval development and metamorphosis in Sipuncula. *Amer. Zool.* 16: 563–571, 1976.

Sawyer, R. T. North American freshwater leeches, exclusive of the Pisciolodae, with a key to all species. *Ill. Biol. Monogr.* 46: 1–154, 1972.

Schroeder, P. C. and Hermans, C. O. Annelida: Polychaeta. In *Reproduction of Marine Invertebrates, Vol. III* (Giese, A. C. and Pearse, J. S., Eds.), pp. 1–205. Academic Press, New York, 1975.

Seymour, M. K. Locomotion and coelomic pressure in *Lumbricus. J. Exp. Biol.* 51: 47, 1969.

Southward, E. C. Horizontal and vertical distribution of Pogonophora in the Atlantic Ocean. *Sarsia* 64: 51–56, 1979.

Stephen, A. C. and Edmonds, S. J. *The Phyla Sipuncula and Echiura.* The British Museum, London, 1972.

Wald, G. and Rayport, S. Vision in annelid worms. *Science* 196: 1434–1439, 1977.

Wells, G. P. Worm autobiographies, *Sci. Am.* 200: 132–141, 1959.

9

The Mollusks

The best known phylum in the invertebrate world is the Mollusca. Laypersons and scientists alike collect the shells of these often beautiful and fascinating animals. Today, with 80,000 living and 35,000 fossil species, the mollusks rank as the largest nonarthropod group. Of all living creatures, only birds and mammals have been classified more thoroughly. Mollusks have been used by humans through the centuries as tools, jewelry, and even money; to this day, few of us are without some souvenir of the phylum in our homes.

Most beginning biology students can compile a fairly long list of mollusks: chitons, snails, limpets, slugs, clams, oysters, octopods, squid, etc. We usually recognize mollusks on sight, but nevertheless have difficulty listing distinct characters common to the entire phylum. Collectively, these animals may be called "the shells," because many of them possess a calcareous shell. Yet shells are by no means unique to the mollusks (foraminiferans, barnacles, ostracods, and brachiopods all have shells), and some mollusks, such as slugs and octopods, have none at all. Nevertheless, mollusks are a very distinct phylum whose members are recognizable by certain generalized molluscan characters. Not all mollusks display all the general characters, but each displays some of them and is related to the other members of the phylum by a long and complex evolutionary history. The generalized molluscan characters are: (1) a dorsal, shieldlike shell secreted by an underlying **mantle;** (2) a ventral, muscular foot; (3) a large visceral hump; (4) a **mantle cavity** formed by the overhang of the mantle relative to the main body and commonly occupied by respiratory gills; (5) a distinct toothlike organ called the **radula;** and (6) **trochophore** and **veliger** larvae.

Living mollusks are distributed among seven classes, which may be grouped into three subphyla. The subphylum Aculifera contains the single aberrant class Aplacophora. These wormlike, shell-less mollusks are often called solenogasters. The second subphylum, the Placophora, also contains only one class, the Polyplacophora. Polyplacophorans commonly are called chitons. The remaining molluscan classes compose the subphylum Conchifera. These classes include the primitive Monoplacophora, the Gastropoda (snails), the Bivalvia (clams and their bivalved relatives), the Scaphopoda (tusk shells), and the Cephalopoda (squid and octopods).

Mollusks live in nearly all environments. They are numerous on the ocean floor, where they range from great depths to coastal and intertidal zones. Pelagic and planktonic mollusks are also known. Freshwater bivalves and gastropods are well established, and certain snails have invaded the terrestrial environment. Parasitism within the phylum is infrequent.

Molluscan Unity

The mollusks, like the annelids and other higher invertebrates, are the products of a long evolutionary history. Molluscan evolution has taken quite diverse courses, as evidenced by the varying forms and life styles within the group. Such diversity complicates a discussion of the unifying features of the phylum. Indeed, there is no single character that is both unique to the Mollusca and common to all its members. As we said previously, an animal is recognized as a mollusk only because of its display of some combination of generalized molluscan characters. It has been fashionable to use these characters to construct a hypothetical, generalized mollusk (Fig. 9.1). Such a creature is not like any known animal, living or fossil, but conveniently introduces the basic features of the phylum. The generalized mollusk often is designated as the "ancestral mollusk," and then it is assumed that the seven modern classes evolved independently from it. Such assumptions are far overdrawn. The generalized mollusk is created by combining several traits which have stabilized over millions of years of animal evolution. This form is a "least common denominator" for the modern members of the phylum, a hypothetical type possessing fully developed, and not protomolluscan, characters. As such, the generalized mollusk is a very poor candidate for the stem form of one of the largest and most diverse phyla in the invertebrate world. However, the hypothetical creature depicted in Figure 9.1 will serve the present purpose of introducing molluscan maintenance, activity, and continuity systems.

MAINTENANCE SYSTEMS

General Morphology

Our generalized mollusk is bilaterally symmetrical, with a flat ventral surface and an arched, shell-covered dorsal surface. Anteriorly, a head region bears the sensory elements and a terminal mouth. The mollusk's visceral hump is covered by a convex dorsal

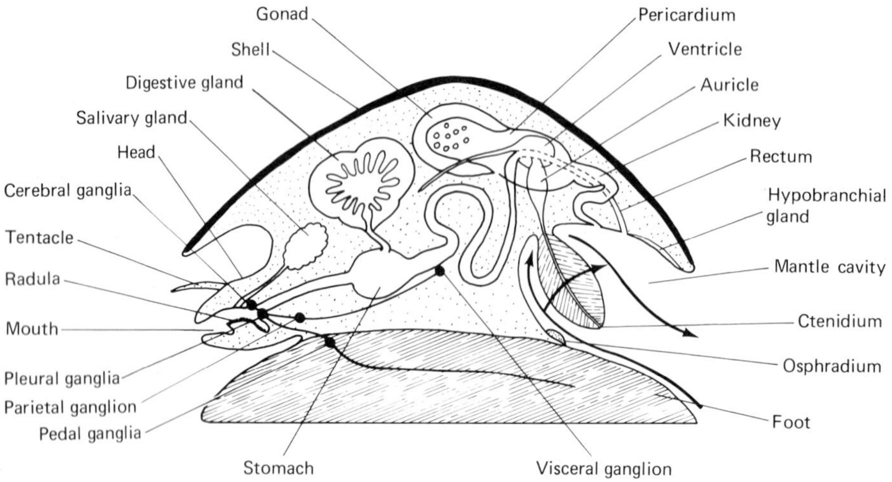

Figure 9.1 Sagittal section through a hypothetical mollusk. (From Grove and Newell, *General Biology*, 1969, University Tutorial Press.)

shell. This protective structure consists of an outer organic layer called the **periostracum** and two inner strengthening sheets of calcium carbonate crystals deposited on an organic matrix (Fig. 9.2). Of the latter, the innermost nacreous layer contains less organic material than the more massive prismatic layer. The periostracum, as well as the organic matrices of the inner shell, is dominated by a protein called **conchiolin.** The entire shell is secreted by the outer layer of the epidermis. Called the mantle, this epidermis is richly supplied with mucous glands and is separated from the shell by a shallow fluid-filled space; this space is the site of conchiolin and crystal formation, and thus the shell can be thickened at any point. Direct contact between the epidermis and the shell occurs only at the folded edge of the mantle, which is the site of periostracum formation and thus the only growing edge of the shell. Beneath the outer, secretory lobe of the mantle are two inner folds. The middle fold may bear sensory elements, and the innermost mantle fold is muscular. Several pairs of retractor muscles fasten the molluscan shell to the foot. Contraction of these muscles clamps the shell down tightly around the body.

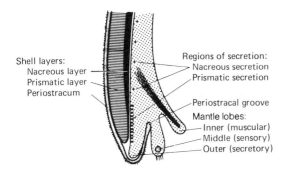

The mantle and shell overhang the central body and foot. Hence a lateral and/or posterior cavity exists about the base of the animal (Fig. 9.1). This space, called the **mantle cavity,** is one of the most important structures in the phylum. It has been modified for respiration, food gathering, and locomotion. Its most common function is respiratory, and in our generalized mollusk, the mantle cavity contains one or more pairs of gills. Mollusks have unique gills, and although various gill forms exist within the phylum, a generalized structure is described here.

The generalized molluscan gill, or **ctenidium,** consists of a flattened central axis from which thin, ciliated filaments project laterally (Fig. 9.3a). The central axis houses blood vessels, muscles, and nerves; it is suspended by membranes connecting with the mantle roof and lateral wall. Lateral filaments alternate along the axis. Each wedge-shaped filament may be supported ventrally by a chitinoid rod. Ctenidial filaments are covered with cilia which draw water currents through the mantle cavity. The use of such currents in respiration and other maintenance systems is discussed subsequently under the appropriate section.

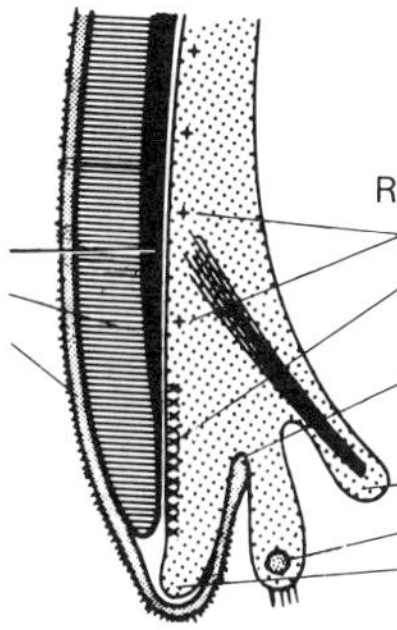

Figure 9.2 Cross section through the shell and mantle margin of a generalized mollusk. (From W.D. Russell-Hunter, *A Biology of Lower Invertebrates*, 1968, Macmillan.)

The molluscan visceral hump houses the digestive, circulatory, excretory–osmoregulatory, and reproductive systems. This area is rather soft and unmuscularized; connective tissues and blood sinuses compose most of its bulk. The tubular digestive tract includes a buccal region, an esophagus, a complex stomach, paired digestive glands, the intestine, and an anus. The anus commonly occupies a dorsomedial position within the rear of the mantle cavity. The circulatory system includes a heart, a single anterior aorta, assorted secondary blood vessels, and an open system of sinuses where blood bathes the tissues directly. A fluid-filled space surrounds the heart; this pericardial cavity is considered a coelom, albeit a rather technically defined one. One or more pairs of metanephridia connect with the coelom of the generalized mollusk. These excretory–osmoregulatory organs are complex enough to be called kidneys. Molluscan gonads usually occupy a large area anterior to the pericardial cavity. They are drained through metanephridial tubules. The cavities of the kidneys and gonads are commonly considered to be of coelomic origin.

Digestion and Nutrition

Molluscan diversity is based largely on an extensive adaptive radiation in feeding styles. A multitude of diets and feeding habits are displayed within the phylum and have influenced the biology of the digestive tract. Nevertheless, the molluscan gut has remained relatively stable in structure. Such stability is attributable to the fact that most mollusks swallow only small particles of food.

The mouth opens into a buccal cavity. There we encounter the **radula,** an organ unique to the Mollusca. The radula (Fig. 9.4) consists of numerous, recurved chitinoid teeth arranged in rows along a membranous belt. This belt is stretched across a cartilaginous rod called the **odontophore.** Both odontophore and radula are associated with protractor and retractor muscles. These muscles not only extend the odontophore from the mouth but also move the

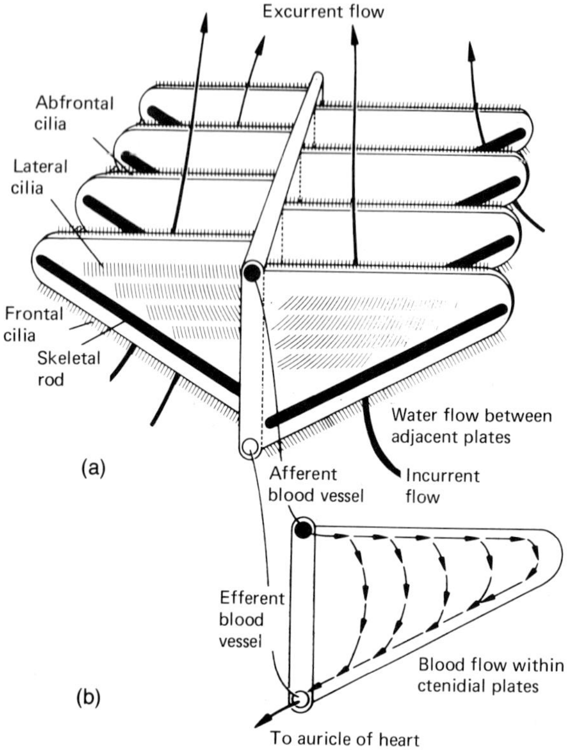

Figure 9.3 A generalized molluscan gill or ctenidium. (*a*) Stereogram showing the ventral to dorsal water current between adjacent gill-plates; (*b*) A single ctenidial filament showing the counter-current flow of blood. (From Russell-Hunter, *A Biology of Lower Invertebrates*.)

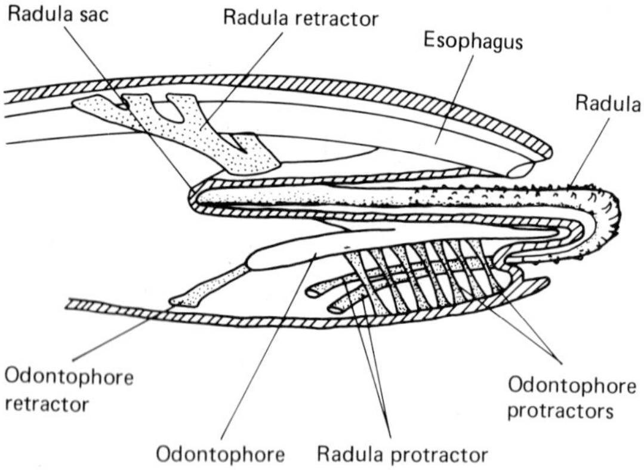

Figure 9.4 Lateral view of a section through the buccal region of *Busycon*, a gastropod, showing the relation of muscles to the radula and odontophore. (From F.A. Brown, *Selected Invertebrate Types*, 1950, John Wiley.)

radular teeth over the odontophore base. Teeth are applied against the substratum, where they scrape particles loose and convey them back into the mouth. Because such scraping rapidly erodes the teeth, new ones are secreted continually at the posterior end of the organ.

Salivary glands discharge mucus into the buccal cavity, where food particles are bound onto a mucous string which is conveyed through the esophagus to the stomach. The generalized molluscan stomach is cone shaped, with its tapered end toward the intestine (Fig. 9.5). The food-laden mucous string is pulled from the esophageal opening at the large end of the stomach cone. The pulling force is generated by cilia in the narrow posterior region of the stomach wall. This region is called the **style sac.** Style sac cilia beat at right angles to the tapered slope of the organ and create the circular current which pulls and winds the mucous string. The rotating mass so formed contacts a ciliary sorting field at the anterior end of the stomach. When small digestible particles are liberated from the mucous mass, they are sorted and then shunted into one of two paired digestive glands. Opening laterally from the stomach, these glands accomplish most of the digestion and absorption. While some extracellular

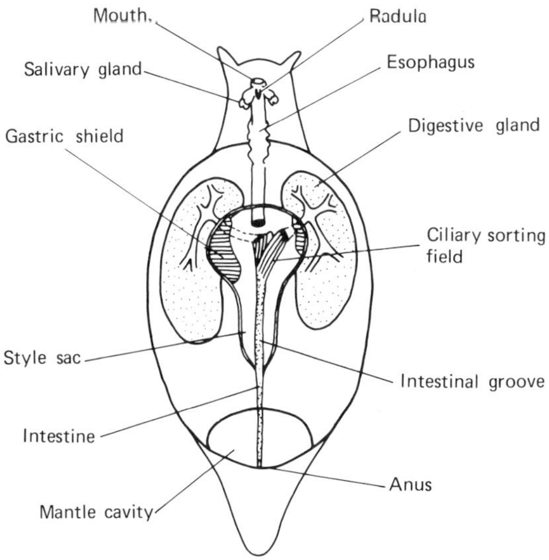

Figure 9.5 Dorsal view of the alimentary canal and associated organs in a generalized primitive mollusk. (From G. Owen, In *Physiology of Mollusca, Vol. II*, K.M. Wilbur and C.M. Yonge, Eds., 1966, Academic Press.)

digestion may occur in the stomach or within the lumen of the digestive glands, much digestion occurs intracellularly within the epithelium of the glandular wall. Meanwhile, food particles too large to enter the digestive gland are swept against the anterior walls of the stomach. Here a chitin-lined gastric shield helps break the particles into sizes appropriate for digestion. Indigestible matter passes along a ciliated groove to the intestine. The latter forms compact fecal pellets. These pellets are voided at the anus, where they are swept away by the excurrent flow of mantle cavity waters. The compact nature of molluscan feces ensures that they do not foul the gills.

Respiration

An animal with a thick dorsal shell and a habit of close ventral contact with the substratum cannot conduct respiratory exchange over its general surfaces. Rather, the generalized mollusk possesses a special respiratory gill. This gill, usually called a ctenidium, was described previously. A respiratory current is drawn through the ctenidium by the beating of lateral cilia on the thin, wedge-shaped filaments (Fig. 9.3). The current enters ventrolaterally, passes over the vascularized surfaces of the gill, and exits dorsolaterally or posteriorly from the mantle cavity. (The dorsal, posterior location of the anus and nephridiopores thus places these waste-releasing openings in the path of the excurrent flow.) Other ctenidial cilia may remove excess particulate matter from the gill and thus prevent its fouling by dirty water. These cleansing cilia are located on the front and rear margins of the filaments. The so-called frontal cilia are the first to contact incurrent waters; they remove particles from the water and transport them upward over the filament to the abfrontal cilia of the rear margin. Along this route, the particles are wrapped in mucus which facilitates their passage. Mucus is secreted not only by the gill filaments but also by two large **hypobranchial glands** (Fig. 9.1). The excurrent flow then removes mucus-bound, unwanted materials from the mantle cavity.

The direction of blood flow through the ctenidium is opposite that of water flow (Fig. 9.3b). Blood enters the molluscan gill through an afferent vessel located near the abfrontal margin of the central axis. As the blood percolates across the broad surface of each filament, it absorbs oxygen from mantle cavity waters. Oxygenated blood drains into an efferent vessel near the frontal margin of the central axis and then flows back to the heart. The opposing directions of blood and water flow allow well-oxygenated water to contact well-oxygenated vascular surfaces, while water with less oxygen meets ctenidial areas also low in oxygen. Such a counterflow system maximizes gas exchange.

Oxygen uptake and transport are facilitated by respiratory pigments. Although hemoglobin is present in some mollusks, the phylum's most common blood pigment is hemocyanin—a copper-containing protein dissolved in the blood plasma. This rather heavy molecule may form aggregations with a collective molecular weight running into the millions. Its concentration in the blood varies with the activity level of the species, active mollusks possessing relatively more of the pigment. While the oxygen-carrying capacities of hemocyanin are generally inferior to those of hemoglobin, the molluscan pigment functions well in a phylum whose members are generally rather slow-moving.

Circulation

The generalized mollusk has an open circulatory system comprising a heart, a few blood vessels, and an extensive network of blood sinuses called the **hemocoel.** The molluscan heart lies dorsally within the pericardial cavity (Fig. 9.1) and features two types of muscularized chambers: the auricle and the ventricle. One or more pairs of auricles are associated with the ctenidia. The efferent blood vessel from each ctenidium drains into an auricle; each auricle then pumps oxygenated blood into the single, medial ventricle. The ventricle pumps blood through a single anterior aorta, which may branch into secondary vessels. Eventually, blood is delivered to a network of sinuses where tissues are bathed directly. Nutrients, absorbed through the outer walls of the digestive gland, also are distributed by the blood. In time, spent blood is collected within the ctenidia, where it begins another cycle through the system.

Because of the open nature of the circulatory system, internal transport in mollusks is not highly efficient. Blood pressure is characteristically low, and circulatory routes are often poorly defined. But once again, such a system suffices for these metabolically slow animals.

Excretion and Osmoregulation

In the generalized mollusk, excretion and osmoregulation involve a complicated metanephridium, often called a kidney. Molluscan kidneys commonly exist in one or more pairs (Fig. 9.1). A nephrostomal region drains the pericardial cavity and discharges into a long, usually coiled, renal tubule. This tubule terminates at a nephridiopore near the anus. Pressure within the heart and the pericardial regions of the aorta filters blood fluids into the pericardial cavity. These fluids enter the renal tubule, whose walls are convoluted and glandular. There, salvageable materials are resorbed and additional waste products are secreted into the forming urine. Upon exiting through the nephridiopore, the urine is swept from the animal with the excurrent flow of mantle cavity waters.

Glands within the walls of the pericardial cavity also play an important excretory role. These pericardial glands secrete wastes into the cavity, thus providing additional input to the kidneys.

Most mollusks eliminate ammonia. As in other animals, the exact form and relative concentration of various nitrogenous wastes are influenced by the habitat. Our generalized mollusk is a small marine creature for which the excretion of toxic, soluble ammonia is no problem. Freshwater mollusks similarly eliminate ammonia. Moreover, mollusks seem to tolerate relatively high concentrations of ammonia in their blood—up to 1000 times as much as mammals can. However, terrestrial species must excrete their nitrogenous wastes in a drier form. This phylum apparently lacks the metabolic pathways for urea formation; uric acid is its water-conserving alternative to ammonia. Relative concentrations of ammonia and uric acid depend on environmental conditions. Terrestrial snails, for example, excrete 10 to 300 times as much uric acid as marine species do, while intertidal forms excrete intermediate amounts.

ACTIVITY SYSTEMS

Mollusks are not a very active group. Except for the cephalopods, they are typically slow-moving or sedentary creatures. Thus, among the noncephalopod classes, we find rather simple activity systems. The cephalopods, however, are quite a spectacular exception. This molluscan class includes some of the largest, fastest, and most intelligent creatures in the invertebrate world. An account of cephalopod activity systems is reserved for the discussion of this specialized class. The following section applies to mollusks in general and thus, specifically, to the idealized form (Fig. 9.1).

Muscles

Among mollusks, the musculature does not play the definitive structural role that it does in the Annelida. The dorsal shell largely determines a mollusk's external form, and the protection and immobility afforded by the shell allow a reduction in dorsal body wall muscles. Only the ventral foot remains well muscularized. Here, as elsewhere in the phylum, muscle fibers are not arranged in distinct longitudinal or circular layers. Rather, they are interwoven with connective tissue and distributed in complex patterns. The loose, spongy nature of molluscan connective tissue and its relationship to the open circulatory system play critical roles in the locomotion of these animals.

Other muscles in the generalized mollusk include the retractor and protractor muscles of the radula and the odontophore, the tentacle and head retractor muscles, and the retractor muscles of the foot. Foot retractor muscles appear in one or more pairs along each side of the mollusk. These strong fibers join the inner surface of the shell with both sides of the foot. Their contraction withdraws the foot and clamps the shell down tightly against the substratum. Thus maximum protection can be provided for the soft viscera.

Nerves

The generalized mollusk has a rather simple nervous system with relatively few ganglia (Fig. 9.1). A pair of cerebral ganglia overlies the esophagus and joins via circumenteric nerve rings with a pair of pleural ganglia and a pair of pedal ganglia. The pleural ganglia give rise to paired nerve cords, which communicate with the viscera and the mantle. Visceral nerves may join posteriorly at paired, visceral ganglia. The pedal ganglia dispatch paired fibers which innervate the foot.

A generalized molluscan nervous system thus comprises a ganglionated, circumenteric region and two nerve cord systems, one pleural–visceral and the other pedal. An extensive subepidermal nerve plexus is also present. In higher mollusks, the circumenteric region constitutes a complex brain, and body nerve organization is modified for an efficient conduction of nervous impulses.

Sense Organs

Molluscan sense organs may include paired tentacles, eyes, statocysts, osphradia, and epipodia. The tentacles are tactile and perhaps also chemosensory. Molluscan eyes are of the pigment cup, retinal type; these organs are developed mostly among gastropods and cephalopods. Most mollusks receive key environmental information in chemical and tactile form. The **osphradium** (Fig. 9.1) is a chemosensory patch located at the base of each gill. This organ detects fouled and/or sediment-bearing water and may serve in food location. Finally, the **epipodium** is a delicate extension of the foot. Its fine tentacular surfaces contact the substratum, where they may respond to chemical and/or tactile stimuli.

Behavior

Although characteristically low-key, molluscan behavior is quite variable and is best discussed in the accounts of the individual classes. Only a general description of the principles of molluscan movement is presented here.

Locomotion is primarily a function of the ventral, muscular foot. The sole of the foot is ciliated in some smaller species and, in all mollusks, is well supplied with mucus-secreting pedal glands. The smallest mollusks move by ciliary action over a slime trail laid down by their pedal glands. (Recall that such locomotion is general among the turbellarians.) Larger mollusks are dependent on muscular action alone. Characteristically, fine waves of muscular contraction sweep anteriorly over the foot. At the height of the sweeping wave, a section of the foot is lifted from the substratum; as the wave passes, the lifted section is lowered a short distance forward from its original position. Thus the mollusk advances. Single waves may extend completely across the foot or may occur alternately and occupy only about half of the pedal width.

Many mollusks use their feet for a different kind of locomotion. In some classes (the Scaphopoda and the Bivalvia), the foot is adapted for burrowing. Such a foot commonly is flattened and/or pointed; its movement involves a special antagonistic relationship between the pedal retractor muscles and the molluscan hemocoel (Fig. 9.51). Mollusks can manipulate local blood pressures within their open body sinuses. When retractor muscles withdraw the foot, head, or tentacles, hemocoelic pressure restores the precontraction state of the foot/head/tentacle. The loose spongy nature of molluscan connective tissue contributes to this antagonism. Valves may exist within these tissues. Their closures retain blood within erect tissues, thus ensuring the continuous extension of a tentacle, head, or foot.

CONTINUITY SYSTEMS

Molluscan continuity involves systems somewhat reminiscent of those encountered in the Annelida. Most marine species are diecious, but hermaphroditism is strongly established among terrestrial and freshwater forms. Reproductive structures include gonads and gonoducts which often are associated with the pericardial coelom. Molluscan embryogeny is typically protostomate and features a **trochophore** larva remarkably similar to that seen in the Annelida. In mollusks, however, a second larval stage, the **veliger,** often occurs. Also in contrast to the situation in many annelids, the phenomena of asexuality and regeneration are insignificant in molluscan continuity.

Reproduction

Primitively, molluscan gonads occurred singly or in pairs along the dorsolateral wall of the pericardial cavity (Fig. 9.1). Indeed, the lumen of the gonadal sac may represent a portion of the coelom. Gametes were released into the pericardial cavity and passed outward through the metanephridial system. Modern mollusks display various modifications of this simple reproductive system. In many types, gametes continue to pass through the coelom and excretory tubules, while in other species, discrete reproductive outlets have evolved. The latter include the more advanced members of the phylum. In separating the reproductive and nephridial systems, these mollusks have developed complex reproductive organs. Such organs may be specialized for gamete storage, egg nutrition and encapsulation, copulation, and the brooding of developmental stages.

Development

In the generalized mollusk, a simple reproductive system releases its gametes into the sea. As fertilization is external, no heavy envelope protects the female gametes, and their yolk supply is minimal. Such metabolically inexpensive gametes are released in enormous numbers. Spiral determinate cleavage produces a blastula; gastrulation follows. The molluscan embryo rapidly develops into a trochophore larva (Fig. 9.6a), a stage almost identical to the annelid trochophore (see Fig. 8.10). A stomodeal foregut, proctodeal rectum, and endodermal midgut form the complete digestive tract. Mesoderm proliferates along either side of the larval gut, and the animal elongates. However, there is no evidence of segmentation in any system during this mesodermal elongation. In many mollusks, the elongated post-trochophore stage develops further into a veliger larva (Fig. 9.6b). In this stage, the foot, shell, and other molluscan characters begin to assume their adult forms. A veliger's most conspicuous organ is its multilobed, ciliated **velum,** which consists of semicircular extensions of the original prototroch. This swimming organ allows many mollusks to play an active, ephemeral role among the zooplankton. The velum degenerates after the young mollusk settles to the substratum for growth and maturation.

A Review

The preceding discussion of molluscan unity necessarily focused on a very generalized hypothetical animal. Some characters of the creature depicted in Figure 9.1 undoubtedly were present in the first mollusks. We reemphasize, however, that this generalized mollusk is *not* the ancestral form of the phylum. The Mollusca have undergone a very long and divergent evolution, as the following account of their diversity demonstrates. At this point, we review the generalized characters on which the unity of the phylum is based. Basically, mollusks are:

(1) bilateral, coelomate, probably unsegmented invertebrates;

(2) animals with a dorsal shell secreted by an underlying mantle, a visceral hump, and a ventral muscular foot;

(3) animals with a complete digestive tract featuring a radula and paired digestive diverticula;

(4) animals with an open circulatory system, respiratory gills within a mantle cavity, and complex metanephridia;

(5) animals whose movements depend on muscular waves and an antagonistic relationship between the muscles and a hemocoelic skeleton;

(6) typical protostomes with a trochophore and often a veliger larva in the life cycle.

Molluscan Diversity

Mollusks obviously are a diverse group. However, the evolutionary pathways along which the phylum's present diversity was produced are less clear. Molluscan taxonomy is thus a rather difficult study. Changes continue to be made in the definition of subphyla, classes, and subclasses, and there is much disagreement among authorities. The present account of three subphyla and seven classes is a relatively new scheme. A rationale for this taxonomy is presented in the discussion of molluscan evolution.

SUBPHYLUM ACULIFERA, CLASS APLACOPHORA

Aplacophorans, commonly known as solenogasters, are a curious group of 130 molluscan species. They are rather short (commonly less than 5cm), vermiform creatures which crawl on or through muddy bottoms or over colonial coelenterates in deeper marine waters throughout the world. Solenogasters have no shell, but their outer surface is studded with calcareous spicules. Their cylindrical shape is produced by the ventral enrollment and closure of the mantle margins. In burrowing types, such as *Crystallophrisson* (Fig. 9.7a, b), the mantle forms a complete tube around the animal; the mantle cavity is posterior and contains a pair of gills. Surface-dwelling solenogasters, such as *Pruvotina* (Fig. 9.7c), have a narrow, longitudinal groove along their ventral surface.

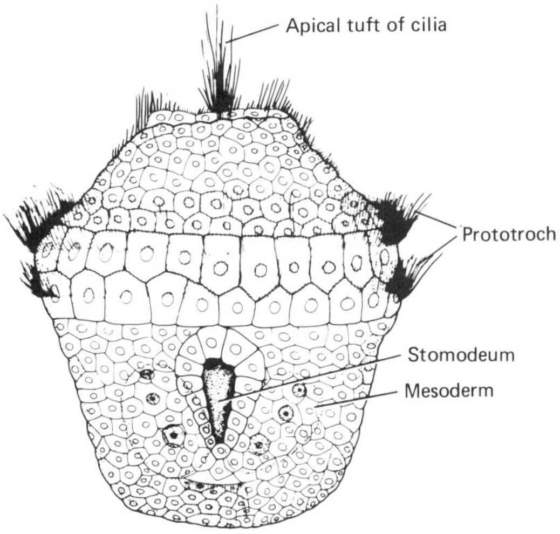

(a)

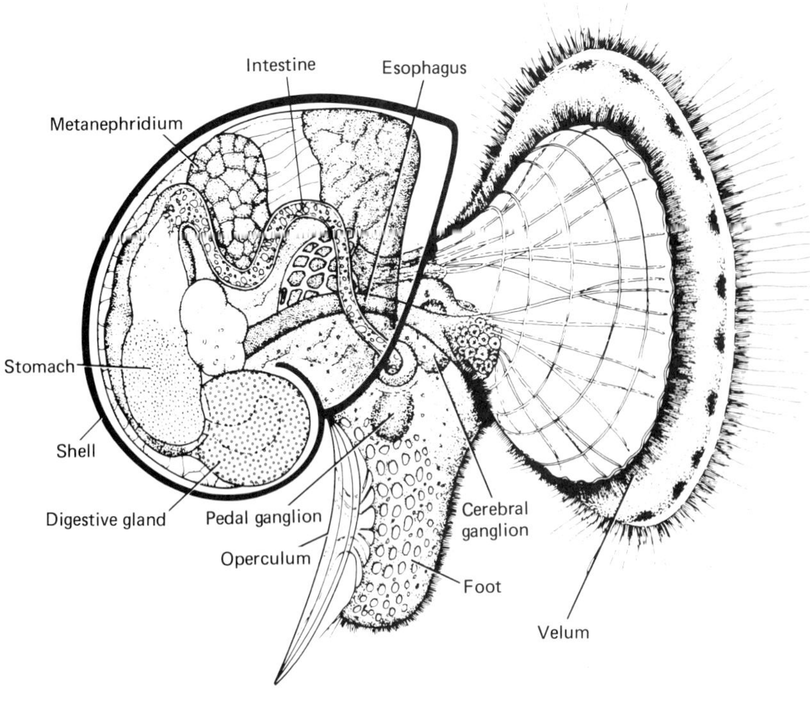

(b)

Figure 9.6 Molluscan larvae. (*a*) The trochophore larva of *Patella*, a primitive gastropod; (*b*) The veliger larva of *Crepidula*, also a gastropod. (From I.W. Sherman and V.G. Sherman, *The Invertebrates: Function and Form: A Laboratory Guide*, 2nd Ed., 1976, Macmillan.)

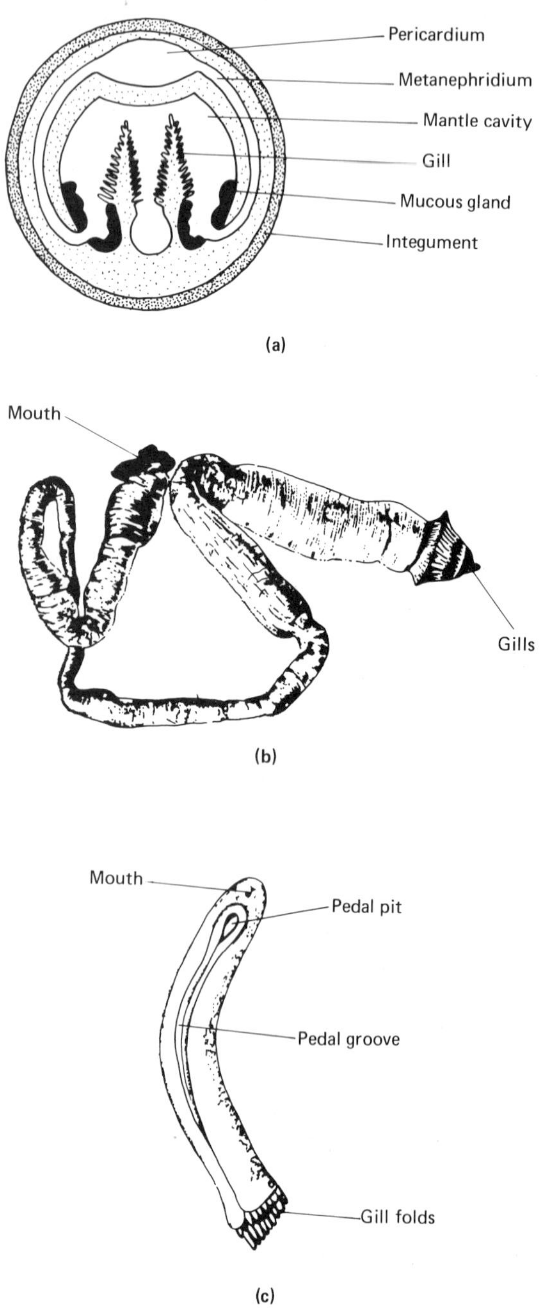

Figure 9.7 Aplacophorans (*a*) Cross section through *Crystallophrisson* taken near the posterior end showing the mantle cavity and gills; (*b*) A burrowing *Crystallophrisson indicum;* (c) *Pruvotina impexa*, a surface dwelling form. (*a* and *b* from E. Fischer-Piette and A. Franc, In *Traité de Zoologie, Vol. V, Part Two*, P.P. Grassé, Ed., 1960, Masson et Cie; c from L.H. Hyman, *The Invertebrates, Vol. VI, Mollusca*, 1967, McGraw-Hill.)

This groove, which represents an interruption in the mantle tube, may contain gills and a vestigial foot.

The aplacophoran digestive tract is straight and tubular, and a radula occurs in most species. Typically, a single gonad communicates with the pericardial cavity, which is drained by excretory tubules. In characteristically molluscan fashion, these tubules and the aplacophoran anus discharge into the mantle cavity. The nervous system in these animals includes cerebral and pleural ganglia with their associated commissures and main body nerves. Such a system may represent the most primitive nervous organization in the phylum.

The Aplacophora are an evolutionary puzzle. Historically, most researchers have considered them an aberrant, highly specialized group. Increasingly, however, it is suggested that these mollusks demonstrate primitive traits. They may be closely related to protomolluscan forms, at least more so than any other living members of the phylum. Their lack of a shell deprives us of a fossil record to study; thus we are often limited to mere speculation. The concluding discussion of molluscan evolution explores some consequences of such speculations.

SUBPHYLUM PLACOPHORA, CLASS POLYPLACOPHORA

The mollusks of the Polyplacophora commonly are known as chitons. Numbering 600 species, these marine animals display many primitive features of their phylum, but also are quite specialized for life along rocky shores. Chitons maintain a low profile in their predominantly subtidal and intertidal environments. They are usually dorsoventrally flattened and always have a greatly reduced head region; their broad ventral foot is applied very closely to the substratum. Their most conspicuous external structure is a shell composed of eight transverse plates (Fig. 9.8), while their internal organ systems are simple and evidence some pseudometamerism.

The eight plates of the chiton shell are arranged along the creature's dorsal midline. Each plate is partially overlapped by the plate lying ahead of it. In the region of overlap, muscle fibers connect the two shell plates and thus permit some independent movement.

The chiton mantle extends over the dorsal surface, where it is called the **girdle.** In *Chaetopleura* (Fig. 9.8a), the girdle is small, but in many chitons, such as *Katharina* (Fig. 9.8b), it covers the lateral edges of the plates. In *Cryptochiton,* the girdle entirely overlies the shell. This dorsal mantle may be covered with calcareous spicules and other ornamentation, as in *Chiton*

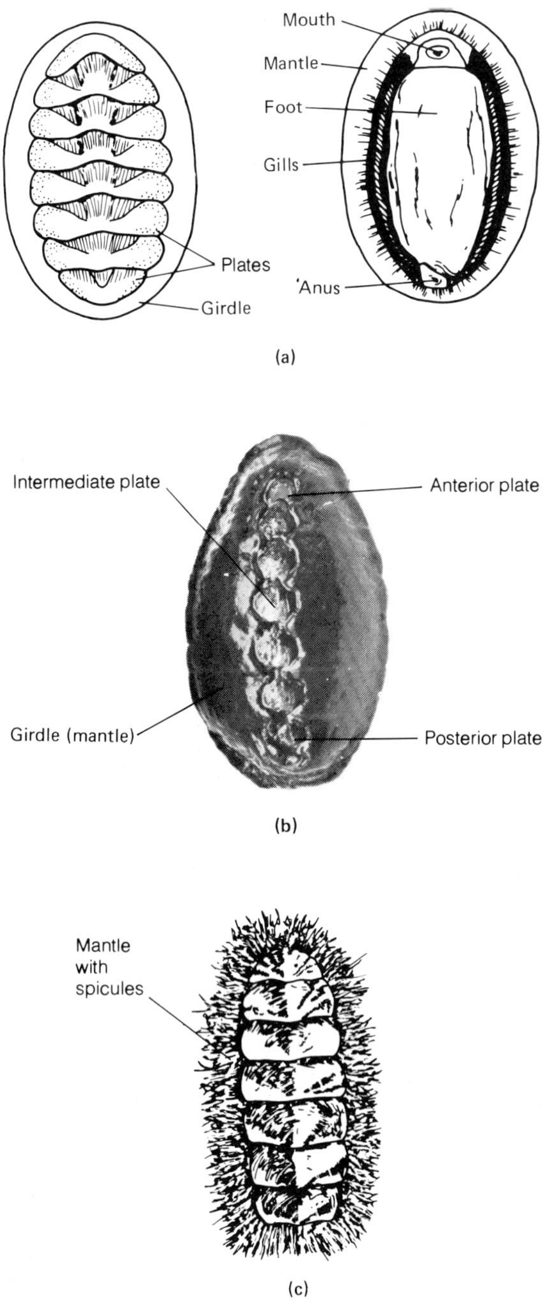

Figure 9.8 Chiton external anatomy. (*a*) Dorsal and ventral aspects of *Chaetopleura apiculata*; (*b*) Dorsal view of *Katharina tunicata*; (*c*) Dorsal view of *Chiton spinosus*. (*a* from Brown, *Selected Invertebrate Types*; *b* from E.F. Ricketts, et al., *Between Pacific Tides*, 4th ed., 1968, Stanford University; *c* from A.H. Cook, In *Cambridge Natural History, Vol. III*, A.E. Harmer and S.F. Shipley, Eds., 1959, Macmillan.)

(Fig. 9.8c). Such a spiny surface is reminiscent of the external cover of the aplacophorans.

Ventrolaterally, the chiton mantle joins the foot. Between the foot and the lateral margin on both sides is a distinct groove (Fig. 9.9). This groove represents the mantle cavity, the site of respiratory exchange across numerous gills. Chiton gills, which may number in the dozens, are similar to the generalized molluscan ctenidia (Fig. 9.3), but commonly display rounded rather than wedge-shaped filaments (Fig. 9.10) and lack supportive rods.

The chiton head is indistinct, a mere zone around the anterior ventral mouth. The entire region is roofed over by and fused with the mantle. No sensory organs are evident, except perhaps a few palps.

Most chitons are microphagous feeders. Their mouth opens into a buccal cavity containing several jaws, a strong radula, and a subradular organ. The magnetite-capped teeth of the radula scrape algae and other small organisms off rocks. In chitons, 17 of these iron-strengthened teeth occupy each transverse row along the radular membrane. The **subradular organ** (Fig. 9.11a) is a chemosensory patch beneath the radula. This patch is protrusible through the mouth, whereupon it tests rock surfaces for the presence of food. When food is located, the organ is withdrawn and the radula goes to work. Feeding behavior in chitons usually involves alternate protrusions of the subradular organ and the radula as the mollusk tests and eats its way along.

Salivary glands discharge nonenzymatic mucus into the buccal cavity. This mucus lubricates food which enters the mouth and passes into the esophagus. A ventral ciliated groove transports food through the esophagus, where sugar glands apply amylases. The esophagus enters an irregularly shaped stomach (Fig. 9.11b). The chiton stomach lacks a ciliary sorting field and is a primary site of extracellular digestion and absorption. Paired ducts from a large digestive gland discharge proteases into the stomach, where food and enzymes are churned by peristaltic action. Food passes into the anterior intestine, where it is retained by a valve until digestion and absorption are completed. This anterior intestinal region may be homologous with the style sac of other mollusks. The opening and closing of the intestinal valve forms fecal pellets which are compacted further by water resorption within the posterior intestine. The digestive tract terminates in a posterior anus. Feces are swept away by mantle cavity currents.

The mantle cavity current, which is central to the physiology of nearly all mollusks, passes along both sides of the chiton body (Fig. 9.9). Anteriorly, the

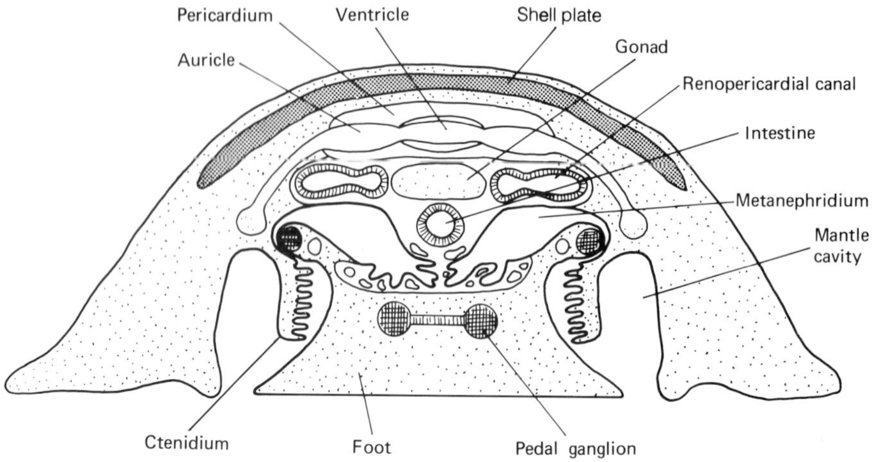

Figure 9.9 Cross section through a chiton showing major anatomical features. (From H. Simroth, In *Klassen und Ordnungen des Tierreichs, Vol. III, Part One*, Bronn, Ed., 1892.)

girdle is raised to permit water intake. Water flows first ventrolaterally, then passes through the ciliated ctenidia and continues along a dorsomedial path. The respiratory current exits from the mantle grooves through one or two posterior gaps.

The chiton circulatory system is essentially that described for the generalized mollusk. A single pair of auricles drains all the gills, and a ventricle pumps blood through the aorta. Blood flows through the open sinuses of the chiton body, returns to the gills for aeration, and completes its circuit to the heart. The

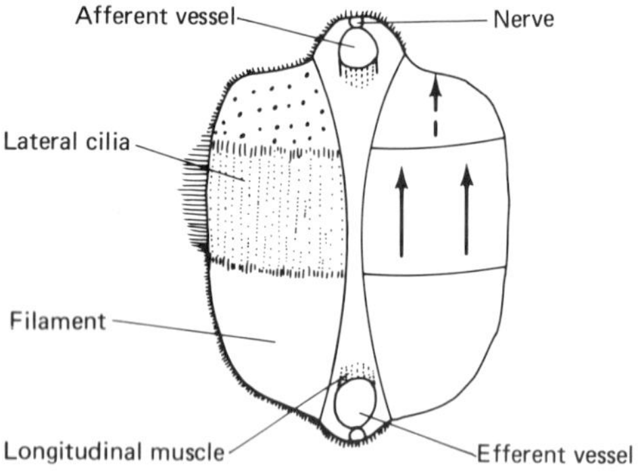

Figure 9.10 A pair of ctenidial filaments from a gill of the chiton *Lepidochitona cinereus*. Compare with Figure 9.3 and note that chiton gill filaments are rounded rather than wedge-shaped. Arrows within the figure denote the direction of water flow. (From Fischer-Piette and Franc, In *Traité de Zoologie*, Grassé, Ed.)

heart itself lies within a relatively spacious pericardial cavity beneath the last two shell plates.

Excretion and osmoregulation are handled by a fairly elaborate metanephridial system. This system includes two large horseshoe-shaped tubules which ramify through the hemocoel (Fig. 9.12). Paired nephrostomes drain the pericardial cavity and open into renopericardial canals. The latter join nephridial tubules which course anteriorly and then recurve to a posterior nephridiopore on the mantle cavity wall. Along the entire length of each tubule, many diverticula terminate blindly within hemocoelic tissues and sinuses. These diverticula are probable sites of secretion and resorption.

There is very little centralization in the chiton nervous system (Fig. 9.13a), ganglia being poorly defined in all members of the class. A circumenteric nerve ring gives rise to nervous loops surrounding the radula and the buccal cavity. Paired pedal nerves also emerge from this anterior ring; along the length of the foot, these nerves are joined by ladderlike connectives. Paired palliovisceral cords innervate the mantle and gut. These nerve trunks meet posteriorly as a single large loop. Lateral connectives join both palliovisceral nerves along their lengths and also connect these trunks with pedal nerves.

Chiton sense organs are rudimentary, as is appropriate for the sedentary life-style of these mollusks. Specialized sensory equipment in the class includes only the subradular organ, osphradialike patches in the mantle cavity, and the **aesthetes.** Aesthetes are clusters of sensory cells which occupy vertical canals

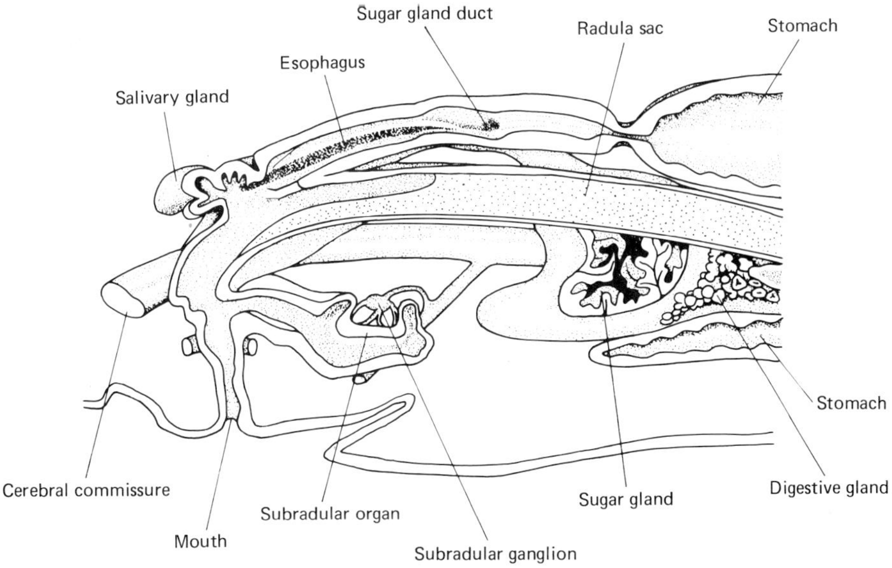

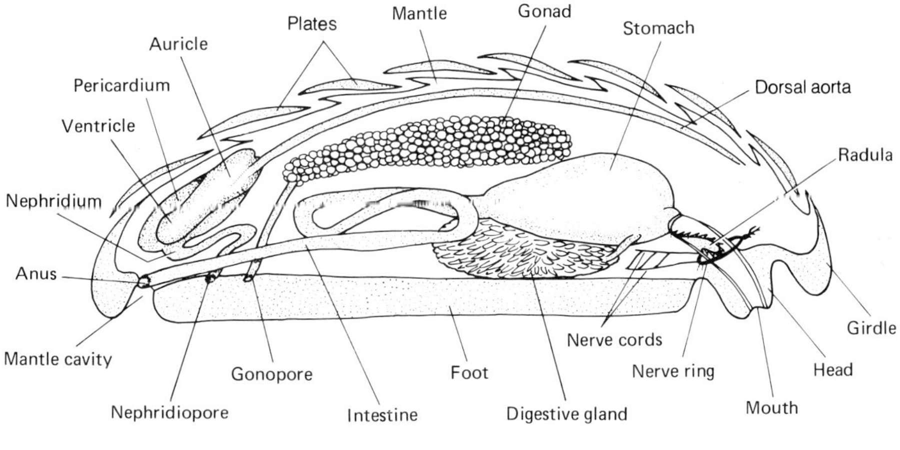

Figure 9.11 (*a*) Sagittal section through the anterior part of *Lepidochitona cinereus* featuring structures of the alimentary canal; (*b*) A longitudinal section through a generalized chiton exhibiting the major features of internal anatomy. (*a* from V. Fretter, Trans. Roy. Soc. Edin., 1937, *59:* 122; *b* from Sherman and Sherman, *The Invertebrates: Function and Form.*)

on the shell plates. Different types of these organs probably have different sensory responsibilities. The simpler aesthetes are perhaps mere tactile receptors, while some are photoreceptive and, as in *Acanthopleura* (Fig. 9.13b), may constitute elaborate eyes. In addition to these sense organs, the general mantle surface is populated by sensory cells. These cells respond variously to light and touch and, along with the aesthetes, compensate for the absence of a sensory head.

Chitons are quite languid animals. They crawl slowly at night and/or when submerged. Crawling involves waves of muscular contraction along the foot. The separate shell plates allow the chiton body to

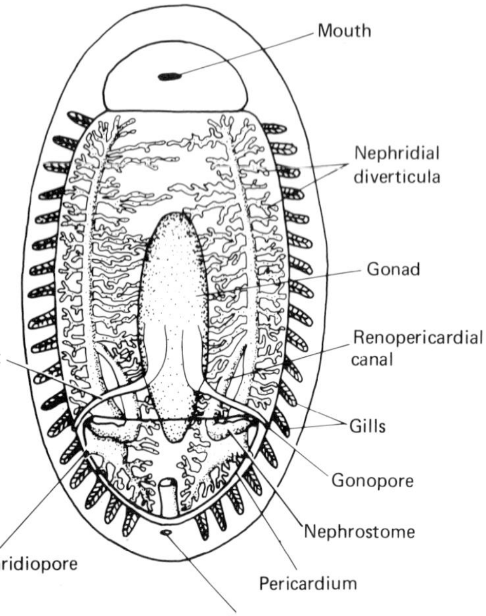

Figure 9.12 Dorsal schematic of excretory and reproductive systems of *Chiton* (After Simroth, In *Klassen und Ordnungen des Tierreichs, Vol III, Part One*, Bronn, Ed.)

conform to uneven substrata and thus to maintain close ventral contact. When a chiton is threatened, it clamps down tightly with its girdle and retracts its foot. Retraction of the foot creates a powerful suction beneath the animal, which further strengthens its hold on the substratum. If dislodged, a chiton may enroll like a pill bug, thus protecting its soft ventral parts.

Certain chitons may establish permanent residences at sites conforming to their body contours. After feeding expeditions, these animals apparently follow chemical or tactile trails to relocate their homes.

Reproduction in this class is a relatively simple process resembling the pattern described for the generalized mollusk. Commonly diecious, most chitons possess a single large gonad beneath the middle shell plates (Fig. 9.12). This gonad is drained by two gonoducts which terminate at discrete gonopores. Such pores are located just anterior to the nephridiopores. External fertilization follows spawning, and the presence of sperm in surrounding waters may stimulate the release of eggs. Gregariousness is general in the class and maximizes opportunities for external union between sperm and eggs. In some species, internal fertilization may occur within the mantle cavity or even within the gonoducts. In the latter case, very yolky embryos are brooded within the reproductive tubules. More commonly, however, the externally formed zygote develops into a free-swimming trochophore (Fig. 9.14). Chitons have no veliger stage, as the trochophore develops directly into a juvenile mollusk. A shell gland on the dorsal surface of the trochophore soon develops distinct, transverse grooves. As the larva elongates, the eight shell plates are secreted within these grooves. The chiton trochophore has eyes which may function in substratum selection. These sense organs degenerate after the mollusk has settled.

The position of the Polyplacophora within molluscan evolution is uncertain. Some consider their rather simple biology to be primitive and ancestral; others argue that the class is specialized and that its seemingly primitive traits reflect an adaptation to a simple, sedentary life-style. In either case, it appears that the class experienced an early separation from the rest of the Mollusca. Finally, the pseudometameric nature of the chiton shell, gills, and nervous system has been the object of much debate. To what extent does this character associate the chitons with the turbellarians, the annelids, and the monoplacophorans? This question and others of its kind are reviewed at the end of this chapter.

SUBPHYLUM CONCHIFERA

The Conchifera represent the evolutionary mainstream of their phylum. The vast majority of mollusks belong to this group, which includes the familiar snails, clams, squid, and octopods. In contrast to solenogasters and chitons, these animals are characterized by a structural emphasis on the dorso-

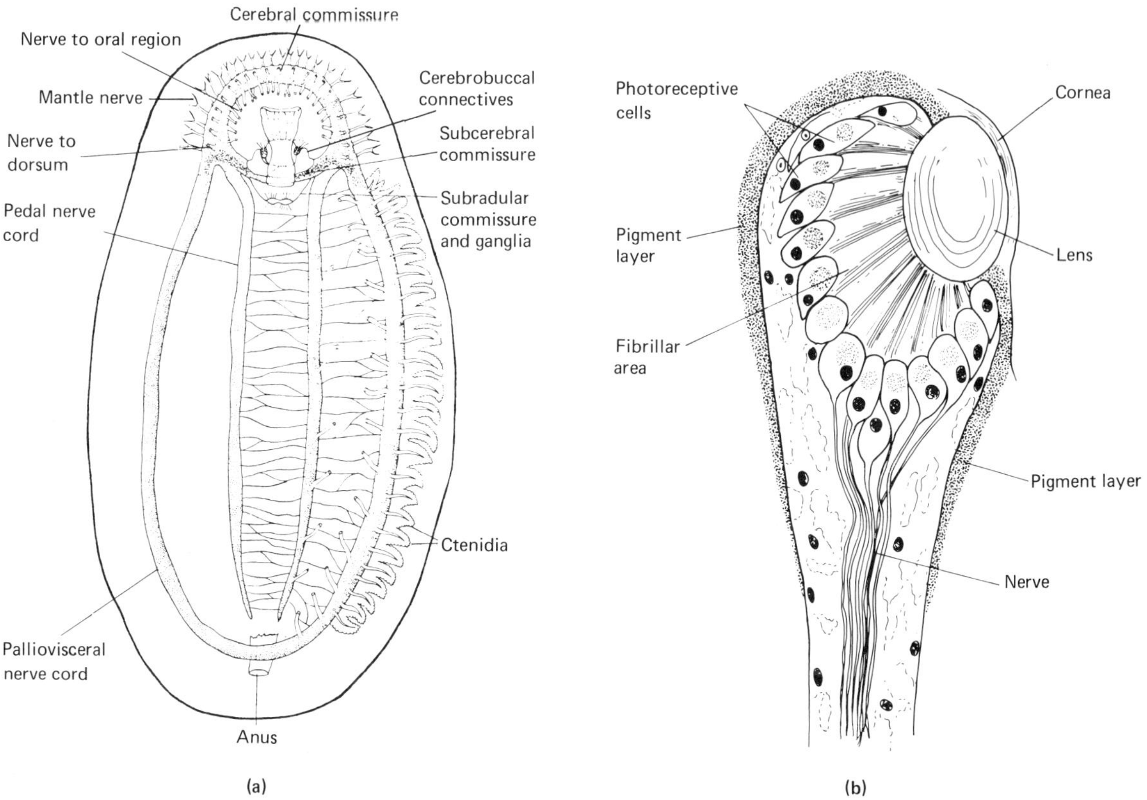

Figure 9.13 (*a*) Dorsal view of the nervous system of a chiton; (*b*) Longitudinal section through an eye-bearing aesthete from *Acanthopleura*. (*a* from J.E. Smith, et al., *Invertebrate Panorama*, 1971, Weidenfeld and Nicholson; *b* from Hyman, *The Invertebrates, Vol. VI*.)

ventral axis. Most conchiferans are not "long and flat," like the members of the two previously discussed subphyla, but rather are "short and high." The increase in height is correlated with an extension of the free eaves of the mantle and a corresponding deepening of the mantle cavity (Fig. 9.16). Mantle cavity structures—gills, nephridiopores, and gonoducts—seldom display any pseudometameric character among the conchiferans; commonly only a single pair remains.

Five classes are recognized within the subphylum. The Monoplacophora, known only from fossils and one living genus, are considered the ancestral conchiferans. The other four classes have followed quite divergent evolutionary pathways. The Gastropoda, the largest molluscan class, are characterized by a coiled shell and the phenomenon of torsion. The Bivalvia are mostly filter-feeding mollusks with bivalved shells. The Scaphopoda are a small group of burrowing mollusks, while the Cephalopoda pursue active, predatory life-styles.

Class Monoplacophora

The Monoplacophora are primitive mollusks, once known only from fossils. They were thought to be extinct since the Devonian until 1952, when living specimens were dredged from a deep trench off the Pacific Coast of Costa Rica. These contemporary monoplacophorans were classified within the genus *Neopilina* (Fig. 9.15). *Neopilina* is probably a specialized monoplacophoran, or it would have suffered the same fate as the other members of its class. Nevertheless, by combining evidence from the fossil record with observations on *Neopilina*, investigators have revealed much of monoplacophoran biology.

Neopilina apparently belongs to the Tergomya, a monoplacophoran line that continued a structural

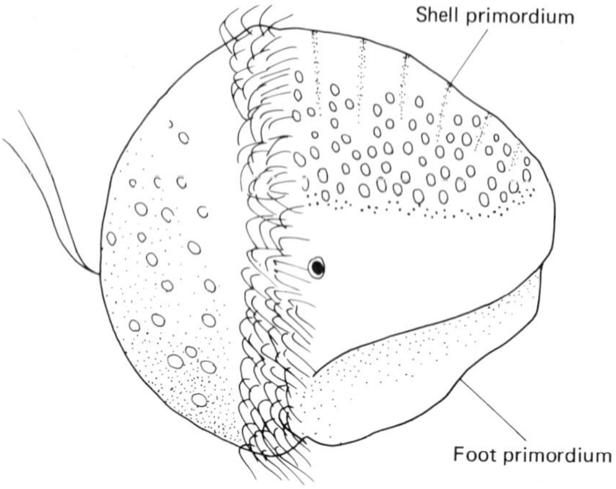

Figure 9.14 Trochophore larva of *Ischnochiton magdalensis.* (From A. Kaestner, *Invertebrate Zoology, Vol. I,* 1967, John Wiley.)

emphasis on the longitudinal axis. Thus these low, flat mollusks resembled the chitons in general form. The tergomyan shell was dome shaped and univalved; in *Neopilina,* it is a mere 3 cm long and bears an anteriorly directed apex. A broad, flat foot forms the ventral surface of the animal. Eight pairs of retractor muscles connect the foot and the shell in *Neopilina,* and paired scars on fossil specimens attest to the general occurrence of such paired retractors throughout the class. Between the overhanging shell and the foot is a shallow mantle cavity. In fossil species, this cavity contained several pairs of gills; *Neopilina* has five or six pairs. In the living genus, the gills are unipectinate; that is, their filaments occur singly, not in lateral pairs as in most mollusks. The respiratory current probably is similar to that of the chitons.

The head of *Neopilina* is roofed over by the mantle. However, it differs from the chiton head zone in its possession of small postoral tentacles and lateral flaps about the mouth. Fossils reveal little of the internal design of the Monoplacophora, so a continuing description of their organ systems must be based on *Neopilina* alone. The simple digestive tract features a radula, a subradular organ, and a stomach with a style sac. The circulatory system includes a paired heart. Two pairs of auricles drain the gills, while a single pair of ventricles drive blood through the aorta and into the open blood sinuses. A pair of pericardial cavities lies behind two large dorsal sacs. Perhaps coelomic in origin, these sacs occupy the position commonly reserved for molluscan gonads. The sacs are evidently sterile, and their function remains a mystery. The nephridial system in *Neopilina* consists of six paired kidneys. These organs have diverticulated tubules reminiscent of those in the Polyplacophora; they discharge into the mantle cavity between the gills. The kidneys may drain either the dorsal sacs or the pericardial cavity.

Nervous organization in *Neopilina* resembles the polyplacophoran pattern, with ganglia restricted to the head region. Although no members of the genus have been observed in their natural habitat, it is assumed that their movement is slow.

Neopilina is diecious, and two pairs of gonads occur laterally in the midbody region. Ducts from these reproductive structures join the third and fourth renal tubules, through which gametes are discharged into the mantle cavity. Development in *Neopilina* has not been observed, but a veliger larva may be present.

Controversy focuses on the serial repetition of body parts in *Neopilina.* The genus displays 16 retractor muscles, 2 ventricles, 4 auricles, 10 or 12 gills, 12 metanephridia, and 4 gonads; all of these organs are paired. Many argue that *Neopilina* is a metameric animal; moreover, some say that the genus illustrates a basic ancestral form and that mollusks are essentially a segmented group. However, much evidence discourages such arguments. *Neopilina* is a specialized mollusk. It has survived in a deep-water habitat, whereas all of its closest relatives have long since perished. Also, the organs of *Neopilina* appear to be merely repetitious; they are not correlated into true metameric units. We cannot define a "segment" in this relict genus. Thus it seems more appropriate to list *Neopilina* among the many pseudometameric invertebrates.

A second, distinct group within the Monoplacophora appears far removed from *Neopilina.* This group is entirely extinct; however, before its demise it may have made some of the most important evolutionary advances in molluscan history. These other monoplacophorans constitute the Cyclomya. Cyclomyans structurally emphasized the dorsoventral axis and thus assumed a much taller body form than their tergomyan relatives (Fig. 9.16). Some spiral coiling of the shell also occurred in these animals.

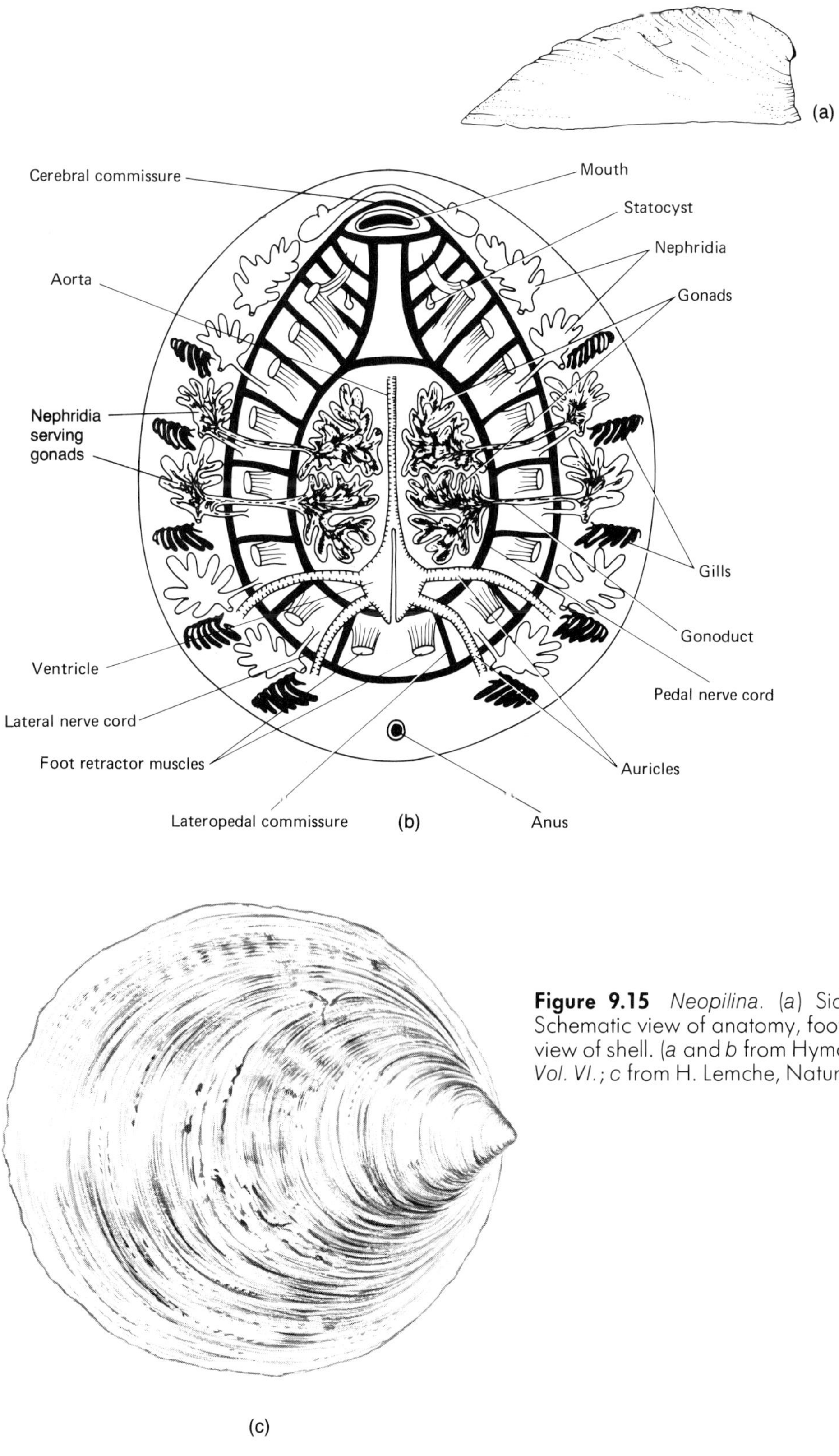

Figure 9.15 *Neopilina*. (*a*) Side view of shell; (*b*) Schematic view of anatomy, foot removed; (*c*) Dorsal view of shell. (*a* and *b* from Hyman, *The Invertebrates*, *Vol. VI.*; *c* from H. Lemche, Nature, 1957, *179*:413.)

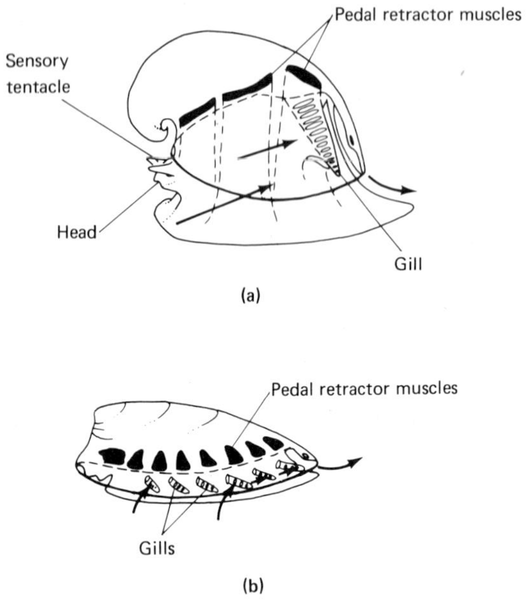

Figure 9.16 Generalized monoplacophorans. (*a*) A cyclomyan emphasizing the dorso-ventral axis, a deep mantle cavity with few but large gills, and a limited number of pedal retractor muscles. Presumably, cyclomyans had a well developed head; (*b*) A tergomyan, such as *Neopilina*, with a more elongate body, many retractor muscles, a poorly developed head, and a shallow mantle cavity with many small gills. (From C.R. Stasek, Mollusca, In *Chemical Zoology, Vol. VII*, M. Florkin and B.T. Scheer, Eds., 1972, Academic Press.)

Retractor muscle scars in the Cyclomya are removed from the shell perimeter. A more medial location for these muscles suggests the presence of a deeper mantle cavity, which would be consistent with a size increase in the gills and, with increasing respiratory efficiency, a probable reduction in gill number. A spacious cavity also would provide a retreat for the vulnerable molluscan head. The head thus could increase in size, sensory specialization, and flexibility. Monoplacophorans of this sort probably became far more active and increased in overall size and complexity. They may have given rise to the largest of the phylum's classes, the Gastropoda.

CLASS GASTROPODA

The snails and slugs of the Gastropoda constitute the largest molluscan class. Indeed, they represent the second largest class in the Animal Kingdom—surpassed only by the insects. Approximately 50,000 modern species have been described, and a rich fossil record adds another 15,000 members. These mollusks pursue a diversity of life-styles in many different environments. They live wherever adequate moisture is present, including many terrestrial habitats. Indeed, with the obvious exception of the arthropods, gastropods include some of the invertebrate world's most successful land animals.

Many gastropods resemble the generalized mollusk described earlier (Fig. 9.1). They have a univalved shell and a broad creeping foot. Internal systems, however, present higher levels of structural and physiological sophistication, and the gastropod shell and body have undergone extraordinary reorganizations.

A general account of the maintenance, activity, and continuity systems of this largest molluscan class precedes descriptions of the major subclasses and orders.

Gastropod Unity

MAINTENANCE SYSTEMS

General Morphology

The first major structural change by the Gastropoda involved the coiling of their shells. Fossil studies suggest that shell coiling began among cyclomyan monoplacophorans and continued with the emergence of the present class. Such coiling was probably an adaptation for growth and balance. Protogastropods grew by adding unequally to the open end of their shells. Balance was maintained because the body steadily shifted beneath the shell's center of gravity during growth. Snails that did not produce coiled shells are known only from fossils; apparently, their tall, conical shells (Fig. 9.17a) were too difficult to balance.

The first coiled shells were bilateral, with each new coil occurring in the same plane and lying entirely outside the previous one (Fig. 9.17b). This planospiral coiling pattern produced a very large shell with associated portage problems. Another adaptation then took place. Helicospiral coiling produced asymmetrical spirals around a central axis called the **columella** (Fig. 9.18). Each new spiral was laid down only slightly outside and below the preceding one, and the shell thus assumed a more compact, asymmetrical form. Asymmetrical shells presented their own portage problems. The body mass was restricted to the open

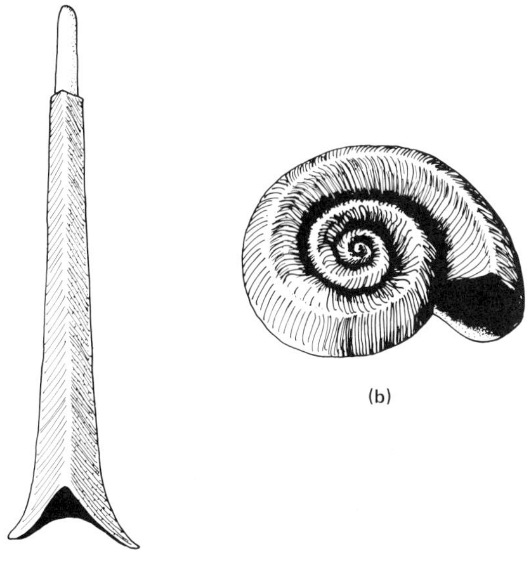

Figure 9.17 Fossils of primitive gastropod shells. (*a*) A tall conical uncoiled shell; (*b*) A planospiral shell. (From R.C. Moore, Ed., *Treatise on Invertebrate Paleontology, Part I: Mollusca 1*, 1960, Geological Society.)

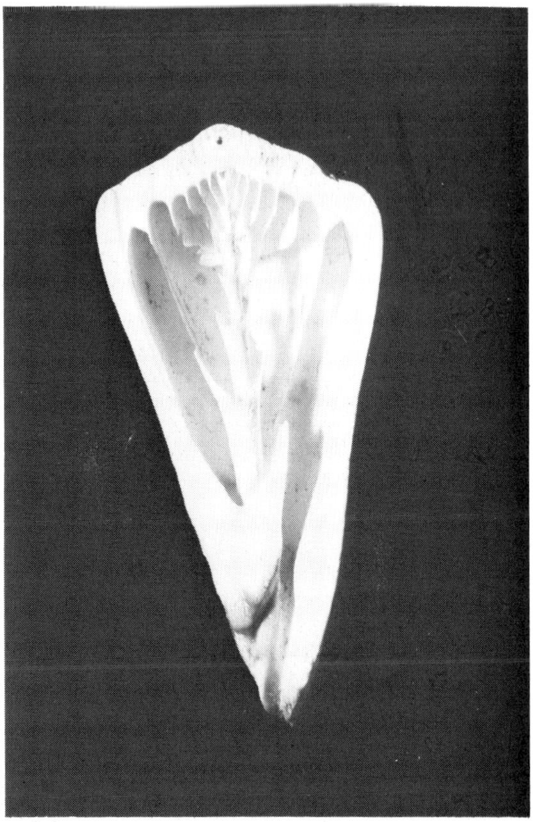

Figure 9.18 Longitudinal section through a helicospirally coiled gastropod shell showing the central axis or columella. (From J. Andrews, *Sea Shells of the Texas Coast*, 1971, University of Texas.)

right side of the spiral, and the shell could hang heavily on the closed left side. In maintaining proper balance, another general reorganization occurred, placing the central axis of the shell (the columella) at an oblique angle to the longitudinal axis of the snail body (Fig. 9.19). This adaptation established the modern gastropod shell: a compact shell, asymmetrically coiled and carried obliquely to the long axis of the body.

The oblique carriage of the shell caused the viscera to bulge along the right side. Such bulging reduced the entire right half of the mantle cavity. The right-hand members of organ pairs associated with the cavity—the right gill, nephridium, and auricle—became reduced or eliminated accordingly. Thus in many modern gastropods, only the left side of the mantle cavity is present, and only one gill, nephridium, and auricle are fully functional.

The second major reorganization of the snail body involved a phenomenon known as **torsion.** Torsion is unrelated to the coiling of the shell. Rather, it refers to the 180° rotation of the main body mass (the viscera, mantle, and mantle cavity) relative to the head and foot. During torsion, the mantle cavity and associated structures are displaced to an anterior position (Fig. 9.20). Thus the digestive tract becomes U shaped as the anus is brought forward; nephridiopores likewise are displaced anteriorly, and the gills and auricles lie ahead of the ventricle. The gastropod nervous system is twisted into a figure eight.

Torsion is a curious process for which no universally accepted explanation has been offered. All gastropods undergo torsion during the larval stage; hence the phenomenon must have some definite adaptive significance. One theory argues that torsion is largely a larval adaptation. The shell supposedly afforded little protection to ancestral planktonic larvae, particularly to their tender head–velum region. Torsion placed the mantle cavity behind the velum; thus this delicate structure could be withdrawn and then covered by the foot. Problems with this theory include the facts that the larval foot is also delicate and that torsion typically occurs quite late in development. Moreover, other molluscan larvae survive their planktonic periods without resorting to anything as traumatic as torsion.

Another theory suggests that torsion brought os-

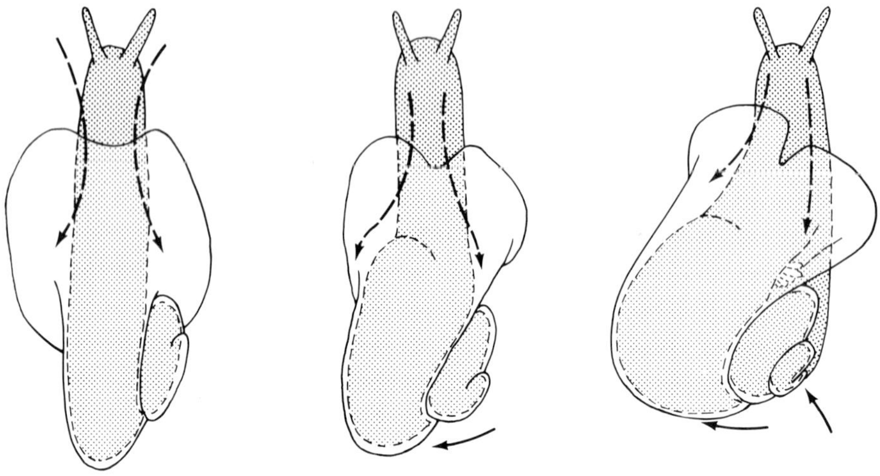

(a) (b) (c)

Figure 9.19 Three drawings showing stages in the adjustment of the helicospiral shell for maintenance of proper balance. (*a*) A helicospiral shell carried as a planospiral one is obviously unbalanced; (*b* and *c*) The adjustment of the shell for proper balance involving a clockwise rotation and upward tilt of the shell. Dashed arrows indicate the pathway of water flow; arrows beneath figures *b* and *c* indicate the clockwise rotation of the shell and the upward tilt of the spire. (From A. Solem, *The Shellmakers: Introducing Mollusks*, 1974, John Wiley.)

phradia forward to monitor cleaner respiratory currents. This theory argues that a posterior mantle cavity receives sediment-heavy water caused by the snail's local disturbance of the bottom during locomotion. An anterior mantle cavity receives cleaner, undisturbed water; also, anterior osphradia can detect and orient to undesirable waters sooner and more efficiently than posterior sensors can. This theory is discounted because snails probably emerged on hard, sediment-free substrata. Mucous secretions and the characteristically slow gait of these mollusks minimize local fouling in any case. Also, incurrent openings to the mantle cavity of early gastropods may have been anteriolateral rather than posterior; a tendency to restrict all mantle cavity openings among modern species represents a more conservative adaptation to the fouling problem.

Other theories suggest that torsion is an adult provision for withdrawing a large head into the mantle cavity, or that it allows the snail to balance its coiled shell during larval settlement.

While its precise causes remain unknown, considerable evolutionary study has focused on the consequences of torsion. Torsion created certain problems for the early gastropods, and diverse solutions to these problems account in part for the present diversity of the class. The most problematic result of torsion is fouling. The anus and nephridiopores could discharge their wastes upon the head of the torted gastropod. No animal can tolerate such a situation! Several adaptations alter the direction of current flow through the mantle cavity (Fig. 9.23). These trends are reviewed in our account of the various gastropod groups.

Nutrition and Digestion

Nutrition and digestion among the gastropods involve nearly every conceivable method of food collection and processing. Herbivores, carnivores, omnivores, scavengers, deposit and suspension feeders, and parasites are all well represented within this class.

The buccal region contains the radula. Radular types vary widely in the class and are useful in taxonomic studies. Gastropod radulae are modified for scraping, grasping, chopping, and conveying. Salivary glands discharge mucus into the buccal cavity, where incoming food is bound into a slimy string. The mucus-bound food is passed through the esophagus, where digestive enzymes may be added from glandular pouches. Due to torsion, the gastropod esophagus may enter the stomach from the rear (Fig. 9.21), although in advanced members of the class, the esophageal opening has migrated anteriorly. The account of the stomach in the generalized mollusk (Fig.

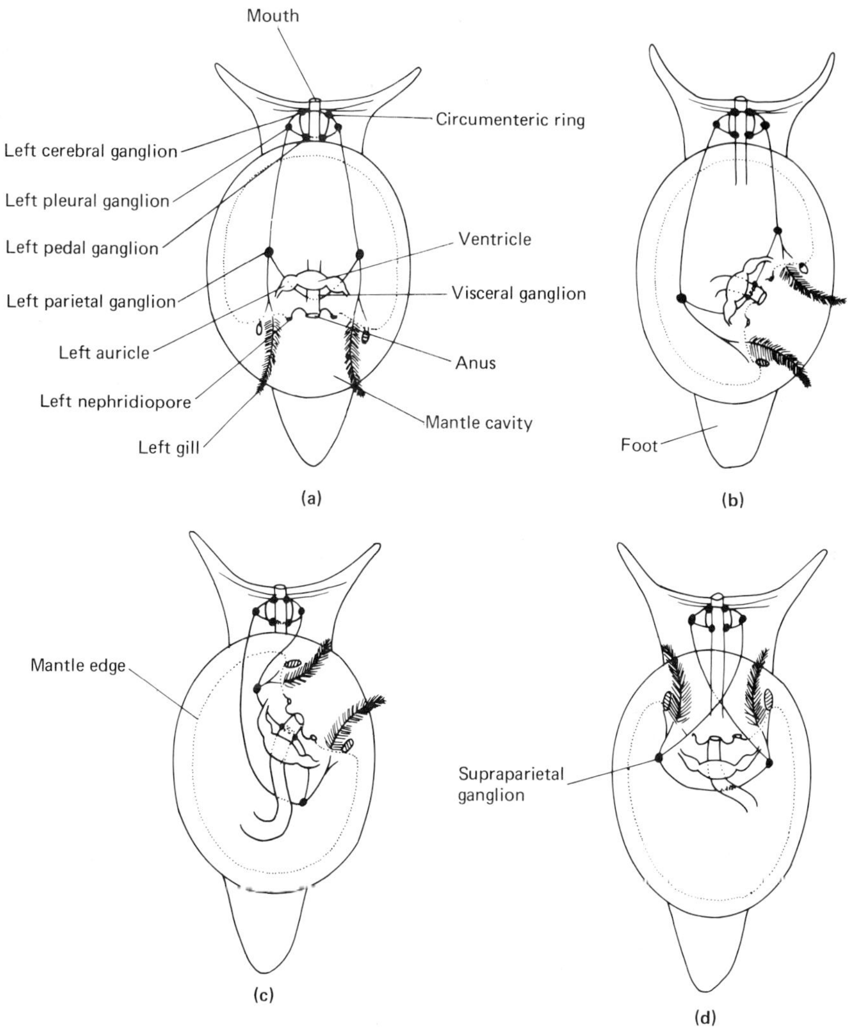

Figure 9.20 Dorsal view of the process of torsion. (*a*) The hypothetical untorted gastropod; (*b* and *c*) Stages in torsion, the counterclockwise rotation of the main body mass; (*d*) The torted gastropod. (From Hyman, *The Invertebrates, Vol. VI.*)

9.5) describes the organ in lower snails. A style sac rotates the food–mucus string while acid secretions loosen nutrients for sorting on a ciliary field. Commonly, this field extends into a long, coiled cecum (Fig. 9.21). Finer digestible particles emerge along the ridges of the cecal wall and pass into one of the paired digestive glands. There digestion is completed intracellularly, and nutrients are transferred to the hemocoel. Large, undigestible particles leave the cecum along a prominent fold called the **typhlosole.** Within the stomach, the typhlosole is continuous with a groove leading to the intestine. The latter forms feces. The gastropod anus opens dorsally within the anterior mantle cavity. As we mentioned previously, its presence has been an important influence in the structural evolution of the class.

Higher gastropods (particularly carnivorous ones) tend toward extracellular digestion. The stomach is usually a simple sac without chitinoid shields or a sorting field, and digestive glands only secrete enzymes. The buccal region displays numerous specializations for particular diets and feeding techniques. Such specializations include complex radulas and protrusible proboscises. These traits, as well as other

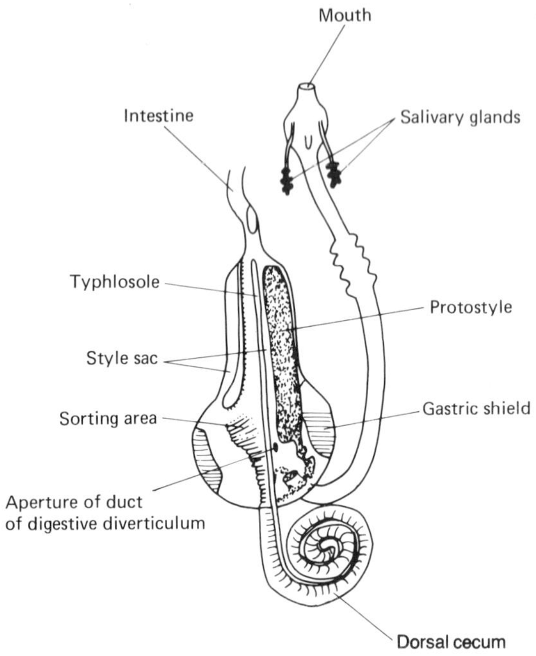

Figure 9.21 The alimentary canal of a prosobranch exhibiting many primitive features. (From Owen, In *Physiology of Mollusca, Vol. II*. Wilbur and Yonge, Eds.)

peculiar aspects of nutrition and digestion among higher gastropods, are described later.

Circulation and Respiration

Characteristic posttorsional right-side reductions in the mantle cavity led to the loss of the right gill and auricle in most higher snails. Blood passes from the remaining auricle to a ventricle which pumps the fluid through an aorta into the arterial sinuses (Fig. 9.22). Due to torsion, the gastropod ventricle is posterior to the auricle, and the main visceral aorta also is directed to the rear. A cephalopedal aorta usually courses anteriorly from the heart. The large cephalopedal sinus collects spent blood for processing through the nephridium and for aeration within the mantle cavity gills or lung. In gastropods with lungs, the mantle cavity wall houses a fine capillary network where oxygen uptake is quite efficient. From the mantle cavity, aerated blood returns to the heart for another passage through the system.

As in all mollusks, respiratory exchange in gastropods involves the mantle cavity. This cavity has been profoundly affected both by torsion and by the asymmetrical snail shell. Both of these developments altered the structure of the cavity and the direction of the respiratory current.

Torsion placed the mantle cavity just above the gastropod head. Therefore, if the respiratory current

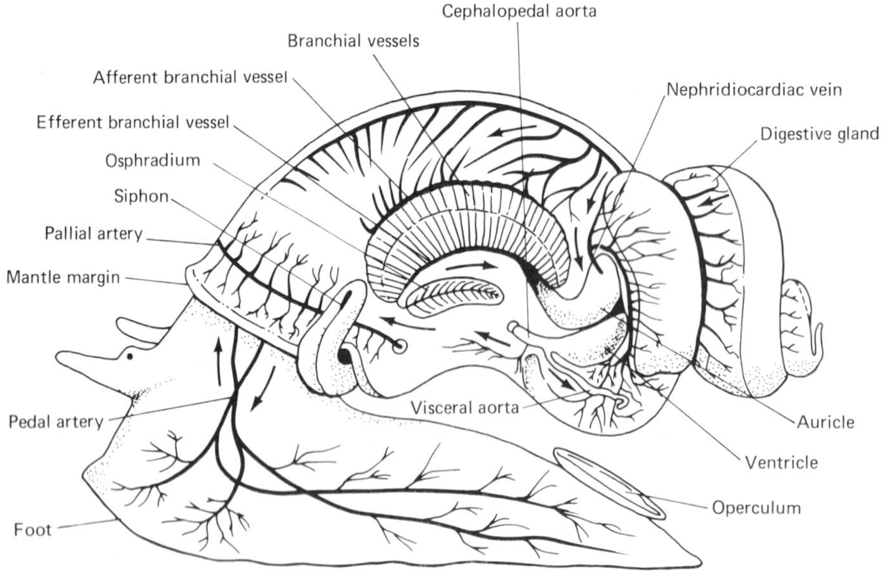

Figure 9.22 The circulatory system of *Buccinum undatum*. Arrows show the flow of blood. (From Moore, Ed., *Treatise on Invertebrate Paleontology, Part I: Mollusca 1*.)

maintained its original course, the head would receive the discharge of the anus and nephridia. Perhaps the earliest adaptation relative to this situation involved a slit or perforation in the shell above the mantle cavity (Fig. 9.23a). The respiratory current exited through this slit and thus waste-laden waters avoided the head entirely. One or more such perforations are seen among limpets, abalones, and other lower gastropods.

Asymmetric growth reduced the right side of the mantle cavity in higher snails and led to the loss of the right-hand gill. With only one gill in the cavity, the mantle current developed a lateral flow (Fig. 9.23b). Water entered from the left side of the snail, passed across the mantle cavity, and exited with digestive and metabolic wastes from the right side.

Finally, terrestrial gastropods lost their gills entirely. The mantle cavity was converted into an air-breathing lung, which has a highly vascularized roof and a very small, adjustable opening called the **pneumostome** (Fig. 9.40a). In land snails, the edges of the mantle cavity are fused to the dorsal surface. This surface forms a muscular floor for the lung. Alternating contractions and relaxations of this floor pump air into and out of the mantle cavity lung as gas exchange occurs.

Excretion and Osmoregulation

Gastropod nephridia comprise highly folded sacs and their associated tubules within the anterior visceral mass. Primitive members of the class possess a pair of these excretory organs, but higher snails show a reduction to a single nephridium. The surviving nephridium is the posttorsional left member, consistent with the surviving left side of the mantle cavity. The nephridial sac communicates with the pericardial cavity either directly or via a short renopericardial canal. Pericardial gland secretions and a blood filtrate pass through the nephridium and are converted into urine by differential secretion and resorption.

Marine gastropods are quite isosmotic with sea water. Accordingly, their osmoregulation is rather slight; moreover, these snails apparently tolerate considerable dilution in brackish waters. Freshwater gastropods maintain a relatively low concentration of body salts, but nonetheless must excrete a copious, hypotonic urine. Finally, land snails conserve their body fluids by excreting uric acid.

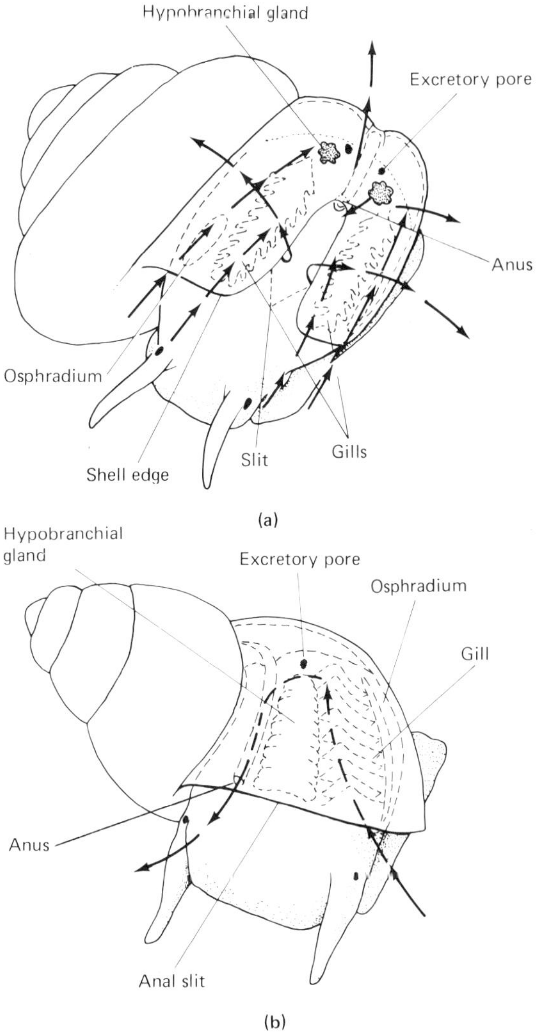

Figure 9.23 Solutions to the problem of fouling resulting from torsion. (*a*) Shell with a slit allowing waste materials to exit from the mantle cavity dorsally and posteriorly; (*b*) Asymmetrical mantle cavity with retention of single left gill; water current enters from the left and exits to the right carrying waste materials. Arrows show direction of water current. (From Solem, *The Shellmakers: Introducing Mollusks*.)

ACTIVITY SYSTEMS

Gastropods are typically molluscan in their activity systems. They possess a generalized molluscan musculature, a nervous system displaying various degrees of centralization, few sense organs, and a proverbially slow behavioral repertoire.

Muscles and Nerves

The foot is the center of muscular activity in gastropods and comprises an array of multidirectional fibers. Retractor muscles of the head, tentacles, radula, odontophore, and shell complete the important muscular elements.

The gastropod nervous system features a central network of ganglia and nerve cords which have undergone torsion. The basic organization of the system prior to torsion is depicted in Figure 9.20a. A pair of cerebral ganglia overlies the esophagus and gives rise to paired fibers innervating the eyes and tentacles; another set of cerebral nerves terminates in a pair of buccal ganglia. The latter mediate the activity of the radula and surrounding structures. The cerebral ganglia also give rise to a pair of ventral cords which connect with pedal ganglia and the nerves of the foot. Laterally, paired cords extend from the cerebral nerve center to a pair of pleural ganglia. The pleural ganglia are associated with the mantle regions; these centers join with the pedal ganglia via pleuropedal connectives. Finally, extending posteriorly from the pleural ganglia is a pair of visceral nerve cords. Each of these cords passes through a parietal nerve center before terminating posteriorly in a visceral ganglion.

Torsion affects the gastropod nervous system along the visceral cords between the pleural and the parietal ganglia (Fig. 9.20b, c, d). The visceral ganglia are twisted forward, and both visceral and parietal ganglial pairs have reversed sides (Fig. 9.20b, c, d). The ganglia that formerly occupied the right side of the gastropod now lie along its torted left side, and vice versa. Additionally, the ganglia on the posttorsional left side are slightly more dorsal than their right-side counterparts. Thus they may be described as the supraparietal and the supravisceral ganglia, respectively.

While primitive gastropods display a torted, figure-eight nervous system, a bilateral arrangement has been restored in most higher snails. This restoration involves a concentration of the several ganglionic pairs and/or a general untwisting of the nerve cords themselves (Fig. 9.24). Concentration of the ganglia results from a shortening of the twisted nerve cords. Commonly, the pedal and pleural ganglia join the cerebral ganglia in a circumenteric nervous assemblage representing the brain. In some gastropods, the visceral ganglia also migrate anteriorly.

Sense Organs

Gastropod sense organs include tentacles, eyes, statocysts, and osphradia. Tentacles commonly are associated with the head, although lateral projections of the foot and mantle may perform similar functions. These organs are primarily tactile and chemoreceptive, but cephalic tentacles may bear the eyes. Gastropod eyes vary from simple, pigment cup ocelli to elaborate visual centers complete with lenses and corneas. Statocysts occur in the foot; osphradia, at the entrance to the mantle cavity. All of these sensory structures are essentially those described for the generalized mollusk. They provide for the rather limited surveillance needs of snails and reflect the generally slow behavior of the class.

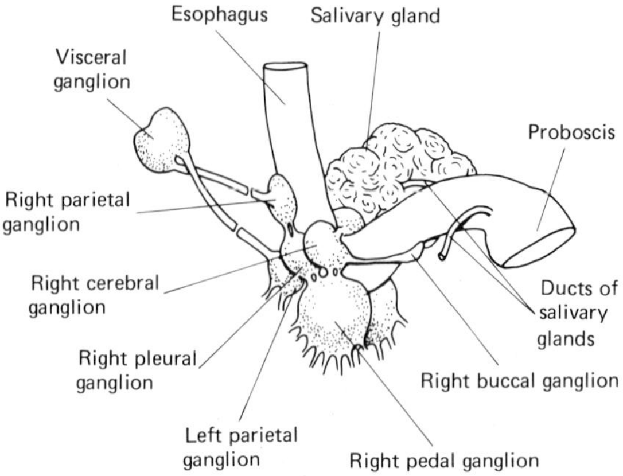

Figure 9.24 Major ganglia in the nervous system of *Busycon*, an advanced prosobranch gastropod. (From Brown, *Selected Invertebrate Types*.)

Behavior

Gastropods are proverbially slow animals. They crawl about, eat, move their tentacles, and withdraw into their shells when threatened. Nonetheless, many exhibit complex defense behavior, territoriality, and homing responses. A few swimming snails demonstrate impressive locomotor abilities. These animals, described under the appropriate subclasses below, possess the most effective gastropod sense organs, and their usually shell-less bodies are streamlined for graceful progress through the water.

CONTINUITY SYSTEMS

Gastropods display a variety of reproductive structures and developmental schemes. Among primitive members of the class, gonads are simple, fertilization is external, and developmental stages include trochophore and veliger larvae. In more advanced gastropods, the gonads are accompanied by accessory glands and structures for the transfer and storage of sperm. These higher snails copulate, and one or both of the larval stages may be suppressed.

Reproduction

Most gastropods are diecious. A single gonad occupies the visceral spire, commonly nestled alongside the digestive gland (Fig. 9.25). Gastropod gonoducts include numerous tubules with considerable variation in form and function. A short gonoduct commonly connects the gonad with the right nephridium. Recall that the right nephridium is often degenerate and, in most gastropods, functions solely to transport gametes. The nephridium discharges into the mantle cavity, where gametes may be released immediately. In many gastropods, however, the mantle wall itself is specialized for gametic transport. A simple ciliated groove occurs in some snails, but in more advanced species, this mantle groove forms a distal, tubular component of the reproductive tract. The mantle tubule has undergone major specializations in higher gastropods.

In the male, the mantle tubule often contains a prostate which secretes seminal fluids (Fig. 9.26a). Sperm are produced in the testis and stored in the highly coiled gonoduct. On copulation, the male gametes pass through the nephridial component of the reproductive tract and along the mantle tubule, where seminal fluid is received. The male system terminates at a penis. The gastropod penis represents an extension of the body wall near the right side of the head.

The female reproductive system comprises a single ovary, a short oviduct, a nephridial zone, and a complex mantle tubule (Fig. 9.26b). The latter is modified for yolk production, egg capsule formation, and sperm reception and storage. Sperm are stored at the nephridial end of the mantle duct. After fertilization occurs at this end, eggs pass into a glandular region of the mantle duct. Albuminal glands add proteinaceous yolk to the newly fertilized eggs before jelly or capsule glands secrete their protective envelopes.

Development

Gastropod zygotes are released in gelatinous strips or masses or in harder capsules which are attached to the

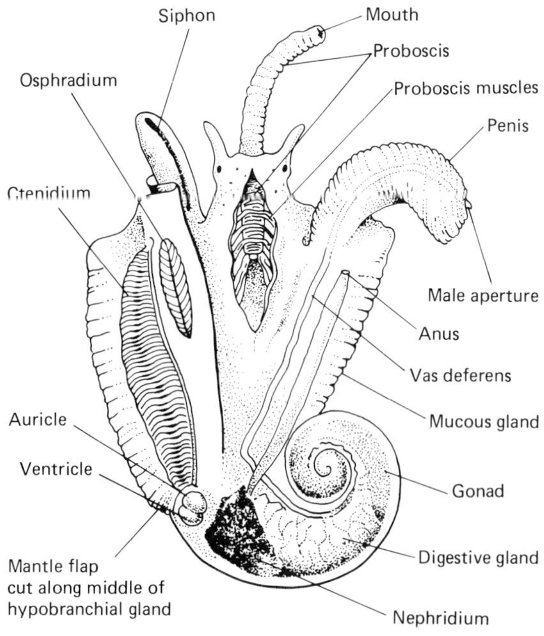

Figure 9.25 Male of *Buccinum undatum* removed from shell. Organs of the mantle cavity have been exposed by making an incision in the mantle dorsally and folding mantle flaps to the side. The proximal end of the proboscis and its muscles have been exposed by an additional incision. (From Moore, Ed., *Treatise on Invertebrate Paleontology, Part I: Mollusca 1*.)

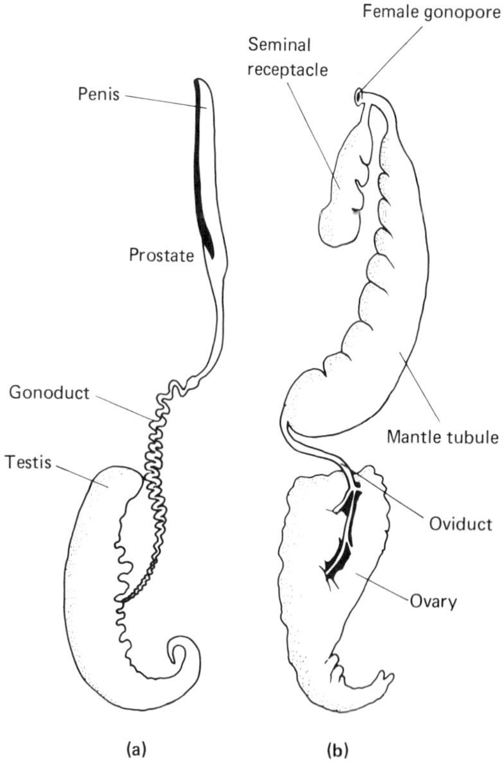

Figure 9.26 (*a*) Male reproductive system of *Charonia*, an prosobranch gastropod; (*b*) Female reproductive system of *Charonia*. (From Hyman, *The Invertebrates, Vol. VI*.)

substratum (Fig. 9.27). Development into the trochophore is rapid, but in most gastropods, this first larval stage is passed within the protective envelope (Fig. 9.28). The shell appears during the late trochophore period, but most adult features do not become apparent until a late veliger stage (Fig. 9.6b). Torsion occurs at the veliger stage, which is free-swimming in many marine species (Fig. 9.6b). The muscles on the right-hand side develop earlier than those on the left; such asymmetrical growth twists the foot and viscera, often in a matter of minutes. Also during the veliger stage, the gastropod shell begins to assume its characteristic, spirally coiled form.

As development proceeds, the larval velum degenerates and the foot becomes capable of movement. The young snail settles to the substratum. Gastropod settling behavior is by no means a random process. Selection, probably involving chemical clues from potential habitats, is important, and completion of adult development may be retarded until a suitable home is found.

Most freshwater and all land snails have no free-swimming larvae. Because of the rigors of their environments, both trochophore and veliger stages are passed in the egg case, from which a small, completely formed snail emerges.

Hermaphroditism is widespread in this class and commonly assumes one of two general forms. A snail may be sequentially hermaphroditic (protandric), meaning that a single individual passes through a male and then a female period. This transition involves the degeneration of the original male reproductive tract and its replacement by a female system. Such transitions maintain a numerical balance of male and female members in a local population. In the second type of hermaphroditism, dual sexual systems are present in the same animal at the same time. Such simultaneous hermaphroditism features a single ovotestis and a complex reproductive tract which divides into separate channels for the independent processing of eggs and sperm.

Gastropod Diversity

The gastropods are divided into three subclasses. The largest of these, the Prosobranchia, includes the more primitive snails. The Opisthobranchia are a marine group characterized by reduced shells. The third subclass, the Pulmonata, is composed almost exclusively of freshwater and terrestrial snails and slugs whose mantle cavity forms a lung.

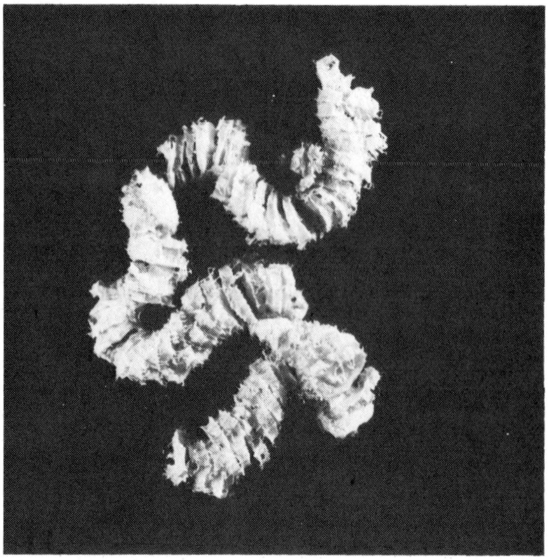

Figure 9.27 A rope of horny egg capsules produced by the female *Busycon spiratum.* (From Andrews, *Sea Shells of the Texas Coast.*)

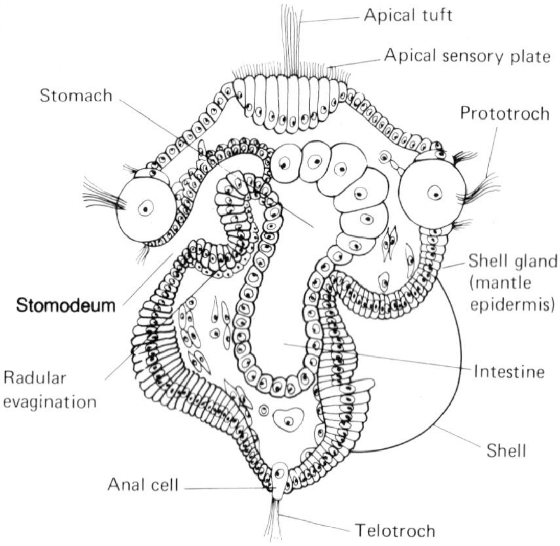

Figure 9.28 A sagittal section through the trochophore larva of *Patella,* a primitive prosobranch. (From Hyman, *The Invertebrates, Vol. VI.*)

SUBCLASS PROSOBRANCHIA

The Prosobranchia are the largest and oldest of the extant gastropod subclasses. Over 30,000 species are represented. Most prosobranchs are marine, but this subclass has members in both freshwater and terrestrial environments. The group is distinguished by an anterior mantle cavity containing one or two gills. A calcareous shield, called an operculum, generally is present on the dorsal, posterior surface of the foot (Fig. 9.29g); this operculum seals the opening of the shell when the foot is retracted. Sexes are usually separate.

Diversity within the Prosobranchia involves adaptive radiation in mantle cavity ventilation mechanisms, feeding styles, locomotor techniques, and habitats. Three large orders are recognized: the Archaeogastropoda, the Mesogastropoda, and the Neogastropoda.

Order Archaeogastropoda

Most archaeogastropods live in clean water on rocky substrata. As their name implies, these snails are an old and relatively primitive group, the majority of which is marine. Both freshwater and terrestrial species are known. They probably resemble the original gastropods, just after asymmetrical coiling and torsion took place. In many archaeogastropods, the mantle cavity and related systems are bilateral. Two bipectinate gills hang from either side of the mantle cavity, and paired auricles and nephridia are usually present (Fig. 9.23a).

Archaeogastropod adaptations to the potential sanitation problems caused by torsion include the most primitive solutions in the class. In some forms, such as *Scissurella* (Fig. 9.29a), the shell is notched above the head (Fig. 9.23a); through this notch, mantle cavity currents discharge digestive and excretory wastes. Other archaeogastropod shells possess a series of holes through which wastes are expelled, as in *Haliotis* (Fig. 9.30a), the common abalone of Pacific coasts. *Fissurella* (Fig. 9.29b), a keyhole limpet, possesses just one hole on its shell. Begun as a notch, this aperture gradually is enclosed and, as a result of differential growth, comes to occupy the apex of the conical shell. In another group of archaeogastropod limpets, the gills are very much reduced. *Acmaea* possesses only a left-side gill. In this genus, water enters anteriolaterally on the left side, passes over the gill, and then flows down the right side of the animal (Fig. 9.29c); water also passes along the left side of the snail without contacting the gill. *Acmaea*'s ventilation system is thus very chitonlike and represents a parallel adaptation for a similar life-style. Other archaeogastropod limpets with this type of current include *Lepeta,* which has no gills, and *Patella,* in which secondary gills form from folds in the mantle wall.

Other archaeogastropods display a single gill and an oblique water current which passes from left to right across the anterior mantle cavity (Fig. 9.23b). *Trochus,* a top snail, is an example. This ventilating system apparently is quite efficient, as it is characteristic of the higher prosobranch orders.

Limpets and abalones are broad, flat archaeogastropods. Their form reflects the relative enormity of the last shell whorl. These prosobranchs cling tightly to hard substrata, where their low postures resist dislodgment by water pressure or would-be predators. Like the chitons, limpets may have domiciles on their rock territories and may exhibit homing behavior. The edge of the limpet shell appears to be an important agent in homesite recognition. Physical abuse of the shell perimeter can completely destroy homing ability in these snails.

Despite the overall primitiveness of the order, certain tropical and subtropical archaeogastropods have colonized both freshwater and terrestrial environments. *Theodoxus* is a common freshwater genus; its right gill has been lost and its mantle current is oblique. *Helicina* is a land snail common in the southeastern United States. The mantle cavity of this archaeogastropod has no gills; rather, it is vascularized and serves as a lung. As a prosobranch, *Helicina* is readily distinguished from the numerous land snails of the Pulmonata by its operculum.

Archaeogastropods are mostly microphagous feeders, scraping algae and other food off rocks in the basic molluscan style. Their digestive tract features a radula with many teeth in each transverse row and a stomach complete with ciliary sorting fields, a style, and paired digestive glands. Reproduction is quite often a simple affair. A penis is usually lacking in these snails, as external fertilization is common. Gametes are discharged through the right nephridium. A free-swimming trochophore larva makes its only gastropod appearance among primitive members of this order.

Order Mesogastropoda

The mesogastropods are the largest prosobranch order and include such familiar snails as conchs and periwinkles. Like the archaeogastropods, they are a predominantly marine group, but exhibit a limited expansion into freshwater and terrestrial environments. Mesogastropods are quite asymmetrical, having lost the right member of several paired organs. Only the left auricle and nephridium remain, and the single gill on the left side of the mantle cavity is unipectinate. This gill is not suspended by membranes; rather, its central axis is attached directly to the dorsal mantle wall. The absence of a suspensory membrane minimizes clogging of the gill and thus allows mesogastropods to exploit muddier habitats.

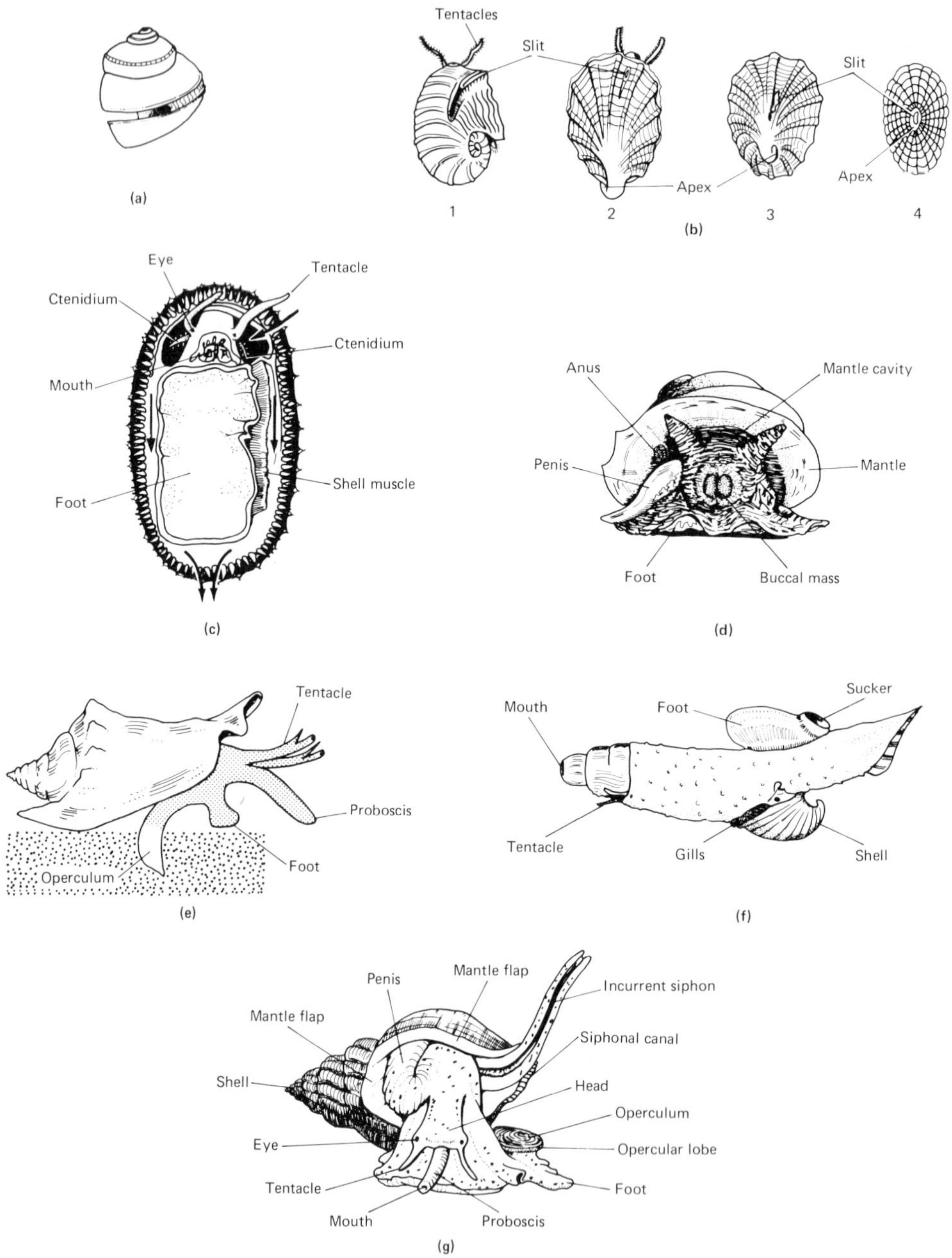

Figure 9.29 Representative prosobranch gastropods (not drawn to scale). (*a*) *Scissurella crispata*, an archeogastropod; (*b*) Four stages in the ontogeny of a keyhole limpet, *Fissurella*, showing the transformation of the shell from coiled to conical and the shift of the notch from the perimeter to the apex; (*c*) Ventral view of *Acmaea virginea* with arrows showing the direction of water flow; (*d*) Anterior view of a male *Littorina littorea*, a mesogastropod, removed from the shell; (*e*) Diagram of a conch, *Strombus*, using its operculum as a claw in locomotion over a sandy substratum; (*f*) An actively swimming pelagic mesogastropod, *Carinaria lamarcki*, in its typical upside-down position; (*g*) *Buccinum undatum*, a neogastropod exhibiting a well developed siphon protected within an extension of the shell, the siphonal canal. (*a* and *e* from Solem, *The Shellmakers: Introducing Mollusks*; *b* from Hyman, *The Invertebrates, Vol. VI*; *c* and *d* from V. Fretter and A. Graham, *British Prosobranch Molluscs*, Ray Society, 1962; *f* from R.T. Abbott, American Seashells, 1954, Van Nostrand *g* from Moore, Ed., *Treatise on Invertebrate Paleontology*, *Part I: Mollusca 1*.)

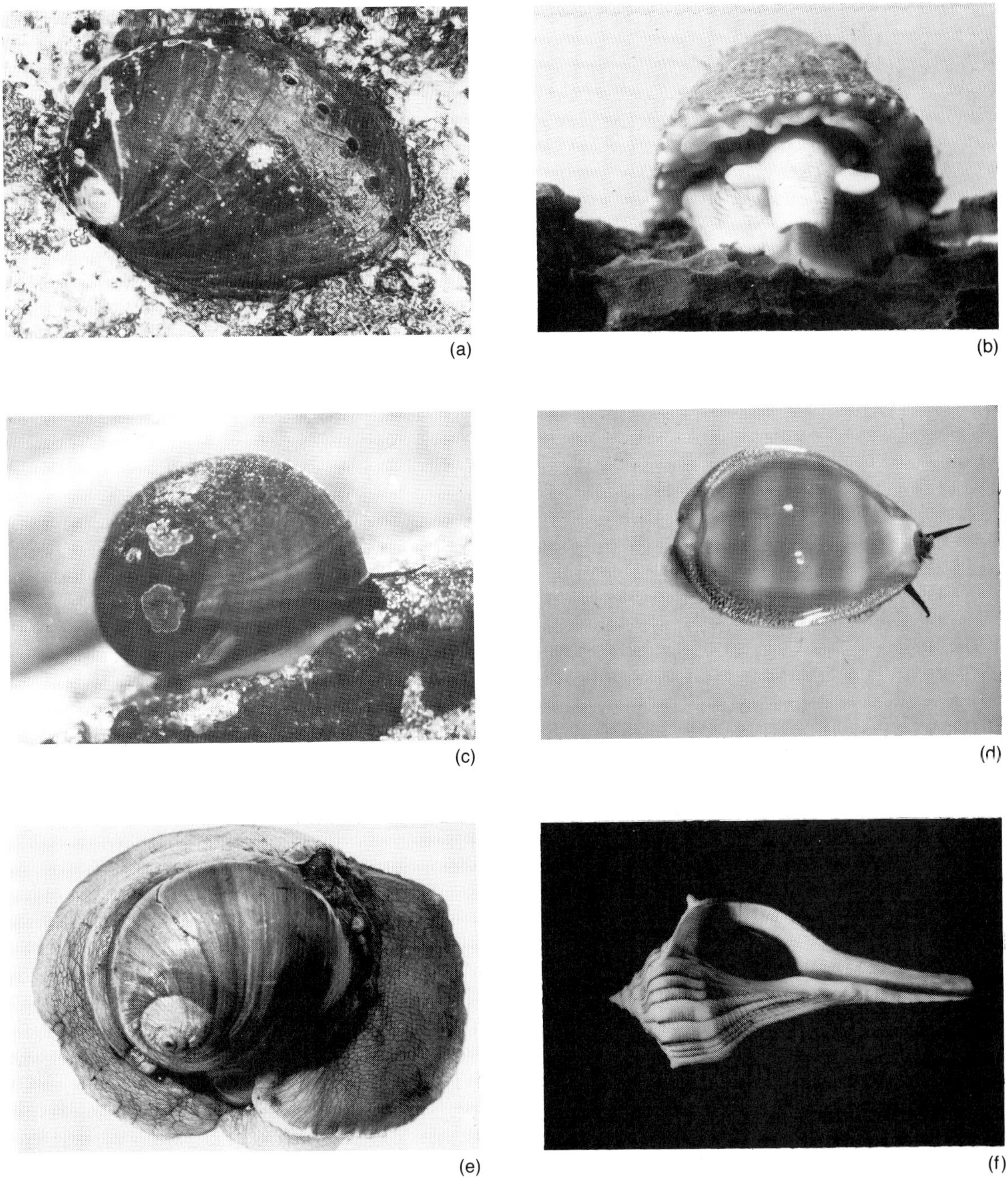

Figure 9.30 Representative prosobranch gastropods. (*a*) The black abalone, *Haliotis cracherodii*, an archeogastropod with a series of holes through which wastes are eliminated from the mantle cavity; (*b*) *Diodora aspera*, a key hole limpet; (*c*) *Norrisia norrisii*, a more advanced archeogastropod; (*d*) A cowrie, *Cypraea*, a mesogastropod browser on sessile colonial invertebrates; (*e*) The giant moon snail, *Polinices lewisi*, using its expanded foot at a plow in moving through soft sandy substrata; (*f*) *Busycon contrarium*, a common large whelk of the South Atlantic and Gulf coasts of the U.S. (*a* and *e* from Ricketts, et al., *Between Pacific Tides*; *b* and *c* from Ward's Natural Science Establishment; *d* and *f* from Andrews, *Sea Shells of the Texas Coast*.)

Mantle cavity organization is fairly uniform throughout the group. Diversity within the Mesogastropoda focuses on feeding styles, locomotor techniques, and habitats.

Although feeding styles are extremely diverse within this order, space permits the description of only a few contrasting types. Marine mesogastropods include algal raspers and cutters, deposit feeders, filter feeders, grazers on sessile animal colonies, and benthic and planktonic predators. *Littorina* (Fig. 9.29d), the common intertidal periwinkle, scrapes algae off the rocks on which it lives. Other marine herbivores, like the conch *Strombus,* may fell and eat larger algal plants. *Strombus* (Fig. 9.29e) also feeds on bottom deposits; its entire buccal region is mounted within a proboscis which can be protruded to pick up food. Such proboscises are most characteristic of carnivorous mesogastropods and neogastropods. Filter feeding in the present order has been studied in the genus *Crepidula.* The mantle cavity of this snail is expanded dorsally to house an enlarged gill. This gill is a food-gathering organ which filters edible particles from mantle waters, wraps them in mucus, and transports them along a ciliated gutter to the mouth. There the radula grasps and ingests them. *Crepidula* seems to be a fairly recent snail; its mode of feeding, although unusual for a gastropod, is well developed among the Bivalvia, and ecological competition with the latter perhaps inhibits gastropod experiments with ctenidial feeding.

Browsing on animal colonies is common among several mesogastropod types, including *Cerithiopsis,* which feeds on sponges, and the cowrie *Cypraea* (Fig. 9.30d), which eats ascidians. Benthic carnivores are represented in this order by *Natica* and *Polinices* (Fig. 9.30e), two genera that bore into other mollusks (commonly bivalves) and extract their tissues. The heteropods are planktonic predacious mesogastropods. As described subsequently, they are modified for fast, observant swimming.

Freshwater mesogastropods (e.g., *Viviparus*) are usually herbivores, as are all terrestrial members of the order. Like the archaeogastropods in such environments, these snails are readily distinguished from their pulmonate neighbors by their operculum.

The Mesogastropoda includes several genera endoparasitic on echinoderms. *Stylifer* lives within the dermis of sea urchins; *Entoconcha,* within the coelom of sea cucumbers. Finally, few parasites in the invertebrate world are more specialized than those of the mesogastropod genus *Enteroxenos.* These snails, which live within sea cucumbers, have no gut and simply absorb nutrients across their body wall, as cestodes do. Their molluscan identity is recognizable by their veliger larva.

Mesogastropod radulae bear seven teeth in each transverse row, and jaws are common in carnivorous forms. The stomach is relatively simple, and digestion is primarily extracellular.

Mesogastropods demonstrate several locomotor techniques. The conch *Strombus,* for example, uses its operculum as a claw (Fig. 9.29e), driving it into the sandy substratum and then contracting the foot to pull the body forward. *Polinices,* the benthic carnivore, uses its foot as a plow. Anteriorly, the foot contains many small pores through which water enters the pedal tissues. Control over water uptake and release enables the foot to expand and contract rapidly as it plows through the sand. Finally, the heteropods (e.g., *Atlanta* and *Carinaria;* Fig. 9.29f) are actively swimming mesogastropods. These snails are compressed laterally, and their foot is modified as a fin. Heteropods possess the most highly developed gastropod eyes, and their visual acuity is comparable to that of fish.

Mesogastropods display more complex sexual systems than do archaeogastropods. A penis is usually present, as are female glands for the production of yolk and gelatinous envelopes. The veliger is free-swimming in most species.

Order Neogastropoda

The third and last prosobranch order, the neogastropods, is virtually all marine carnivores. Neogastropods resemble mesogastropods in many ways; both orders have a single unipectinate gill and only one auricle and nephridium. The neogastropods are distinguished by a mantle siphon, a radula with three or less teeth in each transverse row, and some highly aggressive feeding habits. Neogastropods always have a penis, and their eggs are deposited in capsules. Typically, there are no free-swimming larval stages in this order. Common neogastropods include the whelks, oyster drills, and cone shells.

The neogastropod siphon represents a folded extension of the body wall. This maneuverable entrance to the mantle cavity controls incurrent respiratory flow. Commonly, the shell provides a canal in which the siphon lies, as in *Busycon* (Fig. 9.30f) and *Buccinum* (Fig. 9.29g).

The neogastropod digestive tract is specialized for a carnivorous diet. The radula has fewer teeth in each row, but each tooth is quite strong and frequently multicuspid (Fig. 9.31). A very long proboscis allows the raptorial extension of the buccal mass (Fig. 9.29g). The stomach, a rather simple sac, is the site of extracellular digestion. The ancestral chitinoid shields, ciliary sorting field, and style sac have all disappeared.

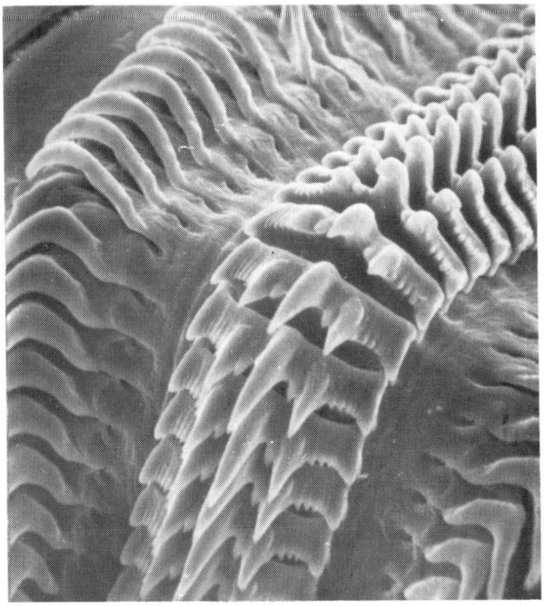

Figure 9.31 A scanning electron micrograph of a portion of the radula of an oyster drill, *Urosalpinx cinerea*. (From M.R. Carriker, Am. Zool., 1969, 9:197.)

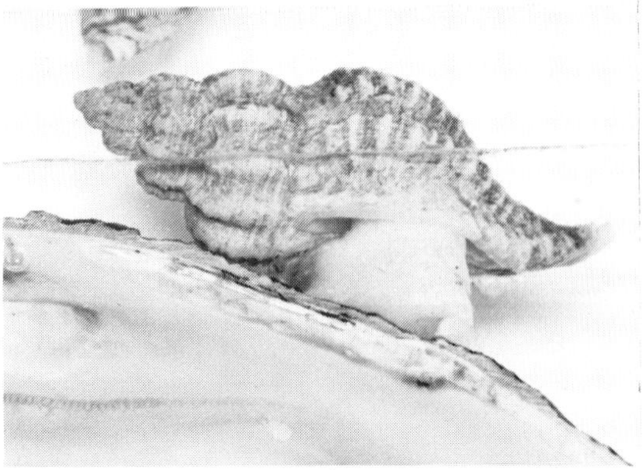

(a)

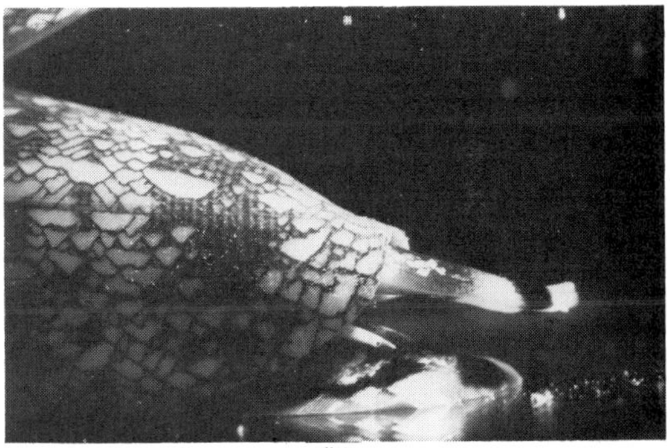

(b)

Figure 9.32 (*a*) The oyster drill, *Urosalpinx cinerea*, a neogastropod; (*b*) A tropical cone shell, *Conus*. (*a* and *b* courtesy of M.R. Carriker.)

Neogastropods inhabit marine bottoms, where polychaetes, other mollusks, and echinoderms are their common prey. The whelk *Buccinum* is a scavenger which tracks decaying flesh over considerable distances. Such ability reflects the highly developed neurosensory system in this order. Many whelks, such as *Melongena, Fasciolaria,* and *Busycon*, prey on living bivalves. *Melongena* quickly slips its proboscis inside a gaping pelecypod shell and extracts soft tissues. Commonly, several individuals feed simultaneously on the same animal. Other whelks use their own shells to abrade the valve margins of prey; the whelk shell and foot then wedge the bivalve open.

Many neogastropods drill holes in their molluscan prey. *Urosalpinx* (Fig. 9.32a), commonly known as the oyster drill, is one such animal. Applying its radula to an oyster shell, *Urosalpinx* grinds at a rate of up to 60 radular strokes per minute. After a while, the snail slides forward over the drilling spot and applies a special glandular area of its foot. This area secretes a softening fluid—probably an acid which dissolves calcium salts or a specific calcium-reducing enzyme. The softening phase lasts a half-hour or so, and then radular drilling resumes. These two phases alternate until *Urosalpinx* penetrates its victim and extracts the latter's soft tissues. The entire process is quite slow, and full penetration usually requires several hours.

The 400 species of the tropical genus *Conus* (Fig. 9.32b) are highly specialized carnivores. In these neogastropods, radular teeth do not form rows on an odontophore; rather, they are secreted as single, arrow-shaped weapons and are stored in a special "quiver" at the base of the proboscis (Fig. 9.33). Radular arrows are released singly from the quiver and are mounted at the tip of the proboscis sac. A long, coiled poison duct discharges into this sac. At the opposite end of this duct, a large contractile bulb injects poison into the hollow tooth and shoots the toxic arrow at a chosen target. Polychaetes, other snails, and even small fish are frequent prey of the cone shells. Certain species, including *Conus geographus,* produce strong neurotoxins which can be lethal to humans.

Before leaving the neogastropods, we mention two aberrant genera. *Columbella,* one of the few herbivores in the order, feeds on brown algae; and *Antemone,* one of the rare freshwater forms, eats snails and small fish.

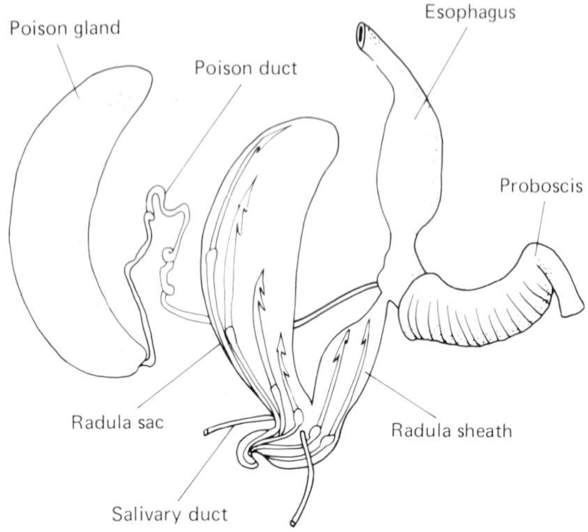

Figure 9.33 Foregut and radular apparatus of the cone shell *Conus striatus*. (For explanation, see text.) (From W.J. Clench and Y. Kondo, Am. J. of Trop. Med., 1943, *23*:105.)

SUBCLASS OPISTHOBRANCHIA

The opisthobranchs are a small but diverse group of snails which have diverged quite dramatically from their prosobranch forebears. This group is characterized by reductions in its shell, mantle cavity, and gills. Such reductions are accompanied by detorsion and the assumption of secondary, bilateral symmetry. Most opisthobranchs are carnivores, and all are hermaphroditic. The subclass contains 1100 species, virtually all of which are marine. Familiar members include the bubble shells, the sea hares, the nudibranchs, and the sea butterflies (pteropods). These very specialized gastropods burrow in soft substrata or crawl among seaweed and encrusted rocks near the shore; some are pelagic.

Up to 12 orders have been described in the subclass. In the following account, however, we focus attention only on three generalized opisthobranch types: (1) the primitive, burrowing opisthobranchs; (2) the sluglike, crawling forms; and (3) the swimmers.

Probably derived from burrowing prosobranchs, the early opisthobranchs included forms closely related to the modern bubble shells. The bubble shells constitute the largest opisthobranch group and initiate a number of adaptive trends seen throughout the subclass. Primitive bubble shells, such as *Acteon* and *Hydatina* (Fig. 9.34a), possess substantial shells, and the former even has an operculum. However, other members of the group display shell reductions. *Haminea* (Fig. 9.34b) has only a thin shell, while the shell of *Philine* is transparent and enclosed by the mantle; finally, in *Runcina* (Fig. 9.34c), the shell has completely disappeared.

Among the bubble shells, the head has expanded to form a broad shield (Fig. 9.34a). It bears two large, chemosensory tentacles called **rhinophores.** As head size increased during the evolution of these mollusks, the mantle cavity retreated along the right side of the body. This migration revoked the torted condition of these snails, restoring a varying level of secondary bilateral symmetry. The opisthobranch nervous system is untwisted, and the auricle lies anterior to the single ctenidium. During its posterior migration, the mantle cavity gradually decreased in size and, indeed, disappeared from higher opisthobranchs.

Reduction of the shell, detorsion, and loss of the mantle cavity involve dramatic changes in opisthobranch biology. Ctenidia are absent, and respiratory responsibilities are transferred to the naked body surface. A general streamlining is evident, as the body assumes a lower, more elongate profile. These changes produce the two other opisthobranch types, the sluglike crawlers (including the sea hares and nudibranchs) and the pelagic swimmers (the sea butterflies).

The sea hares are a primitive group of crawling opisthobranchs. Their considerable length—40 cm in some species—ranks them as the largest members of the subclass. *Aplysia* (Fig. 9.35) is a typical sea hare. It maintains a small mantle cavity and ctenidium along its right side, and a shell remnant lies within the dorsal body wall. Pedal extensions, called parapodia, may join dorsally to enclose the animal. Some species use these parapodia in swimming movements. *A. punctater* flaps them abruptly to lift itself from the substratum; *A. saltater* contracts its parapodia against a dorsal volume of water and thus swims by jet propulsion. *Aplysia,* unlike most opisthobranchs, is a herbivore. It consumes long blades of algae, holding the plant with its jaws while tearing away with its radula.

A related group of crawling opisthobranchs is typified by *Pleurobranchus* (Fig. 9.36). This flat, sluglike gastropod has no mantle cavity, although a gill remains beneath a short flap on the creature's right side. A small internal shell is present, but there are no parapodia.

The shell, mantle cavity, and ctenidia have disappeared completely in the nudibranchs. Their naked dorsal surface has undergone an extensive evolution involving the assumption of sensory, respiratory, and defensive responsibilities. *Tritonia* (Fig. 9.37a), a primitive nudibranch, bears numerous feathery exten-

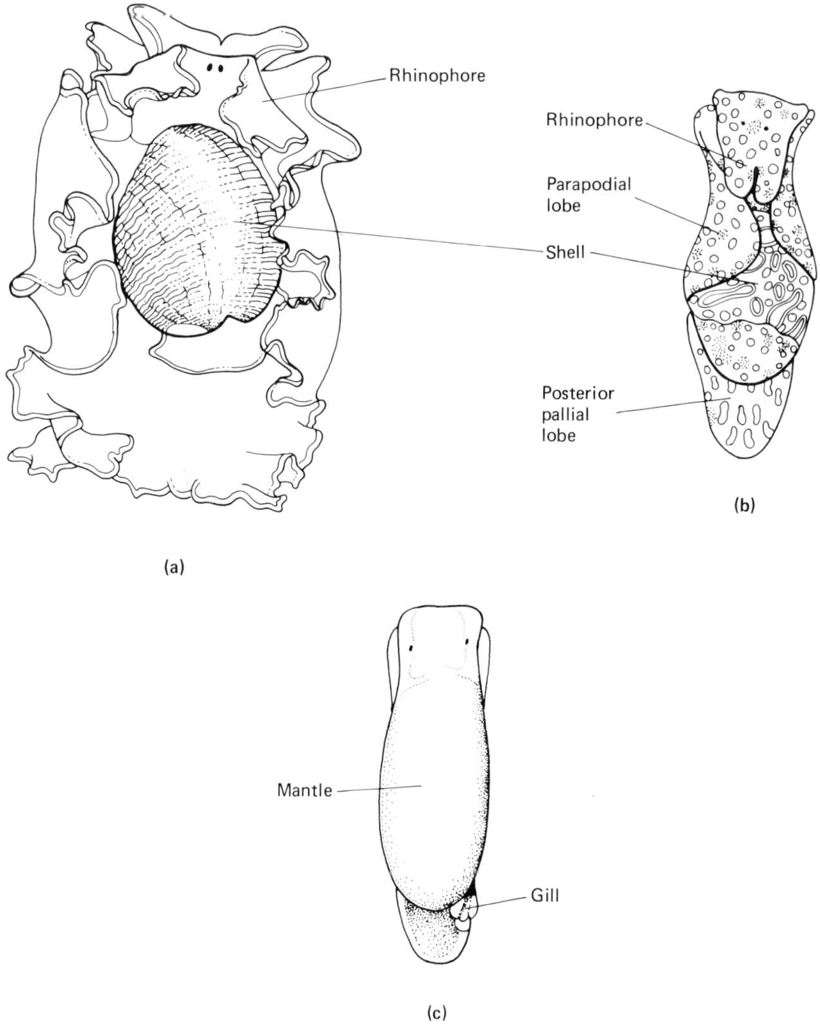

Figure 9.34 Shell reduction in bubble shells. (*a*) *Hydatina* with a well-developed shell; (*b*) *Haminea cymbalum* with a small thin shell; (*c*) *Runcina coronata* lacking a shell. (From T.E. Thompson, *Biology of Opisthobranch Molluscs*, 1976, Ray Society Publishers.)

sions along its back; in the absence of ctenidia, these structures provide ample surfaces for respiratory exchange. The rhinophores are well developed, and their chemosensory activity compensates for the lost osphradium. In many nudibranchs, including *Dirona* (Fig. 9.37b) and *Flabellinopsis (Fig. 9.37c)*, numerous club shaped or branched appendages, called **cerata**, occupy the dorsal surface. The nudibranch hemocoel extends into these appendages, as does the diffuse digestive gland. Those nudibranchs feeding on coelenterates carefully transport undischarged nematocysts to the distal ends of the cerata. There these thread weapons can be fired in defense of the opisthobranch!

Their combinations of bright red, orange, yellow, green, and blue hues rank the nudibranchs among the most colorful creatures in the invertebrate world. Such coloring frequently camouflages these gastropods as they crawl on their prey. Sessile animal colonies, including coelenterates, sponges, barnacles and bryozoans, as well as fish eggs, are favorite food items. The pink *Tritonia plebeia* blends well with the color of the

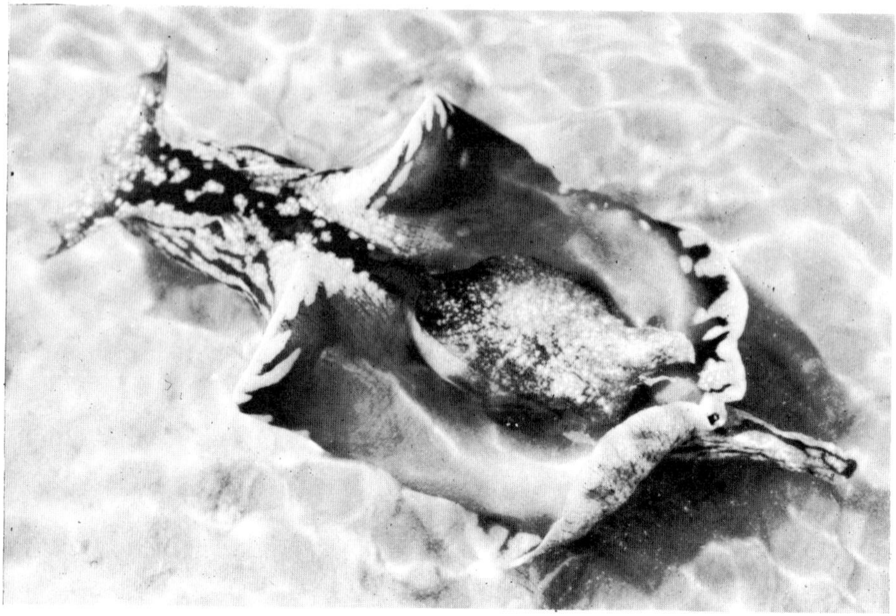

Figure 9.35 Dorsal view of a swimming sea hare, *Aplysia willcoxi*. (From Andrews, *Sea Shells of the Texas Coast*.)

soft corals on which it feeds; the cerata of *Calma glaucoides* match the silver-grays of its fish egg food; and *Rostanga rufescens*' bright red body camouflages it against its common background, a red sponge. Meanwhile, herbivorous nudibranchs favor soft green and brown colors and thus also blend well with their environments.

The best opisthobranch swimmers are the pteropods, or sea butterflies. These mollusks, which include both shelled and naked forms, live among plankton. Shelled pteropods, such as *Limacina* (Fig. 9.38), have long, oarlike parapodia extending anteriorly from the foot. With muscular bases and broad, thin tips, these parapodial oars push the gastropod slowly through the water. *Limacina*, like most shelled pteropods, is a suspension feeder. It traps diatoms, dinoflagellates, and other small plankton on the mucus-coated walls of its mantle cavity. Trapped nutrients are transported along a ciliated groove to the mouth, where they are seized by the radula. Another type of suspension feeding is demonstrated by *Gleba*, a pteropod with a minute shell. *Gleba* fashions a huge external mucous net, from which it hangs by its proboscis. Plankton are caught in this floating trap and conveyed by proboscis cilia to the snail below.

Naked pteropods (e.g., *Clione;* Fig. 9.38b) are the fastest opisthobranch swimmers. These streamlined sea butterflies have no shells, mantle cavities, or gills.

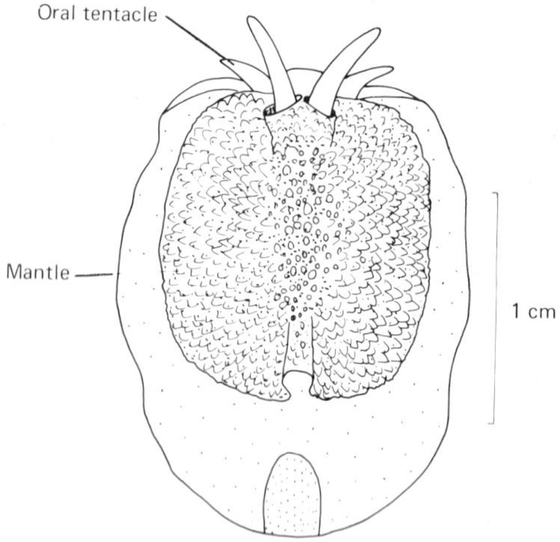

Figure 9.36 *Pleurobranchus membranaceus*. (From Thompson, *Biology of Opisthobranch Molluscs*.)

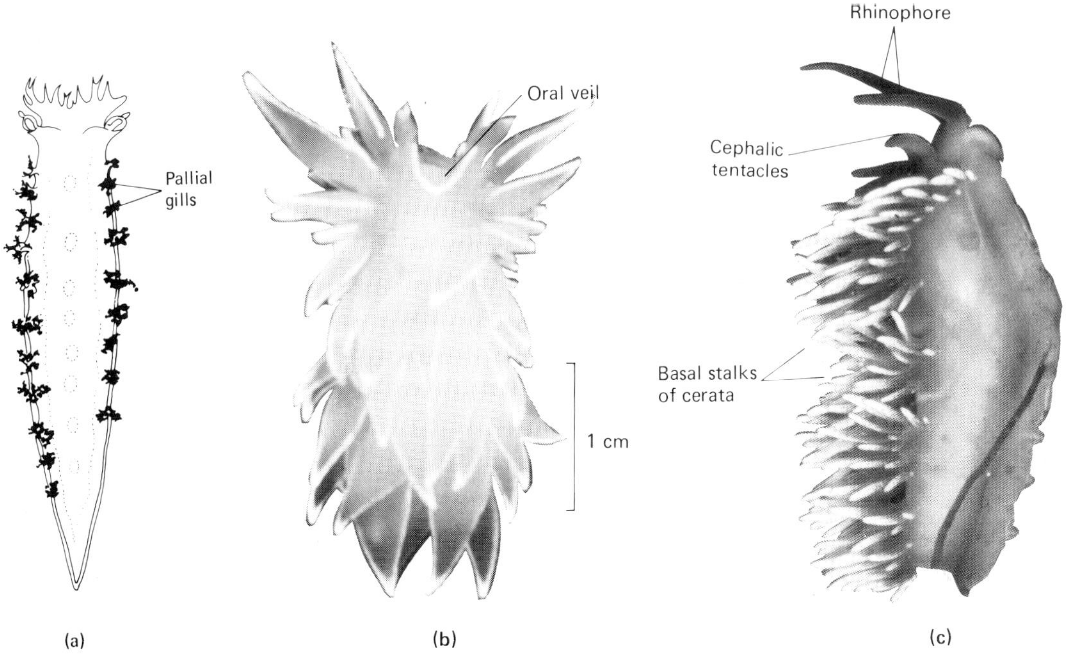

Figure 9.37 Nudibranchs. (*a*) *Tritonia festiva;* (*b*) *Dirona albolineata;* (*c*) *Flabellinopsis iodina.* (*a* from Thompson, *Biology of Opisthobranch Molluscs; b* and *c* courtesy of Ward's Natural Science Establishment.)

Figure 9.38 Pteropods. (*a*) *Limacina helicina;* (*b*) *Clione limacina.* (*a* from Kaestner, *Invertebrate Zoology, Vol. I; b* from Thompson, *Biology of Opisthobranch Molluscs.*)

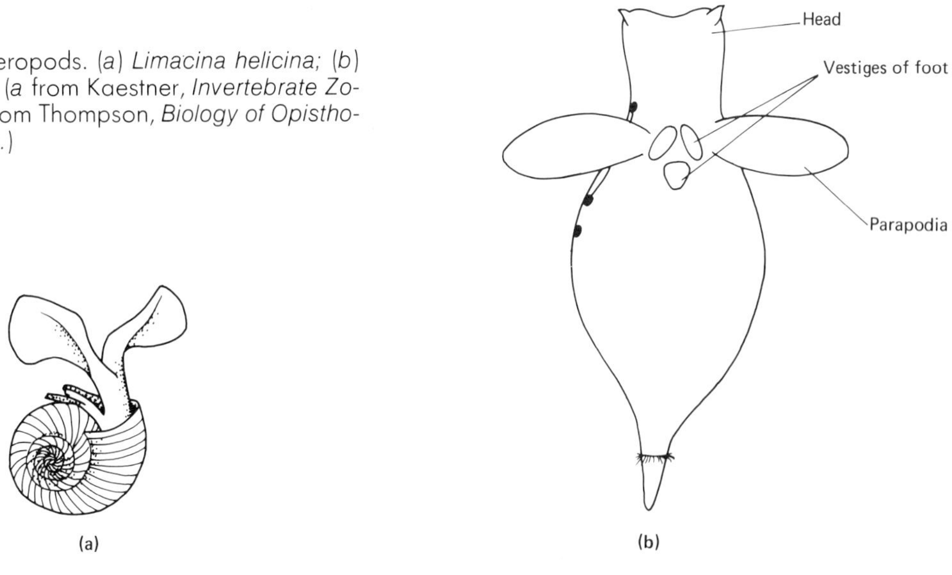

Their parapodia, which are attached ventrally behind the head, produce effective strokes in the dorsoventral plane. Naked sea butterflies are principally carnivorous and often feed on their shelled cousins.

SUBCLASS PULMONATA

The pulmonates represent the third and final gastropod subclass. These snails are adapted primarily for life in freshwater and terrestrial environments. Their mantle cavity forms an air-breathing lung with a restricted opening called the **pneumostome**; there is no gill. The pulmonate shell may be regularly spiraled, or it may be reduced or even absent. With the exception of the marine *Amphibola*, there is never an operculum. Pulmonates are generally herbivorous, and their radulae bear very numerous small teeth in each transverse row. The nervous system is concentrated; all principal ganglia, including those of the viscera, are fused at an anterior brain. Sense organs include eyes, statocysts, and one or two pairs of tentacles. The subclass is entirely hermaphroditic, with complex reproductive systems. Most members lay relatively few, yolky eggs, which complete development within protective gelatinous masses or capsules. A few unusual marine representatives (e.g., *Siphonaria*, a limpet: Fig. 9.39) have a free-swimming veliger, but consistent with their habitat, larval forms are suppressed in most pulmonate species.

Relative to the diverse prosobranchs and opisthobranchs, pulmonates are a rather compact group. Anatomical and gross physiological variations are limited in this subclass, although modifications in shell structure contribute to a diversity of external forms. Traditionally, aquatic pulmonates are considered separately from their terrestrial cousins; indeed, recently some authorities argue that these groups merit two separate subclasses.

Aquatic pulmonates are characterized by a single pair of cephalic tentacles; eyes are located at the tentacle bases. Other distinguishing characters include separate genital openings and egg deposition in gelatinous masses. Only the most primitive pulmonates are marine, and they are confined largely to intertidal habitats along rocky tropical shores. These snails include the limpets *Siphonaria* (Fig. 9.39), *Amphibola*, and *Otina*. The latter resembles an abalone, except for the absence of shell perforations.

The vast majority of aquatic pulmonates live in fresh water. Many, such as *Lymnaea* (Fig. 9.40a), breathe air at the water surface and then submerge for up to an hour at a time. Secondary gills, called **pseudobranchs**, occur near the pneumostome in some species (e.g., *Ferrissia*; Fig. 9.40b). Freshwater pul-

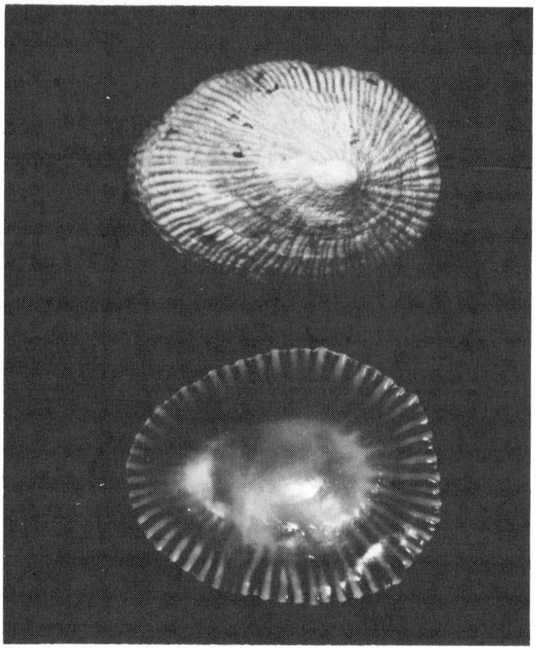

Figure 9.39 The false limpet, *Siphonaria pectinata*, a pulmonate gastropod. (From Andrews, *Sea Shells of the Texas Coast.*)

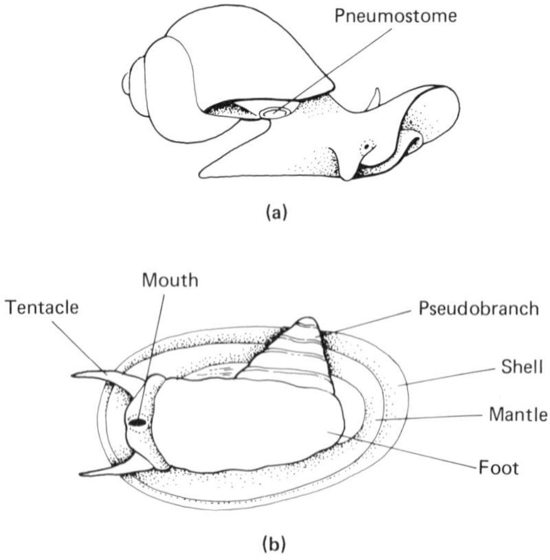

Figure 9.40 (*a*) *Lymnaea*, a typical pulmonate; (*b*) *Ferrissia tarda* showing a pseudobranch. (*a from* Solem, *The Shellmakers: Introducing Mollusks; b from* R.W. Pennak, *Fresh-Water Invertebrates of the United States*, 2nd ed., 1978, Wiley-Interscience.)

monates are important intermediate hosts for many trematode parasites. *Bulinus* and *Physopsis*, for example, harbor the polyembryonic stage of the blood fluke *Schistosoma*.

Land pulmonates far outnumber the aquatic members of the subclass. These snails and slugs have achieved one of the invertebrate world's major colonizations of the terrestrial environment. The proverbial slowness of these creatures favors the geographical isolation of pulmonate populations and thus contributes to the extensive speciation of the group. Both shelled and naked forms are distributed throughout moist temperate and tropical zones.

The land pulmonates possess two pairs of cephalic tentacles; eyes are located atop the second tentacle pair. A common genital opening serves both sexual systems, and eggs are laid in small numbers within calcareous capsules. Shelled pulmonates are typified by the European land snail *Helix pomatia* (Fig. 9.41), which is the edible French escargot. Its turban-shaped spiraled shell is lighter in weight than the shells of some aquatic gastropods. Such lightness may be attributable to a lack of calcium in the diet and also might reflect the increased problem of shell portage on land. The digestive system includes a crop and stomach for extracellular digestion and an absorptive intestine. The pulmonate anus is associated with the pneumostome. The duct from the single nephridium extends along the mantle wall as a ureter and opens near the anus. In the absence of rinse water within the mantle cavity (lung), the ureter allows the snail to void uric acid outside the lung. *Helix* and other land pulmonates must avoid desiccation. Such avoidance may involve confinement to humid environments, nocturnal foraging, and/or estivation. Estivating pulmonates insulate themselves with soil and other debris and then secrete a mucous shield across the shell aperture. This shield, the **epiphragm**, may be fortified with calcium. Considerable tolerance of desiccation is documented in these snails. *Helix* can survive a 50% loss of its fluid weight; the slug *Limax* (Fig. 9.44) tolerates an 80% loss.

Helix, like other pulmonates, is a simultaneous hermaphrodite. Its reproductive structures include a single **ovotestis** and a common duct which bifurcates for the separate processing of sperm and eggs (Fig. 9.42). Mutual copulation often is preceded by extensive foreplay featuring oral, tentacular, and whole body caresses. Curious calcareous darts, produced in a dart sac near the vagina, may be discharged into the partner during courtship.

The land slugs are terrestrial pulmonates whose shells are greatly reduced or absent. In *Testacella* (Fig. 9.43) the shell consists of a small external saddle, but in most slugs it is covered completely by the mantle. These gastropods display very streamlined, nearly bilaterally symmetrical bodies. The pneumostome is often observable along the right side. Reduction of the shell and body streamlining enhance the maneuverability of the group and are associated with the burrowing behaviors and diets of some species. *Limax* (Fig. 9.44), for example, feeds on subterranean roots and tubers, while *Testacella* actively pursues its underground prey of earthworms and other slugs.

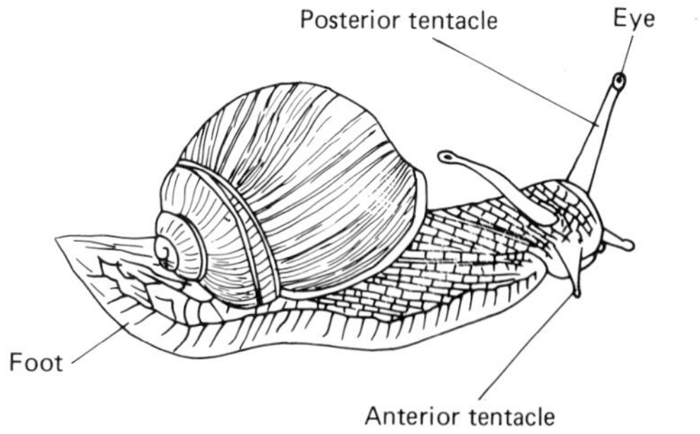

Figure 9.41 The edible French escargot, *Helix pomatia*, a terrestrial pulmonate. (From Hyman, *The Invertebrates, Vol. VI.*)

As in *Helix* and other shelled pulmonates, sexuality among the land slugs includes elaborate foreplay and mutual sperm exchange. In the genus *Limax*, such behavior involves the mutual intertwining of the pair's penes and associated structures (Fig. 9.45). Some of these male organs are longer than the slug body itself, reportedly extending up to 85 cm!

CLASS BIVALVIA

The second largest class in the subphylum Conchifera is the Bivalvia, also known as the Pelecypoda. This group's 28,000 species include the familiar clams, oysters, mussels, and scallops. Bivalves are characterized by a laterally compressed bivalved shell (Fig. 9.46 a,b). Lateral compression has transformed these mollusks into rather extreme demonstrants of the conchiferan plan. The mantle cavity is narrow and deep, and a single pair of very large gills is common. The foot is also compressed, as indicated by the older name Pelecypoda, which means "hatchet foot." The head, having been enclosed completely by the mantle, is lost, and there is no radula. The mantle edge itself assumes sensory responsibilities.

Bivalves are exclusively aquatic, and most taxa are restricted to marine waters. Their diversity is based primarily on feeding style, gill structure, and habitat. Three subclasses are recognized: the Protobranchia,

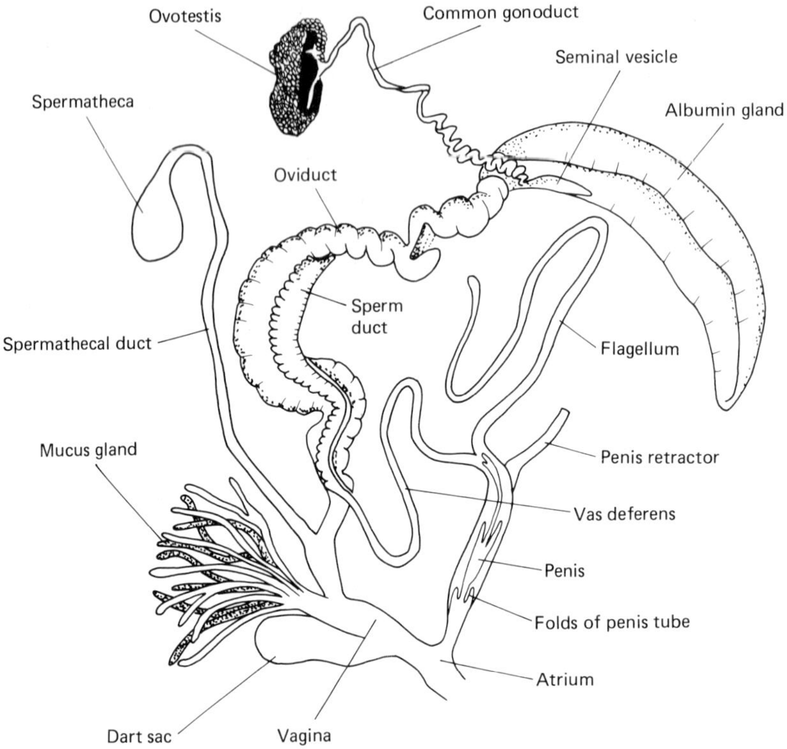

Figure 9.42 The hermaphroditic reproductive system of *Helix pomatia*, dissected out. (From Kaestner, *Invertebrate Zoology*, *Vol. I*.)

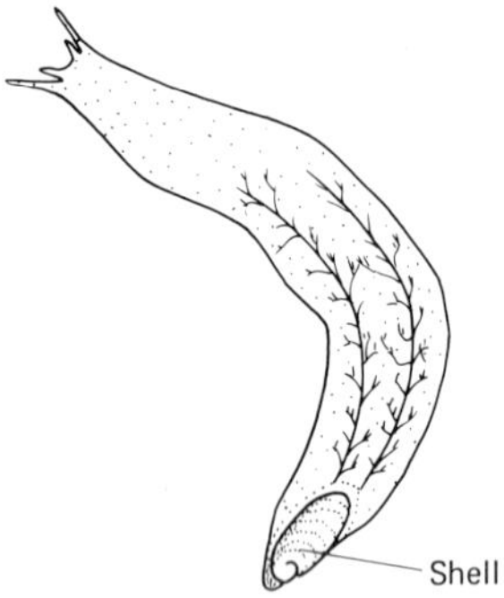

Figure 9.43 The land slug, *Testacella*, with its small external shell. (N.W. Runham and P.J. Hunter, *Terrestrial Slugs*, 1970, Hutchinson.)

the Septibranchia, and the Lamellibranchia. The protobranchs include the most primitive members of the class, characterized by simpler ctenidia and deposit feeding. The septibranchs are highly specialized carnivorous and/or scavenging mollusks with no gills at all. The largest and most familiar bivalve subclass comprises the filter-feeding lamellibranchs. By using their large gills to collect food, these animals have become one of the most successful molluscan groups.

Bivalve Unity

MAINTENANCE SYSTEMS

General Morphology

In no other molluscan class does the shell so dominate external morphology as in the Bivalvia. The hinged, two-part shell completely encloses these animals. In

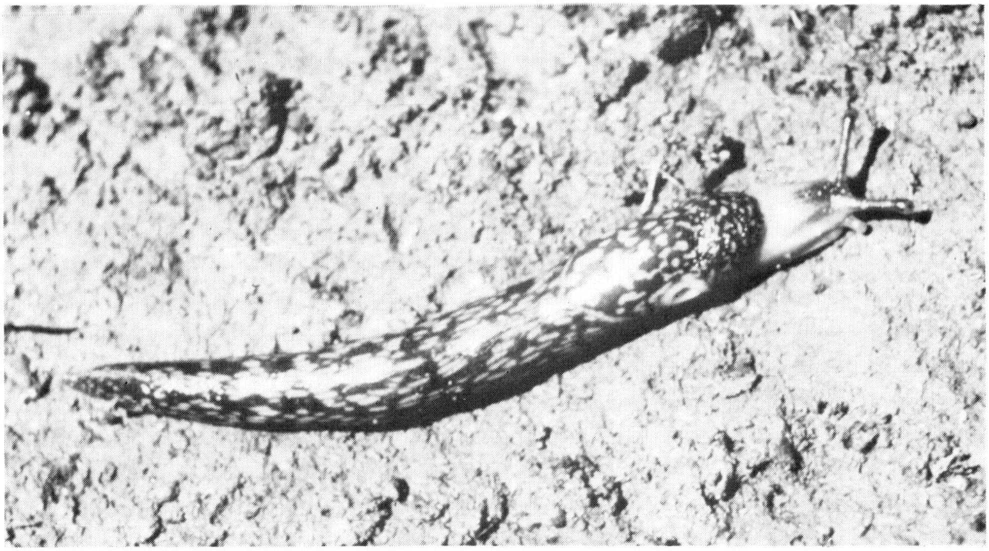

Figure 9.44 *Limax flavus*, a common terrestrial slug. (Carolina Biological Supply.)

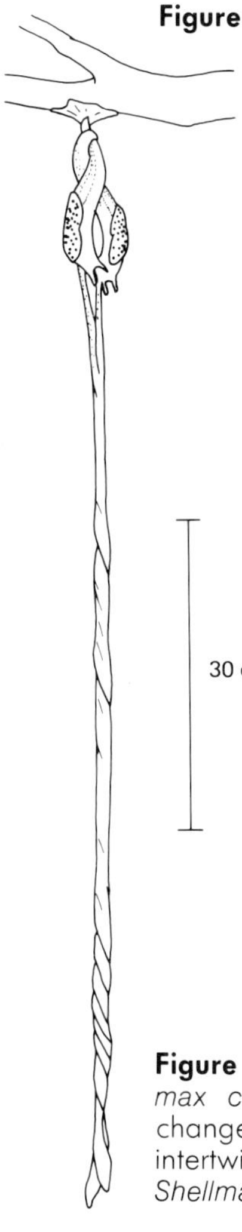

Figure 9.45 Copulating slugs (*Limax cinereoniger*). Sperm are exchanged at the tips of long, intertwined penes. (From Solem, *The Shellmakers: Introducing Mollusks.*)

many bivalves, apertures remain only for incurrent and excurrent water flow and for extension of the bladelike foot. The two valves are connected by a dorsal elastic hinge composed primarily of conchiolin, the periostracal protein (Fig. 9.46a,b). This hinge is located between the umbos of each valve. An **umbo** is the oldest region of the pelecypod valve; it presents a dorsal hump above the hinge and initiates concentric growth lines extending over the surface of the shell. In its resting state, the hinge supports the shell in a slightly open position. Strong anterior and posterior adductor muscles contract against the elastic hinge, closing the valves. The open edges of the shell articulate well, and opposing teeth and/or ridges allow a snug fit during shell closure.

The visceral hump is concentrated dorsally between the shell valves. The mantle extends over this main body mass and forms the walls of a large ventral cavity. Posteriorly, the mantle cavity contains a single pair of gills. Hanging from the visceral mass and extending into the mantle cavity is the hatchetlike foot. This foot is supplied with paired anterior and posterior retractor muscles whose insertions are dorsal to the shell adductor muscles (Fig. 9.47). Anteriorly, a pair of pedal protractor muscles connects the foot with each valve.

The gills dominate the mantle cavity in most bivalves. Gill structure and function are a primary focus of diversity within the group and are discussed under each of the three subclasses.

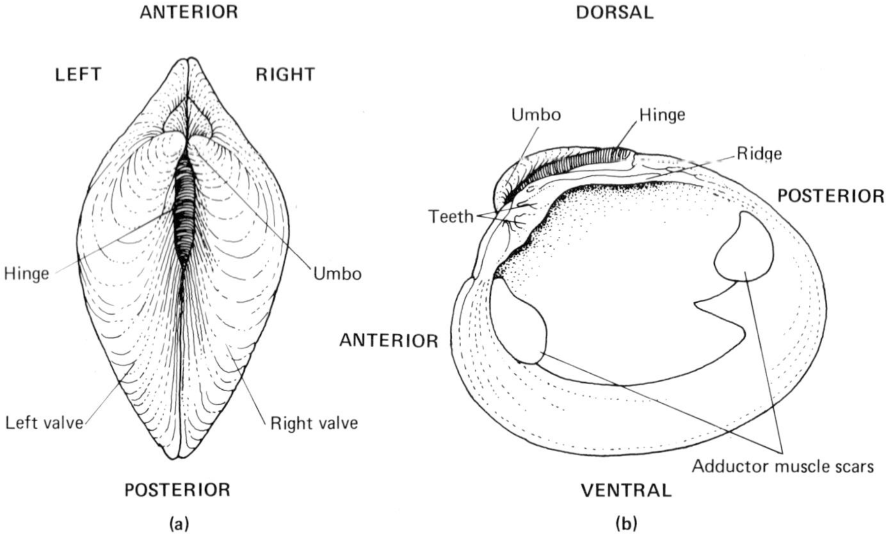

Figure 9.46 Major structural features of bivalve shells. (*a*) Dorsal view of the two valves of the clam shell *Mercenaria campechiensis;* (*b*) Inner surface of the right valve of *Mercenaria.* (From Solem, *The Shellmakers: Introducing Mollusks.*)

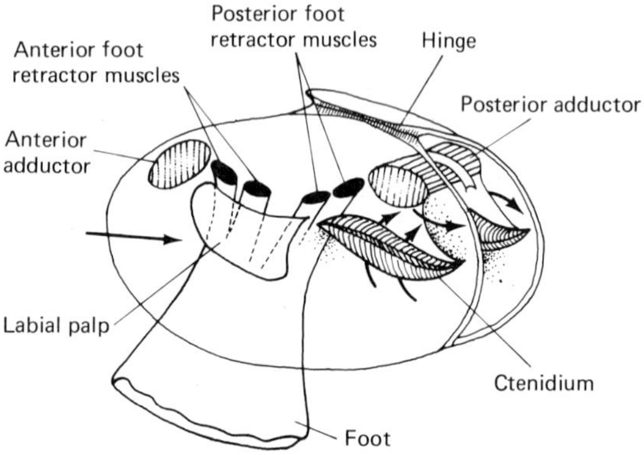

Figure 9.47 Schematic reconstruction of a primitive bivalve showing major muscle masses. (From Wilbur and Yonge, Eds., *Physiology of Mollusca, Vol. I.*)

Nutrition and Digestion

Bivalves have a tubular gut with an anterior mouth and posterior anus (Fig. 9.48). The mouth is flanked by large paired lips, called **labial palps**. Labial palps are involved generally with food collection and sorting. No radula is present in this class. The digestive tract includes an esophagus, a stomach with digestive glands, a rather long intestine, and an anus. The stomach is the center of most digestive activity. Often a complex organ, its structure and function vary among the subclasses.

Respiration and Circulation

Gas exchange and internal transport involve the gills and/or the mantle wall and a typically molluscan, open circulatory system. Bivalves extract a lower percentage of oxygen from surrounding waters than do most other mollusks. Gastropod abalones, for example, remove 48 to 70% of the available oxygen, while scallops extract a mere 2.5 to 6.8%. A general absence of respiratory pigments within the class is largely responsible for this low efficiency. Hemoglobin and/or myoglobin are present in some bivalves, but most

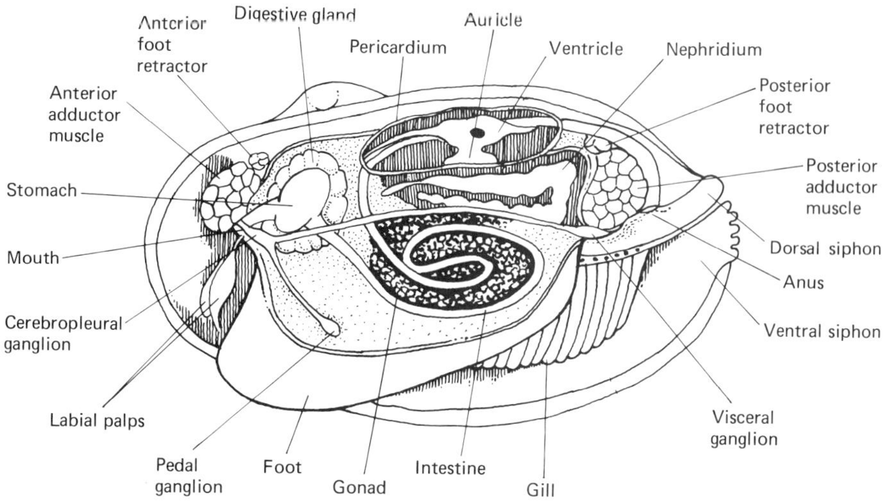

Figure 9.48 Representation of the internal anatomy of a freshwater mussel after removal of left valve, mantle, and gills. (From Brown, *Selected Invertebrate Types.*)

manage without these pigments. However, the enormity of the gills, particularly among the lamellibranchs, more than compensates for the poor gas exchange rate. Such large gills evolved primarily for food-gathering purposes. In developing, however, they provided so well for respiratory needs that blood pigments and efficient gas exchange mechanisms became unnecessary. Moreover, bivalves are a rather sluggish group whose low metabolic rate directs little selective pressure toward the development of more efficient respiratory and circulatory systems.

Internal transport involves the heart, a few blood vessels, and an open network of tissue sinuses (Fig. 9.49). Paired auricles drain the gills, and a single ventricle pumps oxygenated blood through an anterior aorta. In some bivalves, a smaller, posterior aorta may be present. The ventricle is folded around the posterior intestine or rectum in many groups (Fig. 9.48). In such bivalves, the pericardial cavity surrounds both the heart and a part of the gut.

The pelecypod pulse is rather slow. Blood flow is typically molluscan: heart, aorta, tissue sinuses, nephridia, gills, heart. In some bivalves, deoxygenated blood may flow directly from the kidneys to the heart; another route bypassing the gills directs the blood along the thin mantle wall, where some oxygenation may occur. That bivalves function with a poor gas exchange rate, a lack of respiratory pigments, a slow pulse, and a blood circulatory route that may bypass the gills attests to the group's low metabolic rate and the hypertrophied state of the gills relative to the respiratory needs of these animals.

Excretion and Osmoregulation

Water and waste control in the Bivalvia involves a pair of horseshoe-shaped nephridia (Fig. 9.48). These organs occupy the ventral wall of the pericardium and discharge into the upper mantle cavity. Pericardial glands secrete waste products into the pericardial cavity, and the heart delivers a filtrate that, combined with the glandular discharge, passes into the nephridium. In primitive bivalves (protobranchs), the nephridia are glandular throughout their lengths, but in more advanced members of the class, a distal bladder has developed. Nitrogenous wastes are in the form of ammonia. A few bivalve families are abundant in fresh water. The usual osmoregulatory adaptations have occurred, including tubule salt resorption and production of a copious dilute urine.

ACTIVITY SYSTEMS

Bivalves are seldom very active animals. Like many other members of their phylum, they are mostly sluggish, sedentary creatures. Many are sessile. Some bivalves are effective burrowers, usually in soft substrata, but simple water circulation through the mantle cavity remains the dominant activity of the class.

Muscles and Nerves

Important pelecypod muscles include the anterior and posterior adductor muscles, the pedal protractors and retractors, and the fibers of the foot itself (Fig. 9.47). Particularly interesting are the adductor muscles,

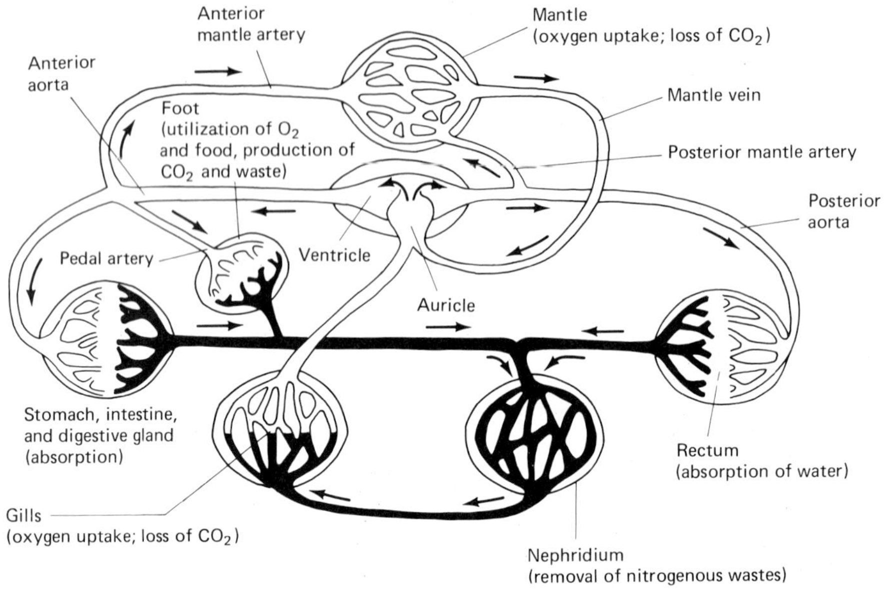

Figure 9.49 Representation of the circulatory system of a generalized bivalve. Oxygenated blood is drawn clear; deoxygenated blood is black. (From R.C. Moore, Ed., *Treatise on Invertebrate Paleontology, Part N,* 1969, Geological Society.)

where two types of contractile fibers are present. One type is smooth and contracts slowly to close the shell for extended periods. The second variety is striated; these adductor muscles contract quickly to shut the shell in emergency situations.

The bivalve nervous system is expectedly simple. Three pairs of ganglia and two major paired nerve cords are present (Fig. 9.48). A cerebropleural ganglion occurs on either side of the esophagus. Each of these nerve centers is connected with a pedal ganglion in the anterior foot. Also connecting with the cerebropleural ganglia via longer nerve cords is a pair of visceral ganglia. The latter are located on the anterioventral surface of the posterior adductor muscle.

The cerebropleural ganglia coordinate the action of the valves and the foot. Pedal and esophageal nervous centers control the anterior adductor muscle and the foot, while the visceral ganglia monitor the posterior adductor muscles and the entrance to the mantle cavity.

Sense Organs

No cephalic sense organs exist in the Bivalvia. Sensory equipment is concentrated in the only tissue area with immediate access to the environment, the edge of the mantle. There, tentacles include tactile and chemosensory cells (Fig. 9.50a). These mantle projections are most common along the route of the respiratory–feeding current. Osphradia usually are present along this route but, surprisingly, often occur within the excurrent path. Such placement discounts their utility in monitoring incoming waters for food, toxins, and sediments; perhaps these sensory patches have a different and, as yet, unknown function in the class.

Other bivalve sense organs include a pair of statocysts in the foot and eyes of various description. Visual centers occupy the mantle edge, where they usually consist of simple pigment cup ocelli. Some bivalves, however, such as the swimming scallop *Pecten*, possess well-developed eyes complete with lenses, corneas, and retinas (Fig. 9.50b).

Behavior

Most locomotor activities within this class are limited to burrowing. A number of burrowing styles are exhibited, but all follow a basic scheme (Fig. 9.51). With the valves gaping slightly, the contraction of pedal protractor muscles extends the foot. Adductor muscles then contract, expelling water forcibly from the mantle cavity. This blast of water loosens local substrata, thus facilitating further penetration by the foot. The bivalve foot penetrates more deeply by hemocoelic pressure (Fig. 9.51b). When fully extended, the foot may anchor itself by manipulating short fibers at its distal end.

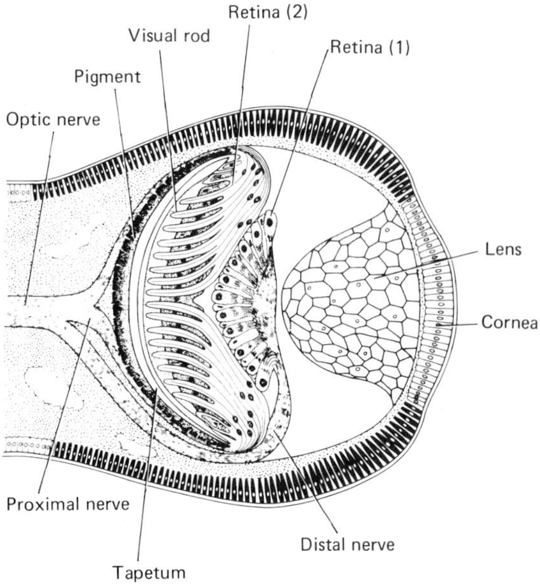

Figure 9.50 (*a*) Photograph of *Aequipecten amplicostatus* showing the mantle edge with eyespots and sensory tentacles; (*b*) Section through an eye of a scallop (*Pecten*). (*a* from Andrews, *Sea Shells of the Texas Coast;* *b* from Sherman and Sherman, *The Invertebrates; Function and Form.*)

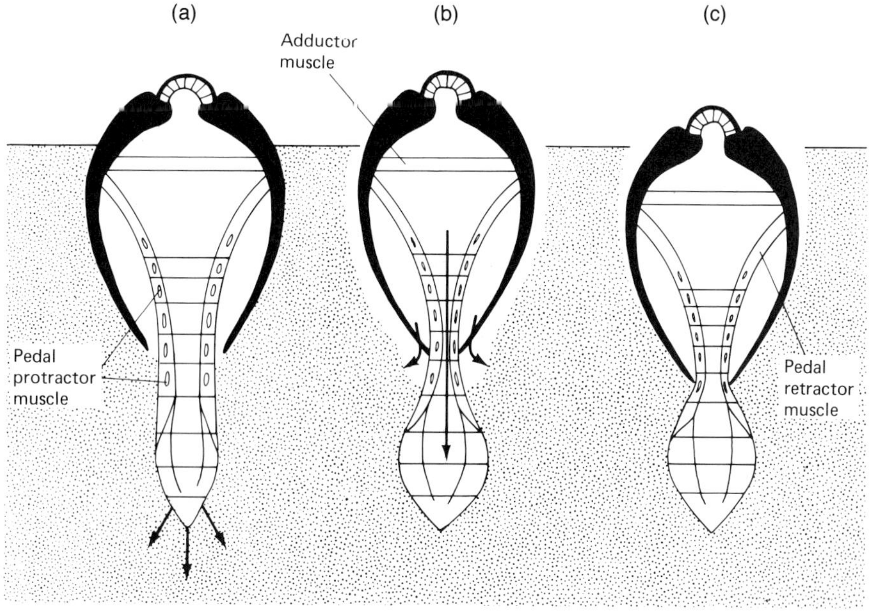

Figure 9.51 Successive stages in the burrowing of a generalized bivalve. (*a*) Valves opened by elasticity of hinge ligament and pressing against sand with foot extended; (*b*) Valves closed by adductor muscles. Water ejected from mantle cavity loosens sand around valves and high pressure in the hemocoel dilates the foot; (*c*) Contraction of foot retractor muscles pulls shell down into loosened sand. (From E.R. Trueman, Zool. Soc. Lond. Symposium, 1968, *22*:167.)

Pedal retractor muscles then draw the entire animal downward toward its anchored foot (Fig. 9.51c). Often, the anterior pedal retractor muscles act before their posterior counterparts, and the rocking motion thus elicited allows the mollusk to penetrate more effectively. By combining this rocking behavior with a highly sculptured, abrasive shell, some bivalves are able to burrow in hard substrata.

CONTINUITY SYSTEMS

Bivalve continuity systems are quite rudimentary. Typical of many sedentary, aquatic invertebrates, numerous gametes are released from simple gonads, fertilization is external, and larval forms are free-swimming. Some modifications of this pattern occur, including the occasional brooding of fertilized eggs within the mantle cavity and gills. Freshwater bivalves in particular often brood their young, whose larval stages are nonmotile.

Reproduction and Development

Most bivalves are diecious. A pair of large gonads occupies the looping zone of the intestine (Fig. 9.48), and a simple gonoduct connects each organ with the mantle cavity. In primitive bivalves, gametes pass through the nephridia, but in higher forms, each gonoduct connects directly with the mantle cavity. Eggs may be fertilized either in the mantle cavity or in external waters. Fertilization potential is enhanced by synchronous gamete release in a local population. This release may be dependent on such environmental cues as tides, water temperature, photoperiods, and/or chemical secretions by other gametes.

A trochophore larva develops in marine pelecypods and is followed by a veliger stage. Most bivalve veligers display a circular, not a bilobed, velum (Fig. 9.52). The bivalved shell originates when the single dorsal plate of the shell gland abruptly folds during its growth. A **byssal gland** on the foot aids the young bivalve during its settlement and attachment to the substratum. After settlement, ventral additions to the shell are made in concentric arcs.

Some marine and nearly all freshwater bivalves brood their young within the mantle cavity. Early developmental stages are passed within a suprabranchial chamber or among the gill filaments. The young emerge prepared to cope with an independent existence.

Bivalve Diversity

The three bivalved subclasses—the Protobranchia, the Septibranchia, and the Lamellibranchia—differ in the structure and function of their gills and in feeding habits, stomach biology, and habitats. Each subclass is discussed in turn.

SUBCLASS PROTOBRANCHIA

The most primitive bivalves belong to the subclass Protobranchia. *Nucula* (Fig. 9.53a) is a standard rep-

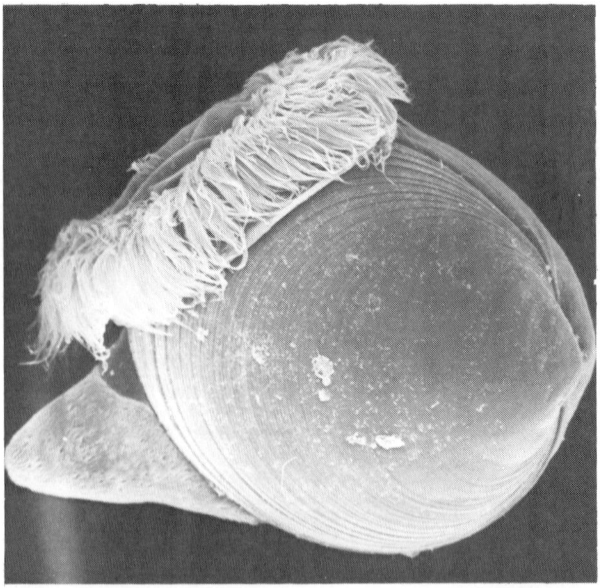

Figure 9.52 Scanning electron micrograph of a bivalve veliger, *Lyrodus pedicellatus*. (From R.D. Turner and P.J. Boyle, Bulletin of the American Malacological Union, 1974.)

resentative. Protobranchs have a single pair of simple, unfolded bipectinate ctenidia whose main function is respiratory. Food collection by ctenidial cilia may occur incidentally, but seldom contributes significantly to the diet. The protobranch foot often has a flattened ventral surface, or **sole**. Members of this subclass burrow in soft bottoms, and their flattened sole is an effective anchor during locomotion.

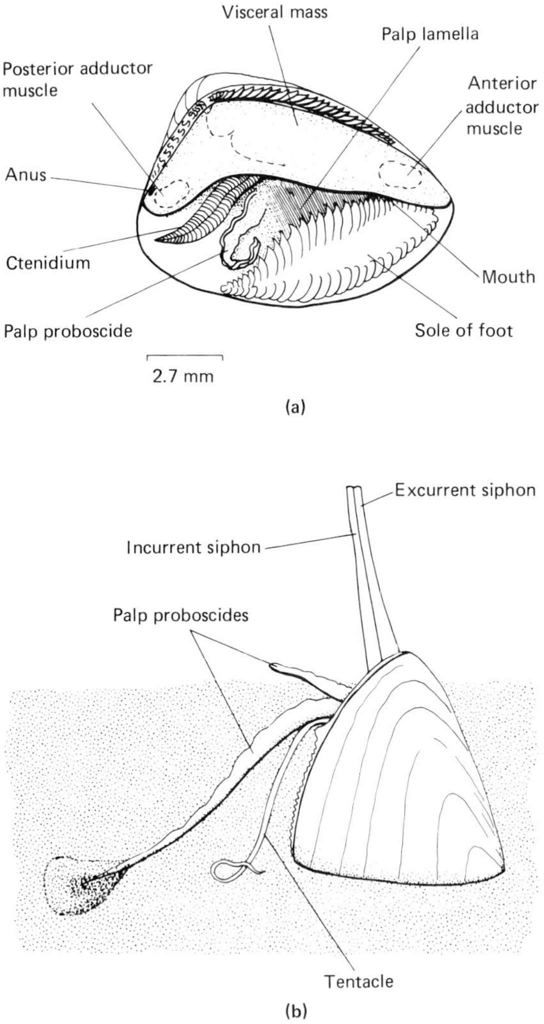

Figure 9.53 (*a*) *Nucula*, a protobranch. The right valve and right mantle fold have been removed to show mantle cavity structures; (*b*) *Yoldia limatula* in its typical feeding position, partially buried in the mud. (*a* from Fretter and Graham, *British Prosobranch Molluscs*; *b* from P. Pelseneer, In *A Treatise on Zoology, Vol. V.* E.R. Lankester, Ed., 1906, Stechert-Hafer Service Agency, Inc.)

The most distinctive aspect of protobranch biology is the mode of food collection. Unlike other bivalves most protobranchs are deposit feeders. Each of their labial palps bears a proboscide of considerable length. These proboscides extend from the shell and penetrate the substratum (Fig. 9.53b), where potential food particles adhere to their mucous surfaces. Cilia propel adhered particles back to the palps, where sorting by size occurs. The sorting field comprises a series of ciliated ridges. Small particles are swept over the larger folds, while heavier matter is trapped within the troughs between adjacent ridges. These troughs transport rejected particles to the edge of the mantle, where they are expelled as **pseudofeces**. Pseudofeces refer to any materials removed from the digestive system before passage through the gut. Particles acceptable as food are shunted into the mouth and through a ciliated esophagus to the stomach. The protobranch stomach (Fig. 9.54) displays many of the characters encountered in archaeogastropods. A dorsal region includes the openings of the esophagus and digestive glands, a ciliary sorting field, and walls lined with protective chitinoid shields. The ventral portion of the protobranch stomach is modified as a style sac. Food-laden mucus enters from the esophagus and is rotated by the cilia of the style sac. In *Solemya* and many other protobranchs, fragments of the rotating mass are tossed onto the ciliary sorting field, where particles of proper size are routed to the digestive

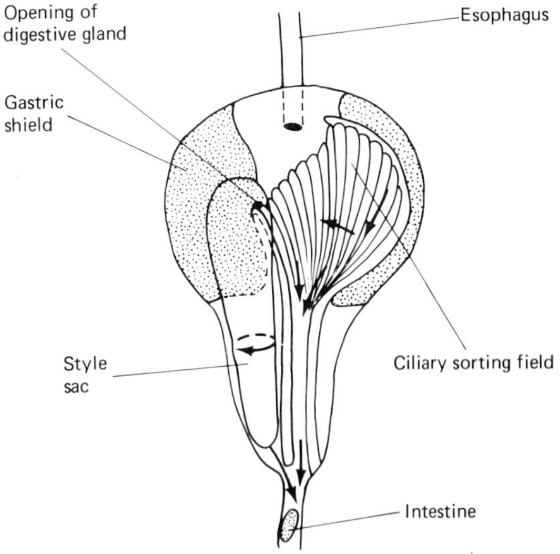

Figure 9.54 The stomach of the protobranch *Nucula*. (From S. Owen, In *Physiology of Mollusca, Vol. II*. Wilbur and Yonge, Eds.)

gland. Digestion may be both extracellular and intracellular, the latter occurring within phagocytotic cells of the digestive glands. In *Nucula* (Fig. 9.53a) and *Yoldia* (Fig. 9.53b), these glands release enzymes into the stomach, where digestion is wholly extracellular. Absorption then occurs in the stomach and intestine.

Undigestible wastes are removed along ciliary tracts which join a ciliated gutter passing into the intestine. The protobranch intestine is rather long and coiled. It loops about the stomach, extends beneath the anterior adductor muscle, and then passes dorsally to the posterior end of the animal. Along this route, it may pass through the pericardial cavity and the heart ventricle. The intestine concludes with a long rectum arching above the posterior adductor muscle and an anus near the excurrent opening of the mantle cavity.

Many of the characters described earlier for the more primitive bivalves apply to the protobranchs. Their nephridia, for example, are glandular throughout and do not form a bladder. Discrete gonopores are absent, as protobranch gonoducts connect with the nephridial system. Fertilization is external, and larval forms are free-swimming.

SUBCLASS SEPTIBRANCHIA

The septibranchs are a small group of very specialized marine bivalves. These animals are carnivores and/or scavengers. Their highly modified gills form muscular septa which divide the mantle cavity into paired ventral and dorsal chambers (Fig. 9.55). Septal pumping draws water into the ventral chamber; this water passes upward through special gill perforations and then exits from the dorsal chamber. Carcasses and small living animals swept in by this current are seized by muscular labial palps. The septibranch stomach reflects the diet of this subclass. Shielded with chitin, the stomach's muscular walls act as a gizzard to grind and crush prey. Ground particles are passed into paired glands, where digestion occurs intracellularly. The ventral style sac is greatly reduced.

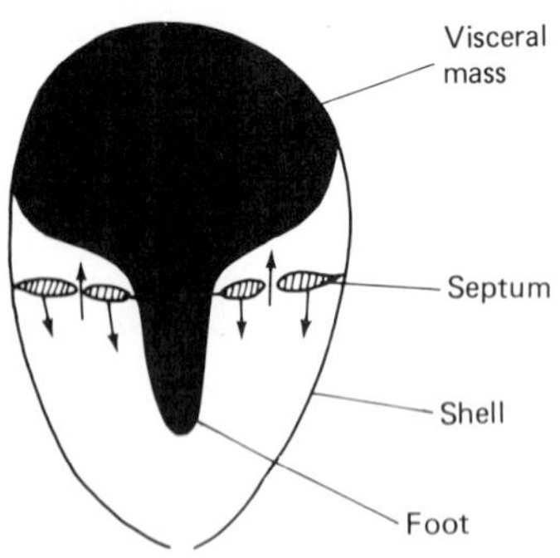

Figure 9.55 Cross section of a septibranch bivalve showing the division by muscular septa of the mantle cavity into paired dorsal and ventral chambers. (From C.M. Yonge, Phil. Trans. Roy. Soc. Lond., 1928, *216*:221.)

Septibranchs probably evolved from protobranch stock. Originally, their septal pump may have developed to ventilate the mantle wall, the site of all gas exchange in this subclass. As this pump became strong enough to suck in prey, the Septibranchia diverged. *Cuspidaria* and the hermaphroditic *Poromya* are representative genera.

SUBCLASS LAMELLIBRANCHIA

The overwhelming majority of bivalves belongs to the subclass Lamellibranchia, and it is on several lamellibranch characters that the overall success of the class is based. Most familiar bivalves—clams, oysters, scallops, and mussels—are lamellibranchs. This class displays the largest and most complex of all molluscan gills. These gills filter plankton and other small edibles from mantle cavity waters. Before the rise of the Lamellibranchia, ctenidial cilia and mucus already served mollusks as means for cleansing particulate matter from the gills. Lamellibranchs, however, adapted to use such particles for food. Their gills elongated and developed complex folds which increased the filtering area (Fig. 9.56a). The rest of the lamellibranch digestive tract adapted to the new types of food then available. Also, lamellibranchs experienced an extensive radiation in habitats and life-styles.

Lamellibranch gill filaments are numerous and individually quite long. Typically, the row of filaments along each side extends to the labial palps. Each filament is folded, thus forming a pair of horseshoe-shaped **demibranchs** on either side of a central axis (Fig. 9.56b). A demibranch is composed of descending and ascending arms. Because of their large size and frequently intricate organization, lamellibranch gills are highly efficient in filtering food particles from mantle cavity waters. Lateral cilia on the demibranchs draw water into the posterior ventral chamber (Fig. 9.56b). This water sweeps through the gills and into the suprabranchial chamber, from which it exits through a posterior opening. Laterofrontal cilia, typically organized into bundles, trap suspended particles and transfer them to the gill's frontal cilia. Frontal cilia then transport potential food items to one of several food grooves (Fig. 9.56b). Food grooves are ciliated tracts along the ventral midpoints of the demibranchs; these tracts terminate with the gill filament bordering the labial palps.

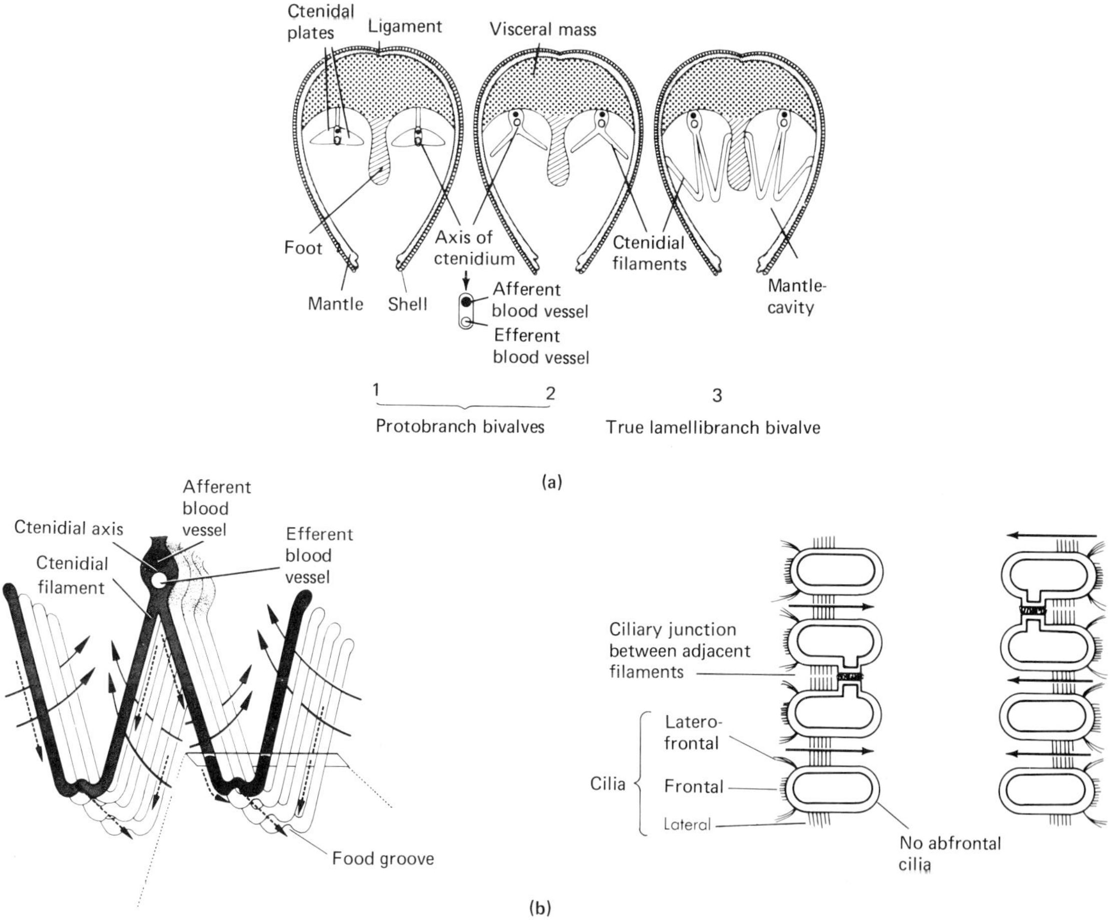

Figure 9.56 (*a*) A comparison of two protobranches and a lamellibranch bivalve in diagrammatic cross section to illustrate the lengthening and ultimate folding of the gill filaments characteristic of lamellibranchs; (*b*) Gill structure in lamellibranchs showing a stereogram of several filaments of one ctenidium and a horizontal section through a demibranch. Solid arrows indicate the direction of water flow; the dashed arrows indicate the pathway of food-laden mucus. (From Russell-Hunter, *A Biology of Lower Invertebrates.*)

Certain structural modifications increase gill support and filtering efficiency. The long ctenidial filaments require reinforcement to maintain their shape under pressure from mantle cavity currents. Tissue connections between the ascending and descending arms of each demibranch are common (Fig. 9.57). Tissue junctions also occur between the mantle wall and the outer arms, as well as between the foot and the inner row of demibranchs. These junctions isolate the ventral, incurrent chamber from the suprabranchial area. Tissue and ciliary associations between adjacent filaments also occur. These associations form the basis of a scheme of gill classification. In **filibranch** gills (Fig. 9.57a), seen in scallops and certain mussels, only ciliary attachments exist between filaments. Most oysters are characterized by **pseudolamellibranch** gills, or gills whose adjacent filaments are connected loosely by infrequent tissue junctions. The most elaborate gill type is the **eulamellibranch** gill (Fig. 9.57b). In these gills, tissue connections between adjacent filaments are quite numerous, so that the entire organ approximates a continuous sheet of tissue. (In fact, "lamellibranch" means "sheet gill.") Water enters the eulamellibranch gill at interfilamental ostia and then passes through vertical tubes to the suprabranchial chamber. Blood vessels no longer associate with each filament, as they do in filibranch and other molluscan gills. Rather in eulamellibranch gills, interfilamental blood vessels are parallel to each of the vertical water tubes.

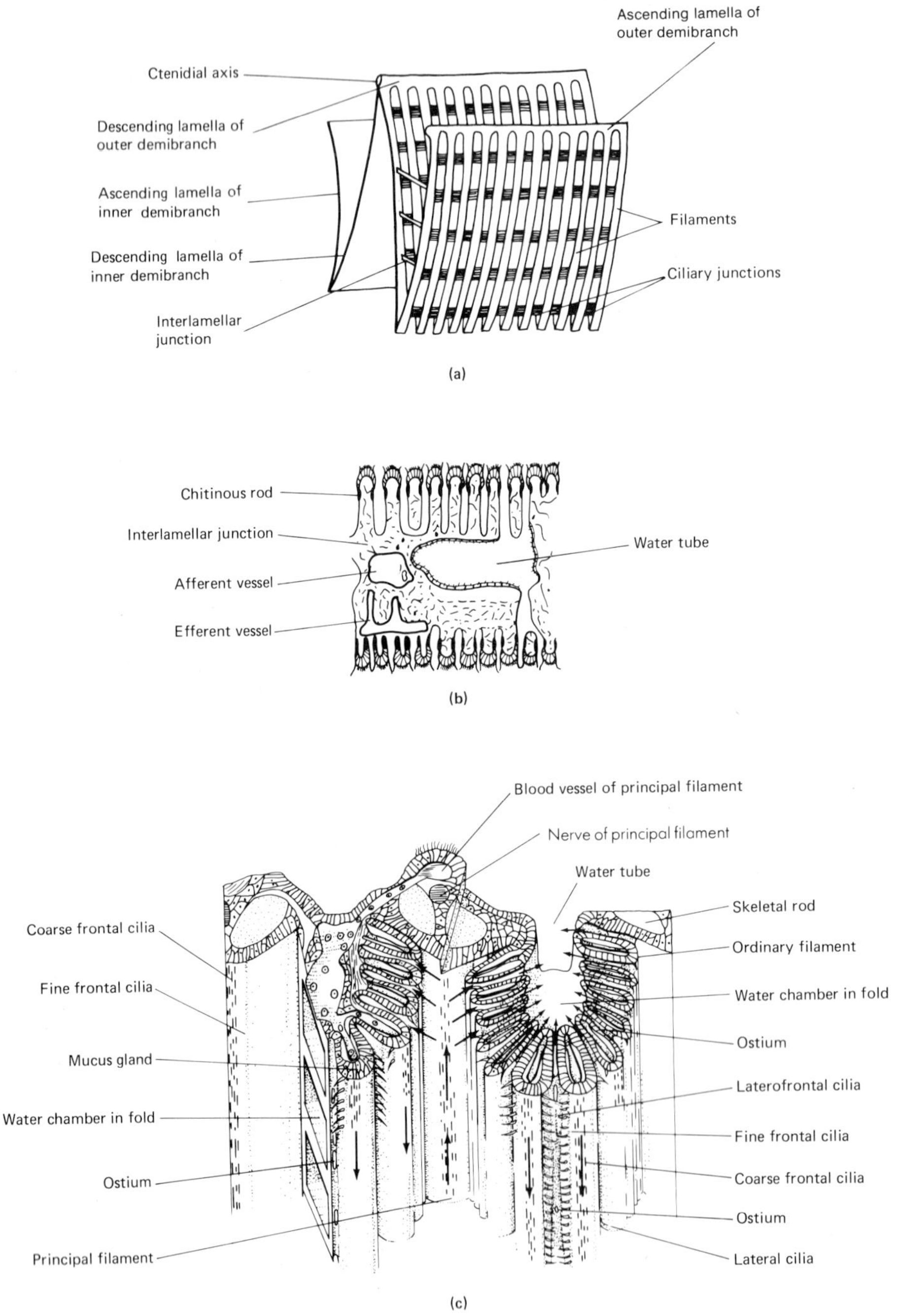

Figure 9.57 Structural modification in lamellibranch gills. (*a*) Part of a filibranch ctenidium from a mussel, *Mytilus edulis*, viewed obliquely. Note the ciliary junctions between the ascending and descending lamellae of each demibranch; (*b*) Horizontal section through a portion of a demibranch of a eulamellibranch gill from a freshwater mussel, *Anodonta;* (*c*) Stereogram of part of a folded gill of an oyster, *Crassostrea virginica*. (See text for a description of folded gills.) (*a* from Moore, Ed, *Treatise on Invertebrate Paleontology*, *Part N; b* from L.A. Borradaile and F.A. Potts, *The Invertebrata*, 1958, Cambridge University; *c* from T.C. Nelson, J. Morphology, 1960, *107*:163.)

Many gills are folded further to increase their surface area. Such secondary folding occurs along the longitudinal row of filaments (Fig. 9.57c). The filament lying deepest within each fold is called a principal filament. Principal filaments can be expanded or withdrawn by muscular action, a process that regulates the exposed surface area of the gill. When currents are clean, expansion of the principal filaments allows maximal ciliary activity, and wide food grooves channel potential nutrients to the labial palps. When sediment-heavy waters enter, the principal filaments withdraw, the ciliary fields are retracted, and the narrowed food grooves accommodate only the smallest particulate matter.

Lamellibranchs regulate the passage of currents and food matter over their gills in still other ways. Two-way ciliary tracts are common; each tract usually handles particles within a certain size range. Heavy particles may be transported posteriorly (away from the mouth) for disposal as pseudofeces. Meanwhile, the food grooves themselves only accept particles of a certain size. The fixed capacity of these ventral transport lanes also governs the total volume of particulate matter.

Particles not rejected by the gills are conveyed to the labial palps, where further sorting occurs. A series of ciliated ridges segregates particles by size and weight, and only the smallest and lightest are passed to the mouth. Rejected particles drop onto the mantle or foot where, with the matter refused by the gills, they are expelled from the animal. Expulsion occurs either through the pedal gaps or through the incurrent aperture. In the latter case, a ciliated groove along the mantle floor channels the pseudofeces posteriorly to the incurrent opening, where a reverse muscular blast discharges them.

Because of their diet, marked changes have occurred in the lamellibranch stomach (Fig. 9.58). Food particles are always small; hence there is no chitinoid gastric lining. Only a small shield remains in association with the tip of the **crystalline style**. The lamellibranch crystalline style is a remarkable organ which may occupy one-third of the body length of these mollusks. It is a large rod whose protein matrix is covered with mucus and adsorbed enzymal secretions. These enzymes, as well as the style framework itself, are produced within the style sac. The rod is rotated by style sac cilia. This rotation draws a mucus—food string from the esophagus into the stomach. The tip of the rotating style is abraded by the small gastric shield, a process that continuously releases amylases, lipases, and cellulases. Thus some digestion of starches and fats occurs extracellularly within the stomach; gastric digestion of proteins is impossible, as proteases would devour the style matrix. As food particles are released from the spinning gastric mucus, they are thrown against a ciliary sorting field. Heavy particles are rejected into a deep groove leading to the intestine. Fine particles and fluids are drawn into one of several openings of the digestive glands, where absorption and intracellular digestion occur. Both incurrent and excurrent tracts are present in these glands; one-way flow through the organ is maintained because only the excurrent tracts are ciliated.

The waste-bearing tracts of the digestive glands join a typhlosole leading to the intestine (Fig. 9.58). Sometimes structured on a tube-within-a-tube plan, this groove prevents digestive wastes from mixing with other stomach contents.

Lamellibranchs exploit a wide range of habitats, where they pursue various life-styles. There are five generalized lamellibranch types: (1) surface-attached lamellibranchs, (2) shallow-burrowing forms, (3) deep-burrowing types, (4) freshwater representatives, and (5) commensals and parasites.

Surface-Attached Lamellibranchs

The oldest members of the Lamellibranchia attach themselves to shallow marine bottoms. Attachment is

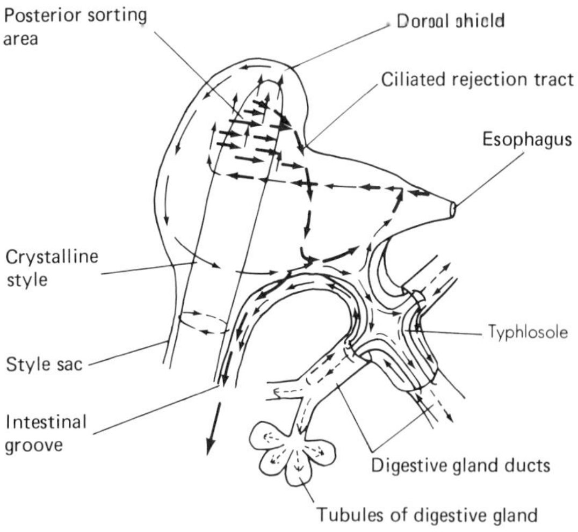

Figure 9.58 The lamellibranch stomach with associated structures. Arrows indicate the probable circulation of particles in the stomach and digestive gland ducts. Heavy arrows represent the path taken by coarse particles; fine arrows, the path followed by fine particles; dotted arrows represent absorption by digestive gland cells. (From G. Owen, Quart J. Micros. Sci., 1955, 96:533.)

by byssal threads or by direct cementation of the right valve. Byssal threads are produced by the byssal gland of the foot. Secretions from this gland flow out along a groove in the foot and contact the substratum; as these secretions harden, the foot's groove molds them into a strong thread. When the foot is removed, the new thread remains to anchor the animal. Other characteristics of surface-attached lamellibranchs include filibranch or pseudolamellibranch gills and a relatively open mantle cavity, the absence of siphons, the reduction of the foot, and a trend toward asymmetry. Four subgroups are discussed here: the Noah's ark shells, the mussels, the scallops, and the oysters.

Noah's ark shells and their relatives are among the most ancient lamellibranchs. These mollusks commonly attach to tropical coral formations, pilings, etc. The group preserves bilateral symmetry; the shell is bulky, and anterior and posterior adductor muscles are evenly developed. The common *Arca* (Fig. 9.59) is representative. This mollusk is stationary, its ventral surface being attached by many short byssal threads. The related genera *Glycymeris* and *Limopsis*, however, burrow shallowly in soft substrata.

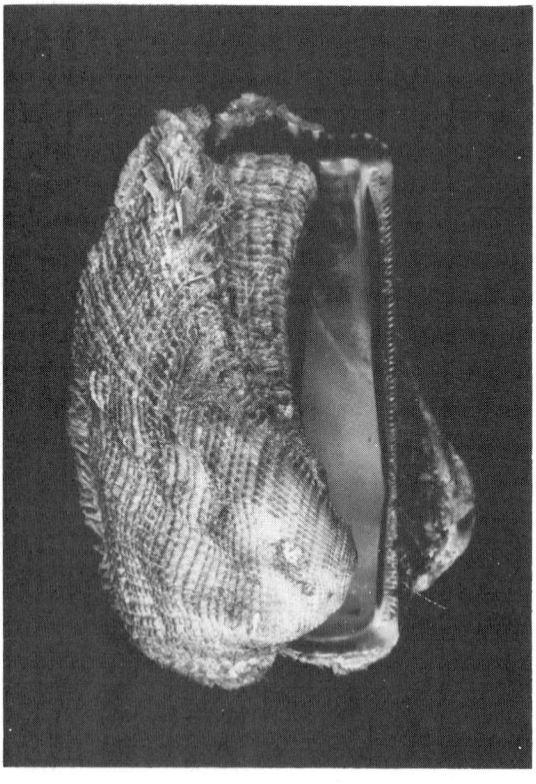

Figure 9.59 Shell of *Arca zebra*. (From Andrews, *Sea Shells of the Texas Coast*.)

The mussels rank among the most familiar and widely distributed bivalves. These mollusks attach by byssal threads to hard substrata off marine coasts throughout the world. Mussel shells are symmetrical, but anteriorly, the valves are often reduced. The byssus and foot have shifted to the anterior end, where the adductor muscle is diminished (Fig. 9.60). Concomitantly, the posterior adductor muscle has migrated forward to a medial position. These changes produce a mollusk with an enlarged, elevated posterior, which may ameliorate respiratory and feeding problems on crowded surfaces.

Mussels have radiated extensively and currently occupy a number of specialized niches. *Mytilus* (Fig. 9.60), the common edible mussel on both American coasts, is often moored on hard substrata in the intertidal zone. Members of the genus *Modiolus* may bury themselves in intertidal grasses or in a nest of their own byssal threads. *Botula*, one of several boring mussels, lives in a burrow in soft rocks.

The oyster is a highly specialized surface-attached lamellibranch. After its initial byssal attachment, it cements its entire right valve to the substratum. An oyster thus lies on its side, the left half uppermost, and differential modifications of the two valves are common. Frequently, the upper valve forms a lid for the lower one, as in the jewel box *Chama* (Fig. 9.61a). Entirely sessile, oysters are a rarity among mollusks in that, as adults, they have no foot. Their anterior end, including the adductor muscle, is reduced. A large crescent-shaped, pseudolamellibranch gill occupies the entire circumference of the mantle (Fig. 9.61b). Oysters usually live on hard substrata, where fouling is seldom a serious problem. However, the group has "quick" adductor muscles whose contractions expel

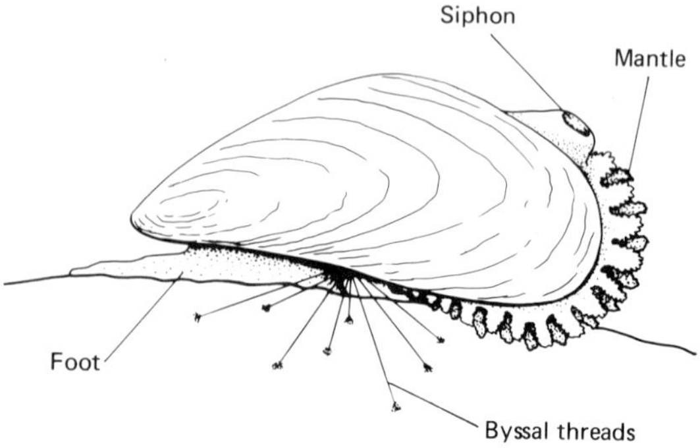

Figure 9.60 *Mytilus edulis*, the common edible mussel showing attachment by byssal threads. (From Solem, *The Shellmakers: Introducing Mollusks*.)

(a)

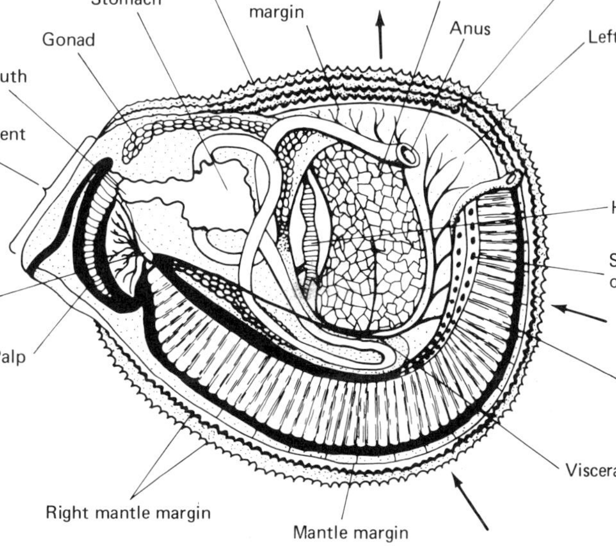

(b)

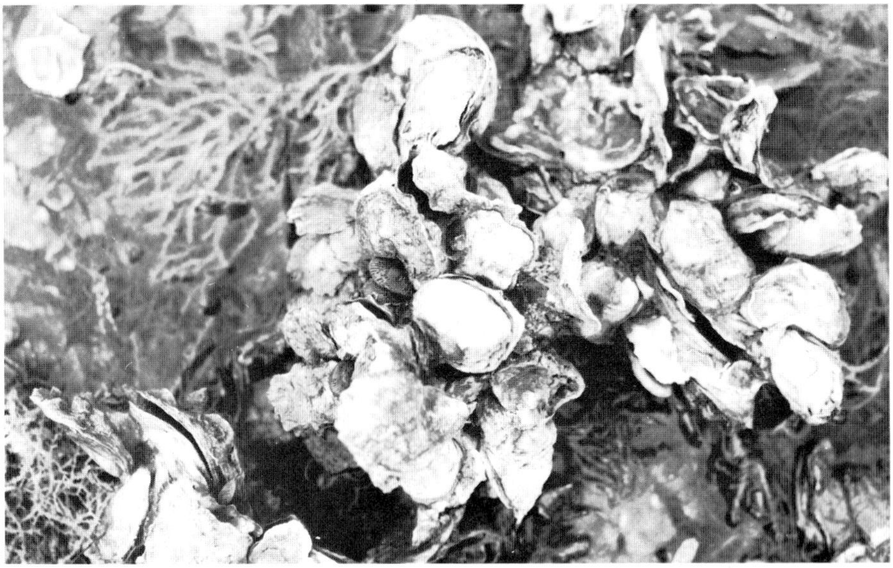

(c)

Figure 9.61 (a) *Chama pellucida;* (b) Internal anatomy of *Ostrea edulis* with left valve, mantle, and gill removed. Arrows indicate water currents; (c) A large clump of oysters, *Crassostrea virginica*, in Aransas Bay, Texas. *Ischnadium recurvum*, a small mussel with a ribbed shell, can be seen in the center of the clump. (a from Ricketts, et al., *Between Pacific Tides;* b from Kaestner, *Invertebrate Zoology*, *Vol. I;* c from Andrews, *Sea Shells of the Texas Coast.*)

sediment. Included among these lamellibranchs are *Crassostrea* (Fig. 9.61c), the American edible oyster, and *Ostrea* (Fig. 9.61b), the European delicacy. *Ostrea* often inhabits gravelly substrata, where the high silt concentration renders its quick "cleaning" muscles most advantageous. This oyster is extraordinary in that it incubates its young within the mantle cavity during part of their larval period.

Pearl formation may occur in any shelled mollusk, but the pearls produced by certain "oysters" of the genus *Pinctada* are the most beautiful and commercially valuable. A pearl begins when some alien object, commonly sand or a parasite, wedges between the mantle and the shell. The mantle then secretes calcareous layers about the object. Frequently, the "pearl" is cemented irretrievably to the shell; however, it may lie within a small mantle fold, where its rotation and the addition of many calcareous layers produce a spherical or ovoid bead.

The last major group of the surface-attached lamellibranchs includes the scallops and the file shells. Many of these mollusks are moored by byssal threads, but related unattached species are the only bivalves with significant swimming capabilities.

Pecten and other scallops have bilaterally symmetrical shells (Fig. 9.50a). A single adductor muscle (representing a migrated posterior adductor muscle) occupies a central position just below the hinge. Slow and quick fibers within this muscle control valve movement. Contraction of the slow fibers closes the valves during sustained nonmobile periods. The quick muscles clap the valves together rapidly, expelling a jet of water from the mantle cavity. By manipulating its muscular mantle lobe, the scallop controls the excurrent jet and thus guides itself through the water. Water may exit from the wide, ventral edge of the mantle or may pass on either side of the hinge.

In conjunction with their more active life-style, *Pecten* and other swimming scallops have evolved sophisticated sense organs along the middle mantle lobe. These organs include numerous tentacles and many small eyes. In some *Pecten* species, beautiful blue eyes comprise a well-developed retina, cornea, and lens (Fig. 9.50b).

Notwithstanding their swimming ability, scallops and file shells are not highly active invertebrates. They are, ultimately, bivalves and are prevented both anatomically and physiologically from assuming a truly pelagic life-style. Their swimming usually represents an escape response. Scallops also spray water to excavate a settling area in soft substrata; *Lima* (Fig. 9.62) and other file shells rest in natural crevices. The latter seldom swim unless threatened.

Many other surface-attached lamellibranchs exist, but space permits description of only a few of the more common forms. The pearl oysters attach by a short byssus, which emerges through a notch on the lower, right-hand valve. These mollusks also display a reduction in the anterior adductor muscle and a medial migration by its posterior counterpart. A trend toward hinge elongation is seen in the group, as typified by *Pinctada*. Further hinge elongation occurs in the wing shells (e.g., *Avicula;* Fig. 9.63a) and culminates in the hammer oyster, *Malleus* (Fig. 9.63b). *Pinna* (Fig. 9.63c) and other pen shells anchor themselves in sand. These mollusks place their anterior ends downward, while their elongate, wing-shaped valves extend above the substratum. The foot and other anterior structures are reduced. Byssal threads from the buried end are secured on individual sand grains, where they provide anchorage. Most of the mantle edge in *Pinna* is free from the shell and thus can be retracted against the viscera to protect internal structures. It may also expel broken shell particles and other debris.

Shallow-Burrowing Lamellibranchs

The second general group of lamellibranchs includes those members of the subclass that burrow shallowly in soft substrata. Actively burrowing types, as well as those that occupy semipermanent or permanent burrows, are represented. These bivalves display eulamellibranch gills and a modified mantle margin which protects the gills from fouling. Burrowing lamellibranchs obviously have potential problems with incoming sediment. However, the mantle is sealed along most of the shell perimeter, leaving apertures only for the foot and for incurrent and excurrent mantle cavity flow. Around the latter openings, the mantle

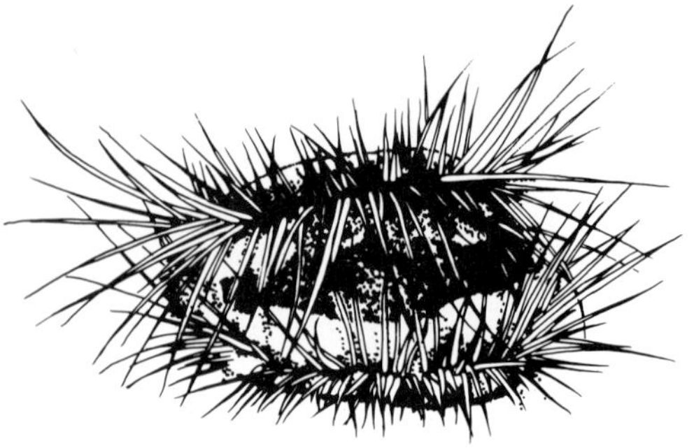

Figure 9.62 *Lima*, a bivalve with long sensory tentacles. (From A.J. Marshall and W.D. Williams, Eds., *Textbook of Zoology: Invertebrates*, 1972, Elsevier Publishing.)

margin is usually drawn out into siphons. A siphon may be formed only from the inner, muscular lobe of the mantle; alternatively, one or both of the other two lobes may be included, in which case siphonal sensory structures and/or shell-like encasements are present. Burrowing lamellibranchs commonly possess rather streamlined shells. Streamlining involves lateral compression, very smooth shell surfaces, and pointed ends. The thin, bladelike foot of these mollusks is their principal burrowing organ. Both freshwater and marine species exist.

The most primitive burrowing lamellibranchs include the cockles and Venus shells. The cockle *Cardium* (Fig. 9.64) is typical in its possession of relatively large and heavy valves. Much of its mantle margin is unsealed. Although the siphons are short, they include sensory tentacles with complex eyes. *Cardium* burrows very shallowly in soft substrata, leaving much of its posterior surface exposed to the water. When attacked, this cockle executes a rather dramatic escape response (Fig. 9.64). Its long muscular foot can be folded underneath the shell; when quickly extended, this foot hurtles the entire animal free from the substratum.

Related to the cockles are the giant clams, typified by *Tridacna* (Fig. 9.65). In *Tridacna,* reportedly the largest of all bivalves, the shell may be a meter in length and weigh well over a hundred kilograms.

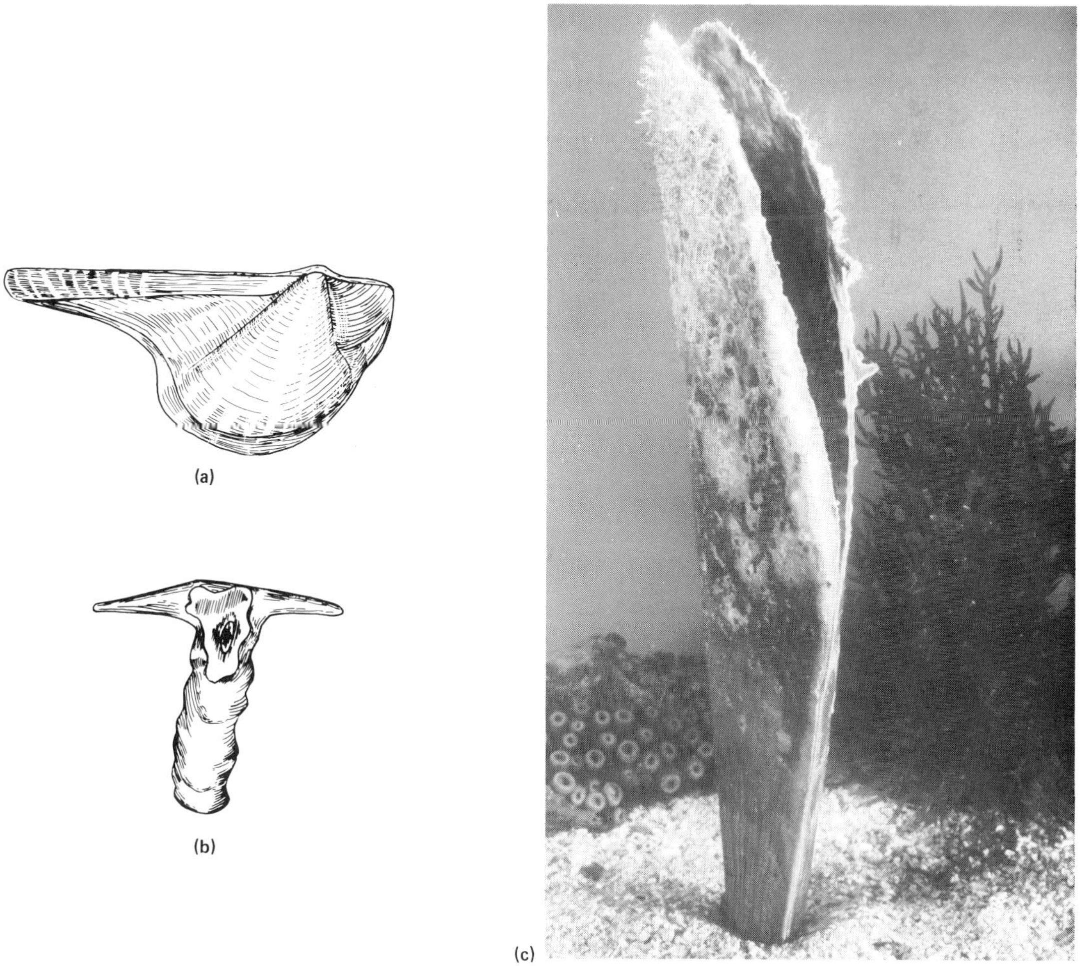

Figure 9.63 (*a*) Lateral view of a wing shell, *Avicula heteroptera;* (*b*) *Malleus vulgaris,* the hammer shell; (*c*) A sea pen shell, *Pinna carnea.* (*a* from A.H. Cooke, In *Cambridge Natural History, Vol. III; b* from M.F. Woodward, *A Manual of the Mollusca,* 1880; *c* courtesy of Grant Heilman.)

Figure 9.64 A cockle, *Cardium echinatum*, shown escaping from a starfish, *Asterias rubens*. (From H. Knudsen, Sarsia 29:371.)

These giant lamellibranchs live among corals in the Indo-Pacific region, where they feed on plankton. *Tridacna* and related genera have undergone a highly unusual reorganization. Primitively, the shell was attached to the substratum by a ventral byssus. Subsequently, the shell hinge and umbos migrated 180° from the traditional dorsal position to a ventral one. Thus, relative to the stationary viscera, the shell of *Tridacna* opens along its dorsal side (Fig. 9.65b). Gills curve around the posterior end of the clam and terminate dorsally below an extensive siphonal complex. The pigmented lips of these siphons present a swirling array of color. Also within the siphonal tissues are symbiotic algae whose photosynthetic products may supplement the diets of these giant clams.

Most burrowing lamellibranchs completely bury

(a)

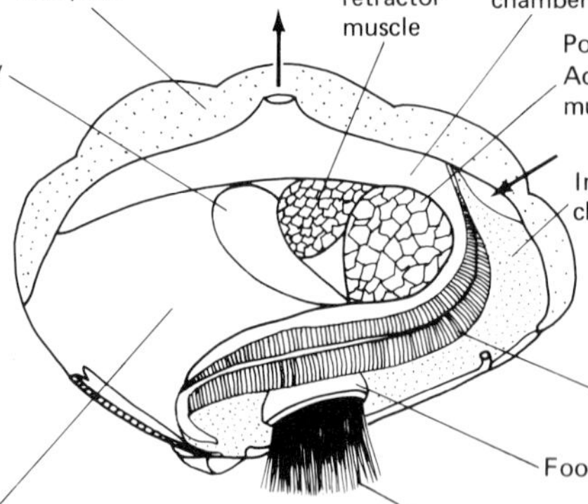

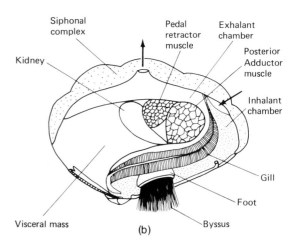

Figure 9.65 (*a*) *Tridacna gigea*, the giant clam; (*b*) *Tridacna crocea*, semi-diagrammatic view of the mantle cavity after removal of the left valve and mantle lobe. The ventral inhalent chamber is drawn stippled, the dorsal exhalent chamber is shown clear. (*a* courtesy of Grant Heilman, photo by George H. Harrison; *b* from C.M. Yonge, Proc. Zool. Soc. Lond., 1953, *123*:553.)

themselves in the substratum, with only their siphons extending to the water above. *Donax* (Fig. 9.66a), a tiny clam occurring on surf beaches, has a pointed anterior end and a wedge-shaped foot. These lamellibranchs ride incoming waves and then burrow rapidly when the surf recedes. At its distal end, the incurrent siphon of *Donax* is bordered by frilly tentacles which prevent the entrance of sand grains. *Siliqua* (Fig 9.66b) and other razor clams are well adapted for rapid burrowing. These mollusks possess long, thin shells. The hinge and umbos are anterior, and the shell length represents an extension of the posterdororsal line. Razor clams have a long, tapered foot which is well muscularized. This strong foot and the streamlined shell account for the speed of these bivalves, which often burrow as quickly as clam diggers can pursue them. The semipermanent nature of razor clam burrows also facilitates a rapid passage through the substratum. In *Tagelus,* long siphons remain exposed for extended periods of time; in contrast, *Ensis* has short siphons and must make periodic trips to the surface to feed and respire.

(a)

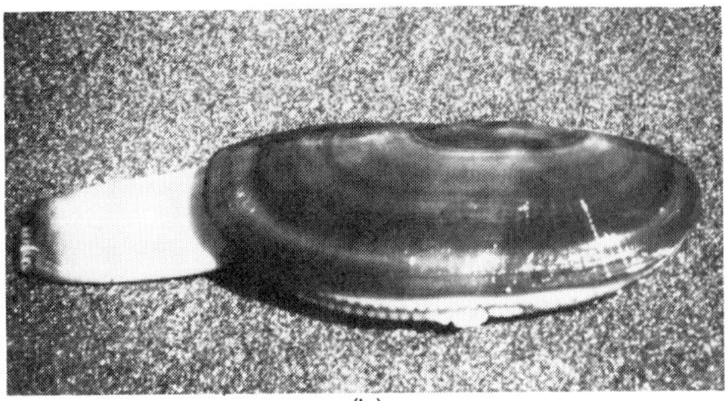

(b)

Figure 9.66 (*a*) Living *Donax variabilis;* (*b*) A razor clam, *Siliqua patula.* (*a* courtesy of Jean Andrews; *b* courtesy of Ward's Natural Science Establishment.)

Deep-Burrowing Lamellibranchs

Many lamellibranchs, such as the edible clam *Mya* (Fig. 9.67a), are quite immobile within permanent burrows. Their assumption of a nearly sessile, adult life-style involves several structural modifications. The mantle margin is fused throughout except for the pedal and siphonal apertures. The immobile shell has lost its streamlining and may be thin and fragile. The hinge is weak and teeth along the edge of the valves are degenerate; the foot, likewise, is reduced.

The siphons of immobile, deeply burrowed lamellibranchs are usually encased in periostracum or even calcium carbonate. These mantle extensions may be quite long, as in *Panope* (Fig. 9.67b), where their extreme length makes retraction between the valves impossible. If the burrow is so deep that the siphons cannot reach the surface, its walls are plastered with mucus to minimize fouling.

Some lamellibranchs burrow not in sand or mud but in hard substrata such as rock and wood. Rock borers include *Pholas* (Fig. 9.68) and *Barnea.* These bivalves attach the foot and then rock the shell back and forth to abrade a hard surface. The anterior half of the shell is sculptured into sharply cutting teeth. Rocking of the shell requires antagonistic contractions of the adductor muscles. The anterior adductor muscle has migrated dorsal to the hinge, so its contraction opens the valves; the posterior adductor closes them.

Wood-boring lamellibranchs include the shipworms of the genus *Teredo* (Fig. 9.69a). These bivalves are quite bizarrely formed. Large siphons extend nakedly from the minute valves. The incurrent siphon is continuous with an extension of the mantle cavity and contains the gill. This siphon applies a calcareous lining to the shipworm's burrow, whose open end may be sealed by paired calcareous plates, the **pallets.** *Teredo* bores by opening and closing its small anterior valves, which are pressed tightly against the blind end of the burrow. Ingested cellulose may be broken down by bacteria within the digestive gland. Shipworms may live for several years, and in some genera, they grow to 2 m in length. A colony of these mollusks can devastate the wooden ships and pilings in which they live (Fig. 9.69b); thus they rank as economically serious pests. However, their activity is beneficial in recycling wood debris in the sea.

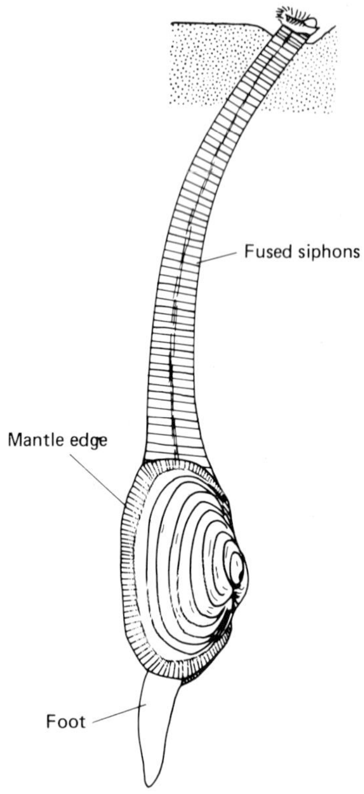

(a)

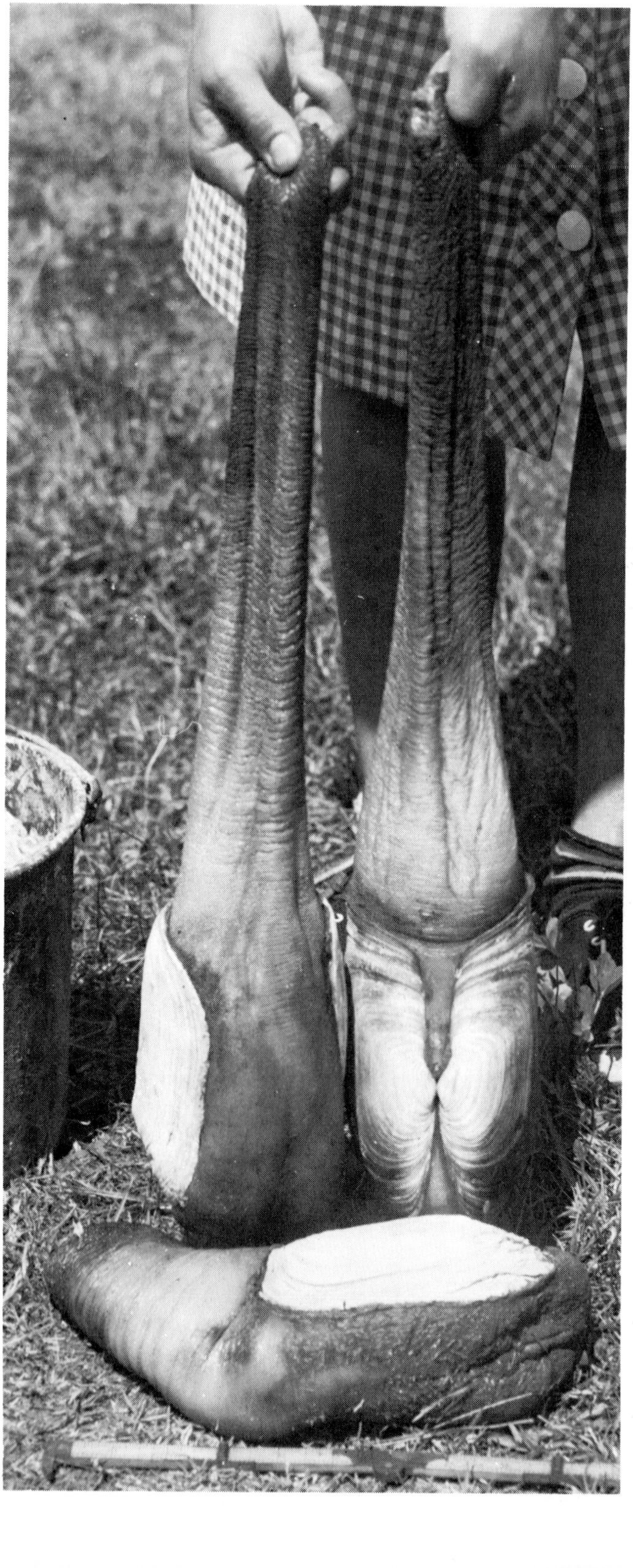

(b)

Figure 9.67 (*a*) *Mya arenaria*, a deep burrower with fused siphons; (*b*) The geoduck, *Panope generosa*, a large burrowing lamellibranch of the Pacific coast whose body and siphons cannot be enclosed within the valves of the shell. (*a* from Moore, *Treatise on Invertebrate Paleontology*, Part N; *b* courtesy of L. Milne and M. Milne.)

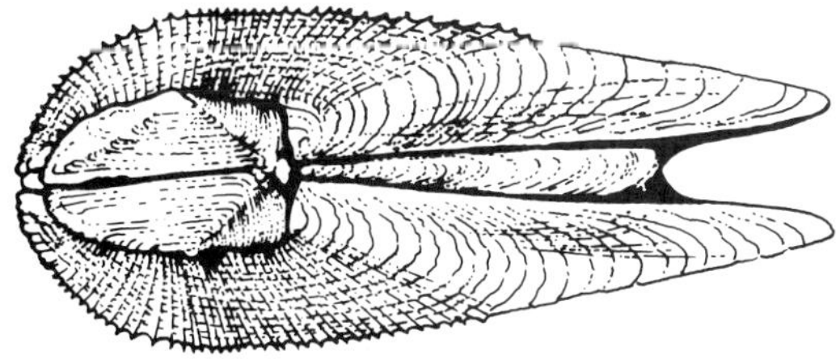

Figure 9.68 *Pholas dactylus*, a rock-boring lamellibranch. (From Woodward, *A Manual of the Mollusca.*)

Freshwater Lamellibranchs

Over 1000 lamellibranch species live in fresh water. Freshwater colonization has occurred in numerous lines, but the family Unionidae has been most successful. Many familiar freshwater clams, including *Unio, Lampsilis,* and *Anodonta,* are members of this group. Lamellibranch adjustments to freshwater life include modifications of their continuity systems. Fertilization and larval incubation are common in the mantle cavity, often within the gills themselves.

Parasitism among freshwater lamellibranch larvae also is common. Released in great numbers, modified veliger larvae called **glochidia** attach to the external surfaces of fish. On some glochidia, such as those of *Anodonta* (Fig. 9.70), hooked valves grasp the host. The attached glochidium causes fish tissues to form a cyst around the parasite. Within this cyst, the larva feeds on host tissues and fluids, gradually assuming adult characters. At some stage in its maturation, the young lamellibranch drops from the fish and burrows into the substratum, where it proceeds with adult life.

Certain species of the North American clam *Lampsilis* ingeniously ensure the transfer of their glochidia to a fish host. Females of this genus often bear a ribbonlike outgrowth of the mantle margin near the excurrent opening (Fig. 9.71). The clam directs this outgrowth toward the river current. At the ribbon's forward end, an eyelike spot is present, and the rear end resembles an undulating tail. Variously colored stripes may occur along the structure. Evidently, the *Lampsilis* mantle ribbon is a fish mimic. Fish approach these clams and hover over their mantle ribbons, apparently inspecting them as potential prey. The shadow of a hovering fish may stimulate the release of up to 300,000 glochidia, which attach parasitically to the fish's gills.

Adult parasitism is rare among the Bivalvia. *Entovalva* lives in the digestive tract of certain sea cucumbers. Commensals are only slightly more common. The latter also tend toward echinoderm hosts, such as sea urchins, brittle stars, and sea cucumbers.

CLASS SCAPHOPODA

A fourth conchiferan class is the Scaphopoda. About 350 recent species have been described; those belonging to the genera *Cadulus* and *Dentalium* (Fig. 9.72) are the best known. Scaphopods commonly are called tusk shells because of their long, slightly curved cylindrical bodies. Elongation has occurred along the anterioposterior axis. The tubular scaphopod shell tapers from its larger anterior aperture to a smaller, posterior opening. The visceral mass is confined to the dorsal side of the shell, while the mantle cavity extends ventrally along its tapered length. Scaphopods burrow in sandy ocean bottoms, often at considerable depths. Characteristically, they bury the larger anterior end of the shell in the sand; the smaller posterior aperture remains exposed just above the substratum.

Scaphopods are selective deposit feeders with unique food-gathering tentacles called **captacula** (Fig. 9.72). Captacula extend from the base of the head into the substratum, where they capture food (often foraminiferans) on their adhesive, spoon-shaped tips. Small food particles are transported back to the mouth along ciliated grooves; if large food items are captured, the captacula themselves may bend to place their catch in the mouth. The scaphopod mouth occupies the end of a proboscislike extension of the body wall. There is no distinct head in these mollusks, and no cephalic sense organs are present. A buccal cavity does contain a strong radula, however. The simple digestive tract (Fig. 9.72) includes an unmodified stomach with large, branching digestive glands. Digestion is largely extracellular within the stomach. A coiled intestine produces feces, which are expelled through the mantle cavity.

Scaphopods have no gills and their circulatory system is rudimentary. Water is drawn slowly into the mantle cavity by ciliary action and/or the suction caused by pedal extension. Respiratory exchange occurs across the mantle wall, whose surface area is increased by folding. Deoxygenated water is expelled by a sudden retraction of the foot. Scaphopods have no heart, and any blood vessels are poorly defined. Blood, circulating through a simple system of tissue sinuses, receives its impetus from the muscular activity of the body wall. Scaphopods have a pair of nephridia, which terminate in pores near the anus (Fig. 9.72).

Tusk shells burrow much as bivalves do. Their sturdy, pointed foot bears lateral elaborations for an-

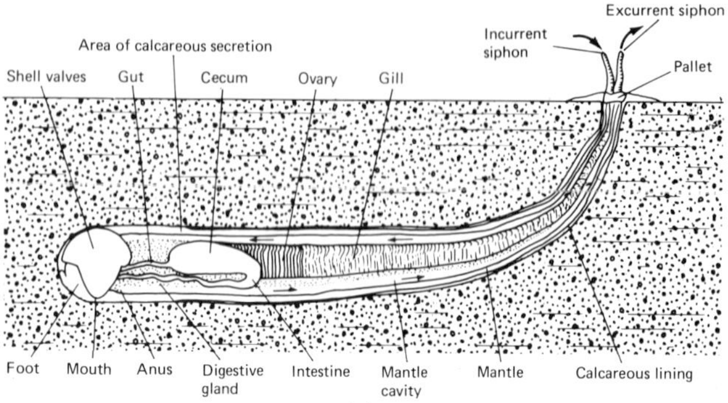

(a)

Figure 9.69 (*a*) Longitudinal section through a wood-boring lamellibranch, *Teredo*, in situ in its burrow; (*b*) Wood broken apart to expose the burrows of the shipworm *Teredo*. (*a* from Moore, *Treatise on Invertebrate Paleontology, Part N*; *b* from Andrews, *Sea Shells of the Texas Coast*.)

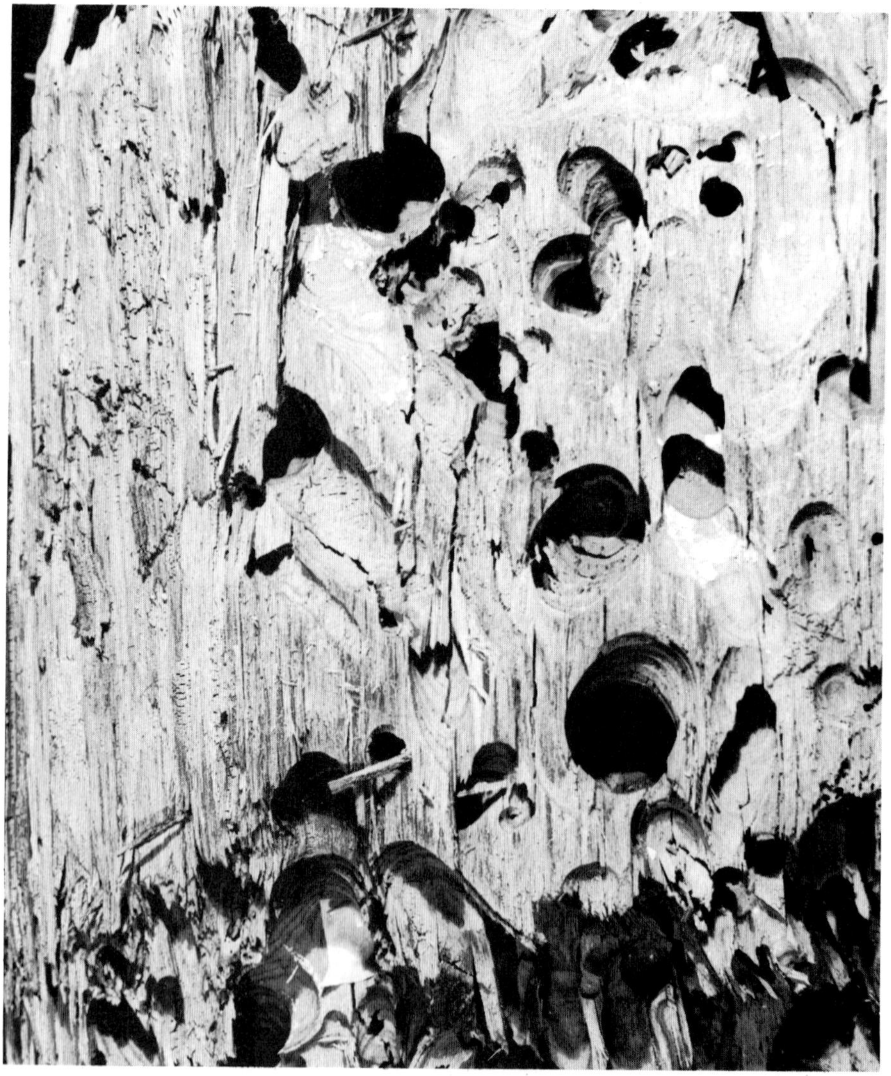

(b)

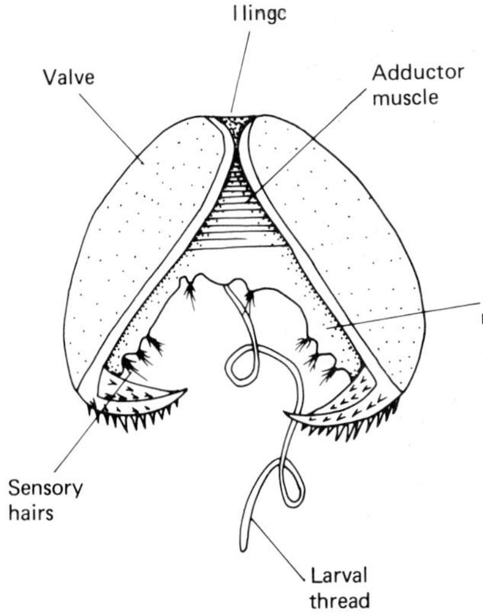

Figure 9.70 Glochidium of *Anodonta imbecillis.* (From Pennak, *Fresh-Water Invertebrates of the United States.*)

Figure 9.71 The edge of the mantle in females of the freshwater clam, *Lampsilis ovata*, is modified to mimic a fish. As a predatory fish snaps at the apparent prey, the clam blows glochidia out of its excurrent siphon. These larval forms attach onto the fish's gills. (From W. Wickler, *Mimicry*, 1968 McGraw-Hill.)

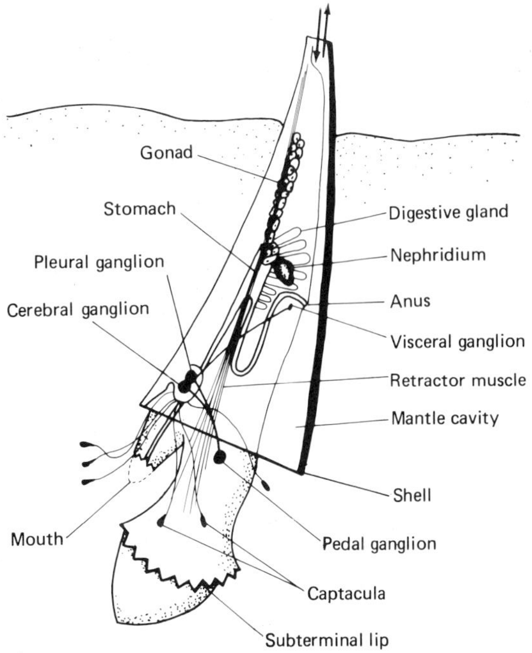

Figure 9.72 Structure of a scaphopod, *Dentalium.* This mollusk lies buried in sand except for its posterior tip through which inhalent and exhalent water currents pass. (R.W. Hegner and J.G. Engemann, *Invertebrate Zoology*, 1968, Macmillan.)

chorage. A terminal disk or a subterminal lip (Fig. 9.72) can be extended to hold the foot firmly in place. Thrusts of the foot followed by pedal anchorage and retraction of the body and shell pull the burrowing animal downward.

The scaphopod nervous system is rudimentary, resembling that of primitive bivalves. The four typical ganglia—cerebral, pleural, pedal, and visceral—occur in pairs, along with their associated nerve cords. The system remains uncentralized, as might be expected for such sedentary mollusks. No sense organs are present on the reduced head, and they are poorly developed elsewhere. A subradular organ occurs, as do statocysts in the foot. In addition, the captacula bear tactile receptors which may transcribe information important for burrowing.

Continuity systems within the Scaphopoda are rather simple. The archetypal molluscan plan is apparent in external fertilization and free-swimming larvae. Scaphopods are diecious, and their single large gonad occupies much of the posterior body (Fig. 9.72). This gonad drains through the right nephridium into the mantle cavity. Scaphopod development resembles that of marine bivalves. Both trochophore and veliger stages occur, and the mantle and shell originally are bilobed (Fig. 9.73). In scaphopods, the two mantle lobes fuse ventrally, producing a tube. The mantle–shell tube elongates to produce the characteristic scaphopod form.

The relationship of scaphopods to other conchiferan groups is uncertain. Some authorities consider them intermediate between gastropods and bivalves. Certainly, scaphopods resemble the latter class in many ways. Their symmetry, reduced heads, burrowing ways, and, perhaps most strikingly, development all point to the bivalves as relatives of some degree. Proponents of the scaphopod—bivalve linkage even suggest homology between the captacula and the palps proboscides of many protobranchs.

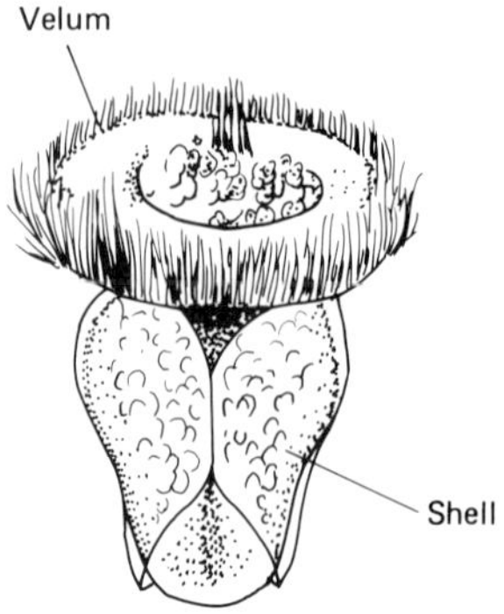

Figure 9.73 Veliger of a scaphopod. (From Bronn, Ed., *Klassen und Ordnungen des Tierreichs, Vol. III, Part One.*)

CLASS CEPHALOPODA

Nearly all of the mollusks discussed to this point are rather sluggish creatures, maintaining low profiles within protective shells. Such is the general nature of life in this phylum; indeed, the slow but efficient craft of the mollusks is the basis of their evolutionary success. We turn now to an adaptive line that has produced quite untraditional mollusks. This line is represented by the Cephalopoda, the fifth and final conchiferan class. The class, which is entirely marine, contains many familiar members, including cuttlefish, squid, and octopods. About 650 cephalopod species are extant, but fossils representing over 7500 additional forms have been described.

Cephalopods are basically pelagic organisms, although some secondary adaptations to bottom crawling have occurred, particularly among the octopods. They are carnivores, and most swim actively in pursuit of prey. Their predaceous life-style involves specializations that, for mollusks, are nothing short of magnificent. Cephalopod evolution has altered all of their systems by varying degrees to support their present way of life. In almost all living forms, the shell is greatly reduced or absent. The functional axes of the cephalopod body have shifted, and the head, foot, and mantle cavity are modified for a unique locomotion. The circulatory system has become closed, and the coelom has expanded. Nowhere, however, are cephalopods more specialized then in their nervous and sensory structures. The class boasts the most sophisticated brain and eyes of any invertebrate group. The cephalopods also deserve recognition as some of the fastest, largest, and most intelligently active members of the invertebrate world. Mollusks may owe their numerical success to a certain adaptive dullness, but in the living cephalopods, the phylum displays some of this planet's most adventuresome animals.

Cephalopod Unity

MAINTENANCE SYSTEMS

General Morphology

Figure 9.74 depicts a generalized modern cephalopod. The cephalopod body form can be derived from that of the generalized mollusk by shifting the original anterioposterior axis (the long axis of the head and foot) into line with the dorsoventral axis (represented by an elongated visceral hump). This shift profoundly alters the cephalopod head and foot. The area surrounding the head (derived in part from the foot) is divided into numerous arms arranged around the mouth. These arms bear adhesive muscular cups or suckers, which are used in prey capture and locomotion. The remaining parts of the foot are rolled into a funnel, which forms a tubular outlet for the mantle cavity. The mantle wall is a highly muscular pumping organ. Mantle muscles pump water into the cavity, where gill aeration occurs. Water is expelled through the funnel, producing a locomotor force.

As we mentioned previously, the shell is reduced or absent in almost all living cephalopods. Ancestral members of the class did have large shells, however, and some consideration of these fossil forms is necessary for an understanding of the group as a whole.

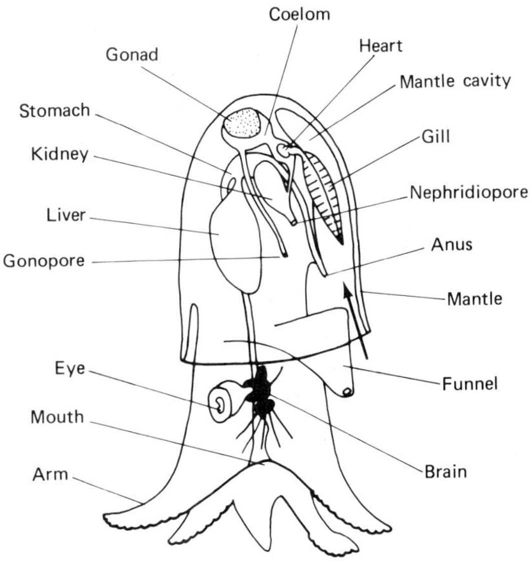

Figure 9.74 Lateral view of a generalized modern cephalopod. (From Lankester, Ed., *A Treatise on Zoology, Vol. V.*)

The earliest known cephalopods were the nautiloids, a group surviving today only in the genus *Nautilus* (Fig. 9.75). Originally, these early mollusks had straight conical shells, but curved and then coiled forms developed later (Fig. 9.89). Nautiloid shells had a number of chambers; the living animal occupied the last chamber, the only one open to the outside world. Successive chambers were divided from one another by transverse septa. Each time the nautiloid outgrew a chamber, it secreted a new one and then sealed off the old chamber with a septum. All the septa were perforated by a single calcareous tube called the **siphuncle** (Fig. 9.75). The siphuncle contained an extension of the mantle cavity and allowed materials to be transported among the shell compartments. Filled with gas and/or fluid, older compartments served as buoyancy devices.

The wholly extinct ammonoids originated from nautiloids with coiled shells. Ammonoid shells were more streamlined than those of the nautiloids (Fig. 9.90). Their compressed angled surfaces minimized friction during locomotion, and ventral keels were common.

All the modern cephalopods except *Nautilus* are members of a third group, the coleoids. These animals—the squid, the cuttlefish, and the octopods—have greatly reduced shells or no shells at all. The coleoid condition can be traced to the oldest known members of the group, the fossil belemnoids. These cephalopods had a reduced shell which was surrounded by the mantle. This internal shell had a straight chambered region, thick lateral shelves, and a broad, forwardly projecting shield (Fig. 9.76). The shield was the attachment site for the muscles of the mantle wall. From the belemnoid shell, the forms of the modern coleoids can be derived (Fig. 9.76). The deep-water cuttlefish *Spirula,* for example, has lost the projecting lateral walls and shield, while the chambered parts of its shell have coiled. *Sepia,* the European cuttlefish, has a relatively straight septal area, but only short shelves and no shield. *Loligo,* the common squid, has lost the main body of the shell entirely; only a portion of the shell remains as a dorsal, horny strip called the pen. Finally, in *Octopus* and related genera, the shell has disappeared altogether.

Nutrition and Digestion

Cephalopods, as we mentioned, are raptorial carnivores. Their arms grasp and entangle prey such as fish, crustaceans, and other mollusks. Suckers adhere tightly as the victim is forced against strong beaklike jaws (Fig. 9.77a). These muscular jaws lacerate prey tissues, which are transported into the buccal cavity by a radula. Salivary glands discharge powerful proteases which reduce some prey tissues prior to ingestion. In some cephalopods, salivary poisons subdue struggling victims.

Contrary to the situation in most mollusks, gut ciliation in the cephalopods is minimal. Food is transported through the esophagus by muscular contractions and enters the stomach (Fig. 9.77b). The cephalopod stomach is a muscular churn which is attached to a larger cecum. Digestion is entirely extracellular. Two glands discharge digestive enzymes through a common duct into the stomach and cecum. These glands are often quite distinct morphologically. The smaller, but less compact gland commonly is called a pancreas; the larger, more solid member of the pair is called the liver. Digestion begins in the stomach, but at intervals, partially digested gastric contents, mixed with additional enzymes from one of the two glands, are discharged into the cecum. There the digestive process is completed. Undigestible particles are trapped on ciliated folds at the cecal opening and returned to the stomach for elimination through the intestine. Absorption usually occurs across the cecal or intestinal wall in *Loligo*, but may take place within the liver in *Octopus* and *Sepia*. The cephalopod intestine discharges at an anus near the exhalent funnel (Fig. 9.74).

(a)

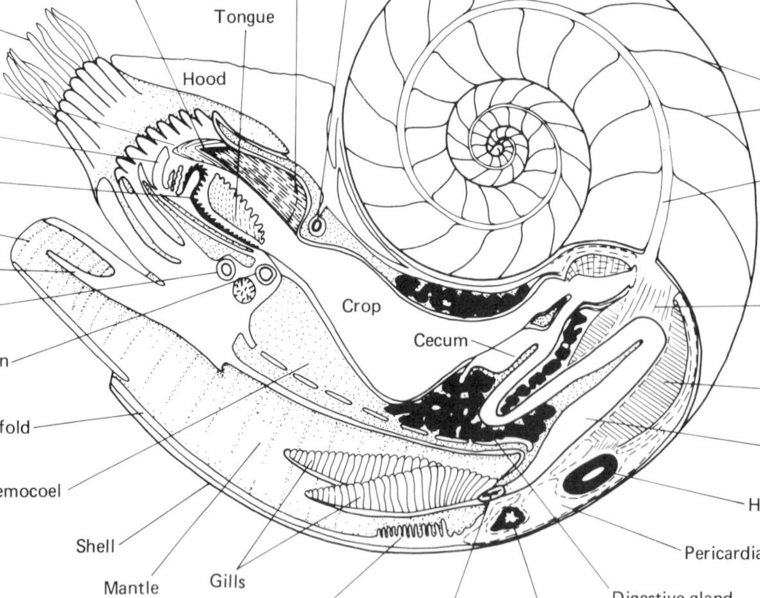

(b)

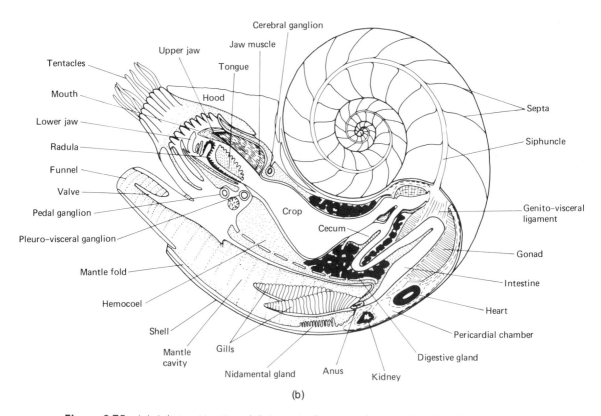

Figure 9.75 *(a)* A living *Nautilus*; *(b)* A sagittal section through *Nautilus* showing major anatomical features. (*a* courtesy of the California Academy of Sciences, photo by Lloyd Ullberg; *b* from R.D. Barnes, *Invertebrate Zoology*, 3rd ed., 1974, W.B. Saunders.)

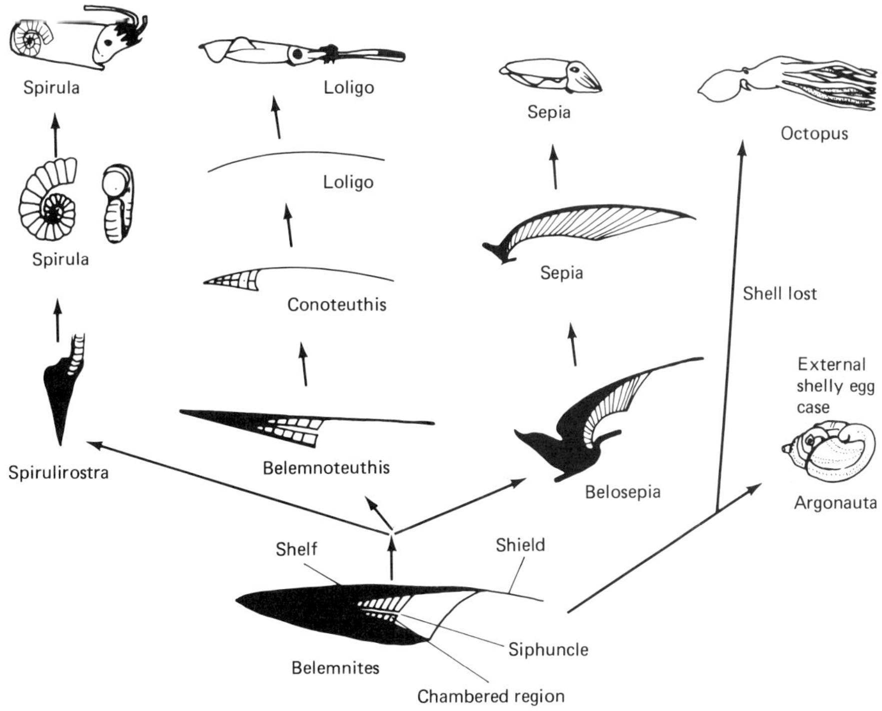

Figure 9.76 Lines of shell modifications within coleoid cephalopods. (For further details, see text.) (From R.R. Shrock and W.H. Twenhofel, *Principles of Invertebrate Paleontology*, 1953, McGraw-Hill.)

Circulation and Respiration

Like other mollusks, cephalopods exchange oxygen and carbon dioxide across their gills and distribute these gases through a blood transport system. Cephalopods, however, have evolved quite specialized gills and circulatory equipment in association with their faster life-styles. Because of their increased metabolic and behavioral activity, such typically molluscan features as gill ciliation and an open circulatory system have been eliminated in the cephalopods. Cephalopod ctenidial-filaments are fan shaped and bear no cilia. Since these animals are basically pelagic, cilia are no longer required to remove sediment from the gills. Perhaps a more direct reason for their absence, however, is the development of the mantle wall as a pumping unit. The muscularization of the mantle represents an adaptation both for locomotion and for increased aeration of the gills. Cephalopods are active creatures, and cilia alone cannot produce respiratory currents strong enough to meet their higher oxygen requirements.

Respiratory flow through the cephalopod mantle cavity is reversed from that seen in other members of the phylum (Fig. 9.78a). Because the ventral foot forms an excurrent siphon, water necessarily exits the cavity ventrally. Water enters the mantle cavity dorsally and laterally. Cephalopod gill filaments possess supporting rods on their dorsal rather than their ventral surfaces, an alteration consistent with the dorsal entrance of the respiratory current. However, blood vessels within the gill axis maintain their original molluscan positions; that is, the afferent vessel lies above the efferent vessel. Cephalopods thus lack the counterflow relationship exploited by other mollusks. (It is interesting to note, however, that the excurrent openings of the digestive, excretory, and genital tracts have shifted to the ventral side of the gills, in line with the excurrent flow of water from the mantle cavity; cephalopods thus continue to exploit the mantle current as a sewerage mechanism.) Despite the loss of the counterflow relationship, respiratory efficiency remains quite high because of the rapid water flow

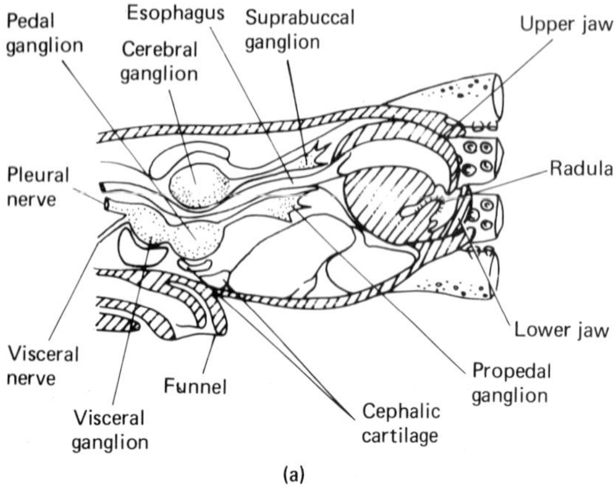

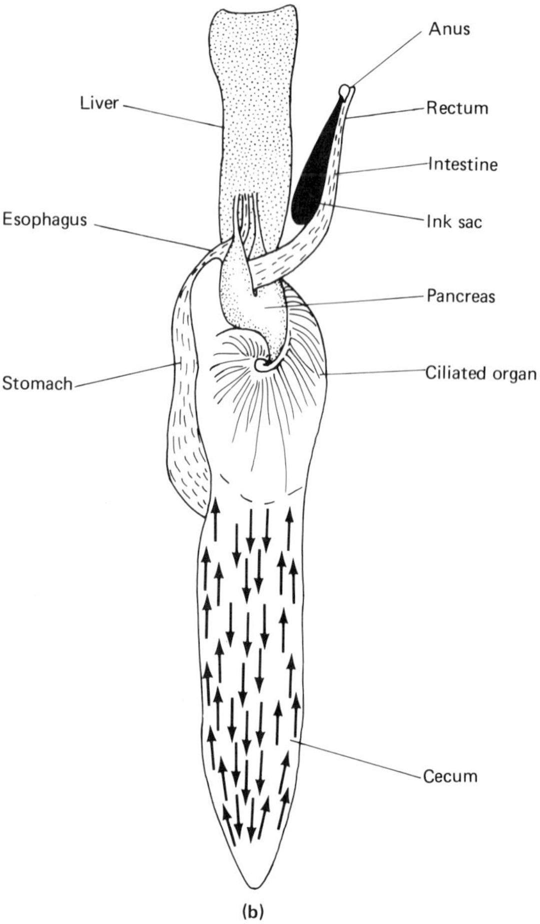

Figure 9.77 (*a*) Longitudinal section through the head region of a squid, *Loligo*, showing structures in the buccal region, cephalic cartilages, and major ganglia in the esophageal region; (*b*) Ventral view of the digestive system of *Loligo*. Arrows indicate directions of ciliary currents in the cecum. (*a* from Brown, *Selected Invertebrate Types; b* from Moore, Ed., *Treatise on Invertebrate Paleontology, Part I,* 1960.)

through the mantle cavity and the closed nature of the circulatory system.

Cephalopod blood contains hemocyanin which, as in other mollusks, is dissolved in the blood plasma. The closed circulatory system comprises a central systemic heart, a well-developed system of major and minor arteries, numerous capillary beds, a venous system, and unique auxiliary pumping stations associated with each gill (Fig. 9.78b). The latter, known as **branchial hearts**, generate high blood pressures within the gills. Blood returning from the gills travels to the central heart, which is surrounded by a large pericardial cavity. This cavity approximates a true coelom; commonly, it surrounds considerable portions of the digestive tract and may communicate with the gonads. This large coelom cushions internal organs against the disruptive pulsations of the mantle wall and contributes to the larger body size potential of the cephalopods.

From the central heart, blood is pumped both anteriorly and posteriorly through major arteries. Minor arteries join capillary beds, where material exchanges with the body tissues occur. Capillaries drain into one of several main venous systems. A large vena cava serving the head region passes posteriorly and bifurcates near the paired nephridia. Each branch traverses the nephridium on its side and enters a branchial heart. Paired anterior and posterior mantle veins join branches of the vena cava near the branchial hearts; the right branch alone also receives a gonadal vein. After the blood is oxygenated under high pressure within the capillaries of the gills, it is returned to the central heart for another circulation.

Excretion and Osmoregulation

Cephalopod nephridia comprise two paired sacs along the venous route to the branchial hearts (Fig. 9.78b). As the vein passes across the renal sac, it gives rise to large appendages which extend into the nephridium.

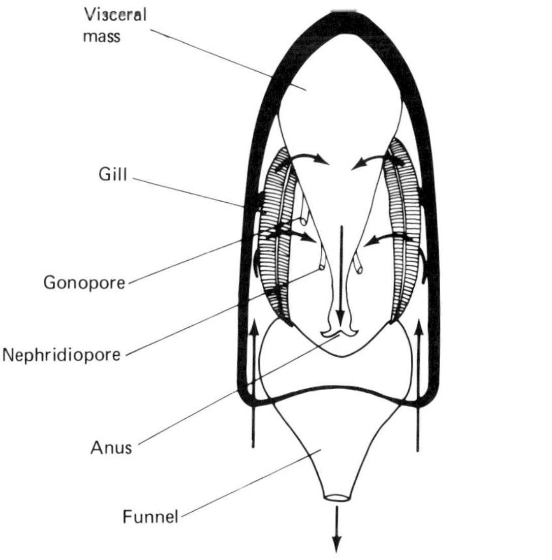

(a)

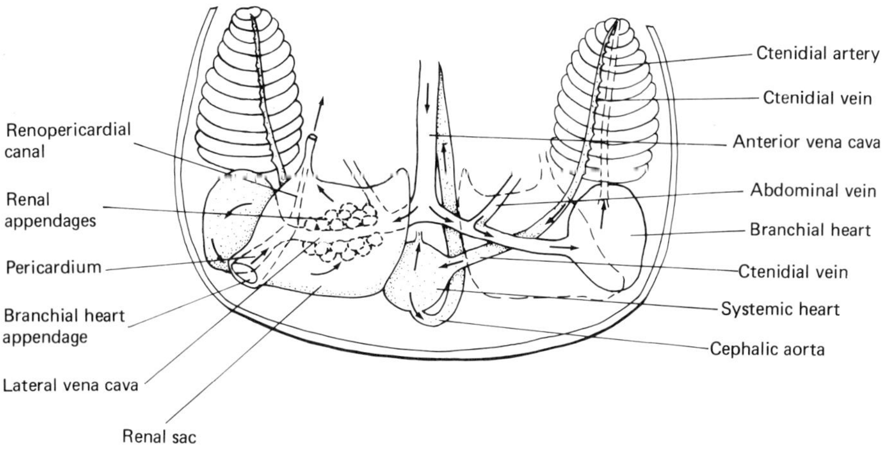

(b)

Figure 9.78 (*a*) Mantle cavity of a cuttlefish, *Sepia officinalis.* Arrows indicate the direction of respiratory current; (*b*) Major blood vessels and excretory structures in *Octopus dofleini*, (*a* from C.M. Yonge, Phil. Trans. Roy. Soc. Lond., 1947, *232:*443; *b* from W.T.W. Potts, Biol. Rev., 1967, *42:*7.)

Coincident with the pumping of the branchial hearts, venous blood ebbs and flows through these renal appendages. Waste materials are secreted across the appendage walls and into the nephridium. The cephalopod nephridium likewise receives wastes from the pericardial coelom, with which it is connected via a renopericardial canal. Also, special appendages of the branchial hearts possibly contribute a waste-laden filtrate to the pericardial fluids. Some resorption of these fluids occurs along the renopericardial canal. The nephridium itself reaches the outside world through a nephridiopore located near the funnel (Fig. 9.78a).

Cephalopods are exclusively marine animals; hence their osmoregulatory problems are not intense. Control over ionic concentrations within body fluids is important, however, in the maintenance of buoyancy among some groups. Squid, in particular, can be rather dense animals, and to avoid sinking, some types maintain a high proportion of lighter ions within an expanded pericardial coelom.

ACTIVITY SYSTEMS

In their nervous and muscular organization, sense organs, and behavior, the cephalopods have diverged radically from their molluscan relatives. Typically, mollusks smell and feel their way slowly through their environments and withdraw into their shells when danger threatens. Cephalopods, in contrast, depend primarily on visual stimuli and a sophisticated behavioral repertoire. These atypical mollusks make critical decisions in their complex brains and then fight, flee, or respond in various other ways with a host of effector systems.

Muscles

Muscles play a more central role in cephalopod activity than in other molluscan classes. Ciliary action is of little significance in this group, as muscles control food capture, digestive transport, production of respiratory currents, and all locomotor activity. Important cephalopod muscles include those of the arms, the funnel, and the mantle cavity wall.

Cephalopod arms are very muscular appendages. They extend and withdraw rapidly, and their ability to entangle and adhere to prey is highly developed. Adherence involves numerous suction cups. These cups have muscularized walls which may bear hooks or other hard reinforcements. Contraction of the wall creates a partial vacuum within the cup when it is applied tightly to a surface. This vacuum grips prey and also provides locomotor traction for bottom-crawling cephalopods.

The muscles of the mantle wall and funnel produce a respiratory current and locomotor jet propulsion. The mantle wall contains both radial and circular fibers. When radial muscles contract, the mantle cavity increases in volume and water is inhaled. When circulars contract, the mantle closes tightly around the head. Placed under considerable pressure, mantle waters are released in a jet through the funnel. The force of the water jet propels the cephalopod in the direction opposite that of the funnel opening. Orientation of the funnel is under muscular control, and a wide range of forward, backward, and lateral movements is possible.

Nerves

Cephalopods possess the most highly organized central nervous system in the invertebrate world. Only the highest arthropods compete with the quality and centralization of cephalopod equipment. In terms of the size of the system, competition with these mollusks comes only from the higher vertebrates.

All of the typical molluscan ganglia have fused in a very large brain around the esophagus (Fig. 9.79a). In *Octopus*, this organ may comprise over 150 million neurons. (By comparison, a typical crustacean has only 30,000 brain cells; we humans have 5 billion in our cerebrum alone.) The supraesophageal part of the brain is derived from the cerebral ganglia. This area gives rise to buccal nerves which join a nervous ring surrounding the buccal cavity. The buccal ring contains both superior and inferior buccal ganglia. Large optic lobes arising within the cerebral ganglia receive numerous optic neurons. These lobes may comprise one-half of the total brain mass. Below the esophagus are the brain centers derived from the pedal ganglia. There nerves are dispatched to the funnel and arms. Visceral and pleural ganglia are fused posteriorly to the cerebral and pedal centers. From the former, three pairs of fibers innervate the main body (Fig. 9.79a). One pair of these nerve cords serves the central organs, including the gills where branchial ganglia are located. A second pair innervates the stomach area; these cords join a gastric ganglion at the junction between the stomach and the cecum. Finally, the third and largest pair of visceral–pleural nerves serves the mantle.

On either side of the mantle, a **stellate ganglion** (Fig. 9.79b) is involved with the ordinary pumping of the mantle wall. When extraordinary mantle activity is required, as in escape swimming movements, a separate system of giant axons overrides the commands of the stellate ganglia. Cephalopod giant axons are the world's largest known nerve fibers. In *Loligo*, these fibers are organized into three descending orders (Fig. 9.79b). First-order fibers within the visceral ganglia are excited by sensory nerves. These giant axons communicate with a second order of large fibers which lead to the stellate ganglia. There third-order fibers of variable size innervate the muscles of the mantle wall. Longer fibers are larger in diameter and thus relay impulses at a faster rate. Shorter fibers are smaller; hence they conduct more slowly. This design ensures that, regardless of the distance to be traveled, all im-

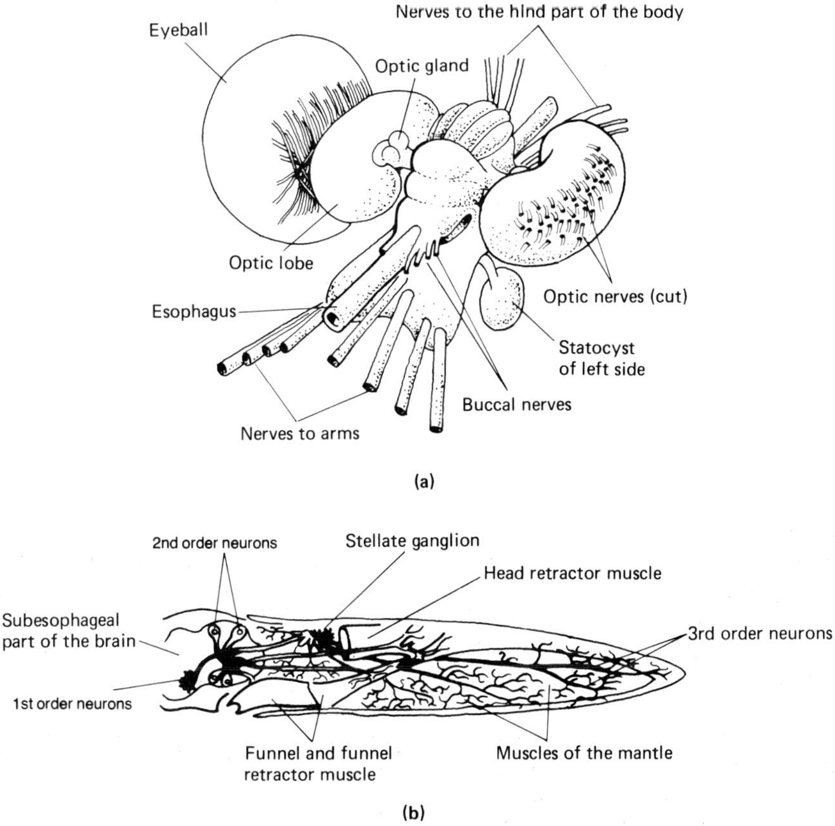

Figure 9.79 (*a*) A diagram of the brain and associated structures in *Octopus;* (*b*) The giant nerve fiber system of *Loligo* showing the location of the first, second, and third order neurons. (For further details, see text.) (*a* from M. Wells, *Lower Animals*, 1968, McGraw-Hill; *b* from M.J. Wells, *Brain and Behavior in Cephalopods*, 1962, Heinemann Educational Books.)

pulses from the third-order giant axons arrive at their associated mantle muscles at the same time. Thus, all of the circular muscles of the mantle wall contract simultaneously, and the cephalopod jets away at a very high speed.

Sense Organs

Cephalopod sense organs include cephalic eyes, statocysts, tactile sites, and chemoreceptors. The latter two are concentrated on the tentacles, where the suckers are particularly well endowed. Osphradia are present only in *Nautilus*.

The major feature of cephalopod sensory biology is their remarkable eyes. Cephalopods possess the most sophisticated visual organs of any invertebrates and depend on visual stimuli as their primary source of environmental information. Cephalopod visual acuity is matched or surpassed only by the higher vertebrates. Indeed, the eyes of *Octopus* (Fig. 9.80a) are remarkably similar to vertebrate eyes. Each eye rests within a cartilaginous socket, one on either side of the head, and comprises a cornea, an iris, a lens, and a retina. Principal differences between cephalopod and vertebrate eyes include their nervous organization and means of focusing and light adjustment. The cephalopod retina receives light directly, and the optic nerves exit from the rear of the retinal cells. In contrast, vertebrate eyes are of the "indirect" type; the retina is turned away from the lens and the optic nerves must pass through the photoreceptor layer, thus creating a "blind spot" within the eye. The cephalopod retina is packed very densely. In *Sepia*, for example, an area corresponding to the human fovea may contain nearly twice as many photorecep-

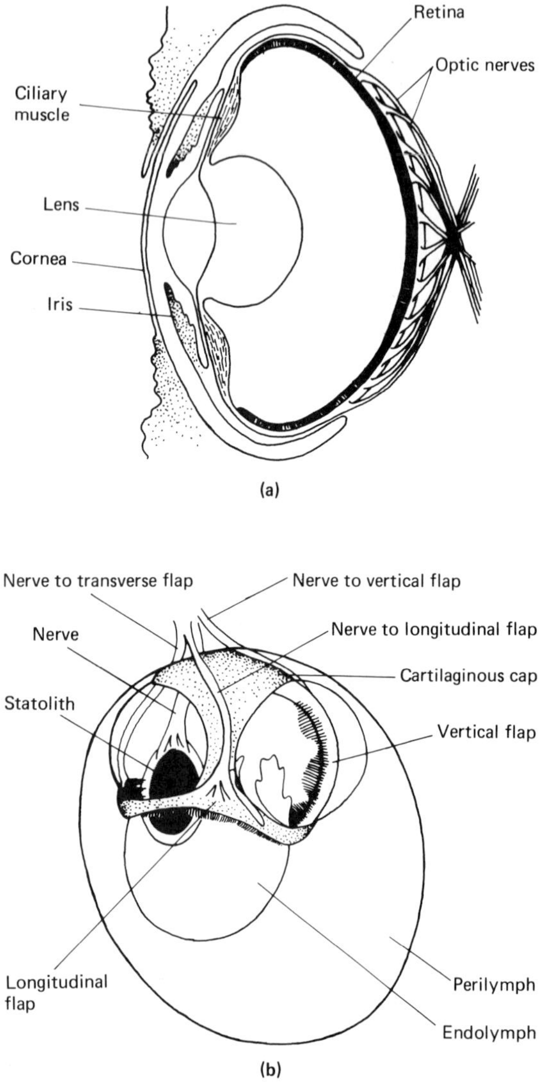

Figure 9.80 (*a*) A vertical section through the eye of *Octopus;* (*b*) Structure of the left statocyst of *Octopus.* (From M.J. Wells, In *Physiology of Mollusca, Vol. II,* Wilbur and Yonge, Eds.)

tive cells per unit area as our fovea does. Obviously, the resolving power of the cephalopod eye is quite high.

The cephalopod lens is spherical and rigid. Thus its focal depth cannot be adjusted by changes in lens shape, as in vertebrates. Cephalopods focus by contracting and relaxing the muscles that surround the eyeball, thus squeezing the lens forward or backward. The amount of light entering the eye is regulated by an iris diaphragm. Also, migrations of the retinal pigments may influence the quantity of light impinging on the photoreceptive cells.

Cephalopod statocysts are more elaborate than any balancing centers we have described thus far. These sense organs occupy cartilaginous cavities on either side of the brain (Fig. 9.79a). Inside the statocyst, special ciliated flaps extend in horizontal, vertical, and transverse planes (Fig. 9.80b). Differential stimulation of these flaps provides a cephalopod with precise information concerning its posture. Such precision is critical to the execution of complex locomotor patterns by these animals.

Behavior

Cephalopods demonstrate the most flexible behavioral repertoire of any invertebrates. These mollusks are intelligent and often personable creatures, which display more intricate behavioral responses than any other invertebrates. Also, cephalopods possess an array of specialized effectors, such as chromatophores, ink sacs, and luminescent structures.

Chromatophores are tiny elastic cells scattered over the body surface of cephalopods without external shells. Containing pigments of various colors, these cells are responsible for the changing hues of many of these mollusks. Color changes involve the expansion and contraction of differently colored chromatophores. Radial muscles are attached to each of the pigment cells; their contraction enlarges the chromatophore and thus its particular pigment is displayed (Fig. 9.81). When these muscles relax, the cell's elasticity restores its original small size and its pigment is unnoticeable.

As cephalopod chromatophores are under muscular control, the nervous system necessarily coordinates color displays. In the brain of *Octopus*, over half a million neurons are associated with chromatophore muscles. These brain cells relate chromatophore activity to environmental stimuli. Precise patterns of coloration require the coordination of thousands of pigment cells. Such patterns may involve simple camouflaging. For example, an octopod in its lair commonly assumes the color of surrounding rocks. While crawling along the bottom, an octopod may change its color continuously to match the tones of each local rock or patch of sand. The cuttlefish *Sepia* can blend with any number of backgrounds and reportedly gives a creditable imitation of a checkerboard!

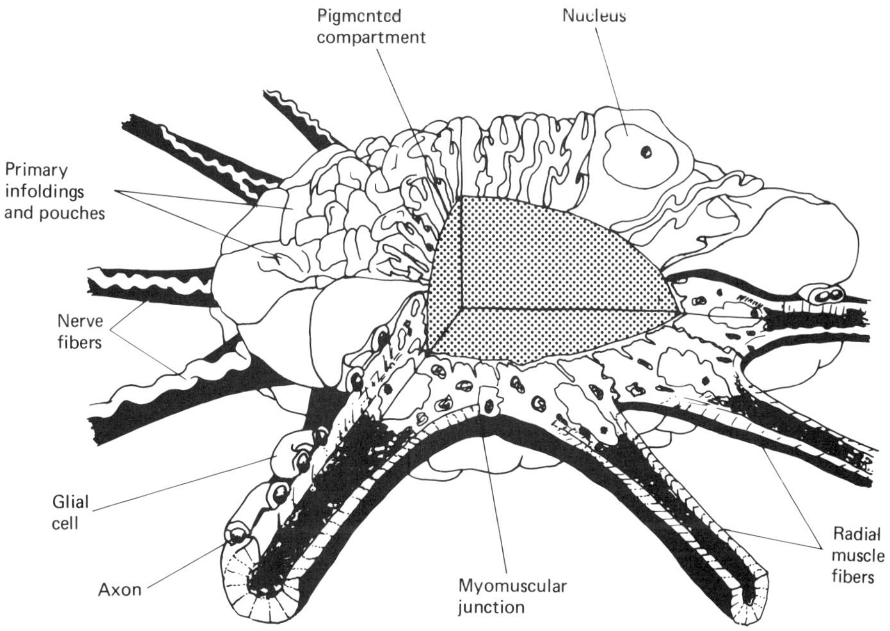

Figure 9.81 Cut-away illustration of a retracted squid chromatophore showing the arrangement of nerve fibers, radial muscle fibers and the folding of the cell membrane around the pigment containing compartment. (From R.A. Cloney and E. Florey, Zeit. f. Zellforsch., 1968, *89*:250.)

Most color changes in cephalopods are correlated with specific behavioral patterns (Fig. 9.82). Chromatophore action in alarmed cephalopods may frighten would-be predators. Both *Sepia* and *Octopus* blanch their bodies and then simulate huge dark eyespots (Fig. 9.82f). By far the most elaborate color displays are associated with cephalopod males during courtship. *Sepia* males exhibit a "zebra pattern," a complex display of purple–brown with yellow or white stripes (Fig. 9.82a). During courtship, the squid *Loligo* sports distinctive short white flecks and a long ventral stripe of the same color.

Because it is such a brilliant cephalopod, *Sepia* has been the object of extensive chromatophore research. Its bright colors are attributable in part to a layer of **iridocytes**; these reflective cells lie below the chromatophores and intensify their colors.

The cephalopod ink sac is a relatively large vesicle lying near the intestine (Fig. 9.77b). It opens through a muscularized duct into the rectum. From the walls of the sac, an ink gland secretes a brownish or black fluid; the dark color is attributable to concentrated melanin pigments. Under appropriate environmental stimulation, commonly the presence of threatening predators, this ink is ejected through the anus. Dispersing ink may blacken the local area, thus allowing the cephalopod to escape unseen in the dark. Certain alkaloids in the ink may temporarily desensitize fish chemoreceptors and so further aid the flight to safety. Some cephalopods, such as the cuttlefish *Sepia*, eject compact puffs of ink to fool would-be predators. When threatened by a potential predator, *Sepia* expands its dark chromatophores and then ejects a compact volume of ink; in contour, this ink mass resembles the darkened mollusk itself. *Sepia* then blanches its body by expanding only its lightest chromatophores. The predator, intent on capturing what it believes to be a black cuttlefish, attacks the ink mass while *Sepia* escapes! This entire series of events may take place in 1 second.

Deep-water cephalopods usually lack ink sacs, and their chromatophores are often modified. In their very dark world, luminescent devices are more practical. These devices include symbiotic, bioluminescent bacteria, as well as structures called **photophores** (Fig. 9.83). The latter lie beneath the chromatophores in certain regions of the body. Within the photophore, a chemical reaction produces light. The overlying

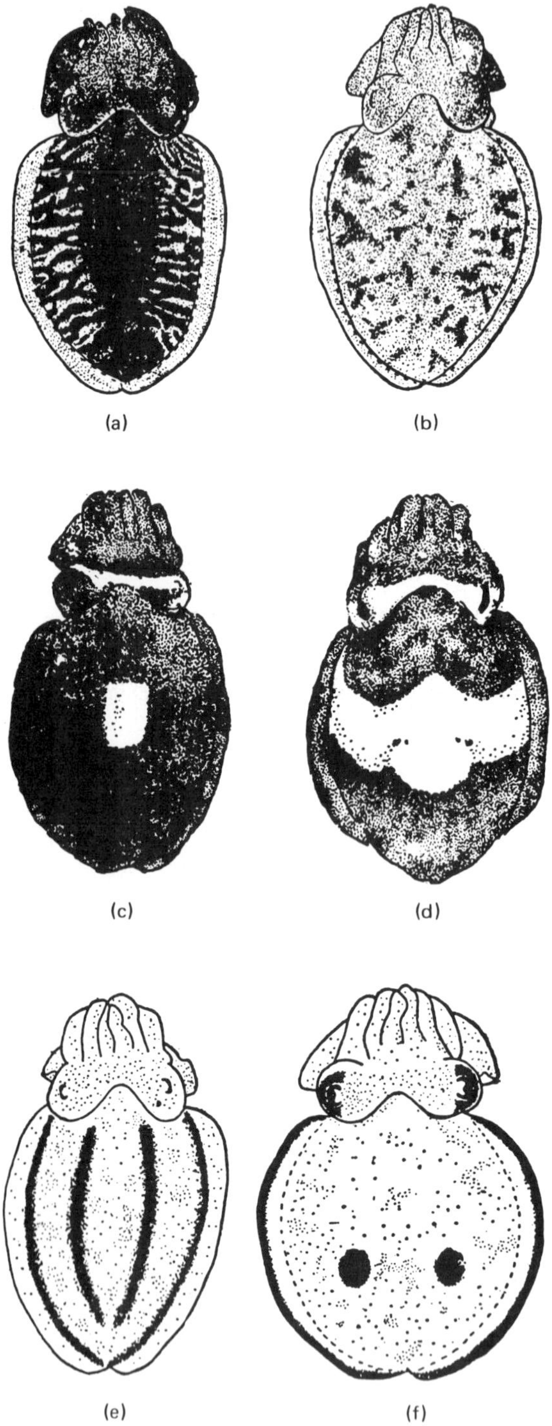

Figure 9.82 Patterns of chromatophore display by a cuttlefish, *Sepia officinalis.* (*a*) Striped pattern accentuated in males in breeding condition; (*b*) The "light mottle" pattern shown by an animal on sand; (*c* and *d*) Patterns displayed in the presence of strong contrasts in the background; (*e* and *f*) Patterns shown by *Sepia* when it is pursued or cornered. (From W. Holmes, Proc. Zool. Soc. Lond., 1940, *110*:17.)

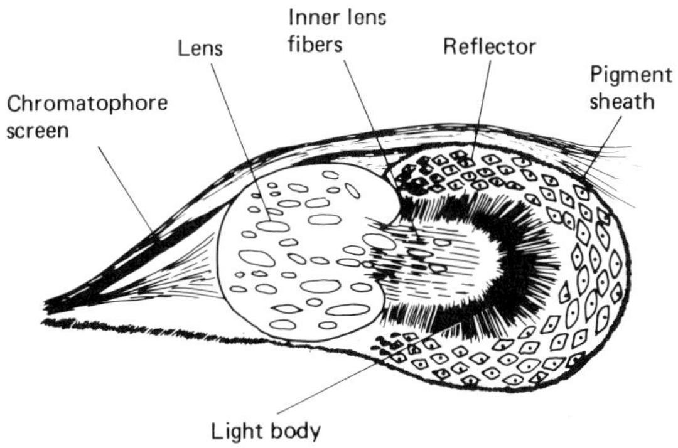

Figure 9.83 Section through a photophore from a deep-sea squid, *Calliteuthis reversa*. From J.A.C. Nicol, *The Biology of Marine Animals*, 1967, John Wiley.)

chromatophore serves as a diaphragm, allowing light emission when its radial muscles contract. Photophores may bear accessory structures, including reflectors, lenses, and color screens. Among deep-sea squid and *Spirula*, a deep-water cuttlefish, luminescence evidently plays a role in species recognition and courtship analogous to the role of chromatophores among shallow-water cephalopods.

Cephalopod learning has been an exciting topic of research for many years. Because it is very adaptable to aquarium life, *Octopus* is the common subject of learning experiments. The following observations demonstrate the ability of these animals to modify their behavior as a result of experience. At the same time, an examination of what octopods consistently fail to learn reveals the limits of their sensory perception and the peculiar organization of their nervous system.

Visual learning in *Octopus* can be demonstrated quite easily. A standard experiment involves the presentation of food accompanied by geometrical figures. Only a few electrical punishments are required to teach the creature that it may not eat in the presence of certain figures. Such experiments also reveal the animal's ability to discriminate among stimuli of various shapes, textures, and weights. Octopods perform quite well over a wide range of visual discriminations. For example, they learn to eat or not depending on whether a circle or a square is presented with their food. Similar discriminations occur between horizontal and vertical rectangles, but such ability declines with obliquely oriented figures or other shapes of increased complexity (Fig. 9.84). Visual discrimination in this group reflects the arrangement of retinal cells, optic nerves, and complex analysis centers of the brain.

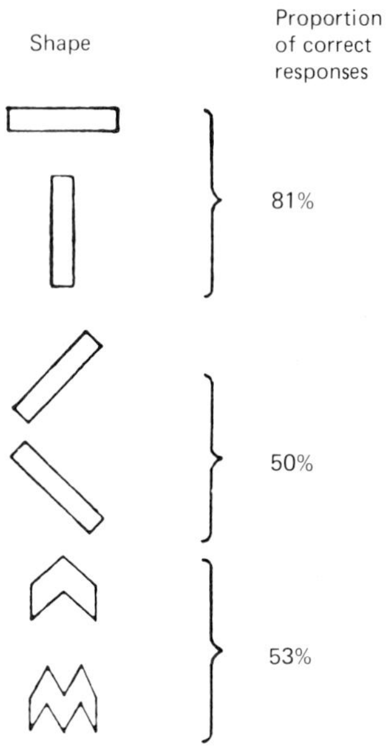

Figure 9.84 Visual discrimination in *Octopus*. Pairs of shapes which *Octopus* can and cannot distinguish. (From M.J. Wells, *Brain and Behavior in Cephalopods*.)

Some textural qualities are distinguished by *Octopus*, but apparently only the general degree of roughness can be analyzed. Thus, the cephalopod discriminates between a smooth cylinder and a grooved one, but fails to distinguish a transversely grooved cylinder from one with vertical grooves. *Octopus* apparently makes no weight discriminations at all.

The failure of this group to discriminate textural quality and weight reflects the lack of centralization of mechanotactile information in the octopod nervous system. Tactile stimuli from the arms and suckers are processed by local nerve centers only. Individual suckers possess ganglia (Fig. 9.85) that receive and transmit signals quite independently. Overall, sucker distortion is recorded centrally—thus the ability to discriminate degrees of roughness. But positional information from the suckers and arms never reaches the brain and hence cannot be integrated into the learning process. The lack of centralization in interpreting mechanotactile phenomena is associated with the extreme flexibility of these mollusks. Octopods would require an enormous number of proprioceptors to

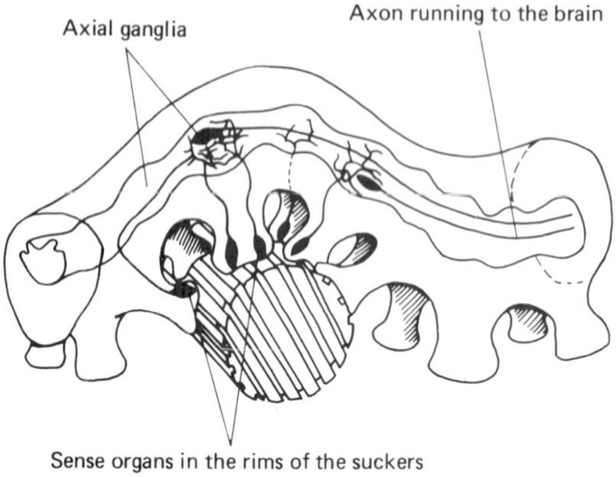

Figure 9.85 Part of an arm of an octopus touching a metal cylinder. Only sensory nerves are shown. (For further details, see text.) (From M.J. Wells and J. Wells, J. Exp. Biol., 1957, *34*:131.)

report to the brain the exact posture of each arm. Likewise, a huge brain center would be required to analyze such information. Octopods simply have not evolved a brain of that size. (It should be noted, however, that octopods have assumed a bottom-crawling life only recently in their evolutionary history. Most members of the class are pelagic, and swimming requires little higher brain analysis of texture, weight, and arm posture.) As we discuss in the following chapters, one of the major advantages of arthropods is the restricted movement of their limbs. Unlike cephalopods, arthropods can monitor the exact positions of their legs with relatively few proprioceptors. Positional stimuli are easily processed in the brains of these invertebrates.

CONTINUITY SYSTEMS

Reproduction

Almost all cephalopods are diecious, and their posterior gonad is large and singular. In males, sperm are produced in the wall of a sacular testis and released into its lumen, a cavity that may have originated from the coelom (Fig. 9.86a). A ciliated vas deferens extends to an anterior seminal vesicle, where mucilaginous secretions from the spermatophoric gland enroll sperm into dense packets surrounded by a complex chitinoid capsule. Such a sperm package is called a **spermatophore** (Fig. 9.86c). Cephalopod spermatophores are shaped like policemen's clubs and are quite complex internally. The sperm mass is surrounded by an inner tunic. A dense fluid occupies the space between this tunic and the outer capsule wall. At the tapered end of the sperm mass, a cement body is attached to a spirally coiled ejaculatory filament. The spermatophore is topped with a chitinoid cap whose apex is drawn out into a long thread. After their production, spermatophores are stored within **Needham's sac**, a large pouch on the left side of the mantle cavity.

Like the testis, the cephalopod ovary is a posterior sac (Fig. 9.86b). Eggs are released into the gonadal lumen, and one or two oviducts lead to the oviducal gland. The oviducal gland secretes albumin around each egg.

Cephalopods copulate. Copulation almost always is preceded by elaborate sexual displays by the male. Chromatophoric and swimming stunts identify the willing male to his partner and discourage her other suitors. Among pelagic forms, such as *Loligo*, copulation usually involves the swarming of an entire community. One of the male's arms is modified as an intromittent organ. This arm, called the **hectocotylus**, may bear smaller suckers which grasp the spermatophores, as in *Sepia* and *Loligo*; alternatively, its distal spoon-shaped end may carry the sperm capsules, as in *Octopus*. In some other cephalopods, such as the octopod *Argonauta*, a special sperm chamber exists within the hectocotylus.

On copulation, a male and female commonly entwine their arms together (Fig. 9.87). With his hectocotylus the male withdraws spermatophores from his funnel or directly from Needham's sac. These sperm packages usually are applied to the mantle cavity wall of the female, near the opening of her oviduct. *Octopus* deposits its spermatophores directly into the oviduct, while *Argonauta* actually abandons its hectocotylus within the mantle cavity of its partner. *Loligo* may receive and store sperm within a special seminal receptacle under the mouth, while a buccal membrane harbors spermatophores in *Sepia* females. Sperm are released from a spermatophore when its cap thread is pulled by the hectocotylus. As the cap is removed, the ejaculatory filament everts, thus drawing out the sperm mass. The cement body holds the sperm against the female until her eggs are released. The **nidamental gland** of the female (Fig. 9.86b) secretes an egg covering which, upon contact with water, hardens into a protective capsule. A number of these yolky egg capsules are released in gelatinous masses which may be attached to rocks or submerged vegetation.

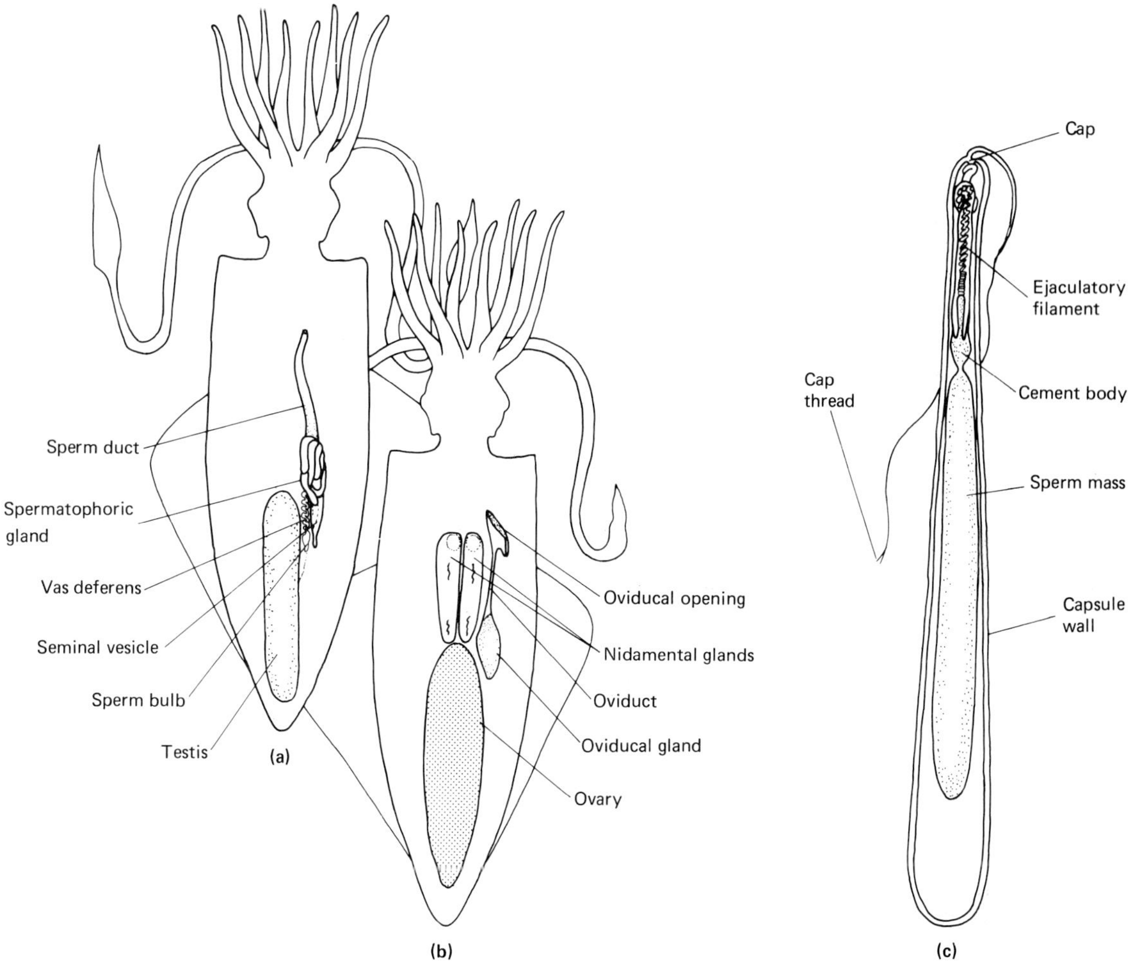

Figure 9.86 Cephalopod reproductive systems. (*a*) Male reproductive system of a squid, *Loligo pealeii;* (*b*) Female reproductive system of *Loligo pealeii;* (*c*) A spermatophore of *Loligo.* (*a* and *b* from D.E. Beck and L.F. Braithwaite, *Invertebrate Zoology Laboratory Workbook*, 3rd ed., 1968, Burgess; c from Brown, *Selected Invertebrate Types.*)

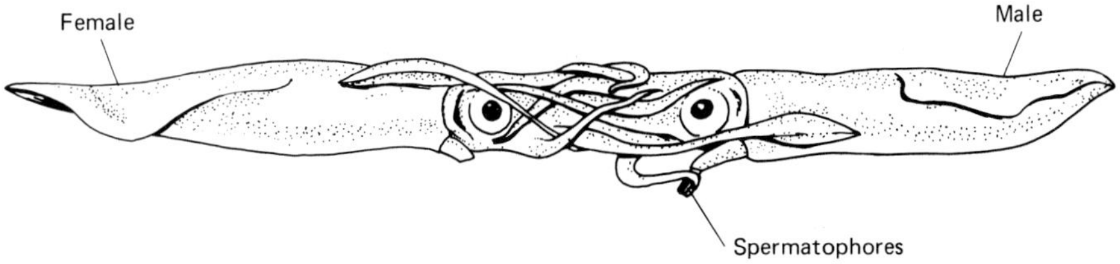

Figure 9.87 A mating pair of *Loligo.* The hectocotylus is withdrawing spermatophores from the male's funnel. (From G. Moment, *General Zoology*, 1967, Houghton Mifflin.)

Development

The female *Octopus* provides elaborate parental care during the embryonic period. Eggs are attached within protective rock crannies and are guarded continuously by the octopod mother. Periodically, she cleans them with water sprayed from her funnel. In *O. vulgaris*, the brooding female does not feed, and soon after the young hatch, she dies. *Argonauta* (Fig. 9.88), the paper nautilus, broods its eggs within an external calcareous shell secreted by two of the female's arms. This thin shell is not only a nursery but also a protective retreat for the female. It may house the diminutive *Argonauta* male as well.

Because of the high yolk content of their eggs, cephalopod embryogeny is radically different from that described for other mollusks. A germinal disk, representing the future embryo, develops at the animal pole. This disk grows along its margins and gradually envelopes the large yolk mass. The typical molluscan larval stages—the trochophore and the veliger—are not recognizable in the class. Eventually, tiny but complete cephalopods hatch and swim away to grow into adults.

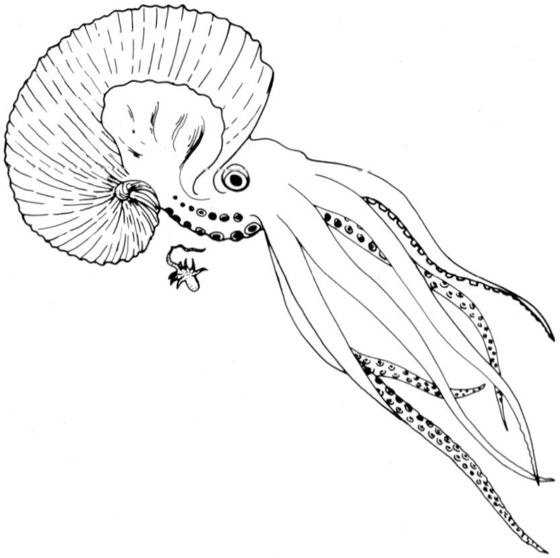

Figure 9.88 Female of the paper nautilus, *Argonauta*. The tiny male is shown beneath. (From N.B. Marshall, *Ocean Life*, 1971, Macmillan.)

Cephalopod Diversity

As we mentioned, most cephalopods are known only from fossils. The class arose in Cambrian seas and subsequently contributed some of the most dominant pelagic animals the invertebrate world has known. Over 7500 fossil species, mostly belonging to the subclasses Nautiloidea and Ammonoidea, have been described. Because of limited recovery conditions, we can only conjecture the total number of extinct forms. All that remains of this great pelagic class are 650 species of the subclass Coleoidea and a single nautiloid genus.

SUBCLASS NAUTILOIDEA

The nautiloids were the first cephalopods. They flourished in Paleozoic oceans, but then declined; this subclass is represented today by some 2500 fossils and *Nautilus*, the single living genus. Nautiloidea are characterized by a straight or coiled external shell (Fig. 9.89). This shell is relatively smooth, and internal sutures on fossil specimens suggest straight or only slightly curved septa. Numerous shell types have existed. Sizes ranged from *Plectronoceras*, which measured a mere 1 cm in diameter, to forms like *Endoceras*, a giant nautiloid whose cone-shaped shell was 4 m long.

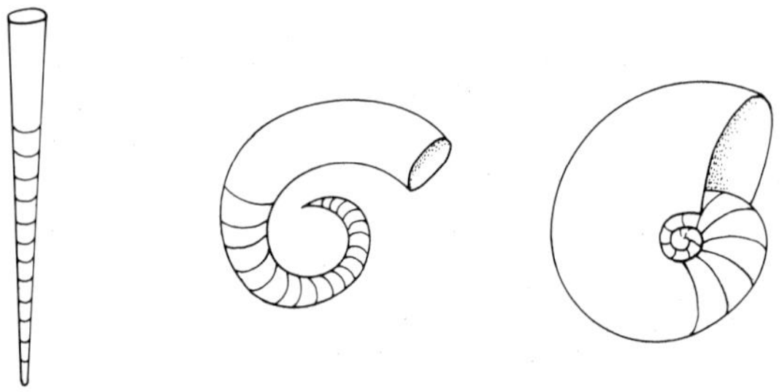

Figure 9.89 Three types of nautiloid shells. (From Moore, Ed., *Treatise on Invertebrate Paleontology, Part K: Mollusca 3*, 1964.

The internal anatomy of the nautiloids can be surmised only from the extant *Nautilus* (Fig. 9.75). We caution, however, that not all scientists agree that *Nautilus* is typical of its subclass. Indeed, its survival suggests that it may be a rather specialized genus. *Nautilus* lives at moderate depths in Indo-Pacific seas. It is the only living cephalopod that has retained the ancestral, external shell. The *Nautilus* shell is about 25 cm across, is bilaterally symmetrical, and forms a planospiral over the head. Each shell whorl is laid over previous ones, so that in the adult mollusk only the last two whorls are visible. The body occupies the last of as many as 30 shell chambers (Fig. 9.75b). When *Nautilus* withdraws into this last chamber, a unique leathery hood blocks the shell aperture (Fig. 9.75a). As many as 90 tentacles are arranged in two concentric whorls around the head. *Nautilus* tentacles have no suckers, and each can be withdrawn into a sheath.

Internal organs often occur in double pairs. There are, for example, four gills, four osphradia, four branchial hearts, and four kidneys. *Nautilus* is not as dramatic behaviorally as other living cephalopods. The genus lacks an ink sac and chromatophores. The mantle wall is not well muscularized, because of its attachment to the external shell. Respiratory and locomotor currents are produced by pulsations of the funnel. *Nautilus* is thus a slow-moving animal, and in most emergencies, it simply withdraws into its shell. It spends much of its time around depths of 600 m, but at night it migrates upward to feed, often on shrimp. Vertical migrations subject the shell to a wide range of hydrostatic pressures. These pressure changes can be endured because *Nautilus* regulates the internal pressure of each of its shell chambers. As the animal migrates, fluid and gas are shuttled along the siphuncle in such a way that internal pressure always balances external water force. Similar regulation of gas and fluid concentrations affects the buoyancy of these mollusks and thus influences their vertical travels.

Nautilus has paired gonoducts. Details of its fertilization and development, however, are not fully known.

SUBCLASS AMMONOIDEA

The ammonoids are an entirely extinct group of externally shelled cephalopods. The group evidently arose from coiled nautiloids in the Silurian period. Like the previous subclass, the ammonoids became the primary carnivores of the open sea. By the end of the Mesozoic, however, all 5000 known species had disappeared.

Figure 9.90 displays two types of ammonoid shells. These shells differed from those of the Nautiloidea in several respects. Lighter in weight and more streamlined, they probably contributed to greater swimming efficiency. As we mentioned earlier, the typical ammonoid shell is quite narrow, has angular surfaces (nautiloids are usually blunt-ended), and may bear a keel-like ventral extension. Ammonoid septa and siphuncles were also complex, as evidenced by the intricate sutural patterns in all fossil specimens (Fig. 9.90). These cephalopods also attained considerable size; the largest ammonoid, *Pachydiscus*, was 3 m in diameter.

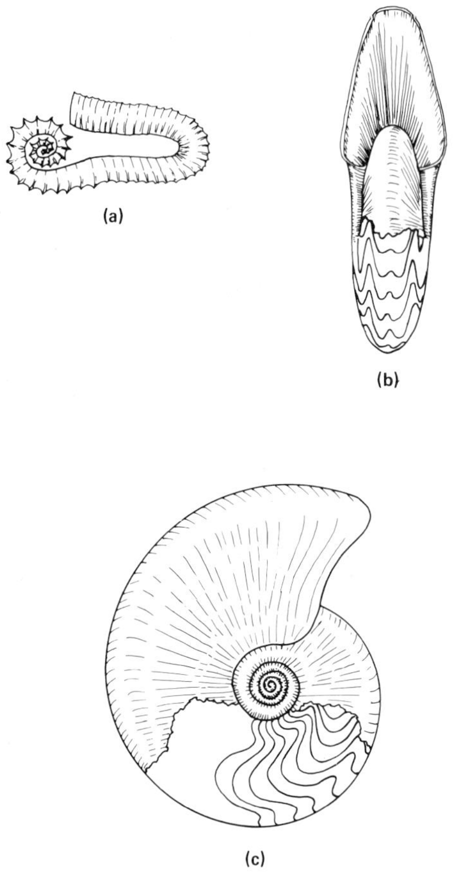

Figure 9.90 Ammonoid shells. (*a*) *Ancyloceras;* (*b* and *c*) Ventral and lateral views of *Manticoceras.* (*a* from Woodward, *Manual of the Mollusca; b* and *c* from Moore, Ed., *Treatise on Invertebrate Paleontology, Part L: Mollusca 4* (*Ammonoidea*), 1957.)

SUBCLASS COLEOIDEA

All contemporary cephalopods except *Nautilus* are coleoids. This subclass contains the internally shelled and shell-less forms, such as the familiar cuttlefish, squid, and octopods. Coleoids have fossil representatives too, including the belemnoids (Fig. 9.76) of the late Paleozoic. In addition to its shell characters, the subclass is distinguished by a few sucker-bearing tentacles, single pairs of gills and nephridia, and highly active life-styles. Our discussion of cephalopod unity focused on coleoid attributes; thus only a general review of the major animals in the group is necessary here. Three orders are recognized: the Decapoda, the Octopoda, and the Vampyromorpha.

Order Decapoda

The Decapoda includes the cuttlefish and squid, as well as the extinct belemnoids. Modern members of the order have two long tentacles and eight shorter arms. These organs bear suction cups reinforced with horny rings and/or hooks. The suckers are distributed generally over the arms, but are restricted to the spatulate tips of the tentacles. Lateral fins function primarily in postural maintenance during locomotion.

Cuttlefish are shorter, blunter decapods. A relatively large internal shell accounts for the rigidity of their body. About 80 species are known. *Sepia* and *Spirula* have been investigated most thoroughly. Many aspects of the biology of *Sepia* were described previously. This decapod lives in shallow coastal waters of Old World subtropical and tropical zones. It is a rather large cuttlefish, with certain species ranging over 1m in length. *Sepia officinalis* (Fig. 9.82), the common European cuttlefish, has a maximum mantle length of 40cm; however, its two tentacles may extend an additional 0.5m from the body. These raptorial organs usually are withdrawn into pockets near the eyes, but can lash out quickly to seize prey. During the day, *Sepia* swims in search of prey. It is not a rapid swimmer and commonly waves its lateral fins for extra locomotor power. Vertical migrations involve a regulation of fluid and gas concentrations within the chambers of the internal shell.

Sepia eats small crustaceans and fish. Particularly interesting is its predation on a sand shrimp. This shrimp buries itself in sand, where its sandy coloration provides camouflage. *Sepia* hunts its prey by spraying jets of water onto the sandy substratum; when a shrimp is uncovered, the cuttlefish seizes it with its tentacles.

Spirula spirula (Fig. 9.91) is widely distributed in tropical and subtropical seas at depths of 200 to 900 m. This unique cuttlefish seldom measures more than 10cm in total length. *Spirula* has a tightly coiled posterior shell and a pair of external fins at its rear end. Between these fins, a discoid luminescent organ emits a steady yellow–green light. *Spirula* is rare among decapods, in that it usually swims vertically rather than horizontally. In ascending, it moves with its blunt, lighted end forward.

Other cuttlefish genera include *Sepiola*, a small round creature which conceals itself on sandy bottoms and awaits its crustacean prey, and *Idiosepius*, which is only 15mm long. The latter lives in tide pools, where a unique mantle sucker attaches it temporarily to green algae.

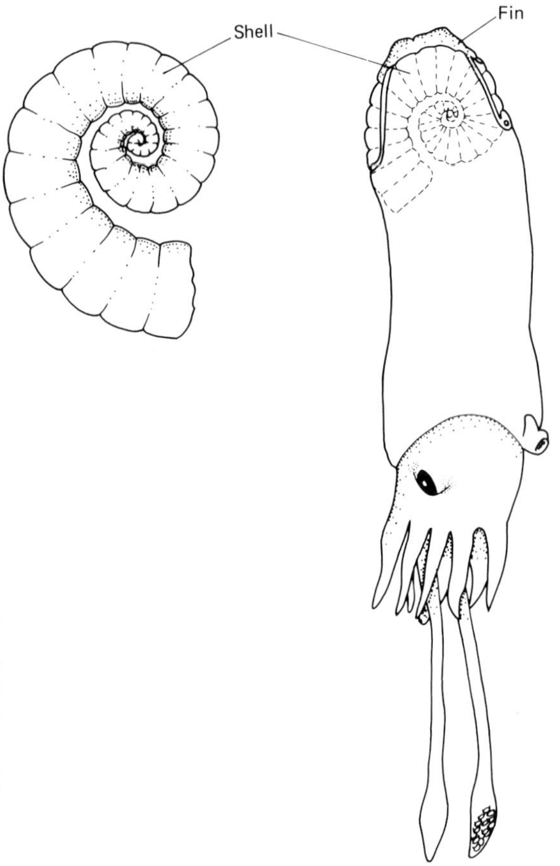

Figure 9.91 The cuttlefish, *Spirula*, and its chambered shell. (From Marshall, *Ocean Life*.)

Squid are generally long and phallus-shaped decapods, whose reduced shell consists merely of an internal, flattened plate, often called a pen. They are distributed throughout the oceans and range from surface waters to depths of 2000m. About 350 species have been described.

Squid present an enormous range in body size. The smallest are only a few centimeters in length, while *Architeuthis* (Fig. 9.92), the famous giant squid, may measure over 15 m from posterior to tentacular tips. This decapod is the largest invertebrate. *Architeuthis* swims along the continental slopes, where it feeds mostly on fish. It may wage awesome sea battles with

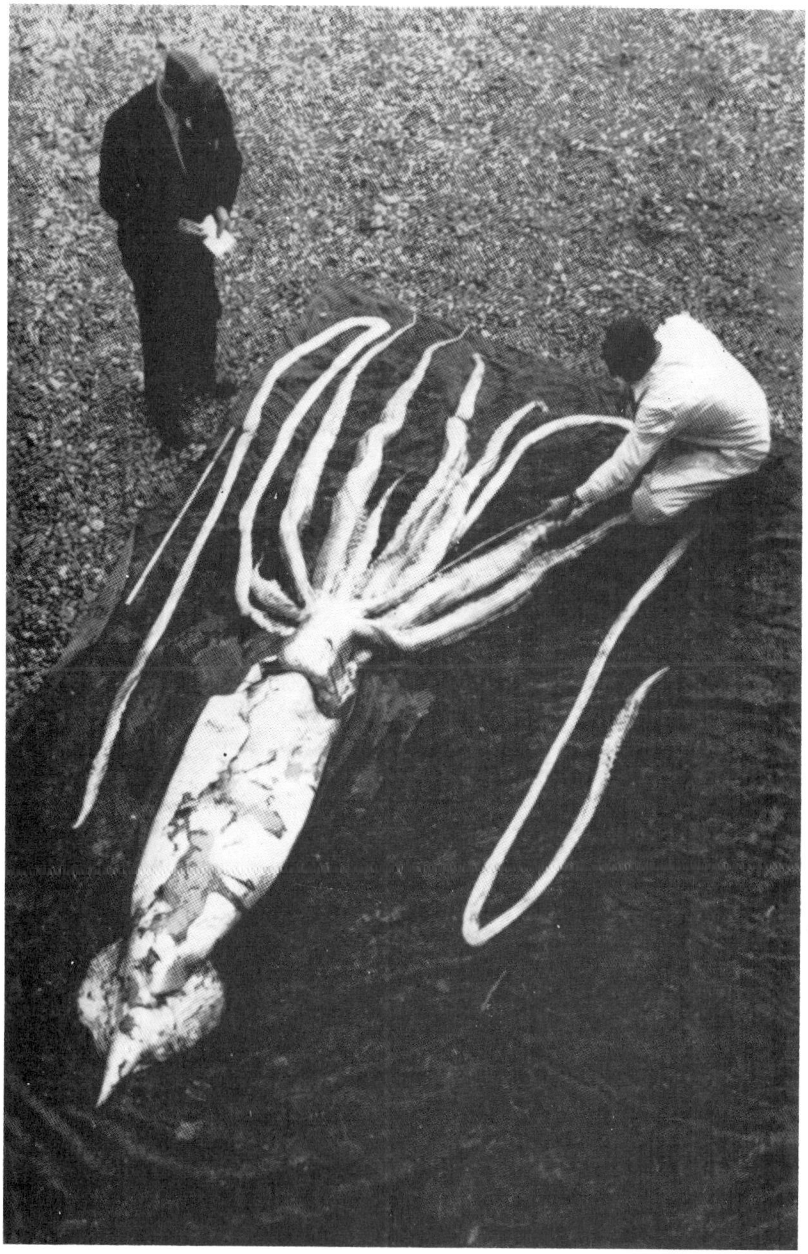

Figure 9.92 A giant squid of the genus *Architeuthis* stranded on the Norwegian coast in 1954. (From M.R. Clarke, Adv. Mar. Biol., 1966, *4*:103.)

sperm whales, and according to a few eyewitness accounts, *Architeuthis* sometimes wins.

Loligo (Fig. 9.93), the common squid, has a more conventional length of approximately 30cm, depending on the species. It lives in deeper offshore waters, where fish and shrimp are its major food. *Loligo* is an accomplished swimmer and often travels in schools of 10 to 100 members. Along with the other good swimmers in this order, *Loligo* bears special valves and cartilaginous articulations along the mantle collar.

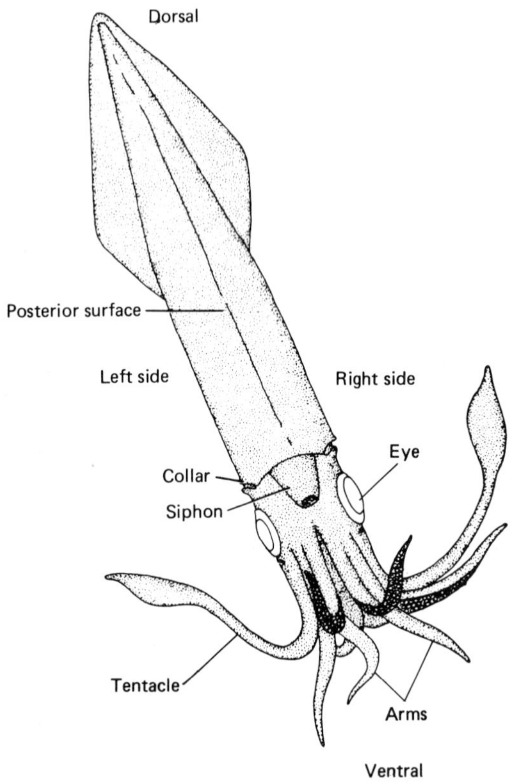

Figure 9.93 External anatomy of *Loligo.* (From Sherman and Sherman, *The Invertebrates: Function and Form.*)

When closed, these structures ensure that water leaves the mantle cavity only through the funnel.

One of the fastest decapods is *Onychoteuthis*, the fabled "flying squid." Its thin body supports broad posterior fins. Repeated strong bursts of water from the mantle funnel may propel *Onychoteuthis* into the air, where its fins allow the squid to glide for a brief time. Such behavior most likely is an escape response from fish predators. Some individual squid have even deposited themselves on the high decks of ships.

Other squid types include *Cranchia*, a small but plump animal feeding on plankton; *Chiroteuthis* (Fig. 9.94a), a deep-water squid whose narrow body and long whiplike tentacles distinguish it as one of the more bizarrely formed cephalopods; and *Histioteuthis* (Fig. 9.94b), another deep-water squid whose webbed arms are studded with photophores.

Order Octopoda

Some cephalopods pursue more sedentary life-styles. They crawl over the ocean bottom, cruise along the water surface, or swim only feebly or for brief periods of time. These cephalopods are members of the Octopoda. Octopods are characterized by globular bodies without lateral fins. Incurrent openings to the mantle cavity are small and ventral. The ancestral cephalopod shell may persist as a tiny internal plate or rod, but often is completely absent. As their name implies, the octopods have eight arms, all of equal length; the suckers on these organs are never hardened, as in squid. About 150 octopod species have been identified.

This order includes, of course, the well-known *Octopus* (Fig. 9.95). In contrast to its image in film and fiction, *Octopus* is never a large animal. Some species are only a few centimeters wide, while the largest, *Octopus hongkongensis*, has a central body less than half a meter across. (The arms, however, extend a considerable length beyond the central body mass.) Most of these animals live in shallow water, where they inhabit natural or self-constructed lairs. The lair serves as a home base, from which *Octopus* forages for food and to which it returns to eat and rest. Coral reefs, rock crevices, and sunken vases provide suitable, ready-made homesites, and *Octopus* improves such domiciles with imported stones and shells. Building materials may be carried for some distance by these creatures; shells, usually representing the remains of crustacean prey, commonly litter the homesite. *Octopus vulgaris* may use stones for protection in still another way. One filmed sequence records this cephalopod holding stones as defensive shields upon attack by a moray eel. Such tool use suggests an intelligence unparalleled by other invertebrates.

Octopus is not a social animal, and lairs are usually well separated. Even during an act as intimate as copulation, individuals commonly maintain as much distance as possible from their mates.

A second octopod genus, *Argonauta* (Fig. 9.88), includes pelagic surface dwellers of warm seas. *Argonauta* owes its buoyancy to a large, thin external shell secreted by the dorsal arms of the female. This secondary shell serves as a brood chamber, as a retreat for the female, and as an occasional domicile for the diminutive male argonaut. Females of the genus commonly measure 30 to 40cm, while the body proper of the male is less than 2cm long. The hectocotylus is well developed, however, and may be several times as long as the rest of the male organism.

Cirrothauma is a deep-water member of the order. Like the squid *Histioteuthis* and the Vampyromorpha described below, this animal has webbed arms and swims by slow, jellyfishlike pulsations. *Cirrothauma* is the only known octopod without eyes.

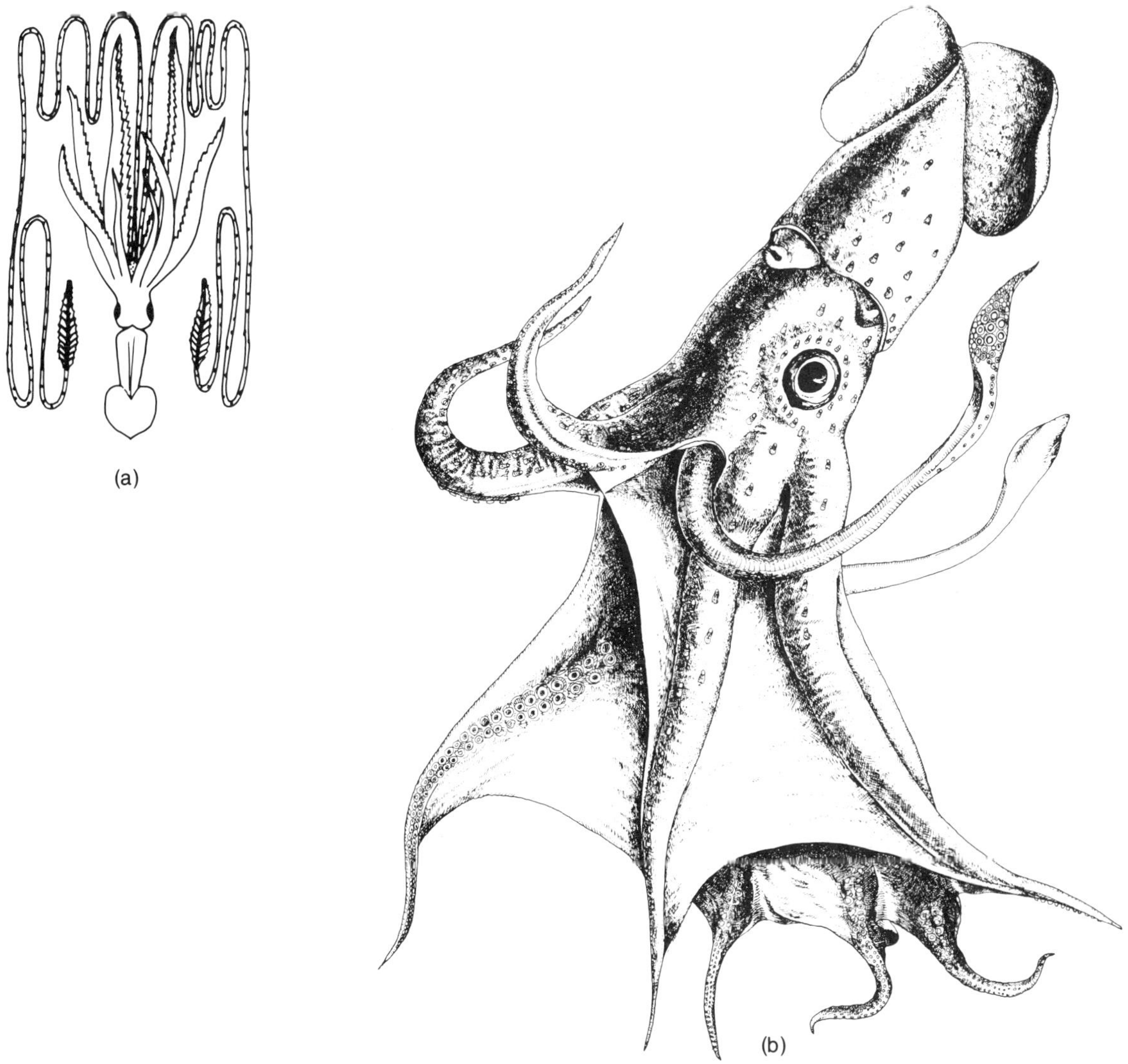

(a) (b)

Figure 9.94 (*a*) The deep-water squid *Chiroteuthis veranyi.* (*b*) *Histioteuthis bonellii*, a deep-water squid with nearly 200 photophores. (*a* from Lankester, Ed., *A Treatise on Zoology*, *b* from F.W. Lane, *Kingdom of the Octopus: The Life-History of the Cephalopoda*, 1974, Sheridan House.)

Order Vampyromorpha

A third cephalopod order comprises only the deep-sea creature *Vampyroteuthis infernales* (Fig. 9.96). This mollusk, called the vampire squid, displays characters common to both the Decapoda and the Octopoda. It has 10 webbed arms, 2 of which are very slender and can be retracted deeply within pockets at the base of the web. *Vampyroteuthis* lives in tropical and subtropical waters at depths of 500 to 3000 m.

Cephalopod Evolution

Despite the fact that they traditionally appear last in any account of the phylum, cephalopods are an ancient group which diverged quite early from the main molluscan line. They may have originated from monoplacophoranlike mollusks which developed a

Figure 9.95 A young *Octopus bimaculoides* in an aquarium. (From Ricketts, et al., *Between Pacific Tides*.)

capacity to leap briefly from the ocean floor, perhaps to escape predators or capture food. The anterior foot became a grasping organ, and sense organs developed rapidly. Leaping gradually could evolve into swimming with associated guidance and propulsive mechanisms.

In terms of species number, cephalopods have dwindled since the Mesozoic, and today less than one-tenth of their historical membership remains extant. The modern coleoids, however, are not mere relicts; rather, they represent a new episode in cephalopod evolution and perhaps have not reached their evolutionary peak. Yet, for all the glamour of the group, some definite limitations on the adaptive potential of modern cephalopods exist. These limitations involve the essence of their molluscan identity. First, as we have seen, the molluscan digestive system is designed to accommodate a continuous uptake of small food particles. While cephalopods have modified the basic design somewhat, their blind cecum can accept materials only on an installment basis. Such an arrangement necessitates the storage of a large volume of food in the upper digestive tract, clearly an encumbrance for a raptorial carnivore. Second, cephalopod nephridia are located near the end of the venal system, where blood pressure lacks the filtration power necessary for life in a dilute medium. Hence, the location of the cephalopod nephridia helps restrict the class to marine waters. Third, severe limitations result from their use of hemocyanin as a respiratory

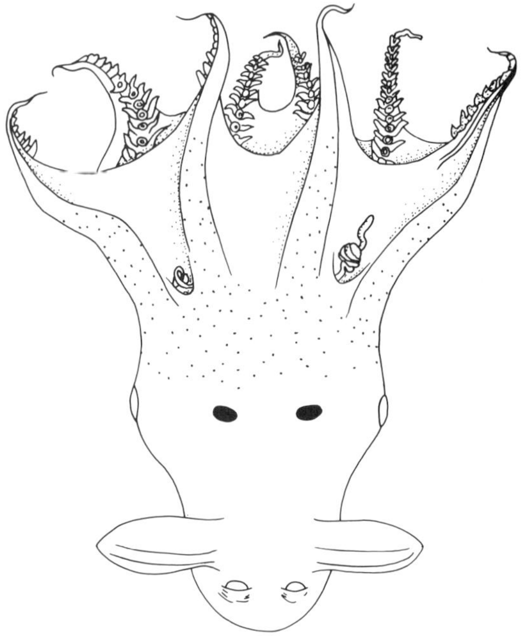

Figure 9.96 The deep-water vampire squid *Vampyroteuthis infernalis*. (From Lane, *Kingdom of the Octopus: The Life-History of Cephalopoda*.)

pigment. Hemocyanin always occurs freely in the blood plasma, and therefore its concentration is limited lest the blood become too sluggish. For all the refinements of the circulatory system—the auxiliary hearts and the many contractile vessels—cephalopod blood has an oxygen-carrying capacity less than one-fifth that of the hemoglobin-containing blood of fishes. With one of the highest metabolic rates in the invertebrate world, cephalopods need all the oxygen that their blood possibly can carry; they cannot migrate into fresh water, where oxygen supply levels may fluctuate. Finally, modern cephalopods lack the traditional molluscan protective device, the shell. They may be agile, extremely fast, and intelligent creatures, but the constant exposure of their delicate bodies still often leads to their undoing.

Molluscan Evolution

The origin of the Mollusca, the relationships among its classes, and the phylum's ties with other animal groups are obscure. Mollusks appeared and the several classes diverged well before the Cambrian; hence few fossils document their earliest history. Considerable debate has centered on their relationship to the Annelida, including the possibility of a close common ancestor for the two phyla. Supporters of such a close relationship emphasize the similarities in the protostomate embryogeny of both groups. The early trochophore larvae of the two phyla are highly similar. Some investigators consider mollusks to be segmented animals; they cite the serial repetition of body parts among the chitons and *Neopilina* and argue for a close bond with the annelids. Others consider mollusks to be primarily unsegmented animals. Chitons and *Neopilina*, they say, are pseudometameric; that is, these animals exhibit repetitive but uncorrelated organ systems. Our discussions of the polyplacophorans and *Neopilina* support the latter opinion. Mollusks almost certainly have some significant relationship to the annelids, but the two groups diverged quite long ago, probably before true metamerism was established in the annelid line.

A scenario by C. R. Stasek derives the Mollusca from turbellarianlike ancestors. This derivation is depicted in Figure 9.97. A proposed ancestral form (Fig. 9.97a) demonstrates several traits common to primitive members of both modern groups: abundant cilia and mucus, locomotion by ventral muscular waves, a diverticulated gut with intracellular digestion, a primitive nerve ring with a few longitudinal nerve cords, and simple excretory and reproductive tracts. The first step toward the modern Mollusca came when the mucus-coated dorsal surface of the stem form developed first a spiculate and then a calcified covering (Fig. 9.97b). The dorsal shell had a number of immediate consequences. First, as respiratory surfaces were reduced, ventral regions evolved ciliary ventilation mechanisms. A hemocoelic network replaced the solid parenchyma, providing more efficient internal circulation for respiratory gases and nutrients. Dorsoventral muscles were consolidated into strong retractor units which joined the shell and foot regions. Second, the eaves of the dorsal shell extended laterally over the protomolluscan body (Fig. 9.97c). This extension better protected the internal organs and increased the locomotor independence of the foot. Eave extension also created the mantle cavity, and evaginations of the cavity wall soon developed into gills. Probably, there were numerous gills within the shallow mantle cavities of early mollusks (Fig. 9.97c), but adaptive trends led to deeper cavities with fewer, larger ctenidia (Fig. 9.97d).

The first members of the phylum were probably no more than 1cm in length. However, dorsal protection,

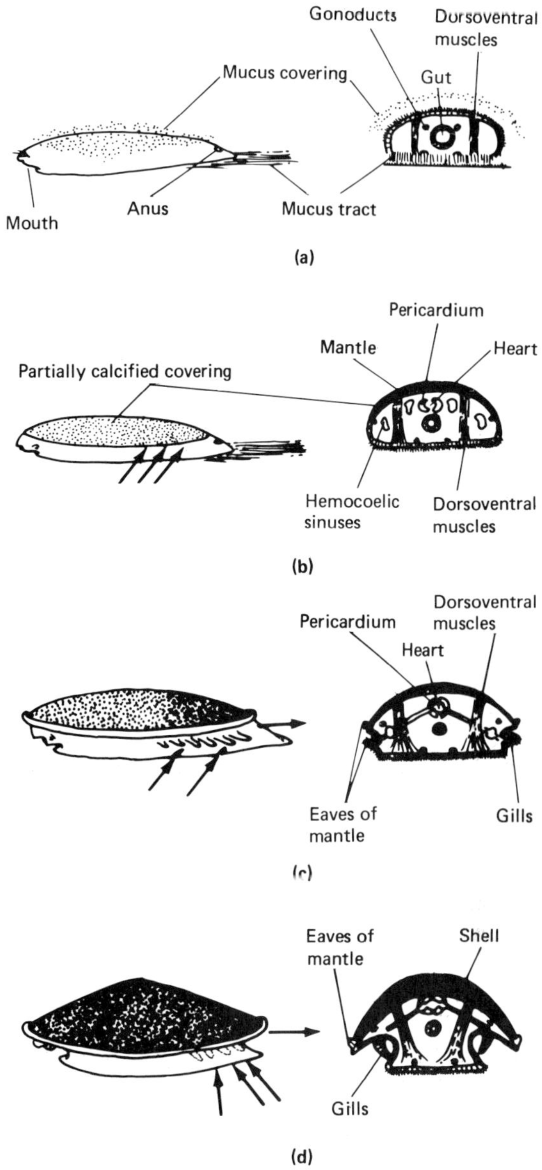

Figure 9.97 Theoretical stages in the origin of mollusks from turbellarian ancestors. All shown in lateral and cross sectional views. (*a*) Ancestral form with ability to secrete mucus as a protective measure and as a locomotory track; (*b*) Transitional stage with radula and partially calcified dorsal covering; (*c*) Later stage with gills contained in an incipient mantle cavity underlying the eaves of the mantle; (*d*) Advanced molluscan stage with well-developed shell, deep mantle cavity and folded gills. Arrows show direction of water currents. (From Stasek, In *Chemical Zoology, Vol. III*, Florkin and Scheer, Eds.)

freedom of movement, and respiratory efficiency with gills and an internal blood transport system led to considerable size increases among later mollusks. At the same time, adaptive radiation within the group established the present subphyla and classes (Fig. 9.98).

According to the scheme outlined above, the subphylum Aculifera (class Aplacophora) would have diverged very early from the main molluscan line (Fig. 9.98). Its divergence would have occurred prior to the establishment of the fully calcified shell and the spacious mantle cavity. If this interpretation is correct, the solenogasters display primitive, not secondarily reduced or specialized, characters. The creature depicted in Figure 9.97b would be a possible ancestor.

The subphylum Placophora (class Polyplacophora) would represent another molluscan group that departed early from the phylum's mainstream. Chitons emphasized pseudometamerism along an anterioposteriorly dominant body line. Figure 9.97c depicts the hypothetical polyplacophoran ancestor.

Finally, the five classes of the Conchifera apparently radiated after the establishment of certain primary characters by the cyclomyan line of the Monoplacophora. Figure 9.97d depicts these characters, which include a univalve shell, a deep mantle cavity housing a single large pair of gills, and a structural emphasis on the dorsoventral axis. Conchiferans also experienced an increase in size and ecological diversity. The gastropods underwent torsion and the spiral coiling of the shell; the class then radiated to fill a multitude of ecological niches. The bivalves adopted filter feeding and burrowing, while the gill-less scaphopods burrowed and fed with captacula. Finally, the cephalopods assumed active, pelagic life styles.

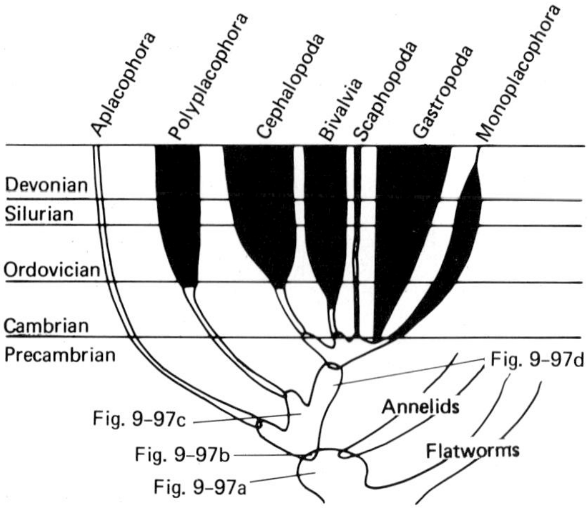

Figure 9.98 Phylogeny of the molluscan classes during Precambrian and Paleozoic times. Darker areas indicate known fossil record; lighter areas indicate inferred early history. Figure numbers refer to proposed ancestral forms as shown in Figure 9.97. Width of the lines of descent roughly indicates the relative size of each class. (From Stasek, In *Chemical Zoology, Vol. III*, Florkin and Scheer, Eds.)

The phylogenetic scheme outlined here is only one of several and is subject to careful scrutiny. Much more study, hopefully aided by some additional hard evidence from comparative biochemistry and paleontology, is vital to our understanding of these animals and their history. The mollusks—perhaps the most diverse of all invertebrate phyla—have a fascinating past, as well as an exciting present and future. They will be objects of biological curiosity for as long a time as there are students to appreciate them.

for Further Reading

Proceedings of the Malacological Society of London, the Journal of Conchology, Nautilus, Veliger, and *Malacologia* are journals devoted exclusively to mollusks.

Abbott, R. T. *American Seashells,* 2nd ed., Van Nostrand Reinhold Co., New York, 1974.

Andrews, J. *Sea Shells of the Texas Coast.* University of Texas Press, Austin, 1971.

Bayer, F. M. and Voss, G. L. (Eds.) *Studies in Tropical American Mollusks.* University of Miami Press, Coral Gables, Florida, 1971.

Bayne, B. L. *Marine Mussels: Their Ecology and Physiology.* Cambridge University Press, New York, 1976.

Bequaert, J. C. and Miller, W. B. *The Mollusks of the Arid Southwest.* University of Arizona Press, Tuscon, 1973.

Boyle, P. R. The physiology and behavior of chitons. *Ann. Rev. Oceanog. Mar. Biol.* 15: 461–509, 1977.

Brace, R. C. Anatomical changes in nervous and vascular systems during the transition from prosobranch to opisthobranch organization. *Trans. Zool. Soc. Lond.* 34: 1–26, 1977.

Brown, K. M. The adaptive demography of four freshwater pulmonate snails. *Evolution* 33: 417–432, 1979.

Browne, R. A. and Russell-Hunter, W. D. Reproductive effort in molluscs. *Oecologia* 37: 23–28, 1978.

Burch, J. B. *How to Know the Eastern Land Snails.* W. C. Brown, Dubuque, Iowa, 1962.

Burch, J. B. *Freshwater Unionacean Clams of North America.* Biota of Freshwater Ecosystems. Ident. Man. No. 11, U.S. Government Printing Office, 1973.

Choat, J. H. and Black, R. Life histories of limpets and the limpet–laminarian relationship. *J. Exp. Mar. Biol. Ecol.* 41: 25–50, 1979.

Clarke, M. A. A review of the systematics and ecology of oceanic squids, *Adv. Mar. Biol.* 4: 91–300, 1966.

Deaton, L. E. Ion regulation in freshwater and brackish water bivalve mollusks. *Physiol. Zool.* 54: 109–121, 1981.

Feder, H. M. Escape responses in marine invertebrates. *Sci, Am.* 227: 92–100, 1972.

Foster-Smith, R. L. The function of the pallial organs of bivalves in controlling ingestion. *J. Molluscan Stud.* 44: 83–99, 1978.

Franz, D. R. and Merrill, A. S. The origins and determinants of distribution of molluscan faunal groups on the shallow continental shelf of the northwest Atlantic. *Malacologia* 19: 227–248, 1980.

Fretter V. and Graham, A. *British Prosobranch Molluscs.* Ray Society, London, 1962.

Fretter, V. (Ed.) *Studies in the Structure, Physiology and Ecology of Mollusks.* Academic Press, New York, 1968.

Fretter, V. and Peake, J., (Eds.) *Pulmonates, Vol. I.* Academic Press, New York, 1976.

Gainey, L. F. The use of the foot and captacula in the feeding of *Dentalium. Veliger* 15: 29–34, 1972.

Galtsoff, P. S. Physiology of reproduction in mollusks. *Am. Zool.* 1: 273–289, 1961.

Galtsoff, P. S. The American oyster. *U.S. Fish Bull.* 64: 1–480, 1964.

Ghiselin, M. T. The adaptive significance of gastropod torsion. *Evolution* 20: 337–348, 1966.

Giese, A. C. and Pearse, J. S. *Reproduction of Marine Invertebrates. Vol. IV, Molluscs: Gastropods and Cephalopods. Vol. V, Pelecypods and Lesser Classes.* Academic Press, New York, 1977 and 1979.

Graham, A. *British Prosobranch and Other Operculate Gastropod Molluscs.* Academic Press, New York, 1971.

Hadfield, M. G. and Hopper, C. N. Ecological and evolutionary significance of pelagic spermatophores of vermetid gastropods. *Mar. Biol.* 57: 315–326, 1980.

Haven, N. The ecology and behavior of *Nautilus pompilius* in the Philippines. *Veliger* 15: 75–80, 1972.

Hyman, L. H. *The Invertebrates, Vol. VI: Mollusca.* McGraw-Hill, New York, 1967.

Jeppesen, L. L. The control of mating behavior in *Helix pomatia. Anim. Behav.* 24: 275–290, 1976.

Johnson, R. I. Zoogeography of North American Unionacea (Mollusca: Bivalvia) north of maximum Pleistocene glaciation. *Bull. Mus. Comp. Zool.* 149: 77–189, 1980.

Judd, Warren. The secretions and fine structute of bivalve crystalline style sacs. *Ophelia* 18: 205–234, 1979.

Kandel, E. R. *Behavioral Biology of Aplysia.* W. H. Freeman, San Francisco, 1979.

Keen, A. M. and Coan, E. *Marine Molluscan Genera of Western North America: An Illustrated Key.* Stanford University Press, 1974.

Keen, A. M. and McLean, J. H. *Sea Shells of Tropical Western America.* Stanford University Press, 1971.

Kohn, A. J. and Riggs, A. C. Catalog of recent and fossil *Conus. J. Molluscan Stud.* 45: 131–147, 1979.

Lemche, H. *Molluscan Phylogeny in Light of Neopilina.* Proceedings, 15th International Congress on Zoology, London, 1959.

Linsley, R. M. Shell formation and the evolution of gastropods. *Am. Sci.* 66: 432–441, 1978.

Marcus, E. and Marcus, E. *American Opisthobranch Mollusks.* University of Miami Press, Coral Gables, Fla., 1968.

Moore, R. C. (Ed.) *Treatise on Invertebrate Paleontology Vols. I-N.* Geological Society of American and University of Kansas Press, Lawrence, 1957–1971.

Morton, B. Feeding and digestion in ship worms. *Ann. Rev. Oceanogr. Mar. Biol.* 16: 107–144, 1978.

Morton, J. E. *Molluscs,* 4th ed. Hutchinson University Library, London. 1967.

Nixon, M. and Messenger, J. B., (Eds.) *The Biology of Cephalopods.* Academic Press, New York, 1977.

Packard, A. Cephalopods and fish: the limits of convergence. *Biol. Rev.* 47: 241–307, 1972.

Pender, W. F. The origin and evolution of the Neogastropoda. *Malacologia* 12: 295–338, 1973.

Pennak, R. W. *Fresh-Water Invertebrates of the United States,* 2nd ed. Wiley-Interscience, New York, 1978.

Portmann, A., Fischer-Piette, E., Franc, A., Lemche, H., Wingstrand, K. G., and Manugault, Pl. Introduction to the Mollusks, the Chitons, the Monoplacophorans, and the Peleycpods. In *Traité de Zoologie* (P. P. Grassé, Ed.), Vol. 5, pp. 1625–2164. Masson et Cie, Paris, 1960.

Potts, W. T. W. Excretion in molluscs. *Biol. Rev.* 1–41, 1967.

Purchon, R. D. *The Biology of Mollusca,* 2nd ed. Pergamon Press, New York, 1977.

Runham, W. W. and Hunter, P. J. *Terrestrial Slugs.* Hutchinson University Library, London, 1971.

Runnegar, B. and Pojeta, J. Mulluscan phylogeny: The paleontological viewpoint. *Science* 186: 311–317, 1974.

Scheltema, A. H. Position of the class Aplacophora in the phylum Mollusca. *Malacologia* 17: 99–109, 1978.

Smith, R. I. and Carlton, J. T. *Lights Manual: Intertidal Invertebrates of the Central California Coast.* Third Edition. University of California Press, Berkeley, 1975.

Solem, A. *The Shell Makers: Introducing Mollusks.* Wiley-Interscience, New York, 1974. (A well-done semi-technical introduction to molluscan biology)

Spight, T. M. On a snail's chances of becoming a year old. *Oikos* 26: 9–14, 1975.

Stasek, C. R. The molluscan framework, In *Chemical Zoology* (M. Florkin and B. T. Scheer, Eds.), Vol. III, pp. 1–44. Academic Press, New York, 1972.

Taylor, J. D. The structural evolution of the bivalve shell. *Paleontology* 16: 519–534, 1973.

Taylor, J. D., Morris, N. J. and Taylor, C. N. Food specialization and the evolution of predatory prosobranch gastropods. *Paleontology* 23: 375–410, 1980.

Thiriot-Quievreux, C. Heteropoda. *Ann. Rev. Oceanogr. Mar. Biol.* 11: 237–261, 1973.

Thompson, F. G. *The Aquatic Snails of the Family Hydrobiidae of Peninsular Florida.* University of Florida Press, Gainesville, 1968.

Thompson, T. E. *Biology of Opisthobranch Mollusca, Vol. I.* Royal Society, London, 1976.

Turner, R. D. *A Survey and Illustrated Catalogue of the Teredinidae.* Museum of Comparative Zoology, Harvard University, Cambridge, 1966.

Tutschulte, T. and Connell, J. H. Reproductive biology of 3 species of abalones *(Haliotis)* in southern California. *Veliger* 23: 195–206, 1981.

Vagvolgyi, J. On the origin of molluscs, the coelom, and coelomic segmentation. *Syst. Zool.* 16: 153–168, 1967.

Vermeij, G. J. and Covich, A. P. Coevolution of freshwater gastropods and their predators. *Am. Nat.* 112: 833–844, 1978.

Vogel, K. and Gutmann, W. F. The derivation of pelecypods: Role of biomechanics, physiology, and environment. *Lethaia* 13: 269–275, 1980.

Von Salvini-Plawen, L. A. Reconsideration of systematics in the Mollusca (phylogeny and higher classification). Malacologia 19: 249–278, 1980.

Wagner, F. J. E. Distribution of pelecypods and gastropods in the Bay of Fundy and eastern Gulf of Maine, Canada. *Proc. N.S. Inst. Sci.* 29: 447–464, 1979.

Wells, M. J. *Octopus: Physiology and Behavior of an Advanced Invertebrate.* Chapman and Hall, London, 1978.

Wilbur, K. M. and Yonge, C. M. (Eds.) *Physiology of Mollusca, Vols, I and II.* Academic Press, New York, 1964 and 1966. (An excellent collection of review chapters on aspects of molluscan physiology)

Yochelson, E. L. An alternative approach to the interpretation of the phylogeny of ancient mollusks. *Malacologia* 17: 165–191, 1978.

Yonge, C. M. Giant clams. *Sci. Am.* 232: 96–105, 1975.

Yonge, C. M. and Thompson, T. E. *Living Marine Molluscs.* Collins, London, 1976.

Young, J. Z. *The Anatomy of the Nervous System of Octopus vulgaris.* Oxford University Press, New York, 1972.

10

Introduction to the Arthropods: Arthropod Unity and the Trilobites

If universities were evenhanded in their zoological curricula, they might offer major courses in "Arthropod Biology" and "Nonarthropod Biology" rather than support the vertebrate–invertebrate dichotomy. Indeed, the animal kingdom is dominated by the arthropods, whose ranks now boast about 800,000 described species. Approximately three of every four animal types are arthropod. The key members of the phylum—the insects, the crustaceans, the spiders, and the mites—are at least somewhat familiar to all of us.

Arthropods are well represented in every major habitat. From the ocean floor to the cracks in kitchen walls, from Arctic regions to the insides of other animals, these creatures form an integral part of the biological economy of the planet. Crustaceans contribute prominently both to the zooplankton and to the benthos. Insects, arachnids, and myriapods, meanwhile, constitute the Invertebrata's most successful land dwellers.

The success of the Arthropoda has resulted largely from three structural characters: segmentation, jointed appendages, and a hard exoskeleton. Arthropods enjoy the same metameric advantages that annelids do, and some researchers believe that polychaete parapodia are related (however distantly) to the ventrolateral appendages of the present group. It is their exoskeleton, however, that enables arthropods to surpass other phyla by all measures of evolutionary success. The arthropod exoskeleton is a hard, multilayered body cover composed of chitin, proteins, and other substances. The skeleton shelters and supports the body, and its extensions allow the appendages to function as powerful levers during locomotion. Accordingly, arthropods are the most mobile invertebrates, a factor accounting for their wide distribution and active life-styles. An additional consequence of the exoskeleton is a reduction in the arthropod coelom. No longer functional as a hydrostatic skeleton, the secondary body cavity of most arthropods persists only as the lumen of reproductive and/or excretory organs.

With the possible exception of cephalopod mollusks, arthropods occupy the zenith of protostomate evolution. The phylum probably had its origins in primitive polychaetelike, segmented worms.

MAINTENANCE SYSTEMS

General structural features and other aspects of maintenance biology in the Arthropoda indicate an annelid heritage. Major differences between the two phyla are attributable to various features of the exoskeleton and to the tendency of many arthropods to reduce metamerism.

General Morphology

A very generalized view of an arthropod is presented in Figure 10.1. The body of a generalized arthropod is a bilaterally symmetrical, segmented tube, with paired appendages located more ventrally than those of the annelids. Primitively, arthropods were rather small and wormlike, but modern members of the phylum display a wide range of sizes and shapes. Small arthropods, such as mites and copepod crustaceans, are often less than a millimeter long, while the largest crab has a leg spread of 3 meters. Because the arthropod body is supported by an external skeleton, there is a definite maximum size limit in the phylum. The size-strength of an exoskeleton increases as the square of the animal's linear dimensions, while the body volume-weight increases as the cubed power. After a certain point, the body simply becomes too large to be supported by an exoskeleton. Aquatic arthropods enjoy greater size potentials because of the buoyancy of their medium; however, giant spiders and insects are strictly Hollywood in origin. In contrast, endoskeletons can support much more weight than exoskeletons. Thus, relative to terrestrial arthropods, vertebrates can attain gigantic sizes on land.

A high degree of cephalization is typical of this phylum. The arthropod head comprises anterior body segments that have fused with an anterior nonsegmental region, the **acron,** which is homologous with the annelid prostomium. Appendages in this area have sensory and/or food-gathering functions. Cephalic eyes are the rule, and the arthropod brain is often well developed. Frequently, segments that are postoral in origin migrate in front of the mouth during development. This migration displaces the oral cavity to a subterminal, ventral position. The midbody com-

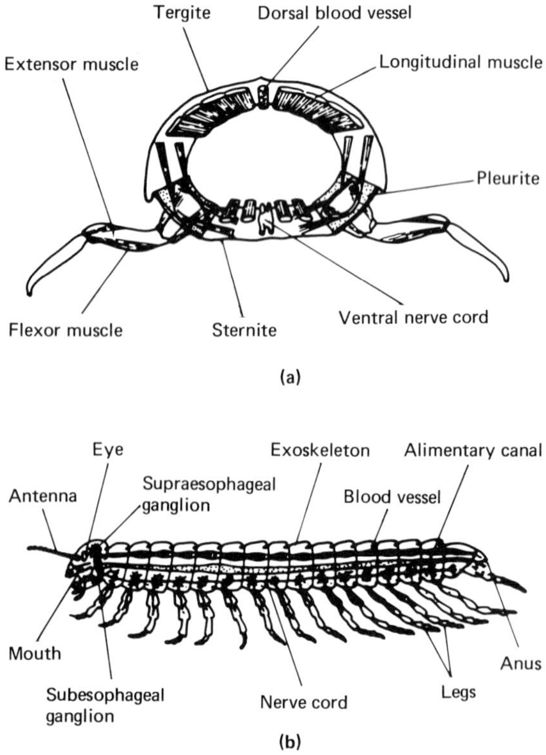

Figure 10.1 (*a*) Cross section through an arthropod segment showing the location of the legs and basal leg muscles in relation to the body wall; (*b*) Representation of a generalized arthropod in lateral view. (*a* from R.E. Snodgrass, *Principles of Insect Morphology*, 1935, McGraw-Hill; *b* from R.H. Wolcott, *Animal Biology*, 1933, McGraw-Hill.)

prises a highly variable number of segments. Segmentation is evident in the exoskeleton and to a variable extent in the nervous, circulatory, and excretory systems. Midbody appendages are important in support and locomotion; they may also be modified for respiratory or reproductive purposes. At the posterior of the generalized arthropod is a nonsegmental pygidium, which bears the anus.

The embryos of all arthropods are metameric, but there is a trend in many adult arthropods to reduce the number of segments. Such metameric reductions reflect an increasing degree of regional specialization of body parts and the development of fewer, more efficient organs. Reductions in metamerism involve one of several processes. Adjacent segments may combine by fusing to varying degrees. In most arthropods, such fusion imposes a supersegmental organization, such as the head–thorax–abdomen body regions of the insects and the prosoma–opisthosoma structure of the chelicerates. This process is known as **tagmatization**, and the individual body regions are referred to as **tagmata**. Alternatively, adjacent segments may become specialized as different selective pressures shape distinct structures on each of them. Such divergence is evident in various cephalic segments and their appendages. Some appendages near the mouth are specialized for food detection, others for food capture, and still others for food maceration. Finally, some arthropod segments may be lost altogether. Evidently, their presence became more of a liability than an asset in the evolution of the species involved.

Fossil evidence suggests that similar appendages were present on each body segment in primitive arthropods. These appendages were used mainly for locomotion. Thus, ancient marine arthropods, such as trilobites, had similar appendages on the head and trunk. The evolution of the arthropods has been accompanied by the specialization of certain appendages for functions besides locomotion and the loss of appendages from many other segments. The claws of a lobster, the fangs of a spider, and the mouthparts of a grasshopper—all of these are modified leg appendages. The immense diversity that arthropods have attained probably depended as much on the phenomenal versatility of their appendages as on anything else.

The wide diversity in the appendages of modern arthropods raises important questions of homology. **Serial homology** refers to the relationship between elements with the same segmental origin but with different structures and functions in divergent contemporary species. Such homology is one of the most useful tools in the study of arthropod evolution. We refer to it often in the following chapters.

The arthropod exoskeleton is one of the more fascinating structures in the invertebrate world. Its composition deserves close attention. This exoskeleton, or cuticle, is a jointed tube which surrounds the main body of the animal and each of the appendages. In each segment, the skeletal tube is divided into a number of plates. In the primitive condition, there are four plates in each segment (Fig. 10.1a): a single dorsal plate, the **tergite**; two lateral plates, each called a **pleurite**; and a ventral plate, or **sternite**. This primitive design is often obscured in modern arthropods whose skeletal plates are fused or secondarily divided. Supersegmental dorsal shields composed of the fused tergites of several segments are common in some groups. Between the plates of adjoining segments, the cuticle is thin and flexible; this flexible area, called the **articular membrane**, allows skeletal movements (Fig. 10.2).

Like the other, simpler cuticles we have studied, this one is noncellular and is secreted by an underlying hypodermis composed of a single layer of cells (Fig. 10.3). The cuticle itself is composed of several layers. Outermost is a thin **epicuticle**, a lipoprotein layer which may be impregnated with wax. Below the epicuticle lies the **procuticle**. This thick layer is subdivided into an outer **exocuticle** and an inner **endocuticle**, where chitin and proteins are joined into highly organized glycoprotein. In the exocuticle, this glycoprotein is strengthened by internal cross-linkages. The process by which these cross-linkages are formed is called **tanning**. The tanned glycoproteins of the exocuticle are the primary source of the skeleton's strength. The absence of this tanned layer from the articular membrane accounts for that region's flexibility, as does the presence of a rubberlike protein called resilin. An additional source of procuticle strength for some arthropods (notably the crustaceans) is the deposition of calcium salts among the

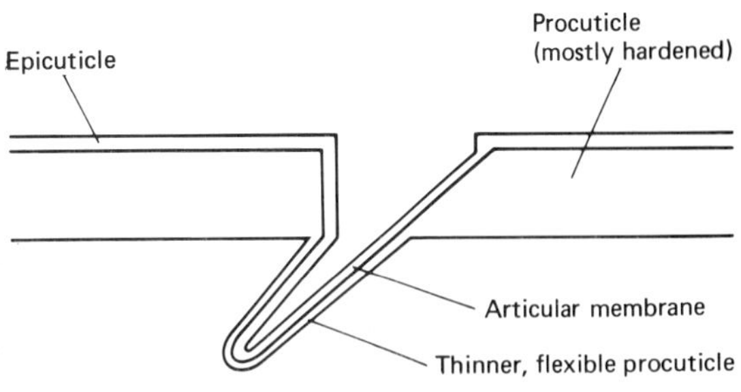

Figure 10.2 View of the structure of an articulation between the skeletal plates of adjoining segments in an arthropod. (From W.D. Russell-Hunter, *A Biology of Higher Invertebrates*, 1969, Macmillan.)

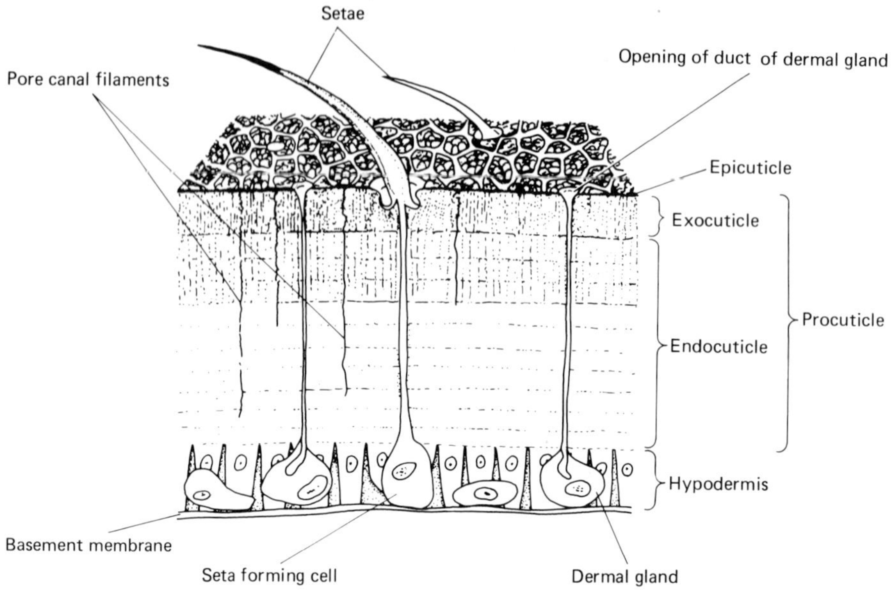

Figure 10.3 Representation of the integument (cuticle plus hypodermis) showing major structural features. (From R.H. Hackman, In *Chemical Zoology, Vol. VI*, M. Florkin and B.T. Scheer, Eds., 1971, Academic Press.)

glycoprotein elements. Regardless of the strength and thickness of the skeleton, the hypodermal cells have access to the body surface via pore canals (Fig. 10.3). Sensory crevices are also present.

In addition to its role in protection and body support, the arthropod cuticle often is specialized regionally for various functions. Lips, jaws, beaks, ploughs, food mills, lenses, wings, pincers, and copulatory organs are among the many accessory structures elaborated from the cuticle.

Note that arthropods are devoid of motile cilia—a situation not unexpected in animals with a hard external covering and a well-developed musculature.

Nutrition and Digestion

The tremendous diversity of the Arthropoda is mirrored in the diverse diets and food-gathering techniques of the phylum. Herbivores, carnivores, scavengers, and parasites are all well represented. The arthropod gut, however, is remarkably uniform (Fig. 10.1b). It consists of a tube extending from the anterior, ventral mouth to the pygidial anus. Both ends of the gut are lined with cuticle and are called the foregut, or stomodeum, and hindgut, or proctodeum, respectively. The foregut is concerned with the intitial processing of foodstuffs. Anteriorly, it aids oral appendages in food acquisition, while more posterior parts of the region are specialized for preliminary digestion and/or food storage. A pharyngeal pump is frequently present. The exact character of the foregut depends, of course, on diet and feeding style. The midgut, an endodermal derivative, digests and absorbs nutrients. Deep folds in its glandular walls often form digestive pouches. Finally, the cuticle-lined hindgut produces feces by withdrawing water from undigested wastes.

Circulation

Internal transport mechanisms in the Arthropoda differ from those of the Annelida. These differences are associated with the reduction in size of the arthropod coelom. Fatty tissues occupy much of the area between the body wall and the gut. Within these tissues, there is a diffuse network of interconnected spaces and sinuses. Called the **hemocoel**, this blood-filled network forms an integral part of the phylum's open blood circulatory system.

The central elements of the circulatory system are a dorsal heart and its associated tubes (Fig. 10.4). The arthropod heart apparently evolved from an annelid-like, contractile dorsal blood vessel. Indeed, in primitive arthropods, the heart is long and tubular, but certain highly evolved members of the phylum display a more compact blood pump. There are no veins. One or more pairs of lateral openings called **ostia** admit blood to the heart. After ostial valves are sealed, the contraction of heart muscles sends blood pulsing outward through the arteries. The arterial system displays various levels of complexity. It may involve only a single, anterior vessel or an elaborate network of continually branching tubes through which blood flows directly to all major body regions. At strategic points, accessory pumps may be formed from muscularized arterial walls. Regardless of arterial development, each vessel eventually discharges its blood into the hemocoel. Blood flows through the sinuses and bathes all tissues directly. In many higher arthropods, the path of blood flow is not random but rather follows a route defined by membranous walls. If specialized respiratory organs such as gills or lungs are present, these structures are bathed immediately prior to the blood's return to the heart.

Arthropod blood, commonly called hemolymph, is essentially the same as the tissue fluid. Dissolved respiratory pigments, most commonly the blue protein hemocyanin, may be present.

Respiration

Recall that annelid respiration usually involves the diffusion of gases across the general body surface. Such a mechanism is seldom suitable for arthropods, owing to their impermeable exoskeleton. Water and gases may diffuse across a very thin, nonwaxy cuticle, but most arthropods possess a more efficient respiratory system. Their more active life-styles, compared with those of the annelids, are associated with specialized methods of gas exchange.

Arthropods were confronted with these respiratory problems very early in their history, and the long and divergent evolution of the phylum has resulted in a variety of novel respiratory adaptations. Among aquatic arthropods, gills have evolved from (or in association with) the trunk appendages. Most terrestrial members of the phylum possess a tracheal system, a network of chitin-lined tubes (**trachea**) which ramify deep within the body (Fig. 14.7). External openings (**spiracles**) are controlled by valves; the internal branches of the system terminate beside or within the cells with which respiratory gases are exchanged.

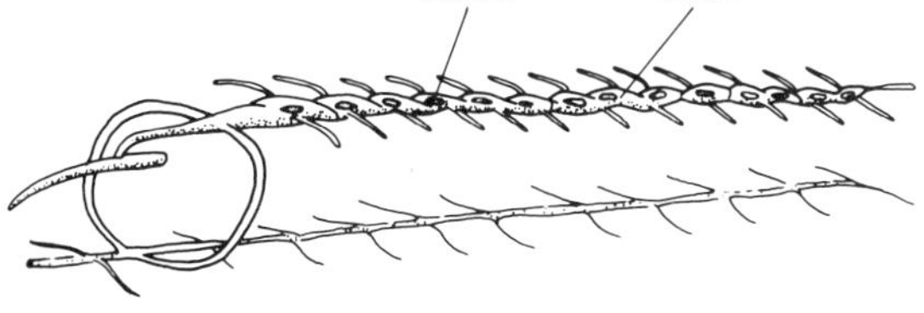

Figure 10.4 Outline of a rather generalized arthropod circulatory system. (From K.U. Clarke, *The Biology of Arthropods*, 1972, Edward Arnold.)

Excretion and Osmoregulation

Arthropods have no spacious coelom and thus no nephridia in the annelid sense. Excretory and osmoregulatory structures reflect the needs of each class. Among the more common organs in the phylum are coxal glands, maxillary or antennal glands, and Malpighian tubules (Fig. 10.5).

Both the **coxal gland** (in arachnids) and the **maxillary** or **antennal gland** (in crustaceans) consist of a thin-walled sac and its associated tubule. The

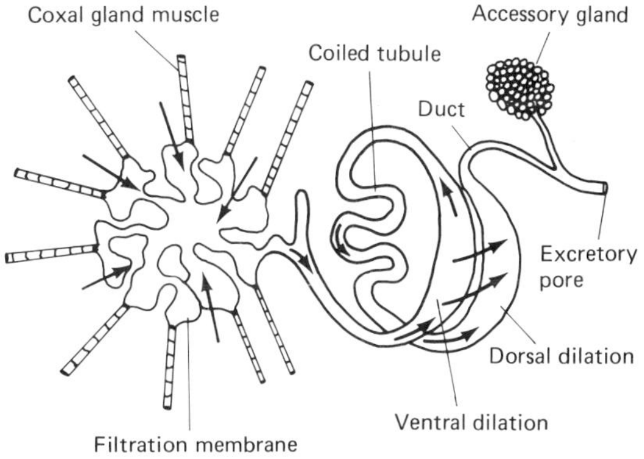

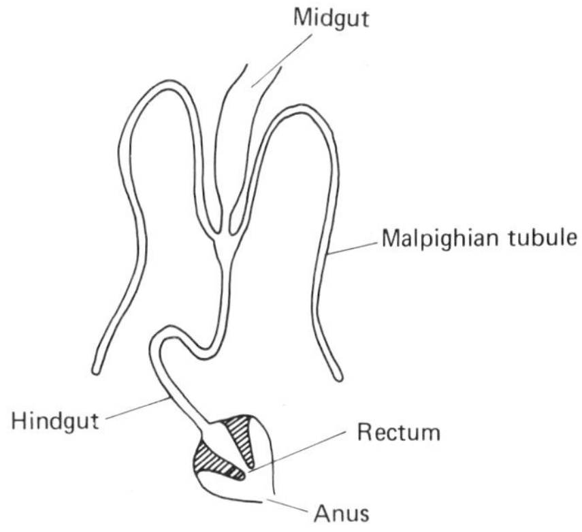

Figure 10.5 (*a*) A coxal gland from the tick *Ornithodorus moubata*. Arrows indicate the direction of fluid movement; (*b*) A diagram showing the origin of Malpighian tubules from the junction of the midgut with the hindgut or proctodeum. (*a* from M.J. Berridge, In *Chemical Zoology, Vol. V*, Florkin and Scheer, Eds., 1970, Academic Press; *b* from E. Bursell, *An Introduction to Insect Physiology*, 1970, Academic Press.)

sac lies within the hemocoel. Waste materials diffuse across the thin walls of the gland and pass through the tubule, where water resorption may occur. Other active resorptions and secretions of salts and organic compounds may take place. The excretory pore lies at the base of an appendage. These glands, which are always paired, may represent coelomic remnants and the tubules may be persistent coelomoducts. They are comparable to metanephridia and could be highly modified derivatives of those organs.

The **Malpighian tubule** is a unique structure which produces the concentrated wastes characteristic of terrestrial arthropods. Arachnids, insects, and myriapods possess one to many pairs of these organs. A Malpighian tubule arises from the junction of the midgut with the proctodeum. This slender, often convoluted tubule ramifies throughout the anterior hemocoel. Blood-borne wastes pass across its thin walls and are transported to the hindgut, where water is resorbed.

Other methods of waste control are described in this phylum. Certain strategically placed cells called **nephrocytes** accumulate and sometimes breakdown waste material. Crustaceans remove some wastes across their gills, and insects may deposit solid wastes in their old cuticle prior to its being shed.

The nitrogenous waste excreted by a particular arthropod reflects its environment and peculiar biochemistry. Being aquatic, most crustaceans simply eliminate ammonia; freshwater species eliminate a hypotonic urine while absorbing necessary salts through the gills. Arachnids are terrestrial, and their predominant nitrogenous waste is dry guanine crystal. Finally, insects evidence their adaptation to land by excreting uric acid.

Many terrestrial arthropods can withstand extreme, prolonged desiccation. This physiological talent is particularly well developed in immature stages, including the arthropod egg.

ACTIVITY SYSTEMS

The arthropods are the most active phylum in the invertebrate world. Their locomotor abilities and behavioral complexity reflect their skeletal innovations, a reorganized musculature, a relatively concentrated nervous system, and sophisticated sensory structures, including compound eyes.

Skeleton and Muscles

Arthropods may not have been the first invertebrates to evolve legs, but they certainly exploit the principle of limbed movement to its maximum advantage. Animals with legs enjoy several advantages over their limbless cohorts. Legs exploit the principle of leverage; provided that a leg resists bending along its length, a considerable amount of movement can take place with relatively small muscular contractions at the leg's pivotal end. When locomotor forces are concentrated in the legs, the central body can maintain forward progress with minimal upset to its housed organs. Such sustained motion conserves energy that otherwise might be spent in continual acceleration and deceleration efforts. Compare this situation with that of the annelids: The peristaltic gait of the annelids distorts the central body. Individual segments cannot be kept in continuous motion (i.e., each segment must come to a complete stop before it can start again), and thus annelids must repeatedly summon the extraordinary muscular effort associated with combatting inertia. Clearly, the arthropods have the more efficient system.

The arthropod skeleton extends along the legs (and all other appendages), where it is variously divided into articles. Articular membranes provide the flexibility necessary for movement (Fig. 10.2). At strategic joints, the cuticle forms condyles and sockets (Fig. 10.6). These joints ensure adequate leg strength without sacrificing mobility. Such leg structure is responsible for the phylum's name, which means "jointed feet." While the legs remain jointed and flexible, the central body of many higher arthropods has become increasingly rigid. Such rigidity, especially among the leg-bearing segments, ensures that the body is pushed forward rather than merely flexed by the leverage forces of the legs and also provides that the work of every leg is applied to the entire body.

The arthropod musculature is organized in relationship to the exoskeleton (Fig. 10.1a). Instead of circular and longitudinal muscle layers, there are distinct bun-

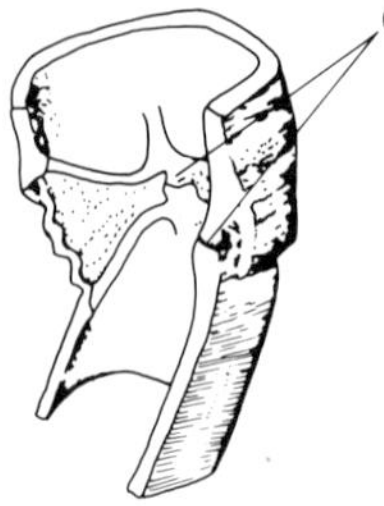

Figure 10.6 Cutaway view of an articulation between two skeletal articles of an arthropod showing coandyles. (From A. Vandel, In *Traité de Zoologie, Vol. VI*, P.P. Grassé, Ed., 1968, Masson et Cie.)

dles of striated fibers which connect strategic points on the inner wall of the skeleton. Commonly, the procuticle is folded inwardly at the site of muscular attachment (Fig. 10.7). Such a specialized attachment point is called an **apodeme.** Muscle fibers are concentrated near the proximal end of the appendages, where the central body supports most of their weight. This arrangement allows the muscle to exert maximum leverage while the muscle weight itself remains in continuous motion with the body. Because arthropod muscles can attach to only one side of the exoskeleton, the development of antagonistic muscle pairs can be a problem. Paired extensors and flexors operate some joints (Fig. 10.1a), but many arthropods must rely on hemolymph pressure to reextend their limbs.

Nerves

The arthropod nervous system is essentially annelid in character, but the present phylum exhibits an increasing concentration of central elements, including the development of a complex brain (Figs. 11.8 and 12.7). A primitive arthropod system includes a dorsal anterior nerve mass, circumenteric connectives, and a doubled ventral nerve cord with paired ganglionic swellings in each segment—all of which are quite annelid. Higher arthropods display a consolidated ventral cord whose ganglia have migrated forward. In the most advanced members of the phylum, important ganglia are concentrated in the head.

The arthropod brain integrates the activities of the whole animal. Its development parallels the evolution of increasingly elaborate sense organs on the head. Each of the major regions of the brain—the protocerebrum, the deutocerebrum, and the tritocerebrum—monitors major cephalic structures (Fig. 10.8). The anteriormost **protocerebrum** receives input from the optic nerves. As we describe later, arthropod eyes can be very complex and provide these animals with some of their most important environmental data. Accordingly, much of the complex behavior of the arthropods involves output from the protocerebrum, particularly from its paired mushroom bodies. The **deutocerebrum,** or midbrain, receives input from important sensory appendages called antennae; it also serves as an integrative center. Finally, the posterior **tritocerebrum** contains the nerves that innervate the mouthparts and the foregut.

Neurologists disagree on the origin of these brain parts. The tritocerebrum, whose circumenteric commissures are postoral, probably represents a forwardly migrated segmental ganglion. The protocerebrum and deutocerebrum also may have their origin in segmental ganglia, or they may have arisen from the nonsegmental acron.

In arthropods, muscle fibers receive input from more than one neuron, and each neuron innervates several muscle fibers. (Compare the vertebrate "one neuron–one muscle fiber" system.) In addition, there are several different types of neurons. **Phasic neurons** transmit impulses that stimulate rapid, short-term muscle contractions. These neurons are fired when fast movements are required. Giant axons, similar to those encountered among the annelids and mollusks, are extreme examples of phasic neurons. Stimulation of muscles by **tonic neurons** produces

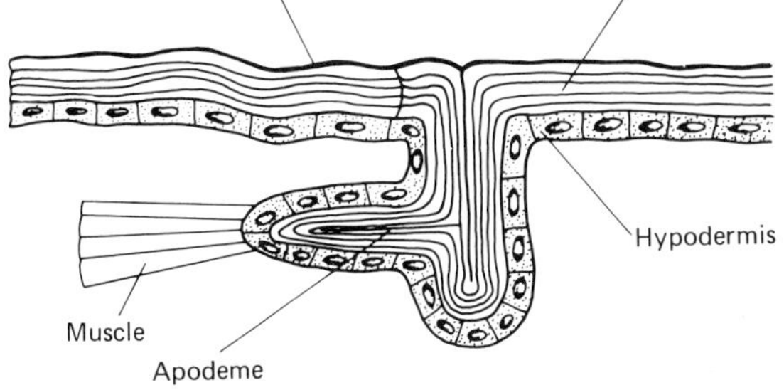

Figure 10.7 Diagram showing the inward folding of the hypodermis and associated cuticle to form an apodeme for muscle attachment. (From Vandel, In *Traité de Zoologie Vol. VI*, Grassé, Ed.)

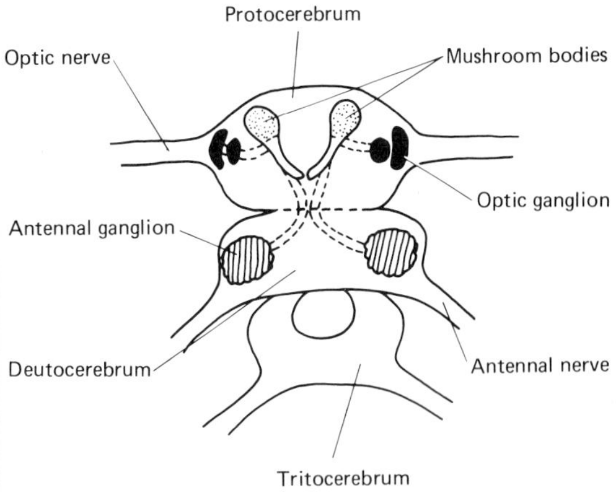

Figure 10.8 A generalized arthropod brain illustrating major brain regions and the sensory inputs to each. (From Vandel, In *Traité Zoologie Vol. VI*, Grassé, Ed.)

slow, prolonged contractions. Their activity regulates posture and very slow movements. **Inhibitory neurons** send impulses to muscles which block contractions.

Such a system permits a wide range of responses by relatively few motor units. Depending on the type of nervous stimulation received, most muscle fibers are capable of variable contraction speed and duration. In addition, some arthropods possess muscle fibers that are themselves inherently phasic (fast) or tonic (slow). This entire situation contrasts sharply with the vertebrate neuromuscular system, in which most muscle fibers respond to individual neurons only in an all-or-none fashion.

Sense Organs

In shielding an arthropod from the outside world, the exoskeleton isolates the animal from environmental stimuli. Those stimuli are necessary, however, to inform the arthropod about the surrounding world. It is not surprising, therefore, that the cuticle has certain areas specialized for sensory input. Tiny canals lined with sensory cells are common, and many permit passage of a sensory hair or bristle to the surface (Fig. 10.3). Such sensory sites are concerned primarily with tactile and chemical stimulation. They are distributed generally over the body surface, but reach their greatest concentration about the joints of the appendages. The common cephalic appendages called **antennae** are always well endowed with sensory cells. Other cuticular sensory structures include balancing and auditory organs.

In most arthropods, eyes are well developed. A great many eye types occur in this phylum, and some are quite complex. The most elaborate are called **compound eyes** (Fig. 10.9). Compound eyes are composed of many long cylinders called ommatidia. Each **ommatidium** contains a full complement of photoreceptive equipment: an outer cornea which functions as a lens, a crystalline cone, light-screening pigments, and a photosensitive retinula and its associated nerves.

The typical ommatidium is long and cylindrical. Its outer end is covered with translucent cuticle which forms a cornea and lens. The cuticular lens, or **facet**, is immobile; thus the focus of the eye is fixed. Below the lens is a long cylindrical body called the **crystalline cone**. The crystalline cone forms a secondary lens (also immobile), and its multilamellar internal structure may screen oblique light rays from the ommatidium. The proximal end of the ommatidium houses the photoreceptive element, the **retinula**, which comprises a rosette of cells around a central translucent chamber. The inner borders of these cells are folded into numerous tubules which project into the central chamber (Fig. 10.9c). These tubular borders are the immediate site of photoreception. When stimulated, the retinula responds as a single unit; impulses travel over a bundle of axons (one per retinular cell) to inform the brain that the ommatidium has experienced a single point of light.

The information received by the brain describes a pattern of light corresponding to an environmental scene. The quality of the image depends largely on the "grain" of the eye; that is, on the density of the ommatidia. Compound eyes possess from 2 or 3 to more than 10,000 ommatidia. On the whole, the arthropod eye does not produce a very distinct image. Its design, however, is well suited to detect movement. A slight change in the position of an object against its background is recorded several times, owing to a corresponding change in the number and distribution of ommatidia that are stimulated.

Arthropods may control the amount of light entering individual ommatidia. Such control is invested in mobile pigments. The distal end of the ommatidium is surrounded by specialized pigment cells, while pigment granules occupy the retinular cells themselves. When light is intense, pigments from both ends migrate centrally. The ommatidial walls become opaque, and any light that reaches a retinula must enter through its associated facet. When light is dim, the pigments retreat and the retinula receives light not only through its own facet but through neighboring units as well.

Arthropods tend to be either diurnal or nocturnal, and their compound eyes are adapted accordingly. Animals active during the day have well-developed screening pigments; commonly, their crystalline cones terminate near their retinulas (Fig. 10.9d). Both of these factors isolate each ommatidium and increase the sharpness of the total image. Experiments reveal that, in such eyes, less than 1% of the light reaching each retinula enters from a neighboring ommatidium. Screening pigments are reduced in nocturnal arthropods, and a longer space commonly separates the crystalline cone from the retinula (Fig. 10.9e). With such an arrangement, each retinula receives stimulation from surrounding units. Indeed, the retinulas of nocturnal arthropods may depend on neighboring ommatidia for as much as half of their light.

Behavior

Arthropods display some of the most interesting behavior in the invertebrate world. Among the other invertebrates, only the cephalopods rival them in

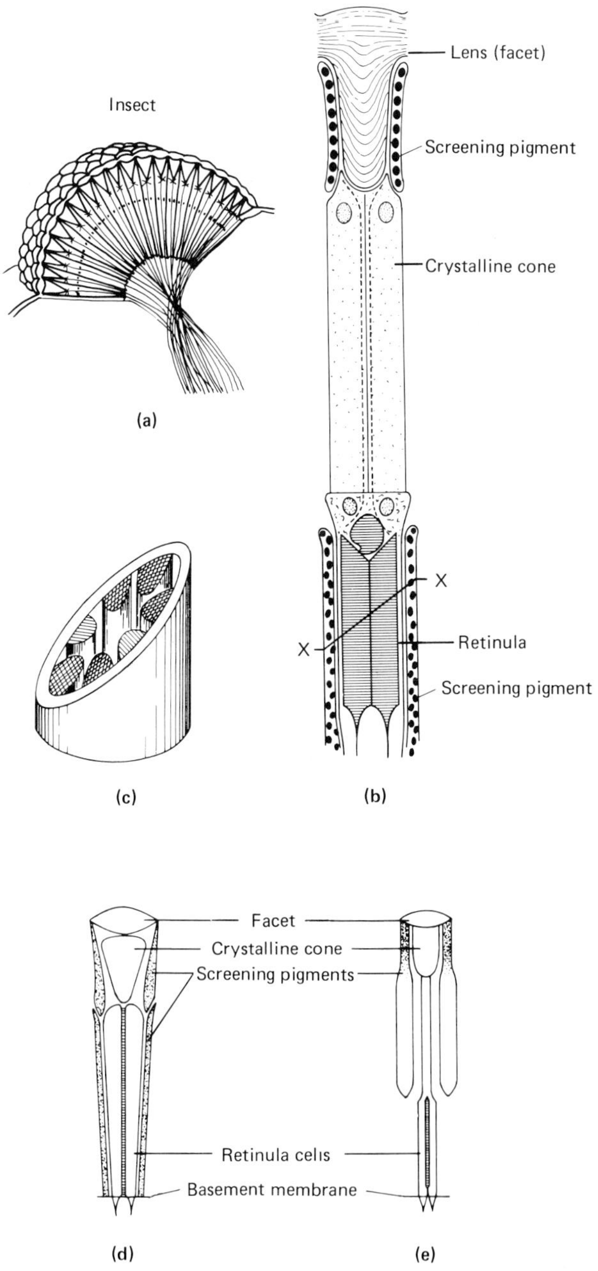

Figure 10.9 (*a*) A portion of the compound eye of an insect; (*b*) Longitudinal view of a single ommatidium from the eye of a firefly; (*c*) Section through the retinula of an ommatidium along the x-------x plane shown in Fig. 10.9b; (*d*) An ommatidium of the apposition type from a diurnal arthropod; (*e*) An ommatidium of the superposition type from a nocturnal arthropod. (For additional explanation, see text.) (*a* from M. Wells, *Lower Animals*, 1968, McGraw-Hill; *b-e* from J.J. Wolken, *Invertebrate Photoreceptors*, 1971, Academic Press.)

terms of locomotor ability and behavioral complexity. Like the squid and the octopus, arthropods confront their environments in an active manner. On the whole, they are highly mobile creatures which range widely in search of food and mates. Many construct shelters from environmental raw materials. Arthropod mobility is a key factor in the phylum's colonization of the terrestrial environment. On land, arthropods have more available oxygen and less mechanical resistance from the surrounding medium—two factors favoring active life-styles. The terrestrial environment presents widely scattered sources of food, shelter, and mates; these attributes also contribute to selection for more active behaviors by the land members of the phylum.

A tremendous diversity of life-styles is seen in the Arthropoda. Most arthropods are mobile, but there are planktonic and even sessile forms, such as the barnacles. Active arthropods, themselves a diverse lot, include crawlers, burrowers, swimmers, and most dramatic of all, flyers. The true flying ability demonstrated by most insect orders is a marvel achieved only by bats and birds in the vertebrate branch of the animal kingdom.

The social systems of bees, ants, and their relatives are some of the most intriguing phenomena in the invertebrate world. These animals communicate and cooperate among themselves in a highly organized fashion. Sexual behavior in this phylum is also a subject of continuous research. With so many arthropods, fascinating courtship procedures have evolved for species identification and mate selection.

Owing to the enormous complexity and specificity of arthropod behavior, further discussion is reserved for the accounts of each group.

CONTINUITY SYSTEMS

Typical arthropod continuity involves separate sexes, gonads and their accessory glands and tubules, copulation, modified cleavage, and serial growth periods defined by the secretion of a new cuticle. Details vary widely among the many groups; thus only the briefest outline is presented here.

Reproduction and Development

Almost all arthropods are diecious. Paired gonads with mesodermal walls occupy the body trunk and connect with accessory glands via tubule systems, some of which are quite elaborate. The gonadal cavity may be a remnant of the coelom; the gonoducts would then represent persistent coelomoducts. The reproductive system terminates ventrally in one or more genital pores. Around these pores, the cuticle may be sculpted for sexual activities. Male arthropods, for example, may possess a cuticular penis. In addition, certain body appendages often are modified for courtship displays and sperm transfer.

Classical arthropod development, although basically protostomate in character, features a unique cleavage. Called superficial cleavage, this arthropod phenomenon reflects the unusual distribution of yolk in the zygote. A fertilized arthropod egg consists of a yolky sphere, at whose center lies a small mass of yolkless cytoplasm surrounding the nucleus (Fig. 10.13a). Early mitotic divisions involve only the nucleus so that in time a syncytial mass occupies the egg center. Each nucleus then migrates to the cytoplasm at the egg surface, where cell membranes form. The embryo then comprises a yolky sphere covered by a thin cellular layer (Fig. 10.10b and c). This stage is the stereoblastula. Along the ventral side of the stereoblastula, a germinal band develops (Fig. 10.10d). Gastrulation occurs along the main axis of the band, transporting future endodermal and mesodermal tissue beneath the outer ectoderm. Paired coelomic spaces appear within the mesoderm. Meanwhile, endodermal cells rapidly surround and absorb yolk nutrients. Segmental proliferation occurs, with each somite receiving one pair each of coelomic pouches, ganglionic precursors, and appendage buds. The growing embryo gradually envelops the entire sphere (Fig. 10.10e–i). Mesodermal growths enclose the blastoel, meeting middorsally to form the heart. In contributing to other mesodermal organs and connective tissues, the coelomic walls disintegrate and the coelom merges with the blastocoel.

Newly hatched arthropods commonly lack many characters that distinguish their species as adults. Entire segments may be missing, and other structures, notably sexual organs, are often absent. Indeed, larval and juvenile arthropods frequently differ so markedly from the adult form that their entire life-style may be distinct. The profound differences between the caterpillar larva and the adult butterfly testify to the divergency that can exist in the life cycle of a single species. Associated with the extensive posthatching development exhibited by arthropods are their peculiar requirements for growth.

The arthropod exoskeleton cannot grow. Thus, once secreted and hardened, an exoskeleton imposes definite size limits on its bearer. Further growth requires that the exoskeleton be shed. The process by which an arthropod removes its old skeleton and forms a new one is called molting (Fig. 10.11).

Molting begins when the epidermis separates from the old exoskeleton. At this time, cell division often occurs, and the epidermis secretes a new epicuticle. Following this, the epidermal cells secrete chitinase and protease, which pass through the new epicuticle and gradually digest the old untanned endocuticle. Most of the products of this digestion are reabsorbed by the epidermal cells and are reused to synthesize the new cuticle. Thus, unlike the keratin of hair and of skin in vertebrates, which are not part of the body's metabolic pool, most of the cuticle of an arthropod can be recycled and reused by the animal. The digestion of the old cuticle and the secretion of the new cuticle occur simultaneously. The process of digestion continues until the old endocuticle is completely gone. The old exocuticle that is tanned and the epicuticle are

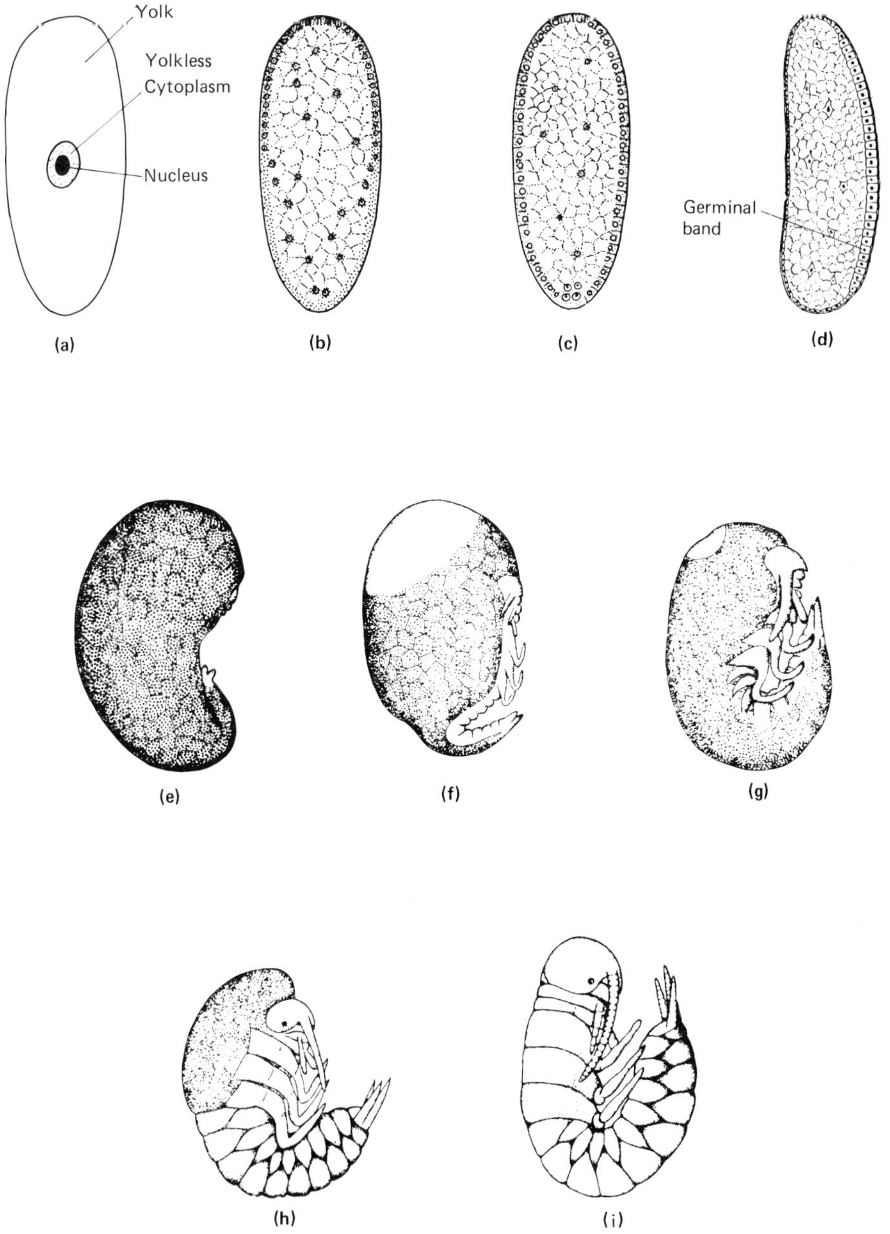

Figure 10.10 Arthropod embryonic development. (*a*) A fertilized insect egg prior to cleavage; (*b*) Nuclei migrating to the periphery; (*c*) Formation of uninucleate cells at the periphery yielding a stereoblastula; (*d*) Formation of the germinal band from the blastoderm on the ventral side of the stereoblastula; (*e-i*) Successive stages in the development of the embryo including segmental proliferation and the development of appendage buds, and the gradual overgrowth of the yolk by the embryo. (*a-d* from Snodgrass, *Principles of Insect Morphology*; *e-i* from A.G. Sharov, *Basic Arthropodan Stock*, 1966, Pergamon Press.)

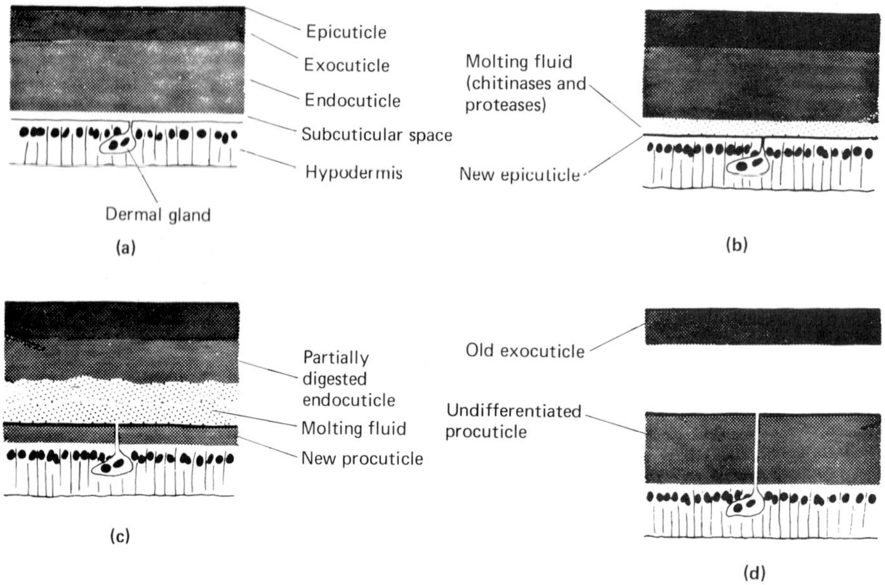

Figure 10.11 Representation of the changes occurring in the integument during the molt cycle. (*a*) Hypodermis separates from the overlying cuticle; (*b* and *c*) Chitinases and proteases are secreted into the intervening space and begin to digest the old endocuticle. New epicuticle and procuticle are secreted; (*d*) Enzyme containing molting fluid is reabsorbed and old exocuticle is cast off exposing the new untanned cuticle. (From R.F. Chapman, *The Insects: Structure and Function*, 1971, Elsevier Publishing.)

not digested. The new epicuticle, which is also resistant to digestion, protects the newly synthesized cuticle from being digested.

Until the very last moment, the muscles and nerves remain attached to the old epi- and exocuticle so that the animal is able to move and receive stimuli from the environment during the process of molting. A few hours before the old cuticle is shed, the epidermal cells usually secrete waxes which waterproof the surface of the new cuticle. Finally, what is left of the old cuticle is shed in a process called **ecdysis**. The old cuticle is ruptured along predetermined lines, and the arthropod crawls free from its old cover. At each ecdysis, the entire cuticle—including those parts forming the stomodeum, the proctodeum, and passages of the reproductive and respiratory tracts—is shed. After ecdysis, the epidermis secretes various chemicals and enzymes which tan and harden the new exocuticle. Also, the epidermis continues to secrete more endocuticle so that the exoskeleton thickens.

Molting is hormonally controlled. This hormonal control has been studied best in insects. A steroid hormone called **ecdysone** is secreted by prothoracic glands in the prothorax of the insect. This hormone acts directly on the epidermal cells and stimulates molting. The secretion of ecdysone by the prothoracic glands is in turn controlled by neurosecretory hormones produced by specialized nerve cells in the insect's brain. The secretion of these neurosecretory hormones is controlled by various environmental and nutritional stimuli. Other arthropods appear to use similar hormonal mechanisms, and an injection of ecdysone causes molting in all arthropods in which it has been tested, including horseshoe crabs, spiders, crustaceans, and insects. The fact that the process of molting is similar in all arthropods and is regulated by the same steroid hormones strongly suggests that all arthropods shared a common ancestor that molted and used the steroid hormone ecdysone.

Because tanning does not occur until after ecdysis, a new exoskeleton is always soft and flexible. It cannot provide maximal support and protection, and thus the immediate postmolt period is a vulnerable time for an arthropod. Many groups have evolved secretive behaviors for this stage. The formation of protective retreats, such as cocoons, is also common. The new exoskeleton stretches to accommodate the tension built up in growing cells prior to ecdysis and expands still further when the arthropod imbibes air and/or water following the molt. Muscle contractions and blood pressure also expand the new exoskeleton. This expansion provides room for tissue growth during the intermolt period.

Ecdysis occurs throughout the lifetime of some arthropods, but in many groups the number of molts is limited. The period between molts is called an **instar**. Larval instars exploit the metabolic upheavals associated with ecdysis to add new body parts. In some groups, new segments may be added with each progressive instar. In those groups with a fixed number of instars, sexual maturity is attained with the final molt.

A Review

As the largest phylum in the animal world, the Arthropoda presents diverse adaptations in its maintenance, activity, and continuity systems. The next four chapters chronicle that diversity, as expressed in the group's various classes and orders. The phylum does, however, owe its success to certain shared characters. Basically, arthropods are:

(1) bilaterally symmetrical, segmented invertebrates, whose hard exoskeleton bears jointed appendages;
(2) animals that often have abandoned classical metamerism in developing specialized structures;
(3) well-cephalized creatures with complex (often compound) eyes and an advanced brain;
(4) animals with an open hemocoel and a dorsal heart;
(5) the most mobile phylum in the invertebrate world and its most successful land dwellers;
(6) protostomes with superficial cleavage and a growth plan involving molting.

Arthropod Diversity

Much remains unsettled in arthropod taxonomy. Not only are relationships among the higher taxa unclear, but questions remain whether the group is even monophyletic. The arthropod assemblage could include convergent animal types which should rightfully constitute two or more separate phyla. For the present organizational purposes, we use the taxonomy listed below. A justification for this scheme, as well as alternative classification systems and speculations on the origin of this super group, is given in the discussion of arthropod evolution (see end of Chap. 14).

Phylum Arthropoda

Subphylum Trilobitomorpha
Class Trilobita: the fossil trilobites

Subphylum Chelicerata
Class Merostomata: the horseshoe crabs and their fossil relatives
Class Arachnida: the spiders, ticks, mites, scorpions *et al.*
Class Pycnogonida: the sea spiders

Subphylum Mandibulata
Class Crustacea: the crabs, shrimp, lobster, copepods, barnacles et al.
Class Chilopoda: the centipedes
Class Diplopoda: the millipedes
Class Pauropoda: the pauropodans
Class Symphyla: the symphylans
Class Insecta (Hexapoda): the insects

SUBPHYLUM TRILOBITOMORPHA

We begin our discussion of arthropod diversity with an account of the phylum's oldest major group, the trilobites. For millions of years these animals dominated the floors of ancient seas, but they became extinct at the end of the Paleozoic era. Today trilobites provide a picture of early arthropod organization and a valuable reference point for evolutionary studies.

Nearly 4000 trilobite species have been identified. In most cases, only the dorsal exoskeleton has been fossilized. This skeleton, which was often reinforced with calcium carbonate, defined the flattened, oval form of these animals (Fig. 10.12a). Often it was enrolled laterally to flank the thinner cuticle of the ventral surface, which bore paired, segmental appendages (Fig. 10.12b). Trilobites were generally small, most not exceeding 10 cm in length. Some species were less than a millimeter long; the exceptionally large species were over half a meter.

The main axis of the trilobite body was divided into an anterior cephalon, a thorax, and a pygidium. At right angles to these divisions, a pair of dorsal longitudinal grooves extended the entire length of the animal, thus defining a medial section and two lateral lobes. This dorsal trilobation is responsible for the subphylum's name.

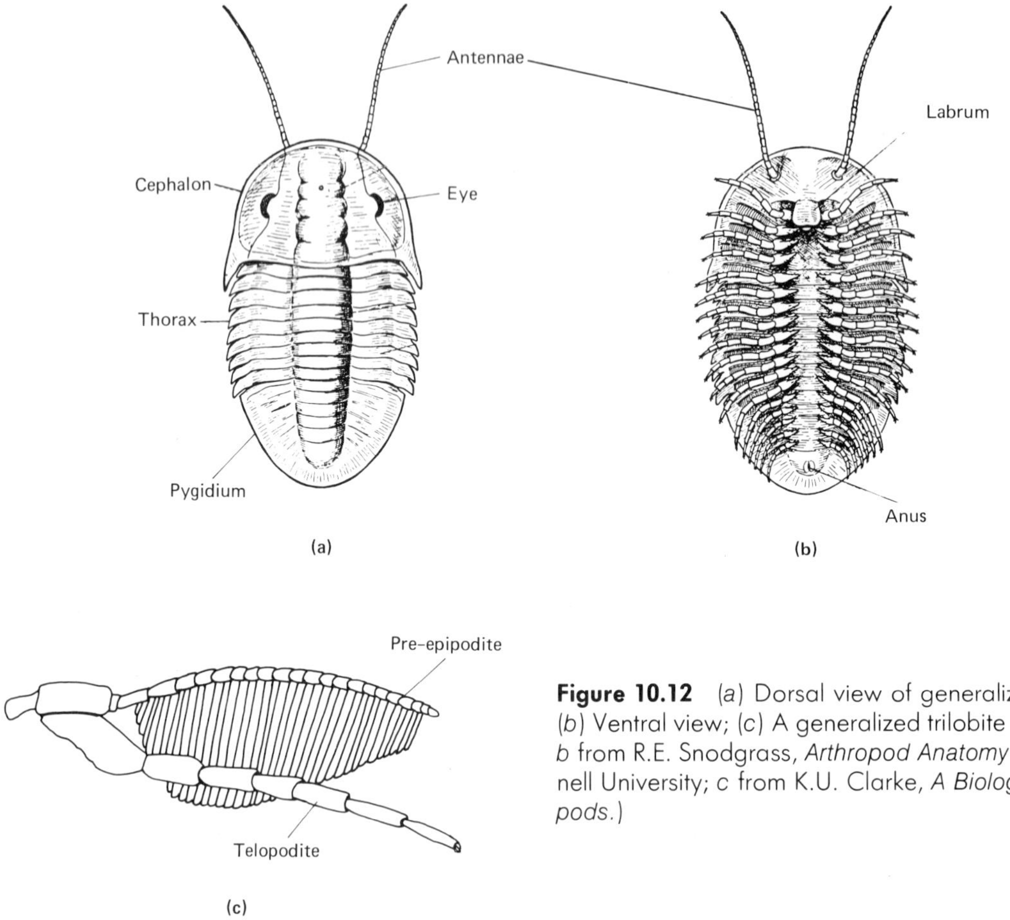

Figure 10.12 (*a*) Dorsal view of generalized trilobite; (*b*) Ventral view; (*c*) A generalized trilobite limb. (*a* and *b* from R.E. Snodgrass, *Arthropod Anatomy*, 1952, Cornell University; *c* from K.U. Clarke, *A Biology of Arthropods*.)

The fused cephalon included the first four postoral segments. A solid cuticular carapace covered this body region dorsally. A pair of compound eyes occupied the lateral lobes of the carapace; the mouth and several paired appendages were located ventrally. The mouth lay behind a liplike extension called the **labrum** (Fig. 10.12b). Sensory antennae (located on the preoral segment) sprouted from either side of the labrum; these appendages may be homologous with insect antennae and crustacean antennules. The four postoral segments bore paired appendages. Trilobite limbs were **biramous;** that is, they had two branches (Fig. 10.12c). The inner branch, called a **telopodite,** was a walking leg. The outer branch, or **preepipodite,** was an attachment site for numerous long filaments, which extended posteriorly in overlapping fields and may have been gills or food-collecting devices, or both.

The trilobite thorax and pygidium comprised a variable number of like segments. (The trilobite pygidium is thus not homologous with the nonsegmental, terminal pygidium of the annelids and most arthropods.) Ventrally, each body segment bore a pair of appendages similar in structure to, but larger than, those of the cephalon. Appendage size decreased toward the posterior end of the pygidium. In this terminal body zone, fused segments were covered dorsally by a solid cuticle.

Fossil specimens reveal very little about the internal biology of the trilobites. Organ systems, evidently arranged along the major body axis, probably resembled those of modern arthropods.

Most of these creatures probably crawled about the sea bottom, where they may have fed on primitive polychaetes and other small invertebrates. Their flattened bodies and dorsal eyes suited the trilobites well

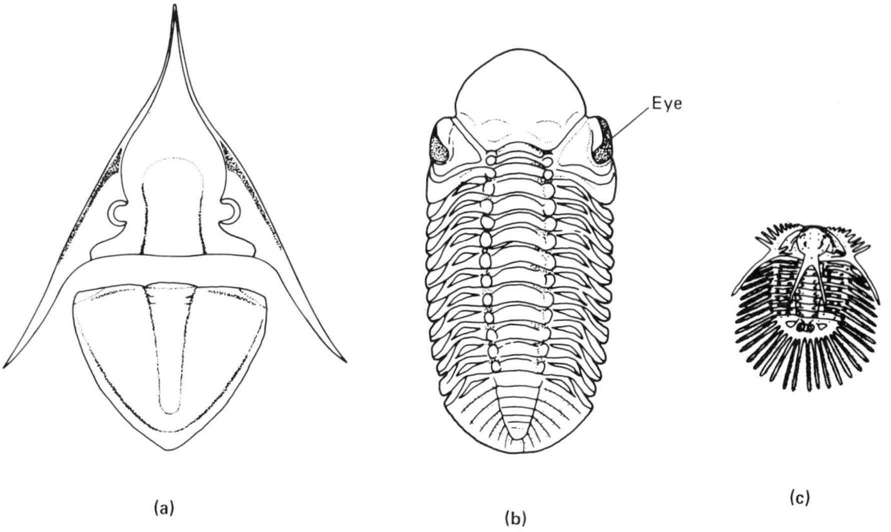

Figure 10.13 (*a*) *Megistaspidella*, a trilobite with a plow-like cephalon. (*b*) A swimming trilobite, *Reedops;* (c) *Radiaspis*, a planktonic trilobite. (*a* and *b* from L. Störmer, In *A Treatise on Invertebrate Paleontology, Part O: Arthropoda 1*, R.C. Moore, Ed., 1959, Geological Society; c from Grassé, Ed., *Traité de Zoologie, Vol. VI.*)

for such a life-style. Those specimens with a plowlike cephalon may have burrowed (Fig. 10.13a); perhaps their food was organic debris ingested with the substratum. Swimming trilobites were narrower and more streamlined, with marginal eyes (Fig. 10.13b). Most likely, their appendages presented broader, paddling surfaces for increased water resistance while swimming. Finally, certain members of the subphylum were planktonic. This group includes the smallest trilobites, some of which bore cuticular spines to aid in flotation (Fig. 10.13c).

Trilobite continuity involved three developmental stages: the protaspis, meraspis, and holaspis larvae (Fig. 10.14). The newly hatched trilobite was a **protaspis** larva, a tiny planktonic form consisting only of the acron and the segments destined to form the head. Several instars elapsed before (with the addition of more segments) a **meraspis** larva was produced. There, a true pygidium began to proliferate thoracic segments from its anterior edge. Several molts later, the **holaspis** larva was formed. This last developmental stage resembled the adult in basic structure. Succeeding instars witnessed growth and maturity to full adulthood.

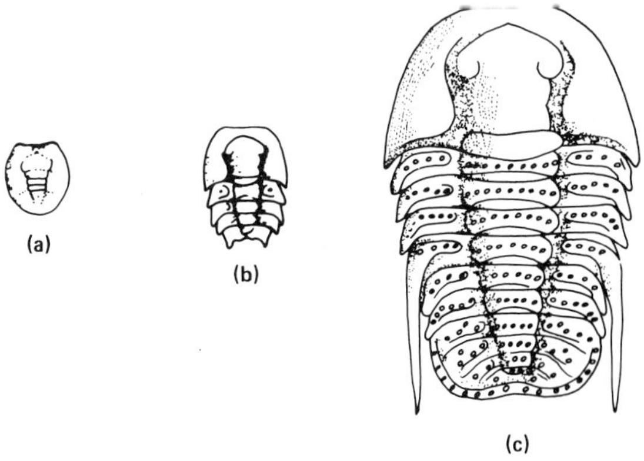

Figure 10.14 Sequential larval stages in trilobite development. (*a*) Protaspis larva; (*b*) Meraspis larva; (c) Holaspis larva. (From Moore, Ed., *A Treatise on Invertebrate Paleontology, Part O: Arthropoda 1.*)

for Further Reading

Anderson, D. T. *Embryology and Phylogeny in Annelids and Arthropods.* Pergamon Press, Oxford, 1973.

Carthy, J. D. *The Behaviour of Arthropods.* W. H. Freeman, San Francisco, 1965.

Cisne, J. L. Trilobites and the origin of arthropods. *Science* 186: 13–18, 1974.

Clarke, K. U. *The Biology of Arthropoda.* American Elsevier, New York, 1973.

Florkin, M., and Scheer, B. T. (Eds.) *Chemical Zoology, Vols. V and VI.* Academic Press, New York, 1971. (Volume V concerns a general introduction to the arthropods and aspects of their nutrition, digestion, metabolism, osmoregulation, growth, a development. Volume VI covers arthropod integument, hemolymph, respiration, excretion, and endocrines.)

Krishnakumaran, A. and Schneiderman, H. A. Chemical control of molting in arthropods. *Nature* (London) 220: 601–603, 1968.

Manton, S. M. Arthropod phylogeny—a modern synthesis. *J. Zool.* 171: 111–130, 1973.

Manton, S. M. *The Arthropoda: Habits, Functional Morphology and Evolution.* Clarendon Press, Oxford, 1977.

Martinsson, A. (Ed.) *Evolution and Morphology of the Trilobita, Trilobitoidea and Merostomata.* Universitetsforlaget, Oslo, 1975.

Moore, R. C. (Ed.) *Treatise on Invertebrate Paleontology, Part O, Arthropoda 1.* Geological Society of America and University of Kansas Press, Lawrence, 1959. (Deals with trilobites)

Schram, F. R. Arthropods: A convergent phenomenon. *Fieldiana Geol.* 39: 61–108, 1978.

Sharov, A. G. *Basic Arthropod Stock.* Pergammon Press, New York, 1960. (Some interesting views on arthropod phylogeny which have come under increasing criticism by S. M. Manton and others)

Shaw, S. R. Optics of arthropod compound eyes. *Science* 165: 88–90, 1969.

Snodgrass, R. E. *Arthropod Anatomy.* Cornell University, Ithaca, New York, 1952.

Wolken, J. J. *Invertebrate Photoreceptors.* Academic Press, New York, 1971.

An assemblage of the Trilobite *Phacops* from the Devonian shells of Ontario, Canada. (Courtesy of the Smithsonian Institution.)

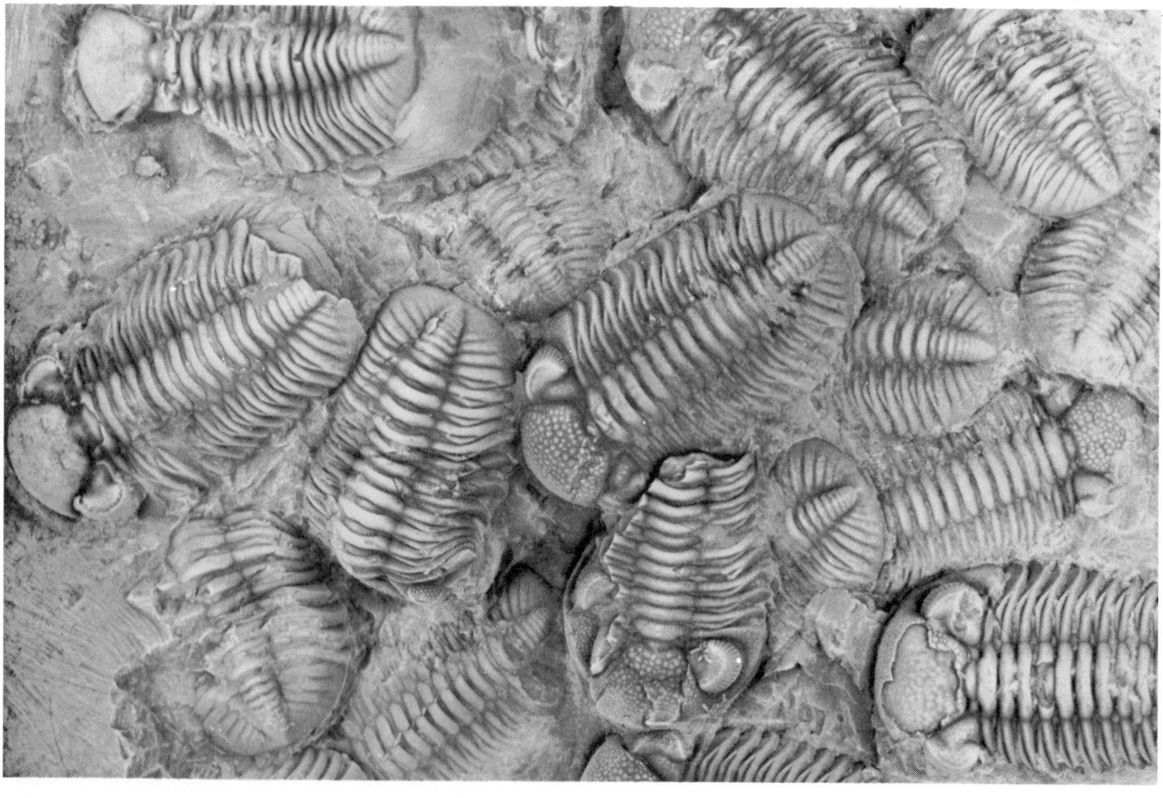

11

Arthropod Diversity: The Chelicerates

Some of the most dreaded animals in the invertebrate world belong to the subphylum Chelicerata. Over 65,000 species have been identified, including the familiar horseshoe crabs, spiders, scorpions, ticks, and mites. The vast majority of these arthropods are completely harmless to humans; however, a few can be quite dangerous, especially to infants.

The chelicerate body consists of an anterior **prosoma** and a posterior **opisthosoma.** The fused prosoma comprises the acron and appendage-bearing segments involved with food acquisition and locomotion. The opisthosoma usually maintains some external segmentation, but its appendages are extremely reduced. Digestion, reproduction, and respiration are important opisthosomal functions.

Distinguishing appendages in the subphylum include the **chelicerae** and the **pedipalps.** The former are immediately anterior to the mouth and function in food acquisition. The latter are postoral and are variously specialized in each of the chelicerate groups. There are no antennae and thus there is never a deutocerebrum in the chelicerate brain. Also lacking are the jawlike, chewing mandibles that characterize crustaceans, insects, and myriapods. Most chelicerates are land animals; indeed, the subphylum probably represents the first line in arthropod evolution to successfully colonize the terrestrial environment.

Three chelicerate classes are recognized. The most primitive group is the Merostomata, an aquatic class composed of many extinct species and the living horseshoe crabs. Nearly all modern chelicerates belong to the Arachnida; here we find the spiders, scorpions, mites, and ticks. The third chelicerate class, the Pycnogonida, contains the puzzling sea spiders.

CLASS MEROSTOMATA

The merostomes are relatively large and primitive aquatic chelicerates. Certain of their opisthosomal appendages form respiratory gills, and a long caudal spine extends from the posterior body (Figs. 11.1 and 11.2). Nearly all merostomes are extinct. Only four species survive in the subclass Xiphosura; the other subclass, the Eurypterida, is composed solely of fossil forms.

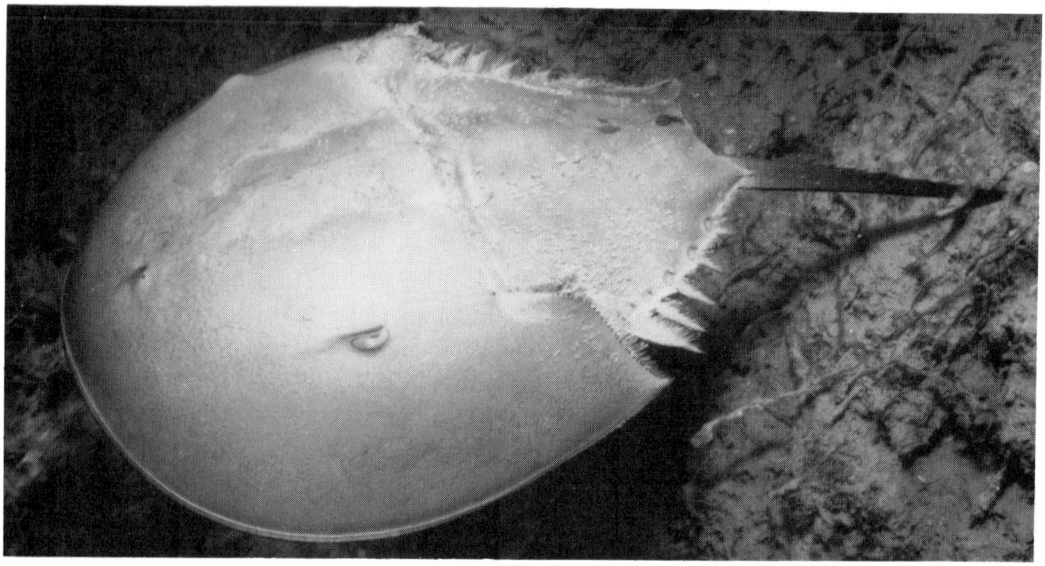

(a)

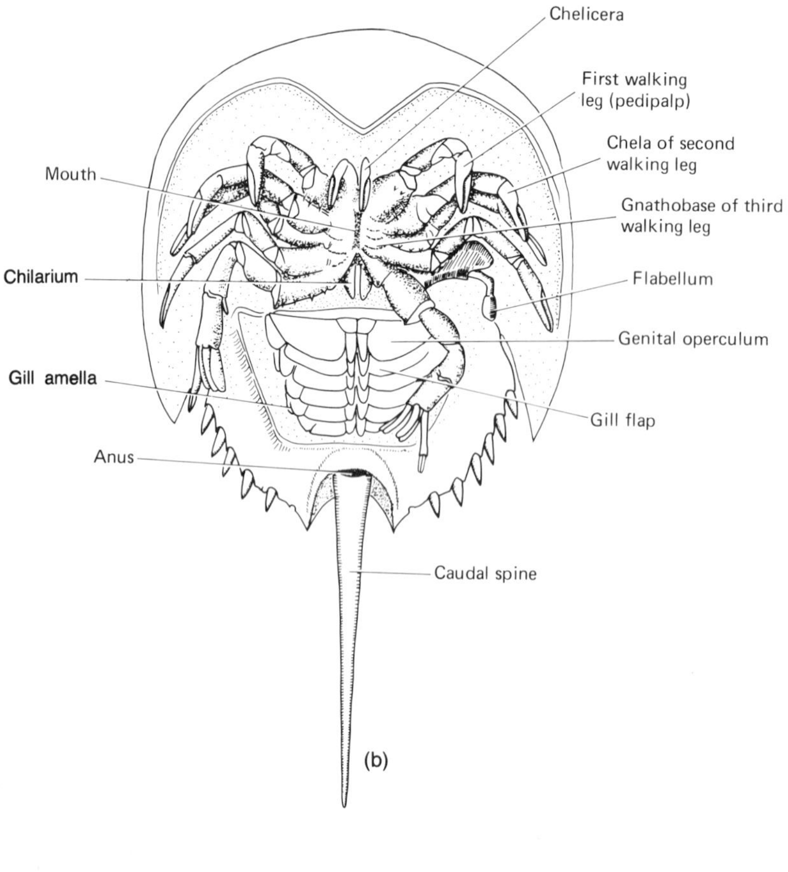

(b)

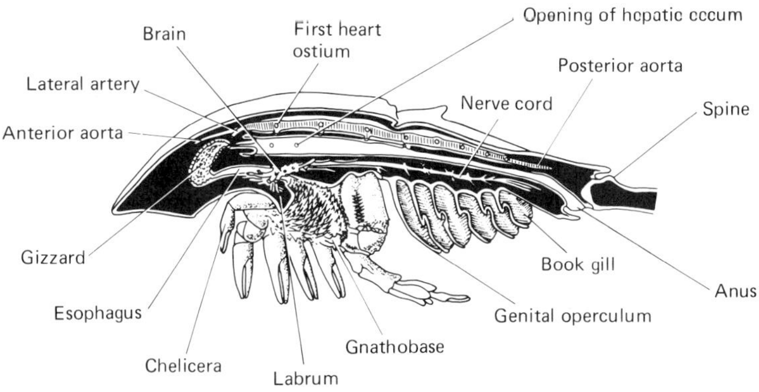

(c)

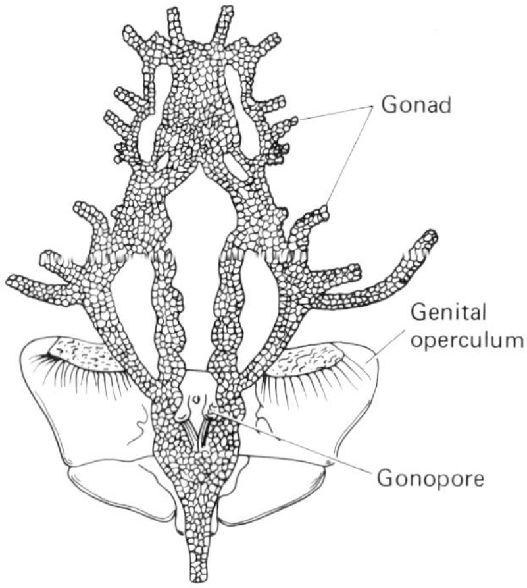

(d)

Figure 11.1 The horseshoe crab, *Limulus polyphemus*. (*a*) Dorsal view; (*b*) Ventral view showing major features of external structure; (*c*) Sagittal section illustrating major anatomical features; (*d*) Reproductive organs; (*e*) Trilobite larva of *Limulus*, ventral view. (*a* courtesy of Charles R. Seaborn; *b*, *c* and *e* from A. Kaestner, *Invertebrate Zoology, Vol. II*, 1968, Wiley-Interscience; *d* from L. Fage, In *Traité de Zoologie, Vol. VI*, P.P. Grassé, Ed., 1949, Masson et Cie.)

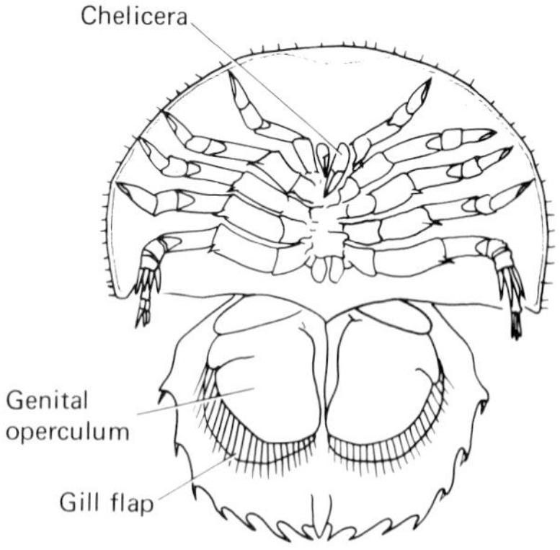

(e)

Subclass Xiphosura

Limulus polyphemus (Fig. 11.1), the horseshoe crab of the Eastern American coast, is the largest living merostome. Three other contemporary species inhabit Asian waters. Additional members of the Xiphosura include numerous extinct forms, most of which date from the Ordovician and Silurian. The biology of *Limulus* is fairly well known, so our present discussion will focus on this animal alone.

Horseshoe crabs have been called living fossils. Their common name describes the horseshoe-shaped, dorsally convex carapace that covers the prosoma. The posterior edges of this carapace form a three-sided slot into which the rear body is fitted (Fig. 11.1a). Both the carapace and the dorsal cover of the opisthosoma are fused shields comprising all tergal plates in their respective regions. A long caudal spine extends from the rear of the opisthosoma. This spine functions in locomotion and righting responses, but is not an offensive or defensive weapon. *Limulus* is one of the largest living arthropods, attaining overall lengths of up to 60 cm.

The ventral body surface bears paired segmental appendages (Fig. 11.1b). Anteriormost are the **chelicerae**, the food-gathering limbs that flank the mouth's upper lip, or **labrum**. A chelicera is composed of three articles, the last two of which form a pair of pincers. Common in several arthropod groups, such clawlike appendages are said to be **chelate**. Ten walking legs are arranged about the narrow, elongated mouth. The first four pairs of these legs are chelate; their spiny basal articles are called **gnathobases**. The fifth pair of walking legs is not chelate. Each of their basal articles includes a gill-cleaning extension known as a **flabellum**, while the distal article on each leg bears four leaflike processes. These processes are swept over the body surface to dislodge mud and other foreign matter. The prosoma ends with a pair of diminutive appendages called **chilaria**, which may be degenerate gnathobases.

Note that horseshoe crabs lack differentiated pedipalps. The first pair of walking legs is serially homologous with the pedipalps of other chelicerates, and its lack of differentiation is clearly a primitive trait.

The appendages of the horseshoe crab opisthosoma are adapted for reproduction and respiration. Each pair is flattened and has fused medially to form a wide membranous flap (Fig. 11.1b). The anteriormost flap is the **genital operculum.** This structure covers the paired gonopores on the eighth body segment. The following five flaps provide surfaces for respiratory exchange. Each of these appendages bears

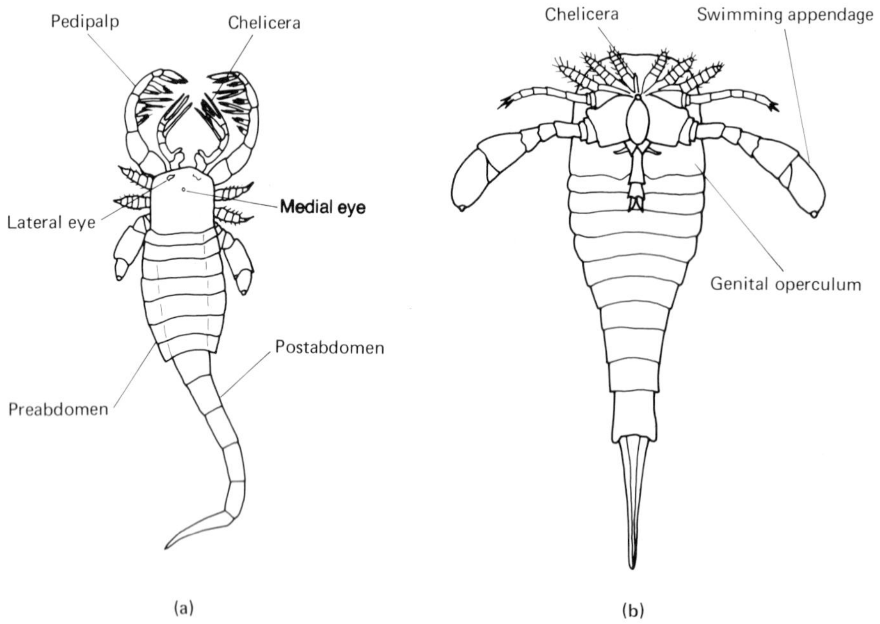

Figure 11.2 (*a*) Dorsal view of a eurypterid, *Mixopterus kiaeri;* (*b*) The ventral surface of a eurypterid (genus *Eurypterus*). (From R.E. Snodgrass, *A Textbook of Arthropod Anatomy*, 1952, Cornell University.)

numerous lamellar folds and is called a **book gill** (Fig. 11.1c).

Horseshoe crabs eat polychaetes, small mollusks, and other soft invertebrates which they encounter on the ocean floor. The chelicerae and other chelate appendages transport food onto the gnathobases. Although the last pair may fragment shells, these spiny articles do not chew prey as they pass it into the mouth. The foregut of *Limulus* includes an esophagus and a gizzard (Fig. 11.1c). These organs, which are lined with cuticle, are arranged along a short anterior loop. In the gizzard, hard chitinous teeth backed by a strong musculature chew foodstuffs. A valve between the gizzard and the midgut remains closed until shell fragments and other undigestible materials are removed through the mouth. The midgut comprises a stomach, paired hepatic ceca, and an intestine. Ducts from the hepatic ceca enter the stomach, where some digestion takes place. Further digestion and absorption occur within the ceca themselves. The intestine leads to a proctodeal rectum and anus, the latter located before the ventral base of the caudal spine.

A long, tubular heart courses dorsally through both body tagmata. Its openings include eight pairs of ostia, three anterior arteries, and four pairs of lateral arteries (Fig. 11.1c). The large size and complexity of these circulatory elements are typical of larger arthropods which possess localized respiratory organs such as book gills. Blood bathes the tissues directly before pooling in two major ventral sinuses. From these sinuses, blood flows into the book gills, then returns to the heart.

Respiratory exchange occurs across the book gill lamellae, and is facilitated by hemocyanin in the blood and by gill movement. Such movement ventilates the outer surface of the gills and also stimulates blood flow through the lamellae. *Limulus* and all other arthropods lack cilia and always use muscles for ventilation and circulation.

Coxal glands near the gizzard handle excretion in the horseshoe crab. Four of these glands are present on each side of the foregut. Blood-borne wastes absorbed across their walls eventually are collected within a single chamber on each side of the body. From this chamber, a long convoluted tubule leads to a muscular bladder. Excretory pores occur at the base of the last pair of prosomal legs.

The horseshoe crab brain forms a ring around the esophagus (Fig. 11.1c). A large protocerebrum and tritocerebrum are anterior to the looping foregut. Connectives join a subesophageal nerve mass formed by the fusion of all remaining prosomal ganglia. From this mass, a ventral nerve cord passes into the opisthosoma, where it links five segmental ganglia. Lateral nerves also communicate between the rear body and the brain collar. All of these nerves are peculiar in that they are housed within arteries.

Sense organs in *Limulus* include a pair of lateral and medial eyes and the ventral frontal organ. In addition, spines on the gnathobases contain cells that respond to food juices. The distal articles of the walking legs also bear chemosensitive centers. Compound lateral eyes are located dorsally, just outside lateral ridges on the carapace (Fig. 11.1a). No other compound eyes are known among the chelicerates. Their ommatidia are grouped rather loosely, perhaps indicating an ability to detect movement but hardly any image formation. The medial eyes are invaginated cups lined with retinal cells; a single pair flanks the central ridge of the carapace. The chemoreceptive frontal organ lies anterior to the labrum on the ventral reflection of the carapace. A third pair of eyes, degenerate in adult horseshoe crabs but probably functional in larval forms, is adjacent to the frontal organ.

Limulus crawls over or through soft bottoms. Its dorsal eyes, low body contour, and rounded anterior end represent adaptations for such movements. Certain smaller individuals swim with their ventral surfaces directed upward, in which case the book gills operate as paddles.

Sexes are separate. Both male and female reproductive systems comprise a series of interconnecting ducts and tubules which ramify symmetrically through the prosoma and the opisthosoma (Fig. 11.1d). (This situation is unusual for chelicerates, whose gonads commonly are confined to the rear body.) Paired gonopores are guarded by the genital operculum.

A male horseshoe crab mounts the opisthosomal shield of the larger female as she prepares to lay her eggs. The female excavates several small nests in the sand, into which she deposits up to 1000 ova. As they are laid, the eggs are fertilized by the male and then covered with sand.

Embryogeny features total cleavage and the formation of a stereogastrula. Two centers of mesodermal proliferation develop. One contributes the first four prosomal segments; the second, posterior center then produces the remaining parts. Very early in development, the first seven segments are isolated as the future prosoma.

The horseshoe crab hatches as a **trilobite larva** (Fig. 11.1e). This stage resembles a trilobite, particularly in the similarity between its prosoma and the trilobite cephalon. At first, the caudal spine is extremely short and several abdominal appendages are missing. A young *Limulus* swims and burrows, and with each succeeding instar it gradually adds adult segments. The spine, too, grows to adult length.

Horseshoe crabs require about 10 years to reach full maturity, and their total life span may be twice that long.

Subclass Eurypterida

Perhaps the largest arthropods the world has known belonged to the subclass Eurypterida. These merostomes, all of which are now extinct, flourished in Paleozoic seas, where one species *(Pterygotus)* attained a length of almost 3 m. The subclass made some advances into fresh water and possibly even onto land. Two hundred species are known.

Eurypterids resembled the modern horseshoe crab. A prosoma, opisthosoma, and caudal spine were present, and the ventral appendages of the two groups were similar (Fig. 11.2). The eurypterid carapace, however, was more compact, as it covered a relatively smaller anterior body. Dorsally, it housed lateral and medial eyes. The opisthosoma was unique in that it was divided into two zones—a preabdomen with seven appendage–bearing segments and a tapering postabdomen of five limbless segments. In some species, the postabdomen may have borne a terminal stinger.

Eurypterids crawled and/or swam. Well-developed walking legs carried them over the ocean floor or along freshwater bottoms. Certain species whose book gills were protected by ventral extensions of the preabdominal skeleton may have ventured onto land. In some eurypterids, such as the scorpionlike *Mixopterus kiaeri* (Fig. 11.2a), the anterior pair of legs formed raptorial pedipalps. In swimming species, the fifth (and sometimes the fourth) pair of prosomal appendages were expanded distally to form paddles. Preabdominal appendages included a genital operculum and book gills. Eurypterid larvae apparently resembled those of the horseshoe crab.

The merostomes obviously are a class whose past is more important than its present. Similarities between this group and the Trilobita encourage phylogenetic schemes that describe the trilobites as the immediate ancestors of the merostomes. Indeed, certain fossil forms appear to bridge the gap between the two groups. At the more recent end of merostome evolution, it is likely that the eurypterids gave rise to the largest of the chelicerate classes, the Arachnida.

CLASS ARACHNIDA

The Arachnida is the first group of arthropods that makes a major contribution to the phylum's dominance of the animal world. The class includes about 60,000 known species of spiders, scorpions, mites, ticks, and their close relatives (Fig. 11.3). Many more thousands of species no doubt await identification. Some of these creatures inspire fear in the most courageous people. However, while poison glands are present in certain arachnids, very few members of the class can permanently harm humans.

Arachnids are land animals. A terrestrial history dating from the Devonian probably qualifies them as the arthropod pioneers of that environment. On land, the class underwent an extensive adaptive radiation which saw the exploitation of different subhabitats and life-styles. Later, a few arachnids returned to aquatic environments. Accordingly, several distinct orders exist today. Each of these subgroups is described following a brief account of the unifying features of the class.

Arachnid Unity

Most arachnid features are associated with life on land. In adapting to this most inconstant of environments, arachnids have evolved rather water-tight bodies and aggressive behaviors.

MAINTENANCE SYSTEMS

General Morphology

The arachnid body consists of a prosoma and a variously metameric opisthosoma. The seventh body segment forms the junction between these two regions; it may be reduced to a narrow stalk called the **pedicel**. In the most primitive members of the class, both a segmented preabdomen and a postabdomen are present. Most arachnids, however, have undergone some fusion of the posterior body. Advanced orders feature a wholly unsegmented opisthosoma, while in the most specialized arachnids (the mites), the entire body is fused (Fig. 11.3c).

A solid carapace covers the dorsal surface of the prosoma. Sternal plates and/or the bases of the appendages constitute the ventral surface. In the opisthosoma, skeletal design depends on the metamerism of the rear body. Arachnid epicuticle is impregnated with wax and thus forms a water-resistant barrier.

Prosomal appendages include a pair each of chelicerae and pedipalps and four pairs of walking legs. The chelicerae are forwardly directed, grasping limbs whose size varies among the different orders. Arachnids with small chelicerae usually have large, chelate pedipalps. Alternatively, the pedipalps may

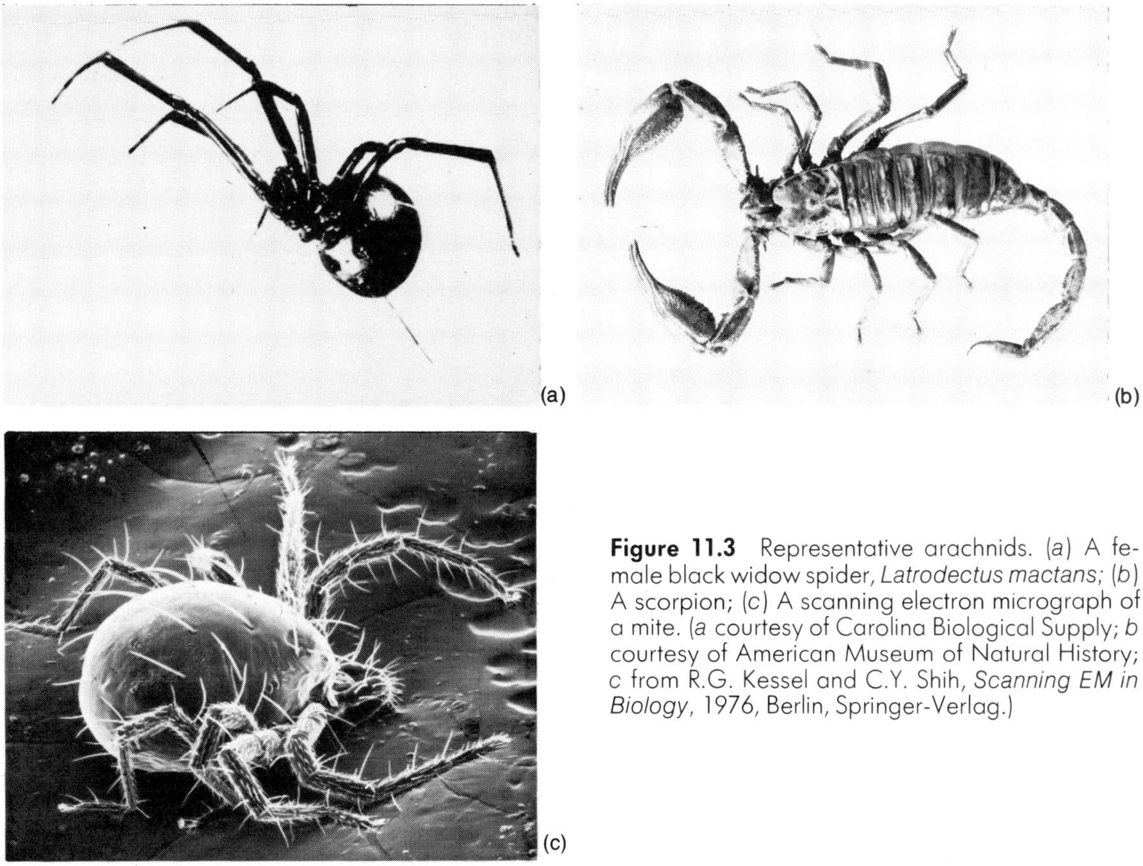

Figure 11.3 Representative arachnids. (*a*) A female black widow spider, *Latrodectus mactans;* (*b*) A scorpion; (*c*) A scanning electron micrograph of a mite. (*a* courtesy of Carolina Biological Supply; *b* courtesy of American Museum of Natural History; *c* from R.G. Kessel and C.Y. Shih, *Scanning EM in Biology*, 1976, Berlin, Springer-Verlag.)

serve as walking legs and/or sensory limbs. Opisthosomal appendages, if present, are modified as respiratory structures or as spinnerets.

Nutrition and Digestion

The typical arachnid is a raptorial carnivore feeding on smaller arthropods. The chelicerae, pedipalps, or legs capture and manipulate prey. The foregut is a relatively narrow tube which processes only small, moist particles of food. Therefore, because arachnids (like other chelicerates) have no real jaws, prey tissues must be reduced prior to their arrival at the mouth. This reduction can be a rather slow process; thus arachnid prey often are immobilized shortly after capture. Immobilization commonly results from toxins discharged by poison gland ducts on the chelicerae, the pedipalps, or a terminal stinging apparatus. When the victim is subdued, the chelicerae tear it into smaller pieces.

Digestion begins outside the body, when midgut enzymes flow from the mouth and penetrate prey tissues. Also, saliva may be discharged from glands in the chelicerae or pedipalps. The labrum and pedipalpal bases flank a prebuccal chamber in which this initial digestion occurs. The hairy inner surfaces of the pedipalps or first legs may channel food to the mouth. A muscularized pharyngeal pump then sucks the partially digested food through the foregut and into the stomach area. Many midgut diverticula branch through the body (Fig. 11.4). Secretory cells in the midgut wall discharge digestive enzymes, and nutrients are taken up by absorptive cells. Most arachnids store excess nutrients in interstitial cells adjacent to the midgut diverticula. Such storage enables some species to withstand starvation for long periods. The arachnid digestive tract concludes in a short proctodeal intestine and an anus at the rear end of the body.

Circulation

The arachnid circulatory system has characteristic arthropod features (Fig. 11.5). A tubular heart usually lies within a dorsal chamber in the anterior opis-

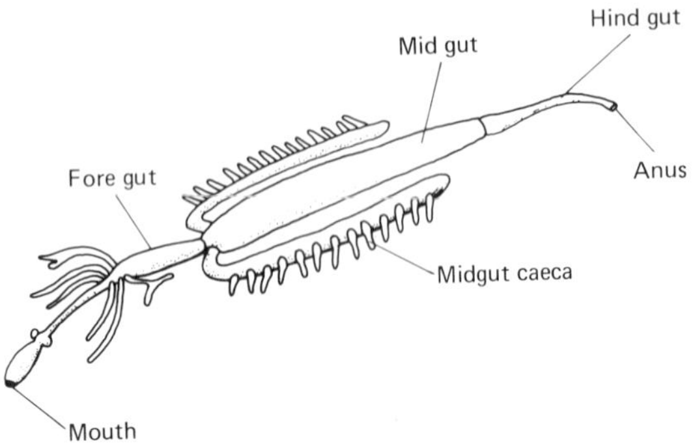

Figure 11.4 Diagram of a generalized arachnid gut. (From K.U. Clarke, *The Biology of Arthropods*, 1972, Edward Arnold.)

thosoma. In those arachnids whose abdomen is metameric, the heart conforms in segmentation. Paired ostia form the incurrent openings to this organ, whose muscularized walls pump blood through several arterial systems. A large aorta supplies the prosoma. A smaller, posterior vessel communicates with rear tissues, and a variable number of paired lateral arteries service the midbody. Among higher arachnids there is a reduction in the heart size, and in some groups the heart occupies only a single segment. In the mites, the tiniest arachnids, it is often absent. This reduction and eventual disappearance of the heart are correlated not only with decreasing body size but also with the evolution of efficient tracheal respiration in the advanced members of the class.

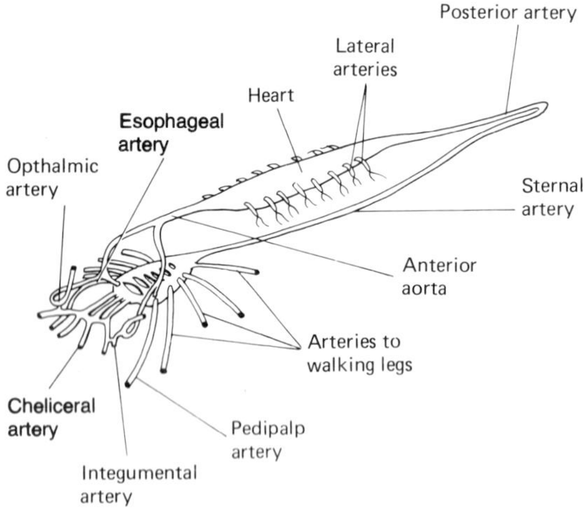

Figure 11.5 Diagram of the circulatory system of a scorpion, an arachnid with an extensive arterial system. (From Clarke, *The Biology of Arthropods*.)

Respiration

Arachnid respiration involves **book lungs** and/or **tracheae**. The former seem to represent book gills adapted to function on land. Book lungs occur in pairs along the ventral surface of the opisthosoma. Each consists of a small, invaginated air chamber which opens externally through a narrow slit called a **spiracle** (Fig. 11.6). One side of the lung chamber is folded into numerous lamellae held apart by cuticular bars. Air enters the book lung through the spiracle, and respiratory exchange occurs across the thin walls of the lamellae. Muscular control over spiracle size and chamber volume regulates water loss and may influence the respiratory rate. Those arachnids depending primarily on book lungs for gas exchange usually possess hemocyanin. Their well-developed heart and arterial systems reflect the respiratory responsibility of their blood.

Certain arachnids (many spiders) possess book lungs and tracheae, but tracheae alone suffice in some groups. Two types of tracheal systems have been described. The more primitive **tracheal lungs** consist of cuticle-lined tubules which extend into the hemocoel (Fig. 11.7). These tubules arise from an invaginated chamber which opens through a single surface spiracle. Gas exchange occurs with the hemolymph. In the more modern tracheal system, tubules descend deeper into the body and terminate in fluid-filled cavities adjacent the tissues. Respiratory exchange takes place directly with the body tissues, thus bypassing the circulatory system. In arthropods with such advanced tracheae, the heart and arterial system are reduced and hemocyanin may be absent. Each trachea may operate independently, but in more advanced arachnids, intertracheal networks are common. Definite routes of air flow develop, while individual spiracles assume inhalent or exhalent roles. The efficiency of the entire system is thus enhanced.

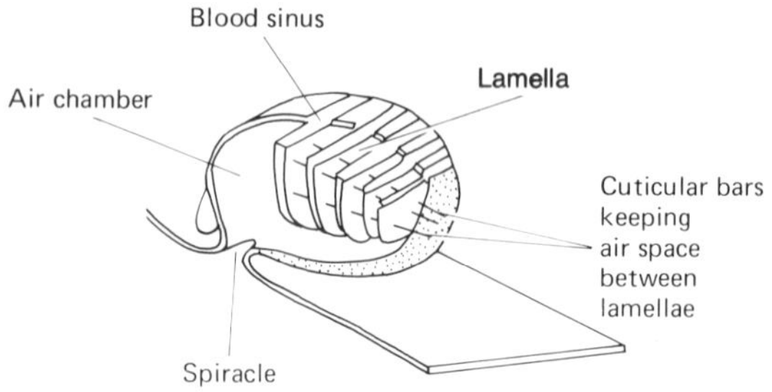

Figure 11.6 A book lung of a spider. (From Clarke, *The Biology of Arthropods*.)

Tracheae probably evolved from book lungs in animals under selective pressure to conserve water. These tubules are particularly effective among smaller arachnids, whose water-retention problems are most acute. (However, tiny mites living in areas of high humidity have no such problems, and their tracheae often have disappeared.)

Excretion and Osmoregulation

Excretion and water regulation in the Arachnida involve coxal glands and Malpighian tubules. Nephrocytes have also been described. Coxal glands are paired lateral sacs in the prosoma. Depending on the arachnid group, up to four pairs are present. These nephridialike organs may enclose remnants of the protoarthropod coelom. They absorb blood-borne wastes and transport them through a long coiled tubule to an excretory pore. Each pore is located on a **coxa,** or basal article, of a prosomal appendage.

One or two pairs of Malpighian tubules are located in the opisthosoma. These tubules originate from the junction of the midgut and intestine (Fig. 10.5b). Winding among the digestive diverticula, they absorb hemocoelic wastes across their own thin walls. These wastes pass through the tubules and into the hindgut, where water is resorbed. Most arachnid waste is excreted as crystalline guanine. This very dry nitrogenous product illustrates the terrestrial adaptation of the class.

ACTIVITY SYSTEMS

Most arachnids are aggressive animals which actively seek prey, mates, and shelter. Their central nervous systems are concentrated, and their sense organs—especially the tactile ones—are well developed. A number of specialized effector organs are involved in some of the more unusual behaviors of the class.

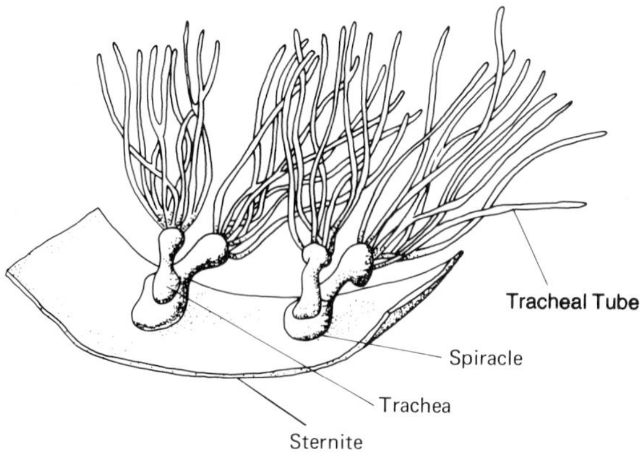

Figure 11.7 Tracheal lungs of *Ricinoides*, an arachnid with a primitive tracheal system. (From Clarke, *The Biology of Arthropods*.)

Muscles and Skeleton

Arachnid appendages are rather long and comprise many articles. The midportions of the legs often describe an arc somewhat higher than the central body. This posture seems to elicit a curiously unsettling response in many people.

Arachnid appendages are specialized for various activities. One pair of raptorial pincers is almost always present. These aggressive limbs are the chelicerae in spiders and mites. Scorpions, pseudoscorpions, and related smaller orders have small chelicerae, but their pedipalps are large and chelate. Regardless of which pair is so developed, the raptorial appendages are important in many activities, including prey capture, defense, digging, sexual display, and sometimes sperm transfer.

If the prosoma and opisthosoma are joined by a narrow pedicel, flexion between the two regions is possible. Such movement allows posterior effector organs to be brought forward or otherwise aimed. Thus, vinegaroons can spray their enemies with acetic acid from posterior glands and spiders can aim silk threads from their terminal spinnerets. Scorpions, although lacking a pedicel, have a long, narrowly segmented opisthosoma that can be flexed to aim a sharp posterior stinger.

Flexor muscles at arachnid joints are well developed. In the absence of extensor muscles, however, extremely high hemolymph pressures cause the reextension of contracted appendages.

Nerves

The arachnid nervous system has undergone considerable concentration. Several stages in that concentration are still displayed in the class (Fig. 11.8). Primitive arachnids (the scorpions) possess a supraesophageal brain, circumesophageal connectives, and a ventral nerve cord linking single ganglia in the opisthosomal segments (Fig. 11.8a). The brain comprises a protocerebrum, which serves the eyes, and a tritocerebrum, which innervates the chelicerae. In higher arachnids, opisthosomal ganglia have fused with the brain collar (Fig. 11.8b). In such animals, all major nerves emanate from the esophageal complex, and highly integrated behaviors are possible.

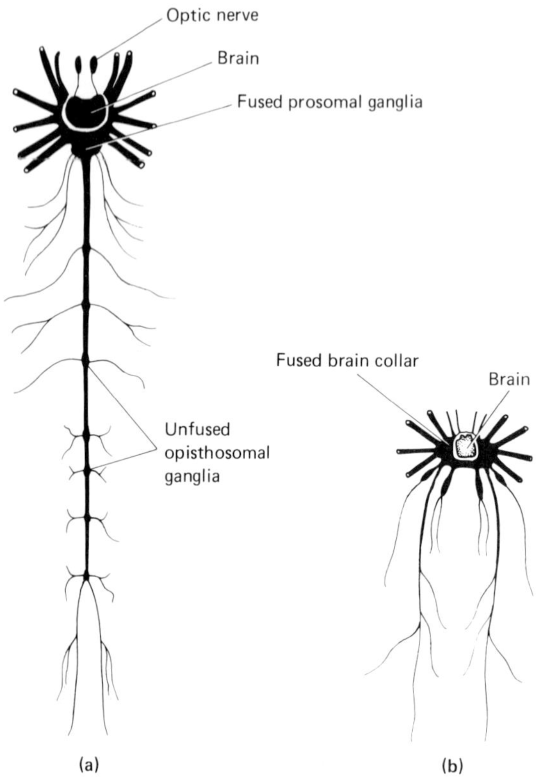

Figure 11.8 Arachnid nervous systems. (*a*) Primitive nervous system of a scorpion with a well-developed ventral nerve cord; (*b*) Advanced nervous system of an opilionid. (From J. Millot, In *Traité de Zoologie, Vol. VI*, Grassé, Ed.)

Sense Organs

Arachnid sensory devices include eyes, complex sensory hairs, and slit sense organs. Arachnid eyes are never compound, and their number and arrangement vary among the different orders. Commonly, each eye consists of a thick cuticular cornea and lens, a hypodermal layer, and a retina (Fig. 11.9). The retina, which contains the photoreceptive cells, faces either the light source or the inner surface of the eye. In the latter case, a postretinal membrane, the **tapetum,** reflects light into the retina (Fig. 11.9b). Image formation depends on the density of the retinal cells and is poorly developed in most arachnids. Certain spiders, however, have good image-forming eyes.

For many arachnids, the most important environmental information is transmitted through numerous, variously modified sensory hairs. These structures are especially abundant on the appendages. Usually either the chelicerae, the pedipalps, or the first pair of walking legs are specialized as the major tactile organs. Functionally, such appendages are comparable to the antennae of other arthropod groups.

Some hairs on the tip of anterior appendages are hollow and may be olfactory. Others, called **trichobothria,** are exceedingly long and thin. The base of a trichobothrium occupies a complexly designed socket (Fig. 11.9c). Neurons associated with the socket respond to the slightest movements of the hair.

Slit sense organs are narrow cuticular crevices covered externally by a membrane (Fig. 11.9d). A sensory hair originating in a cell below the cuticle touches the inner surface of this membrane. Any movement by the exoskeleton is detected by the hair. Slit sense organs also respond to sound vibrations. These organs may occur singly or in parallel series.

Behavior

Arachnids tend to be rather secretive animals. Many are nocturnal, and most live in ground litter or in crevices. These attributes reflect the small size of arachnids in a world full of would-be predators. Size certainly influences habitat selection. Very small species, because of their large surface-to-volume ratios, occupy humid areas, where water-retention problems are less acute. Some larger arachnids, such as scorpions and mygalomorph spiders (the "tarantulas"), can live in the desert.

Remarkable arachnid talents include poisoning, silk spinning, and ritual dancing. Poisoning, as we discussed, reflects the need to subdue prey while digestion begins outside the mouth. Silk may also be used against struggling prey. Many spiders wrap their victims in threads before attempting to eat them. The use of silk in other capacities, ranging from hunting traps to home construction and locomotion, testifies to the adaptive ingenuity of these animals. Finally, this class displays unending variations in courtship dancing and copulatory behaviours. Because they are such an aggressive, raptorial group, sex is often a risky act for an arachnid who must convince a potential mate that its intentions are reproductive and not nutritional. Sexual acts and other specific behaviors are described in the accounts of each of the arachnid orders.

CONTINUITY SYSTEMS

We have emphasized the rigors of the terrestrial environment, particularly its pressures on reproduction and development. Developmental stages are always small and thus suffer from a large surface-to-volume

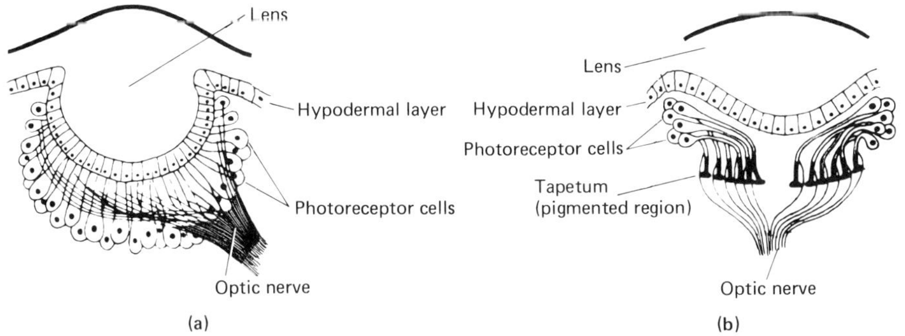

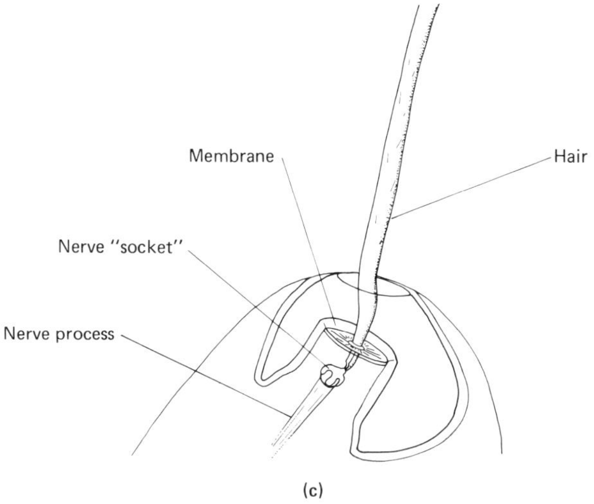

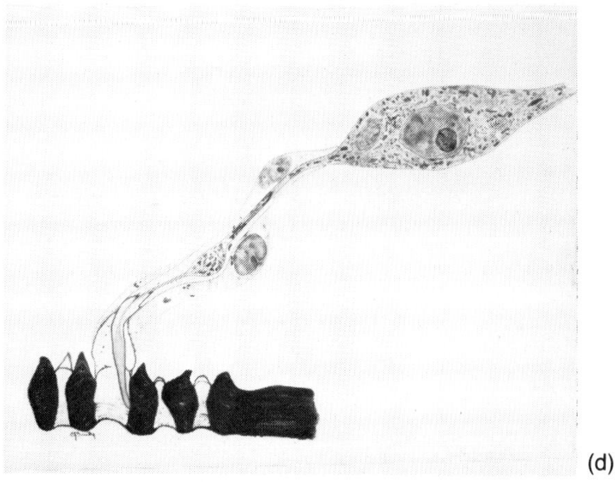

Figure 11.9 Arachnid sense organs. (*a*) Section through a direct eye of a spider; (*b*) Section through an indirect eye of a spider; (*c*) Cutaway view of a trichobothrium of the spider *Tegenaria;* (*d*) A two-dimensional reconstruction of an arachnid slit sense organ. (*a* and *b* from K.R. Snow, *The Arachnids: An Introduction*, 1970, Columbia University; *c* from P. Görner, Cold Spring Harbor Symp. Quant. Biol., 1965, *30*:69; *d* from M.M. Salpeter and C. Walcott, Exp. Neurol., 1960, *2*:237.)

ratio. They are highly vulnerable to water loss and generally lack the resources for sophisticated maintenance systems. Accordingly, on land they are usually sheltered until such systems can be developed.

All terrestrial invertebrates, arachnids included, wrap their eggs in protective envelopes. Prior to such packaging, the eggs are fertilized and given a yolk supply adequate for the prehatching period. Thus females possess well-developed yolk and envelope-producing glands, as well as receptacles for sperm storage. Males, meanwhile, must transfer their sperm to the female in good condition. All of these requirments are reflected in the diverse and always interesting phenomena of terrestrial sex.

All arachnids are sexual creatures. In both males and females, opisthosomal gonads connect via ductule systems with a genital pore on the eighth body segment. These organs may be paired or single.

Several methods of sperm transfer exist in this class. A few arachnids copulate. In most groups, however, the ventral midbody gonopores are inaccessible for direct sexual intercourse. The production of spermatophores is one solution to this physical problem. Originally, these insulated sperm packages simply were deposited on the ground and, if found by a female, picked up. The male may have laid chemical and/or silken trails to his spermatophores. Eventually, however, the male assumed a more active role in attracting the female. Today complex courtship rituals are common. These rituals identify willing and suitable partners, as they assume postures that facilitate spermatophore deposition and uptake.

In some arachnids without spermatophores, various limbs deliver the sperm. Spiders use their pedipalps, while certain mites use the chelicerae or first pair of walking legs to ladle sperm into the female gonopore.

Fertilized eggs may develop within the female reproductive tract, but most groups deposit their developing ova in a protected external site. Often an insulated brood chamber or cocoon is fashioned from silk. Cocoons may be attached to rocks or vegetation or may be carried about by the mother. Postnatal care is common. Females guard their young, and nursing with regurgitated food may occur. The juveniles of some species are transported on their mother's back.

Arachnid Diversity

Arachnids were probably the first arthropods to live on land. Once they had undergone the basic physiological adjustments necessary for terrestrial life, an extensive adaptive radiation ensued. Arachnids everywhere took advantage of their previously little-exploited environment. With the coming of the insects, the carnivorous nature of the class became more defined and the arachnids diversified still further. Today we recognize 10 different orders with living members.

ORDER SCORPIONES

The most primitive arachnids are the scorpions, an ancient group whose terrestrial history dates from the Devonian. The earliest scorpions were probably aquatic, but they may have been the first arthropods to achieve any real success on land. Among laypersons, the scorpions are notorious for their venomous stinging apparatus. About 800 species have been identified in tropical, subtropical, and temperate zones, where both jungle and desert habitats are exploited. Scorpions are nocturnal animals which spend their daylight hours in seclusion beneath stones and plant matter or in burrows. Not uncommonly, they reside within buildings. The primitive character of the order is evidenced by the persistent metamerism of several organ systems.

Scorpions are large by arachnid standards. Some late Paleozoic individuals measured almost a meter in length, but most modern members of the order are 3 to 9 cm long. General body form is reminiscent of the euryterids (cf. Fig. 11.2). The relatively small prosoma is covered dorsally by a fused carapace; its ventral surface is formed by the leg coxae and a fused sternum (Fig. 11.10). Scorpions possess multiple pairs of eyes. Large medial eyes are mounted atop middorsal tubercles, while up to five pairs of smaller eyes occur about the anterior carapace margin. Prosomal appendages include small chelicerae, mammoth raptorial pedipalps, and four pairs of walking legs. The rear body is divided into a preabdomen and postabdomen. The preabdomen consists of seven segments, the first of which bears the genital opercula. The second abdominal segment bears the **pectines,** unique comblike appendages which probably are tactile or chemosensory. Also on this segment, paired slitlike spiracles open into book lungs. Similar respiratory structures are located on the third, fourth, and fifth preabdominal segments. The scorpion postabdomen comprises five narrower, ringlike metameres. The last of these units bears the anus and the stinging apparatus.

Scorpions are carnivores. Their prey, usually insects, is seized with pedipalpal claws and, if large, is immobilized by the stinging apparatus. The chelicerae tear apart prey tissues, which are predigested and sucked into the mouth. The scorpion gut is typically arachnid, featuring a pumping pharynx and paired endodermal ceca in both the prosoma and the opisthosoma. Other maintenance organs include a seven-segmented heart (corresponding in metamerism to the preabdomen), four pairs of book lungs, one pair of coxal glands, and two pairs of Malpighian tubules.

The nervous system is primitive in that the preabdomen retains a ventral cord with seven distinct ganglia (Fig. 11.8a). Besides eyes and pectines, scorpion sensory structures include trichobothria and slit sense organs. Trichobothria on the pedipalps are instrumental in food location.

The most interesting aspect of scorpion activity systems is their poisonous stinging for prey immobiliza-

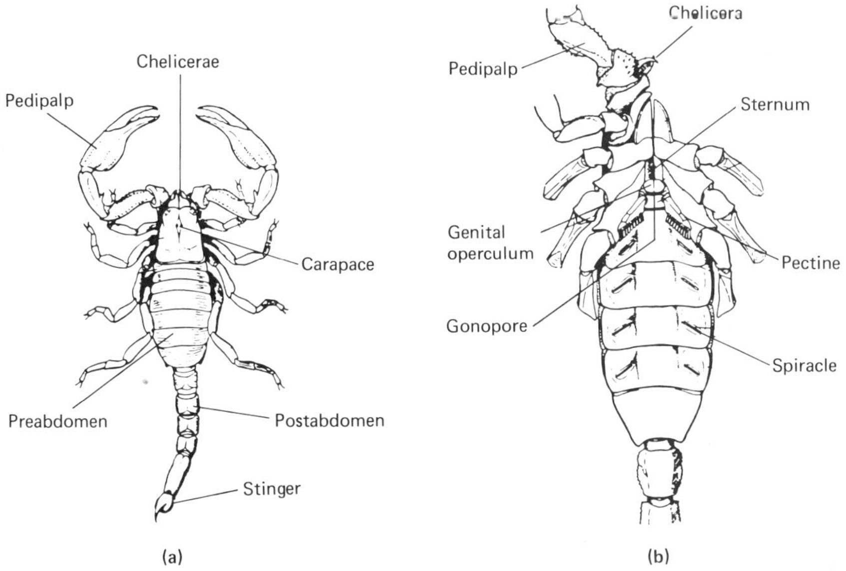

Figure 11.10 (*a*) Dorsal view of a scorpion, *Chactas*, showing external anatomical features; (*b*) Ventral view of a scorpion, *Pandinus*, showing external features. Anterior appendages on the left side have been removed. (From Snodgrass, *A Textbook of Arthropod Anatomy*.)

tion and defense. Located in the last body segment, the scorpion stinging apparatus consists of a slightly curved barb and expanded base (Fig. 11.11). Within this base is a pair of poison glands. Upon stimulation mediated by giant axons, the muscles surrounding these glands contract and venom is squirted out through a common duct, which terminates at an opening just below the tip of the barb. Typically, the stinging apparatus is brought forward over the head to puncture a victim being held by the pedipalps.

Scorpion venoms are neurotoxic and kill or paralyze small invertebrates. For humans, the attack of most species resembles a serious bee sting. Dangerous exceptions must be noted, however. Venom from the Saharan *Androctonus australis* can kill a person within 7 hr. Several species of *Centruroides* in Mexico, Arizona, and New Mexico can also be quite toxic.

Scorpion continuity involves separate sexes, courtship with indirect sperm transfer, and ovoviviparous or viviparous development. Both male and female gonads are tubular and ramify preabdominally among the midgut diverticula. In the female, paired seminal receptacles flank a single genital atrium (Fig. 11.12a). The male system includes glands for spermatophore production (Fig. 11.12b). Near the male atrium, paired tubules mold the two halves of the sperm package. In both sexes, the atrium opens through a single gonopore between the genital opercula on the first preabdominal segment (Fig. 11.10b).

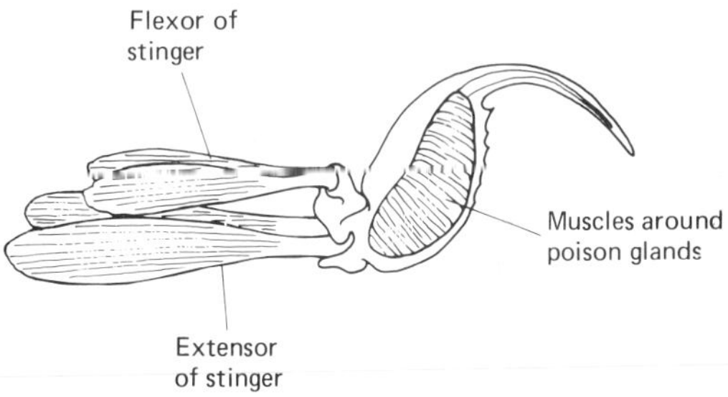

Figure 11.11 A section of a scorpion stinger with associated musculature. (From P.A. Meglitsch, *Invertebrate Zoology*, 2nd ed., 1972, Oxford Univ.)

A repetitious courtship dance determines mating suitability. Commonly, the scorpion pair faces one another while walking in a circle with their abdomens raised. Eventually, the male seizes the female with his pedipalps, and a forward and backward walking routine is begun. This routine may continue for many hours until a suitable surface for spermatophore depostion is found. The pectines seem important in site selection. In time, a male attaches his spermatophore to the ground. This spermatophore comprises a basal stalk, the sperm mass, an opening apparatus, and a

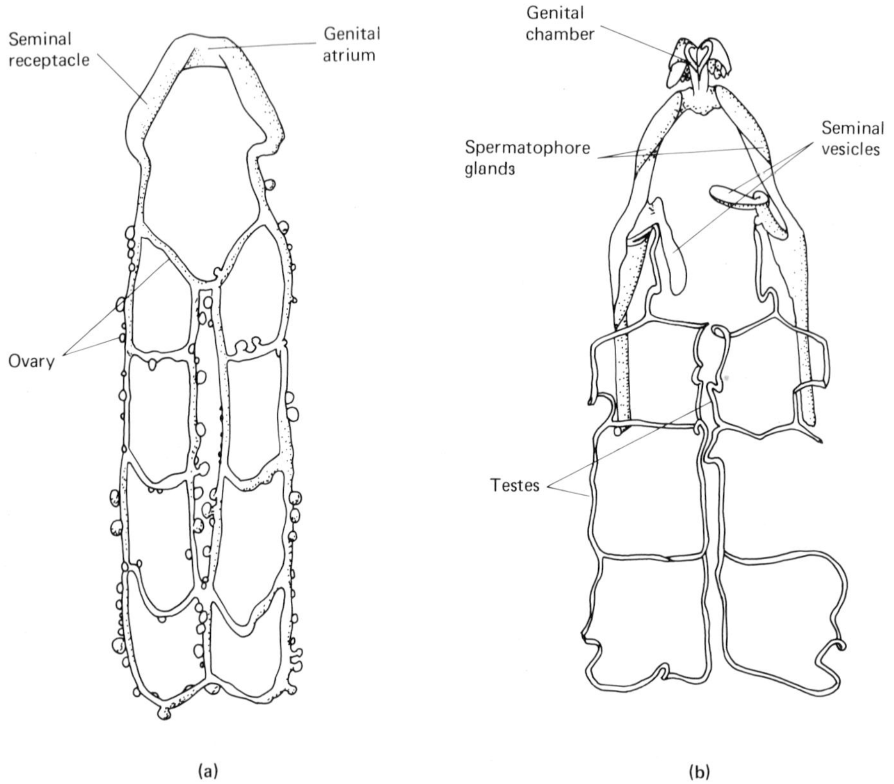

Figure 11.12 Reproductive structures in scorpions. (*a*) Female system from *Parabuthus;* (*b*) Male system from *Heterometrus.* (From J. Millot and M. Vachon, In *Traité de Zoologie, Vol. VI, Grassé*, Ed.)

trigger (Fig. 11.13a, b). The male positions the female's genital segment over the spermatophore trigger. Pressure from the female body stimulates the release of sperm into her gonopore (Fig. 11.13 c, d, e).

Development occurs within the female reproductive tract. The eggs of ovoviviparous species are brooded within ovarian tubules, while those of viviparous scorpions develop within ovarian diverticula. Ovoviviparous eggs are large and yolky. In contrast, the small eggs of viviparous species have very little yolk. Truly remarkable nutritional arrangements are made for these eggs. A column of tissue joins each ovarian diverticulum with the mother's midgut wall (Fig. 11.14). Absorptive cells acquire nutrients from the maternal digestive tract and pass them through the column to the developing embryo.

Many months usually elapse before up to 90 baby scorpions hatch and crawl onto their mother's dorsal surface. There they pass the first instar. Additional molts on the ground lead to sexual maturity.

ORDER UROPYGI

Whip scorpion and vinegaroon are names commonly applied to the arachnids of the order Uropygi. One hundred and thirty species have been described from warm, usually humid parts of the world. These animals are wholly nocturnal and pass daylight hours beneath ground debris. Tunneling is common.

The whip scorpion body comprises a prosoma covered by a single or bipartite carapace and a segmented abdomen with a terminal flagellum (Fig. 11.15). The latter appears to be photosensitive. Central body lengths vary from a few millimeters to almost 7 cm in the large American species *Mastigoproctus giganteus.* Prosomal appendages include small chelicerae, much larger raptorial pedipalps, and four pairs of legs. The first pair of legs is antenniform and functions solely as sensory appendages; the other legs are locomotor.

Whip scorpions process their food in the typical arachnid fashion. Powerful pedipalps capture and crush prey (probably small insects), and the chelicerae

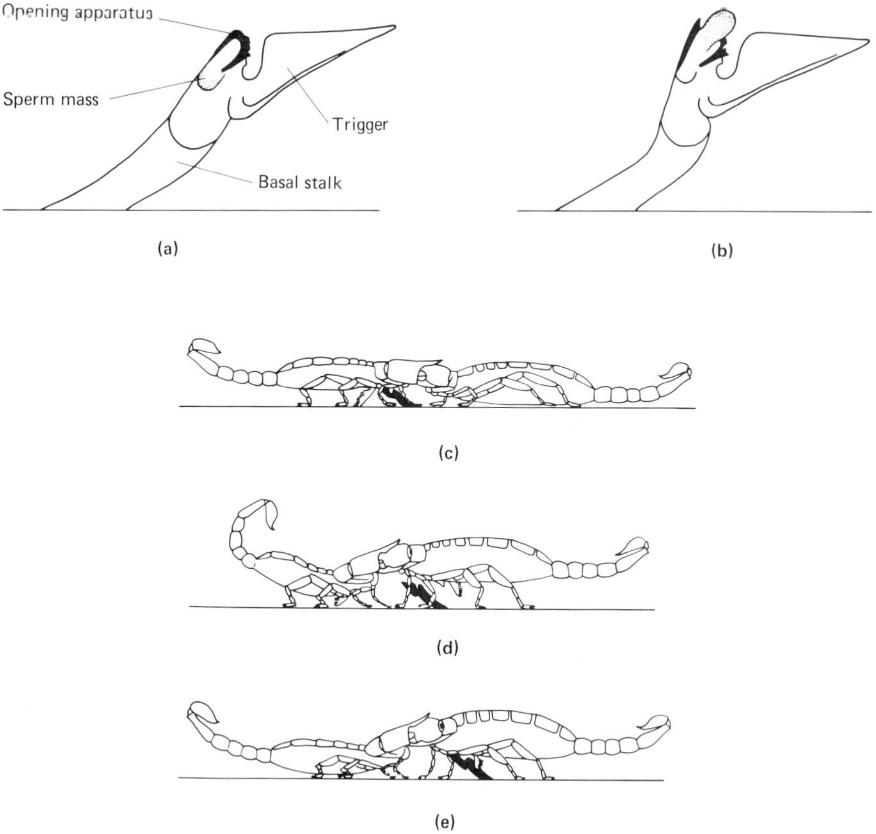

Figure 11.13 (*a*) Lateral view of a scorpion spermatophore with opening apparatus closed; (*b*) Trigger open with sperm mass protruding; (*c–e*) Three stages in the courtship sequence of a scorpion by which the male positions the female's genital segment over the trigger of the spermatophore. Male is on the left. (From H. Angermann, Zeit. f. Tierpsychol., 1957, *14*:276.)

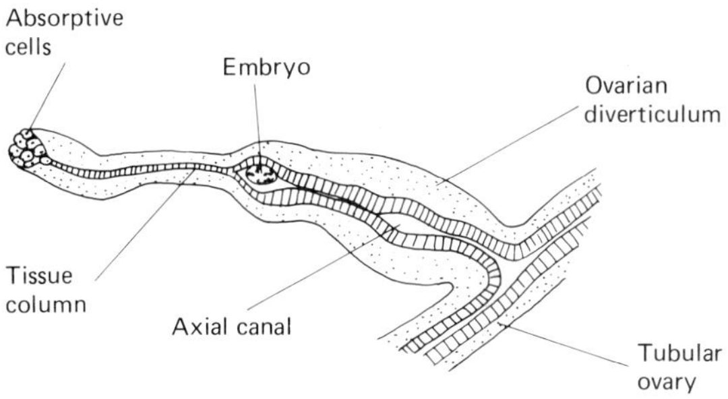

Figure 11.14 An embryo within an ovarian diverticulum from the scorpion *Hormurus australasiae.* The embryo receives nutrients from the mother's midgut via the absorptive cells and the tissue column. (From C. Dawydoff, In *Traité de Zoologie, Vol. VI*, Grassé, Ed.)

mash victims during ingestion of their tissues. The gut, which has numerous ceca, terminates in an anus below the flagellum. Flanking the anus are the openings of paired anal glands. These glands produce a fluid of concentrated acetic acid which is sprayed on would-be adversaries. Some caprylic acid also is present in the fluid and can erode the cuticle of arthropod enemies.

The uropygid circulatory system is well developed centrally, in conjunction with the presence of two pairs of book lungs. Coxal glands and Malpighian tubules are the excretory organs. The nervous system is only partially concentrated, and a large fused abdominal ganglion occupies the 14th segment.

Sex among the whip scorpions involves spermatophore transfer. Courtship rituals may last for

Figure 11.15 Dorsal view of *Mastigoproctus giganteus,* the whip scorpion or vinegaroon. (Courtesy of the American Museum of Natural History.)

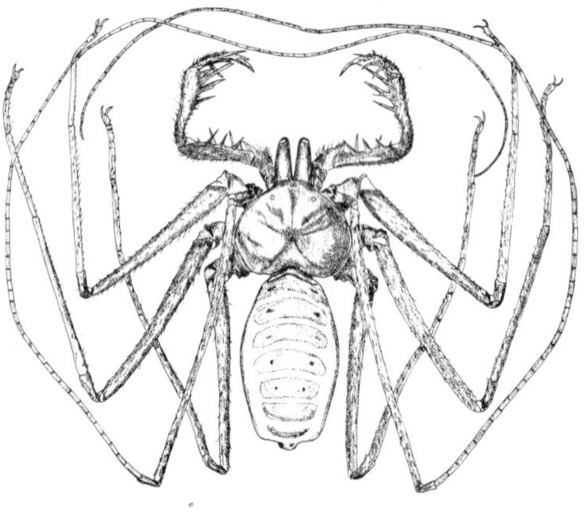

Figure 11.16 Dorsal view of an amblypygid, *Charinus milloti.* (From J. Millot, In *Traité de Zoologie, Vol. VI,* Grassé, Ed.)

days. The spermatophore may be deposited by the dancing male, who then maneuvers the female's gonopore over his sperm package. This method is similar to that described for most true scorpions. Alternatively, as in the American *Mastigoproctus,* male chelicerae may stuff the spermatophore directly into the female orifice.

The female selects a secluded site at which to lay up to 40 large eggs. Some species tunnel underground, where they excavate brood chambers. The uropygid mother remains with her offspring, as they hatch and pass their first few instars attached to her back. After the young whip scorpions dismount, their mother dies.

ORDER AMBLYPYGI

Amblypygids resemble uropygids in distribution, behavior, and average body size. Indeed, the two groups formerly were united within a common order, the Pedipalpi. Amblypygids have been called tailless whip scorpions and whip spiders. About 60 species are known.

Amblypygids are distinguished by a flattened central body and the absence of a terminal flagellum (Fig. 11.16). Anal glands are also missing. The first pair of amblypygid legs is extremely long and whiplike; as in the whip scorpions, these appendages are tactile organs. The pedipalps are strong and spiny. They seize insect prey and then pass it to the chelicerae, whose distal fangs pull apart the victim's tissues. The digestive system, as well as other aspects of internal biology, shows some affinity with the spiders. Both a pumping pharynx and a postpharyngeal pumping stomach are present.

There are two pairs of book lungs, while both coxal glands and Malpighian tubules provide for excretory needs. Fused opisthosomal ganglia have migrated below the esophagus.

Amblypygid courtship rituals involve gentle, exploratory stroking by the sensory legs. Sperm transfer is once again rather scorpionlike. The male deposits a spermatophore stalk and applies two drops of sperm to its top. He then maneuvers the female into a proper position for uptake.

As the eggs are laid, they are retained within a membranous pouch on the female's ventral abdomen. A product of special reproductive glands, this pouch houses the amblypygid young through their hatching and first molt. The second instar is passed on the mother's back, and third-instar juveniles carefully dismount from her rear end. Should one fall off anteriorly or laterally, it is likely to be eaten by its mother.

ORDER PALPIGRADI

Little is known about the tiny arachnids called palpigrades. Possibly, they represent dwarfed Uropygi. Fifty species have been identified from warm, humid environments, where they live beneath ground litter and in caves. No individuals are over 3 mm long. The palpigrade body comprises a bipartite prosoma, a pedicel,

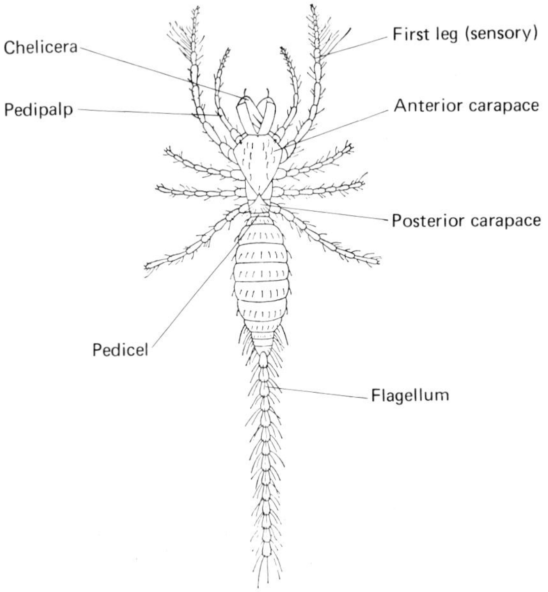

Figure 11.17 Dorsal view (greatly enlarged) of a palpigrade, *Koenenia mirabilis*. (From T.H. Savory, *Arachnida*, 1964, Academic Press.)

and an opisthosoma with a long terminal flagellum (Fig. 11.17). Prosomal appendages include a pair of stocky, forwardly directed chelicerae, pedipalps, and four leg pairs. The pedipalps are primarily locomotor limbs, while the first pair of legs is sensory. There are no eyes: Sensitive hairs and trichobothria are the only known sense organs.

Book lungs may occur in certain species, but respiratory as well as circulatory organs are usually absent. Palpigrades are very small, and their warm, humid habitats allow informal maintenance systems. A very thin cuticle permits diffusion to play an important physiological role. There are no Malpighian tubules, but one pair of coxal glands may be present. Continuity systems are undescribed, but some species apparently have extremely limited male populations.

Order Araneae

The most common arachnids are the spiders of the order Araneae. Over 35,000 species have been described, and new ones are identified monthly. Spiders

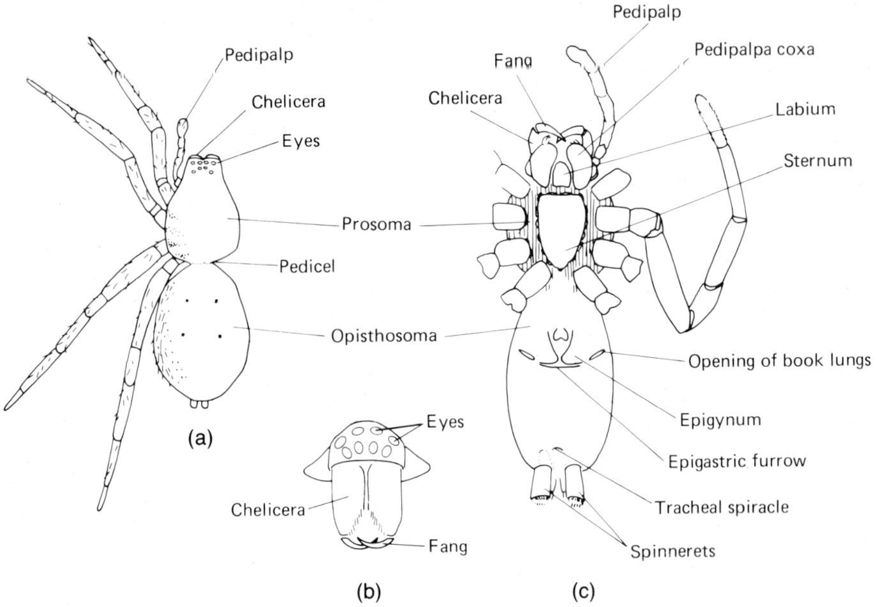

Figure 11.18 External anatomy of a spider. (*a*) Dorsal view, appendages on right omitted; (*b*) Frontal view of face and chelicerae; (*c*) Ventral view with most appendages omitted. (From W.J. Gertsch, *American Spiders*, 2nd ed., 1979, Van Nostrand.)

live on every continent except Antarctica and exploit virtually all terrestrial niches. They are found in jungles and temperate-zone forests, in deserts, meadows, and caves, and along freshwater shores and marine coast lines. One species, the Eurasian *Argyroneta aquatica,* lives in fresh water. Not only are spiders widely distributed, but in favorable areas their populations can be enormous. An acre of grassy meadow may support over 2 million individuals.

Several factors contribute to the success of the Araneae. Venom is important in their defense and prey capture. Some hunting spiders are very mobile and possess outstanding eyes; others are attuned primarily to vibrational phenomena. The group's mastery of silk use is perhaps most directly responsible for its success. This scleroprotein thread plays a major role in many aspects of spider life. Finally, the order exhibits a complex array of sexual behaviors that support extensive speciation.

MAINTENANCE SYSTEMS

The spider body consists of a well-defined prosoma, pedicel, and opisthosoma (Figs. 11.3a and 11.18). Central body size varies from half a millimeter to nearly 10 cm in mygalomorphs (the tarantulas). The appendages may have a much wider span. Dorsally, the prosoma is covered by a single carapace with up to eight anterior eyes. Ventrally, a fused sternum separates the appendage bases (Fig. 11.18c). Except in very primitive forms, the spider opisthosoma is unsegmented; however, some individuals display colored abdominal bands which recall the original body segments.

Prosomal appendages include chelicerae, pedipalps, and four pairs of walking legs. Spider chelicerae are short and heavy and bear a distal fang (Fig. 11.18c). Ducts from a poison gland open on the fang into the prosoma (Fig. 11.19). The leglike pedipalps basal article of the chelicerae and may extend deeply into the prosoma (Fig. 11.19). The leglike pedipalps often have sensory responsibilities. In the male, they also serve as gonopods for indirect sperm transfer. The chelate walking legs are often hairy.

The opisthosoma lacks appendages, but several of its structures may develop from embryonic appendage buds. Such structures include one or two pairs of book lungs and terminal spinnerets. The latter contain the openings of the silk glands; usually three pairs are present. Each **spinneret** consists of a raised cone bearing the openings of numerous ducts (Figs. 11.19 and 11.20). Each opening is called a **spool** or **spigot.** The ducts arise from silk glands in the posterior abdomen. Up to six types of silk glands produce strands of various thicknesses and adhesive properties. Several strands, each exiting from a single spool or spigot, combine to form each silk thread. The thread is quite elastic at first, but its proteins harden when stretched. We do not know how the silk is cut, but enzymes from oral secretions may be involved.

An **epigastric furrow** crosses the ventral surface of the opisthosoma on the eighth body segment (Fig.

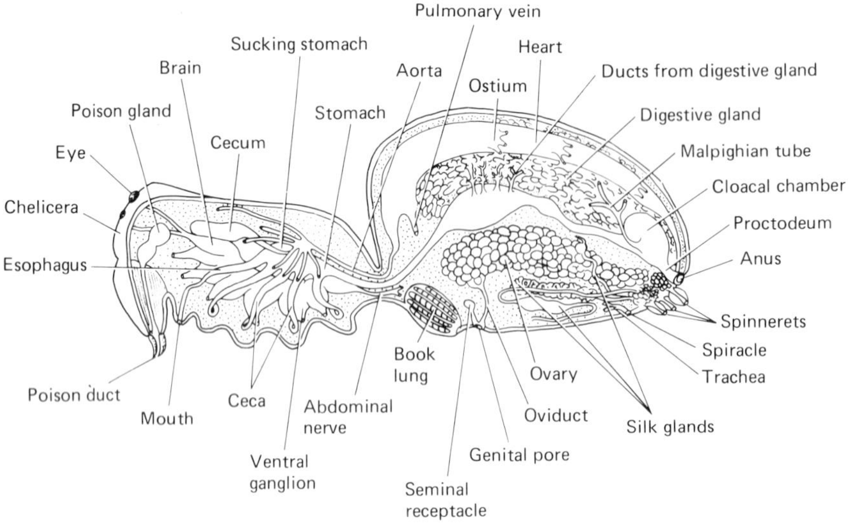

Figure 11.19 Longitudinal section through a spider illustrating internal organs. (From R.W. Hegner and J.G. Engemann, *Invertebrate Zoology*, 1968, MacMillan.)

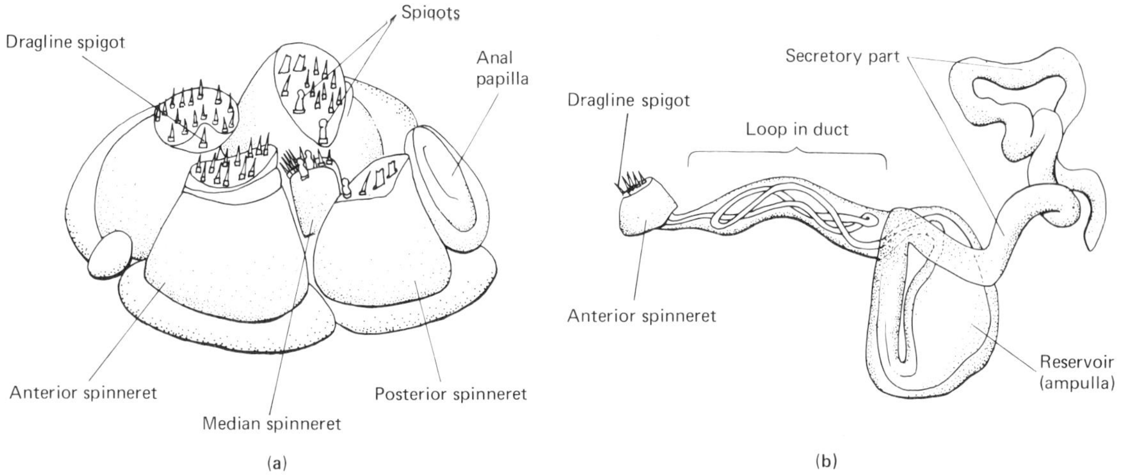

Figure 11.20 The orb weaver, *Araneus diadematus*. (*a*) The arrangement of spinnerets and their spools and spigots on the posterior tip of the abdomen; (*b*) Diagram of the major silk gland. (From R.S. Wilson, Am. Zool., 1969, 9:104.)

11.18c). This transverse groove is associated with reproductive and respiratory pores.

Like most arachnids, spiders are carnivores. Insects provide their staple diet, but larger species occasionally take a small vertebrate. Hunting forms—including the mygalomorphs and the trapdoor, wolf, and jumping spiders—swiftly seize prey with the pedipalps and chelicerae. Others trap their meals in silken snares (Fig. 11.18c). The spider foregut includes both a immobilize the victim, whose tissues are predigested and sucked into the foregut. Setae surrounding the mouth prevent oversized particles from entering the digestive tract. These hairs are located on the pedipalpal coxae, the labrum, and the **labium.** The latter is a postoral lip attached to the anterior end of the sternum (Fig. 11.18c). The spider foregut includes both a pharyngeal pump and a muscularized, sucking stomach (Fig. 11.19). The midgut walls open into numerous paired ceca, some of which penetrate the head region and the basal articles of the limbs. A multibranched digestive gland occupies much of the opisthosoma. The posterior midgut widens as a cloacal chamber, and a short protodeum leads to the terminal anus.

Primitive spiders possess two pairs of opisthosomal book lungs. The anterior pair opens through spiracles at either end of the epigastric furrow (Fig. 11.18c). In most members of the order, the second pair has been replaced by tracheae. Their spiracles have migrated posteriorly and often have fused into a single opening near the spinnerets. In some spiders, both pairs of book lungs have been replaced by tracheae. This condition occurs in many minute species and in the aquatic *Argyroneta,* which has remained an air-breathing animal.

Central circulatory elements reflect the nature of the respiratory organs. Book lungs require a stronger heart and well-defined arteries. When posterior tracheae are present, the posterior aorta is diminished.

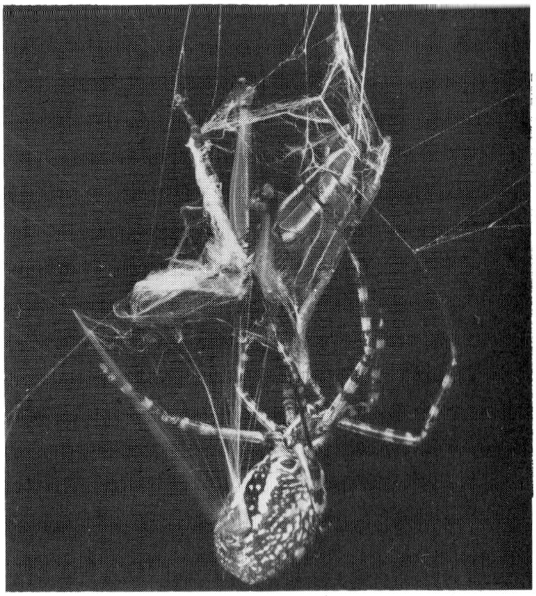

Figure 11.21 Garden spider, genus *Argiope*, with prey. (Photo by Walker Van Riper, courtesy of the University of Colorado Museum.)

In those spiders relying entirely on tracheal respiration, the heart and all associated vessels are reduced.

Excretion is handled by the prosomal nephrocytes, coxal glands, midgut wall, and Malpighian tubules. Primitive spiders have two pairs of coxal glands, but one or both pairs are reduced or missing in most higher members of the order. There is some evidence that silk and poison glands have evolved from coxal gland precursors. Malpighian tubules discharge wastes into the cloacal chamber, whose walls also release guanine and other nitrogenous acids into the lower digestive tract.

ACTIVITY SYSTEMS

Up to 31 flexor muscles are present in the leg of a large spider. There are no extensor muscles, however; limb extension is achieved by hemolymph pressure. Many spiders have a resting blood pressure as high as that of humans, and closing an aortal valve can double that pressure during rapid movements or ecdysis. The narrow pedicel confers considerable mobility to the opisthosoma, allowing the terminal spinnerets to be aimed.

The central nervous system has consolidated around the esophagus (Figs. 11.19 and 11.22). A supraesophageal brain contains the protocerebrum and tritocerebrum; the former is expanded in hunting species with well-developed eyes. All prosomal and opisthosomal ganglia have fused with the subesophageal nerve mass. Sense organs include the dorsal eyes, tactile and chemosensitive hairs, trichobothria, and slit sense organs. The arrangement of up to eight eyes on the anterior carapace is important taxonomically.

Hunting types, such as the wolf spiders (Fig. 11.23a), have two rows of four eyes each. The anterior, medial pair is direct; the rest are indirect. These spiders can be detected at night when flashlight beams reflect off their white tapetum. Image formation is minimal in most spiders, whose retinas are not densely packed. Jumping spiders, however, have very large eyes with good resolving power. They see images clearly, as evidenced by the importance of vision in

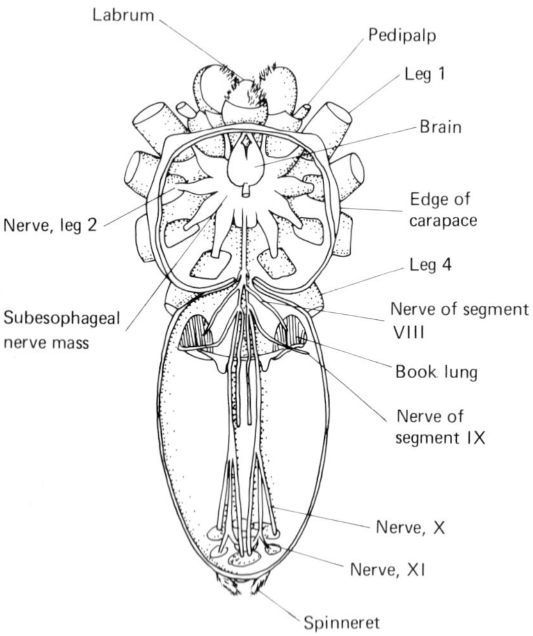

Figure 11.22 Dorsal view of the nervous system of the spider, *Tegenaria*. (From Kaestner, *Invertebrate Zoology, Vol. III*.)

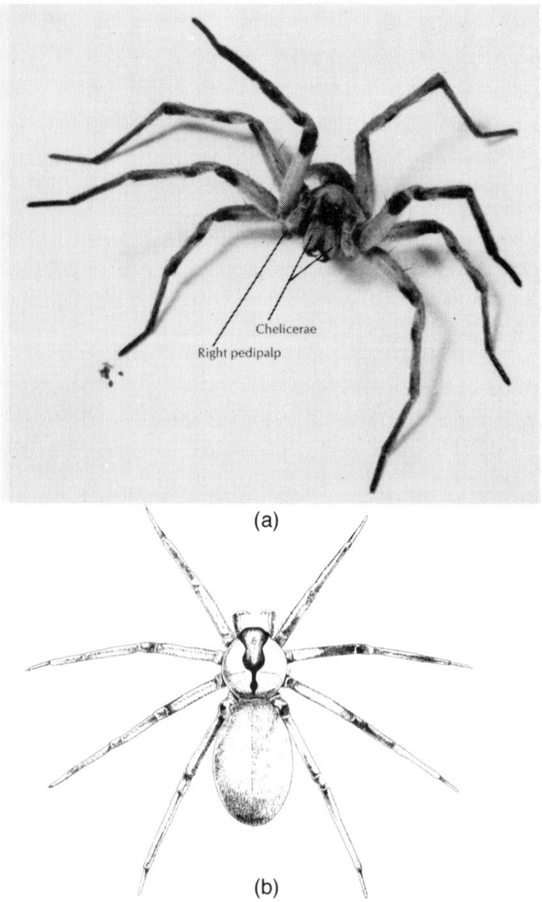

Figure 11.23 (*a*) *Lycosa*, a wolf spider; (*b*) The brown recluse, *Loxosceles reclusa*. (*a* from Hickman and Hickman, *Integrated Principles of Zoology*; *b* from R.D. Barnes, *Invertebrate Zoology*, 4th ed., 1980, Saunders College Publishing.)

complex courtship displays. In web-building spiders, vibrations are the most important source of environmental information. Parallel series of slit sense organs **(lyriform organs)** occupy strategic points near leg joints. By vibrations sensed with its trichobothria, a spider discerns the presence of prey. Hollow olfactory setae at the tips of the pedipalps and anterior legs are important in the recognition of prey, mates, and young.

Spider behavior features the use of two substances—venom and silk. Most spider venom is neurotoxic and quickly subdues invertebrate prey. Danger to humans is negligible because most species cannot insert their chelicerae through human tissue or because their toxin is too weak. At least, such is the case for *most* species. *Lactrodectus,* a genus that includes the black widow spiders (Fig. 11.3a), produces a venom that can cause extreme pain and other symptoms akin to appendicitis in humans. Death is rare, but may occur by respiratory paralysis. In *Loxosceles reclusa* (Fig. 11.23b), the American brown recluse, the venom kills human tissue around the bite. Pain and slow healing aggravate the attack. Contrary to popular belief, North American tarantulas are not poisonous. However, sharp stinging bristles brush off their abdomen if they are handled disrespectfully. A few South American and Australian mygalomorphs are highly venomous and can capture small vertebrates.

Silk is the glory of the order. This spider thread is the strongest natural fiber in the invertebrate world. Its uses vary from locomotion and prey capture to reproduction and development.

Silk commonly acts as a dragline during locomotion. Indeed, this may have been its original role. A line of dry silk is deposited continuously as a spider walks (Fig. 12.24b). Adhesive silk attaches the dragline at strategic points for vertical travel and other complex movements. Capturing prey with this line is a relatively simple adaptive step. Gnaphosid spiders run around their victims, releasing thread continuously until their prey is bound and immobile. Many spiders insulate their nests with silk. Trap-door spiders, for example, lurk within a silk-lined tunnel (Fig. 11.24e). Prey trips over the silk threads radiating from the tunnel's trapdoor, whereupon the hungry spider leaps onto its victim (Fig. 12.24d). Complex traps have developed from such signal lines. Conical webs surround the underground nests of the grass funnel weavers *Agelenopsis* and *Agelena.* These spiders emerge from a tunnel at the center of their silken funnel and harvest any insects caught in the trap.

Silken webs that catch flying prey rank among the most ingenious arachnid products. Simple irregular webs are built by *Dictyna* (Fig. 11.24c), which throws threads over stones and plant stems in a haphazard but evidently successful manner. Cobweb spiders, such as *Theridion* (Fig. 11.24f), anchor their diffuse traps with taut vertical threads coated with adhesive drops. An insect running or flying into the web causes these tight sticky threads to break and spiral into the web center. As the insect struggles, more adhesive threads are loosed to entangle him. The bolas spiders have perfected a unique system of prey capture. They hang from a single horizontal thread and dangle an additional thread whose free end carries a drop of glue (Fig. 11.24a). The spider swirls this "bolas" thread to entangle insects. The American *Mastophora* lurks quietly as a moth approaches, then suddenly casts its bolas at the insect. In some species, the bolas may be coated with a pheromone attractive to male moths!

The most complex silk traps are the orb webs (Fig. 11.25). Orb weaving is an unlearned behavior which has evolved more than once in the Araneae. The web is begun when the spider stretches a horizontal thread between two fixed surfaces. A second thread is dropped at right angles to the first and is pulled tight to form a Y-shaped frame. Additional threads may border and support this frame. With the frame providing a general pattern, additional radii are laid. Spacing between these radii is determined by the weaver's leg span. When the radii are completed, the spider lays a temporary spiral of dry silk from the center to the perimeter of the web. The spider then retraces its route, taking up the first spiral and replacing it with adhesive silk. The distance between succeeding passes of the spiral reflects the size of potential prey. Spiders that take larger prey have webs with wider meshes, and vice versa. When an unwitting insect flies into the web, it creates a pattern of vibrations which the spider's lyriform organs and trichobothria translate. The spider may pluck its web to test the influence of the prey load on the web's inherent vibrational properties. Edible captives are bound with additional silk before or after being poisoned by cheliceral secretions.

When the adhesiveness of the spiraled threads diminishes, the old threads are eaten by the spider and new ones are spun. Most spiders reweave all or part of their web each day or night. Radioactive labeling demonstrates that the proteins of the eaten thread rapidly reappear in the silk glands.

Spiders that spin orb webs have poorly developed eyes. However, their elaborate sense of touch more than compensates for their visual deficiencies. In comparison with the hunting spiders, web builders have

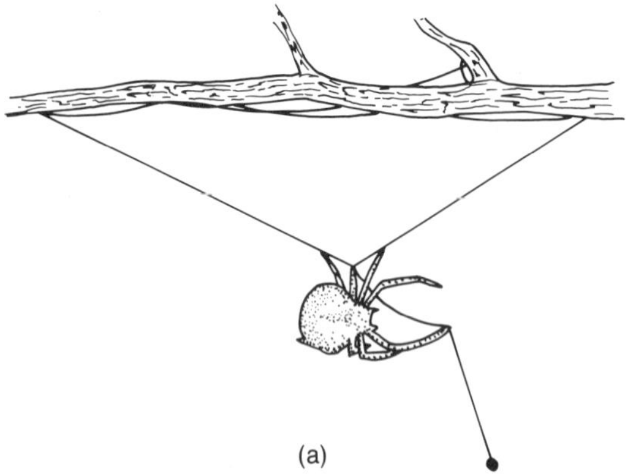

(a)

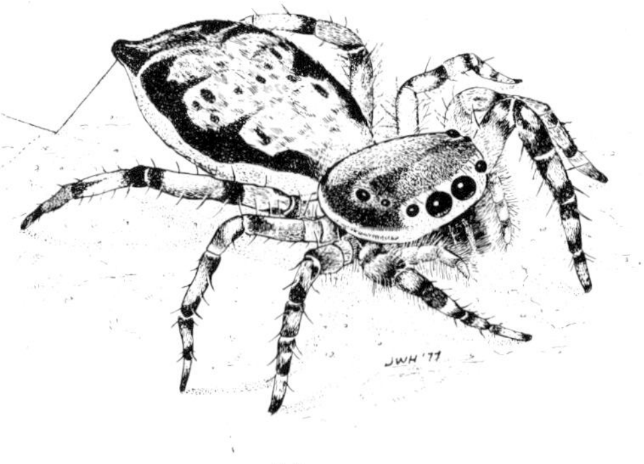

(b)

Figure 11.24 (*a*) Bolas spider holding thread; (*b*) Jumping spider with dragline; (*c*) *Dictyna* web; (*d*) Trapdoor spider capturing prey; (*e*) Underground web of trapdoor spider. Note above-ground strands that alert spider of approaching prey; (*f*) The cobweb of *Theridion*. (See page 349 for more details.) (*a*, *e*, and *f* from Barnes, *Invertebrate Zoology*, *b* from Barnes, after L.R. Levi, *A Guide to Spiders and their Kin*, 1968, Golden Press; *c* and *d*: photos by Walker Van Riper, courtesy of the University of Colorado Museum.)

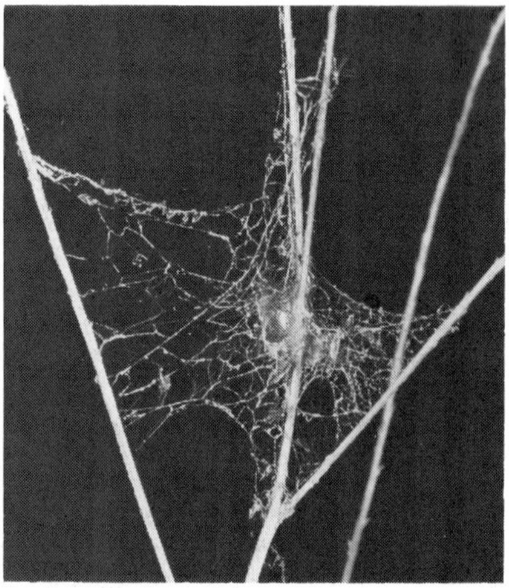

(c)

(d)

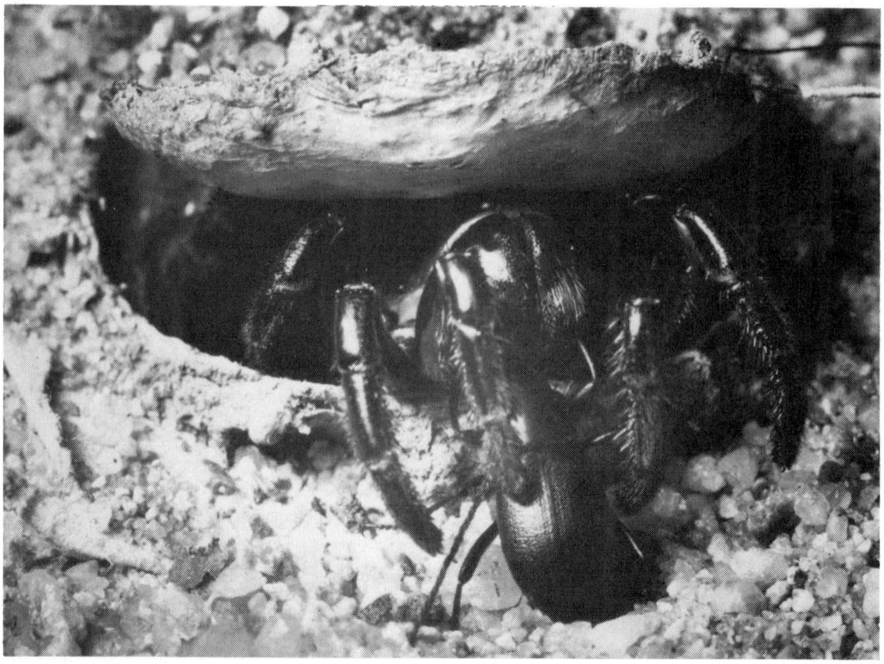

(e)

(f)

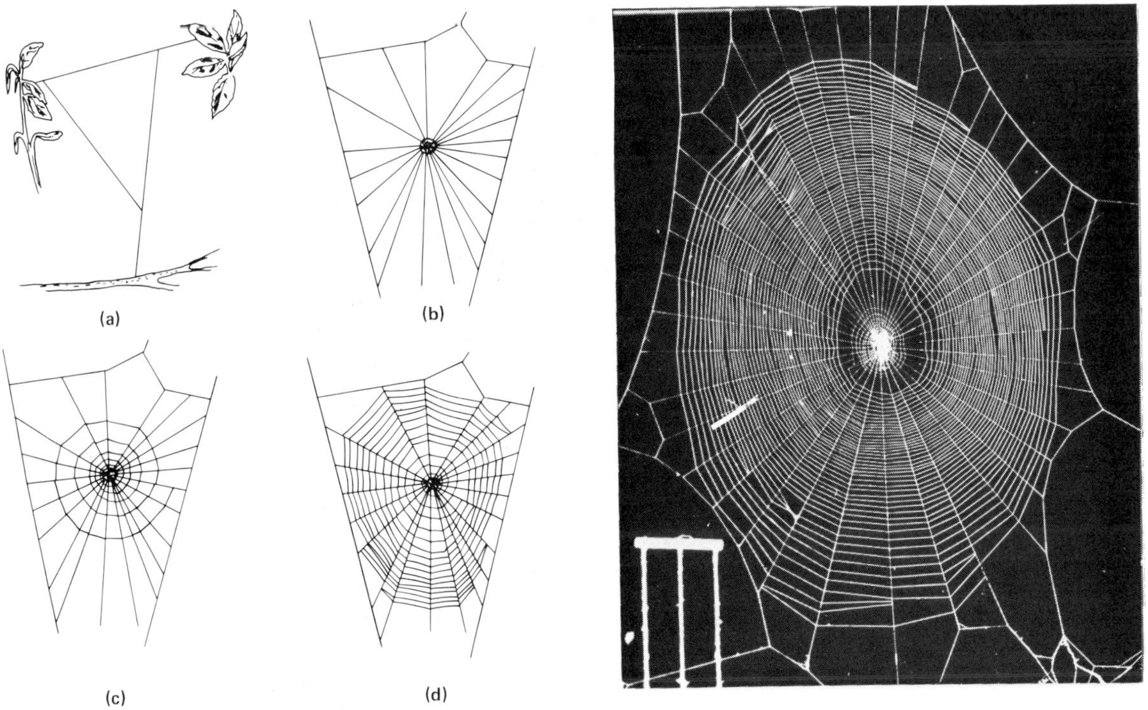

Figure 11.25 *(a–d)* Stages in the construction of an orb web. (For explanation, see text.); *(e)* Completed orb web of the orb weaver *Araneus diadematus*. *(a–d* from Gertsch, *American Spiders; e* courtesy of P.N. Witt.)

smaller, narrower limbs. An extra claw on the walking legs helps them to crawl freely over the web. How they avoid sticking to their own adhesive silk is not known.

CONTINUITY SYSTEMS

In male spiders, a pair of tubular testes unites in a common duct and the single gonopore lies on the eighth body segment. The female has two ovaries which, when ripe, fill most of the abdomen. A fused oviduct leads to a cuticular genital atrium and a gonopore on the middle of the epigastric furrow. In primitive spiders, a pair of seminal receptacles branches from the sides of the female atrium. Most modern members of the order have separate copulatory openings for these reservoirs (Fig. 11.26a). Such openings lie anterior to the epigastric furrow on a sculptured plate known as the **epigynum.**

Indirect sperm transfer involves the male pedipalps as intromittent organs. The penultimate article of these appendages are modified to receive sperm and inject them into the female's copulatory opening. Basic palpal structure includes a sperm reservoir and a coiled ejaculatory duct, which extends into a cuticular penis-like horn called the **embolus** (Fig. 11.26b). The organ is often complicated by species-specific accessory structures that grip the female opening. Typically, the male spider spins a minute web, onto which he ejaculates a small quantity of semen. Silk for this web emanates from special glands near his gonopore. The pedipalps are dipped into the sperm mass, filling their reservoirs. The male then seeks out a mate.

Spider males are usually smaller than their female counterparts. Their legs, however, are often longer and may bear additional chemosensors for the detection of willing females. Draglines are important guides for the mate-seeking male, as spiders recognize the silk threads of their own species. Apparently, a male can tell if a conspecific web is occupied by a mature female. The webless hunting spiders use their advanced visual powers in the search for sexual partners. Pheromones may also be involved.

Because araneids are aggressive carnivores, the smaller male must exercise extreme caution when advertising his intentions to the female. Elaborate courtship expresses sexual desire and tests mating suitability. The highly visual jumping spiders dance for prospective partners, waving their conspicuously striped abdomens (Fig. 11.27). Web-dwelling females

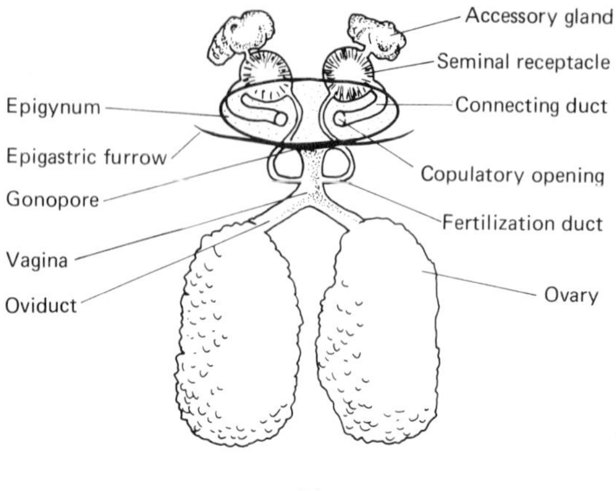

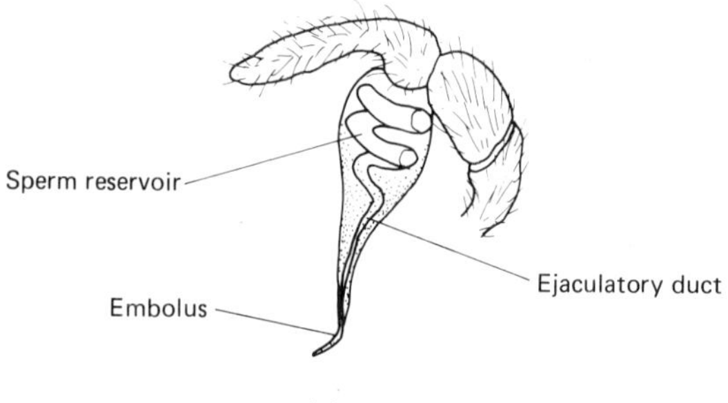

Figure 11.26 (*a*) Female reproductive system of a spider; (*b*) Simple pedipalp of a male spider, *Segestria senoculata*. (From Kaestner, *Invertebrate Zoology, Vol. II.*)

may have their threads plucked by the male, as he sends harplike messages of his sexual interest. Some males present the female with a fly in silk wrappings; if she accepts the gift, mating may occur while she eats it. Certain spiders court more aggressively, and either sex may wrap the other in silk prior to mating. In a few species, the female may bite her partner while mating; in others, she may eat her lover if he fails to escape following copulation. Male wolf spiders leap atop their mates and hold them with the chelicerae.

After pairing is established, the male pedipalps become engorged with hemolymph and each embolus is pushed out. This structure then is inserted into one of the female's copulatory openings. Part of the embolus may break off inside the female, thus preventing further mating by both individuals.

Sperm is stored in the seminal receptacles of the female. Fertilization occurs immediately prior to egg laying. The mother spider spins a layer of silk, deposits her fertilized eggs, and then covers them with an additional silk layer. The entire package is bound with more silk until a protective cocoon has been fashioned. Spiders hang their cocoons in their webs or tunnels, attach them to stones or vegetation, or carry them about. Developing young are usually well guarded.

Most young spiders emerge from the cocoon as second instars. They may remain with their mother for a time and share her food. Like most carnivorous animals, spider siblings disperse to ensure an adequate food supply for all. Ballooning is a common method of travel. A young spider crawls to the top of a plant or rock and casts a long thread of silk into the wind. If the wind is strong enough, the spiderling is swept away. Great distances can be covered by ballooning spiders; some have reached ships several hundred miles at sea.

A variable number of molts occur during a spider's lifetime. Large animals molt most often (up to 15 times

Figure 11.27 Courtship position of a male jumping spider, *Gertschia noxiosa*. (From Kaestner, *Invertebrate Zoology, Vol. II.*)

in some species), and females commonly have more instars than conspecific males. Many spiders reproduce only once and then die. Some tarantulas, however, have survived in laboratory cages for 25 years (Fig. 11.28).

ORDER RICINULEI

Ricinuleids are small, heavily armored arachnids that live in dark, moist places. Caves are favorite habitats. Only 25 species have been identified, with no individual measuring over a centimeter in length. Diagnostic features include a thick exoskeleton with an anterior hood, or **cucullus,** which drops over the mouth and chelicerae (Fig. 11.29). The chelicerae are small and grasping; the pedipalps resemble the legs. Owing to reduction of its posterior segments, the opisthosoma is short.

Termites and insect larvae are common ricinuleid prey. Simple tracheae provide respiratory exchange, while excretory organs include a single pair each of Malpighian tubules and coxal glands. All ganglia have fused with the brain, and sense organs are limited to a few scattered hairs.

During the short courtship ritual, the male mounts the female's back and then transfers a wad of sperm with one of his third legs. Prior to hatching, the eggs may be transported beneath the female cucullus. Young ricinuleids spend much of their time underground, where molting usually occurs.

Figure 11.28 *Brachypelma smithi,* a mygalomorph spider. These spiders are commonly called "tarantulas" in the United States. (Courtesy of American Museum of Natural History.)

ORDER PSEUDOSCORPIONES

The next group of arachnids owes its name to a superficial resemblance to the order Scorpiones. These so-called pseudoscorpions number about 2000 species worldwide. Humid environments, commonly those of plant litter, often support huge populations of these animals. Pseudoscorpions resemble scorpions in their possession of outsized raptorial pedipalps (Fig. 11.30). However, the opisthosoma, although segmented, is simple and flattened, with a rounded posterior and no stinging apparatus. Pseudoscorpions also are much smaller than their namesakes. Rarely do any individuals exceed 1cm in length; most measure less than half that. This order is quite advanced, however, as evidenced by the complexity of the appendages and various organ systems.

Pseudoscorpion chelicerae comprise two articulating fingers with interesting accessories (Fig. 11.30c). The distal tip of one finger is raised into a **galea**, a set of horny processes which contain the openings of silk gland ducts. The silk glands themselves lie within the prosoma and are homologous with the cheliceral poison glands of spiders. The serrated borders of the fingers are used as combs for grooming. The formidable pedipalps house venomous glands within their distal articles. Small insect and arachnid prey is grasped, poisoned, and passed to the chelicerae. These appendages tear apart the exoskeleton; ex-

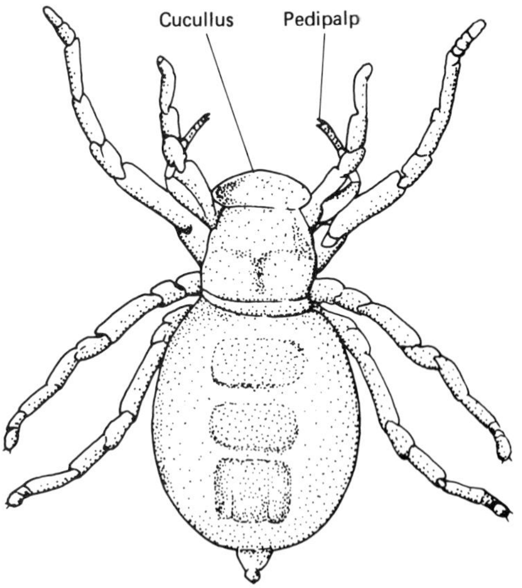

Figure 11.29 Dorsal view (enlarged) of a juvenile ricinuleid, *Ricinoides feae.* (From Millot, In *Traité de Zoologie, Vol. VI,* Grassé, Ed.)

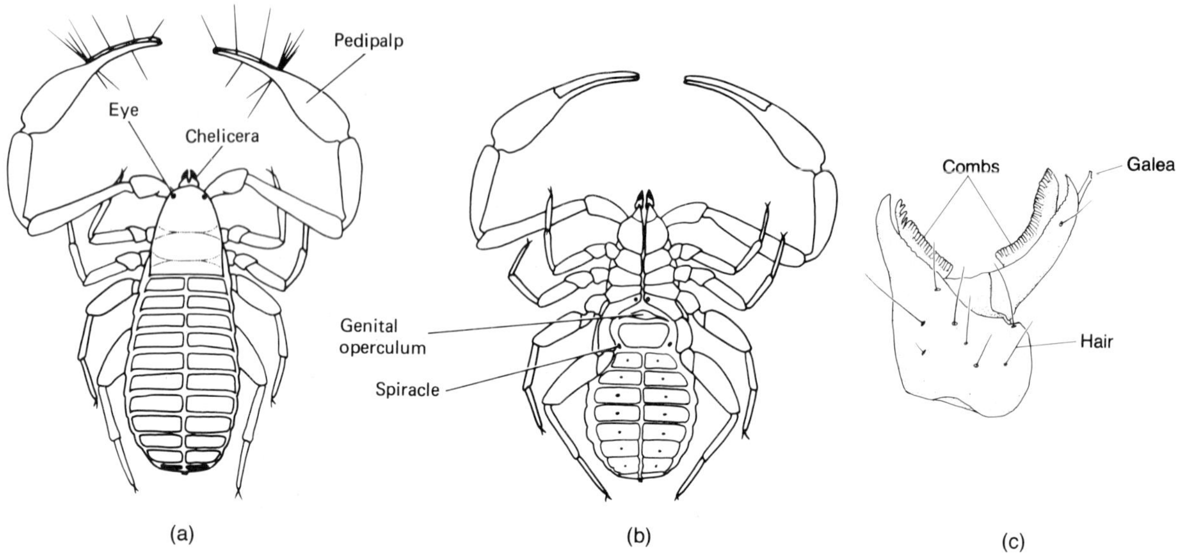

Figure 11.30 Dorsal (*a*) and ventral (*b*) views of external morphology of a pseudoscorpion based on the male of *Chelifer cancroides;* (c) External surface of a cheliceral finger from the pseudoscorpion *Microcreagris sequoiae.* (*a* and *b* from P. Weygoldt, *The Biology of Pseudoscorpions*, 1969, Harvard University; c from Vachon, In *Traité de Zoologie, Vol. VI*, Grassé, Ed.)

posed inner tissues are predigested and then sucked into the gut in the typical arachnid fashion. When the oral regions become clogged, cheliceral hairs remove trapped particles. These hairs and other feeding parts are wiped clean by the cheliceral combs.

The small pseudoscorpion heart gives rise to a single, anterior artery. Well-developed tracheae deliver oxygen directly to the tissues. No Malpighian tubules are present. The order survives with a single pair of coxal glands and prosomal nephrocytes.

All ganglia are concentrated in the pseudoscorpion brain. Tactile and vibrational phenomena dominate the sensory world of these animals, but chemical clues may be important in sexual behavior. Eyes, if present, are lateral and indirect. Other sensory structures include trichobothria, which are common on the pedipalps, slit sense organs, and simple sensory cells.

Sperm transfer is always indirect and, as in many other arachnids, involves spermatophore deposition and uptake. Some pseudoscorpion males apparently deposit their spermatophores at random, so that the female has few clues to their whereabouts. In other species, the male lays silken guide lines in a radiating pattern from each sperm package. These silk threads are manufactured in rectal glands. Finally, higher pseudoscorpions undergo pairing courtship rituals. The male may seize the female with his pedipalps and then pull her over his attached spermatophore. The most sophisticated pairing rituals, however, involve very little physical contact. The male of *Chelifer cancroides* a species found in human homes throughout the world, evaginates a pair of soft tubes from his genitalia. These tubes, sometimes called "ram's horn organs," attract the female, who dances in unison with her partner. After spermatophore deposition, the female straddles the sperm mass and is shaken by the male until sperm enter her gonopore. *Chelifer* males apparently are territorial and will not court outside their home range.

Silk glands play important roles in pseudoscorpion development. A mated female gathers nesting materials (plant matter, stones, food scraps), which she binds with silk from her galeal openings. The female remains in this nest while her young develop. Glandular cells along the genital atrium of the nesting mother secrete an external brood pouch (Fig. 11.32) where the pseudoscorpion embryos develop. Additional glands secrete "milk" on which the young are nourished. Often, the third instar is attained before juvenile pseudoscorpions emerge, and succeeding molts are passed within temporary silken retreats. Temperate-zone species may overwinter in cocoons.

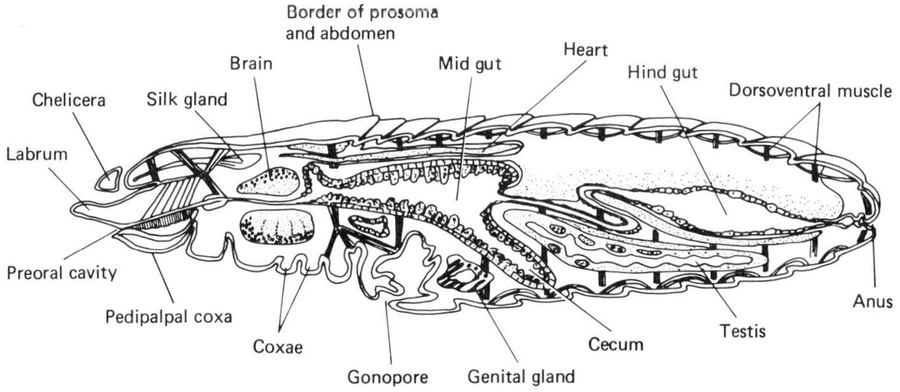

Figure 11.31 Longitudinal section through a male pseudoscorpion, *Chelifer*, showing major features of internal anatomy. (From Kaestner, *Invertebrate Zoology*.)

ORDER SOLIFUGAE

There are 800 species of arachnids called solifugids. These animals are found in the tropics and subtropics, where they inhabit both humid and dry areas. Often active in daylight, solifugids sometimes are called sun spiders.

A length of up to 7cm ranks the solifugids among the largest arachnid types. Their prosomal carapace comprises two movable parts (Fig. 11.33). The larger anterior part bears two medial eyes and a huge pair of chelicerae. Each chelicera is longer than the entire prosoma itself and consists of two elements which articulate as pincers in the vertical plane. The chelicerae are raised and lowered by flexions between the anterior carapace and its smaller posterior division. Other prosomal appendages include long, slender pedipalps and four pairs of legs. Solifugid pedipalps bear terminal adhesive pads which aid in food acquisition. The first legs are small and tactile. The remaining three pairs generate the solifugid's rapid speed, a trait that has earned these arachnids another title, "wind spiders." The solifugid opisthosoma is large and segmented.

Like most arachnids, solifugids are raptorial carnivores. Prey caught by the sticky pedipalps is transferred to and crushed by the chelicerae. Termites lead the list of favorite insect prey, while the largest solifugids may devour small lizards. The midgut has four pairs of prosomal pouches, and large, spiderlike ceca occupy the opisthosoma. Central circulatory elements are reduced. Solifugids possess one of the Arachnida's most sophisticated tracheal systems. Both prosomal and opisthosomal spiracles are present, and a tubular network branches extensively throughout the body. Excretory responsibilities are discharged by single pairs of Malpighian tubules and coxal glands.

The central nervous system includes a brain collar and a concentration of opisthosomal ganglia in the eighth body segment. Large medial eyes are present, and sensitive hairs are concentrated on the pedipalps and the first pair of walking legs.

Solifugid sex involves relatively short courtship rituals, and sperm transfer is usually indirect. Typically, the male grasps the female and, while stroking her with his pedipalps and first legs, turns her over and exposes her genital area with his chelicerae. Sperm may be dropped on the ground and then picked up and stuffed into the gonopore by the male chelicerae. In most American species, however, the male drops his sperm directly into the female orifice. Following fertilization within the female reproductive tract, dozens of eggs are laid in underground nests. Solifugid young remain immobile through one or more instars as adult characters develop.

ORDER OPILIONES

Most of us call the opilionids "daddy longlegs" or harvestmen. About 3200 species have been identified. Humid, organically rich environments among forest litter are their favorite dwelling places. The rather wide speciation of this order in part reflects the diverse temperature and humidity ranges and the development of either nocturnal or diurnal habits by neighboring groups.

The opilionids are often mistaken as spiders, but the absence of a pedicel clearly distinguishes them from

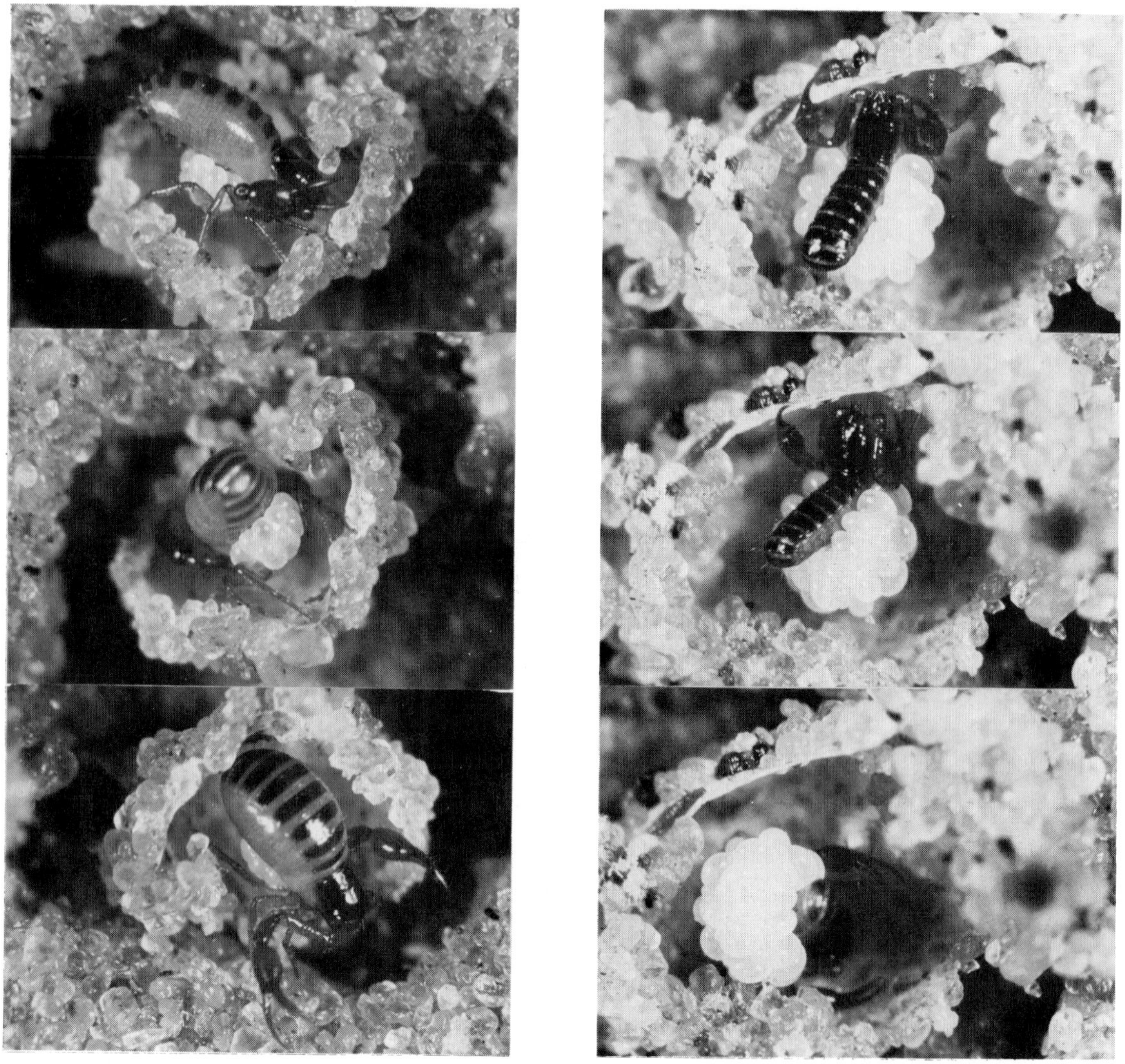

Figure 11.32 A series of photographs showing a female pseudoscorpion, *Neobisium muscorum*, in her nest with her brood in an external pouch. *Left:* Female carrying brood sac with early embryonic stages. *Right:* Female carrying brood with late embryonic stages. (From Weygoldt, *The Biology of Pseudoscorpions.*)

the Araneae. The broadly fused prosoma and opisthosoma produce a rather oval-shaped central body (Fig. 11.34a and c). A pair of direct, medial eyes flanks a middorsal tubercle (Fig. 11.34b). Opilionid legs are often very long. While the central body is usually less than a centimeter in diameter, typical leg spans are 3 to 40 times that length. The legs may be prehensile, wrapping about plant stems as the opilionid crawls along the ground or up into trees and shrubs. The chelicerae comprise three articles, the last two of which form a pair of forwardly directed pincers. The short pedipalps are leglike and bear small distal chelae.

Opilionids are unusual arachnids in that some are omnivorous. Animal prey includes small insects and even snails, while dead organisms, fruits, and vegetation are eaten at times by various species. The pedipalps bring foodstuffs to the chelicerae, which tear them apart. Predigestion occurs, but opilionids apparently ingest more particulate matter than do other arachnids.

The heart and arterial network are reduced, while

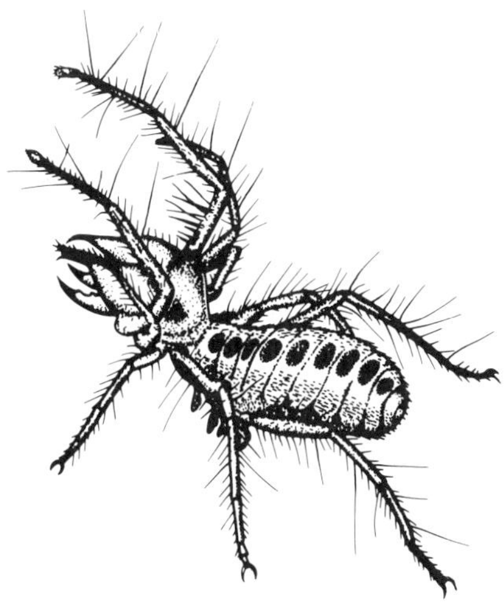

Figure 11.33 A large North African solifugid, *Galeodes arabs*, in a defensive posture. (From R.D. Barnes, *Invertebrate Zoology*, 3rd ed., 1974, W.B. Saunders.)

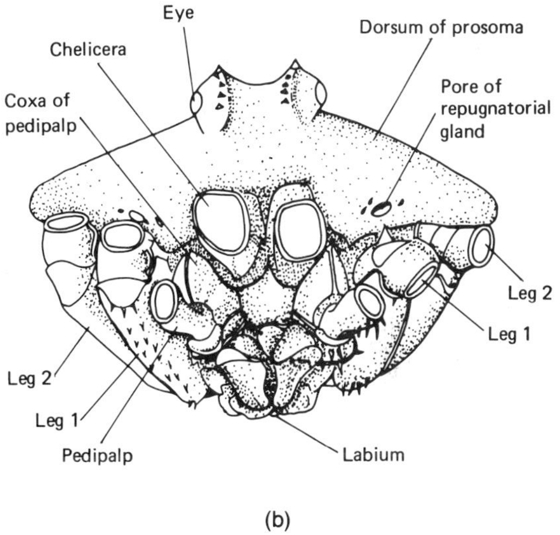

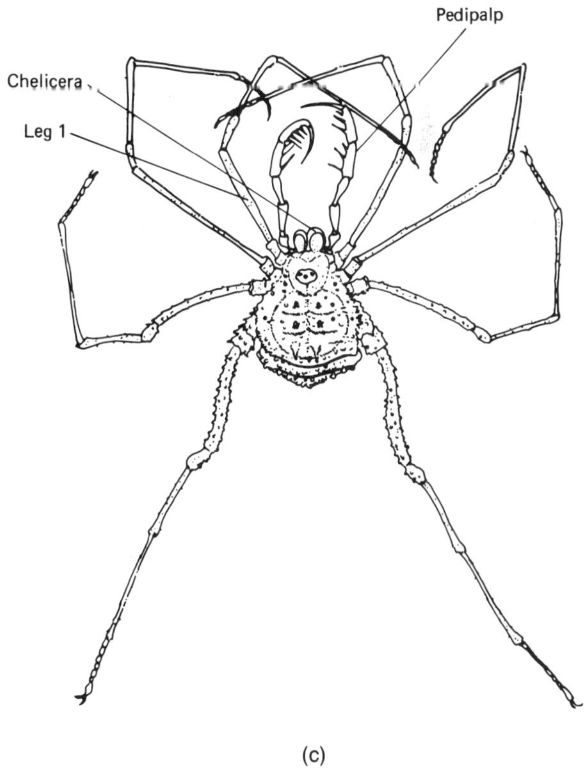

Figure 11.34 (*a*) An opilionid (Family Phalangiidae) feeding on a dead moth; (*b*) Anterior view of an opilionid, *Platybunus bucephalus*. Appendages are cut off near their bases; (*c*) Dorsal view of an opilionid showing external features. (*a* from Milne and Milne, *Invertebrates of North America*—photo courtesy of E.S. Ross; *b* from Kaestner, *Invertebrate Zoology*; *c* from G.W. Krantz, *A Manual of Acarology*, 2nd ed., 1978, Oregon State University.)

the tracheal systems are well-developed and include secondary spiracles on the legs. Tubules from a single pair of coxal glands ramify into the opisthosoma. In addition to the eyes, opilionid sensory devices include scattered tactile hairs and slit sense organs. These structures are most common on the long second legs.

Opilionids defend themselves with **repugnatorial glands**, autotomy, and mime. The repugnatorial glands are located near the bases of the pedipalps. Their acrid secretion may be sprayed directly on enemies, or a drop of the repulsive substance may be picked up by a leg and thrown at an attacker. The wide leg span of the harvestmen provides a natural perimeter of defense. Intruders may capture a single appendage, which can be autotomized. Regeneration does not occur. Finally, threatened opilionids may withdraw the legs over their central body and "play dead" for many minutes.

Opilionid gonads are typical tubular structures, but in both sexes, the genital atrium is modified distally. In the male, it forms an extensible penis; in the female, an ovipositor. Harvestmen are short on courtship, as the male simply thrusts his penis into the female gonopore. Following fertilization, the female lays her eggs in abandoned snail shells, behind bark, in plant stems, or in the ground (Fig. 11.35). Abdominal pumping extends her ovipositor, which, having penetrated to an appropriate site, discharges numerous ova. In temperate-zone species, the eggs often overwinter before hatching. The long legs of the opilionids can be troublesome during ecdysis. Most individuals must hang from branches while their chelicerae and pedipalps pull their long limbs from the old cuticle.

ORDER ACARINA

Acarines are the little animals commonly called mites and ticks. There may be more of them than of all other arachnids combined. Only about 25,000 species have been identified, but current estimates of the total number of acarine types run well over 100,000. As with the spiders, individual populations are often enormous. A few mites and ticks pose health and economic threats to humans: They eat our crops and infect our livestock, and some species are parasites of human blood and tissues. We are relatively knowledgeable about these acarines, but the rest of the order has been neglected. The destruction of mite habitats (tropical rain forests in particular) threatens to annihilate many species before they can be described.

Mites live wherever adequate moisture is available. Rich humid soil and organic litter support large populations of many species. Arid deserts and polar zones

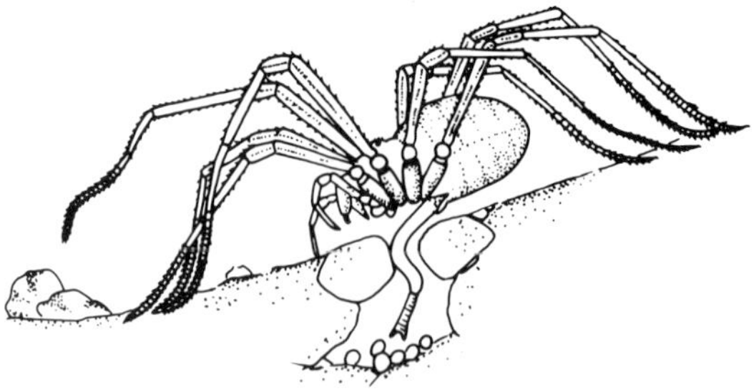

Figure 11.35 A female daddy long legs ovipositing. (From Kaestner, Invertebrate Zoology, Vol. II.)

have mite fauna as well. Nearly 3000 species from aquatic habitats are known.

The extensive distribution and other successes of the Acarina are attributable in part to their ability to package sophisticated organ systems in a minute body. Most grown mites are less than a millimeter long. Such a small size allows them to occupy a diversity of microhabitats. Not only inherently small areas (e.g., between the barbs of bird feathers) but also regions of low food concentration may be exploited. Their size makes many mites extremely difficult for predators to detect and capture. Reproductive potentials are also high.

Another factor may contribute to the diversity of this order. Many acarologists (as experts on this group are called) believe that the Acarina represents a composite group whose similarities result from convergent evolution; however, further evidence is necessary to confirm or refute this theory. The following discussion describes only the basic characters of the order, without regard to their origin. Although acarologists use a specialized vocabulary in their literature, we continue, for convenience, with the same terms employed for the other arachnid groups.

Maintenance Systems

In nearly all acarines, the prosoma and opisthosoma have fused, and the body is covered by a single dorsal carapace (Fig. 11.36a). This fusion creates a rather ovoid shape, although vermiform mites also occur (Fig. 11.36b). The cuticle of the central body and the appendages may be sculptured, and many hairs populate its surface. These elaborations are often unusual; many are bizarre (Figs. 11.36c and 11.39). The form and arrangement of acarine hairs are keys to the taxonomy of the order.

The area constituting the acarine head is called the **capitulum** (or **gnathosoma**). The capitulum bears

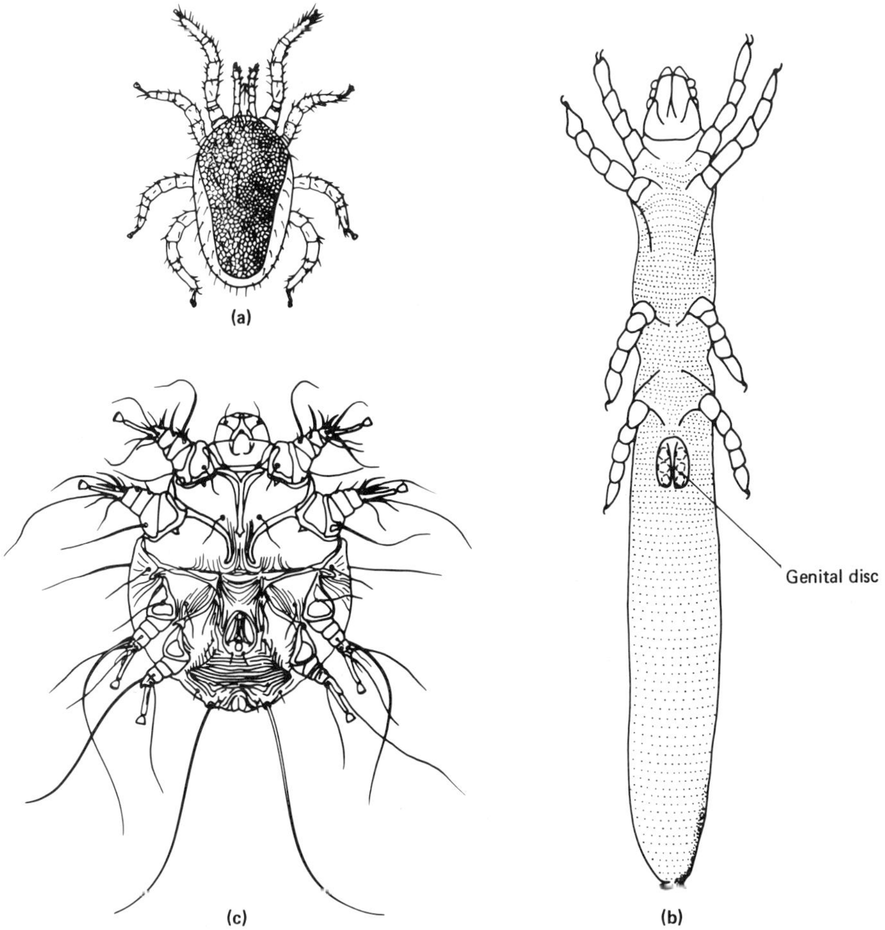

Figure 11.36 Mites. (*a*) Dorsal view of a male *Dermanyssus gallinae;* (*b*) Ventral view of a female gall mite, *Nematalycus;* (*c*) Ventral view of a male *Knemidokoptes mutans.* (*a* and *c* from G.O. Evans and E. Browning, *Some British Mites of Economic Importance,* 1955, British Museum; *c* from Krantz, *A Manual of Acarology.*)

the complex mouth parts, the chelicerae, and the pedipalps (Fig. 11.37). It is roofed over by the **rostrum**, a dorsal projection of the carapace. Fused pedipalpal coxae form a trough, which drops from the rostrum and encloses the oral region. Chelicerae extend between this trough and the rostrum. A folded joint membrane connects the cheliceral bases with the central body (Fig. 11.37b), so these appendages can be extended and withdrawn at will. Below the chelicerae and just above the mouth, a labrum joins the pedipalpal coxae to enclose a prebuccal chamber. The walls of this chamber constitute the **buccal cone**. The buccal cone also may be extensible, owing to folded membranes that attach it to the anterior body.

Acarines compare with nematodes in their diversity of feeding habits. Herbivores, carnivores, scavengers, commensals, and parasites of both plants and animals are represented in this order. The mouth parts, digestive tract, and general body morphology reflect the nutritional ecology of each species.

Prominent herbivores include spider mites and gall mites, as well as some marine algae eaters. Spider mites (Tetranychidae) are a serious pest on cotton and fruit trees. As in most herbivorous mites, each chelicera is modified distally to form a single, piercing stylet, with which the spider mite punctures a leaf and then sucks out its fluids. Gall mites (Tetrapodili) tunnel through plant tissues and extract fluids from individual cells. These vermiform acarines could be considered parasites, but their manner of feeding only reflects

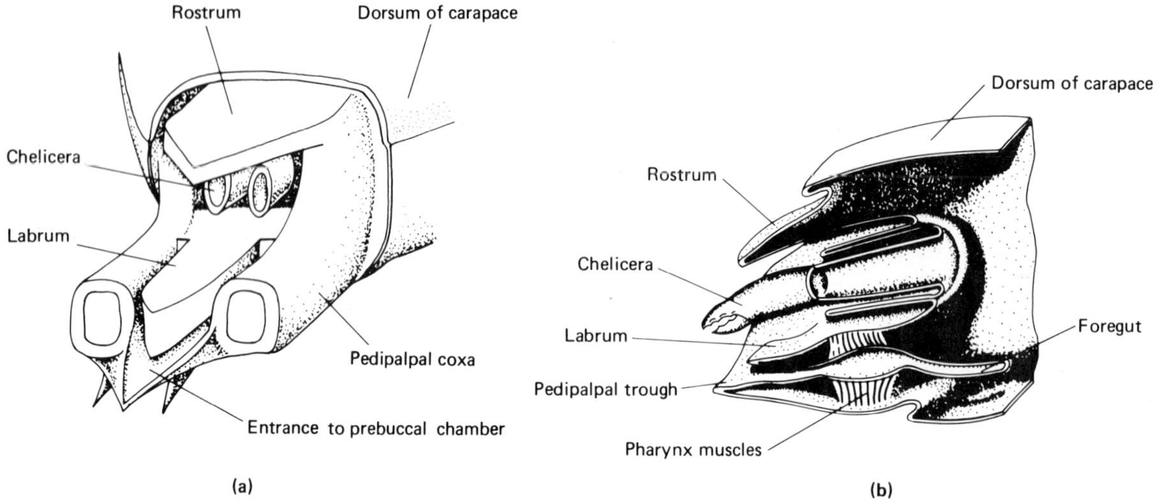

Figure 11.37 Diagrams of the capitulum of a mite. (*a*) An oblique frontal perspective with chelicerae and pedipalps cut off near the base; (*b*) Longitudinal section. (From Kaestner, *Invertebrate Zoology, Vol. II.*)

their small size relative to their food source. These and other crop-eating mites become increasingly problematic when their natural predators and competitors from the insect world are reduced by insecticides.

Carnivorous mites eat nematodes and small arthropods. Immature stages are common and often more easily acquired prey. Many mites feed on insect eggs, while aquatic species (Hydracarina) take small crustaceans. Frequently, the larvae of water mites parasitize the gills of crustaceans or aquatic insects. Predatory mites have chelate pedipalps, and teeth often border the margins of their cheliceral pincers. Most carnivorous mites suck out prey fluids in the traditional arachnid manner, but some rip their victims apart with the chelicerae. Prey fragments are brought into the prebuccal cavity, where enzymes reduce them.

Scavenger mites include a diverse collection of feeders on corpses, plant products, and decaying organic matter. Entomological collections must be protected from *Threophagus entomophagus*, which thrives on dead insects. Storage mites (Acaridae) are common in such places as old mattresses and food supply centers. One of them, the flour mite *Acarus siro*, lives in grain mills, where enormous populations may flourish. The oribatid mites (Fig. 11.38a) eat organic remains in leaf mold. *Dermatophagoides* lives among the dust in our homes.

The feather mites are commensals on birds. Their appendages and/or body surfaces may be greatly modified for anchorage on feather barbs. *Analges* (Fig. 11.38b) for example, has huge third walking legs for this purpose. These mites feed on skin oils and broken feather parts. Other commensals eat dead mammalian skin and fur. Some Mesostigmata live in bee or wasp nests, where they consume feces and carrion.

Many mite groups parasitize invertebrates and vertebrates. Ectoparasitism is most common, but endoparasitism has developed through the penetration of host respiratory tracts. Mites incorporate parasitism into their life cycles in a variety of ways. Often only the larval stages are parasitic. This type of life cycle is demonstrated by *Leptotrombidium* (Fig. 11.39) and related forms, whose young are the aggravating little beasts known as chiggers. The chigger egg hatches in soil and the young crawl onto vertebrate bodies. They feed on dermal tissues, where their presence causes severe and prolonged itching. Following their meal, the larvae drop off. Subsequent molts produce a free-living adult whose main food is insect eggs.

The ticks are a good example of a periodic parasite. These largest of all acarines parasitize nearly every land vertebrate, including humans, dogs, cattle, and sheep. Most ticks attach to their host only while feeding. First, their closed chelicerae stab through the host integument. As the chelicerae are opened, lateral teeth widen the wound and sharp barbs on the pedipalpal bases are inserted into the host tissues. With the aid of anticoagulant secretions, the tick then sucks out blood.

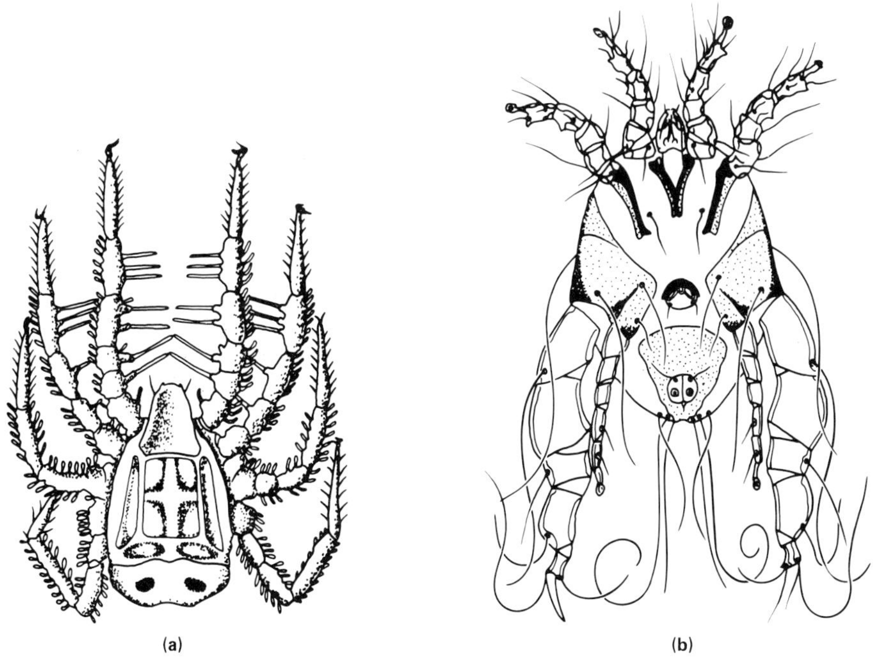

Figure 11.38 (*a*) The bizarrely-shaped oribatid mite, *Caeculus echinipes*. (*b*) The feather mite *Analges*, a commensal of birds. Ventral view of a male. (*a* from W. Kühnelt, *Soil Biology*, 1976, Michigan State University; *b* from Kaestner, *Invertebrate Zoology*, *Vol. II*.)

Figure 11.39 A chigger mite, *Leptotrombidium akamushi*, greatly enlarged. (From Vercammen-Grandjean and Langston, *Chigger Mites of the World*.)

Most ticks have extensive midgut ceca and a flexible exoskeleton, which allow huge blood meals to be taken. Females of *Amblyoma clypeolatum* can swell to 3 cm in length, and in many species, a gorged tick can survive over a year between feedings. Numerous disease organisms are transferred by these acarines. *Dermacentor* (Fig. 11.40a and b), for example, carries the spirochaete bacterium that causes Rocky Mountain spotted fever.

Sarcoptes scabiei (Fig. 11.40c) is a mange mite that lives within human skin. Infestation by this animal is called the 7-year itch. These tiny arachnids (males are a quarter-millimeter long; females are only twice that) tunnel through the epidermis, reproducing for generations on end. Their compact circular bodies enable them to turn within narrow passageways. Transfer from one human host to another occurs during direct contact with infested skin.

Highlights of the mite digestive tract include a pumping pharynx and up to seven pairs of midgut ceca (Fig. 11.41). Intracellular digestion occurs in many mites, as evidenced by the lack of an anus in one large group (the Trombidiformes). The uptake of nutrients in preprocessed fluid form minimizes undigested wastes.

Some mites have tracheal systems with anterior spiracles. Appendage coxae commonly bear these pores, thus causing speculation that they are not homologous with the spiracles of other arachnid groups. Many mites, however, lack specialized respiratory organs. These animals are small, and direct diffusion across their relatively thin cuticles provides sufficient gas exchange. Mites that depend on the general body surface for respiration are confined to humid areas. Because of their small size, mites have a greatly reduced heart and arterial system. General

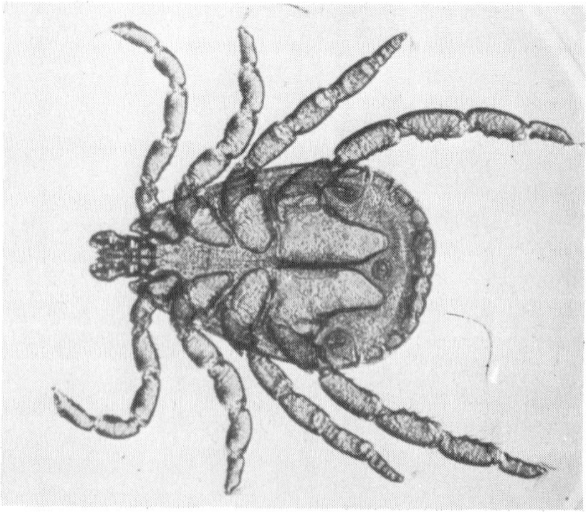

(a)

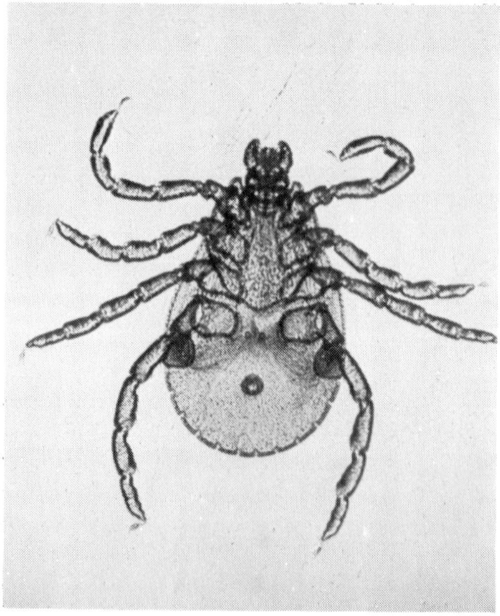

(b)

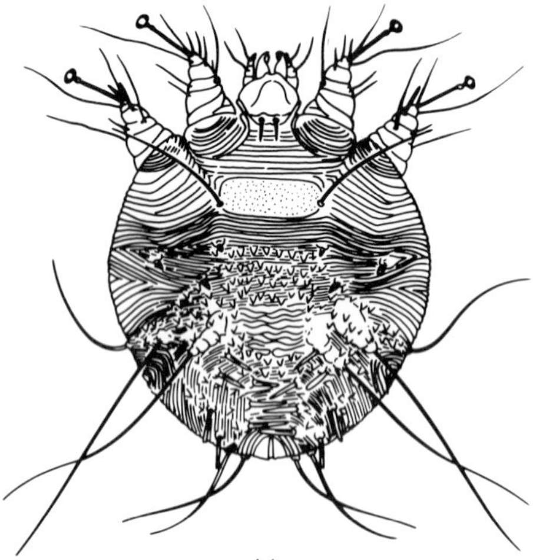

(c)

Figure 11.40 (*a*) Male and (*b*) female of *Dermacentor andersoni*, a tick vector of Rocky Moutain spotted fever; (c) The mange mite, *Sarcoptes scabiei*. Dorsal view of a female. (*a* and *b* courtesy of Ward's Natural Science Establishment; c from E.W. Baker, *et al.*, *A Manual of Parasitic Mites of Medical and Economic Importance*, 1956, National Pest Control Association.)

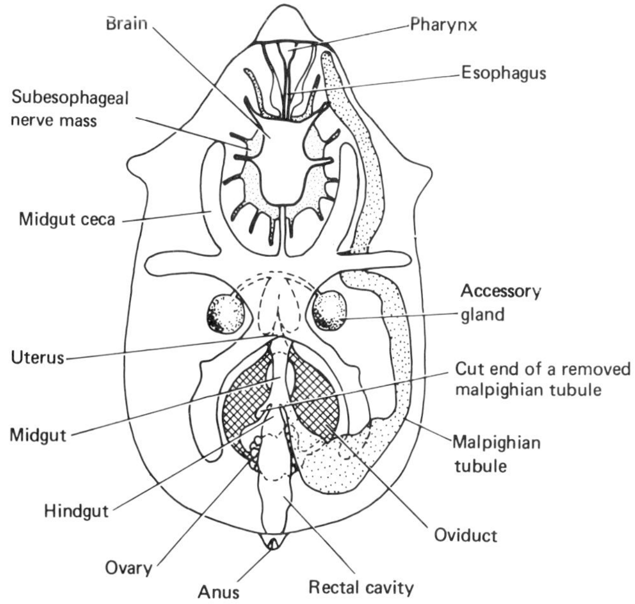

Figure 11.41 A dorsal view of the interior anatomy of the mite *Caminella peraphora.* (From Krantz, *A Manual of Acarology.*)

muscular contractions apparently circulate the hemolymph.

One to four pairs of coxal glands are usually present, with or without one or two paired Malpighian tubules. Wastes collected by the midgut wall are discharged into the digestive tract and excreted with Malpighian products through the hindgut. In mites with a blind midgut (the Trombidiformes), this pattern is interrupted. A unique medial pouch collects wastes, which are discharged through a cuticularized tubule and posterior medial pore. This organ may have evolved from the separated hindgut; if so, its excretory pore is homologous with the anus.

Mites are especially subject to water loss. Many species burrow more deeply into moist soil during drying winds. Other mites, such as the flour mite *Acarus*, absorb water from the air.

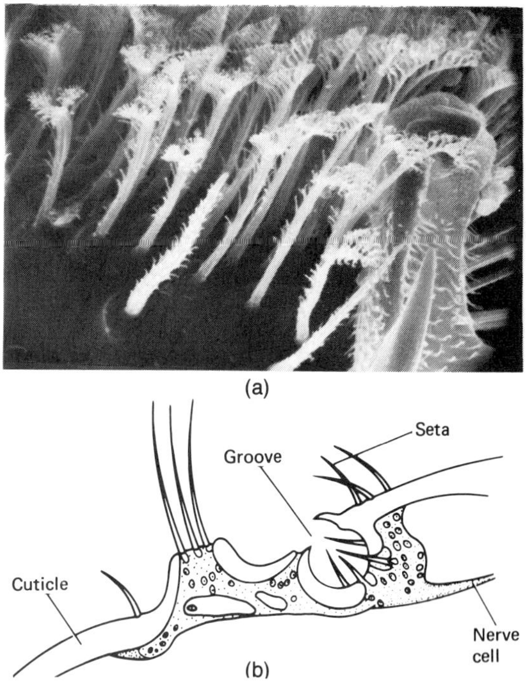

Figure 11.42 (*a*) A scanning electron micrograph showing tenent hairs from the tarsus of a mite. These hairs have expanded filiform tips which are provided with a glandular secretion. This allows adhesion to smooth surfaces; (*b*) A longitudinal section through Haller's organ of the mite *Ixodes reduvius.* (*a* from Kessel and Shih, *Scanning EM in Biology; b* from M. André, In *Traité de Zoologie, Vol. VI,* Grassé, Ed.)

Activity Systems

In mites, the nervous system is concentrated, all ganglia having fused around the esophagus. Touch and smell are the dominant senses. Numerous tactile hairs are scattered over the body surface, where their various structural modifications provide fascinating viewing in a scanning electron micrograph (Fig. 11.42). These hairs are concentrated on the first pair of walking legs. Upon encountering an object, a water mite grasps it with these legs; if the object moves, the mite attempts to eat it. Acarine trichobothria (often called **pseudostigmatic organs**) detect vibrations and wind. **Haller's organ,** apparently a chemosensitive structure, is located on the first legs. It consists of a wide groove lined with setae and nerve cells. Structures resembling slit sense organs are usually present. Vision is not well developed in this order, and indeed, many mites have no eyes at all. The oribatids have a single medial eye, and five eyes are present on some water mites. There is evidence that the brain itself responds directly to light passing through the translucent body surface.

Acarines are very sensitive to humidity. In ticks, this sensitivity is localized in the anteriormost legs. Temperature recognition is an important source of host information for many ticks, and heat-sensitive centers have been identified on the posterior legs. Chigger larvae may respond to the carbon dioxide in human breath or to amino acids on the skin. These stages are negatively geotactic and thus tend to climb host bodies.

The last three leg pairs are modified for locomotion. The limbs of aquatic mites have long hairs which increase water resistance during their effective swimming stroke (Fig. 11.43). Legs are often short in burrowing forms (Fig. 11.40b), and in most parasites they are adapted for clinging to the host. Some mites attach to beetles or bees and are transported by these insects. Such nonparasitic travel is called **phoresy** and was encountered earlier among certain nematode groups.

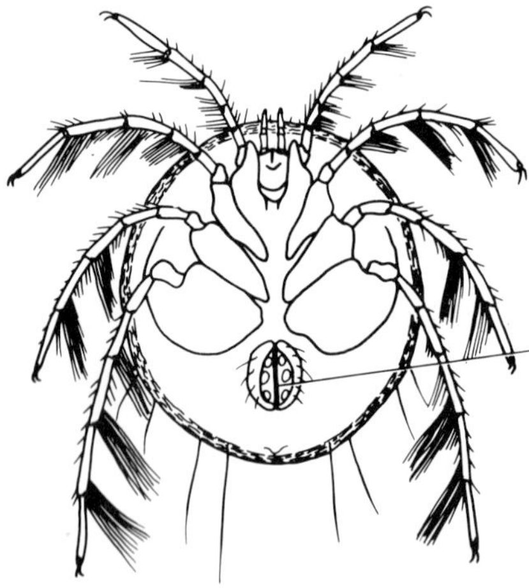

Figure 11.43 A water mite, *Mideopsis orbicularis*. Ventral view showing the characteristic long hairs on the legs. (From R.W. Pennak, *Fresh-Water Invertebrates of the United States*, 2nd ed., 1978, Wiley-Interscience.)

Continuity Systems

The gonads and their accessory glands occupy the central body of the mite. The testes are usually paired, but the ovary is often single. Seminal receptacles are present in the female, and many male mites have a cuticular penis. In both sexes, a sternite bears the medial gonopore (Fig. 11.43). Sperm transfer may be direct or indirect. Indirect mating involves spermatophore transfer by one of the appendages. Chelicerae usually are the intromittent organs, but water mites use their third pair of swimming legs. Alternatively, a stalked spermatophore is deposited on the ground, where a passing female takes up its apical sperm mass. Such a mechanism is reminiscent of pseudoscorpion sex. Direct sperm transfer involves the insertion of the penis into the female gonopore. The mite couple is held together by male appendages, which may be specialized for the copulatory embrace. Some mites have ventral, posterior suction cups that increase adherence during mating.

Eggs may be laid in humid soil or ground litter. Some mites attach their ova to other individuals, and a few are ovoviviparous. Mite juveniles have only six legs (Fig. 11.44). During successive molts, adult characters—including the missing leg pair—are added. Life spans range from about a week up to 3 years.

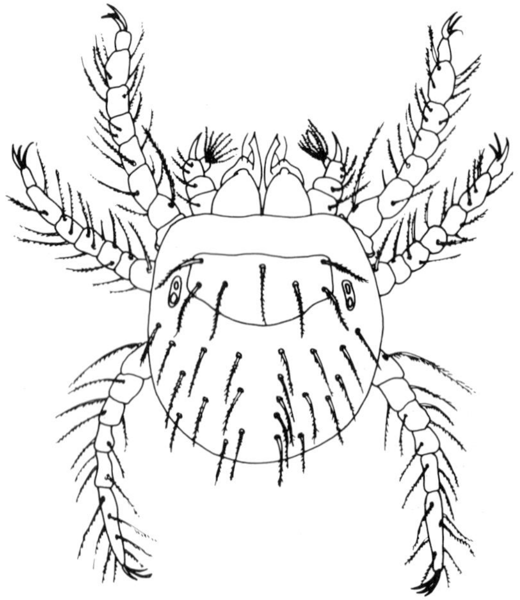

Figure 11.44 Dorsal view of the larva of a chigger mite, *Trombicula autumnalis*. (From Savory, *Arachnida*.)

CLASS PYCNOGONIDA

The pycnogonids are extremely peculiar arthropods, which are usually grouped with the chelicerates. Some authorities, however, place them in a unique subphylum. These animals resemble other chelicerates in brain structure and in the presence of chelicerae. Pedipalps are usually present, and four pairs of walking legs are common. But several other characters are unique to the pycnogonids and suggest that they are highly aberrant chelicerates, if not an independent group.

Pycnogonids are called sea spiders because all 600 species are marine. They live in all oceans and inhabit littoral regions as well as the deep sea. Frequently, sea spiders are collected on bryozoan, hydrozoan, or soft coral colonies. The pycnogonid body plan emphasizes surface area over volume. (Fig. 11.45). The central trunk is very narrow and seldom exceeds a centimeter in length. The legs can be very long, however, and often constitute half or more of the animal's bulk. The largest sea spider, the benthic *Dodecolopoda mawsoni,* has a trunk 6 cm long and a leg span of almost half a meter.

The trunk represents the prosoma. In most pycnogonids, the opisthosoma is extremely reduced and may exist only as a posterior knob with an anus (Fig. 11.45). Anteriorly, a trunk proboscis bears the mouth and the slender lateral chelicerae. The chelicerae are reduced in those sea spiders whose proboscis is large and flexible. Behind these appendages is a smaller, often reduced pair of sensory pedipalps. The third pair of pycnogonid limbs is particularly unusual. These small appendages are called **ovigerous legs** because they are used by the male to carry developing eggs. In females, they are often degenerate. Ovigerous legs also may be utilized for grooming.

Between the proboscis and the trunk, a dorsal tubercle bears up to four indirect eyes. The trunk is distinctly segmented. Lateral projections on each segment precede the walking legs, which bear one or more terminal claws. Most sea spiders have only four pairs of walking legs, but five or six pairs are present in some groups. These additional legs are puzzling. It is not known whether segmental subdivision has occurred or whether the last walking appendages have developed on opisthosomal metameres.

Pycnogonids consume coelenterates, bryozoans, sponges, and sometimes soft mollusks and sea cucumbers. The pedipalps find an appropriate surface for the chelicerae to attack. Smaller prey is torn apart by these appendages. The whole proboscis may be inserted into larger prey, such as the sea anemone, whose tissues are extracted in a semiparasitic manner. The pycnogonid mouth opens into a strong, hard-surfaced pharynx (Fig. 11.46). This organ is a powerful pump, and spines along its cuticularized walls reduce and strain incoming particulate matter. From the midgut, deeply branched ceca penetrate far into the appendages. Digestion is largely intracellular. Cells lining the midgut lumen absorb partially digested fluids

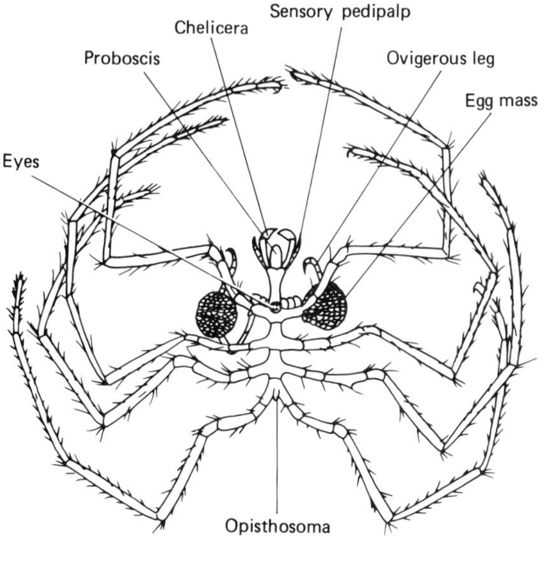

Figure 11.45 Adult male of a pycnogonid, *Nymphon rubrum.* (From L. Fage, In *Traité de Zoologie, Vol. VI,* Grassé, Ed.)

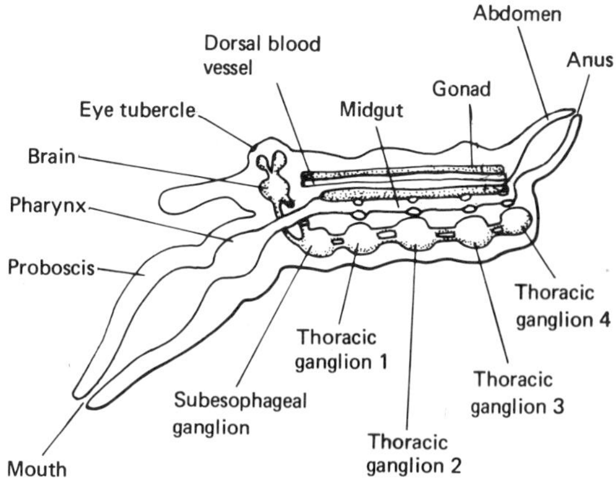

Figure 11.46 Longitudinal section through the body of a pycnogonid showing the alimentary canal and other major internal organs. (From P.E. King, *Pycnogonids,* 1973, St. Martin's Press.)

and then detach to wander about the cavity. Eventually they contact absorptive cells, which transfer their digested nutrients to the hemolymph. The digestive cells then exit through the anus, carrying waste products with them.

A dorsal heart courses the length of the trunk, pumping blood through an anterior aorta into the proboscis area. The circulatory fluid then drops below a horizontal membrane which extends throughout the body. Underneath this membrane, blood flows posteriorly into the rest of the trunk and the appendages, then passes up through the horizontal membrane and back to the heart.

Specialized organs for respiration and excretion are not present. Evidently, the high surface-to-volume ratio allows direct diffusion to play a major role in these maintenance activities. Also, the unusual migrations of the digestive cells may have an excretory purpose. Moreover, unlike the arachnids, pycnogonids benefit from the relatively benign environment of the sea.

Their nervous system features a brain collar and a doubled ventral cord (Fig. 11.46). A protocerebrum innervates the dorsal eyes, and a tritocerebrum associates the cheliceral nerves. The subesophageal ganglion handles the pedipalps and ovigerous legs. A laterally fused ganglion occupies the ventral cord in each appendage-bearing trunk segment. Sense organs include the eyes and numerous sensitive hairs: The latter attain their greatest concentration on the pedipalps and ovigerous legs. Sea spiders are rather slowpaced, and their terminal claws grasp almost anything they touch. A few can tread water.

Pycnogonid gonads lie above the gut and branch into the appendages (Fig. 11.46). Ripe gametes collect in the legs, and gonopores are located on all the coxae. (These multiple gonopores further distinguish the pycnogonids from the arachnids and merostomes.) As the female discharges several hundred ova, the male fertilizes them externally. Using secretions from leg glands, he cements the eggs into a large spherical mass, which is brooded against his ovigerous legs (Fig. 11.45). Setae on these legs help to hold the eggs in place.

The first posthatching pycnogonid stage is a **protonymphon** (Fig. 11.47). Chelicerae, pedipalps, and ovigerous legs are the only appendages present, and the trunk is not developed. Protonymphons may remain on their father's body or may become temporary ectoparasites on coelenterates, bivalves, or nudibranchs. Subsequent molts witness the serial addition of leg-bearing segments and the gradual assumption of an adult life-style.

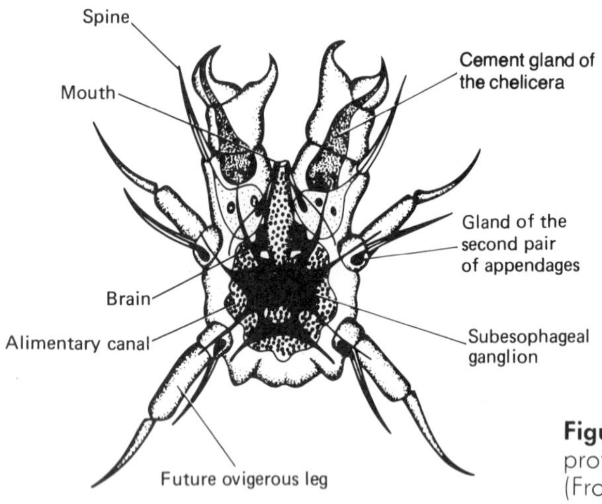

Figure 11.47 Ventral view illustrating the structure of the protonymphon larva of the pycnogonid *Achelia echinata*. (From King, *Pycnogonids*.)

for Further Reading

Anderson, J. F. Energy content of spider eggs. *Oecologia* 37: 41–58, 1978.

Arthur, D. R. *Ticks and Disease.* Pergammon Press, London, 1962.

Babu, K. S. Anatomy of the central nervous system of arachnids. *Zool. Jahrb. Abt. Anat.* 82: 1–154, 1965.

Baerg, W. *The Tarantula.* University of Kansas Press, Lawrence, 1958.

Balogh, J. *The Oribatid Genera of the World.* Akademiai Kiado, Budapest, 1972.

Barth, F. G. and Picklemann, P. Lyriform slit sense organs—Modeling an arthropod mechanoreceptor. *J. Comp. Physiol.* 103: 39–54, 1975.

Brach, V. *Anelosimus studiosus* and the evolution of quasisociality in theridiid spiders. *Evolution* 31: 154–161, 1977.

Bristowe, W. S. *The World of Spiders.* New Naturalist Series, London, 1971.

Buchli, H. H. R. Hunting behavior in the Ctenizidae. *Am. Zool.* 9: 175–198, 1969. (An account of a family of trapdoor spiders)

Burgess, J. W. Social Spiders. *Sci. Am.* 234(3): 101–106, 1976.

Campbell, A. and Harris, D. L. Reproduction of the American dog tick, *Dermacentor variabilis,* under laboratory and field conditions. *Environ. Entomol.* 8: 734–739, 1979.

Cloudsley-Thompson, J. L. Some aspects of physiology and behavior of *Galeodes arabs. Entomol. Exp. Appl.* 4: 257–263, 1961. (Good account of this North African solifugid)

Cloudsley-Thompson, J. L. *Spiders, Scorpions, Centipedes and Mites.* Pergammon Press, New York, 1968. (A popular account of most arachnid groups)

Cook, D. *Studies on Neotropical Water Mites.* American Entomological Institute, 1980.

Cooke, J. A. L. The biology of Ricinulei. *Zool.* 151: 31–42, 1967.

Dondale, C. D. and Redner, J. H. Revision of the wolf spider genus *Alopecosa* in North America. *Can Entomol.* 111: 1033–1056, 1979.

Edgar, W. D. Prey and feeding behavior of adult females of the wolf spider *Pardosa amentata. Neth. J. Zool.* 20: 487–491, 1970.

Eisner, T., Meinwald, J., Munro, A. and Ghent, R. Composition and function of the spray of the whip scorpion, *J. Insect Physiol.* 6: 272–298, 1961.

Eisner, T. Survival by acid defense. *Nat. Hist.* 71: 10–19, 1962. (An account of uropygid defense mechanisms)

Gardner, B. T. Observations on three species of jumping spiders. *Psyche* 72: 133–147, 1965.

Gertsch, W. J. *American Spiders,* 2nd ed. Van Nostrand Reinhold, New York, 1979.

Foil, L. D., Coons, L. B., and Norment, B. R. Ultrastructure of the venom gland of the brown recluse spider, *Loxosceles reclusa. Int. J. Insect Morphol. Embryol.* 8: 325–334, 1979.

Francke, O. F. Spermatophores of some North American scorpions *J. Arachnol.* 7: 19–32, 1979.

Fry, W. G. A classification within the Pycnogonida. *Zool. J. Linn. Soc.* 63: 35–78, 1978.

Hadley, N. F. Adaptional biology of desert scorpions. *J. Arachnol.* 2: 11–23, 1974.

Harwood, R. H. Predatory behavior of *Argiope aurantia. Am. Midl. Nat.* 91: 130–138, 1974.

Kaestner, A. *Invertebrate Zoology, Vol. II.* Wiley-Interscience, New York, 1968. (Describes all chelicerates in somewhat more detail than other invertebrate zoology texts.)

Kaston, B. J. Spiders of Connecticut. *State Geol. Nat. Hist. Surv. Bull.* 70: 1–874, 1948. (Good taxonomic study including most Eastern U.S. spiders)

Kaston, B. J. The evolution of spider webs. *Am. Zool.* 4: 191–207, 1964.

Kaston, B. J. Comparative biology of American black widow spiders. *Trans. San Diego Soc. Nat. Hist.* 16: 33–82, 1970.

Kaston, B. J. Supplement to spiders of Connecticut. *J. Arachnol.* 4: 1–72, 1977.

Kaston, B. J. *How to Know the Spiders,* 3rd ed. W. C. Brown, Dubuque, Iowa, 1978. (Basic taxonomic guide)

King, P. E. *Pycnogonids.* St. Martin's Press, New York, 1973.

Kovoor, J. Silk and the silk glands of Arachnida. *Annee Biol.* 16: 97–172, 1977.

Krantz, G. W. *A Manual of Acarology,* 2nd ed. Oregon State University, Corvallis, 1978.

Kraus, O. On the phylogenetic position and evolution of the Chelicerata. *Entomol. Ger.* 3: 1–12, 1976.

Kullman, E. J. Evolution of social behavior in spiders. *Am. Zool.* 12–419, 1972.

Legg, G. Sperm transfer and mating in *Ricinoides hanseni. J. Zool.* 182: 51–61, 1977.

Levi, H. W. Notes on the life history of the pseudoscorpion *Chelifer cancroides. Tran. Am. Microsc. Soc.* 67: 290–299, 1948.

Levi, H. W. Adaptations of respiratory systems of spiders. *Evolution* 21: 571–583, 1967.

Levi, H. W. and Levi, L. R. *A Guide to Spiders and Their Kin.* Golden Press, New York, 1968.

Manton, S. M. Habits, functional morphology and evolution of pycnogonids. *Zool. J. Linn. Soc.* 63: 1–22, 1978.

McCrone, J. D. 1950. Spider venoms: biochemical aspects. *Am. Zool.* 9: 153–156, 1969.

McDaniel, B. *How to Know the Mites and Ticks.* Wm. C. Brown, 1979.

Merrett, P. (Ed.) *Aracnology.* Symp. Zool. Soc. Lond. No. 42, Academic Press, New York, 1978.

Milne, L. J. and Milne, M. J. *The Crab that Crawled Out of the Past.* Athenieum, New York, 1965.

Moore, R. C. (Ed.) *Treatise on Invertebrate Paleontology,* Part P. Arthropoda 2. Geological Society of America and the University of Kansas Press, Lawrence, 1955. (An account of fossil chelicerates)

Mumma, M. J. A synoptic review of North American, Central American and West Indian Solipugida. *Arthropods of Florida,* Vol. V, Contribution No. 154. Bureau of Entomology, Florida Department of Agriculture and Consumer Services, 1970.

Pennak, R. W. *Fresh-Water Invertebrates of the United States.* 2nd ed. Wiley-Interscience, New York, 1978. (Includes an account of freshwater mites)

Platnick, N. I. The evolution of courtship behavior in spiders. *Bull. Brit. Arch. Soc.* 2: 40–47, 1971.

Polis, G. A. and Farley, R. D. Population biology of a desert scorpion *(Paruroctonus mesanensis):* Survivorship, microhabitat, and the evolution of life history strategy. *Ecology* 61: 620–629, 1980.

Pollock, J. Life of the ricinulid. *Animals* 8: 402–405, 1966.

Robinson, M. H. Predatory behavior of *Argiope argentata. Am. Zool.* 9: 161–173, 1969.

Robinson, M. H. and Robinson, B. Comparative studies of the courtship and mating behavior of tropical araneid spiders. *Pac. Insects Monogr.* 36: 1–218, 1980.

Sankey, J. H. P. and Savory T. H. *British Harvestmen.* The Linnean Society of London, Academic Press, London, 1974.

Savory, T. H. "Daddy Longlegs." *Sci. Am.* 207: 119, 1962.

Savory, T. H. *Introduction to Arachnology.* F. Muller, London, 1974.

Savory, T. H. *Arachnida,* 2nd ed. Academic Press, London, 1977.

Schaefer, M. Winter ecology of spiders. *Z. Angew Entomol.* 83: 113–134, 1977.

Sekiguchi, K. and Sugita, H. Systematics and hybridization in the four living species of horseshoe crabs. *Evolution* 34: 712–718, 1980.

Shuster, C. N. Natural history of *Limulus. Contrib. Woods Hole Oceanogr. Inst.* 564: 18–23, 1950.

Snodgrass, R. E. The feeding organs of Arachnida, including mites and ticks. *Smithson. Misc. Collect.* 110: 1–93, 1948.

Snow, K. R. *The Arachnids: An Introduction.* Columbia University Press, New York, 1970.

Stormer, L. Arthropod invasions of land during late Silurian and Devonian times. *Science* 197: 1362–1364, 1974.

Tiegs, O. W. and Manton, S. M. The evolution of the Arthropoda. *Biol. Rev.* 33: 255–337, 1958.

Turnbull, A. L. Ecology of true spiders. *Annu. Rev. Entomol.* 18: 209–229, 1949.

Tuttle, D. M. and Baker, E. W. *Spider Mites of the Southwestern United States.* University of Arizona Press, Tucson, 1968.

Vachon, M. The biology of scorpions. *Endeavor* 12: 80–89, 1953.

Van der Hammen, L. The evolution of the coxa in mites and other groups of Chelicerata. *Acarologia* 19: 12–19, 1977.

Weygoldt, P. *The Biology of Pseudoscorpions.* Harvard University Press, Cambridge, 1969.

Williams, D. S. and McIntyre, P. The principal eyes of a jumping spider have a telephoto component. *Nature* 288: 578–580, 1980.

Williams, S. C. Scorpions of Baja California, Mexico, and adjacent islands. *Occas. Pap. Calif. Acad. Sci.* 135: +–127, 1980.

Wilson, R. S. Control of drag-line spinning in certain spiders. *Am. Zool.* 9: 103, 111, 1969.

Witt, P. N. The web as a means of communication. *Biosci. Commun.* 1: 7–23, 1975.

Witt, P. N., Redd, C. F., and Peakall, D. B. *A Spider's Web.* Springer-Verlag, New York, 1968. (A study on the techniques of web making)

Yoshikura, M. Development of a whip scorpion *Typopeltis. Acta Arachnol.* 17: 19–24, 1961.

12

Arthropod Diversity: The Crustaceans

The most delicious invertebrates, at least as judged by most humans, belong to the class Crustacea. This group comprises the popular shrimps, lobsters, and crabs, as well as a host of smaller, lesser known arthropods. About 27,000 species have been described. This number is low in comparison with other major arthropod classes; however, the crustaceans express in population sizes what they may lack in speciation. These animals swarm through the oceans much as insects swarm on land. They dominate marine zooplankton, where crustacean populations often reach tens of thousands of individuals per cubic meter. Crustaceans are the primary consumers of oceanic plant life and in turn are prey for the sea's many carnivores. Thus they represent essential links in the marine food chain. Crustaceans are also important in many freshwater environments, and a few have established themselves on land.

Traditionally, crustaceans are grouped with insects and myriapods in a single enormous subphylum, the Mandibulata. Today many biologists are rather skeptical of this arrangement, and some modern classification systems isolate the Crustacea from the rest of the phylum. Certainly, crustaceans are unique in several ways. No other arthropod class is as aquatic. Crustacean head appendages include two pairs of antennae, a pair of mandibles, and two pairs of maxillae. Typically, crustacean appendages are **biramous**; that is, they bear two distal branches (Fig. 12.1b). Except for the trilobites, all other arthropods have uniramous appendages.

Early in the nineteenth century, the Crustacea were classified into two large subclasses, the Entomostraca and the Malacostraca. The former included all of the smaller crustaceans, such as fairy shrimps, water fleas, copepods, and barnacles. Today these animals compose seven distinct subclasses, but the term "entomostracan" still describes them collectively. The Malacostraca, meanwhile, maintains its integrity as a

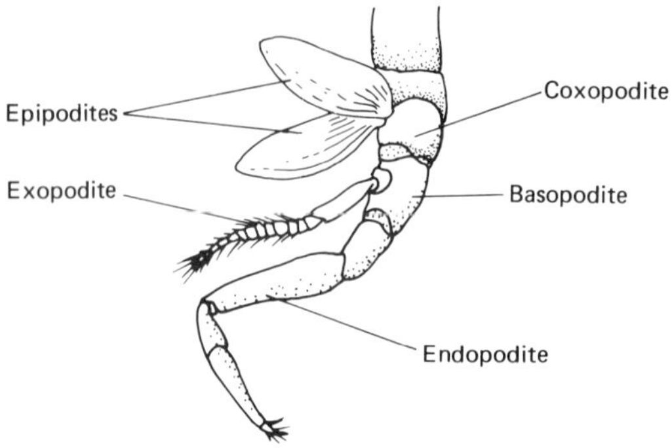

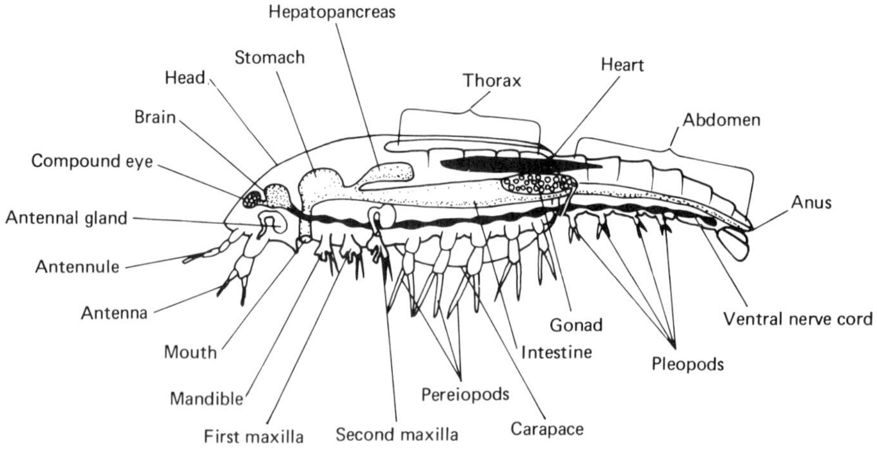

Figure 12.1 (*a*) A basic crustacean appendage. The two major branches of the biramous appendage are the exopodite and the endopodite; (*b*) Diagram showing the organization of a generalized crustacean. (*a* from R.E. Snodgrass, *A Textbook of Arthropod Anatomy*, 1952, Cornell University; *b* from A. Kühn, *Grundriss der Allgemeinen Zoologie*, 1926, George Thieme Verlag.)

group. This eighth subclass includes the larger crustaceans, such as shrimps, lobsters, crayfish, and crabs.

Crustacean Unity

General crustacean characters reflect the aquatic habitats of these animals and their presumed descent from a common ancestor. That ancestor probably possessed a head with five appendages and a long trunk with uniform, paired limbs. Specialization of trunk regions and their appendages has occurred throughout the evolution of the class.

MAINTENANCE SYSTEMS

Crustaceans vary in external form, but their internal organization is often rather standardized. Much of their biology reflects adaptation for life in water. Size also plays an important role in several maintenance systems. Accordingly, it will be helpful to recall the old entomostracan–malacostracan divisions.

General Morphology

Crustaceans vary tremendously in size. Many adult entomostracans are microscopic, while some lobsters are half a meter long. Although increased size typifies the more advanced crustaceans, recall that higher chelicerates are often among the smaller members of their subphylum. These opposite trends reflect the distinct habitats of the two groups. Higher chelicerates live on land, where support problems are acute. Crustaceans, on the other hand, enjoy the buoyancy of aquatic habitats and their relatively greater concentration of food. Accordingly, in the malacostracan line, they have produced the largest modern arthropods.

Crustacean form emphasizes diverse adaptations of the group's appendages. The basic structure of these limbs has been modified for divergent functions, both in different crustaceans and in different segments of the same animal. A basic crustacean appendage is depicted in Figure 12.1a. It consists of a proximal **protopodite** and two distal branches. The protopodite itself has two parts: the **coxa** or **coxopodite,** which is attached to the body proper, and the **basopodite**, from which the distal elements hang. Those two elements are an inner **endopodite** and an outer **exopodite.** Appendage structure and nomenclature are complicated by specialized extensions on the above named parts. Coxopodite extensions are called **epipodites**. Those attached to an endopodite are called **endites**; to an exopodite, **exites.** Epipodites, endites, and exites play key roles in the maintenance, activity, and continuity systems of most crustaceans.

The crustacean body comprises a head, thorax, abdomen, and terminal telson (Fig. 12.1b). As mentioned previously, the head bears five paired appendages. Anteriormost are the first antennae, also known as **antennules**. The antennules are presegmental in origin and are homologous with the antennae of insects and myriapods. They are often uniramous. The second antennae, or simply the **antennae**, may be homologous with arachnid chelicerae. The next appendages are the jawlike **mandibles**. Between the chewing bases of these limbs is the crustacean mouth. An upper lip, the **labrum**, extends from the body wall anterior to the oral cavity; and a lower lip, the **labium**, drops from the foregut at the posterior border of the mouth. Behind the labium are two paired appendages, the **first** and **second maxillae**. The mandibles and both pairs of maxillae tend to be short, flattened limbs; their function in food manipulation is described below.

Originally, the crustacean trunk was probably a uniform region with paired, ventrolateral appendages. This condition is barely preserved among the most primitive contemporary forms. The evolutionary history of the class has witnessed the specialization of various trunk areas and their appendages. As part of this adaptive specialization, fusions, reductions, and functional isolation of trunk regions have resulted in discrete body tagmata. Thus a thorax and abdomen are often distinguished, or a cephalothorax may be formed. In most crustacean groups, tagmatization includes the formation of a **carapace**. A carapace is a cuticular shield which originates from cephalic tergites. These plates extend back over the body and may fuse with skeletal elements on the trunk. In most malacostracans, the carapace also extends along the sides of the animal to enclose a ventral respiratory chamber. In some groups, such as ostracods and barnacles, it envelops the entire body.

The earliest function of trunk limbs probably combined locomotion, respiration, and feeding. As various types of locomotion have been refined, legs have specialized for swimming, crawling, and/or burrowing. Trunk appendages nearest the mouth may acquire and process foodstuffs. Such limbs, many of which are chelate, join the mandibles and maxillae as part of an expanded head zone. Sensory centers occur on many appendages and are most numerous on both antennal pairs. Other limbs are adapted for respiratory chores or for various roles in reproduction. The latter include legs specialized for courtship, copulation, and egg brooding. Appendage adaptation may involve atrophy or hypertrophy of any of the basic elements described above. For example, in mandibles, only the protopodite remains well developed as a grinding or cutting gnathobase. Endopodites dominate walking legs, while exopodite structures are emphasized on swimming appendages. The epipodites of respiratory limbs are usually expanded.

Nutrition and Digestion

Herbivores, carnivores, scavengers, and parasites are all well represented in the Crustacea. Food-gathering techniques include raptorial predation, browsing, endo- and ectoparasitism, and suspension filter feeding. The latter is highly developed among planktonic entomostracans, which harvest diatoms, other phytoplankton, and small zooplankton. Detritus kicked into suspension is also important in the diet of many filtering forms. Recall that all arthropods lack cilia. Suspension-feeding crustaceans filter their food with tiny body hairs, or setae. Often multibranched, these setae form intricate nets which trap suspended particles of an appropriate size. Other setae groom the filtering net and brush trapped food particles toward the mouth. Filtering areas are common on the appendages, particularly on the endites of anterior limbs. Originally, the feeding current was probably powered by routine swimming movements, but in many contemporary suspension feeders, modified appendages maintain a nutritional current independent of locomotor waves.

In raptorial crustaceans, specialized trunk limbs seize and tear prey: The claws of lobsters, crayfish, and crabs are familiar examples. In both carnivores and herbivores, maxillar endites move food fragments toward the mandibles. Each mandibular endite bears hard spiny surfaces which cut or grind against those on the opposite side of the mouth. Such opposing jaws are called **gnathobases**. Food is crushed, softened, and chewed before entering the digestive tract. Crustacean scavengers eat in much the same way. Omnivorous feeding is not uncommon in the class, and various combinations of suspension filter feeding, scavenging, and eating live plants and animals are exhibited by many species.

In free-living crustaceans, the digestive tract comprises a complex stomodeal foregut, a pouched midgut, and a proctodeal hindgut (Fig. 12.2). Although crustaceans chew with their mandibles, oversized particulate matter usually enters the foregut, which further grinds and strains the food. In the foreguts of larger crustaceans, cuticular ridges lined with stiff setae remove oversized food. A triturating area, or **gastric mill**, is also common, where opposing cuticular plates with hard-toothed surfaces grind together to reduce food fragments (Fig. 12.31). Some small entomostracans can handle immediately almost all of the tiny particles that they ingest, and their foreguts are rather simple.

The crustacean midgut is the traditional site of digestion and absorption. One to several pairs of ceca are present. Never arranged by segments, these diverticula commonly are clustered about the anterior midgut, and in malacostracans they form a large **hepatopancreas** (Fig. 12.56). Crustacean ceca comprise a maze of tubules whose walls are composed of secretory and absorptive cells. The tubules also store digested nutrients. The proctodeal hindgut concludes with an anus on the telson.

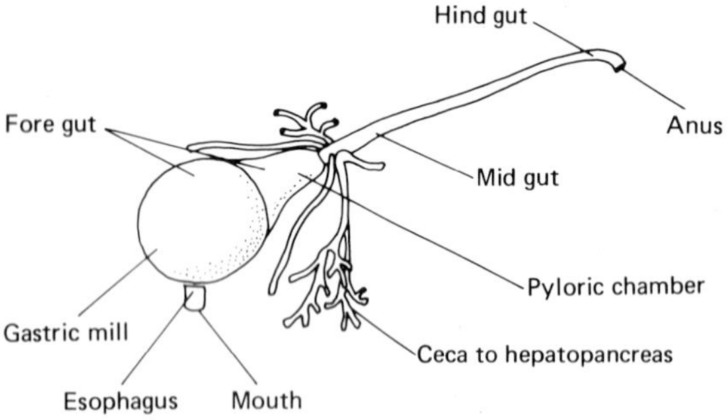

Figure 12.2 Diagram of a crustacean digestive tract. (From K.U. Clarke, *The Biology of Arthropods*, 1972, Edward Arnold.)

Circulation

Like other arthropods, crustaceans have a spacious hemocoel and their blood bathes the tissues directly. A heart lies within a dorsal pericardial sinus and opens through variously organized vessels. The development of central circulatory elements reflects the size and conformation of the crustacean involved. Some trends demonstrated by arachnids are apparent also in the present class. Primitive, elongate crustaceans have a tubular heart with numerous segmental ostia (Fig. 10.4). The heart is more compact among advanced and/or shorter members of the class (Figs. 12.3 and 12.12e) and fewer ostia are present.

Among many entomostracans there is no arterial development. Routine body contractions circulate the hemolymph. In the Malacostraca, arterial systems always include an anterior aorta, while posterior, lateral, and ventral vessels exist in those groups with a strong central heart (Fig. 12.3). In many higher malacostracans, arteries branch extensively before reaching capillary beds in intimate association with the tissues. Thus the circulatory system nearly becomes closed. In every case, however, the hemolymph is discharged into the sinuses, and the tissues are bathed directly.

Membranes partition the sinuses, so that blood follows a well-defined route through the hemocoel.

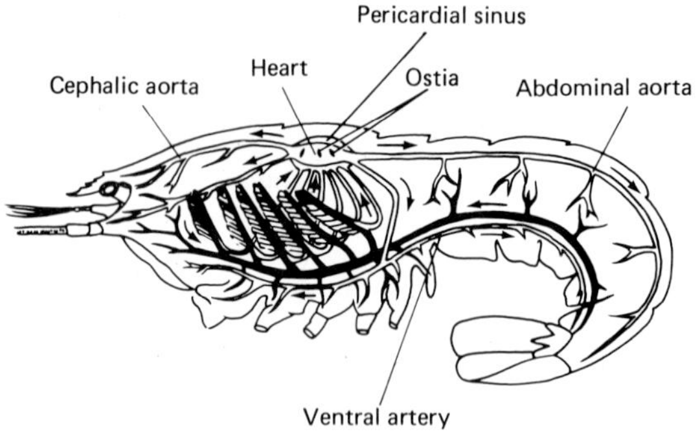

Figure 12.3 The circulatory system of a lobster, an advanced malacostracan. Unshaded vessels carry oxygenated blood, shaded vessels carry deoxygenated blood. (From R.W. Hegner and J.G. Engemann, *Invertebrate Zoology*, 1968, Macmillan.)

Valves prevent back flow and are especially important in entomostracans with no arteries. In larger crustaceans, accessory arterial pumping stations help maintain hemolymph pressure, especially in cephalic regions. Before returning to the pericardial sinus, crustacean blood is aerated in the gills.

Respiratory pigments include dissolved hemocyanin and, less often, hemoglobin; some crustaceans have none at all. A number of wandering cells are also present in the blood. Certain amoebocytes called **explosive cells** are involved in clotting. When a crustacean is injured, these cells release a substance that causes fibrous elements to form and then trap other cells at the wound site.

Respiration

Most crustaceans respire across gills associated with the appendages. They are usually epipodites, arising as broad, flat extensions of the coxopodites (Fig. 12.1a). Apparently, crustacean gills originated from thin flaps of tissue which were ventilated by swimming movements. Primitively, a gill was present on every trunk appendage. Adaptive trends in the class include an increase in gill surface area, a decrease in gill number, and better ventilation mechanisms.

A gill has more surface area if it presents numerous filaments arranged about a central axis. Such feathery gills have evolved in some crustacean groups (Fig. 12.4a). Often, the anterior trunk assumes most of the respiratory workload, as the gills become concentrated behind the mouth. Among filter feeders, the nutritional current and the respiratory current thus reinforce one another. Either the gill-bearing or the neighboring appendages produce this current. When a carapace surrounds the respiratory segments, it provides a narrow channel through which water flow can be controlled (Fig. 12.4b). Often, the inner walls of the carapace itself assume some role in oxygen uptake.

Gills are absent from most entomostracans, whose small size allows respiratory diffusion across thinly cuticularized surfaces. Crustaceans that venture onto land have made minimal respiratory adjustments and are confined to humid environments. (See the following discussion of Isopoda and Decapoda.)

Respiratory pigments notwithstanding, crustacean blood has an oxygen-carrying capacity below that of cephalopods and far below that of fishes. Some crustaceans, such as the lobsters, are oxygen conformers whose metabolism simply drops when oxygen supplies diminish. Others, such as crayfish, are oxygen regulators. In response to a drop in oxygen concentration, these crustaceans ventilate more rapidly to maintain a constant gas supply to the tissues.

Excretion and Osmoregulation

Among aquatic crustaceans, the elimination of nitrogenous wastes and the control of osmotic balance prove less demanding than among terrestrial arthropods. Direct diffusion at the body surface contributes significantly to these processes. The gills in particular are an important site of excretion and osmoregulation, where both diffusion and active absorption and secretion occur. Crustaceans never have Malpighian tubules, but excretory organs resembling

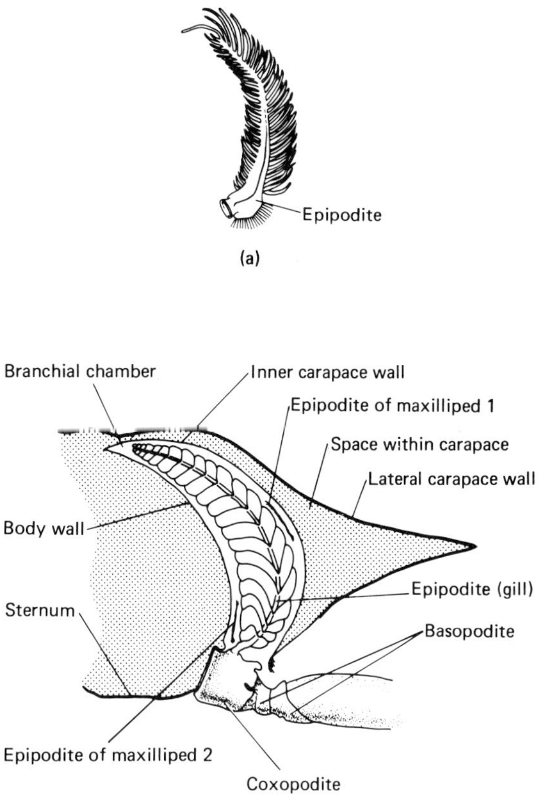

Figure 12.4 (*a*) Feathery gill derived from the epipodite of the third walking leg of a crayfish, *Cambarus longulus;* (*b*) Cross section through the right half of the cephalothorax of a crab at the level of walking leg showing the branchial or gill chamber. (*a* from Snodgrass, *A Textbook of Arthropod Anatomy; b* from A. Kaestner, *Invertebrate Zoology, Vol. III: Crustacea*, 1970, Wiley-Interscience.)

coxal glands are present in the head. These paired structures are known as **antennal glands** or **maxillary glands**, depending on which appendage base houses their excretory pores. Larval crustaceans may possess both types, but one pair usually degenerates before adulthood. Each gland consists of an end sac and a tubule, where resorption and secretion determine the nature of the urine (Fig. 12.5). Larger crustaceans, particularly freshwater species, have a longer and more convoluted tubule. A complex labyrinth may dominate its anterior end, as in crayfish. Distally, the excretory tubule often expands as a bladder.

Crustaceans are primarily ammonotelic. Ninety percent of the nitrogenous waste of the marine members of the class is excreted as ammonia. Freshwater and even terrestrial crustaceans also eliminate this waste. The high solubility of ammonia allows its release across the gills. Even if the antennal or maxillary glands are plugged, ammonia discharge continues at about the same rate. Other, less soluble wastes, such as uric acid, may exit the body with undigested matter through the anus. Nephrocytes, often clustered in the coxopodites, also remove solid metabolic wastes.

Most crustaceans are osmotic conformers. If placed in a medium that is not isosmotic to their body fluids, they lose or absorb water until a balance is achieved. Normal body volume is restored by regulating internal salt content. Some crustaceans, however, live in aquatic environments with which they cannot conform and still survive. Many brackish-water species, including amphipods, isopods, and decapods, osmoregulate over a relatively wide range of salinity. The brine shrimp *Artemia* can live in a saturated solution of sodium chloride by continuously secreting salt from its gills and by producing a hypertonic urine. Freshwater crustaceans may form a hypotonic urine, and their gills actively absorb salt. The antennal glands of the crayfish produce a copious, dilute urine. Also known

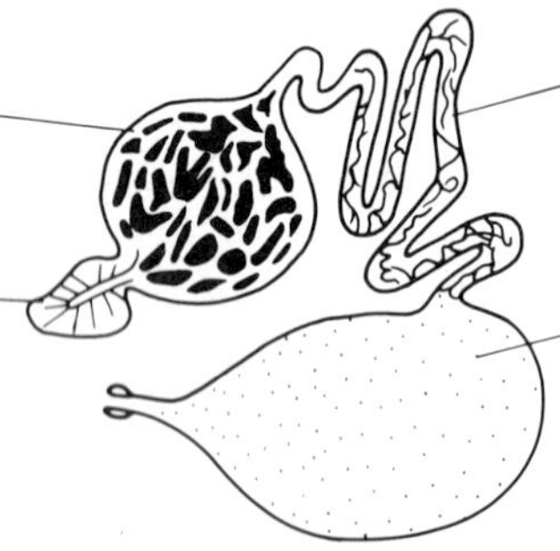

Figure 12.5 Diagram of a crustacean antennal or maxillary gland. (From G. Kümmel, Zool. Beitrag., 1964, *10*:227.)

as green glands because of the green color of their labyrinth, these organs are irrigated by a rich capillary network. Finally, excretory and osmoregulatory systems in terrestrial crustaceans are not highly developed. No hypertonic urine is produced, and most water-retention problems are met on a behavioral rather than a physiological level.

ACTIVITY SYSTEMS

Crustaceans include some of the most active invertebrates in the sea. Depending on local conditions, freshwater species are sometimes quite aggressive, and terrestrial members of the class, although often rather secretive, exhibit complex display behaviors. The crustaceans possess the skeletal and neuromuscular facilities typical of the Arthropoda and have adapted these systems with considerable success. An array of specialized sense organs has evolved. With such complex activity systems, these animals have developed a versatile behavioral repertoire.

Skeleton and Muscles

The crustacean exoskeleton is unusual in the phylum in that it is strongly reinforced with calcium carbonate. This mineral salt is deposited within the epicuticle and the procuticle (Fig. 12.6). In the latter, three distinct layers are formed: an outer tanned and calcified exocuticle; an intermediate untanned but calcified endocuticle; and the innermost endocuticle, which is untanned and uncalcified. Pigment granules in the exocuticle account in part for the brilliant colors of many crustaceans.

The density of skeletal calcite varies throughout the class. In smaller entomostracans, the cuticle is often very thin and flexible owing to low calcium concentrations, but the outer cover of larger malacostracans is usually thick and crusty. Indeed, the name Crustacea was inspired by such exoskeletons.

Crustacean muscles are highly developed. Flexors and extensors maneuver the jointed appendages for swimming, crawling, jumping, burrowing, food procurement, respiration, and reproduction. In the larger malacostracans, these fibers and the intersegmental trunk musculature provide our tasty "shellfish" dinners.

Nerves

Modern crustaceans display various stages in the evolution of a concentrated central nervous system. These stages are seen throughout the annelid–ar-

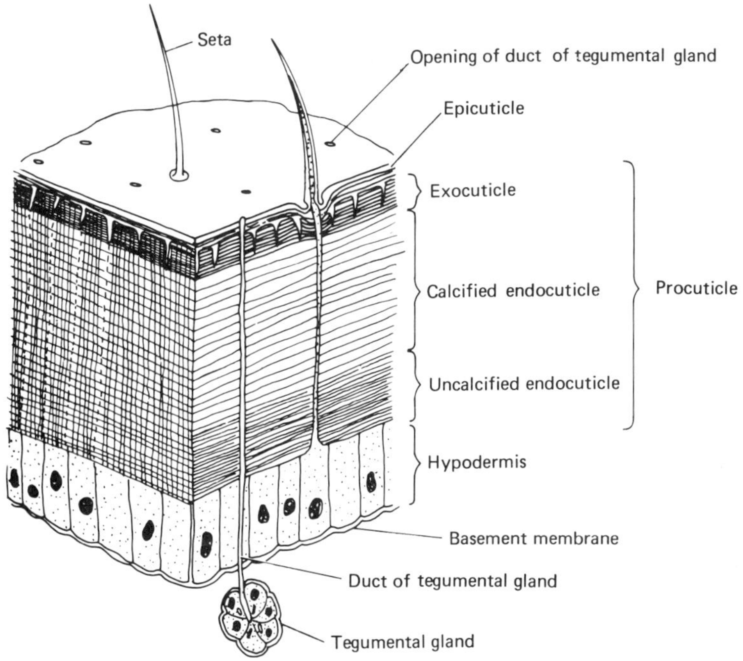

Figure 12.6 Cross section of a crustacean cuticle. (From R. Dennell, In *Physiology of Crustacea, Vol. 1: Metabolism and Growth*, T.H. Waterman, Ed., 1960, Academic Press.)

thropod line. More primitive crustaceans have a ladderlike system, in which paired ventral cords join double ganglionic swellings in each trunk segment (Fig. 12.7a). In stomatopods, medial fusion has occurred (Fig. 12.7b). This situation is common in other malacostracans whose abdomens are long and well developed and also occurs in many entomostracan groups. Finally, longitudinal fusion in the crab nervous system unites all trunk ganglia in a single ventral mass (Fig. 12.7c). In other members of the class, a wide range of intermediate fusion has been described.

There is some evidence that the crustacean brain operates primarily as an inhibitory center. For example, complex behaviors associated with feeding and reproduction continue unimpaired in many brainless animals. The crustacean brain is tripartite. The protocerebrum innervates the compound eyes and a medial eye, if present. The antennules are served by the deutocerebrum, and the tritocerebrum handles the antennae. Connectives surround the esophagus and join a subenteric nerve mass. Also arising from the brain are dorsal giant axons, which extend the length of the trunk in elongate crustaceans such as crayfish. Dorsolateral giant axons may connect each trunk ganglion. These giant axons mediate coordinated wholebody responses such as the leaping of shrimp and crayfish. Impulses along the dorsal fibers are initiated from the brain, but dorsolateral axons can be influenced by any of the ganglia which they connect.

Sense Organs

Crustaceans possess simple and compound eyes, chemoreceptors, and a host of varied tactile centers. In larval stages, a **nauplius eye** is located medially on the head (Fig. 12.8a). This structure represents a small cluster of inverse pigment cup ocelli. The nauplius eye degenerates in most adult crustaceans, although in some entomostracans it persists as the primary visual organ. Compound eyes are typical of most adult members of the class. These organs are located laterally on the head and may occupy optical stalks called peduncles (Fig. 12.8b). The **peduncle** widens the visual field of the eye, as does the convexity of the cornea. Some higher malacostracans see over a 200° arc. In most crustaceans, image formation is poor owing to a paucity of ommatidial units. In lobsters, however, up to 14,000 ommatidia allow considerable differentiation of size and shape. Color vision evidently occurs in certain crustacean groups.

Chemoreceptors are concentrated on the antennules, antennae, and other appendages near the mouth. Of these receptors, the **esthetascs** are probably the most important. Esthetascs normally consist of compact rows of sensory hairs filled with nerve cell processes (Fig. 12.8c). Also, various cuticular crevices and pores are lined with dendritic endings which may respond to chemical stimulation. These structures are common on the gnathobases.

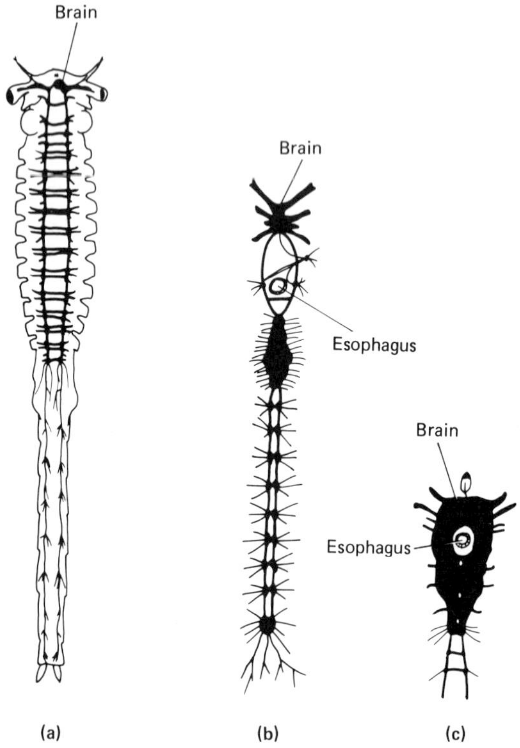

Figure 12.7 Crustacean nervous systems. (*a*) Nervous system of a primitive crustacean, *Branchinecta*, a fairy shrimp. Body is outlined; (*b*) Isolated nervous system of a mantis shrimp, (Stomatopoda); (*c*) Isolated system of a crab, (Brachyura), showing fusion of all trunk ganglia into a single ventral mass. (*a* from R.W. Pennak, *Fresh-Water Invertebrates of the United States*, 2nd ed., 1978, Wiley-Interscience; *b* and *c* from W. Kükenthal and T. Krumbach, *Handbuch der Zoologie, Band III*, 1926, De-Gruyter.)

Touch plays a critical role in many aspects of crustacean behavior. Tactile bristles are scattered over the body surface and reach their greatest density about the appendage joints and tips. The location, stiffness, and basal innervation of these hairs allow the crustacean to interpret any stimulation received. Water currents, passing enemies or prey, and the nature of the substratum are detected. Proprioceptors—specialized tactile receptors within the musculature—monitor the position and tension of the muscle fibers. Proprioceptors respond in a phasic or tonic manner. Phasic receptors are stimulated by faster, short-term muscle events, while tonic receptors are involved with basic posture and orientation to the substratum.

Statocysts are another tactile organ. These invaginated ectodermal chambers contain a solid body, the statolith, which is secreted or is formed from cemented sand grains (Fig. 12.8d). Gravitational forces on the statolith result in the differential stimulation of nerve cell processes lining the inner wall of the statocyst. Thus the crustacean is informed regarding its posture, acceleration, and vibration.

Temperature, sound, and pressure recognition are important for some crustaceans. A sense of water pressure apparently plays a critical role in the vertical migrations of many planktonic members of the class.

Behavior

Like most arthropods, crustaceans depend heavily on their behavioral repertoire for day-to-day survival. They move about, acquire food, construct homes, and communicate with potential mates. When faced with danger, crustaceans hide or run away, threaten or fight. Few other invertebrates are as flexible in their responses or as interesting to observe.

The first crustaceans were probably small, epibenthic swimmers whose hairy trunk appendages were uniformly adapted for paddling. As some members increased in size and complexity, crawling and burrowing became more efficient. Trunk endopodites differentiated as heavier, more muscularized walking legs. Other crustaceans remained small, and many became prominent constituents of the zooplankton. A very few ventured onto land. These different crustacean types have evolved various behavioral programs to sustain their life-styles.

The life of a planktonic entomostracan is rather passive. By definition, planktonic movements are dominated by water currents. Vertical migration, however, is a controlled movement characteristic of many groups. Such migrations ensure that a population occupies the most favorable vertical level at any given time of day. The usual pattern involves passive sinking

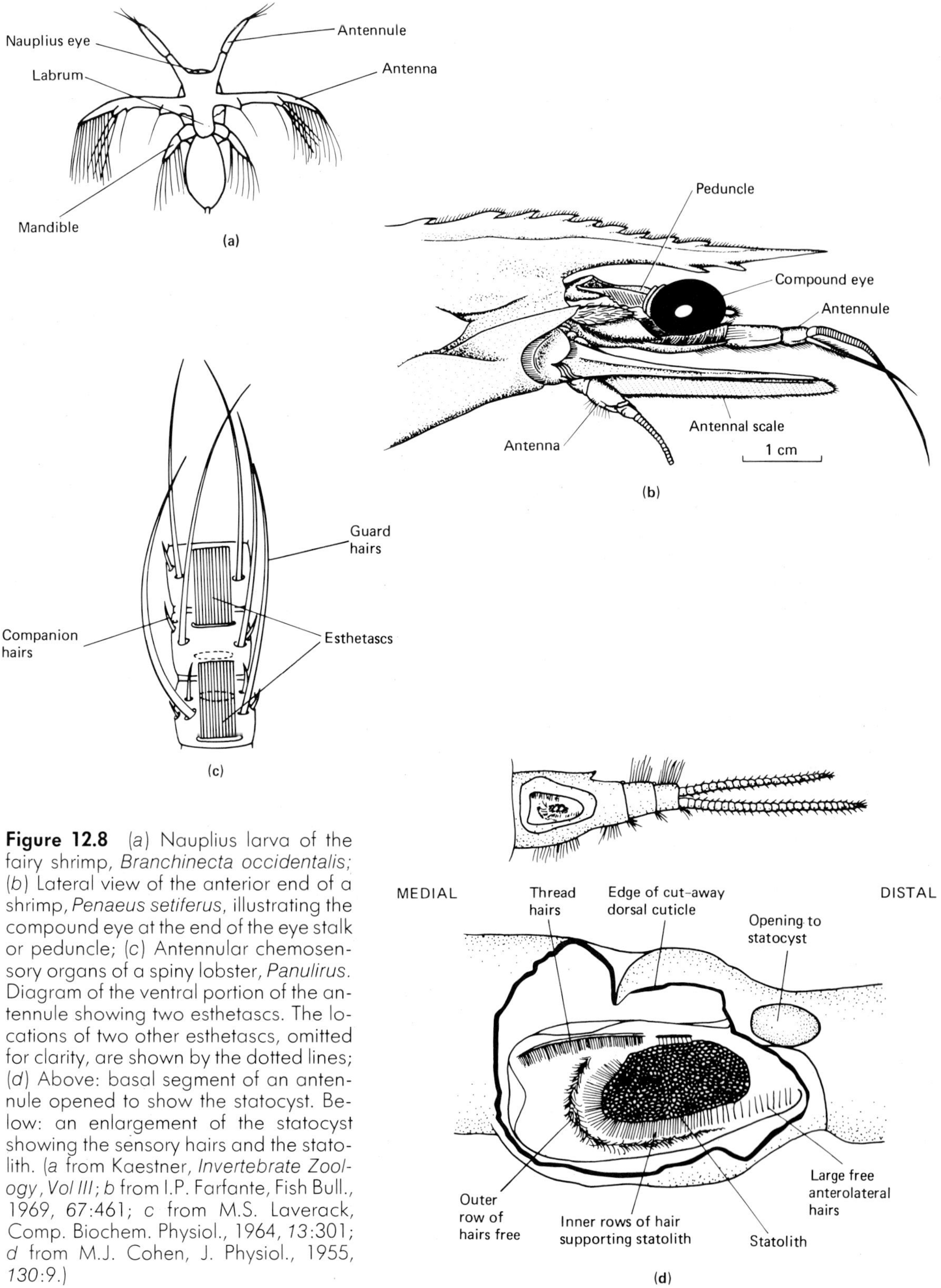

Figure 12.8 (*a*) Nauplius larva of the fairy shrimp, *Branchinecta occidentalis;* (*b*) Lateral view of the anterior end of a shrimp, *Penaeus setiferus*, illustrating the compound eye at the end of the eye stalk or peduncle; (*c*) Antennular chemosensory organs of a spiny lobster, *Panulirus*. Diagram of the ventral portion of the antennule showing two esthetascs. The locations of two other esthetascs, omitted for clarity, are shown by the dotted lines; (*d*) Above: basal segment of an antennule opened to show the statocyst. Below: an enlargement of the statocyst showing the sensory hairs and the statolith. (*a* from Kaestner, *Invertebrate Zoology, Vol III*; *b* from I.P. Farfante, Fish Bull., 1969, 67:461; *c* from M.S. Laverack, Comp. Biochem. Physiol., 1964, *13*:301; *d* from M.J. Cohen, J. Physiol., 1955, *130*:9.)

during the daylight hours when sunlight and predators are most concentrated in upper waters. With sunset, the population rises to enjoy abundant surface food in the relative security of darkness. Apparently, water pressure is the sense by which these little animals recognize their depth; swimming orientation depends on the detection of light from above.

Similar environmentally coordinated rhythms are seen in benthic and coastal crustaceans. Along the coast, some animals organize their activities around tidal cycles. Fiddler crabs, for example, live within burrows which are covered at high tide. At low tide, they walk and feed along the exposed beach.

Crustaceans are often very colorful animals. In many forms, integumental pigments are concentrated within branched **chromatophores** (Fig. 12.9). The distribution of chromatophores and the controlled dispersal of their pigments determine the animal's color. Thus, at one point, a shrimp may be nearly transparent, but later may be red, blue, or black. Pigment dispersal often follows diurnal rhythms; fiddler crabs, for example, are dark during the day and pale at night. Environmental backgrounds also may influence pigmentation. *Ligia,* an isopod which lives along marine coasts, is light colored when lying on sand, but darkens against seaweed or black rocks. Many shrimps, likewise, undergo camouflaging color changes.

The mechanism that controls crustacean chromatophores is very different from that controlling similar structures in the cephalopods. Recall that the latter are muscularized sacs whose size is influenced by nervous input. In contrast, crustacean chromatophores are under hormonal control. The hormones involved are manufactured within a neurosecretory complex called the **X organ.** The X organ lies near the eyes and contains numerous nerve cell bodies (Fig. 12.10). Produced within these nerve cells, hormones travel down neurosecretory axons to a **sinus gland,** where they are discharged into the hemolymph. Different hormones have varying effects on the chromatophores and their pigments. By its nature, hormonal control is much slower than nervous control; thus crustacean color changes are far more gradual than those of cephalopods. The crusty exoskeleton of the present group obviously affords a greater margin of security during the color-changing phase.

greater margin of security during the color-changing phase.

The location of the X organ and sinus gland suggest that visual clues are involved in chromatophore commands. Such is probably the case in protective coloration. Diurnal rhythms of color change, however, such as those exhibited by fiddler crabs, persist even in blinded animals.

Among crustaceans that live on land, behavior is a key element in maintenance biology. No terrestrial crustacean compares favorably with arachnids and insects in terms of its physiological adaptation to land. This condition reflects the absence of a waxy epicuticle. Accordingly, the wanderings of most species are restricted to humid areas. Burrowing is common, as is nocturnality. During dry periods, some crustaceans, such as the common pill bug, enroll their bodies to protect the ventral respiratory surface.

Some crustaceans demonstrate learning ability. Lobsters, crabs, and pill bugs have been taught simple mazes, and crabs kept in aquariums rapidly associate human presence with the arrival of food. In nature, the routine memory of a burrow's location is no small talent. This process may involve the perception of polarized light.

CONTINUITY SYSTEMS

The methods that ensure the continuity of crustacean species are somewhat different from those seen in other arthropods. Most other members of the phylum are terrestrial and provide the utmost care for gametes and developmental forms. Crustaceans are not subject to the same selective pressures during their early stages. Compared with their marine neighbors from other phyla, however, crustaceans have rather sophisticated continuity systems. Fertilization is internal and brooding is common. Free-living larval stages are the rule, and complex molting control mechanisms are in effect.

Reproduction and Development

Sexes are usually separate; however, hermaphroditism occurs sporadically in several subclasses and

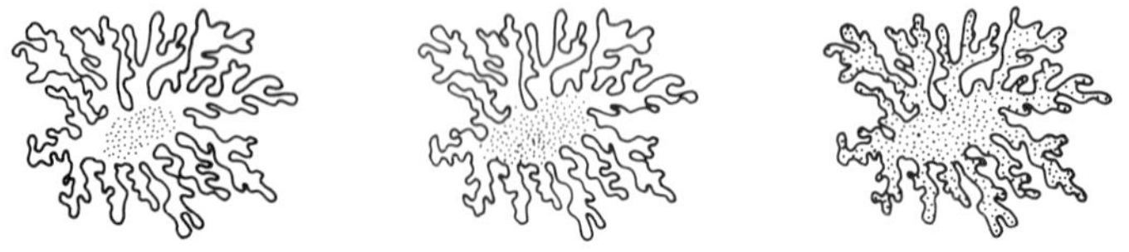

Figure 12.9 Three crustacean chromatophores. Pigment granules are concentrated on left, dispersed on right. (From M. Wells, *Lower Animals,* 1968, McGraw-Hill.)

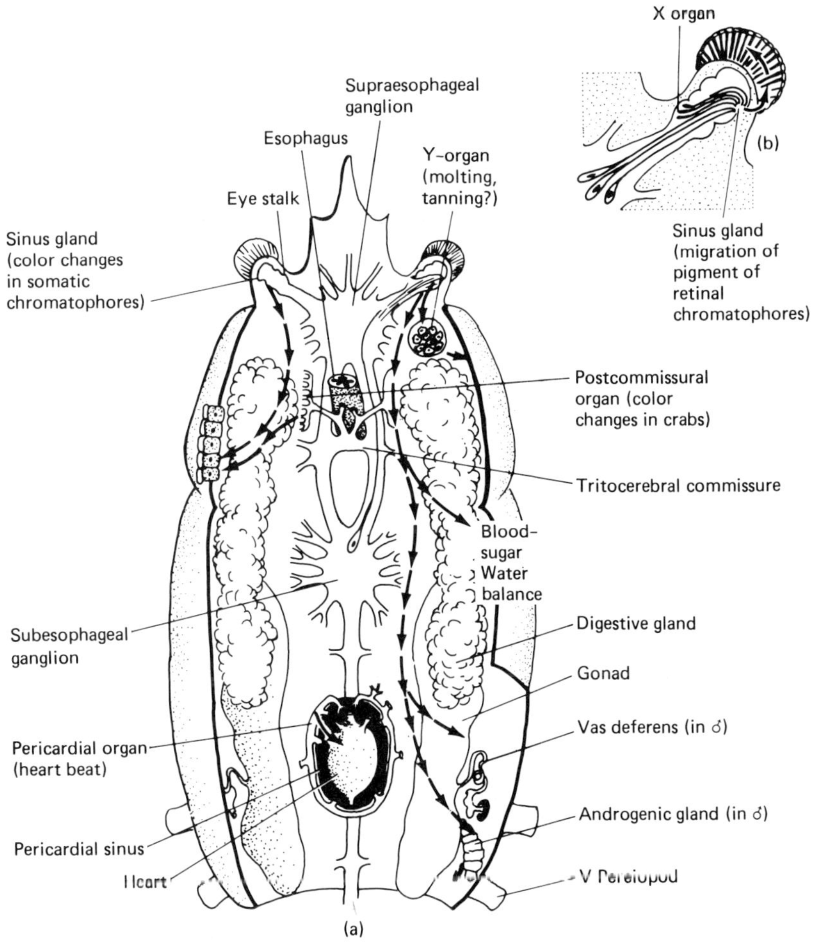

Figure 12.10 Schematic representation of the neuroendocrine system in decapod crustaceans. (*a*) Cephalothorax of a generalized decapod with carapace removed to show sites of hormone production and principal target organs. Arrows indicate transport of hormones from sources to target organs; (*b*) Eyestalk and sinus gland detail showing effect of sinus gland secretion on retinal chromatophores. (From M.S. Gardiner, *Biology of the Invertebrates*, 1972, McGraw-Hill.)

is most notable among the barnacles. Crustacean gonads are long paired organs which lie above the midgut. Their ducts descend to the ventral surface, where the gonopores are located on the protopodites or a sternal plate. Some crustaceans have a penis, but sperm transfer by a specialized trunk appendage **(gonopod)** is common. Other male appendages may hold the female during copulation. Prior to fertilization, sperm usually are stored within seminal receptacles at the lower end of the female reproductive tract.

Most often, the fertilized eggs are brooded. Brooding sites and procedures vary from subclass to subclass and even within smaller taxonomic groups. Some crustaceans, such as the water flea *Daphnia*, have an internal brood pouch. In many other entomostracans, eggs are retained by the mother in an external ovisac secreted during egg deposition. Pericarids are distinguished by a thoracic brooding shelf, while decapods cement their developing eggs to the abdominal appendages.

The typical crustacean passes through numerous larval stages. These stages and their names vary among the different subclasses and orders. At this point, we will introduce only a very generalized larval progression. Later discussions define the particular scheme of each subgroup.

The first larval stage is the **nauplius** (Fig. 12.8a). The nauplius larva is characterized by a single medial eye (the nauplius eye) and three pairs of appendages, which represent the future antennules, antennae, and mandibles. Setae increase the swimming efficiency of these larval limbs. The trunk arises when new segments are proliferated at successive molts. When several trunk segments are formed, the young crustacean has reached the **zoea** stage. In the zoea, trunk appendages are functional in swimming. Further molts witness the addition of posterior body parts until, with the completion of a full set of segments and appendages, the **postlarva** stage is attained. Most postlarvae resemble the adult form of their species. In some crustacean groups, however, an elaborate metamorphosis precedes adulthood. The crab postlarva, for example, has a long abdomen which is reduced and flexed under the thorax at the adult molt. The barnacle postlarva undergoes profound changes, as it abandons a free-swimming life-style in favor of a sessile adult existence. Parasitic crustaceans (especially copepods) also experience drastic alterations in form during their development.

Any or all of these developmental phases may be suppressed in higher crustaceans. No malacostracan has a feeding nauplius, and in the crabs and lobsters this stage is passed prior to hatching. Freshwater crayfish pass all of their larval stages in the egg and emerge as juveniles.

Growth

Hormonal control of molting and sexual maturity involves the X organ–sinus gland neurosecretory complex (Fig. 12.10). Hormones from this complex act on a second pair of glands, the Y organs. Each **Y organ** is located in an antennal or maxillary segment. (They occupy the segment lacking an excretory gland.) Apparently, the X organ hormone inhibits the manufacture and/or release of a molt-promoting hormone (ecdysone) from the Y organ. Under the influence of the Y organ hormone, the epidermis separates from the old cuticle, secretion of a new skeleton begins, and calcium and other substances are resorbed. Resorption of calcium is most intense along rupture lines of the old skeleton. Water uptake stretches the new cuticle, expanding its size. A quiescent period follows, during which the new cuticle is hardened. Calcium salts, having been stored in the hepatopancreas, are redeposited.

The hormonal events that control molting are initiated by any of several environmental and metabolic factors, including day length, temperature, and nutritional status. Some crustaceans, such as lobsters, grow and molt throughout the life span, but the number of instars is limited in most other groups. Many crabs stop molting after they mature sexually.

Hormonal control over gamete production and secondary sexual characters is less well described. The X organ–sinus gland complex may elaborate a hormone that inhibits egg production in the female. When the mating season nears, concentrations of this substance decline and other hormones secreted elsewhere in the central nervous system stimulate ovarian activity. As the eggs develop, ovarian hormones stimulate preparations in the brooding area.

Sexual characters in crustacean males are determined by an **androgenic** gland (Fig. 12.10). This gland usually lies at the base of the sperm duct. Its activity promotes the development of testes, copulatory appendages, and secondary sexual features. If the androgenic gland is removed from a young crustacean male, successive molts witness a gradual sexual conversion. The testes assume ovarian characters and the male-type appendages are lost. If the androgenic gland is transplanted into a young female, she gradually assumes a male identity and even attempts to copulate with another female. Such efforts are always unsuccessful, however, owing to the failure of sperm duct development in these transsexuals.

A Review

The remainder of this chapter discusses the diversification of the Crustacea. We pause now to review the general characters on which the progress of the class has been based. Basically, crustaceans are:

(1) bilaterally symmetrical, metameric invertebrates with a calcified exoskeleton and biramous appendages;

(2) arthropods whose cephalic appendages include paired antennules, antennae, mandibles, and first and second maxillae;

(3) aquatic arthropods which respire across gills;

(4) active invertebrates, usually with well-developed visual and tactile organs, a concentrated nervous system, and giant axons;

(5) arthropods with highly integrated behaviors often based on diurnal and tidal rhythms;

(6) predominantly diecious protostomes whose indirect development features several larval stages.

Crustacean Diversity

Although only 27,000 crustacean species have been identified, classification systems for the group list no less than 8 subclasses and over 30 orders. Obviously, the Crustacea are a varied lot. Their diversity involves differences in appendage adaptation and regional specialization of the trunk, nutritional methods, habitats, and development. Following the classification outline below, we discuss each of the major groups in turn. Note that the old group Entomostraca comprised all of the subclasses except the Malacostraca.

Class Crustacea
- Subclass Cephalocarida
- Subclass Branchiopoda
  - Order Anostraca
  - Order Notostraca
  - Order Diplostraca
    - Suborder Conchostraca
    - Suborder Cladocera
- Subclass Ostracoda
- Subclass Copepoda
- Subclass Mystacocarida
- Subclass Branchiura
- Subclass Cirripedia
- Subclass Malacostraca
  - Group Phyllocarida
    - Order Leptostraca
  - Group Eumalacostraca
    - Superorder Syncarida
      - Order Anaspidacea
      - Order Bathynellacea
      - Order Stygocaridacea
    - Superorder Hoplocarida
      - Order Stomatopoda
    - Superorder Peracarida
      - Order Mysidacea
      - Order Cumacea
      - Order Tanaidacea
      - Order Isopoda
      - Order Amphipoda
    - Superorder Eucarida
      - Order Euphausiacea
      - Order Decapoda
        - Suborder Natantia
          - Section Penaeidea
          - Section Caridea
          - Section Stenopodidea
        - Suborder Reptantia
          - Section Macrura
          - Section Anomura
          - Section Brachyura

Subclass Cephalocarida

The most recently discovered crustacean subclass comprises only four species and is called the Cephalocarida. Less than half a centimeter long, these tiny animals live in the mud–water interface of soft coastal bottoms. Their body is composed of a rounded head, a thorax with nine paired appendages, and a gradually tapering limbless abdomen (Fig. 12.11a). A

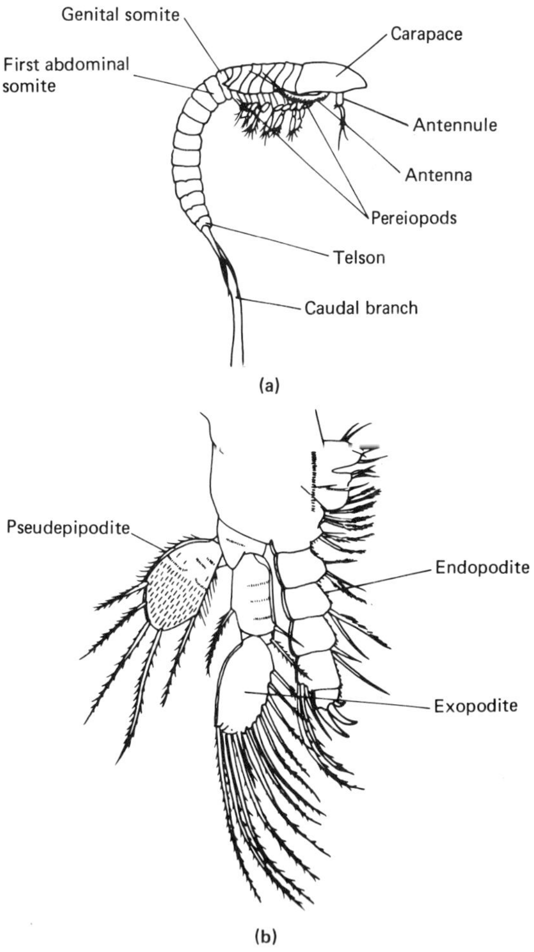

Figure 12.11 Cephalocaridans. (*a*) *Hutchinsoniella macracantha;* (*b*) A maxilla from *Lightiella incisa.* (*a* from T.H. Waterman and F.A. Chace, In *Physiology of Crustacea, Vol. 1: Metabolism and Growth*, T.H. Waterman, Ed.; *b* from R.U. Gooding, Crustaceana, 1963, 5:293.)

pair of caudal branches hangs from the telson. Cephalocarids have no eyes, probably because vision is of little value in their muddy environment. Their antennules and antennae are short; their mandibles simple. There is little or no differentiation between the maxillae and the thoracic legs. These partially flattened appendages are considered triramous, owing to a large, lateral branch on each coxopodite (Fig. 12.11b). This branch is called a **pseudepipodite** and has been likened to the trilobite preepipodite (Fig. 10.12c). All thoracic appendages bear endites which border a ventral food groove. The legs stir up detritus, which is swept along this groove and into the mouth.

Little is known about the internal biology of the cephalocarids. They are peculiar among free-living crustaceans in that they are hermaphroditic. Eggs are brooded within an external ovisac carried against the first abdominal segment. A late naupliar form, the **metanauplius**, is the hatching stage.

Following their discovery in 1955, cephalocarids became the object of considerable phylogenetic debate. The issue was whether their simple organization represented a primitive or a reduced state. Today the consensus is that these crustaceans are indeed primitive and that their ancestors played a central role in the early history of the class.

Subclass Branchiopoda

The branchiopods are another rather primitive group of small crustaceans which includes fairy shrimps, tadpole shrimps, clam shrimps, and water fleas. With few exceptions, these are freshwater organisms. Their distinguishing physical traits include a relatively uniform, segmented trunk and numerous flattened appendages (Fig. 12.12). Fringed with setae (Fig. 12.12a), these limbs are important in nutrition, respiration, and locomotion. In all but the fairy shrimps, a prominent carapace covers much of the body. Branchiopods exhibit a reproductive versatility that allows them to exploit ephemeral environments such as temperate-zone puddles and ponds. Eight hundred species are distributed among three orders: the Anostraca, the Notostraca, and the Diplostraca. The diplostracans compose two suborders, the Conchostraca and the Cladocera.

Anostracans are the fairy shrimps. With lengths of up to 10 cm, these animals are the largest, if also the most primitive, branchiopods. They have no carapace. The brine shrimp *Artemia* (Fig. 12.12b) is the most studied form. The notostracans, or tadpole shrimps, have a large dorsal carapace over the an-

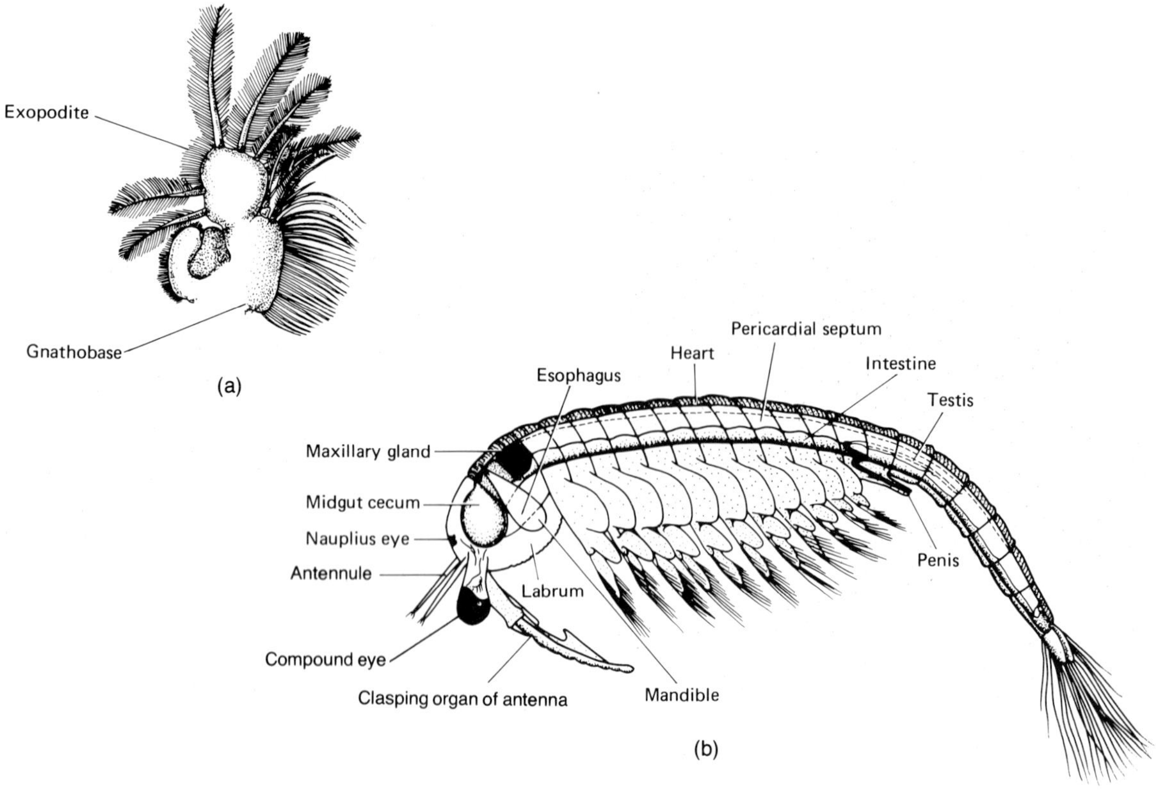

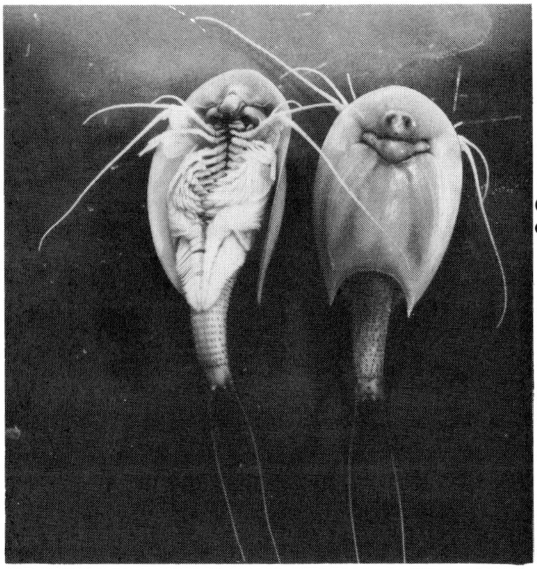

(c)

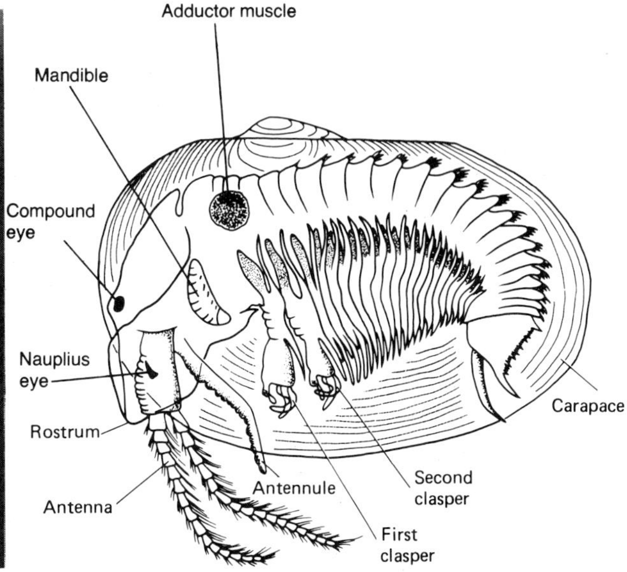

(d)

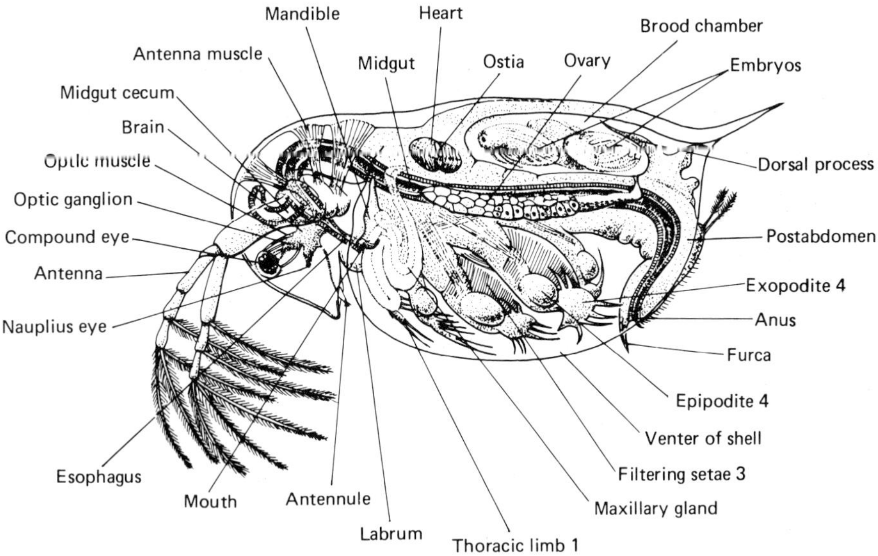

(e)

Figure 12.12 (*a*) Third thoracic appendage of the cladoceran *Simocephalus vetulas; (b)* Lateral view of the brine shrimp, *Artemia salina*, an anostracan. *Artemia* normally swims on its back; (*c*) The tadpole shrimp, *Triops longicaudatus*, a notostracan; (*d*) *Cyzicus*, the clam shrimp, a conchostracan; (*e*) Lateral view of the water flea, *Daphnia pulex*, a cladoceran. (*a* from W.F.R. Weldon, In *Cambridge Natural History*, *Vol. IV.*, S.F. Harmer and A.E. Shipley, Eds., 1959, Macmillan; b and e from Kaestner, *Invertebrate Zoology*, *Vol. III;* c from Pennak, *Fresh-Water Invertebrates of the United States; d* from W.T. Edmondson, H.B. Ward and G.C. Whipple, Eds., *Fresh-water Biology*, 1959, John Wiley.)

terior body. Beneath this carapace, up to 70 pairs of appendages occur on superannulated segments. *Triops* (Fig. 12.12c) is a common notostracan. The conchostracans are called clam shrimps because a bivalved carapace encloses most of their body. No true hinge is present, however. *Cyzicus* (Fig. 12.12d) is typical of this suborder. The water fleas of the suborder Cladocera are the most advanced branchiopods. Their body is very compact: No more than six trunk appendages are present, and except for the head region, all body parts are enclosed within a carapace. Cladocerans include the only marine branchiopods, as well as the only species that prosper in large freshwater bodies such as lakes and slow-moving streams. *Daphnia* (Fig. 12.12e) is the best-known genus.

Almost all branchiopods are filter feeders. The backward stroke of thoracic appendages draws water over the trunk's lateral surfaces. Setae along the appendage margins cull organic material (phytoplankton, zooplankton, and/or suspended detritus) from the water current (Fig. 12.13). Bound in mucus, this material is swept forward along a ventral food groove. Anterior propulsion is provided by the appendages' forward stroke. Gnathobase setae also transport food toward the mouth.

Other feeding methods characterize browsing and predacious branchiopods. Herbivores scrape plant surface tissues, while certain cladocerans, such as *Leptodora* (Fig. 12.14a), seize small animals with their enlarged first trunk appendages. In such crustaceans, sharp mandibles chew the prey. *Triops* lacks filtering setae, but feeds on accumulated sediment, algae, and larger protozoans. The branchiopod gut is a simple tube composed of an esophagus, a midgut with ceca, and an intestine (Fig. 12.14b,e).

A long tubular heart with numerous paired ostia extends through the trunk of most branchiopods. Because of their compacted body, however, cladocerans exhibit a relatively short ovoid heart with only a single pair of ostia (Fig. 12.12e). Arterial systems are not well developed in this subclass, but hemocoelic membranes often establish a definite route of blood flow. Well-irrigated trunk appendages are the principal respiratory sites. (Indeed, branchiopod means "gill foot.") The epipodites on each limb are called gills, but the broad flat branchiopod appendage is suited overall for gas exchange. In filter-feeding species, the nutritional current ventilates these respiratory areas. Those species living in stagnant waters are more apt to have hemoglobin. Some branchiopods, including *Daphnia,* apparently lack this pigment when oxygen supplies are adequate, but manufacture it when oxygen becomes scarce.

Maxillary glands handle branchiopod excretion. In notostracans and diplostracans, excretory tubules coil through the carapace before opening through a pore on the second maxillae. As in most freshwater crustaceans, the gill is a site of active salt uptake. At the other environmental extreme, the brine shrimp *Artemia* flourishes in Utah's Great Salt Lake, where salt concentrations approach the saturation point. Survival there requires salt excretion from the gills and anus and production of a hyperosmotic urine by the maxillary glands.

The branchiopod nervous system remains relatively unconcentrated and ladderlike (Fig. 12.7a). Sense or-

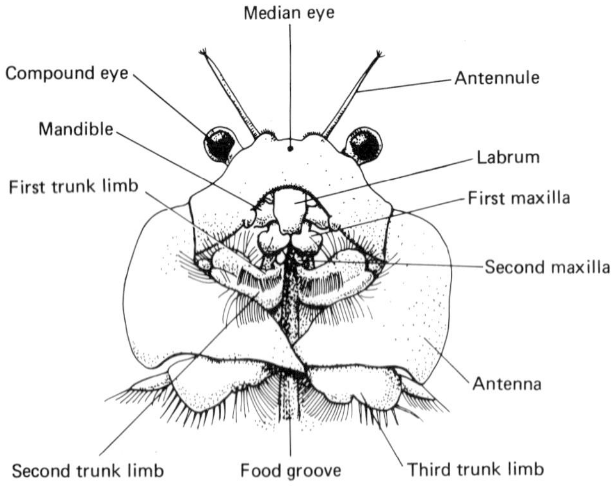

Figure 12.13 Ventral view of the anterior end of a male *Artemia salina* (order Anostraca) showing appendages used in filter feeding. (From J. Green, *A Biology of Crustacea*, 1961, Witherby.)

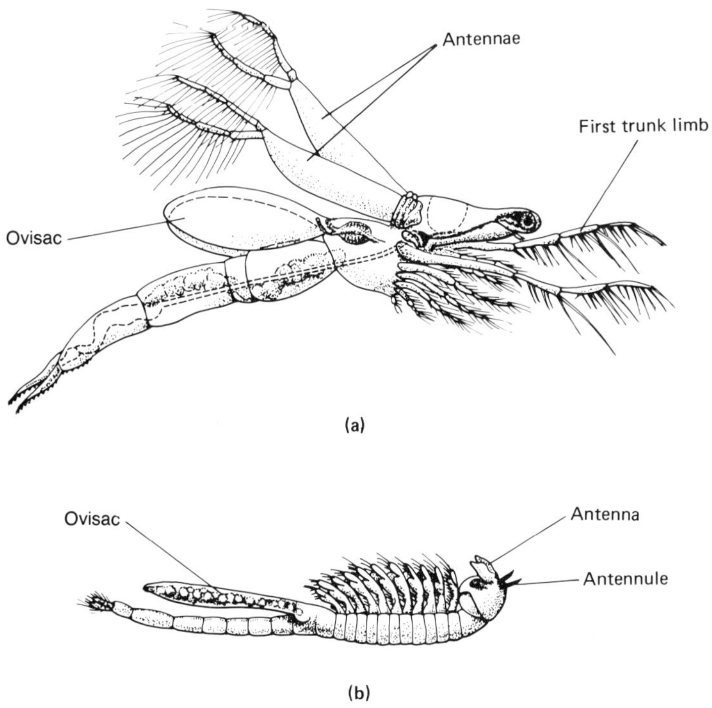

Figure 12.14 (*a*) A female *Leptodora*, a cladoceran, with ovisacs; (*b*) Female fairy shrimp, *Branchinecta*, an anostracan. (*a* from Pennak, *Fresh-Water Invertebrates of the United States; b* from W.T. Calman, In *A Treatise on Zoology, Part VII,* E.R. Lankester, Ed., Stechert-Hafer Service Agency.)

gans include compound and nauplius eyes, numerous tactile sites, and chemoreceptors. Anostracans have stalked compound eyes (Fig. 12.12b), while in the cladocerans these organs have fused into a single medial complex (Fig. 12.12e). The **frontal organ** on the head of many branchiopods is probably chemosensitive. In all branchiopods, the antennules are reduced. The second antennae, however, are greatly expanded in diplostracans, where they are important in locomotion. The powerful stroking of these appendages propels clam shrimps and water fleas through the water in a series of jerks. Other branchiopods move under power of their thoracic limbs. In such animals, the feeding, respiratory, and locomotor currents may be combined. Fairy shrimps swim on their backs (Fig. 12.14b); tadpole shrimps swim dorsal side up. Bottom crawling occurs in some members with tubular endopodites, while certain notostracans use their carapaces to plow through sediment.

Paired branchiopod gonads in the trunk usually open through simple ducts onto a genital sternite. Once again, however, cladocerans are exceptional in that the gonopores are subterminal in the male and dorsal in the female. In males, either the second antennae or a pair of anterior trunk appendages is modified for grasping during copulation. Fairy shrimps have a pair of penes; males wrap their long bodies around the female until each penis contacts one of her gonopores. In other branchiopods, sperm are released beneath the female's carapace.

Brooding is general in the subclass. The fairy shrimp mother secretes an ovisac which she fills with developing eggs (Fig. 12.14b). Tadpole shrimps retain embryos against abdominal appendages, and clam shrimps have a brooding space within their bivalved carapace. In water fleas, the female's dorsal gonopore opens into a brood chamber enclosed by the carapace. Cladoceran young remain within this dorsal chamber until they emerge as miniature adults. In other branchiopods, a nauplius larva is often the first free-living stage.

Additional aspects of branchiopod continuity adapt the subclass well for freshwater life. Perhaps because branchiopods are potentially easy prey for fish and other aquatic carnivores, they are often most prosperous when isolated in temporary freshwater bodies. Living in such ephemeral environments, however, requires tolerance toward recurrent droughts and seasonal temperature cycles. Rapid reproduction is also necessary during the brief life of the puddle or pond, and some mechanism for dispersal is needed.

Branchiopod adaptations to these conditions are

reminiscent of those described for some rotifers (see Chap. 7). Parthenogenesis is common during optimal growth periods. Parthenogenetic eggs are thin walled and require very little incubation. According to the species and its local conditions, huge parthenogenetically produced populations may be established during the summer months, during two peak periods in the spring and fall, or whenever the environment permits. When conditions become less favorable, sexual branchiopods are produced. Their fertilized eggs are surrounded by thicker shells which retard development. In cladocerans, the entire dorsal brood chamber may be covered by a thick envelope. When the young mother molts, this protective chamber detaches with its load of embryos.

Dormant branchiopod eggs are extremely resistant to drying, freezing, and baking. They may be transported by wind or in patches of mud stuck to the feet of insects, birds, and mammals. Some eggs reportedly survive ingestion by birds and later hatch in the latter's feces. When favorable conditions recur, rapid development produces a new generation of branchiopods which reproduce by parthenogenesis. This capacity for dispersal and rapid population growth explains the sudden appearance of these crustaceans in isolated, temporary pools. Before this process was fully understood, such magical appearances caused the anostracans to be called fairy shrimps.

Subclass Ostracoda

The ostracods, or mussel shrimps, are entomostracans that superficially resemble bivalve mollusks. Like the conchostracans, the ostracods have a bivalved carapace which encloses the entire body (Fig. 12.15a). Fossilized remains of these carapaces have allowed the identification of about 10,000 extinct species, from the Cambrian onward. Ostracod fossils are often associated with oil deposits, a factor that motivates their study. Today the subclass lists 2000 living species, two-thirds of which are marine. Shallower, organically rich waters are preferred by the ostracods, although some are planktonic or bathyal. Freshwater species are widely distributed, and a few ostracods in New Zealand and southern Africa inhabit moist forest floors.

Several adaptive trends typical of the evolution of higher crustaceans are seen in the Ostracoda. The body has shortened through the loss and/or fusion of ancestral segments, and considerable differentiation of the appendages has occurred (Fig. 12.15b). Fused skeletal plates form the all-encompassing carapace.

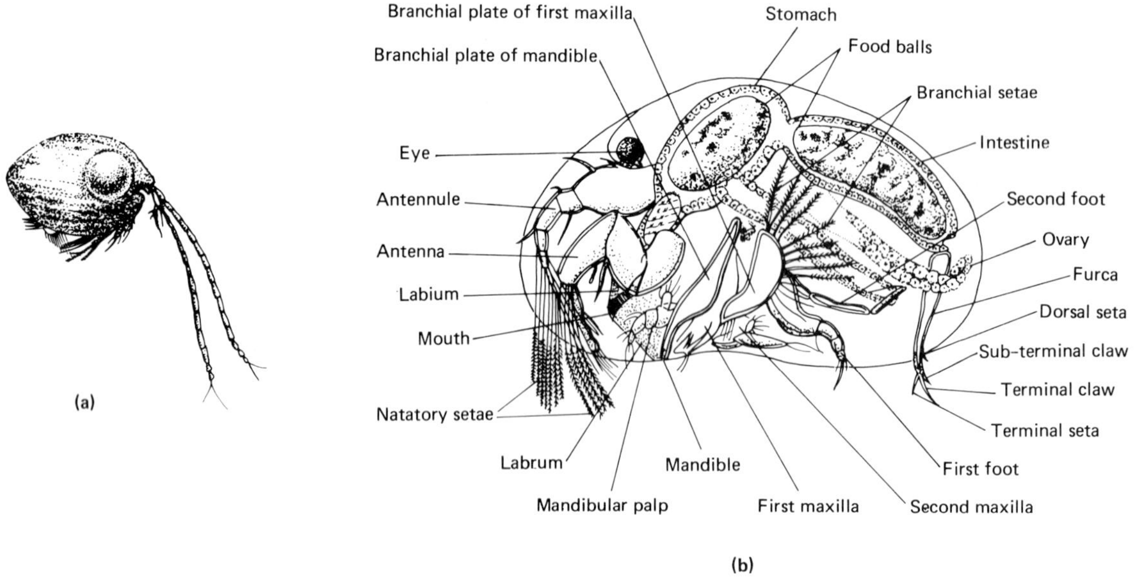

Figure 12.15 (*a*) A myodocopid ostracod; (*b*) Anatomy of *Cypris virens*, and ostracod. (*a* from W. Schmitt, *Crustaceans*, 1965, University of Michigan; *b* from W.T. Edmondson, H.B. Ward, and G.C. Whipple, Eds, *Freshwater Biology*, 1959, John Wiley.)

This bivalved shield has a true dorsal hinge of uncalcified cuticle. Adductor muscles close the carapace valves and stretch the hinge. Interlocking ridges along the carapace margins help maintain the closed position, but when adductor muscles relax, the elasticity of the hinge reopens the valves. The carapace has both inner and outer walls, with an intermediate hemocoelic space.

The ostracod head dominates the body. Both pairs of antennae are large and often plumose. In some ostracods, these appendages can be extended through carapace notches even when the valves are closed (Fig. 12.15a). Grinding mandibles surround the mouth and are followed by paired maxillae. The second maxillae may have thick setae, but frequently they are leglike and resemble the thoracic limbs. The ostracod trunk is extremely reduced. Segmentation is diminished, and no more than two pairs of appendages are present. The second maxillae and trunk appendages are modified for locomotion, food procurement, valve maintenance, and sexual embrace. A pair of caudal branches hangs from the posterior body.

Many mussel shrimps are filter feeders. Maxillary and thoracic appendages draw a current forward over the oral region, where setae on the mandibles, maxillae, and/or first trunk legs harvest suspended organic material. Plant and animal matter, both living and dead, is coated with mucus and then transferred to the mandibles for chewing. In the course of feeding, a good deal of mud is often swallowed.

In some ostracods, the antennae or mandibles grasp small invertebrates, which then are chewed and eaten. Most members of the subclass are only about 1 mm long, so ostracod prey is always rather tiny. An exception is *Gigantocypris,* a pelagic form 30 mm in length; this ostracod may feed on small fish. Depending on the diet, the foregut may include a grinding chamber or chitinous ridges which strain particulate matter. The midgut comprises a large stomach often bearing digestive ceca.

In one primitive marine order (Myodocopa), both gills and a dorsal heart are present. Most ostracods, however, lack specialized respiratory and circulatory structures. Integumental gas exchange usually suffices in the subclass, while blood circulation between the carapace walls provides adequate internal transport. The nutritional current in filter feeding ostracods aids both of these processes. Excretion involves antennal or maxillary glands, or both.

Despite their expanded heads, the ostracod central nervous system is not greatly concentrated. Sense organs include eyes and numerous, scattered hairs. Ostracods may have a single nauplius eye, a pair of simple ocelli, or compound eyes; some have no visual organs at all. One or both pairs of antennae usually provide locomotor power. In swimming ostracods, the antennal surface area is increased by long lateral hairs. Burrowing species have stocky first antennae, which they plunge into the substratum. Such ostracods include terrestrial forms that plow through humus. Kicking thoracic legs augment antennal efforts in crawling forms, and contractions of the posterior trunk also push the animal along. The carapace is quite thick in bottom-dwelling ostracods, but swimming species have lighter, thin-walled shields. Some ostracods climb on plants by using antennal hooks modified from setae. Their thoracic limbs may also be clawed.

In certain marine members of the subclass, labral glands produce a bluish, bioluminescent secretion. When water is added to dehydrated minced *Cypridina,* light strong enough for reading may be produced.

Ostracod females have a pair of trunk ovaries, tubules with dilated seminal receptacles, and usually paired gonopores. These openings lie between the last pair of trunk appendages. The male system includes paired testes, each of which may comprise four long tubules. These tubules discharge into a common sperm duct, which leads to an ejaculatory apparatus. Paired cuticular penes are located at the trunk posterior. Only one penis is present in species with a single female gonopore (Conchoeciidae).

During copulation, the male holds the female with his second antennae or with specially modified claws on his second maxillae. Some ostracods produce enormous sperm, which may be longer than the adult male himself. Indeed, the largest sperm in the world belong to the Australian *Platycypris.* These male gametes are 10 mm long; in comparison, human sperm measure 0.06 mm.

Philomedes has an interesting copulatory schedule. The short-lived males of this genus are planktonic, while females spend most of their lives on the seafloor. *Philomedes* females do not acquire natatory setae until the sexual molt. Then they join the males in an orgiastic swarm. Following copulation, the female's second maxillae cut off her swimming hairs, and she returns to the bottom.

After fertilization, ostracod eggs are usually attached to plants or other bottom material. Freshwater mussel shrimps are brooded within the parental valves and a modified nauplius larva with a protective bivalved carapace is the hatching stage. Freshwater ostracods are shorter lived than their marine cousins. As among the branchiopods, both parthenogenesis and overwintering eggs are common.

Subclass Copepoda

Most entomostracans belong to the Copepoda, a subclass that contains 4500 species. Copepods are extremely abundant in both marine and freshwater habitats, and a few terrestrial forms live among moss and moist humus. One-fourth of the subclass is parasitic. According to some estimates, there are more individual copepods than all other organisms combined. The different species of the nearly cosmopolitan genus *Calanus* (Fig. 12.16a) could be the most populous animals in the world. Copepods are the most important primary consumers of the open sea and are themselves consumed by larger invertebrates and fish. Freshwater copepods, many of which belong to the genus *Cyclops* (Fig. 12.16b), play an analogous role in the food chains of their environment.

Free-living members of the subclass are seldom more than 2 mm long. Their rounded heads include five segments bearing the typical cephalic appendages and one or two trunk somites (Fig. 12.16b). Long antennules are modified for locomotion, food capture, and copulatory embrace. Sensory second antennae are often reduced. Mandibles flank the mouth, and both pairs of maxillae have flattened gnathobases. The next segment bears the **maxillipeds,** paired appendages modified for feeding. Three to five thoracic segments have locomotor legs, some of which may be specialized for sexual activity. The limbless abdomen tapers posteriorly through four segments and concludes with two long caudal branches.

Rotating cephalic appendages establish a feeding current which swirls about the oral region (Fig. 12.16c). These limbs may beat hundreds of times per minute in actively feeding copepods. As the current eddies anteriorly across the second maxillae, a fine setose net filters out phytoplankton and some protozoans (Fig. 12.16d). The endites of the maxillipeds and first maxillae transfer trapped food particles to the mandibles and mouth. Some bottom-dwelling copepods stir detritus into suspension and then filter it for food. The few predacious species capture small invertebrates with their first maxillae. Aided by the antennules and posterior mouthparts, these spiny appendages transfer food (often oligochaetes) to the chewing mandibles. Copepods that eat large algae feed in a similar way.

The foregut consists simply of a muscular, swallowing esophagus. Midgut ceca are rare. Such alimentary simplicity is expected in animals that consume small, finely filtered food.

Over 1000 copepod species are parasitic. These animals tend to be larger and more vermiform than their free-living cousins (Fig. 12.17a). One species of *Pennella,* a whale parasite, ranges up to one-third of a meter in length. Various parasitic styles are seen in this subclass. Ectoparasitism on the gills of fish is particularly common (Fig. 12.17b). Such parasites wound host gills with their sharp head appendages. Their labrum and labium form a tubular oral lip which is inserted into the wound; within this lip, the mandibles cut away at the gill tissues. In other ectoparasites, the maxillae and/or hooked second antennae grasp the host. Copepod endoparasitism is less common, but involves numerous hosts, including polychaetes, echinoderms, mollusks, and ascidians. Thoracic limbs may be reduced in ectoparasites, while all appendages have atrophied in endoparasitic species. The most highly adapted of these copepods have no mouthparts at all, and nutrients are absorbed across their integument.

Other aspects of copepod maintenance are rather simplified. A small heart is present in one order, but most copepods lack a formal circulatory system. Gut peristalsis enhances hemolymph circulation. Gills are absent, as gas exchange takes place across the integument. Anal respiration has been described in a few genera. Antennal glands are active during the larval stages, but in adult copepods maxillary glands handle excretory functions.

In most members of the subclass, the ventral nerve cord and its segmental ganglia have fused medially, and shortening of the trunk has produced some longitudinal consolidation (Fig. 12.18). The most concentrated nervous systems occur in parasitic forms. Sense organs include tactile sites and chemoreceptors on the appendages and a medial eye. Derived from the larval nauplius eye, the latter comprises a cluster of three ocelli. Compound eyes are unknown among the copepods. Esthetascs are located on the first antennae.

Copepods float, swim, and crawl. Planktonic species swim with their setose thoracic appendages. The plumose antennules aid in flotation, as do feathery extensions of the caudal branches (Fig. 12.19). In addition, the cephalic feeding current has some locomotor effect. Diurnal vertical migrations are common. Copepods visit surface waters during safer nocturnal hours, then sink from harsh sunlight and predators during the daytime. Apparently, water pressure is the key factor in determining depth, while light provides the orienting stimulus. Benthic copepods, whose antennules are always smaller and less elaborate than those of swimming species, walk on thoracic legs. Some tiny vermiform copepods undulate through the spaces between sand grains.

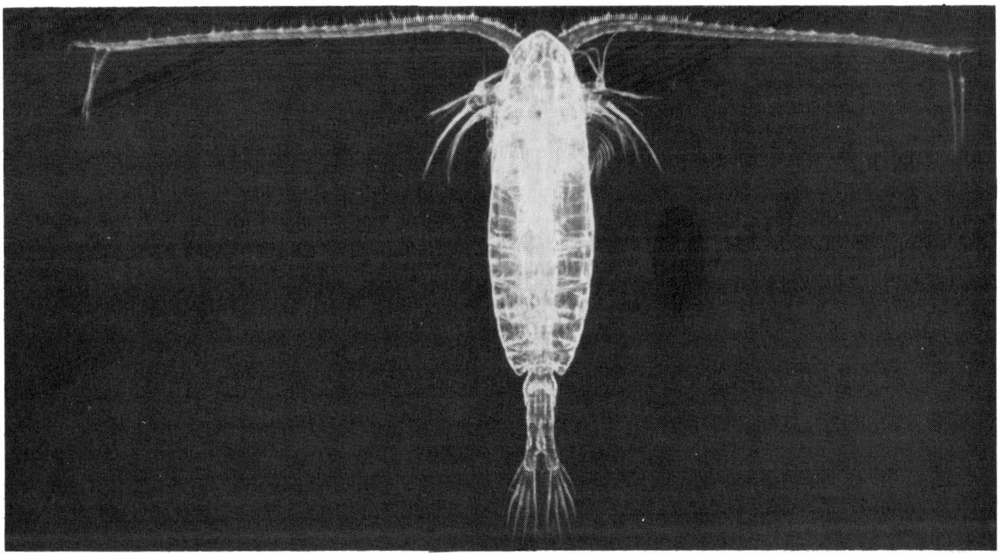

(a)

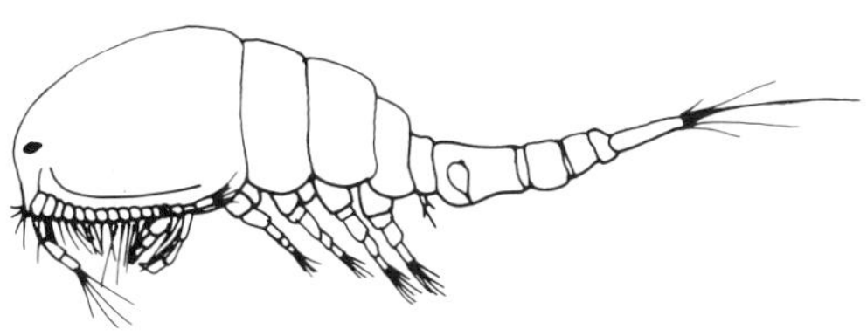

(b)

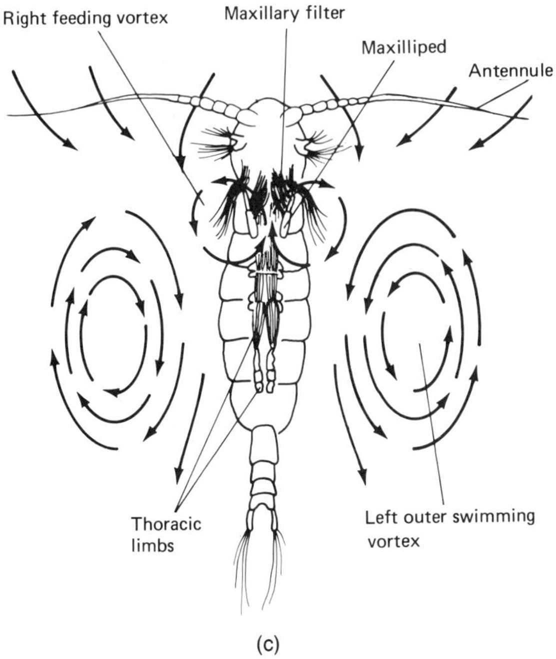

(c)

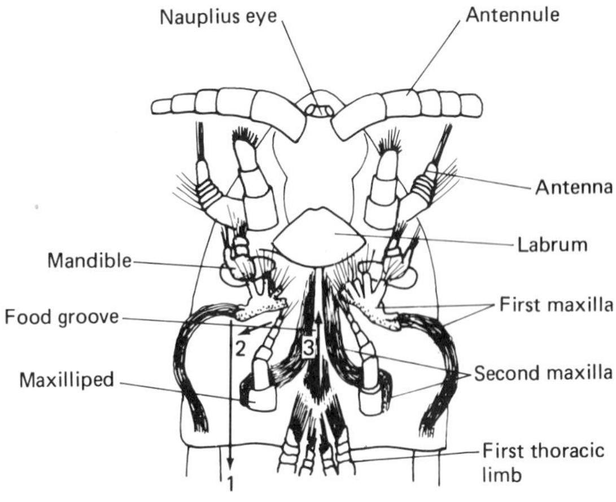

(d)

Figure 12.16 Copepods. (*a*) *Calanus helgolandicus*. Adults are about the size of rice grains; (*b*) Lateral view of female *Cyclops;* (*c*) Diagram of the vortices produced by a swimming *Calanus*. The vortices bring food particles into the midline and through the maxillary filter; (*d*) Anterior portion of *Diaptomus* in ventral view. The second maxillae have medially directed filter nets overlaid by the maxillipeds. Arrows indicate major water currents: 1 = main current; 2 = filtered water drawn toward main current; 3 = food current drawn toward food groove. (*a* courtesy of D.P. Wilson; *b* from Pennak, *Fresh-Water Invertebrates of the United States; c* from W.D. Russell-Hunter, *A Biology of Lower Invertebrates*, 1968, Macmillan; *d* from Kaestner, *Invertebrate Zoology*, *Vol. III.)*

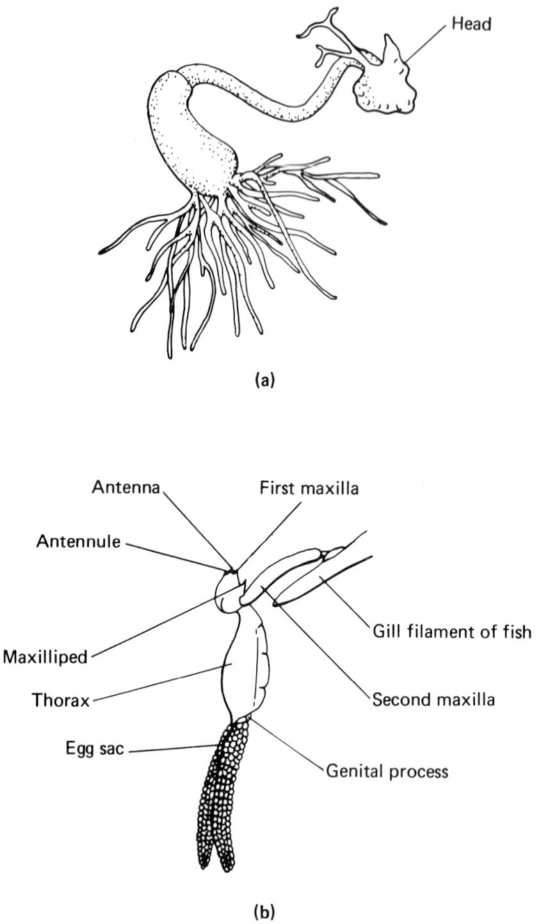

Figure 12.17 Parasitic copepods. (*a*) *Lernaeolophus sultans*, an ectoparasite of marine fish; (*b*) *Salmincola salmonea*, a fish gill parasite. (*a* from S. Yagamuti, *Parasitic Copepoda and Branchiura of Fishes*, 1963, John Wiley; *b* from Waterman and Chace, In *Physiology of Crustacea, Vol. I.*, Waterman, Ed.)

Copepod ovaries are single or paired. Oviducts bear lateral diverticula, where eggs are prepared for fertilization; paired gonopores occupy the first abdominal segment. This segment also bears the openings of the seminal receptacles, paired pouches connecting internally with the rest of the female system. Most copepod males have a single testis. Near the male gonopore, which also occupies the first abdominal segment, the sperm duct is dilated and glandular. This region is the site of spermatophore production.

Male copepods seize female partners with their enlarged antennules and/or maxillipeds. In addition, claws on the last thoracic appendages may grasp the female abdomen. These thoracic limbs then transfer a

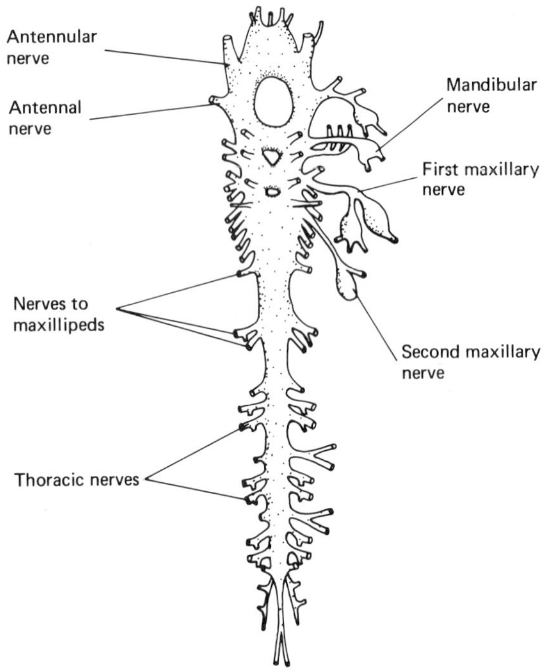

Figure 12.18 Central nervous system of the copepod *Calanus*. (From S.M. Marshal and A.P. Orr, *The Biology of a Marine Copepod: Calanus finmarchicus*, 1952, Springer-Verlag.)

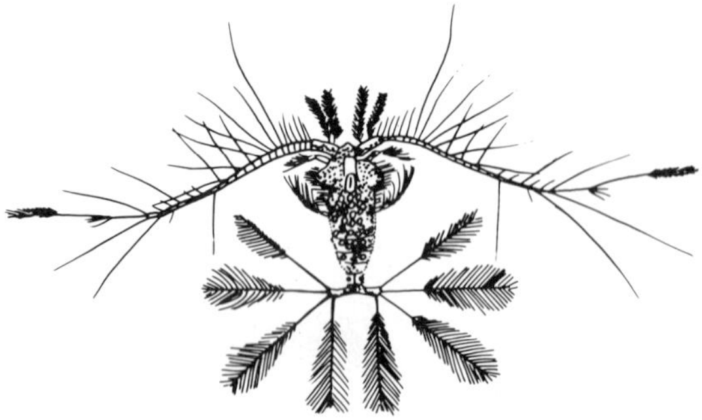

Figure 12.19 *Calocalanus pavo*, a planktonic copepod with well-developed furcal branches. (From T.J. Parker and W.A. Haswell, *Textbook of Zoology, Vol. I*, 1921, Macmillan.)

spermatophore to the openings of the female's seminal receptacles. Adhesive secretions hold the package in place as sperm enter her body.

Fertilization and egg deposition may be delayed for several days. Many planktonic species simply discharge their eggs into the water, but others secrete a brooding ovisac. The egg-filled ovisac may be released or may be carried by the female's thoracic limbs (Fig. 12.20a). A nauplius larva (Fig. 12.20b) hatches and eventually develops into a juvenile **copepodid** (Fig. 12.20c). The copepodid is adultlike, but its trunk is not fully formed. Usually the nauplius passes six instars before the copepodid stage, and five more molts precede adulthood. External segmentation becomes apparent at the first copepodid molt, and sexual dimorphism is evident by the third instar. Adult copepods do not molt.

Freshwater adaptations include thick-walled eggs, which are dormant during periods of environmental severity. Also, copepodids and adults may survive decision in an encysted state.

Parasitism always involves adjustments in continuity systems, and the copepods are no exception. Marked sexual dimorphism is common, because the female of a species often becomes more parasitic than the male. In ectoparasitic copepods, the male may roam over the host body, while the female remains permanently attached. In some species, the male is free-living except during breeding periods; in others, he may be diminutive and live on the female body. Such dimorphism facilitates sexual reproduction and thus parallels the hermaphroditism common among other parasitic groups. In most copepod parasites, free-living larval stages allow species dispersal. Often the copepodid is the hatching stage. In one order (Monstrilloidea), however, the first nauplius and the adult are free-living, while all intermediate stages are endoparasites in polychaetes.

Subclasses Mystacocarida and Branchiura

Two small entomostracan subclasses resemble the copepods in some ways. Whether this resemblance reflects a common history or merely evolutionary convergence is not clear. Mystacocarida comprises only three species, which live among coastal sand grains. The Branchiura are an ectoparasitic group with 75 species.

The vermiform mystacocaridan body is less than half a millimeter long (Fig. 12.21a). Large and setose antennules resemble those of the swimming copepods, mandibles are elongate, and the first trunk appendages are maxillipeds. The remaining four pairs of trunk legs are extremely short and flattened. The limbless posterior body terminates in a telson with short caudal extensions. Mystacocaridans eat detritus

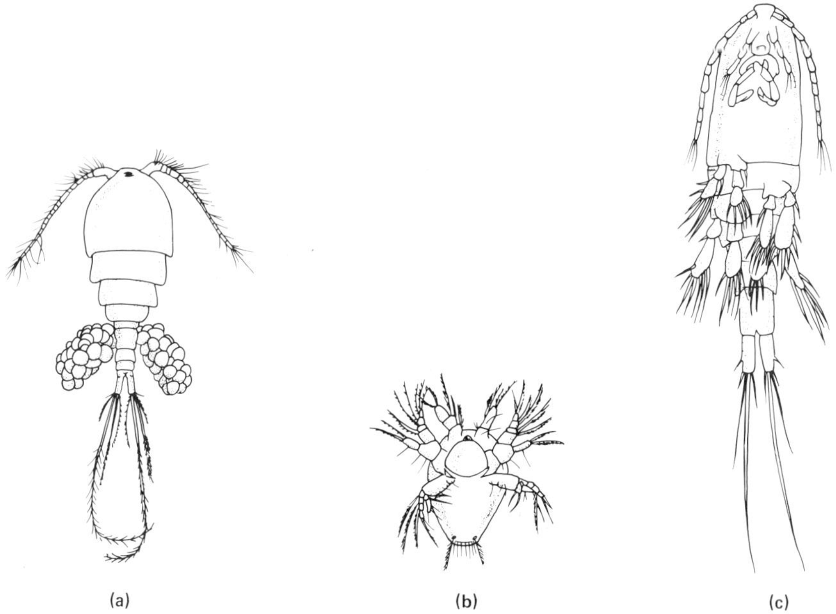

Figure 12.20 (*a*) Female *Macrocyclops ater* with ovisacs; (*b*) A nauplius larva of *Cyclops*; (*c*) Copepodid stage of *Cyclops*. Only alternate legs are shown on each side. (From Pennak, *Fresh-Water Invertebrates of the United States*.)

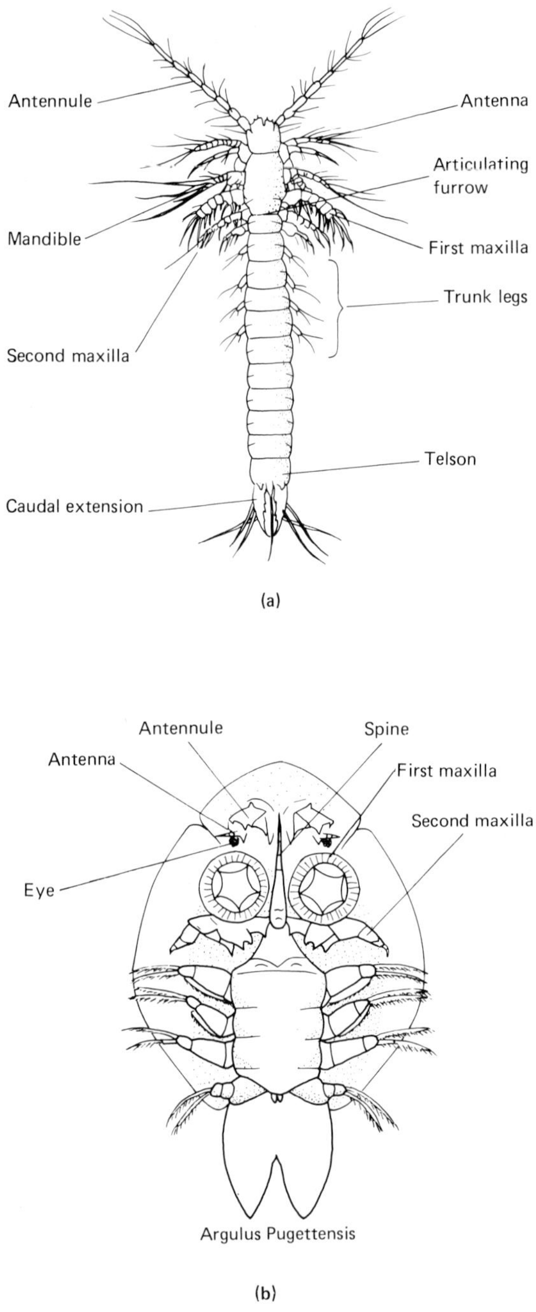

Figure 12.21 (*a*) Dorsal view of *Derocheilocaris remanei*, a mystacocaridan. (*b*) Ventral view of *Argulus pugettensis*, a branchiuran. (*a* from Kaestner, *Invertebrate Zoology, Vol. III*; *b* from S. Yamaguti, *Parasitic Copepoda and Branchiura of Fishes*, 1963, Wiley-Interscience.

and microorganisms, which their mouthpart setae brush from sand grains. Little is known of their internal biology. Apparently, they have no heart and no midgut ceca. The cuticle is thin, and diffusion must contribute to respiration and excretion. Curiously, the second antennae and the mandibles are walking limbs. A nauplius eye persists in the adult. Mystacocarids are diecious, and their eggs hatch as nauplius larvae.

The branchiurans are ectoparasites on fish and amphibians. They share several characters with the parasitic copepods, but are distinguished by their compound eyes and a large carapace, which encloses most of their dorsoventrally flattened body (Fig. 12.21b). Both pairs of antennae are reduced and, like other cephalic appendages, may be modified for parasitic attachment. Terminal claws are common on the antennules and the second maxillae. The first maxillae may form suckers, while the fused labrum and labium form a tubular entrance to the digestive tract. Some *Argulus* (Fig. 12.21b) have a hollow, reportedly poisonous spine just anterior to the mouth. Branchiurans grip their host's skin or gills and suck out blood and mucus. Large midgut ceca store huge fluid meals. A heart with a single pair of ostia is present, as is an anterior aorta. Branchiuran blood contains hemoglobin.

Maxillipeds are not differentiated, and the four pairs of trunk appendages are large and setose. With these limbs, branchiurans swim from one host to another and make sexual contacts. Sexes are separate, and copulation takes place on the host body. Branchiuran eggs develop on the bottom and usually do not hatch until all appendages are functional. The young crustaceans then swim upward to find a host.

Subclass Cirripedia

Rocky coastlines and harbors throughout the world are inhabited by populations of barnacles. These sessile marine creatures were classified as mollusks until well into the nineteenth century, but today they are recognized as the crustacean subclass Cirripedia. Superficially, barnacles do resemble some mollusks in their possession of a thick calcareous shell. A study of larval stages and internal anatomy, however, clearly reveals their crustacean heritage. Almost 900 species have been described; one-third of them are parasitic.

Barnacles are very unusual crustaceans. Their sessile life-styles are supported by a peculiar body orientation and hermaphroditic reproductive systems. Because many barnacles will attach to living as well as nonliving substrates, parasitism has evolved several times in the subclass. Parasitic modifications can be extreme, so we will treat the freeliving and parasitic barnacles separately.

THE FREE-LIVING BARNACLES

To understand the form of the adult barnacle, a late developmental stage called a **cypris larva** (Fig. 12.22a) must be examined. A bivalved carapace envelops the cypris, which looks very like an ostracod. When this free-swimming larva settles, cement glands near the antennules secrete adhesives which attach its preoral end (Fig. 12.22b). The entire body then bends 90°, until its main axis is nearly parallel with the substratum and the legs are vertical (Fig. 12.22c).

The larval valves define the contours of the adult **mantle.** The inner surface of this mantle borders a spacious cavity around the central body (Fig. 12.23). Externally, the mantle secretes calcareous plates that form the definitive shell. Within this mollusk-like shell, the barnacle body retains its modified crustacean character. The large cephalic area alone is attached to the mantle wall, and the trunk extends freely into the mantle cavity. Cephalic appendages include a pair of mandibles and bladed first maxillae. A labrum is

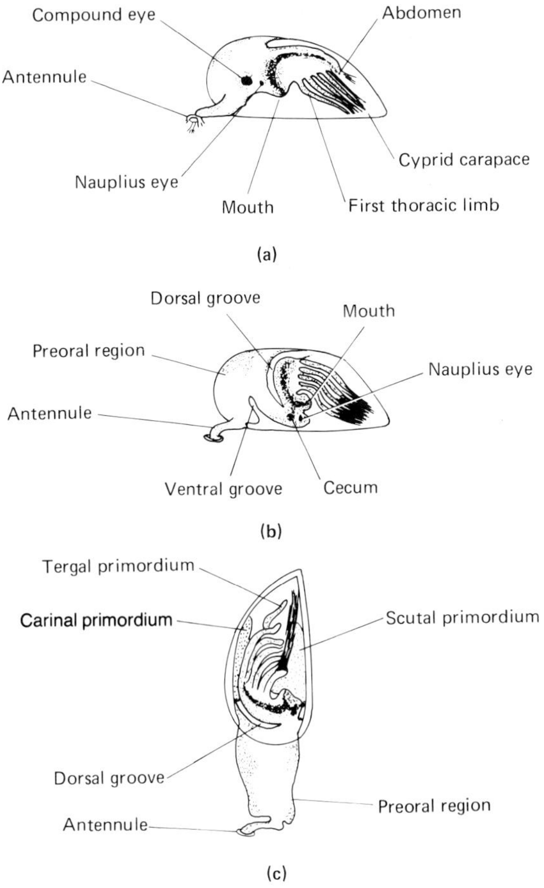

Figure 12.22 View of left side of a gooseneck barnacle, *Lepas*, at various stages of metamorphosis. (*a*) A free-swimming cypris larva; (*b*) An attached cypris undergoing metamorphosis; (*c*) Later stage: The cypris has not yet shed the carapace, but the body has completed its 90° rotation. The preoral region has formed a stalk and some primordial shell plates have appeared. (From Kaestner, *Invertebrate Zoology, Vol. III.*)

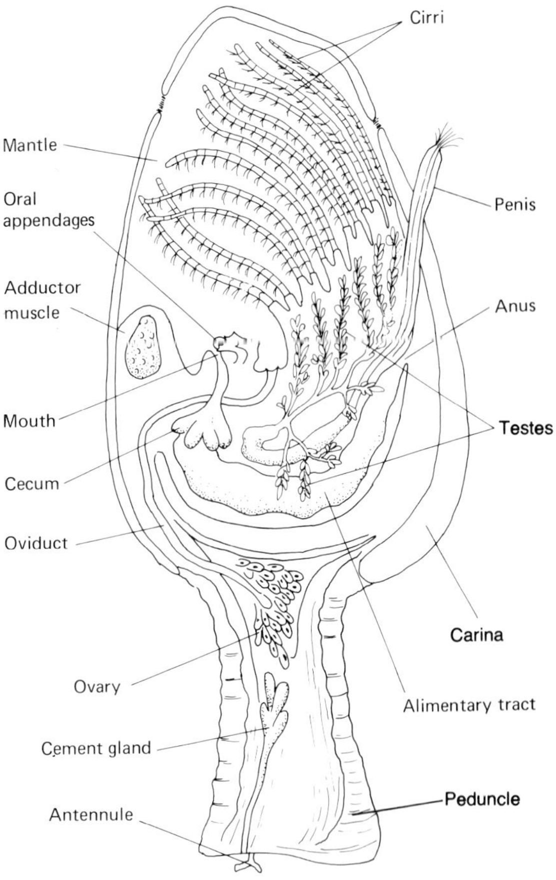

Figure 12.23 The anatomy of a gooseneck barnacle. (From V. Schechter, *Invertebrate Zoology*, 1959, Prentice-Hall.)

prominent, and fused second maxillae form a posterior lip. Six pairs of thoracic legs are well developed and can be extended through an apical opening in the barnacle shell. Each of these limbs comprises two narrow branches, or **cirri,** which bear long sweeping setae. The posterior trunk is degenerate.

A barnacle's external appearance reflects the number and arrangement of its shell plates. A consideration of shell structure requires the learning of several specialized terms. Both stalked and stalkless barnacles occur. The stalked (or gooseneck) barnacles are attached to the substratum by a **peduncle,** which is formed by an elongation of the preoral end (Fig. 12.25a). Distally, this stalk supports a calcareous framework, the **capitulum,** which houses the mantle and central body. Originally, the capitulum probably had paired valves similar to those of the cypris larva. Such gaping valves, however, would make barnacles extremely vulnerable to predation. Accordingly, capitular adaptations have afforded better protection through fine control over the shell aperture.

In one primitive line, the shell is supported by a dorsal keel, the **carina.** Two paired plates, the **terga** and **scuta,** meet apically to flank the shell aperture. The size of this opening is controlled by adductor muscles between the scuta. Barnacles with this type of shell are called lepadids. *Lepas* (Fig. 12.24a), the common goose barnacle, is representative of these crustaceans. Its long, fleshy peduncle commonly attaches to floating objects or fastens commensally to larger nektonic animals. In commensal lepadids, such as *Conchoderma* (Fig. 12.24b) on whales and *Alepas* on jellyfish, the capitular plates are often reduced.

In scalpellid barnacles, the peduncle is armored with whorls of calcareous plates. These plates also surround the base of the capitulum, where they include a large carina and its opposing **rostrum** (Fig. 12.25a). The rostrum lies outside the area of head attachment to the mantle wall. A variable number of paired lateral plates join the carina and rostrum in a calcareous ring at the base of the scalpellid capitulum. This basal ring of plates displaces the terga and scuta to a more apical position. *Pollicipes polymerus* (Fig. 12.25a) is a common scapellid barnacle on our Pacific coast. Other members of this line include *Scalpellum* and *Lithotrya*, a barnacle that bores in coral rock.

(a)

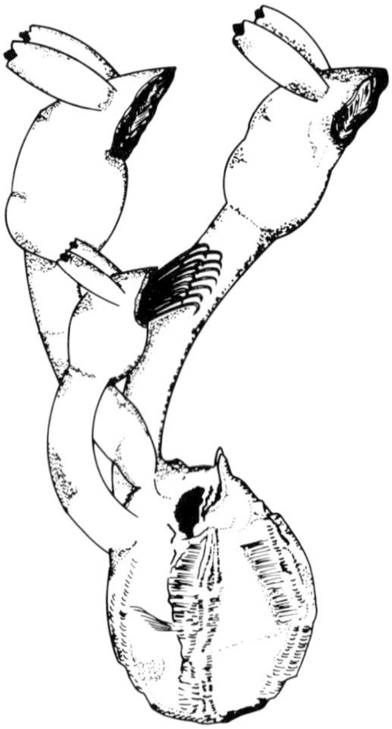

(b)

Figure 12.24 (*a*) A cluster of *Lepas*; (*b*) Three rabbit-eared barnacles, *Conchoderma auritum*, attached to an acorn barnacle, *Coronula diadema*, which attaches in turn to the skin of a whale. (*a* courtesy of S. Arthur Reed; *b* from G.E. MacGinitie and N. MacGinitie, *Natural History of Marine Animals*, 1968, McGraw-Hill.)

The scalpellids probably gave rise to stalkless barnacles. The latter have no peduncle, and a flattened preoral disk (or basis) is attached directly to the substratum. Stalkless barnacles, or balanomorphs, present lower profiles, which better withstand the pounding waters of rocky intertidal and subtidal zones. The balanomorph shell is dominated by a ring of plates including the carina, carinolaterals, laterals, rostrolaterals, and rostrum. Fusion among these plates is common. For example, in the acorn barnacle *Balanus*, both rostrolateral plates have fused with the rostrum (Fig. 12.25b), and in *Creusia* a similar fusion has occurred among the plates flanking the carina. Finally, in *Pyrgoma* the plates form an unbroken ring. In many balanomorphs, including some species of *Balanus* (Fig. 12.26), the walls of the rostrum and carina may extend above the terga and scuta; thus the apical plates occupy a protective central vestibule.

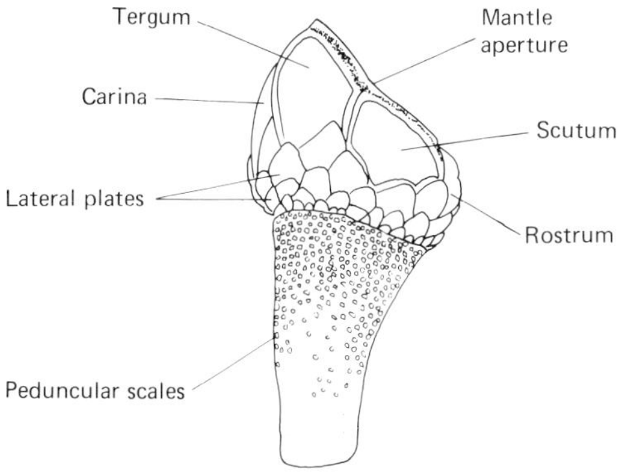

(a)

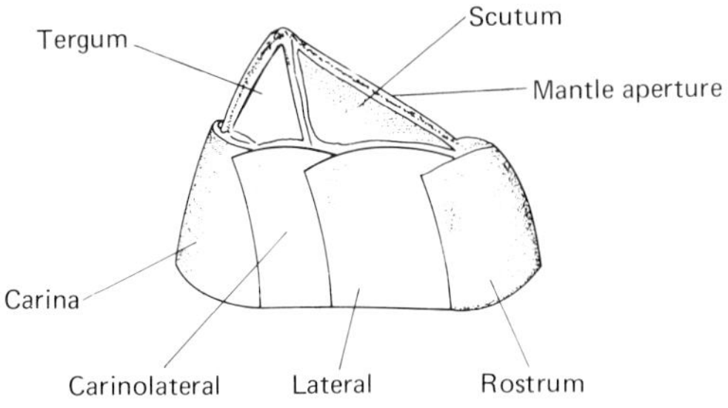

(b)

Figure 12.25 Comparison between scalpellid (*a*) *Pollicipes polymerus*, and balanomorph (*b*) *Balanus*, barnacles. The terga and scuta flanking the aperture are drawn with a double border to facilitate comparison. (From Kaestner, *Invertebrate Zoology, Vol. III.*)

Balanomorphs have adapted along several divergent lines. Some species live on intertidal algae, while others are commensal on fish, sea turtles, and whales. The latter often have modified or reduced shell plates. In *Coronula,* for example, the shell ring is deeply folded and is imbedded in the dermis of the whale host (Fig. 12.24b). Another whale barnacle, the elongated *Xenobalanus,* has almost no skeleton at all. These changes in commensal barnacles foreshadow parasitic adaptations in the subclass.

An aberrant group of cirripedians, the Acrothoracia, lacks all vestiges of a calcareous shell. These unusual barnacles employ enzymes and mantle spines to bore through coral and mollusk shells.

Free-living barnacles gather food with their biramous, setose legs. When the terga and the scuta are withdrawn, these limbs extend and sweep rhythmically, generating a feeding current from which plankton are filtered. Alternatively, the legs may be held motionless against the sea current. In *Balanus,* only the first two legs serve as filters. The four posterior pairs, whose setae are spaced too widely to trap plankton, generate the feeding current. In every case, the anteriormost appendages collect all trapped food particles and transfer them to the mouth. Some larger barnacles, such as *Lepas,* may snare medium-sized crustaceans. Such prey is placed directly into the mouth by the capturing limb. *Conchoderma* is a stalked barnacle which attaches to another barnacle, *Coronula,* a commensal on whales (Fig. 12.24b). *Conchoderma* faces the anterior end of the whale. The whale's swimming movements generate a current that enters the barnacle's mantle, passes over the filtering legs, and exits through two ear-shaped carapace lobes.

The barnacle foregut is usually simple, although boring species have a gastric mill. A pair of ceca branches from the anterior end of the midgut.

Circulatory and respiratory mechanisms are informal. Barnacles have no heart or arterial system, but hemocoelic membranes maintain orderly blood flow through the central sinuses and the mantle wall. The muscularized borders of a large anterior sinus propel the blood. Respiratory exchange occurs across the mantle walls and along the thoracic legs. In some barnacles, such as *Lepas,* the exchange area is increased by folds in the mantle wall near the shell aperture. Intertidal species protect their respiratory surfaces by tightly sealing the shell when the tide is out.

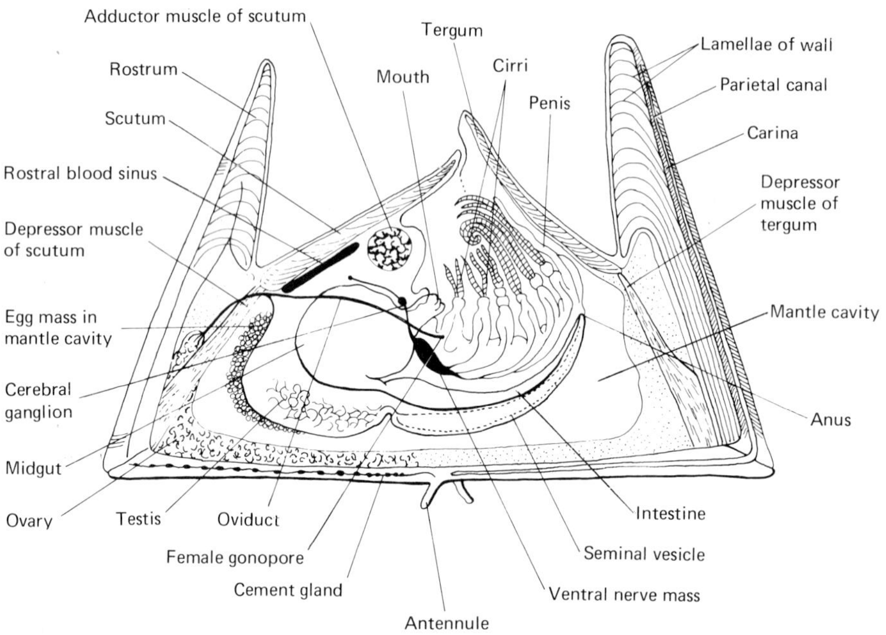

Figure 12.26 Vertical section showing major anatomical features of an acorn barnacle, *Balanus*. (From Calman, In *A Treatise on Zoology, Part VII*, Lankester, Ed.)

Such informal methods of circulation and gas exchange are expected in most barnacles, whose diameters are less than a few centimeters. Metabolic rates are also low. However, even the largest members of the subclass are no better equipped. Some species of *Lepas* have a peduncle 75 cm long, and one species of *Balanus* is 12 cm wide. This situation suggests that barnacles evolved from crustaceans in which hearts, arteries, and gills were never developed.

Excretion in this subclass involves maxillary glands. Large end sacs may be compartmentalized, and excretory tubules are long and extensively coiled.

Reductions in activity systems are typical of sessile animals. Barnacle muscles include those that move the legs and control the shell aperture. In stalked forms, the peduncle has its own musculature. The barnacle nervous system consists of a circumesophageal collar and a variously concentrated ventral network. In balanomorphs, all trunk ganglia have fused below the foregut, but distinct thoracic ganglia are present along a ventral cord in the stalked groups. A nauplius eye persists in some adults.

Coincident with their sessile ways, most barnacles have hermaphroditic reproductive systems. Hermaphroditism ensures the possibility of breeding among all neighboring individuals. Self-fertilization may also occur. Ovaries occupy the peduncle of stalked barnacles and the basis and/or mantle wall of other forms. Paired oviducts open beneath the first thoracic appendages. The male system consists of cephalic testes and ducts leading to a very long penis (Fig. 12.23). Where possible, each barnacle extends its penis into the mantle cavity of a neighbor.

Complemental male dwarfs occur in some free-living barnacles, including certain species of the large genera *Balanus* and *Scalpellum*. These diminutive forms attach to a hermaphroditic individual. In certain species, complemental males represent little more than sperm-producing machines. In these barnacles, the conventional hermaphroditic form becomes a female, and the species is diecious. *Trypetesa* is a diecious barnacle which bores in gastropod shells. The tiny male with his gonadal supply systems superficially resembles a human penis (Fig. 12.27). He has no gut, and antennules are his only appendages. The large female *Trypetesa* may support up to a dozen or more males on her mantle wall.

In most barnacles, brooding occurs within an ovisac which is retained inside the mantle cavity. Over 10,000 nauplius larvae may hatch from a single parent. The barnacle nauplius has a three-sided carapace and paired cephalic appendages (Fig. 12.28a). Six

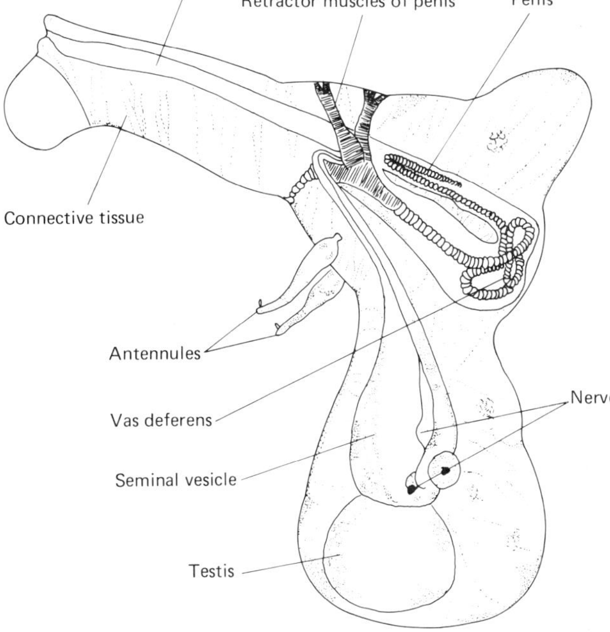

Figure 12.27 Enlarged view of a male barnacle, *Trypetesa*. (From Calman, In *A Treatise on Zoology, Part VII*, Lankester, Ed.)

instars later, the cypris larva appears (Fig. 12.28b). The cypris never feeds; its sole duty is to select an attachment site. Each coastal species has a characteristic vertical distribution, which depends on its resistance to desiccation and predation and on interspecific competition for space. Dense settling has reproductive advantages for most barnacles. Aggregations develop in part from naupliar swarms produced by simultaneous hatching and from protein attractants secreted by established individuals. Barnacles that settle densely on ships and harbor installations cause costly fouling. A barnacle-laden hull slows a ship's speed considerably.

The anterior end of the cypris is fastened to the substratum by cement gland secretions. These secretions include the strongest glues in the invertebrate world; no manmade product seals more effectively. The preoral tip, including the antennules and cement glands, expands to form the peduncle in stalked barnacles or the basis in stalkless groups. Rotation of the body and lengthening of the cirri produce the adult condition. Postsettlement ecdysis reveals the first shell plates. Continual mantle secretions add to the height and thickness of the shell throughout the life span. Ecdysis occurs periodically in adult barnacles. Molting involves the mantle lining and the cuticle of the central body and its appendages, but does not affect the shell.

THE PARASITIC BARNACLES

Body form in parasitic cirripedians ranges from the unusual to the bizarre (Fig. 12.29). In one large order (Rhizocephala), all traces of the calcareous shell, metamerism, and appendages are gone. Another parasitic group (Ascothoracia) has a bivalved carapace and a long abdomen (Fig. 12.29a).

Most parasitic barnacles feed on decapod crustaceans. Host nutrients are absorbed across the walls of an expanded peduncle. This region houses a network of digestive processes which ramify throughout the host. One such barnacle, *Sacculina* (Fig. 12.29b), parasitizes crabs. If the host is a male, its androgenic glands are among the first tissues to be digested. Destruction of this gland may cause the atrophy of sexual structures and the eventual castration of the crab.

Other barnacle parasites (Ascothoracia) live on echinoderms and soft corals. Ectoparasitic forms suck out host tissues along a gutter formed by the first maxillae. Endoparasites have degenerate mouthparts, as their expanded mantle surfaces directly absorb host nutrients.

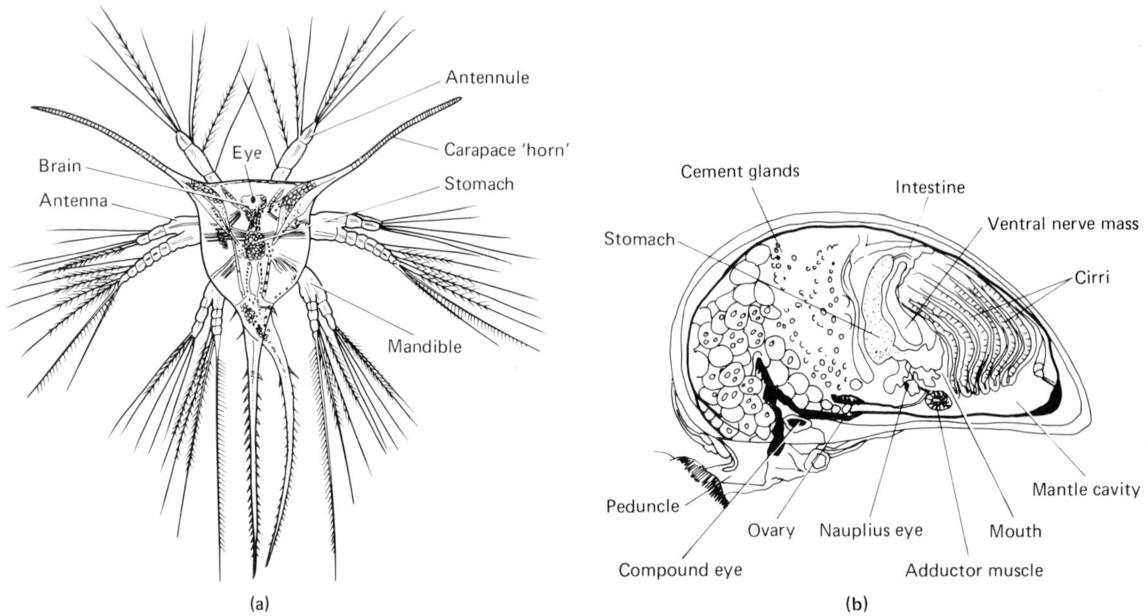

Figure 12.28 Barnacle larvae. (*a*) Nauplius larva of *Lepas fascicularis;* (*b*) Cypris larva of *Lepas australis.* (From G. Smith, In *Cambridge Natural History, Vol. IV,* Harmer and Shipley, Eds., 1920).

Parasitic barnacles usually are diecious, and their complex life cycles include free-living larval stages. One of the more interesting barnacle parasites is *Peltogasterella* (Fig. 12.29c). Upon penetrating a hermit crab host, the female cypris dedifferentiates and wanders among the crab tissues as a mass of simplified cells. Absorptive processes spreading from the barnacle mass take up nutrients. When the female parasite is ready to reproduce, she extends a large developmental chamber through the abdominal surface of her host. A tiny male cypris swims into this chamber and differentiates into a functional testis. Sexual reproduction follows, and hatching nauplii develop into the male and female cypris larvae of the next generation.

Subclass Malacostraca

Most of the larger and more familiar crustaceans belong to the subclass Malacostraca, including the crabs, lobsters, crayfish, shrimps, and pill bugs. Over 18,000 species, accounting for 70% of all known crustacean types, have been described. Malacostracans have considerable economic importance: They provide more human food than all other invertebrate groups, and smaller members of the subclass are eaten by many commercially valuable fish. These animals live in marine and fresh water throughout the world, and some of them occupy moist terrestrial niches.

Malacostracans are large by crustacean standards, although considerable size variation exists. The smallest syncarids and peracarids are less than a millimeter long, while the American lobster *Homarus* can extend 60 cm. Conventionally the malacostracan body plan is introduced in terms of a generalized, hypothetical animal, the **caridoid facies.** The caridoid facies is to the Malacostraca what the "generalized mollusk" (see Chap. Nine) is to its phylum. It is *not* an ancestral form, but simply a convenient model of basic design.

The caridoid facies has a head, thorax, and abdomen (Fig. 12.30). The head and thorax are covered by a large dorsal carapace, which extends laterally and projects from the anterior end as a rostrum. Cephalic appendages include paired biramous antennules, antennae, mandibles, and two pairs of maxillae. The antennae bear a prominent scalelike exopodite. Mandibular gnathobases have both grinding and cutting surfaces. Palps extend from both the mandibles and the maxillae. In addition to these appendages, the head bears a pair of stalked compound eyes.

The thorax of the caridoid facies comprises eight segments, all of which bear paired appendages called

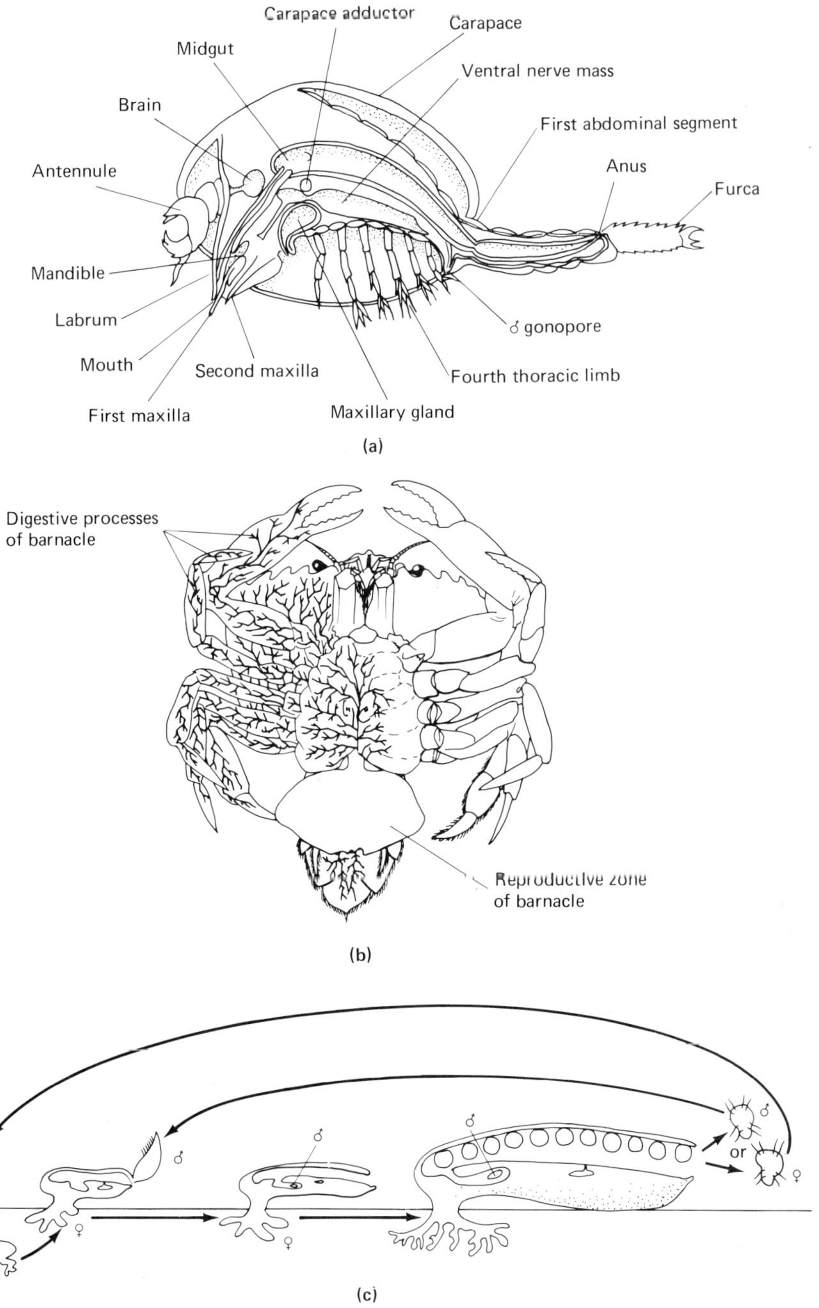

Figure 12.29 Left longitudinal section through the male of a parasitic barnacle, *Ascothorax ophiocetenis;* (*b*) Ventral view of a green crab, *Carcinus maenas*, infected with *Sacculina carcini.* The right side of the crab is shown as if transparent to illustrate the extensive root system of *Sacculina;* (*c*) Diagram of the life cycle of *Peltogasterella*, a barnacle parasitic on hermit crabs. (See text for further details.) (*a* after V.L. Wagin, Acta Zool., 1946, *27*:155, from Kaestner, *Invertebrate Zoology*, *Vol. III*; *b* from Kaestner, *Invertebrate Zoology*, *Vol. III;* *c* from R. Yanagimachi, Crustaceana, 1961, 2:184.)

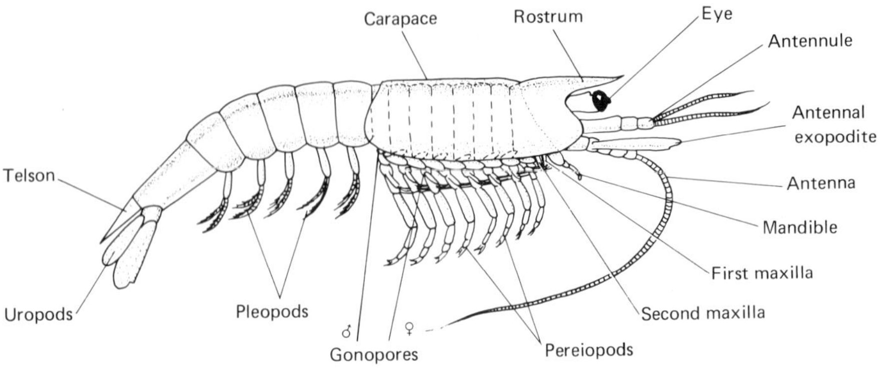

Figure 12.30 Diagram of the "caridoid facies," the generalized malacostracan body plan. (From Calman, Crustacea, In *A Treatise on Zoology, Part VII*, Lankester, Ed.)

pereiopods. Their endopodites usually are developed as walking legs. Small thoracic exopodites may be important for swimming, but these branches are absent from many groups. Up to three pairs of the anteriormost pereiopods may be modified as feeding **maxillipeds.** Other limbs may be large and chelate for defense and prey capture. These appendages, such as the claws of crabs and crayfish, are called **chelipeds.** Thoracic gills represent flattened epipodites or extensions of the inner carapace wall. In larger malacostracans, gill surfaces are increased by elaborate branching.

The generalized malacostracan abdomen consists of six segments, the first five of which bear paired **pleopods.** Pleopods, also known as swimmerets, are swimming appendages. Spines on each flattened exopodite and endopodite may interlock to form an unbranched paddling limb. Wide, well-ventilated pleopod surfaces are often exploited for respiratory exchange. In female malacostracans, the pleopods may be modified for egg brooding; males often use the anterior pairs as copulatory structures. The last abdominal segment is flattened and bears a pair of **uropods.** These broad limbs join the posterior **telson** to form a swimming tail.

The subclass contains both filter feeders and raptorial carnivores, as well as members that eat plants, carrion, and detritus. The digestive tract includes a complex foregut, a pouched midgut, a tubular intestine, and a posterior anus. The foregut is modified for grinding and straining. Its cuticular gastric mill, or stomach, comprises two regions (Fig. 12.31). An anterior **cardiac stomach** grinds food particles between opposing cuticular ossicles, which are gnashed by the powerful foregut musculature. The **pyloric stomach** forms a straining chamber. There, folded cuticular ridges and fine rows of setae prevent larger particles from entering the midgut ceca. The malacostracan midgut may form the rear part of the pyloric stomach, where one to four pairs of digestive diverticula are present. These pouches constitute the large **hepatopancreas**, the site of final digestion and absorption. Large undigestible matter is removed through the intestine and anus.

A large dorsal tubular heart is common in this subclass, and arterial systems are well developed, particularly among the larger forms. Malacostracans respire across their gills and any other appropriate surfaces, such as the broad pleopods. Maxillary or antennal glands are the excretory organs. The gills also are an important site of waste removal and osmotic control.

Most malacostracans have rather uncentralized nervous systems. Limited ganglionic concentration reflects the elongate body and the retention of abdominal segmentation and appendage pairs. In some groups, however, fusion has occurred among thoracic ganglia. Long malacostracans possess giant axons which mediate escape responses.

Malacostracan sense organs include stalked compound eyes, statocysts, muscle receptor organs, other proprioceptors, and assorted tactile and chemosensitive hairs, including esthetascs. The eyes of some lobsters contain over 10,000 ommatidia and thus compare favorably with those of the most visual insects. Other sense organs are concentrated near the joints of the appendages and within muscle tissues, where they respond to changes in the orientation of the cuticle.

Behaviorally, malacostracans are the most complex

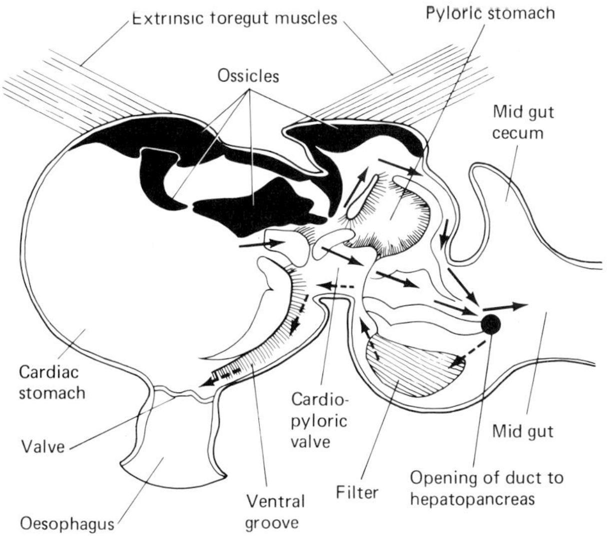

Figure 12.31 The foregut structures typical of advanced malacostracans. Solid arrows indicate the pathway of particles to midgut, broken arrows indicate the pathway of fluids to and from the hepato-pancreas. (From Grove and Newell, *Animal Biology*, 1969, University Tutorial Press.)

crustaceans. Anyone who has tried to catch crabs or shrimps or to play with a crayfish has experienced firsthand both the environmental awareness of these creatures and their ability to respond rapidly to external stimuli. They possess a host of specialized effector systems, including chromatophores and diversely modified appendages.

Malacostracan sex follows patterns established within the lower crustacean subclasses. Gonads extend through the trunk, and gonopores open on the eighth thoracic segment in males and on the sixth segment in females. Courtship commonly involves an embrace by specialized male appendages. Sperm transfer is either direct or utilizes spermatophores. Usually, the female broods developing eggs against her ventral surface or, as in the peracarids, a preconstructed brooding shelf is present. Malacostracan development features a parade of larval forms. The nauplius is passed prior to hatching in most orders.

Adaptive radiation within the Malacostraca has produced some 12 orders. Several distinct lines differ in metameric reduction, carapace and gill structure, feeding styles, and habitats. Specialization of the appendages is also a central factor in such radiation. The remainder of this chapter is devoted to the malacostracan groups. We begin with the Phyllocarida, a small and isolated series of crustaceans unlike all other malacostracans. The rest of the subclass composes the Eumalacostraca. This series includes four superorders: the Syncarida, the Hoplocarida, the Peracarida, and the Eucarida.

GROUP PHYLLOCARIDA (NEBALIACEA):

Order Leptostraca

Only 10 species of phyllocarids are known, and all of them are in a single order, the Leptostraca. These small marine creatures live among littoral mud and seaweed. One genus, *Nebaliopsis*, is planktonic.

Phyllocarids are unique among the malacostracans in that their abdomen is partially covered by a bivalved carapace (Fig. 12.32a). An adductor muscle can close the valves beneath the ventral surface. Anteriorly, a hinged rostrum protrudes over the head, where both pairs of antennae are long and setose. The mandibles and the maxillae bear long palps. The second maxillae and thoracic legs are uniformly broad and flat, resembling branchiopod appendages (cf. Figs. 12.12a and 12.32b). The beating of these limbs produces a current from which food particles are filtered. This current also ventilates epipodite gills and the inner wall of the carapace, which are the principal respiratory areas. Phyllocarids have a long tubular heart with seven paired ostia. Both antennal and maxillary glands are present. Their abdomen bears four pairs of swimming pleopods. The fifth and sixth abdominal segments have reduced appendages, and none are present on the seventh metamere. The eighth segment is modified as a telson and has a forked tail.

Phyllocarid females brood developing ova against their thoracic legs. Early larval stages are passed within

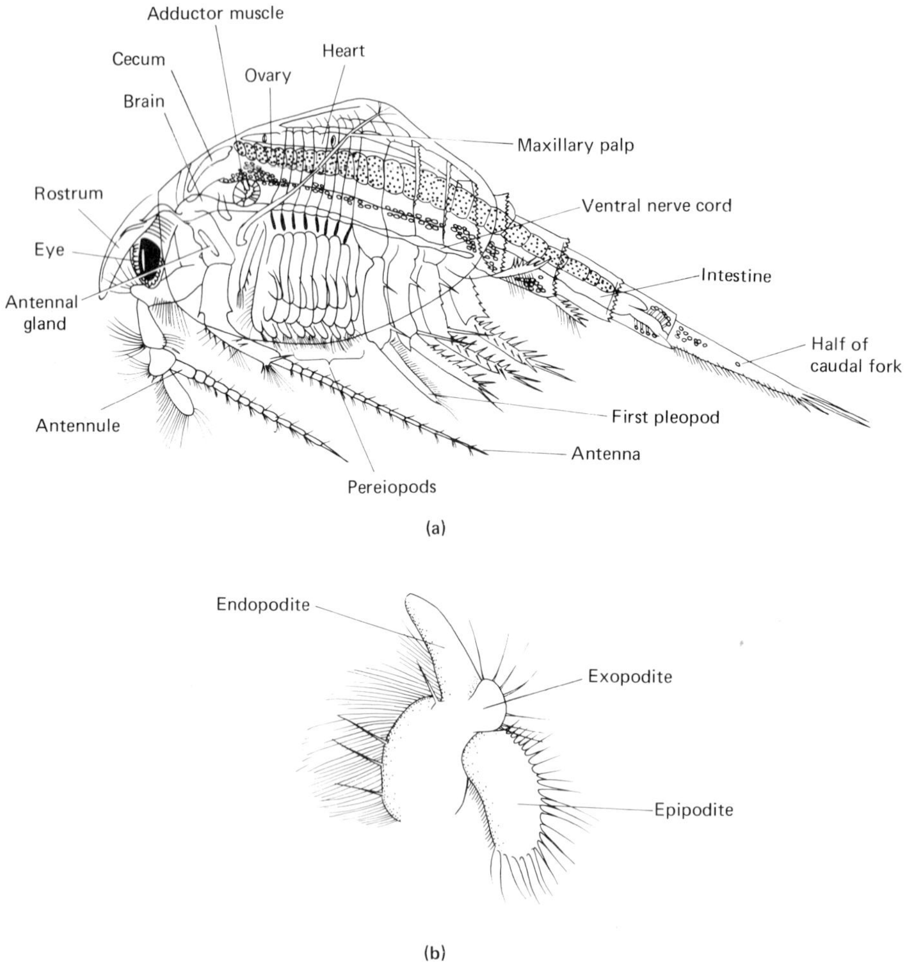

Figure 12.32 (*a*) *Nebalia geoffroyi*, a phyllocaridan; (*b*) First thoracic appendage of *Nebaliopsis typica*. (*a* from G. Smith, In *Cambridge Natural History, Vol. IV*, Harmer and Shipley, Eds., *b* from Calman, Crustacea, In *A Treatise on Zoology, Part VII*, Lankester, Ed.)

the egg, and the hatching juvenile, or **manca,** possesses a partially formed carapace.

Phyllocarids are probably the most primitive malacostracans. The lack of differentiation among their thoracic appendages and the "extra" abdominal segments are primitive traits. A fossil history dating from the Silurian supports their placement near the origin of the malacostracan branch.

GROUP EUMALACOSTRACA

SUPERORDER SYNCARIDA

The Eumalacostraca includes all malacostracans except the phyllocarids. Some of its most primitive members are the syncarids, a now small superorder with about 30 sporadically distributed freshwater species. The group was once more extensive, however, and many of its fossil members were marine. Syncarids lack the thoracic carapace so common among most malacostracans (Fig. 12.33). Tagmatization and differentiation of trunk appendages are not advanced.

The syncarid thorax bears uniform walking legs, and swimming pleopods distinguish the abdomen. Most modern species eat detritus, which they gather with their maxillae and anterior thoracic limbs. Others eat small plants and animals seized with their legs; most fossil syncarids apparently fed in this way. A dorsal tubular heart extends through most of the

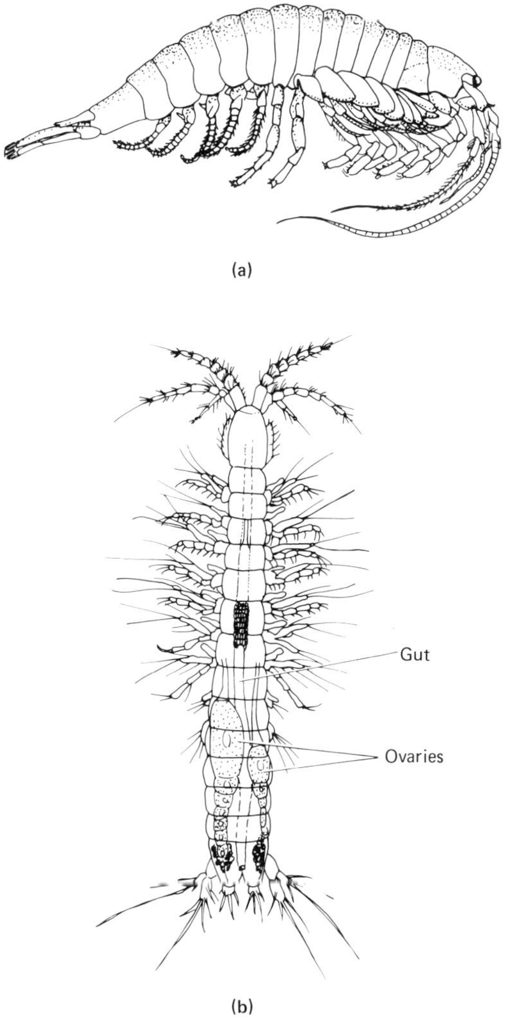

Figure 12.33 (*a*) A syncaridan, *Anaspides tasmaniae;* (*b*) A cave-dwelling syncaridan, *Bathynella natans.* (*a* from W.P. Pycraft, *The Standard Natural History*, 1931, Frederick Warne; *b* from H. Jakobi, Zool. Jahrb. Abt. Syst., 1954, *83*:1.)

trunk. Respiratory exchange occurs across epipodite gills on the thorax, and maxillary glands handle excretion.

Some syncarids have compound eyes, but many species are blind. Both antennae bear numerous sensory processes, including proximal statocysts in the second pair. Syncarids walk on their thoracic endopodites and/or swim with their pleopods. Pleopod exopodites are fringed with natatory setae, and the uropods join the telson to form a powerful tail. Body flexions are also important in locomotion.

Pleopod endopodites are developed only on the first two abdominal segments of syncarid males. These limbs are copulatory structures. Fertilized eggs packaged in resistant shells are laid externally. Larval stages are passed prior to hatching. The young syncarid enters its freshwater world in essentially adult form.

Three syncarid orders are recognized. Order Anaspidacea contains larger forms which live in Australian lakes. This group includes *Anaspides* (Fig. 12.33a), a raptorial feeder 5 cm long; another anaspidacean, *Paranaspides*, is a filter feeder. In this order, the first thoracic segment is always fused with the head; its appendages are maxillipeds. The orders Bathynellacea and Stygocaridacea comprise similar blind crustaceans living in cave pools and underground streams. These syncarids, typified by the European *Bathynella* (Fig. 12.33b), are only 1 or 2 mm long. Their first trunk segment is free from the head.

SUPERORDER HOPLOCARIDA

Mantis shrimp, so named because of their resemblance to the insect preying mantis, belong to the superorder Hoplocarida. Only one order, the Stomatopoda, is defined in this very independent line in malacostracan evolution. Stomatopods are characterized by a short carapace, raptorial thoracic appendages, abdominal gills, and, in males, a penis. They are entirely marine, dwelling in burrows or crevices on littoral bottoms.

The stomatopod body is 4 to 40 cm long and flattened dorsoventrally (Fig. 12.34a). Triramous antennules and stalked compound eyes extend in front of the carapace. This dorsal shield covers the head and no more than four thoracic segments. Each of the first five thoracic legs ends in a movable finger which folds back onto the leg's penultimate article. Such "jackknife" pincers are called **subchelae**. The second legs are much enlarged and their fingers have a sharp, often serrated edge. The last three pairs of thoracic appendages are walking legs. All six abdominal segments bear pleopods, and large uropods join a sculptored telson to form an effective swimming tail.

Stomatopods are raptorial carnivores. Invertebrates, including other crustaceans, and small fish compose their diet. Some stomatopods hunt actively, but many simply lurk within their burrows or crevices and pounce upon prey that wanders by. The subchelate second legs can slice a shrimp cleanly in half. The stomatopod heart is a metameric tube extending through most of the trunk. Thirteen ostia are present, and paired lateral arteries branch into each segment

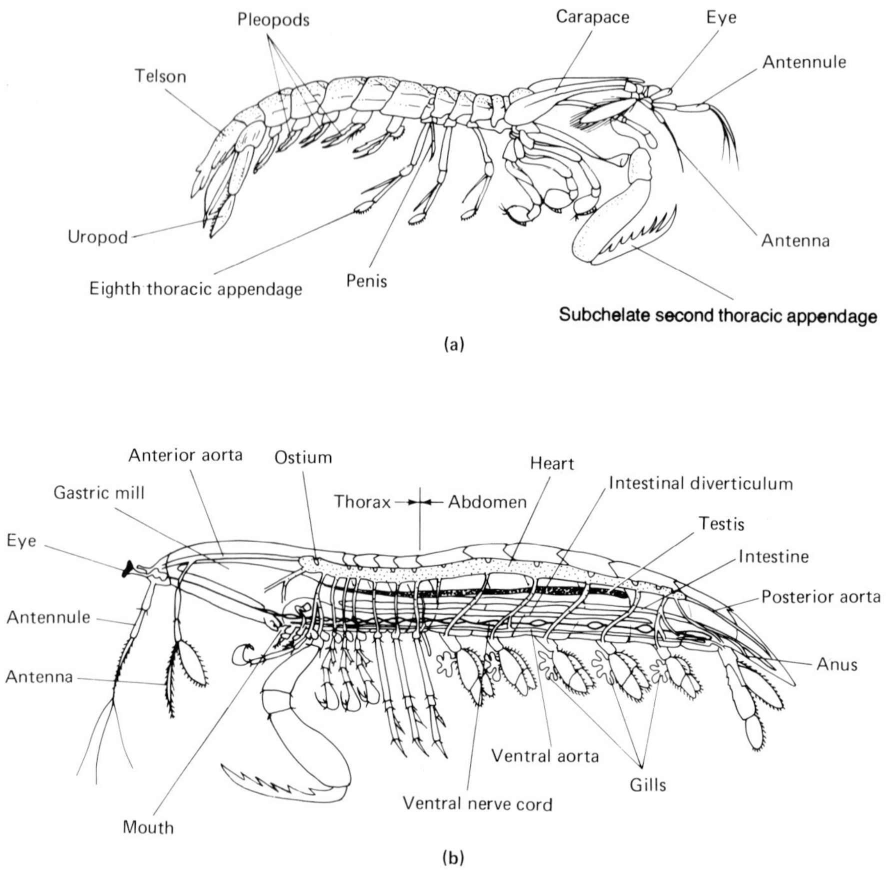

Figure 12.34 (*a*) A stomatopod, *Squilla mantis;* (*b*) Schematic showing the internal anatomy of a stomatopod. (*a* from Calman, Crustacea, In *A Treatise on Zoology, Part VII*, Lankester, Ed.; *b* from R. Siewing, In *Phylogeny and Evolution of Crustacea*, H.B. Whittington and W.D.I. Rolfe, Eds., 1968, Museum of Comp. Zool. Symposium, Harvard University.)

(Fig. 12.34b). Blood, which often contains hemocyanin, is aerated within gills on the pleopods. Abdominal (rather than thoracic) gills are an unusual character of mantis shrimps. Maxillary glands are the excretory organs.

The brain communicates via long circumesophageal connectives with the fused ganglia of the mouthparts and subchelate limbs. Other trunk ganglia are spaced along a ventral nerve cord. Highly developed compound eyes are critical for prey location. Between these large stalked organs is a single nauplius eye. Tactile sites and chemoreceptors are generally distributed, and esthetascs occupy the distal branches of the antennules.

Mantis shrimp crawl and/or swim. Their uropods and antennal scales are effective rudders. Many stomatopods occupy ready-made crevices in stone or coral or the abandoned burrows of other animals; sand-dwelling species dig their own homes. The short stomatopod carapace allows considerable body flexibility inside the burrow. The telson or the raptorial appendages may guard the entrance to the lair, or the opening may be sealed with collected debris.

Stomatopod continuity involves copulation, egg brooding, and conspicuous larval forms. Elongate ovaries empty through a single gonopore on the sixth thoracic segment. A seminal receptacle opens nearby. In males, tubular abdominal testes discharge through a jointed cuticular penis between the last thoracic legs. When mating, stomatopods appress their ventral surfaces, the female lying on her back and the male crouching over her.

Cement glands on the posterior thoracic segments of the female secrete adhesives which bind fertilized eggs in a single, brooding mass. In *Squilla*, this mass comprises up to 50,000 developing embryos; held by the subchelate legs, it is rotated and cleaned continuously. Some mantis shrimp, such as the coral-inhabiting *Gonodactylus*, protect a smaller egg mass at the rear of their burrows. Early free-living stages resemble a decapod zoea. In the larva, the carapace covers more of the head and thorax than in the adult (Fig. 12.35).

SUPERORDER PERACARIDA

About half of all malacostracans belong to the superorder Peracarida. Several evolutionary trends are evident in this large and diverse group. Primitive peracarids are shrimplike filter feeders with a prominent dorsal carapace. They exhibit little differentiation in thoracic and abdominal appendages. In the higher members of the superorder, the carapace has been reduced. Direct feeding on larger animals and plants is common, and trunk appendages are often quite specialized.

In all peracarids, at least the first thoracic segment is fused with the head; appendages on this somite are maxillipeds. The last four thoracic legs are always free from the carapace. Also general in the superorder is a ventral brooding shelf, or **marsupium** (Fig. 12.36). Horizontal plates extending inwardly from several thoracic coxopodites form the floor of the shelf, while thoracic sternites constitute the marsupial roof. Within this brooding area, peracarid young pass most of their larval stages.

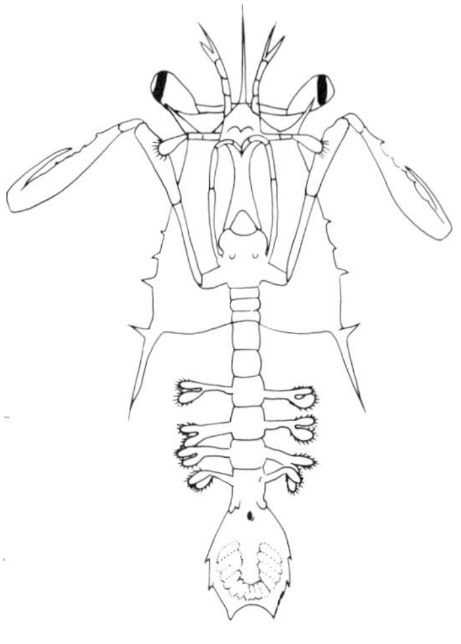

Figure 12.35 Larval stage of a stomatopod, *Squilla*. (From W.T. Calman, *The Life of Crustacea*, 1911, Methuen and Co.)

About 9000 species have been described from marine, freshwater, and terrestrial habitats. Five major orders are recognized. The Mysidacea probably are the most primitive peracarids. A single adaptive line leading from this order apparently is responsible for the Cumacea, Tanaidacea, and Isopoda. A separate line leads to the order Amphipoda.

Order Mysidacea

The mysids are simple shrimplike peracarids. Most of the 450 described species live in the ocean, where both bottom-dwelling and pelagic types are known. Migration into fresh water apparently is a recent phenomenon. Tidal basins and estuaries are frequent habitats, and a few species have invaded inland lakes. Mysids have an indirect commercial importance, as they are common prey for shad and flounder. In the Great Lakes, they constitute over half the diet of some trout.

Mysids are peracarids of moderate size. Most species are 1 to 3 cm long; the giant of the order, the pelagic *Gnathophausia*, can measure over one-third meter. Their most conspicuous, primitive trait is their large carapace. This dorsal shield encloses most of the head and thorax. A pointed rostrum projects from its anterior end, and posteriorly it covers (but is not fused with) the last four thoracic segments. Both pairs of antennae are long, and the second pair supports flattened scales. One or two thoracic segments have fused with the head; their shortened appendages constitute maxillipeds. The remaining thoracic legs are similar pereiopods with swimming setae. The mysid abdomen bears segmental pleopods, which are sometimes reduced.

Filter feeding is common in this primitive order. Exites on the second maxillae, often aided by maxillipeds and thoracic exopodites, create the feeding current. Plankton and detritus are strained by maxillary

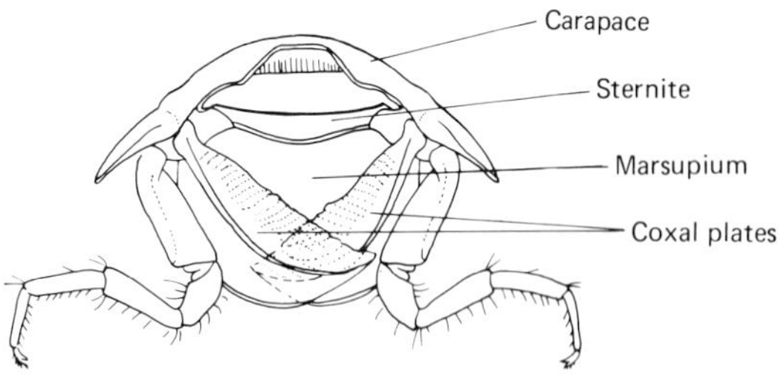

Figure 12.36 A female isopod, *Ligia exotica*. Cross section through the second thoracic segment showing the marsupium. (From Snodgrass, *A Textbook of Arthropod Anatomy*.)

endites (Fig. 12.37b). Some mysids also handle larger food with their thoracic endopodites. Filter feeding is not practiced in deep-water environments, where most mysid species are adapted for scavenging.

Other maintenance systems illustrate the primitive nature of the Mysidacea. An elongate heart has paired segmental arteries (Fig. 12.37c), and the gills are thoracic epipodites. Significant respiratory exchange also occurs across the inner walls of the carapace. Ventilation currents are produced by the epipodites of the first maxillipeds. Some mysids possess both antennal and maxillary glands, but the latter are absent from most species.

The mysid nervous system is organized segmentally, with distinct trunk ganglia. Sense organs include large compound eyes on movable stalks and statocysts embedded in the endopodites of the uropods. Mysids crawl on their pereiopods or swim with their pleopods. The flattened uropods and telson form an effective tail paddle. Setose thoracic exopodites may be important in swimming as well. Species that swim with these limbs combine their feeding, ventilation, and locomotor currents. This is an advanced condition, as the more primitive mysids operate three distinct currents for these processes.

Mating follows the sexual molt of the female, at which time her marsupium first appears. The male swims beneath her and sheds his sperm into her brood pouch, where fertilization occurs. The marsupium shelters all early developmental stages, and young mysids assume an independent life only after most adult characters are complete. The tendency of adult mysids to swarm contributes to their importance as prey for fish.

Order Cumacea

Cumaceans are marine peracarids which live in mud or sand burrows. Most inhabit shallower bottoms, where populations may be as dense as several thousand individuals per square meter. The order's 800 species include the well-described *Diastylis* (Fig. 12.38).

Cumaceans are rather tiny animals, most of which are measured in millimeters. The head and thorax are enlarged, in contrast to the long but narrow posterior body. A dorsal carapace extends to the third or fourth thoracic segment. Posteriorly, this shield is fused to the trunk, and its two long sides almost join beneath the ventral surface. The antennules are small, and in females the antennae are missing; however, the male antennae are long grasping appendages. The trunk limbs show considerable specialization. Three pairs of maxillipeds have differentiated, and the fourth thoracic legs are prehensile. The midthoracic segments contribute to the marsupium, and the posterior legs dig the cumacean burrow. Pleopods may be developed only in males. The uropods are peculiar tubelike appendages, which extend upward from the buried abdomen (Fig. 12.38a).

The first pair of maxillipeds is involved with nutrition and respiration. Their gill-bearing epipodites draw a feeding and ventilation current over the ventral, anterior surface. Maxillary setae filter food in *Diactylis*, while in some other cumaceans the mouthparts scrape the organic coating from sand grains held by the maxillipeds. Gas exchange occurs across the maxilliped gills and the carapace walls. The feeding-ventilation current then is deflected dorsally and anteriorly by long exopodites on the first maxillipeds. The heart is small, but is well muscularized above the respiratory zone (Fig. 12.38b). Maxillary glands handle cumacean excretion.

The ventral nerve cord has fused laterally, but little if any longitudinal concentration has occurred. The protocerebrum is small, as are the unstalked eyes. Males have better eyes then females and are generally more active. Cumaceans dig burrows with their rear thoracic legs. As these limbs excavate a hole, the body is backed into the position shown in Figure 12.38a. When it abandons its burrow, a cumacean swims with its thoracic exopodites until finding a new homesite. Males also may swim with their pleopods, especially during the nocturnal swarming of some species.

Swarming probably coincides with the mating period. Early cumacean instars are passed in the marsupium, and immature juveniles emerge at the **manca** stage. The manca lacks burrowing legs and sexual characters. Thus it tends to disperse before settling.

Order Tanaidacea

A third peracarid order comprises the tiny marine animals called tanaids. Tanaids continue some of the same trends established by the cumaceans and foreshadow certain isopod adaptations. Many investigators believe them to be intermediate between these two groups. About 350 species have been described, most of which live on muddy littoral bottoms.

Tanaids are usually a few millimeters long. Their shortened carapace extends only to the first two thoracic segments (Fig. 12.39). Only one pair of maxillipeds is present; the larger, second thoracic legs are chelate and raptorial. Other thoracic appendages may be chelate, and in some tanaids they spin mucous tubes. These limbs commonly lack exopodites. Abdominal pleopods are missing in some females.

A water current generated by the maxillipeds flows

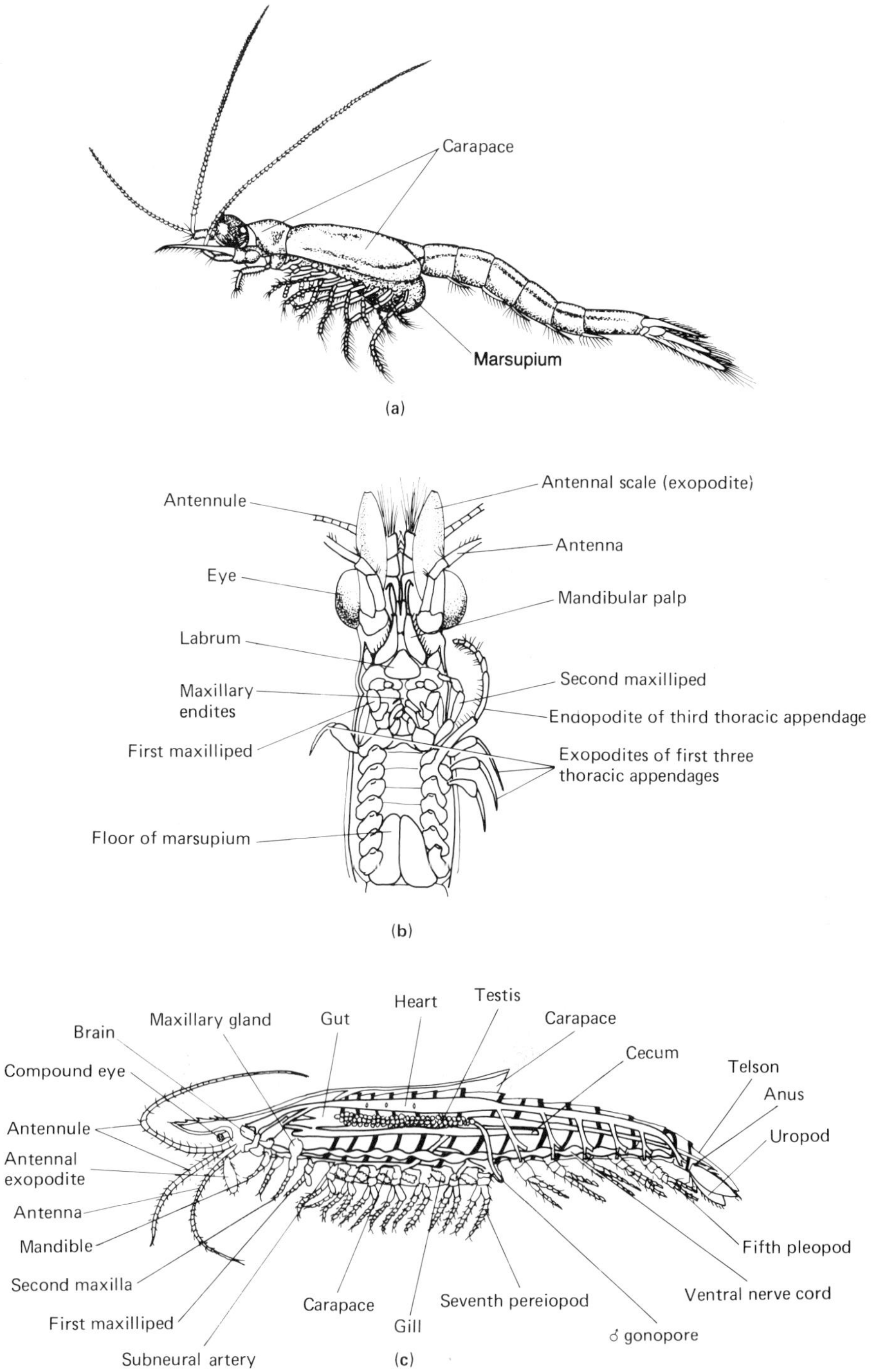

Figure 12.37 (*a*) *Mysis relicta*, a freshwater mysidacean; (*b*) Ventral view of the cephalothoracic region of *Mysis relicta*. Most thoracic appendages have been removed; (*c*) The internal anatomy of a male mysidacean. (*a* from Pycraft, *The Standard Natural History*; *b* from Calman, Crustacea, In *A Treatise on Zoology*, *Part VII*; *c* from Kaestner, *Invertebrate Zoology*, *Vol. III*.)

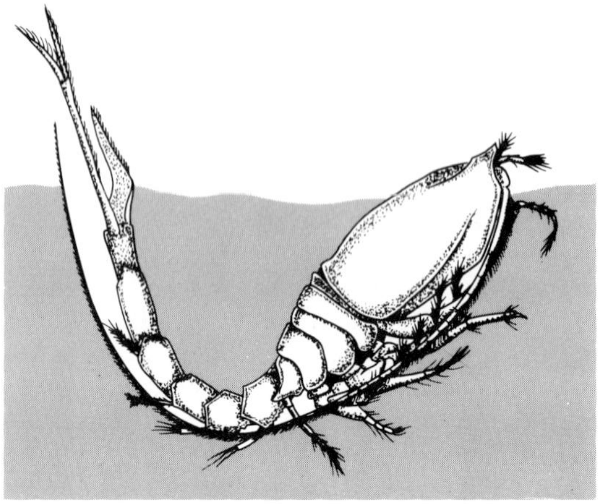
(a)

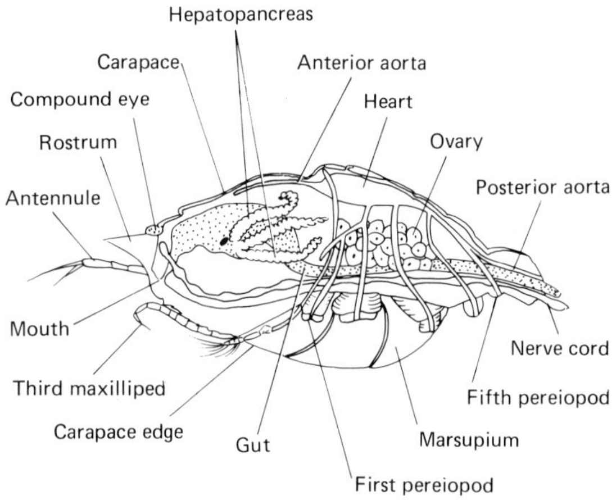

(b)

Figure 12.38 (*a*) *Diastylis stygia*, a cumacean; (*b*) The internal anatomy of a female cumacean. The legs and most of the abdomen have been removed. (*a* from Parker and Haswell, *A Textbook of Zoology, Vol. I;* and Green, *A Biology of Crustacea*, 1961, H.F. and G. Witherby; *b* from Kaestner, *Invertebrate Zoology*, *Vol. III.*)

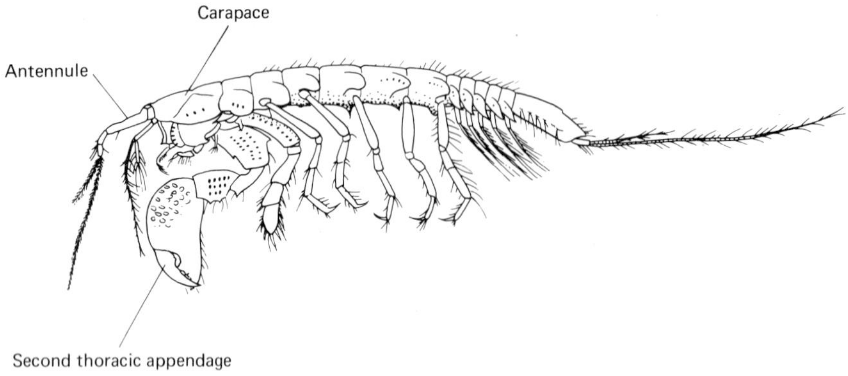

Figure 12.39 A tanaidacean, *Apseudes spinosus*. (From G. Smith, In *Cambridge Natural History*, *Vol. IV*, Harmer and Shipley, Eds.)

anteriorly, where maxilliped and maxillary setae may filter potential food particles. However, a decided trend within the order leads away from filter feeding. Clumps of detritus may be passed to the mouth by chelate limbs, and in many species the second legs snatch nematode prey. The water current is retained, however, as it ventilates the inner respiratory walls of the carapace.

Small compound eyes occupy a pair of lateral processes on the head. These organs are better developed among male tanaids, which also have more effective esthetascs and stronger pleopods. Many tanaids burrow in mud or live among bottom plants and debris.

Heterotanais oerstedi is one of the more interesting species. As in numerous other tanaids, mucous glands open on the distal articles of certain pereiopods. Their secretions are pulled into threads, which are then spun into a dwelling tube. Diatoms and various bottom debris are incorporated into the tube's walls. When a male *Heterotanais* finds a tube housing a mature female, he opens it with his large chelipeds. Both sexes then begin a courtship ritual which may last several hours. After fertilization, the male is ejected and the female seals her tube tightly. Emerging from the marsupium at the manca stage, the young eat phytoplankton trapped in the walls of their mother's tube. Eventually, they construct tubular side compartments which their mother seals off behind them. *Heterotanais* and other tanaids display a peculiar sexual plasticity. Young raised near adult females become males, and vice versa. After brooding, females may change into males, but this transformation does not occur if the female is kept in an aquarium with a male. Such patterns ensure a sexual balance in tanaid populations and may foreshadow hermaphroditic adaptations in the order.

Order Isopoda

The isopods rank among the most successful crustaceans. Only the Decapoda surpass them in number of species, and in terms of habitat diversity they are without peer in their class. Most isopods are marine, but nearly every suborder has at least a few freshwater members. Terrestrial isopods (the pill bugs) are some of the most successful crustaceans on land. Finally, there are many interesting parasitic forms. Four thousand species have been described.

Most isopods are distinguished by a dorsoventrally flattened body (Fig. 12.40). Their low profile is highlighted by lateral extensions of the tergites and sternites; thoracic coxopodites may fuse with these skeletal plates and thus further dramatize the flattened appearance. Several aspects of isopod biology represent the culmination of trends seen among the cumaceans and tanaids: These include the loss of the carapace, the abandonment of filter feeding, and a reduction of the abdomen.

Overall body lengths average around 1 cm, but some isopods are much larger. Among the giants of the order is *Bathynomus giganteus,* a bathypelagic species over 40 cm long and 15 cm wide. The typical isopod head is rounded anteriorly and includes one or two fused thoracic segments. Uniramous antennules are reduced and may be absent from terrestrial forms. The second antennae are better developed. Except for the anteriormost pair, which forms maxillipeds, the thoracic appendages are crawling pereiopods. The first few trunk legs may be prehensile. Dorsally, the isopod abdomen often is undifferentiated from the thorax, and fusion among its segments is common. In many freshwater groups, for example, most of the rear body has fused with the telson (Fig. 12.41). Abdominal pleopods are specialized for swimming and respiration.

Exceptions to typical isopod form occur in *Calathura* (Fig. 12.42a) and related genera, whose trunks are long and tubular. Other aberrant isopods

Figure 12.40 The rock louse, *Ligia pallasi,* an isopod common on rocky shores above the high tide line. (Courtesy of Ward's Natural Science Establishment.)

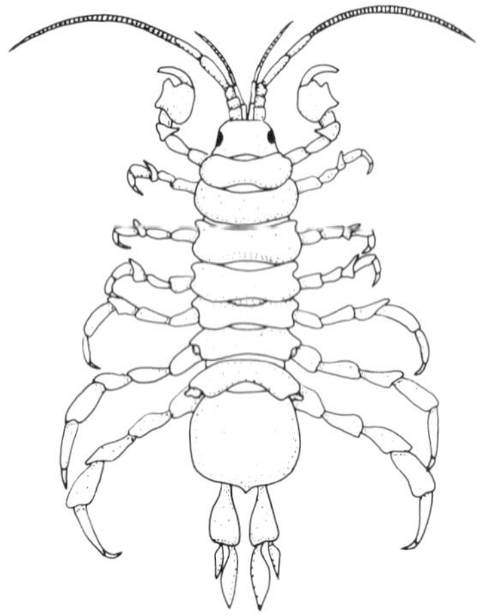

Figure 12.41 A freshwater isopod of the genus *Asellus*—bristles omitted. (From R.W. Pennak, *Fresh-Water Invertebrates of the United States*.)

include the insectlike gnathids (e.g., *Gnathia;* Fig. 12.42b) and the laterally compressed *Neophreatoicus* (Fig. 12.42c). Parasitic isopods are extremely deviant in form (e.g., *Portunion;* Fig. 12.42d).

Isopods have abandoned the filter feeding of lower peracarids. With their reduced maxillae, maxillipeds, and prehensile legs, they seize and hold food, to which the mouthparts are applied. A variety of plants and animals, both living and dead, is consumed. *Astacilla* (Fig. 12.43) ambushes unsuspecting crustaceans. This predacious isopod lurks motionless atop a coelenterate or bryozoan colony and then snatches prey with its long chelate antennae. Some isopods eat wood (e.g., *Limnoria;* Fig. 12.44), while omnivorous diets are common in many groups. The labrum and maxillipeds closely surround the mouth and the strong chewing mandibles. Parasitic isopods have piercing mandibles and their lips form a sucking cone. The absence of filter feeding preadapts the order for terrestrial life, and among land species the oral region is remarkably insectlike.

The isopod foregut includes a gastric mill, whose complexity reflects the texture of principal foodstuffs. A grinding cardiac stomach occurs in many forms. Up to three pairs of midgut ceca are the sites of final digestion and absorption. Wood-eating isopods may possess mutualistic gut bacteria which digest cellulose.

Isopods respire across pleopod gills, thus joining the stomatopods as the only crustaceans with abdominal respiratory organs. In aquatic species, both branches of the pleopods are flattened and transverse ridges often increase their surface areas. In some isopods, the anteriormost pleopods cover the remaining gills; in

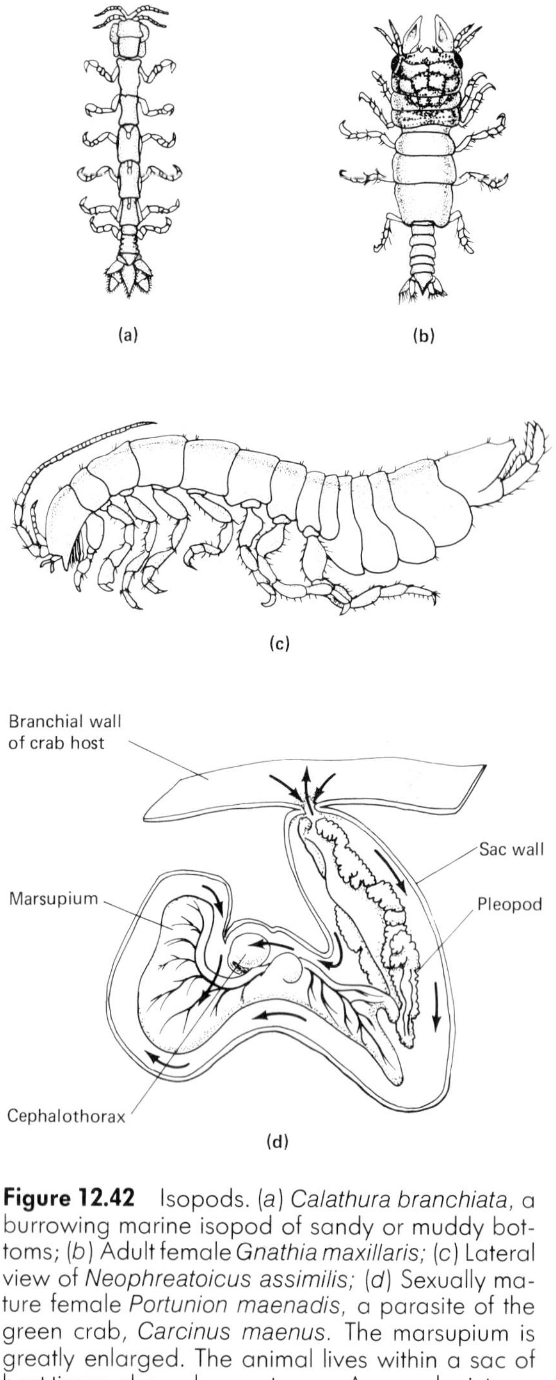

Figure 12.42 Isopods. (*a*) *Calathura branchiata*, a burrowing marine isopod of sandy or muddy bottoms; (*b*) Adult female *Gnathia maxillaris;* (*c*) Lateral view of *Neophreatoicus assimilis;* (*d*) Sexually mature female *Portunion maenadis*, a parasite of the green crab, *Carcinus maenus*. The marsupium is greatly enlarged. The animal lives within a sac of host tissue, shown here cut open. Arrows depict respiratory water current. Return current is behind animal and thus is not visible in this view. (*a* from G.A. Schultz, *How to Know the Marine Isopod Crustaceans*, 1969, Brown; *b* from Calman, Crustacea, In *A Treatise on Zoology, Vol. 8*, Lankester, Ed.; *c* and *d* from Kaestner, *Invertebrate Zoology, Vol. III*)

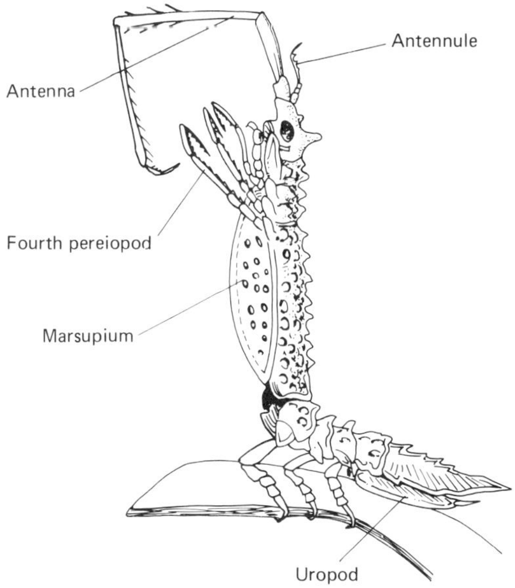

Figure 12.43 A predacious isopod, *Astacilla pusilla*, in ambush position. (From Kaestner, *Invertebrate Zoology, Vol. III.*)

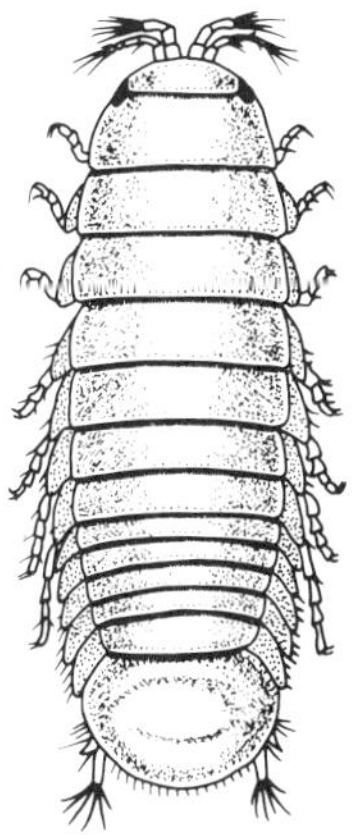

Figure 12.44 The gribble, *Limnoria lignorum*, a wood-boring isopod. (From Calman, *The Life of Crustacea.*)

others, elongated uropods play a similar role. Such protection of the respiratory surface represents another preadaptation for life on land. Terrestrial isopods themselves further shield their gills from the drying air. Their abdominal exopodites form a series of gill opercula, as endopodites alone are involved in gas exchange. The chamber thus formed is kept moist at all times. Uropods may transport water into the gill chamber, and in some terrestrial isopods, tracts along the tergites channel all available moisture to the ventral surface. In the most highly adapted species, the gills are replaced by **pseudotracheae.** These respiratory tubules ramify beneath the pleopod surfaces. Despite all these measures, however, terrestrial isopods remain limited to humid environments. Forest litter and the undersides of stones are common retreats. Pill bugs often roll up into a ball to reduce respiratory water loss (Fig. 12.45b).

Like the gills, the isopod heart occupies an abdominal position (Fig. 12.46). Maxillary glands are the excretory organs, and ammonia is the dominant nitrogenous waste. Even terrestrial species are ammonotelic, although much of their ammonia is released as a gas across the body wall. Their failure to excrete a drier waste product and the absence of a waxy epicuticle limit the terrestrial range of isopods and weigh against them in their ecological competition with insects.

The isopod nervous system, like that of most malacostracans, is not highly concentrated: It is ladderlike, and thoracic ganglia are quite distinct (Fig. 12.47). Sense organs include unstalked compound eyes and statocysts in the telson. The tips of the antennae usually bear chemosensitive esthetascs.

Most isopods are crawlers. Some move very rapidly both forward and backward, and many can walk vertically or upside down on the undersurface of rocks and logs. Each leg of the shoreline *Ligia* (Fig. 12.40) steps up to 16 times per second! Burrowing is well developed in some groups. Usually, the anterior legs shove excavated sand and mud to either side of the body. Wood-burrowing isopods such as *Limnoria* (Fig. 12.44) are serious pests on marine docks. Almost all aquatic isopods can swim, although swimming effectiveness varies widely. Some isopods, such as *Munnopsis,* paddle with flattened thoracic legs, but most members of the order swim with their pleopods. Frequently, the anterior abdominal appendages are specialized entirely for swimming, leaving only the posterior limbs for respiratory exchange.

Terrestrial isopods have a simple behavioral mechanism which concentrates them in dark, damp places: When exposed to drying air, they walk quickly and do not slow down until a damp area is reached. They are photonegative as well. A tendency toward increased activity at night inspires their nocturnal wanderings.

Male isopods possess paired testes, whose ducts open through genital papillae on the last thoracic segment. As the first pleopods bend these papillae backward, sperm are ejaculated into a chamber on the second abdominal segment. The female system in-

(a)

(b)

Figure 12.45 *Armadillidium vulgare*, the common terrestrial pillbug or sowbug. (*a*) Normal active posture. (*b*) Protective rolled-up posture. (Photos courtesy of Grant Heilman.)

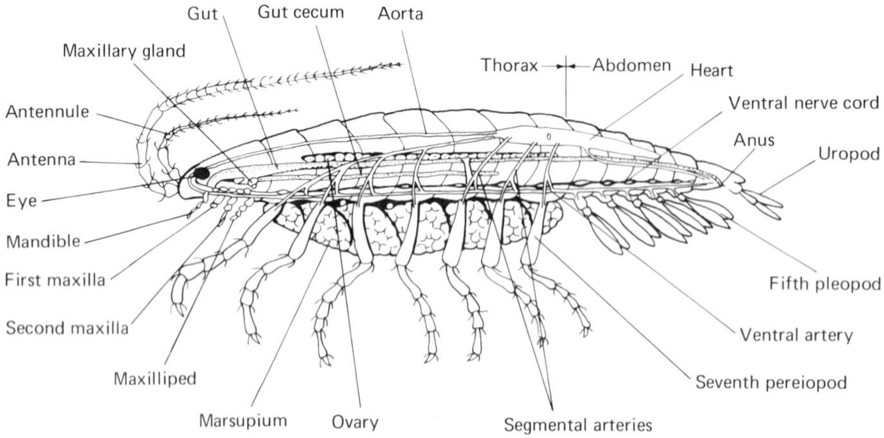

Figure 12.46 The major internal and external anatomical features of a generalized isopod. (From Siewing, In *Phylogeny and Evolution of Crustacea*, Whittington and Rolfe, Eds.)

cludes paired ovaries and oviducts and usually semi nal receptacles. Pheromones secreted by an isopod female inform the male of her presence and sexual capacity. He confirms identification by touching her with his second antennae. Insemination occurs when the male vibrates his second pleopods against the female's gonopores.

As other peracarids, isopods usually incubate their young in a marsupium. In exceptional species, brooding occurs within invaginations of the thoracic sternites. Maxilliped epipodites frequently stir the water inside the marsupium, thus increasing ventilation for the developing young. Even in terrestrial species, the brood chamber is filled with fluid, an indication of the order's marginal adaptation to land. Young hatch as postlarval manca (Fig. 12.48).

Parasitic isopods are often hermaphroditic, and complex life cycles involving one or more hosts are common. A typical history features an independent larva that becomes ectoparasitic (or commensal) on a copepod. After several instars elapse, it detaches from the copepod and crawls on the bottom. If a crab is encountered, the isopod parasitizes its gills or brooding area. The first individual to infest a crab is often female. Subsequent parasites may attach not to the crab but to the body of an established comrade. The late arriver becomes a male.

Order Amphipoda

The second major line in peracarid evolution extends from the Mysidacea to the Amphipoda. While the Amphipoda are not as diverse as the Isopoda, there is much convergent evolution between the two orders. In both groups, there is no carapace and one or two thoracic segments have fused with the head (Fig. 12.49a). A single pair of maxillipeds is present. Thoracic exopodites are missing, and the abdomen is seldom set off from the anterior body. The primary differences between the two orders involve body compression and gill location. Most amphipods are compressed laterally, whereas isopods are dorsoventrally flattened. Amphipods have thoracic gills; but recall that, in isopods, the respiratory organs are abdominal.

Most of the 3600 amphipod species are marine, but freshwater and semiterrestrial forms are also known. Marine amphipods include both pelagic and bottom-dwelling forms. Freshwater species are often quite abundant among pond and stream vegetation. The common beach fleas represent the order's most successful semi-terrestrial group, and there have been limited migrations into humid forest litter.

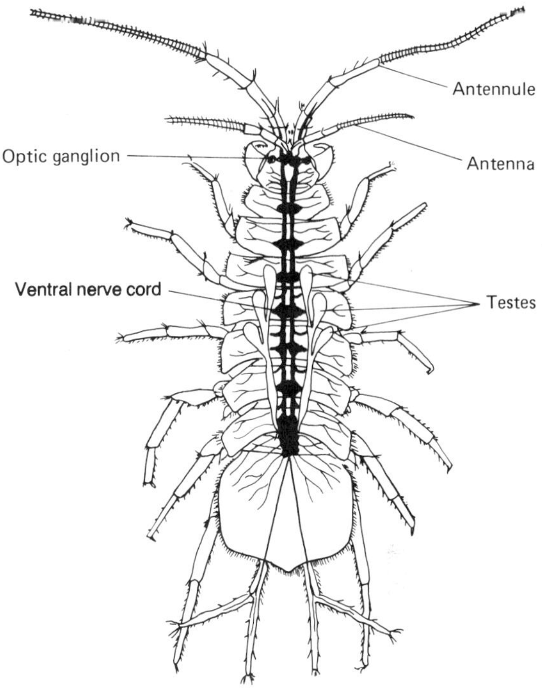

Figure 12.47 Nervous system of the freshwater isopod, *Asellus aquaticus*. (From L.A. Borradaile and F.A. Potts, *The Invertebrata*, 1961, Cambridge University.)

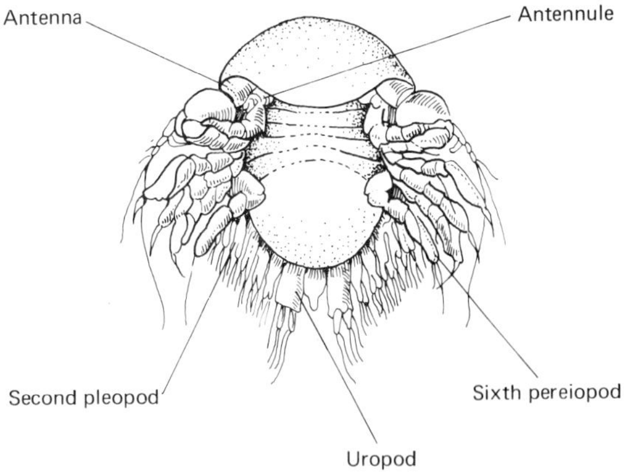

Figure 12.48 Postlarval manca stage of the isopod, *Clypeoniscus meinerti*. (From Kaestner, *Invertebrate Zoology, Vol. III.*)

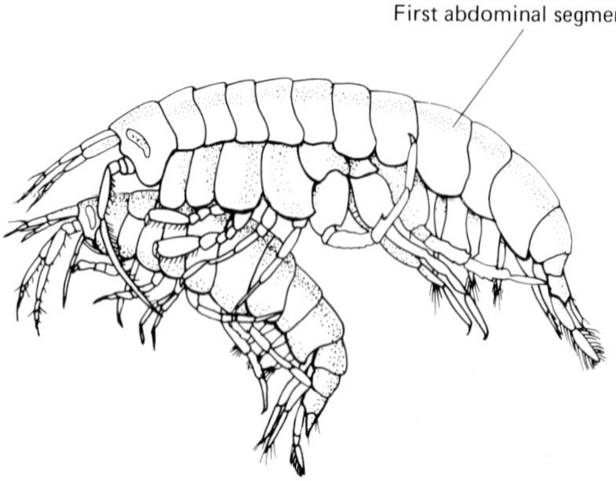

(a)

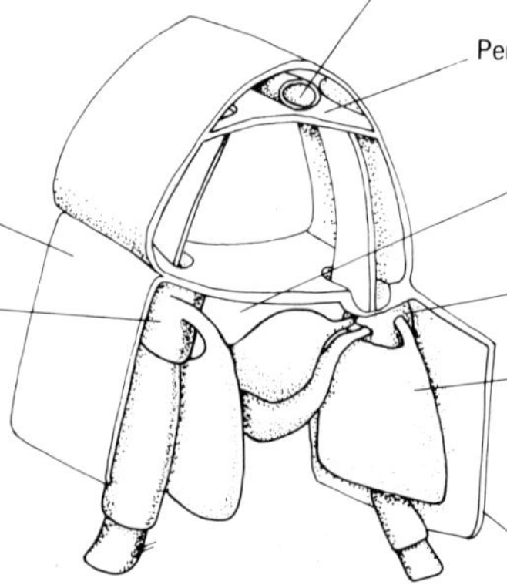

(b)

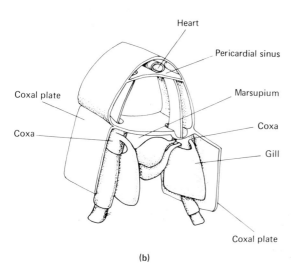

Figure 12.49 Copulating amphipods, *Gammarus locusta.* Male above, female beneath; (*b*) Cross section through a thoracic segment of an amphipod showing the typical lateral compression of the body. (*a* from G. Smith, In *Cambridge Natural History, Vol. IV*, Harmer and Shipley, Eds; *b* from Kaestner, *Invertebrate Zoology, Vol. III*.)

Amphipods share a common size range with isopods. The largest species live in the deep sea, where one specimen was measured at almost 30 cm. The lateral compression of the amphipod body is enhanced by a convex dorsal surface and flattened coxal plates (Fig. 12.49b). Although typically uniramous, both pairs of antennae are well developed. The maxillipeds have fused at their bases, and the second and third thoracic appendages are modified as **gnathopods.** These chelate or subchelate limbs seize and hold food and other objects. The remaining thoracic legs are similar pereiopods. Abdominal pleopods usually include three anterior pairs specialized for swimming and three posterior pairs which resemble uropods.

Exceptional amphipods include pelagic forms such as *Rhabdosoma* and *Mimonectes* (Fig. 12.50a,b). Their extended and/or inflated regions aid in flotation. Predacious caprellids, including *Caprella* (Fig. 12.50c), have a narrow elongate thorax, while the abdomen is greatly reduced in the whale lice (Fig. 12.50d). A few dorsoventrally flattened amphipods superficially resemble isopods.

Most amphipods eat detritus and/or dead plants and animals. The gnathopods pick up food from the substratum and, with the help of the maxillipeds, pass it to the strong chewing mandibles. Organically rich mud adhering to the antennae or other appendages may be scraped clean by the maxillipeds and mouthparts. Other amphipods are filter feeders. Anterior pleopods, sometimes aided by the pereiopods, draw a feeding current, and setae on the anterior thoracic appendages and mouthparts cull particles of an appropriate size. Predation is limited in the order. Caprellids commonly crawl over seaweed, coelenterates, or bryozoans, where they feed on encrusted debris. Some species lurk motionless against their sessile background and await unsuspecting prey (Fig. 12.50c). When a small crustacean or polychaete chances by, it is seized by the caprellid's gnathopods. Parasitism is not common in this order. A few amphipods live in sponge channels or within tunicate gills, and some are ectoparasites on fish. Whale lice are commensal on cetaceans and eat the organic matter that gathers on the surface of their host.

The tubular amphipod heart has up to three pairs of ostia and a simple arterial system (Fig. 12.51) and lies directly above the thoracic gills. Amphipod gills include two to six pairs of lamellar epipodites, which may be folded or branched. In some groups, accessory gills occupy the first abdominal segment, thoracic sternites, or appendages. Broad, elongated coxopodites flank a protective respiratory channel along the ventral surface. Anterior pleopods push a ventilating current through this channel. In tube-dwelling or burrowing species, the antennae may generate the respiratory current. Gas-exchange mechanisms are little different on land, where the entire body surface may be active in respiration. This condition restricts terrestrial amphipods to the most humid areas.

The excretory organs are antennal glands. In brackish or fresh water, amphipods exhibit some osmotic regulation. Several species of *Gammarus* can

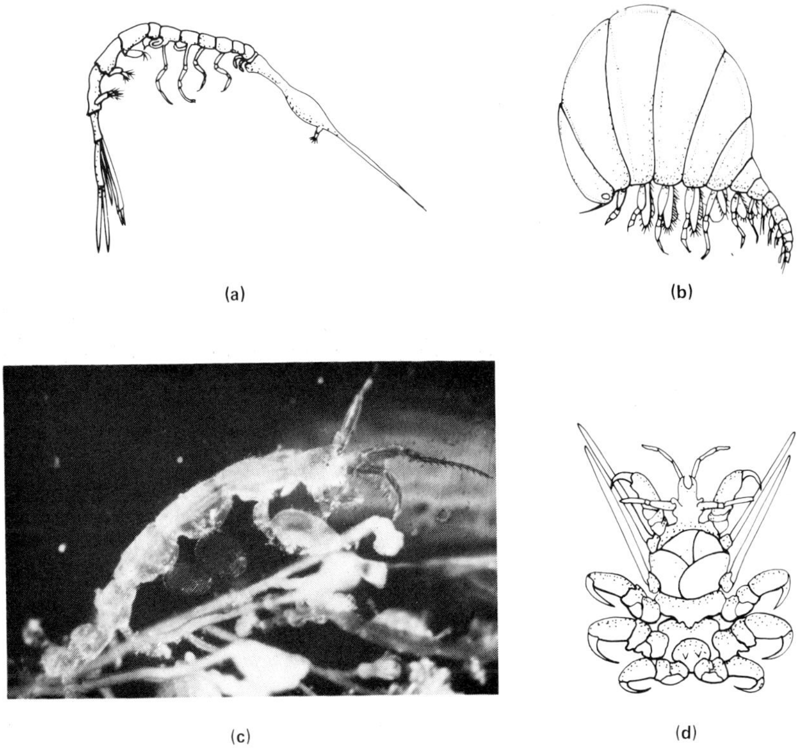

Figure 12.50 (*a*) *Rhabdosoma piratum* and (*b*) *Mimonectes loveni*, pelagic amphipods; (*c*) *Caprella equilibra*, a predacious amphipod, clinging to seaweed; (*d*) A whalelouse, *Cyamus erraticus*. Ventral view of a female with marsupium. (*a* from Calman, Crustacea, In *A Treatise on Zoology*, *Vol. VIII*, Lankester, Ed.; *b* from Calman, *The Life of Crustacea*; *c* courtesy of D.P. Wilson; *d* from Kaestner, *Invertebrate Zoology*, *Vol. III*.)

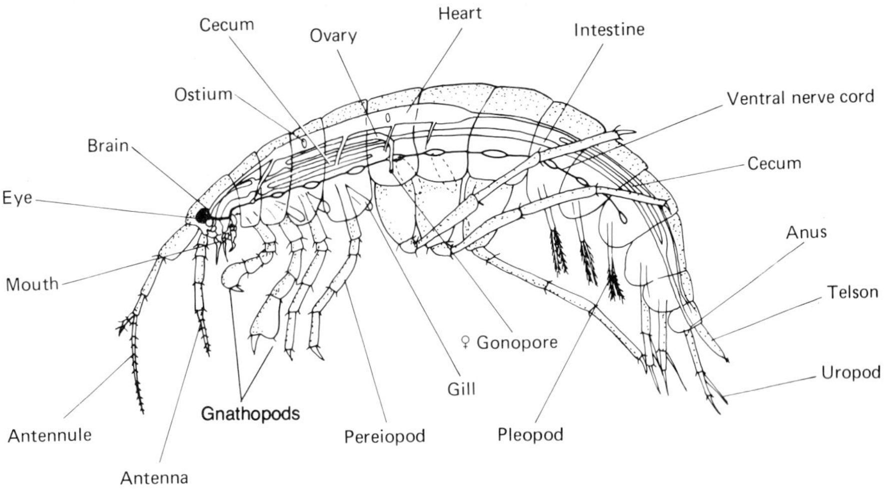

Figure 12.51 The major internal and external anatomical features of a generalized amphipod. (From Siewing, In *Phylogeny and Evolution of Crustacea*, Whittington and Rolfe, Eds.)

excrete a dilute urine while absorbing salts through the gills. Terrestrial species have made little if any excretory modifications.

In most amphipod nervous systems, the ganglia of the mouthparts and maxillipeds have fused. Sensory structures include compound eyes and a typical assortment of sensory hairs. As among the isopods, the eyes are never stalked. In most amphipods, the ommatidia do not possess individual facets; rather, the entire eye is covered by a continuous layer of corneal cuticle. Pelagic amphipods frequently have enormous eyes, while those of cave-dwelling species are reduced or absent. The latter have highly developed sensory hairs.

Amphipod habits are varied. Some marine species spend their time either swimming or attached to larger pelagic organisms, such as jellyfish. Most, however, live on the seafloor, where they crawl or burrow and occasionally swim. The abdomen can be flexed beneath the body and then suddenly straightened, thus flipping the animal forward. Caprellids are effective climbers; their clawed thoracic legs help them scramble over plants and sessile invertebrates. Various appendages are employed by burrowing amphipods. The gnathopods are particularly effective in digging sand and mud, while some groups rely heavily on the antennae for this work. Active burrowers may pass excavated material posteriorly from one appendage to the next until the load is flipped away by the uropods and telson. Some amphipods plaster their burrow walls with cement produced by glands which open on thoracic sternites. Similar adhesives join the sand particles and other debris that form the homes of tube-dwelling species. *Siphonoecoetes* drags its tube while crawling over the bottom; its antennae are used in locomotion. Some amphipods burrow in wood. Those living within ascidian tunics or sponge walls may be on their adaptive way toward a parasitic existence.

Terrestrial amphipods burrow on moist sandy beaches or through humid forest litter. The common beach fleas feed on organic debris washed in by the sea. On cloudy days, these amphipods often hop actively around stranded algal masses or broken crab shells. Beach fleas like *Talorchestia* can leap an incredible 50 times their body length, a feat ranking them among the champion jumpers of the invertebrate world.

Amphipod reproductive systems include paired tubular gonads and gonopores on the typical malacostracan segments. The male has a pair of penis papillae. Female pheromones stimulate the male, who may grasp his mate's marsupium with his uropods. Some prospective fathers anticipate mating by riding the backs of females preparing to undergo the sexual molt. Sperm are liberated into the female ventilation current and swept into the marsupium. The female then releases her eggs into the brooding chamber, where fertilization occurs. Amphipod young are incubated until all adult segments are formed. When they are ready to assume independent lives, the mother may undergo a molt during which her marsupium is cast off.

Superorder Eucarida

The last superorder of malacostracans, and on several counts the most important of all crustacean groups, is the Eucarida. Eucarids are the largest and most advanced animals in their class. They are distinguished by a combination of characters: A very prominent dorsal carapace is fused with all thoracic segments, the eyes are always stalked, and the gills are thoracic. There is never a marsupium. Indirect development is complicated by several larval stages. Two orders are recognized: the Euphausiacea and the Decapoda.

Order Euphausiacea

Only about 90 species of euphausiaceans have been identified, but they include some of the most numerous and trophically significant crustaceans of the sea. These animals are commonly known as krill. An average body length of 3 cm, a marine pelagic range, and a tendency to swarm combine to make euphausiaceans important prey for whales and large fish. Indeed, 5 million individuals of *Euphausia superba* have been counted in the stomach of a single blue whale.

Euphausiaceans are rather primitive, shrimplike eucarids. In fact, they bear a strong resemblance to the most primitive peracarids, the mysids (cf. Figs. 12.37 and 12.52), which suggests that the two largest malacostracan superorders are closely related.

The euphausiacean body is laterally compressed. A carapace has fused with the entire cephalothorax, but its short sides never enclose a ventral gill chamber. A dorsal rostrum extends over the head, where both antennae are well developed. The second pair has prominent scales. Thoracic legs always include exopodites, but none are differentiated as maxillipeds. Heavily setose pleopods extend from the large abdomen, which may represent over half the body length. The telson is elongate.

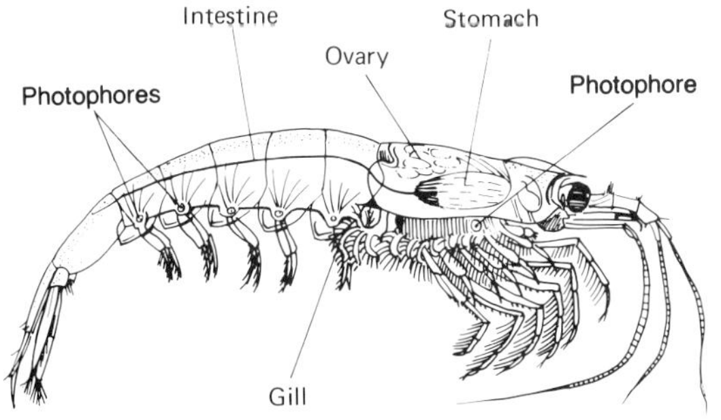

Figure 12.52 Female of *Euphausia pellucida*. (From Smith, In *Cambridge Natural History, Vol. IV*, Harmer and Shipley, Eds.)

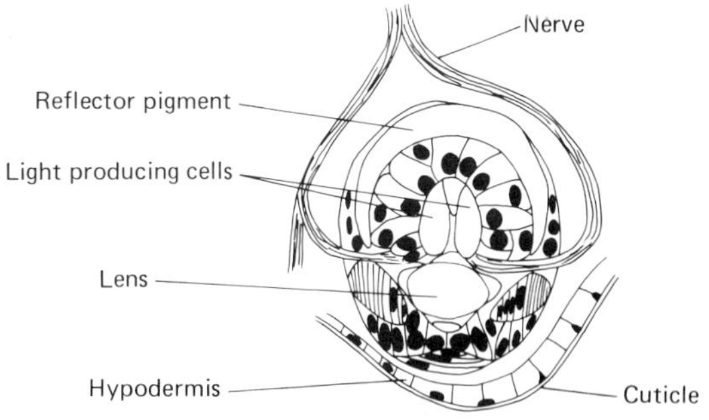

Figure 12.53 Section through a photophore from a thoracic coxopodite of *Nematoscelis*, a euphausiacean. (From Calman, Crustacea, In *A Treatise on Zoology, Part VII*, Lankester, Ed.)

Most euphausiaceans feed on plankton filtered by their thoracic appendages. Setae along the inner margins of the endopodites form a large central net. Swimming itself drives plankton-rich water through the net; strained food particles are passed over the small, flattened maxillae to the mandibles. *Stylocheiron* is one of the few predacious euphausiaceans. Its large, chelate third legs seize arrowworms, copepods, and other small prey. The cardiac stomach is well developed in predacious forms, less so among filter feeders. A complex midgut gland (the hepatopancreas) occupies most of the cephalothorax.

Respiratory gas exchange occurs across thoracic epipodites (Fig. 12.52). Often highly branched, these gills are ventilated by setose exopodites on the same appendages. The heart lies in the rear half of the cephalothorax and pumps blood into an extensively branched arterial system. Antennal glands are the excretory organs.

Large stalked compound eyes may be augmented by a persistent nauplius eye, and esthetascs are present on the antennae. Mouthpart ganglia are fused, but nerve centers of the trunk are spaced along a ventral cord. Under power of their pleopods, euphausiaceans swim freely through the open sea. Varying depth preferences are expressed by different species. Many live in surface waters, but some forms occur as deep as 3500 m. Vertical migration is common, as is swarming. Densities of 50,000 individuals per cubic meter are typical in some species. A swarm usually contains animals of about the same age and may represent a lifelong association. Most euphausiaceans are luminescent. (The Greek "phausis" means "shining light.") Photophores are located on the eyestalks, certain thoracic coxopodites, and the anterior abdominal sternites. These small organs house a lens, a reflector, and a battery of cells containing the light-producing material (Fig. 12.53). Most likely, luminescence plays a role in intraspecific communication. Blind euphausiaceans always lack photophores.

In both sexes, branched gonads discharge through coiled gonoducts. A euphausiacean male transfers his spermatophore to a female with his anterior prehensile pleopods. Fertilized eggs may be brooded briefly within the thoracic feeding net or may be cemented to the sternites. Many species, however, simply shed their zygotes into the open sea. Nauplius larvae hatch, but do not feed. At least in *Euphausia superba*, the developmental stages occupy deeper waters than the adult. This species lives in the Antarctic, where surface water is colder than that several meters deep.

Order Decapoda

Most of us began our appreciation of decapod crustaceans at an early age, either along beaches or at the dinner table. This group includes the familiar shrimps, lobsters, crayfish, and crabs. These animals are the largest and most complex members of the Crustacea and, with about 8400 known species, constitute its most populous order. Most decapods are marine, but freshwater environments also have limited shrimp and crab fauna. Crayfish are particularly successful in streams, and a few families have ventured onto land.

Decapod Unity

Within the Decapoda, several adaptive trends in crustacean evolution are culminated. Foremost among these trends are tagmatization and the specialization of appendages. Decapod limbs display numerous modifications and are a primary focus of the order's behavioral complexity. Finally, decapod continuity encompasses some of the invertebrate world's most interesting larval forms.

MAINTENANCE SYSTEMS

Although the decapods vary widely in body form, maintenance systems are fairly standard throughout the order. Of course, some modifications occur in freshwater and terrestrial groups.

General Morphology

As malacostracans, decapods have a thorax with eight segments and an abdomen with six segments. Pereiopods and pleopods are the appendages. Female gonopores occupy the sixth thoracic segment; male pores, the eighth. As eucarids, their prominent dorsal carapace is fused with the entire thorax; the eyes are stalked, the gills are thoracic, and there is no marsupium. Decapods differ from the euphausiaceans in that their first three thoracic limbs are modified as maxillipeds. This adaptation leaves a total of 10 walking legs, a condition that accounts for the name of the order. Thoracic exopodites are seldom present. The first walking leg on each side is often chelate and much larger than the others. This limb, the **cheliped,** bears the prominent claw of crabs, lobsters, and crayfish. Laterally, the decapod carapace extends below the coxopodites to form the sides of a branchial chamber.

There are two general types of decapods, each with a characteristic body form. The first group, the Natantia, includes the shrimps (Fig. 12.54a). These decapods are specialized for swimming. They have a laterally compressed body, a thin exoskeleton, narrow legs, and an extended, well-developed abdomen. The second group, the Reptantia, includes walking forms such as crabs, crayfish, and lobsters (Fig. 12.54b, c). These animals may be dorsoventrally flattened, with a rather thick cuticle. Heavy walking legs and powerful chelipeds characterize the reptant decapods. In lobsters and crayfish, the abdomen is well developed, but the posterior body is reduced in true crabs.

Decapods vary enormously in size. Small porcelain crabs are only several millimeters wide, while the Japanese spider crab may span over 3 m. A wide range of colors is possible in the many species with integumental chromatophores. Some decapods manipulate their body color for camouflage. Other

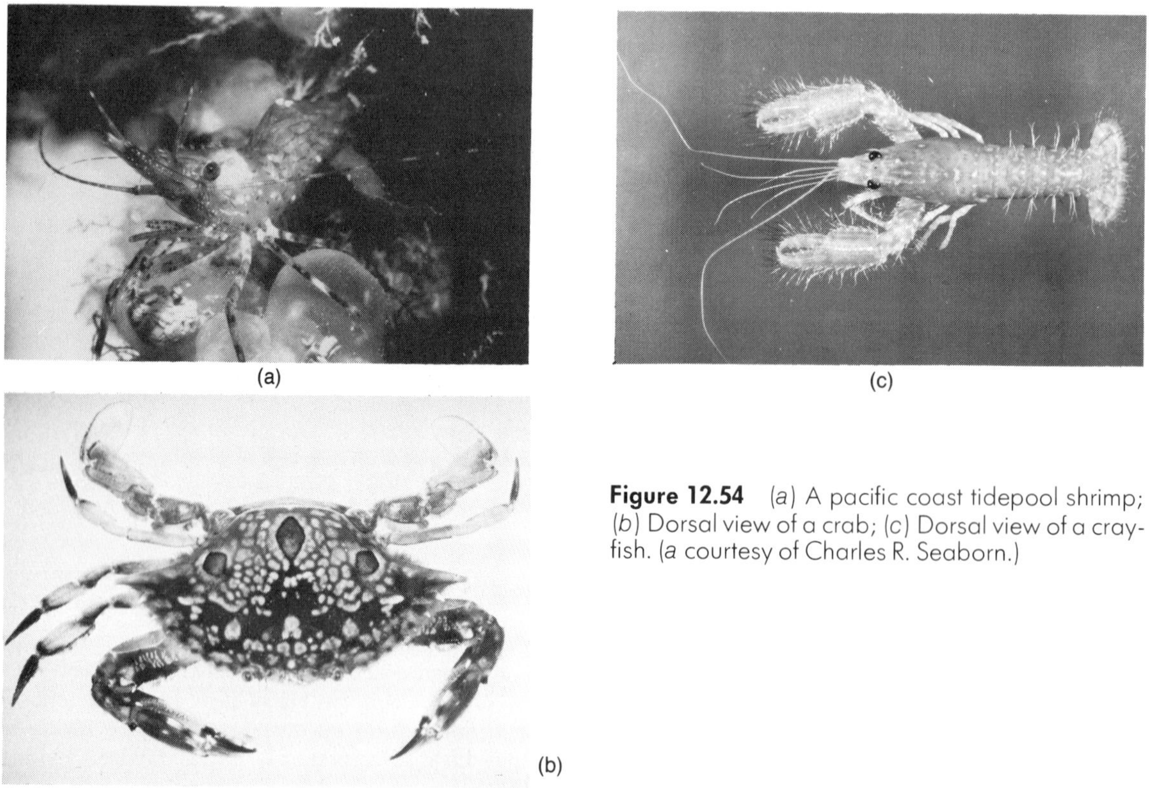

(a) (b) (c)

Figure 12.54 (*a*) A pacific coast tidepool shrimp; (*b*) Dorsal view of a crab; (*c*) Dorsal view of a crayfish. (*a* courtesy of Charles R. Seaborn.)

means of camouflage include complex cuticular ornaments and the growth of sessile organisms on the body surface.

Nutrition and Digestion

Decapods seldom have specialized diets. Most species feed on a wide range of plant and animal matter, both living and dead. Freshwater and terrestrial groups, however, tend to be more herbivorous. Most decapods seize food with their chelipeds or smaller chelate legs and then transfer it to the mouthparts. Characteristically, the chewing mandibles, maxillae, and maxillipeds are stacked one atop the other. The third maxillipeds are outermost and, in crabs, bear flattened plates which cover the other appendages. This entire region occupies a ventral body wall depression called the **buccal frame** (Fig. 12.55). The buccal frame is bordered by an anterior plate (the **epistome**) and by the lateral walls of the carapace. Sharp, biting maxillipeds and maxillae hold food material securely while the mandibles chew.

Some decapods stuff detritus between the feeding appendages. Among these types, the borders of the maxillipeds may bear filtering setae. In some burrow-dwelling decapods, setae on the walking legs extract food particles from a current drawn by the pleopods. In other species, plumose second antennae filter out plankton.

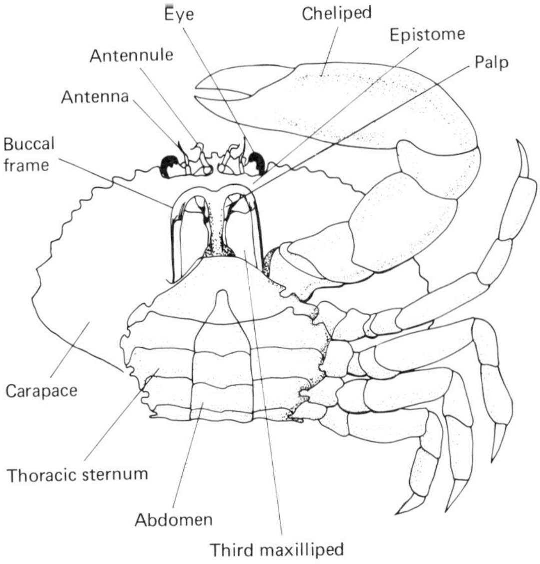

Figure 12.55 A brachyuran crab in ventral view showing the buccal frame. Right cheliped and pereiopods have been removed. (From Kaestner, *Invertebrate Zoology, Vol. III.*)

The decapod alimentary canal features a complex gastric mill, a midgut with an hepatopancreas, and a hindgut with its anus on the telson (Fig. 12.56). The gastric mill comprises an anterior cardiac stomach followed by a smaller pyloric stomach. In the cardiac stomach, food is ground and some digestion occurs when enzymes pour forth from the midgut. Variations in the musculature and cuticular teeth of the cardiac stomach reflect the diet and the chewing ability of the mandibles. Plankton and detritus feeders have a much simpler gastric mill, as do most shrimps, whose mandibles are well developed. Ground food is strained within the pyloric stomach (Fig. 12.31). The walls of this chamber are deeply folded and setose. The pyloric musculature regulates the size of the organ's channels while squeezing the food posteriorly. Final straining occurs within a gland filter, an area with exceedingly fine setae. Only the tiniest particles are admitted to the hepatopancreas. Oversized matter is deflected for direct passage through the intestine.

The decapod hepatopancreas comprises the tubular ceca, which arise from the anterior midgut. A single cecum or several paired ceca may be present, and the entire organ is often quite large. Enzymes are produced and released by its glandular walls. Digestion, which begins in the gastric mill, is completed in the midgut. Absorption and food storage occur within the hepatopancreas.

Circulation

The decapod heart is a rather compact, polygonal pump with three or five pairs of ostia (Fig. 12.56). Arterial systems attain their highest development in this crustacean order. Typically, the heart gives rise to an anterior aorta, two hepatic vessels, an abdominal artery, and a sternal artery. All of these vessels branch numerous times, and in larger decapods, capillary beds are formed. Eventually, blood is discharged into the tissue sinuses. A large ventral sinus collects spent hemolymph for aeration in the gills. Hemocyanin dissolved in the plasma facilitates oxygen uptake, and then the blood is returned to the heart. At an average rate of blood flow, a round trip through the decapod body requires up to 1 minute.

Respiration

Decapods respire across thoracic gills, which are housed within a branchial chamber formed by a deep lateral fold of the carapace (Fig. 12.4b). A decapod gill consists of a central axis and its lateral projections. The central axis houses afferent and efferent blood channels. The lateral projections assume one of three

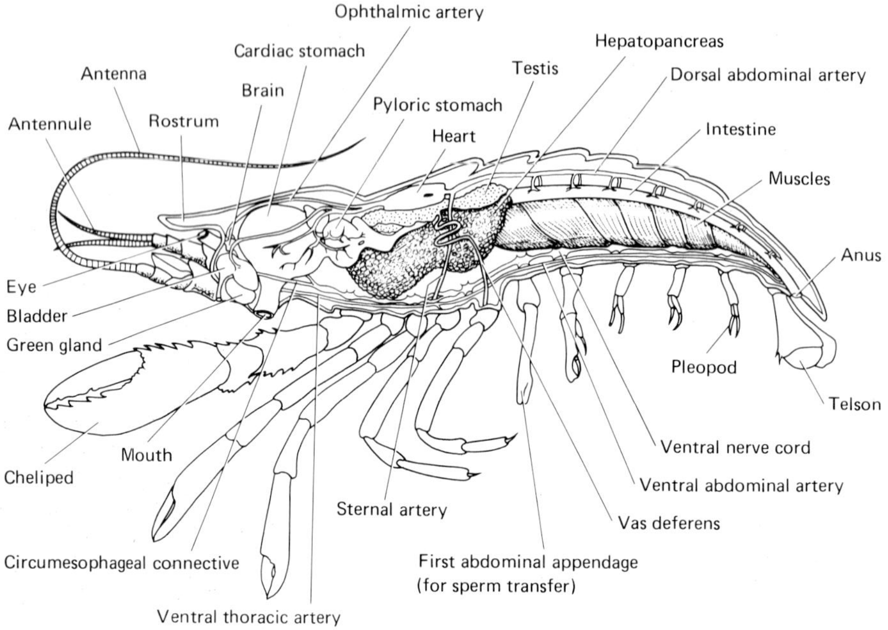

Figure 12.56 Diagram showing the internal anatomy of a lobster in lateral view. (From Hegner and Engemann, *Invertebrate Zoology*.)

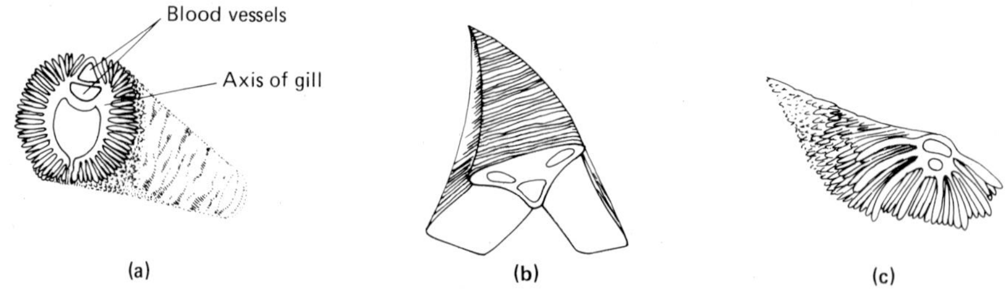

Figure 12.57 Decapod gill types. (*a*) Dendrobranchiate; (*b*) Phyllobranchiate; (*c*) Trichobranchiate. (From P.A. Meglitsch, *Invertebrate Zoology*, 2nd ed., 1972, Oxford University.)

basic forms, which are useful in taxonomic studies. The **dendrobranchiate** gill is typical of primitive shrimps; it bears paired lateral lobes, each of which has many branches (Fig. 12.57a). Most higher shrimps and crabs possess simple lamellar, or **phyllobranchiate,** gills (Fig. 12.57b). Lobster and crayfish have **trichobranchiate** gills, which bear numerous but unbranched lateral projections (Fig. 12.57c).

A ventilation current is produced by the beating of an extension of the second maxillae. This extension is called a **scaphognathite** or, more commonly a gill bailer. In shrimps, the current enters along the ventral and posterior margins of the carapace, flows across the gills, and exits by the head. The carapace is fitted more closely in lobsters and crayfish, and ventral incurrent openings are restricted to the region immediately surrounding the bases of the appendages. Among crabs, water enters the branchial chamber primarily around the chelipeds. Water flows through the gills and then exits anteriorly near the mouth (Fig. 12.58). Water flow near the buccal frame is important in the nutrition of filter-feeding groups.

Fouling of the respiratory surface is a potential

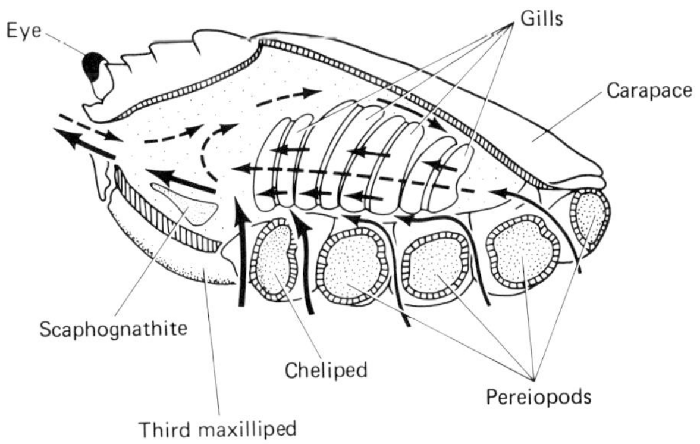

Figure 12.58 Water flow through the gill chamber of a brachyuran crab, *Carcinus*. Solid arrows indicate forward currents; dotted arrows indicate reversed currents. (From K.D. Arudspragasam and E. Naylor, J. Exp. Biol., 1964, *41*:306.)

problem for many bottom-dwelling decapods, particularly for burrowing species. Several mechanisms ensure that the gills remain clean. The general trend toward limiting the area of incurrent water flow minimizes fouling. Also, some decapods hold their antennae together to form a siphon along which clean water can be imported, while coxopodite setae strain any suspended particulate matter. Nevertheless, dirt still enters the branchial chamber. The gill bailer then may reverse its beat, causing a change in current direction which has a cleansing effect. In crabs, setose epipodites on the maxillipeds are brushed along the gills to remove clogged particles (Fig. 12.4b).

Decapods that spend all or part of their lives on land have made various respiratory adjustments. Some amphibious crabs flood their branchial chambers while in the water. When they crawl out onto land, maxilliped flaps seal the chamber openings and the animals continue to respire in an aquatic style. In some groups, the contained water can be reaerated: The gill bailer drives the branchial water out over cuticular troughs on the body surface, where the water absorbs fresh oxygen and then returns to the branchial chamber through its incurrent openings.

Terrestrial decapods have smaller and fewer gills then aquatic forms do. This situation reflects the selective pressure to minimize evaporative water loss and the superior availability of oxygen on land. Other respiratory adaptions recall those of terrestrial gastropods. Openings to the branchial chamber are smaller, and highly vascularized chamber walls are an important site of respiratory exchange. The gill bailer fans air in and out of the chamber.

Behavioral factors are critical in decapod respiration on land. Most land decapods are crabs. Terrestrial species tend to be nocturnal, and daytime burrowing is very common. During dry periods, crabs burrow toward the water table.

Excretion and Osmoregulation

Decapod antennal glands consist of a saccule, a labyrinth, an excretory tubule, and a bladder (Fig. 12.5). Because the folded glandular walls of the labyrinth are often green in color, the entire organ sometimes is called the green gland. The dominant nitrogenous waste is ammonia, which exits through the antennal pores and over the gill surfaces. Antennal glands regulate internal fluids and ions. Most marine decapods are osmotic conformers and, when placed in a hyposmotic environment, passively absorb water. The subsequent increase in internal pressure causes more filtration through the antennal gland and the passage of more urine. Because most decapod urine is essentially isosmotic with the hemolymph, internal osmotic levels drop to those of the environment. Truly marine decapods then die, but those adapted to brackish or freshwater environments begin some form of osmoregulation. This regulation involves the resorption of ions by the antennal gland and the uptake of salts by the gills. For example, freshwater crayfish excrete a copious urine that may be 10 times less concentrated than their hemolymph (Fig. 12.59). This dilute urine results from an active resorption of salts in the tubular portions of the green glands. Crayfish gills also take up salts. In these animals, the gills and body surface are less permeable to water than in marine decapods.

Terrestrial decapods have made few excretory modifications. Some species apparently produce less urine than comparable aquatic decapods, but there is little difference in the composition of the urine itself. Although small amounts of uric acid may be produced, ammonia remains the primary nitrogenous waste. The behavior of terrestrial decapods reflects their potential dehydration problems. As mentioned previously, they may burrow in the daytime and drink

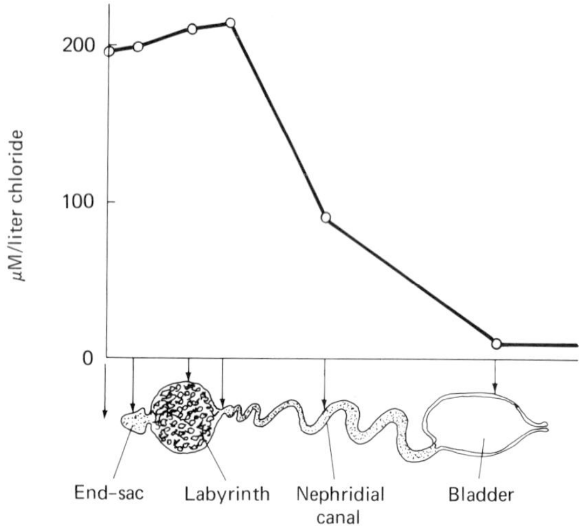

Figure 12.59 Diagram of an antennal gland of a crayfish, *Astacus astacus*, together with data on the chloride content of the excretory fluid. (From G. Parry, Excretion, In *The Physiology of Crustacea*.)

often, and many species periodically immerse themselves in water.

ACTIVITY SYSTEMS

Muscles and Nerves

The decapods are the largest crustaceans, and their motile abdomens and limbs are well muscularized. Everyone who has eaten crab legs and lobster tail can attest to the substantial fibers in these body regions. Decapods have a well-developed brain, whose protocerebral lobes extend into the eyestalks. The ganglia that service the mouthparts and all three pairs of maxillipeds have fused below the esophagus. Other thoracic and abdominal ganglia are often distinct along a single or doubled nerve cord (Fig. 12.56). In true crabs, however, all trunk ganglia have fused in a large nerve mass in the anterior thorax (Fig. 12.7c). Giant axons are common throughout this order.

Sense Organs

Stalked compound eyes lead the list of decapod sense organs. These structures are supported by jointed, movable peduncles, which allow scanning in several directions. Ommatidia are often numerous. Gravity centers are as important as eyes for many decapods, especially the burrowing forms. Statocysts are almost always present at the base of the antennules (Fig. 12.8d). In crabs, these organs often include a fluid-filled labyrinth whose nerve cells respond to rotation, acceleration, and other complex changes in motion. Such statocysts are similar to our own semicircular canals.

Other sense organs in the Decapoda include stretch receptors between the trunk segments, intramuscular tension-sensitive cells, and **myochordotonal organs.** The latter are nerve cell clusters on the joints of the walking legs and are probably proprioceptors. Esthetascs are present on the antennules, and the mouthparts are also well endowed with chemosensitive hairs.

Behavior

Decapods swim, jump, walk, and burrow with the most agile invertebrates. Shrimps have strong swimming pleopods fringed with natatory setae. Their laterally compressed body, pointed rostrum, and antennal scales are adaptations for swimming. Subtle abdominal flexions are important in steering, while rapid flexions result in sudden backward leaps. Lobsters and crayfish jump backward in the same manner. These primarily walking decapods swim only to escape danger. In some crabs the last thoracic legs are flattened as swimming paddles (Fig. 12.76a).

Most decapods walk on their thoracic legs, which are sturdy and well muscularized among the non-shrimp members of the order. When large chelipeds are present, they are not employed in locomotion. In crabs, the reduction and ventral flexion of the abdomen have shifted the body's center of gravity forward, so that the walking legs efficiently support the entire weight of the animal. This condition makes crabs the fastest and most agile members of the Reptantia. They can run forward or backward, and many move sideways.

Burrowing is practiced throughout this order. Some shrimps use their pleopods to excavate homes in soft sand or mud. Reptantia usually dig with their thoracic appendages, although mole crabs use their abdomens. Burrowing is particularly important for freshwater and terrestrial forms. These decapods escape the periodic severities of their environments by withdrawing underground. Most land species remain in burrows during daylight hours. By night (or anytime during heavy rains) they run about quickly; some tropical species climb trees.

Many decapods exploit natural shelters among rocks, vegetation or invertebrate colonies. Several species are commensals in polychaete tubes in coral heads, or in the cavities of mollusks and echinoderms. Others, such as the hermit crabs, live inside abandoned gastropod shells. This tendency to hide is very strong among decapods and must reflect their edibility. Chelipeds and other appendages, such as the spiny antennae of certain lobsters, are effective weapons, but camouflage is often a more successful means of protection. Decapod chromatophores are well developed, and many species (especially shrimps) assume a body color that blends well with the local background. Other individuals attach various objects to their outer surfaces (Fig. 12.60). A crab encrusted with stones, algae, anemones, bryozoans, sponges, and the like is very difficult to identify when motionless (Fig. 12.74d). One crab (*Dromia*) cuts out a small piece of sponge and places it over its carapace (Fig 12.74b).

Figure 12.60 (*a*) A decorator crab, *Stenocionops furcata*, detaches an anemone from a stone; (*b*) The crab lifts the anemone with its right cheliped and brings it down to the carapace for attachment. (From C. Cutress, D.M. Ross, and L. Gutton, Can. J. Zool., 1970, *48*:371. Photos courtesy of D.M. Ross.)

CONTINUITY SYSTEMS

Reproduction

Male decapods possess a pair of gonopores on or near the coxopodites of the last thoracic segment. Ducts lead from these pores to paired, tubular testes. Distally, the ducts produce spermatophores or mold outgoing sperm into compact cords. An ejaculation musculature is present, and in true crabs there may be a penis. The ovaries resemble the testes, and oviducts open on the female's sixth trunk segment. When the male has a penis, the female oviduct is modified distally as a vagina and seminal receptacle. Otherwise, an independent sperm storage area may be present on the thoracic sternites.

Sexually mature decapods locate mates by sight, sound, and/or smell. Pheromones are particularly important among marine and freshwater crabs. Often chelipeds are more highly developed in the male, who uses them for sexual display and precopulatory caresses. In certain crabs, pairing occurs prior to the sexual molt of the female; the male faithfully attends her until her "coming out" makes copulation possible.

The mating couple maneuvers until their ventral regions are applied (Fig. 12.61). The first two pairs of male pleopods are modified for sperm transfer. Semen is pumped out along these limbs, which are grooved or enrolled as conducting cylinders. When the pleopod tips are inserted into the openings of the seminal receptacle, sperm flow into the female. If there is no seminal storage area, sperm are molded into spermatophores, which are cemented near the female's genital pores.

Figure 12.61 Copulation in the crayfish, *Orconectes propinquus*. Male above. (From D.W. Crocker and D.W. Barr, *Handbook of the Crayfishes of Ontario*, 1968, Life Science Misc. Publications, Royal Ontario Museum.)

Development

Although terrestrial crabs mate on land, they must return to water to release their young. Similarly, many marine species whose adults migrate into brackish areas require normal sea water to reproduce; their larvae cannot survive in a dilute medium.

Certain primitive shrimps release their fertilized eggs into the ocean, but brooding occurs in all other decapods. Embryos are cemented against the pleopods until they hatch (Fig. 12.62). This order presents an array of larval types. Each major group displays its own developmental program, reflecting the adult form, larval ecology, and evolutionary history of its members. Characteristic stages include the nauplius, protozoea, zoea, and postlarva.

As among other crustaceans, the decapod **nauplius** is an unsegmented phase with three pairs of appendages (Fig. 12.63a). This stage is free-living only in the nonbrooding shrimps; in all other decapods, it is passed in the egg. The **protozoea** follows the nauplius by several molts. There, thoracic and abdominal segments appear and the carapace takes shape (Fig. 12.63b). There is no fusion between the carapace and the thorax, however, and cephalic appendages still

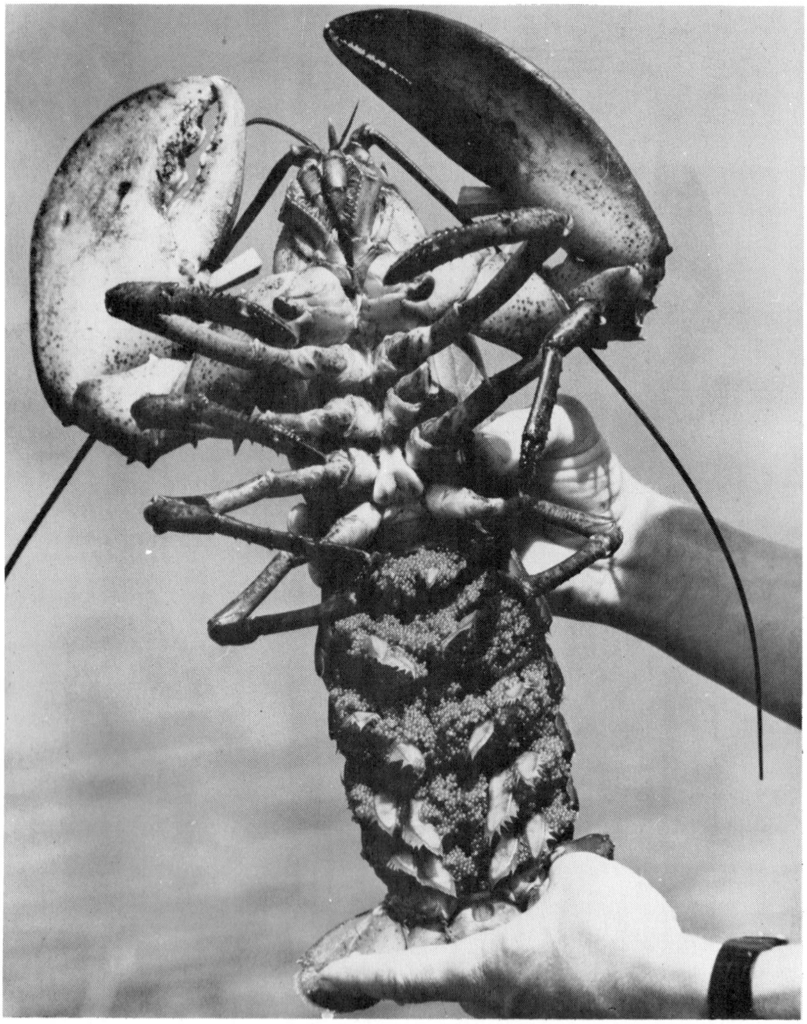

Figure 12.62 Ventral view of female lobster, *Homarus americanus*, with eggs attached to pleopods. (From J.E. Bardach, et al., *Aquaculture: The Farming and Husbandry of Freshwater and Marine Organisms*, 1972, Wiley-Interscience. Photo courtesy of D.M. Ross.)

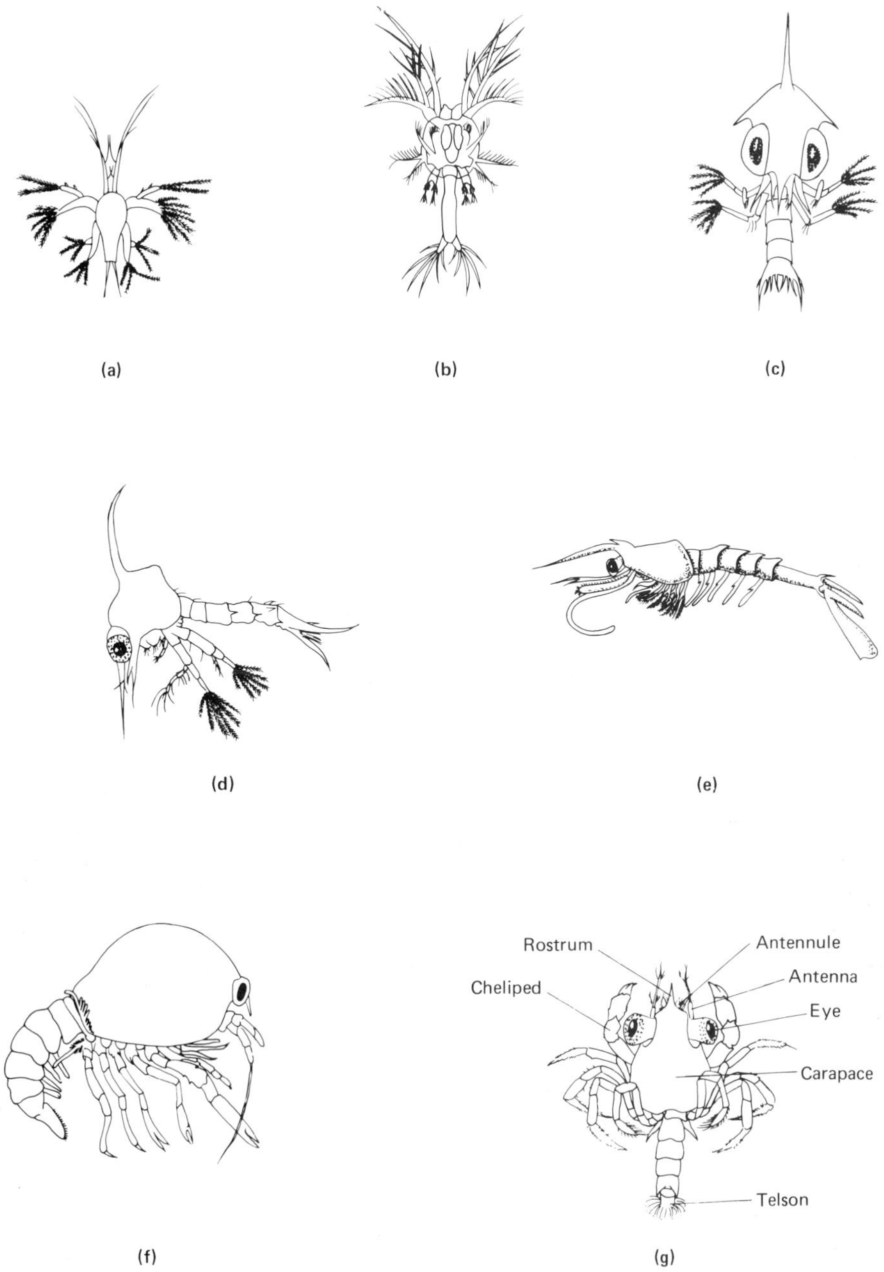

Figure 12.63 Decapod larval forms (not to scale). (*a*) Nauplius of *Penaeus*, a shrimp; (*b*) Protozoea of *Penaeus;* (*c*) First zoea of the blue crab, *Callinectes sapidus*, ventral view; (*d*) Lateral view of third zoea of *Callinectes;* (*e*) Postlarval mysis stage of *Penaeus;* (*f*) Newly hatched post-larva of a crayfish, *Astacus fluviatilis;* (*g*) Megalops of *Callinectes*. (*a*, *b*, and *e* from Hegner and Engemann, *Invertebrate Zoology; c* and *d* from E.P. Churchill, Chesapeake Biol. Lab. Publ., 1942, *49*:3; *f* from Calman, *The Life of Crustacea; g* from Waterman and Chace, In *The Physiology of Crustacea, Vol. 1*, Waterman, Ed.)

control locomotion. Brooding shrimps hatch as protozoea larvae, but in the Reptantia this stage also is passed in the egg. Most lobsters and crabs hatch as **zoea** larvae. During this stage, the remaining trunk segments and their limbs are added and the carapace fuses with the thorax (Fig. 12.63c, d). A prominent anterior rostrum is common, and stalked eyes appear. At the zoea stage, locomotion becomes the domain of trunk appendages. Several zoeal molts give rise to the **postlarva** (Fig. 12.63e, f, g), a stage characterized by a complete set of segments and their functional appendages. In most decapods, the postlarva resembles the adult; only sexual structures may be missing. Among true crabs, however, the postlarva or **megalops** has an extended abdomen (Fig. 12.63g).

Adult lobsters grow and molt throughout their lives. Their considerable size—large ones weigh over 5 kilograms (kg)—is thus possible. In contrast, many other decapods stop growth and molting once sexual maturity is attained.

Decapod Diversity

The almost 8400 known species attest to the diversity of the Decapoda, which is expressed by divergent body form and variations in habitat and life-style. The swimming Natantia and the crawling Reptantia define the principal dichotomy of the group. Within these divisions, further classification is based largely on structural and behavioral differences.

SUBORDER NATANTIA

This suborder includes all the shrimps. Its members are adapted primarily for swimming, although bottom crawling and even burrowing are pursued by some species. Actually, few shrimps are truly pelagic, and most spend a considerable portion of their time on the seafloor. There are some freshwater forms and one known semiterrestrial species. Shrimps are 1 to 20 cm long. The body is compressed laterally, as is the anterior rostrum. Thoracic legs are relatively slender, and the pleopods are well developed and fringed with natatory setae. An antennal scale provides a steering rudder. About 2000 species are listed among three sections: the Penaeidea, the Caridea, and the Stenopodidea.

The Penaeidea

Only about one-sixth of all shrimp species are penaeids, which are among the most primitive decapods. Distinguishing characters include chelate third legs and dendrobranchiate gills. Penaeids never brood their young, and they are the only decapods that hatch as nauplius larvae. Littoral habitats are common.

Penaeus (Fig. 12.64) is a large genus typical of the group. These cosmopolitan shrimps live on shallow sandy bottoms, where several species dominate the catch of Atlantic and Gulf Coast fishermen. Another penaeid genus, the pelagic *Sergestes*, contains many luminescent species. Light production occurs within photophores located on the body surface or within the hepatopancreas. Luminescence is characteristic of several shrimp groups which swim through deeper waters.

The Caridea

Most shrimps belong to the Caridea. This group is distinguished by phyllobranchiate gills and nonchelate third walking legs. Usually, the abdomen is sharply bent (Fig. 12.65d). Like other higher decapods, carideans brood their young against the pleopods; hatching occurs at the protozoea stage. The caridean membership is primarily marine, but includes important freshwater families.

The largest group of carideans includes the snapping or pistol shrimps. These animals inhabit burrows or crevices; using sharp legs as needles, *Alpheus pachychirus* (Fig. 12.65a) sews a tubular home from a sheet of filamentous algae. Pistol shrimps possess a single grossly enlarged cheliped (Fig. 12. 65b). When this claw snaps shut, a sharp cracking sound is produced. The noise may stun small fish that wander too near the shrimp's retreat. Cheliped snapping also may discourage predators, but its primary purpose proba-

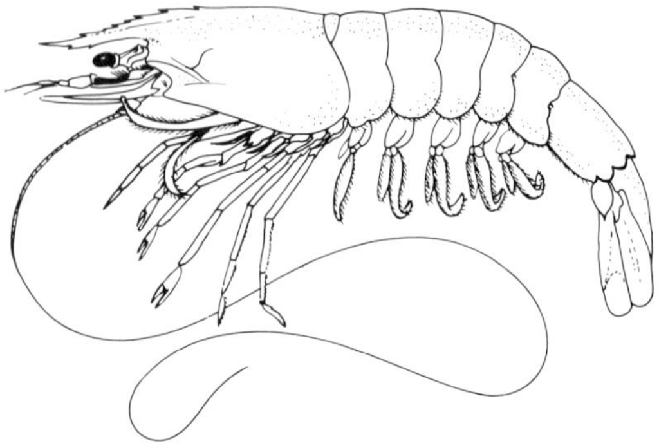

Figure 12.64 Lateral view of female *Penaeus setiferus*, a commercially important shrimp. (From I.P. Farfante, Fisheries Bull., 1969, 67:461)

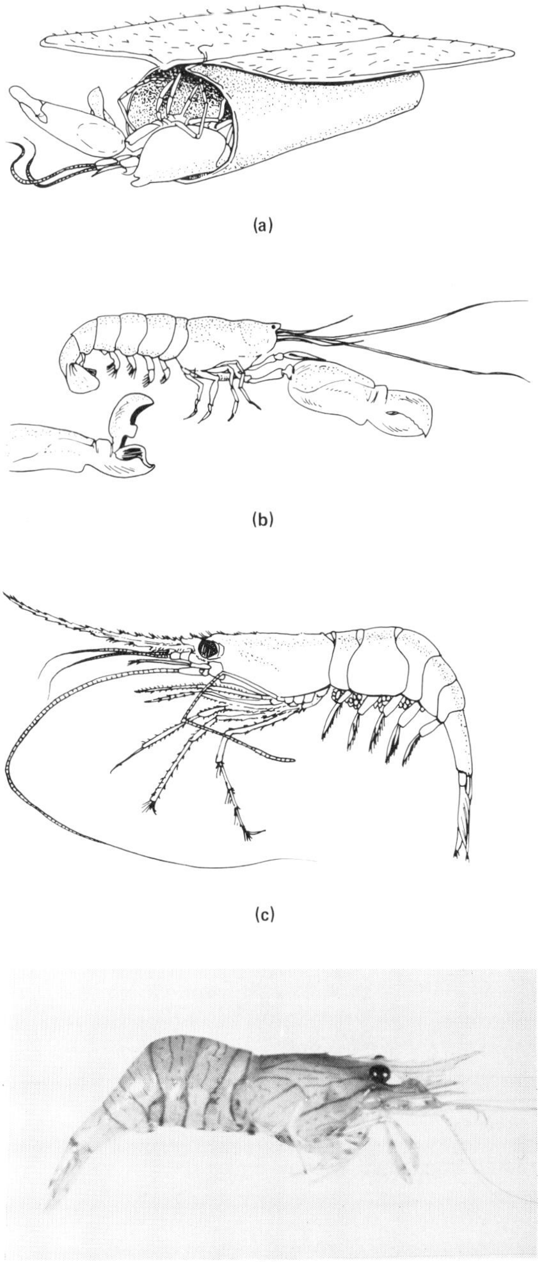

Figure 12.65 (*a*) *Alpheus pachychirus* in its algal tube; (*b*) *Alpheus californiensis*, a pistol shrimp. Below is a "cocked" pistol cheliped; (*c*) The Norwegian deep water water shrimp, *Pandalus borealis;* (*d*) A swimming brackish water shrimp, *Palaemon macrodactylus*. (*a* from W.L. Schmitt, *Crustaceans*, 1965, University of Michigan; *b* from G.E. MacGinitie and N. MacGinitie, *Natural History of Marine Animals*, 2nd ed., 1968, McGraw-Hill; *c* from Calman, *The Life of Crustacea; d* from J.A.L. Cooke.)

bly involves intraspecific communication. Apparently, pistol shrimps sound off to warn potential neighbors against locating too close to their own territory.

Sand shrimps of the genus *Crangon* burrow in soft bottoms during the day. Only the eyes of the entrenched animals protrude above the ground. At night, they emerge to feed on small invertebrates, algae, and detritus. The Baltic *Crangon crangon* is caught for livestock feed. Members of this genus are frequent subjects of physiological experiments, including chromatophore research. Another caridean shrimp whose chromatophores have been studied is *Hippolyte. Hippolyte varians* schools along the Atlantic coast of Europe, where it finds protection among green, red, and brown algae. The shrimps may locate near an algal bed of matching color, or their chromatophores may adapt over several days to resemble the plant background. Even mottled surroundings can be emulated. At night, these animals turn a translucent blue. *H. acuminatus* lives on sargassum.

Pandalus (Fig. 12.65c) is a protandric hermaphrodite. To the age of 2 years, most individuals are male; then they become females. *Pandalus danae* is the so-called coon-striped shrimp of the Northwest Pacific.

*Palaemon (*Fig 12.65d) and *Palaemonetes* include many common prawns. Numerous species inhabit marine coasts and brackish and fresh waters. *Macrobrachium rosenbergii* lives in freshwater and may grow to 24 cm; it is the object of extensive aquaculture research. *Atyaephyra desmarestii* is populous among river and stream vegetation in Europe. Commensal carideans include *Trypton*, which inhabits sponge channels, and *Anchistus custos*, which lives in pen shells. Only one semiterrestrial species has been described. This shrimp, the tropical *Merguia rhizophorae*, leaps among mangrove roots along Central and South American beaches.

The Stenopodidea

The last small group of natant decapods is the Stenopodidea. Its members prefer warm marine waters, where they often are commensal on various colonial invertebrates. Gills are trichobranchiate, and large third chelipeds are common. *Stenopus* (Fig. 12.66) is a beautiful shrimp, whose iridescent body may display all the colors of the rainbow. These animals inhabit coral reefs, where they obtain much of their food by cleaning the body surfaces of fish. *Spongicola* is one of several members that live within sponges. Sexual pairs of *S. venusta* enter a sponge, whose growth eventually entraps them for life.

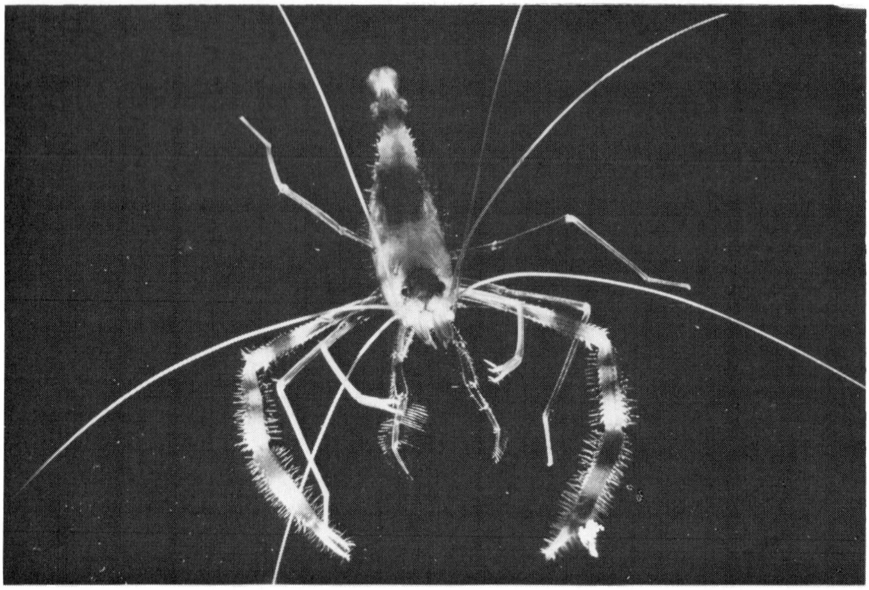

Figure 12.66 *Stenopus hispidus*, the barberpole or banded coral shrimp. (Courtesy of Kent W. Allen.)

SUBORDER REPTANTIA

This suborder features the lobsters, crayfish, and crabs. These decapods are adapted primarily for walking. Dorsoventral compression is common, and the rostrum may be reduced. The pereiopods are heavy and include an anterior pair of chelipeds. The abdomen, which may be reduced, never has swimming pleopods, and antennal scales are not developed. No member of the group hatches before the zoea stage. About 6400 species have been described among three major sections: the Macrura, the Anomura, and the Brachyura.

The Macrura

Macrurans are the more primitive members of the Reptantia. These crustaceans have a powerful elongate abdomen, whose uropods and telson form an effective tail fan. The most prominent macrurans are the marine lobsters and freshwater crayfish. Spiny lobsters, burrowing "shrimps," and some deep-water macrurans complete the list of 700 species.

Lobsters live among the crannies of rocky marine bottoms. Their tendency to maneuver into tight places facilitates their capture in fishing pots. *Homarus americanus* is the most commonly trapped lobster off the Maine coast. A few females have grown to over half a meter long and have weighed 20 kg. Such large animals may be 50 years old. The rostrum is prominent, and the chelipeds are strong and raptorial. These limbs are developed asymmetrically. On one side, the claw is small but sharp and cutting, while the opposite side supports an expanded pincer with rounded teeth used for crushing shells. Scavenging is common, although small fish, crabs, and snails also are eaten live and some lobsters dig out clams. Slow crawling over the bottom is the usual locomotion; *Homarus* ranges over many square kilometers of the seafloor. For escape, the animal depends on quick contractions of its strong abdominal musculature to spurt it backward through the water.

Crayfish are closely related to lobsters, but are smaller than their marine cousins, most species averaging about 10 cm in length. The giant of the group, Australia's *Euastacus armatus*, grows up to 50 cm long. Crayfish are the most successful decapods in fresh water, owing in part to their well-developed antennal glands. These excretory organs produce a copious, dilute urine while the gills actively absorb essential salts. Most of the 500 crayfish species are burrowers. Some of their homes extend to underground water and may serve as shelters during winter freezes or summer droughts. Certain crayfish, which rely primarily on an underground water supply, eventually might give rise to terrestrial species.

The European *Astacus* has a 20-year life span in

streams and ponds. When a fungus in the late nineteenth century decimated these edible crayfish, the North American *Orconectes* was introduced. This genus now flourishes in the Old World. It is hardly edible, however, and its success probably inhibits the recovery of *Astacus*. Many of the 66 species of *Orconectes* (Fig. 12.61) remain common in the Eastern United States and Canada. *O. limosus* does not burrow and may be active by day; it feeds throughout the year, surviving in deeper water in the wintertime. Another major North American genus is *Cambarus* (Fig. 12.67c), whose numerous species include burrowing and cave-dwelling forms. The burrowers may damage crop areas, as they excavate large quantities of moist soil and build mud chimneys around their burrow entrances. Some of these chimneys may be 30 cm high.

Typical of most freshwater invertebrates, crayfish larval stages are passed in the egg. For a brief time after hatching, juveniles may remain attached to maternal pleopods.

Spiny lobsters belong to a discrete macruran group. These animals have rather cylindrical carapaces and lack the heavy chelipeds of true lobsters. All pereiopods are similar and nonchelate; the rostrum is small. Instead of raptorial legs, spiny lobsters have long, sharply pointed second antennae. When these appendages are shaken, they rattle loudly and can also inflict painful wounds. Such talents are important in intraspecific aggression and defense from potential predators, such as octopods. Edible spiny lobsters include *Palinurus* in Europe and *Panulirus* (Fig. 12.68a) along both American coasts.

The related Spanish, or shovelnose, lobsters have flattened carapaces with sharp lateral edges. Their second antennae are short and bear leaflike processes with toothed margins. The edible *Scyllarus* (Fig. 12.68b) often is associated with coral reefs.

The burrowing shrimps include *Upogebia* and *Callianassa* (Fig. 12.69). These decapods, which may be classified as macrurans or anomurans, have flattened chelipeds and pleopods, which may be useful for occasional swimming. Their primary means of locomotion, however, is tunneling with the thoracic legs. Often, mazelike burrows are formed in the mud off both American coasts. In areas of high population density, burrowing shrimps may be important geologic agents. *Callianassa* feeds primarily on the minute organic matter in its excavated material and occasionally takes a polychaete. Beating pleopods circulate water through the U-shaped burrow of *Upogebia pugettensis*. This creature feeds on suspended detritus and plankton filtered by its setose legs. The tunnels of burrowing shrimps provide homes for many invertebrate commensals, including polychaetes, clams, copepods, and pea crabs.

A final macruran group comprises several deep-sea forms called eryonids. *Polycheles* (Fig. 12.70) is the best-known genus. Dorsoventral flattening is pronounced in these animals, and their first legs are often long, although narrow and chelate. Eryonids are ranked among the most ancient macrurans.

The Anomura

Anomurans include 1300 species which, because of their limited economic importance, are not as well known as the lobsters and true crabs. The most famous of these decapods are the hermit crabs, the remarkable little crustaceans that inhabit old gastropod shells. Other members of the group include the king crabs, coconut crabs, porcelain crabs, and sand or mole crabs.

Anomurans have a peculiar abdomen (Fig. 12.71a). Its structure is somewhat intermediate between the well-developed tail of the lobster and the small, ventrally flexed posterior of the true crab. The soft anomuran abdomen is long, but usually narrow and curved. Its locomotor value is negligible, and the pleopods are greatly reduced or absent. Asymmetry is common, but appears curiously late in larval development. The third thoracic legs always lack chelae, and the fifth pair is reduced and flexed dorsally. Scavenging and detritus feeding are the usual modes of nutrition. Other anomuran traits are highly variable and contribute to suspicions that the group is polyphyletic.

The habit of residing within gastropod shells has evolved at least twice within the group. One line includes a hundred species of *Pagurus* (Fig. 12.71), many of which live off European and North American coasts. A second line describes the land hermits and their relatives (e.g., *Birgus*; Fig. 12.72d). Abandoned shells are used, as there is almost never aggression against a living snail. Hermit crabs, however, may battle one another for a potential home (Fig. 12.71b). Apparently, there is little specificity as to shell origin; any home of an appropriate size, weight, and spirality is acceptable. The hermit crab inserts its asymmetrical abdomen into the mollusk shell and grips the columella with modified uropod hooks. Contraction by longitudinal muscles along the inner side of the spiral applies the abdomen tightly against the shell. The absence of pleopods from this side allows a close fit. Thoracic legs may brace against the shell, thus further increasing the hermit crab's grip. When the animal withdraws, large chelipeds usually block the shell opening.

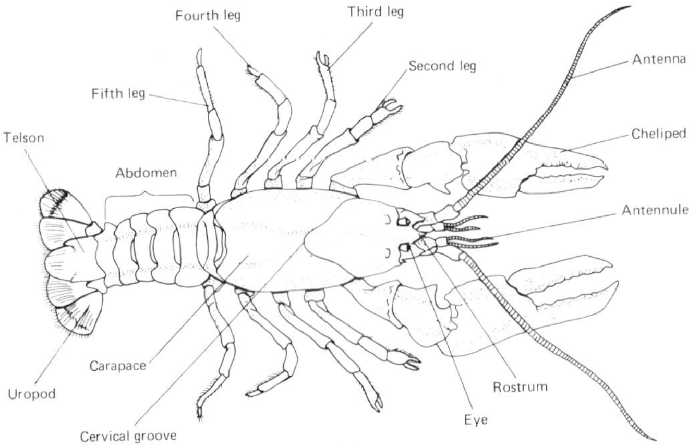

(a)

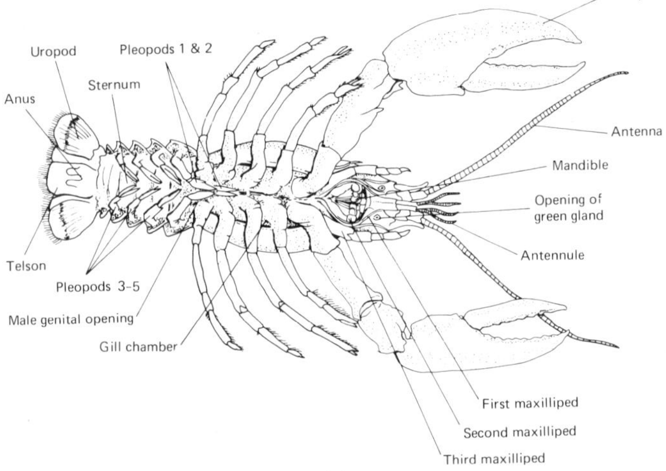

(b)

(c)

Figure 12.67 Dorsal (*a*) and ventral (*b*) views of the external anatomy of a crayfish, *Cambarus;* (*c*) Ventral view of a female crayfish, *Procambarus*, with juveniles attached to the pleopods. (*a* and *b* from I.W. Sherman and V.G. Sherman, *The Invertebrates: Function and Form: A Laboratory Guide*, 2nd ed., 1976, Macmillan; *c* courtesy of Carolina Biological Supply.)

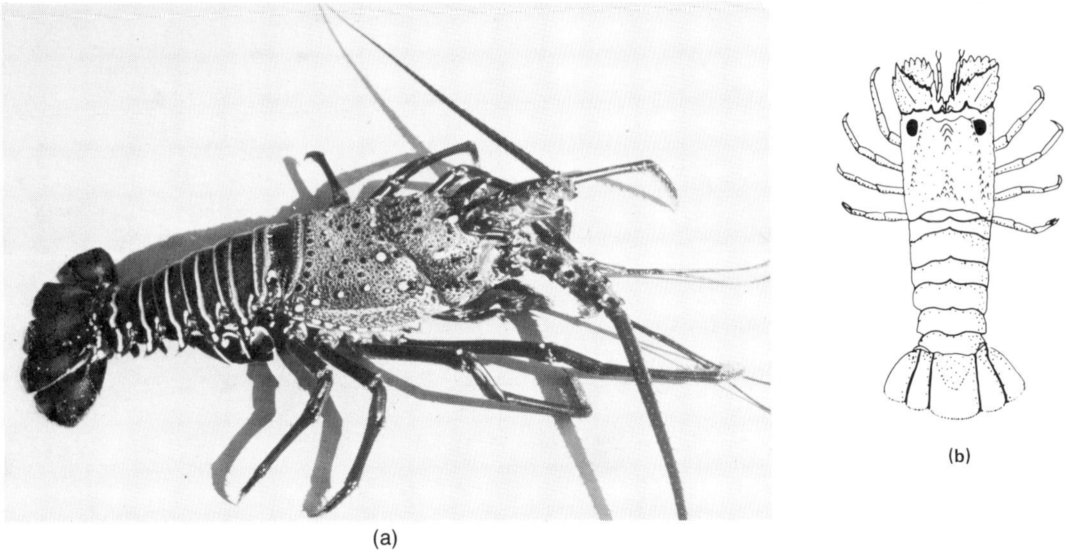

Figure 12.68 (*a*) The Atlantic coast spiny lobster, *Panulirus argus;* (*b*) A shovel nose lobster, *Scyllarus arctus.* (*a* courtesy of Jack Zbar and Coral Reef Photographers, Inc.; *b* from G. Smith, In *Cambridge Natural History, Vol. IV*, Harmer and Shipley, Eds.)

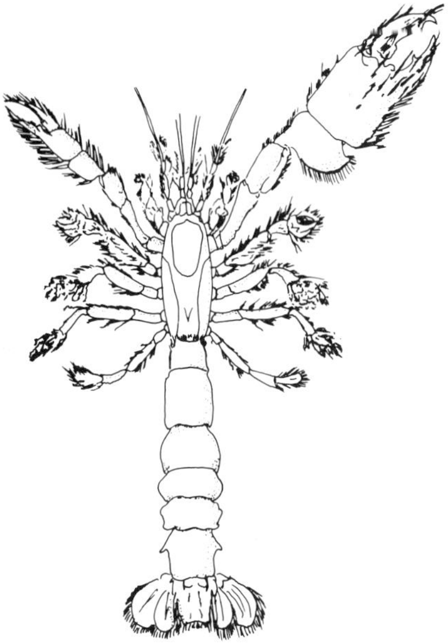

Figure 12.69 A burrowing shrimp, *Callianassa goniophthalma.* Note that one cheliped is larger than the other. (From Kaestner, *Invertebrate Zoology, Vol. III.*)

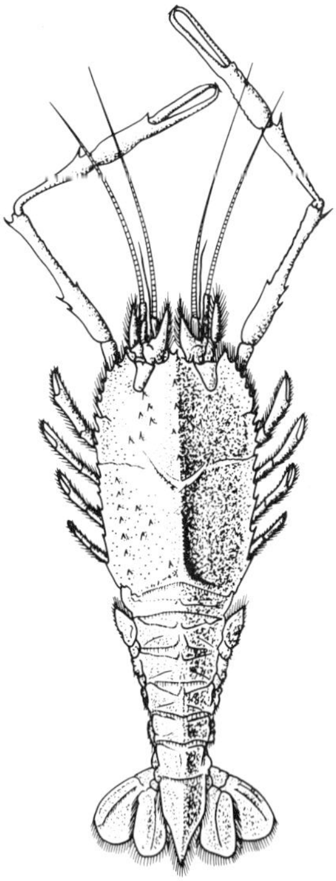

Figure 12.70 *Polycheles phosphorus*, a deep sea eryonid from the Indian Ocean. (From Calman, *The Life of Crustacea.*)

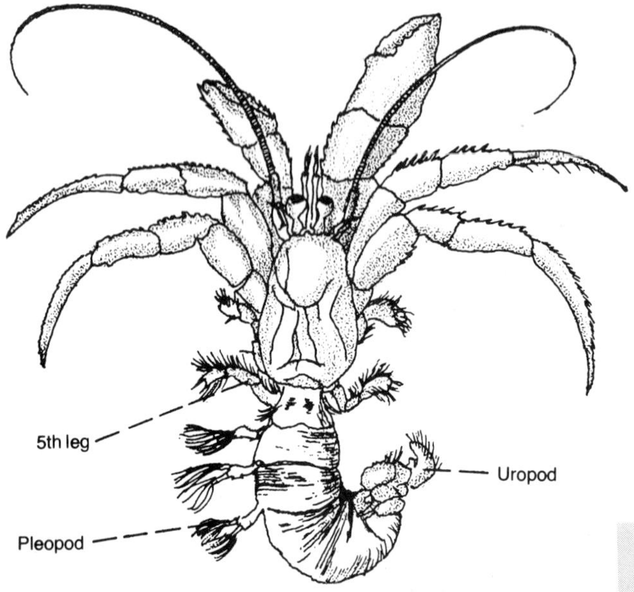

(a)

(b)

Figure 12.71 Hermit crabs. (*a*) *Pagurus* removed from shell; (*b*) *P. armatus* battling over a shell. One crab (right) is trying to evict its opponent; (*c*) *P. dalli;* (*d*) A crab inspecting a potential new home. (*a* from Barnes, *Invertebrate Zoology; b* courtesy of Charles R. Seaborn; *c* courtesy of Ward's Natural Science Establishment; *d* courtesy of New York Aquarium/ Osborn Laboratories of Marine Sciences.)

(c)

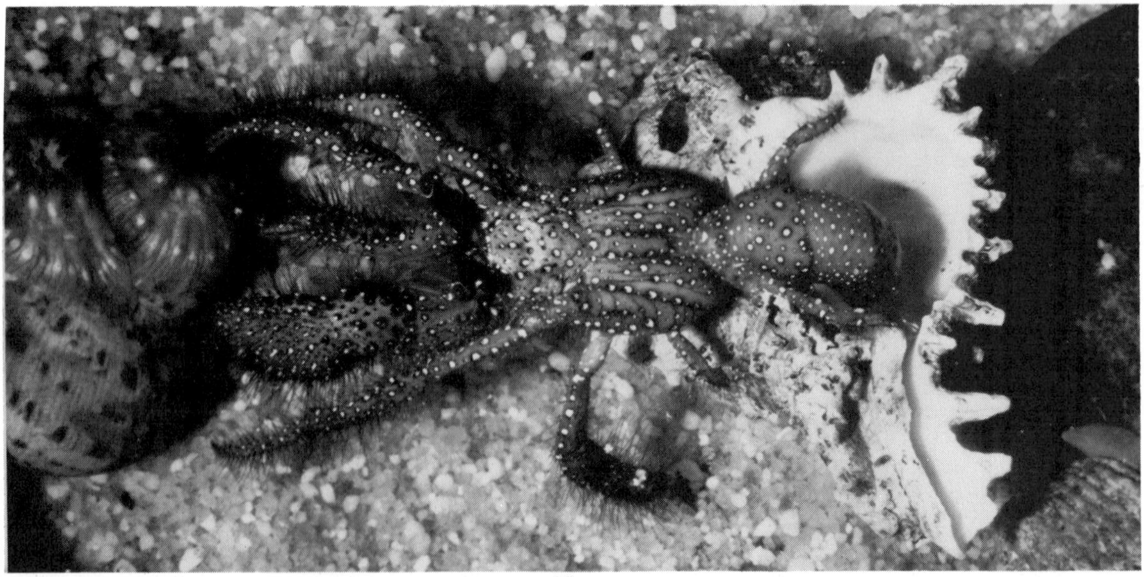

(d)

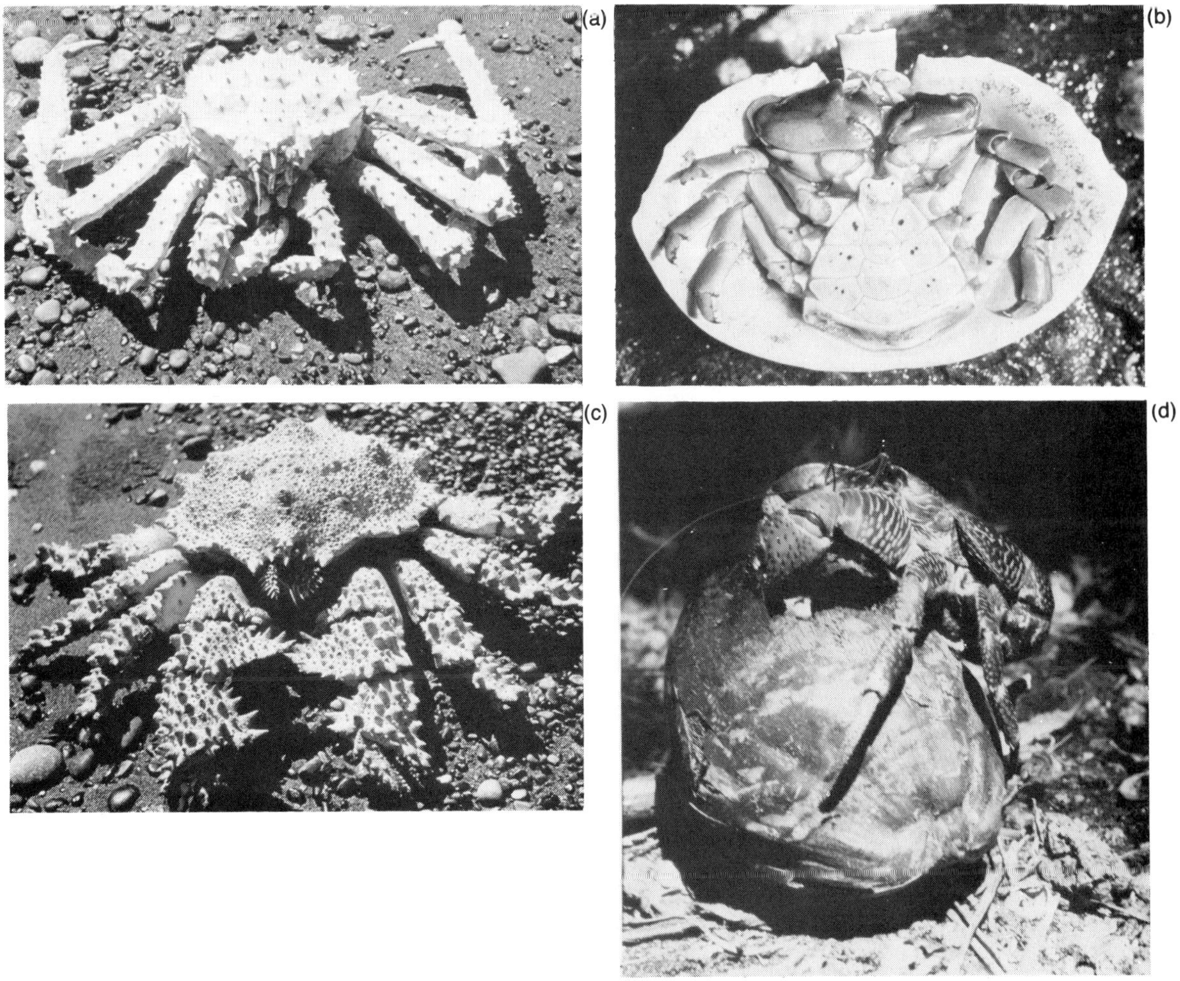

Figure 12.72 (*a*) A Pacific coast spider crab, genus *Lithodes;* (*b*) Ventral view of *Cryptolithodes sitchensis*, the umbrella-backed crab; (*c*) *Lopholithodes foraminatus*, a Pacific coast species with an unusual form; (*d*) The semi-terrestrial coconut crab of the Indo-Pacific region, *Birgus latro*. (*a, b* and *c* from Ward's Natural Science Establishment; *d* courtesy of S. Arthur Reed.)

As a hermit crab grows, periodic transfers to larger shells are necessary. The new shell is selected before the old one is abandoned (Fig. 12.71d), and the transfer takes place with discreet swiftness.

Obviously, the gastropod shell affords protection for the soft anomuran abdomen. This protection is enhanced when bryozoans, sponges, barnacles, and/or anemones attach to the shell surface (Fig. 12.71c). Encrusting bryozoans may dissolve the shell adopted by *Pylopagurus*. As a result, this hermit crab acquires a protective bryozoan cloak whose growth keeps pace with its own. Certain relatives of the hermit crab insert themselves in a variety of other retreats, including sponges, roots, and scaphopod shells.

King crabs and coconut crabs apparently evolved from anomurans whose abdomens were sheltered in shells or similar retreats. King crabs, whose cuticle is heavily calcified, include *Paralithodes camtschatica,* the commercial king crab of the North Pacific. Pleopod asymmetry indicates its anomuran heritage. The semiterrestrial coconut crab *Birgus* (Fig. 12.72d) lives in the Indo-Pacific. Larvae develop in the sea, and when they first crawl onto land, juvenile *Birgus* occupy gastropod shells. Adults, which may grow to lengths of 30 cm, have reduced abdomens and are unhoused. They climb high into coconut trees to feed on the fruit and to obtain husks for burrow insulation. Closely related to *Birgus* is the land hermit *Coenobita*. This anomuran lives in coastal burrows. It retains a snail shell about its abdomen and may go wading periodically to fill its shell with water.

Porcelain crabs are small anomurans which live among littoral rocks (Fig. 12.73a). They are filter feeders, collecting food on maxilliped combs. Owing to their ventrally flexed abdomen, porcelain crabs are often mistaken for brachyurans; however, uropods and asymmetrical pleopod development confirm their anomuran identity. A related group includes *Aegla,* the only known freshwater genus. Its 20 species inhabit temperate areas of South America.

Emerita (Fig. 12.73b) is representative of the sand or mole crabs. These anomurans burrow in sand along the wave line. Their abdomen is bent ventrally and their pereiopod tips are flattened for greater digging efficiency. Chelipeds are absent. Mole crabs extend their plumose second antennae into receding surf, from which they filter plankton and detritus. The first antennae may form siphons which draw respiratory water.

The Brachyura

Brachyurans are true crabs. With 4500 known species, they constitute the largest section of their class. Brachyurans are arguably the most highly developed crustaceans and, certainly, culminate several trends established by less complex forms. True crabs have a broad, flattened carapace, which unites most of their body as a cephalothorax. Anteriorly, the carapace is fused with the epistome (Fig. 12.55). The buccal frame is well developed, and broad third maxillipeds cover the other feeding appendages. Lateral to each second antenna is a stalked compound eye. Gills are phyllobranchiate.

The abdomen is reduced and flexed beneath the cephalothorax (Fig. 12.55). Female pleopods are retained for egg brooding, but in males only the first two pairs are present as copulatory appendages. Uropods are usually lacking. These changes in the brachyuran abdomen have shifted the body's center of gravity. The weight of these animals is balanced over the walking legs, a condition increasing their locomotor efficiency. Crabs are highly mobile for crustaceans and often walk sideways. In this type of movement, leading legs pull while trailing legs push. During an extended walk, periodic 180° turns allow the legs to alternate roles. The first pereiopods are raptorial chelipeds, which function in prey capture and defense but play little role in locomotion. Brachyurans hatch as zoea larvae. Their life cycle includes a distinctive **megalops** stage whose abdomen is extended in the macruran fashion (Fig. 12.63g).

True crabs have diversified in structure, habits, and habitats. Architectural details of the carapace, buccal frame, third maxillipeds, and pereiopods are important taxonomic keys. Crabs range from the ocean depths to the intertidal zone, into fresh water, and even onto land. Whether crawling or burrowing, hiding or threatening attack, these clever animals rank among the most fascinating members of the invertebrate world.

Apparently, the most primitive brachyurans are burrowers on deeper littoral bottoms. These crabs have a long carapace and flattened pereiopods for digging in

(a)

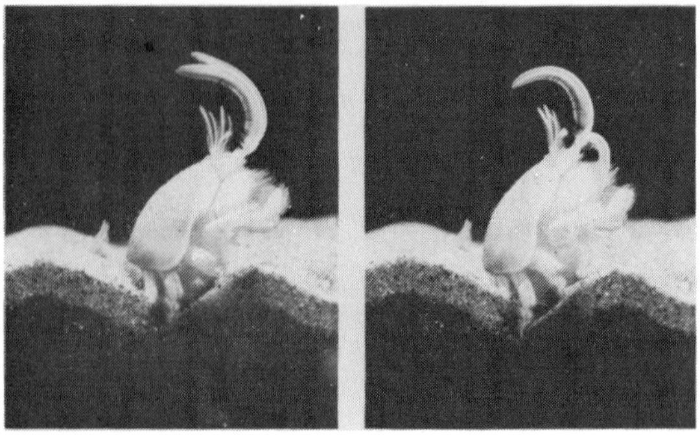

(b)

Figure 12.73 (*a*) *Petrolisthes eriomerus*, a porcelain crab; (*b*) The mole crab, *Emerita analoga*, feeding. At left the antennae are extended into the current; at right accumulated food particles are scraped from one antenna by specialized appendages. (*a* courtesy of Ward's Natural Science Establishment; *b* from Ricketts, et al., *Between Pacific Tides*.)

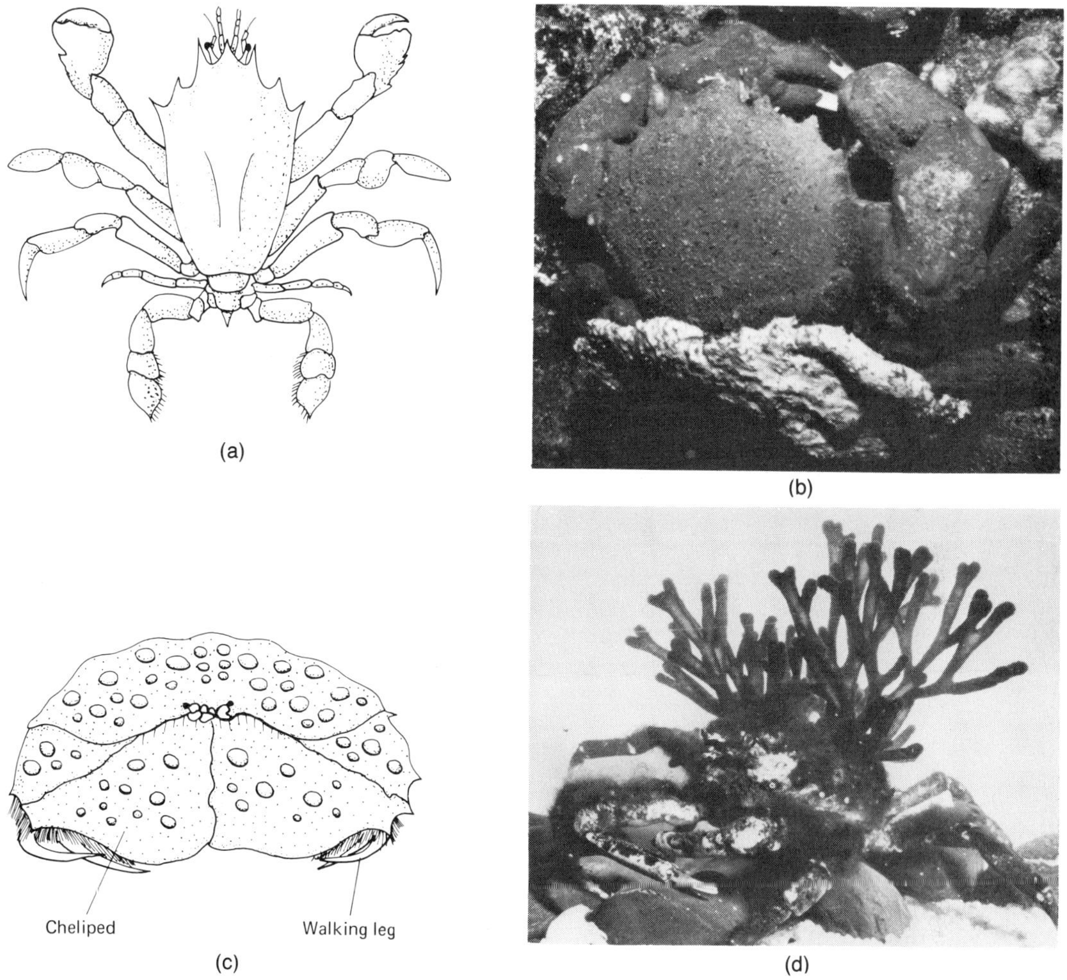

Figure 12.74 (*a*) A primitive burrowing brachyuran of the Indian Ocean, *Lyreidus channeri;* (*b*) *Dromia* with sponge on dorsal surface; (*c*) Frontal view of *Calappa granulata* showing the enormous chelipeds hiding the face; (*d*) *Libinia dubia*, camouflaged with a covering of algae, hydroids, and bryozoans. (*a* from Kukenthal and Krumbach, *Handbuch der Zoologie, Band III; b* courtesy of Charles R. Seaborn; *c* from *Cambridge Natural History*, Harmer and Shipley, Eds; *d* courtesy of New York Aquarium/Osborn Laboratories of Marine Sciences.)

soft substrata (Fig. 12.74a). Another primitive group includes *Dromia* (Fig. 12.74b), a small crab that wears a sponge cap for camouflage. In *Dromia* and its close relatives, the last two pairs of pereiopods are flexed upward to hold various disguises over the carapace. In *Dorippe*, for example, these limbs hold mollusk shells, ascidians, or even a severed fish head for camouflage. Other primitive types include the box crabs of the genus *Calappa* (Fig. 12.74c). Germans call these animals "shame-faced crabs" because they hold enlarged chelipeds over their eyes and mouthparts. These strong pincers can cut open a gastropod shell.

Disguises are especially common among the spider crabs. These marine decapods have a triangular carapace, which often is encrusted with sponges, colonial coelenterates, bryozoans, and algae (Figs. 12.60 and 12.74d). Spider crabs include the world's largest living arthropod, the Japanese *Macrocheira kaempferi* (Fig. 12.75).

Crabs are not known for their swimming ability. In one group, however, the portunids, flattened fifth pereiopods form effective swimming paddles. Some of these crabs can even catch fish. Portunids include the Atlantic blue crab, *Callinectes sapidus* (Fig. 12.76a), which is superbly edible. In Europe, most edible crabs belong to the genus *Cancer*, a nonswimming group.

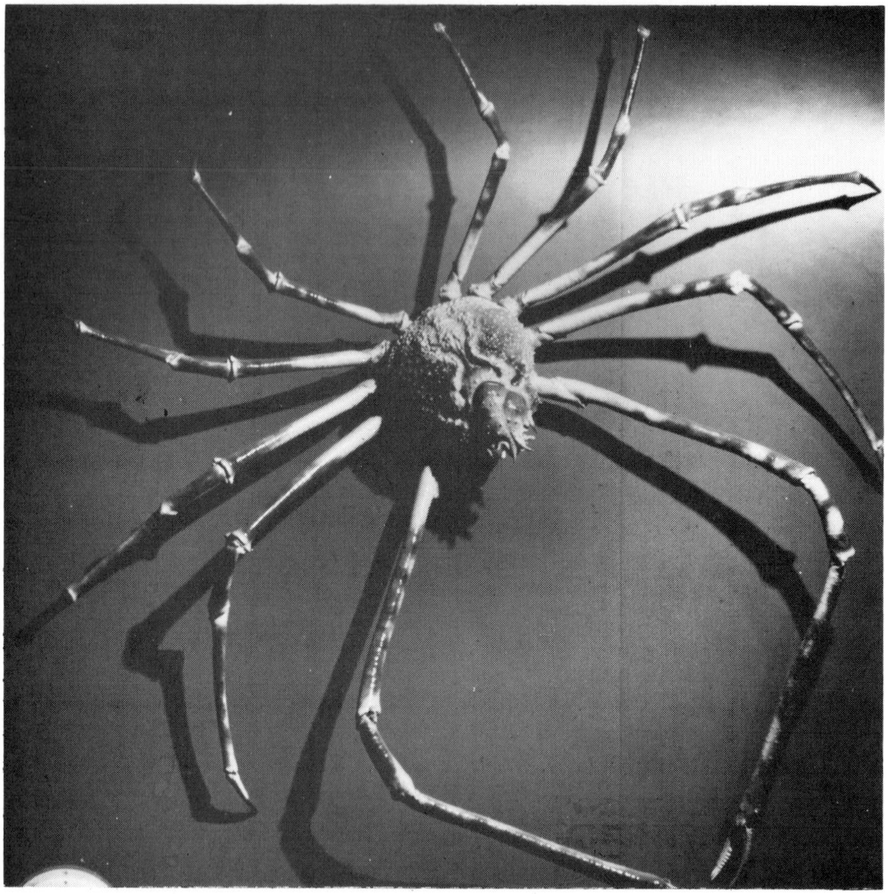

Figure 12.75 A model of a Japanese spider crab, *Macrocheira kaempferi,* the world's largest living arthropod, on exhibit at the American Museum of Natural History. (Courtesy of American Museum of Natural History.)

Cancer magister (Fig. 12.76b), the Dungeness crab, is fished along our Pacific coast.

Most commensal brachyurans are small pea crabs. Many share the homes of burrowing or tube-dwelling invertebrates, including polychaetes and burrowing shrimps. The mantle cavities of mollusks and the pharynxes of tunicates are inhabited by some pea crab species. *Pinnixa chaetopterana* is a filter feeder which lives with the polychaete *Chaetopterus.*

The coral gall crab *Hapalocarcinus* is one of the most interesting commensals in the invertebrate world. The female larva settles on branching coral. The young crab's feeding and ventilation currents influence the pattern of coral growth until eventually the decapod is encaged by bars of calcium carbonate. Between these bars, planktonic food and the minute male of the species easily enter; however, the female remains imprisoned for life.

Several brachyuran groups live in brackish and fresh water. Many, such as the Chinese mitten crab *Eriocheir,* return to the sea to reproduce. *Eriocheir* is a fair swimmer and is common in Asian rivers and rice paddies. Incidental transport in ballast water has introduced growing *Eriocheir* populations in the navigable rivers of Northern Europe. Crabs that spend their entire lives in fresh water are mostly tropical and subtropical. One such species, the edible river crab *Potamon fluviatile,* is found in the Mediterranean area.

Because of their superior mobility and the relative tightness of their branchial cavity, brachyurans are preadapted for amphibious and terrestrial life. Crabs are never vigorous land dwellers, however, and most must drink copiously or often immerse themselves in water. Breeding always requires a return to the ancestral aquatic environment. Among the most successful

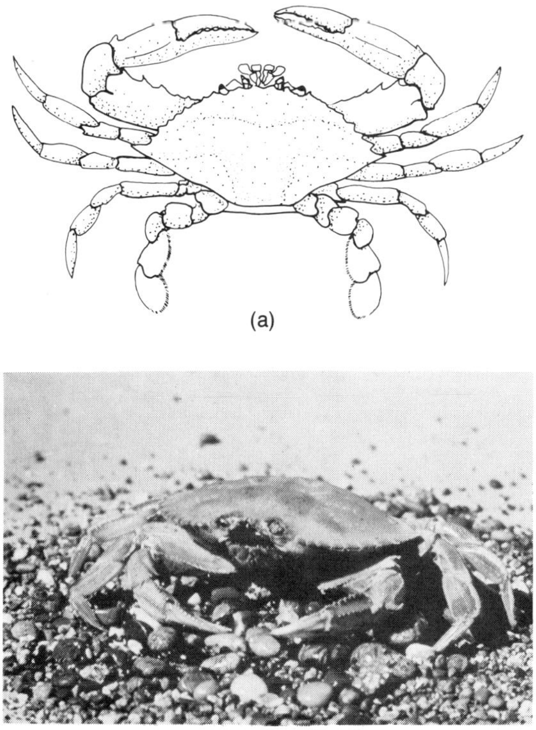

(a)

(b)

Figure 12.76 (*a*) The blue crab, *Callinectes sapidus*, an abundant portunid crab along the Atlantic and Gulf coasts of North America; (*b*) *Cancer magister*, the Dungeness crab of the Pacific coast. (*a* from Kaestner, *Invertebrate Zoology, Vol. III; b* courtesy of Ward's Natural Science Establishment.)

amphibious invertebrates are the fiddler crabs *(Uca)* and the ghost crabs *(Ocypode)*.

Uca (Fig. 12.77a) burrows in intertidal mud or sand. When the tide is out, it emerges to feed on detritus and carrion. Marked sexual dimorphism typifies this genus. *Uca* females have small symmetrical chelipeds, but in the male, one of these limbs is enormous. During courtship, the male brandishes his enlarged cheliped or thumps it on the beach surface (Fig. 12.77 b, c, d, e). Copulation may occur after the male has lured a sexual partner into his burrow. Cheliped displays are also important in other behavioral areas, including intraspecific aggression and defense.

Ocypode (Fig. 12.78a) can burrow in sand above the high-tide mark. At speeds of over 1.5 m/sec, ghost crabs are among the fastest decapods afoot. Other semiterrestrial brachyurans include *Grapsus,* which lives on coastal rocks above the splash zone, and the soldier crab *Mictyris.* Large populations of the latter march in formation along beaches. *Gecarcinus* may be the crab best adapted to land. Living in the American and West African tropics, it burrows shallowly in meadows and forests. A related genus, *Cardisoma* (Fig. 12.78b), ranges into southern Florida, where it burrows down to groundwater.

Crustacean Evolution

Crustacean evolution is a muddled study. As paleontology and comparative anatomy reveal more of the history of these invertebrates, our phylogenetic ideas undergo an evolution of their own. Old theories are discarded, and new ones slowly are formulated. The major crustacean groups date from the Precambrian, whose fossil record is poor. Accordingly, the origin of the class is obscure. Nevertheless, it appears that crustaceans are not closely related to insects and probably not to trilobites either (more on these topics at the end of Chap. 14).

Regarding relationships within the class, there is more agreement. The Cephalocarida have been accepted by most authorities as very primitive crustaceans. The uniformity of their trunk and paired appendages is a key trait thought to be possessed by the ancestors of the entire class. From cephalocarid limbs, the biramous appendages of all other crustaceans can be derived. Higher crustaceans may constitute four distinct lines (Fig. 12.79): the Branchiopoda, the Os-

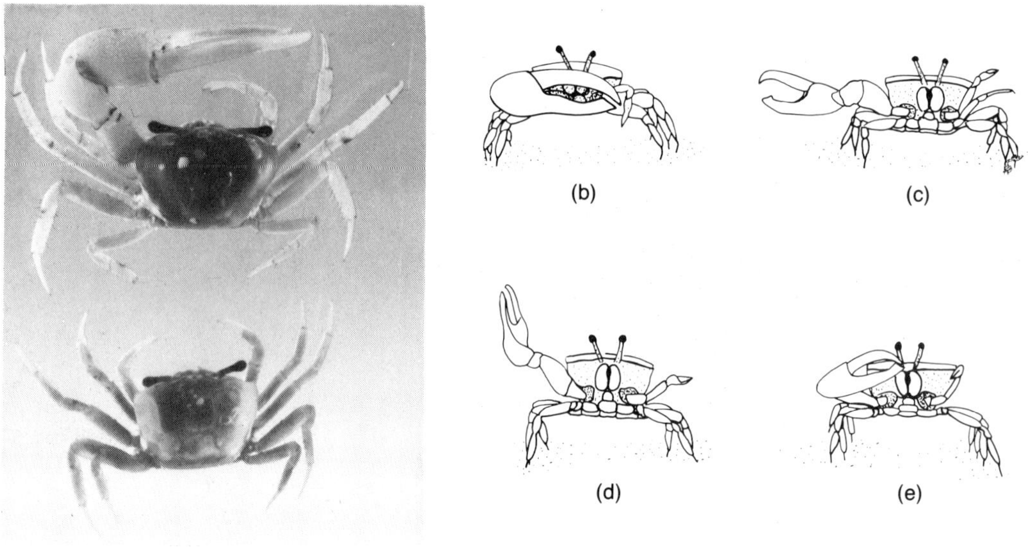

(a) (b) (c) (d) (e)

Figure 12.77 Fiddler crabs. (*a*) *Uca pugilator:* Male (*above*), female (*below*); (*b–e*) A courtship display in a male *Uca lactea* known as the lateral circular wave. The cheliped starts from the flexed position (*b*), is unflexed outwards (*c*), raised upwards (*d*), and finally returned to the starting point (*e*). (*a* courtesy of Ward's Natural Science Establishment; *b–e* from J. Crane, Zoologica, 1957, *42*:69.)

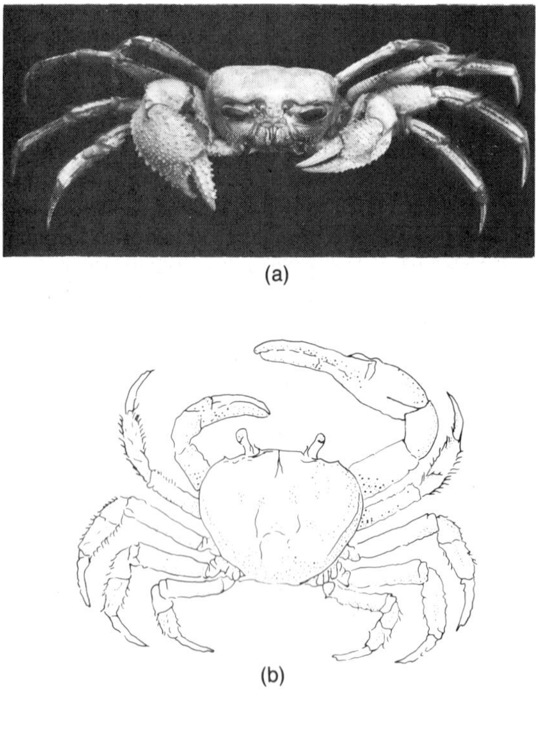

(a) (b)

Figure 12.78 (*a*) Model of a ghost crab, *Ocypode albicans;* (*b*) A land crab, *Cardisoma armatum.* (*a* courtesy of the American Museum of Natural History; *b* from Kaestner, *Invertebrate Zoology*, *Vol. III.*)

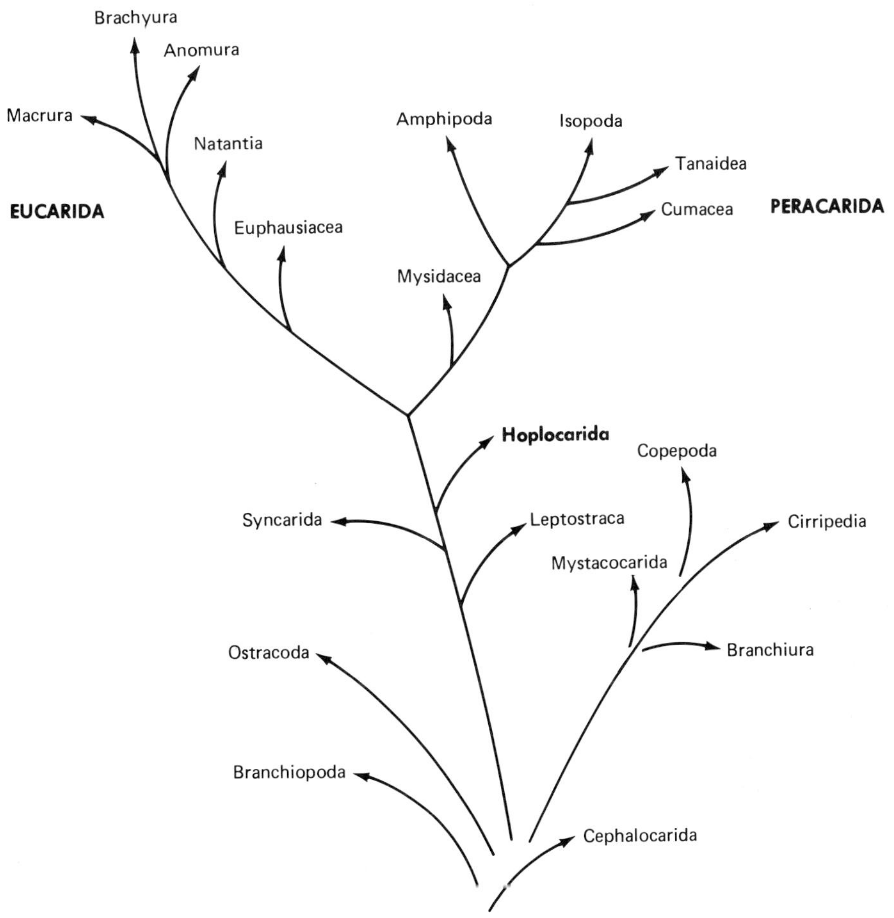

Figure 12.79 Presumed phylogenetic relationships within the Class Crustacea.

tracoda, the Cirripedia–Copepoda, and the Malacostraca. With the exception of the ostracods, each of these lines has primitive members with a highly metameric trunk and rather uniform appendages. The history of the Ostracoda remains puzzling, and efforts to ally these entomostracans with other subclasses have been frustrated. Meanwhile, barnacles and copepods (along with the mystacocarids and branchiurans) are assembled in a supergroup known as the Maxillopoda. Finally, the large but well-defined Malacostraca culminates in two branches, the Peracarida and the Eucarida.

for Further Reading

Adiyodi, K. G. and Adiyodi, R. G. Endocrine control of reproduction in decapod Crustacea. *Biol. Rev.* 45:121–165, 1970.

Aiken, D. E. Photoperiod, endocrinology, and crustacean molt cycle. *Science* 164:149–155, 1969.

Allen, J. A. Recent studies on the rhythms of post-larval decapod Crustacea. *Annu. Rev. Oceanogr. Mar. Biol.* 10: 415–436, 1972.

Anderson, D. T. On the embryology of the cirripede crustaceans and some considerations of crustacean phylogenetic relationships. *Phil Trans. Roy. Soc. London* 256:138, 1969.

Atwood, H. L. An attempt to account for the diversity of crustacean muscles. *Am. Zool.* 13:357–378, 1973.

Barker, P. L. and Gibson, R. Observations on the feeding mechanism, structure of the gut and digestive physiology of the European lobster *Homarus gammarus J. Exp. Mar. Biol. Ecol.* 26:297–324, 1977.

Barnard, J. L. The families and genera of marine gammaridean Amphipoda. *U.S. Nat. Mus. Bull.* 271:1–535, 1969.

Berkes, F. Some aspects of feeding mechanisms of euphausiids. *Crustaceana* 29:266–270, 1975.

Bourget, E. and Crisp, D. J. Factors affecting deposition of the shell in *Balanus balanoides J. Mar. Biol. Ass. U.K.* 55: 231–249, 1975.

Bousfield, E. L. *Shallow-water Gammaridean Amphipoda of New England.* Cornell University Press, Ithaca, 1973.

Bousfield, E. L. A revised classification and phylogeny of amphipod crustaceans. *Trans. Roy. Soc. Can. (Ser IV) 16:*343–390, 1978.

Browne, R. A. Reproductive pattern and mode in the brine shrimp *(Artemia salina) Ecology* 61:466–470, 1980.

Butler, T. H. Shrimps of the Pacific coast of Canada. *Can. Bull. Fish. Aquat. Sci.* 202: 1–280, 1980.

Caine, E. A. Habitat adaptations of North American caprellid Amphipoda. *Biol. Bull.* 155: 288–296, 1978.

Chace, F. A. and Hobbs, H. H. The freshwater and terrestrial decapod crustaceans of the West Indies. *Smithson. Inst. U.S. Nat. Mus. Bull.* 292:1–258, 1969.

Christy, J. H. Adaptative significance of reproductive cycles in the fiddler crab *Uca pugilator:* A hypothesis. *Science* 199: 453–455, 1978.

Crane, J. *Fiddler Crabs of the World.* Princeton University Press, 1975.

Crocker, D. W. The crayfishes of New England, U.S.A. *Proc. Biol. Soc. Wash.* 92:225–252, 1979.

Crocker, D. W. and Barr, D. W. *Handbook of the Crayfishes of Ontario.* University of Toronto Press, 1968.

Diaz, H. The mole crab *Emerita talpoida:* A case of changing life history pattern. *Ecol. Monogr.* 50:437–456, 1980.

Diaz, H. and Rodriguez, C. The branchial chamber in terrestrial crabs: a comparative study. *Biol. Bull.* 153:485–504, 1977.

Fincham, A. A. Eyes and classification of malacostracan crustaceans. *Nature* 287: 729–731, 1980.

Foster, B. A. Desiccation as a factor in the intertidal zonation of barnacles. *Mar. Biol.* 8:12, 1971.

Fox, R. S. and Bynum, D. H. The amphipod crustaceans of North Carolina estaurine waters. *Chesapeake Sci* 16:223–237, 1975.

Green, J. *A Biology of Crustacea.* Quadrangle Book, Chicago, 1961. (A concise, 180-page overview of the class, organized by life processes)

Hartnoll, R. G. Mating in Brachyura. *Crustaceana* 16:161–181, 1969.

Hartnoll, R. G. and Smith, S. M. Pair formation in the edible crab, *Cancer pagarus. Crustaceana.* 36:23–28, 1979.

Herring, P. J. and Locket, N. A. The luminescence and photophores of euphausiid crustaceans. *J. Zool.* 186:431–462, 1978.

Hessler, R. R. and Newman, W. A. A trilobite origin for the Crustacea. *Fossils and Strata* 4:437–459, 1975.

Hobbs, H. H. *Crayfishes (Astacidae) of North and Middle America.* Biota of Freshwater Ecosystems Identification Manual No. 9. EPA, U.S. Gov. Printing Office, 1972.

Ivanov, B. G. On the biology of the Antartic krill *Euphausia superba. Mar. Biol.* 7:340, 1970.

Johnson, P. T. *Histology of the Blue Crab, Callinectes sapidus.* Praeger Publishers, New York, 1980.

Kaestner, A. *Invertebrate Zoology, Vol. III:* Crustacea. Wiley-Interscience, New York, 1970. (A 423 page review of the class, well organized, and illustrated)

Leathem, W. and Maurer, D. Decapod crustaceans of the Delaware Bay area, USA. *J. Nat. Hist.* 14:813, 828, 1980.

Lockwood, A. P. M. *Physiology of Crustacea.* W.H. Freeman, San Francisco, 1967.

Manning, R. B. *Stomatopod Crustacea of the Western Atlantic.* University of Miami Press, Miami, 1969.

Marshall, S. M. Respiration and feeding copepods. *Adv. Mar. Biol.* 11:57–120, 1973.

Marshall, S. M. and Orr, A. P. *The Biology of Calanus finmarchicus.* Oliver and Boyd, London, 1955. (Now available from Springer-Verlag, New York, 1972)

Mauchline, J. and Fischer, L. R. The biology of euphausiids. *Advances in Marine Biology, Vol. VII.* Academic Press, New York, 1969.

McLaughlin, P. A. The hermit crabs of northwestern North America. *Zool. Verh.* 130:1–396, 1974.

McLaughlin, P. A. *Comparative Morphology of Recent Crustacea.* W.H. Freeman, 980.

McMahon, B. R. and Burggren, W. W. Respiration and adaptation to the terrestrial habitat in the land hermit crab *Coenobita clypeatus J. Exp. Biol.* 79:265–282, 1979.

Menzies, R. J. and Frankenberg, D. *Handbook of the Common Marine Isopod Crustacea of Georgia.* University of Georgia Press, Athens, 1966.

Modlin, R. F. The life cycle and recruitment of the sand shrimp, *Crangon septemspinosa,* in the Mystic River estuary, Connecticut, USA. *Estuaries* 3: 1–10, 1980.

Molenock, J. Evolutionary aspects in the courtship behavior of anomuran crabs. *Behavior 53:1*–30, 1975.

Moore, R. C. (Ed.) *Treatise on Invertebrate Paleontology,* Part R, Arthropoda 4, Vols. I and II. Geological Society of America and University of Kansas Press, Lawrence, 1969. (Describes all fossil crustaceans except ostracods, which are handled in Part Q, Arthropoda 3; also contains a widely accepted taxonomy of the class)

Naylor, E. *British Marine Isopods.* Academic Press, New York, 1972.

Nelson, W. G. Reproductive patterns among gammaridean amphipods. *Sarsia* 65:61–72, 1980.

Nolan, B. A. and Salmon, M. The behavior and ecology of snapping shrimp. *Forma Function* 4:289–335, 1970.

Omori, M. The biology of pelagic shrimps in the ocean. *Adv. Mar. Biol.* 12:233–324, 1974.

Palmer, J. D. Biological clocks of the tidal zone. *Sci Am.* 232: 70–79, 1975.

Reese, E. S. The behavior mechanisms underlying shell selection by hermit crabs. *Behavior* 21:78–126, 1963.

Roger, C. Feeding rhythms and trophic organization of a population of pelagic crustacean. *Mar. Biol.* 32:365–378, 1975.

Rosenfield, A. Structure and secretion of the carapace in some living ostracods. *Lethaia* 12:353–360, 1979.

Salmon, M. Signal characteristics and accoustic detection by fiddler crabs: *Uca rapaz and Uca pugilator. Physiol Zool.* 44: 210, 1971.

Sanders, H. L. The Cephalocarida. *Mem. Conn. Acad. Arts. Sci.* 15:1–80, 1963.

Schmitt, W. L. *Crustaceans.* University of Michigan Press.

Schultz, G. A. *How to Know the Marine Isopod Crustaceans.* W.C. Brown, Dubuque, Iowa, 1969.

Spight, T. M. Availability and use of shells by intertidal hermit crabs. *Biol. Bull.* 152:120–133, 1977.

Steel, C. G. H. Mechanisms of coordination between molting and reproduction in terrestrial isopod Crustacea. *Biol. Bull.* 159:206–218, 1980.

Stevcic, Z. The main features of brachyuran evolution. *Syst. Zool.* 20: 331–340, 1971.

Tomlinson, J.T. The burrowing barnacles (Cirripedia: Order Acrothoracia.) *Bull. U.S. Nat. Mus.* 296:1–62, 1969.

VanWeel, P. B. Digestion in Crustacea. In *Chemical Zoology* (M. Florkin and B. T. Scheer, Eds.), Vol. V, pp. 97–115. Academic Press, New York, 1970.

Warner, G. F. *The Biology of Crabs.* Van Nostrand Reinhold, New York, 1977.

Waterman, T. H. (Ed.) *The Physiology of Crustacea, Vols. I–II.* Academic Press, New York 1960–1961. (Volume I covers metabolism and growth; it has an excellent introduction. Volume II treats crustacean activity systems. The work includes a valuable taxonomic system)

Webb, P. W. Mechanics of escape responses in crayfish *(Orconectes virilis)* J. Exp. Biol. 79:245–264, 1979.

Whittington, H. B. and Rolfe, W. D. I. (Eds.) *Phylogeny and Evolution of Crustacea.* Harvard University Press, Cambridge, 1963. (A collection of papers and discussions at the 1963 conference dealing with major problems in crustacean relationships)

Wickins, J. F. Prawn biology and culture. *Ann. Rev. Oceanogr. Mar. Biol.*

Wittman, K. J. Adoption, replacement and identification of young in marine Mysidacea. *J. Exp. Mar. Biol. Ecol.* 32: 259–274, 1978.

Yamaguti, S. *Parasitic Copepoda and Branchiura of Fishes.* Wiley-Interscience, New York, 1963.

13

Arthropod Diversity: The Myriapods

Many-legged and secretive, the myriapods compose a relatively small but interesting branch of the Arthropoda. Centipedes and millipedes account for most of the 11,000 described species. Myriapods, like most insects, are land animals, and the two groups are known collectively as terrestrial mandibulates. Moreover, it is hypothesized that insects arose from protomyriapod stock.

Myriapods are characterized by a short head and a very elongate, segmented trunk. All of their limbs are uniramous. Cephalic appendages include a single pair of antennae, mandibles, and variously modified maxillae. Paired legs occupy the many trunk segments. Metamerism is quite persistent in several internal organs, including the heart, tracheae, and nerves. Never as physiologically adapted to land as insects, most myriapods are limited to humid, organically rich environments, such as forest soils and humus. Tropical habitats are common, but many species range through the temperate zones.

Early classification systems conceived the myriapods as a single class. As later study underscored the diverse biologies of these creatures, four separate classes became recognized: the Chilopoda, the Diplopoda, the Pauropoda, and the Symphyla. Today most biologists use the term myriapod in a nontaxonomic sense to refer collectively to these four classes. There have been some recent arguments, however, to reestablish Myriapoda as a single class.

Class Chilopoda

The 3000 species of chilopods are better known as centipedes. For various reasons, these many-legged creatures both attract and repel us; indeed, as Cloudsley-Thompson writes, they "seem to exert a weird fascination on the morbid appetites of the hysterical and insane." At least in the myriapod group, centipedes are quite unusual in that they are active, aggressive carnivores. They have poisonous limbs called **prehensors**, and their dorsal gonads connect with gonopores at the posterior end of the body. Humid soil, forest litter, and undersurfaces of stones are their common habitats. Some chilopods, such as *Scutigera* (Fig. 13.1a), frequent the basements and bathrooms of Europe and North America.

Average temperate-zone centipedes are less than a centimeter long. Body size increases in warmer latitudes, where some species reach 20 cm or more in length. The chilopod head, like that of other myriapods, has a single pair of long antennae and forwardly directed mouthparts (Fig. 13.2a). A labrum drops from the rounded anterior body to form a forelip. Behind this structure, a preoral cavity houses the mandibles and a recessed mouth. Short and semicircular, the mandibles bear sharp, biting teeth. Flattened first maxillae form a posterior lip. Second maxillae simply overlie the first, or they may be elongate, as in Scutigersa (Fig. 13.2b) The first trunk appendages, which are usually held over the mouthparts, are called prehensors in centipedes. Ducts from venomous glands open on the terminal claws of the prehensors, thus rendering them effective offensive and defensive limbs. The spiny coxae of the prehensors are fused ventrally to mark the posterior border of the head (Fig. 13.2c). The remaining trunk appendages are similar walking legs. From 15 to 173 pairs of these limbs are present; curiously, there is always an uneven number of leg-bearing segments. The last two segments are legless, and a pygidium forms the posterior end of the body.

Centipedes capture and eat small terrestrial invertebrates such as worms, snails, and other arthropods. Detritus figures in the diet of some burrowing forms. A large centipede like *Scolopendra gigantea* (over 25 cm in length) may feed on small lizards and toads; one individual in a zoo reportedly thrived on baby mice! The prehensors seize and poison prey. Some of the larger members of the class may rear up on their hind legs to snatch flying insects. The victim is passed through the mouthparts, where the mandibles, often aided by salivary secretions, chew its flesh. In some centipedes with weaker mandibles, additional enzymes flow from the mouth and arachnid-style predigestion occurs. The alimentary canal is long and tubular (Fig. 13.2d). In the lengthy foregut, long cuticular spines strain oversized particles. Midgut digestion and absorption follow, and the short hindgut terminates through a pygidial anus.

Like most other myriapods, centipedes possess a dorsal, tubular heart, whose paired segmental ostia alternate with lateral arteries. Respiration is through tracheae. In most chilopods, segmental spiracles lie

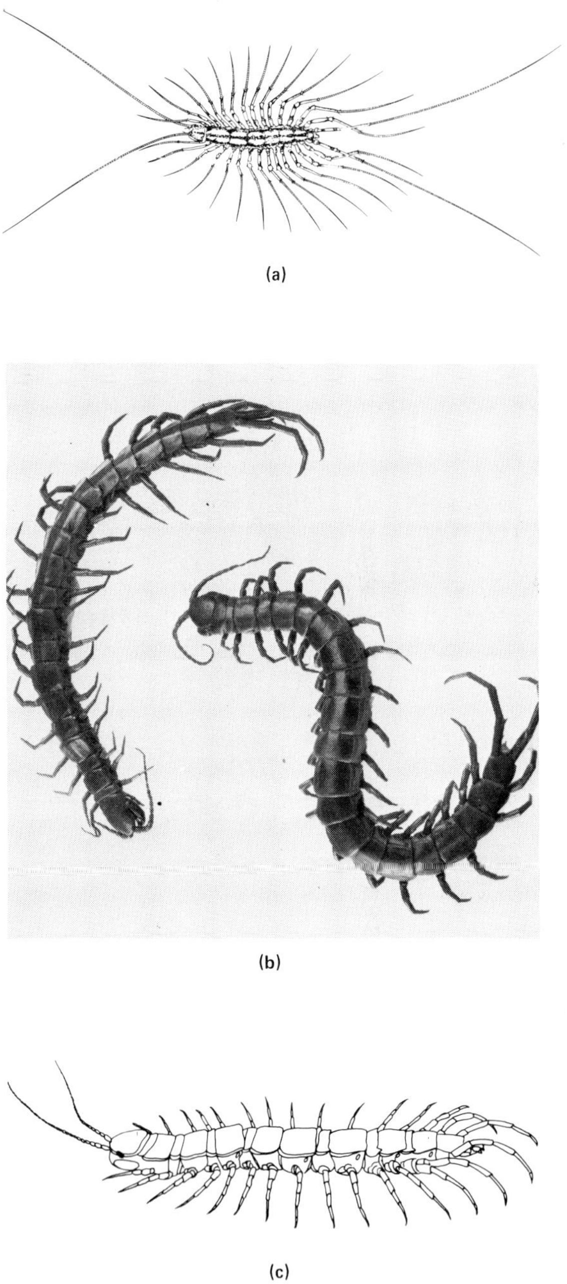

Figure 13.1 Representative Chilopoda. (*a*) *Scutigera coleoptrata*, a centipede frequently found in human habitations; (*b*) *Scolopendra*, which may reach 25 cm in length; (*c*) *Lithobius forficatus*, a centipede whose overlapping tergites alternate in size. (*a* from R.E. Snodgrass, *A Textbook of Arthropod Anatomy*, 1952, Cornell University; *b* courtesy of American Museum of Natural History; *c* from G. Rilling, Zool. Jahrb. Abt. Anat., 1960, *78*:39.)

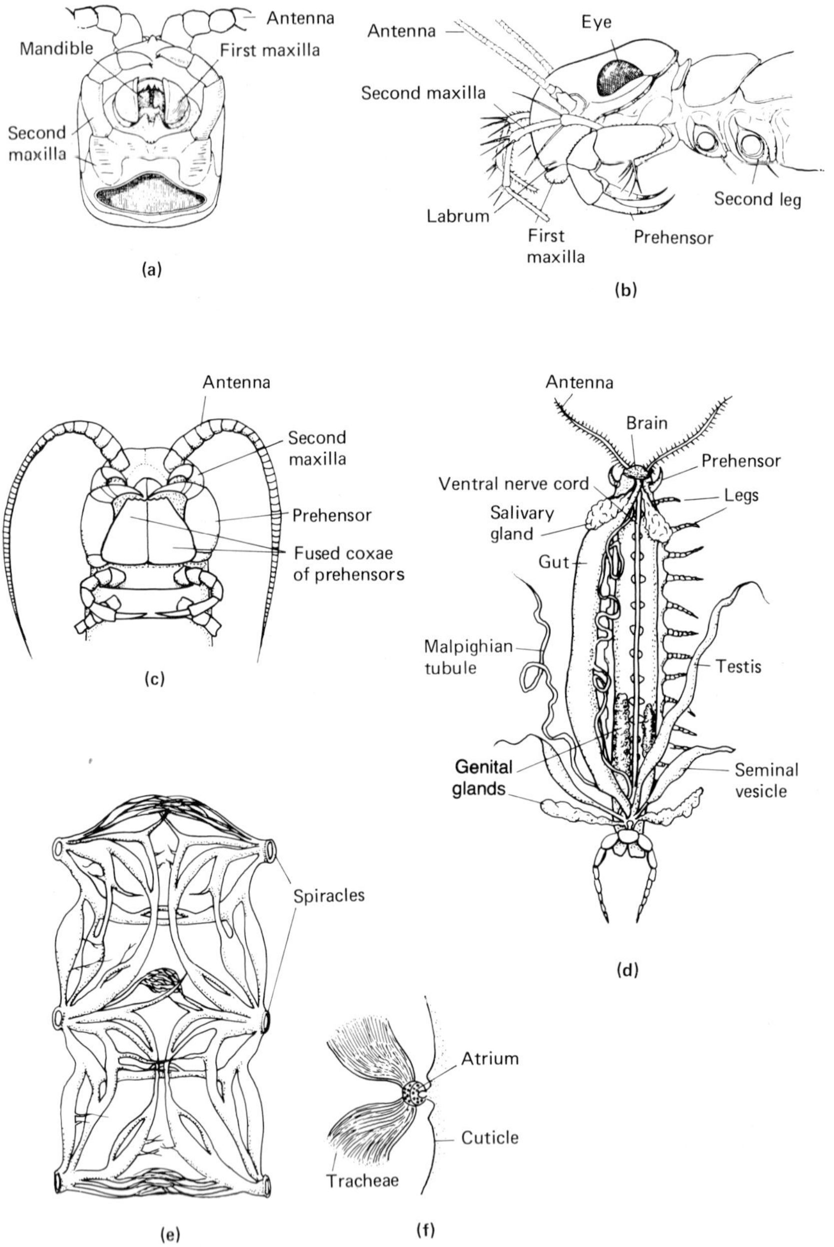

Figure 13.2 (*a*) Ventral view of the head and mouth parts of a centipede, *Otocryptops sexspinosa;* (*b*) Lateral view of the head and anterior body segments of *Scutigera coleoptrata;* (*c*) Ventral view of the head and anterior body segments of *Lithobius americanus* showing the large coxae of the prehensors; (*d*) Internal anatomy of *Lithobius;* (*e*) A portion of the tracheal system of *Scolopendra cingulata* including three pairs of spiracles; (*f*) One of several pairs of tracheal clusters making up the tracheal system of *Scutigera coleoptrata.* (*a* and *b* from Snodgrass, *A Textbook of Arthropod Anatomy; c* from A.S. Packard, *Zoology*, 1904; *d* from F.G. Sinclair, In *Cambridge Natural History, Vol. V*, 1959, Macmillan; *e* and *f* from K.W. Verhoeff, In *Klassen und Ordnungen des Tierreichs, Vol. V*, H.G. Bronn, Ed., 1925.)

near the bases of the legs, and interconnecting respiratory tubules ramify extensively throughout the body tissues (Fig. 13.2e). *Scutigera* and its relatives (the scutigeromorphs) are unusual in that their hemolymph serves as a respiratory medium. In these centipedes, middorsal spiracles open into an atrium from which numerous short tubules descend to the pericardium (Fig. 13.2f). Blood is aerated as it enters the powerful scutigeromorph heart. This arrangement reflects the more active life-style of these centipedes.

Water retention is critical for centipedes and the other myriapods. These creatures lack the waxy epicuticles possessed by arachnids and insects and also cannot close their spiracles. Water is readily lost over the general surface and from the respiratory tract. Chilopods and their relatives require very humid environments; many die of desiccation when the humidity drops below 70%. Another factor restricting their range is their excretory physiology. Despite the presence of a pair of Malpighian tubules, well over half of centipede nitrogenous wastes is voided as ammonia.

The chilopod central nervous system expresses the typical myriapod pattern (Fig. 13.2d). A protocerebrum innervates the eyes, a deutocerebrum services the antennae, and the tritocerebrum joins the fused ganglia of the mouthparts. A doubled ventral cord links distinct segmental ganglia throughout the trunk. The chilopod world is dark and moist, and sense organs responding to tactile and chemical stimuli are most developed. If present, eyes usually occur as lateral clusters of ocelli on the head. In the scutigeromorphs, these clusters are dense and resemble compound eyes. Vibration-sensitive structures called **organs of Tömösvary** often occupy the antennal bases. Tactile and chemosensitive hairs are concentrated along the antennae and the distal elements of the legs. Long and sensitive, the last trunk appendages of *Scutigera* serve as posterior antennae. Temperature and humidity sensors are also critical to chilopod behaviors; these sense organs probably lie along the antennae.

Centipedes are specialized for rapid running over the substratum. Such movements usually are restricted to nocturnal hours, when lower temperatures and higher humidities describe the least stressful conditions.

Certain mechanical difficulties are incumbent on any animal that walks quickly on 300, or even only 30, legs. First, there is a coordination problem: The legs must not trip over one another. While running, centipedes stretch their bodies, thus widening the spaces between adjacent legs. Every leg begins stepping just before the leg behind it, so that a wave of step activation passes from anterior to posterior over the body. Because each leg tends to be longer than its anterior neighbor, distal entanglement is not likely (Fig. 13.3a). Second is the problem of stability under high speed. Centipedes run faster by decreasing the duration of their legs' effective strokes; thus, in a rapidly moving animal, fewer legs are in contact with the substratum at any given time. Stability is maintained by keeping paired legs in opposite phases of the locomotor cycle. Surface contact is increased by the presence of many distal articles on each appendage. Often, the distal third of the leg rests against the substratum. Finally, centipedes have a rather rigid trunk, whose overlapping tergites may alternate in size (Fig. 13.1c). Also, tergites and sternites may be staggered. This arrangement minimizes lateral undulations and thus increases the efficiency of forward movement. Centipedes, like other myriapods and the arachnids, lack extensor muscles in the legs. Their trunk segments are unfused, and thus local hydrostatic extension of the appendages is possible. The fastest centipedes are the scutigeromorphs: A combination of the longest legs and shortest trunk (only 15 segments) in this class facilitates speeds of up to 50 cm/sec.

Some centipedes (the geophilomorphs) are specialized for burrowing. Seldom seen above ground, they tunnel through loose soil to depths of 40 cm. Their locomotion is remarkably similar to that of annelids. Alternate contractions of powerful longitudinal and circular muscles push the body through cracks in the ground; short legs provide anchorage. Trunk flexibility results in part from wide articular membranes and accessory skeletal plates (Fig. 13.3b).

Centipedes defend themselves with their poisonous prehensors, spiny hind legs, and repugnatorial glands. Contrary to popular fears, chilopod venoms are seldom strong enough to harm people seriously, and the prehensors of temperate-zone groups usually cannot pierce human skin. The southern California *Scolopendra heros*, however, can prick human skin with its crawling legs. If agitated, it drops a noxious secretion into each of the tiny wounds, leaving a painful trail of swollen tissue. Against invertebrate and small vertebrate prey, the prehensors are highly effective weapons. Using these limbs, large centipedes have successfully battled scorpions. The bright colors of many tropical species evidently advertise them as animals that cannot be attacked with impunity. Another line of defense, the repugnatorial glands, also involves chemical warfare. Ducts from these glands open on the legs or trunk sternites and discharge either noxious secretions or entangling adhesives. Burrowing chilopods unlease these substances against

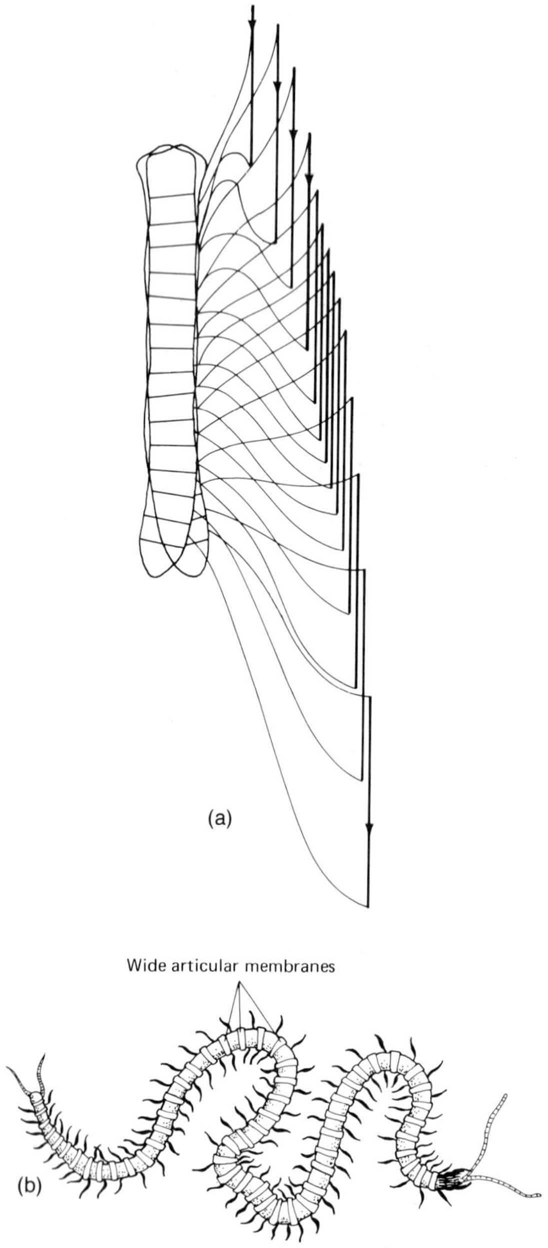

Figure 13.3 (*a*) The field of movement of the legs in a running *Scutigera*. The heavy vertical lines represent the movements of the tip of each leg relative to the body during the propulsive backstroke. Each leg is drawn twice: at the beginning and at the end of each backstroke. Recovery strokes for individual limbs are not shown; (*b*) A geophilomorph centipede, *Geophilus longicornis*. (*a* from S.M. Manton, J. Linn. Soc. Zool., 1952, *42*:100; *b* from Sinclair, In *Cambridge Natural History*, *Vol. V*, Harmer and Shipley, Eds.)

ants, and surface dwellers may trap inimical wolf spiders with their sticky repugnatorial threads.

Most centipedes, however, are not easily provoked. Secretive ways and escape responses remain their most common defense. If all else fails, centipedes may rely on yet another factor: Judged by the reactions of many birds, they apparently rank among the most unpalatable members of the invertebrate world.

Pregenital and genital segments are just anterior to the pygidium. A single ovary lies above the gut, and its duct terminates on the genital segment. Surrounding the female gonopore is a small pair of grasping appendages (Fig. 13.4a). In the male, a similar pair of gonopods flanks the penis. The latter houses the aperture of a sperm duct which drains 1 to 14 testes in the dorsal trunk (Fig. 13.4b and c). Distally, the male duct contains glandular areas which produce spermatophores. Genital silk glands are also present in all male centipedes except scutigeromorphs. The male spins an irregular platform of threads, onto which he deposits his sperm package. Some courtship surrounds this act, as the mating pair strokes each other with their antennae and maneuvers about the spermatophore net (Fig. 13.4d). Eventually, the female takes up the sperm with her gonopods. One scutigeromorph, *Thereuopoda decipiens*, reportedly grasps his spermatophore in his maxillipeds and then inserts it into the female orifice.

Eggs are laid in the ground. The female coats them with moisture and fungicides, and some parental guarding of the developmental stages is common. Characteristically, a centipede mother wraps her body around the brood (Fig. 13.4e) Hatchlings display various levels of maturity. Some species already possess a full set of adult segments, but others, including the scutigeromorphs, must await posthatching molts before the trunk is completed. Four to six years is a typical life span.

Class Diplopoda

Diplopods, also known as millipedes, are the largest myriapod group. Over 7500 species have been identified worldwide. In testimony to the diversity of these animals, the class is organized into at least 14 orders. Moreover, it is estimated that less than 20% of the actual number of diplopod species are known. Like the centipedes, millipedes are often long and many-legged (Fig. 13.5), and most crawl through moist, dark soils and humus. Millipedes are distinguished by herbivorous feeding, slower gaits, and unusual metameric organization.

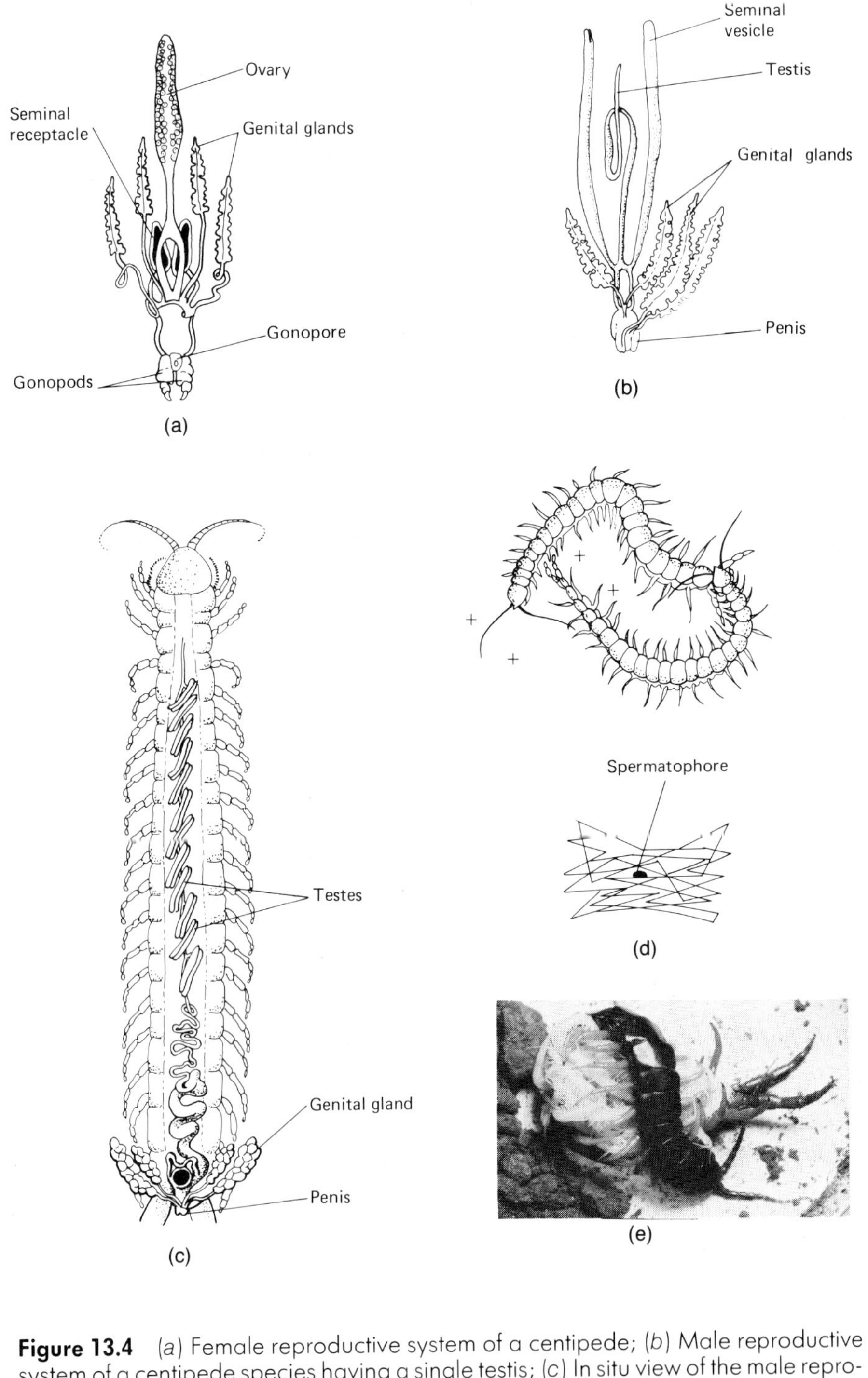

Figure 13.4 (*a*) Female reproductive system of a centipede; (*b*) Male reproductive system of a centipede species having a single testis; (*c*) In situ view of the male reproductive system of a centipede having twelve testes. (*d*) *Above*: male and female *Scolopendra cingulata* near their sperm web. Male is on the left; web limits are indicated by crosses. *Below:* sperm web of *S. cingulata;* (*e*) Female with brood. (*a–c* from Verhoeff, In *Klassen und Ordnungen des Tierreichs, Vol. V*, Bronn, Ed.; *d* from H. Klingel, *Naturwissenschaften*, 1957, *44*:338; *e* from H. Klingel, Zeit, f. Tierpsychol., 1960, *17*:11.)

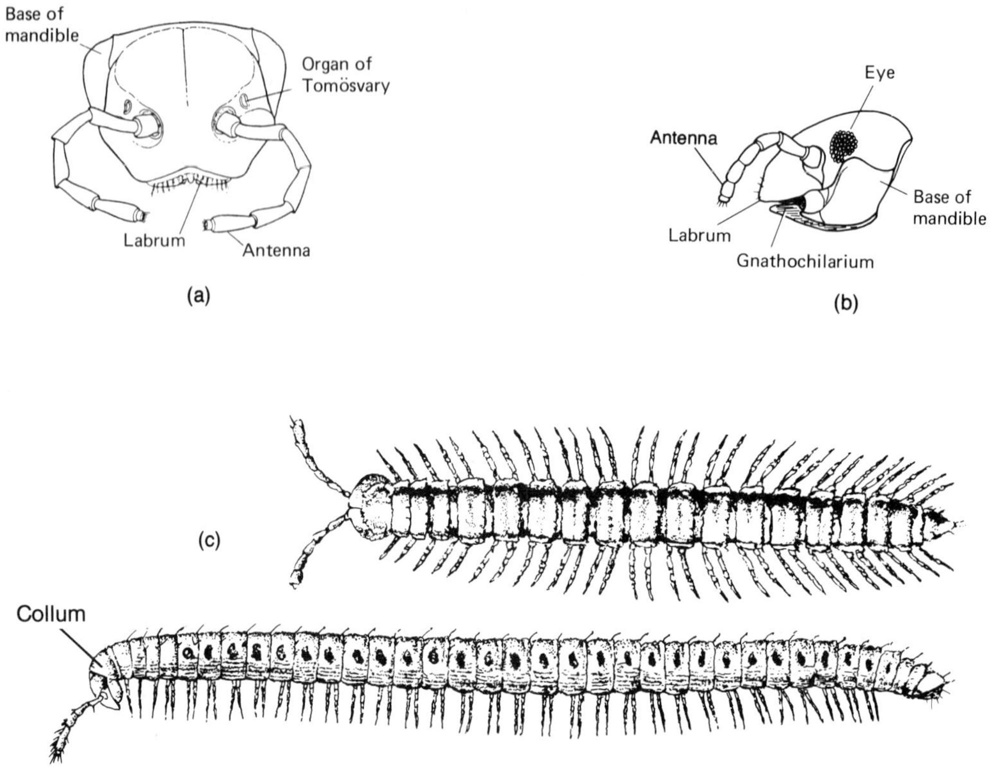

Figure 13.5 (*a*) Anterior view of the head of *Apheloria coriacea*, a millipede, class Diplopoda; (*b*) Lateral view of the head of *Arctobolus marginatus*; (*c*) General body form in millipedes. *Above:* dorsal view of *Stronglyosoma italicum*. *Below:* a lateral view of *Nopoiulus palmatus*. Both views show the collum, the anterior single body segments and the more numerous diplosegments. (*a* and *b* from Snodgrass, *A Textbook of Arthropod Anatomy*; *c* from Verhoeff, In *Klassen und Ordnungen des Tierreichs, Vol. V.*, Bronn, Ed.)

Diplopod body size approximates that of the chilopods. The largest diplopod, the African *Archispirostreptus*, is 28 cm long; the smallest (a pselaphognath), only 2 mm. The head bears a pair of short antennae and large mandibles. A preoral chamber is bounded by the labrum, the lateral mandibular bases, and the **gnathochilarium** (Fig. 13.5b). The latter represents the flattened and fused first maxillae, which form a floor beneath the mandibles. All vestiges of second maxillae are incorporated into the gnathochilarium, as the third postoral segment is essentially limbless. Called the **collum**, this segment forms a conspicuous collar (Fig. 13.5c). The next three body units bear single pairs of legs and other metameric structures. Along the rest of the trunk, however, fusion among pairs of tergites produces a chain of diplosegmented rings (Fig. 13.5c). Each of these rings expresses a doubled complement of metameric organs, including two pairs of legs. Some posterior diplosegments may lack appendages, and the trunk ends with a pygidium.

Most millipedes are vegetarians, partial to dead and decaying plant matter and to leaves with a rich calcium content. Millipedes play an important role in the breakdown of leaf litter in deciduous forests, where their recycling activities have been compared with those of earthworms. Although the mandibles are often quite strong, food usually is softened with saliva inside the preoral chamber. Some millipedes are specialized feeders on plant juices, and in *Siphonophora* the labrum and gnathochilarium form a piercing suctorial beak. *Blaniulus guttulatus* (Fig. 13.6a), the "spotted snake millipede," is a pest in greenhouses and in fields of potatoes, oats, wheat, and strawberries. Deep-burrowing species may ingest soil, while scavenging and omnivorous feeding occurs within some rock-dwelling groups. The diplopod gut is a simple tube. As in other myriapods, there are no

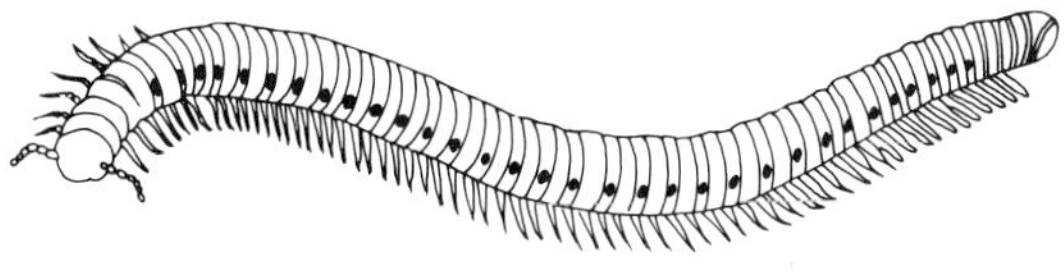

(a)

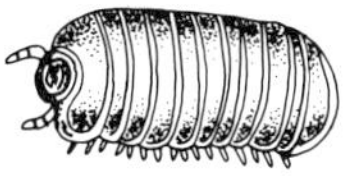

(b)

(c)

Figure 13.6 (*a*) *Blaniulus guttulatus*, the "spotted snake" millipede, a pest in greenhouses; (*b*) *Glomeris marginata*, one of the so-called pill millipedes; (*c*) *Orthoporus ornatus*, a large desert millipede. This female is about 12 cm long. (*a* and *b* from Sinclair, In *Cambridge Natural History, Vol. V*, Harmer and Shipley, Eds; *c* courtesy of Clifford S. Crawford.)

ceca in the long midgut region, and the anus occupies the ventral side of the pygidium.

The heart extends through the trunk, and a single aorta penetrates the head region (Fig. 13.7a). Two pairs of ostia and unbranched lateral arteries are present in each diplosegment, and one pair of each occupies the single anterior somites. The tracheal network is similarly arranged with four ventral spiracles on each diplosegment. Gas exchange occurs either with the hemolymph or directly with the tissues. There is a single pair of Malpighian tubules. Like centipedes, most millipedes inhabit moist areas because of their inability to close the spiracles and their lack of a waxy epicuticle. The few types that tolerate dry air apparently absorb and retain water within invaginated sacs on their coxopodites. Desert millipedes spend most of their lives underground in an inactive state. They emerge seasonally when adequate moisture is present.

The millipede nervous system features a brain with large olfactory centers and a subesophageal mass incorporating the fused ganglia of the mandibles and the gnathochilarium. The remaining body ganglia are spaced along a ventral nerve cord. One ganglionic swelling occurs in each of the first four trunk segments; double swellings are present in the diplosegments. Sense organs include tactile and chemosensitive hairs, innervated cones on the antennae and gnathochilarium, organs of Tömösvary, and assorted ocelli. Many millipedes have no eyes, but their general integument may be light sensitive. Like other myriapods, these creatures shun illuminated areas. Most environmental data are translated by tactile and chemosensitive centers on the antennae and mouth parts. As a millipede travels about, its antennae constantly explore the surrounding terrain.

Diplopod architecture can be explained largely in terms of locomotor needs. The majority of millipedes has a tough exoskeleton reinforced with calcium carbonate. Commonly, the legs are small and are attached quite ventrally (Fig. 13.7b). Muscles tend to be few in number, short in length, and thick in diameter. These qualities allow millipedes to push their bodies slowly but powerfully through soil and humus. Unlike burrowing centipedes, whose trunk muscles alternately extend and contract the body, millipedes depend entirely on their legs for locomotor force. Their burrowing power is directly proportional to the number of legs pushing at any given time. Dozens of adjacent limbs may move simultaneously as the millipede forces its way through packed substratum. Diplosegments provide an unparalleled density of powerful legs along the trunk and also minimize inefficient undulations of the body.

Millipedes may be strong burrowers, but they are slow and lack agility. Clearly, this condition is compatible with herbivorous feeding — unlike centipedes, millipedes never chase their food. But what can a millipede do when pursued by a would-be predator? For the most part, millipedes avoid this danger by remaining inaccessible beneath stones, soil, and humus. If trapped above ground, however, some species enroll their bodies like pill bugs. These so-called pill millipedes include some of the shorter members of the class; their enlarged posterior tergites protect the head when the trunk is rolled up. Longer millipedes coil the body spirally to shield its delicate ventral surface. These behaviors also aid in water retention.

Millipedes also possess repugnatorial glands which open through ducts on the tergites. Noxious secretions containing hydrogen cyanide may ooze from these openings, but some millipedes spray the fluids over considerable distances. The large West Indian *Rhinocricus lethifer* can shower birds and small mammals half a meter away and has caused blindness in chickens.

In both diplopod sexes, a single pair of long, tubular gonads lies below the gut (Fig. 13.7a). The gonopores occupy the third trunk segment. In the female, each oviduct opens into a genital atrium which contains one or more seminal receptacles. The male may possess a single penis or a pair of these structures. Nevertheless, sperm transfer involves modified legs as intromittent organs. The mandibles transfer sperm in pill millipedes, but more commonly the limbs of the seventh trunk segment serve as gonopods. Specific modifications of these limbs are valuable in diplopod taxonomy. Males bend their anterior segments back under the trunk until the penes contact the gonopods. These limbs then take up spermatophores. Copulating millipedes embrace with their legs and mandibles and may coil their trunks together (Fig. 13.8). Sometimes a couple remains intact for up to 2 days. During mating, an enzyme dissolves the spermatophore and sperm move into the seminal receptacles of the female.

Eggs are laid in the ground. There may be a brood-

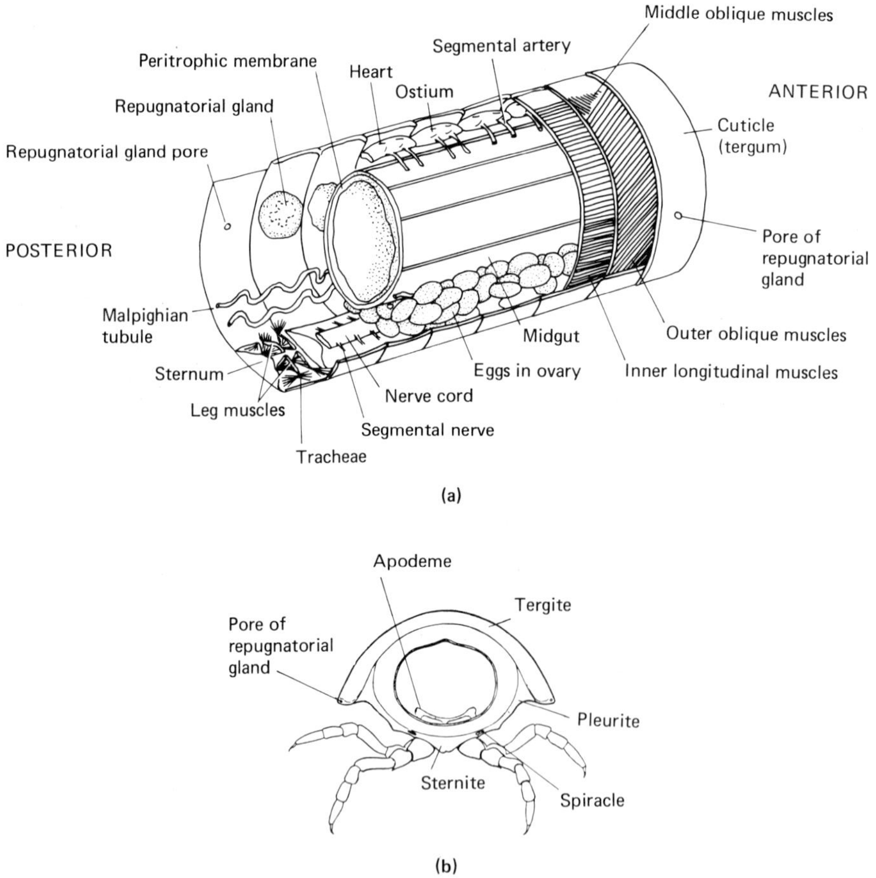

Figure 13.7 (*a*) Dorso-lateral stereodiagram of midbody segments of a female *Spirobolus marginatus;* (*b*) Cross section through the fifth body segment of a millipede, *Apheloria coriacea.* (*a* from F.A. Brown, *Selected Invertebrate Types*, 1950, John Wiley; *b* from Snodgrass, *A Textbook of Arthropod Anatomy.*)

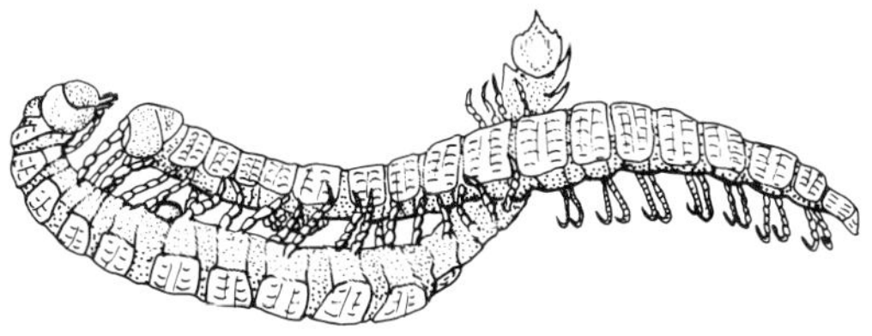

Figure 13.8 Two polydesmid millipedes, *Brachydemus superus*, copulating. The male is beneath. (From J.L. Cloudsley-Thompson, *Spiders, Scorpions, Centipedes, and Mites*, 1968, Pergamon.)

ing nest of soil and humus walls reinforced with parental feces. Some millipedes, such as *Polydesmus*, construct the entire nest from excrement. First, a small depression is made in some firm substrate. The female then defecates in repeated circles, raising a domed chamber with successive spirals of her long, stringlike feces. The eggs are dropped through the chamber roof just before it is completed, and the entire nest is then camouflaged with debris.

Hatching millipedes typically possess no more than seven segments and only three paired legs. Developmental molts witness the proliferation of segments from the anterior edge of the pygidium. Ecdysis may occur within a shelter similar to the brooding nest described above. Commonly, the old cuticle is eaten for its calcium content. Life spans range from 1 to 7 years.

One group of diplopods is quite distinct from the rest of the class: the pselaphognaths, of which almost 100 species have been described. *Polyxenus* (Fig. 13.9) is the best-known genus. Pselaphognaths are tiny millipedes with a soft but bristly cuticle. Mandibular teeth, repugnatorial glands, and gonopods are missing, while leglike second maxillae are present on the collum. Some species climb high into trees. Parthenogenesis is widespread among pselaphognaths, but an interesting heterosexuality has been described for *Polyxenus lagurus*. In this species, genital glands produce silk threads, which the male fashions into a bridge across a narrow crevice. Sperm are dropped onto the bridge, and long signal threads are laid down at right angles to it. If another male encounters one of these threads, he follows it to his rival's sperm, eats them, and deposits his own fresh gametes in their place. When a suitable female finds the bridge, she takes the sperm into her gonopore. Pselaphognaths may protect their brood within tree bark chambers swept out by body setae. Broken parental setae adhere to the surfaces of eggs, affording them protection against mites and other predators.

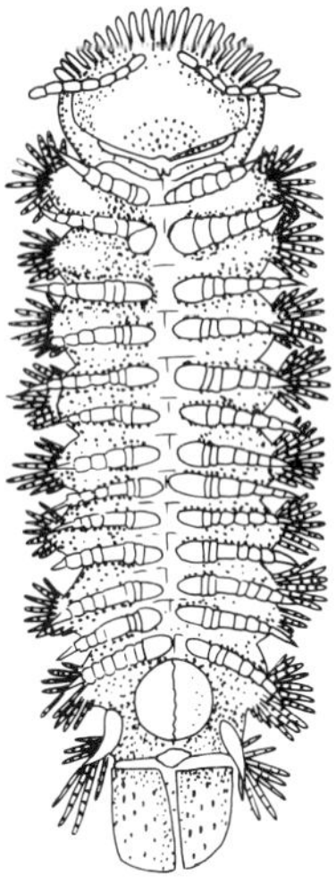

Figure 13.9 Ventral view of *Polyxenus lagurus* (greatly enlarged). (From Verhoeff, In *Klassen und Ordnungen des Tierreichs, Vol. V*, Bronn, Ed.)

Class Pauropoda

The pauropods are minute myriapods which resemble millipedes in some respects. About 380 species have been identified. Pauropods never exceed 2 mm in length and usually have no more than 11 trunk segments. Six tergites dominate the dorsal surface and produce segmental coupling, as in the millipedes (Fig. 13.10a). Such an arrangement minimizes undulations during pauropod locomotion. Pauropod antennae are unusual for the myriapod group, in that they are branched. Terminal flagella and other sensory lobes highlight these appendages. The mouthparts include a pair of curved mandibles and a gnathochilarium. As in the millipedes, a collum is present just behind the head.

Pauropods apparently eat fungus and decaying organic matter, which they encounter while crawling through soil and humus. Salivary glands discharge enzymes near the mouth, and a pumping pharynx draws food into the unbranched midgut. Because of the small size of these creatures, other maintenance systems are rather informal. Hearts and tracheae are seldom present, and direct diffusion over the thin exoskeleton must play a major physiological role. A single pair of Malpighian tubules joins the hindgut. Besides the receptor-studded antennae, sensory structures include tergal trichobothria and cephalic disks, which may resemble organs of Tömösvary. No eyes are present.

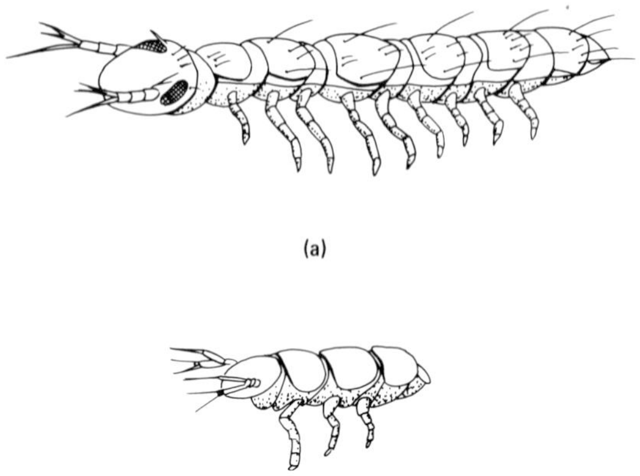

(a)

(b)

Figure 13.10 A pauropod, *Pauropus silvaticus*. (*a*) Dorsolateral view of an adult; (*b*) A newly-hatched specimen. (From K.U. Clarke, *The Biology of Arthropods*, 1972, Edward Arnold.)

The ovaries are ventral, but during development the testes migrate above the gut. As in the millipedes, the gonopores occupy the third trunk segment. Without courtship, the female takes up a spermatophore, which the male has suspended over a crevice with two threads. Pauropods lay their yolky eggs in decaying wood. Hatchlings lack many adult segments and usually have only three pairs of legs (Fig. 13.10b). Populations as great as 2 million individuals per acre may thrive on moist forest floors.

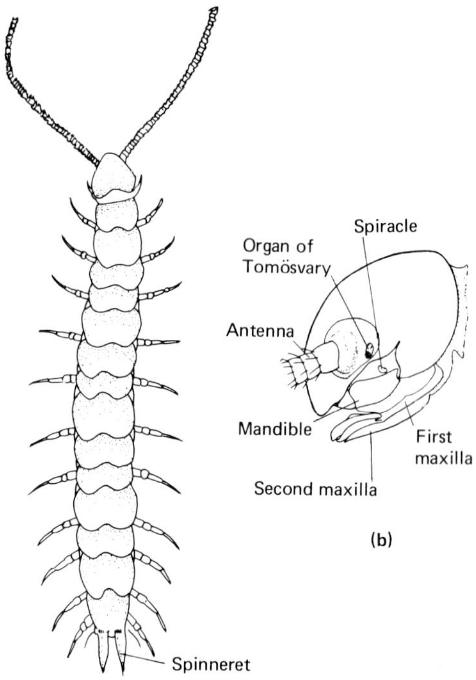

(a)

(b)

Figure 13.11 Symphylans. (*a*) Dorsal view of an adult *Scutigerella immaculata*, a frequent pest in nurseries and greenhouses; (*b*) Lateral view of the head of *Hanseniella agilis*. (*a* from A.E. Michelbacher, Hilgardia, 1938, *11*:82; *b* from Snodgrass, *A Textbook of Arthropod Anatomy*.)

Class Symphyla

The last and smallest myriapod class comprises the symphylans, another group of tiny blind arthropods that live in soil and humus. The 120 known species have been the focus of considerable phylogenetic speculation, as they may share close common ancestors with the insects.

Symphylans never exceed 1 cm in length (Fig. 13.11a). Their cephalic appendages resemble those of insects. The antennae are long, the mandibles are well developed, and both pairs of maxillae are prominent. The first maxillae extend beneath the jaws, and the bases of the second pair have fused as a posterior lip (Fig. 13.11b). Twelve paired legs occupy the trunk, but curiously, there are up to 22 tergites. These supernumerary dorsal plates confer great flexibility to the trunk and enable symphylans to writhe through decaying vegetation. The function of a pair of spinnerets on the penultimate segment is unknown.

Symphylans eat soft green plants. The best-known species, *Scutigerella immaculata* (Fig. 13.11a), is a serious pest in plant nurseries and flower gardens. One report estimates that 90 million of these myriapods may infest an acre of soil in large commercial greenhouses. The heart and a branching arterial network are well developed in this class. Spiracles are limited to a single pair near the mandibular bases. The posterior body depends on oxygen transport by the hemolymph; respiratory diffusion across the cuticle is also important. Malpighian tubules are the excretory organs. Symphylans have a rather large brain, but no eyes. Sensitive hairs, including posterior trichobothria and organs of Tömösvary, are the primary sensory structures.

In both sexes, paired gonads discharge through gonopores on the third trunk segment. Between molts, a mature male symphylan erects numerous stalks topped with spermatophores. When a female encounters one of these elevated sperm packages, she bends the stalk over with her antennae and bites off the spermatophore. Sperm are stored within her preoral cavity. When her eggs are ripe, the female removes them from her gonopore with her mouthparts and cements them to moss or other substrates. Fertilization occurs after she coats each egg with the stored sperm. *Scutigerella* hatches with only half of its trunk segments and appendages. Over 50 molts may occur during the 4-year life span of this myriapod.

for Further Reading

Auerbach, S. I. The centipedes of the Chicago area with special reference to their ecology. *Eco. Mongr.* 21: 97–124, 1951.

Binyon, J. and Lewis, J. Physiological adaptations of two species of centipedes (Geophilomorpha) to life on shore. *J. Mar. Biol. Assoc.* 43: 49–55, 1963.

Blower, B. Chilopod and diplopod cuticle. *Quart. J. Microsc. Sci.* 92: 141–161, 1951.

Blower, J. G. (Ed.) *Myriapoda.* Academic Press, London, 1974.

Burrt, E. Exudate from millipedes, with particular reference to its injurious effects. *Trop. Dis. Bull.* 44: 7, 1947.

Camatini, M. (Ed.) *Myriapod Biology.* Academic Press, New York, 1979.

Cloudsley-Thompson, J. L. The behavior of centipedes and millipedes. *Annu. Mag. Nat. Hist.* 5: 417–434, 1952.

Cloudsley-Thompson, J. L. *Spiders, Scorpions, Centipedes and Mites.* Pergamon Press, New York, 1968. (Three chapters are devoted to the mryiapod classes, with emphasis on their natural history.)

Crawford, C. S. Feeding-season production in the desert millipede *Orthoporus ornatus. Oecologia* 24: 265–276, 1976.

Crawford, C. S. and Matlack, M. C. Water relations of desert millipede larvae, larva-containing pellets, and surrounding soil. *Pedobiologica* 19: 48–55, 1979.

Edwards, C. A. The ecology of Symphyla. *Entomol. Exp. Appl.* Amsterdam 2: 257–267, 1959.

Eisner, T. and Meinwald, J. Defensive secretions of arthropods. *Science* 153: 1341–1350, 1966.

Filka, M. E. and Shelley, R. M. Millipede fauna of the Kings Mountain region of North Carolina, USA. *Brimleyana* 4: 1–42, 1980.

Hoffman, R. L. and Payne, J. A. Diplopods as carnivores. *Ecology* 50: 1096–1098, 1959.

Kaestner, A. *Invertebrate Zoology, Vol. II.* Wiley-Interscience, New York, 1968. (Once again, a detailed systematic account with extensive bibliography. Six chapters are involved with the myriapods.)

Lewis, J. G. E. Food and reproductive cycles of the centipede *Lithobius variegatus* and *L. forficatus. Proc. Zool. Soc. London* 144: 269–283, 1965.

Manton, S. M. The locomotion of the Chilopoda and Pauropoda. *Linn. Soc. Zool.* 42: 118–167, 1952.

Manton, S. M. The structure, habits and evolution of the Diplopoda. *Linn. Soc. Zool.* 42: 229–368, 1954.

Manton, S. M. Body design in Symphyla and Pauropoda. *J. Linn. Soc. Zool.* 46: 103–141, 1966.

Meske, C. Sense physiology of Diplopoda and Chilopoda. *Z. Physiol.* 45: 61–77, 1961.

Moore, R. C. (Ed.) *Treatise on Invertebrate Paleontology,* Part R, Arthropoda 4, Vol. 2. Geological Society of America and University of Kansas Press, Lawrence, 1969. (Covers the fossil myriapods)

Schaller, F. *Soil Animals.* University of Michigan Press, Ann Arbor, 1968.

Shear, W. Studies in the millipede order chordeumida (Diplopoda). A revision of the family Cleidogonidae and a reclassification of the order in the New World. *Bull. Mus. Comp. Zool.* 144: 151–352, 1972.

Shelley, R. M. Revision of the millipede genus *Pleuroloma* (Polydesmida: Xystodesmidae) Can. J. Zool. 58: 129–168, 1980.

Starling, J. H. Ecological studies on the Pauropoda of the Duke Forest. *Ecol. Mongr.* 14: 291–310, 1972.

Tiegs, O. W. The embryology and affinities of the Symphyla, based on a study of *Hanseniella agilis. Quart. J. Microsc. Sci.* 82: 1–225, 1940.

Toye, S. A. Effect of desiccation on millipedes. *Entomol. Exp. Appl.* 9: 369–377, 1966.

Woodring, J. P. and Blum, M. S. Anatomy, physiology and comparative aspects of the repugnatorial glands of *Orthocricus arboreus* (Diplopoda: Spirobolida). *J. Morphol.* 116: 99–208, 1965.

14

Arthropod Diversity: The Insects and Arthropod Evolution

Flying ability is the cornerstone of success in the largest animal class, the Insecta. While other land invertebrates creep slowly over or through the ground, insects wing swiftly through the air. Thus they better exploid the rather unevenly distributed resources of the terrestrial environment which, otherwise so limting for invertebrate organisms, supports over 750,000 insect species. This number exceeds the combined total of all other animal groups. Moreover, estimates of the number of undescribed insect species range to 10 million. We will never identify all of these creatures; hence most of our efforts are concentrated on the few groups directly affecting the human economy.

Some 20,000 insect species occur in aquatic habitats. Larval stages, in particular, are common in lakes and ponds. Even glacial streams, oil swamps, and hot springs support some insect life, however. Of all the habitats that invertebrates occupy, only the open sea has not been invaded by insects. This situation reflects the terrestrial origin of these animals and the crustacean dominance of marine arthropod niches.

No other invertebrates are as intimately involved with human lives as the insects. They pollinate over half of all flowering plants, including most of our food crops. Such service is hardly free, however, as other insects yearly consume about one-third of the potential harvest. Insects affect our lives in other ways as well. Butterflies and moths are creatures of unusual beauty. Bees provide us with honey, silkworms produce silk, and mosquitoes give us itching bites and transmit the organisms that cause malaria and yellow fever. Among other disease organisms transmitted by insects are those responsible for sleeping sickness, typhus, bubonic plague, and typhoid fever. Yearly, we broadcast thousands of tons of poisons on our planet to try to kill these creatures. Success in such ventures is often limited and always must be weighed against their long-term ecological consequences. Insects are the only animals that seriously compete with humans for food, but in our struggle with insects, we must be careful lest we become our own worst enemies (see Special Essay, this chapter).

To accord the insects the same depth of treatment given the other invertebrate classes would require a twofold increase in the size of this book. Therefore, we limit the present chapter to an account of the general biology of the class, with the intention of providing a view of the insects against which the other arthropod groups can be compared. To that end, the distinctive vocabulary used by insect specialists is largely avoided. For further reading, the interested student should consult the reference list at the end of this chapter.

Insect Unity

Insects are the culmination of a number of evolutionary trends in the Arthropoda. Their adaptations reflect the selective pressures of the terrestrial environment and include a hard waterproof skeleton, variable feeding habits, dry feces, tracheal respiration, uric acid excretion, outstanding mobility, keen sense organs, internal fertilization, and mechanisms to protect immature stages.

MAINTENANCE SYSTEMS

Insect maintenance biology addresses the problems of life on land. Adequate body support and water conservation are its central themes. While internal systems among the various insect groups are fairly uniform in dealing with these problems, body architecture and feeding methods are not. The latter represent areas of diversity on which the extensive speciation of the class is based.

General Morphology

The insect body comprises three distinct tagmata: the head, the thorax, and the abdomen (Fig. 14.1). The head segments are tightly fused, and in adults most of their boundaries cannot be distinguished. Large compound eyes are situated laterally on the head, and a single pair of antennae sprouts from the anterior end (Figs. 14.2 and 14.4). In most insects, ocelli are located between the compound eyes. The segment that bears the second antennae in crustaceans has no appendages in the insects. Although a few predacious forms have forwardly projecting jaws (Fig. 14.10), the mouthparts of most insects are directed downward (Fig. 14.2a). A labrum is present as an upper lip.

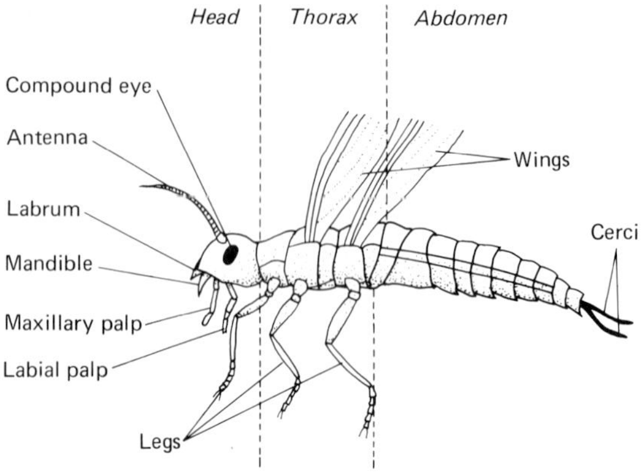

Figure 14.1 Lateral view of a generalized insect showing major external anatomical features. (From D. Sharp, *Cambridge Natural History, Vol. V*, 1901, Macmillan.)

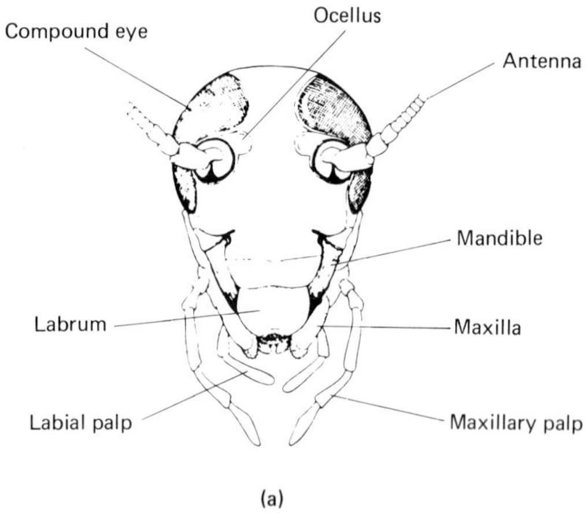

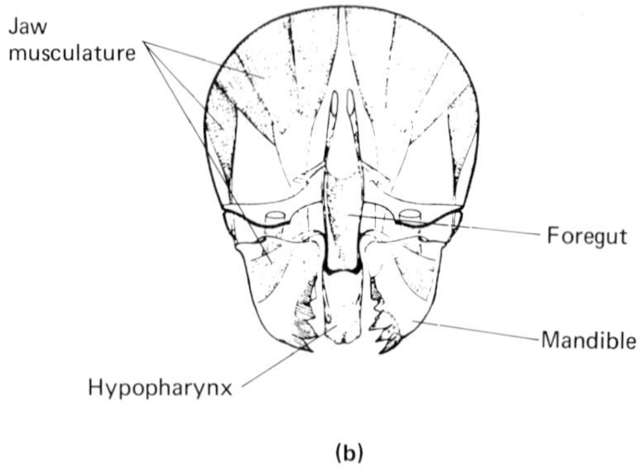

Figure 14.2 (*a*) Anterior view of the head of a cockroach, *Periplaneta americana;* (*b*) Section through the head showing the mandibles and associated musculature. (From R.E. Snodgrass, *A Textbook of Arthropod Anatomy*, 1952, Cornell University.)

There are three pairs of feeding appendages which are derived from ancestral legs; these include paired mandibles and first and second maxillae. Insect mandibles are formed from entire limbs and grind or cut with their tips (Fig. 14.2b). (Recall that crustaceans have gnathobasic mandibles.) Typically, the first maxillae are short but leglike, although both these appendages and the mandibles have been modified in insects with specialized feeding styles. As described in the following section on nutrition, these paired appendages may form stylets, sucking tubes, absorptive sponges, or any of several other structures for acquiring food. The second maxillae are fused as a labium and commonly bear palps. Between the labium and the first trunk segment, a wide articular membrane allows considerable flexibility in head movements.

The thorax consists of three segments, which bear all of the adult locomotor structures. These segments are called the prothorax, the mesothorax, and the metathorax. Each bears a pair of walking legs, which typically are long, clawed, and well muscularized. Laterally, the walls of the mesothorax and metathorax are drawn out as paired wings (Fig. 14.3). Along strategic lines, the wing is supported by thickened tubes called veins, which contain extensions of insect circulatory, respiratory, and nervous systems. Their organization is a key to species identification.

Wings are absent from some insect groups. The most primitive members of the class (the subclass Apterygota) apparently diverged before wings originated in the main insect line. Other groups, such as parasitic and burrowing insects, have lost their wings secondarily. Functional wings are also absent from the immature stages of all insects.

The abdomen consists of up to 11 segments, whose appendages are reduced or absent in most adult insects. Highly modified limbs are associated with the openings of the reproductive tract in both sexes, and a single pair of sensory appendages called **cerci** may project from the last segment. Abdominal leg rudiments are present, however, in the embryo and in some developmental stages.

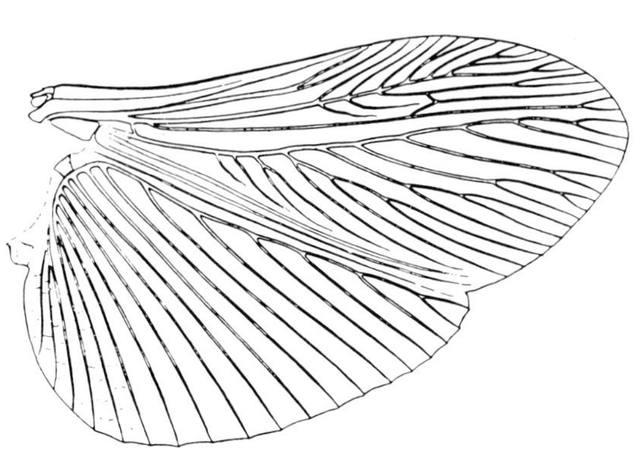

Figure 14.3 The right hind wing of *Periplaneta* fully spread. (From Snodgrass, *A Textbook of Arthropod Anatomy*.)

Insects are small creatures. The longest species, the walking stick *Pharanacia serratipes*, is a third of a meter in length, but the rhinocerus beetle *Titanus giganteus* has the greatest body volume. Among the tiniest insects are thrips, feather-wing beetles, and parasitic wasps. Some of these creatures are less than a tenth of a millimeter long. As mentioned in Chapter Ten, the small size of insects reflects in part the limited supportive capacity of an exoskeleton on land. Smallness offers distinct advantages also, such as the ability to exploit microhabitats. For example, the parasitic wasps mentioned above develop *inside* the eggs of other insects.

Nutrition and Digestion

A major factor in insect speciation is their ability to adapt to a wide range of diets and feeding styles. Herbivores, carnivores, scavengers, and parasites are all well represented in this class. Some insects are omnivorous, but others exhibit highly specific food preferences.

The mouthparts reflect the manner in which food is acquired. Insects that simply chew and swallow solid food have the least modified oral structures. Their short but leglike maxillae hold foodstuffs as sharp mandibles cut them to an ingestible size.

In other insects, the mouthparts are adapted for piercing, sucking, and/or sponging. These creatures usually consume liquid nutrients. In butterflies and moths, for example, distal elements of the maxillae form a long tube through which nectar is sucked. Between meals, this tube coils beneath the head (Fig. 14.4a). The mandibles may be entirely lost in butterflies and moths, but jaws are retained by bees for wax manipulation during hive construction. Bees feed on nectar drawn up by their elongated maxillae and labium (Fig. 14.4b). In flies, the labium expands distally as a fleshy sponge; absorbed nutrients are transported along surface capillaries to the mouth (Fig. 14.4c). Houseflies have reduced mandibles and maxillae, but cutting jaws are well developed in the horseflies. These insects wound their food source and then sponge up blood. Aphid and mosquito mouthparts also penetrate living food and extract nutrients. In these insects, all oral structures—the labrum, mandibles, maxillae, and labium—are drawn out as a piercing beak (Fig. 14.4d). Channels along this beak

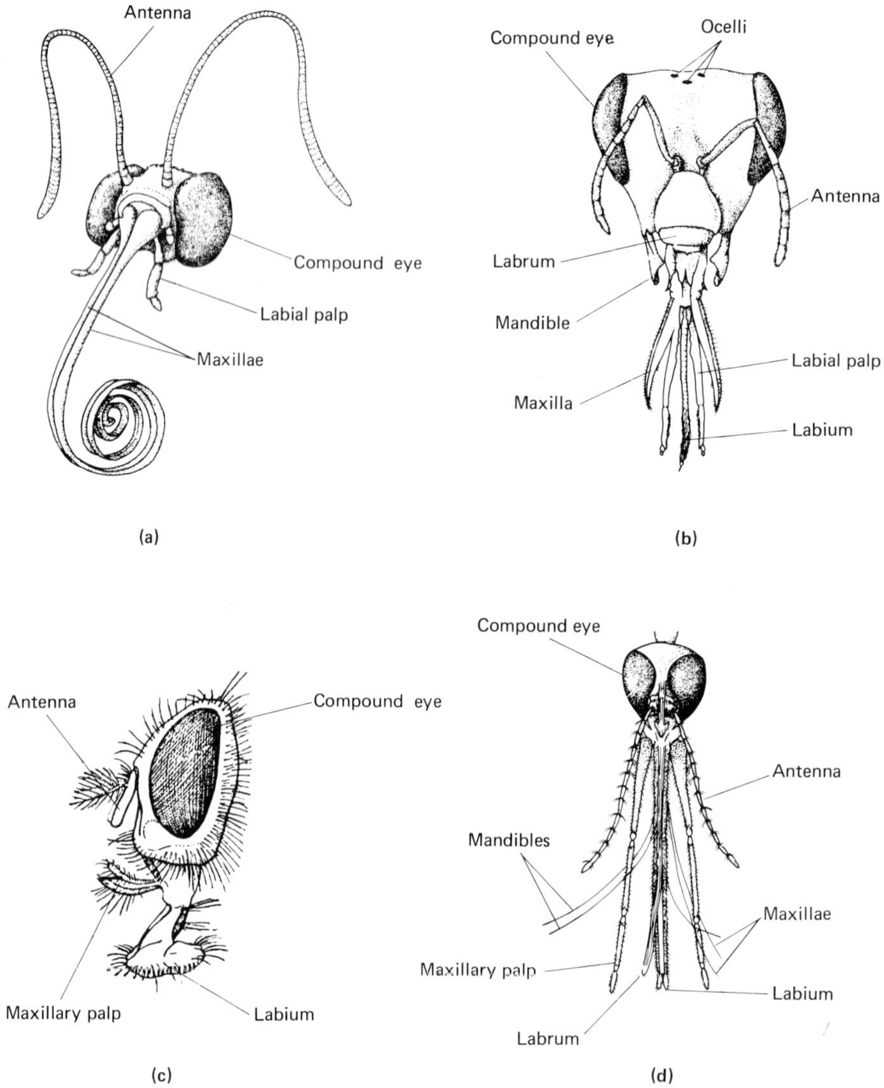

Figure 14.4 Head and mouth parts of various insects. (*a*) Butterfly. (*b*) A honeybee, *Apis mellifera*, anterior view; (*c*) A housefly, *Musca domestica*, lateral view; (*d*) A mosquito, anterior view. (*a*, *b*, and *d* from I.W. Sherman and V.G. Sherman, *The Invertebrates: Function and Form: A Laboratory Guide*, 2nd ed., 1976, Macmillan; *c* from R.E. Snodgrass, *Principles of Insect Morphology*, 1935, McGraw-Hill.)

allow both the outward passage of enzymes (including anticoagulants in mosquitoes) and the inward sucking of liquid food.

Mandibular and salivary glands discharge enzymatic fluids into the buccal cavity and over the mouthparts. These secretions soften and suspend incoming nutrients and may initiate digestion. Their activity also allows sponging and sucking insects to absorb some solid food.

The insect foregut includes an esophagus, crop, and proventriculus (Fig. 14.5a). The anterior esophagus is muscularized, and in sucking insects, it forms a powerful pump. The crop is a storage center, whose walls are highly extensible in insects that consume infrequent but large meals. Also, some preliminary digestion may occur in the crop. The **proventriculus** regulates food passage into the midgut. Among insects whose nutrients are already in dissolved or suspended

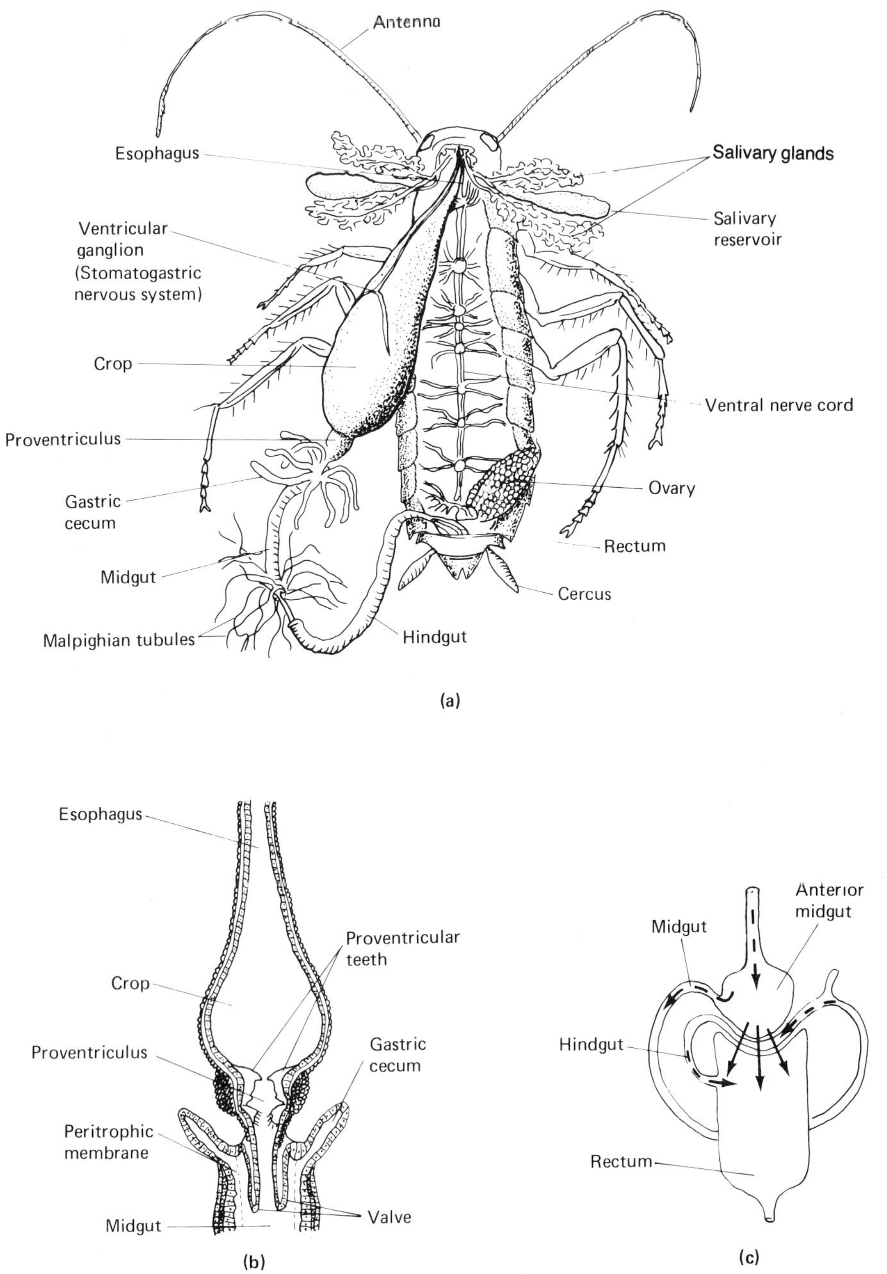

Figure 14.5 Internal anatomy of the oriental cockroach, *Blatta orientalis.* The gut has been pulled out so that its parts are easily visible; (*b*) Section through the crop, proventriculus, stomodeal valve, and anterior midgut of *B. orientalis;* (*c*) Midgut modifications in a fluid feeding insect (a coccid, Order Homoptera). Solid arrows show the movement of water from the anterior midgut directly to the hindgut; the dashed arrows represent the movement of concentrated nutrients through the digestive loop of the midgut. (*a* from G. Rolleston and W.H. Jackson, *Forms of Animal Life*, 1888, Oxford University; *b* from Snodgrass, *Principles of Insect Morphology;* *c* from M.J. Berridge, In *Chemical Zoology*, *Vol. V*, Florkin and Scheer, Eds., 1970, Academic Press.)

form, this area may be merely a valved straining chamber. In insects that consume bulkier food, a complex proventriculus serves as a gastric mill. Its cuticular walls bear strong opposing teeth and grinding surfaces which are gnashed together by powerful proventricular muscles. When the food is reduced sufficiently, a valve opens the passage to the midgut (Fig. 14.5b).

The tubular midgut is lined with columnar cells, some of which manufacture and secrete enzymes into the midgut lumen, where digestion is completed. Most absorption also occurs in this area. A rapid, continuous replacement of the midgut lining suggests a very high level of activity. Most insects have gastric ceca, usually near the anterior end of the midgut (Fig. 14.5a, b). These pouches increase the region's working surface area and also may serve as breeding grounds for mutualistic bacteria. The insect gut provides a home for numerous protozoans and microorganisms. Most renowned among these symbionts are zooflagellates that digest cellulose in the hindguts of wood-eating insects, such as the termites.

A **peritrophic membrane** often borders the midgut lumen of insects that eat solid food. This thin cuticular lining is secreted by the cells of the midgut wall (Fig. 14.5b). Its production is continuous, and in the hindgut it may envelop feces. The peritrophic membrane evidently protects the delicate inner lining of the midgut. Enzymes pass freely through it, while the membrane's grid limits the size of absorbed particles.

Insects that imbibe fluid nutrients rarely possess a peritrophic membrane. This situation reflects the less abrasive nature of their food. The potential midgut problems of such insects are quite different. If a large quantity of liquid is consumed, concentrations of midgut enzymes may drop below their level of optimum efficiency. Such dilution seldom occurs, however, owing to the existence of complexly looped digestive tracts (Fig. 14.5c). Excess water passes directly from the anterior midgut into the hindgut. Meanwhile, concentrated fluids are channeled through the midgut loop, where effective digestion occurs. Such a mechanism permits survival on fluids with very low concentrations of nutrients.

While some absorption of nutrients may occur in the insect hindgut, this region is concerned primarily with water conservation. Rectal glands (Fig. 14.9), composed of stacks of elongate cells, are the principal sites of water resorption. Both Malpighian discharge and fluids associated with undigested wastes are treated. Insects living in arid environments pass a dry, powdery feces.

Circulation

Insect circulatory systems conform to the basic arthropod pattern. A dorsal tubular heart courses through the abdomen and narrows as an aorta in the thorax (Fig. 14.6). Blood is pumped into the head and then flows posteriorly through perivisceral and ventral sinuses. Circulation through the legs may be directed by diaphragms which isolate incurrent and excurrent pathways. If the legs are especially long, as in grasshoppers, booster hearts give the blood added impetus. These accessory pumps are formed by pulsating membranes. Similar structures force blood through the wing veins. In the posterior body, hemolymph flows dorsally into a pericardial cavity. Segmentally arranged ostia readmit blood to the heart.

Hemolymph pressure is rather low in insects. Nutrients, wastes, and hormones are circulated by the blood, but the latter bears minimal responsibility for respiratory transport. The routine muscular activity of the legs, wings, and abdomen contributes to blood flow through the sinuses. On occasion, this flow may be exploited for hydraulic power. In this way, butterflies and moths unroll their maxillar feeding tubes. Hemolymph pressure also is necessary for body flexions associated with ecdysis, defecation, and aeration of the tracheal system.

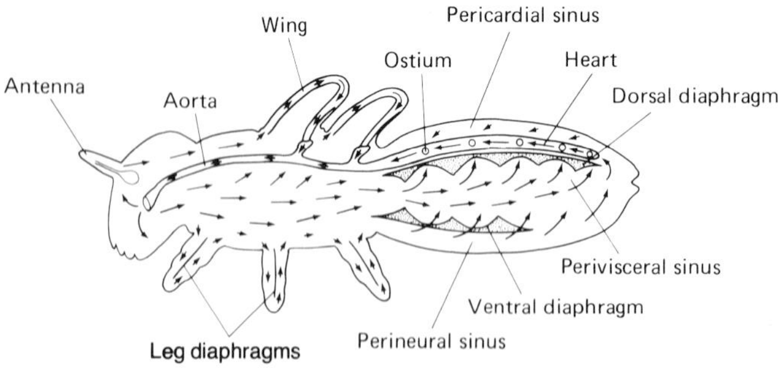

Figure 14.6 Side view of a generalized insect circulatory system. (From H. Weber, *Lehrbuch der Entomologie*, 1933, Gustav Fischer Verlag.)

Respiration

Gas exchange by direct diffusion over the body surface is the only respiratory method among some tiny, primitive insects and during embryonic stages in most groups. While a majority of adult insects relies on tegumental exchange to a slight degree, their respiratory needs are met primarily by a tracheal system. This system includes numerous valved spiracles and an anastomosing network of internal tubules through which oxygen diffuses or is pumped directly to the tissues (Fig. 14.7a). Carbon dioxide likewise diffuses

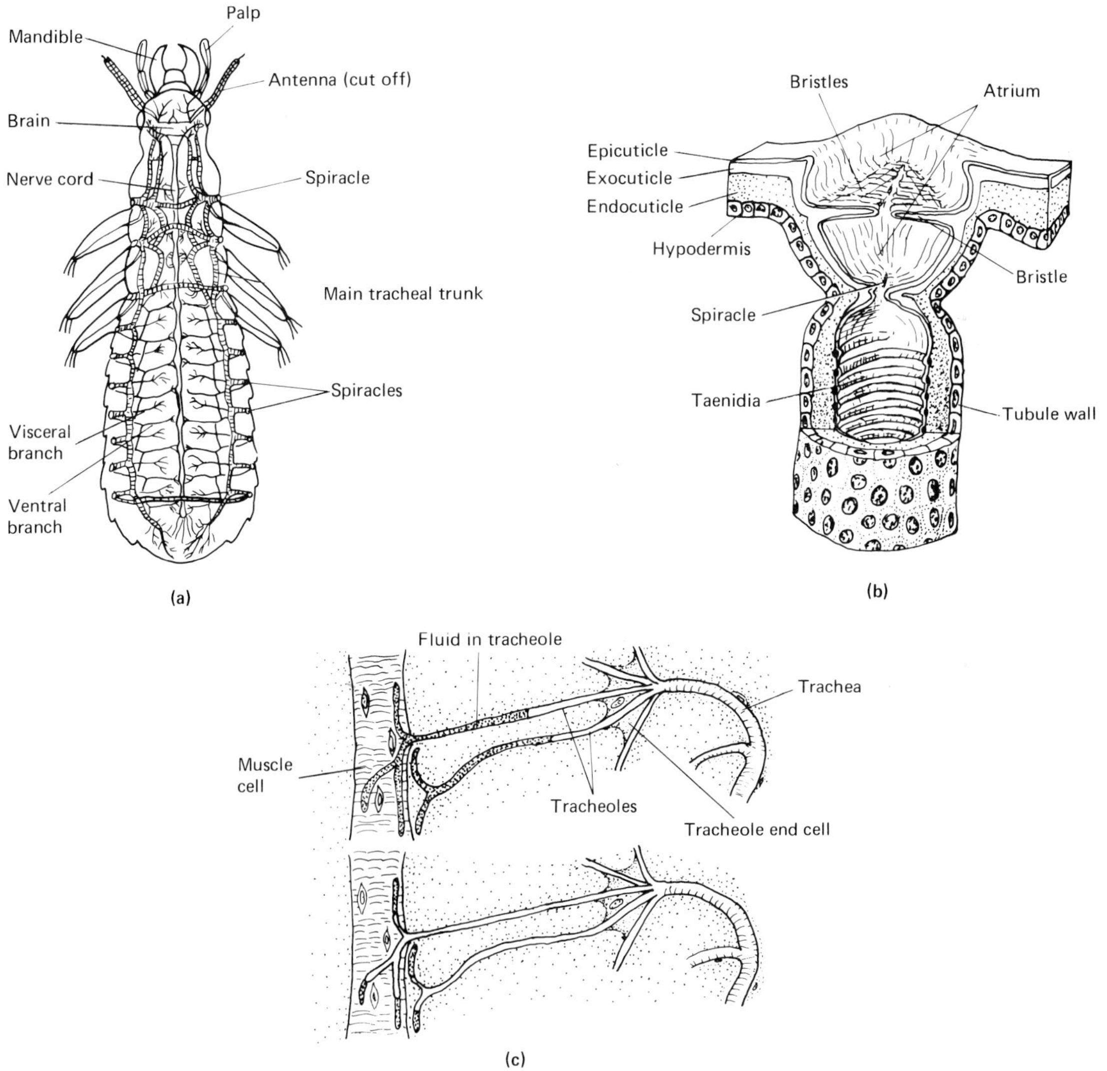

Figure 14.7 (*a*) Tracheal system of a generalized insect; (*b*) Longitudinal section through a spiracle; (*c*) Diagrams showing the movement of fluid in the tracheoles. *Above:* the tissues are well oxygenated and there is considerable fluid in the tracheoles. *Below*: the tissues are oxygen deficient and the fluid volumes in the tracheoles decrease, allowing the tissues increased access to oxygen. (*a* from H.H. Ross, *A Textbook of Entomology*, 3rd ed., 1965, John Wiley; *b* from G. Siefert, *Entomologisches Praktikum*, George Thieme Verlag; *c* from Snodgrass, *Principles of Insect Morphology*.)

out through these tubules. The direction and rate of air flow through the tracheal system may be precisely controlled.

Paired spiracles occupy up to eight abdominal segments, the metathorax, and sometimes the mesothorax. These openings often are recessed within an atrium whose walls are lined with setae (Fig. 14.7b). Such hairs inhibit parasites and debris from entering the respiratory tract. The spiracle itself is guarded by a muscularized valve. Because it arises as an invagination of the body wall, each trachea is lined with cuticle. Where the tracheal lumen is large, the tubule walls are reinforced with spiraling skeletal rings called **taenidia**.

Tracheal organization varies. Often each primary tubule gives rise to three branches, which serve the dorsal, visceral, and ventral body, respectively. Each of these branches repeatedly subdivides until the smallest respiratory tubules, the **tracheoles**, terminate within special cells. These cells, called **tracheole end cells**, secrete and house the final offshoots of the system (Fig. 14.7c). Less than a micrometer in diameter, these offshoots penetrate and service adjacent tissues. Incoming oxygen is dissolved in fluid at the distal end of the tubule. The volume of fluid in the tubule is related to the respiratory condition of local cells. When surrounding cells are low in oxygen, tracheolar fluid levels drop, thus allowing oxygen-rich air to penetrate closer to its target.

Insect tracheal systems are more elaborate than those of other arthropods. Longitudinal trunks often connect the primary tubules and their major branches (Fig. 14.7a). Likewise, commissures may link paired elements in each segment. These connections ensure that no area of the body will suffocate if the closest spiracle becomes clogged. System-wide ventilation is also possible. In many insects, spiracles are specialized as inhalent or exhalent pores and air flow through the tracheal network follows a well-defined route. Tracheal pouches called air sacs store excess air which, upon release, accelerates tracheal gas flow. Contraction of the air sacs is caused by hemocoelic pressure initiated by muscular activity in the legs or abdomen. Once begun, ventilation waves are sustained by the elasticity of the tracheal walls.

Aquatic insects retain tracheal systems, although accessory gills have developed in some species (Fig. 14.8a). Spiracles may be reduced or absent, however, as oxygen may diffuse across the integument and enter a capillarylike network of tracheal tubules (Fig. 14.8b). Anal respiration occurs in some aquatic groups. These insects pump water in and out of their rectums, whose walls contain subsurface tracheae (Fig. 14.8c). Aquatic insects with functional spiracles

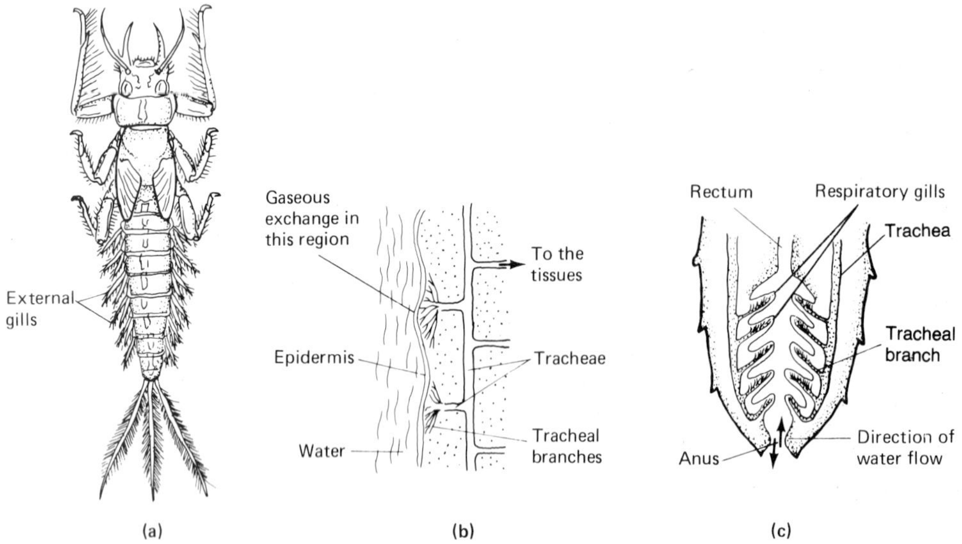

Figure 14.8 (*a*) Dorsal view of a mayfly nymph, *Polymitarcys albus*, an aquatic insect with external gills; (*b*) Diagram of integumentary respiration in an aquatic insect; (*c*) The rectal gills of a dragonfly nymph. (a from D.J. Borror and D.M. Delong, *An Introduction to the Study of Insects*, 1964, Holt, Rinehart and Winston; *b* and c from Ross, *A Textbook of Entomology*.)

hold a bubble of air over each of these openings. This bubble is secured by wax and hydrophobic hairs; the latter may be as dense as 2 million per square millimeter. Oxygen from the water continuously diffuses into the bubble.

Excretion and Osmoregulation

Excretory physiology among insects is dominated by a single theme—the need to conserve water. Malpighian tubules are almost universal. These paired structures extend through the hemocoel, and their bases are attached to the midgut–hindgut junction (Fig. 14.9). If only a single pair is present, the tubules are highly coiled. Some insects possess over a hundred pairs of Malpighian tubules, although size and convolution decrease when these structures are numerous.

Malpighian tubules are muscularized and move within the hemocoel as they absorb products from the blood. Most nitrogenous wastes are absorbed as uric acid formed in the tissues. Amino acids, salts, and water also pass into the Malpighian lumen and are transported by peristaltic contractions toward the hindgut. Some resorption may occur in the lower part of the tubules, but the rectum is the primary site of such activity. Essential molecules and ions are returned to the hemolymph, and the withdrawal of water from both excretory and digestive effluents produces a dry, compact waste. Uric acid is voided in crystalline form.

Excretion is a different problem for aquatic species, whose urine is usually copious. Following large meals, insects that feed on blood or plant juices also pass rather dilute wastes.

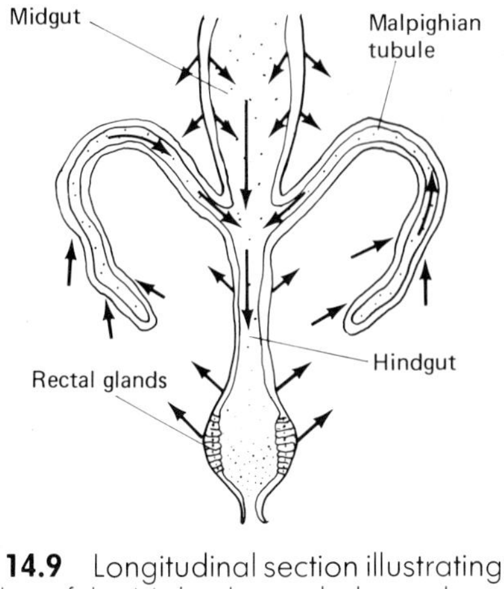

Figure 14.9 Longitudinal section illustrating the relationship of the Malpighian tubules to the mid and hindguts. Arrows indicate water flow. (From Ross, *A Textbook of Entomology*.)

Insects exhibit several auxiliary methods of waste control. The walls of the Malpighian tubules and those of the midgut are similar in structure, and the latter probably transfer some excretory products into the digestive lumen. Storage excretion is common and utilizes fat bodies, pericardial nephrocytes, and the cuticle itself. Wastes deposited in the exoskeleton are voided conveniently at each molt.

While resorption in the hindgut minimizes water loss from the digestive and excretory tracts, a waxy epicuticle limits evaporation from the general body surface.

ACTIVITY SYSTEMS

The unparalleled success of the insects is attributable largely to their outstanding locomotor powers, which reflect an ability, almost unique among the invertebrates, to pursue the advantages of terrestrial life. Because their maintenance systems have overcome problems associated with support and water supply, insects range extensively over the terrestrial horizons. Compared to aquatic habitats, dry air offers far more oxygen and less mechanical resistance to movement. Such factors greatly enhance locomotor potential. The requisites of life—food, water, shelter, and mates—are less evenly distributed on land, and this factor also contributes to selection for more sophisticated activity systems. Among the invertebrates, only spiders, crabs, and cephalopods approach the behavioral complexities of the insects.

Skeleton and Muscles

Insects possess the hard exoskeleton and striated muscle fibers typical of the Arthropoda. Locomotor appendages are limited to the thorax, much of which is filled with leg and wing muscles. Insects, having only three pairs of relatively long legs, are capable of fast, easily coordinated movements. The compact thoracic box ensures that the force exerted by each leg moves the entire body. A characteristic gait features alternate stepping by leg triads (Fig. 14.10). The foreleg and hindleg on each side are grouped with the opposing midleg. One three-leg set advances and then supports the body as the opposite set steps forward. Even at high speeds, the animal thus remains quite stable.

Insect legs may be modified in a number of ways (Figs. 14.11a–g): The jumping hindlegs of the grasshopper, for example, are greatly elongated. The raptorial forelegs of the praying mantis are used for food capture, but not for locomotion. The legs of honeybees bear several specialized areas, including

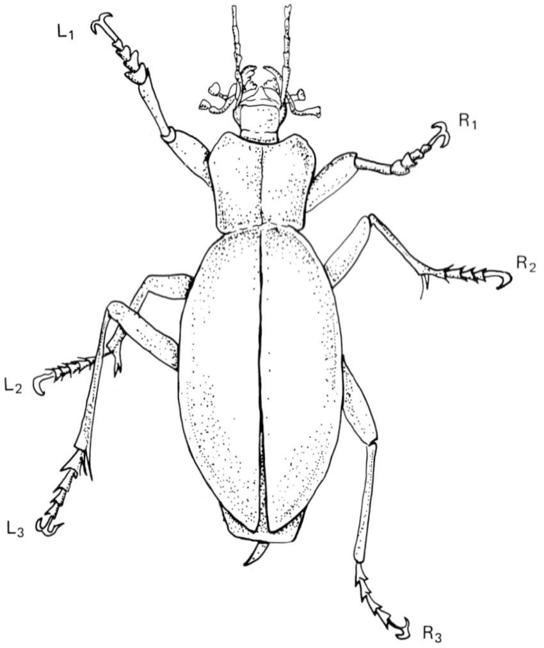

Figure 14.10 A beetle walking or running. Three legs (L_1, R_2, and L_3) are directed forward while the other three (R_1, L_2, and R_3) are directed backward. (From A.S. Packard, *Zoology*, 1904, Holt.)

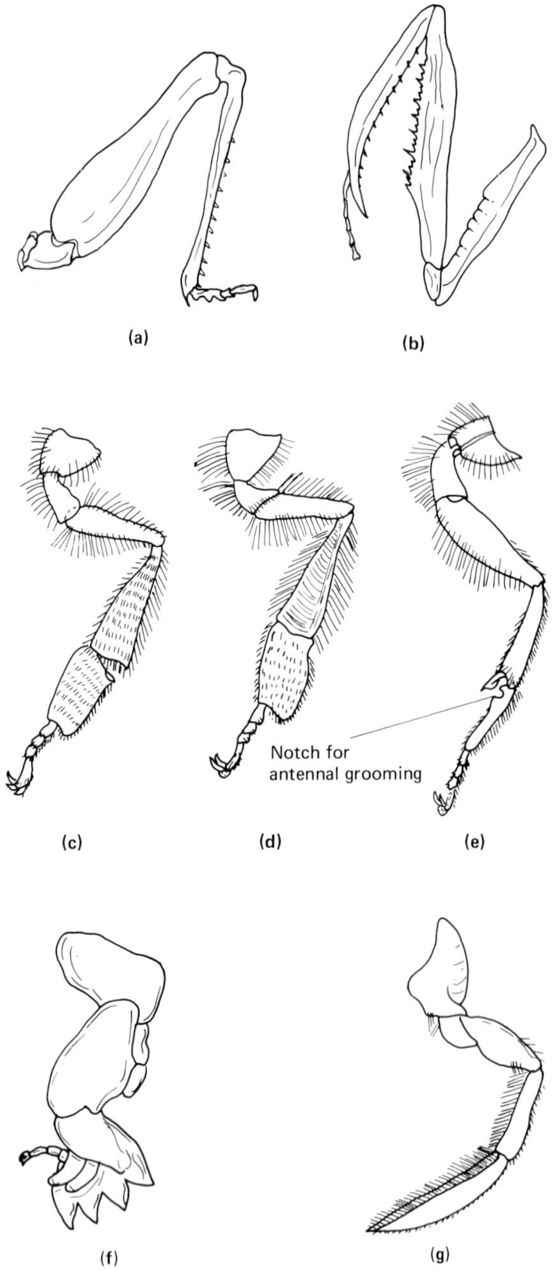

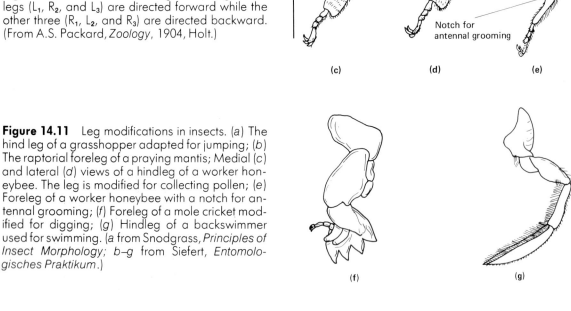

Figure 14.11 Leg modifications in insects. (*a*) The hind leg of a grasshopper adapted for jumping; (*b*) The raptorial foreleg of a praying mantis; Medial (*c*) and lateral (*d*) views of a hindleg of a worker honeybee. The leg is modified for collecting pollen; (*e*) Foreleg of a worker honeybee with a notch for antennal grooming; (*f*) Foreleg of a mole cricket modified for digging; (*g*) Hindleg of a backswimmer used for swimming. (*a* from Snodgrass, *Principles of Insect Morphology; b–g* from Siefert, *Entomologisches Praktikum*.)

brushes for collecting pollen and cleaning the compound eyes and a semicircular notch through which the antennae are groomed. Digging insects have short but strong limbs, while the legs of swimming species are often flat and hairy.

The evolutionary history of insect wings is not well understood. These remarkable locomotor structures may have arisen from rigid lateral extensions of the thoracic tergites. Such extensions might have enabled jumping insects to remain airborne for brief periods. Alternatively, wing precursors may have been articulated structures which originally served as protective covers for the thoracic spiracles or as thermoregulatory organs.

The wing, an extension of the dorsal wall, rests upon a pleurite fulcrum (Fig. 14.12). The location of this fulcrum so near the proximal end of the wing affords optimum leverage; that is, very short muscular contractions produce wide movements of the wing. Up-and-down beating involves the contraction of both indirect and direct flight muscles. Direct flight muscles are attached to various points on the sclerotized base of the wings. These fibers contribute to downward beating, control the tilting of the wings, and fold the wings back over the abdomen when not in use. Indirect flight muscles include dorsoventral and longitudinal fibers which manipulate each winged tergite. Contraction by dorsoventral muscles lowers the dorsal plates, resulting in the elevation of the wings (Fig. 14.12a, b). Longitudinal contractions force the tergites upward and the wings downward (Figs. 14.12c, d, e). Slowflying insects like butterflies beat the wings about 20 times per second. At such modest speeds, individual muscle contractions account for each wing stroke. Most bees and flies, however, beat their wings about 200 times every second, while gnats and midges may exceed 1000 strokes per second! Nerve cells cannot conduct action potentials at such high frequencies. Evidently, the wings are beating several times for each arriving impulse.

In fact, each muscular contraction causes several wing beats. This phenomenon results from the elasticity of the tergite–wing hinge. This hinge is stable at two positions: one when the wings are fully elevated; the second, when they are fully lowered (Fig. 14.12b, e). The middle position is inherently unstable, and the hinge always snaps upward or downward from this position into a more stable one. When either set of indirect flight muscles contracts, the hinge is forced into its unstable middle position (Fig. 14.12a). Elastic forces then overcome muscular ones, causing the wings to snap upward or downward. During this snap, the contracting muscles briefly relax. Meanwhile, the opposing set of indirect flight muscles, which has been extended by the wing movement, contracts. Its contraction returns the hinge to midline instability, causing the hinge to snap in the opposite direction. The first set of muscles then resumes contracting, another oscillation occurs, and so on. In this way, the thorax and wings move up and down in response to dual, repeatedly interrupted contractions of each muscle set. Nervous input simply initiates this cycle and ensures that the oscillations persist.

The mere beating of the wings, however, cannot produce flight. The wings must be held at a slight positive angle into the wind to generate lift. This angle is maintained throughout the beating cycle by turning the wing over during the upstroke (Fig. 14.13). During this cycle, the wings are also moved forward and backward. The airfoil quality of the wing is enhanced by venation patterns that create a convex upper surface. Air flows more rapidly over the curved top of the wing and spills from its trailing edge, thus producing both lift and thrust forces.

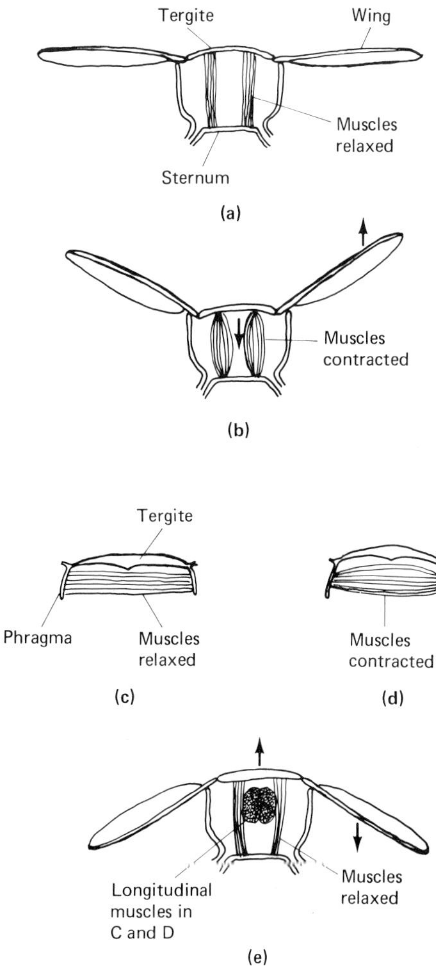

Figure 14.12 Diagrams of the muscles and wing movements producing insect flight. (*a*) Cross section of a thoracic segment showing an unstable wing position; (*b*) Dorsoventral muscles contract, pulling down the tergite and elevating the wings; (*c*) Longitudinal section of a thoracic segment showing longitudinal muscles at rest; (*d*) Longitudinal muscles contracted causing an upward bowing of the tergite and (*e*) a corresponding downward movement of the wings. (From Ross, *A Textbook of Entomology*.)

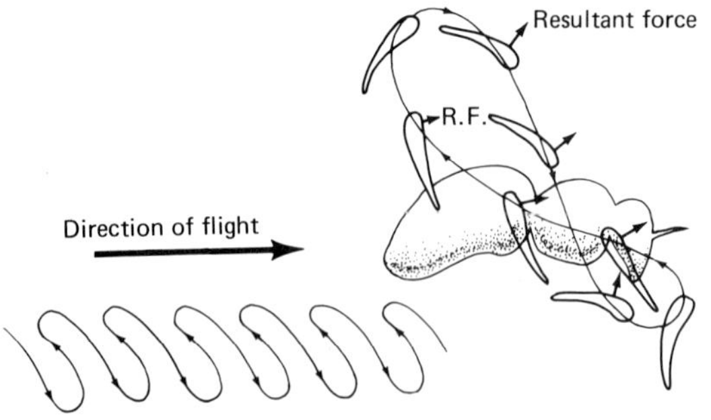

Figure 14.13 The path followed by a fly's wing in flight. Righthand figure shows the path of the wing tip of a suspended insect. The wing turns over during the upstroke to maintain a positive angle of attack. Arrows represent the resultant of both lift and thrust. Lefthand figure shows the path traced by the wing tip of a fly in flight. (From M. Wells, *Lower Animals*, 1968, McGraw-Hill.)

The two pairs of wings differ among various insect groups. In butterflies, both flap in unison. Each pair of damselfly wings, however, is always at an opposite phase in the beating cycle; with this arrangement, the hindwings avoid the air turbulence created by the fore wings. Flies and mosquitoes have only one pair of wings. On the metathorax, they have been reduced to short knobs, the **halteres,** which have gyroscopic power. In contrast, among beetles, only the second pair of wings is used in flight. Very tiny insects are scarcely denser than the air itself. Their bristly wings (Fig. 14.14) more or less paddle along, not unlike the locomotion of planktonic arthropods. Some of the largest insects, such as dragonflies and horseflies, are the fastest flyers. These arthropods have been clocked at up to 50 km/hr.

Insect flight muscles rank among the world's strongest contractile fibers. Their mitochondria are enormous, and tracheoles deliver oxygen deep into the fibrils. Nearby fat bodies provide high energy reserves.

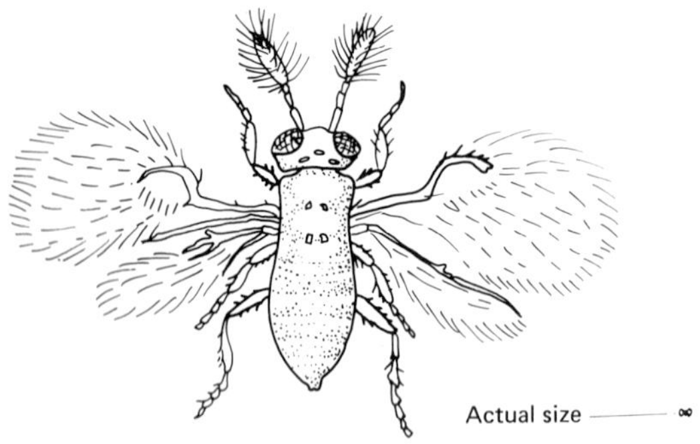

Figure 14.14 The bristly wings of a very small insect, *Trichogramma;* a wasp that develops as a parasite in the eggs of other insects. (From Wells, *Lower Animals.*)

Nerves

Insects have not evolved large complex nervous systems like those of vertebrates. After all, only so much can be crammed into a small package. Instead, insects have developed miniaturized systems of programmed, or instinctive, behavior. Such systems appear to require fewer nerve cells and less space than systems that permit learning and complex behavior of the kind seen in higher vertebrates.

The insect central nervous system, like that of other arthropods, features a brain and a ventral nerve cord connecting variously fused ganglia (Fig. 14.15). The insect brain comprises a large protocerebrum, which innervates the eyes, a deutocerebrum with control over the antennae, and a small tritocerebrum, which is associated with the limbless postantennal segment. The subesophageal nerve mass contains the fused ganglia of the mouthparts. Thoracic nerve centers may

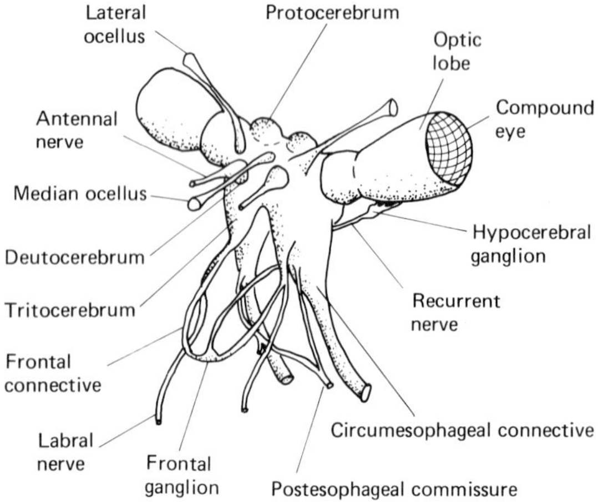

Figure 14.15 Latero-frontal view of the brain of a generalized insect. (From K.U. Clarke, *The Biology of Arthropods*, 1972, Edward Arnold.)

also be fused, and some forward migration by reduced abdominal ganglia is common.

A stomatogastric network is also present. Similar to the human autonomic nervous system, this network arises from the anterior brain to innervate the glands and muscles of the foregut (Fig. 14.5). Other important elements in insect nervous equipment are the neurosecretory cells. These neurons also function as endocrine glands and secrete hormones into the blood. Neurosecretory hormones regulate molting, sexual maturity, and water balance.

Sense Organs

Insect sense organs include compound and ocellar eyes and a host of tactile and chemosensitive structures. The latter feature highly specialized associations of cuticular and nervous elements, most of which are concentrated on the appendages.

Insect compound eyes resemble those of the crustaceans. Although their visual acuity is poor compared with the powers of the human eye, at least some insects distinguish colors and certain geometric forms (Fig. 14.16). The sharpest insect eyes probably belong to the dragonfly, which has up to 30,000 ommatidia. One quality in which insect eyes excel our own is measured by a flicker-fusion test. A light flickering at about 50 times per second is perceived as a continuous beam by the human eye. Fast-flying insects, however, discriminate up to 300 separate flickers per second. Such ability enables them to monitor a rapidly changing environment while flying. Simple medial eyes are present in most insect larvae and persist in many adult species. Up to three of these ocelli are present between the compound eyes, where they detect general levels of illumination. Ocellar stimulation apparently heightens the overall sensitivity of the nervous system and its sensory outposts.

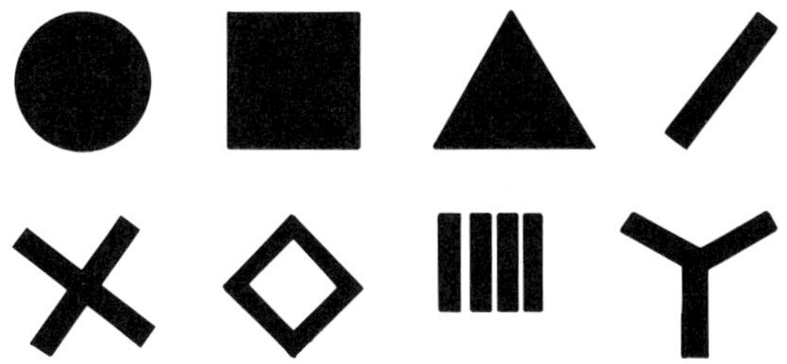

Figure 14.16. Honeybees do not readily discriminate the figures in the top row, one from another, but they can easily discriminate any top row figure from any bottom row figure. (From J.D. Carthy, *An Introduction to the Behavior of Invertebrates*, 1958, Hafner.)

Chemoreceptors are concentrated along the antennae and on the mouthparts. Often, sensory endings are presented through tiny cuticular canals atop peglike elevations (Fig. 14.17a). In many insects, these receptors have extremely low response thresholds. Some moths, for example, respond to the sex attractants (pheromones) secreted by potential mates located over 4 km away.

Tactile receptors include simple hairs and complex centers called **chordotonal organs,** which are aggregations of subcuticular processes and their associated nerve fibers (Fig. 14.17b). These organs respond to changes in muscular and skeletal tensions and are common near all jointed areas of the body. Specialized chordotonal structures called **tympanal organs** are present in sound-producing insects (Fig. 14.17c). Constructed in part from tracheal pouches, these vibration-sensitive organs are involved in the intraspecific communication of grasshoppers, crickets, and related forms.

During flight, insects rely on a number of sensory systems. Air speed probably is translated from the stimulation to the eyes and antennal hairs. Muscular activity is under local control in the thorax. By keeping the dorsal ommatidia under maximum illumination, upright posture is maintained. As soon as the tips of the legs touch something, the activity of all flight muscles is inhibited.

Behavior

In this limited space, it is difficult to make generally useful comments on the behavior of over three-quarter million animal species. Accordingly, we present a mere overview of some of the principles involved and then turn briefly to one of the most fascinating areas of invertebrate behavior: insect societies.

Having evolved sophisticated mechanisms to detect and respond to environmental situations, insects exhibit complex behavioral patterns. Alone among the invertebrates, insects range widely over the terrestrial scene, seeking out optimal sites for food, shelter, and mates. Each species performs diverse, highly integrated acts by which it maximizes its access to environmental benefits. Most of these responses relate directly to basic orientations to light, gravity, humidity, temperature, touch, and certain chemicals. Specific taxes vary from group to group and even among

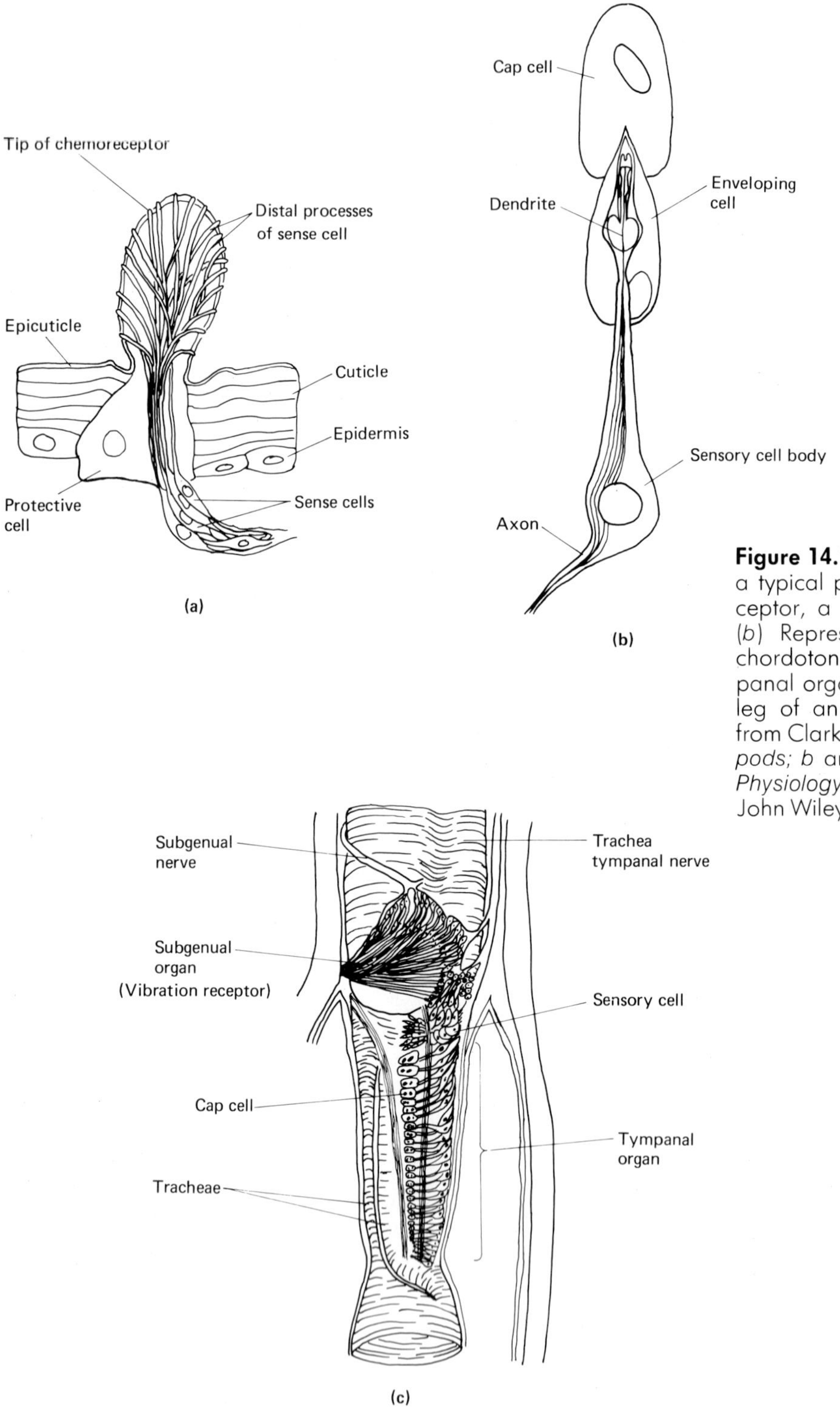

Figure 14.17 (*a*) Section through a typical peglike insect chemoreceptor, a sensillum basiconicum; (*b*) Representation of an insect chordotonal sensillum; (*c*) Tympanal organ from the prothoracic leg of an orthopteran insect. (*a* from Clarke, *The Biology of Arthropods; b* and *c* from V.G. Dethier, *Physiology of Insect Senses*, 1963, John Wiley.)

various stages in the life cycle of a single species. Most burrowing larvae, for example, are negatively phototactic and positively geotactic. Meanwhile, larvae that hatch on the ground but feed on plant leaves are attracted to light and climb against gravitational forces. When such insects mature, these patterns may change dramatically, depending on the ecology of the adult form.

While simple taxes explain much of insect behavior, the real situation is typically much more complicated. Land is a complex environment, and insects must often assimilate conflicting orientational demands. Internal physiological conditions, such as hunger or sexual state, also complicate the expression of simple, innate responses.

Integration occurs in the brain. Insects with their brains removed still perform a variety of acts, but fail to adapt their behavior to changing environmental conditions. They become hyperactive, often stumbling about until they drop from fatigue or entanglement. Apparently, stereotyped responses are programmed in local neuromuscular organization. The brain routinely inhibits these prepackaged responses until environmental conditions warrant them. Extended behavioral patterns, such as those associated with courtship and mating, involve the serial release of numerous stereotyped acts. Evidently, each act releases a response in the mate, whose behavior then stimulates another act by the first animal, and so on (Fig. 14.22). Similar feedback series unfold in many behavioral areas as insects interact with their animate and inanimate surroundings.

Learning has been demonstrated in this class. Cockroaches have been taught to run simple mazes (Fig. 14.18), and bees learn to associate feeding opportunities with certain colored dishes or specific times of day.

Insects interact with other organisms on a variety of levels. Of these interactions, none are more intimate than those among members of the same species of termites, bees, wasps, and ants. Like humans, these social insects are dependent on each other for survival.

Insect societies are rigidly organized. Each member is totally dependent on the group and cannot function except in relationship to it. Societal organization involves polymorphism, a division of labor and elaborate communication systems.

We encountered polymorphism as far back as certain colonial coelenterates whose polyps are differentiated as reproductive and feeding zooids. A similar type of polymorphism occurs among the social insects (Fig. 14.19). Typically, there is a large worker caste whose members are sterile. These insects perform the bulk of the society's chores, such as home construction, food acquisition, and brood care. In some groups (notably the termites), a second sterile caste, the soldiers, exists. These insects bear specialized offensive and defensive structures, such as oversized mandibles and glands that discharge noxious chemicals. One or more queens and a few fertile males are the only reproductive members of most insect societies. These individuals are usually quite large. In termites, for example, the queen may weight a thousand times as

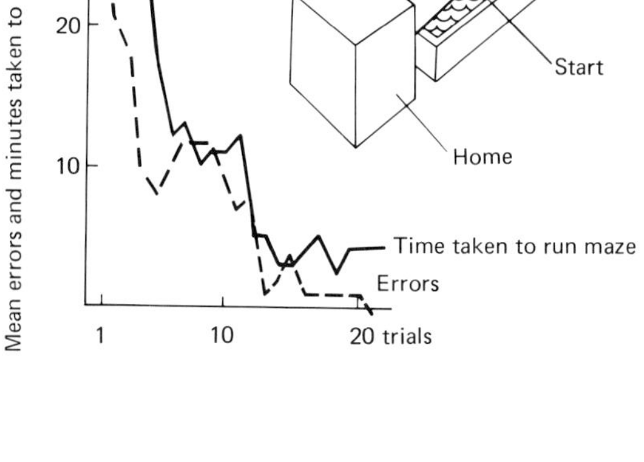

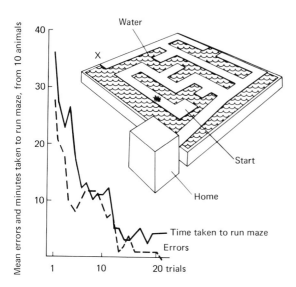

Figure 14.18 An experiment on maze learning in a cockroach. The graph shows the mean number of errors (occasions on which the animal took a wrong turn or retraced its steps towards the start) and the total time taken to run to the shelter in successive trials. (From Wells, *Lower Animals*.)

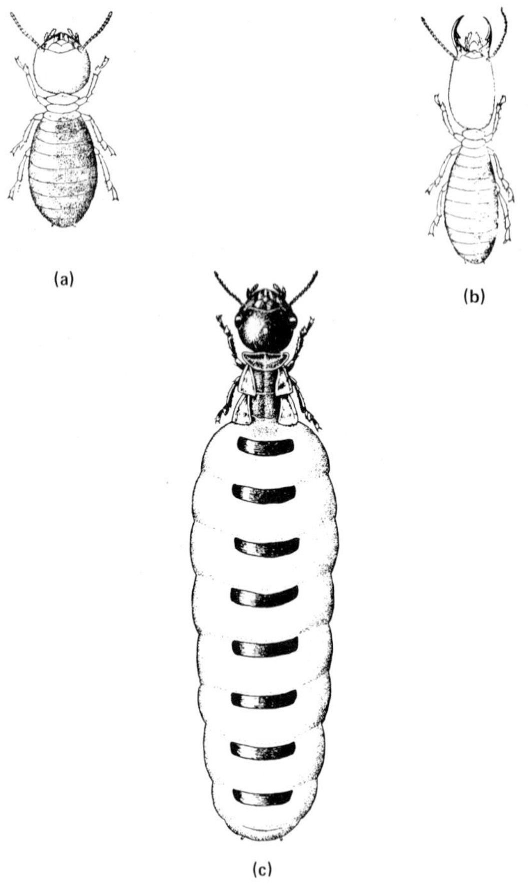

Figure 14.19 Three castes from a colony of a termite (*Amitermes hastatus*) drawn to the same scale. (*a*) Worker; (*b*) Soldier; (*c*) Primary queen with greatly swollen abdomen. (From E.O. Wilson, *The Insect Societies*, 1971, Harvard University.)

much as a worker: Totally preoccupied with egg laying she may produce 50 million young in her lifetime. The reproducers are the only winged individuals in ant societies. Wings allow sexual forms to disperse and initiate new colonies when the old one is overpopulated or otherwise threatened.

Of the many eggs laid each day in a thriving insect colony, only a few each season develop into reproductive adults. Among bees, larval nourishment is the critical development factor. Larvae fed exclusively on **royal jelly,** a nutritious secretion from the esophageal glands of a nursing worker, grow into queens. Strictly genetic factors determine the appearance of male bees, which are always haploid and arise from unfertilized eggs. Termite reproductives and soldiers exude a substance that inhibits the development of additional members of their castes. Under this inhibition, all immature termites become workers. When an increase in colony size, the death of a queen, or some other factor reduces the concentration of an inhibitory pheromone, the appropriate individuals develop.

Communication is vital to the success of all cooperative ventures, insect societies included. Members of each colony definitely recognize their comrades; if a worker strays from its home group, it seldom is accepted by another colony. Among termites and ants, most communication is chemical. Ants lay down odor trails as they forage; thus they can find their way back to the nest and lead other workers to any discovered food source. Mutual grooming and antennal caressing occur within the nest. Termite young lick the anal regions of adult members of the colony. Among other functions, this act transfers a starter supply of cellulose-digesting zooflagellates to the gut of the young animal.

The remarkable communication systems of honey bees have been described by Karl von Frisch. When a scout bee discovers a food source, it returns home and dances on the wall of its hive. If the source is nearby, the bee performs a simple **round dance,** alternately circling in clockwise and counterclockwise directions (Fig. 14.20a). The other bees are excited by the dancing scout. Soon following it outside, they find the food by orienting to chemical signals present on the scout's body. If the food source is more than 80 m from the hive, the scout expresses in its dance the exact distance and direction of the source. A **waggle dance** traces two semicircles with a straight run between them (Fig. 14.20b). The food's distance is described by sounds and wagging movements executed during the straight run. The further away the food lies, the longer the sounds last and the more slowly the dancing bee waggles its abdomen. The angle of the straight run describes the direction of the food source in relation to the sun. A run straight up the hive wall denotes a location directly toward the sun. When food exists at an angle to the left or right of the sun, the bee runs at that same angle to the left or right of the vertical (Fig. 14.20c). Even on cloudy days these dances are effective, because bees detect the sun's location by their analysis of polarized light. These remarkable creatures also allow for the changing position of the sun through the day, so that their directional information remains valid over many hours. This talent enables bees to forage several kilometers from the nest.

Many more facts can be told of the strange and

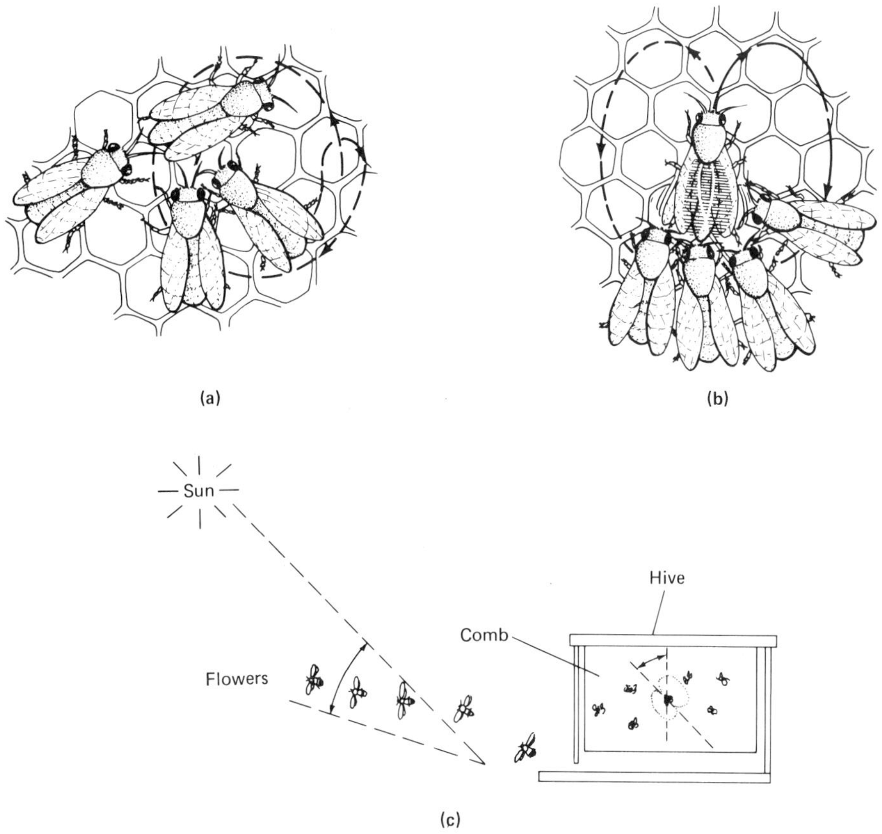

Figure 14.20 (*a*) The round dance of the honeybee. The form of the dance is indicated by the dashed arrows. Three workers follow the dancer; (*b*) The waggle dance of the honeybee. The form of the dance is indicated by the arrows. During the straight run the dancer's abdomen waggles from side-to-side. Four workers follow the dancer. (*c*) A diagram illustrating how a dancing bee within a dark hive substitutes gravity for the sun's position in communicating the direction of a food source in relation to the sun. (*a* and *b* from K. von Frisch, *The Dance Language and Orientation of Bees*, 1967, Harvard University; *c* from R.E. Hutchins, *Insects*, 1966, Prentice-Hall.)

sophisticated ways of the social insects. The slaveholding and gardening of various ants, the efficient architecture of the termites, and the swarming of bees are remarkable events in the invertebrate world. For further reading, the interested student should consult the list at the end of this chapter.

CONTINUITY SYSTEMS

Terrestriality and molting dominate the reproductive and developmental patterns of the Insecta. Owing to pressures from the physical environment, survival on land demands that gametes and embryonic stages be well protected. This situation has contributed to the evolution of complex gonadal systems, direct sperm transfer, and parental care. Growth necessitates molting, which in turn makes possible relatively dramatic structural and functional changes at regular intervals in the life cycle. The insects that most exploit these changes actually switch ecological niches within a single lifetime. It is important that we understand the principles of insect continuity, as it is these systems that afford some of our best opportunities for effective pest control.

Reproduction

Insect sexes are separate. The female has a pair of ovaries, whose ductules unite in a common oviduct (Fig. 14.21a). Insect ovaries comprise a cluster of tubular ovarioles. Oogenesis begins distally in each ovariole, and ova gradually mature as they migrate toward the base of the gonad. The common oviduct leads to a genital chamber, as does the **bursa copulatrix**, or vagina. Ducts from accessory glands and spermathecae (seminal receptacles) also open into the chamber. The genital orifice is located on the eighth abdominal segment in most species (Fig. 14.21b).

The male reproductive system is similar (Fig. 14.21c). Each of the two testes comprises a number of sperm tubes. Paired vasa deferentia dilate as seminal vesicles and then unite in a single ejaculatory duct. Near this duct, accessory glands discharge seminal fluids into the reproductive tract. The lower end of the ejaculatory duct is housed within a penis, which extends ventrally from the ninth abdominal segment (Fig. 14.21d).

Each insect species has its own recognition methods, which may include complex courtship rituals. Mature males and females attract one another by chemical, tactile and/or visual clues; some insects, such as the crickets, "sing" for their prospective mates. Courtship often involves a series of behaviors, in which a response by one partner stimulates the next act of its mate (Fig. 14.22). Such rituals may last many hours before they are consummated.

In most insects, sperm are transferred while the male holds his penis inside the genital chamber of the female. Special abdominal claspers on the male augment his sexual grip (Fig. 14.23). Other cuticular modifications in the phallic area may contribute to the copulatory embrace. Such modifications are highly species specific; thus they aid the insects in determining mating suitability and guide entomologists in their taxonomic work.

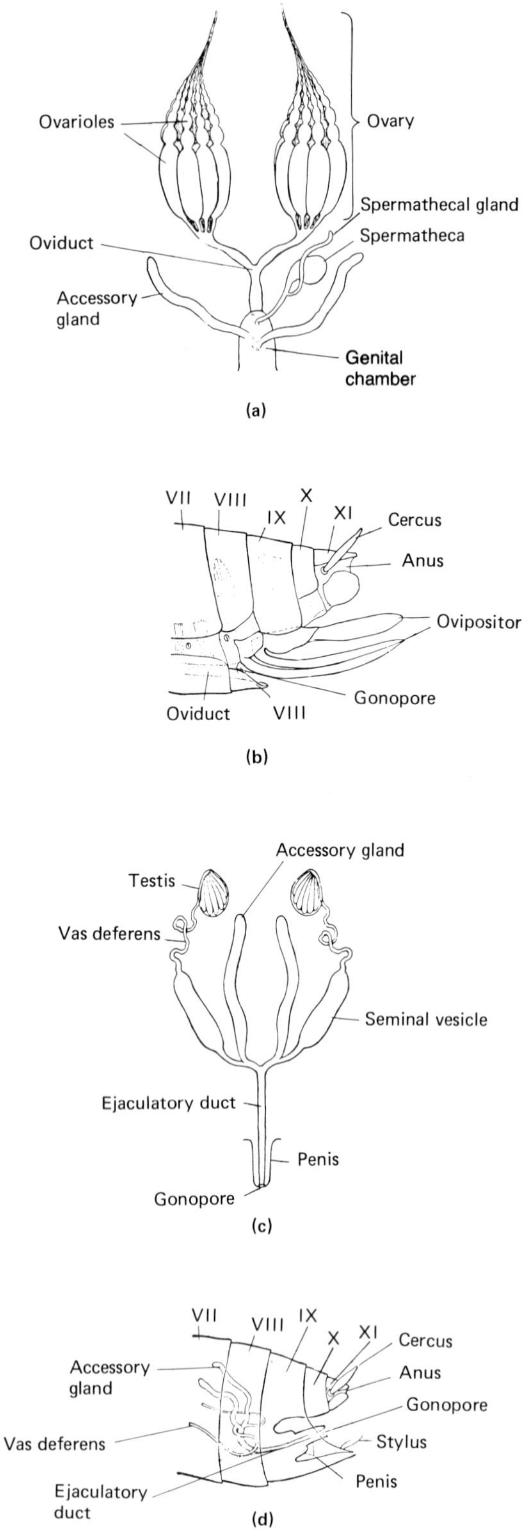

Figure 14.21 Generalized insects. (*a*) Female reproductive organs; (*b*) Lateral view of female external genitalia; (*c*) Male reproductive organs; (*d*) Lateral view of male external genitalia. (From Snodgrass, *Principles of Insect Morphology*.)

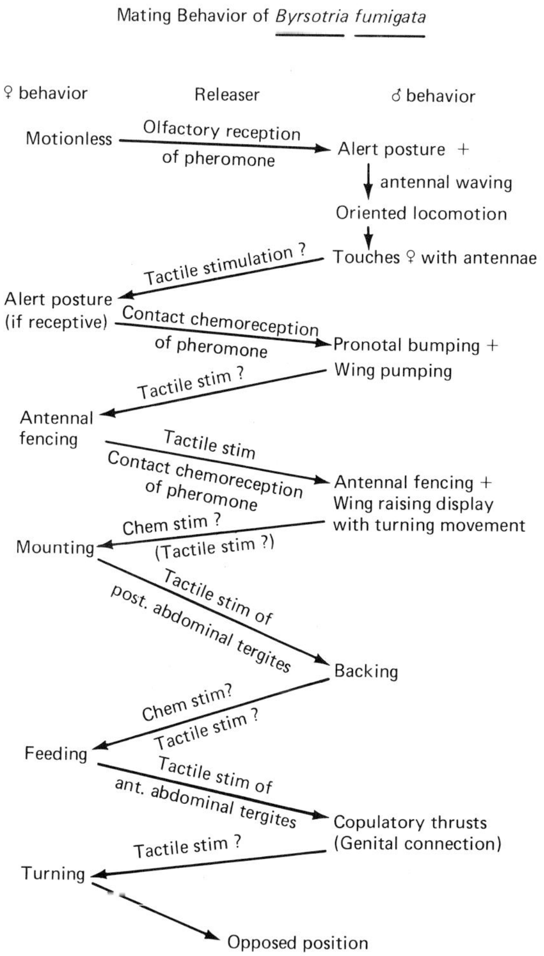

Figure 14.22 A summary of the mating behavior of a cockroach, *Byrsotria fumigata*, in which a series of behavior ultimately leads to copulation. (From R. H. Barth, *Behaviour*, 1964, *23:* 19.)

Sperm simply may be suspended in accessory gland secretions, or these secretions may harden about the gametes to produce spermatophores. Spermatophores are produced by primitive, wingless insects. Males leave these sperm packages on the ground, where they are found and taken up by females.

The insect female stores large quantities of sperm within her spermathecae. The storage from a single mating may be sufficient to fertilize her eggs for an entire lifetime. Insect eggs are protected by a thick membrane produced within the ovary itself. Accessory glands contribute adhesives and/or secretions which harden over the fertilized ova. In many species, cuticular extensions around the female gonopore form an ovipositor (Fig. 14.21b). This structure places the eggs within a protective brooding site, such as a shallow underground chamber or plant stem (Fig. 14.24).

Development

Although almost all species develop a full complement of adult segments in the egg, newly hatched insects may differ in their maturity. Only among the most primitive, wingless members of the class (e.g., silverfish) do the young closely resemble their adult counterparts. In the winged groups, metamorphosis accompanies juvenile development. In general, there are two types of metamorphic development among insects: The more gradual type is called incomplete metamorphosis, or **hemimetabolous development.** Cockroaches, bugs, and grasshoppers are hemimetabolous insects. In this group, the young resemble adults in many ways—they have compound eyes and antennae, and their feeding styles are usually similar. Functional wings and sexual structures however, are always lacking. (Fig. 14.25). In hemimetabolous insects, immature stages are sometimes called **nymphs.** The preadult nymph has a pair of wing rudiments which have developed from external buds. At the adult molt, these rudiments expand to form the wings. Dragonflies and mayflies have aquatic young which undergo a modified hemimetabolous development. Aquatic nymphs, often called **naiads,** are peculiar in those groups with land-dwelling adults. Typically, the naiads bear specialized respiratory organs (gills) and their wing development is retarded. Wings have no adaptive value for the naiad and do not appear until the last instar leaves its aquatic home. The final molt thus involves a rather dramatic change, as respiratory and locomotor specializations of the cuticle are modified for terrestrial life.

Such dramatic changes are common in the second type of insect development, called complete metamorphosis, or **holometabolous development.** Holometabolous insects, such as ants, bees, flies, moths, and beetles, hatch as wormlike larvae (Fig. 14.26). These immature stages bear little resemblance to the adult form. They lack compound eyes and antennae, and their entire ecological niche is usually distinct. Larval mouthparts may be wholly unlike those of the adult, and their manner of feeding is almost always different. Larval life may be quite prolonged, as the juvenile consumes enormous quantities of food. Often, the greater part of an insect's lifetime is spent as a larva, and this stage attains a body volume

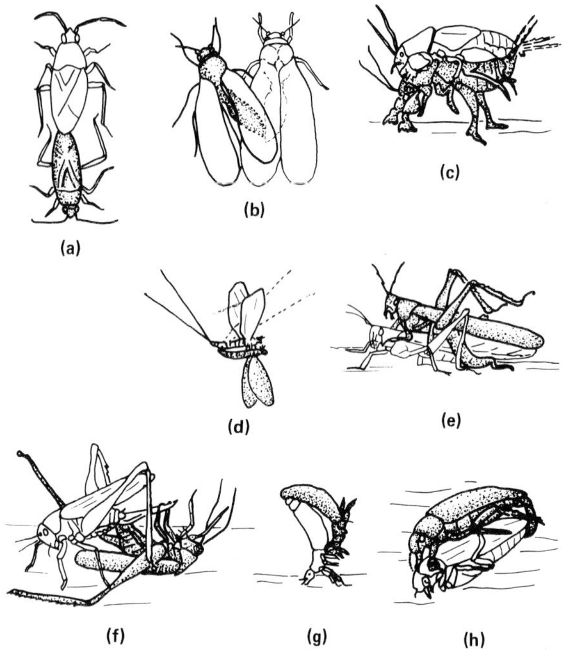

Figure 14.23 Copulatory positions in various insects. Males shown in black. (*a*) and (*b*) Dorsal views; (*c–h*) Lateral views. (From Weber, *Lehrbuch der Entomologie.*)

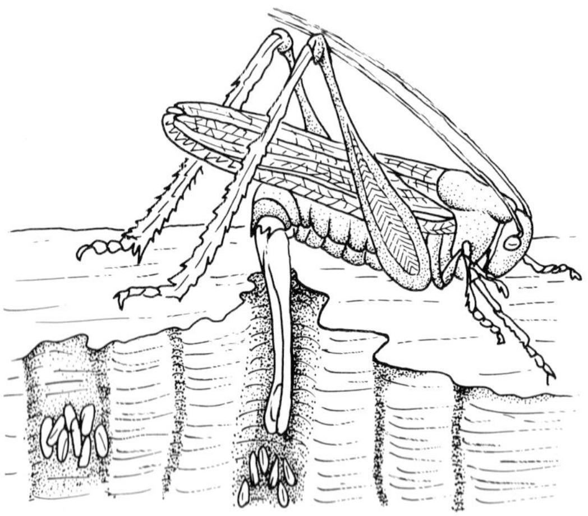

Figure 14.24 A female grasshopper has a well developed ovipositor for laying eggs in underground burrows. (From Peabody, *Cecil's Book of Insects*, 1868.)

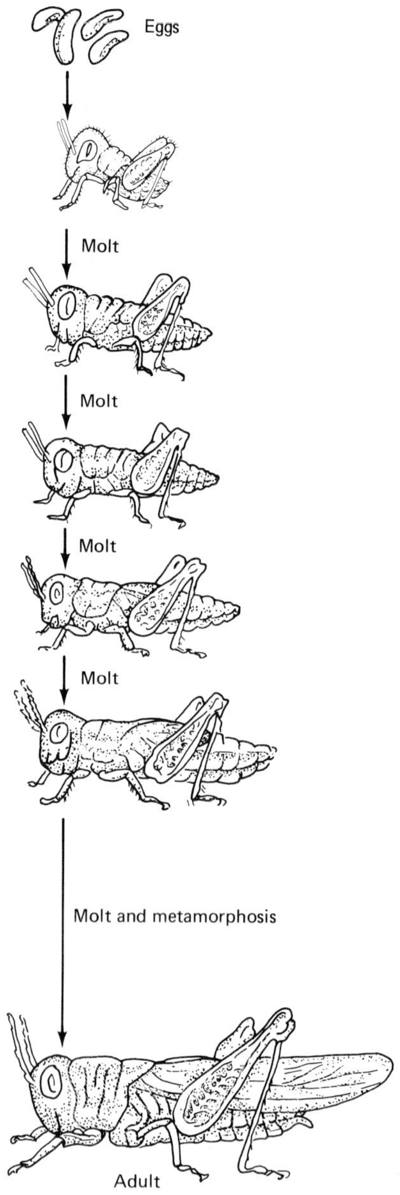

Figure 14.25 Post-embryonic development in a grasshopper, *Melanoplus differentialis*, a typical hemimetabolous insect. Each immature stage is known as an instar or nymph. (From C.D. Turner, *General Endocrinology*, 1955, W.B. Saunders.)

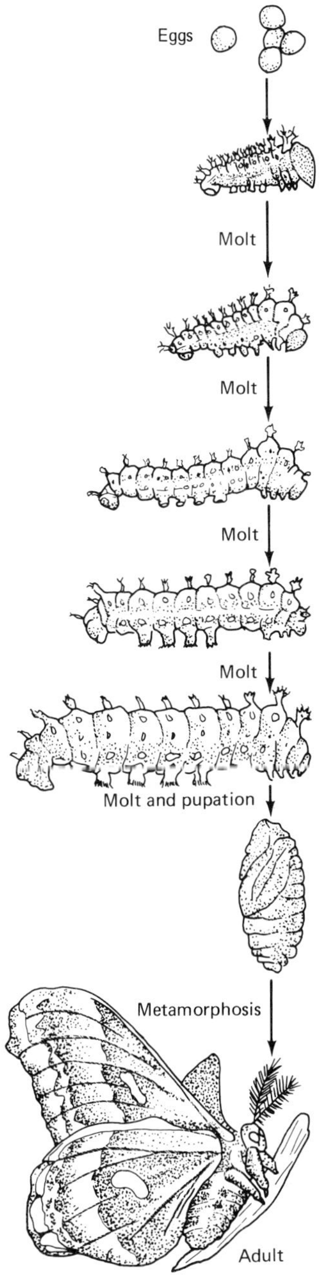

Figure 14.26 Post-embryonic development in a giant silkworm moth, *Hyalophora cecropia*, a typical holometabolous insect. (From Turner, *General Endocrinology*.)

larger than that of the adult. Eventually, such feeding halts, and a single molt initiates **pupation.** During pupation, the insect, now called a **pupa,** is quiescent and does not feed. It can do little to defend itself; thus some protection, such as a cocoon or an underground burrow, is usually prearranged. The pupa's energy is applied to the wholesale transformation of its body. Many structures are broken down and reorganized according to adult needs. For the first time, external wings and sexual organs are formed. Although the reorganization itself occurs fairly rapidly, the pupal stage may endure for several months. In the temperate zone, the pupal stage is a convenient form in which to pass the winter. When postpupal ecdysis occurs, the winged adult emerges to assume its own niche.

Holometabolous insects outnumber hemimetabolous species by a ratio of 10 to 1. Their adaptive advantages evidently include the fact that larvae seldom compete directly with adults for food and other resources.

The role of ecdysone in initiating molting is described in Chapter 10. This hormone works in conjunction with a second endocrine product in controlling the sequence of events in insect metamorphosis. This second product, **juvenile hormone,** is manufactured and released by the **corpora allata,** a pair of endocrine glands associated with the brain (Fig. 14.27). When ecdysone initiates a molt in an early larval instar, the accompanying concentration of juvenile hormone is high. Such a high concentration ensures a larva-to-larva molt. After the last larval instar is reached, the corpora allata cease to secrete juvenile hormone. Low concentrations of juvenile hormone result in a larva-to-pupa molt. Finally, when the pupa is ready to molt, juvenile hormone is absent from the hemolymph, and at ecdysis the adult form appears. Curiously, juvenile hormone may be active in adult insects, whose cyclical reproduction reflects hormonally controlled fluctuations in gonadal activity.

A Review

Before a brief, illustrated survey of the various insect orders, we recount the basic characters of the Insecta. Inherited in their general forms from arthropod ancestors, these characters have been perfected by insects as the class radiated along several adaptive lines. Basically, insects are:

(1) the largest group of invertebrates, accounting for well over half of all animal species;

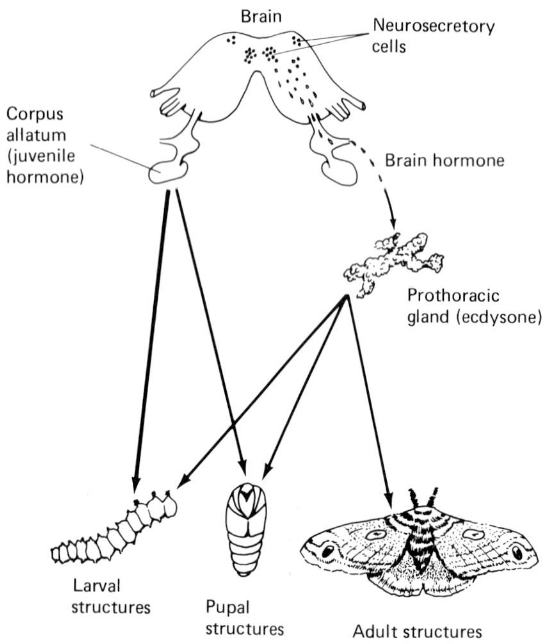

Figure 14.27 Schematic diagram showing the principal endocrine organs of an insect (the giant silkworm moth, *Hyalophora cecropia*) and the interactions of their chief hormonal products in postembryonic development (molting and metamorphosis). (From H.A. Schneiderman and L.A. Gilbert, Science, 1964, *143*:327.)

(2) terrestrial arthropods with three distinct tagmata, antennae, mandibles, maxillae, six legs, and usually four wings;

(3) feeders on a wide range of plant and animal matter, with mouthparts adapted to the diet and mode of food uptake;

(4) animals with a dorsal tubular heart, an open hemocoel, a sophisticated tracheal system, and Malpighian tubules;

(5) the only truly flying invertebrates with keen sense organs and elaborate behavioral programs;

(6) protostomes with a complex sex life and a metamorphic development that often includes ecologically distinct stages within a single life cycle.

Insect Diversity

Three-quarters of a million known species of insects are distributed among 28 extant orders. It is quite impossible to describe them in detail in a book of this length. Therefore, we provide only the following gallery of selected insect types (Figs. 14.28 through 14.55). We hope that the illustrations will kindle increased appreciation for these fascinating creatures. For the curious student, further reading is suggested at the end of this chapter.

Arthropod Evolution

The Arthropoda occupy a high point of invertebrate evolution in the protostomate line. Lower protostomes exhibit bilateralism, metamerism, and the secondary body cavity. Building on these characters, the arthropods have evolved a chitinous exoskeleton with jointed legs and organ systems of increasing complexity. While this general history is rather clear, controversy reigns in several areas of arthropod evolution. Particularly controversial are the details of the origin of the phylum and the relationships among its member groups.

Most zoologists agree that arthropods arose from some protoannelid stock. Selective pressures involving locomotion no doubt played a major role in the evolution of the arthropod skeleton and in the related hemocoelic developments. The onychophorans (see the following) may preserve an early stage in the transition from protoannelid to protoarthropod form.

The emergence of the arthropod subphyla and classes is shrouded in Precambrian obscurity. Paleontology reveals little about this period, and by the time fossils were formed in the Cambrian, most of the major groups were already established. In the absence of hard data, we must speculate with evi-

Text continued on page 483.

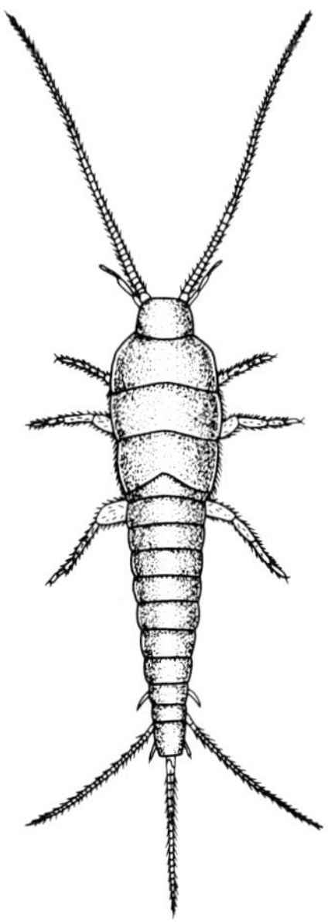

Figure 14.28 A silverfish, *Lepisma saccharina* (Order Thysanura). (From A.D. Imms, *A General Textbook of Entomology*, 1957, Methuen and Co.)

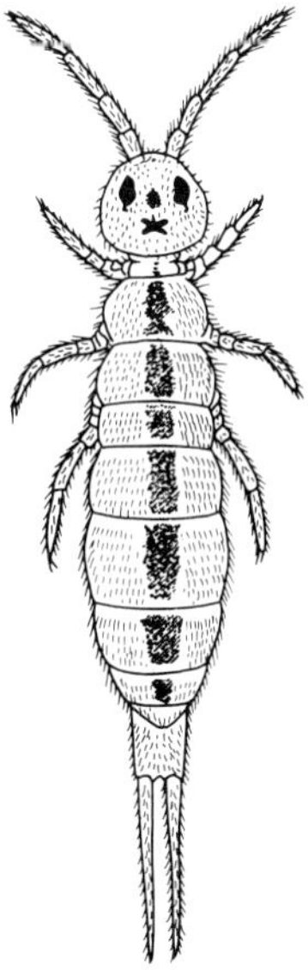

Figure 14.30 A spring-tail, *Isotomurus palustris* (Order Collembola), greatly enlarged. (From R.W. Pennak, *Fresh-Water Invertebrates of the United States*, 2nd ed., 1978, Wiley-Interscience.)

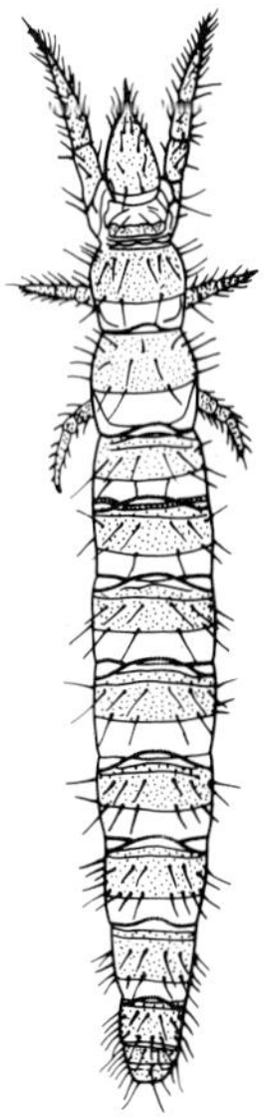

Figure 14.29 *Acerentomon doderoi* (Order Protura), greatly enlarged. (From J.H. Comstock, *An Introduction to Entomology*, 9th ed., 1940, Cornell University.)

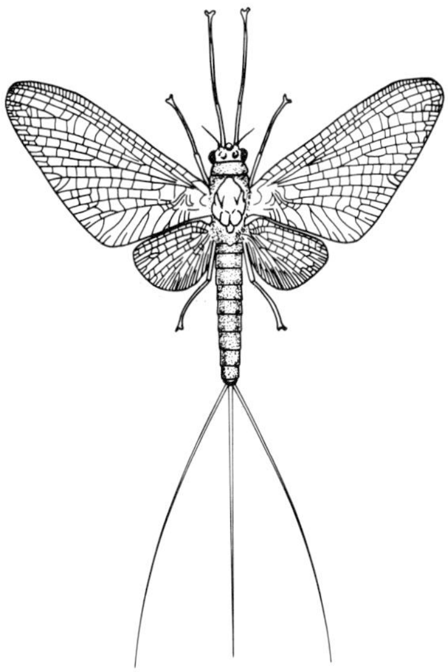

Figure 14.31 An adult mayfly, *Potamanthus* (Order Ephemeroptera). (From Pennak, *Fresh-Water Invertebrates of the United States*.)

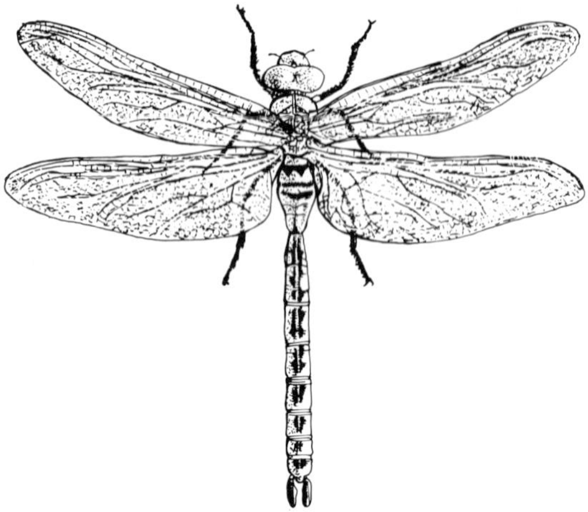

Figure 14.32 A dragonfly, *Anax formosus* (Order Odonata). (From Sharp, In *Cambridge Natural History, Vol. V.*)

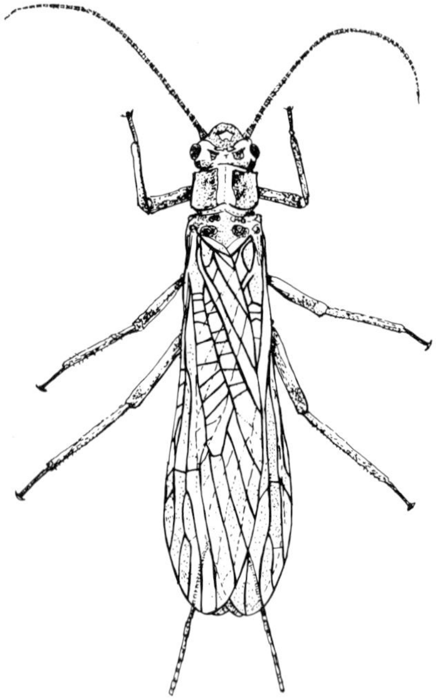

Figure 14.33 A stonefly, *Isoperla confusa* (Order Plecoptera). (From Pennak, *Fresh-Water Invertebrates of the United States.*)

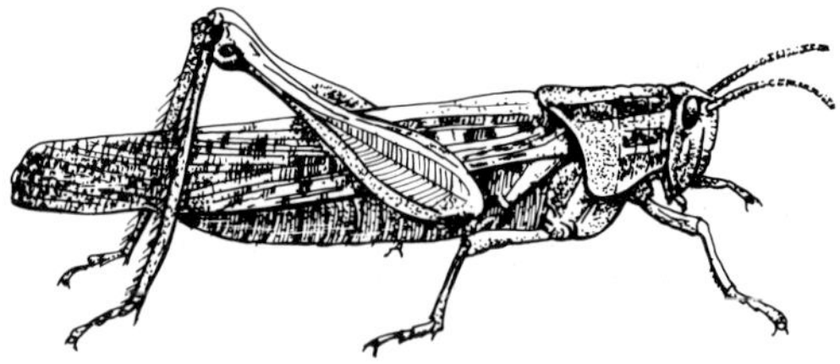

Figure 14.34 A grasshopper, *Schistocerca americana* (Order Orthoptera). (From Comstock, *An Introduction to Entomology.*)

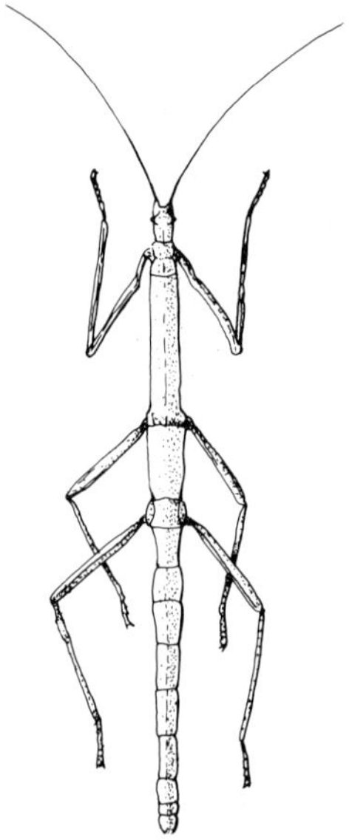

Figure 14.35 A walking stick, *Carausius morosus* (Order Phasmida). (From Imms, *A General Textbook of Entomology.*)

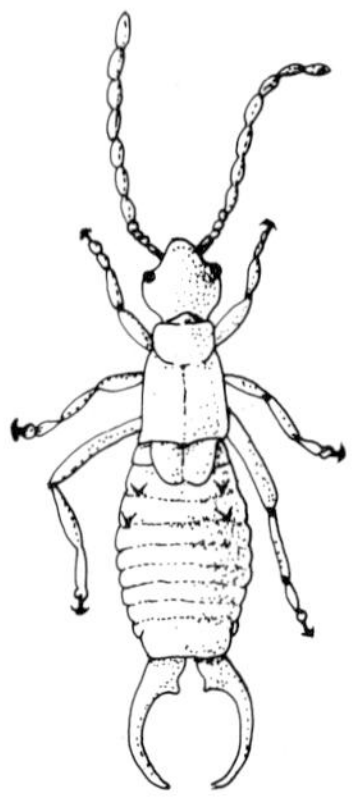

Figure 14.36 An earwig, *Forficula auricularia* (Order Dermaptera). (From Comstock, *An Introduction to Entomology.*)

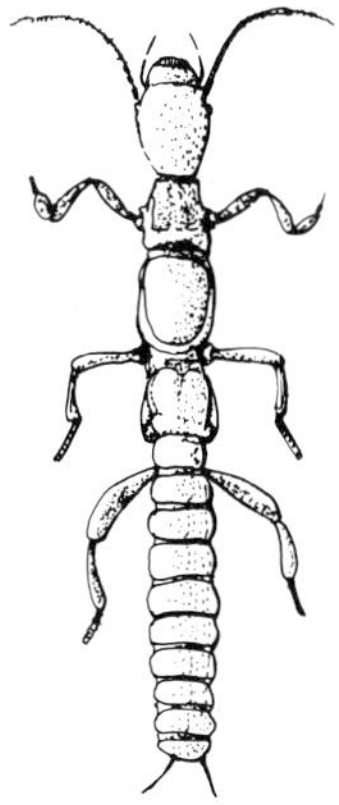

Figure 14.37 *Embia sabulosa*, female (Order Embioptera). (From Comstock, *An Introduction to Entomology*.)

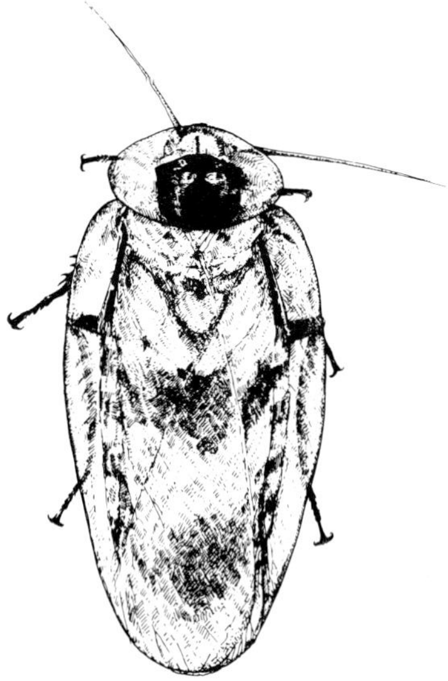

Figure 14.38 A cockroach, *Blaberus giganteus* (Order Dictyoptera). (Drawing by Salley Blakemore from a photo by Jack Salmon in L.M. Roth and E.R. Willis, The Biotic Associations of Cockroaches, Smithsonian Miscellaneous Collections, 1960, *141*:1–470.)

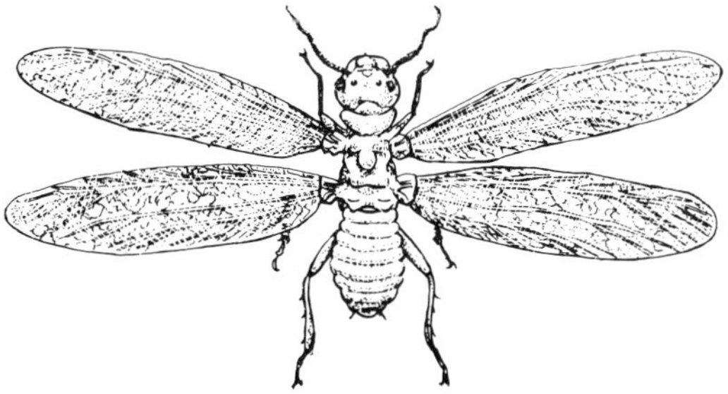

Figure 14.39 A termite, *Hodotermes mossambicus*. Winged adult (Order Isoptera). (From Sharp, In *Cambridge Natural History*, *Vol. V*.)

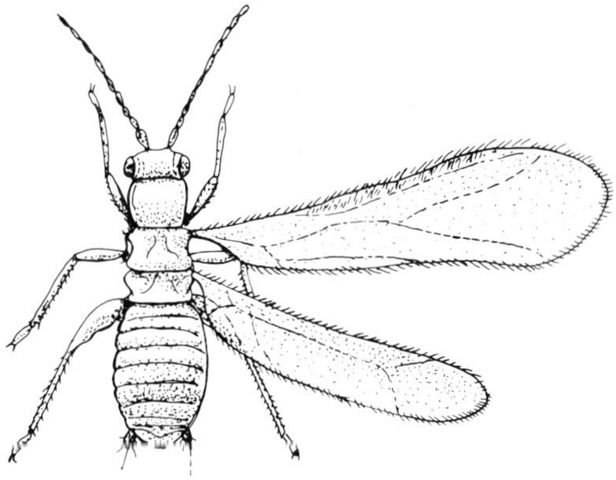

Figure 14.40 *Zorotypus hubbardi*, winged adult female (Order Zoraptera). Left fore and hind wings removed. (From Comstock, *An Introduction to Entomology*.)

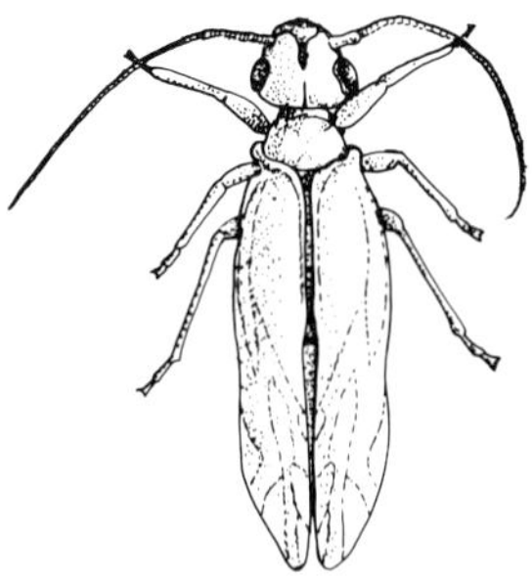

Figure 14.41 A bark louse, *Caecilius manteri* (Order Psocoptera), greatly enlarged. (From Borror and Delong, *An Introduction to the Study of Insects*.)

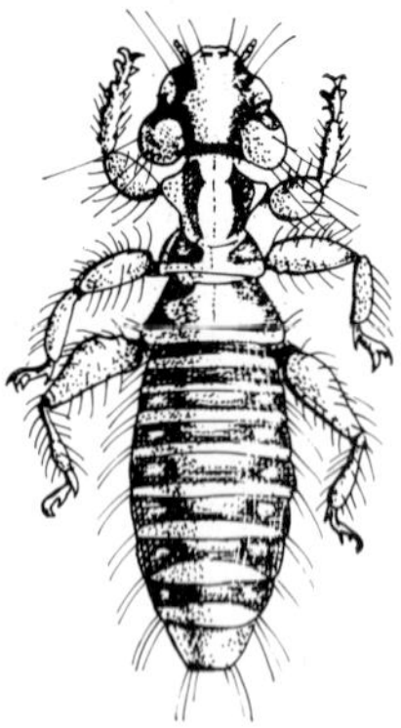

Figure 14.42 A biting louse, *Trinoton luridum*, an ectoparasite of ducks (Order Mallophaga), greatly enlarged. (From Sharp, In *Cambridge Natural History, Vol. V.*)

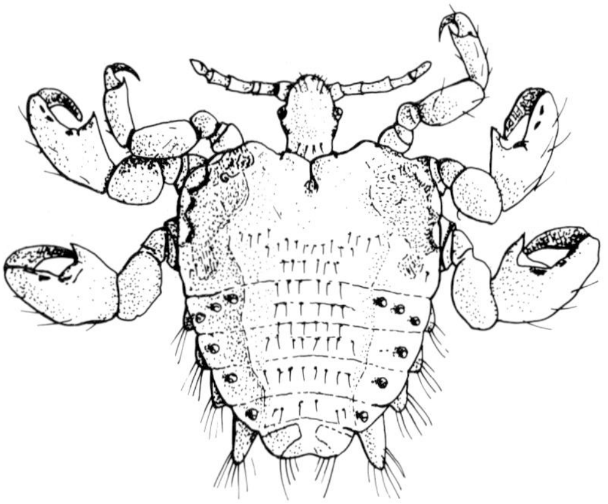

Figure 14.43 The human pubic louse, *Phthirus pubis* (Order Anoplura), greatly enlarged. (From J. Smart, *Lice*, 1954, Brit. Mus. Pub.)

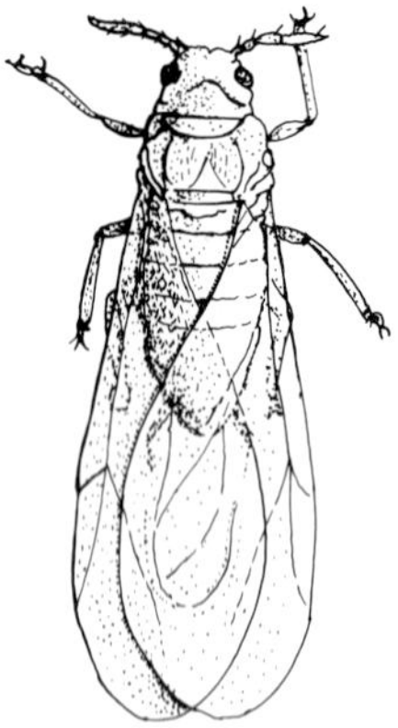

Figure 14.44 A plant louse, *Phylloxera*, a notorious pest in vineyards. Winged female (Order Homoptera). (From Comstock, *An Introduction to Entomology*.)

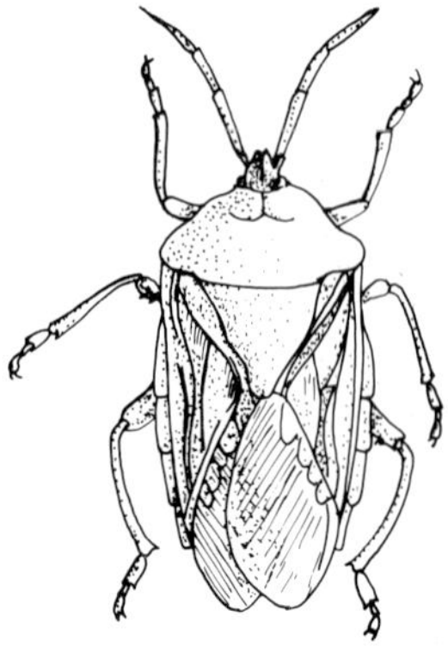

Figure 14.45 A stink bug, *Eusthenes pratti* (Order Hemiptera). (From Sharp, In *Cambridge Natural History, Vol. VI.*)

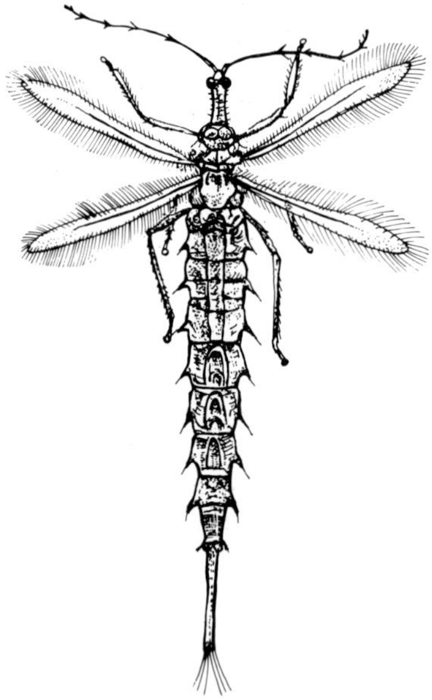

Figure 14.46 A thrip, *Idolothrips spectrum* (Order Thysanoptera), greatly enlarged. (From Sharp, In *Cambridge Natural History, Vol. VI.*)

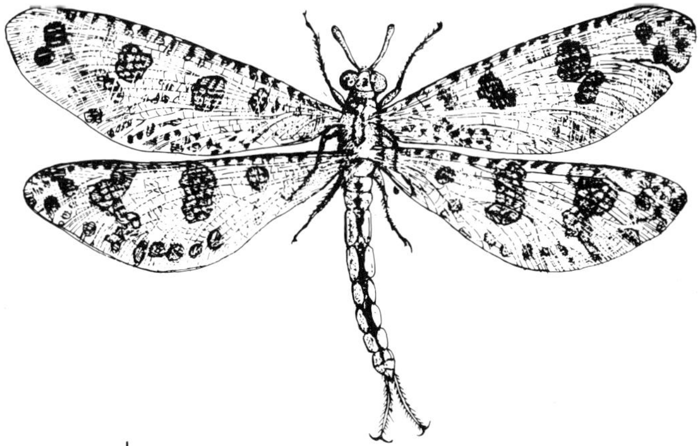

Figure 14.47 An ant-lion, *Myrmeleon libelluloides* (Order Neuroptera). (From Peabody, *Cecil's Book of Insects.*)

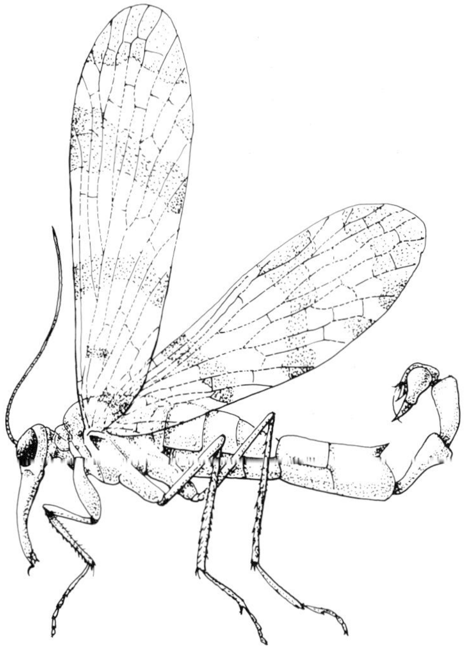

Figure 14.48 A male scorpionfly, *Panorpa venosa* (Order Mecoptera), showing appendages of the right side only. (From Borror and Delong, *An Introduction to the Study of Insects.*)

Figure 14.49 An adult caddisfly, *Rhyacophila fenestra* (Order Trichoptera). (From Pennak, *Fresh-Water Invertebrates of the United States.*)

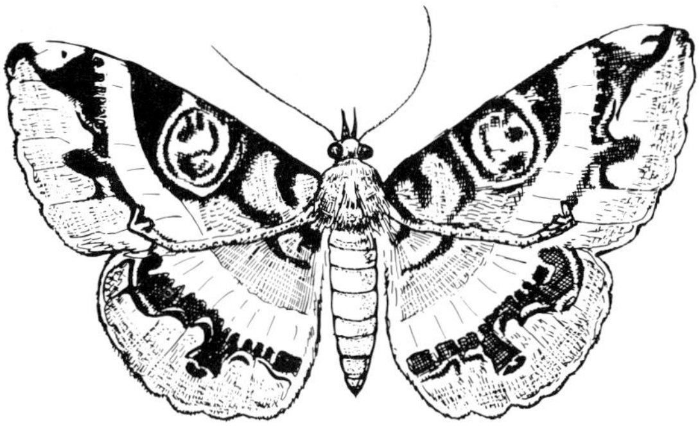

Figure 14.50 A moth. (Order Lepidoptera). (From Kirby, *Elementary Textbook of Entomology*, 1892.)

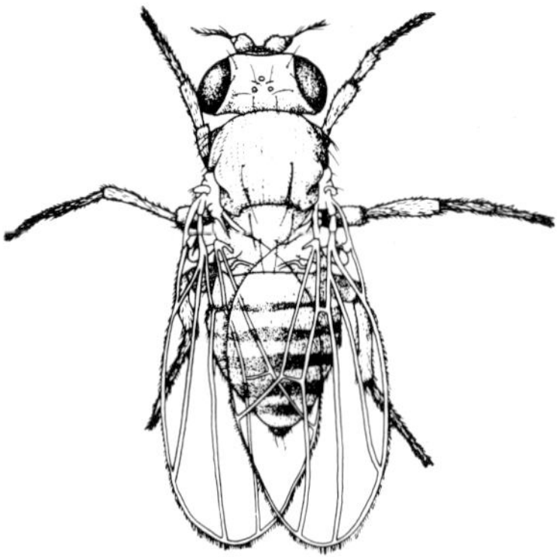

Figure 14.51 Adult female fruitfly, *Drosophila melanogaster* (Order Diptera). (From Sherman and Sherman, *The Invertebrates: Function and Form.*)

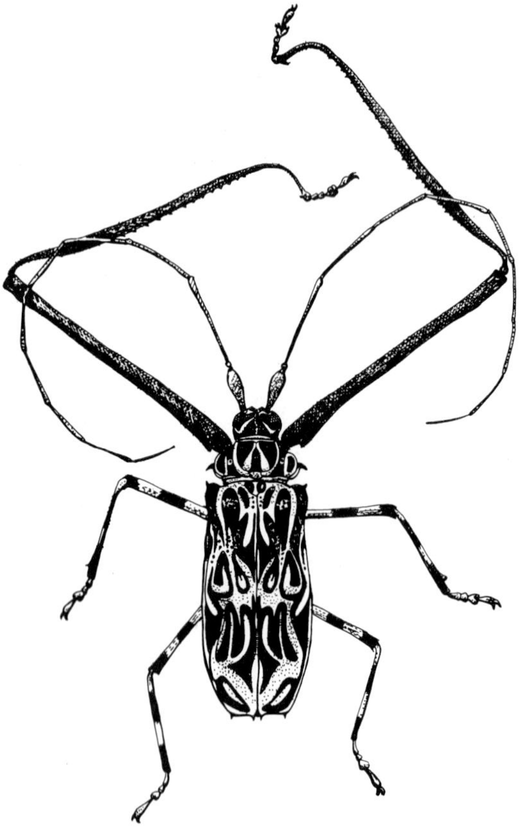

Figure 14.52 A harlequin beetle, *Acrocinus longimanus* (Order Coleoptera). (From Kirby, *Elementary Textbook of Entomology.*)

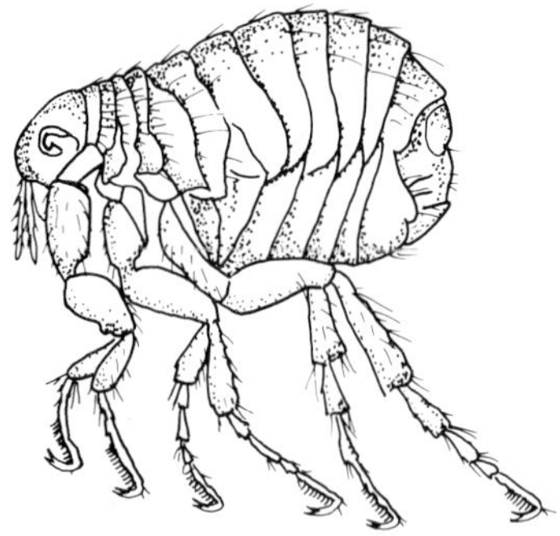

Figure 14.53 The human flea, *Pulex irritans* (Order Siphonaptera), greatly enlarged. (From Borror and Delong, *An Introduction to the Study of Insects.*)

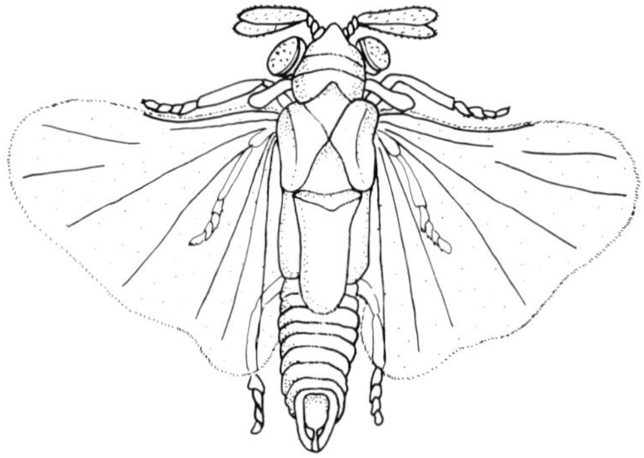

Figure 14.54 A wing male stylopid, *Opthalmochlus duryi* (Order Strepsiptera), greatly enlarged. Stylopids are parasites of other insects for most of their life cycle. (From Comstock, *An Introduction to Entomology.*)

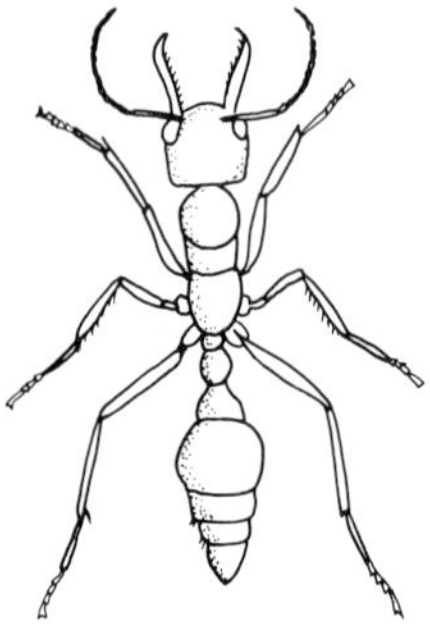

Figure 14.55 An Australian bull-dog ant, *Myrmecia pyriformis* (Order Hymenoptera). (From *Cambridge Natural History, Vol. VI.*)

dence gleaned from embryology, comparative anatomy, and biochemical studies. Such speculations generally support one of two major theories. The first proposes a unified, monophyletic Arthropoda comprising three great subphyla: Trilobitomorpha, Chelicerata, and Mandibulata. The second supports a polyphyletic origin for the Arthropoda and describes fundamental differences among its several groups.

A popular monophyletic argument derives the entire Arthropoda from prototrilobites (Fig. 14.56a). These early arthropods allegedly founded both the chelicerate and the mandibulate lines while their immediate progeny, the trilobites, flourished in the early Paleozoic. Supporters of this view stress the unity of the Mandibulata, pointing to such shared features as antennae, mandibles, and compound eyes.

Not so, claim the supporters of a polyphyletic origin. These zoologists argue that several protoarthropod lines arose from the early annelids and that each major arthropod group has a long and independent evolutionary history (Fig. 14.56b). While recognizing some affinities between the insects and myriapods, the polyphyletic theory isolates the crustaceans from the terrestrial mandibulates. Among reasons cited for this isolation are the biramous appendages, second antennae, gnathobasic mandibles, and prominent digestive glands of the Crustacea. Some polyphyletic schemes argue for a loose relationship among trilobites, crustaceans, and merostomes. Certain paleontological evidence favors this arrangement as curious fossil forms called pseudocrustaceans (Fig. 14.57) bear possible affinities with all three groups.

The polyphyletic view is compatible with the absence of fossil forms linking crustaceans with insects and myriapods. Under the polyphyletic scheme, however, convergent evolution must account for all similarities between the two groups, including the almost identical compound eyes and hormonal control of molting. Moreover, convergent evolution must explain other widespread features in the phylum, such as the existence of tracheal systems and Malpighian tubules in both arachnid and insect–myriapod lines.

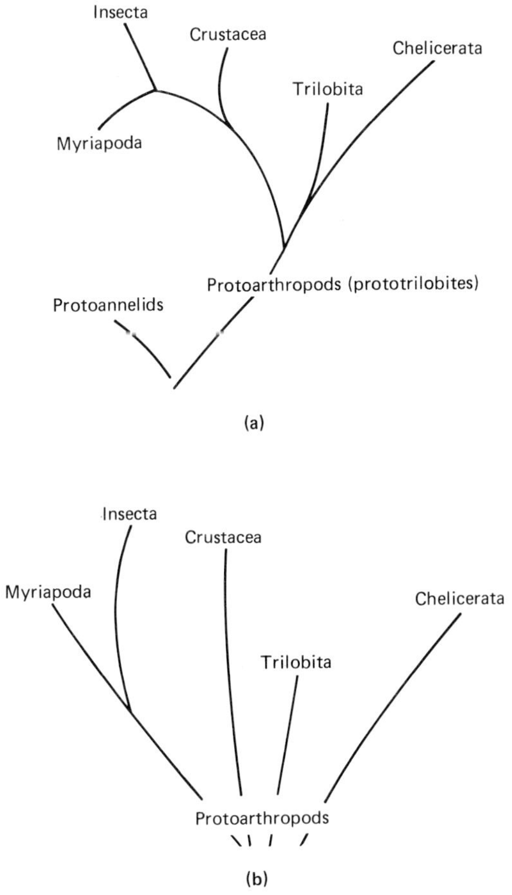

Figure 14.56 (*a*) An arthropod phylogenetic tree assuming a monophyletic origin for the phylum. (*b*) A phylogenetic tree assuming a polyphyletic origin for the arthropods.

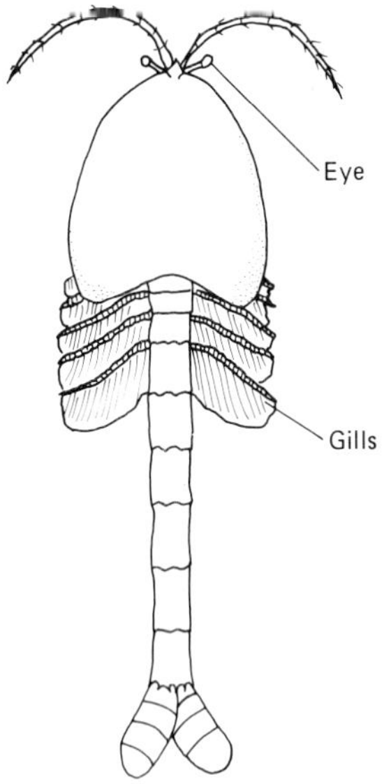

Figure 14.57 A fossil pseudocrustacean, *Waptia fieldensis*. (From L. Stormer, In R.C. Moore, Ed., *A Treatise on Invertebrate Paleontology, Part O*, 1959, Geological Society.)

Clearly, there are positive and negative attributes to both the monophyletic and the polyphyletic points of view. Further evidence, preferably of a paleontological nature, is necessary before firmer conclusions can be drawn. It is important, however, to reflect on the late Precambrian world in which this controversial drama took place. The protoannelids and protoarthropods most likely represented a large assemblage of invertebrates. Arthropod ancestors may constitute at least an order or class of animals in the modern sense. Even from the most rigid monophyletic view, it is unlikely that the entire Arthropoda would be traceable to a single founder species. On the other hand, an ardent polyphyletic advocate should recognize that some relationship existed among the stem animals in all "independent" arthropod lines. Ultimately, the monophyletic–polyphyletic debate may require a compromise based on such recognitions.

Upon leaving the arthropods, we mention three groups whose phylogenetic relationships are not well established. Some of these animals exhibit unique combinations of characters individually associated either with arthropods or with annelids. Accordingly, many authorities see them as old groups that diverged from segmented worms near the time of the first arthropods. Because of our uncertainty about their past, each group tentatively is considered a separate phylum.

PHYLUM ONYCHOPHORA

In the minds of many zoologists, the onychophorans bridge the gap between annelids and arthropods. Their annelid characters include a very similar body wall and unjointed legs, while a chitinous cuticle, a hemocoel, tracheal respiration, and mandiblelike appendages suggest close arthropod affinities. About 65 species have been described from tropical and southern temperate zones, where sheltered, humid environments are preferred. Curiously, members of the same genus may inhabit both Africa and South America. This unusual distribution supports the geological theory that the southern continents were once connected.

Onychophorans are wormlike animals, 1 to 15 cm long (Fig. 14.58a). At the anterior end, they have a single pair of large ringed antennae and paired oral papillae (Fig. 14.58b). Fleshy lobes border a ventral depression which contains the mouth. Flanking the oral opening is a pair of jaws, or mandibles. The long body trunk displays no obvious segmentation other than its paired ventrolateral legs. Up to 43 pairs of these unjointed, stubby appendages are present; each bears distal walking pads and a pair of terminal claws (Fig. 14.58b). Small scaly tubercles populate the entire external surface.

The onychophoran body wall comprises a very thin chitinoid cuticle, a simple epidermis and dermis, and muscle fibers in circular, diagonal, and longitudinal layers. The body cavity is a hemocoel.

Onychophorans eat small invertebrates and some plant matter. The mandibles seize and cut prey, which is softened preorally by salivary gland secretions. These glands apparently represent modified nephridia. A muscular, chitin-lined pharynx sucks reduced food into the foregut. A narrow esophagus precedes the long, wide intestine, where digestion and absorption are completed (Fig. 14.59a). A short hindgut passes to the ventral, posterior anus.

Onychophoran circulation is remarkably similar to that of arthropods. A dorsal tubular heart with paired

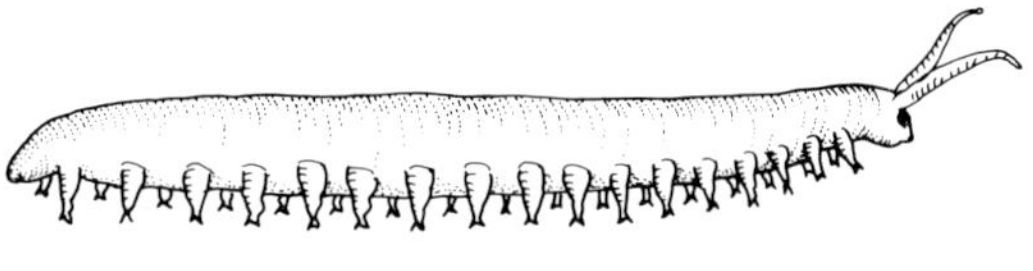

(a)

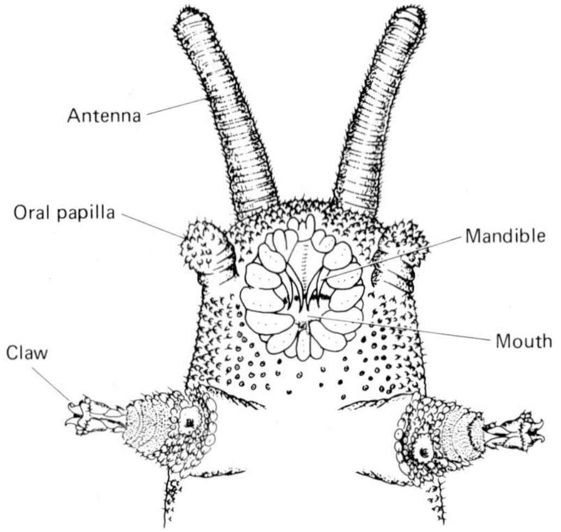

(b)

Figure 14.58 (*a*) Lateral view of *Peripatus capensis;* (*b*) Ventral view of the anterior end of *Peripatus* showing antennae, oral papillae, jaws and the first pair of ventrolateral legs. (From T.J. Parker and W.A. Haswell, *Textbook of Zoology*, 1921, Macmillan; leg detail from A. Sedgwick, In *Cambridge Natural History, Vol. V.*)

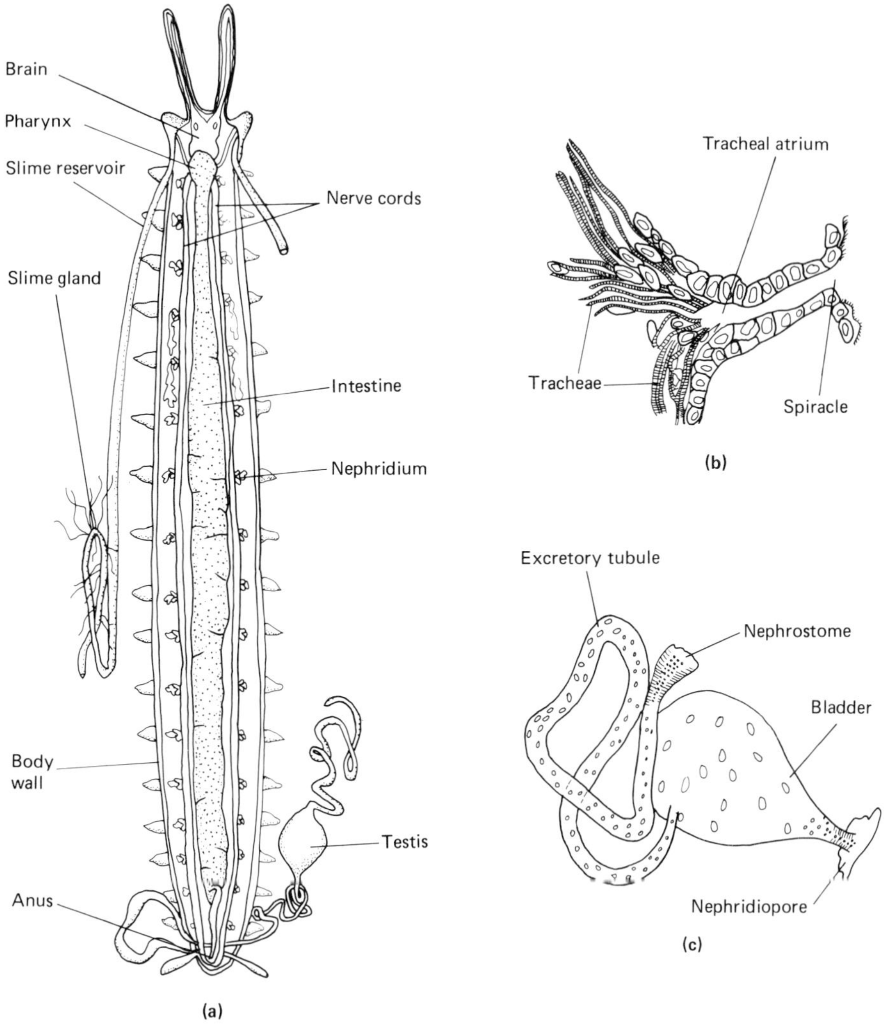

Figure 14.59 *Peripatus.* (*a*) Internal anatomy of a male with one testis and one slime gland removed; (*b*) Section through a tracheal atrium; (*c*) Diagram of a nephridium. (*a* and *c* from V. Schechter, *Invertebrate Zoology*, 1959, Prentice-Hall; *b* from Sedgwick, In *Cambridge Natural History, Vol. V.*)

segmental ostia pumps the colorless blood forward for general distribution through the hemocoel. Although less well defined than in higher arthropods, hemocoelic sinuses help route blood through the body. Very numerous minute spiracles populate the body surface. Each opening leads to a tracheal atrium located within the muscle layers of the body wall (Fig. 14.59b). From this atrium, unbranched tracheae penetrate the tissues. Like myriapods, onychophorans cannot close their spiracles. This fact, coupled with their very thin cuticle, makes them quite vulnerable to water loss. Accordingly, onychophorans are restricted to the most humid environments and are active only at night.

Except for those at the extreme ends, each body segment houses a pair of nephridia. Thought to enclose remnants of a true coelom, these excretory organs comprise the familiar nephrostome, tubule, bladder, and nephridiopore (Fig. 14.55c). The nephridiopores occupy the bases of the legs. Nitrogenous wastes probably are eliminated as ammonia.

The onychophoran nervous system includes a suprapharyngeal brain, circumenteric commissures, and two ventral nerve cords (Fig. 14.59a). Although commissures join the ventral cords in each segment, trunk ganglia are not conspicuous. Sense organs include antennae, eyes, and numerous sensory sites on the general body surface. The eyes are located at the antennal bases. These direct, pigment-cup ocelli help to orient the animal away from desiccating sunlight. Tactile and chemosensitive cells are common on the larger body tubercles. Certain antennal and tubercular sites respond to humidity levels.

Onychophorans crawl very slowly. Like annelids, their locomotion depends in part on a hydrostatic skeleton. Like spiders, their legs contain flexor but not extensor muscles. Onychophorans extend the body lengthwise, place a leg forward on the ground, and then push the leg back. Although waves of activity spread posteriorly over the body, onychophoran leg stepping is never as regular as that of myriapods. Apparently, intersegmental reflexes are not as well developed. Because of their flexible cuticles, however, these animals can enter very tight places. Onychophorans defend themselves with adhesive secretions. Produced within large anterior glands, these slimy secretions are sprayed up to half a meter from openings on the oral papillae.

Onychophoran females are somewhat larger than their male counterparts. In both sexes, paired, posterior gonads connect with ducts which unite before the single gonopore near the anus (Fig. 14.59a). Males form large spermatophores near the end of their reproductive tracts. Females have paired seminal receptacles and uteri. Sperm transfer is not well described, but in at least one species, the spermatophore penetrates the female body wall. Two genera lay their eggs externally, but in most other onychophorans, embryos develop within their mother's uterus. Viviparous species nourish their young on uterine secretions and may form placentalike connections with the embryo. Molting is necessary to accommodate growth.

PHYLUM TARDIGRADA

The tiny animals called tardigrades (alias water bears) are among the most curious invertebrates. Various aspects of their biology suggest affinities with aschelminthes, deuterostomes, and annelid–arthropod stem forms. About 350 species have been identified. Most tardigrades are terrestrial, living on the moist surfaces of mosses and lichens. A few species dwell interstitially on marine or freshwater bottoms; some live on freshwater plants.

Tardigrades are seldom more than half a millimeter long. Four pairs of short, clawed legs extend ventrolaterally from the stocky body (Fig. 14.60a). The proteinacious cuticle may be smooth or sculptured and bristly; it contains no chitin. The tardigrade epidermis is distinguished by a species-specific constancy of cell number. (Recall that many aschel-

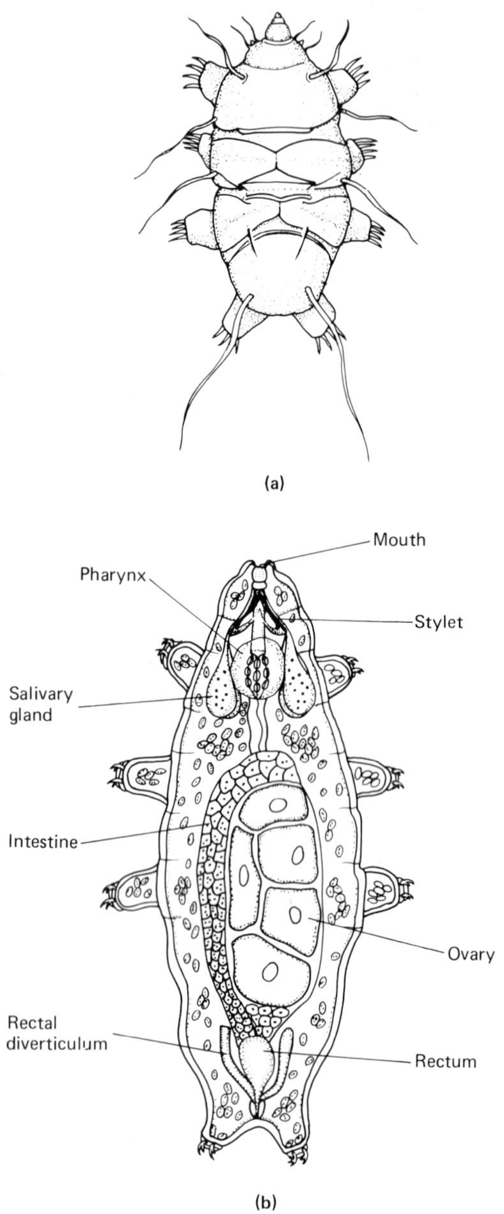

Figure 14.60 (*a*) Dorsal view of *Echiniscus testudo*, a tardigrade; (*b*) Internal anatomy of a female tardigrade, *Macrobiotus schultzei*. (From A.E. Shipley, In *Cambridge Natural History, Vol. IV.*)

minthes display a similar constancy.) The body wall musculature is not layered, but rather consists of individual strands connecting strategic subcuticular sites. The secondary body cavity has been called both a hemocoel and a pseudocoel.

Most tardigrades feed on plant cells. They have no jaws, but a pair of sharp stylets occupies sheaths on the walls of a buccal tube (Fig. 14.60b). These stylets can be extended through the mouth to pierce individual plant cells. A powerful pharynx then sucks the cell contents into the gut. Digestion and absorption occur in the wide intestine, and wastes are eliminated through the short hindgut and anus. A few tardigrades impale small aschelminthes on their stylets; however, carnivorous feeding in the phylum is not well understood.

No circulatory and respiratory organs exist in these minute animals. Excretion may involve three glands which merge with the midgut–hindgut junction (Fig. 14.60b). Indeed, these glands sometimes are called Malpighian tubules, although their excretory role has not been demonstrated conclusively. The intestinal wall may eliminate some wastes through the gut, and other unwanted materials may be deposited in the old cuticle prior to a molt. Although most tardigrades live under fluctuating osmotic situations, their osmoregulatory devices are unknown. If their moss or lichen homes dry out, as frequently happens, tardigrades shrivel up but do not die. They survive periods of dryness and extreme cold in a cryptobiotic state (see Special Essay, Chap. 7).

The tardigrade nervous system is similar to that of annelids and arthropods. A large multilobed brain joins a subpharyngeal nerve mass, and a double ventral nerve cord connects four prominent body ganglia. Sensory structures include paired anterior eyespots and a modest assortment of spines and bristles. Tardigrades walk on their legs, clinging to the substratum with their terminal claws. Their movements are often rather jerky, because each limb is operated by only five or six muscle bands. Each band may consist of a single muscle cell.

Tardigrade sexes are separate, although in some species, females predominate and parthenogenesis may occur. Both the testis and the ovary are large sacs lying above the posterior gut (Fig. 14.60b). In females, the gonopore may open into the rectum. The aquatic male injects sperm into the anus or gonopore of a female just before she molts. Fertilization occurs between her old and new cuticles, and embryos may develop in the shed skeleton. Some aquatic tardigrades produce thin-shelled, rapidly developing eggs when growth conditions are optimal, but switch to hard-shelled eggs when the environment deteriorates. (A similar situation was encountered among rotifers.) Perhaps reflecting their unstable environment, terrestrial species always produce thick-shelled eggs; fertilization occurs in the ovary.

Tardigrade development is direct and features the formation of five pairs of coelomic pouches. Curiously, these pouches arise from evaginations of the gut, and their walls eventually form the body musculature. Tardigrades molt to accommodate posthatching growth.

The relationship of tardigrades with other phyla is most uncertain. Their cell constancy, parthenogenesis, and variable egg shell types are reminiscent of rotifers. Their segmented nervous system and embryonic pouches indicate annelid–arthropod affinities, although enterocoely compromises their protostomate qualities. The uniqueness of these animals reflects their very specialized way of life and continues to frustrate our understanding of their history.

PHYLUM PENTASTOMIDA

There are about 70 species of pentastomids. Also known as tongue worms, these animals are parasitic within the respiratory tract of vertebrates. All vertebrate classes are parasitized, but tropical lizards are especially common hosts.

Pentastomids measure up to 15 cm in length. An anterior, medial snout bears the mouth, and there are two pairs of anteriolateral protuberances with sharp claws (Fig. 14.61a). These five projections account for the phylum's name. The rest of the body is long and annulated; a chitinous cuticle covers the entire animal.

Pentastomids grasp the lung or nasal tissues of the host with their claws, which are secreted by hook glands. The mouth is applied to the wound, and a muscular pharynx sucks blood into the straight gut (Fig. 14.61b). Anticoagulants may be secreted by a cephalic gland. Circulatory, respiratory, and excretory organs are absent. A simple ladderlike nervous system is present, but the circumenteric nerve ring lacks a prominent dorsal ganglion.

While the above-mentioned systems are rather rudimentary, pentastomids, like most other parasites, have sophisticated reproductive organs and a complex life cycle. Males are typically more motile than females and crawl within the primary host in search of mates. Fertilization is internal. Each female may pass hundreds of thousands of eggs, which exit with host feces. When eaten by an intermediate host (also a vertebrate, commonly a fish or rodent), the eggs hatch into tiny mitelike larvae (Fig. 14.61c). This stage encysts

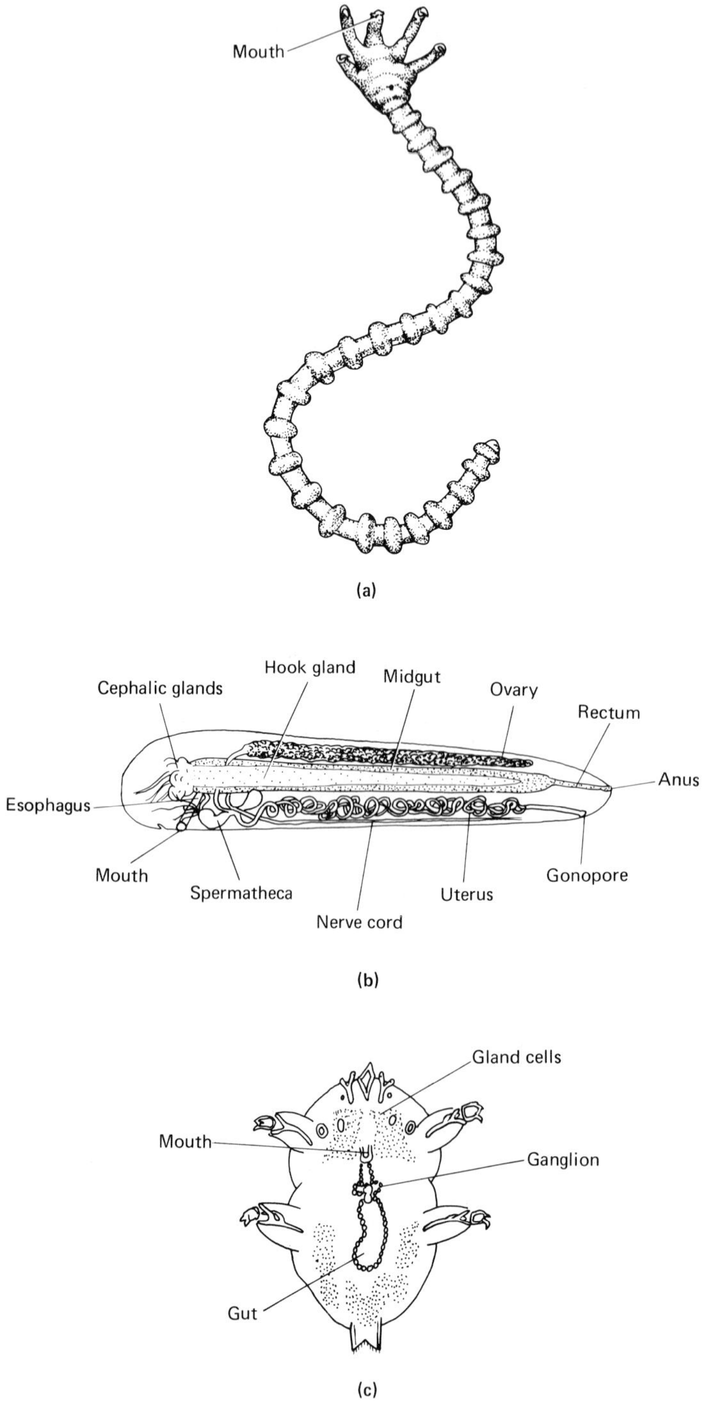

Figure 14.61 Pentastomids. (*a*) External anatomy of *Cephalobaena tetrapoda;* (*b*) The internal anatomy of a female *Porocephalus teretiusculus*, lateral view; (*c*) Ventral view of a larval *Porocephalus proboscideus*. (*a* and *b* from Shipley, In *Cambridge Natural History, Vol. IV;* c from Comstock, *An Introduction to Entomology*.)

until it is eaten by the primary host. The adult parasite migrates up the esophagus to the respiratory tract.

Because parasitism has so specialized the pentastomids, their phylogenetic relationships are difficult to trace. Few researchers doubt their arthropod affinities, but the group has been allied with such diverse classes as mites and myriapods. Tardigrade relationships have also been proposed.

Competing with the Invertebrate World

Like no other invertebrates, certain insects challenge the human niche on earth: Some insects eat our crops, while others transmit diseases both to us and to our livestock. It is wise to remember that harmful species represent less than 1% of all insects, but equally wise to combat these creatures when their interests threaten our own. The final measure of wisdom, however, involves the manner in which we struggle against our insect foes: They must be combated intelligently, lest we do more harm than good to our own life support systems.

Responsible people recognize that human population sizes must be controlled. An effective yet unthinkable means of limiting human population growth would be to dump tons of poisons on our cities each year. While such an indiscriminate method of human population control is abhorrent to our basic values, it has seemed proper in dealing with insects. For years, poisons like DDT (dichlorodiphenyltrichloroethane) have been poured over the planet to destroy insect pests. In some places and for certain periods of time, these poisons were largely successful. They killed inimical insects—but also beneficial ones, which pollinate crops and prey on destructive species. Insecticides, especially fat-soluble DDT, accumulate in other animals and have killed birds and fishes. In recognition of its ultimate threat to human health, the use of such poisons is increasingly restricted.

There are more sophisticated ways in which to continue this struggle. Research on the behavior and physiology of insects has opened possible avenues of effective, ecologically sane population control. One of the simplest of these methods pits insects against insects. A grape field infested with leaf hoppers, for example, may be protected by releasing thousands of tiny wasps. These wasps parasitize the leaf hoppers, killing their hosts. Ladybugs have been imported to prey on destructive aphids and scale insects.

Other biological control methods focus on insect reproduction and developmental phenomena. Some species, such as screwworm adults, mate only once. Millions of sterile males can be cultured and then released in an infested area. They copulate with potent females, sating them sexually for a lifetime but fathering no offspring. Some insects, notably the moths, use pheromones to attract mates over long distances. Analogous compounds can be synthesized and set in traps to lure would-be mates. Some very interesting work has been done using juvenile hormone analogs as possible insect control agents. When consumed by an insect, such compounds disrupt larval development and sexual maturity in the adult. Juvenile hormone analogs have no known deleterious effects on other animals. Only a few grams per acre may be sufficient to control a pest population.

for Further Reading

Anderson, D. T. *Embryology and Phylogeny in Annelids and Arthropods.* Pergamon Press, New York, 1973.

Atkins, M. D. *Introduction to Insect Behavior.* Macmillan, New York, 1980.

Beck, S. D. *Insect Photoperiodism.* 2nd ed. Academic Press, New York, 1980.

Borror, D. J., De Long, D. M., and Triplehorn, C. A. *An Introduction to the Study of Insects.* 5th ed. Saunders College Publishing, 1981.

Boudreaux, H. B. *Arthropod Phylogeny with Special Reference to Insects.* Wiley-Interscience, New York, 1979.

Chapman, R. F. *The Insects: Structure and Function.* 2nd ed. Elsevier, New York, 1976.

Cisne, J. L. Trilobites and the origin of arthropods. *Science* 186: 13–18, 1974.

Counce, S. J. and Waddington, C. H. (Eds.) *Developmental Systems: Insects.* Academic Press, New York, 1972.

Daly, H. V. and Doyen, J. T. *An Introduction to Insect Biology and Diversity.* McGraw-Hill, New York, 1979.

Danks, H. V. Canada and its insect fauna. *Mem. Entomol. Soc. Can.* 108: 1–573, 1979.

Douglas, M. W. Thermoregulatory significance of thoracic lobes in the evolution of insect wings. *Science* 211: 84–86, 1981.

Eisner, T. and Wilson, E. O. *The Insects: Readings from Scientific American.* W. H. Freeman, 1977.

Emden, H. F. van (Ed.) Insect plant relationships. *Symp. Roy. Entomol. Soc. London* 6–215, 1973.

Friedlander, C. P. *The Biology of Insects.* Universe Books, New York, 1977.

Frisch, K. von. Bees: Their Vision, Chemical Senses and Language. Cornell University Press, Ithaca, New York, 1971.

Gould, T. Communication of distance information by honey bees. *J. Comp. Physiol.* 104: 161–174, 1976.

Gupta, A. P. (Ed.) *Arthropod Phylogeny.* Van Nostrand Reinhold, New York, 1979.

Hepburn, H. R. (Ed.) *The Insect Integument.* Elsevier, New York, 1976.

Horn, D. J. *Biology of Insects.* Holt, Rinehart and Winston, New York, 1976.

Horridge, G. A. (Ed.) *The Compound Eye and Vision of Insects.* Clarendon Press, Oxford, 1974.

Kapoor, V. C. *Taxonomic Approach to Insecta.* Advent Books, New York, 1980.

Kukalova-Peck, J. Origin and evolution of insect wings and their relation to metamorphosis, as documented by the fossil record. *J. Morph.* 156: 11–126, 1978.

Little, V. A. *General and Applied Entomology.* Harper and Row, New York, 1972.

Locke, M. and Smith, D. S. (Eds.) *Insect Biology in the Future.* Academic Press, New York, 1980.

Manton, S. M. The evolution of arthropodan locomotory mechanisms. Part 1: Locomotory habits, morphology and evolution of the hexapod class. *Zool. J. Linn. Soc.* 51: 203–400, 1972.

Manton, S. M. Arthropod phylogeny–A modern synthesis. *J. Zool.* 171: 11, 1973.

Manton, S. M. *The Arthropoda: Habits, Functional Morphology and Evolution.* Oxford University Press, 1978.

Merritt, R. W. and Cummins, K. W. *An Introduction to the Aquatic Insects of North America.* Kendall-Hunt Publishing, Dubuque, Iowa, 1978.

Morgan, C. I. and King, P. E. *British Tardigrades.* Synopses of the British Fauna No. 9 Academic Press, London, 1976.

Oster, G. F. and Wilson, E. O. *Caste and Ecology in the Social Insects.* Princeton University Press, 1978.

Pollock, L.W. *Tardigrada.* Marine flora and fauna of the northeastern U.S. NOAA Technical Reports NMFX Circular 394. U.S. Govt. Printing Office, 1976.

Price, P. W. *Insect Ecology.* Wiley-Interscience, New York, 1975.

Richards, O. W. and Davies, R. G. *Imm's Outlines of Entomology, Vols. I and II.* Chapman and Hall, London, 1977.

Riley, J., Banaja, A. A., and James, J. L. The phylogenetic relationships of the Pentastomida: the case for their inclusion with the Crustacea. *Int. J. Parasit.* 8: 245–254, 1978.

Rockstein, M. (Ed.) *The Physiology of Insecta, Vols. I–VI.* 2nd ed. Academic Press, New York, 1973.

Sharov, A. G. *Basic Arthropodan Stock.* Pergammon Press, New York, 1966. (Sharov argues for the monophyletic origin of the arthropods from protrilobites.)

Snodgrass, R. E. *Principles of Insect Morphology.* McGraw-Hill, London and New York, 1935.

Tiegs, O. W. and Manton, S. M. The evolution of the Arthropoda. *Biol. Rev.* 33: 255–337, 1958.

Wigglesworth, V. B. *Insect Hormones.* Oliver and Boyd, Edinburgh, 1964.

Wigglesworth, V. G. *The Principles of Insect Physiology.* Chapman and Hall, London, 1972.

Wilson, E. O. *The Insect Societies.* Belknap Press of Harvard University, Cambridge, 1971.

15

The Lophophorates

Curious embryogenies typify the phyla known collectively as lophophorates. These animals include the phoronids, the bryozoans, and the brachiopods. Lophophorates cannot be categorized neatly as either protostomes or deuterostomes. Rather, their varied assortment of developmental characters suggests that they may be intermediate between the two major phylogenetic lines.

The mouth forms from or near the blastopore in most lophophorates, and phoronids and bryozoans have Trochophore-like larvae. These are clearly protostomate traits. Deuterostomate traits include modified radial, indeterminate cleavage in at least some members of all three groups. Such cleavage may be the only pattern among brachiopods, a phylum that undergoes modified enterocoely and has no trochophore-like larva. The lophophorate coelom further allies these animals with the deuterostomate branch of the invertebrate world. Like most larval echinoderms and hemichordates, lophophorates possess a compartmentalized coelom. Among deuterostomes, the secondary body cavity is divided into three areas: an anterior **protocoel,** a midbody **mesocoel**, and a posterior **metacoel.** Many lophophorates lack a protocoel, but a mesocoel and metacoel are usually quite distinct. The lophophorate protocoel may have been lost because of the pronounced head reduction in these animals.

The most characteristic structure of these phyla is the **lophophore,** a crown of ciliated tentacles that surrounds the mouth. The lophophore represents a fold in the anterior body wall. The tentacles are hollow outgrowths of this fold, and each receives an extension of the mesocoel. Coelomic pressure and a specialized musculature operate the lophophore, which can be extended for filter feeding. Other characteristics common to the lophophorates include a U-shaped digestive tract, sessile life-styles, and protective external secretions. These secretions vary among the phyla and contribute to their diversity of appearance.

Phylum Phoronida

The smallest and probably the most primitive group of lophophorates is the phylum Phoronida. Only 16 species in two genera (*Phoronis* and *Phoronopsis*) have been identified; no higher taxonomic levels have been established. Phoronids are solitary tube dwellers along marine coastlines, where most species are subtidal. Their cylindrical bodies lie freely within secreted chitinous tubes, which are buried on sandy bottoms or attached to various hard substrata. A few species penetrate calcareous shells and rocks.

Ranging from 0.5 to 25 cm in length, the phoronid body has relatively few external features (Fig. 15.1a). It expands posteriorly for anchorage within the tube, and its anterior end bears the lophophore. This crescent-shaped structure encompasses two parallel ridges, which flank the ventral mouth and often spiral dorsally (Fig. 15.1b). Each ridge supports numerous tentacles. The latter arise from a medial break in each ridge, and their number increases throughout the lifetime of the phoronid. The long, narrow mouth is located along a groove between the two lophophoral ridges; it is covered by a flap called the **epistome.** The phoronid anus lies outside the dorsal spirals of the lophophore. This opening is flanked by a pair of nephridiopores.

Perhaps because their tubes provide protection, phoronids have a rather thin body wall. It comprises an epidermis, weak circular muscles, a strong longitudinal muscle layer, and a coelomic peritoneum. Numerous mucus-secreting cells populate the lophophore. The phoronid coelom displays the typical lophophorate divisions: a trunk metacoel and an anterior mesocoel that penetrates the tentacles. A small protocoel occupies the epistome.

Phoronids collect plankton and other organic matter with their lophophores. Cilia on the tentacles draw a water current toward the groove between the lophophore ridges, where mucus traps suspended particles. Food is ushered to the mouth by groove cilia. A U-shaped digestive tract includes a descending esophagus, a stomach, and a long ascending intestine (Fig. 15.2). Intracellular digestion may occur in the stomach.

Phoronid circulation involves a closed blood vessel system. No heart is present, but the major blood vessels are contractile and include a dorsal vessel through which blood flows anteriorly and a ventral vessel for posterior transport. The dorsal tube contributes smaller vessels to each lophophore tentacle, and the ventral vessel gives rise to numerous capillary ceca which

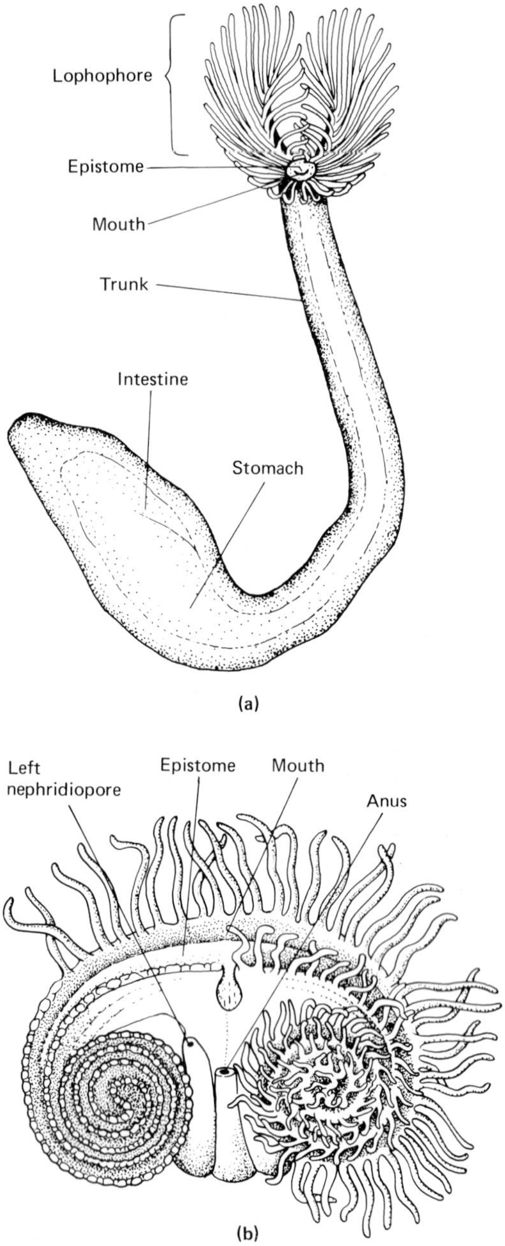

Figure 15.1 (*a*) External features of a phoronid worm, *Phoronis architecta;* (*b*) Dorsal view of the anterior end of *Phoronis australis*. The tentacles of the lophophore are cut away on the left side, and those of the inner ridge are shortened on the right side to show their arrangement—actually all tentacles are the same length. (*a* from L.H. Hyman, *The Invertebrates, Vol. V: Smaller Coelomate Groups*, 1959 McGraw-Hill; *b* from A.E. Shipley, In *Cambridge Natural History, Vol. II*, S.F. Harmer and A.E. Shipley, Eds., 1959, Macmillan.)

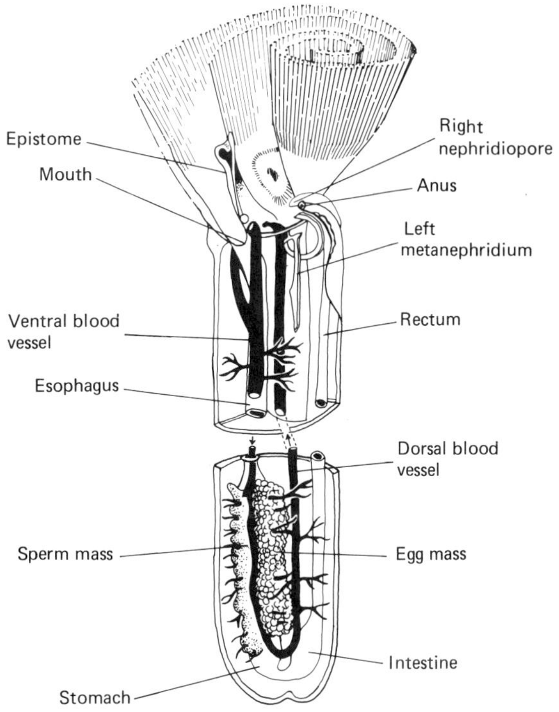

Figure 15.2 A view of the internal anatomy of a phoronid worm. The body has been sectioned longitudinally; most of the trunk has been omitted. (From Shipley, In *Cambridge Natural History, Vol. II*, Harmer and Shipley, Eds.)

serve the internal organs. Posteriorly, a network of gastric sinuses connects the ventral and dorsal vessels. No specialized respiratory structures exist, as the tentacles provide ample gas-exchange surfaces. Phoronid blood contains cell-bound hemoglobin, as does the coelomic fluid. A single pair of metanephridia opens into the metacoel; as mentioned previously, nephridiopores flank the anus (Fig. 15.1b).

In keeping with their sessile existence, phoronid muscular and nervous systems are rather simple. Longitudinal muscles in the tentacles allow gentle swaying, and the stronger longitudinal body wall muscles permit swift withdrawal of the animal. Circular muscles are weakly developed in the trunk and are absent from the tentacles. An epidermal nerve ring occupies the base of the lophophore and dispatches nerves to the tentacles and the body wall. A single giant fiber passing along the left side of the trunk coordinates quick withdrawal movements. Scattered sensory cells are linked by a subepidermal nerve plexus. No specialized sense organs are present.

Most phoronids are hermaphroditic. Sperm and eggs are produced from coelomic peritoneum and exit from the parent via the nephridial system. Fertilization is probably internal in most species, and brooding may occur between the ridges of the lophophore. One species reproduces by transverse fission and budding. Phoronid cleavage is radial, but the blastopore may form the mouth. The larval stage is called an **actinotroph** (Fig. 15.3). Resembling an elongate trochophore, this larva bears an oral collar of ciliated tentacles, a complete gut, and a posterior telotroch. The latter is a ciliated locomotor ring. The actinotroph

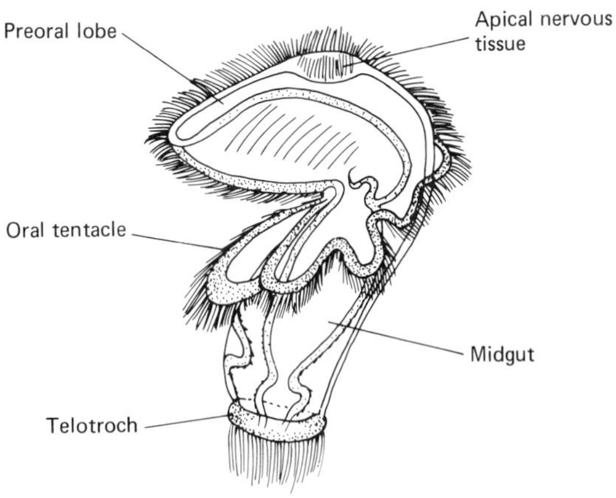

Figure 15.3 Actinotroch larva of a phoronid worm. (From Hyman, *The Invertebrates, Vol. V.*)

pursues a planktonic existence for a month or so before it settles to the bottom, assumes a more adult form, and secretes a tubular home.

Phylum Bryozoa

The bryozoans, also known as moss animals (Ectoprocta) or ectoprocts, are by far the most abundant and diverse group of modern lophophorates. There are about 4000 living and 16,000 fossil bryozoan species. Colonial and sessile, bryozoans live mostly along marine coastlines, although deep-sea species and freshwater forms are also known. Bryozoan colonies are common on rocks, pilings, kelp, and the hulls of ships, where they can cause major fouling problems. Their encrusting or arborescent growth patterns often resemble those of hydroid coelenterates (cf. Figs. 5.10, 5.14, and 5.16). Bryozoans are readily distinguished, however, by the cilia on their tentacles, and closer inspection reveals a more complex, eucoelomate body plan.

Bryozoans are individually tiny animals. Possibly, they have evolved from ancient phoronid stock; if so, their unique characters represent adaptations associated with smaller body size and colonial life.

Bryozoan Unity

Our discussion of bryozoan unity will focus on the individual zooid. Colonial organization in the phylum varies widely and is treated later.

MAINTENANCE SYSTEMS

General Morphology

The typical bryozoan zooid is less than a millimeter long. Its body form varies from tubular to cuboidal (Fig. 15.4). The zooid secretes a protective, chitinous and often calcareous exoskeleton called a **zoecium.** Unlike the chitinous tubes of phoronids, this encasement forms a part of the body wall. An orifice in the zoecium allows the extension of the lophophore. Commonly, the skeleton is invaginated at this opening, forming an inner chamber or atrium. When the lophophore is retracted, an operculum may seal the orifice.

The body wall is unmuscularized in the many bryozoan species with unmovable calcareous exoskeletons. In such animals, there is merely a secretory epidermis and the coelomic peritoneum. Circular and longitudinal muscles lie between these layers in freshwater bryozoans, whose skeletons are pliant. The spacious coelom comprises the typical mesocoel and metacoel; a protocoel is present in the phylactolaemates. Crossing the large metacoel are sets of lophophore retractor muscles and a cord of tissue called the **funiculus** (Fig. 15.4), which may be involved in nutrient transport between zooids and in statoblast production (see page 497). The bryozoan body proper is called the **polypide.** The polypide consists of the lophophore and the digestive tract, both of which resemble the corresponding structures in phoronids. In marine bryozoans, however, the lophophore has only one ridge of tentacles. Over a hundred tentacles are present in some freshwater species; marine forms have between 8 and 34. The body wall extends into a tentacular sheath at the base of the lophophore (Fig. 15.4). This muscularized sheath encloses the tentacles when they are withdrawn and may drape around the orifice when the lophophore is extended.

As in other colonial organisms, many bryozoans display polymorphic zooids. Standard feeding morphs are called **autozooids,** while various **heterozooids** are specialized for defense, sanitation, and reproduction.

Nutrition and Digestion

Bryozoans feed much as phoronids do. With the lophophore extended, tentacular cilia sweep phytoplankton and other minute edibles into the mouth. Food movement is enhanced by ciliary action in the pharynx and by the muscular pumping of the foregut. The rest of the U-shaped digestive tract includes an esophagus, a glandular stomach, the ascending intestine, and a rectum (Fig. 15.5). The gut is muscular throughout, and the posterior stomach is ciliated. Ciliary action rotates gastric mucus and food. Intracellular digestion of fats may occur in marine bryozoans, but in freshwater species, digestion is probably entirely extracellular. Feces are bound in mucus and are voided through the anterior anus.

Circulation, Respiration, and Excretion

Perhaps because bryozoan zooids are so minute, they survive without special organs for internal transport, gas exchange, and waste removal. Coelomic fluids distribute food and respiratory gases; the latter are exchanged across the body surfaces, principally those

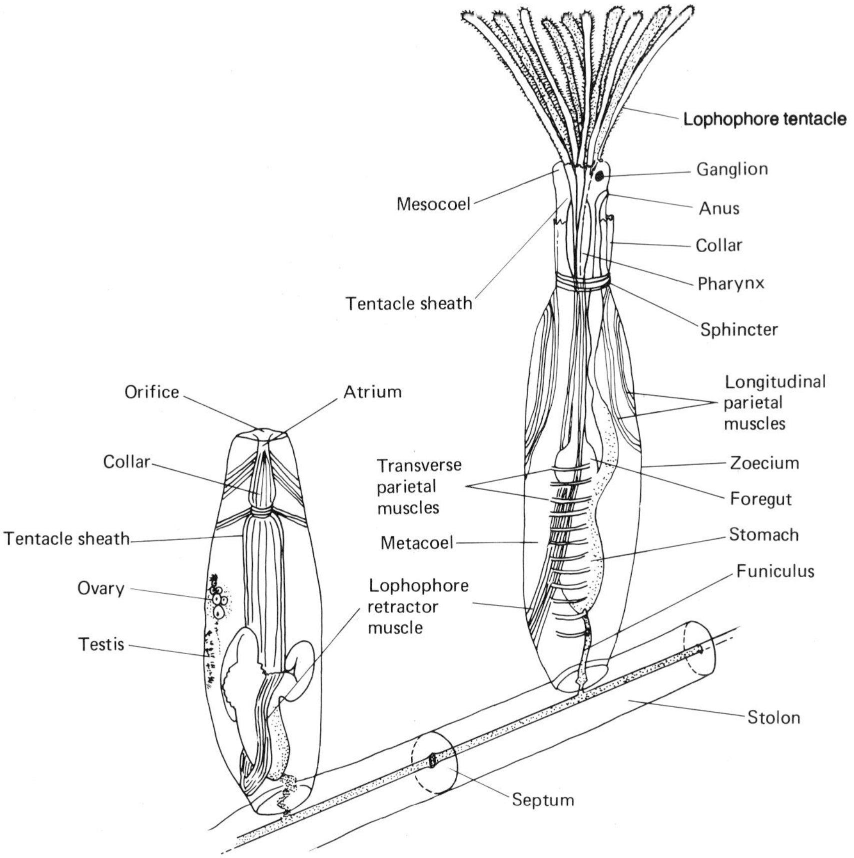

(a)

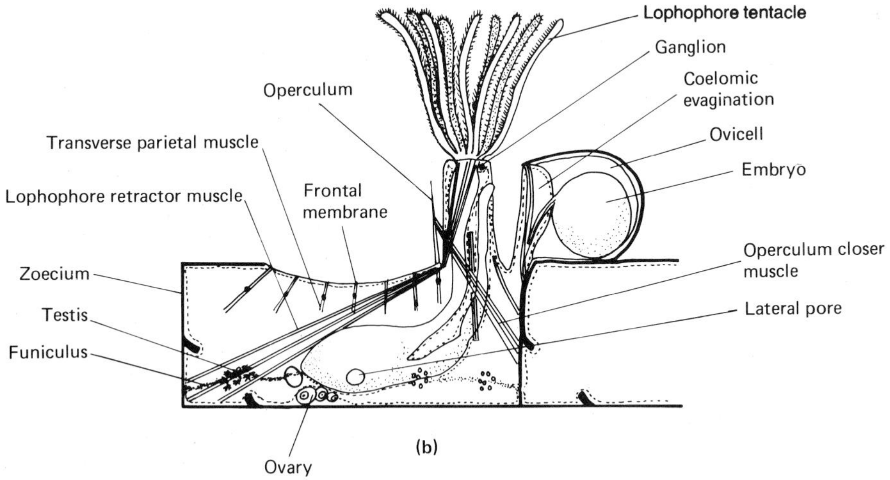

(b)

Figure 15.4 (*a*) A small portion of a colony of the bryozoan *Bowerbankia* showing two tubular zooids attached to a stolon. The left-hand zooid is fully retracted; the tentacles of the right-hand zooid are expanded; (*b*) Diagrammatic section through a box-like cuboidal zooid of an encrusting bryozoan. (From J.S. Ryland, *Bryozoans*, 1970, Hutchinson.)

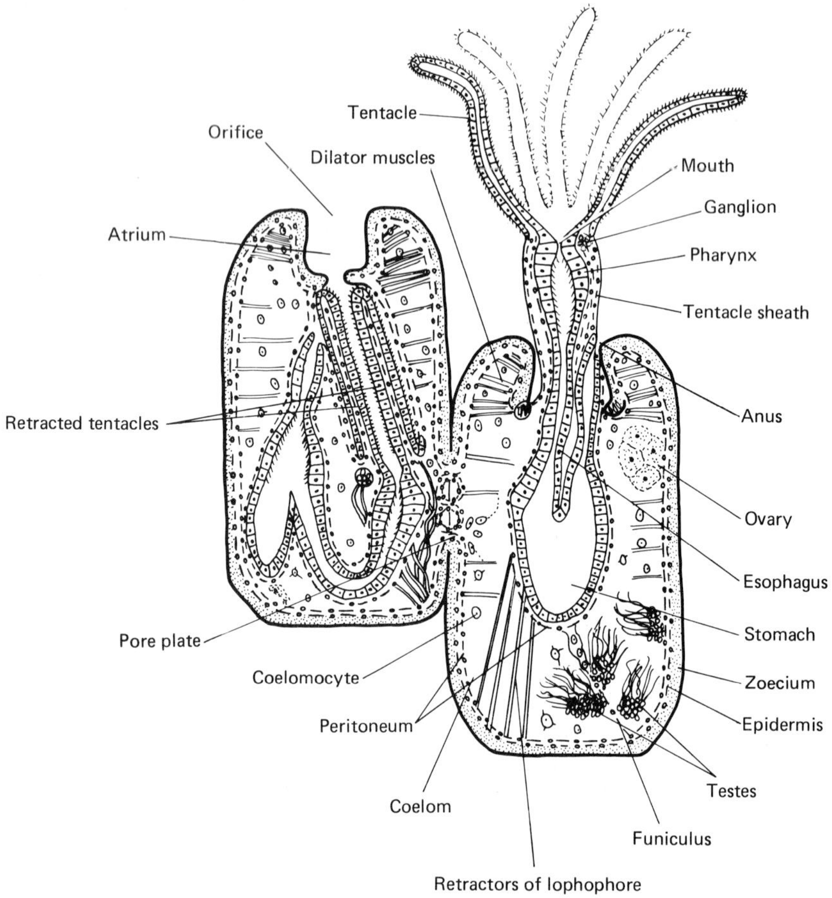

Figure 15.5 Longitudinal section illustrating the internal anatomy of two adjacent zooids. *Left*: lophophore retracted. *Right*: lophophore expanded. (From Hyman, *The Invertebrates, Vol. V.*)

of the lophophore. Soluble wastes simply diffuse away, while coelomocytes may engulf particulate matter. Degenerated polypides called brown bodies (see next page) may be involved in storage excretion.

ACTIVITY SYSTEMS

As sessile organisms, bryozoans exhibit limited muscular development. Muscular activity focuses on the operation of the lophophore. Extension mechanisms vary widely, but always involve coelomic (mesocoelic) fluid pressure. In freshwater species with a pliant exoskeleton, the contraction of the body wall musculature produces the necessary hydrostatic force. Most marine species have a rigid zoecium; thus other elements must act against the coelomic fluid. These elements are discussed later. However it is extended, the lophophore is withdrawn by two large pairs of retractor muscles (Figs. 15.4 and 15.5).

A major ganglion lies near the bryozoan mouth and gives rise to a pharyngeal nerve ring, whose outlets communicate with the tentacles, tentacular sheath, and gut. A subepidermal nerve plexus extends over the body. Known sensory devices are limited to scattered tactile and chemosensitive cells.

CONTINUITY SYSTEMS

Bryozoans are subject to the same reproductive and developmental pressures as other sessile colonial invertebrates, and their adaptations are similar. The phylum is almost entirely hermaphroditic, and reproductive polymorphism is common. The life cycle includes dispersive larvae and the production of resistant bodies when environmental conditions deteriorate. Asexual reproduction leads to the development of colonies.

Reproduction

Bryozoan gametes are produced from the lining of the metacoel. Fertilization is not well understood. In a few groups that have been studied, sperm exit via two of the tentacles and subsequently adhere to the lophophoral regions of other zooids. Eggs may be released through an opening on the lophophore called a **coelomopore.** Sometimes this pore is elevated on the **intertentacular organ** (Fig. 15.11c). In some species, sperm enter the coelomopore for internal fertilization.

Development

Most bryozoans brood their eggs. Brooding may occur within the coelom, the tentacular sheath, or the atrium. In one large group (the cheilostomes), a special chamber called the **ovicell** houses the developing embryo (Figs. 15.4b and 15.11f). The polypides of brooding zooids commonly degenerate, and their remnants form dark cellular clusters called **brown bodies.** New polypides arise from the body wall after brooding. Often, the brown body is surrounded by the new rectum and is expelled.

In nonbrooding species, a feeding trochophore-like **cyphonautes** larva drifts for several months among the marine zooplankton (Fig. 15.6). Brooding bryozoans produce short-lived, nonfeeding larvae. Upon settlement, a posterior adhesive sac fastens the larva to a substratum and adult metamorphosis occurs. The lone zooid, or **ancestrula,** produces similar individuals by budding. Budding patterns vary widely between species, as evidenced by the range of colonial forms in the phylum. Arborescent growths feature budding stolons or zooids with apical reproduction, while encrusting colonies feature lateral budding zones.

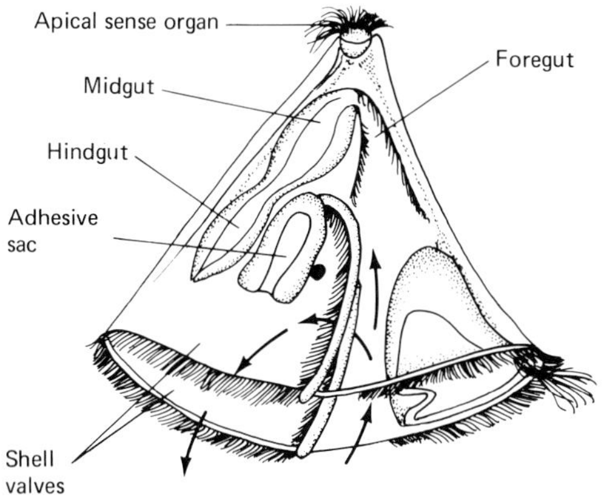

Figure 15.6 Cyphonautes larva of the bryozoan *Membraniporal*. *Arrows* indicate the feeding current. (From Ryland, *Bryozoans*.)

Statoblasts are common in freshwater bryozoans. Formed from buds along the funiculus (Fig. 15.7a), these resistant bodies house peritoneal and epidermal cells as well as food reserves. They secrete protective valves and may bear spines that aid in flotation or adhere to traveling animals (Fig. 15.7b). Produced in enormous numbers, statoblasts help a bryozoan population to survive hard times, such as droughts and winter. Under proper environmental conditions, dormant statoblasts may disperse widely before germinating (Fig. 15.8a).

A Review

Basically, bryozoans are:

(1) sessile colonial lophophorates, whose exoskeleton is attached to the body wall;

(2) colonies of tiny zooids, each having a U-shaped gut and a muscular lophophore, but no circulatory, respiratory, or excretory organs;

(3) hermaphrodites that usually brood their young and form complex colonies by asexual means.

Bryozoan Diversity

Although not well known by the layperson, bryozoans are a sizable and diverse phylum. The 4000 living species currently are grouped into three classes: the Phylactolaemata, the Stenolaemata, and the Gymnolaemata. Their members differ in terms of habitat, skeletal construction, polymorphism, colonial growth patterns, and lophophore extension techniques.

CLASS PHYLACTOLAEMATA

The phylactolaemates include 50 species, all of which live in fresh water. Although their habitat is often associated with higher invertebrates, these animals are probably the most primitive modern bryozoans. Like the phoronids, phylactolaemates have cylindrical zooids, an uncalcified exoskeleton, an epistome, and a large horseshoe-shaped lophophore (Fig. 15.7a). Contraction of the body wall musculature results in the hydraulic extension of the lophophore. Like many

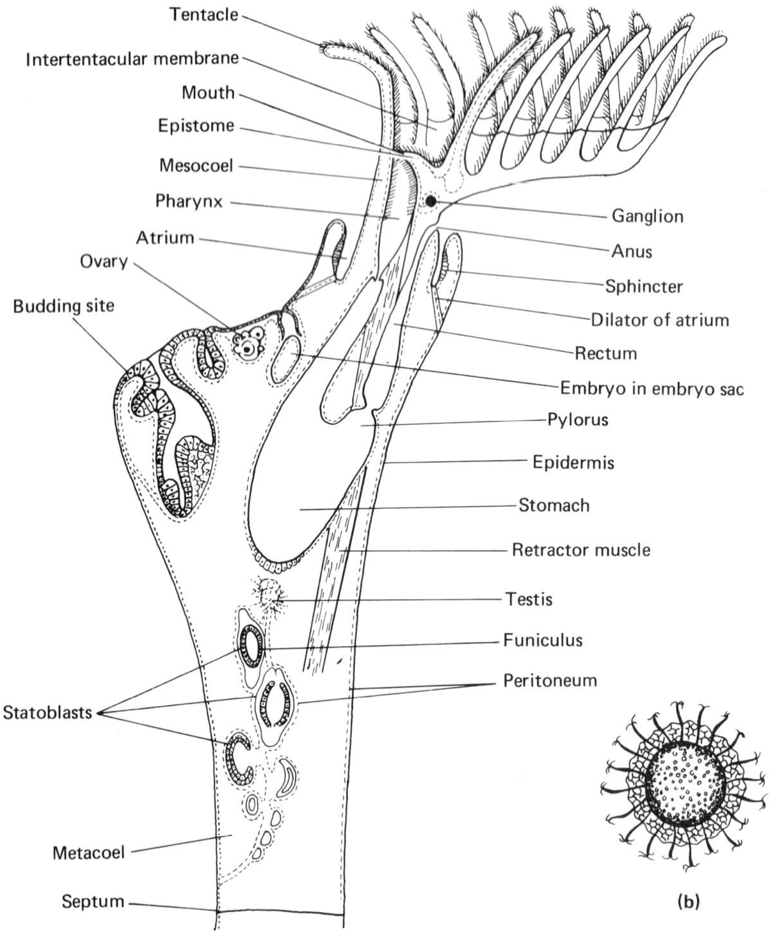

Figure 15.7 (*a*) Section through a zooid of a generalized phylactolaemate bryozoan. Note the development of new polypides at the budding site and the development of statoblasts from the funiculus; (*b*) A spiny statoblast from the freshwater bryozoan, *Cristatella.* (From Ryland, *Bryozoans.*)

freshwater invertebrates, phylactolaemates brood their young. Fertilization occurs in the metacoel, and an embryo sac develops near the site of egg cell formation. Statoblasts are formed for dispersion and in preparation for winter or drought. Because statoblasts can be transported so easily, many phylactolaemates enjoy a cosmopolitan distribution.

Phylactolaemate colonies are unique in several respects. There is no polymorphism, all individuals being similar autozooids. The exoskeleton never completely isolates any member, so that the colony shares a continuous coelom. Colonial growth patterns include both lophopodid and plumatellid types. Lophopodid colonies, such as *Lophopus* (Fig. 15.8a), comprise zooids embedded in a soft, gelatinous mass. In *Cristatella,* the base of this mass is muscularized, allowing the colony to creep very slowly (Fig. 15.8b). *Plumatella* and *Fredericella* form plumatellid colonies in which serially arranged zooids extend in a plantlike manner (Fig. 15.8c).

CLASS STENOLAEMATA

Stenolaemate bryozoans dominated in the Paleozoic, but today are mostly extinct. Approximately 550 fossil genera are described, but only a few marine families have living members. Only these, all of which belong to the order Cyclostomata, are discussed here.

Cyclostomate autozooids are tall and tubular with a heavily calcified skeleton and no body wall muscles (Fig. 15.9a). A circular orifice occupies the distal tip of the body; an epistome is never present. Although cyclostomes cannot distort their body wall, their lophophore nevertheless is extended by coelomic fluid pressure. Distal extensions of the mesocoel surround the anterior atrium, whose borders are bound to the

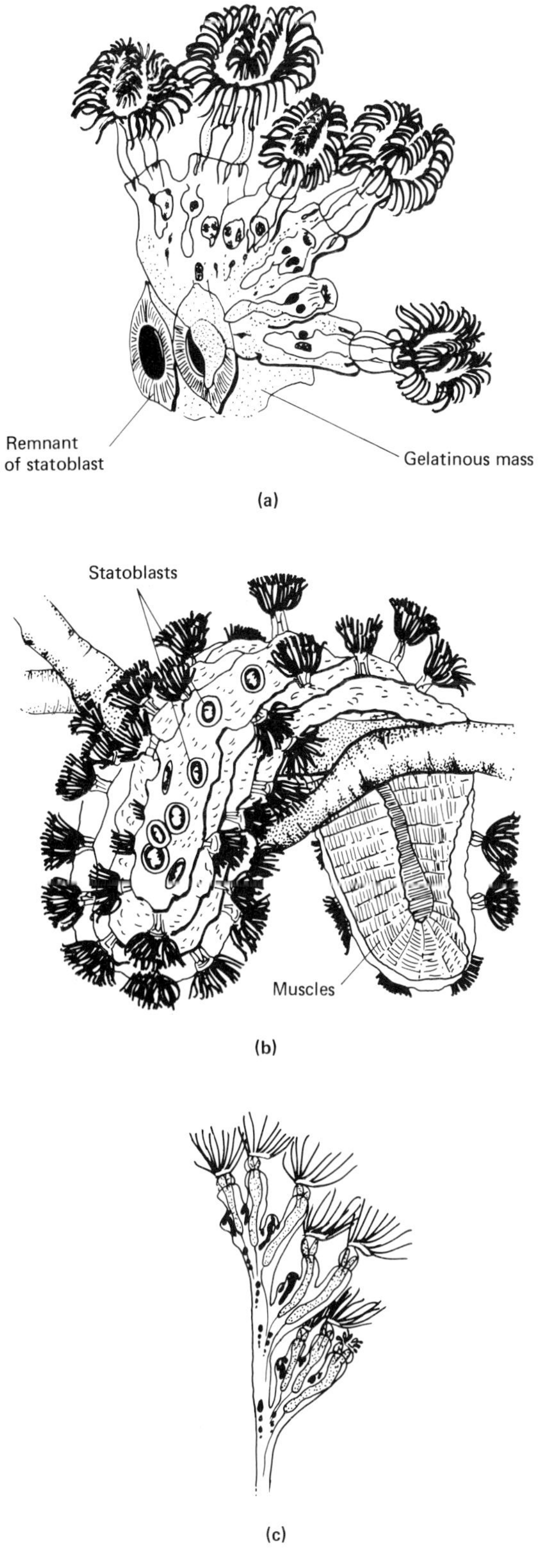

Figure 15.8 (*a*) Colony of the freshwater bryozoan *Lophopus crystallinus* developing from a germinated statoblast; (*b*) A colony of *Cristatella mucedo* creeping along plant stems—zooids face outward on both sides; (*c*) One branch of *Plumatella fungosa*. (*a* and *c* from P. Brien, In *Traité de Zoologie, Vol. V*, P.P. Grassé, Ed., 1960, Masson et Cie, *b* from Hyman, *The Invertebrates, Vol.V*.)

body wall by dilator muscles. Contraction of these muscles widens the atrium while forcing the surrounding fluid into the rest of the coelom. As a result, the lophophore itself is engorged with fluid and extends.

Colonial form varies, as illustrated by *Crisia, Tubulipora*, and *Stomatopora* (Fig. 15.9b–d). Pores between adjacent zooids allow some exchange of coelomic fluids. Heterozooids in this order are limited to a few anchoring individuals and reproductive morphs. Cyclostomate embryogeny is unusual in that a single zygote commonly gives rise to scores of embryos.

CLASS GYMNOLAEMATA

The vast majority of modern bryozoans belongs to the Gymnolaemata, a class with two orders. Almost all of these animals are marine. Their distinguishing characters include cylindrical or cuboidal zooids, a circular lophophore, and polymorphism. Gymnolaemates have no epistome and no body wall musculature. Pores exist in the interzooidal walls, but are filled with epidermal cells.

Order Ctenostomata

Ctenostomes compose the smaller and more primitive gymnolaemate group. This order exhibits an uncalcified skeleton, often stoloniferous colonial growth, and limited polymorphism. No operculum covers the anterior orifice. Bands of muscle span the ctenostomate coelom and insert on the flexible body wall (Fig. 15.4a). Their contraction causes coelomic pressure to extend the lophophore.

In many ctenostomate colonies, such as *Bowerbankia* (Figs. 15.4a and 15.10a), the autozooids are borne along a creeping and/or branching stolon. The stolon itself represents a fused series of reduced morphs called **kenozooids.** The funiculus of each autozooid merges with a similar cord of tissue in the stolon; thus zooids within the colony are linked.

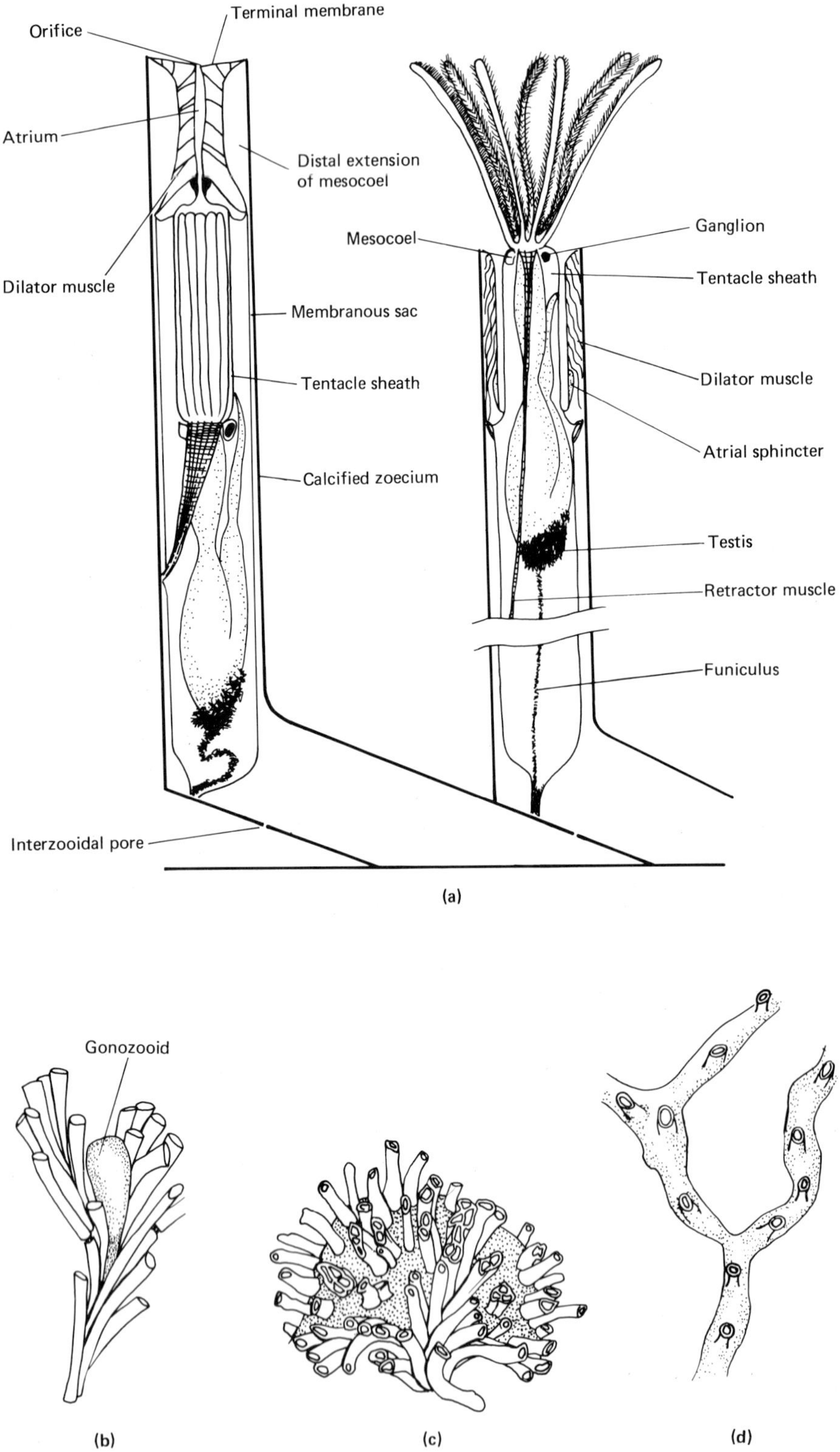

Figure 15.9 Cyclostomate bryozoans. (*a*) Zooid structure. *Left:* lophophore retracted. *Right:* lophophore expanded; (*b*) *Crisia*, a colonial form, possessing many separate tubular zooids; (*c*) A colony of *Tubulipora flabelaris;* (*d*) *Stomatopora*, a colonial form consisting of branched rows of fused zooids. (*a* from Ryland, *Bryozoans; b* and *d* from Hyman, *Invertebrate Zoology, Vol. V; c* from Brien, In *Traité de Zoologie, Vol. V.* Grassé, Ed.)

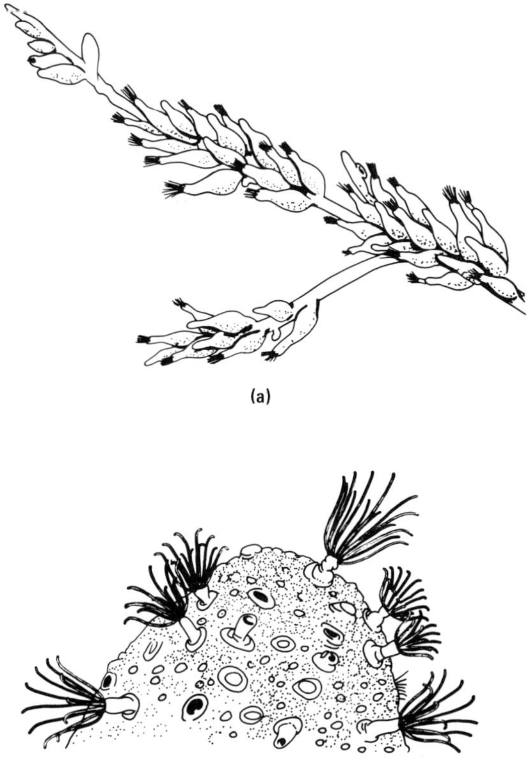

Figure 15.10 (*a*) A portion of a colony of *Bowerbankia imbricata,* a ctenostomate bryozoan; (*b*) *Alcyonidium polyoum,* a nonstoloniferous ctenostome. (From M. Prenant and G. Bobin, *Faune de France, Vol. 60,* 1956, Paul Lechevalier.)

Nonstoloniferous colonies include *Nolella, Alcyonidium* (Fig. 15.10b), and the freshwater *Paludicella.* Some trends toward flattened zooids occur in these genera, and individuals may be joined laterally. These trends are developed more extensively in the second gymnolaemate order.

Order Cheilostomata

The cheilostomes are by far the largest and most widely distributed group of modern bryozoans. Their distinguishing features include flattened, boxlike zooids, usually with calcified walls and an operculum. Polymorphism is well developed in this order. Most species brood their young in an ovicell.

Bugula (Fig. 15.11a) is a common genus, whose colonies are upright and plantlike in form. Most cheilostomes, however, grow in encrusting sheets (e.g., *Membranipora* and *Electra;* Fig. 15.11b, c). Encrusting zooids are fused along their lateral and end walls, and the orifice occupies a broad frontal surface. In the more primitive cheilostomes, a frontal membrane stretches across the exposed surface of each zooid (Fig. 15.4b). The lateral walls are completely rigid, so muscles insert instead on this frontal membrane. Their contraction depresses the membrane, creating the hydrostatic pressure necessary for lophophore extension. Less primitive cheilostomes provide increasing protection for their frontal surfaces. A recessed membrane is surrounded by spines in *Callopora* (Fig. 15.11d) and in *Cribrilina* (Fig. 15.11e) these spines have fused to form a sievelike cover. Finally, the ascophorans shelter their frontal membrane beneath a nearly solid skeletal surface (Fig. 15.11f). The membrane itself lines a sac called the **ascus.** The contraction of muscle bands inserted on the lower side of the membrane expands the ascus, thus elevating coelomic fluid pressure and extending the lophophore. An operculum covers both the opening to the ascus and the orifice above the lophophore.

Cheilostomes are the most polymorphic of the bryozoans. Their heterozooids include defensive avicularia, sanitational vibracula, and reproductive morphs. In an **avicularium,** most structures are reduced except for the operculum, which is modified as a jaw (Fig. 15.12a). These morphs are either sessile or mounted atop movable stalks. Abductor and adductor muscles snap the avicularium open and shut, as it defends the colony against fouling organisms. The **vibraculum** also features a modified operculum (Fig. 15.12b). In this morph, the cap forms a long bristle which sweeps across the colony, apparently cleaning it. Reproductive morphs develop from autozooids. In most species, the distal body wall bulges to form an **ovicell** (Fig. 15.11f). A coelomic evagination fills part of the ovicell, and the embryo is incubated in the remaining space. In brooding zooids, polypides commonly degenerate into brown bodies.

Phylum Entoprocta

Sessile, mostly colonial animals called entoprocts resemble bryozoans in many ways. Indeed, they were long classified among the Bryozoa and then were

(a)

Figure 15.11 (*a*) A small part of a colony of *Bugula*, a cheilostomate bryozoan featuring typical feeding autozooids; (*b*) An encrusting colony of *Membranipora*, a common cheilostome; (*c*) Lateral view of a typical box-like zooid of an encrusting cheilostome, *Electra pilosa*; (d) Frontal view of two zooids of *Callopora*, in which spines protect the frontal membrane; (*e*) In *Cribrilina* the spines have fused to form a protective cover over the frontal membrane; (*f*) Section through a zooid of an ascophoran cheilostome, in which the frontal membrane is protected by a nearly solid skeletal surface. (*a* Courtesy of Ward's Natural Science Establishment; *b* Courtesy of Carolina Biological Supply; *c* from E. Marcus, In *Die Tierwelt der Nord und Ostsee, Part VII*, G. Grimpe and E. Wagler, Eds., 1926, Akademische Verlagsges; *d–f* from Ryland, *Bryozoans*.)

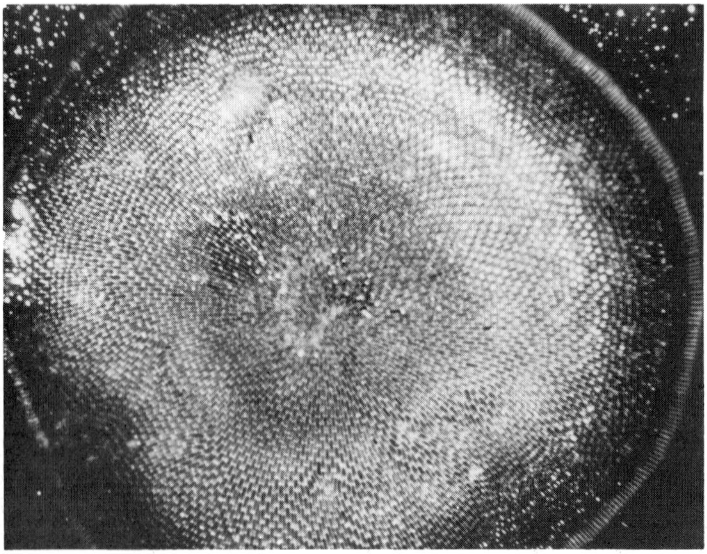

(b)

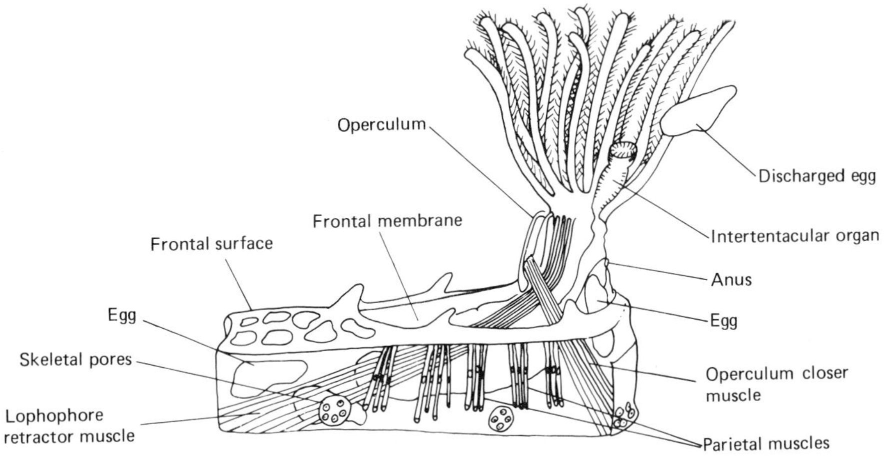
Operculum
Discharged egg
Frontal membrane
Intertentacular organ
Frontal surface
Anus
Egg
Egg
Skeletal pores
Operculum closer muscle
Lophophore retractor muscle
Parietal muscles

(c)

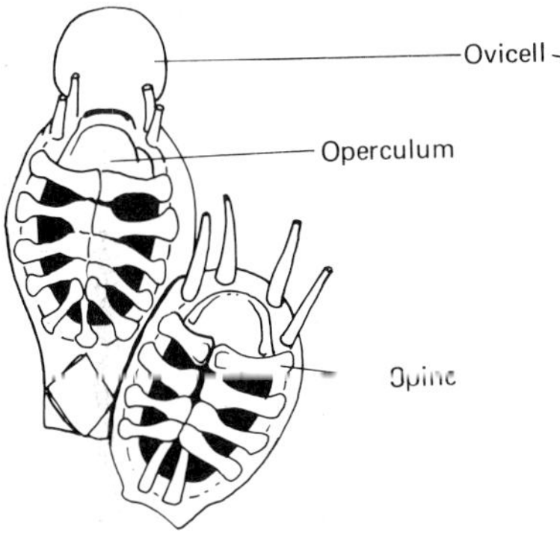
Ovicell
Operculum
Spine

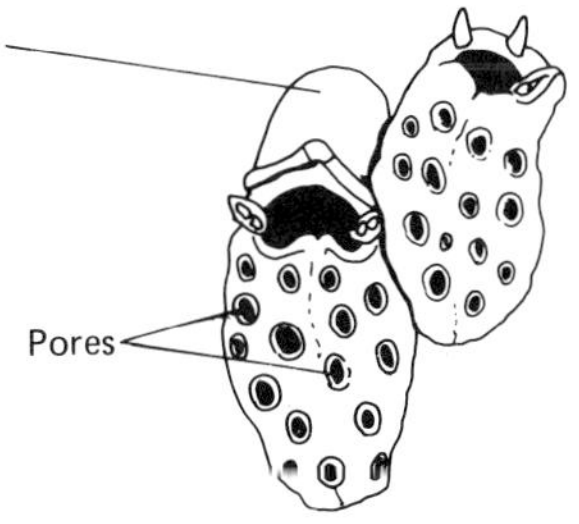
Pores

(e)

(d)

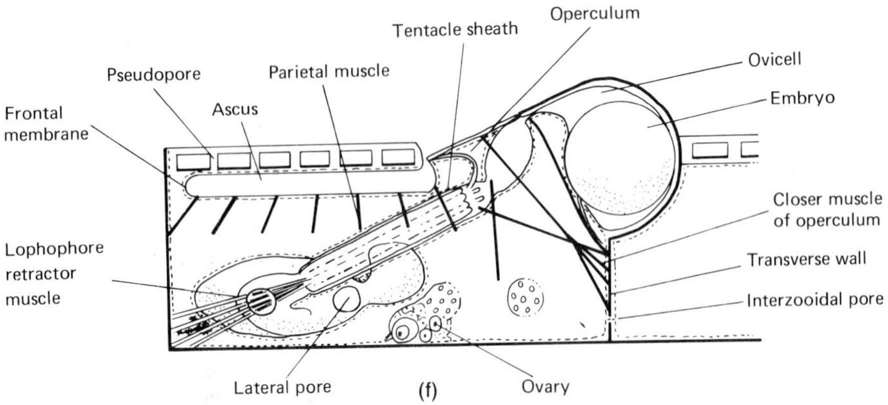
Operculum
Tentacle sheath
Ovicell
Pseudopore
Parietal muscle
Embryo
Frontal membrane
Ascus
Closer muscle of operculum
Lophophore retractor muscle
Transverse wall
Interzooidal pore
Lateral pore
Ovary

(f)

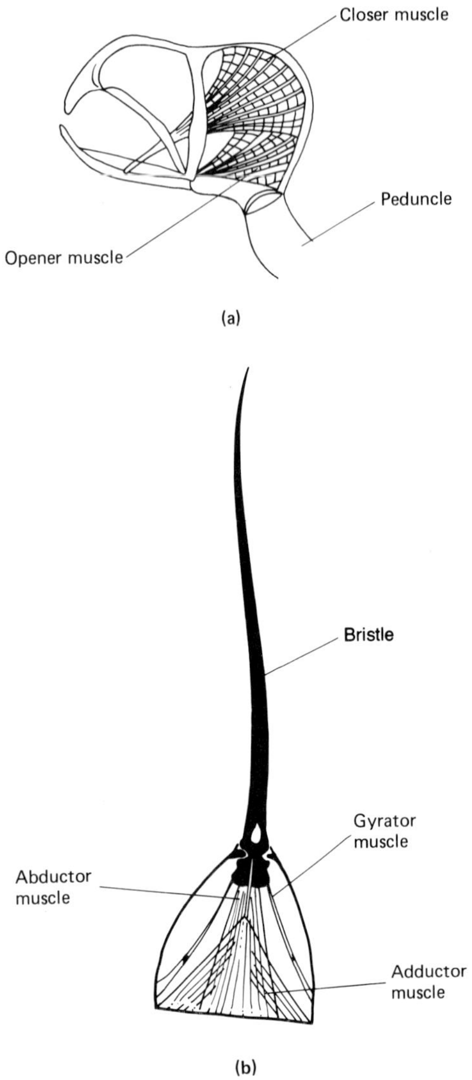

Figure 15.12 Cheilostome heterozooids. (*a*) A pedunculate avicularium of *Bugula;* (*b*) Transverse section through a vibraculum; (*a* from R.W. Hegner and J.G. Engemann, *Invertebrate Zoology*, 1968, Macmillan; *b* from Ryland, *Bryozoans*.)

allied with the pseudocoelomate phyla. Recently, a close entoproct–bryozoan relationship has been expounded again, and some authorities consider entoprocts as a bryozoan subphylum. We tentatively list them as a separate phylum and discuss them in this chapter for comparative purposes.

About 90 entoproct species have been described. Except for one small genus, they are limited to marine environments, where they attach themselves to rocks, shells, and other hard substrata in shallow coastal areas. Some species are commensal.

Individual entoprocts are less than half a centimeter long. Their body consists of a central, boat-shaped **calyx** and a slender stalk. The calyx houses the internal organs and supports a ring of hollow tentacles (Fig. 15.13). Within this ring, a depression called the vestibule contains the mouth and anus. Thus, unlike the arrangement in bryozoans, the entoproct anus is surrounded by the ring of tentacles. A U-shaped digestive tract comprises a short muscular esophagus, a bulbous stomach, an ascending intestine, and a rectum. Often, the anus is perched atop an anal cone. Entoprocts are ciliary feeders. Tracts of cilia along the tentacular grooves route organic particles and small planktonic organisms to the mouth. Ciliary action in the digestive tract circulates nutrients, and stomach glands initiate extracellular digestion. Absorption takes place in the stomach and the intestine.

The small area between the vestibule and the digestive loop is the site of the protonephridial, nervous, and reproductive systems. Paired protonephridia, each with a single flame bulb, terminate through a common nephridiopore near the mouth. A large ganglion in this region innervates the stalk and calyx and dispatches nerves to the tentacles. Bristles on the tentacles and at the outer margins of the calyx are the major sensory structures.

Most entoprocts are diecious. Simple, paired gonads above the stomach are connected by ducts to a gonopore beside the nephridial outlet. Fertilization occurs in the ovary, and developing embryos may be protected briefly within a part of the vestibule called the brood chamber (Fig. 15.13). Cleavage is spiral and probably determinate. A ciliated trochophore-like larva hatches and, after a brief free-swimming existence, attaches and undergoes adult metamorphosis.

Entoproct colonies arise by asexual reproduction. New individuals usually bud from a creeping stolon, as in *Mysoma* (Fig. 15.14a). Solitary entoprocts, such as *Loxosomella* (Fig. 15.14b), reproduce asexually by budding from the calyx.

Entoprocts were isolated from the Bryozoa because of their cleavage pattern, the location of their anus, and their pseudocoel. Recent studies of development, including the identification of a trochophore-like larva, suggest that entoprocts may be closely related to the bryozoans after all.

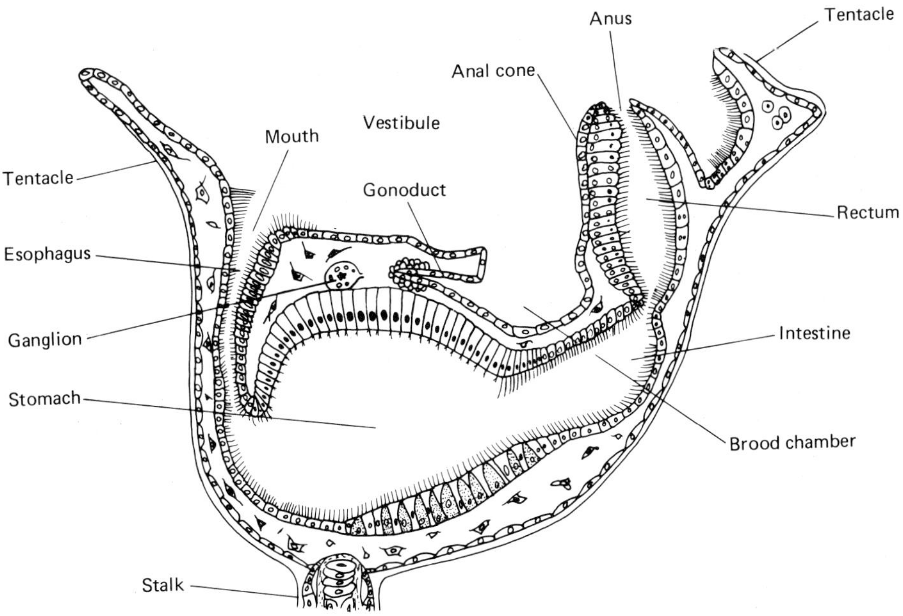

Figure 15.13 A sagittal section through an entoproct (*Pedicellina*) featuring the histology of the digestive tract. (From L.H. Hyman, *The Invertebrates, Vol. III: Acanthocephala, Aschelminthes and Entoprocta*, 1951, McGraw-Hill.)

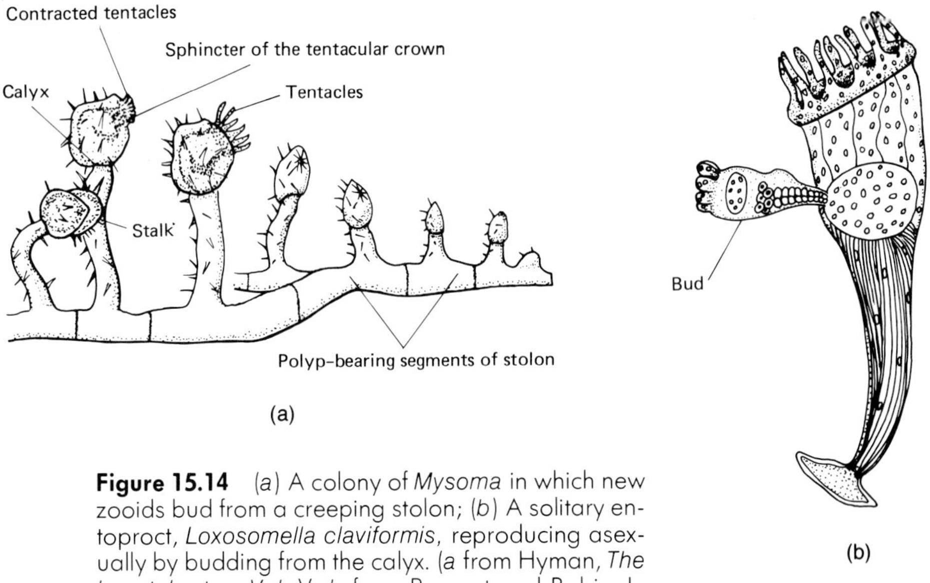

Figure 15.14 (*a*) A colony of *Mysoma* in which new zooids bud from a creeping stolon; (*b*) A solitary entoproct, *Loxosomella claviformis*, reproducing asexually by budding from the calyx. (*a* from Hyman, *The Invertebrates, Vol.* V; *b* from Prenant and Bobin, In *Faune de France, Vol. 60.*)

Phylum Brachiopoda

The brachiopods, also known as lamp shells, compose a phylum of lophophorates with calcareous bivalved shells. Because of their external appearance, they once were classified with the mollusks, whom they only superficially resemble. Brachiopod valves are dorsal and ventral (rather than lateral, as in the bivalve mollusks) and enclose a much different body form. This phylum is wholly marine, and most of the 280 living species inhabit colder waters along the continental slopes. Most brachiopods, including the New England *Terebratulina septentrionalis* and the Pacific coast *Terebratalia transversa,* attach themselves to hard bottoms. Some genera, such as *Glottidia* and *Lingula* (Fig. 15.15), live in vertical tunnels in soft substrata.

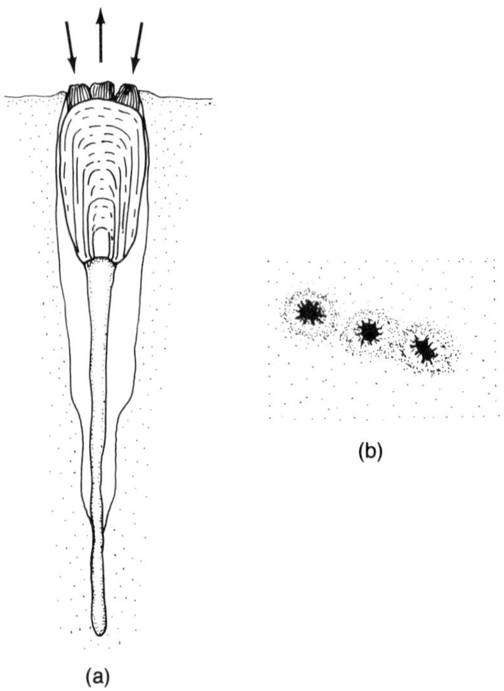

Figure 15.15 (*a*) The brachiopod *Lingula* in feeding position in its burrow. Arrows show direction of water currents. When disturbed the shell is withdrawn by contraction of the pedicle. (*b*) A *Lingula* burrow with its typical tripartite opening as seen from the surface of the substratum. (From M.J.S. Rudwick, *Living and Fossil Brachiopods,* 1970, Hutchinson.)

Thirty thousand fossil species, many of which lived in warmer seas, are known from the Paleozoic and Mesozoic eras. The decline of the brachiopods has not been explained, but some zoologists theorize that they simply have not competed well with the bivalve mollusks.

Brachiopod shells are usually rather drab in color, but sculpted surfaces, often with concentric growth rings, are common (Fig. 15.16a). The ventral valve frequently is larger than the dorsal valve and, at the posterior hinge line, may curve upward as a lamplike "spout" (Fig. 15.16b). Two types of hinges occur in this phylum. Inarticulate brachiopods, such as *Lingula* (Fig. 15.15a), lack a true hinge, and their valves are bound together only by muscles. Articulate brachiopods possess interlocking teeth and sockets which form a true hinge (Fig. 15.16c).

Some brachiopods cement their shells directly to the substratum, but in most species, a stalk, or **pedicle,** intervenes. The pedicle may be long and muscular, as in *Lingula,* which can withdraw deep into its burrow. Other brachiopods, including most articulate forms, possess shorter stalks of connective tissue. In some species, such as *Laqueus californicus* (Fig. 15.16a), the stalk arises through the spout of the ventral valve.

Brachiopod valves are secreted by an underlying mantle. This body wall layer surrounds the internal organs, including the anterior lophophore (Fig. 15.17). Along the gape of the valves, the mantle often bears spines, which may protect against fouling. The lophophore fills most of the anterior two-thirds of the shell chamber. In brachiopods, this organ comprises two lobes, each of which bears a row of tentacles and a brachial groove along its base (Fig. 15.18a, b, c). Brachiopod evolution has witnessed adaptive trends toward increased looping and/or spiraling of the lophophoral lobes (Fig. 15.18d). Such lobes present greater surface areas for food acquisition and respiration.

Like other lophophorates, brachiopods are filter feeders. With the valves open, tentacular cilia draw water across the lophophore along well-defined incurrent and excurrent paths (Fig. 15.18a and c). Small organic particles are transferred to the brachial grooves for transport to the mouth. Inarticulate brachiopods have a complete gut featuring an esophagus, a stomach with a digestive gland, an intestine, and an anus. Articulates have a blind intestine and thus no anus (Fig. 15.17a); their lophophores are highly developed filters allowing ingestion of only the tiniest food particles.

A heart over the stomach gives rise to anterior and

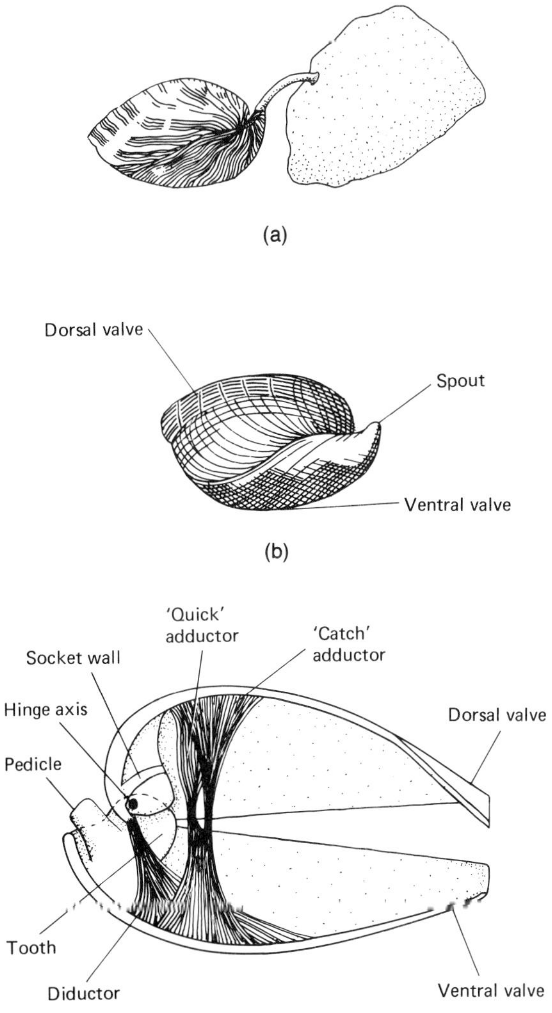

Figure 15.16 (*a*) *Laqueus californicus*, a brachiopod whose shell shows the typical concentric growth rings; (*b*) Side view of *Hemithyris psittacea* showing the curving spout of the ventral valve; (*c*) Sagittal section of *Waltonia*, an articulate brachiopod, featuring the shell hinge and its musculature. Most of the soft parts have been omitted. The "quick" adductor snaps the shell shut in response to various sensory stimuli. The "catch" adductor reacts more slowly, but can hold the shell tightly closed for long periods. (*a* from G.E. MacGinitie and N. MacGinitie, *Natural History of Marine Animals*, 1968, McGraw-Hill; *b* from Hyman, *The Invertebrates, Vol. V*; *c* from Rudwick, *Living and Fossil Brachiopods*.)

posterior channels that branch into the body sinuses. In this open system blood circulation is slow, so coelomic fluid probably plays a major role in oxygen transport. Certain coelomocytes containing hemerythrin aid in oxygen delivery. Both the mesocoel within the lophophore and the metacoel within the mantle are well suited for respiratory exchange. Ciliated mantle canals branch from the metacoel (Fig. 15.17b) and contribute to internal transport. Brachiopods have one or two pairs of metanephridia. Metacoel wastes, often packaged in coelomocytes, are drained by nephrostomes (Fig. 15.17a); nephriopores flank the mouth.

Like other sessile and sedentary invertebrates, brachiopods have rudimentary muscular and nervous systems. Paired diductor and adductor muscles open and close the valves (Fig. 15.16c and 15.17); they include both fast and slow fibers. Pedicle muscles allow some rotation of the animal, and burrow-dwelling types like *Lingula* can withdraw deeply into protective retreats. Ventral ganglia dominate a nerve ring surrounding the esophagus. Nerves lead outward to the body organs and a subepidermal plexus. At least one species of *Lingula* has a statocyst, but, in other brachiopods, known sensory sites are limited to spines and sensitive cells along the mantle margins.

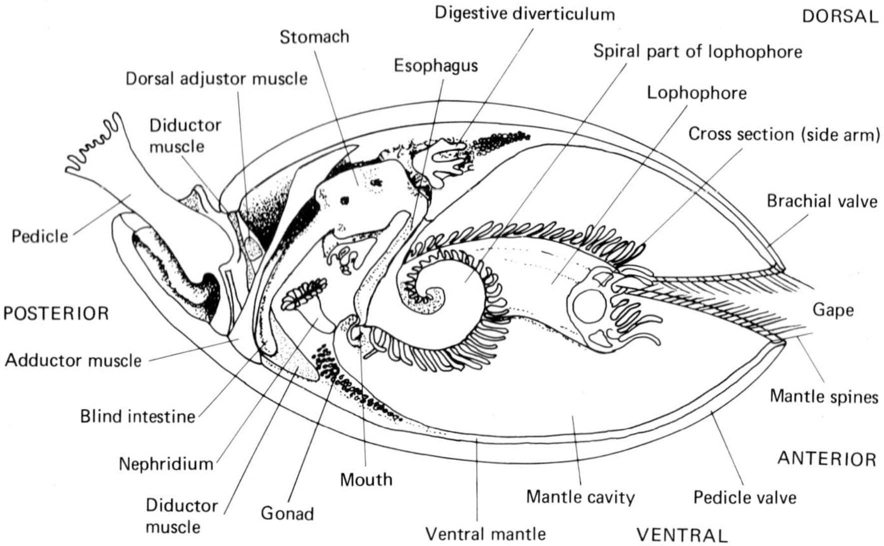

(a)

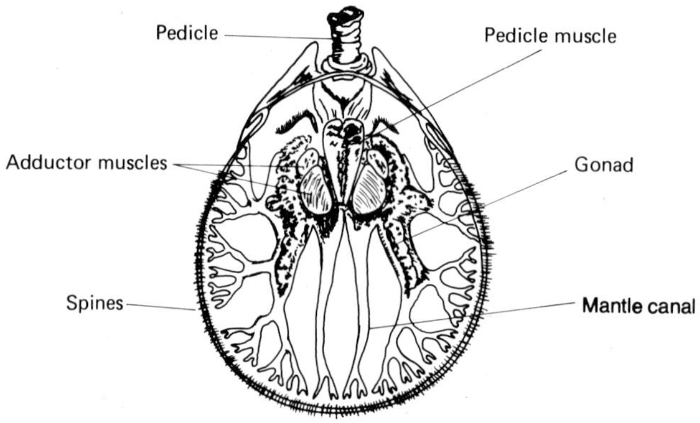

(b)

Figure 15.17 (*a*) Major anatomical features of *Terebratulina*, an articulate brachiopod; (*b*) Ventral valve of *Macandrevia*. (*a* from A. Williams and A. J. Rowell, In *Treatise on Invertebrate Paleontology, Part H*, R. C. Moore Ed., 1965, Geological Society; *b* from Hyman, *The Invertebrates, Vol. V.*)

Brachiopods are diecious. Gametes are released from gonadal areas along the peritoneum of the metacoel (Fig. 15.17) and exit through the nephridia. Fertilization is generally external, and brooding is rare. Brachiopod embryogeny displays definite deuterostomate characters, including radial cleavage and modified enterocoely. The larvae of inarticulate brachiopods possess a locomotor lophophore and a shell-secreting mantle (Fig. 15.19a). Their free-swimming existence ends when, because of their growing shells, they sink to the bottom. The larvae of articulate species have three lobes: one each for the developing lophophore, mantle, and pedicle (Fig. 15.19b). When settlement occurs, the mantle lobe surrounds the lophophoral area and shell secretion begins.

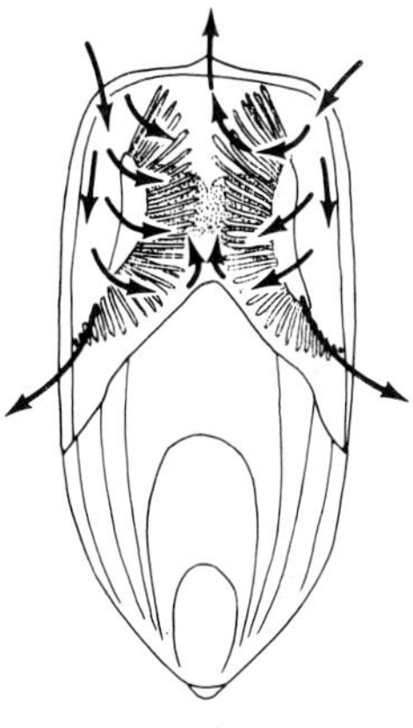

(a)

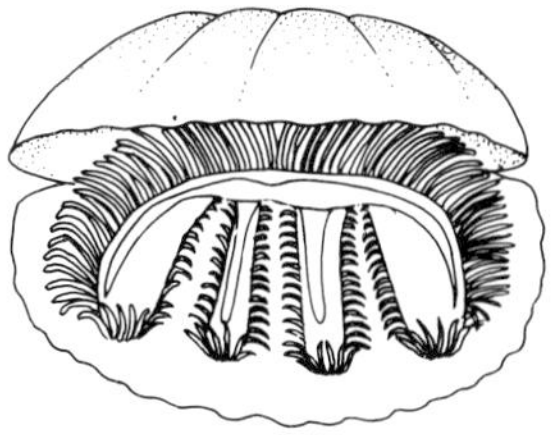

(b)

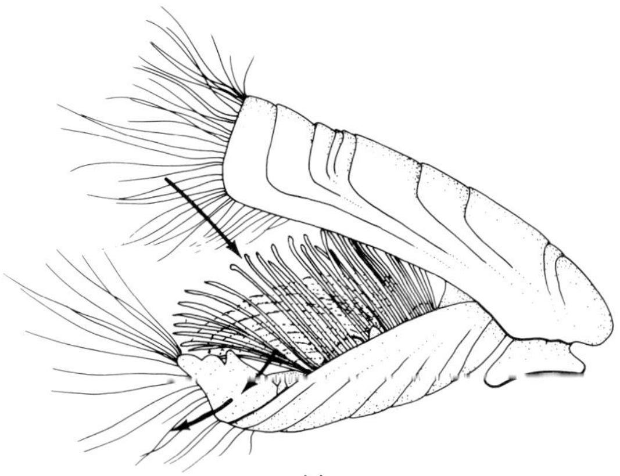

(c)

Figure 15.18 (*a*) Dorsal view of *Lingula*; (*b*) Dorso-frontal view of the quadrilobed lophophore of *Megathiris*; (*c*) Lateral view of *Pumilus antiquatus* in feeding position. Arrows show the direction of the water current; (*d*) The pattern of lophophore evolution in brachiopods shown schematically. Tentacles are sown as circles. Smaller circles represent younger tentacles. The primitive type (1) gains additional tentacles by becoming bilobed (2). Further modification in this direction produces type (3) from which have evolved both the highly spiraled lophophore (4) and the horseshoe-shaped and spiraled lophophore (5). (*a* and *d* from P. de Bcauchamp, In *Traitó dc Zoologic, Vol. V*, Grassé, Ed; *b* from Williams and Rowell, In *Treatise on Invertebrate Paleontology, Part H*, Moore, Ed; *c* from D. Atkins, after P. de Beauchamp, In *Traité de Zoologie Vol. V*, Grassé, Ed.)

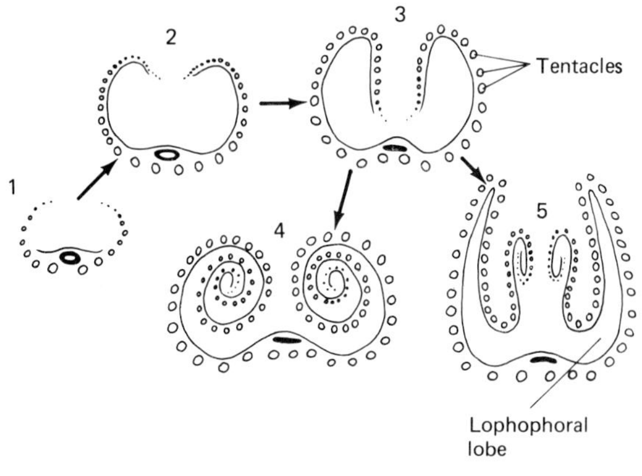

(d)

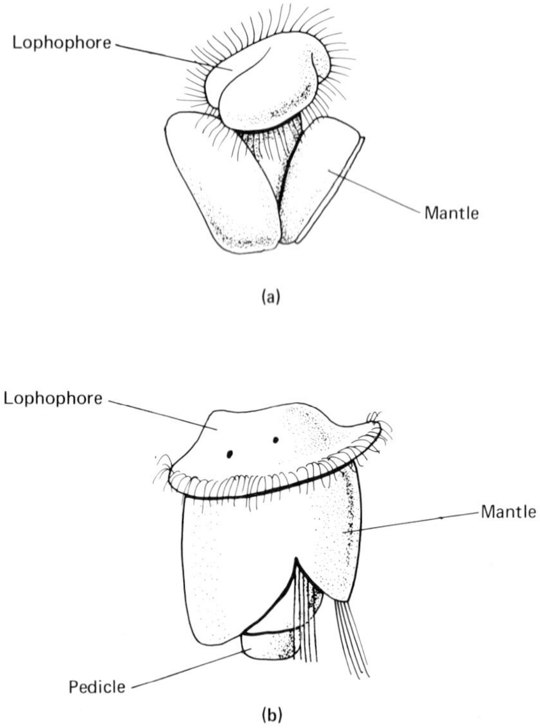

Figure 15.19 (*a*) Larva of *Lingula*, an inarticulate brachiopod; (*b*) Larva of *Argyrothera cordata*, an articulate brachiopod. (*a* from N. Yatsu, J. Coll. Sci., Univ. of Tokyo, 1902, *17*(4); *b* from H. Plenk, Arbeit. Zool. Inst. Univ. Wein, 1913, *20*.)

Lophophorate Evolution

Lophophorates are curious animals, whose phylogenetic position is not known. Even the relationships among the lophophorate phylum are not well defined, although the lophophore and the divided coelom strongly suggest a common ancestry for the entire group. Many authorities consider the phoronids ancestral to the bryozoans, while brachiopods probably developed independently. Phoronids, in turn, could have arisen from wormlike creatures which gradually assumed a tube-dwelling, filter-feeding life. With the perfection of this life-style, the head became reduced, some body systems were simplified, and a protective covering developed. Similar adaptations have occurred in many sessile and sedentary invertebrates. Differences between the lophophorate phyla can be explained in part by their development of discrete types of exoskeletons.

Embryological factors remain a key focus of lophophorate research. The unusual development of these animals suggests possible relationships with echinoderms, hemichordates, and other deuterostomes. Perhaps lophophorates and deuterostomes arose from similar stocks of primitive eucoelomates. Some authorities have even suggested that ancient lophophorates gave rise directly to the first echinoderms (see Chap. 16) and the pterobranch hemichordates (see Chap. 17).

for Further Reading

Banta, W. C. Body wall morphology of Reteporellina evelinae. *Am. Zool.* 17: 75–91, 1977.

Banta, W. C., and Carson, R. Bryozoa from Costa Rica *Pac. Sci.* 31: 381–424, 1977.

Boardman, R. S. and Cheetham, A. H. Skeletal growth, intracolony variation, and evolution in Bryozoa: a review. *J. Paleontology* 43: 205–233, 1969.

Cook, P. L., and Chimonides, P. J. Morphology and systematics of some rooted cheilostome Bryozoa. *J. Nat. Hist.* 15: 97–134, 1981.

Cooper, G. A. *Brachiopods from the Caribbean Sea and Adjacent Waters.* Studies in Tropical Oceanography No. 14. University of Miami Press, Coral Gables, 1977.

Emig, C. C. The systematics and evolution of the phylum Phoronida. *Z. Zool. Syst. Evolutionsforsch.* 12: 128–151, 1974.

Emig, C. C. Embryology of Phoronida. *Amer. Zool.* 17: 21–37, 1977.

Farmer, J. D., Valentine, J. W., and Cowen, R. Adaptive strategies leading to the ectoproct groundplan. *Syst. Zool.* 22: 233–239, 1973.

Fishcher, A. G. Brackish oceans as the cause of the Permo-Triassic marine faunal crisis. In *Problems in Paleoclimatology* (A.E.M. Nairn, Ed.), pp. 566–574, 1964. (Discusses reasons for the rapid decline of the brachiopods and other marine invertebrates)

Gilmour, T. H. K. Ciliation and function of the food-collecting of lophophorates. *Can. J. Zool.* 56: 2142–2155, 1978.

Gosner, K. L. *Guide to Identification of Marine and Estuarine Invertebrates: Cape Hatteras to the Bay of Fundy.* Wiley-Interscience, New York, 1971. (Pages 222–248 describe the lophophorates in this region)

Gutman, W. F., Vogel, K., and Zorn, H. Brachiopods: biochemical interdependencies governing their origin and phylogeny. *Science* 199: 890–893, 1978.

Hayward, P. J. Cheilostomata (Byozoa) from the South Atlantic. *J. Nat. Hist.* 14: 701–722, 1980.

Hughes. R. L., Jr., and Woollacott, R. M. Photoreceptors of bryozoan larvae. Zool. Scr. 9: 129–138, 1980.

Hyman, L. H. *The Invertebrates, Vol. V: Smaller Coelomate Groups.* pp. 228–609. McGraw-Hill, New York, 1959.

Jebram, D. Laboratory diets and qualitative nutritional requirements for bryozoans. *Zool. Anz.* 205: 333–334, 1980.

Larwood, G. P. *Living and Fossil Bryozoa: Recent Advances in Research.* Academic Press, New York, 1973.

MacGintie, G. E., and MacGinite, N. *Natural History of Marine Animals.* McGraw-Hill, New York, 1968.

MacKay, S., and Hewitt, R. A. Ultrastructural studies on the brachiopod pedicle. *Lethaia* 11: 331–339, 1978.

McCammon, H. M. and Reynolds, W. A. (Organizers) Symposium: Biology of the Lophophorates. Am. Zool. 17: 3–150, 1977. (Papers from a 1975 symposium on many aspects of lophophorate biology, including embryogeny life cycles, body wall morphology, and entoproct-bryozoan relationships)

Mukai, H., and Oda, S. Comparative studies on the statoblasts of higher phylactolaemate bryozoans. *J. Morphol.* 165: 131–156, 1980.

Nielsen, C. Entoproct life cycles and the entoproct/ectoproct relationship. *Ophelia* 9: 209–341, 1971.

Nielsen, C. The relationship of Entoprocta, Ectoprocta and Phoronida. *Am. Zool.* 17: 149–150, 1977.

Pennak, R. W. *Fresh-Water Invertebrates of the United States,* 2nd ed. Wiley-Interscience, New York, 1978.

Rider, J., and Cowen, R. Adaptive architectural trends in encrusting ectoprocts. *Lethaia* 10: 29–41, 1977.

Rogick, M. D. Bryozoa. In *Freshwater Biology* (W. T. Edmondson, H. B. Ward, and G. C. Whipple, Eds.), 2nd ed., pp. 495–507. John Wiley and Sons, New York, 1959.

Rudwick, M. J. S. *Living and Fossil Brachiopods.* Hutchinson University Library, Hutchinson, London, 1970.

Ryland, J. S. *Bryozoans.* Hutchinson University Library, Hutchinson, London, 1970.

Ryland, J. S. Physiology and ecology of marine bryozoans. *Adv. Mar. Biol.* 14: 285–443, 1976.

Steele-Petrovic, H. M. Brachiopod food and feeding processes. *Paleontology* 19: 417–436, 1976.

Winston, J. E. Polypide morphology and feeding behavior in marine ectoprocts. *Bull. Mar. Sci.* 28: 1–31, 1978.

Woollacott, R. M., and Zimmer, R. L. (Eds.) *Biology of Bryozoans.* Academic Press, New York, 1977.

16

The Echinoderms

Patently symbolic of marine life, the echinoderms represent some of the most celebrated and curious denizens of the invertebrate world. Approximately 6500 living species of sea stars, sea urchins, sea cucumbers, and their relatives belong to this fascinating phylum. With the echinoderms, we begin our account of the deuterostomate line in animal phylogeny. Recall that deuterostomes are distinguished by certain peculiarities in their embryogeny (see Chap. Three); the group includes the phylum Chordata, which embraces *Homo sapiens* and all other vertebrates. How and to what extent echinoderms share a common evolutionary history with humans remains one of the great puzzles of invertebrate zoology.

Echinoderms are also famous and puzzling in their own right, for their pentamerous radial symmetry and curious coelomic systems are unique among invertebrates. Their unusual symmetry caused historical confusion regarding their relationship with the coelenterates and ctenophores, and not until the middle of the past century were echinoderms properly distinguished from the radiate phyla.

To be sure, echinoderms are quite different from coelenterates. The present phylum is more complexly organized. Its members possess a true mesodermal coelom and some well-developed organ systems. Echinoderms, however, do display a lack of specialization in certain aspects of their biology. Circulatory, osmoregulatory, sensory, and reproductive organs are rudimentary. Much of our discussion focuses on this rather peculiar assortment of evolutionary advancement. The echinoderm coelom likewise receives major attention, as it both illustrates the deuterostomate character of the phylum and contributes to a unique water–vascular, or **ambulacral** system. Other outstanding features of the group include a calcareous endoskeleton and an array of bilaterally symmetrical larval forms.

Echinoderms are distributed throughout the oceans

and are particularly abundant in the Indo-Pacific. They have never invaded freshwater, terrestrial, or parasitic habitats.

Echinoderms present a long and auspiciously detailed fossil history. Consideration of space will limit our discussion of the 20,000 fossil members of the phylum, as we concentrate instead on the five major groups of living echinoderms. These groups are the class Crinoidea (sea lilies and feather stars); the subclasses Asteroidea (sea stars) and Ophiuroidea (brittle stars), both of which belong to the class Stelleroidea; the class Echinoidea (sea urchins and sand dollars); and the class Holothuroidea (sea cucumbers).

Echinoderm Unity

Echinoderms display an array of shapes and postures, as well as some diversity of internal anatomy and physiology. A sea cucumber does not advertise its kinship with the sea stars; the closely allied sea urchins and sand dollars likewise disguise their relationship. Therefore, our account of the unity of the Echinodermata is necessarily quite generalized and frequently requires mention of the many exceptions to the rules. Nevertheless, echinoderms remain a discrete group, demonstrating several unique characters upon which their phyletic unity is based. Recognizing that these characters are modified to suit the specialized biologies of the individual classes and subclasses, we outline them here as a general portrait of echinoderm life.

MAINTENANCE SYSTEMS

General Morphology

Compared to most invertebrates, echinoderms are rather large animals. As adults, they are never microscopic, ranging upward from several millimeters in diameter, and a few members have achieved great size. Certain Paleozoic crinoids had stalks 25 m in length, but meter-long sea cucumbers and sea stars are the largest modern forms.

The most striking aspect of echinoderm morphology is their pentamerous radial symmetry, which has been acquired secondarily. Early echinoderms no doubt were bilaterally symmetrical, as are the present-day larvae. As we shall see, modern echinoderm larvae undergo a metamorphosis that transforms their original bilateralism into the radial symmetry of the adult. Such a transformation is unparalleled in the animal world. Many fossil echinoderms were sessile animals; their radial symmetry apparently was an adaptation to a sessile or semi-sessile life-style. Radial symmetry, however, was never perfectly developed in the phylum, and the more active modern echinoderms (such as sand dollars) tend toward the restoration of adult bilateralism.

While the radial symmetry of the echinoderms is easily explained, their pentamerous nature is more puzzling. Why do key structures appear so consistently in fives or multiples of five? Why not threes, fours, or sixes? Currently, the best answer to this question involves a brief lesson in skeletal architecture. One of the first evidences of radial symmetry in a developing echinoderm is the appearance of a ring of skeletal plates about the anus (Fig. 16.1). Sutures between these plates present very weak structural lines; accordingly, the skeleton is strongest when the number of skeletal plates is small. Moreover, with an odd number of plates, opposing sutures never lie along a single structural line. A final requirement dictates that enough plates be present to create a radial form. Apparently, the number best fulfilling all of these requirements is five; thus, the pentamerous radial symmetry of the echinoderms.

As other radially symmetrical animals, echinoderms have no head and their major body parts are arranged concentrically about an oral–aboral axis. Characteristically, the echinoderm body is divided into five specialized areas (Fig. 16.2). These areas may be defined by distinct arms or rays, as in the stelleroids and crinoids, or may represent tracts along the body surface, as in echinoids and holothuroids. Each zone is

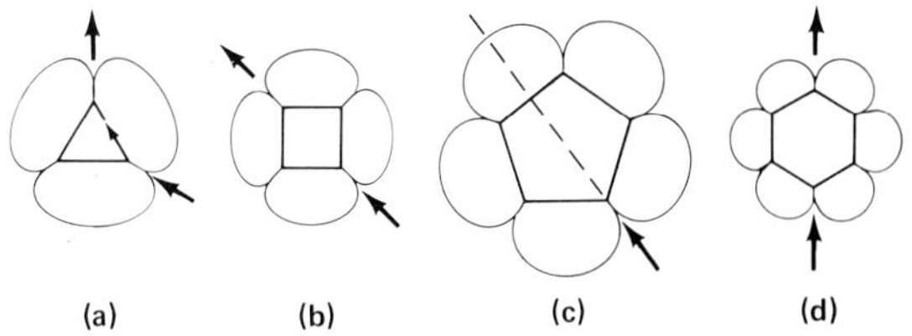

Figure 16.1 Diagrams suggesting a reason for the establishment of pentamerous symmetry in echinoderms. (*a*) With three skeletal plates a line of weakness exists (shown by the arrows) and the number of plates is insufficient to create a radial form; (*b*) With four plates, opposing sutures are in line creating two planes of weakness; (*c*) With five plates, no suture is in line with any other, hence there is no existing plane of weakness; (*d*) With six plates, opposing sutures are in line creating three planes of weakness. (For additional explanation, see text.) (From D. Nichols, *Echinoderms*, 1966, Hutchinson.)

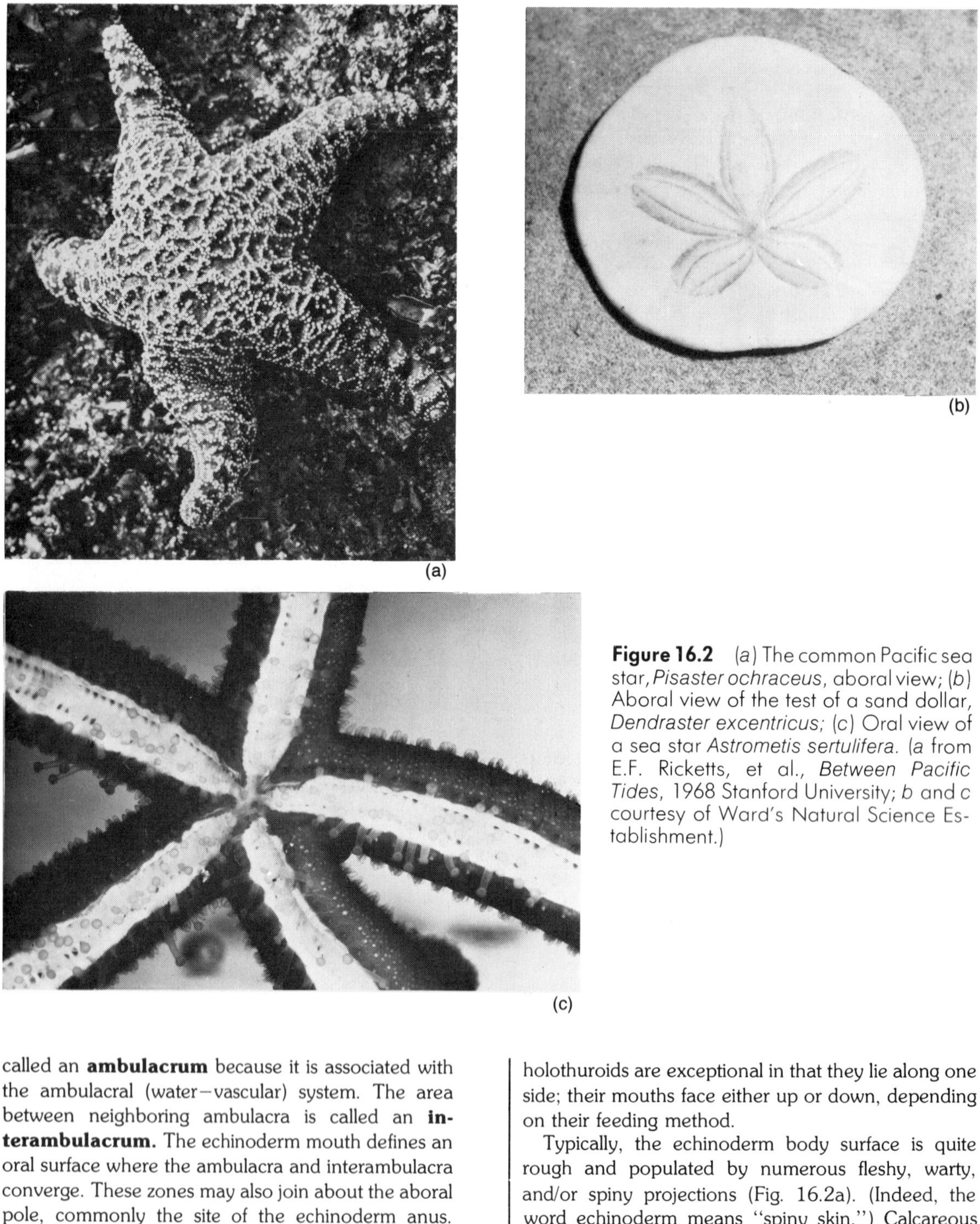

Figure 16.2 (*a*) The common Pacific sea star, *Pisaster ochraceus*, aboral view; (*b*) Aboral view of the test of a sand dollar, *Dendraster excentricus*; (*c*) Oral view of a sea star *Astrometis sertulifera*. (*a* from E.F. Ricketts, et al., *Between Pacific Tides*, 1968 Stanford University; *b* and *c* courtesy of Ward's Natural Science Establishment.)

called an **ambulacrum** because it is associated with the ambulacral (water–vascular) system. The area between neighboring ambulacra is called an **interambulacrum.** The echinoderm mouth defines an oral surface where the ambulacra and interambulacra converge. These zones may also join about the aboral pole, commonly the site of the echinoderm anus. Crinoids orient with their oral surfaces directed upward, but most other members of the phylum apply the mouth directly to the substratum. The holothuroids are exceptional in that they lie along one side; their mouths face either up or down, depending on their feeding method.

Typically, the echinoderm body surface is quite rough and populated by numerous fleshy, warty, and/or spiny projections (Fig. 16.2a). (Indeed, the word echinoderm means "spiny skin.") Calcareous extensions of the endoskeleton may occur as short entrenched knobs or long movable spines. **Papulae** and **podia** are fleshy appendages with a variety of functions; the latter represent the most distal elements of the ambulacral system and, as outlined subsequently, participate in several aspects of echinoderm physiology. Perhaps the most specialized external projection is a **pedicellaria** (Fig. 16.3). Typically, this small structure comprises calcareous jaws mounted atop a muscularized extension of the body wall. The jaws articulate on the distal end of the organ, which can bend and grasp local debris and/or food. Thus the pedicellariae help to maintain a clean body surface and may be important in food acquisition as well. Certain sea stars possess unstalked pedicellariae, which act as pincers (Fig. 16.3b).

The echinoderm body wall comprises a usually flagellated epidermis, a thick connective dermis which secretes and houses the endoskeleton, variable muscular layers, and a peritoneum bordering the coelom (Fig. 16.4). The epidermis is often overlaid by a thin cuticle which extends over the entire external surface, including the spines and other body projections. Situated among the epidermal cells are mucous gland cells and neurosensory elements. Epidermal flagella produce currents for sanitation, ventilation, and other functions. A subepidermal nerve plexus extends beneath the entire layer.

Within the dermis, the echinoderm endoskeleton consists of numerous calcareous units called ossicles. Apparently, each **ossicle** is a complex of tiny, uniformly arranged crystals of magnesium calcite ($CaCO_3$ + $MgCO_3$) and other mineral salts. Echinoderm ossicles exhibit a variety of sizes and shapes, including rods, disks, and irregular forms (Fig. 16.5). They may occur singly within the connective matrix of the dermis or may be variously bound. Binding by collagen fibers is pronounced in the central disk of the ophiuroids and among the echinoids, whose plate-shaped ossicles are united in a rigid theca, or test.

The body wall musculature reflects the nature of the endoskeleton, particularly the articulation of its major ossicular units. In echinoderms with rigid tests (the echinoids), this musculature may be absent. The rather flexible asteroids possess outer circular fibers and inner longitudinal ones, while in ophiuroid arms, specialized longitudinal muscles join large, linearly arranged ossicles. Finally, holothuroids possess microscopic ossicles in a pliant body wall; as we might expect, their muscle layers are well developed.

The echinoderm coelom comprises four systems of internal tubes, sinuses, and cavities, each of which is specialized for one or more functions: (1) the perivisceral coelom, (2) the ambulacral system, (3) the hemal system, and (4) the perihemal system.

The perivisceral coelom is the largest of the four systems and is comparable with the secondary body cavity of other eucoelomate animals. It surrounds the digestive and reproductive tracts and extends into the arms of the asteroids and crinoids. Echinoderm coelomic fluid is virtually identical to sea water, but contains phagocytotic coelomocytes which serve a variety of maintenance functions.

The ambulacral system represents this phylum's major adaptive innovation. Basically, this system is a network of ciliated or flagellated canals through which fluid is transported to hydraulically active tube feet, or podia. The following description is based primarily on the best-known sea stars, but is generally applicable to the phylum. The system's external opening occurs at the **madreporite**, a sievelike plate which occupies an

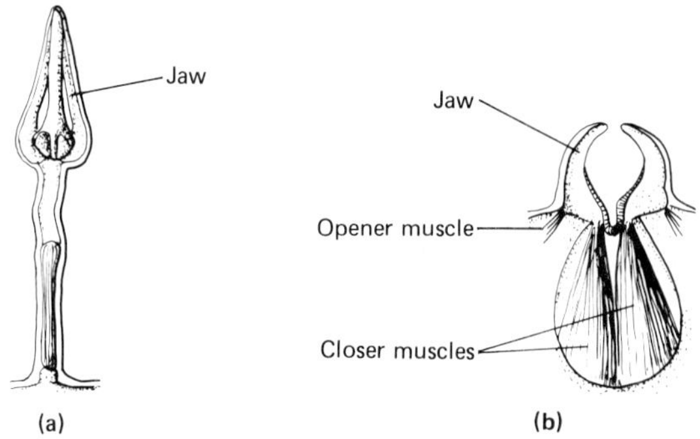

Figure 16.3 (*a*) A stalked pedicellaria from the sea urchin, *Echinus esculentus;* (b) Section through a stalkless pedicellaria from a sea star, *Hippasteria plana.* (From L. Cuénot, In *Traité de Zoologie, Vol. XI,* P.P. Grassé, Ed., 1948, Masson et Cie.)

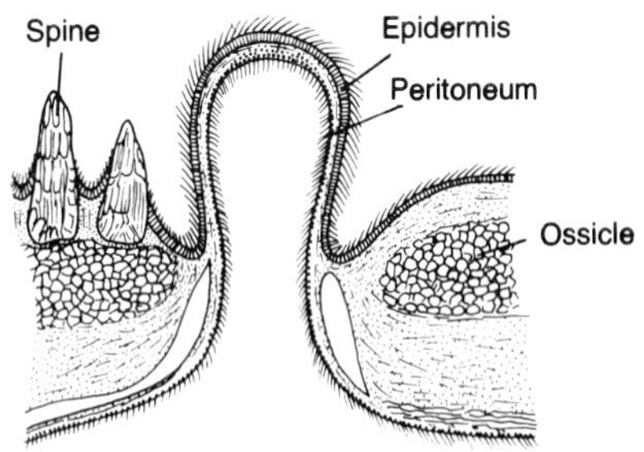

Figure 16.4 A section through the body wall of a sea star passing through a papula. (From Cuénot, In *Traité de Zoologie, Vol. XI,* Grassé, Ed.)

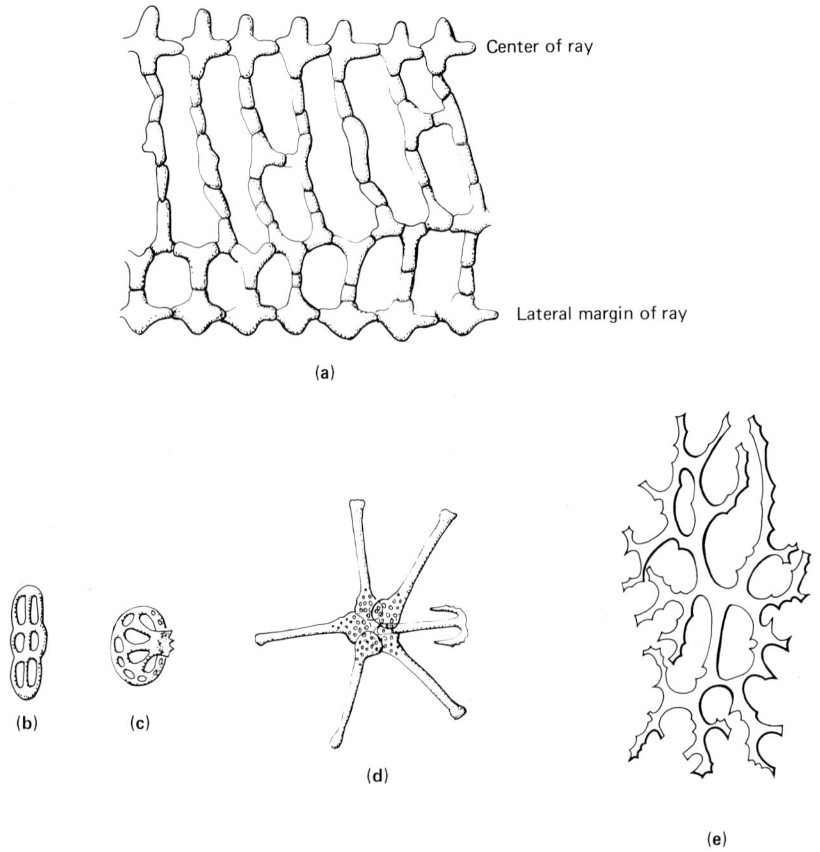

Figure 16.5 (*a*) Arrangement of the endoskeletal ossicles in the arm of a sea star; (*b–e*) Various ossicle types from sea cucumbers (Holothuroidea). (*a–d* from L.H. Hyman, *The Invertebrates, Vol. IV: Echinodermata*, 1955, McGraw-Hill; *e* from D.M. Raup, In *Physiology of Echinodermata*, R.A. Boolootian, Ed., 1966, Wiley-Interscience.)

aboral interambulacrum (Fig. 16.6). (The exact position of the madreporite varies throughout the phylum; see the following discussion of the various classes.) Sea water percolates down through the flagellated furrows of the madreporite and is collected in numerous tiny pore canals which join to form a single **stone canal.** The stone canal, named because of its calcareous reinforcement, descends to an oral **water ring.** The water ring surrounds the esophagus. Extending from this ring, five **radial canals** define the ambulacral zones. Smaller extensions of the water ring include four or five pairs of folded outgrowths called **Tiedemann's bodies** and one to five (or more) larger, muscularized sacs called **polian vesicles.** The former, which are peculiar to the sea stars, may produce coelomocytes. Polian vesicles are general in this phylum and may be reservoirs for excess fluid within the ambulacral system.

Radial canals give rise along their lengths to short lateral canals bearing the tube feet, or podia. These distinctive organs are composed of a proximal muscularized bulb called an **ampulla** and the extensible foot proper (Figs. 16.6 and 16.7). Characteristically, the entire structure can be isolated from its lateral canal by a closed valve. So isolated, the podium represents an independent hydraulic unit. The several functions of this hydraulic organ—respiratory, locomotor, nutritional, sensory—are described subsequently in the appropriate sections.

Two parallel networks of channels and sinuses apparently function in internal transport: the so-called hemal and perihemal systems. Neither is well described for any member of the phylum, and much controversy concerns their activities. Further discussion is reserved for the account of echinoderm circulation.

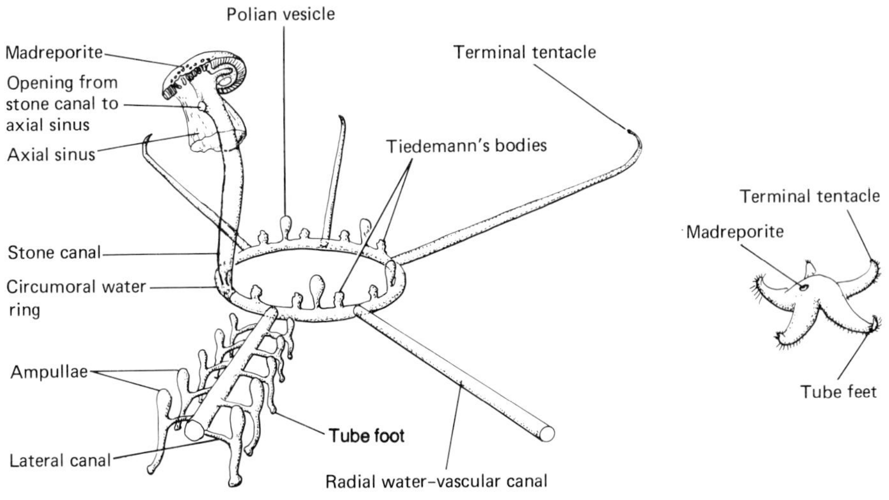

Figure 16.6 Diagram of the water-vascular of a sea star. The small sea star on the right shows those parts of the system visible externally. (From D. Nichols, In *Physiology of Echinodermata*, Boolootian, Ed.)

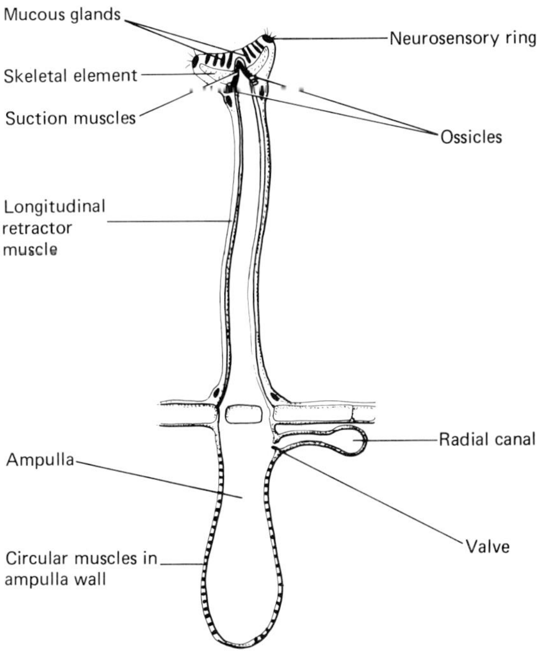

Figure 16.7 Longitudinal section through a sucker-bearing tube foot of a sea urchin (*Echinus*). (From Nichols, *Echinoderms*.)

Nutrition and Digestion

Echinoderm diets and feeding styles rival those of the mollusks in terms of their diversity. Suspension feeders, deposit feeders, grazers, raspers, scavengers, and predators are all well represented in this phylum. Generally speaking, most echinoderms eat anything they can ingest. Ravenous appetites have been described for many members of the phylum; apparently, an echinoderm that consumes a significant percentage of its tissue weight daily is quite typical.

Suspension feeding is probably this phylum's most primitive method of nutrition. Indeed, the entire ambulacral network may have evolved as a food-gathering mechanism. Crinoids and stelleroids still channel food to the mouth along ambulacral tracts. Such food transport involves the muscular activity of the podia, mucous entrapment, and flagellar currents. In all echinoderms, oral podia are involved in food tasting. In holothuroids, these tube feet form tentacles, which may trap suspended organic material or may penetrate soft substrata to gather detritus. The most sophisticated food-gathering system in this phylum belongs to the echinoids. Many members of this class possess a chewing apparatus known as **Aristotle's lantern** (Fig. 16.37), which effectively rasps organic material off hard surfaces and may fell tall kelp plants.

The echinoderm digestive tract is rather simple. It is a complete tube, except in the ophiuroids, which have no intestine or anus. The mouth usually occupies the center of the oral surface; there, the ambulacra and interambulacra converge. A muscular peristomial membrane surrounds the oral aperture, which leads into a short esophagus. In most echinoderms, the rest of the digestive system includes a long, frequently coiled stomach–intestine, a rectum, and the anus.

Circulation

Echinoderms have no heart and no true blood vessel system. Internal transport needs are met by their extensive and complex coelom, particularly by its hemal and perihemal components. Ciliary or flagellar action continuously circulates coelomic fluids, as does contraction of the peristomium and body wall muscles.

The hemal system itself varies from class to class, but its essential features include oral and aboral hemal rings, which communicate via a spongy mass of tissue known as the **axial gland** (Figs. 16.6 and 16.8). This gland is surrounded by the **axial sinus.** Authorities cannot agree on the function of the axial gland, but reports of a pulsating vessel within the structure suggest a circulatory role. The axial gland parallels the stone canal of the ambulacral system. Along this gland, hemal elements arise in association with the digestive tract. From the oral ring, hemal branches extend radially along the ambulacra, and comparable branches of the aboral ring communicate with gonadal sinuses.

The perihemal system, including the axial sinus, surrounds the hemal system and may augment its circulatory function.

The sinuses and channels of the hemal and perihemal systems are unwalled and communicate indirectly with the walled chambers of the perivisceral coelom. Their fluids, although sometimes referred to as blood, do not seem distinct from coelomic fluid. Hemal and perihemal fluids contain a variety of wandering coelomocytes. These multipurpose maintenance cells are implicated in numerous physiological processes, including the phagocytosis and transport of nutrients and particulate wastes, food storage, oxygen transport, enzyme release, clot formation, the synthesis of various pigments and connective tissues, and the secretion and deposition of calcium carbonate.

Respiration

The first echinoderms were rather small, sessile or sedentary animals, whose respiratory rates were quite low. These ancestral forms established the basic body plan of the phylum, including its calcareous endoskeleton. Such a plan minimized surface respiratory exchange and impeded the internal circulation of oxygen and carbon dioxide. For echinoderms to increase in size and pursue more active life-styles, special adaptations were necessary to ensure adequate gas transport. Today, these adaptations vary from class to class and thus attest to the relatively late and makeshift nature of this phylum's respiratory program.

Gas exchange occurs wherever internal body fluids are brought near the surrounding sea water, in areas such as body wall evaginations, sunken pouches, and respiratory tubules. The most common respiratory surface is the podium. This evagination of the ambulacral system probably did not evolve primarily as a respiratory organ, but its thin wall and fluid-filled lumen are exploited in gas exchange by all echinoderms. Podia represent the major respiratory surfaces of crinoids and handle up to 50% of the gas exchange in sea stars. Sea stars also possess numerous coelomic evaginations called **papulae** (Fig. 16.4). These structures, which occur at interruptions in the skeletal framework, present only a thin layer of peritoneum and its overlying epidermis to the surrounding sea. Both layers are ventilated by flagellar action.

Other echinoderms have evolved more sophisticated respiratory structures. Brittle stars possess 10 respiratory sacs called **bursae** (Fig. 16.30), and sea

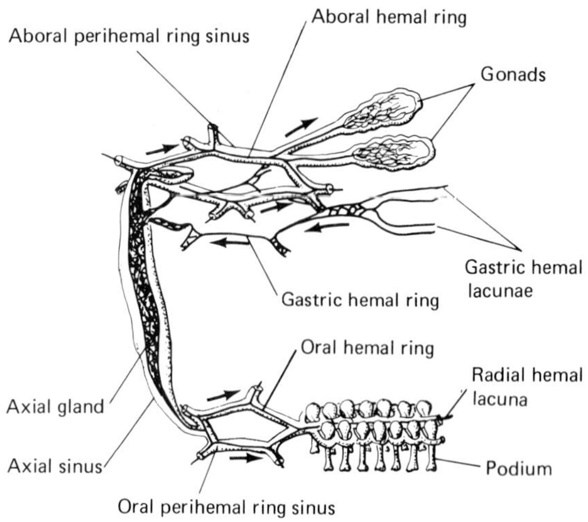

Figure 16.8 Schematic diagram showing the major portion of the hemal and perihemal systems of a sea star in isolation from other body structures. (From G. Ubaghs, In *Treatise on Invertebrate Paleontology, Part S,* R.C. Moore, Ed., 1967, Geological Society.)

urchins have bushy peristomial gills (Fig. 16.33a). These thin-walled structures surround the mouth and are ventilated by flagellar action and/or the pumping of the oral musculature. Finally, the most complicated respiratory system in this phylum belongs to the holothuroids. These animals pump water through paired, branching tubules which ramify throughout their bodies (Fig. 16.43). Called **respiratory trees**, these tubules are associated with the hemal system. Holothuroid hemal fluid contains **hemocytes**, specialized hemoglobin-containing coelomocytes which transport oxygen.

Most echinoderms demonstrate a rather passive respiratory attitude. They simply gear their oxygen consumption to local conditions, metabolizing more rapidly when well supplied with oxygen and more slowly when oxygen is scarce. Only the holothuroids strive to maintain a consistent concentration of internal oxygen. Under stress from low oxygen tensions in the immediate environment, these animals accelerate the pumping of their respiratory trees and recruit the surfaces of oral tentacles and the general body itself for respiratory exchange.

Excretion and Osmoregulation

Echinoderms are quite unspecialized in their excretory and osmoregulatory physiology. They are largely ammonotelic, but may excrete urea, uric acid, and various other organic wastes as well. Soluble wastes are removed across any convenient, thin-walled surface—podia, papulae, bursae, gills, respiratory trees, and/or the madreporite. Insoluble wastes are engulfed by coelomocytes. These wandering cells then exit across any of the aforementioned surfaces. Also, some solid wastes may pass into the gut lumen and leave with the feces.

Echinoderms have virtually no powers of osmoregulation. No nephridium regulates the osmotic composition of their coelomic fluids. Hence (with few ionic variations) these fluids are identical to sea water. Obviously, osmotic responsibility devolves upon the individual tissues, which expend a great deal of energy to maintain osmotic balance. Because of their inability to osmoregulate on the coelomic level, echinoderms are absolutely restricted to marine habitats.

ACTIVITY SYSTEMS

The functional radial symmetry of the echinoderms precludes their development of a head and the neurosensory and behavioral complexity that cephalization normally involves. Accordingly, echinoderms are rather sedentary or only slow-moving animals, whose responsiveness is limited both in kind and amount. Muscular development reflects body form, but nervous and sensory elements always remain rudimentary.

Muscles

The echinoderm musculature includes a number of independently operating units: the muscles of the body wall, the podial fibers, and the many short muscles which operate specialized effectors, such as spines and pedicellariae. Circular and longitudinal body wall muscles are well developed in asteroids and holothuroids. These animals produce general body flexions, and some holothuroids can even burrow. Such body wall fibers are absent from the rigid echinoids, which depend primarily on their spines for locomotion. Echinoid muscles include an elaborate system which operates Aristotle's lantern (Fig. 16.37). In both ophiuroids and crinoids, the arms are composed of articulating ossicles. These ossicles are bound by elastic ligaments and a system of antagonistic muscles which produce serpentine arm movements.

The most characteristic effector organ in this phylum is the podium, an ambulacral appendage which operates by an antagonistic musculature and a contained volume of watery fluid (Fig. 16.7). As described previously, the podium and its proximal bulb, the ampulla, are separated from a lateral ambulacral canal by a valve. The closing of this valve isolates a volume of fluid within the podium. The contraction of circular muscles in the walls of the ampulla forces the fluid into the tube foot proper. The tough dermal walls of the podium prevent radial expansion, so the organ elongates. Longitudinal muscles in the podium contract to shorten the foot, forcing water back into the ampulla and thus stretching the latter's circular fibers. Differential contraction of longitudinal muscles on opposite sides of the foot causes bending movements, while muscles on the terminal disk may produce suction. The tube feet may pull the echinoderm along by alternate contractions of their circular, terminal, and longitudinal muscles, or the podia may step forward when longitudinal muscles alternately contract along an extended foot.

Nerves

Echinoderms have a diffuse nervous system which, although more highly organized, is somewhat reminiscent of the nervous apparatus of the coelenterates. Many neurons are associated with the epidermis, where sensory and motor pathways direct the podia, spines, pedicellariae, and other surface effectors. A radial central nervous system comprises up to three nerve rings and their associated nerve cords (Fig.

16.10). These rings never approximate a brain and may represent condensations of the peripheral nerve plexus. The oral nerve ring dominates in most echinoderms. This circular or pentagonal center lies immediately beneath the oral epidermis, where it surrounds and innervates the foregut. From the oral ring, five radial nerves extend out along the ambulacra. These radial nerves include both sensory and motor elements; they contact the subepidermal plexus and service the tube feet. As elaborations of the body wall, the tube feet also are innervated by the subepidermal plexus and commonly bear terminal nerve rings (Fig. 16.7).

A deep oral nerve ring is present in most members of the phylum. Its predominantly motor fibers parallel those of the oral system and innervate the body wall musculature as well as the podia. The third, or aboral, nerve ring usually is limited to a diffuse network associated with the anus and gonads. In the crinoids, however, the aboral ring is very well developed.

Sense Organs

Rather slow and tolerant creatures, echinoderms lack elaborate sensory structures. In these headless animals, true sense organs are limited to clusters of photosensitive elements and certain balance centers. Most worldly data are collected by scattered sensory cells, which communicate with the subepidermal nerve plexus. These cells are concentrated on the podia, the ambulacral margins, the spines, and the pedicellariae. Chemosensitive cells are particularly numerous around the mouth and anus. In addition, the subepidermal neurons themselves may be sensitive to outside events, particularly to the impingement of light on the general body surface.

At the end of each asteroid arm is a small tentacle with a cluster of photosensitive cells at its base (Fig. 16.9). Often, these cells occupy a pigment cup, and they may be associated with a lens. The only other sense organs currently described for the echinoderms are the statocysts of certain holothuroids and the echinoid sphaeridia, both of which are gravity receptors.

Behavior

With few exceptions, echinoderms are inhabitants of the ocean floor, where they sit or slowly crawl about. Their behavioral repertoires vary from class to class and are dominated by responses to touch, gravity, and light.

The echinoderm body surface is tactile. With the notable exception of the crinoids, most members of the phylum maintain oral contact with the substratum; tactile elements on the podia and other surface structures ensure that proper posture is maintained. If the animal is overturned, these oral sensors respond to the absence of surface contact. Muscles are signaled and various righting responses follow. In some upsidedown sea stars, for example, one or more arms are flexed aborally until their outermost podia contact the substratum. Podial steps pull the flexed arms further beneath the central body until the animal flops over into its normal stance.

Touch plays a greater role than gravity in determining posture. The majority of echinoderms will orient in any position, as long as maximum surface

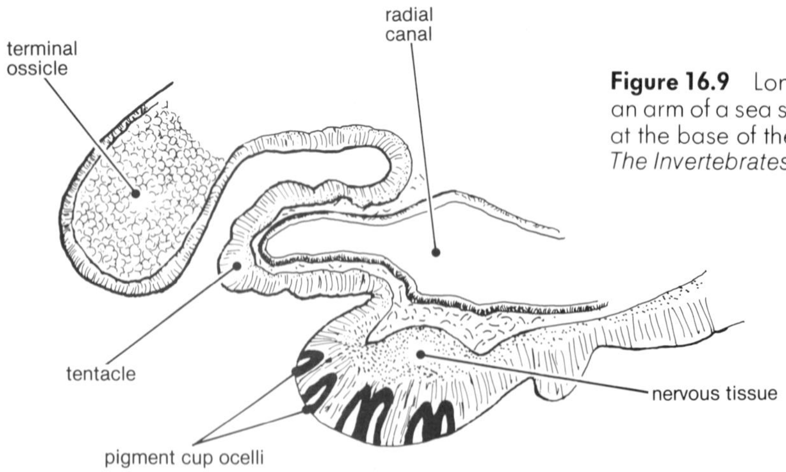

Figure 16.9 Longitudinal section through the tip of an arm of a sea star showing the pigment cup ocelli at the base of the terminal tentacle. (From Hyman, *The Invertebrates, Vol. IV.*)

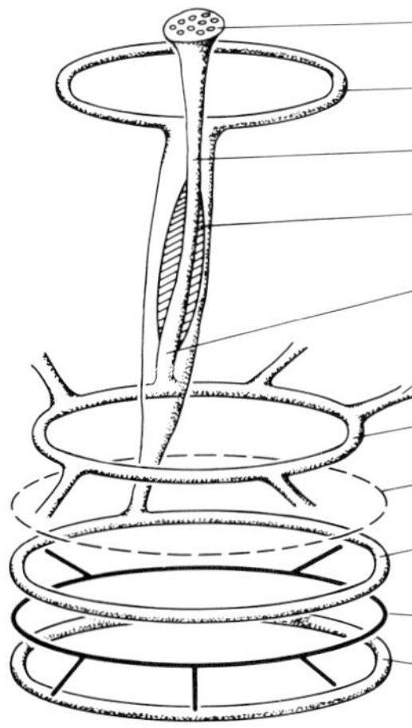

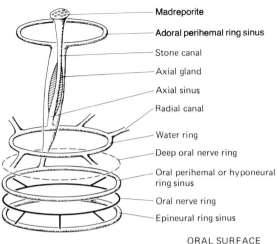

Figure 16.10 A diagram of the relative positions of coelomic ring sinuses (perihemal ring sinuses), water ring, and nerve rings in an echinoderm. (The hemal system has been omitted.)

contact is made. Indeed, it is not unusual to observe feather stars attached upside down to overhanging rocks or even projecting from a rock at right angles to the line of gravity (Fig. 16.11). Gravity does play a major orientational role, however, in the lives of burrowing echinoderms. These individuals, most of which are irregular echinoids (sand dollars and heart urchins) and holothuroids, depend on gravitational clues to direct their digging through soft substrata.

Echinoderm responsiveness to light is variable. Generally, these animals avoid strong illumination, but some species may be attracted by low-intensity light.

Because most echinoderms are radially symmetrical but not sessile, there may be problems regarding the direction of their movements: For example, how does a sea star determine which of its several arms will lead in locomotion? How do its hundreds of podia, distributed in separate fields, step in unified fashion? Complete answers for these questions are not available, but certain observations suggest a nervous hierarchy directing the locomotion of these animals. In isolated asteroid arms, all podia move in a single direction. If the arm contains some fragment of the oral nerve ring, the podia step toward the arm's distal tip. If no part of the nerve ring remains attached, the podia walk toward the oral end. While no in-phase marching occurs, podial movement nonetheless is coordinated and unidirectional. Such coordination apparently involves the radial nerve. Severing this nerve isolates the podia within an arm and destroys their ability to act in unison. The coordination of all arms in an intact animal is the responsibility of the oral nerve ring. By a process as yet unknown, one arm gains dominance over the others and leads the entire asteroid in its outward direction. Such dominance apparently is localized in the junction of the oral nerve ring and the radial nerve of the dominant arm. In most asteroids, any arm can dominate temporarily; thus most sea stars need not turn before changing direction. (Some reports describe a constantly favored arm in certain individuals; however, any such dominance appears to be a phenomenon of individuals rather than species.) Finally, if the oral nerve ring is cut, coordination among the arms is destroyed. Individuals having undergone these neurotomies may pull themselves apart as their arms step in opposing directions.

Large aggregations of echinoderms occur in many areas of the world. Feather stars, sand dollars, and brittle stars, in particular, are noted for their gregarious behaviors. Such phenomena raise the question of sociality among echinoderms. Do these animals, in fact, respond directly to one another in a social sense? Little

Figure 16.11 A feather star or comatulid (*Comanthus panicinna*) clinging to a vertical rock surface. (Photo courtesy of Dave Hayes.)

current evidence supports the existence of true social behavior in this phylum. Some territorial activity, including actual fighting, has been reported in one sea urchin species. However, aggregational behavior by echinoderms often is interpreted simply as multiple individualistic responses to favorable environmental conditions. Certain aspects of echinoderm reproductive behavior may indicate a prelude to social interaction. These aspects, which include synchronous spawning, are described in later sections.

In the first half of this century, researchers employed a variety of experimental techniques to test whether echinoderms modify their behavior by experience. This research involved escape behaviors, righting movements, and avoidance conditioning. In many experiments, some evidence of learning surfaced: After repeated trials, individual ophiuroids did extricate their arms from traps with greater efficiency; asteroids learned to use previously unfavored arms in righting movements; and various echinoderms discriminated among surface textures and shades. All of the experiments, however, did not support echinoderm learning ability, and the entire question remains unresolved. Acceptance of the idea of learning ability in echinoderms, or any other invertebrates, depends largely on a personal definition of learning and the experimental design approved to demonstrate its existence.

CONTINUITY SYSTEMS

Echinoderm continuity systems illustrate both the simplicity and the sophistication of the phylum. Reproductive structures are simple, as are gamete production and fertilization. Echinoderm development, however, is a complex process, which involves a deuterostomate embryogeny, numerous and phylogenetically significant larval forms, and a remarkable metamorphosis.

Echinoderms have served as subjects for countless research projects involving early developmental phenomena. Much of our knowledge about the basic principles of fertilization and early cleavage has been furnished by work on these remarkable invertebrates. This phylum no doubt will continue to be a source of experimental animals, and thus its continuity systems should be considered carefully.

Reproduction

Almost all echinoderms are diecious, although sexual dimorphism is rarely evident in the phylum. Only a few sea cucumbers and some brittle stars are hermaphroditic. The primitive crinoids have no sexual organs; their gametes are simply produced by specialized areas of the coelomic peritoneum. Most echinoderms, however, possess radially arranged gonads; interambulacral gonopores commonly surround the oral or the aboral pole. Sea cucumbers have only one sexual organ.

Most echinoderms simply release their gametes for external fertilization in the sea. Hundreds of thousands may be discharged by each individual. In addition to sheer numbers, several mechanisms ensure adequate fertilization, including the simultaneous maturation of gametes in a local population and aggregational behavior. Generally, temperate-zone echinoderms spawn in the spring and summer months. The synchrony of maturation probably reflects the interaction of neurosecretory complexes with such environmental factors as day length, lunar periodicity, and temperature. Aggregational behavior has been reported for some echinoderms, particularly the echinoids and ophiuroids. Aggregation typically coincides with the spawning period, as numerous individuals congregate in warm shallow waters.

Some echinoderms, notably certain sea cucumbers, brood their fertilized eggs. Holothuroid brooding may occur along the animal's lower surface or even within its coelom. Brooding also has been described in each of the other classes, where ambulacral grooves, bases of spines, or specialized chambers shelter developing young.

Development

Echinoderm developmental programs are definitively deuterostomate, and each class has one or more characteristic larval types. Radial and indeterminate cleavage produces a hollow blastula. Gastrulation by invagination follows. As the archenteron advances across the embryo, lateral pouches separate from its distal end. The cavities of these pouches represent the future echinoderm coelom, and the pouch walls are the precursors of all mesodermal tissues. Hence, the secondary body cavity has an enterocoelous origin and the mesoderm is an entomesoderm. The archenteron eventually fuses with the stomodeum. As in all deuterostomes, the echinoderm mouth arises independently of the archenteron. The anus is formed from, or near the original site of, the blastopore.

Development of the several coelomic systems varies from class to class. Certain common features, however, suggest the embryogeny of ancestral forms. One scheme relates the developmental story as follows: Lateral pouches separate from the archenteron and then elongate parallel to the gut. As these pouches constrict further, they may form up to three separate, paired compartments. Such compartments are called the left and right axocoels, hydrocoels, and somatocoels (Fig. 16.12). The **somatocoels** surround most of the gut, where they give rise to the perivisceral coelom; the borders of these paired cavities join to form visceral mesenteries. The **axocoels** and **hydrocoels** on one or both sides may remain together, thus forming elongate axohydrocoels. Alternatively, they may join apically to form a single horseshoe-shaped anterior coelom.

In some echinoderms, the right axocoel contributes to the madreporite. In most members of this phylum, however, the right-side coeloms are diminutive and indeed may never be fully formed. The left axohydrocoel always dominates in the formation of the ambulacral system. A stone canal develops between the axocoel and the hydrocoel components on the left side (Fig. 16.12b). The left axocoel forms a hydropore on the dorsal surface of the embryo, while the left hydrocoel forms a scalloped loop around the esophagus (Fig. 16.12b,c). This loop represents the beginning of the ambulacral water ring, and its scallops define the sites of radial canal formation.

Concurrent with the development of coelomic sys-

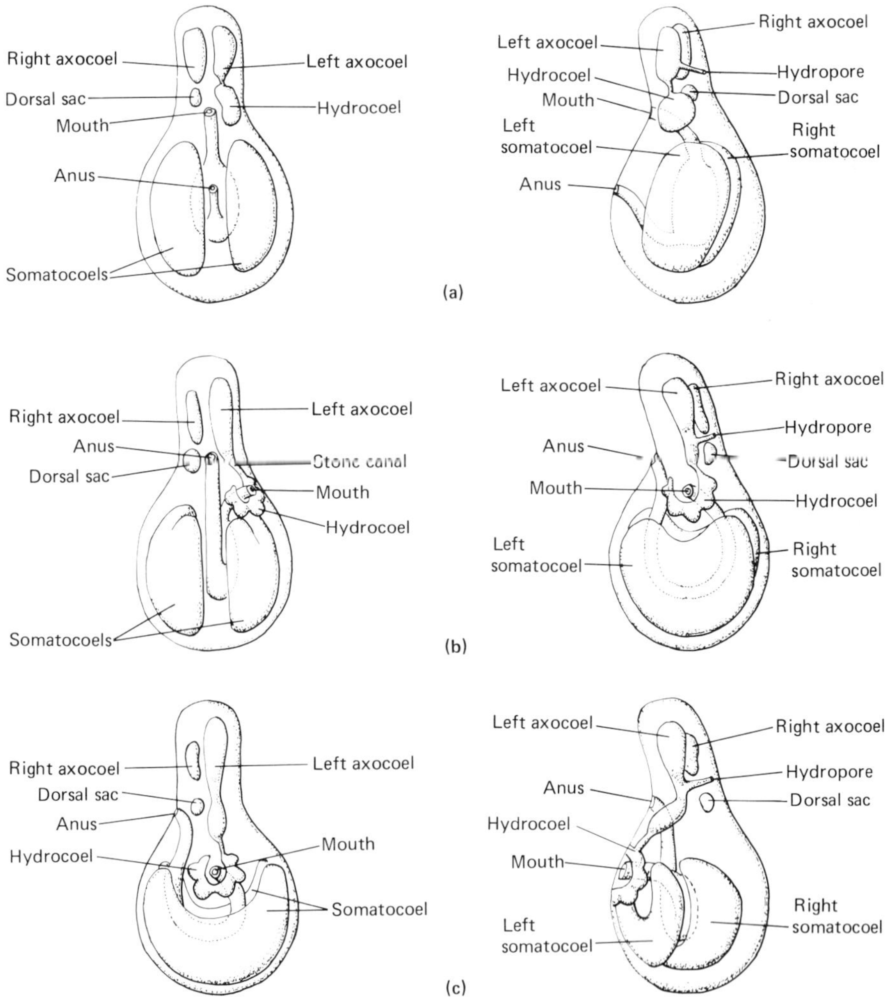

Figure 16.12 (*a–c*) Three stages in the formation of coelomic compartments in an echinoderm larva. The left-hand figures are views from the oral surface, the right-hand figures are views from the left side. (From Ubaghs, In *Treatise on Invertebrate Paleontology, Part S*, Moore, Ed.)

tems, other changes transform the echinoderm embryo into a bilaterally symmetrical, free-swimming larva, the **dipleurula** (Fig. 16.13a). These changes include the completion of a functional gut and the consolidation of surface flagella into distinct locomotor bands. Characteristically, the ventral mouth is located within a depression surrounded by flagellar bands. The larval anus is also ventral, but lies outside this oral depression. In most echinoderms, projections of the body wall form long larval arms, along whose lengths the flagellar bands extend. These larval arms are ephemeral structures and do not contribute to the arms of adult crinoids and stelleroids. They provide buoyancy and locomotion during the planktonic phase of the life cycle. Echinoderm larvae represent an important element in marine zooplankton.

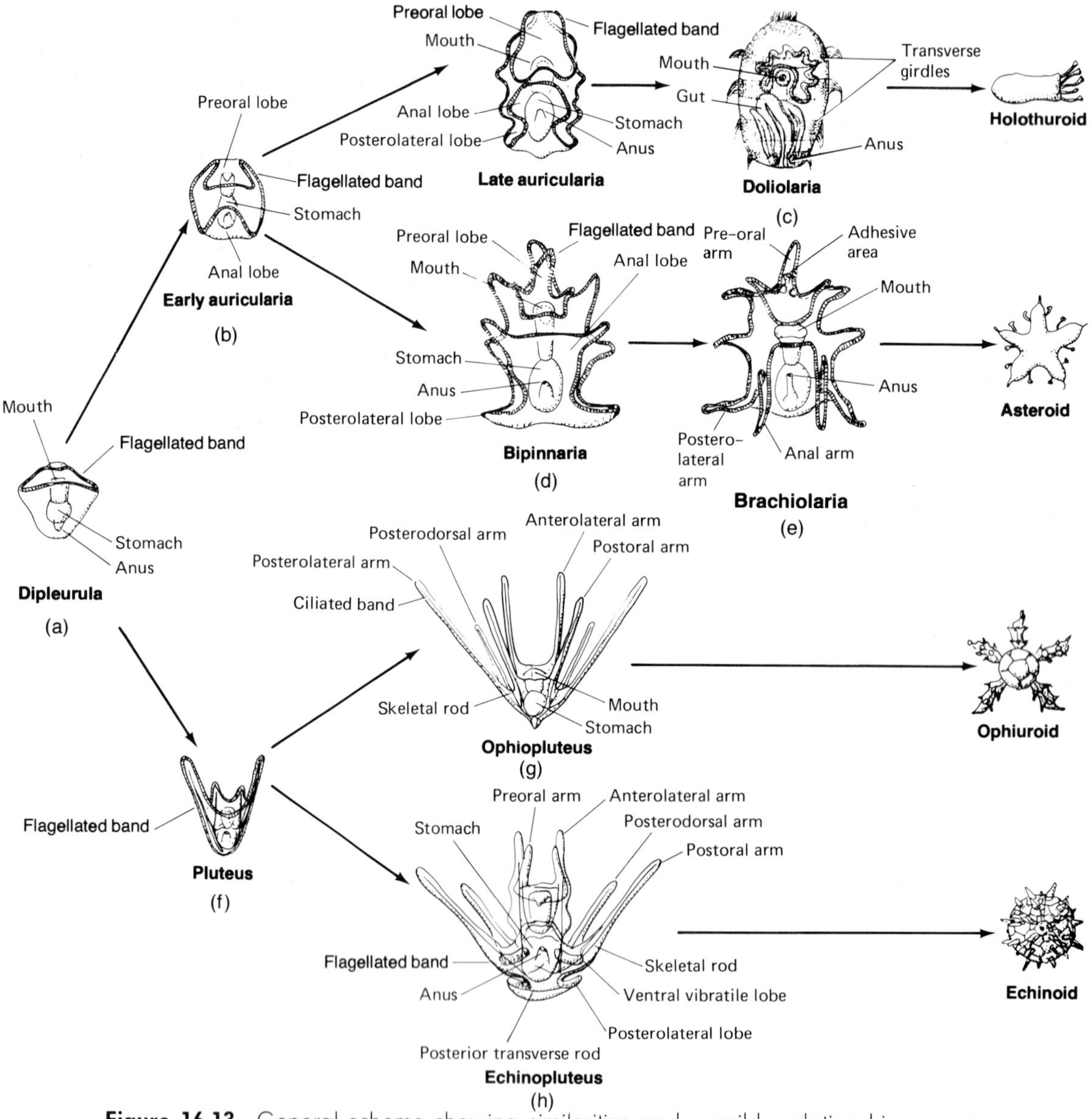

Figure 16.13 General scheme showing similarities and possible relationships among echinoderm larvae. Stylized adult forms much reduced in size. Crinoids are omitted. Letters in figure refer to information in text. (From Ubaghs, In *Treatise on Invertebrate Paleontology, Part S*, Moore, Ed; doliolaria from S. Runnström, Bergens Mus. Aarbok, 1927,*1*)

Larval morphology, especially the arrangement of locomotor tracts along the arms, is peculiar in each of the echinoderm classes. Echinoderm larval stages have been the object of considerable phylogenetic interest, as they may present clues to the relationships among the various echinoderm groups as well as to the phylum's overall place in the animal world.

The most primitive larval form may be the **vitellaria,** which appears in the crinoids and in some holothuroids. This stage is nonfeeding and barrel-shaped and has no arms. Transverse bands of flagella encircle the plump, compact body. In holothuroids, an **auricularia** (Fig. 16.13b) is often the first larval stage; in this form, a single flagellar band surrounds both the mouth and the anus. The holothuroid auricularia becomes a barrel-shaped **doliolaria** (Fig. 16.13c) when the flagella reorganize along three to five transverse girdles.

The first definitive asteroid larva is the **bipinnaria** (Fig. 16.13d). A young bipinnaria resembles the holothuroid auricularia, but within several weeks, it sprouts arms. Asteroids then pass through another larval stage, the **brachiolaria** (Fig. 15.13e). The brachiolaria is characterized by three extra arms on the anterior, ventral surface. These rather stubby projections surround an adhesive area which attaches the asteroid to the substratum during its adult metamorphosis. Ophiuroids and echinoids have similarly shaped, long-armed **pluteus** (Fig. 16.13f) larvae. It is debated whether such similarity evidences a close kinship between the two groups or represents evolutionary convergence. Both larvae are flattened along their future oral surfaces. The **ophiopluteus** (Fig. 16.13g) has four pairs of long flagellated arms; the **echinopluteus** (Fig. 16.13h) has four to six pairs. In both groups, calcareous rods provide skeletal support for these longest of echinoderm larval arms. The pluteus larva requires a developmental period as long as several months.

All echinoderms undergo a rather elaborate metamorphosis, during which larval bilateralism gives way to the radial symmetry of the adult. As the process begins, crinoid vitellaria and asteroid brachiolaria attach themselves to suitable substrata. In contrast, members of other echinoderm groups remain free-swimming throughout most of their metamorphosis, settling only as the growth of the adult skeleton renders them too heavy to remain afloat.

The pattern of metamorphosis varies with the adult body form of the animal involved. Generally, the anterior larval end degenerates, while the middle and posterior regions reorganize in radial fashion (Fig. 16.14). In most echinoderms, the larval left side becomes the adult's oral surface; the right side contributes the aboral surface. Typically, the gut and other internal systems are almost entirely reformed. The greater development of coelomic structures on the left side of the larva effects a temporary asymmetry (Figs. 16.12 and 16.14). Gradually, the left hydrocoel forms the water ring and radial canals grow outward to define the pentamerous ambulacral fields. Other organ systems follow the structural lead of the ambulacral system, and their composite growth determines the adult form. The adult endoskeleton begins with five ossicles which surround the aboral pole. As in other systems, skeletal elements are added radially during the growth of the animal.

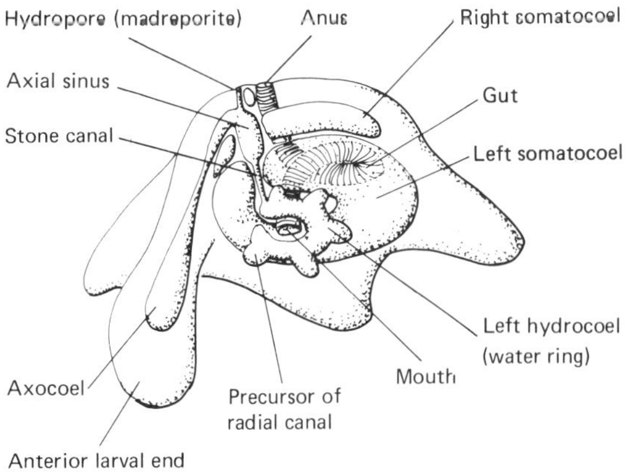

Figure 16.14 A late stage in the metamorphosis of a brachiolaria larva into a juvenile sea star. (From K. Heider, Verh. Deutsch. Zool. Gesel., 1912, *22*.)

Regeneration

Echinoderms are well known for their regenerative powers. Damaged arms, spines, podia, and other structures often are autotomized and replaced by new ones. Coelomocytes evidently play an active role in regeneration. Many of these cells migrate to ruptured areas, where they deliver nutrients and phagocytose damaged tissues.

Specific regenerative capacities vary from class to class and even from genus to genus. Sea stars can duplicate whole arms, and many of these creatures can reform completely from a single arm and as little as one-fifth of their central disk (Fig. 16.15). If the disk fragment bears the madreporite, it can be even smaller and still direct the regeneration of a whole animal. Brittle stars have more limited regenerative powers. Commonly, one or more arms and the entire central disk must remain intact for complete regeneration to

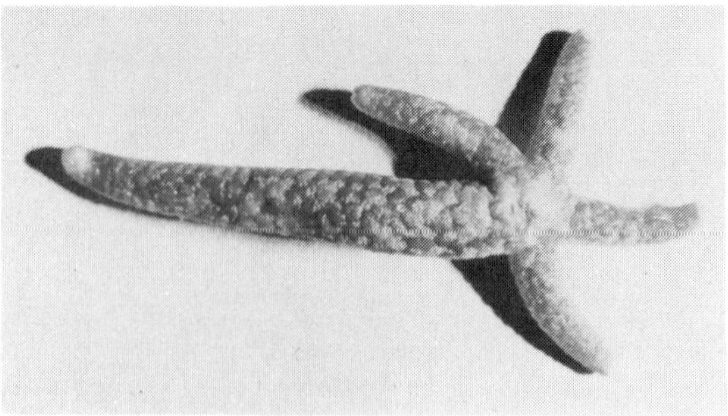

Figure 16.15 A sea star, regenerated from a single arm and one-fifth of the central disk. (Photo courtesy of Ward's Natural Science Establishment.)

occur. Members of both stelleroid subclasses may reproduce by **fissiparity.** Fissiparity involves the active self-division of an echinoderm; the subsequent regeneration of its halves produces two whole animals. Division occurs along a predetermined line between the major ossicles of the central disk.

Holothuroids also regenerate quite well. In at least some types, the cloacal region is a regenerative center. In many species, regeneration is impossible from fragments that do not include part or all of this posterior zone. Echinoids repair spines and broken areas of their tests, but whole-body regeneration has not been described in this class. Finally, the crinoids, as the stelleroids, can replace their arms quite readily. In most forms, however, virtually all of the central disk and at least a pair of arms must remain intact to ensure survival. The crinoid aboral nerve center is essential to the reformation of internal organs.

A Review

Echinoderms display a combination of characters most unlike any other invertebrate phylum. While diversity within its ranks is also quite pronounced, the following traits are general in the group and distinguish these curious creatures from all other animals. Basically, echinoderms are:

(1) marine invertebrates with secondary pentamerous radial symmetry;

(2) animals with a calcareous endoskeleton and a complexly partitioned coelom which includes an ambulacral system;

(3) noncephalized animals, whose nervous and sensory equipment remains quite rudimentary;

(4) predominantly diecious invertebrates with simple sexual systems;

(5) deuterostomes with an indirect development featuring numerous bilaterally symmetrical larval forms and a complex metamorphosis.

Echinoderm Diversity

Echinoderms are an ancient group. Their long and relatively well-documented evolution has produced a diversity of animal forms within the four modern classes. Such diversity centers on variations in morphology (in large, a function of diverse skeletal organization), feeding ecology, behavior, and habitat. Respiratory structures and continuity systems also exhibit interesting variations throughout this phylum. The present account is limited to the extant groups. The more numerous fossil echinoderms are discussed briefly in concluding remarks on the evolution of the phylum.

Beyond the level of class, echinoderm taxonomy is a complex study which concentrates primarily on skeletal detail. Because of limited space, we will not attempt a systematic review of the various orders: The interested student should consult the suggestions for further reading at the end of this chapter.

CLASS CRINOIDEA

The oldest and most primitive living echinoderms belong to the class Crinoidea. Crinoids arose and flourished in the Paleozoic, from which era over 5000 fossil species have been described. During their Paleozoic heyday, sessile crinoids formed dense gardens rising from shallow ocean floors (Fig. 16.16). Today's sessile crinoids, however, include only about 80 species called sea lilies (Fig. 16.17b,c). Most modern members of the class are known as comatulids or feather stars (Fig. 16.18). These sedentary and/or swimming crinoids include about 700 species. Crinoid distribution, as that of many other echinoderm groups, is centered in the Indo-Pacific.

Whether attached or free-swimming, all crinoids exhibit a similar morphological plan. Their pentamerous central body comprises a cuplike, aboral **calyx** and its oral roof, the **tegmen** (Fig. 16.17a). The calyx, which houses the internal organs, is defined by one or more whorls of skeletal ossicles. Aborally, a central ossicle forms the calyx base, where some sort of stem is attached. Among sessile sea lilies, this stem is a

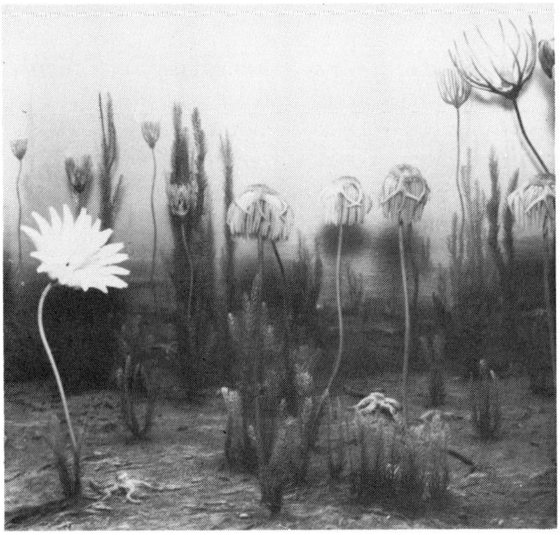

Figure 16.16 Reconstruction of a sea-bottom from Lower Carboniferous times when stalked crinoids were conspicuous. (Photo courtesy of the Smithsonian Institution.)

long-flexible stalk, whose ossicles are stacked in vertical series. Along the stalk, whorls of appendages called **cirri** may be present (Fig. 16.17b). Stalk lengths average several centimeters in modern sea lilies, but certain fossil forms stood an astonishing 25 m above the seafloor. The comatulids have dramatically reduced stems. Typically, only a single whorl of large cirri anchors these crinoids during their stationary periods (Figs. 16.11 and 16.18).

The membranous tegmen may be reinforced with scattered ossicles (Fig. 16.17a). Five ambulacral fields pass across this oral roof and converge at its central mouth. Thus, in contrast to the situation in other living echinoderms, the crinoid mouth occupies the upper, exposed body surface. The anus also is present on this surface and frequently is mounted atop a raised anal cone. Crinoid arms arise from the border of the calyx and tegmen. Primitively, five arms are present, as in the living *Ptilocrinus* (Fig. 16.17c). In other modern forms, however, forking of the arms is common; in most species, the arm number represents a multiple of five. The sea lily *Metacrinus* for example, has 60 arms, while the comatulid *Comanthina* bears as many as 200. Open ambulacral grooves extend from the tegmen into each of the arms. Crinoid arms have numerous lateral appendages called **pinnules,** which receive branches of the ambulacral system. Crinoid podia usually are fused in groups of three (Fig. 16.19). Each podial group is sheltered beneath a movable plate called a **lappet.**

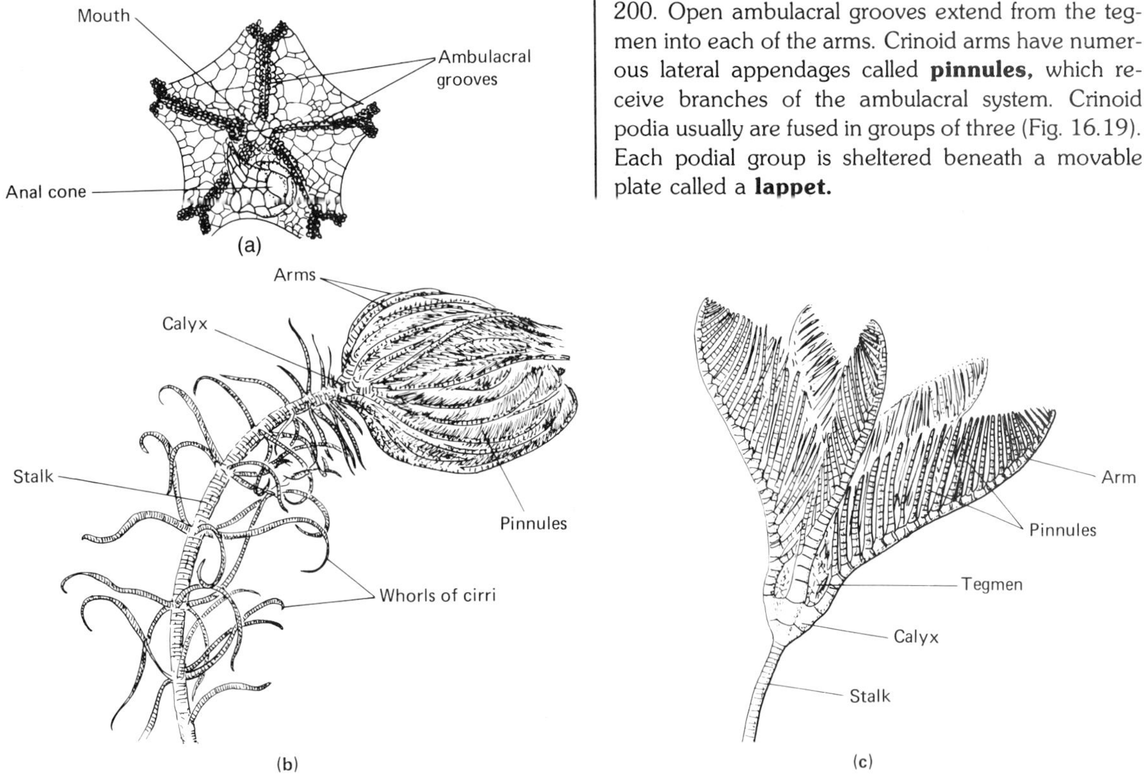

Figure 16.17 (*a*) Oral view of the tegmen of a crinoid; (*b*) Drawing of a living sea lily, *Cenocrinus asteria*, bearing many arms and numerous whorls of cirri along the stalk; (*c*) A living sea lily, *Ptilocrinus pinnatus*, with five heavily pinnulated arms. (From Hyman, *The Invertebrates*, Vol. IV.)

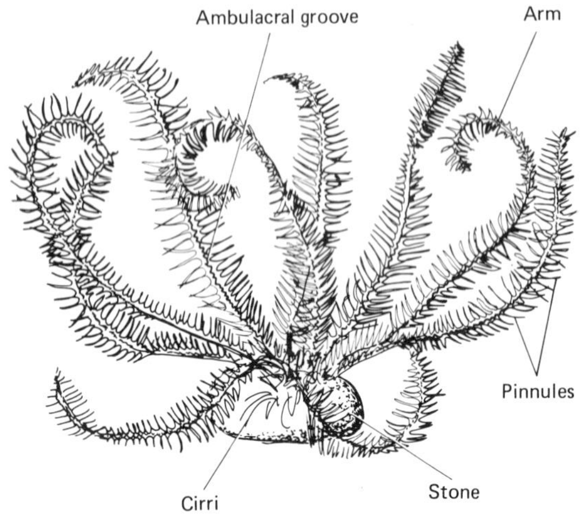

Figure 16.18 Drawing of a living comatulid, *Antedon bifida.* (From Hyman, *The Invertebrates, Vol. IV.*)

Crinoid ossicles are rather large and thick by echinoderm standards and typically occupy almost the entire dermal area. Their size and serial organization lend a jointed appearance to the arms and stalks. Strong elastic ligaments connect the ossicles of the arms and stalk and allow some bending. These ligaments are antagonized by longitudinal muscles.

Crinoids eat plankton and detritus, which precipitate on their arms. Trapped in mucus, potential food is passed orally by ambulacral podia and flagella. A specialized ring of oral podia tastes arriving foodstuffs and admits favorable items to the digestive tract. Crinoid podia thus have their primary role in food acquisition. It is widely believed that food gathering was the original function of the ambulacral system and that the ancestral process is preserved in this ancient class.

The crinoid digestive system comprises a mouth, an esophagus, a long intestine, a rectum, and an anus (Fig. 16.20). The intestine may give rise to diverticula as it circles the inner calyx wall. The rectum then ascends through the anal cone, where fecal pellets are compacted with mucus and discharged away from incoming food. In many crinoids, the ambulacral grooves in the anal region are conspicuously detoured to prevent their fouling. Crinoids commonly draw water into and out of their rectums, an activity that may represent an enema or a respiratory process.

Crinoid coelomic systems display several specializations. The primary, perivisceral coelom is reduced by connective tissues. This coelom extends into each arm as four or five distinct canals (Fig. 16.21). At the base of the calyx, the perivisceral system is isolated as a **chambered organ** (Fig. 16.20). The prominent aboral nervous system surrounds this region. Five coelomic canals descend from the chambered organ into the stalk and/or cirri.

A hemal system traverses the connective tissues of the perivisceral region. Closely associated with the gut, this system also extends into the arms, where its canals

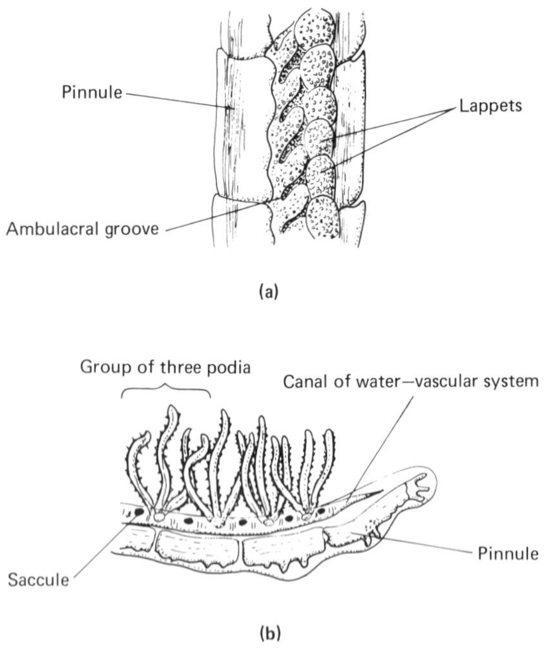

Figure 16.19 (*a*) Portion of a pinnule of a crinoid showing lappets and other skeletal elements. Podia omitted. (*b*) Section through the tip of a crinoid pinnule showing groups of podia and saccules. (From Hyman, *The Invertebrates, Vol. IV.*)

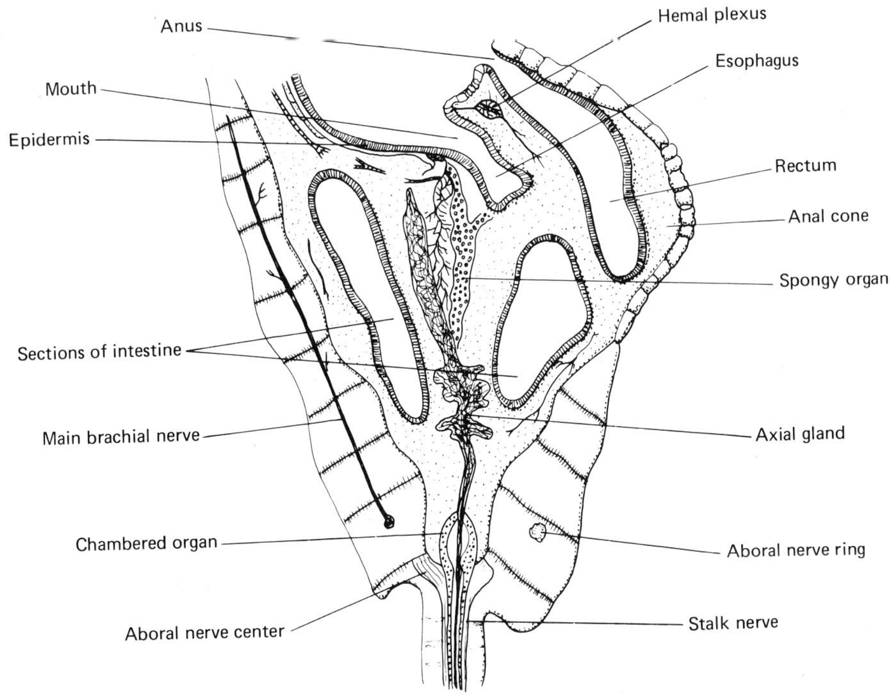

Figure 16.20 Internal anatomy of a sea lily, *Neocrinus decorus.* (From Hyman, *The Invertebrates, Vol. IV.*)

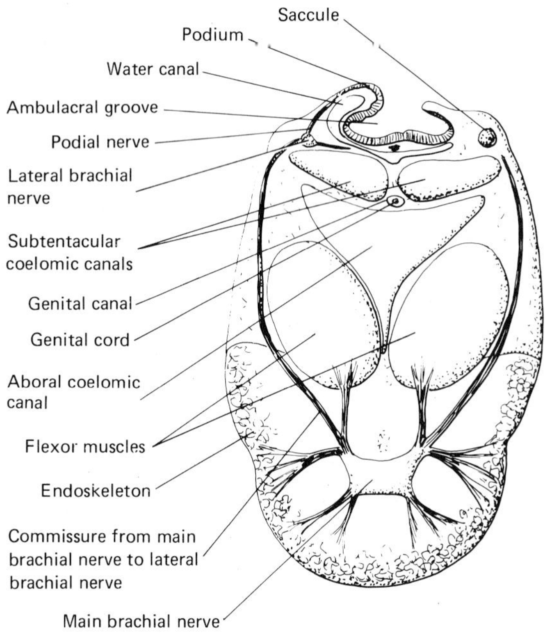

Figure 16.21 Section through an arm of a crinoid. (From Hyman, *The Invertebrates, Vol. IV.*)

participate in reproduction (see the following). A so-called **spongy organ** of uncertain function is related to the axial gland.

The crinoid ambulacral system lacks a madreporite. Hundreds of tiny ciliated funnels in the tegmen wall apparently control the passage of sea water into and out of the body. The central water ring issues numerous stone canals; these short, interradial tubes absorb fluid directly from the coelom. Radial canals lead from the water ring into each of the arms, where they fork into the pinnules. Within the pinnules, a lateral canal connects to each tripodial unit. Because crinoid podia lack ampullae, their extension depends on the contraction of radial canal muscles.

Coelomocytes transport materials through the various canals and sinuses, while open ambulacral surfaces provide amply for respiratory exchange. Total ambulacral area in the crinoids can be quite large. In one multiarmed sea lily, *Metacrinus rotundus,* a composite groove length of 80 m has been recorded. Such remarkable lengths represent adaptations for food acquisition, but their incidental respiratory contributions spare the crinoids the necessity to develop specialized gas-exchange structures. Ambulacral surfaces also represent waste removal sites. Insoluble wastes are stored within small saccules along the ambulacra (Figs. 16.19 and 16.21). These saccules may void themselves periodically.

The crinoid nervous system differs from that of other echinoderms in that its aboral components are predominant. This situation relates to the peculiar orientation of the crinoids and to the activity of their stalks, cirri, and arms. The aboral network surrounds the chambered organ (Fig. 16.20). This cup-shaped nervous mass dispatches fibers down the coelomic canals of the stalk and cirri. Brachial nerves extend into each of the arms, where they terminate in podial nerves (Fig. 16.21). Branches of the brachial nerves also service the aboral sides of the arms and the ossicular musculature. A deep oral system comprises a nerve circle outside the water ring. Its radiating nerves communicate with ambulacral structures and the tegmen. Finally, the diminutive oral system consists of a circumesophageal loop and a subambulacral nervous network.

The only sense organs in this class are small podial papulae. These projections and many scattered sensory hairs respond to touch, chemicals, and light.

The sea lilies are limited behaviorally because of their sessile condition. Only slight bending movements of the stalk and arms are possible. When nutrients appear, the arms may wave about to maximize their food-capturing potential. Feather stars (comatulids) are rather more active. These stalkless crinoids can walk and swim, although they spend most of their time standing still. Walking involves the clawlike cirri and, occasionally, the arms. The cirri are strongly thigmotactic. Feather stars stand at any angle as long as their cirri maintain ground contact; they quickly right themselves if their oral end is forced against a surface.

The brief swimming periods of the comatulids provide one of the more graceful spectacles in the animal world. Feather stars swim by alternately raising and lowering sets of five arms. In the 10-armed *Antedon* (Fig. 16.18), for example, every other arm belongs to an opposite locomotor set; thus each arm passes downward as its two neighbors sweep upward, and vice versa. In many-armed species, the sequential sweeping of numerous sets of arms is quite inspiring. In addition, the unusual coloration of many comatulids, particularly the tropical species, enhances their beauty while swimming.

Although difficult to distinguish, crinoid sexes are always separate. Indistinct germinal tissues are scattered through tubular extensions of the perivisceral and hemal coeloms in the arms and pinnules. When the pinnule wall ruptures, gametes are either released to the sea or retained on the sticky surfaces of the pinnules. In the latter case, hatching may occur at a larval stage. In *Notocrinus* and other cold-water crinoids, a brood pouch may house later developmental stages.

Crinoid embryogeny features a nonfeeding **vitellaria.** Settlement is rapid as the young echinoderm applies an apical adhesive pit against the substratum. In both sea lilies and comatulids, a stalk is formed, and the first few weeks offer only a sessile existence. Eventually, however, the comatulids separate from the juvenile stalk.

CLASS STELLEROIDEA; SUBCLASS ASTEROIDEA

The best-known echinoderms are, of course, the sea stars. These animals, which are also known by the misleading name "starfish," have been studied more intensively than any other group in the phylum. Many asteroid characters are rather primitive and generalized and thus served as the bases for much of the discussion of echinoderm unity. Sea stars live along most of the world's coastlines, where they crawl over both rocky and sandy surfaces or occasionally burrow in soft substrata. Coral reefs are common habitats. About 1800 living asteroid species have been identified.

The asteroid body comprises a central disk and its radiating arms. Typically, five arms are present, as in *Pisaster, Astropecten, Asterias,* and *Oreaster* (Figs. 16.2a and 16.22a–c). Many sea stars, however, have more than five arms. In *Crossaster* (Fig. 16.22d), for example, 7 to 14 arms are present, while in *Solaster* the number may reach 50.

The relationship of the arms to the central disk distinguishes the asteroids from the brittle stars of the subclass Ophiuroidea. Asteroid arms generally are short in relationship to the disk diameter. In most species, they taper gradually from the central body. Exceptions occur, however, as in the longer-armed *Freyella* (Fig. 16.22f) and the short-armed *Patiria* (Fig. 16.22g). In the latter, the arms are hardly distinct from the central disk. Asteroids average 12 to 24 cm in diameter, although large forms such as *Acanthaster* (Fig. 16.22h) may span an entire meter.

The central disk houses the major organs, although branches of the digestive, ambulacral, nervous, and reproductive systems extend into each arm (Fig. 16.23). The oral surface is held against the substratum. At the center of this surface, the mouth is surrounded by a peristomial membrane and a circle of protective spines. The aboral surface of the disk bears a central anus and a conspicuous madreporite, which is situated asymmetrically within an interambulacral zone.

Asteroids are rather crusty animals. Their external surfaces bear an assortment of skeletal appendages, including spines, tubercles, paxillae, and pedicellariae. Spines and tubercles are hard protective extensions of the dermal skeleton. A **paxilla** (Fig. 16.24a) is a specialized ossicle associated with certain burrowing asteroids (e.g., *Astropecten;* Fig. 16.32a). Exposed above the general body surface, this calcareous structure bears movable spines along its edges. When lowered, the spines of adjacent paxillae enclose a trough for respiratory and feeding currents; thus these currents flow freely even while the animal is buried.

Pedicellariae, the jawed guardians of the external surface, are common in the subclass. Both stalked and sessile types are present (Fig. 16.3), often in association with delicate respiratory papulae.

The central disk and arms are supported by the dermal skeleton. Figures 16.24 and 16.25 depict generalized cross sections of a sea star arm. Typically, the arm skeleton includes numerous marginal ossicles on the aboral and lateral sides. Orally, paired ambulacral ossicles and a smaller pair of superambulacral ossicles form the ambulacral groove. These skeletal elements lie above (aboral to) the radial canals of the ambulacral system. Only the ampullae of the tube feet are within the skeletal framework. Thus, like the crinoids, asteroids display open ambulacra. In all other echinoderms, the radial canals are enclosed within the endoskeleton.

The asteroid ambulacral system follows the general outline described earlier. Its central network includes the aboral madreporite, a stone canal, and a water ring complete with Tiedemann bodies and polian vesicles; radial canals, lateral canals, and two or four rows of podia extend into each arm. At the end of each arm, the radial canal terminates in a single tentacle (Fig. 16.23b). Asteroid podia are highly effective locomotor organs. They also capture and convey food and serve as respiratory surfaces.

The primary respiratory surface in this subclass is provided by the papulae (Fig. 16.4 and 16.25). These evaginations of the coelomic peritoneum occur at gaps in the asteroid dermis and allow proximity between the sea and the fluids of the perivisceral coelom. This main body coelom is more spacious in the asteroids than in other echinoderm groups, occupying most of the internal disk and extending broadly into each of the arms (Figs. 16.23a and 16.25).

Asteroids are basically carnivorous, and their diverse diet is rich in small invertebrates and even some fish. Mollusks and colonial animals are common food items. Some sea stars are scavengers, and a few supplement their diets with detritus and/or suspended food particles, which are trapped on ambulacral mucus. Their digestive system comprises a sphinctered mouth, a short and sometimes diverticulated esophagus, a complex double stomach, a short intestine, and an anus (Fig. 16.23). The asteroid stomach consists of an initial oral chamber called the **cardiac stomach** and a second, aboral **pyloric stomach** with paired, branching ceca in each of the arms. Mesenteries suspend each cecum within the arm's perivisceral coelom. Most digestion occurs extracellularly within the cecal lumen, whose walls house secretory, absorptive, and storage cells. A flagellated epithelium helps to maintain an organized current through the digestive tract.

Many asteroids evert their cardiac stomach by applying pressure on the coelomic fluid of the central disk. The organ is protruded through the mouth, whereupon it engulfs prey and may initiate external digestion. Bound to the body wall by 10 paired mesenteries, the cardiac stomach is withdrawn by retractor muscles. Some asteroids, such as the bat star *Patiria miniata,* sweep their everted stomachs along organically rich substrata. Many stars feed on bivalves, and in certain areas they may threaten oyster beds. These asteroids can insert their cardiac stomach through a slight gap between the shell valves of their prey. The inserted stomach may digest the shell's adductor muscles, and thus the sea star obtains easy access to the whole mollusk. An asteroid may take its entire meal, however, through a shell opening no more than 0.1 mm wide.

Arm muscles are well developed among the asteroids. Short fibers connecting ambulacral and superambulacral ossicles control the depth of the ambulacral grooves (Fig. 16.24). In the body wall, strong circular and longitudinal muscles produce arm movements within the limits of ossicular flexibility. In most species, arm flexion is limited to aboral bending, as in righting movements. Podial stepping accounts for the slow foraging of these echinoderms. Extended podia are used as levers on horizontal surfaces; during vertical climbs, they adhere tightly to the substratum and then contract longitudinally to pull the animal upward. Podia are strongly thigmotactic, a property that determines asteroid orientation. Light reception is localized in numerous pigment cup ocelli, which form an eyespot at the distal tip of each arm. Certain photopositive asteroids, including *Asterias rubens* and *Astropecten irregularis,* can be attracted by shining light at these sense organs.

(a)

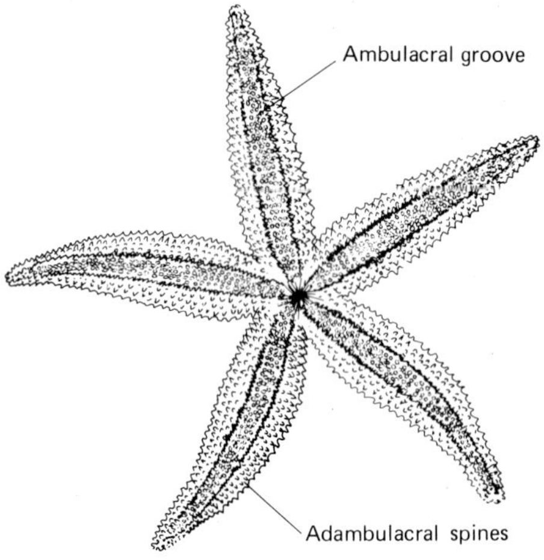

(b)

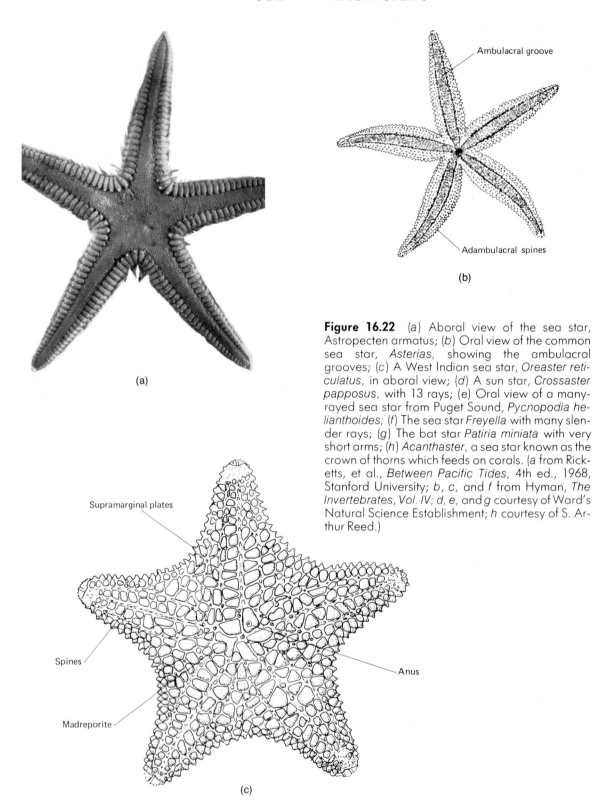

Figure 16.22 (*a*) Aboral view of the sea star, Astropecten armatus; (*b*) Oral view of the common sea star, *Asterias*, showing the ambulacral grooves; (c) A West Indian sea star, *Oreaster reticulatus*, in aboral view; (*d*) A sun star, *Crossaster papposus*, with 13 rays; (e) Oral view of a many-rayed sea star from Puget Sound, *Pycnopodia helianthoides;* (*f*) The sea star *Freyella* with many slender rays; (*g*) The bat star *Patiria miniata* with very short arms; (*h*) *Acanthaster*, a sea star known as the crown of thorns which feeds on corals. (*a* from Ricketts, et al., *Between Pacific Tides*, 4th ed., 1968, Stanford University; *b*, *c*, and *f* from Hyman, *The Invertebrates, Vol. IV; d*, *e*, and *g* courtesy of Ward's Natural Science Establishment; *h* courtesy of S. Arthur Reed.)

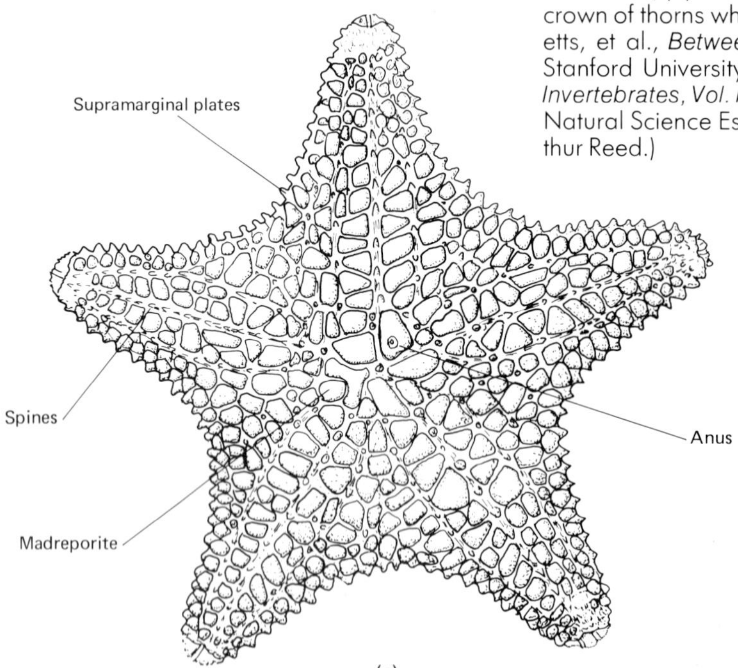

(c)

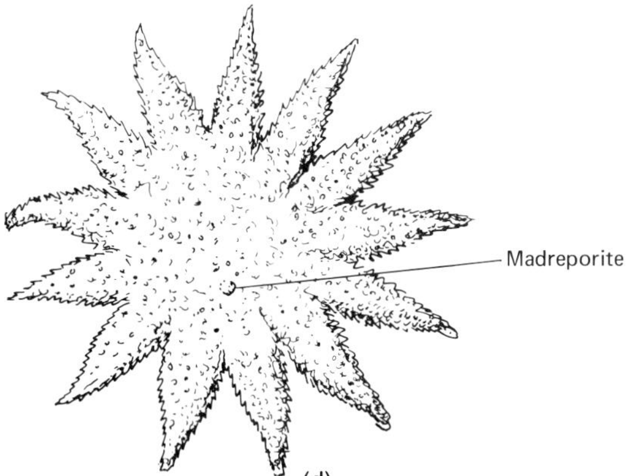

(d)

(g)

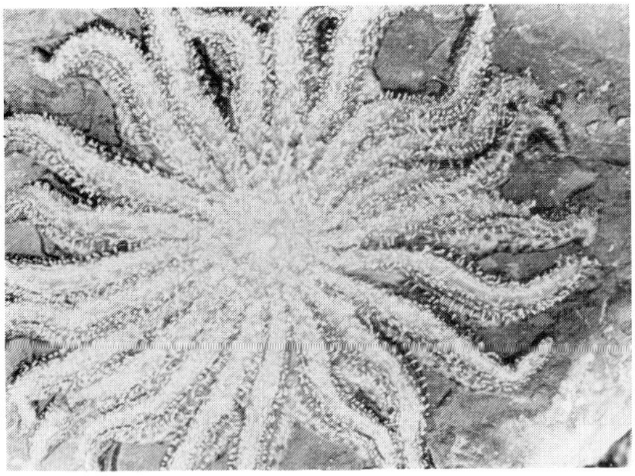

(e)

(h)

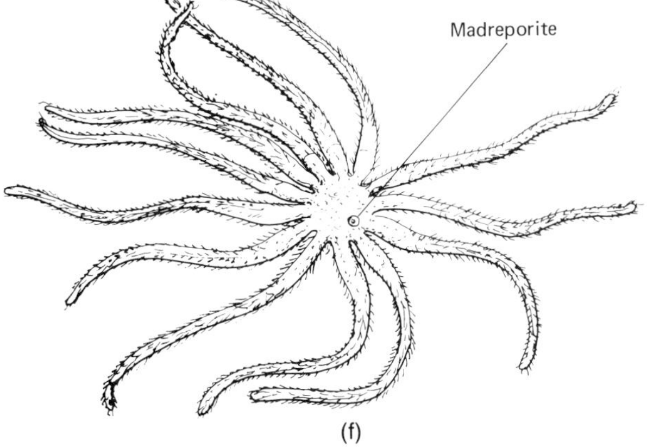

(f)

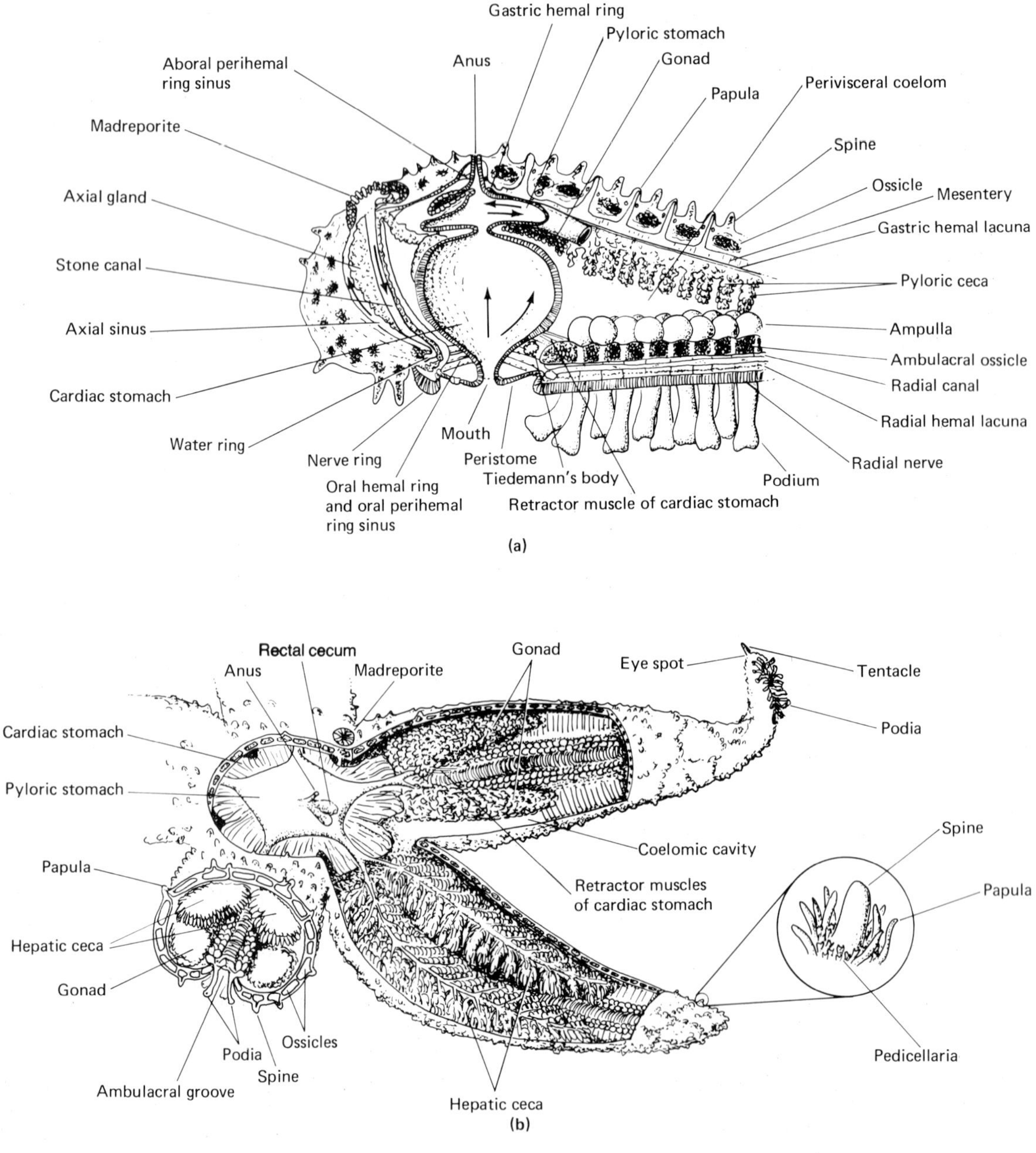

Figure 16.23 (*a*) Sea star anatomy: a vertical section cut through the madreporite-bearing interradius and the proximal part of the opposite arm of *Asterias rubens* (*b*) Cut-away diagram showing the anatomy of a sea star in aboral view. Three arms are cut off, one is seen in cross section. The aboral surface of the disk and two arms have been removed. The ceca have been removed from the upper arm. The enlarged insert shows a spine, papulae and pedicellariae. (*a* from V. Fretter and A. Graham, *A Functional Anatomy of Invertebrates*, 1976, Academic Press, and D.E. Beck and L. F. Braithwaite, *Invertebrate Zoology Laboratory Workbook*, 1968, Burgess; *b* from T.I. Storer, et al., *General Zoology*, 1972, John Wiley.)

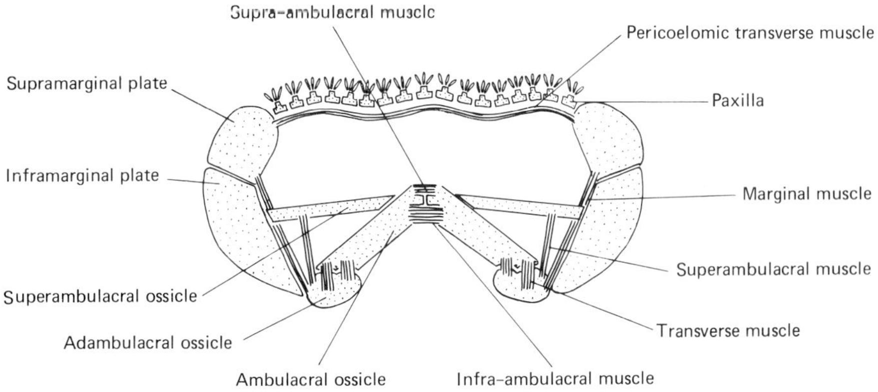

Figure 16.24 Cross section of an arm of *Astropecten irregularis* showing skeletal elements and musculature. Note the paxillae covering the aboral surface. (From D. Heddle, In *Echinoderm Biology*, N. Millot, Ed., 1967, Academic Press.)

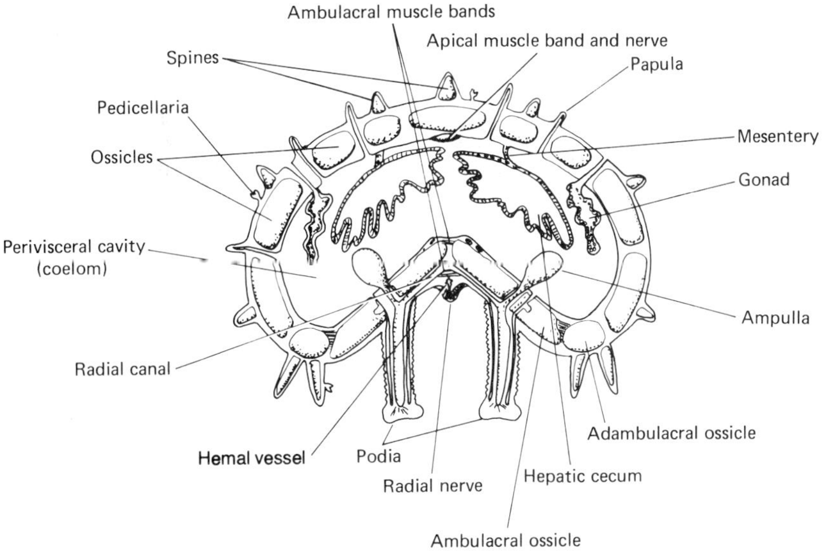

Figure 16.25 Cross section through an arm of a generalized sea star. (From Beck and Braithwaite, *Invertebrate Zoology Laboratory Workbook.*)

A few protandric hermaphrodites are known in this subclass, but the vast majority of asteroids is diecious. Their gonads are paired treelike structures which occupy the coelomic extensions in the arms (Figs. 16.23 and 16.25). When ripe, the gonads fill these cavities. Gametes are discharged through interambulacral gonopores located between the bases of adjacent arms. External fertilization is common and implies synchronous spawning. According to some authorities, the axial gland may chemically recognize conspecific gametes. This recognition might mediate general gamete release by aggregated individuals. An efficient form of gamete release is displayed by *Archaster typicus:* The spawning male of this species lies atop the aboral surface of the female; his arms alternate with hers and thus their gonopores are brought together.

Brooding is common among asteroids of high latitudes. Most of these sea stars shelter developing young between their oral surface and the substratum. Others brood within specialized aboral depressions or within baskets formed by clustered spines.

Asteroid development features the multiarmed **bipinnaria** larva and its successor, the **brachiolaria.** Like the crinoids, asteroids adhere to the substratum as metamorphosis begins. Later, when the podia develop enough strength to lift the body, the young sea star—typically no more than 1 mm in diameter—crawls away.

CLASS STELLEROIDEA; SUBCLASS OPHIUROIDEA

The second stelleroid subclass includes the ophiuroids, also known as brittle or serpent stars. Such common names describe the snakelike arms of these animals and their tendency to autotomize if disturbed. Two thousand species have been listed, thus making this group the largest in the phylum. Indeed, ophiuroids are considered to be the most successful modern echinoderms. Their success is attributed in part to their size—small by echinoderm standards—and their highly flexible and mobile bodies.

Structurally, ophiuroids are not a diverse lot. Two subgroups can be distinguished, however, on the basis of arm morphology: the common brittle stars (Fig. 16.26), whose arms are simple; and the deeper-water basket stars (Fig. 16.27), whose arms are coiled and often branched. Basket stars include the largest ophiuroids, whose disks may reach 12 cm in diameter. In most other members of this subclass, the central disk is only 1 to 3 cm across, although the arms are always several times longer.

Ophiuroids live on most marine bottoms; however, their tendency to conceal themselves beneath stones, vegetation, and invertebrate colonies makes them difficult to observe and collect. In offshore waters, ophiuroids may be so populous that they blanket the ocean floor. Densities as high as several hundred individuals per square meter have been reported.

Ophiuroids are readily distinguished from asteroids by their narrow flexible arms, which are sharply set off from the flattened central disk (Figs. 16.26 and 16.27). The central disk houses the major organ sys-

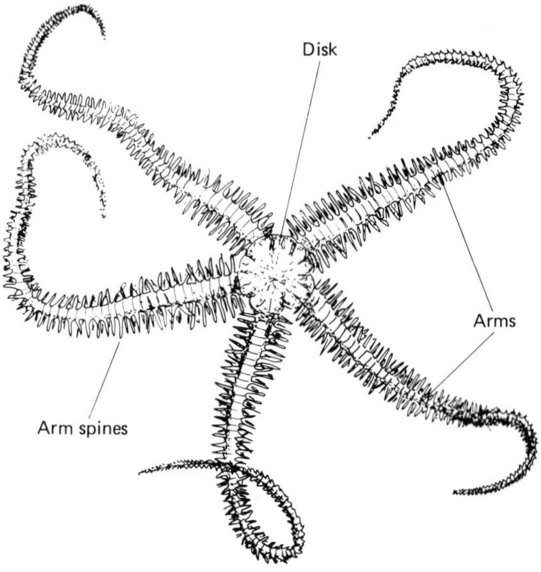

(a)

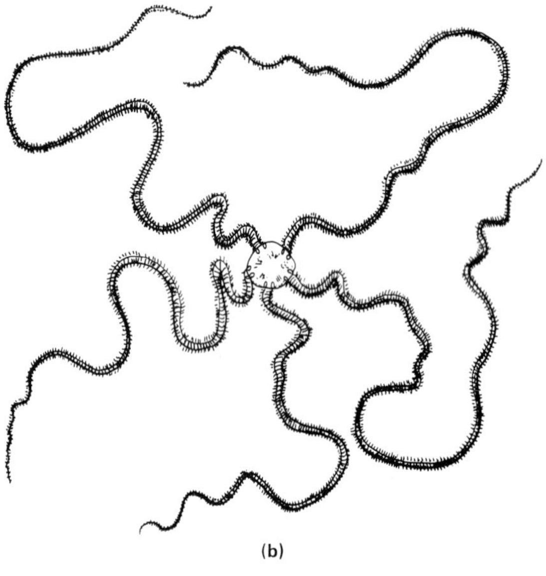

(b)

Figure 16.26 (*a*) Aboral view of the West Indian brittle star, *Ophiocoma*, a typical ophiuroid with spiny arms. (*b*) *Orchasterias columbiana*, a Pacific Coast brittle star with exceptionally long arms. (*a* and *b* from Hyman, *The Invertebrates, Vol. IV;*)

Figure 16.27 *Gorgonocephalus caryi*, a basket star. Oral view. (Photo courtesy of Ward's Natural Science Establishment.)

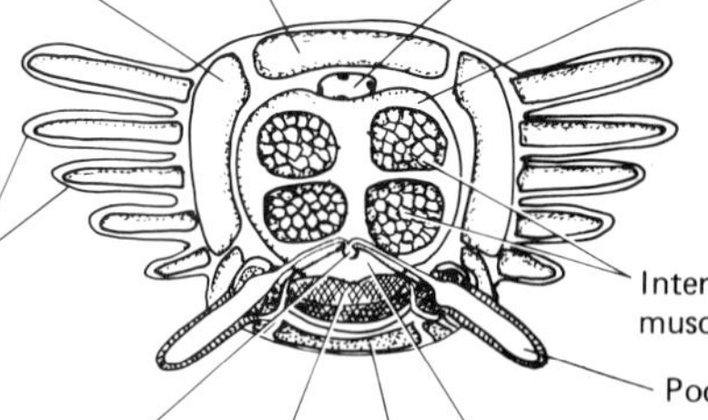

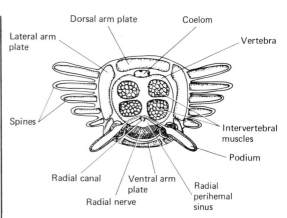

Figure 16.28 Cross section of an ophiuroid arm cut between two vertebrae so that the intervertebral muscles are visible. (From E.W. MacBride, In *Cambridge Natural History, Vol. I*, S.F. Harmer and A.E. Shipley, Eds., 1959, Macmillan.)

tems, and few radial elements extend into the arms. Aborally, the disk comprises a central apical plate surrounded by concentric rings of ossicles. Spines are scarce and usually are confined to the lateral margins of the arms and disk. The ophiuroid epidermis is reduced as dermal and skeletal elements extend to the body surface. Ossicles are rather flat and tightly fitted. In the arms, the endoskeleton is dominated by large central disks called **vertebrae** (Fig. 16.28), so named because of their resemblance to the bones of the vertebrate spinal column. Ophiuroid vertebrae articulate in linear series (Fig. 16.30). Oral and aboral pairs of intervertebral muscles manipulate these large plates. Arm flexibility depends on intervertebral articulation. In most brittle stars, the articulation is such that only lateral swaying is possible. Basket stars, however, move their arms in both lateral and vertical planes.

The vertebrae are surrounded by a sheath of surface ossicles called arm plates (Fig. 16.28). Usually four in number, these plates occupy dorsal, ventral, and lateral positions about the circumference of each vertebra. Serial repetition of the arm plates and vertebrae creates the jointed appearance of the arms. Ophiuroid arms extend across the oral surface of the disk and terminate around the central mouth (Fig. 16.29a). There, the arms contribute to five interradial jaws which project as spiny triangles into the oral aperture. At the base of one or more of these triangles is a madreporite.

Within the central disk, small external openings occur alongside the arm margins. Called **bursal slits**, these openings lead to five pairs of bursae (Figs. 16.29a and 16.30). An ophiuroid **bursa** is an invagination of the oral surface floor. It extends upward into the central disk, where its functions are both respiratory and reproductive. The bursal epithelium is usually flagellated. Flagellar action, often combined with muscular pumping by the disk wall, draws sea water into and out of the bursae. Respiratory gases are exchanged during the pumping cycle, and soluble wastes may be voided. In addition, ophiuroid gonads discharge into the bursae; gametes may be released to the sea immediately or, especially in many Antarctic species, the bursae may serve as brood chambers.

The ophiuroid coelom is not nearly as spacious as its asteroid counterpart. The digestive tract and bursae dominate the central disk, while radial portions of the coelom are reduced to mere canals between the vertebrae and the arm plates (Fig. 16.28).

Ophiuroids possess a closed ambulacral system; that is, there is no ambulacral groove and only the podia extend outside the endoskeleton. Radial canals pass through perforations in the arm vertebrae, and lateral canals present tube feet between the ventral and lateral arm plates (Fig. 16.28). Ophiuroid podia lack ampullae, but a valve isolates them from their lateral canal. Muscular contractions of the podial wall, and sometimes of the radial canal, account for podial movement.

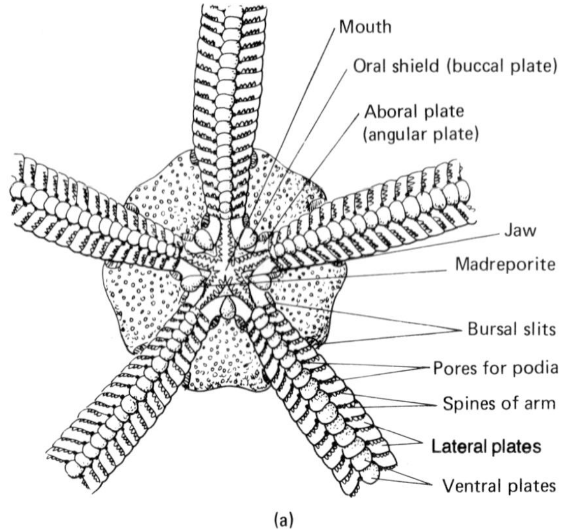

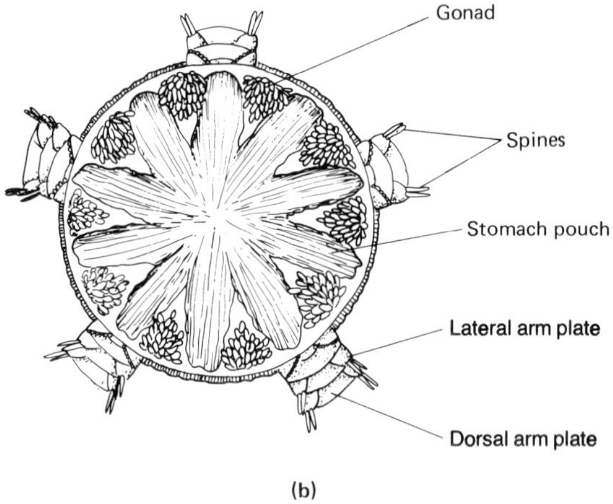

Figure 16.29 (*a*) Oral view of the central disk and arm bases of a brittle star, *Ophioderma;* (*b*) Aboral view of the viscera of an ophiuroid. (From Beck and Braithwaite, *Invertebrate Zoology Laboratory Workbook.*)

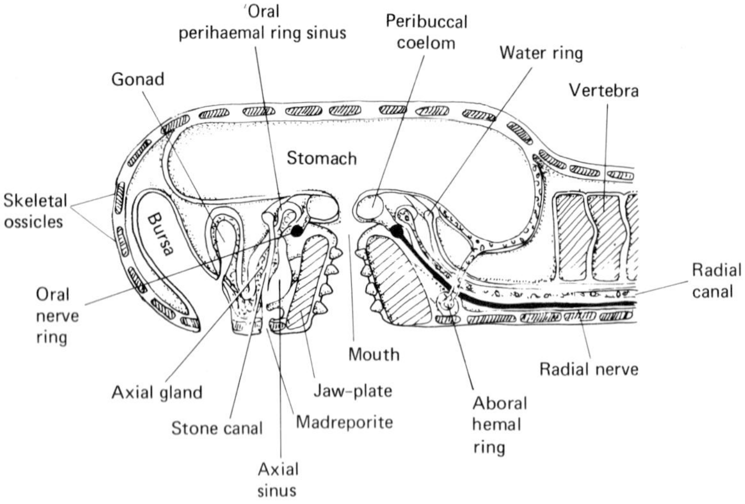

Figure 16.30 A vertical section through the central disk and the base of one arm of an ophiuroid, passing through one genital disk, one gonad, the axial complex, and one ambulacrum. (From Nichols, *Echinoderms.*)

Ophiuroids augment their basic diet of bottom material with occasional larger prey. Detritus and protozoans of the marine floor are staple items, which are captured by mucous nets erected between spines. These nets are harvested by the podia. Podial and flagellar action then conveys food to the mouth. Larger ophiuroid prey include polychaetes, small crustaceans, and mollusks, as well as smaller ophiuroids. Victims are entangled by the arms and then looped beneath the central disk for ingestion. Ophiuroids cannot capture highly mobile prey, and carrion constitutes an important element in their diet. The diets of individual species display a considerable range of specificity. Some ophiuroids, such as *Ophionereis reticulata,* reportedly eat only vegetable matter. Others, such as *Ophiura texturata* and *Ophiothrix fragilis*, may restrict their intake to bottom material. Meanwhile, *Ophiocomina nigra* is probably most typical of the subclass, in that it feeds on a variety of items, including detritus, suspended matter, live prey, and carcasses.

The ophiuroid digestive system is rather simple (Fig. 16.30). A prebuccal cavity lies behind the jaws. At the cavity's upper end, a peristomial membrane surrounds the mouth and a short esophagus leads to the stomach. The ophiuroid stomach is suspended by aboral mesenteries. Extending from the stomach margins, 10 pouches alternate in position with the bursae and gonads (Fig. 16.29b). Ophiuroids have no intestine or anus. The stomach is the site of digestion and absorption, while undigestible materials are voided through the mouth.

Ophiuroids are the liveliest echinoderms. They travel at modest speeds (up to 3 cm/sec) along the ocean floor and apparently have no arm preference. Characteristically, leading and trailing arms lift the central disk, while lateral arms propel the animal forward in a series of leaps. The serpentine arms may coil around objects (Fig. 16.31). To right themselves, overturned brittle stars extend two opposing limbs and then lift and twirl the central body with the remaining arms. Most members of this subclass conceal themselves by day within rock crevices or among vegetation and animal colonies. Some ophiuroids are burrowers; *Amphioplus*, for example, uses its podia to construct underground channels, in which it lives. Podia are seldom used in locomotion, but do play an important role in chemoreception. Podial receptors probably detect food at a distance, and generally distributed photosensitive cells mediate the animals' negative response to light. There are no specialized sense organs in the Ophiuroidea.

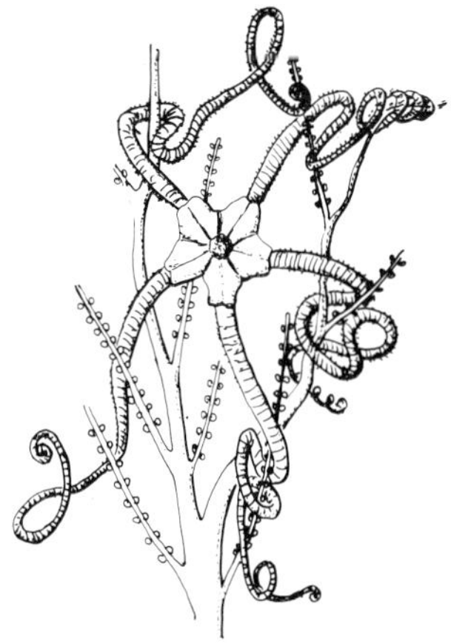

Figure 16.31 A brittle star, *Asteronyx excavata* from Panama with its arms coiled around the branches of a black coral. (From Hyman, *The Invertebrates, Vol. IV.*)

One to several saccular gonads are associated with each bursa (Figs. 16.29b and 16.30). They lie within the central coelom and discharge near the bursal slits. Sexual dimorphism is limited in this subclass. An exception, however, is *Amphilycus androphorus* (Fig. 16.32); the diminutive male of this species is transported by the female, who holds him against her mouth. While most ophiuroids are diecious, the subclass does claim more hermaphroditic members than

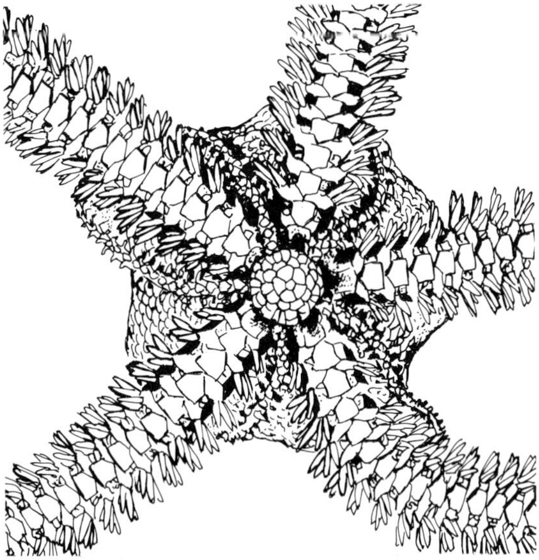

Figure 16.32 A female of the brittle star *Amphilycus androphorus* carrying the dwarf male mouth to mouth. (From Hyman, *The Invertebrates, Vol. IV.*)

most other echinoderm groups. *Amphipholis squamata* is a hermaphroditic serpent star of almost cosmopolitan distribution. *A. squamata* embryos attach to the bursal wall of their parent and actually receive nourishment there; later, juvenile stars are born alive. Other brooding ophiuroids release their offspring at an early larval stage. Most species, however, do not brood, but simply wager the survival of the next generation on the fortunes of external fertilization and larval development in the open sea. The long-armed **ophiopluteus** larva gradually undergoes adult metamorphosis. The young brittle star continues to swim until its adult skeleton forces it to sink to a bottom home.

Fissiparity has been observed in some ophiuroids, but this asexual process usually is limited to smaller, six-rayed groups. The genus *Ophiactis* is particularly renowned for its asexuality. Fissiparity probably plays a major role in the success of *O. savignyi*, a commensal living throughout the tropics.

The ophiuroids include most of the commensal echinoderms. Sponges are favorite homes for many species (50 to 75 *O. savignyi* individuals have been recovered from a large sponge mass), and other echinoderms also serve as ophiuroid residences. *Ophiomaza*, for example, lives on a feather star and *Amphilycus androphorus* occurs beneath a certain sand dollar.

CLASS ECHINOIDEA

The third modern class of echinoderms is the Echinoidea, a group that includes the familiar sea urchins, heart urchins, and sand dollars. Unlike the echinoderms previously discussed, most echinoids are globular or oval in form and have no arms. Numerous spines on the body surface account for the class name, which means "hedgehoglike." Echinoids are further distinguished from all other echinoderms by their endoskeletons; the flattened skeletal ossicles of these animals are joined into a solid, immobile test. About 900 species have been identified from littoral and benthic areas throughout the world.

Traditionally, the class is divided into two semiofficial groups. The more primitive sea urchins, whose bodies display the secondary radial symmetry general to the phylum, constitute the "regular echinoids." The more advanced heart urchins and sand dollars have experimented with tertiary bilateral symmetry. These specialized creatures are called "irregular echinoids." Because of the differences in structure, we discuss the two groups separately.

The Regular Echinoids

Regular echinoids, or sea urchins, live on hard surfaces, typically near the low-tide mark. Most of these animals are 6 to 20 cm in diameter, and some may be quite colorful. Sea urchins bear long moveable spines, which function variously in locomotion, defense, and burrowing.

The spherical urchin test comprises alternating ambulacral and interambulacral fields, which join at the upper aboral pole and the lower oral center. The mouth is surrounded by 10 buccal podia and 10 peristomial gills (Fig. 16.33a). The latter are body wall evaginations which present surfaces for respiratory exchange. Conventional tube feet continue aborally along the five ambulacral fields, which consist of double rows of flattened ossicles. Perforations in the ambulacral ossicles allow the extension of podia from the closed water–vascular system. The five interambulacral fields likewise are defined by double rows of ossicles. All 10 fields converge aborally in a ring of plates about the anus. Five circumanal ossicles associated with the interambulacral fields are called genital plates (Fig. 16.33b). As their name implies, these large plates bear the echinoid gonopores. One of the genital plates houses the madreporite.

Most echinoid ossicles bear one to several raised knobs called **tubercles**, which are the sites of spine attachment. A concave socket at the spine's base fits over the convex tubercle, and two circular sheaths of muscle allow spine movement about the tubercle–socket joint (Fig. 16.34). Sea urchins employ their long spines in locomotion, burrowing, and defense. Spines cooperate with the podia in producing basic travel power; characteristically, they push off hard surfaces. Certain sea urchins, including the European *Paracentrotus lividus* and the North American *Strongylocentrotus purpuratus* (Fig. 16.35a), burrow by rotating their spines against rock walls. Burrowing urchins live in rough coastal waters. In such areas, even nonburrowing species lodge themselves within rock crannies, where spines act as wedging devices. The short, broad spines of *Colobocentrotus* (Fig. 16.35b) are well adapted for anchorage against strong wave action.

Other sea urchins use their spines as defensive structures. Not only are these instruments often sharply pointed, but many are hollow and may contain poisons. The urchin *Asthenosoma*, for example, has poisonous spines which are dangerous to humans. *Lytechinus*, a common urchin off southeastern waters of the United States, camouflages itself by piling stones, shells, and other debris over its aboral surface. This pe-

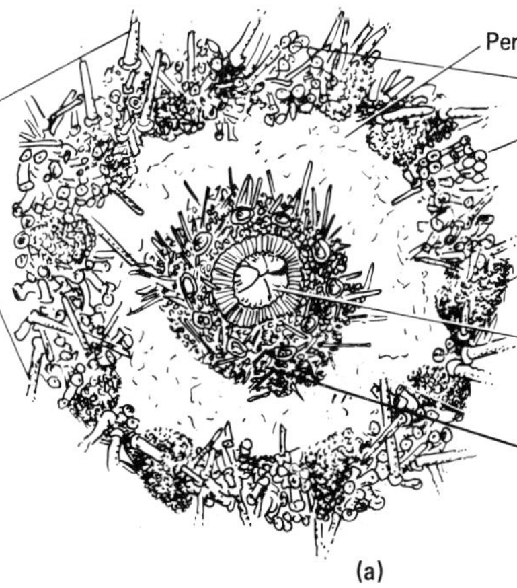

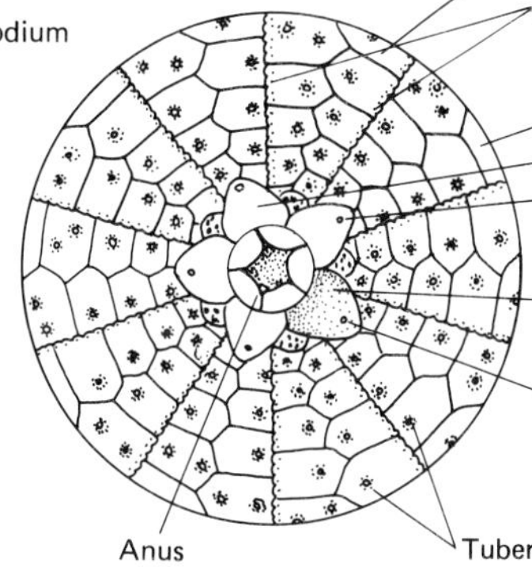

Figure 16.33 (*a*) The peristomial region of a sea urchin, *Echinus esculentus;* (*b*) A representation of the test of a sea urchin (*Arbacia*). Aboral view. (*a* from MacBride, In *Cambridge Natural History, Vol. I,* Harmer and Shipley, Eds; *b* from Beck and Braithwaite, *Invertebrate Zoology Laboratory Workbook.*)

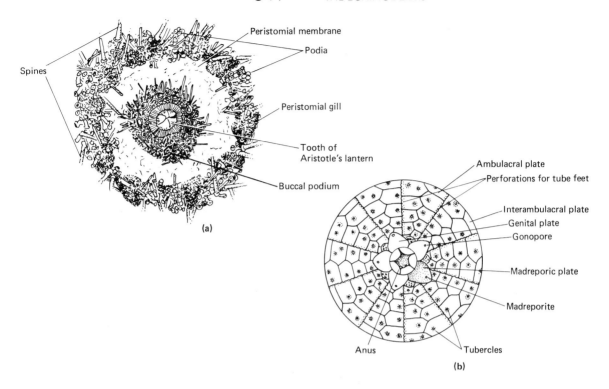

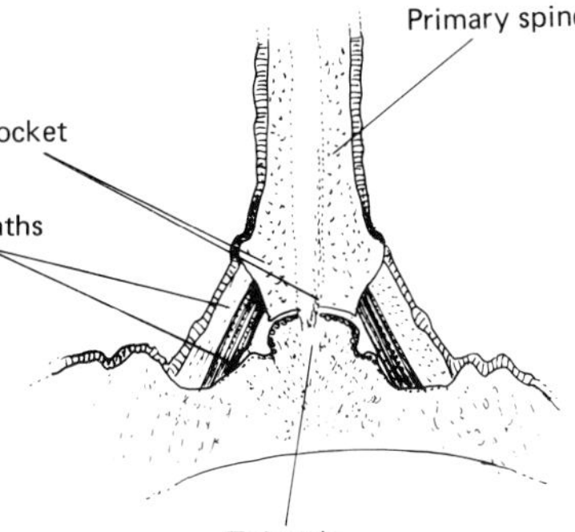

Figure 16.34 The base of a primary spine of a sea urchin. (From Cuénot, In *Traité de Zoologie, Vol. XI,* Grasse, Ed.)

culiar activity may represent a reaction against strong light. Urchins, in general, are negatively phototactic. Their tube feet contain photosensitive cells.

Sea urchins also have pedicellariae. These stalked appendages are distributed over the general body surface for defense and sanitation. Echinoid pedicellariae differ from those of the asteroids, in that a skeletal rod supports the stalk and three articulating jaws are present apically. When an urchin is threatened, its spines bend down to protect the body surface, while the pedicellariae rise to meet the oncoming foe (Fig. 16.36). Pedicellarial jaws open when touched on their outer surfaces and close rapidly if stimulated inside. Poison glands occur on the pedicellariae of some urchins, including *Lytechinus*. Such glands discharge onto the teeth of the jaws; thus their bites are quite toxic. Smaller animals are paralyzed by pedicellarial poisons, while larger creatures may be forced to withdraw. Spines, podia, and pedicellariae create an active forest on the echinoid body surface.

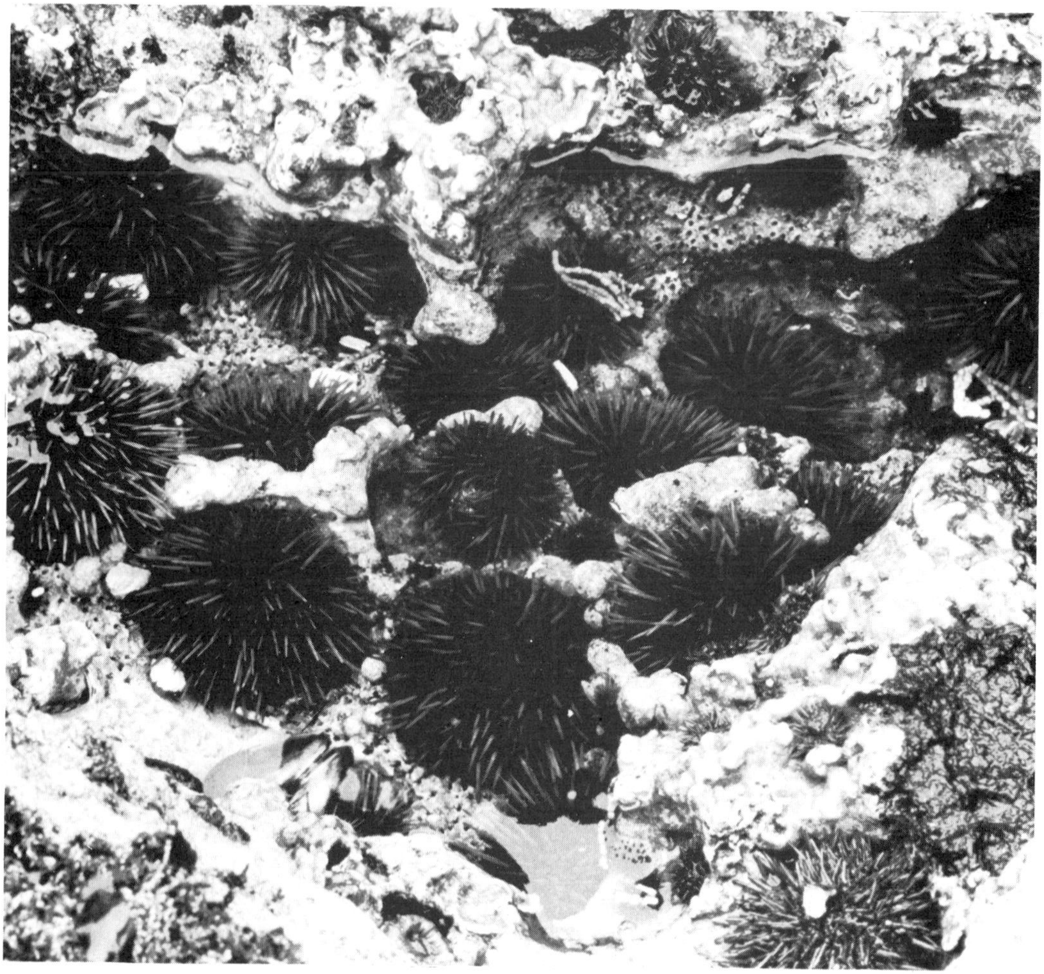

(a)

(b)

Figure 16.35 (*a*) Sea urchins (*Strongylocentrotus purpuratus*) nestled in pits in a rocky surface; (*b*) A Hawaiian sea urchin, *Colobocentrotus*, with short broad spines adapted for anchorage. (*a* from Ricketts, et al., *Between Pacific Tides*; *b* courtesy of S. Arthur Reed.)

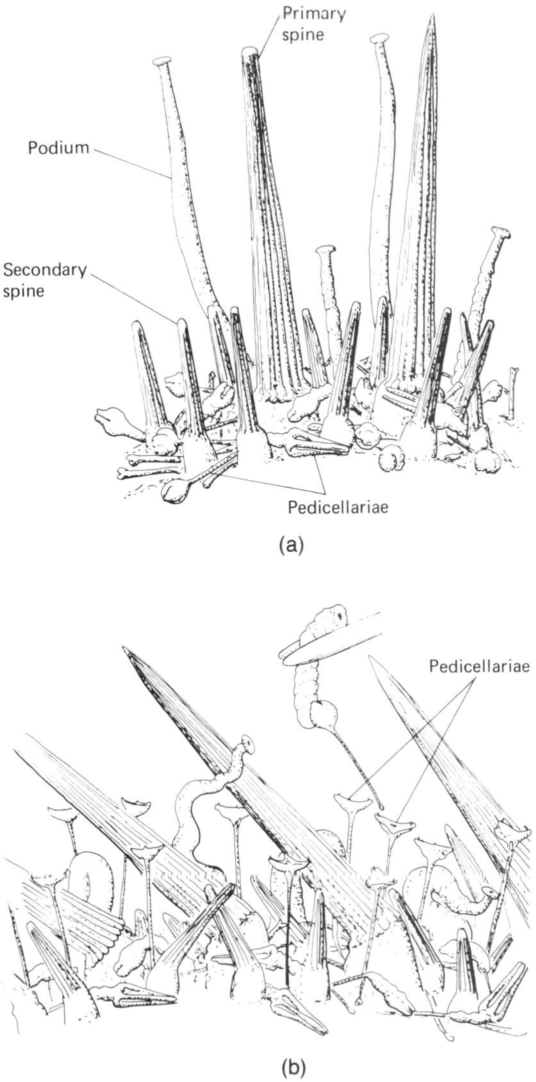

Figure 16.36 (*a*) Portion of the surface of an undisturbed sea urchin, *Psammechinus miliaris* showing primary and secondary spines, podia and several types of pedicellariae; (*b*) The same surface when stimulated by a predator's podium (held in forceps); spines bend down, podia retract, and pedicellaria extend. (From M. Jensen, Ophelia, 1966, 3:209; drawn by Kai Olsen.)

The often colorful, graceful bending and swaying of these appendages is a delightful spectacle, best appreciated beneath a dissecting microscope.

Sphaeridia complete the list of body wall appendages in the regular echinoids. These small, transparent structures are mounted atop stalks within the ambulacral fields. Often, they are limited to the oral hemisphere; in some urchins (e.g., *Arbacia*), they are localized about the peristome. Sphaeridia contain a statocyst for balance maintenance.

The most remarkable organ in the Echinoidea is the chewing apparatus known as **Aristotle's lantern** (Fig. 16.37). Composed of numerous calcareous plates and their associated nerves, muscles, and ligaments, the lantern surrounds both the buccal cavity and the pharynx. Five large plates called pyramids form its central structure. These pyramids are more or less triangular in shape, and their sharpest angles bear strong rasping teeth, which are exposed through the mouth. Each tooth is produced within an aboral dental sac on its respective pyramid. New teeth are transported orally along a calcareous band and replace worn ones as required. Various muscles and ligaments bind the pyramids together and also attach the entire lantern to the body wall. The teeth also are associated with muscles. Thus, not only can the lantern be raised and lowered through the mouth, but the teeth themselves can be opened and shut.

Aristotle's lantern contributes to the respiration of sea urchins. Certain muscles within the lantern's aboral end pump coelomic fluid past peristomial gills. Such ventilation enhances gas exchange between the gills and the outside medium.

Sea urchins eat marine plants, dead animals, and sessile invertebrates. Algae and ectoprocts are the favorite foods of many species. The digestive tract comprises a buccal cavity and pharynx (both located within Aristotle's lantern), an esophagus, a long intestine, a rectum, and an anus (Fig. 16.38a). At the esophageal–intestinal junction, a blind cecum may produce digestive enzymes. The intestine loops between Aristotle's lantern and the test wall, then doubles back and circles in the opposite direction. A small tube, called the siphon, branches from the anterior end of the intestine, then courses parallel to the gut before discharging back into the intestinal lumen. Such a siphon is unique to the echinoids. It probably withdraws intestinal water, thus concentrating enzymes and food during the initial stages of digestion.

The gonads of regular echinoids occupy the aboral hemisphere of the test (Fig. 16.38a). Their location along the interambulacra coincides with the genital plates around the anus. Sexes are separate, but rarely dimorphic. Most sea urchins release their gametes for external fertilization in the sea. Some brooding occurs among cold-water species, whose spines may cradle developing eggs around the mouth and/or anus.

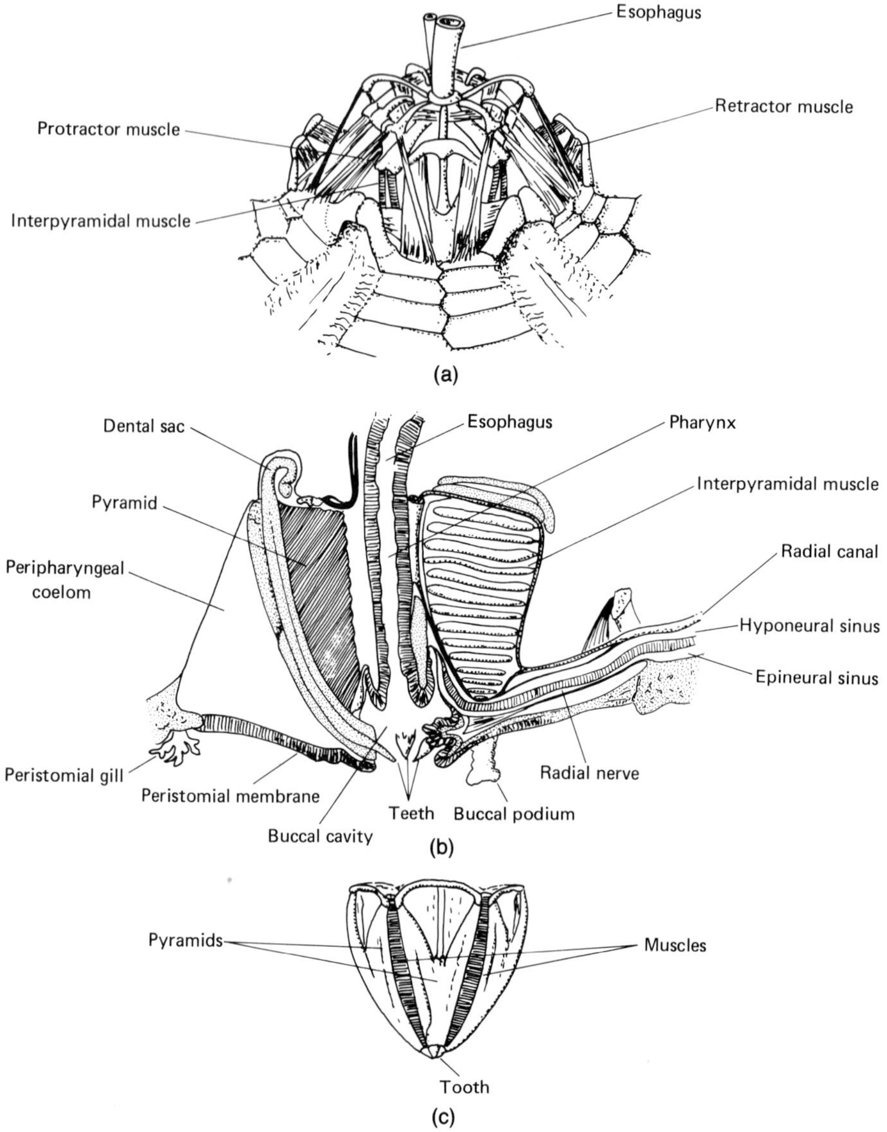

Figure 16.37 (*a*) Aristotle's lantern of the sea urchin, *Echinus esculentus* in situ as viewed from within the test; (*b*) Sagittal section of the sea urchin, *Echinus esculentus*, through the region bearing Aristotle's lantern; (*c*) Side view of Aristotle's lantern removed from the sea urchin, *Tripneustes ventricosus*. (*a and b from* Cuénot, In *Traité de Zoologie*, *Vol. XI*, Grassé, Ed; c from Hyman, *The Invertebrates*, *Vol. IV*.)

Echinoid embryogeny features the 10- or 12-armed **echinopluteus** larva. This planktonic stage may endure for several months before the weight of its growing skeleton causes it to sink to the ocean floor. No attachment period occurs. Echinoid adult metamorphosis may be completed in a single hour.

The Irregular Echinoids

The more advanced group of the Echinoidea displays some tertiary bilateralism. These so-called irregular echinoids, which include the heart urchins and the sand dollars, have definite anterior and posterior ends

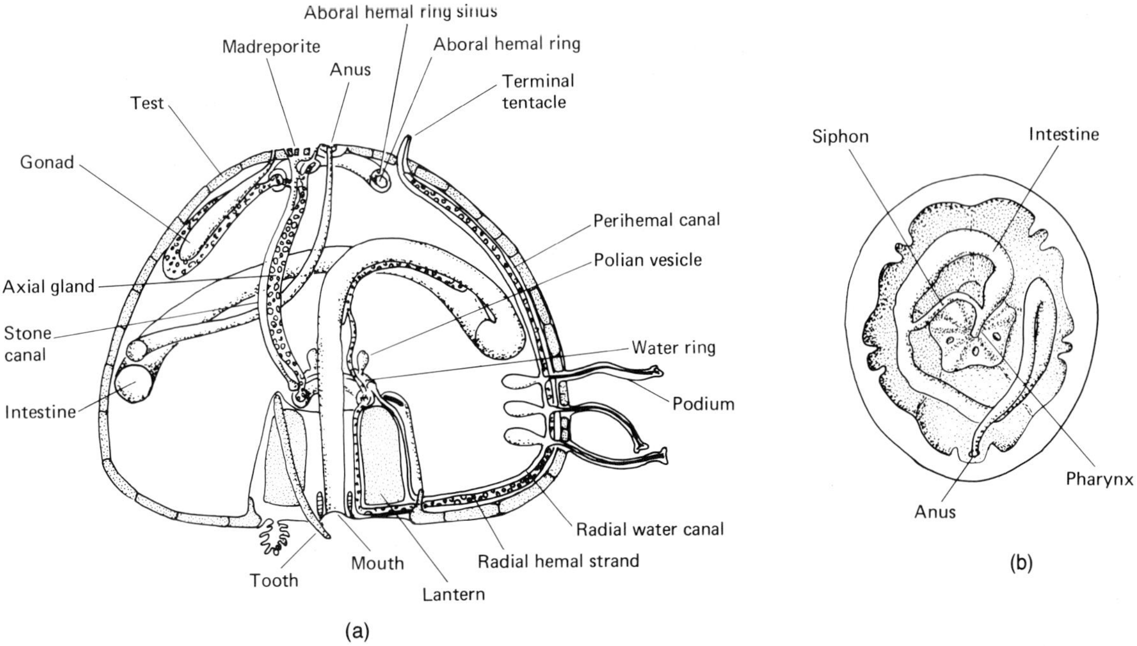

Figure 16.38 (*a*) Vertical section through a sea urchin; (*b*) The digestive tract of the sand dollar, *Echinocyamus pusillus*, dorsal view. (*a* from Nichols, Echinoderms; *b* from Cuénot, In *Traité de Zoologie Vol. XI*, Grassé, Ed.)

and move in a single forward direction. Their burrowing existence in soft substrata involves several adaptations. In all irregular echinoids, small but numerous spines are the primary tools for burrowing. Burrowing procedures vary from species to species, but normally emphasize the lateral displacement of sand by anterior spines. Ambulacral systems typically exist as isolated oral and aboral fields. The podia have almost no locomotor responsibility and instead are specialized for respiratory, nutritional, and sanitational roles.

The heart urchins normally construct and occupy semipermanent chambers beneath the sand surface (Fig. 16.39), where they are relatively quiescent, stirring mainly to eat small organic particles and to maintain their burrow walls. The heart urchin body is oval in contour, with a flat oral surface and a rounded aboral one. The mouth occupies an anterior position on the lower surface, while the aboral anus has migrated into the posterior interambulacrum. Heart urchin podia are restricted to two distinct areas, one aboral and one oral. The five aboral ambulacra are called **petaloids** because of their flowerlike arrangement. Thin-walled aboral podia provide respiratory surfaces. In some heart urchins, such as *Echinocardium*, a specialized anterior petaloid maintains the burrow surfaces. Its podia remove obstructions from the burrow opening and coat the walls with mucus, which apparently is produced within specialized spines. Modified podia in the posterior body construct a feces drain at the rear end of the burrow.

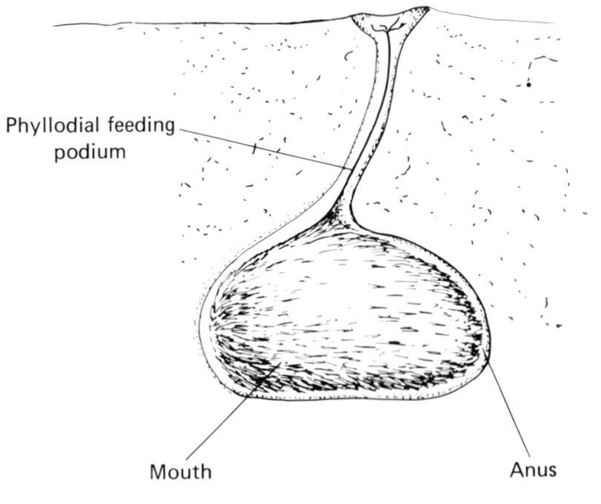

Figure 16.39 A heart urchin, *Echinocardium cordatum*, in its burrow. (From Cuénot, In *Traité de Zoologie, Vol. XI*, Grassé, Ed.)

The oral ambulacra are called **phyllodes**. Arranged florally about the mouth, the phyllodial podia gather and convey food. Some of these tube feet are very extensile and bear terminal branches (Fig. 16.39). Heart urchins eat minute organic particles in the sand. Such a diet does not require a chewing apparatus, so there is no Aristotle's lantern.

Sand dollars are the more active members of the irregular Echinoidea. These echinoderms burrow within the upper few centimeters of sandy ocean bottoms. The sand dollar body is flattened and more or less circular in contour (Fig. 16.40). The mouth occupies the center of the oral surface, but the anus has migrated from its traditional aboral position to a ventral (oral) position within the posterior interambulacrum (Fig. 16.40a). Radially arranged notches occur about the edges of the body disk in the keyhole sand dollars (e.g., *Mellita*; Fig. 16.41). Called **lunules**, these notches apparently allow sand to pass to the opposite side of the disk during burrowing.

Sand dollar podia form aboral petaloids and flank oral food grooves which converge at the central mouth (Fig. 16.40a). No phyllodes are present. These echinoids feed on organic particles mixed in sand. Specialized spines hold sand and debris aloft while food particles precipitate onto the mucus-coated aboral surface. Flagella transport potential nutrients over the disk margin or through the lunules to the oral surface. The podia of the oral grooves then convey foodstuffs to the mouth. Sand dollars possess an immobile Aristotle's lantern.

Irregular echinoids have no peristomial gills. Respiratory exchange occurs across the petaloid podia. Finally, continuity systems among these animals display only slight variations from the sea urchin norm. Typically, only four gonads are present, as the anus usurps a gonadal site within the posterior interambulacrum. Brooding has been described in some cold-water heart urchins and in a single sand dollar species. Specialized pits on the petaloids serve as brood chambers.

CLASS HOLOTHUROIDEA

The fourth modern echinoderm class, is the Holothuroidea. Holothuroids are the marine animals commonly called sea cucumbers. About 1100 species have been identified from ocean floors throughout the world. Sea cucumbers are common in littoral regions and also rank among the most successful colonizers of the marine abyss. Indeed, in certain deep-sea areas, they may constitute as much as 90% of the total animal mass.

In several respects, sea cucumbers are quite unlike other echinoderms. Many of their unusual features are associated with their unique posture. The oral–aboral axis of the body is markedly elongated; the holothuroid lies along this side (Fig. 16.42). With the

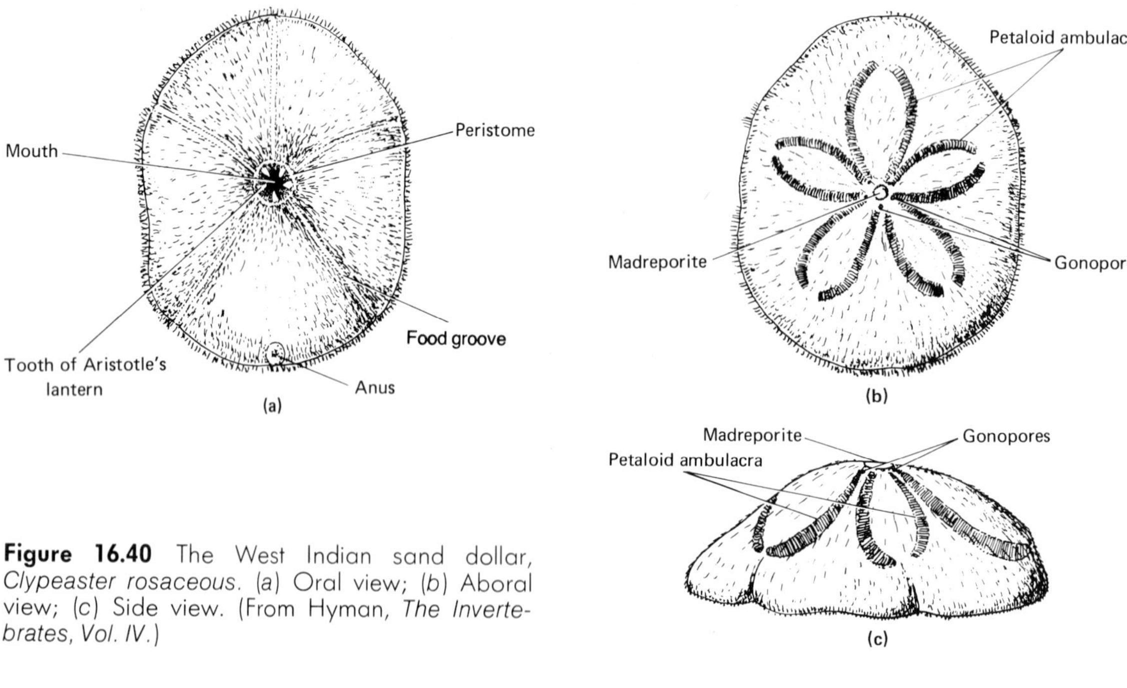

Figure 16.40 The West Indian sand dollar, *Clypeaster rosaceous*. (*a*) Oral view; (*b*) Aboral view; (*c*) Side view. (From Hyman, *The Invertebrates, Vol. IV.*)

Figure 16.41 A keyhole sand dollar (*Mellita quinquiesperforata*) burrowing in sand. (Photo courtesy of Lorus J. and Margery Milne.)

inevitable differentiation of lower and upper surfaces of the body trunk, sea cucumbers display a definite tertiary bilaterialism. The mouth and anus occupy anterior and posterior positions, respectively. Other characters distinguishing this class include its microscopic ossicles, unique respiratory trees, and large oral podia, which form a crown of tentacles around the mouth.

Sea cucumbers display a variety of shapes, sizes, and colors. A few are rather spherical, such as *Ypsilothuria* (Fig. 16.42a); others are long and thin, such as *Leptosynapta* (Fig. 16.42b). For most members of the class, however, the word "cucumber" accurately describes the body form. The most common genera, including *Cucumaria, Holothuria,* and *Thyone* (Fig. 16.42c, d), approximate the shape of their vegetable namesake. Their popular name is also descriptive of a common size range of 10 to 30 cm. Exceptional sea cucumbers include short species of no more than 3 cm, while some species of *Stichopus* (Fig. 16.42e) may be a full meter in length and 25 cm in diameter. Drab coloration (blacks, browns, and pale greens) is most typical of the class, but some types (notably poisonous ones) sport brilliant coats of red and orange.

As in other echinoderms, five ambulacral fields extend across the holothuroid surface from mouth to anus. Three of these fields are located on the lower, or ventral, surface, which is called the **sole** (Fig. 16.42e). The podia of the sole are always better developed than those of the dorsal surface. Indeed, in many holothuroids, dorsal podia have degenerated altogether. Another trend diminishes the orderly arrangement of podia along holothuroid ambulacra. In *Thyone,* for example, the tube feet are distributed quite randomly. These trends culminate in such holothuroids as *Euapta, Synapta,* and *Leptosynapta,* whose surfaces are entirely devoid of podia. This reduction and ultimate loss of podia reflects the assumption of locomotor and respiratory responsibilities by other organs.

The holothuroid body wall is dominated by a thick connective dermis, which houses numerous microscopic ossicles. Holothuroid ossicles display a wide and interesting range of shapes (Fig. 16.5) and are important taxonomically. Their presence imparts a tough, leathery character to the body surface. Beneath the dermis are a layer of circular muscles and five banded regions of longitudinal fibers, the latter underlying the ambulacral fields. These muscles produce the peristaltic movements of some sea cucumbers.

Much larger ossicles form a calcareous bulb around the holothuroid mouth and esophagus (Fig. 16.43). This ring supports the buccal region and provides an attachment site for the longitudinal fibers of the trunk wall and the retractor muscles of the tentacles. Holothuroid tentacles form a bushy ring around the mouth. These structures, which represent specialized buccal podia, are extended by pressure from coelomic fluids. They collect food and may aid in burrowing. The tentacles, along with the entire oral region, can be withdrawn into the anterior end of the animal (Fig. 16.42e).

Sea cucumbers feed on bottom deposits and/or suspended matter. Their tentacles are coated with food-trapping mucus and sweep along the surface or

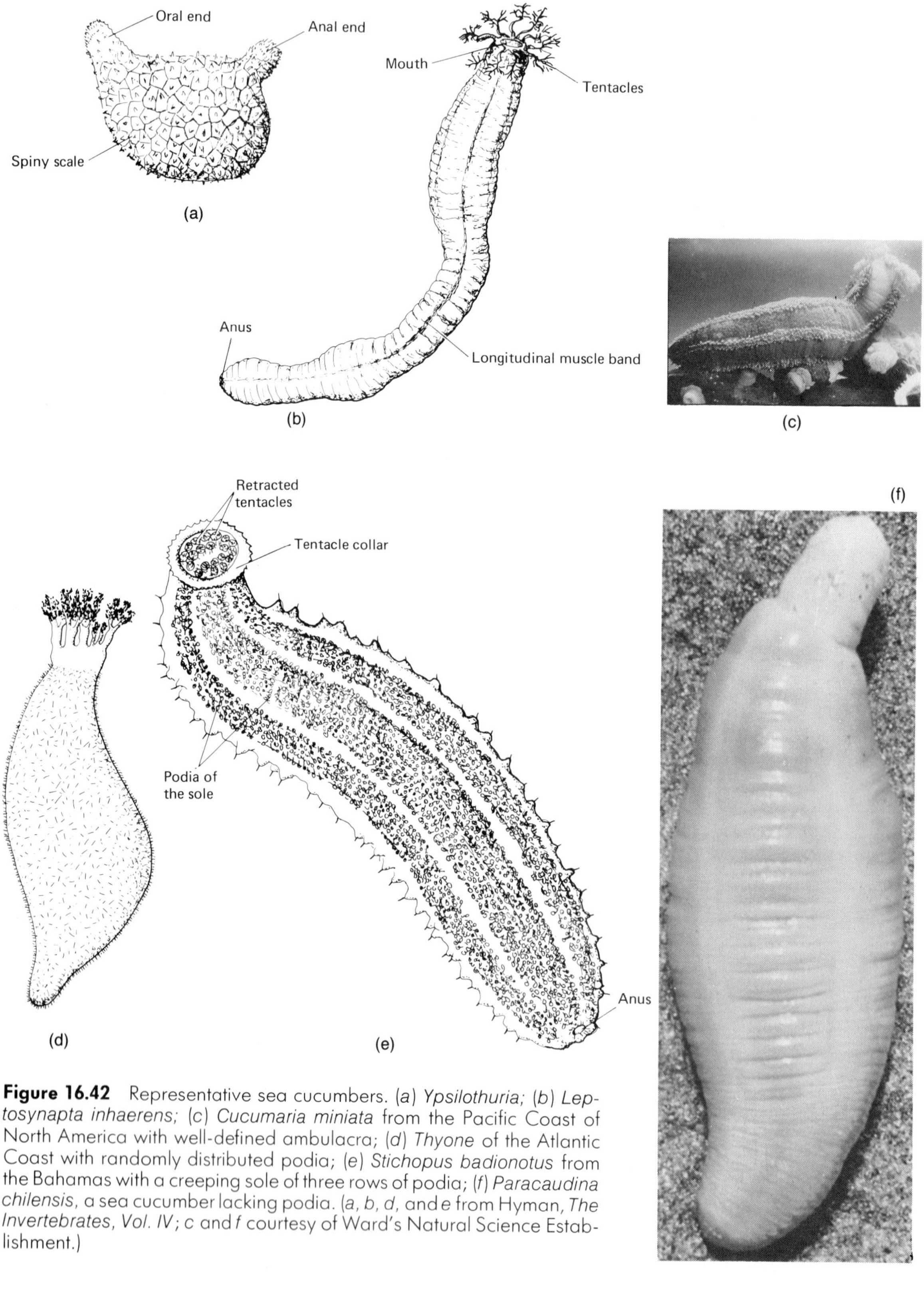

Figure 16.42 Representative sea cucumbers. (*a*) *Ypsilothuria;* (*b*) *Leptosynapta inhaerens;* (*c*) *Cucumaria miniata* from the Pacific Coast of North America with well-defined ambulacra; (*d*) *Thyone* of the Atlantic Coast with randomly distributed podia; (*e*) *Stichopus badionotus* from the Bahamas with a creeping sole of three rows of podia; (*f*) *Paracaudina chilensis*, a sea cucumber lacking podia. (*a*, *b*, *d*, and *e* from Hyman, *The Invertebrates, Vol. IV*; *c* and *f* courtesy of Ward's Natural Science Establishment.)

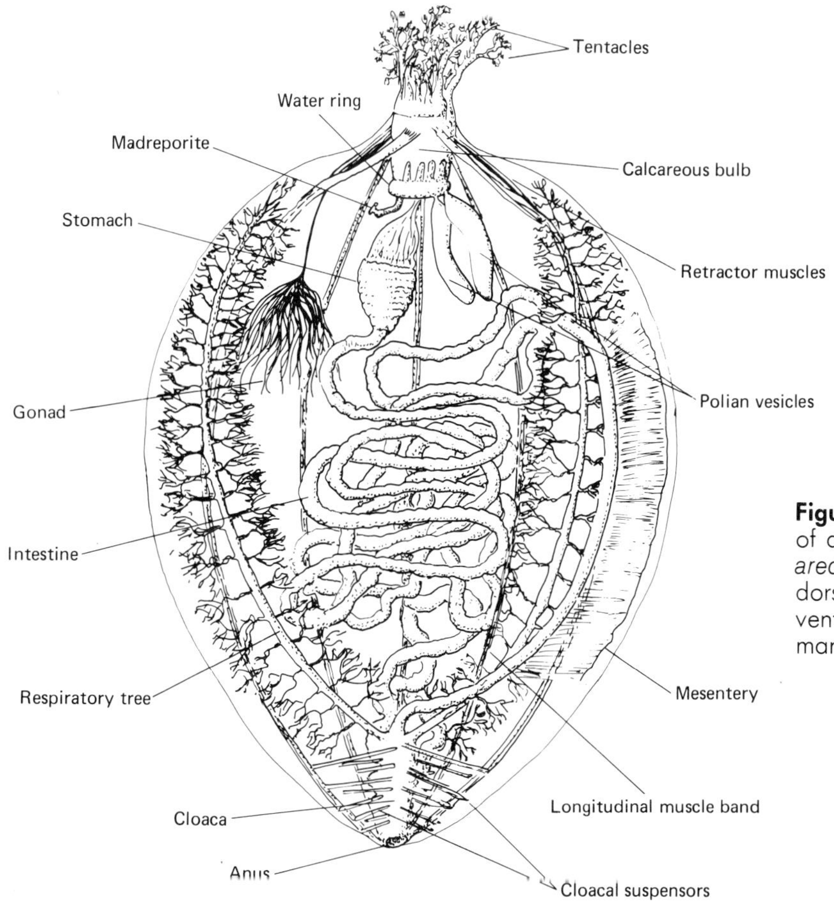

Figure 16.43 Internal anatomy of a sea cucumber, *Thyone briareous* from the side; functional dorsal surface at left, functional ventral surface at right. (From Hyman, *The Invertebrates, Vol. IV.*)

plunge into the substratum. *Cucumaria elongata* extends its tentacles into the water above its semipermanent burrow and traps plankton. Holothuroids insert their food-laden tentacles one at a time into their mouths. Oral insertion may refurbish the mucous coating. An active burrower like *Euapta* consumes the material through which it passes. In areas of high population density, these holothuroids contribute significantly to the mixing of sediment.

The holothuroid digestive system is dominated by a lengthy, coiled intestine (Fig. 16.43), which loops three times about the spacious coelom before terminating in a posterior cloaca. Mesenteries attach the intestine to the body cylinder. The hemal system is highly developed in the gut region. Apparently, enzyme-bearing hemal coelomocytes traverse the intestinal wall and participate in food breakdown. When coelomocytes later migrate from the intestinal lumen, they carry digested nutrients for general distribution.

Gas exchange is handled by unique holothuroid organs, the **respiratory trees** (Fig. 16.43). These paired branched tubules originate from the cloaca and extend on either side of the digestive tract. The trees are pumped full of water by contractions of the cloacal musculature. This process involves: (1) dilation of the cloaca to draw in water; (2) closure of the anal sphincter; and (3) contraction of the cloacal muscles to force water into the respiratory trees. From the distal branches of the trees, dissolved oxygen diffuses into the coelomic systems. Certain coelomocytes (called hemocytes) bear hemoglobin and thus transport oxygen. Sea water within the tree tubules receives ammoniacal wastes. As these important physiological exchanges are completed, the system is flushed by a single strong contraction of the tubule muscles.

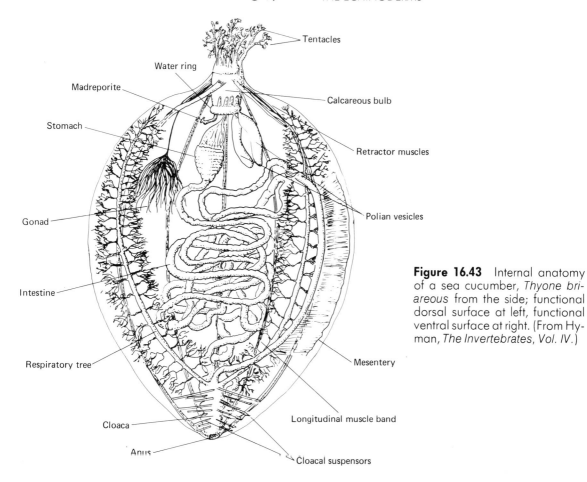

Leptosynapta and other long, burrowing sea cucumbers do not have respiratory trees. Oxygen uptake occurs across their general body surfaces, and excretion may involve ciliated urns. The latter are funnel-shaped coelomic organs associated with the body wall at or near the sites of mesentery attachment. Coelomocytes carry wastes to the ciliated urns and may exit the body through these organs.

The holothuroid ambulacral system is typically echinoderm in form. In most sea cucumbers, however, the madreporite is not at the body surface, but rather lies fully within the coelom. The central water ring, with its variable number of polian vesicles, surrounds the base of the calcareous bulb (Fig. 16.43). Radial canals extend to the buccal tentacles, as well as to the ambulacra of the body trunk.

Holothuroid locomotion involves either podial creeping or peristaltic contractions of the body wall musculature. Some species employ both methods. Certain deep-sea holothuroids (e.g., *Scotoplanes*) walk on a few large and highly specialized podia. Other peculiar sea cucumbers include the pelagic *Pelagothuria* (Fig. 16.44), in which an anterior ring of webbed papillae contracts to produce jellyfishlike swimming movements.

The holothuroid gonad is singular and comprises a moplike cluster of tubules and their common gonoduct (Fig. 16.43). The organ occupies a dorsal, anterior region of the coelom. Its gonopore discharges at the base of the buccal tentacles. Most sea cucumbers are diecious, although a few protandric hermaphrodites have been described. Fertilization is usually external, and development proceeds through **auricularia** and **doliolaria** stages. Holothuroid metamorphosis occurs gradually during the planktonic phase. The early appearance of buccal tentacles defines a preadult stage known as a **pentactula larva** (Fig. 16.45). Growth and continued maturation are accompanied by settlement and the assumption of adult life.

Brooding is not uncommon in the Holothuroidea and is well developed among Antarctic species. In such forms, the tentacles usually collect the eggs and tuck them beneath the sole. *Thyone rubra* and *Leptosynapta* are unusual in that their eggs are incubated within the coelom. Fertilization has not been described for these sea cucumbers, whose offspring exit from the parent through a rupture in the cloacal region.

Holothuroids are known for their evisceration. This process refers to the active discharge of internal body organs. Evisceration may take place through the anus, in which case the respiratory trees, digestive tract, and gonads are voided; or it may occur through the oral opening, where it involves the anterior gut and its associated structures. Evisceration is observed most often among laboratory animals, which are subjected to stressful conditions (fouling, overcrowding, etc.), but the phenomenon occurs in nature as well. Regeneration follows evisceration and thus the process may have a rejuvenating effect.

Sometimes confused with evisceration is a defense mechanism employed by *Holothuria* and a few other sea cucumbers. In such echinoderms, a mass of sticky tubules can be discharged from the cloacal region. These so-called **tubules of Cuvier** entangle predators and thus allow the holothuroid time to escape.

Echinoderm Evolution

The history of the Echinodermata presents another of the invertebrate world's great puzzles. Despite a rich fossil record, the origin of the phylum and its proper place within the animal kingdom remain rather mysterious. Echinoderms are an ancient group, much

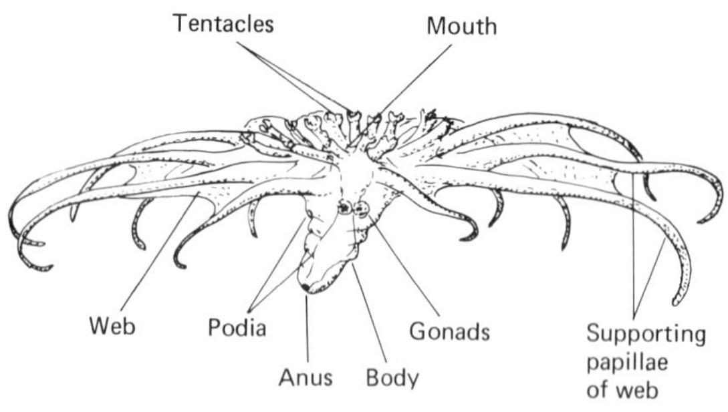

Figure 16.44 A pelagic sea cucumber, *Pelagothuria ludwigi.* (From Hyman, *The Invertebrates, Vol. IV.*)

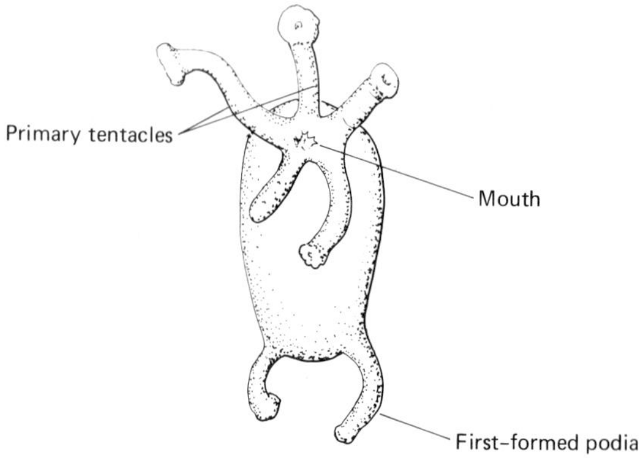

Figure 16.45 Pentactula larva of a sea cucumber, *Cucumaria frondosa,* with 5 buccal tentacles and 2 podia along the developing midventral ambulacrum. (From Hyman, *The Invertebrates, Vol. IV.*)

older than all 20,000 fossil species. Lacking clues to the formative Precambrian history of these animals, we can only guess at their origin and relationships.

Sixteen extinct classes have been described from the Paleozoic era. Even at that distant point in time, echinoderms were a diverse phylum whose modern characters were at various stages of development. Most fossil species shared certain features with the modern sea lilies. The majority were stalked, sessile animals, whose oral surfaces were directed upward to collect suspended nutrients. Usually, a solid theca enclosed both oral and aboral sides of the body. The ambulacral system occupied oral grooves; branches of this definitive echinoderm network often extended into narrow appendages called **brachioles.**

Three fossil groups vie for recognition as the oldest echinoderm types: the Helicoplacoidea, the Carpoidea, and the Eocrinoidea. All were well developed and quite distinct by the onset of the Cambrian period.

The helicoplacoids were either asymmetrical or bilaterally symmetrical echinoderms. Their bodies were spindle-shaped, and a single ambulacrum wound about the spirally pleated skeleton (Fig. 16.46). These ancient animals may have rested vertically in soft substrata. Their skeletal pleats made it possible to extend the body for feeding and to contract for protection. Such a lifestyle would be reminiscent of that of many minor coelomates, including lophophorates, sipunculids, and hemichordates.

Helicoplacoids were not the only echinoderms in which radial symmetry was lacking. The diverse carpoids (now recognized as four classes) were also more or less bilateral in form. Typically, the central body stretched horizontally along the substratum, to which it was attached by a short stalk (Fig. 16.47a). A crown of tentacles—similar to those of the lophophorates—may have surrounded the upper, central mouth, and in later carpoids, one or two brachioles were present (Fig. 16.47b).

The eocrinoids resembled the modern crinoids, except that, like most fossil echinoderms, the central body was enclosed entirely by the skeletal theca (recall the membranous tegment of modern crinoids). Arms and pinnules were not present on the eocrinoid body, but pentamerous brachioles did extend from the oral

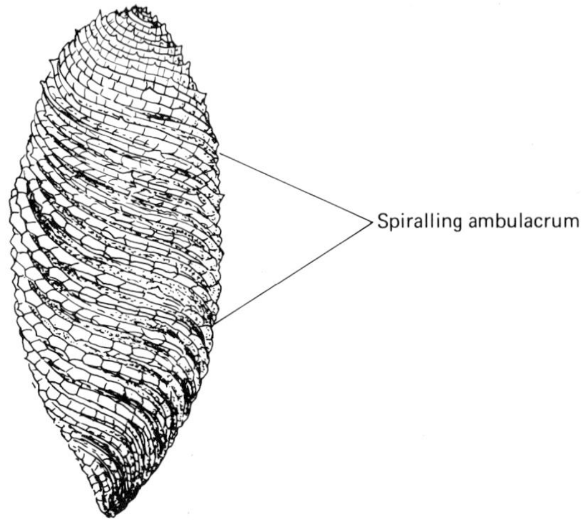

Figure 16.46 Reconstruction of a helicoplacoid, *Helicoplacus curtisi*, from the Lower Cambrian, in lateral view. (From R.C. Moore, ed., *Treatise on Invertebrate Paleontology, Part U.*)

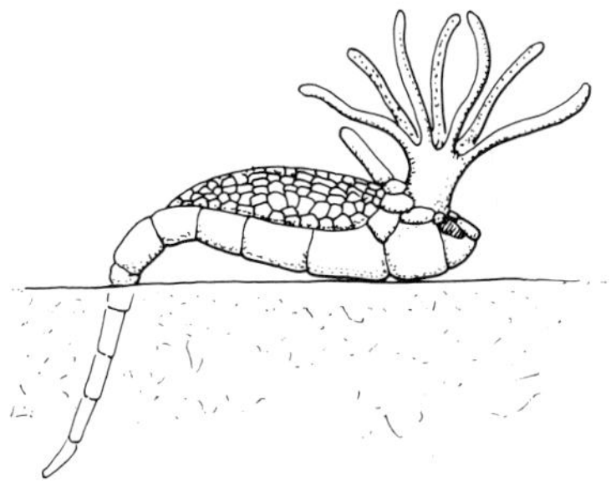

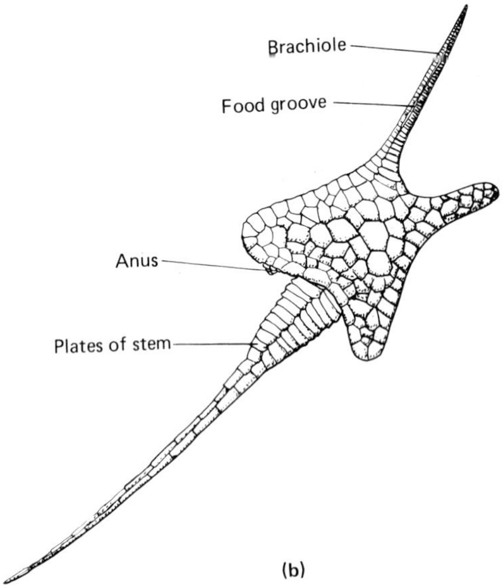

Figure 16.47 (*a*) Reconstruction of a carpoid, *Gyrocystis*, in its probable feeding position; (*b*) A carpoid, *Dendrocystites*, from the Ordovician. (*a* from Nichols, *Echinoderms; b* from Hyman, *The Invertebrates, Vol. IV.*)

surface (Fig. 16.48). Resembling the crinoids, and probably related to them, are two other groups of fossil echinoderms; the Cystoidea and the Blastoidea. Both cystoids and blastoids had an oval theca, which was attached to the substratum either directly or through a stalk (Fig. 16.49). Pentamerous ambulacra extended from the mouth out over the thecal surface; in the blastoids, the water–vascular fields often were mounted atop skeletal ridges (Fig. 16.49b).

Later fossil forms anticipated the free-living existence of most modern members of the phylum. The edrioasteroids had an oval or spherical theca and were probably the first echinoderms to possess spines. Sessile edrioasteroids were stalked (Fig. 16.50a), while unattached species bore an aboral sucker for adherence during resting periods. The latter probably moved by flexing their thecal plates. Pentamerously arranged ambulacra extended across the thecal surface. Some edrioasteroids with sharply defined am-

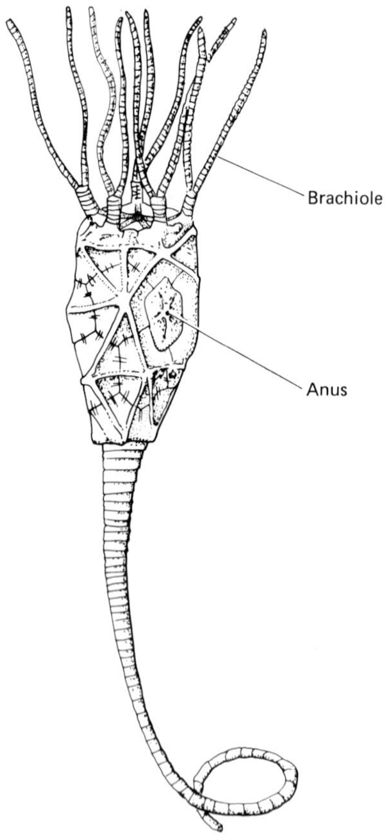

Figure 16.48 An eocrinoid, *Macrocystella mariae* from the upper Cambrian. (From Cuénot, In *Traite de Zoologie, Vol. XI*, Grassé, Ed.)

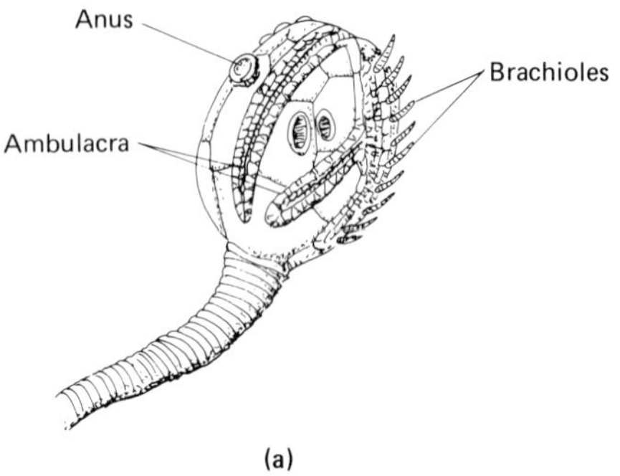

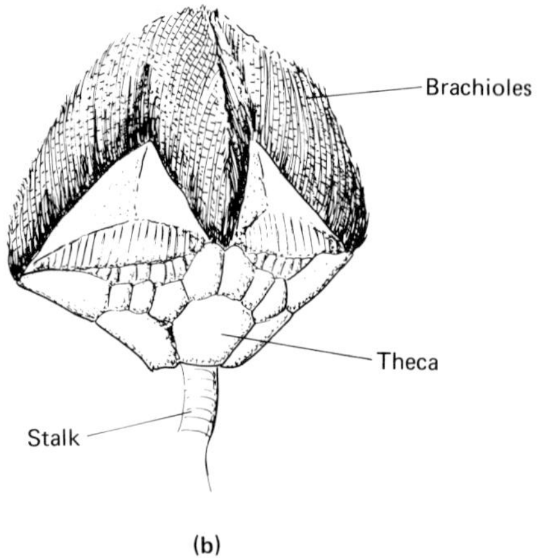

Figure 16.49 (*a*) A cystoid, *Callocystis juvetti* from the Silurian; (*b*) A blastoid, *Blastoidocrinus*. (*a* from Cuénot, In *Traité de Zoologie, Vol. XI*, Grassé, Ed; *b* from Hyman, *The Invertebrates, Vol. IV*.

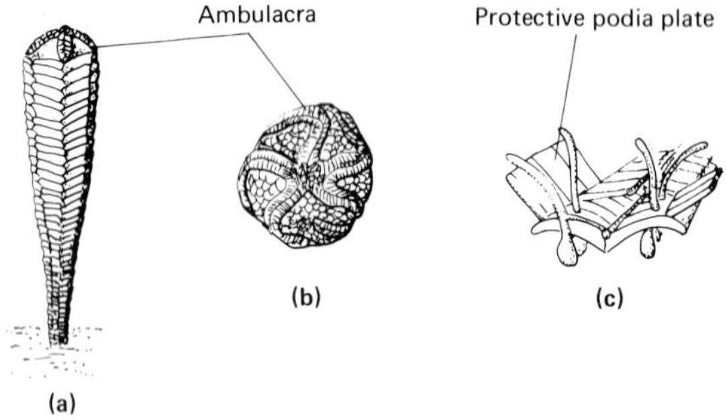

Figure 16.50 Edrioasteroids. (*a*) *Pyrgocystis*, a sessile stalked form from the Ordovician and Silurian periods with ambulacra restricted to the oral surface; (*b*) *Edrioaster*, a stalkless form existing from the Ordovician into the Devonian; (*c*) Reconstruction of part of one ambulacrum of *Edrioaster*. The first two coverplates on each side are open, the rest are nearly closed. The length of the tube feet is unknown. (From Nichols, *Echinoderms*.)

bulacra resembled a modern sea star clutching a plated ball (Fig. 16.50b). The podia projected through holes in ambulacral ossicles which themselves were flanked by protective, movable plates (Fig. 16.50c).

The ophiocistioids apparently were quite mobile echinoderms which applied a flattened oral surface to the substratum (Fig. 16.51). The aboral surface was dome shaped. At the center of the lower surface, five ambulacra converged at a mouth which was armed with chewing plates (Fig. 16.51b). Thus the ophiocistioids foreshadowed the modern posture of the phylum, and in their oral architecture, they resembled the echinoids. Ophiocistioid ambulacra were unusual in that each contained three pairs of giant tube feet. These outsized podia almost surely had locomotor powers.

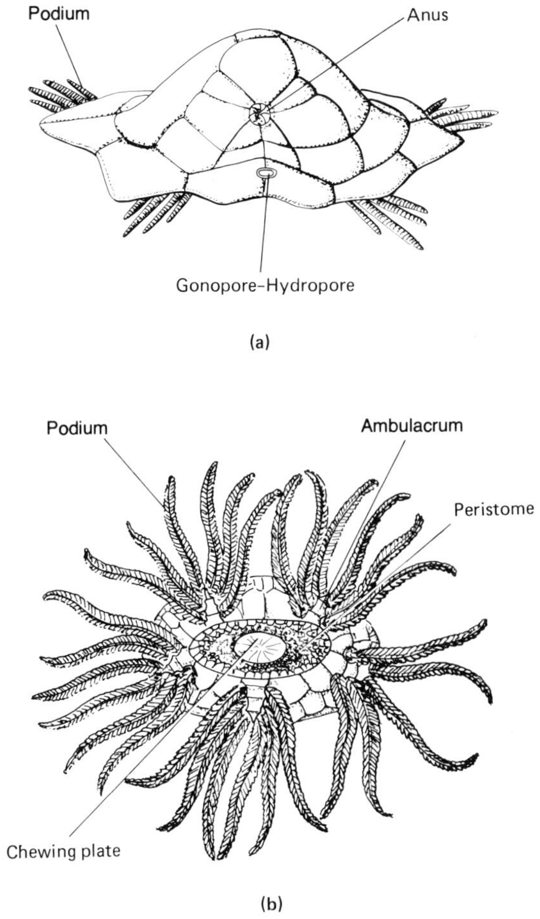

Figure 16.51 Ophiocistioids. (*a*) Aboral surface of *Volchovia*; (*b*) *Sollasina*, oral surface. (From Hyman, *The Invertebrates, Vol. IV*.)

Authorities continue in their attempts to interpret the fossil record and other clues to the evolution of the Echinodermata, but differences of opinion abound. The oldest interpretation of echinoderm history established two subphyla, the Pelmatozoa and the Eleutherozoa. Pelmatozoan echinoderms were attached animals and included the crinoids and most fossil forms. According to this theory, their sessile habits were associated with the phylum's assumption of both secondary radial symmetry and a protective endoskeleton. The pelmatozoan mouth was directed upward, and the ambulacral system apparently evolved as a food-gathering mechanism. Later, the eleutherozoan, or free-living, echinoderms arose from some pelmatozoan stock. These animals reversed the orientation of the mouth, applying it against the substratum, and adapted for a more mobile life-style. Podia became involved in locomotion, and several independent trends toward bilateralism ensued.

The pelmatozoan–eleutherozoan dichotomy came under increasing criticism, largely because of the absence of adequate transition stages between the two subphyla. As we mentioned previously, some of the oldest echinoderms were not sessile, radially symmetrical animals. Mobile echinoderms may well have a continuous evolutionary lineage.

A more recent classification system recognizes four echinoderm subphyla (Fig. 16.52). The first, the Homalozoa, includes all of the carpoids; this extinct group apparently is not ancestral to any living forms. The second subphylum is the Crinozoa. The cystoids and blastoids are peripheral classes within this group, as are the eocrinoids. The central class Crinoidea is represented by both fossil and living members. The third subphylum, the Asterozoa, comprises the modern stelleroids. And the fourth subphylum, the Echinozoa, includes the fossil helicoplacoids, edrioasteroids, and ophiocistioids, as well as the modern echinoids and holothuroids. This new system recognizes the limits of our knowledge of echinoderm origin and provides a framework for further research and speculation. Currently, the helicoplacoids are indicated as possible ancestors for the modern Echinozoa; and the ophiocistioids certainly show echinoid affinities. The origin of the Asterozoa remains more obscure. Two lines of descent—one from the crinoids and the second from the helicoplacoids via the edrioasteroids—are plausible.

There are, however, serious objections to this four-subphyla system. In denying a sessile ancestry for all echinoderms, the classification does little to explain the radial symmetry and skeleton of most free-living modern forms. The wide separation of the echinoids

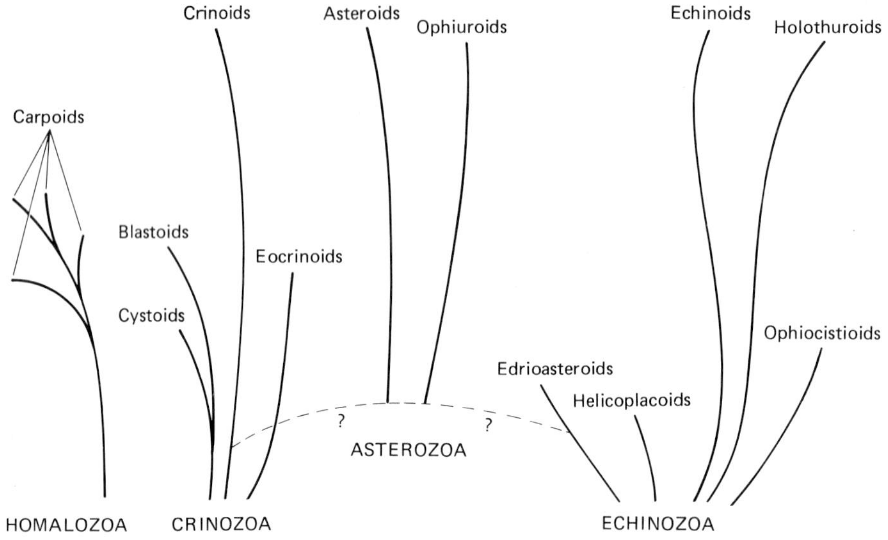

Figure 16.52 A possible phylogenetic scheme for the Phylum Echinodermata.

and ophiuroids completely discounts resemblances between the pluteus larvae of the two classes. As we mentioned, however, such a resemblance might be explained in terms of evolutionary convergence.

Because echinoderms are such unusual invertebrates, their relationships with other phyla are not well defined. As deuterostomes, they are allied with hemichordates and chordates, and some affinity with the lophophorate phyla is implied. More specific theories have traced echinoderm origins to various larva-like forms and even to the phylum Sipuncula.

A classical theory of echinoderm origin describes a bilaterally symmetrical, swimming or crawling animal called a **dipleurula** (Fig. 16.53a). This creature represents a composite of the initial larval stages in the modern classes. Three coelomic compartments are present in the dipleurula. According to this theory, the dipleurula's evolutionary progression is recapitulated in modern echinoderm embryogeny. In adults, coelomic entities gradually were lost on the right side, and the left compartments surrounded the gut to initiate the phylum's radial symmetry. The water–vascular system then evolved from the middle coelom.

Related to the dipleurula theory is another historical view which describes a pentactula ancestor (Fig. 16.53b). The theoretical **pentactula** was a bilateral animal with a ring of hollow tentacles about the mouth. These tentacles supposedly were operated by a hydraulic system emanating from an isolated middle coelom. Thus the tentacles would have borne a strong resemblance to a lophophore. Echinoderms might have developed from the pentactula when the tentacles became reinforced with calcite and thin-walled extensions sprouted along their lengths (Fig. 16.53c). Each tentacle would have become a radial canal, and its extensions would have formed podia.

Another theory suggests that sipunculans may be ancestral to both the lophophorates and the echinoderms. This idea relates to the pentactula theory in that it derives the ambulacral system from an oral ring of hydraulically active tentacles (Fig. 16.53d, e). The sipunculan theory argues that calcification of the tentacles was a defensive adaptation and that podia may have originated as respiratory organs. Podial muscularization also represented a defensive adaptation (they could thus be withdrawn from predators) and made possible later roles in food capture and locomotion. Although it has advantages, the sipunculan theory is criticized most frequently because of the distinct protostomate nature of the sipunculan group. Once again, we confess ignorance on a central evolutionary question and resign ourselves to more puzzlement and further research on this most curious phylum.

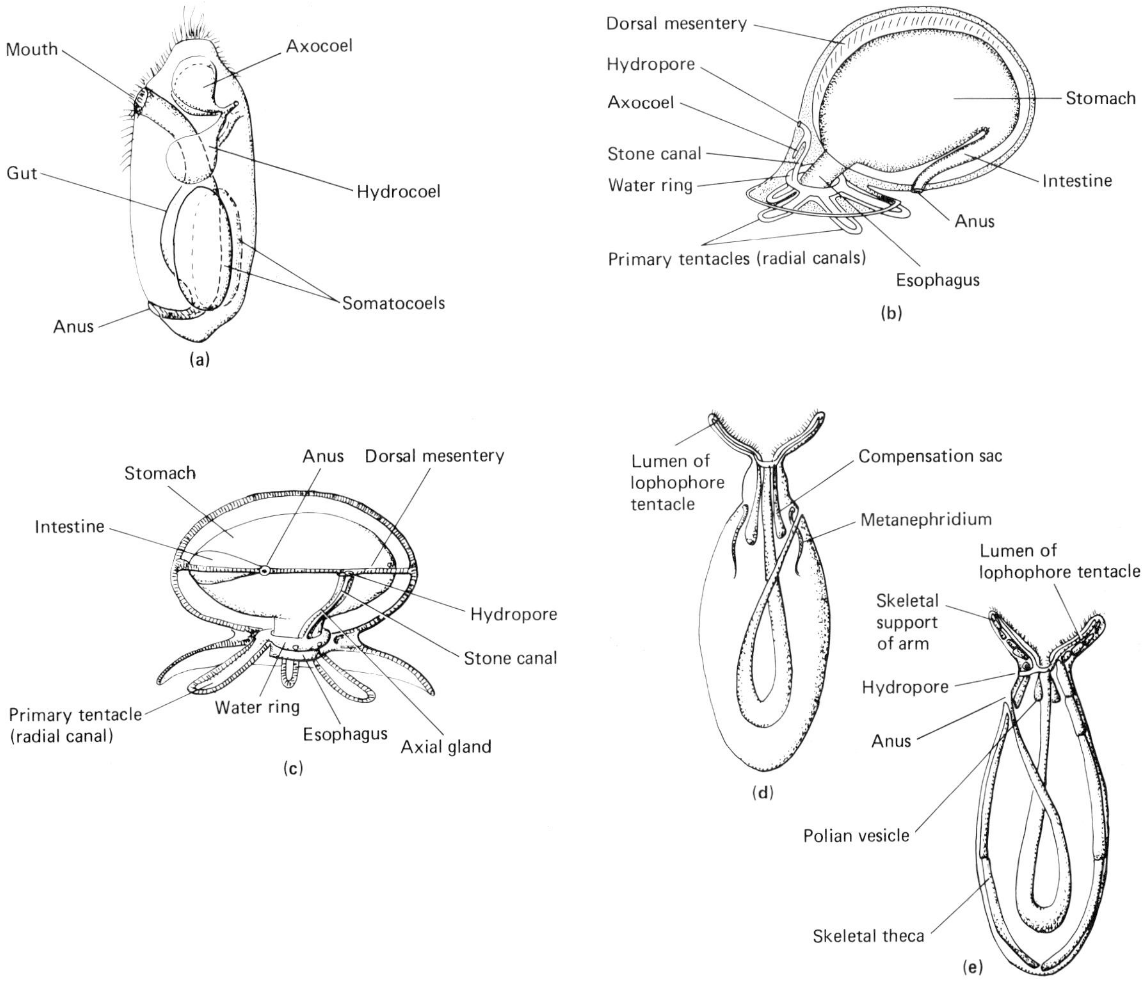

Figure 16.53 (*a*) The hypothetical dipleurula larva; (*b*) The hypothetical pentactula ancestor in its original bilateral condition; (*c*) The pentactula after assuming radial symmetry; (*d* and *e*) Diagrams to show the transformations necessary to derive a primitive echinoderm from a sipunculan-like animal; (*d*) Basic anatomy of a sipunculan-like animal; (*e*) Basic anatomy of a primitive echinoderm. (*a* from Cuénot, In *Traité de Zoologie*, *Vol. XI*, Grassé, Ed; *b* and *c* from Hyman, *The Invertebrates*, *Vol.IV*; *d* and *e* from Nichols, In *Echinoderm Biology*, Millot, Ed.)

for Further Reading

Anderson, J. M. Studies on functional morphology in the digestive system of *Oreaster reticularis Biol. Bull.* 154: 1–14, 1978.

Binyon, J. *Physiology of Echinoderms.* Pergammon Press, Oxford, 1972.

Boolootian, R. A. (Ed.) *Physiology of Echinodermata.* Wiley-Interscience, New York, 1966.

Broom, D. M. Aggregation behavior of the brittle star *Ophiotrix fragilis. J. Mar. Biol. Assoc. U. K.* 55: 191–197, 1975.

Burke, R. D. Podial sensory receptors and the induction of metamorphosis in echinoids. *J. Exp. Mar. Biol. Ecol.* 47: 223–234, 1980.

Carey, A. G. Food sources of sublittoral, bathyal and abyssal asteroids in the northeast Pacific Ocean. *Ophelia* 10: 35–47, 1972.

Chia, F., and Whiteley, A. H. (Eds.) Developmental biology of the echinoderms. *Am. Zool.* 15: 483–775. (Papers presented at a symposium in 1973).

Christensen, A. M. Feeding biology of the sea star *Astropecten irregularis. Ophelia* 8: 1–134, 1970.

Clark, A. M. *Starfishes and their Relations.* British Museum, London, 1962.

Coe, W. R. *Starfishes, Serpent Stars, Sea Urchins,and Sea Cucumbers of the Northeast.* Dover Publications, New York, 1972.

Downey, M. E. *Starfishes from the Caribbean and the Gulf of Mexico.* Smithsonian Contributions to Zoology 126: 1–158, 1973.

Eakin, R. M., and Brandenburger, J. L. Effects of light on ocelli on sea stars. *Zoolmorphologie 92:* 191–200, 1979.

Feder, H. M. Organisms responsive to predatory sea stars. *Sarsia* 29: 371–394, 1967.

Fell, H. B. The phylogeny of sea stars. *Phil. Trans. Roy, Soc. London Ser. B.* 246: 381–485, 1963.

Fenner, D. H. The respiratory adaptations of the podia and ampullae of echinoids. *Biol. Bull.* 145: 323–339, 1973.

Fish, J. D. Biology of *Cucumaria elongata. J. Mar. Biol. Assoc. U.K. 47:* 143, 1967.

Florkin, M., and Scheer, B. T. (Eds.) *Chemical Zoology Vol. III: Echinodermata, Nemotoda and Acanthocephala.* Academic Press, New York, 1969.

Haugh, B. N. and Bell, B. M. Fossilized viscera in primitive echinoderms. *Science* 209: 653–657, 1980.

Herreid, C. F., LaRussa, V. F., and Defesi, C. R. Blood vascular system of the sea cucumber, *Stichopus moebii. J. Morphol.* 150: 423–451, 1976.

Hotchkiss, F. H. Case studies in the teratology of starfish. *Proc. Acad. Nat. Sci. Phila.* 131: 139–157, 1979.

Houk, M. S., and Hinegardner, R. T. The formation and early differentiation of sea urchin gonads. *Biol. Bull.* 159. 280–294, 1980.

Hyman, L. H. *The Invertebrates, Vol. IV: Echinodermata.* McGraw-Hill, 1955. (The chapter "retrospect" in Vol. V 1959 for this series summarizes the literature on echinoderms from 1955 to 1959)

LaTouche, R. W. The feeding behavior of the feather star *Antedon bifida. J. Mar. Biol. Assoc. U.K.* 58: 877–890, 1978.

Lawrence, J. M. On the relationship between marine plants and sea urchins. *Ann. Rev. Oceanogr. Mar. Biol.* 13: 213–286, 1975.

Laxton, J. H. Aspects of the ecology of the coral-eating starfish, *Acanthaster planci. Biol. J. Linn. Soc.* 6: 19–45, 1974.

Macurda, D. B., and Meyer, D. L. Feeding posture of modern stalked crinoids. *Nature* 247: 394–396, 1974.

Matsumuara, T. Hasegawa, M., and Shigel, M. Collagen biochemistry and phylogeny of echinoderms. *Comp. Biochem. Physiol. B. Comp. Biochem.* 62: 101–1–6, 1979.

Millot, N. (Ed.) *Echinoderm Biology.* Academic Press, New York, 1967.

Millot, N. The photosensitivity of echinoids. *Adv. Mar. Biol.* 13: 1–52, 1975.

Nichols, D. *Echinoderms,* 4th ed. Hutchinson University Library, London, 1969.

Paine, R. T. Size-limited predation: An observational and experimental approach with the *Mytilus-Pisaster* interaction. *Ecology* 57: 858–873, 1976.

Pentreath, R. J. Feeding mechanisms and the functional morphology of podia and spines in some New Zealand ophiuroids. *J. Zool.* 161: 395–429, 1970.

Seilacher, A. Constructional morphology of sand dollars. Paleobiology 5: 191–221, 1979.

Stephenson, D. G. Pentamerism and the ancestral echinoderm. *Nature (London)* 250: 82–83, 1974.

Strathmann, R. R. Larval feeding in echinoderms. *Am. Zool.* 15: 717–730, 1975.

Strathmann, R. R. Echinoid larvae from the northeast Pacific with a key and comment on an unusual type of planktotrophic development. *Can. J. Zool.* 57: 610–616, 1979.

Timko, P. Sand dollars as suspension feeders: A new description of feeding in *Dendraster excentricus. Biol. Bull.* 151: 247–259, 1976.

Wilkie, I. C. Arm autonomy in brittle stars. *J. Zool.* 186: 311–330, 1978.

Woodley, J. D. The behavior of some amphiurid brittle stars. *J. Exp. Mar. Biol. Ecol.* 18: 29–46, 1975.

17

The Other Deuterostomes

We conclude our phyletic series with an account of the remaining deuterostomate groups, which include the urochordates, hemichordates, and chaetognaths. Such animals rank among the most specialized of all invertebrates. Because of their relationship with the largest deuterostomate group, the Vertebrata, these creatures occupy a unique position in the invertebrate world. Among their ancestors, we seek the origins of our own branch of the animal kingdom.

The deuterostomate line is an ancient one, fully as old as the protostomate lineage; indeed, many authorities suggest that deuterostomes were the first coelomates. In any case, a number of deuterostomate types must have come and gone during the long history of the group, and the extant forms represent a diverse assemblage. There is no obvious phylogenetic relationship among the various modern deuterostomes, so relatively high taxonomic levels are assigned to the major groups. Both the hemichordates and the chaetognaths are recognized as phyla, while the urochordates are considered a subphylum of the Chordata. Phylum Chordata also includes the subphyla Cephalochordata and Vertebrata, two groups discussed in vertebrate biology texts.

Despite their diversity, deuterostomes do share certain common traits. Their embryogenies demonstrate the characters outlined in Chapter Three: radial, indeterminate cleavage; an anus derived from the blastopore; and enterocoely. Furthermore, many of these animals possess a well-defined tripartite coelom. The three coelomic compartments include an anterior **protocoel**, a middle **mesocoel**, and a posterior **metacoel**. These cavities are compared with the axocoel, hydrocoel, and somatocoel of the echinoderms. All of the animals described in this chapter are marine.

Phylum Chordata—Subphylum Urochordata

The deuterostomes of the subphylum Urochordata commonly are known as tunicates. Adult tunicates generally do not look like chordates (Fig. 17.1). The majority are secondarily simplified, filter-feeding animals, which live on the shallow ocean floor. Only the tunicate larva clearly displays the requisite chordate structures: the pharyngeal clefts, the dorsal hollow nerve cord, and the notochord. Most of the 1500 known species belong to the class Ascidiacea. Two other tunicate classes, the Thaliacea and the Larvacea, contain highly specialized planktonic forms.

CLASS ASCIDIACEA

The ascidians, or sea squirts, are sessile urochordates. Throughout the oceans of the world, they are attached to hard substrata and even to sand and mud. While certain species live on deep-sea bottoms, littoral habitats are favored by most members of the group. Coral reefs often abound with ascidian forms, and sea squirt colonies may blanket underwater rocks and pilings.

The adult body is derived through a complex larval metamorphosis. By tracing the fate of larval parts, it is possible to describe the structural axes of the adult animal. Ascidians are attached at their posterior ends. Attachment is direct on hard substrata, but usually is mediated by a stalk in species that inhabit sandy or muddy bottoms. At the free end of the animal are two openings, the **buccal siphon** and the **atrial siphon** (Fig. 17.2). The apical buccal siphon occupies the ascidian anterior, while the atrial siphon defines the animal's dorsal side. These two openings play vital physiological roles.

The ascidian body wall comprises a thin epidermis, a thick connective dermis, and multidirectional smooth muscle fibers. The outermost surface, however, is an epidermal secretion called the **tunic,** which consists of a fibrous matrix (composed largely of a celluloselike material called **tunicin**) and its associated salts and proteins. In addition, amoeboid cells usually migrate within this outer cover, and in a few ascidians, blood vessels are present. Attachment is largely a function of the tunic, whose posterior is modified to bind to the substratum. Rootlike **stolons** extend from the attached surfaces of some ascidians (e.g., *Clavelina*; Fig. 17.3). These structures provide anchorage and also may serve as budding sites. The tunic invaginates at the buccal and atrial siphons, forming a lip about both openings. The base of the buccal opening is circled by tentacles.

Internally, the ascidian body is dominated by a large pharyngeal basket (Fig. 17.2). This rather cylindrical organ is attached to the ventral body wall. Surrounding the pharynx dorsally and laterally is a cavity known as the atrium, which communicates with the outside world via the atrial siphon. Along its free margins, the ascidian pharynx bears numerous slits called **stig-**

(a) (b)

Figure 17.1 (*a*) A solitary tunicate or sea squirt, *Herdmania momus* (Class Ascidiacea); (*b*) A colonial ascidian (*Botrylloides* sp.) encrusting a rocky substratum. (Photos courtesy of S. Arthur Reed).

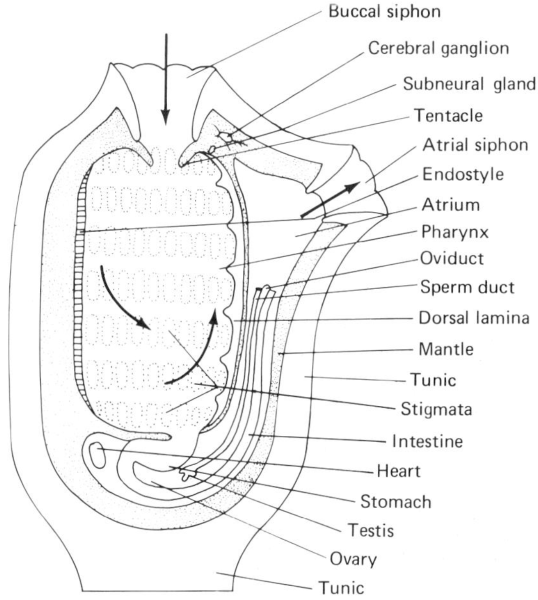

Figure 17.2 Longitudinal section through an ascidian illustrating basic anatomy. Arrows show direction of water flow (From H. Harant and P. Vernieres, *Faune de France, Vol. 27,* 1933, Paul Lechevalier.)

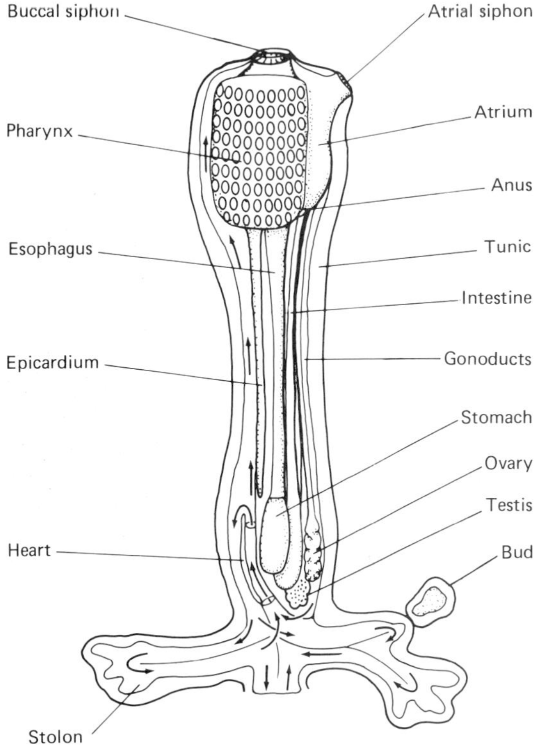

Figure 17.3 The structure of the ascidian *Clavelina lepadiformis* including the root-like stolons. Arrows indicate the pattern of blood circulation.(From P. Brien, In *Traité de Zoologie, Vol. XI,* P.P. Grassé, Ed., 1948, Masson et Cie.)

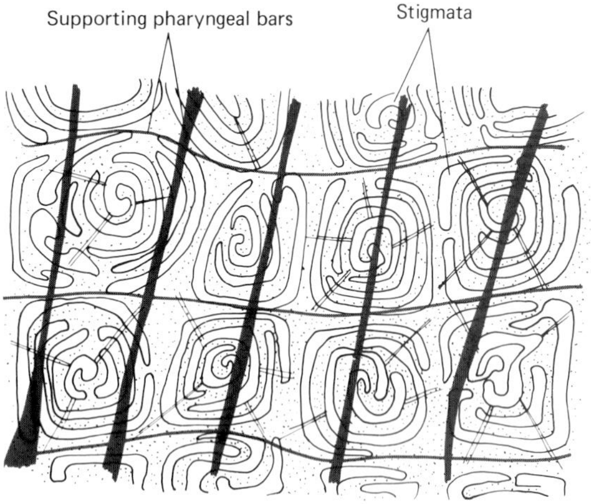

Figure 17.4 A portion of the pharyngeal basket of the ascidian, *Corella parallelogramma,* viewed from within showing spiraled stigmata. (From M. de Sélys-Longchamps, Ann. Soc. Roy. Zool. et Malaco-logiques de Belgique, 1914, *48:* 127.

mata. In many ascidians, the stigmata and their supportive structures form a complex grid network, and spiraled slits are common in some species (Fig. 17.4). These modifications, which are often useful in taxonomy, increase the pharyngeal surface area. The stigmata are lined with cilia which produce a steady current of water through the animal. Sea water enters the pharynx through the buccal siphon, passes through the stigmata into the atrium, and then exits through the atrial siphon. This current is involved in ascidian nutrition, respiration, excretion, and reproduction. Large quantities of water—up to several thousand times the body volume per day—are passed by all ascidians.

A sea squirt feeds on plankton, which it filters from the water passing through its pharynx. Along the ventral side of the pharynx, a vertical groove known as the **endostyle** discharges large quantities of mucus (Figs. 17.2 and 17.5). This mucus moves as a continuous sheet across the pharyngeal walls. In passing over the stigmata, the mucous sheet collects suspended food particles and then continues toward the dorsal midline. There, a tissue ridge called the **dorsal lamina** or a vertical series of tentacles called dorsal **languets** enrolls the pharyngeal mucus within a gutter (Fig. 17.5). A mucus–food cord is delivered posteriorly to

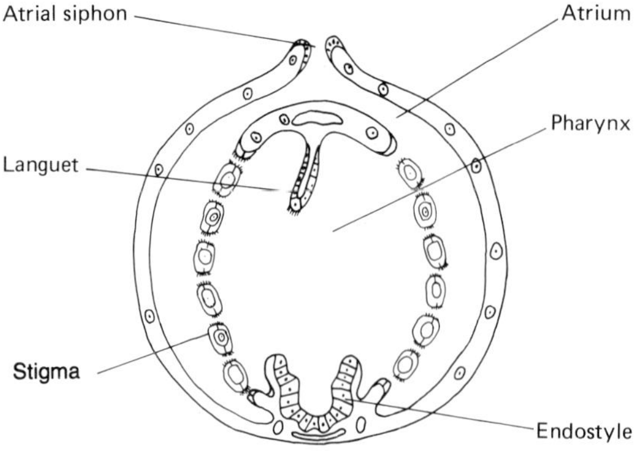

Figure 17.5 Cross section of *Clavelina lepadiformis* at the level of the atrial siphon. (From O. Seeliger, In *Klassen und Ordnungen des Tierreichs, Vol. III*, H. G. Bronn, Ed., 1911, C. F. Winter.)

the esophageal opening on the dorsal floor of the pharynx.

The rest of the digestive tract comprises a ciliated, horseshoe-shaped tube, which descends into the ascidian posterior and then ascends to the atrial floor (Fig. 17.2). This tube includes a narrow esophagus, a sphinctered stomach, and a rather long and sometimes coiled intestine. Digestion occurs extracellularly within the stomach; absorption is largely an intestinal event. Feces are discharged through the anus, which is located just below the atrial siphon. The atrial current flushes waste matter from the body.

Other ascidian maintenance systems are rather rudimentary. The open circulatory system features a pericardial region and a somewhat ill-defined heart, which is simply an unwalled muscular fold of the pericardium (Figs. 17.2 and 17.3). It lies near the digestive loop, with its two open ends pointed ventrally and dorsally. The heart pumps blood through an informal system of vessels and open sinuses. From the ventral side of the heart, a large subendostylar vessel passes anteriorly along the ventral wall of the pharynx. This vessel joins numerous blood channels, which

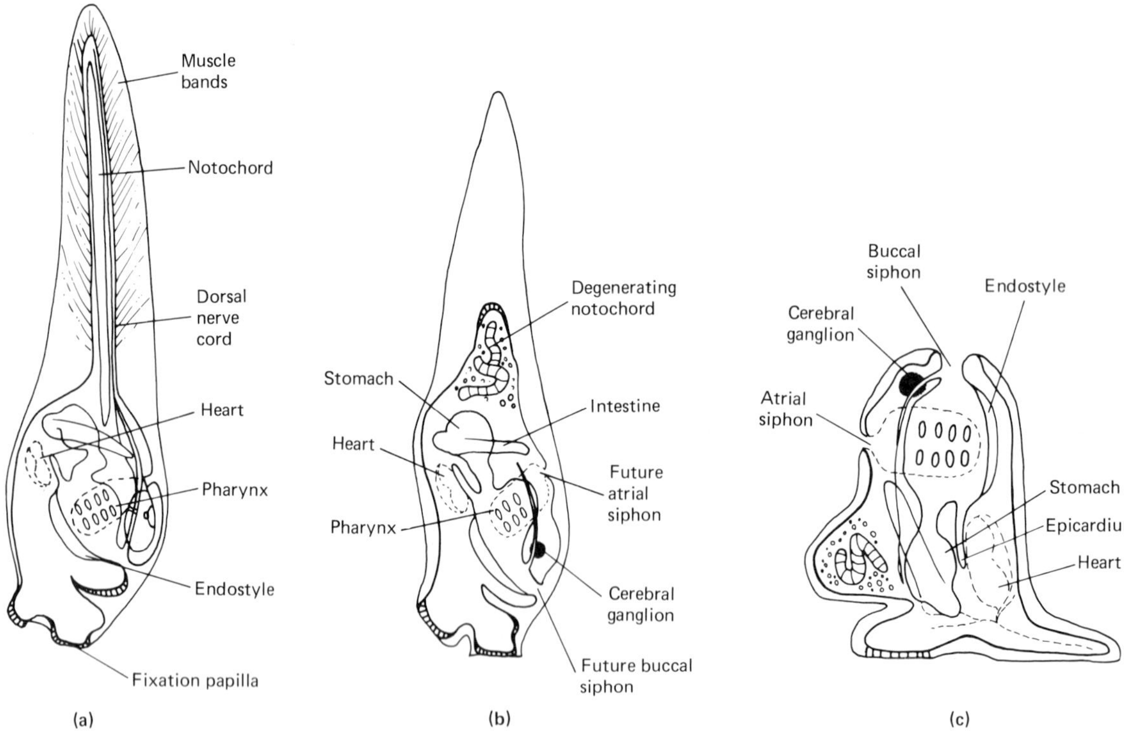

Figure 17.6 (*a*) Planktonic larva of the ascidian *Clavelina;* (*b*) An early stage in the metamorphosis of the "tadpole" larva of *Clavelina;* (*c*) A late stage in metamorphosis. Most adult structures apparent. (From Seeliger, In *Klassen und Ordnungen des Tierreichs, Vol. III*, Brown, Ed.)

saturate the pharyngeal grid. Pharyngeal blood collected within a medial dorsal sinus is passed to the viscera. An abdominal sinus returns blood to the dorsal end of the heart. Oddly, blood may pass in either direction through the ascidian circulatory system. Periodic reversals in blood flow occur, as the heart pumps blood first from its ventral and then from its dorsal end. This phenomenon involves a fluctuation in dominance by dual excitation centers at each end of the heart.

Oxygen uptake occurs within the blood channels of the pharyngeal grid. In some species, additional respiratory exchange may take place across the external surface. Reflecting their sessile condition, ascidians have a very low metabolic rate. Typically, they extract no more than 10% of the available oxygen from the surrounding sea water. Such inefficiency is all the more tolerable because of the hypertrophied surface area of the pharynx. Ascidian blood contains no respiratory pigment, but numerous amoebocytes are present. These multipurpose maintenance cells include **vanadocytes** and excretory bodies. Vanadocytes occur in only a few species. These green cells house large quantities of vanadium, an element extracted from sea water by pharyngeal mucus. Vanadial concentrations half a million times that of sea water may be established, but why ascidians accumulate this rare metal remains a mystery. Excretory amoebocytes absorb uric acid and other solid wastes. When full, these cells migrate to certain areas of the body (usually the outer walls of the digestive and reproductive organs), where they are deposited for the duration of the life span. Ammonia represents 90½ of the nitrogenous waste of ascidians and is removed by diffusion across any convenient surface.

Ascidians do not have a true coelom. An obvious relationship with other deuterostomes indicates that the ascidian coelom has been lost in the course of their rather specialized evolution. The pericardium and an unusual epicardium, however, may represent coelomic remnants. The ascidian **epicardium** is an endodermally derived tube, or system of tubes, which parallels most of the digestive loop (Fig. 17.3). It often participates in asexual budding, but its exact function remains unknown.

As sessile creatures, adult ascidians do not possess well-developed activity systems. Their body wall muscles include only minor fibers, whose contractions are limited by the flexibility of the tunic. Sphincter muscles surround the buccal and atrial siphons and can seal these openings when the animal is threatened. As we described, ciliary action plays a major role in feeding and respiratory processes, and the importance of muscular activity is correspondingly low.

The ascidian nervous system features a (so-called) cerebral ganglion which lies within the body wall between the two siphons (Fig. 17.2). From this ganglion, several nerves communicate with the major body structures. Removal of the cerebral ganglion seldom interferes with basic physiological processes, a fact that testifies to the decentralization of nervous control in these animals. Below the ganglion lies a mass of tissue known as the **subneural gland** (Fig. 17.2). A duct from this gland leads into a ciliated funnel within the pharyngeal wall; posteriorly, the organ connects with a length of tissue representing the larval nerve cord. The subneural gland may be homologous with the vertebrate pituitary gland.

Specialized sense organs are absent from adult members of this class. However, widely scattered epidermal sensory cells are numerous about the siphonal openings, within the atrium, and on the buccal tentacles. They transduce tactile and chemical stimuli and connect with local neuromuscular elements which control the water current.

Almost all ascidians are hermaphrodites. Typically, a single testis and an ovary are present near the stomach and separate gonoducts parallel the intestine before discharging below the atrial siphon (Figs. 17.2 and 17.3). In some species, both gonads and gonoducts are paired. Because the atrium serves both the digestive and reproductive systems, it may be called a cloaca.

Solitary ascidians usually produce eggs with minimal yolk content. These ova are fertilized in the open sea. Radial cleavage and gastrulation by epiboly or invagination produce an embryo which soon elongates. Subsequent developmental stages express the chordate identity of the class. The archenteron elaborates a supportive rod, the notochord, and an ectodermal evagination enrolls to form the dorsal hollow nerve cord. Both of these structures occupy a caudal extension of the larva (Fig. 17.6a). For this reason, the ascidian larva often is called a "tadpole." The ascidian tadpole never forms a coelom. Anteriorly, it bears a pigment cup ocellus and a single statocyst. As the pharynx begins to form, only a few stigmata are associated with a tiny cavity. A mouth, representing the future buccal siphon, usually remains closed throughout the larval period. The ascidian larva never feeds; indeed, its body is wholly enclosed by the epidermally secreted tunic. The tunic elaborates a swimming fin in the tail region, where the contraction of muscular bands provides locomotor power.

Anteriorly, the epidermis and tunic form three **fixation papillae.** After the planktonic period—a stage that often lasts no more than several hours—these papillae attach to a suitable substratum (Fig. 17.6b). Adult metamorphosis follows. Ascidian metamorphosis involves the resorption of the notochord and the dorsal nerve cord and a considerable expansion of the area around the mouth. That expansion displaces the mouth from its original position near the fixation papillae. Eventually, the entire digestive tract and other organ systems are rotated 180° toward the upper, free end of the body (Fig. 17.6c). The pharynx increases in size. With the addition of more stigmata, an enlargement of the atrium, and the establishment of siphonal water flow, adult life begins.

Asexual reproduction contributes to the formation of numerous colonial forms. Individual ascidian zooids tend to be rather small, but whole colonies can attain great size. In the simplest colonial species, zooids bud from a common stolon. This stolon may be vinelike, with scattered, globular individuals—as in *Perophora listeri* (Fig. 17.7)—or short, with densely packed, elongate zooids—as in *Clavelina lepadiformis.* Other ascidian colonies display various levels of structural and functional cooperation among their members. In *Cyathocormus* (Fig. 17.8a), a stalked base supports a cuplike arrangement of zooids. Each individual has its buccal siphon pointed toward the outside of the cup, and its atrial siphon discharges into the cup center. *Coelocormus* (Fig. 17.8b) is a colonial ascidian whose doubled wall supports inner and outer layers of zooids. All of the buccal siphons open on the exterior, while the atrial siphons discharge into the lumen separating the two zooid layers. This lumen has its external opening at the base of the cup. The trend toward greater integration of colonial zooids continues within the genus *Botryllus* (Fig. 17.9). In these colonial animals, star-shaped clusters of individuals share a common tunic and a common atrial siphon. Several of these clusters may occur on a single colony.

CLASS THALIACEA

The thaliaceans, or salps, are planktonic urochordates. Only six genera have been identified, most of which are confined to warm seas. Salps resemble ascidians, but their atrial siphons are posterior. The digestive–respiratory current is expelled behind these creatures, jetting them through the water. Both colonial and solitary forms are known, and the class exhibits some very peculiar life cycles. All thaliaceans pass through both sexual and asexual stages. A sexually produced individual, the **oozooid,** undergoes asexual budding to form **blastozooids.** Blastozooids

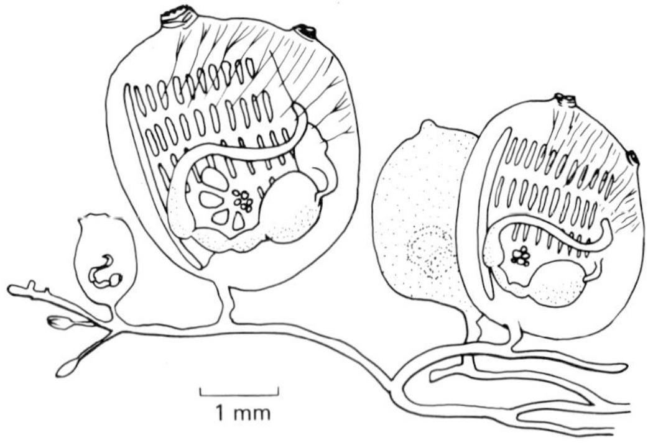

Figure 17.7 Part of a colony of *Perophora listeri*. (From R. H. Millar, *British Ascidians*, 1970, Linnaean Society of London.)

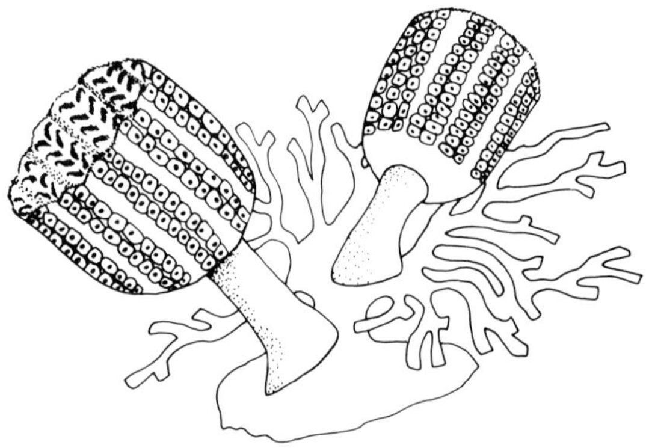

(a)

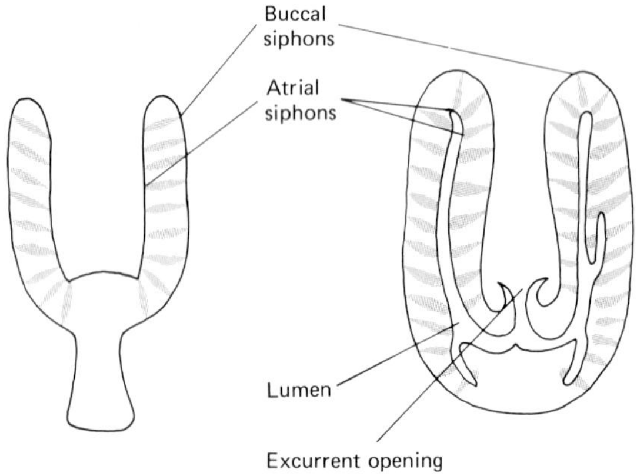

(b)

Figure 17.8 (*a*) Two colonies of *Cyathocormus mirabilis* attached to coral; (*b*) At left a schematic sagittal section through a colony of *Cyathocormus*; at right a schematic sagittal section through a colony of *Coelocormus*. (From Brien, In *Traité de Zoologie, Vol. XI*, Grassé, Ed.)

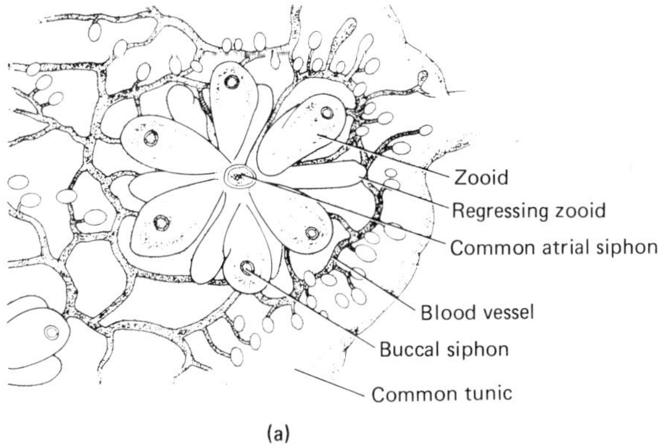

Figure 17.9 (*a*) Part of a colony of *Botryllus schlosseri* showing the external features of a single cluster of individuals; (*b*) Section through a cluster from a *Botryllus* colony showing the arrangement of the zooids. (From I.W. Sherman and V. G. Sherman, *The Invertebrates: Function and Form: A Laboratory Manual*, 2nd ed, 1976, Macmillan.)

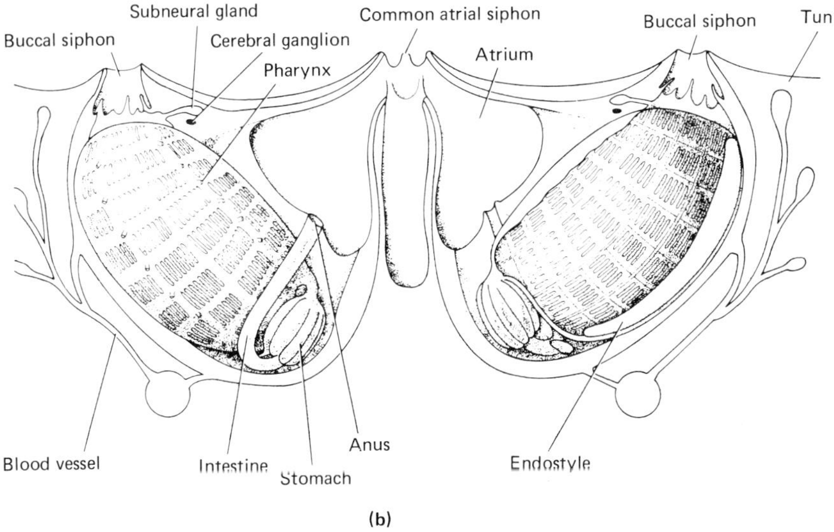

mature into sexual individuals whose fertilized gametes develop into the oozooids of the next generation.

Pyrosoma (Fig. 17.10), a colonial thaliacean, presents a raft of zooids organized about a central cloaca. Such organization is reminiscent of certain colonial ascidians (Fig. 17.8). All of the buccal siphons open externally, while the atrial siphons discharge into the central cavity. Water jets from a single cloacal opening at the rear of the colony. *Pyrosoma* colonies can become quite large—some are over 2 m long—and all are brightly luminescent. Pharyngeal glands apparently produce light, and the cerebral ganglion houses a photoreceptive organ.

Each member of the *Pyrosoma* colony produces a single egg, which develops within the common atrium. The resulting oozooid elaborates a stolon with four blastozooid buds (Fig. 17.10b). This embryonic complex then exits from the parent. The oozooid delivers nutrients to the blastozooids, which grow to surround the oozooid atrium. Blastozooid budding then produces a new *Pyrosoma* colony.

Solitary thaliaceans include *Salpa* and *Doliolum*. Muscle bands partially encircle the cylindrical body of Salpa (Fig. 17.11a); their contraction draws water through the animal. The life cycle of this thaliacean includes an oozooid which buds many blastozooids along a stolon. This stolon extends from the tip of the oozooid endostyle (Fig. 17.11b). As the stolon lengthens, sexually mature blastozooids detach from its distal tip.

Doliolum (Fig. 17.12) demonstrates one of the most complex life cycles in the invertebrate world. As adults, *Doliolum* individuals are barrel shaped, with distinct muscle bands completely surrounding the body. A typical tadpole larva undergoes metamorphosis to produce an oozooid. Various types of buds originate from a stolon and migrate to one of three

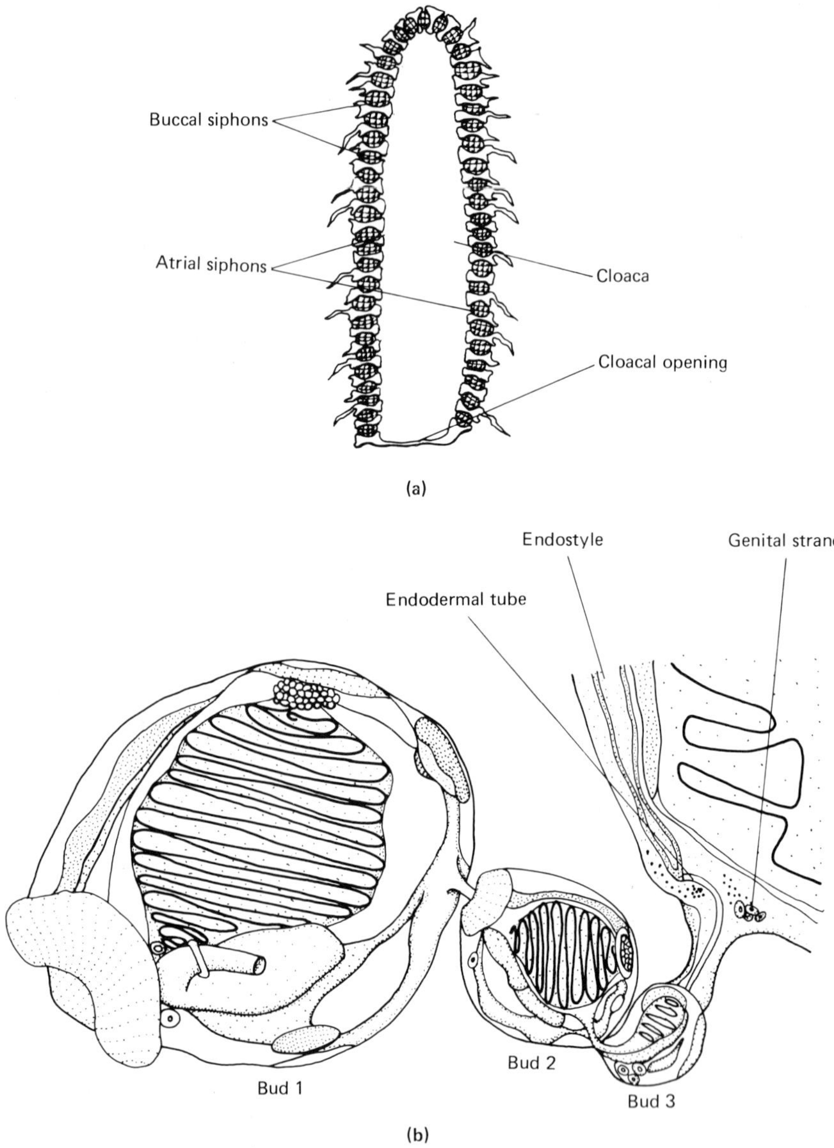

Figure 17.10 (*a*) Sagittal section through a colony of *Pyrosoma*, a thaliacean; (*b*) Chain of blastozooid buds forming from the stolon of an oozoid in *Pyrosoma*. (*a* from Brien, In *Traité de Zoologie, Vol. XI*, Grassé, Ed; *b* from N. J. Berrill, *Growth, Development and Pattern*, 1961, W. H. Freeman.)

rows on a posterior dorsal process (Fig. 17.12b). The parent oozooid resorbs most of its internal organs as it produces these buds. On the two outer rows of the dorsal process, buds develop into nutritive **gastrozooids.** On the middle row, both **phorozooids** and **gonozooids** develop. Phorozooids, which resemble adult oozooids, detach from the parent and swim away, each carrying several gonozooids at its base. Each gonozooid matures sexually and then detaches in turn. Its fertilized gametes become the oozooids of the next generation.

CLASS LARVACEA

The most highly specialized urochordates belong to the Larvacea. As their name indicates, these inverte-

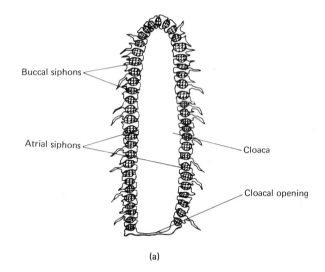

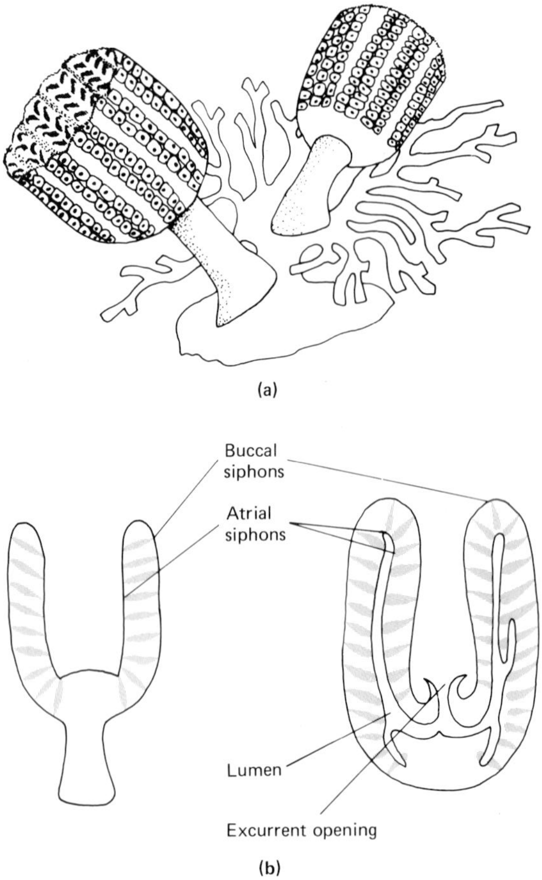

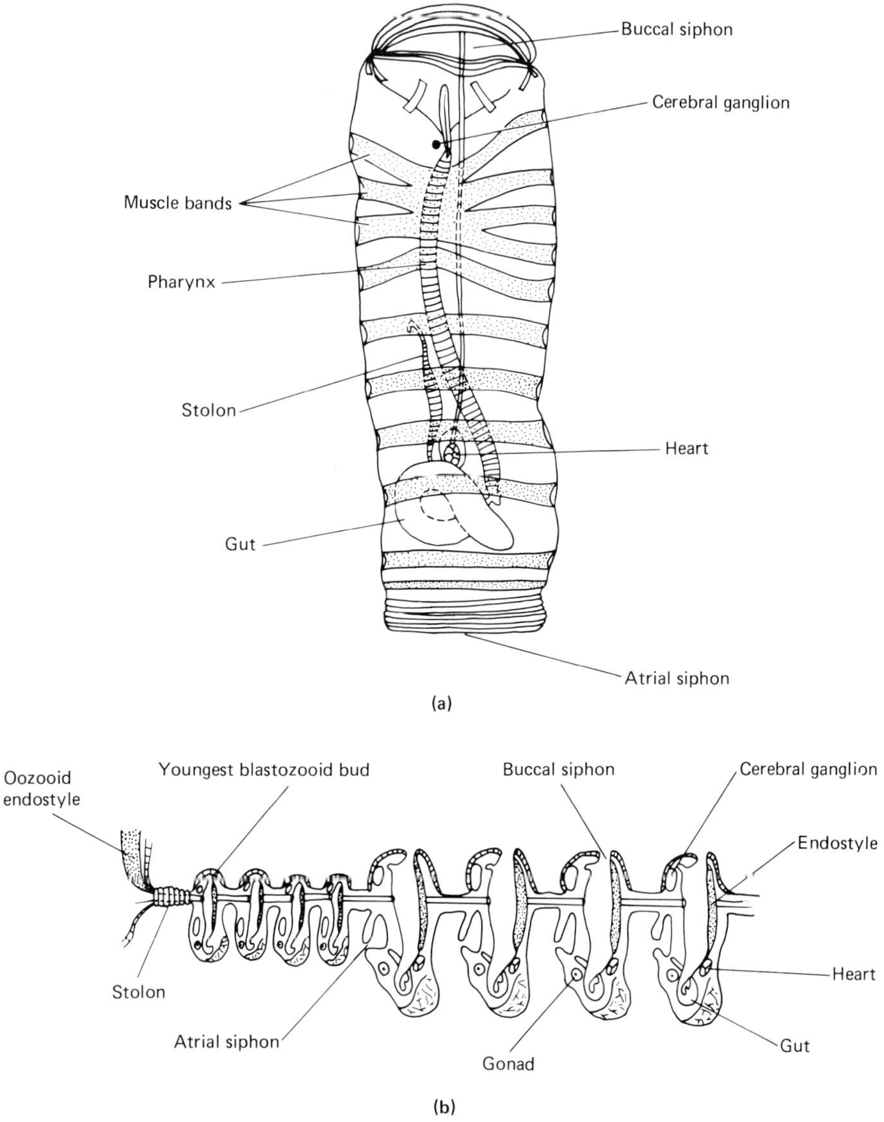

Figure 17.11 (*a*) Dorsal view of an oozoid of the thaliacean, *Salpa cylindrica* (tunic not shown); (*b*) Diagram showing the formation of successive generations of blastozooid buds along the stolon of a salp. Blastozooids are increasingly mature distally. (*a* from J. E. A. Godeaux, In *Chemical Zoology, Vol. VIII*, 1974, Academic Press; *b* from Brien, In *Traité de Zoologie, Vol. XI*, Grassé, Ed.)

brates retain many larval characters throughout adult life. Larvaceans are planktonic organisms with a worldwide distribution. They are always small—only a few millimeters in length—and frequently transparent. Red and violet gonads supply the only color in most species.

The larvacean body resembles an ascidian tadpole whose tail is tucked ventrally (Fig. 17.13). This tail contains the notochord and a dorsal nerve cord. The larvacean digestive system comprises an anterior mouth, a pharynx with an endostyle, a tubular gut, and a midventral anus. These animals are hermaphroditic, and their reproduction is always sexual.

The outstanding feature of the class is the gelatinous "house" secreted by epidermal gland cells. This house may extend as a mere bubble from the mouth or may

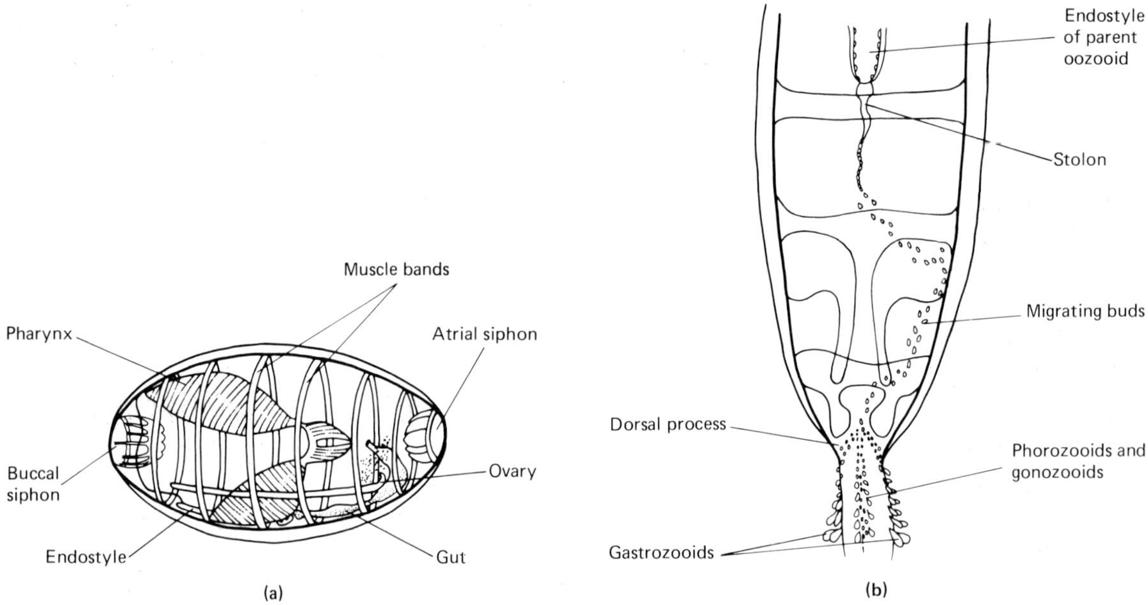

Figure 17.12 (*a*) *Doliolum denticulatum*, mature gonozooid; (*b*) *Doliolum gegenbauri*, posterior part of an oozooid showing the stolon, migrating buds and the dorsal process with three rows of developing buds. (*a* from G. Neumann, In *Handbuch der Zoologie, Vol. 5*, W. Kukenthal and J. Krumbach, Eds., 1934, W. de Gruyter; *b* from H. Harant, In *Traité de Zoologie, Vol XI*, Grassé, Ed.)

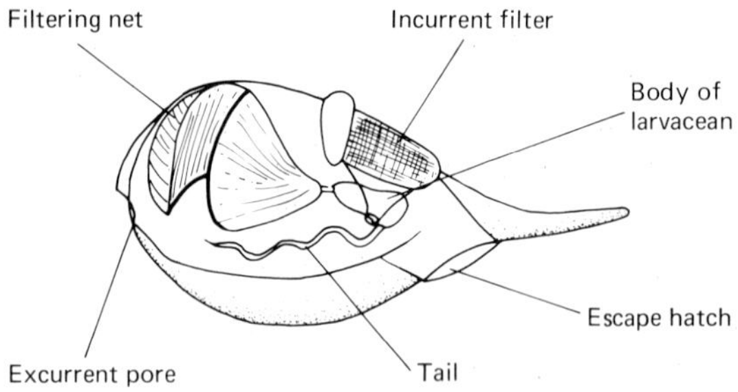

Figure 17.13 A larvacean, *Oikopleura albicans* within its gelatinous "house". (From Lohmann and A. Buckmann in *Handbuch der Zoologie Vol. 5*, Kukenthal and Krumbach, Eds., 1934.)

completely surround the larvacean. In *Oikopleura* (Fig. 17.13), the house is much larger than the body, allowing the animal to move within its gelatinous confines. Lashing movements of the tail draw water through the house. Water enters through a screened orifice which blocks the entry of larger plankton and debris. As the current flows toward the animal's mouth, finer nets strain suspended particles, and only the tiniest plankton are delivered to the pharyngeal filters. Water pressure eventually opens the hinged door of an excurrent pore. On exiting through this pore, the feeding current provides some locomotor force.

In time, the incurrent pore of *Oikopleura* becomes clogged with particulate matter. The animal then abandons its house through an escape hatch. Secretion of a new house follows, a process that may be repeated as often as every 3 hrs.

Phylum Hemichordata

Hemichordates are sedentary wormlike deuterostomes, which once were classified as a subphylum of the Chordata. They are now ranked as a distinct phylum with certain chordate affinities, including the possession of pharyngeal clefts. About 70 species have been identified on marine bottoms. Two classes, the Enteropneusta and the Pterobranchia, are recognized.

CLASS ENTEROPNEUSTA

The hemichordates of the Enteropneusta are commonly known as acorn worms. These animals burrow on soft bottoms or live beneath debris and other cover in shallow marine waters. An acorn worm has a rather large vermiform body with an anterior, conical proboscis (Fig. 17.14). This proboscis is followed by a short collar zone and the elongate, somewhat limp trunk. Overall lengths of 10 to 45 cm are common, but certain species, such as *Balanoglossus gigas,* may extend over 150 cm. The three body zones—the proboscis, collar, and trunk—represent the protosome, mesosome, and metasome, respectively. The proboscis contains a single coelomic cavity, the protocoel. In most enteropneusts, this cavity has been invaded by connective and muscular tissues. A single middorsal pore joins the protocoel with the outside world. The collar overlaps the proboscis stalk and bears the ventral mouth (Fig. 17.15). The collar coelom, or mesocoel, is paired and opens middorsally through a system of canals and pores. Finally, the trunk contains an isolated, paired coelom with no external openings. This long posterior region houses most of the digestive tract and, at its anterior end, bears the external openings of the pharynx and the reproductive system.

Acorn worms include both deposit and suspension feeders. Burrowing forms usually consume the sand and mud through which they tunnel and then digest out any nutritive substances. Above their burrows, coiled castings testify to the relatively large volumes of bottom material that these animals pass through their guts (Fig. 17.18). In suspension feeding, plankton and detritus are captured on the mucus-coated surface of the proboscis. The beating of proboscis cilia sends captured particles toward the mouth. Enroute, most potential food items pass through a preoral ciliary organ on the proboscis stalk. This organ is probably chemosensitive, and acceptable food is admitted to the mouth. The enteropneust collar is highly flexible, and its edges withdraw to cover the mouth when the animal stops feeding.

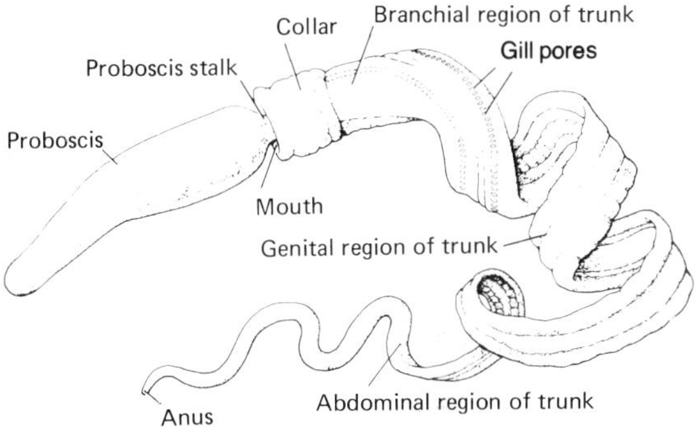

Figure 17.14 An acorn worm, *Saccoglossus* (Class Enteropneusta). (From Sherman and Sherman, *The Invertebrates, Function and Form.*)

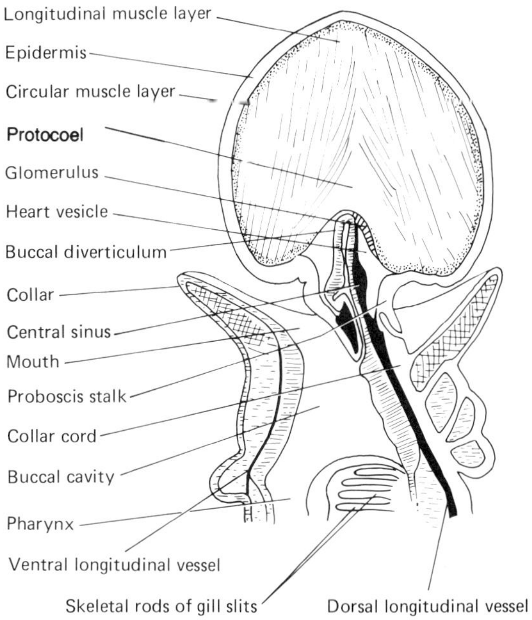

Figure 17.15 Section through the proboscis and collar region of an acorn worm, *Glossobalanus minutus*. (From L. H. Hyman, *The Invertebrates, Vol. V: Smaller Coelomate Groups*, 1959, McGraw-Hill.)

The mouth opens into a large buccal cavity. From this cavity, a diverticulum projects into the proboscis (Fig. 17.15). This peculiar structure formerly was considered a notochord, a misunderstanding that supported the early classification of acorn worms with the Chordata. It appears, however, to be merely an anterior projection of the digestive tract. Posteriorly, the buccal cavity leads into a pharynx whose dorsolateral wall is perforated by numerous clefts (Figs. 17.15 and 17.16). Food passes ventrally through the pharynx and into an esophagus and intestine. The acorn worm esophagus is unusual in that it communicates with the body surface via several pores in its dorsal wall. The anterior half of the organ pumps food through a midesophageal constriction. As that constriction squeezes food and mucus into a narrow cord, the esophageal pores allow excess water to escape through the esophageal pores. Digestion and absorption are intestinal processes, and feces are voided through a terminal anus.

Enteropneust internal transport involves an open blood-vascular system comprising a prominent dorsal vessel, a ventral vessel, and a system of interconnecting sinuses (Fig. 17.15). Colorless enteropneust blood flows anteriorly through the dorsal tube and descends to a central sinus within the proboscis stalk. There, a muscularized heart vesicle pumps blood into several anterior sinuses whose walls evaginate into the protocoel. These sinuses are known collectively as the **glomerulus** and probably have an excretory function. From the glomerulus, blood passes into the ventral vessel, which communicates with an elaborate system of sinuses serving the digestive tract and the body wall. The sinuses then return the blood to the dorsal vessel.

Respiratory activity by acorn worms centers within their pharyngeal clefts. These paired, U-shaped slits on the inner wall of the pharynx open into branchial sacs, which in turn are drained by dorsolateral gill pores (Fig. 17.16). As many as 100 or more of these gills may be present, and their pores commonly occupy a longitudinal groove on the anterior trunk. Pharyngeal cilia establish a respiratory current which enters the mouth, passes through the gills, and exits through the dorsolateral pores. The tissues surrounding the clefts are supported by skeletal rods and are

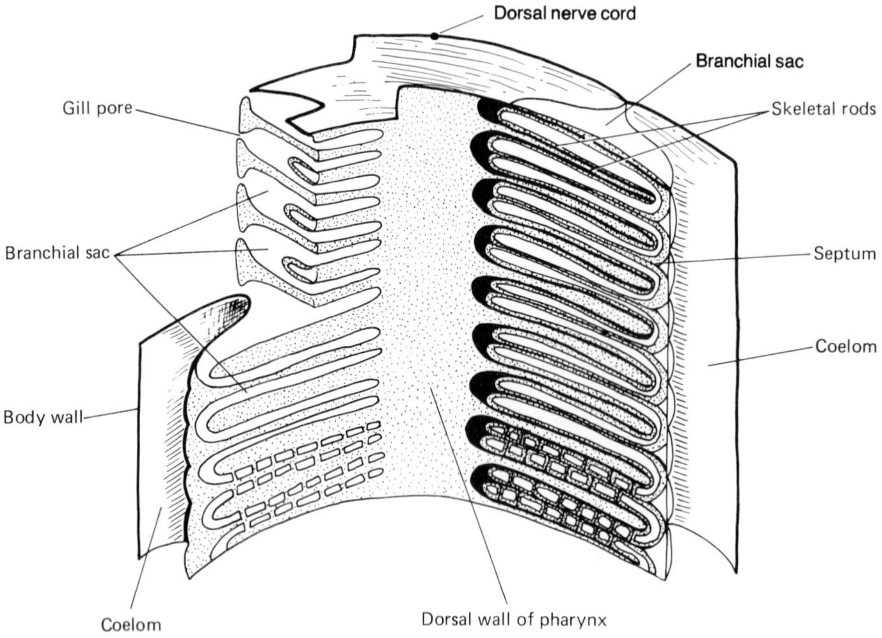

Figure 17.16 A portion of the pharynx of an acorn worm showing the branchial apparatus. The pharynx is opened ventrally and partially cut away at upper left. Skeletal rods (solid black structures) represented on right side only; cross bars shown for the lower two gill slits only. (From Hyman, *The Invertebrates, Vol. V.*)

richly supplied with blood sinuses. Respiratory exchange occurs between the circulating waters and the pharyngeal blood.

The enteropneust nervous system is rudimentary, consisting of a subepidermal nerve plexus which condenses into a few longitudinal nerve cords and two main nerve rings (Fig. 17.17). The most prominent cords are located middorsally and midventrally in the proboscis and trunk regions. An anterior nerve ring within the proboscis and a circumenteric ring at the upper end of the trunk join these long nerve cords. The ventral cord is interrupted at the collar, but the dorsal nerve traverses the collar region as a sunken, hollow tube containing giant fibers. This collar tube may be homologous with the chordate dorsal hollow nerve cord.

Except for the preoral ciliary organ, sensory structures on the acorn worms are limited to generally distributed neurosensory cells.

Enteropneusts are rather inactive invertebrates. Many simply lie beneath vegetation and rocks, and even burrowing species are slow-moving. The latter tunnel by peristaltic contractions of their proboscis, which can be anchored to prevent slippage. Their trunk is extremely fragile. Its external cilia may aid in locomotion, but typically, the entire region must be pulled along by the proboscis. Two common enteropneust genera, *Balanoglossus* and *Saccoglossus*, construct U-shaped burrows whose walls are plastered with mucus (Fig. 17.18). These hemichordates crawl in either direction through their homes and periodically emerge for feeding and defecation.

Acorn worms have separate sexes. Numerous paired gonads occur laterally in the anterior trunk region. In some genera, the sex organs occupy genital wings extending from the trunk wall (Fig. 17.19). Each gonad discharges through a separate gonopore, and gametes are fertilized in the sea. The early stages of enteropneust development are remarkably similar to those of the echinoderms. Typically, the protocoel, mesocoels, and metacoels are derived from archenteron pouches (Fig. 17.20a), and a feeding

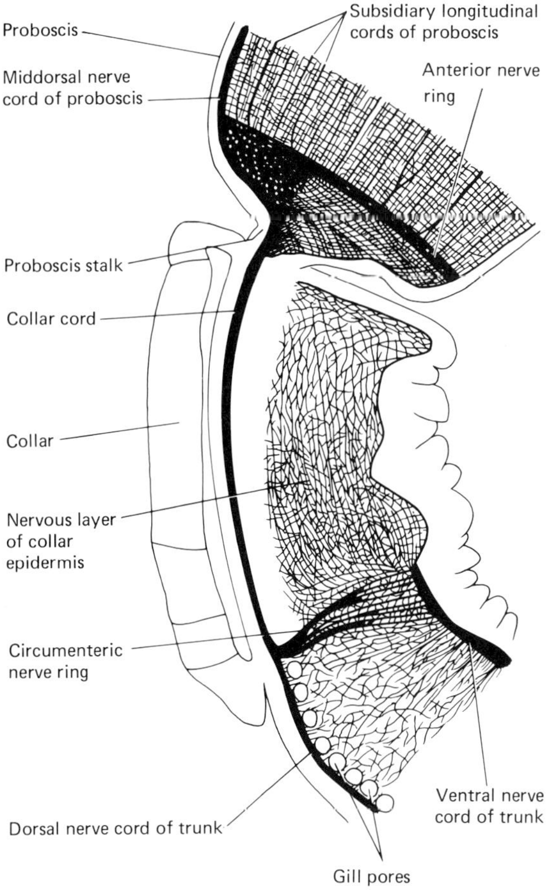

Figure 17.17 The anterior part of the nervous system of an acorn worm, *Saccoglossus cambrensis*. Longitudinal view. (From E.W. Knight-Jones, Phil. Trans. Roy. Soc. Lond., 1952, *236B*:315.)

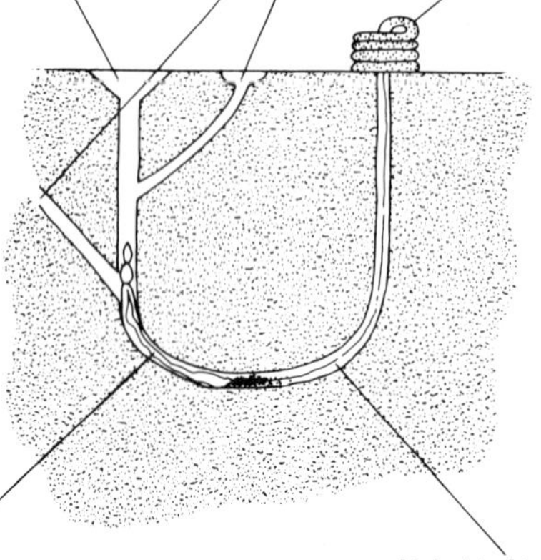

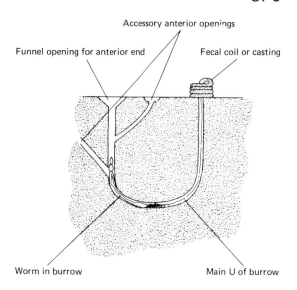

Figure 17.18 Schematic diagram of the U-shaped burrow of an acorn worm, *Balanoglossus clavigerus*. (From Hyman, *The Invertebrates, Vol. V.*)

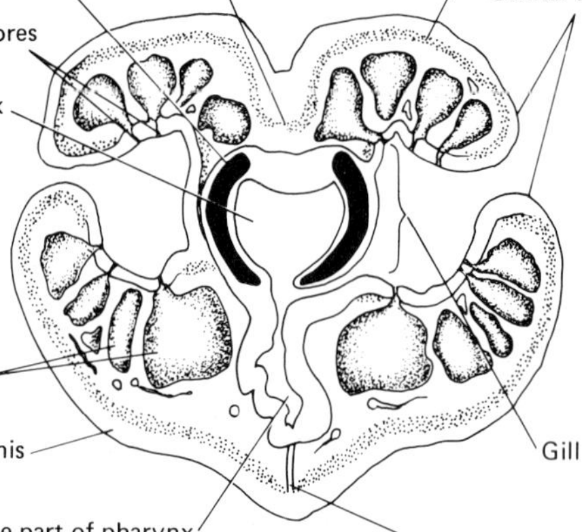

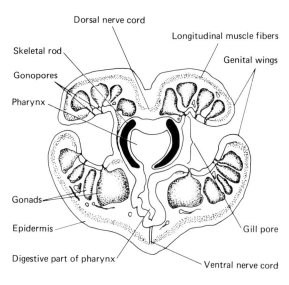

Figure 17.19 Cross section through the pharyngeal region of the enteropneust, *Stereobalanus*, showing the 4 genital wings. (From Hyman, *The Invertebrates, Vol. V.*)

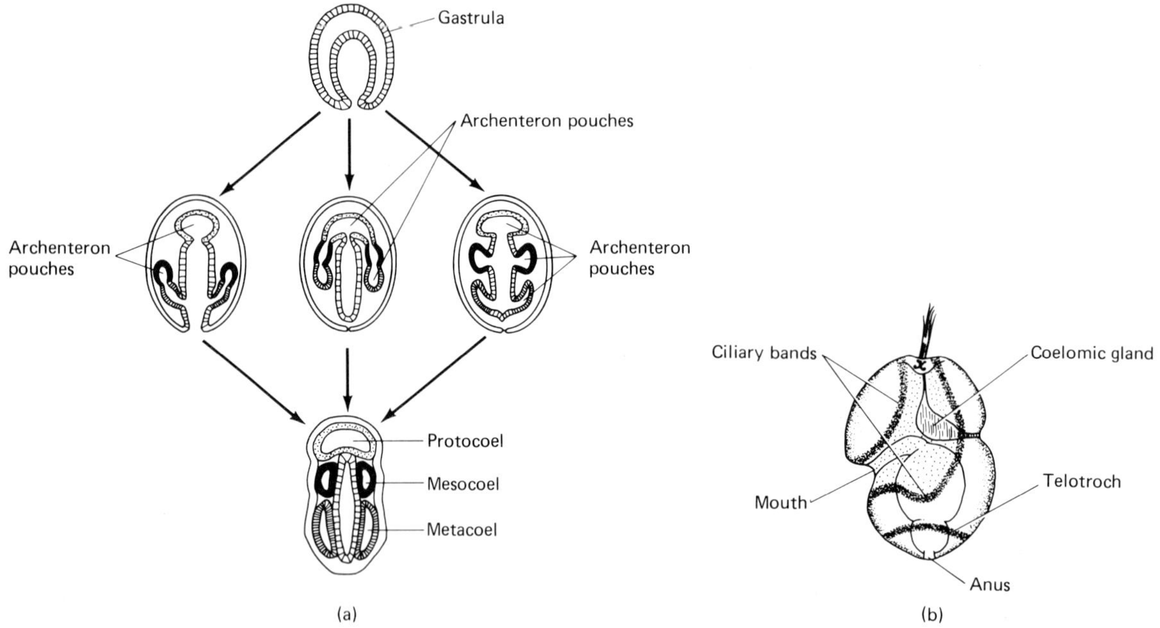

Figure 17.20 (*a*) Different modes of formation of the coelomic cavities in the Enteropneusta; (*b*) Early tornaria larva of an enteropneust. (*a* from A. Remane, In *The Lower Metazoa*, E.C. Dougherty, Ed., 1963, University of California; *b* from G. Stiasny, Zeit. wiss Zool., 1914, *110*.)

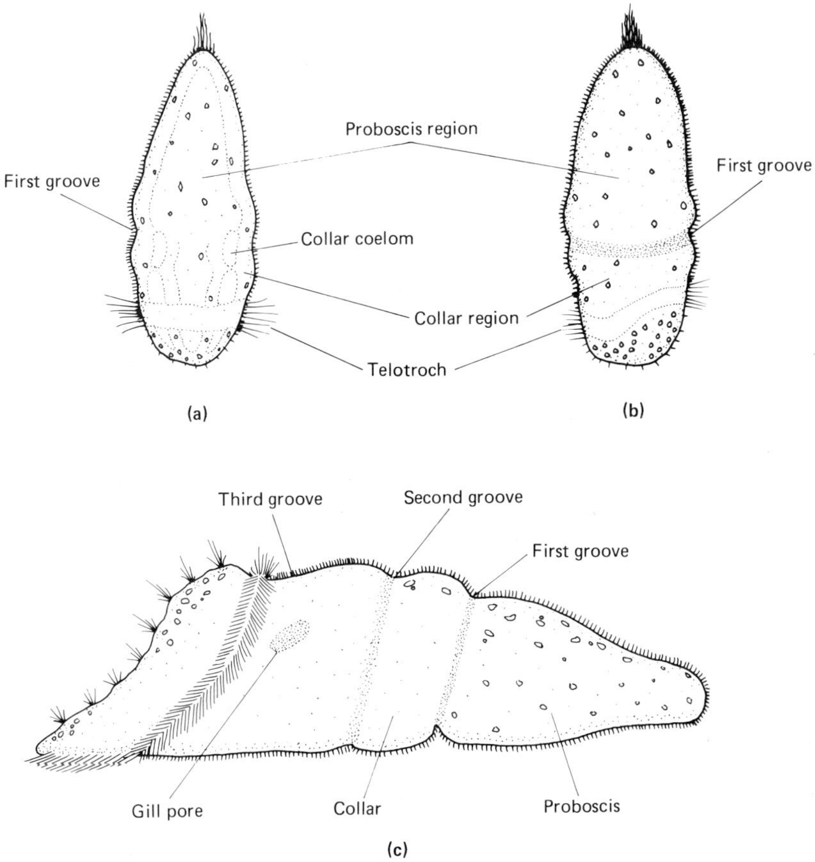

Figure 17.21 Metamorphosis in the acorn worm *Saccoglossus*. (*a*) Late tornaria larva; (*b*) Larva at the moment of settlement; (*c*) Creeping larva taking on adult structures. (From C. Burdon-Jones, Phil. Trans. Roy. Soc. Lond., 1952, *236B*:553.)

larva with ciliary bands soon develops (Fig. 17.20b). The enteropneust larva is called a **tornaria** and rather resembles the asteroid bipinnaria stage. In time, a transverse constriction appears at mouth level on the developing larva, and the future proboscis, collar, and trunk regions begin to differentiate (Fig. 17.21).

The tornaria larva is suppressed among some acorn worms. Hatching may occur at any of several late developmental stages, and in the Atlantic *Saccoglossus kowalevskii*, fully differentiated juveniles emerge from the eggs. Asexual reproduction and regeneration occur in several members of this class.

CLASS PTEROBRANCHIA

A second hemichordate class comprises three genera of tiny sessile animals called pterobranchs. These deuterostomes inhabit deeper seabottoms. One genus, *Atubaria* (Fig. 17.22a), is solitary and lives among colonial hydroids. The other genera, *Cephalodiscus* and *Rhabdopleura* (Fig. 17.22b), dwell in secreted tubes. *Cephalodiscus* occurs in aggregations, but individual animals occupy private tubes. *Rhabdopleura* is a colonial organism whose zooids are distributed along a common, creeping stolon.

Pterobranchs are typically no more than a few millimeters long. Their body includes the three hemichordate regions: the proboscis, the collar, and the trunk. The pterobranch proboscis projects anteriorly as a cephalic shield (Fig. 17.22a). The collar contains the ventral mouth and presents one to nine pairs of large dorsal arms. These collar arms support a field of ciliated tentacles which collect food. Pterobranchs display the same five coelomic compartments

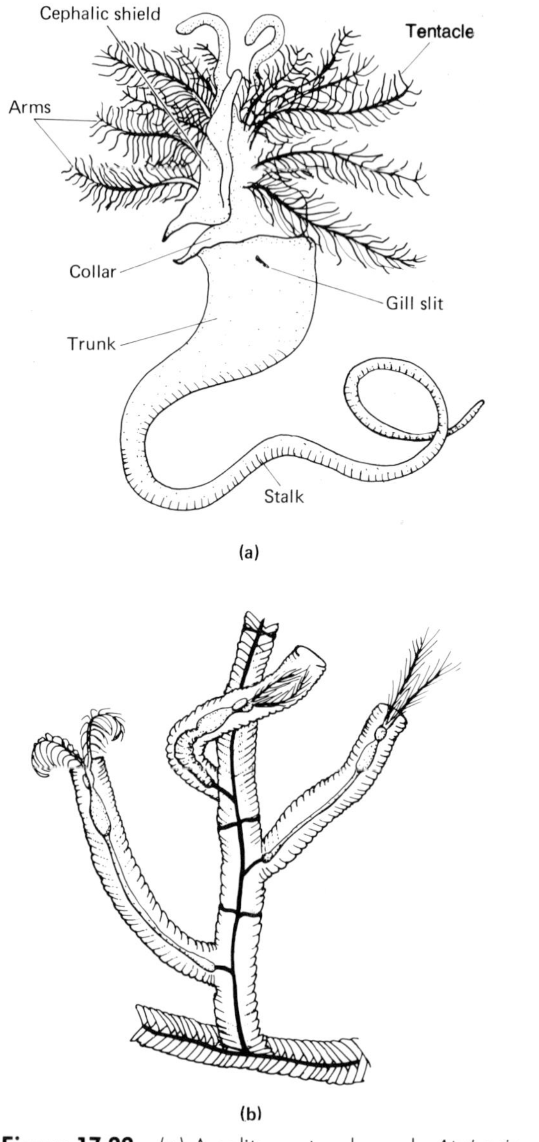

Figure 17.22 (*a*) A solitary pterobranch, *Atubaria*; (*b*) Part of a colony of *Rhabdopleura*. (*a from* Hyman, *The Invertebrates, Vol. V*; *b* from C. Dawydoff, In *Traité de Zoologie, Vol.XI*, Grassé, Ed.)

described for the enteropneusts, and the arms and tentacles receive extensions of the collar's mesocoels. These appendages thus can be likened to a lophophore.

Most other pterobranch systems are similar to those found in the acorn worms. A single pair of pharyngeal clefts occurs in *Cephalodiscus*, whose intestine is looped and discharges through an anus located on the collar. (Such a U-shaped digestive tract is typical of many tube-dwelling invertebrates.) Little is known about pterobranch continuity systems. Both hermaphroditic and diecious species occur, and there is a ciliated planktonic larva. Asexual budding has been described in colonial and aggregational forms.

Phylum Chaetognatha

The last deuterostomate phylum features the chaetognaths, or arrowworms. These invertebrates are common in marine plankton throughout the world; a single genus, *Spadella* (Fig. 17.23), is benthic. About 100 species—most belonging to the genus *Sagitta*—have been identified.

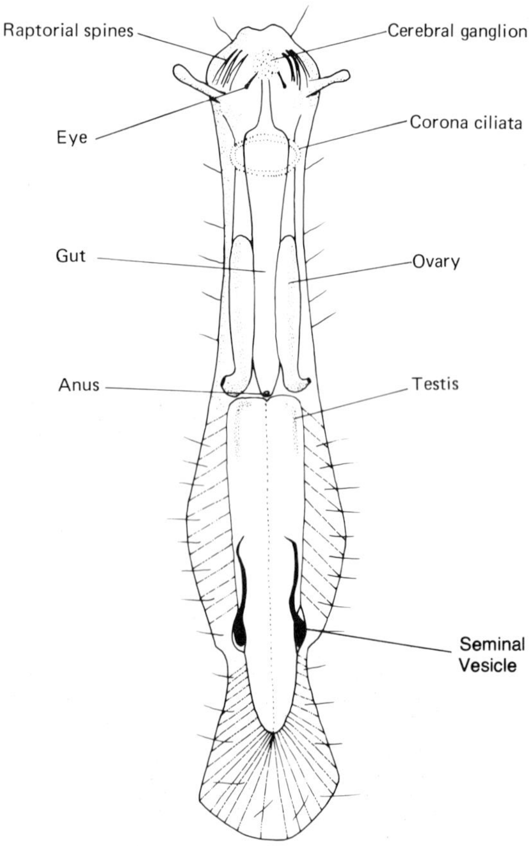

Figure 17.23 An unusual benthic chaetognath, *Spadella cephaloptera*. (From W. Kuhl, In *Klassen und Ordnungen des Tierreichs, Band 4*, Bronn, Ed. (1938))

The arrowworm derives its name from its long, dartlike body (Fig. 17.24). A central transparent fuselage comprises the head, trunk, and tail regions; lateral and caudal fins extend horizontally from the trunk and tail. Arrowworms are small animals, most measuring less than 4 cm in length.

The body wall of these animals is remarkably similar to the outer cylinder of the aschelminthes. A cuticle overlies a multilayered epidermis, whose basement membrane is thickened to support epidermal fins. All body wall muscles are longitudinal, and there is no coelomic peritoneum. Thus the secondary body cavity of these animals is somewhat reminiscent of the aschelminth pseudocoel. After the deuterostomate fashion, however, the chaetognath coelom is divided into compartments. A single compartment occupies the head region, and the trunk displays paired coelomic spaces. One or two tail compartments probably are formed secondarily from the trunk coelom.

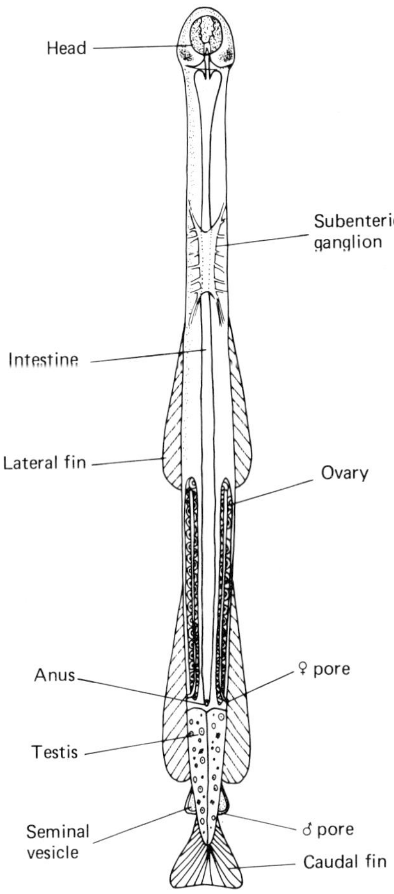

Figure 17.24 *Sagitta elegans*, a more typical planktonic arrowworm in ventral view. The hood covers the head. (From Kuhl, In *Klassen und Ordnungen des Tierreichs, Band 4*, Bronn, Ed.)

Arrowworms are carnivores, whose anterior ends are specialized for raptorial prey capture and ingestion. The head is rounded, and its ventral depression, the vestibule, surrounds the mouth (Fig. 17.25). A variable number of large curved spines borders the vestibule, and shorter teeth occur in rows at its anterior end. Both spines and teeth are made from chitin. Immediately behind the head, a muscular fold in the body wall forms a hood which can be drawn over the entire anterior region. This hood protects and streamlines the head during locomotion. An attacking chaetognath swims rapidly toward prey, withdrawing its hood to expose its sharp spines. These spines grasp the victim (commonly a small fish) and force it through the mouth. A muscular pharynx adds lubrication as it pumps incoming food into the long intestine. The intestine, which is suspended dorsally and ventrally by membranous mesenteries, is the site of extracellular digestion and absorption. The anus occurs just anterior to the trunk-tail junction.

Other chaetognath maintenance systems reflect the simplicity characteristic of smaller invertebrates. No organs are specialized for circulation, respiration, excretion, or osmoregulation. The compartmentalized coelom apparently serves the internal transport needs of these animals, while simple diffusion over the general body surface is adequate for other physiological processes.

Chaetognath activity systems are rather more developed. These active, cephalized animals possess

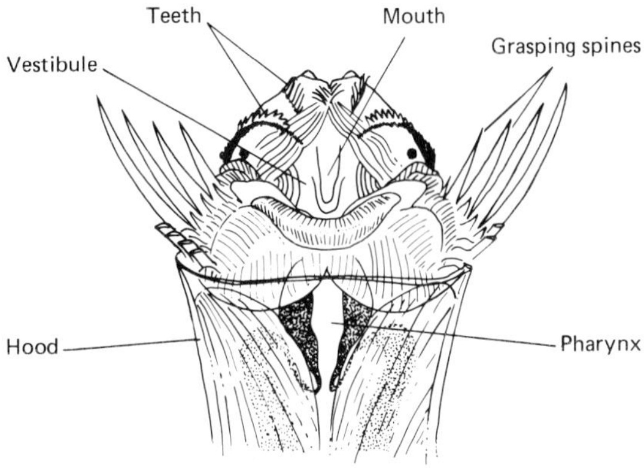

Figure 17.25 The anterior end of an arrowworm *Sagitta elegans*. The hood is retracted exposing the raptorial spines. Some spines have been cut off at the base. (From Hyman, *The Invertebrates, Vol. V.*)

better sense organs and more sophisticated neuromuscular machinery than most deuterostomes. Their muscles include the longitudinal units of the body wall and the finer, manipulatory fibers of the hood, spines, and teeth. The chaetognath nervous system is centered in a circumpharyngeal collar, whose numerous ganglia are associated with various cephalic nerves. A large cerebral ganglion dominates the collar dorsally (Fig. 17.23); from this nerve center, circumenteric connectives extend to a subenteric ganglion (Fig. 17.24), which dispatches paired sensory and motor nerves to the trunk and tail. Chaetognath sense organs include a pair of inverse eyes, a corona ciliata, and rows of tactile bristles along the length of the trunk. The eyes, each representing a cluster of five pigment cup ocelli, are located on the rear dorsal surface of the head (Fig. 17.23). The **corona ciliata** is a tract of cilia that loops between the head and the anterior trunk (Fig. 17.23). Its function is unknown, but chemoreception and/or rheoreception are suggested.

Arrowworms are accomplished swimmers. Quick contractions of the longitudinal musculature propel their body missiles rapidly through the water. Their prominent fins aid in flotation and gliding.

Chaetognaths exhibit well-developed hermaphroditic reproductive systems and a deuterostomate embryogeny. A long pair of testes occupies the tail coelom (Figs. 17.23 and 17.24). Sperm are packaged into spermatophores, which are released from a posterior seminal vesicle. The female system includes a pair of ovaries at the posterior end of the trunk coelom (Figs. 17.23 and 17.24). These gonads discharge through paired gonoducts at the trunk–tail junction. In the benthic *Spadella,* spermatophores are mutually exchanged. Each individual deposits a sperm package on the neck region of its mate. As the spermatophore erodes, sperm emerge, flow posteriorly over the body surface, and enter the gonopores. Planktonic arrowworms seem to be fertilized in much the same manner. In certain species, self-fertilization has been indicated.

Fertilized eggs may be released immediately or may adhere to the parent's outer surface. *Spadella* eggs are often attached to bottom plants. Chaetognath embryogeny includes a definitively deuterostomate cleavage pattern. Invagination produces an archenteron whose anterior end forms the coelomic pouches. The chaetognath coelom, however, arises from invaginations of the archenteron and not from outpockets, as in other deuterostomes. Only two pairs of these pouches are formed. Direct development produces a juvenile arrowworm.

Deuterostomate Evolution

The deuterostomes are diverse invertebrates, whose evolutionary histories are exceedingly problematic. All modern deuterostomes are highly specialized animals. Each of the major groups has pursued an independent adaptive course for some time, so interphyletic relationships are very difficult to define. Unlike the echinoderms, the animals described in this chapter have not left abundant fossils. Accordingly, evolutionary studies have focused on deuterostomate embryogeny and comparative larval anatomy. These disciplines have shed some light on the history of this important branch of the animal world.

The salient characters of deuterostomate embryogeny—indeterminate radial cleavage, a blastopore anus, and enterocoely—suggest a common origin for these animals. The Hemichordata often are considered primitive, and the pterobranchs in particular may closely resemble the ancestral deuterostomes. Pterobranch arms and tentacles (with their mesocoelic extensions) are at least superficially similar both to lophophores and to the water-vascular systems of the Echinodermata. Some authorities hypothesize that early pterobranchs and lophophorates arose from a common stem and that enteropneusts and echinoderms then diverged from pterobranch stock. Similarities between the hemichordate tornaria larva and the asteroid bipinnaria further encourage this hypothesis. The fact that all echinoderm larvae undergo a more drastic adult metamorphosis than does the tornaria might suggest that hemichordates remain more faithful to the ancestral deuterostomate form.

In any case, it is generally agreed that the Hemichordata diverged quite early from the deuterostomate line that led to the Chordata. The presence of pharyngeal clefts in all hemichordates, as well as the (arguably) hollow collar nerve cord in the enteropneusts, suggests chordate affinities, but the two phyla must have been distinct before all of the chordate characters were established.

Soon after their definitive chordate features were established, urochordates diverged in adaptation for an adult sessile life-style. The lack of a coelom within the group surely represents a secondary loss: Urochordate embryogeny too clearly allies them with the eucoelomate, deuterostomate line. Urochordates never display the body segmentation that typifies the cephalochordates and vertebrates. This condition is usually attributed to the divergence of the urochordates before the establishment of metamerism in the higher chordate subphyla.

As the only entirely planktonic deuterostomate phylum, the chaetognaths display a number of unique characters. Their aschelminthlike body wall, their cephalization, and the chitin in their spines and teeth are peculiar for deuterostomes. Yet embryological evidence supports the deuterostomate heritage of these animals, and their unusual traits must be attributed to a long and independent evolutionary history.

In conclusion, it is important to reject any notion that the urochordates, the hemichordates, and the chaetognaths represent mere milestones in a progression toward the Vertebrata. True, their evolution may be related, however distantly, to our own; but these animals pursue quite different modes of existence. Their success cannot be diminished merely because vertebrate chordates also evolved.

for Further Reading

Aldridge, A. Appenicularians. *Sci. Am.* 235: 94–102, 1976. (An account of the larvaceans)

Alvarino, A. Chaetognaths. *Oceanogr. Mar. Biol. Annu. Rev.* 3: 115–194, 1965.

Barrington, E. *The Biology of Hemichordata and Protochordata.* W. H. Freeman, San Francisco, 1965.

Barrington, E. and Jeffries, R. (Eds.) *Protochordates.* Academic Press, New York, 1975.

Berrill, N. J. *The Tunicata.* Ray Society, London, 1950.

Berrill, N. J. *The Origin of the Vertebrates.* Clarendon Press, Oxford, 1955.

Bone, Q. The origin of the chordates. *J. Linn. Soc. Zool.* 44: 252, 1960.

Cloney, R. A. Observations on the mechanism of tail resorption in Ascidians. *Amer. Zool.* 1: 67–87, 1961.

Feigenbaum, D. L. Hair-fan patterns in the chaetognatha. *Can. J. Zool.* 56: 536–546, 1978.

Florkin, M. and Scheer, B. T. (Eds.) *Chemical Zoology, Vol. VIII: Deuterostomians, Cyclostomes and Fishes.* Academic Press, New York and London, 1974.

Ghiradeli, E. Some aspects of the biology of the chaetognaths. *Adv. Mar. Biol.* 6: 271–375, 1968.

Giese, A. C. and Pearse, J. S. (Eds.) *Reproduction of Marine Invertebrates Vol II.* Academic Press, New York, 1975. (Includes chapters on chaetognaths, hemichordates and tunicates.)

Gilmour, T. H. J. Feeding in pterobranch hemichordates and the evolution of gill slits. *Can. J. Zool.* 57: 1136–1142, 1979.

Goodbody, I. The physiology of ascidians. *Adv. Mar. Biol.* 12: 1–149, 1974.

Gosner, K. L. *Guide to Identification of Marine and Estuarine Invertebtrates: Cape Hatteras to the Bay of Fundy.* Wiley-Interscience, New York, 1971.

Hyman, L. H. *The Invertebrates, Vol. V: Smaller Coelomate Groups,* pp. 1–71. McGraw-Hill, New York, 1959.

Jones, J. C. On the heart of the orange tunicate, *Ecteinsacidia turbinata.* Biol. Bull. 141: 130–145, 1971.

King, K. R. The life history and vertical distribution of the chaetognath, Sagitta elegans, in Dabob Bay, Washington, USA. *J. Plankton Res.* 1: 153–168, 1979.

Knight-Jones, E. W. On the nervous system of *Saccoglossus cambrensis. Phil. Trans, Roy. Soc. London Ser. B.* 236: 315–354, 1952.

Knight-Jones, E. W. The feeding of *Saccoglossus. Proc. Zool. Soc. London* 123: 637, 1953.

Komai, T. The homology of the notochord in pterobranchs and enteropneusts. *Am. Nat.* 85: 270, 1951.

Kriebel. M. E. Studies on the cardiovascular physiology of tunicates. *Biol. Bull.* 134: 434–455, 1968.

Millar, R. H. *British Ascidians.* Academic Press, New York, 1970.

Millar, R. H. The biology of ascidians. *Adv. Mar. Biol.* 9: 1–100, 1971.

Monniot, C. and Monniot, F. Abyssal tunicates: an ecological paradox. *Ann. Inst Oceanogr.* 51: 99–129, 1975.

Monniot, C. and Monniot, F. Recent work on the deep-sea tunicates. *Ann. Rev. Oceanogr. Mar. Biol.* 16: 181–228, 1978.

Moreno, I. The grasping spines and teeth of six chaetognath species observed by scanning electron microscopy. *Ant. Anz.* 145: 453–463, 1979.

Plough, H. H. *Sea Squirts of the Atlantic Continental Shelf from Maine to Texas.* Johns Hopkins University Press. Baltimore, 1978.

Smith, J. H. The blood cells and tunic of the ascidian *Halocynthia aurantium. Biol. Bull.* 138: 354–378, 1970.

Strathman, R. R. The evolution and loss of feeding larval stages of marine invertebrates. *Evolution* 32: 894–906, 1978.

18

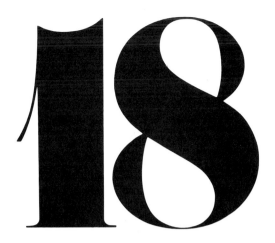

Phyla and Evolution: A Review

"Anything said on these questions lies in the realm of fantasy."

—L. H. Hyman, *The Invertebrates, Vol. V.*

Libbie Hyman was right. Despite our increasing sophistication in dealing with all forms of evolutionary evidence, we must resort, in part, to speculation and fantasy when addressing any of the major questions of animal phylogeny. When did life originate? How were the first metazoans formed? What are the relationships among the modern animal phyla? None of these questions has been answered or is even answerable at our present level of knowledge. Thus, left only to speculate, scientists have done so in abundance. This chapter reviews some of the theories by which zoologists seek to understand the history of the invertebrate world.

Evolutionary Tools

Invertebrate evolution is so puzzling because hard fossil evidence is limited and information from other sources is open to varied interpretations. Almost all phyla originated in Precambrian time; that is, more than 600 million years ago. Unfortunately, few fossils are contained in Precambrian rock. These ancient rocks have been exposed to extreme pressures, which apparently deformed and/or destroyed any contained fossils. Moreover, Precambrian invertebrates were mostly softbodied and could not have fossilized well. Fossils from the Cambrian period are abundant, but as most phyla were already quite distinct by that time, these relics shed little light on interphyletic relationships. Fossil evidence is invaluable, however, in tracing progressions within smaller taxonomic groups. This evidence is compelling when it presents a stepwise fossil series illustrating a transformation from ancient to modern types. These series are rare, however, largely because fossil formation and preservation are influenced by so many random environmental factors. Thus, information from the fossil record usually requires supporting evidence from comparative studies on the anatomy, physiology, biochemistry, and embryology of modern forms.

Comparative anatomy is the oldest tool of the evolutionary biologist. Charles Darwin's work was based almost entirely on such studies, and the discipline continues today as a favored approach to an understanding of animal relationships. In relatively recent times, comparative studies in physiology and biochemistry have assumed a similar importance in evolutionary research. By examining the similarities and differences among modern invertebrate types, we can hope to create a reliable portrait of their common ancestors and to illustrate pathways along which the recent types have diverged. Any such recreations must be predicated on characters central to the animals involved. Such deeply rooted characters as symmetry, germ layers, metamerism, secondary body cavities, locomotion, organ system development, and protein chemistry are most reliable. By focusing on these essential traits, we are less likely to be confused by superficial similarities among unrelated organisms. Convergent evolution is widespread among invertebrates, as unrelated organisms frequently display similar adaptive solutions to common problems. Thus, for example, the ostracods, brachiopods, and bivalves have similar hinged shells, but very discrete ancestries. It is not always easy to determine whether similarities reflect convergent evolution or descent from a close common ancestor. For example, metamerism within both the Annelida and the Arthropoda is viewed by most scientists as evidence of a relationship between the two phyla, yet homology between polychaete parapodia and arthropod appendages is contested. Lively debates concern the occurrence of compound eyes in both crustaceans and insects and the presence of Malpighian tubules in both arachnids and terrestrial mandibulates. Do these phenomena result from convergent evolution or close common ancestry? What about the relationships among the several pseudocoelomate phyla? Are sipunculoid tentacles homologous with bryozoan lophophores and/or echinoderm ambulacral systems? Clearly, such questions can be asked endlessly. By developing our own criteria for judging such matters, each of us can contribute to continuing speculations on the history of the invertebrates.

Perhaps nowhere do phylogenetic theories differ as much as in the degree to which they accept evidence from comparative embryology. Late in the nineteenth century, the celebrated axiom, "Ontogeny recapitulates phylogeny," was introduced. This principle held that the evolutionary history of each species was preserved in its embryonic development. Supposedly, scientists needed only to observe developmental stages in order to review a species' evolution in its entirety! While this principle did stimulate early advances in both evolutionary and developmental biology, its shortcomings are widely recognized by modern scientists. Today we realize that all developmental forms are themselves subject to natural selection and that their biology represents adaptations for embryonic survival as well as preparations for adult life. Because most early stages share similar survival problems (small size, limited locomotor power, relatively simple organization), some convergent evolution is inevitable. Once again, similarities resulting from convergence must be distinguished from those reflecting a phylogenetic relationship. The appearance of a trochophore larva among both annelids and mollusks has long been considered as evidence of a relationship between these two major phyla. A few authorities, however, argue that these larvae are products of convergent evolution. Likewise, relationships among the deuterostomate groups depend heavily on evidence from comparative development, so that even the most popular theories are wisely questioned.

Clearly, many pitfalls exist along all routes of evolutionary research. With proper awareness of potentially misleading evidence, we turn now to the central questions of invertebrate history.

The Origin of Multicellularity

Following a very long series of chemical and biochemical phenomena, the first animals, or protozoans, were formed. Figure 2.69 depicts the most likely relationships among the protozoan subphyla and indicates possible origins for the metazoan world. Two major schools of thought exist concerning the origin of multicellular animals from protozoan ancestors. The classical view derives the first metazoans from colonial flagellates. This so-called colonial theory describes the coelenterates as the most primitive metazoans. A more recent alternative to this view is the plasmodial theory, which derives the entire Metazoa from multinucleate ciliates via the acoel flatworms.

The **colonial theory** was first espoused in the late nineteenth century and remains today as the most popular interpretation of early metazoan history. It begins with a colonial zooflagellate similar to *Sphaeroeca volvox* (Fig. 18.1a). This ancestral colony

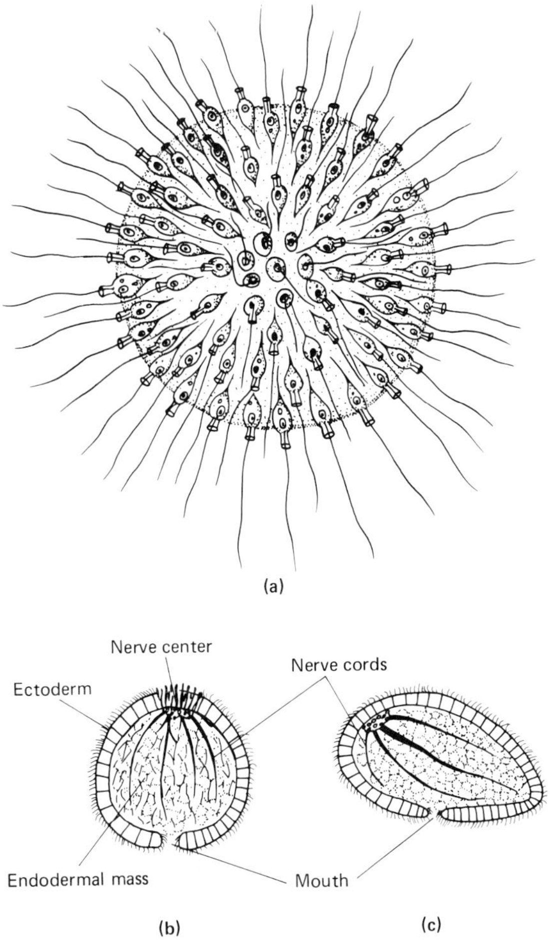

Figure 18.1 (*a*) A colony of the choanoflagellate, *Sphaeroeca volvox;* (*b*) A planuloid organism, the ancestor of the radiate phyla according to the colonial theory; (*c*) A bilaterally symmetrical acoeloid organism derived from the planula and ancestral to the flatworms according to the colonial theory. (*a* from A. Hollande, In *Traité de Zoologie, Vol. II*, P. P. Grassé, Ed., 1952, Masson et Cie, *b* and *c* from L. H. Hyman, *The Invertebrates, Vol. II: Platyhelminthes and Rhynchocoela*, 1951, McGraw-Hill.)

represented a hollow sphere of flagellated cells. According to the theory, groups of cells within the colony became increasingly specialized and interdependent. The anteriormost cells adapted for sensory and locomotor roles, while other cells assumed nutritive and reproductive responsibilities. Gradually, digestive and reproductive cells were displaced to the inside of the colony, producing a solid organism (Fig. 18.1b). The resultant multicellular animal was radially symmetrical and planktonic. It would have resembled the planula larva of the Coelenterata; and indeed, the colonial theory argues that the radiate phyla arose from such a creature. The theory also derives the flatworms and the rest of the Metazoa from a planuloid organism which assumed a bottomcrawling existence (Fig. 18.1c). Bottom-crawling would have evolved in association with bilateralism and the differentiation of dorsal and ventral surfaces.

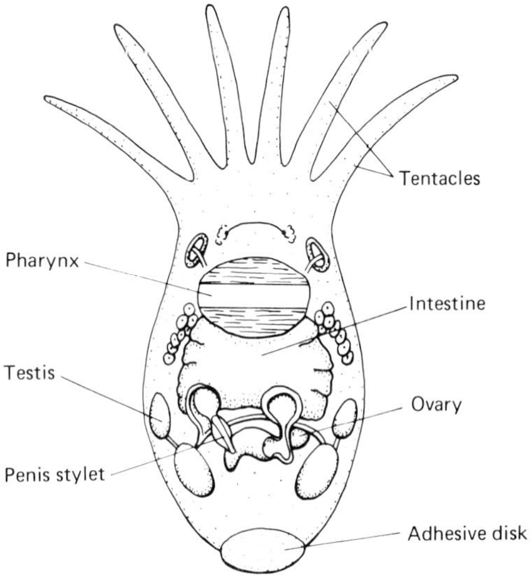

Figure 18.2 *Temnocephala*, a commensal or parasitic turbellarian of the type considered ancestral to anthozoan coelenterates by supporters of the plasmodial theory. (From Hyman, *The Invertebrates, Vol. II.*)

The colonial theory draws much of its support from embryological sources. It sees a definite recapitulation of the origin of multicellularity in the early developmental stages of most metazoans. Thus, a typical hollow blastula would recall an ancestral colonial zooflagellate and the gastrula stage would recapitulate the evolution of a solid, layered organism. Moreover, the gradual cellular differentiation described by the colonial theory is entirely consistent with the embryogeny of all metazoan organisms. Other factors supporting the theory are the widespread occurrence of flagellated gametes throughout the animal world and the presence of flagellated cells in several tissue types among lower metazoans.

As discussed previously, considerable caution must be exercised in evaluating the phylogenetic significance of embryological events. Some scientists totally reject the colonial theory, persuasively arguing that any general themes in metazoan embryogeny reflect convergent adaptations to similar developmental problems. Economical ways to produce a metazoan from a single-celled zygote may be limited; thus the ubiquitous blastulas and gastrulas are not necessarily of any intrinsic importance to the question of metazoan origins.

Evolutionary biologists who argue along these lines are likely to support the **plasmodial theory**. This theory draws heavily from comparative morphology in deriving acoel flatworms from multinucleate ciliates and postulates a bilateral protociliate with many similar nuclei. (Differentiation of macronuclei and micronuclei supposedly occurred later in the evolution of the protozoan group.) Theoretically, this protozoan gave rise to the first metazoans when cell membranes developed around each of the nuclear centers. Supporters of the plasmodial theory emphasize several anatomical similarities between the acoels and the ciliates. Members of both groups are of comparable size and symmetry and are ciliated externally. Acoel nuclei often are not isolated within cell boundaries. The ciliate cytostome and the acoel mouth are somewhat comparable; and some researchers homologize ciliate endoplasm with acoel endoderm, trichocysts with rhabdites, and contractile vacuoles with protonephridia. Ciliate conjugation has even been likened to the mutual copulation of hermaphroditic acoels.

Critics of the plasmodial theory charge that these similarities are very superficial and do not represent homologies. Indeed, it is difficult to homologize protozoan organelles with the tissues and organs of a true metazoan. Critics also point out that some acoels have definite cells with single nuclei and plasma membranes. Moreover, where such cells are not defined, acoels exhibit a syncytial rather than a plasmodial condition. A syncytial condition implies the degeneration of boundaries between preexisting cells, whereas a plasmodial condition, as in the ciliates, describes a *de novo* multinucleate state. The embryogeny of early metazoans never includes a phase at which a multinucleate embryo becomes cellularized. (Superficial cleavage in many arthropod eggs involves a cellularization phase, but this condition is clearly a specialized one relating to extraordinary yolk volume) This lack of recapitulation undermines the plasmodial theory among zoologists who stress comparative embryology. It is no particular problem, however, for the theory's most ardent supporters, who routinely discount the value of embryologically based arguments.

If bilateral flatworms are the most primitive metazoans, the radial symmetry of the coelenterates must be derived secondarily. This requirement places the arguably bilateral anthozoans at a primitive level, while scyphozoans and hydrozoans are interpreted as the more advanced coelenterates. Problems with this progression are substantial. All trends within the Coelenterata—from the increasing definition of a mesodermal layer to the more elaborate nervous system—argue for an evolutionary progression from the hydrozoans upward to the anthozoans. The biradial or bilateral symmetry of the Anthozoa may not be primary and could represent an adaptation to enhance water flow through a more complex coelenteron.

Also vulnerable to criticism is the neorhabdocoel order suggested as an ancestor for the anthozoans. This order, typified by *Temnocephala* (Fig. 18.2), may resemble coelenterates, but certainly in a very superficial way. These flatworms are commensal or parasitic on turtles, crustaceans, and mollusks—a rather specialized mode of living for the proposed stem form of an ancient and rather simple phylum.

Recently, attempts have been made to combine both the colonial and the plasmodial theories in support of a polyphyletic origin for the Metazoa. According to this combined theory, the radiate animals could have evolved from colonial flagellates, while flatworms and higher metazoans could have descended from multinucleate ciliates. This view shares most of the advantages and disadvantages of both theories and is subject to the same criticisms. It does improve on the plasmodial theory, however, by removing the derivation of coelenterates from bilateral flatworms. The combined theory has received minimal support, perhaps in part because either of the monophyletic views offers greater aesthetic appeal. Some scientists, however, have accepted the possibility of polyphyletic origins, at least as far as the sponges are concerned. If the sponges arose independently from other multicellular animals, we might believe that other discrete origins have occurred. Perhaps the polyphyletic view deserves more attention than it has so far received.

The Origins of the Coelom and Metamerism

Solving the riddles of metazoan origin would gain us only a foothold on animal phylogeny. Millions of years of evolution have produced a variety of metazoans, and while we recognize some affinities among certain phyla, no single model relates all of these animals in a manner acceptable to every zoologist. Therefore, we encounter still another speculative arena filled with competing theories.

Theories concerning the evolution of the higher Metazoa focus on two structural-functional phenomena: the coelom and metamerism. By addressing the origins of these conditions, we speculate most profitably on the relationships among the phyletic groups expressing various levels of coelomic and segmental organization.

The **enterocoel theory** derives the coelom from the gastric pouches of early anthozoans. According to this theory, the anterior-posterior axis of these ancient creatures elongated, as a mouth and anus formed from opposite ends of the gastrovascular opening (Fig. 8.43). Meanwhile, four coelenteronic pouches lost their connection with the gut. The anteriormost pouch formed a protocoel. The two lateral spaces contributed paired mesocoels, and the posterior pouch became a pair of metacoels (Fig. 18.3a).

The enterocoel theory is supported by much of deuterostomate embryogeny; enterocoelous formation of the secondary body cavity in these animals supposedly recapitulates the coelom's evolutionary origin. Also, the ciliated coelomic peritoneum of most metazoans does resemble the borders of the coelenteron. The enterocoel theory, however, is objectionable on several counts. First, it is difficult to imagine what adaptive gain occurred by isolating the gastric pouches from the gut. These outpocketings provide valuable surface amplification in modern coelenterates, and their early isolation would have required an immediate and major selective advantage. Second, problems arise upon examination of the theory's phylogenetic implications. Like the plasmodial theory, the enterocoel theory describes the anthozoans as the most primitive coelenterates. General evolutionary trends within that phylum are thus held to be almost entirely retrogressive. Deuterostomes are described as the more primitive eucoelomates, and the protostomes must be derived from them. However, nothing like enterocoely occurs in the protostomate assemblage.

Protostomes may have arisen from deuterostomate ancestors in the following manner (Fig. 18.3): The loss of all coelomic compartments would produce acoelomate animals. If the protocoels and mesocoels were lost, an unsegmented coelomate group, such as the mollusks, would arise (Fig. 18.3a–c). Finally, if the surviving metacoel elongated and developed paired compartments, the annelid-arthropod line would be established (Fig. 18.3d–f). Called the **cyclomerism theory,** this scheme is advocated by most supporters

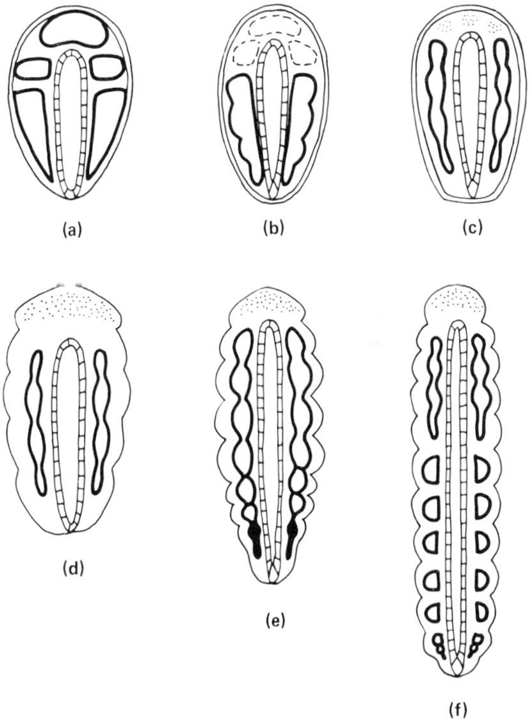

Figure 18.3 (*a–c*) Loss of protocoels and mesocoels to give rise to an unsegmented coelomate group such as the mollusks; (*d–f*) Elongation of the metacoel and its division into paired compartments yields the annelid-arthropod line according to the cyclomerism theory. (From W. D. Hartman, In *The Lower Metazoa*, E. C. Dougherty, Ed., 1963, University of California.)

of the enterocoel theory. The lack of a convincing explanation for the loss of coelomic pouches, particularly among the acoelomates, is the theory's major weakness.

The **schizocoel theory** presents an alternative view, deriving the coelom from ever-widening splits in the mesoderm of acoelomate ancestors. As the enterocoel theory is supported by the development of enterocoelous metazoans, so the schizocoel theory derives sustenance from the embryogeny of schizocoelomates. The acoelomate condition is viewed as primitive, and protostomes would precede the deuterostomes. Problems with the schizocoel theory include its lack of explanation for enterocoely in the deuterostomate group.

A third view, the **gonocoel theory,** states that the coelom arose from the emptied cavities of one or more paired gonads (Fig. 8.42). According to this theory, an acoelomate ancestor produced germ cells from the interior walls of its hollow gonads; gametes ripened within the gonadal cavity and then were released to the outside. The gonadal cavity remained, perhaps improving internal transport or allowing organ growth, and eventually evolved into a coelomic space. The reproductive systems of primitive annelids and mollusks function in essentially this manner, a factor providing the gonocoel theory its best support. The theory requires that acoelomates precede eucoelomates, and thus the evolutionary trend within the higher Metazoa is a progressive one. Because the gonads of many primitive flatworms are arranged pseudometamerically, their transformation into coelomic cavities would originate segmentation. This evolutionary economy also makes the theory attractive.

However, many factors discourage the gonocoel theory. No embryological association exists between the coelom and the gonads. Coelomic development always precedes the appearance of sexual structures, and gonadal tissues never participate in the formation of the coelomic peritoneum. The gonocoel theory suggests that the selective pressures involved in coelomic formation were exerted *after* breeding. Such pressures could not influence the evolution of many lower metazoans, which breed only once in the life cycle. Finally, the phylogenetic consequences of the gonocoel theory are troublesome, particularly when allied with the origin of metamerism. If the coelom and metamerism arose simultaneously, the unsegmented coelomates necessarily derive from annelidlike creatures. This derivation again requires an apparent evolutionary retrogression—not an impossibility, but certainly a less appealing idea.

All of the above theories share a common fault: They are based almost entirely on observations from comparative anatomy and embryology and speculate little on the selective forces that might have shaped the first coeloms and metameric structures. Some contemporary phylogeneticists, championed by R. B. Clark (see this chapter's suggestions for further reading), have concentrated on coelomic and metameric functions in seeking possible origins for these features.

Most of the evidence describes the primary role of the coelom as a conveyor of hydrostatic pressures during locomotion by peristaltic waves. (This type of locomotion involves the alternate contraction of sets of longitudinal and circular muscles, often in a burrowing vermiform animal.) When some form of locomotion other than peristalsis appears, such as the looping movements of the leeches or the limbed locomotion of the arthropods, the coelom regresses. Such regression reinforces the conviction that the coelom's function is primarily (and originally) mechanical rather than physiological. Any ancient group of acoelomates standing to profit from increased burrowing power would have experienced selective pressures favoring a secondary body cavity. The coelom could have arisen from digestive pouches, mesodermal vacuoles, gonadal cavities, or any other source that provided the necessary mechanical advantage. Possibly, the coelom has multiple origins—a view that would undermine the supposed unity of the higher Metazoa.

Clark's theory likewise maintains that metamerism arose as an adaptation for increased locomotor efficiency. The segmented musculature in early burrowing coelomates applied hydrostatic forces with greater precision. Metameric muscles evolved in concert with segmentally arranged nervous systems, and circulatory and excretory elements assumed a similar organization. Metamerism within the chordate line probably also represents a muscular adaptation. Segmentally arranged muscle sets along the notochord produce the finely controlled contraction waves necesssary for effective swimming.

Phylogenetic Trees

Theories on the origins of multicellularity, the coelom, and metamerism can be combined in various ways to construct phylogenetic trees. Products of some of the best scientific minds of our time, these grand schemes nonetheless remain dependent on multiple orders of speculation. None of these schemes can be proven or disproven, but a study of each aids our appreciation of the dynamic history of the invertebrates.

Phylogenetic trees tend to be rather personalized. Four of the more popular schemes are depicted in Figures 18.4 through 18.7 and are named after their principal supporters.

Hyman's phylogenetic tree (Fig. 18.4) is based on the colonial theory of metazoan origin. A planuloid ancestor gives rise to both the radiate phyla and the acoelomates. Eucoelomates evolve from acoelomates, with the pseudocoelomates diverging at an intermediate stage. The higher metazoans are organized along two distinct branches, representing the traditional protostomate and deuterostomate dichotomy. The lophophorates are intermediate between these two great branches. Mollusks diverge from the annelid-arthropod line before the appearance of metamerism in the latter group.

Marcus' scheme of invertebrate phylogeny draws on the colonial theory of metazoan origin and the enterocoel theory of coelom formation (Fig. 18.5). It describes a primitive position for the anthozoans, deriving the other coelenterate classes from them. All higher metazoans arise from coelomate stock. Deuterostomes diverge early from the dominant protostomate lineage. Coelomic regression leads to the acoelomates, while placement of the mollusks so near the annelid-arthropod climax supposes their basic metameric character.

Hanson favors the plasmodial theory; thus the acoelomates on his phylogenetic tree arise from ciliate protozoans (Fig. 18.6). The radiate phyla diverge early from the main course of animal evolution. Higher metazoans occupy a number of divergent lines, reflecting multiple origins of coelomate and metameric conditions. By placing the mollusks between the annelids and the arthropods, Hanson illustrates his view that the mollusks are primitively metameric.

The fourth phylogenetic tree, that of Hadzi (Fig. 18.7), is most unusual. This scheme is based primarily on the plasmodial theory and a peculiar attitude toward the development of metamerism. Ciliates give rise to the acoels, which establish the primitive Ameria, an unsegmented organizational level. Other members of the Ameria—the radiate phyla, the pseudocoelomates, the higher flatworms, and the mollusks—diverge from a single rising line. This line leads to the Polymeria, a highly segmented level shared by the annelids and arthropods. Continuing evolution along the *same* line leads to the Oligomeria, a group characterized by a reduction in segment number. Finally, this single evolutionary line culminates in the Chordata. Hadzi's highly controversial phylogeny totally discounts embryological evidence and coelomic significance. (Note the wide separation between annelids and mollusks and the derivation of deuterostomes from higher protostomes.)

There is little reason to believe that we will ever understand invertebrate evolution in its entirety. Yet, it would be unfortunate indeed to abandon altogether our attempts at phylogenetic reconstruction. Modern biology requires increasing emphasis on detailed studies of single communities and species, making it all too easy to lose an appreciation for the wider historic drama of invertebrate life. All of us, both students and professors, would profit by looking beyond the tedium of exams and experiments to see that drama and to wonder at it.

Ultimately, wonder may remain our best access to the deeper mysteries of the invertebrate world. Phylogenetic arguments can appeal to reason and logic, yet these faculties are inadequate in dealing with subjects that transcend human experience. After all, it is not entirely "logical" that such creatures as sea stars,

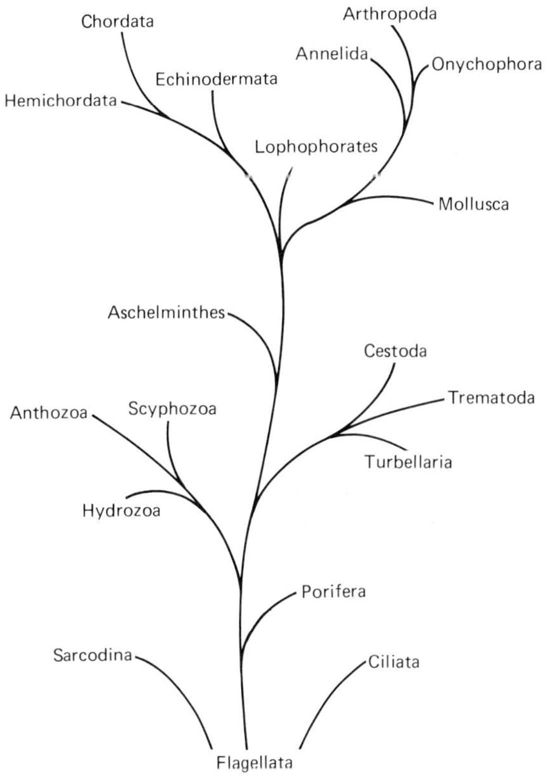

Figure 18.4 A phylogenetic tree of the Animal Kingdom based on the ideas of Hyman.

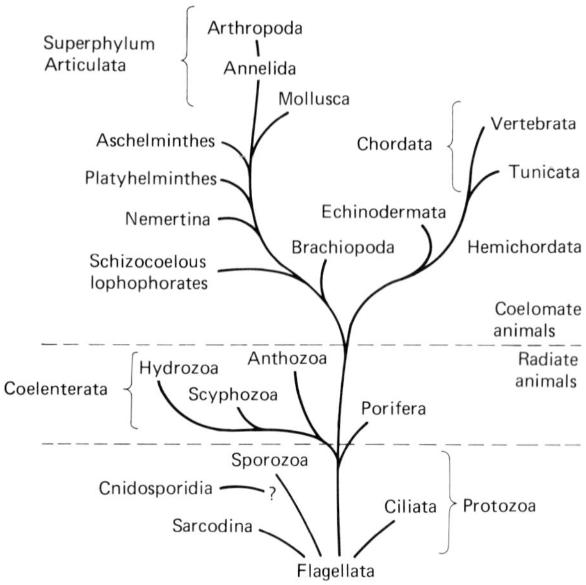

Figure 18.5 Scheme of animal phylogeny as devised by Marcus. (From E. Marcus, Quart. Rev. Biol., 1958, 33: 24)

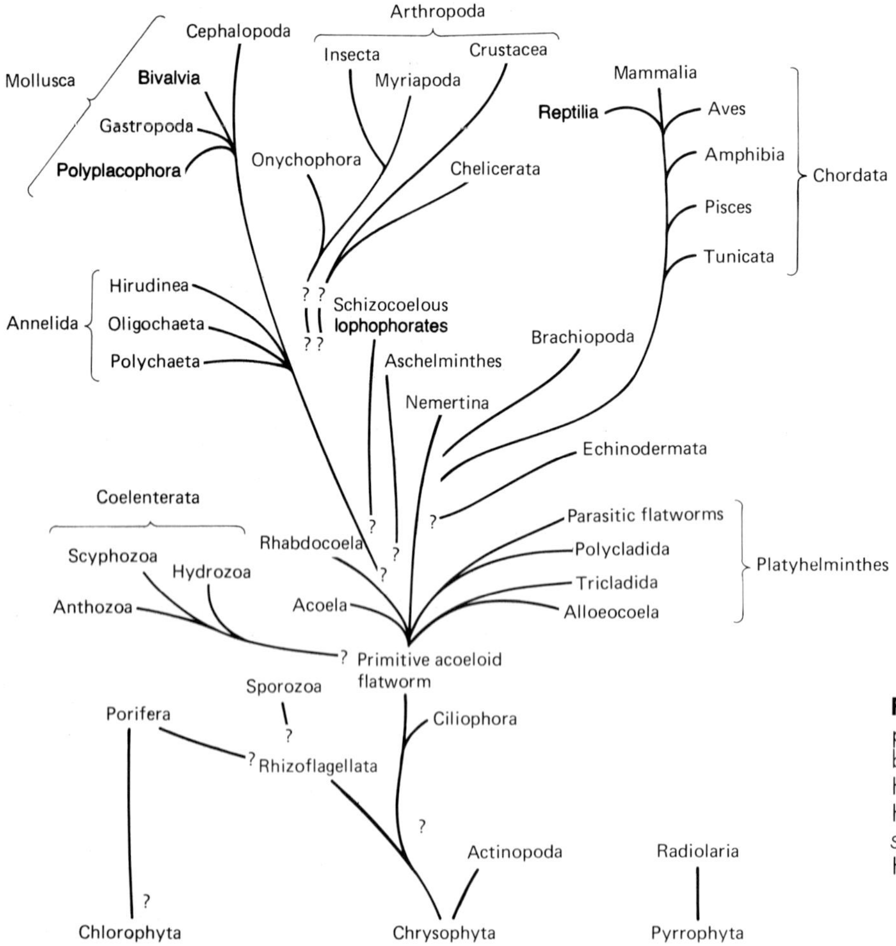

Figure 18.6 A possible phylogeny of animals based on the views of Hanson. (From E.D. Hanson, *Animal Diversity*, 1972, Prentice-Hall.)

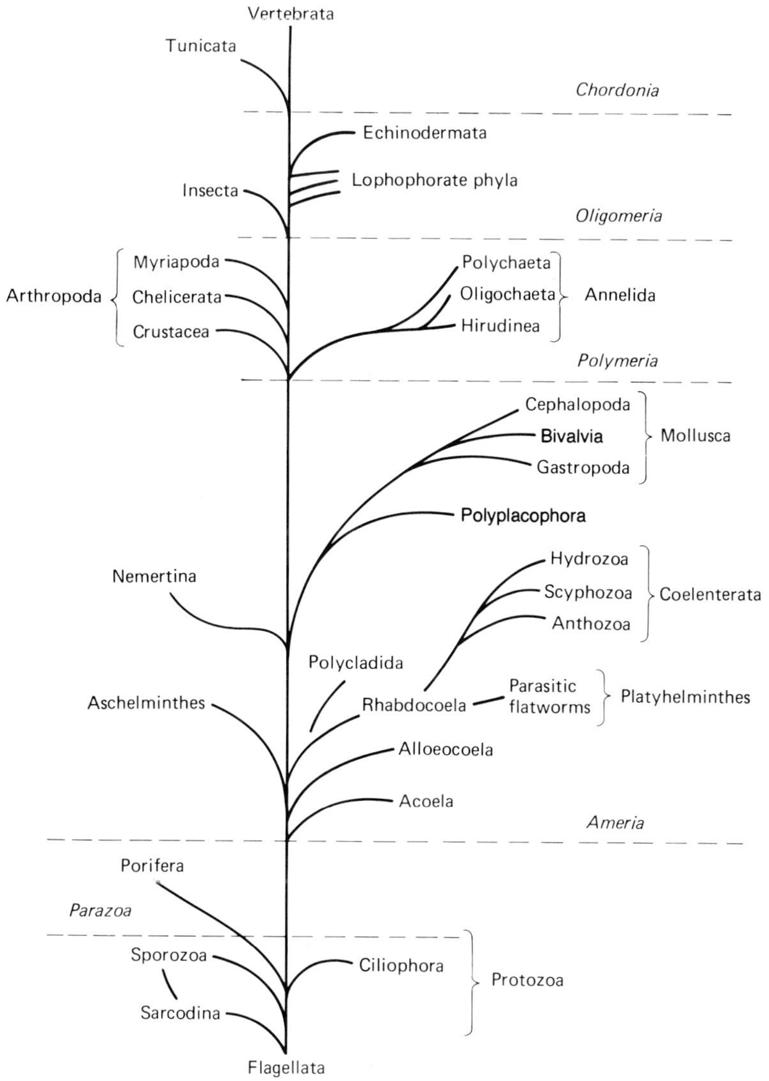

Figure 18.7 A phylogenetic tree for animals as proposed by Hadzi. (From J. Hadzi, *Syst. Zool.*, 1953, 2:145.)

fairy shrimps, and corals should live on earth. Given the principles of molluscan biology, who would have thought it "reasonable" for flying squid to evolve? J. B. S. Haldane has written, "The world is not only stranger than we suppose; it is stranger than we can suppose." And so it is with the invertebrate world. It is a world not only for earnest study and speculation, for intellectual challenge and debate, but also for endless wonder and appreciation of mysteries both past and present.

for Further Reading

Anderson, D. T. *Embryology and Phylogeny in Annelids and Arthropods.* Pergamon Press, New York, 1973.

Clark, R. B. *Dynamics in Metazoan Evolution.* Clarendon Press, Oxford, 1964. (An excellent work that reviews theories of coelomic and metameric origin and then seeks to explain the evolution of these phenomena as adaptations for locomotor efficiency)

Dougherty, E. C. (Ed.) *The Lower Metazoa.* University of California Press, Berkely, 1963. (Papers from a symposium on the phylogeny of the lower metazoans: includes defenses of major theories by their most renowned supporters)

Greenberg, M. J. Ancestors, embryos and symmetry. *Syst. Zool.* 8: 212–221, 1959. (An argument for combining the colonial theory and the plasmodial theory in support of a polyphyletic origin of the Metazoa)

Hadzi, J. An attempt to reconstruct the system of animal classification. *Syst. Zool.* 2: 145–154, 1953.

Hadzi, J. *The Evolution of the Metazoa.* MacMillan, New York, 1963. (Hadzi expounds on his controversial views)

Haeckel, E. The gastraea theory, the phylogenetic classification of the Animal Kingdom and the homology of the germ almellae. *Quart. J. Microsc. Sci.* 14: 142–165 and 223–247, 1974. (The landmark article introducing the colonial theory of metazoan origin)

Hand, C. On the origin and phylogeny of the coelenterates. *Syst. Zool.* 8: 191–202, 1959.

Hanson, E. D. On the origin of the Eumetazoa. *Syst. Zool.* 7: 16–47, 1958.

Hanson, E. D. *The Origin and Early Evolution of Animals.* Wesleyan University Press, 1977.

House, M. R. (Ed.) *The Origin of Major Invertebrate Groups.* Academic Press, New York, 1979.

Hyman, L. H. *The Invertebrates Vol. I: Protozoa through Ctenophora.* McGraw-Hill, New York, 1940. (Hyman describes her support of the colonial theory)

Hyman, L. H. *The Invertebrates, Vol. II: Platyhelminthes and Rhynochocoela.* McGraw-Hill, New York, 1951. (Includes discussion of the origin of bilateralism, the coelom, and segmentation)

Marcus, E. On the evolution of the animal phyla. *Quart. Rev. Biol.* 33: 24–58, 1958. (This author supports the colonial theory and the entercoel theory)

Valentine, J. W. Coelomate superphyla. *Syst. Zool.* 22: 97–102, 1973.

Appendix I

Geological Time Scale

Era	*Period*	*Epoch*	*Beginning of Interval: Millions of years before present*	*Principal events in invertebrate evolution*
CENOZOIC	Quaternary	Pleistocene	1	Increase in cheilostomate bryozoans; spread of gastropod and bivalve mollusks; oligochaetes; great spread of winged insects; first copepods.
	Tertiary	Pliocene	13	Radiolarians and foraminiferans abundant
		Miocene	25	Pulmonate mollusks wide-spread
		Oligocene	36	First centipedes
		Eocene	58	Testate amoebas First moths and butterflies; extinction of belemnoids
		Paleocene	63	Decline of brachiopods; decline of cyclostomate bryozoans
MESOZOIC	Cretaceous	—	135	Belemnoids abundant; great increase in cyclostomate bryozoans; increase of eulamellibranch and prosobranch mollusks; increase of malacostracans, barnacles, centipedes, insects and arachnids; decline of brachiopods; decline and extinction of ammonoids; marked decline of merostomates. First euglenoid protozoans; first gorgonians; first cheilostomate bryozoans; first pteropod and octopod mollusks; first ants, bees, and wasps.
	Jurassic	—	181	Radiolarians abundant; belemnoids abundant; great spread of ammonoids; echinoids abundant; foraminiferans with siliceous tests; invasion of freshwater by gastropods; rock and wood-boring bivalves; first dinoflagellates; first ciliates; first decapod cephalopods (squids).
	Triassic	—	230	Belemnoids widespread; ammonoids flourishing; continued increase of diplopods, insects, and arachnids; first modern crinoids; first modern reef building corals. First barnacles; first modern dragonflies; first flies.

Geological Time Scale — Continued

Era	Period	Epoch	*Beginning of interval: Millions of years before present*	*Principal events in invertebrate evolution*
PALEOZOIC	Permian	—	280	Brachiopods abundant; cockroaches widespread; increase of insects and arachnids. Extinction of eurypterids, trilobites, and blastoids. First crickets, mayflies, beetles, and bugs.
	CARBONIFEROUS: Pennsylvanian	—	320	Brachiopods abundant; millipedes abundant; foraminiferans with calcareous tests. First pulmonate gastropods; first chilopods and winged insects (dragonflies and cockroaches).
	CARBONIFEROUS: Mississippian	—	345	Nematomorphs; belemnoids; holothuroids. First opisthobranch gastropods; first ophiuroids.
	Devonian	—	405	Radiolarians abundant; great glass sponges abundant; eurypterids abundant; great development of brachiopods and increase of arachnids. Decline of nautiloids and rise of ammonoids; decline of merostomates. Extinction of carpoids and cystoids. First siphonophores; first branchiopods and malacostracans; first diplopods and springtails.
	Silurian	—	425	Siliceous sponges abundant; growth of coral reefs widespread; great development of brachiopods; cystoids abundant; increase of ostracods. Decline of trilobites continuing. First scaphopods and first prosobranch gastropods; first air breathing arachnids (scorpions); myriapods; first mandibulate arthropods.
	Ordovician	—	500	Gymnolaemate bryozoans abundant; period of greatest differentiation of brachiopods; period of maximum development of trilobites (early) and of beginning decline (late). Nautiloids abundant, some with coiled shells; chitons; bivalves. Great spread of gastropods. Tubiculous annelids, with calcareous tubes; ostracods, barnacles and malacostracans; blastoids, cystoids, crinoids, asteroids, ophiuroids, and echinoids (without Aristotle's lantern). Extinction of eocrinoids. First foraminiferans; first alcyonarians and zoantharians; first pycnogonids.
	Cambrian	—	600	Representatives of most invertebrate phyla already established. Radiolarians; sponges with siliceous spicules; hydrozoans, scyphozoans, brachiopods; gastropods with uncoiled, symmetrical shells; nautiloids with straight or curved shells; annelids of several types, some resembling polychaetes; onychophorans; trilobites; xiphosurans and eurypterids; branchiopods; chaetognaths (possibly), eocrinoids, carpoids.
Proterozoic	Pre-Cambrian	—	1600	Siliceous sponge spicules and trails of floating and crawling animals or plants.
Archeozoic		—	3600	Evidence of blue-green algae and bacteria.

From Gardiner, M. S., 1972, *The Biology of Invertebrates* (McGraw-Hill), as taken from:
Kulp, J. L. *Science*, 1961, v. 133, pp. 1105-1114 — Geological Time Scale.
Moore, Lalicker, Fisher. 1952, *Invertebrate Fossils* (McGraw-Hill).
Kummel, B. 1961, *History of the Earth* (Freeman).
Shrock, R. R. and Twenhofel, W. H. 1953, *Principles of Invertebrate Paleontology* (McGraw-Hill).

Appendix II

A Classification Outline For The Invertebrates

This classification is limited to genera mentioned in this book. An asterisk * denotes an extinct taxon. Words in *italics* denote representative genera.

SUBKINGDOM PROTOZOA

Phylum Protozoa: Unicellular; solitary or colonial.
Subphylum Sarcomastigophora: With pseudopodia and/or flagella, monomorphic nuclei.
Class Mastigophorea: With flagella; asexual reproduction by longitudinal fission.
Subclass Phytoflagellata: Chromoplasts usually present; with few flagella.
Order Dinoflagellida: With two flagella in grooves of sculptured cellulose envelope; encystment in palmella form common.
Ceratium, Noctiluca, Gymnodium, Gonyaulax

Order Cryptomonadida: With two yellow or brown chromoplasts, two flagella, anterior vestibule.
Chilomonas

Order Chrysomonadida: With one or two yellow or brown chromoplasts, two unequal flagella, siliceous cysts.
Ochromonas, Paraphysomonas, Dinobryon

Order Euglenida: With photoreceptive stigmata; primarily freshwater, large and green.
Euglena, Peranema

Order Volvocida: With two or more flagella, green chromoplasts; primarily freshwater, sexual and colonial.
Chlamydomonas, Gonium, Volvox, Polytomella

Subclass Zooflagellata: Chromoplasts absent; one to many flagella.
Order Choanoflagellida: With single flagellum surrounded by protoplasmic collar; freshwater.
Salpingoeca, Proterospongia, Codosiga, Sphaeroeca

Order Kinetoplastida: With kinetoplasts, undulating membrane; often parasitic.
Trypanosoma, Leishmania

Order Metamonadida: With axostyles; often parasitic (includes hypermastigids).
Trichomonas, Giardia, Trichonympha

Order Rhizomastigida: With pseudopodia and flagella; primarily freshwater.
Mastigamoeba

Class Sarcodinea: With pseudopodia
Subclass Rhizopodia: Amoeboid
Order Amoebida: Test lacking; with lobopodia; mostly freshwater or parasitic; Naked amoebas.
Amoeba, Chaos, Entamoeba

Order Arcellinida: Test present; with lobopodia or filopodia.
Arcella, Euglypha, Difflugia

Subclass Granuloreticulosia: With granular reticulopodia.
Order Foraminiferida: With chambered perforated calcareous test, flagellated gametes; sexual and asexual stages.
Globigerina, Gromia, Rotalulla, Globigerinoides, Tritomphalus

Subclass Actinopodia: Spherical, planktonic; with radiating axopodia.
Order Radiolaria: With central siliceous capsule, outer calymma, axopodia; marine.
Trypanosphaera, Acanthometron, Cyrtocalpis

Order Heliozoida: Without central capsule; primarily freshwater.
Actinophrys, Actinosphaerium, Heterophrys

Class Opalinatea: With cilia; monomorphic nuclei; reproduction by syngamy and/or longitudinal fission; parasitic.
Opalina

Subphylum Sporozoa: Endoparasitic; with complex life cycle, including spores.
Class Telosporea: Reproduction both sexual and asexual.
Subclass Gregarinia: Larger, mature forms extracellular parasites; schizogony lacking; syzygy common.
Gregarina

Subclass Coccidia: Smaller, mature forms intracellular parasites; schizogony common.
Plasmodium, Eimeria

Subphylum Cnidospora: Parasitic, spores with polar filaments.

Class Myxosporidea: Larger, histozoic and coelozoic in fish; up to four polar filaments; spores develop from several nuclei.

Unicapsulina, Myxidium, Henneguya

Class Microsporidia: Minute; mostly cytozoic in insects; single polar filament; spores develop from one nucleus.

Nosema

Subphylum Ciliophora: With cilia, polymorphic nuclei; reproduction by conjugation and transverse fission.

Class Ciliatea: With the characteristics of the subphylum.

Subclass Holotrichia: Primitive; ciliature relatively uniform.

Prorodon, Paramecium, Tetrahymena, Didinium, Pleuronema

Subclass Peritrichia: Mostly sessile; often colonial; cilia concentrated in oral region.

Vorticella, Opercularia, Ophrydium

Subclass Spirotrichia: Oral ciliature spiraling into buccal cavity; body cilia clumped as cirri.

Order Heterotrichida: Large; uniform body cilia.

Stentor, Spirostomum

Order Oligotrichida: Marine; reduced body cilia; sometimes loricate.

Tintinnopsis

Order Hypotrichida: Dorsoventrally flattened; with prominent adoral membranelles.

Euplotes

Subclass Suctoria: Stalked, sessile, devoid of cilia in adult; ingestion through tentacles; mostly ectosymbionts.

Ephelota, Acineta

SUBKINGDOM AGNOTOZOA

Phylum Placozoa†: Body of two flattened epithelioid layers surrounding central mesenchyme.

Trichoplax

Phylum Mesozoa: No organs; body of cellular tube surrounding reproductive cells.

Order Dicyemida: Complex life cycle, including nematogen, rhombogen and infusiform stages; cephalopod parasites.

Pseudicyema

Order Orthonectida: Plasmodial parasitic stage unattached within various marine invertebrates.

Rhopalura

†This newly-recognized phylum consists of a single, puzzling species. Including the Placozoa, there are 33 invertebrate phyla described in this outline.

SUBKINGDOM PARAZOA

Phylum Porifera: Porous body with water canal system lined by choanocytes; incipient tissue formation only; sessile; mostly marine.

Class Calcarea: Skeleton of calcareous spicules; body plan usually asconoid or syconoid; marine.

Leucosolenia, Sycon

Class Hexactinellida: Skeleton of hexamerous, siliceous spicules; deep water; marine.

Euplectella

Class Demospongiae: Skeleton of siliceous spicules and/or sporgin; leuconoid body plan; a few freshwater species.

Spongia, Spongilla, Verongia

Class Sclerospongiae: As Demospongiae, but with outer calcareous covering.

SUBKINGDOM METAZOA
SECTION RADIATA

Phylum Coelenterata: Diploblastic; with modified radial symmetry, gastrovascular cavity, nematocysts, polypoid and medusoid stages; planula larva.

Class Hydrozoa: Both polypoid and medusoid stages common; with non-cellular mesoglea, simple coelenteron, epidermal gonads; medusae with velum.

Order Hydroida: Polypoid stage dominant; often sessile and colonial.

Suborder Calyptoblastea: With athecate zooids, leptomedusae.

Obelia

Suborder Gymnoblastea: With athecate zooids, anthomedusae.

Tubularia, Clava, Pennaria, Hydractinia, Eudendrium, Hydra, Sarsia

Order Trachylina: Primitive; medusa dominant.

Gonionemus, Craspedacusta

Order Hydrocorallina: With massive calcareous skeleton, dactylozooids.

Stylaster, Millepora

Order Siphonophora: Pelagic; colonial; with polymorphic zooids, float or swimming bell.

Physalia, Praya, Stephalia

Order Chondrophora: With highly integrated colonies; multi-chambered chitinoid pneumatophore.

Velella

Class Scyphoza: Medusa dominant; mesoglea partially cellularized; velum absent; with tetramerous symmetry; radial canals, gastrodermal gonads.

Order Stauromedusae: With medusoid bell supported by polypoid column.

Haliclystus

Order Cubomedusae: With cuboidal bell, four rhopalia.
Chironex, Chiropsalmus, Carybdea

Order Semaeostomae: Saucer or bowl-like bell; oral tentacles, 8-16 rhopalia.
Aurelia, Cyanea

Order Rhizostomae: With porous, filter-feeding manubrium; oral arms fused.
Cassiopeia, Stomolophus, Rhizostoma

Class Anthozoa: Medusa absent; gametes from septal endoderm; mesoglea cellular; with complex coelenteron, stomodeum and more than four septa.
Subclass Zoantharia (Hexacorallia): Hexamerous symmetry; with more than eight unbranched tentacles.
Order Actiniaria: Solitary with attached pedal disk; usually two siphonoglyphs. Sea anemones.
Metridium, Calliactis, Corynactis

Order Madreporaria (Scleractinia): With colonial polyps, no siphonoglyphs, massive external calcareous skeleton. Stony corals.
Acropora, Porites, Diploria

Order Ceriantharia: Elongate burrowing anemones; with one siphonoglyph.
Cerianthus

Subclass Alcyonaria (Octocorallia): Octamerous symmetry; with eight pinnate tentacles and outer cover of coenenchyme; almost all colonial.
Order Gorgonacea: Plantlike growth; axial gorgonin rod. Sea fans and sea whips.
Gorgonia, Eunicella, Leptogorgia

Order Stolonifera: Polyps joined by basal stolon or mat. Organ pipe coral.
Tubipora

Order Pennatulacea: Colony with flat, fleshy body. Sea pansies.
Renilla

Phylum Ctenophora: Solitary; radial symmetry; with eight meridional rows of ciliated plates, aboral sense organ, cydippid larva.
Class Tentaculata: Tentacles present.
Order Cydippida: Body round or ovoid.
Pleurobrachia

Order Lobata: Body compressed laterally.
Mnemiopsis

Order Cestida: Body compressed in tentacular plane.
Cestum, Velamen

Order Platyctenea: Body compressed in oral-aboral plane.
Coeloplana

Class Nuda: Tentacles absent.
Beröe

SECTION PROTOSTOMIA

The Acoelomate Bilateria

Phylum Platyhelminthes: Acoelomate; unsegmented; hermaphroditic; triploblastic; dorsoventrally-flattened; with protonephridia; anus absent.

Class Turbellaria: Mostly free-living; with ciliated epidermis, rhabdoids, simple life cycle.

Subclass Archeophora: Primitive; with yolky eggs, spiral cleavage.

Order Acoela: Without permanent digestive cavity, protonephridia or gonads; pharynx simple; marine.

Convoluta

Order Catenulida: With unbranched gut, dorsal male gonopore, simple pharynx; freshwater.

Alaurina

Order Polycladida: Thin, but often broad body; with highly branched gut, plicate pharynx, Muller's larva; marine.

Prostheceraeus, Planocera

Subclass Neoophora: With external yolk cells, yolk glands, modified cleavage.

Order Neorhabdocoela: With bulbous pharynx; simple sac-like gut.

Mesostoma

Order Temnocephalida: Ectocommensal; with reduced cilia, anterior tentacles, posterior suckers.

Temnocephala

Order Seriata: With plicate pharynx, lateral gut diverticula.

Suborder Tricladida: Elongate; gut with three branches.

Bothrioplana, Bipalium, Geoplana, Dugesia, Procerodes

Class Trematoda: Life cycle simple or complex; parasitic; with unciliated tegument, complex gut, one or more suckers. Flukes.

Order Monogenea: Largely ectoparasitic on single host; with posterior opisthaptor; oral sucker small or absent.

Polystoma, Sphyranura, Benedenia

Order Digenea: Endoparasitic with two or more hosts; with polyembryonic, supernumerary larval stages.

Fasciola, Opisthorchis, Leucochloridium, Schistostoma

Order Aspidobothrea: Endoparasitic; with enormous opisthaptor.

Cotylaspis

Class Cestoda: Endoparasitic with multiple hosts; digestive tract lacking. Tapeworms.

Subclass Eucestoda: Strobilated body; anterior scolex with holdfast organs; reproductive system in each proglottid.

Taenia, Dibothriocephalus

Subclass Cestodaria: Nonstrobilated; with single reproductive system.

Gyrocotyle

Phylum Rhynchocoela (Nemertina, or Nemertea): Acoelomate; eversible proboscis within rhynchocoel; digestive tract with anus; closed circulatory system.

Class Anopla: Proboscis unbarbed; mouth posterior to brain; development indirect; marine.

Lineus, Tubularus, Cephalothrix, Cerebratulus, Procarina

Class Enopla: Proboscis barbed; mouth anterior to brain; development direct; marine, freshwater and semiterrestrial.

Prostoma, Geonemertes, Amphiphorus, Malacobdella

The Pseudocoelomates
The Aschelminth Phyla

Phylum Nematoda: Pseudocoelomate; unsegmented; body wall muscles exclusively longitudinal; cilia and protonephridia absent. Roundworms.

Class Aphasmida (Adenophorea): Phasmids absent; mostly free-living.

Trichosomoides, Trichinella, Enoplus

Class Phasmida (Secernentea): Phasmids present; mostly parasitic.

Ascaris, Mesorhabditis, Rhabditis, Ancylostoma, Criconema, Tylenchus, Camallanus, Cephalobellus, Aphelenchoides

Phylum Rotifera: Pseudocoelomate; with anterior ciliated corona, jawed mastax; parthenogenesis common, primarily freshwater.

Class Monogononta: Often sessile; males degenerate; females with single germovitellarium.

Collotheca, Euchlanis, Asplanchna, Limnias, Epiphanes

Class Bdelloidea: Corona with two trochal disks above cingulum; males absent.

Philodina, Rotaria

Class Seisonacea: Body elongate; corona reduced; epizoic; marine.

Seison

Phylum Gastrotricha: Pseudocoelomate; corona lacking; cuticle sculptured into spines and/or scales; marine and freshwater.

Chaetonotus, Lepidodermella

Phylum Kinorhyncha: Pseudocoelomate; body with 13 zonites; protrusible, spiny oral cone; marine.

Echinoderes

Phylum Nematomorpha: Body hairlike; cuticle smooth; larva parasitic.

Gordius

Other Pseudocoelomate Phyla

Phylum Acanthocephala: Anterior spiny proboscis; digestive tract lacking; acanthor larva; adults parasitic.

Macracanthorhynchus, Acanthocephalus, Polymorphus

Phylum Gnathostomulida: Minute; marine interstitial; hermaphroditic; anus, protorephridia, blood vessels all lacking.

Gnathostomula, Austrognathia

The Eucoelomate Protostomes

Phylum Annelida: Eucoelomate; segmented, with closed circulatory system; metanephridia, trochophore larva.

Class Polychaeta: With distinct head, paired lateral parapodia, abundant setae; marine.

Subclass Errantia: Well cephalized; segments numerous and essentially similar; swimmers, crawlers and active burrowers.

Nereis, Platynereis, Aphrodita, Glycera, Alciopa, Syllis, Autolytus

Subclass Sedentaria: Segments specialized; parapodia reduced; tubiculous or sedentary burrowers.

Arenicola, Amphitrite, Sabella, Chaetopterus, Serpula, Polygordius,

"Class" Archiannelida: Small; interstitial; segmentation indistinct; parapodia absent; marine.

Class Oligochaeta: Head indistinct; with clitellum, few setae; hermaphroditic; freshwater and terrestrial.

Order Tubificida: Usually with one pair each of testes and ovaries; mostly freshwater.

Tubifex

Order Haplotaxida: Basically two pairs each of testes and ovaries; mostly terrestrial.

Lumbricus

Class Hirudinea: Usually dorsoventrally flattened; coelom filled with connective and muscle tissue; setae absent; predatory and/or ectoparasitic; hermaphroditic.

Order Rhynchobdellida: With eversible jawless pharynx, colorless blood restricted to blood vessels; aquatic.

Glossiphonia, Theromyzon, Ozobranchus

Order Gnathobdellida: With noneversible, jawed pharynx; red blood circulates in coelomic sinuses; aquatic and terrestrial.

Hirudo

"Class" Branchiobdellida: Epizoic on crayfish; with anterior and posterior suckers; setae lacking.

Stephanodrilus

"Class" Acanthobdellida: Single species parasitic on salmonid fish.

Acanthobdella

Phylum Pogonophora: Body divided into prosoma with cephalic lobe, trunk, and opisthosoma; digestive tract absent; marine.

Spirobrachia

Phylum Echiura: Body cylindrical or ovoid with nonretractable proboscis; deposit feeders; marine.

Urechis, Echiuris, Bonellia

Phylum Sipuncula: Retractable anterior introvert with crown of tentacles around mouth; U-shaped gut; marine.

Dendrostomum, Sipunculus

Phylum Priapulida: Cylindrical body with large anterior proboscis; spiny, tuberculous surface; predaceous; marine.

Priapulus, Halicryptus

Phylum Mollusca: With dorsal shell secreted by mantle, ventral muscular foot, dorsal visceral hump, mantle cavity with gills, radula, trochophore and/or veliger larvae.

Subphylum Aculifera: Vermiform body covered with calcareous spicules; rudimentary foot.

Class Aplacophora: With the characteristics of the subphylum.

Crystallophrisson, Pruvotina

Subphylum Placophora: Shell with eight overlapping dorsal plates; broad flat foot surrounded by groove containing gills.

Class Polyplacophora: With the characteristics of the subphylum.

Chaetopleura, Katharina, Cryptochiton, Chiton, Acanthopleura, Lepidochitona, Ischnochiton

Subphylum Conchifera: Dorsal surface not covered by spicules or eight plates; with structural emphasis on the dorsoventral axis.

Class Monoplacophora: Single domelike shell and multiple pairs of organs; three relict species.

Neopilina

Class Gastropoda: Body torted within spirally coiled shell; head well-developed; foot large and muscular.

Subclass Prosobranchia: Mantle cavity anterior; shell well developed; operculum present; mostly diecious and marine.

Order Archeogastropoda: With paired auricles, gills, metanephridia; many teeth in each transverse row of radula. Limpets and abalones.

Scissurella, Haliotis, Fissurella, Diodora, Acmaea, Lepeta, Patella, Trochus, Norrisia, Theodoxus, Helicina

Order Mesogastropoda: With single auricle, gill, metanephridium; seven teeth in each transverse row of radula. Periwinkles, cowries, and conchs.

Littorina, Strombus, Crepidula, Cerithiopsis, Natica, Polinices, Viviparus, Stylifer, Entoconcha, Enteroxenos, Atlanta, Carinaria, Tonna, Charonia

Order Neogastropoda: Usually carnivorous with reversible proboscis, bipectinate osphradium; three teeth in each transverse row of radula; exclusively marine. Whelks, cone shells, oyster drills.

Busycon, Buccinum, Melongena, Fasciolaria, Urosalpinx, Conus, Columbella, Antemone

Subclass Opisthobranchia: Detorsion with reduction/loss of mantle cavity and shell; single gill, auricle and nephridium; hermaphroditic; mostly marine.

Order Cephalaspida: Shell usually prominent; head expanded as burrowing shield. Bubble shells.
Acteon, Hydatina, Haminea, Philine, Runcina

Order Aplysiacea: Shell reduced; foot with lateral parapodia. Sea hares.
Aplysia

Order Notaspidea: Shell small and internal; gill within mantle cavity remnant.
Pleurobranchus

Order Nudibranchia: Shell, mantle cavity and gills lacking.
Tritonia, Dirona, Flabellinopsis, Calma, Rostanga

Order Thecosomata: With broad flapping parapodia; pelagic. Shelled sea butterflies (pteropods).
Limacina, Gleba

Order Gymnosomata: With parapodial fins; pelagic. Unshelled pteropods.
Clione

Subclass Pulmonata: Respiration across vascularized wall of mantle cavity; gills and operculum lacking; hermaphroditic; freshwater and terrestrial.
Amphibola, Siphonaria, Otina, Lymnaea, Ferrissia, Bulinus, Physopsis, Helix, Limax, Testacella

Class Bivalvia: Shell with two hinged valves; head and radula lacking; large mantle cavity with gills.

Subclass Protobranchia: Protobranchiate gills; labial palps for food collection.
Nucula, Yoldia

Subclass Septibranchia: Gills modified as muscular septa.
Cuspidaria, Poromya

Subclass Lamellibranchia: Filter-feeding gills with elongate, folded filaments.
Arca, Glycymeris, Limopsis, Mytilus, Modiolus, Botula, Chama, Crassostrea, Ostrea, Pinctada, Pecten, Lima, Avicula, Maleus, Pinna, Cardium, Tridacna, Donax, Tagelus, Ensis, Siliqua, Mya, Panope, Pholas, Barnea, Teredo, Unio, Lampsilis, Anodonta, Entovalva, Mercenaria, Aequipecten, Lyrodus, Brachiodontes

Class Scaphopoda: Shell tubular; with food-gathering captacula.
Cadulus, Dentalium

Class Cephalopoda: Head and eyes well-developed; nervous system highly concentrated; pelagic; marine.

Subclass Nautiloidea: External shell straight or coiled with simple sutures.
*Nautilus, *Plectronoceras, *Endoceras*

Subclass *Ammonoidea: External shell coiled with complex sutures.
**Pachydiscus, *Ptychoceras, *Manticoceras*

Subclass Coleoidea: Internal shell much reduced or absent.
Order Decapoda: With lateral fins, two tentacles and eight arms.
Suborder *Belemnoidea: Internal shell with straight chambered region, thick lateral shelves and broad shield.
Suborder Sepioidea: Body short and broad. Cuttlefish.
Sepia, Spirula, Sepiola, Idiosepius

Suborder Teuthoidea: Body elongate; shell remnant as pen. Squid.
Loligo, Architeuthis, Onchoteuthis, Calliteuthis, Cranchia, Chiroteuthis, Histioteuthis

Order Vampyromorpha: Deep water; with ten arms (two of which are minute).
Vampyroteuthis

Order Octopoda: Body globular with eight arms; shell and fins lacking.
Octopus, Argonauta, Cirrothauma

Phylum Arthropoda: Body segmented and covered by a chitinous exoskeleton with jointed appendages; well cephalized with complex eyes and advanced brain; open hemocoel; molting to accommodate growth.
Subphylum *Trilobitomorphaa: Body divided into cephalon, thorax and pygidium; carapace with dorsal trilobation; similar biramous limbs.
Class Trilobita: With the characteristics of the subphylum.
**Phacops, *Megistaspidella, *Reedops, *Radiaspis*

Subphylum Chelicerata: Body divided into prosoma and opisthosoma; appendages include preoral chelicerae and postoral pedipalps; antennae and mandibles lacking; primarily terrestrial.
Class Merostomata: Mouth between bases of prosomal walking limbs with opisthosomal respiratory limbs, caudal spine; aquatic.
Subclass Xiphosura: Prosoma semi-circular; book gills present. Horseshoe crabs.
Limulus

Subclass *Eurypterida: Body elongate; opisthosoma divided into pre- and post adbomen; raptorial pedipalps common. Giant water scorpions.
**Pterygotus, *Mixopterus, *Eurypterus*

Class Arachnida: Prosoma with four pairs of walking legs; opisthosoma without appendages; gonopores on second opisthosomal segment; respiration via book lungs or tracheae; usually terrestrial.
Order Scorpiones: Small chelicerae, large chelate pedipalps; opisthosoma divided into pre- and postabdomen; terminal stinger; comblike pectines.
Androctonus, Centruroides, Chactas, Pandinus, Parabuthus, Heterometrus, Hormurus

Order Uropygi: Raptorial pedipalps; terminal flagellum; acid discharge from anal glands. Vinegaroons.
Mastigoproctus

Order Amblypygi: Flattened body with no flagellum; long and slender tactile first legs. Whip spiders.

Charinus

Order Palpigradi: Minute, with bipartite prosoma; pedipalps as walking legs; respiratory organs lacking.

Koenenia

Order Araneae: Prosoma and opisthosoma joined at narrow pedicel; poison ducts open on cheliceral claws; posterior silk glands. True spiders.

Latroductus, Argyroneta, Loxosceles, Agelenopsis, Agelena, Dictyna, Theridion, Mastophora, Araneus, Tegenaria, Lycosa, Cteniza, Segestria, Brachypelma

Order Ricinulei: Prosoma with anterior cucullus; chelicerae grasping; pedipalps leglike; male third legs modified as copulatory organs.

Ricinoides

Order Pseudoscorpiones: Chelicerae with silk glands; large raptorial pedipalps with poison glands; flattened opisthosoma with rounded posterior.

Chelifer, Microcreagris, Neobisium

Order Solifugae: Bipartite prosomal carapace; large chelicerae; long slender pedipalps; first legs tactile. Sun spiders.

Galeodes

Order Opiliones: Prosoma and opisthosoma broadly fused; slender chelicerae, leglike pedipalps. Harvestmen.

Platybunus

Order Acarina: Dorsal carapace fused over compact ovoid body; complex mouthparts form capitulum; often parasitic. Ticks and mites.

Threophagus, Acarus, Dermatophagoides, Analges, Amblyoma, Dermacentor, Sarcoptes, Dermanyssus, Nematalycus, Knemidokoptes, Caeculus, Caminella, Ixodes, Mideopsis, Trombicula, Leptotrombidium, Ornithodorus

Class Pycnogonida: Prosoma long and narrow; opisthosoma degenerate; males with ovigerous legs. Sea spiders.

Dodecolopoda, Nymphon, Achelia

Subphylum Mandibulata: Oral appendages modified as mandibles.

Class Crustacea: Head with paired antennules, antennae, mandibles, first and second maxillae; biramous appendages; procuticle reinforced with calcium carbonate; predominantly aquatic.

Subclass Cephalocarida: Rounded head; thorax with nine paired triramous appendages; telson with caudal branches; second maxillae undifferentiated, eyes: lacking.

Hutchinsoniella, Lightiella

Subclass Branchiopoda: Uniform trunk with numerous flattened appendages; abdominal appendages lacking; mostly filter-feeding in fresh water.

Order Anostraca: Carapace lacking; eyes stalked. Fairy shrimp.

Artemia, Branchinecta

Order Notostraca: Carapace over anterior body. Tadpole shrimp.

Triops

Order Diplostraca: Carapace laterally compressed; second antennae modified for swimming.

Suborder Conchostraca: Body enclosed by bivalved carapace. Clam shrimp.

Cyzicus

Suborder Cladocera: Compact body with six or less trunk appendages; carapace over all body except head. Water fleas.

Daphnia, Leptodora, Simocephalus

Subclass Ostracoda: Body enclosed by laterally compressed bivalved carapace; head dominant with locomotor antennules and antennae. Mussel shrimp.

Gigantocypris, Cypridina, Platycypris, Philomedes, Cypris

Subclass Copepoda: Small; compound eyes and carapace lacking; with long antennules, naupliar and copepodid larvae; parasitic aberrations.

Calanus, Cyclops, Pennella, Diaptomus, Lernaeolophus, Salmincola, Calocalanus, Macrocyclops

Subclass Mystacocarida: Vermiform; with large setose antennules, elongate mandibles, differentiated maxillipeds.

Derocheilocaris

Subclass Branchiura: Dorsoventrally flattened body; sucking mouth; ectoparasitic. Fish lice.

Argulus

Subclass Cirripedia: Body enclosed by calcareous plates secreted by folded carapace (mantle); thoracic legs slender and setose; head appendages degenerate; cypris larva; adults sessile or parasitic. Barnacles.

Lepas, Conchoderma, Alepas, Pollicipes, Scalpellum, Lithotrya, Balanus, Creusia, Pyrgoma, Coronula, Xenobalanus, Trypetesa, Sacculina, Peltogasterella, Ascothorax

Subclass Malacostraca: Body with 19 segments: head (5) thorax (8) with pereiopods, abdomen (6) with pleopods; carapace covering head and all or part of thorax; male gonopore on eighth thoracic segment, female on sixth.

Group Phyllocarida: Bivalved carapace extending to abdomen; hinged rostrum; pereiopods flattened; abdomen with eight segments.

Order Leptostraca: With the characteristics of the group.

Nebaliopsis, Nebalia

Group Eumalacostraca: With the characteristics of the subclass.

Superorder Syncarida: Carapace absent; sporadic distribution in fresh water.
Order Anaspidacea: First thoracic segment fused to head.
Anaspides, Paranaspides

Order Bathynellacea: First thoracic segment free from head.
Bathynella

Superorder Hoplocarida: Carapace short; antennules triramous; second thoracic legs subchelate; brooding common.
Order Stomatopoda: With the characteristics of the superorder. Mantis shrimp.
Squilla, Gonodactylus

Superorder Peracarida: With at least the first thoracic segment fused to the head; maxillipeds present; brooding within marsupium.
Order Mysidacea: Carapace covering thorax; shrimplike with uropods and telson forming tail fin.
Gnathophausia, Mysis

Order Cumacea: Carapace inflated; abdomen slender; pleopods developed in males only.
Diastylis

Order Tanaidacea: Short carapace fused to first two thoracic segments; single pair of maxillipeds.
Heterotanais, Apseudes

Order Isopoda: Carapace lacking; usually dorsoventrally flattened; abdomen reduced but with pleopod gills.
Bathynomus, Calathura, Gnathia, Neophreatoicus, Portunion, Astacilla, Limnoria, Ligia, Munnopsis, Armadillidium, Asellus, Clypeoniscus

Order Amphipoda: Carapace lacking; usually laterally compressed; second and third legs modified as gnathopods; gills thoracic.
Gammarus, Rhabdosoma, Mimonectes, Caprella, Siphonoecoetes, Talorchestia, Cyamus

Superorder Eucarida: Carapace fused with all thoracic segments; eyes stalked; gills thoracic; marsupium never present.
Order Euphausiacea: Body laterally compressed and shrimplike; gill chamber and maxillipeds lacking. Krill.
Euphausia, Stylocheiron, Nematoscelis

Order Decapoda: Three pairs of maxillipeds; pereiopods uniramous; ventral branchial chamber.
Suborder Natantia: Body laterally compressed; rostrum and antennal scales prominent; pereiopods slender; pleopods developed for swimming.
Section Penaeidea: Third pereiopods slender and chelate; gills dendrobranchiate; young hatch as nauplius larvae.
Penaeus, Sergestes

Section Caridea: Third pereiopods nonchelate; gills phyllobranchiate; eggs brooded, hatch as protozoea.

Alpheus, Crangon, Hippolyte, Pandalus, Palaemon, Palaemonetes, Macrobrachium, Atyaephyra, Trypton, Anchistus, Merguia, Spirontocaris

Section Stenopodidea: Third cheliped enlarged; gills trichobranchiate; often commensal.

Stenopus, Spongicola

Suborder Reptantia: Body not laterally compressed; dorsoventral flattening common; rostrum reduced; pereiopods heavy and modified for walking.

Section Macrura: Abdomen elongate and muscular; uropods and telson modified as swimming fan. Lobsters and crayfish.

Homarus, Euastacus, Astacus, Orconectes, Cambarus, Procambarus, Palinurus, Panulirus, Scyllarus, Upogebia, Callianassa, Polycheles

Section Anomura: Abdomen variable, often asymmetrical; third pereiopods nonchelate; fifth pereiopods reduced and dorsally flexed; eyes medial to antennae.

Pagurus, Birgus, Pylopagurus, Lithodes, Paralithodes, Lopholithodes, Cryptolithodes, Coenobita, Aegla, Emerita, Holopagurus, Petrolisthes

Section Brachyura: Carapace broad and dorsoventrally flattened; carapace fused with epistome; abdomen reduced and flexed beneath cephalothorax; megalops larva; eyes lateral to antennae. True crabs.

Dromia, Dorippe, Calappa, Macrocheira, Callinectes, Cancer, Carcinus, Pinnixa, Hapalocarcinus, Eriocheir, Potamon, Uca, Ocypode, Grapsus, Mictyris, Gecarcinus, Cardiosoma, Stenocionops, Lyreidus, Oregonia

Class Chilopoda: Body dorsoventrally flattened; 15-173 segments with one pair of legs each; poison duct with prehensors; posterior gonopore; carnivorous. Centipedes.

Scutigera, Scolopendra, Thereuopoda, Lithobius, Otocryptops, Geophilus

Class Diplopoda: Body cylindrical with two pairs of legs on each of 20-100 segments; gonopore on third segment; mostly herbivorous. Millipedes.

Archispirostreptus, Siphonophora, Blaniulus, Rhinocricus, Polydesmus, Polyxenus, Orthoporus, Apheloria, Glomeris, Spirobolus

Class Pauropoda: Body cylindrical with nine legs on eleven segments; antennae triramous; gonopore on third segment.

Pauropus

Class Symphyla: Body with twelve or less limbed segments; head insectlike; spinnerets on penultimate segment.

Scutigerella, Hanseniella

Class Insecta: Body divided into head, thorax and abdomen; head with antennae, mandibles, maxillae and labium; thorax with three pairs of walking legs and two pairs of wings; abdomen with posterior gonopores; tracheal respiration; excretion through Malpighian tubules; hemimetabolous or holometabolous development.

Order Thysanura: Silverfish and bristletails.

Lepisma

Order Protura: Proturans.

Acerentomon

Order Collembola: Springtails.

Isotomurus

Order Ephemeroptera: Mayflies.

Potamanthus, Polymitarcys

Order Odonata: Damselflies and dragonflies.

Anax

Order Plecoptera: Stoneflies.

Isoperla

Order Orthoptera: Grasshoppers, locusts, and crickets.

Schistocerca, Melanoplus

Order Phasmida: Walking sticks.

Carausius, Pharanacia

Order Dermaptera: Earwigs.

Forficula

Order Embioptera: Webspinners.

Embia

Order Dictyoptera: Cockroaches.

Blaberus, Periplaneta, Blatta, Eyrsotria

Order Isoptera: Termites.

Hodotermes, Amitermes

Order Zoraptera: Zorapterans.

Zorotypus

Order Psocoptera: Booklice and barklice.

Caecilius

Order Mallophaga: Chewing lice.

Trinoton

Order Anoplura: Sucking lice.

Phthirus

Order Homoptera: Cicadas, leafhoppers and aphids.
Phylloxera

Order Hemiptera: True bugs.
Eusthenes

Order Thysanoptera: Thrips.
Idolothrips

Order Neuroptera: Lacewings and antlions.
Myrmeleon

Order Mecoptera: Scorpionflies.
Panorpa

Order Trichoptera: Caddisflies and water moths.
Rhyacophila

Order Lepidoptera: Butterflies and moths.
Hyalophora

Order Diptera: True flies.
Drosophila, Musca

Order Coleoptera: Beetles.
Acrocinus, Titanus

Order Siphonaptera: Fleas.
Pulex

Order Strepsiptera: Stylopids.
Opthalmochlus

Order Hymenoptera: Ants, bees and wasps.
Myrmecia, Apis, Trichogramma

Phylum Onychophora: Body elongate and unsegmented; legs fleshy and clawed; cuticle chitinous; tracheal respiration; mandibles present.
Peripatus

Phylum Tardigrada: Body short, cylindrical and unsegmented; four parts of stubby clawed legs; circulatory and respiratory organs lacking.
Echiniscus, Macrobiotus

Phylum Pentastomida: Body elongate and unsegmented; anterior oral snout with two pairs of clawed protuberances; parasites of vertebrate respiratory tracts.
Cephalobaena, Porocephalus

The Lophophorate Phyla

Phylum Phoronida: Body cylindrical and enclosed within chitinous tube; lophophore crescent-shaped; metanephridia, blood vessels present; actinotroph larva; marine.
Phoronis, Phoronopsis

Phylum Bryozoa (Ectoprocta): Body enclosed in zoecium; lophophore usually circular; excretory and circulatory organs lacking; colonial; marine and freshwater.
Class Phylactolaemata: Zoecium uncalcified; lophophore horseshoe-shaped; zooids not polymorphic; colonial coelom continous; body wall muscles developed; freshwater.
Lophopus, Cristatella, Plumatella, Fredericella

Class Stenolaemata: Zoecium calcified with circular orifice; lophophore circular; zooids tall and tubular; lophophore extended by septal muscles; marine.
Crisia, Tubulipora, Stomatopora

Class Gymnolaemata: Zooids cylindrical or cuboidal; circular lophophore extended by body wall deformation; zooids polymorphic; marine.
Order Ctenostomata: Zoecium uncalcified; operculum absent; colonial growth usually stoloniferous.
Bowerbankia, Nolella, Alcyonidium, Paludicella

Order Cheilostomata: Zoecium calcified; operculum present; zooids boxlike; brooding in ovicell common.
Bugula, Electra, Membranipora, Callopora, Cribrilina

Phylum Entoprocta: Body with stalked central calyx; mouth and anus within tentacular ring.
Mysoma, Loxosomella, Pedicellina

Phylum Brachiopoda: Body enclosed dorsoventrally within calcareous valves; lophophore spiral; metanephridia present; solitary; marine.
Class Inarticulata: Valves similar; hinge absent; anus present.
Lingula, Glottidia

Class Articulata: Valves dissimilar; hinge present; anus lacking.
Terebratulina, Terebratalia, Laqueus, Hemithyris, Waltonia, Macandrevia, Megathirus, Pumilus, Argyrotheca

SECTION DEUTEROSTOMIA

Phylum Echinodermata: Body with secondary pentamerous radial symmetry, ambulacal system, calcareous endoskeleton; marine.
Subphylum *Homalozoa: Body bilateral with short stalk and perhaps a crown of oral tentacles. Carpoids.
**Gyrocystis, *Dendrocystites*

Subphylum Crinozoa: Body radial and globular with oral surface uppermost.
Class *Eocrinoidea: Skeletal theca; brachioles pentamerous; arms lacking.
**Macrocystella*

Class *Cystoidea: Theca oval or spherical, porous; sessile.

Callocystis

Class *Blastoidea: Theca oval; with aboral stalk and hydrospires; elevated ambulacra bordered by brachioles.

**Blastoidocrinus*

Class Crinoidea: Body with central cuplike calyx and oral tegmen; arms with lateral pinnules; spines and madreporite lacking. Feather stars and sea lilies.

Ptilocrinus, Cenocrinus, Antedon, Comanthus, Neocrinus, Metacrinus, Comanthina, Notocrinus

Subphylum Asterozoa: Radial; generally star-shaped with five arms; oral surface lowermost; non-sessile.

Class Stelleroidea: With the characteristics of the subphylum.

Subclass Asteroidea: Arms relatively wide and housing spacious extensions of coelom; ambulacra open; podia locomotory. Sea stars.

Pisaster, Astrometis, Hippasteria, Astropecten, Asterias, Oreaster, Solaster, Pycnopodia, Freyella, Patiria, Acanthaster, Marthasterias, Archaster

Subclass Ophiuroidea: Arms narrow and flexible; most organ systems confined to central body; ambulacra closed; anus lacking. Brittle stars and basket stars.

Ophiocoma, Ophiactis, Orchasterias, Ophiopholis, Ophioderma, Gorgonocephalus, Asteronyx, Amphilycus, Ophionereis, Ophiura, Ophiothrix, Ophiocomina, Amphioplus, Amphipholis, Ophiomaza

Subphylum Echinozoa: Globoid or cylindrical; arms lacking.

Class *Helicoplacoidea: Body spindle-shaped; ambulacrum singular and spiraled.

**Helicoplacus*

Class *Edrioasteroidea: Mostly sessile; oral surface uppermost; five adoral ambulacra.

**Pyrgocystis, *Edrioaster*

Class *Ophiocistioidea: Tube feet giant and scaly; oral surface lowermost.

**Volchovia, *Sollasina*

Class Echinoidea: Spherical or disk-shaped; spines movable; often with Aristotle's lantern and pedicellariae. Sea urchins and sand dollars.

Echinus, Arbacia, Strongylocentrotus, Colobocentrotus, Psammechinus, Tripneustes, Echinocyamus, Echinocardium, Clypeaster, Mellita, Dendraster, Paracentrotus, Asthenosoma, Lytechinus

Class Holothuroidea: Elongate oral-aboral axis; tertiary bilateral symmetry; oral podia modified as tentacles; ossicles microscopic; respiratory trees. Sea cucumbers.

Cucumaria, Leptosynapta, Ypsilothuria, Thyone, Stichopus, Paracaudina, Pelagothuria, Euapta, Synapta, Scotoplanes, Holothuria

Phylum Hemichordata: Vermiphore; with pharyngeal clefts and dorsal hollow nerve tube; marine particle feeders.

Class Enteropneusta: Anterior conical proboscis; gill slits numerous; tentacles lacking; gut straight; solitary. Acorn worms.

Balanoglossus, Saccoglossus, Stereobalanus, Glossobalanus

Class Pterobranchia: Small to microscopic; anterior tentacles; gill slits sometimes lacking; gut U shaped; after tubiculous, aggregated or colonial.

Atubaria, Cephalodiscus, Rhabdopleura

Phylum Chaetognatha: Body torpedo-shaped with lateral fins; head with raptorial spines; hermaphroditic; marine, mostly planktonic. Arrow worms.

Spadella, Sagitta

Phylum Chordata: With pharyngeal clefts, dorsal hollow nerve cord, and notochord.

Subphylum Tunicata: With chordate characteristics developed in larva only; adults sessile or pelagic, generally marine, hermaphroditic, filter feeding.

Class Ascidiacea: Adults sessile with large central pharynx; outer tunic; incurrent buccal and excurrent atrial siphons. Sea squirts.

Clavelina, Perophora, Cyathocormus, Coelocormus, Botryllus, Herdmania, Botrylloides, Corella

Class Thaliacea: Siphons at opposite ends of barrel-shaped zooid; solitary or colonial; pelagic. Salps.

Pyrosoma, Salpa, Doliolum

Class Larvacea: Adults with persistent larval characteristics; body within secreted "house"; minute; planktonic.

Oikopleura

Glossary

ABYSSAL PLAIN the deep ocean floor.

ACANTHOR the first larval stage of an acanthocephalan.

ACETABULUM a ventral attachment organ in digenetic trematodes.

ACICULUM a skeletal rod which reinforces a parapodium in errant polychaetes.

ACOELOMATE lacking a secondary body cavity.

ACONTIUM a long thread which extends from the middle lobe of a septal filament in anthozoans.

ACRON the anterior pre-segmental tip of the arthropod body.

ACTINOTROCH the larval form of a phoronid.

ACTINULA the immature polypoid stage which follows settlement of the planula in many coelenterate life cycles.

ACTIVITY SYSTEMS structures and processes which underlie the behavior of an animal.

AESTHETE a cluster of sensory cells in vertical canals on chiton shell plates.

ALVEOLUS one of many membrane delineated spaces surrounding the infraciliature of ciliate protozoans.

AMBULACRAL SYSTEM the hydraulic (water-vascular) system which operates the tube feet of an echinoderm.

AMBULACRUM a body zone associated with one of the radial elements of the ambulacral system of an echinoderm.

AMICTIC pertaining to diploid, thin-shelled, parthenogenetically produced eggs or individuals.

AMOEBOCYTE any amoeboid, wandering cell.

AMPHIBLASTULA a sponge larva with anterior, externally flagellated micromeres and posterior macromeres.

AMPHID an anterior sensory and glandular pit in nematodes.

AMPULLA the proximal muscularized bulb of an echinoderm podium.

ANCESTRULA the original, founder zooid of a bryozoan colony.

ANDROGENIC GLAND a gland in male crustaceans that promotes the development of sexual characters.

ANISOGAMETE either one of the two distinct types of sex cells produced by most animals.

ANNULUS a superficial (nonsegmental) ring on the external body surface, as in leeches.

ANTENNA a sensory appendage on an arthropod head.

ANTENNAL GLAND the excretory organ of many crustaceans.

ANTENNULE one of the pair of anteriormost crustacean appendages.

ANTHOMEDUSA the bell-shaped medusa of athecate hydroids.

APICAL SENSE ORGAN a sensory complex at the aboral pole of ctenophores.

APODEME an inward folding of the arthropod procuticle and site of muscle attachment.

APOPYLE the excurrent opening of a choanocyte chamber in a leuconoid sponge.

ARCHENTERON the central cavity of the gastrula and future gut of the adult.

ARCHEOCYTE in sponges, a wandering cell capable of differentiating into any of a number of cell types.

ARISTOTLE'S LANTERN the complex chewing apparatus characteristic of echinoids.

ARTICULAR MEMBRANE the flexible region between the exoskeletal plates of an arthropod.

ASCONOID pertaining to the simplest sponge body plan which features a single spacious spongocoel.

ASCUS an expandable sac lined by the frontal membrane of an ascophoran bryozoan.

ATOKE the immature, asexual form of an epitokous polychaete species.

ATRIAL SIPHON the excurrent opening along the dorsal side or at the posterior end of a tunicate.

AURICULARIA the first larval stage of most holothuroids.

AUTOGAMY the union of nuclei within and derived from a single cell, as in certain ciliate protozoans.

AUTOTROPHIC capable of synthesizing food from nonbiological sources.

AUTOZOOID a feeding individual in a colony, particularly in bryozoans.

AVICULARIUM a bryozoan heterozooid whose operculum forms a prominent, defensive jaw.

AXIAL GLAND a spongy mass of tissue associated with the axial sinus of echinoderms; possibly circulatory in nature.

AXIAL SINUS a sinus that connects the oral and aboral ring sinuses in the echinoderm hemal system.

AXOCOEL the anterior coelomic compartment (protocoel) of an echinoderm embryo.

AXONEME the microtubule bundle which constitutes the core of a cilium or flagellum.

AXOPODIUM a slender ectoplasmic pseudopodium that is reinforced by a central axis of spiraling microtubules.

AXOSTYLE a long bundle of microtubules in metamonad zooflagellates.

BASAL BODY (KINETOSOME) the possibly self-replicating base of a cilium or flagellum.

BASOPODITE the distal part of a crustacean protopodite; attachment site for the endopodite and exopodite.

BENTHOS bottom-dwelling aquatic organisms.

BIPINNARIA the first larval stage of an asteroid.

BIRAMOUS composed of two branches.

BLASTOCOEL the central cavity of the blastula.

BLASTOPORE external opening of the archenteron.

BLASTOSTYLE a coelenterate gonozooid without a mouth or tentacles.

BLASTOZOOID an asexually produced individual that, upon maturation, reproduces sexually in a thaliacean life cycle.

BLASTULA an embryonic sphere of cells formed after numerous cleavages of the zygote.

BOOK GILL one of a series of respiratory flaps derived from the opisthosomal appendages of primitive chelicerates.

BOOK LUNG a small, invaginated air chamber on the ventral surface of the opisthosoma in many arachnids.

BOTRYOIDAL TISSUE loose mesenchymal tissue in the reduced coelom of leeches.

BRACHIOLARIA the second larval stage of an asteroid.

BRACHIOLE narrow appendages containing branches of the ambulacral system in fossil echinoderms.

BRANCHIAL HEART a blood-pumping organ at the base of a cephalopod gill.

BROWN BODY a dark cellular cluster representing the remnants of the degenerated polypide in a brooding bryozoan.

BUCCAL CAVITY a variously modified, preoral or postoral chamber.

BUCCAL CONE the walls of the prebuccal chamber in acarines.

BUCCAL FRAME a ventral body wall depression that contains the mouth and most of the feeding appendages of a crab.

BUCCAL SIPHON the incurrent opening at the anterior end of a tunicate.

BURSA a respiratory invagination of the oral surface of an ophiuroid.

BURSA COPULATRIX a portion of the genital chamber or vagina of a female insect into which the spermatophore is inserted at copulation.

BURSAL SLIT the opening to the bursa in an ophiuroid.

BYSSAL GLAND a gland on the foot of certain bivalve mollusks; its secretions aid in attachment.

CALCIFEROUS GLANDS esophageal glands whose secretions cause the crystallization of inorganic ions in annelids.

CALYMMA a frothy, vacuolated region which surrounds the central capsule in radiolarian protozoans.

CALYX[1] the boat-shaped central body of an entoproct.

CALYX[2] the cuplike aboral component of the central body of a crinoid.

CAPITULUM[1] **(GNATHOSOMA)** the head zone of an acarine.

CAPITULUM[2] the calcareous framework which houses the mantle and body of a barnacle.

CAPTACULA the food-gathering tentacles of scaphopod mollusks.

CARAPACE a cuticular shield which originates from the edges of cephalic tergites and covers a variable portion of the arthropod body.

CARDIAC STOMACH[1] the grinding portion of the gastric mill in a malacostracan crustacean.

CARDIAC STOMACH[2] the oral eversible chamber of an asteroid gut.

CARIDOID FACIES a generalized hypothetical animal with the basic features of malacostracan crustaceans.

CARINA the dorsal keel of a barnacle shell.

CELL ROSETTE a ciliated area surrounding a pore between the gastrodermal canals and the mesoglea of ctenophores.

CELLULAR LEVEL OF ORGANIZATION a multicellular body plan in which the physiological independence of most cells is preserved (said of sponges).

CEPHALIZATION the adaptive concentration of sense organs and nervous tissues in the anterior end of bilaterally symmetrical animals; head formation.

CERATA club-shaped or branched appendages on the dorsal surface of many nudibranchs.

CERCARIA a polyembryonically produced, fluke-like larval stage of digenetic trematodes.

CERCI sensory appendages that project from the last abdominal segment of an insect.

CEREBRAL ORGAN ciliated sensory pits on the dorsal head region of nemertines.

CHAMBERED ORGAN the aboral center of the perivisceral coelom of a crinoid.

CHELATE clawed; pincerlike.

CHELICERA one of the pair of preoral food-gathering appendages typical of chelicerate arthropods.

CHELIPED a large chelate pereiopod, such as the claw of a crab.

CHEMOSYNTHESIS production of food from non-biological raw materials using energy from the oxidation of nonorganic compounds.

CHILARIA diminutive appendages on the last prosomal segment of horseshoe crabs.

CHLORAGOGEN yellow or brownish tissue associated with the intestinal wall of annelids; a site of storage excretion and intermediary metabolism.

CHOANOCYTE a flagellated collar cell characteristic of sponges.

CHOANOCYTE CHAMBER one of the numerous, small, choanocyte-lined spaces in a leuconoid sponge.

CHORDOTONAL ORGAN insect sensory structures that respond to changes in muscular and skeletal tension.

CHROMATOPHORE a pigment-containing cell which contributes to the variable color of an animal.

CHROMOPLAST an organelle containing pigment in phytoflagellates.

CILIUM a short, hairlike locomotory organelle occurring in organized rows or groups.

CINCLIDE a small lateral pore in an anthozoan body wall.

CINGULUM the lower ciliary wheel of the corona in some rotifers.

CIRRUS[1] a group of fused cilia, as in spirotrichs.

CIRRUS[2] an eversible penis in flatworms.

CIRRUS[3] a finger-like projection of a polychaete parapodium.

CIRRUS[4] one of the two setose branches of the thoracic legs of barnacles.

CIRRUS[5] one of the lateral appendages on the stalk or aboral base of a crinoid.

CLITELLUM a glandular swelling over several mid-body segments in oligochaetes and leeches; its secretions produce a cocoon.

CLOACA an exit chamber common to both the reproductive and the digestive systems.

CNIDOCIL a bristly extension of a nematocyte whose stimulation triggers nematocyst discharge.

COELENTERON the primary gastrovascular body cavity of the radiate animals.

COELOM a secondary body cavity completely surrounded by mesodermal tissues.

COELOMOPORE an opening on the bryozoan lophophore through which eggs may be released.

COELOZOIC pertaining to an endoparasite in the body cavity.

COENENCHYME the soft, primarily mesogleal tissue which covers the external surface of soft coral colonies.

COENOSARC the living tissues of the hollow stems in hydroid colonies.

COLLENCYTE an elongate sponge cell type that may have a communicative function.

COLLOBLAST an adhesive cell characteristic of ctenophores.

COLLUM the limbless, third postoral segment which forms a collar around a millipede.

COLONIAL THEORY a phylogenetic view that holds that metazoans arose from colonial zooflagellates.

COLUMELLA the central axis of helicospirally coiled gastropod shells.

COMB ROW a series of parallel ciliary plates that extend over the body surface of ctenophores.

COMPOUND EYE an arthropod visual organ composed of numerous, discrete units called ommatidia.

CONCHIOLIN the principal protein found in mollusk shells.

CONJUGATION the cytoplasmic joining of two ciliates for the exchange of genetic material.

CONTINENTAL SHELF a shallow platform extending outward from the coast.

CONTINUITY SYSTEMS structures and processes which effect the reproduction and development of an animal.

CONTRACTILE VACUOLE a subcellular osmoregulatory organelle which rhythmically collects and expels fluids.

COPEPODID an intermediate larval stage in a copepod life cycle.

COPULATORY BURSA a sperm receiving area in the female reproductive tract.

CORACIDIUM the ciliated first larval stage of many tapeworms.

CORALLITE the skeleton of a single polyp in a stony coral colony.

CORMIDIUM In siphonophore colonies, a cluster of zooids including one phyllozooid, one gastrozooid, one gonozooid and one dactylozooid.

CORONA the anterior ciliary complex in rotifers.

CORONA CILIATA an anterior, sensory, horseshoe-shaped tract of cilia on a chaetognath.

CORPORA ALLATA a pair of endocrine glands associated with the insect brain, which manufacture and release juvenile hormone.

COXA the proximal segment of an arthropod appendage; attaches appendage to the body proper.

COXAL GLAND an excretory organ in arachnids.

COXOPODITE the proximal segment of a crustacean appendage.

CROP an esophageal dilation for food storage.

CRYSTALLINE CONE a long, cylindrical secondary lens within an ommatidium of an arthropod compound eye.

CRYSTALLINE STYLE a mucous and proteinaceous rod which releases enzymes in the stomach of some bivalve and gastropod mollusks.

CRYPTOBIOSIS a state of suspended animation induced by desiccation.

CTENIDIUM the generalized molluscan gill.

CUCULLUS the anterior hood characteristic of ricinuleids.

CYCLOSIS cytoplasmic streaming.

CYDIPPID the characteristic larval form of ctenophores.

CYPHONAUTES the feeding trochophore-like larva of nonbrooding bryozoans.

CYPRIS LARVA the late, settling larva of barnacles.

CYSTICERCUS an encysted preadult stage in some tapeworm life cycles.

CYTOPROCT an area of protozoan cell membrane specialized for exocytosis.

CYTOSTOME an area of protozoan cell membrane specialized for food vacuole formation.

CYTOZOIC pertaining to an intracellular endoparsite.

DACTYLOZOOID a mouthless coelenterate polyp whose single tentacle is loaded with nematocysts.

DELAMINATION gastrulation by tangential cell divisions which isolate an inner portion of the blastula.

DEMIBRANCH a horseshoe-shaped or V-shaped folded gill filament of lamellibranch bivalves.

DENDROBRANCHIATE pertaining to the repeatedly branched gills of primitive decapod crustaceans (e.g., penaeid shrimp).

DERMAL OSTIA the outermost incurrent openings in higher sponges.

DEUTOCEREBRUM the second of the three pairs of ganglia composing the arthropod brain and site of antennal nerve input.

DIAPAUSE a state of developmental arrest, often requiring exposure to low temperature before development can proceed.

DIPLEURULA a theoretical, bilaterally symmetrical ancestor of echinoderms; also, that stage in echinoderm development between the gastrula and an early larval form.

DIRECTIVE a primary septum or mesentery associated with the siphonoglyph in zoantharian anthozoans.

DOLIOLARIA the late larval stage of most holothuroids.

DORSAL LAMINA a pharyngeal ridge of tissue that enrolls food-laden mucus in an ascidian.

ECDYSIS the shedding of the old exoskeleton by an arthropod.

ECDYSONE the hormone which stimulates ecdysis in arthropods.

ECHINOPLUTEUS the larval stage of an echinoid echinoderm.

ECTODERM the outermost embryonic germ layer which forms epidermal, nervous and sensory structures.

ECTOPARASITE a parasite which attaches to its host's external surface.

ECTOPLASM cytoplasm in the gel phase which commonly borders the cell membrane.

ELYTRA the dorsal horizontal plates which form the roof of the respiratory channel in certain polychaetes (e.g., scale worms).

EMBOLUS a copulatory horn on the pedipalp of male spiders.

ENCYSTMENT the formation of a protective outer covering at the onset of a dormant stage in the life cycle.

ENDITE a process borne by the endopodite of a crustacean.

ENDOCUTICLE the inner layer of arthropod procuticle, characterized by untanned glycoproteins.

ENDODERM the innermost embryonic germ layer which forms the lining of the digestive tract.

ENDOPARASITE a parasite which attaches to or otherwise occupies a site within the body of its host.

ENDOPLASM cytoplasm in the sol phase which commonly occupies all but the outer perimeter of the cell.

ENDOPODITE the inner, distal portion of a crustacean appendage.

ENDOSTYLE a mucus-discharging groove along the ventral side of the pharynx of an ascidian.

ENTEROCOEL-CYCLOMETAMERISM THEORY a phylogenetic view which derives the segmented coelom from the gastric pouches of anthozoans.

ENTEROCOELOMATE forming a coelom by enterocoely (said of deuterostomes).

ENTEROCOELY coelom formation by evaginations of the archenteron.

EPHYRA in a scyphozoan life cycle, the stage which is produced by the scyphistoma and which matures into the adult medusa.

EPIBOLY gastrulation by the inward displacement of vegetal pole cells due to the more rapid growth of animal pole cells.

EPICARDIUM an endodermally derived tube, or system of tubes, with an uncertain function in ascidians.

EPICUTICLE the outermost proteinaceous-lipoidal layers of the arthropod exoskeleton.

EPIGASTRIC FURROW a transverse groove on the spider opisthosoma; associated with reproductive and respiratory openings.

EPIGYNUM a sculptured plate bearing the copulatory openings on the anterior opisthosoma of many female spiders.

EPIPHRAGM a mucous, sometimes calcified shield which seals the shell aperture in hibernating or estivating pulmonate gastropods.

EPIPODITE a process borne by the coxopodite of a crustacean.

EPIPODIUM a delicate, sensory extension of the molluscan foot.

EPISTOME[1] the anterior plate which borders the buccal frame of a decapod crustacean.

EPISTOME[2] a flap which covers the mouth of a phoronid worm.

EPITOKE the sexual, swimming form of an epitokous polychaete.

EPITOKY the asexual formation of swarming sexually-reproducing errant polychaetes.

ESTHETASC a compact row of chemosensory hairs common among crustaceans.

EUCOELOMATE possessing a true coelom.

EUKARYOTIC pertaining to organisms whose cells contain membrane-bound organelles.

EULAMELLIBRANCH pertaining to lamellibranch gills with numerous tissue connections between filaments.

EXITE a process borne by the exopodite of a crustacean.

EXOCUTICLE the outermost layer of the arthropod procuticle, characterized by tanned glycoproteins.

EXOPODITE the outer, distal element of a crustacean appendage.

EXPLOSIVE CELLS amoebocytes which aid in the clotting of crustacean blood.

FACET the cuticular lens of an arthropod eye or ommatidium.

FILIBRANCH pertaining to lamellibranch gills with only ciliary attachments between filaments.

FILOPODIUM a slender, pointed pseudopodium.

FISSIPARITY asexual reproduction by fission in stelleroid echinoderms.

FIXATION PAPILLA one of three epidermal projections that attach a settling ascidian larva to the substratum.

FLABELLUM a gill-cleaning, coxal extension of the fifth pair of walking legs in horseshoe crabs.

FLAGELLUM a long, whiplike locomotory organelle usually occurring singly or in small numbers.

FLAME BULB blind, hollow, internally ciliated projections of a flame cell.

FLAME CELL the central cell of a protonephridium.

FRONTAL ORGAN a likely chemosensitive area on the head of many branchiopod crustaceans.

FUNICULUS a cord of tissue, possibly with a transport function, which connects the gut to the body wall in a bryozoan.

GALEA on the chelicerae of pseudoscorpions, a set of horny processes that contain the openings of silk gland ducts.

GAMOGONY the transformation of a trophozoite or merozoite into a gamont and then into gametes in the life cycle of sporozoan protozoans.

GAMONT the sexual (gametic) stage in the sporozoan life cycle, produced by gamogony from a trophozoite or a merozoite.

GASTRIC MILL a triturating area in the crustacean foregut.

GASTRODERMIS the tissue layer which lines the coelenteron in radiate animals.

GASTROVASCULAR CAVITY the central body cavity or coelenteron of radiate animals.

GASTROZOOID[1] nutritive polyps in polymorphic coelenterate colonies.

GASTROZOOID[2] an asexually budded, nutritive individual in the life cycle of *Doliolum* (Thaliacea).

GASTRULA the developmental stage produced by gastrulation and usually characterized by distinct ectodermal and endomesodermal cells and an archenteron.

GASTRULATION the developmental process by which cells move into the blastocoel, thus converting the blastula to a gastrula.

GEL PHASE a colloidal state of cytoplasm in which subcellular particles combine in a clear gelatinous mass.

GEMMULE a mass of undifferentiated cells surrounded by a protective case, produced by freshwater sponges.

GENITAL ATRIUM a chamber into which open the various components of the reproductive system.

GENITAL OPERCULUM a wide membranous flap which covers the gonopores on the eighth body segment of horseshoe crabs.

GERMARIUM the area which produces yolkless eggs in the female reproductive tract of neorhabdocoels and triclads (Turbellaria).

GERMOVITELLARIUM the combined ovary and yolk gland of rotifers.

GIRDLE[1] the transverse groove on the body surface of a dinoflagellate.

GIRDLE[2] the portion of the mantle of a chiton which extends over the dorsal surface.

GIZZARD a foregut region which is modified for grinding food.

GLOCHIDIUM the parasitic veliger larva of freshwater lamellibranch bivalves.

GLOMERULUS the anterior blood sinuses of an enteropneust; possibly excretory in function.

GNATHOBASE the spiny basal article common to many arthropod appendages in the oral region.

GNATHOCHILARIUM the flattened and fused first maxillae of a millipede.

GNATHOPOD one of the prehensile limbs on the second and third thoracic segments of some isopods and amphipods.

GNATHOSOMA see **CAPITULUM.**

GONOCOEL-PSEUDOMETAMERISM THEORY a phylogenetic view which derives the annelid coelom from the gonadal cavities of early turbellarians.

GONOPHORE a medusa bud that produces gametes but that never separates from the colony on which it forms.

GONOPOD a crustacean trunk appendage that is specialized for sperm transfer.

GONOTHECA a transparent case which surrounds a gonozooid in thecate hydroids.

GONOZOOID[1] a reproductive polyp in polymorphic hydrozoan colonies.

GONOZOOID[2] an asexually budded individual that matures into the sexual form of *Doliolum* (Thaliacea).

GORGONIN a horny material which composes the axial rod of a gorgonian colony.

GÖTTE'S LARVA the free-swimming larva of certain polyclad flatworms.

GRAVID filled with fertilized eggs.

HALLER'S ORGAN a chemosensitive groove on the first legs of acarines.

HALTERE a short gyroscopic knob derived from the second wing of dipteran insects; a balancing organ.

HECTOCOTYLUS the arm which a male cephalopod uses to transfer his spermatophores to a female.

HEMIMETABOLOUS DEVELOPMENT the life cycle sequence of insects which undergo a gradual, "incomplete" metamorphosis.

HEMOCOEL an extensive network of blood sinuses which are coelomic in origin.

HEMOCYTE hemoglobin-containing coelomocytes in the hemal fluid of holothuroid echinoderms.

HEPATOPANCREAS a glandular area associated with the anterior midgut of malacostracan crustaceans; a modified cecum.

HERMAPHRODITISM the condition of having both male and female reproductive systems.

HETEROTROPHIC obtaining food in organic form.

HETEROZOOID a nonfeeding, specially modified member of a colony, as in bryozoans.

HIRUDIN an anticoagulant secreted by the salivary glands of blood-sucking leeches.

HISTOZOIC pertaining to an endoparasite occurring within host tissues.

HOLASPIS the final larval stage of a trilobite.

HOLOMETABOLOUS DEVELOPMENT the life cycle sequence of insects which undergo a dramatic, "complete" metamorphosis.

HOLOZOIC pertaining to heterotrophic ingestion of solid food.

HYDROCAULUS the vertical stalk of a hydroid colony.

HYDROCOEL the middle coelomic compartment (mesocoel) of an echinoderm embryo.

HYDRORHIZA the branching root-like portion of a hydroid colony.

HYDROSTATIC SKELETON an enclosed body of fluid which can be manipulated by the surrounding muscles.

HYDROTHECA the transparent glassy cup which surrounds the zooids in thecate hydroids.

HYPOBRANCHIAL GLAND a mucus-secreting gland in the mantle cavity of mollusks.

HYPOSTOME the oral cone of some coelenterate gastrozooids.

INCURRENT CANAL a passageway through the external cortex which precedes the prosopyles of more advanced sponges.

INFRACILIATURE the cilia and all associated structures in a ciliate protozoan.

INFUNDIBULUM the deep, tubular buccal cavity of peritrich ciliates.

INFUSORIFORM LARVA a stage in the life cycle of a dicyemid mesozoan which is produced sexually within infusorigens and which leaves the primary host, probably to enter some intermediate host.

INFUSORIGEN a stage of uncertain origin in the life cycle of dicyemid mesozoans; it occupies the axial cell of a rhombogen and sexually produces infusoriform larvae.

INGRESSION gastrulation by the inward migration of surface cells.

INHIBITORY NEURON a nerve cell whose impulses inhibit other nervous activity.

INSTAR any of the several molt-delineated, larval stages of a metamorphosing arthropod.

INTERAMBULACRUM a body zone between neighboring ambulacra in an echinoderm.

INTERSTITIAL CELL in coelenterates, an undifferentiated formative cell.

INTERTENTACULAR ORGAN an elevated extension of the bryozoan lophophore on which the coelomopore is located.

INTERTIDAL ZONE a coastal area between high and low tide marks.

INTROVERT the anterior, tentacle-bearing body zone of sipunculans.

INVAGINATION gastrulation by the inward migration of cells from the vegetal pole.

IRIDOCYTE a reflective cell associated with the chromatophores of some cephalopods.

ISOGAMETE one of the uniform sex cells produced by some protozoans.

JUVENILE HORMONE an endocrine product of the corpora allata which influences the maturation of insects.

KENOZOOID a reduced bryozoan morph that forms part of the stolon of a colony.

KINETODESMA a bundle of fibrils, each of which originates from a basal body in ciliate protozoans.

KINETOPLAST a large body containing nucleic acids in kinetoplastid zooflagellates (trypanosomes).

KINETOSOME see **BASAL BODY.**

KINETY a longitudinal unit of cilia, basal bodies and kinetodesmata in the infraciliature of ciliate protozoans.

KLINOKINESIS movement in response to stimulus gradients.

LABIAL PALPS[1] large, paired lips flanking the mouth of a bivalve mollusk.

LABIAL PALPS[2] sensory extensions of the labium of insects.

LABIUM a postoral lip typical of many arthropods; in insects it is formed from the second pair of maxillae.

LABRUM a lip-like extension in front of the mouth of many arthropods.

LANGUET one of a vertical series of tentacles along the dorsal wall of an ascidian pharynx.

LAPPET[1] one of numerous lobes in the bell margin of scyphozoans.

LAPPET[2] a movable, skeletal plate that shelters podial groups in a crinoid.

LEPTOMEDUSA the flattened, saucer-shaped medusa of thecate hydroids.

LEUCONOID pertaining to the most complex sponge body plan which features choanocyte chambers and highly branched water channels.

LOBOPODIUM a simple pseudopodium with a blunt tip.

LOPHOPHORE an extensible crown of ciliated tentacles that surrounds the mouth of phoronids, bryozoans, entoprocts and brachipods.

LORICA[1] the external covering of a chrysomonad.

LORICA[2] an armored external case of thickened cuticle in rotifers.

LUCIFERASE an enzyme that catalyses a bioluminescent reaction.

LUCIFERIN the substrate in a bioluminescent reaction.

LUNULE a radial notch through the body of a keyhole sand dollar.

LYRIFORM ORGAN a parallel series of slit sense organs typical of web-building spiders.

MACROGAMONT a sporozoan gamont which produces a single, large nonmotile gamete.

MACRONUCLEUS in ciliates a large polyploid nucleus which governs vegetative processes.

MADREPORITE the sieve-like external opening of the ambulacral system in an echinoderm.

MAINTENANCE SYSTEMS structures and processes which define the boundaries of an organism and maintain internal order.

MALE ATRIUM a chamber which bears the openings of the sperm duct(s) and the male gonopore; often specialized for copulation.

MALPIGHIAN TUBULE a slender, often convoluted tubule which discharges into the hindgut of many terrestrial arthropods.

MANCA the hatching, juvenile stage of several crustacean groups.

MANDIBLE one of the paired, short, jawlike appendages which flank the mouth of a mandibulate arthropod.

MANTLE[1] the outermost tri-lobed layer of the body wall in mollusks.

MANTLE[2] the carapace of a barnacle.

MANTLE CAVITY a sheltered space bounded by extensions of the mantle in mollusks.

MANUBRIUM the tubular, mouth-bearing extension of the subumbellar surface of a medusa.

MARSUPIUM the ventral thoracic brooding chamber of peracarid crustaceans.

MASTAX the complex pharynx of rotifers.

MASTIGONEME a tiny lateral projection along a flagellum.

MAXILLA one of the paired appendages immediately behind the mouth in mandibulate arthropods.

MAXILLARY GLAND the excretory organ of many crustaceans.

MAXILLIPED one of the paired anterior trunk appendages that have become adapted for food gathering in higher crustaceans.

MEDUSA the bell-shaped, swimming, sexually-reproducing stage in the life cycle of many coelenterates; a jellyfish.

MEGALOPS a crab postlarva with an extended abdomen.

MEHLIS' GLAND an accessory gland of uncertain function in the female reproductive tract of trematodes.

MEMBRANELLE an organelle comprising three short rows of cilia which borders the buccal cavity of some ciliates.

MERASPIS the second larval stage of a trilobite during which segmental proliferation began.

MEROZOITE an infective stage in the sporozoan life cycle, produced by schizogony from a trophozoite.

MESENTERY[1] a mesodermal fold which suspends the digestive tract in eucoelomates.

MESENTERY[2] in anthozoans, a gastrodermal and mesogleal ingrowth of the body wall; a septum.

MESOCOEL the midbody coelomic compartment of a deuterostomate or lophophorate animal.

MESODERM the intermediate embryonic germ layer which forms muscles, gonads and the major elements of many organ systems.

MESOGLEA a gelatinous matrix and its associated elements in the body wall of sponges and between the epidermal and gastrodermal layers of radiate animals.

METACERCARIA an encysted preadult stage in the life cycle of digenetic trematodes.

METACOEL the posterior coelomic compartment in a deuterostomate or lophophorate animal.

METAMERE one of a series of homologous body segments in metameric animals.

METAMERISM the segmentation of mesodermal body parts.

METANAUPLIUS a late nauplius larva.

METANEPHRIDIUM (NEPHRIDIUM) an osmoregulatory-excretory organ consisting of a nephrostome, nephridial tubule and nephridiopore.

MICROFILARIA a juvenile filarial (nematode) worm.

MICROGAMONT a sporozoan gamont which produces many small, motile gametes.

MICRONUCLEUS in ciliates a small diploid nucleus involved in reproduction.

MICROPYLE a pore in a sponge gemmule through which germinating cells emerge.

MICROTRICHES extensions of the cestode integument.

MICROTUBULE a hollow proteinaceous tubule occurring in bundles within the core of a cilium or flagellum.

MICTIC pertaining to haploid eggs which can develop parthenogenetically into males or which can be fertilized, thus producing females after dormancy.

MIRACIDIUM the first free-swimming larva of digenetic trematodes.

MONOPODIAL pertaining to a hydroid colonial growth pattern in which new polyps bud below each apical polyp.

MÜLLER'S LARVA the free-swimming larva of certain polyclad flatworms.

MYOCHORDOTONAL ORGAN probably proprioceptive nerve cell clusters in the third segments of the walking legs of decapod crustaceans.

MYOCYTE a long, curved contractile cell that surrounds the pores of sponges.

MYONEME a contractile bundle of microfilaments, common in stalked ciliates.

NAIAD the nymph of an aquatic hemimetabolous insect.

NAUPLIUS the first larval stage of most crustaceans, characterized by three pairs of appendages.

NAUPLIUS EYE a medial, cephalic cluster of inverse pigment-cup ocelli characteristic of larval crustaceans and retained by the adults of some groups.

NECTOPHORE a swimming medusoid morph which forms part of a siphonophore colony.

NEEDHAM'S SAC a spermatophore storage pouch in the mantle of male cephalopods.

NEKTON swimming animals.

NEMATOBLAST an immature nematocyte.

NEMATOCYST a threadlike stinging organelle characteristic of coelenterates.

NEMATOCYTE an ovoid cell which contains a nematocyst.

NEMATOGEN in the life cycle of dicyemid mesozoans, a parasitic stage which asexually produces infective larvae.

NEPHRIDIOPORE the external opening of a protonephridial or metanephridial system.

NEPHROCYTE a cell which functions in storage excretion in some arthropods.

NEPHROSTOME a ciliated funnel opening into the coelom which forms the major collecting region of a metanephridium.

NERVE NET a loose subepidermal network of neurons, as in coelenterates.

NEUROPODIUM the ventrolateral lobe of a polychaete parapodium.

NIDAMENTAL GLAND the organ which secretes the protective egg covering in female cephalopods.

NOTOPODIUM the dorsolateral lobe of a polychaete parapodium.

NUCHAL ORGAN a ciliated chemosensitive pit on the prostomium of annelids and anterior end of sipunculans.

NYMPH an immature stage of a hemimetabolous insect.

OCELLUS a rudimentary eye containing pigmented, photosensitive cells.

ODONTOPHORE the cartilaginous rod which supports the radula in mollusks.

OMMATIDIUM the individual photoreceptive unit of an arthropod compound eye.

ONCHOMIRACIDIUM the larval form of monogenetic trematodes.

ONCOSPHERE the nonciliated first larval stage of certain tapeworms.

OOTYPE the female genital chamber in trematodes and cestodes.

OOZOOID a sexually produced individual that undergoes asexual budding in the life cycle of a thaliacean.

OPERCULUM any lid-like structure, such as those that cover the nematocyst in coelenterates or the shell opening in prosobranch gastropods.

OPHIOPLUTEUS the larval stage of an ophiuroid.

OPISTHAPTOR the posterior attachment organ in monogenetic trematodes.

OPISTHOSOMA the posterior tagma of the chelicerate body.

ORGAN OF TÖMÖSVARY a vibration sensitive structure at the base of the antenna in many myriapods.

ORGAN-SYSTEM LEVEL OF ORGANIZATION a metazoan body plan in which true organ-systems are present.

OSCULUM the large excurrent opening of a sponge.

OSPHRADIUM a chemosensory structure within the molluscan mantle cavity.

OSSICLE the crystalline, calcareous unit of the echinoderm endoskeleton.

OSTIUM a small incurrent opening, as in the water channel system of a sponge or in the heart of an arthropod.

OVICELL an evaginated chamber of the body wall of a bryozoan, which houses a developing embryo.

OVIGEROUS LEG one of the third pair of appendages of male pycnogonids; used to carry developing eggs.

OVOTESTIS a gonad which produces both sperm and eggs.

PALLET a calcareous plate which may seal the burrow opening of shipworms (*Teredo:* Bivalvia).

PALMELLA the encysted non-flagellated form of a phytoflagellate.

PAPULA in an echinoderm body wall, a fleshy, thin-walled area serving respiratory and excretory functions.

PARAPODIA the paired, ventrolateral, segmentally arranged appendages of polychaetes.

PARENCHYMA the solid mesodermal tissue characteristic of acoelomate animals.

PARENCHYMELLA the larval form of a leuconoid sponge.

PARTHENOGENESIS the development of an egg without fertilization.

PAXILLA a specialized ossicle with movable spines in burrowing asteroids.

PECTINES comblike sensory appendages on the second preabdominal segment of scorpions.

PEDICEL a narrow stalk between the prosoma and opisthosoma of some arachnids.

PEDICELLARIA a muscularized, jawed extension of the body wall of an echinoderm.

PEDICLE the attachment stalk of a brachiopod.

PEDIPALP one of the first pair of postoral appendages variously specialized in chelicerate groups.

PEDUNCLE[1] an elongation of the preoral end which forms the stalk of gooseneck barnacles.

PEDUNCLE[2] the optical stalk of some crustaceans.

PELAGIC REGION the open sea.

PELLICLE in ciliates, the complex outer surface composed of layers of cell membrane and other organelles.

PENTACTULA a theoretical, bilaterally symmetrical ancestor of echinoderms.

PENTACTULA LARVA the preadult stage of a holothuroid echinoderm.

PEREIOPOD one of the thoracic limbs of a malacostracan crustacean.

PERIOSTRACUM the outer organic layer of mollusk shells.

PERISARC the thin, protective chitinoid tube which surrounds the coenosarc in hydroid colonies.

PERISTOMIUM the mouth-bearing first segment of annelids.

PERITONEUM the mesodermal lining of the coelom.

PERITROPHIC MEMBRANE a thin cuticular lining in the midgut of insects and certain other terrestrial arthropods.

PETALOID the aboral ambulacra of an irregular echinoid.

PHAGOCYTOSIS holozoic uptake of food particles by the cell membrane to form a food vacuole.

PHASIC NEURON a nerve cell whose impulses stimulate rapid, short-term muscle contractions.

PHASMID a posterior, unicellular gland of uncertain function in nematodes.

PHORESY a nonparasitic behavioral pattern in which one organism is transported by another of a different species.

PHOROZOOID an asexually budded, swimming individual in the life cycle of *Doliolum* (Thaliacea).

PHOTOCYTE a cell in which a bioluminescent reaction occurs.

PHOTOPHORE an organ in which light is produced.

PHOTOSYNTHETIC (PHOTOAUTOTROPHIC) capable of synthesizing food from nonbiological sources by using light energy.

PHYLLOBRANCHIATE pertaining to the simple lamellar gills of certain specialized shrimps and most advanced decapods (e.g., anomurans and brachyurans).

PHYLLODE the oral ambulacrum of an irregular echinoid.

PHYLLOZOOID a nematocyst-studded, tentacle-like structure on siphonophore colonies.

PHYLUM one of the principle divisions of the Animal Kingdom.

PHYTOPLANKTON the autotrophic (photosynthetic) members of the plankton.

PILIDIUM the larval form of nemertine worms.

PINACOCYTE an epidermal cell of a sponge.

PINNULE[1] one of numerous lateral projections on the tentacles of soft corals.

PINNULE[2] one of numerous lateral appendages on the arms of a crinoid.

PINOCYTOSIS saprozoic uptake of nutrients by the cell membrane.

PLANKTON small floating organisms typically too weak to swim against water currents.

PLANULA the free-swimming ciliated larva of many coelenterates.

PLASMODIAL THEORY a phylogenetic view that holds that metazoans arose from multinucleate ciliates.

PLEOPOD one of the abdominal appendages of malacostracan crustaceans; a swimmeret.

PLEROCERCOID the encysted preadult stage in some tapeworm life cycles.

PLEURITE a segmental, lateral plate of an arthropod exoskeleton.

PLUTEUS the long-armed larva of an ophiuroid or an echinoid.

PNEUMATOPHORE the gas float of a siphonophore colony.

PNEUMOSTOME the opening to the lung of terrestrial gastropods.

PODIUM a tube foot; one of numerous distal elements in an echinoderm ambulacral system.

POLAR CAPSULE a body which houses coiled polar filaments in cnidosporan protozoans.

POLAR FILAMENT a spirally coiled structure which is extruded by cnidosporan protozoans for anchorage to their hosts.

POLIAN VESICLE a large muscularized extension of the water ring of an echinoderm.

POLYEMBRYONY the asexual production of larval forms by non-motile embryonic stages, as in the life cycle of digenetic trematodes.

POLYKINETY a spiraling row of cilia which descends into the infundibulum of peritrich ciliates.

POLYP the bottle-shaped, sessile, usually asexually-reproducing stage in a coelenterate life cycle.

POLYPIDE the central body of an individual bryozoan.

POROCYTE in a sponge, a tubular epidermal cell that surrounds an ostium.

POSTLARVA a late larval stage in most crustaceans; characterized by a complete set of adult segments.

PRE-EPIPODITE the outer branch of a trilobite appendage.

PREHENSOR one of the paired, poisonous, chelate first thoracic limbs of a centipede.

PROCERCOID a larval stage occurring within an intermediate host in the life cycle of certain tapeworms.

PROCUTICLE the typically thick, chitinous and proteinaceous layer of an arthropod exoskeleton; the exocuticle and the endocuticle.

PROGLOTTID a unit of the strobila of a cestode.

PROKARYOTIC pertaining to organisms (largely bacteria and blue-green algae) whose cells do not contain membrane-bound organelles.

PROSOMA the anterior tagma of the chelicerate body.

PROSOPYLE a small intercellular space preceding the canal system of syconoid and leuconoid sponges.

PROSTOMIUM the anterior, presegmental tip of annelids.

PROTANDRY the production of sperm before eggs by a hermaphrodite.

PROTASPIS the first larval stage of a trilobite.

PROTOCEREBRUM the first of the three pairs of ganglia composing the arthropod brain and site of optic nerve input.

PROTOCOEL the anterior coelomic compartment of a deuterostomate animal.

PROTONEPHRIDIUM a primitive excretory-osmoregulatory organ featuring tubules blindly ending in flame bulbs.

PROTONYMPHON the first post-hatching stage of a pycnogonid.

PROTOPODITE the proximal region of a crustacean limb; the coxopodite and the basopodite.

PROTOTROCH the equatorial, ciliary ring of a trochophore larva.

PROTOZOEA a decapod crustacean larval stage which follows the nauplius; the hatching form of more advanced shrimp.

PROVENTRICULUS a straining or grinding chamber at the foregut-midgut junction of an insect.

PSEUDEPIPODITE a large, lateral branch on the coxopodite of a cephalocaridan crustacean.

PSEUDOBRANCH a secondary gill of an aquatic pulmonate gastropod.

PSEUDOCOEL a secondary body cavity derived from the blastocoel and located between endodermal and mesodermal tissues.

PSEUDOCOELOMATE possessing a pseudocoel.

PSEUDOFECES any material rejected by the digestive system before passage through the gut.

PSEUDOLAMELLIBRANCH pertaining to lamellibranch gills with infrequent tissue junctions between filaments.

PSEUDOMETAMERISM the serial repetition of body parts without the formation of true metameres.

PSEUDOPODIUM a locomotory projection of the plasma membrane in sarcodine protozoans.

PSEUDOSTIGMATIC ORGAN a type of trichobothrium occurring in acarines.

PSEUDOTRACHEA respiratory tubules which ramify beneath pleopod surfaces in isopods.

PUPA the quiescent preadult stage of a holometabolous insect.

PUPATION the developmental changes which occur at the larva-to-pupa molt of holometabolous insects.

PYGIDIUM the postsegmental end of annelids.

PYLORIC STOMACH[1] a region of the gut of malacostracan crustaceans that strains food material after passage through the gastric mill.

PYLORIC STOMACH[2] the aboral pouched region of the asteroid midgut.

RADIAL CANAL[1] an evaginated, choanocyte-lined pocket of the spongocoel in syconoid sponges.

RADIAL CANAL[2] in the echinoderm ambulacral system, the tubular component that extends from the water ring outward along each ambulacral zone.

RADIAL CLEAVAGE a type of early embryonic cell division in which consecutive cleavage planes are perpendicular.

RADULA a protrusible ribbon of teeth characteristic of the molluscan buccal cavity.

REDIA a polyembryonically produced developmental stage of digenetic trematodes; gives rise to cercariae.

REDUCTION BODY a de-differentiated mass of sponge tissue from which a new animal can regenerate.

RENETTE GLAND a single-celled gland of uncertain function in primitive nematodes; possibly related to the excretory tubules of more advanced nematodes.

REPUGNATORIAL GLAND in opilionids, a gland whose noxious secretion is defensive in nature.

RESPIRATORY TREE one of the pair of branched respiratory tubules originating from the cloaca of a holothuroid.

RETICULOPODIUM a pseudopodial type consisting of a network of anatomosing filopodia.

RETINULA the photoreceptive rosette of cells within a single ommatidium in the arthropod compound eye.

RETROCEREBRAL ORGAN a glandular structure of uncertain function near the brain of rotifers.

RHABDOID a rod-shaped epidermal body in flatworms.

RHAGON an immature stage of a leuconoid sponge.

RHINOPHORE large chemosensory tentacles on the heads of some opisthobranch gastropods.

RHOMBOGEN in the life cycle of dicyemid mesozoans, a stage representing a mature nematogen which produces infusoriform larvae.

RHOPALIUM a compound sense organ in scyphozoans, containing an ocellus, a statocyst, tactile and probably chemoreceptors.

RHYNCHOCOEL a body cavity in nemertines which houses the proboscis apparatus.

ROSTRUM[1] in the capitulum of a barnacle, a calcareous plate adjacent the site of head attachment to the mantle wall.

ROSTRUM[2] an antero-dorsal projection of the carapace in acarines and malacostracan crustaceans.

ROUND DANCE movements by a scout bee which signal the presence of nearby food.

ROYAL JELLY a worker bee secretion which nourishes a developing queen bee.

SAPROZOIC pertaining to the heterotrophic uptake of organic materials in dissolved or suspended form.

SCAPHOGNATHITE an extension of the second maxilla whose beating draws the ventilation current across the gills of a decapod crustacean; the gill bailer.

SCHIZOCOEL THEORY a phylogenetic view that derives the coelom from splits in the mesoderm of acoelomate ancestors.

SCHIZOCOELOMATE forming a coelom by schizocoely, said of protostomes.

SCHIZOCOELY coelom formation by the splitting of a band of mesodermal tissue.

SCHIZOGONY the asexual proliferation of merozoites from a trophozoite in the sporozoan life cycle.

SCLEROBLAST a spicule-secreting cell in the mesoglea of sponges.

SCOLEX the cestode head, which bears the holdfast organs.

SCUTUM one of a pair of apical plates which flank the shell aperture of a barnacle.

SCYPHISTOMA in the scyphozoan life cycle, the polypoid stage which strobilates ephyrae.

SEMINAL RECEPTACLE a sperm storage area in the female reproductive tract.

SEMINAL VESICLE a sperm storage area in the male reproductive tract.

SEPTAL FILAMENT the trilobed free margin of an anthozoan septum.

SEPTUM[1] in anthozoans, a gastrodermal and mesogleal ingrowth of the body wall.

SEPTUM[2] a double layer of peritoneum which separates adjacent segments in annelids.

SERIAL HOMOLOGY the homology pertaining to segmentally repeated structures in metameric animals.

SESSILE pertaining to organisms which are permanently attached to the substratum.

SETA (CHAETA) a chitinous bristle, common to annelids.

SINUS GLAND see **X-ORGAN AND SINUS GLAND COMPLEX.**

SIPHONOGLYPH a ciliated gutter along an anthozoan stomodeum.

SIPHUNCLE a calcareous tube which perforates all shell chambers of *Nautilus* and many fossil cephalopods.

SLIT SENSE ORGAN vibration-sensitive cuticular crevices typical of many arachnids.

SOL PHASE a colloidal state of cytoplasm in which subcellular particles are in suspension and the cytoplasm is relatively fluid and granular.

SOLE[1] the flattened, ventral surface of the foot of a mollusk.

SOLE[2] the ventral surface of a holothuroid.

SOLENIA gastrovascular tubes within colonies of soft coral.

SOMATOCOEL the posterior coelomic compartment (metacoel) of an echinoderm embryo.

SPATIAL SUMMATION the phenomenon whereby stimulus strength is increased to threshold level by simultaneous subthreshold stimulation at different sites.

SPECIES a taxon representing populations of similar organisms capable of mating and producing fertile offspring.

SPERMATOPHORE a protective sperm package which is transferred from one animal to another.

SPICULE a small supportive spike, such as the calcareous or siliceous elements in sponge skeletons.

SPIGOT the external opening of a spider silk gland duct.

SPINNERET a modified abdominal appendage bearing the openings of silk gland ducts in spiders.

SPIRACLE an external opening into an arthropod tracheal system.

SPIRAL CLEAVAGE a type of early embryonic cell division in which consecutive cleavage planes are oblique.

SPONGIN a meshwork of protein fibers in the skeleton of advanced sponges.

SPONGIOBLAST a sponge cell which manufactures spongin.

SPONGOCOEL the main body cavity of a simple sponge.

SPONGY ORGAN an organ of uncertain function associated with the axial gland of crinoids.

SPORE an encysted stage which characterizes the life cycles of sporozoan and cnidosporan protozoans.

SPOROCYST a germinal sac which gives rise to rediae in the life cycle of a digenetic trematode.

SPOROGONY meiotic production of sporozoites from the zygote in a sporozoan life cycle.

SPOROPLASM a cytoplasmic mass with two nuclei which emerges from cnidosporan spores and enters host tissues.

SPOROSAC a highly specialized hydrozoan gonophore which has lost all medusoid structures.

SPOROZOITE an infective stage which emerges from the spore in the life cycle of a sporozoan protozoan.

STATOBLAST a resistant, dispersive, reproductive body formed along the funiculus of freshwater bryozoans.

STATOCYST an equilibrium sense organ consisting of a central cavity, protruding tactile hairs, and a statolith.

STELLATE GANGLION a nerve center associated with the rhythmic muscular activity of the mantle wall in cephalopods.

STEREOGASTRULA a solid gastrula produced by ingression or delamination.

STERNITE a segmental, ventral plate of an arthropod exoskeleton.

STIGMA[1] a photoreceptive eyespot in euglenoids.

STIGMA[2] (Pl: **STIGMATA**) a pharyngeal slit through which the respiratory/feeding current passes in a urochordate.

STOLON a rootlike branch of an animal colony that provides anchorage and budding sites.

STOMODEUM the anterior ectodermally-lined portion of any digestive tract.

STONE CANAL a calcareous tube which drains water from the madreporite into the water ring of an echinoderm.

STROBILA the cestode body trunk, which consists of many proglottids.

STROBILATION the asexual production of ephyrae by a scyphistoma in the life cycle of scyphozoan coelenterates.

STYLE SAC the ciliated, narrow posterior region of the stomach in primitive mollusks.

SUBCHELA a cutting limb whose distal finger folds back onto the penultimate article.

SUBNEURAL GLAND a mass of glandular tissue below the cerebral ganglion of an ascidian.

SUBRADULAR ORGAN a chemosensory patch beneath the radula of chitons.

SULCUS the longitudinal groove along the body surface of a dinoflagellate.

SYCONOID pertaining to a sponge body plan which features radial canals along the spongocoel.

SYMPODIAL pertaining to a hydroid colonial growth pattern in which new polyps bud from the tip of each stem.

SYZYGY the union of two older sporozoan trophozoites which then undergo joint gamogony, encystment, and sporogony.

TAENIDIUM a spiraling cuticular thread which reinforces the wall of an insect trachea.

TAGMA in an arthropod, a body zone comprising several segments.

TAGMATIZATION the adaptive formation of supersegmental body zones in arthropods.

TANNING the hardening of cuticular proteins in arthropods by the formation of cross-linkages.

TAPETUM the post-retinal reflective membrane of an indirect eye.

TEGMEN the membranous oral covering of a crinoid calyx.

TEGUMENT the unciliated, cytoplasmic layer which covers the external surface of trematodes and cestodes.

TELOPODITE the inner, walking leg of a trilobite.

TELOTROCH a small ciliary ring at the posterior tip of a trochophore larva.

TELSON the anus-bearing terminal segment in various arthropod groups.

TEMPORAL POLYMORPHISM the alteration of different morphs in the life cycle, as in the polyp and medusa of coelenterates.

TEMPORAL SUMMATION the phenomenon whereby stimulus strength is increased to threshold level by repeated stimuli of subthreshold intensity.

TERGITE a segmental, dorsal plate of an arthropod exoskeleton.

TERGUM one of a pair of apical plates which flank the shell aperture of a barnacle.

TEST any hard outer case or covering secreted or constructed by an organism.

THECA the skeletal cup secreted by a stony coral polyp.

THIGMOTACTIC responding to touch.

TIEDEMANN'S BODY one of several folded outgrowths of the water ring in an echinoderm ambulacral system; possibly a site of coelomocyte production.

TISSUE LEVEL OF ORGANIZATION a metazoan body plan in which cells are organized into tissues, but few or no true organs are formed (said of radiate animals).

TONIC NEURON a nerve cell whose impulses stimulate slow, prolonged muscle contractions.

TORNARIA the enteropneust larval form that resembles an asteroid bipinnaria.

TORSION the 180° rotation of the main body mass in gastropods.

TRACHEA a chitin-lined respiratory tube which penetrates to the hemocoel in many terrestrial arthropods.

TRACHEAL LUNG a small cluster of trachea in certain primitive arachnids.

TRACHEOLE a small trachea.

TRACHEOLE END CELL the terminal cell of a tracheole.

TRICHOBOTHRIUM a long, thin tactile hair typical of many arachnids.

TRICHOBRANCHIATE pertaining to lobster and crayfish gills, which bear numerous but unbranched lateral processes.

TRICHOCYST an organelle located below the infraciliature of ciliates and capable of discharging striated, barb-tipped shafts.

TRILOBITE LARVA the hatchling stage of the horseshoe crab.

TRIPLOBLASTIC possessing three embryonically formed germ layers.

TRITOCEREBRUM the third of the three pairs of ganglia making up the arthropod brain and site of input from nerves associated with the mouth parts.

TROCHOPHORE the characteristic first larval stage of annelids and mollusks.

TROCHUS the upper ciliary wheel of the corona in some rotifers.

TROPHOZOITE a mature sporozoite; undergoes schizogony to produce merozoites.

TROPHUS a strong masticating jaw within the rotiferan mastax.

TUBERCLE any knoblike process, such as the ossicular attachment site for echinoid spines.

TUBULES OF CUVIER a mass of sticky tubules that can be discharged from the cloacal region of certain holothuroids.

TUNIC the outermost, epidermally secreted covering of a urochordate.

TUNICIN a polysaccharide resembling cellulose that composes the fibrous matrix of the urochordate tunic.

TYMPANAL ORGAN a specialized vibration sensitive (hearing) structure in a sound-producing insect.

TYPHLOSOLE[1] a dorsal infolding of the intestinal wall of an annelid.

TYPHLOSOLE[2] an excurrent trough in the midgut of some mollusks.

UMBO the dorsal hump above the hinge and the oldest region of a bivalve shell.

UNDULATING MEMBRANE a simple long row of fused cilia which borders the buccal cavity in certain ciliates.

URN peritoneal cell complexes involved in excretion in sipunculans.

UROPOD one of the pair of flattened appendages on the sixth abdominal segment of some malacostracan crustaceans; uropods and the telson form a tail fan.

UTERINE BELL a structure which controls the passage of encysted acanthocephalan larvae from the pseudocoel to the female reproductive tract.

VANADOCYTE an ascidian amoebocyte that contains very high concentrations of vanadium.

VAS DEFERENS (Pl: **VASA DEFERENTIA**) a sperm duct.

VELIGER the second larval stage in many mollusks.

VELUM[1] a thin circular flap which extends inward from the bell margin of hydrozoan medusae.

VELUM[2] a ciliated swimming organ of molluscan veliger larvae.

VERTEBRA one of a series of large central disks in the endoskeleton of ophiuroid arms.

VESTIBULE a simple preoral chamber.

VIBRACULUM a bryozoan heterozooid whose opeculum forms a prominent, sanitational bristle.

VITELLARIA a primitive larval form in crinoids and some holothuroids.

VITELLARIUM the area which produces yolk cells in the female reproductive tract of neorhabdocoels and triclads.

WAGGLE DANCE movements by a scout bee which communicate the location of a food source.

WATER RING the central, circular element of the echinoderm ambulacral system.

X-ORGAN AND SINUS GLAND COMPLEX a neurosecretory complex which is associated with chromatophoric display, molting and sexuality in crustaceans.

Y-ORGAN a crustacean gland that manufactures and/or releases a molt-promoting hormone.

ZOEA an intermediate larval stage among many crustaceans; characterized by developing trunk appendages.

ZOECIUM the exoskeleton of an individual bryozoan.

ZONITE one of the thirteen distinct sections of a kinorhynch body.

ZOOID an individual member of a colony, as in coelenterates and bryozoans.

ZOOPLANKTON the animal members of the plankton.

ZOOXANTHELLAE symbiotic dinoflagellates, occurring in an encysted (palmella) form.

Index

Note: Page numbers in italics represent illustrations; *t* indicates tables.